The Ciliated Protozoa

The Ciliated Protozoa

Characterization, Classification, and Guide to the Literature

Third Edition

Denis H. Lynn
University of Guelph
Canada

Springer

Denis H. Lynn
Department of Integrative Biology
University of Guelph
Guelph, Ontario
Canada
ddr@uoguelph.ca

Author's notes:
With the exception of the *Glossary* (Chapter 2) and *The Ciliate Taxa* (Chapter 17) all chapters have been rewritten for the third edition. The *Glossary* and *The Ciliate Taxa*, originally from the second edition by John O. Corliss, have been extensively revised and considerably expanded. Original illustrations of ciliate taxa drawn by Owen Lonsdale and reconstructions of the somatic cortex illustrated by Jennie Knopp are contained throughout the book, except in Chapter 2.
The author acknowledges that there are instances where he was unable to trace or contact the copyright holder for permission to reproduce selected material in this volume. The author has included complete source references for all such material. If notified, the publisher will be pleased to rectify any errors or omissions at the earliest opportunity.

All figures are available for free viewing via the book's homepage at www.springer.com

ISBN 978-1-4020-8238-2 e-ISBN 978-1-4020-8239-9
DOI: 10.1007/978-1-4020-8239-9
Springer Dordrecht Heidelberg London New York

Library of Congress Control Number: 2008923552

Originally published by Pergamon Press, New York, 1979

Cover illustration: A phylogenetic tree of the 11 classes of ciliates
Cover image of marsh © 2007 JupiterImages Corporation

Printed on acid-free paper

9 8 7 6 5 4 3

springer.com

To my parents, Beverley and Blanche Lynn, whose faithful support and unfailing encouragement were only stopped by their tragic and untimely deaths.

"Some come my friend be not afraid, we are so lightly here.
It is in love that we are made, in love we disappear."

Leonard Cohen

Acknowledgements

To my three mentors: Jacques Berger, Eugene B. Small, and John O. Corliss

Jacques Berger (1934–1995)

Jacques was my "academic father", who I met with his research crew in the summer of 1968 while I was investigating the autecology of mosquito larvae in Newfoundland, working for Marshall Laird. Jacques' sense of humor, his strong, well–founded opinions, his broad and deep academic interests in ciliates, baseball and golf, and in later years the history of science, attracted me to the academic life at a time when I was considering a career in medicine.

Eugene B. Small

Gene, who I consider my "academic uncle", has been a colleague, with boundless enthusiasm for the ciliates and life, always willing to engage in discussion on topics as arcane as "the number of ciliated kinetids in Kinety X" to the spiritual elements of Zen. As a mentor, Gene generously shared both his deep intuitive insights into the taxonomic significance of the cortical characters of ciliates and his broad knowledge of ciliate diversity. Over the years, our collaboration has been inspirational and deeply rewarding.

John O. Corliss

John, my "academic grandfather" as Jacques Berger completed his Doctoral dissertation with John in 1964, has always been supportive of my career in ciliatology. John has been a professional role model by demonstrating outstanding devotion to students and colleagues both far and wide, and to one's discipline. His willingness in 1997 to begin our collaboration on a Third Edition demonstrated his enthusiasm for "our wee beasties" and his confidence in our partnership. For his trust, encouragement, and big-hearted generosity, I am sincerely grateful.

Preface to the Third Edition

Soon after I began graduate studies in Protozoology at the University of Toronto in September 1969, Jacques Berger brought in his copy of the First Edition of "The Ciliated Protozoa" as required reading. This book synthesized the "state of knowledge" of ciliate systematics at that time, and it brought the formal study of ciliate diversity, especially in its nomenclatural aspects, to a highly professional level. In the following decade events occurred that set me on the path to pursuing ciliate research. John Corliss, the author of that little-big ciliate book visited Jacques in Toronto, and I met him. John suggested that I visit Gene Small at the Department of Zoology, University of Maryland, USA. In 1971, I met Gene, whose enthusiasm for "these wee bugs" was infectious, and whose intuitive grasp of the systematic significance of particular features was marvellous. I resolved to return to Maryland to work with Gene, taking a "sabbatical" leave from my doctoral thesis research to do so. There was, of course, another wonderful reason for the move to Maryland in September 1972 – I had met Dr. Portia Holt who, at the time, was working as a postdoctoral fellow with Dr. Corliss. So the 1972–1973 period was a rich experience of immersion in ciliate systematics, coupled with immersion in my developing relationship with my future wife, Portia. During this time, John Corliss, Head of the Department of Zoology, provided financial assistance as well as academic support. At that time, John was beginning preparations for the Second Edition of "The Ciliated Protozoa", having just co-authored a major revision of the ciliate macrosystem with his colleagues in France. By 1974, these co-authored and authored papers on the new macrosystem were published, including his paper entitled "The changing world of ciliate systematics: historical analysis of past efforts and a newly proposed phylogenetic scheme of classification for the protistan phylum Ciliophora". This was the "Age of Ultrastructure," as John called it, but the "newly proposed phylogenetic scheme" was only moderately influenced by these new data.

While in Maryland, Gene Small and I became deeply involved in discussing the implications of ultrastructural features, and these discussions lead to my publication of "the structural conservatism hypothesis" in 1976. Applying that idea, Gene and I proposed a radically different macrosystem for the ciliates in 1981, which I supported by a major review of the comparative ultrastructure of ciliate kinetids, demonstrating the conservative nature of these important cortical components. While ultrastructural study still formed an element of my research program in the 1980s, Gene encouraged me to consider moving into molecular phylogenetics to test the robustness of our ideas, which had now been slightly modified with publication of the First Edition of "An Illustrated Guide to the Protozoa". In an ultimately productive sabbatical year in 1986–1987, I worked with Mitch Sogin at the National Jewish Hospital, Denver, to learn the techniques of cloning and sequencing. Mitch and I were finally able to provide one of the first larger comparative datasets on genetic diversity of ciliates based on the small subunit rRNA gene sequences, derived at that time by reverse transcriptase sequencing. On the other side of the Atlantic, our colleagues in France, led by André Adoutte, were using the same approach with the large subunit rRNA gene and generating an even larger dataset. Both approaches demonstrated two things: firstly, confirmation that the ultrastructural approach informed by structural conservatism was providing resolution of the major natural assemblages or clades of ciliates; and secondly, genetic distances between groups of ciliates were as vast as the genetic distances between plants and animals – THE major eukaryotic kingdoms at that time!

I continued to collaborate with Mitch, and in 1991 my first "molecular" Magisterial student, Spencer Greenwood, published an article establishing 1990 or thereabouts as the beginning of the "Age of Refinement" – the period when gene sequencing techniques would deepen our understanding of the major lines of evolution within the phylum. Nearing the end of that decade, I was fairly confident that we had resolved the major lines of evolution, mostly confirming the system that Gene and I had proposed in the mid-1980s. We published a revisionary paper in 1997 in the *Revista* as a tribute to our Mexican colleague, Eucario Lopez-Ochoterena.

As I look back on my correspondence, it was about this time that I approached John regarding writing a Third Edition of "The Ciliated Protozoa". While turning "just" 75 in 1997, John enthusiastically embraced the idea and we began collaborating on a book proposal that travelled with several editors through different publishers. We finally signed a contract with Springer-Verlag in 1999, and the project began.

It was difficult for us working with each other "at a distance" and making the commitment to focus on "The Book" with all the other competing responsibilities and obligations of academic life, especially since John was in retirement. I began work on the "Class" chapters, while John's commitment was to revision of "The Ciliate Taxa", now **Chapter 17** in the Third Edition. In reviewing my correspondence, John's health took a turn for the worse in early 2003. He was busy writing his last major "op ed" piece – "Why the world needs protists" (Corliss, 2004). This took a major joint effort for us to complete, and by the end of 2003 John reluctantly agreed to withdraw from "The Book" project, and assign copyright over to me. It is with deep gratitude that I heartily thank John for this gift, and for his many years of mentoring both me and the protistological community. The Third Edition would have benefited significantly from his deep and careful understanding of taxonomic and nomenclatural practises, and I can only hope that I have achieved to some degree the level of excellence that he established in the first two editions. A major regret has been the omission of figures from **Chapter 17**, "The Ciliate Taxa". There were significant hurdles to obtain copyright permissions for the over 1,000 required illustrations, and I put the publication schedule ahead of this element. There are a number of significant illustrated guides to genera and species that have recently been published. References are made to these throughout the book as sources that readers can consult for this aspect of ciliate diversity. A future project that I am contemplating is an illustrated guide to all the valid ciliate genera.

This book has been a collaborative effort from the beginning. In addition to my indebtedness to John, I have appreciated the support provided by my new contacts at Springer – Dr. Paul Roos, Editorial Director, Environmental/Sciences, and Betty van Herk, his Senior Assistant. Since the Third Edition depended heavily on several sections from the Second Edition – notably the **Glossary** and **The Ciliate Taxa**, I was helped immensely by the secretarial assistance of Lori Ferguson, Felicia Giosa, Irene Teeter, and Carol Tinga, who created electronic files of Chapters 2, 20, and 22 from the Second Edition. Illustrations have been a major component of previous editions, and I have re-used these when appropriate. Ian Smith at BioImage, College of Biological Science, University of Guelph, scanned and "cleaned up" many images from the Second Edition from hard copy files provided to me by John. I am also deeply grateful to Ian for patiently tutoring me in the idiosyncracies and some of the finer points of Adobe Photoshop and Corel Draw, as I constructed the over 100 plates for the Third Edition.

Three students have made major contributions to the project. For the "representative taxa", Owen Lonsdale, a former graduate student in Environmental Biology, University of Guelph, has rendered beautiful schematic drawings of genera based on various literature sources. Since the somatic kinetid has been a major element in our systematic approach, I have worked with Jennie Knopp, a talented University of Guelph Biology Major, to render three–dimensional reconstructions of the somatic cortex of most of the classes. This collaboration has stretched both our imaginations. I thank Jennie for her patience as she worked through many revisions to "get it right"! Finally, I sincerely appreciate the careful and attentive reading that Eleni Gentekaki, my doctoral student, has done of the text. She has identified trouble spots, has been mindful of terms that should be in the **Glossary**, and has found a variety of typographical errors.

Finally, I am deeply indebted to several colleagues with expertise in the taxonomy of various groups and to whom I sent sections of **Chapter 17**, "The Ciliate Taxa". While none of these colleagues can be held responsible for errors or omissions in **Chapter 17** OR the taxonomy that I have ultimately decided to present, since our opinions did differ sometimes substantially, I do wish to thank for their comments the following in alphabetical order: Félix–Marie Affa'a – clevelandellids; Helmut Berger – hypotrichs and stichotrichs; Stephen Cameron – trichostomes; John Clamp – apostomes and peritrichs; Igor Dovgal – chonotrichs and suctoria; Wilhelm Foissner – haptorians; and Weibo Song – scuticociliates. Finally, I am deeply grateful to Erna Aescht who reviewed the entire **Chapter 17** with a degree of care and precision that I could not have expected. She has contributed immeasurably to the accuracy of this chapter, and I cannot thank her enough.

I thank my recent and current academic family – my research associate, Michaela Strüder-Kypke, and my graduate students, Dimaris Acosta, Chitchai Chantangsi, Eleni Gentekaki, Chandni Kher, Megan Noyes, and Jason Rip – for their forebearance as their "boss" excused himself yet again to work on "The Book". Finally, I wish to thank my wife, Portia, who has provided constant support and a shared vision of the completion of this work, even though it has taken much longer than either of us originally anticipated!

Guelph, July, 2007

Contents

List of Figures

List of Tables

Chapter 1
Introduction and Progress in the Last Half Century

Abstract The history of ciliate systematics has been divided into five periods: (1) the Age of Discovery; (2) the Age of Exploitation; (3) the Age of Infraciliature; (4) the Age of Ultrastructure; and (5) the Age of Refinement. Progress in each of these periods arose through an interaction of technology and conceptual views. For example, refined silver staining techniques revealed the law of desmodexy of the ciliate cortex and enabled the development of comparative morphogenetics in the Age of Infraciliature. Electron microscopy was essential for the conceptual notion of levels of organization below the cell and provided the impetus for the structural conservatism hypothesis in the Age of Ultrastructure. In this latter age, the foundations for the current classification system have been laid. Gene sequencing has provided the next technological innovation, which has enabled testing and revising our views on relationships in the current Age of Refinement. Major differences between the scheme presented herein with its two subphyla and 11 classes and other competing schemes are briefly discussed.

Keywords Kinetid, cortex, rRNA gene, molecular phylogeny, organic design

Systematics as a discipline was defined by Simpson (1961) as "the scientific study of the kinds and diversity of organisms and of any and all relationships among them" (p. 7). One aim of modern systematics is to represent these relationships among organisms by natural classifications: these are hierarchical and reflect as closely as possible the true phylogeny of a group of organisms. The approach to establishing a hierarchical classification is influenced by the conceptual views of how significant particular characters are in inferring relationships, and these conceptual views, in their turn, are influenced by the technical approaches in vogue. In this context, Corliss (1974a) discussed the historical development of ciliate systematics in four periods: (1) the Age of Discovery (1880–1930), exemplified by Bütschli; (2) the Age of Exploitation (1930–1950), exemplified by Kahl; (3) the Age of the Infraciliature (1950–1970), exemplified by Chatton, Lwoff, and Fauré-Fremiet, and during which Corliss (1961) published the first edition of "The Ciliated Protozoa"; and (4) the Age of Ultrastructure, whose beginnings around 1970 were summarized in the review chapter by Pitelka (1969). The zenith of the Age of Ultrastructure (1970–1990) was at the time of the second edition of "The Ciliated Protozoa" by Corliss (1979), and its ending might be established around 1990, at the appearance of the first reports on gene sequences of ciliates. Indeed, Greenwood, Sogin, and Lynn (1991a) suggested this criterion as the beginning of a fifth age – the Age of Refinement (1990–present), during which the major lines of evolution and our closest approach yet to a natural classification for the phylum might be possible. It is therefore useful to briefly review this history, especially emphasizing the last 50 years to understand how ciliate systematics has indeed progressed.

1.1 The Ages of Discovery (1880–1930) and Exploitation (1930–1950)

Bütschli (1887–1889) and Kahl (1930–1935), exemplifying the Ages of Discovery and Exploitation, respectively, primarily used light microscopic observations of living ciliates, without the use of sophisticated stains. From the Age of Discovery to the Age of Exploitation, the number of higher taxa doubled as our understanding of diversity exploded (Table 1.1). The conceptual approach focused on the character of the somatic and oral ciliature and on a consideration that evolution proceeded from simpler forms to more complex forms. This is reflected in the characterization of the higher taxa by Bütschli as Holotricha – evenly covered by somatic cilia – and Spirotricha – with a prominent spiralling adoral zone of membranelles (Table 1.1). The suctorians with their bizarre tentacled appearance and absence of external ciliature were given equivalent stature to all other ciliates by both Bütschli and Kahl. Other specialized and "complex" sessile forms, like the chonotrichs and peritrichs, were also segregated to a higher rank by Kahl, equivalent to Holotricha and Spirotricha (Table 1.1). Within these higher taxa, oral features, indicated by the suffix "-stomata", were major characters to indicate common descent (Table 1.1). It is interesting to note that the opalinid "flagellates" were considered "protociliates" during the Kahlian period based on the views of Metcalf (1923, 1940) among others (Table 1.1).

TABLE 1.1. Major systems of ciliate classification popular prior to 1950.[a]

Bütschlian Era[b] (1880–1930)[a]	Kahlian Era (1930–1950)
INFUSORIA	Subphylum Ciliophora
Ciliata	**CILIATA**
Holotricha	*Protociliata*
Gymnostomata	**Opalinata**
Trichostomata	*Euciliata*
Astomata	**Holotricha**
Spirotricha	Gymnostomata
Heterotricha	Prostomata
Oligotricha	Pleurostomata
Hypotricha	Hypostomata
Peritricha	Trichostomata
Suctoria	Apostomea
	Hymenostomata
	Thigmotricha
	Stomodea
	Rhynchodea
	Astomata
	Spirotricha
	Heterotricha
	Ctenostomata
	Oligotricha
	Tintinnoinea
	Entodiniomorpha
	Hypotricha
	Peritricha
	Mobilia
	Sessilia
	Chonotricha
	SUCTORIA

[a] Classes are indicated in bold capital letters; subclasses, in italics; orders, in bold; suborders and "tribes", further indented in Roman type.

[b] It should be noted that Bütschli (1887–1889) originally proposed a scheme that differed slightly from that shown (see Corliss, 1962a; Jankowski, 1967a). Later workers in the period re-arranged it so that it came to resemble the form presented here. In all cases, the number of major groups remained essentially the same.

1.2 The Age of the Infraciliature (1950–1970)

Five scientists – Chatton and Lwoff, Klein, von Gelei, and Fauré-Fremiet – stand out as the pioneers of this period, which Corliss (1974a) suggested extended from about 1950 to 1970. Yet, the roots of this age originated earlier in the 20th century in descriptions of the different technical approaches to using silver to stain the cortex and other structures of ciliates – the dry silver method of Klein (1929) and the wet silver method of Chatton and Lwoff (1930). The observations made by these pioneers culminated in seminal conceptual papers attributing a variety of causal relationships to various infraciliary structures (Chatton & Lwoff, 1935b; Klein, 1928, 1929; von Gelei, 1932, 1934b; von Gelei & Horváth, 1931). Chatton and Lwoff's (1935b) law of desmodexy stands out as one of the "rules" emerging from this period that has stood the test of time: true kinetodesmata and/or kinetodesmal fibrils, when present, lie to or extend anteriad and/or to the organism's right of the kinety with which they are associated (see **Chapter 2**). With this rule, one can not only identify a ciliate, but also one can deduce the polarity of the cell. The developmental autonomy and "genetic" continuity of the infraciliature was summarized at the beginning of this period by Lwoff (1950) in his book entitled "Problems of Morphogenesis in Ciliates".

Fauré-Fremiet and his students applied these conceptual views of the developmental importance of infraciliary patterns to resolving phylogenetic problems within the phylum. Fauré-Fremiet's (1950a) discussion of comparative morphogenesis of ciliates rested on the conceptual presumption that similarities in pattern of the ciliature during division morphogenesis revealed the common ancestry of lineages (see Corliss, 1968). These similarities in division morphogenesis were particularly important in establishing the phylogenetic affinities of polymorphic forms, such as peritrichs, suctorians, and chonotrichs. Using similarities in division morphogenesis and an imagined evolutionary transformation from hymenostome to thigmotrich to peritrich, Fauré-Fremiet (1950a) made the case for the "hymenostome" affinities of the peritrichs (Fig. 1.1). His student, Guilcher (1951), argued that suctorians and chonotrichs ought not to be

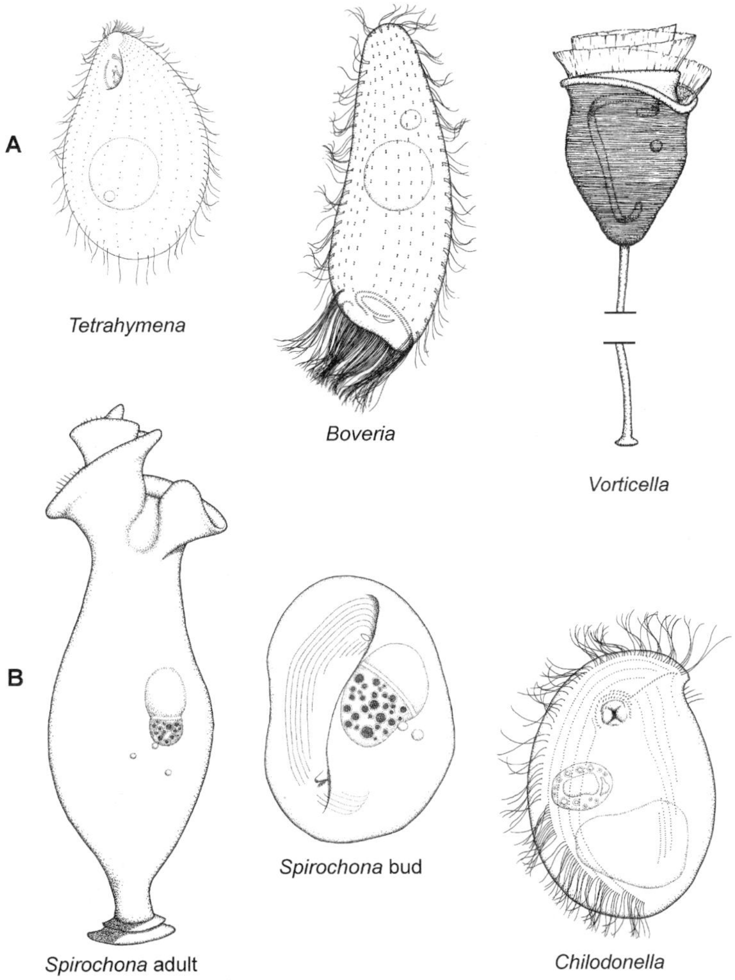

FIG. 1.1. **A** Schematic drawings of the hymenostome *Tetrahymena*, the thigmotrich *Boveria*, and the peritrich *Vorticella*. Fauré-Fremiet (1950a) related these three groups in a transformation series, imagining that evolution of the peritrich form proceeded through a thigmotrich-like intermediate from an ancestral *Tetrahymena*-like hymenostome. **B** Schematic drawings of the cyrtophorine *Chilodonella* and of the mature form and the bud of the chonotrich *Spirochona*. Guilcher (1951) argued that the similarities in pattern between the chonotrich bud and the free-living cyrtophorine suggested a much closer phylogenetic relationship between these two groups although the classification scheme of Kahl suggested otherwise (see Table 1.1)

greatly separated from other ciliate groups, and she claimed that chonotrichs might in fact be highly derived cyrtophorine gymnostomes (Fig. 1.1).

Furgason (1940) in his studies of *Tetrahymena* had imagined a more global evolutionary transformation of the oral apparatus of ciliates, premissed on the assumption that the three membranelles or oral polykinetids of *Tetrahymena* and the hymenostomes preceded the evolution of the many membranelles of the heterotrichs, like *Stentor* (Fig. 1.2).

This view was supported by Fauré-Fremiet (1950a) and Corliss (1956, 1961) who envisioned the hymenostomes as a pivotal group in the evolutionary diversification of the phylum. Corliss (1958a) used this concept of transformation of oral structures from simpler to more complex to argue that the hymenostomes, in their turn, had their ancestry in "gymnostome"-like forms, such as the nassophorean *Pseudomicrothorax*, which itself became another pivotal ancestral type. This led to the rearrangement of higher taxa and the proposal of a "Faurean" classification system by Corliss (1961) (Table 1.2).

This new view still maintained the Holotricha and Spirotricha, but the opalinids had now been removed based on the recognition that they shared many significant features with flagellate groups (Corliss, 1955, 1960a). Considering the work of the French ciliatologists, Corliss (1961) transferred the peritrichs, suctorians, and chonotrichs into the Holotricha, recognizing their probable ancestry from groups placed in this subclass. Oral structures continued to play a dominant role in characterizing orders as indicated by the common suffix "-stomatida" (Table 1.2).

Of course, the underlying assumption of the transformation of oral structures proposed by Fauré-Fremiet, Furgason, Corliss, and others was that the oral polykinetids or membranelles of these different ciliates – *Pseudomicrothorax*, *Tetrahymena*, and *Stentor* – were homologous. It was the invention of the electron microscope, which was just beginning to demonstrate its applicability during the latter part of this period, that was to provide the evidence to refute this assumption and therefore undercut the general application of this concept.

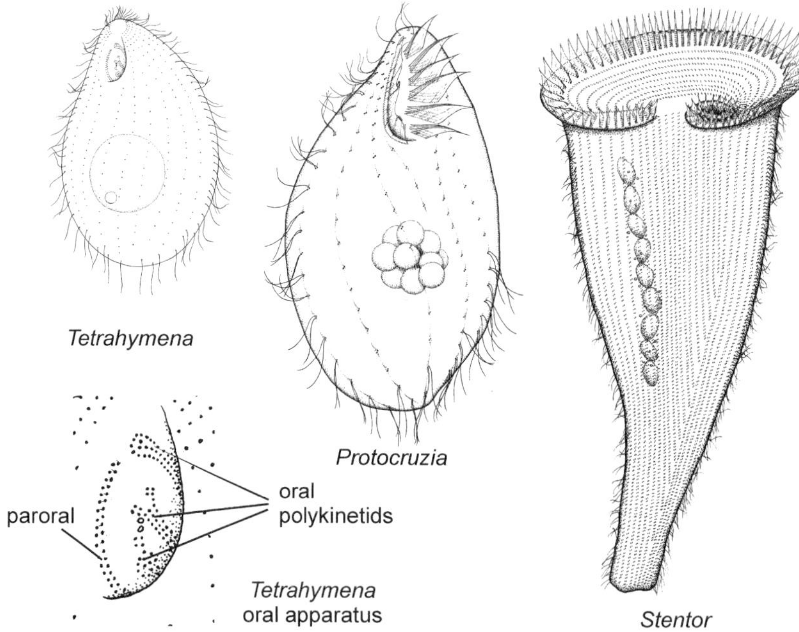

FIG. 1.2. Schematic drawings of three ciliates that have multiple oral polykinetids. The hymenostome *Tetrahymena* has three oral polykinetids and a paroral while the spirotrich *Protocruzia* and the heterotrich *Stentor* have many more than three. Furgason (1940) imagined that evolution proceeded by proliferation of oral polykinetids or membranelles and so the major groups of ciliates could be ordered by this conceptual view into more ancestral-like and more derived

TABLE 1.2. Faurean classification and post-Faurean system adopted by Corliss (1979).[a]

Faurean Era (1950–1970)	Post-Faurean Era (1970–1981)	
Subphylum Ciliophora	Phylum Ciliophora	
CILIATA	**KINETOFRAGMINOPHORA**	**OLIGOHYMENOPHORA**
Holotricha	*Gymnostomata*	*Hymenostomata*
Gymnostomatida	**Primociliatida**	**Hymenostomatida**
Rhabdophorina	**Karyorelictida**	Tetrahymenina
Cyrtophorina	**Prostomatida**	Ophryoglenina
Suctorida	Archistomatina	Peniculina
Chonotrichida	Prostomatina	**Scuticociliatida**
Trichostomatida	Prorodontina	Philasterina
Hymenostomatida	**Haptorida**	Pleuronematina
Tetrahymenina	**Pleurostomatida**	Thigmotrichina
Peniculina	*Vestibulifera*	**Astomatida**
Pleuronematina	**Trichostomatida**	*Peritricha*
Astomatida	Trichostomatina	**Peritrichida**
Apostomatida	Blepharocorythina	Sessilina
Thigmotrichida	**Entodiniomorphida**	Mobilina
Arhynchodina	**Colpodida**	**POLYHYMENOPHORA**
Rhynchodina	*Hypostomata*	*Spirotricha*
Peritrichida	**Synhymeniida**	**Heterotrichida**
Sessilina	**Nassulida**	Heterotrichina
Mobilina	Nassulina	Clevelandellina
Spirotricha	Microthoracina	Armophorina
Heterotrichida	**Cyrtophorida**	Coliphorina
Heterotrichina	Chlamydodontina	Plagiotomina
Licnophorina	Dysteriina	Licnophorina
Oligotrichida	Hypocomatina	**Odontostomatida**
Tintinnida	**Chonotrichida**	**Oligotrichida**
Entodiniomorphida	Exogemmina	Oligotrichina
Odontostomatida	Cryptogemmina	Tintinnina
Hypotrichida	**Rhynchodida**	**Hypotrichida**
Stichotrichina	**Apostomatida**	Stichotrichina
Sporadotrichina	Apostomatina	Sporadotrichina
	Astomatophorina	
	Pilisuctorina	
	Suctoria	
	Suctorida	
	Exogenina	
	Endogenina	
	Evaginogenina	

[a]Classes are indicated in bold capital letters; subclasses, in italics; orders, in bold with the ending "–ida"; suborders, further indented with the ending "–ina".

1.3 The Age of Ultrastructure (1970–1990)

As with other ages, the technological roots of the Age of Ultrastructure began in the 1950s and 1960s. The silver proteinate staining technique of Bodian or protargol staining became established as the light microscopic stain of choice during this period, although it had its technological innovators in the previous age (Kozloff, 1946; Kirby, 1950; Tuffrau, 1967). However, it was electron microscopy, promoted by Pitelka (1969), that gained preference in resolving questions in both the systematics and cell biology of ciliates. These early results, coupled with two seminal papers by

Jankowski (1967a, 1973c), prompted the French group of de Puytorac, Batisse, Bohatier, Corliss, Deroux, Didier, et al. (1974b) and, both with his French colleagues and independently, Corliss (1974a, 1974b) to propose revised classifications. Corliss (1979) used a slightly modified version in his third edition to "The Ciliated Protozoa" (Table 1.2). About this time, Jankowski (1980) proposed a new system, which still placed major emphasis on oral features as indicated by the names of some of his classes – Apicostomata, Pleurostomata, Rimostomata, Synciliostomata, Cyrtostomata, and Hymenostomata.

The major feature of these post-Faurean schemes was the prominent elevation of oral features. The three classes in the phylum were now characterized by the nature of the oral apparatus: small, simple kinetal fragments characterized the Class Kinetofragminophora; typically three oral polykinetids or membranelles characterized the Class Oligohymenophora; and many more than three membranelles characterized the Class Polyhymenophora (Table 1.2). All three names derived from the conceptual vision of Jankowski (1967a, 1973c, 1975), which shared the same assumption as Furgason's: homology was assumed among "oligo"-membranelles and "poly"-membranelles.

Before we return to a refutation of this assumption, it is important to set the conceptual stage, which was being constructed during the early 1960s. A seminal paper of this period was by Ehret (1960) and entitled "Organelle systems and biological organization". Influenced by systems theory, cell biology, and the emerging field of molecular biology, Ehret imagined cells to be constructed of a series of levels of organization – from molecules to macromolecular aggregates to organelles to envelope systems (= cells). He concluded –

> Within this reference frame of understanding, the cell ceases to occupy a central location as a fundamental unit of life. It appears, instead, as a special case among the single- and multiple-envelope systems that comprise all forms of life. (p. 122)

This perspective had a liberating effect for it demanded that we not constrain our view to the importance of cellular characters, but look "below" the cell at features that might be just as significant to an understanding of the common descent of protists. Ehret and McArdle (1974) then imagined the *Paramecium* cell to be constructed of levels, the simpler ones integrating to build more complex levels. In the context of the ciliate cortex, these levels can be imagined as macromolecule (i.e. tubulin), suborganelle or macromolecular aggregate (i.e., microtubule), unit organelle (i.e., kinetosome, cilium, microtubular ribbon), organellar complex (i.e., kinetid), and organellar system (i.e., locomotory system or kinetome) (Lynn, 1981; also see **Chapter 2** for definitions).

A number of scientists had imagined cells and organisms to be built in a series of increasingly complex levels of organization and had concluded that this important property constrained morphological variation, especially at the lower levels of biological organization. In other words, if one constructs something of bricks of a certain shape that are assembled in a precise sequence, changing the ultimate arrangement has less drastic consequences than changing the shape of each brick. Bronowski (1970) had termed this the principle of stratified stability: "the building up of stable configurations does have a direction, the more complex built on the next lower, which cannot be reversed in general" (pp. 242–243). Independently, Lynn (1976a, 1981) called it the principle of structural conservatism: the conservation of structure through time is inversely related to the level of biological organization. Thus, if the ciliate cortex and infraciliature were conceived as being constructed of repeating and highly integrated units, then there should be strong selection on preserving this unit structure (i.e., the kinetid) to construct the cortical system (Fig. 1.3). Lynn and Small (1981) then argued that this principle gave us an approach to examining the comparative ultrastructure of the ciliate cortex and to infer common descent: structurally similar kinetids should be homologues, limited to vary by the "selective forces" of stratified stability or structural conservatism.

In the 21st century, this may all seem self-evident. However, there was one major conceptual problem with it at the time – the idea of 'organic design'. Pantin (1951, 1966) and Grimstone (1959) had argued that microtubules, basal bodies or kinetosomes, and the cilium were of such low complexity that they could conceivably have evolved many times, unlike "the more complex and improbable metazoan organs which, determined by a far more numerous set of genes, appear to have arisen only once" (p. 277, Grimstone, 1959), and "it seems

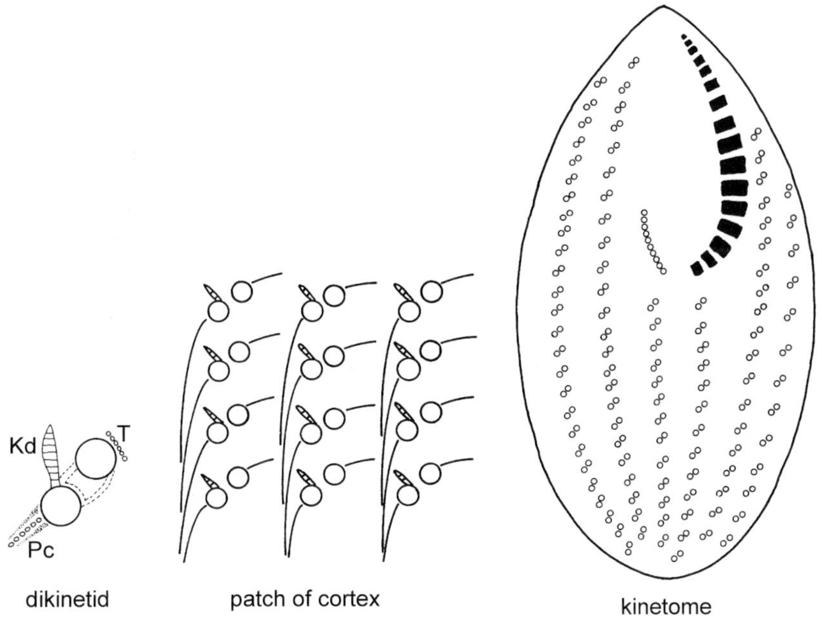

FIG. 1.3. The hierarchical organization of the ciliate cortex. The fundamental component of the cortex is the dikinetid, an organellar complex here composed of seven unit organelles, which are the two kinetosomes, two cilia (not shown), transverse (T) and postciliary (Pc) microtubular ribbons, and the kinetodesmal fibril (Kd). In a patch of cortex, the microtubular ribbons and kinetodesmal fibrils of adjacent kinetids are closely interrelated. The interrelated kinetids comprise the components of the next higher level in the hierarchy, the organellar system called the kinetome. Two major cortical organellar systems are the somatic region or kinetome and the oral region, functioning in locomotion and feeding, respectively. (from Lynn & Small, 1981.)

highly improbable that the unique assemblage of genetic factors which ensures the development of a pentadactyl limb would ever be selected independently on two separate occasions" (p. 144, Pantin, 1951). Thus, from this view, similarities in kinetids would have arisen by a non-adaptive process, rather than as a result of natural selection. Instead, these structures were determined by thermodynamics and "by physical and spatial properties of matter rather than by functional needs … of a transcendental rather than adaptive origin" (p. 4, Pantin, 1966). Yet, a little over a decade later, the flagellum of *Chlamydomonas* was reported to have at least 170 polypeptides (Huang, Piperno, & Luck, 1979) and the cilium of *Paramecium* to have at least 125 polypeptides (Adoutte et al., 1980), and this picture has become even more complex in the intervening decades. Thus, these organelles are clearly not simple, but indeed are extremely highly ordered complexes. It is therefore reasonable to conclude that their structural complexity is as much a result of natural selection as the organs of metazoa or the pentadactyl limb.

With this conceptual perspective, Small and Lynn (1981) applied structural conservatism to make sense of the diversity of ciliate kinetids. They also relied on the notion that somatic structures are more highly conserved than oral ones (Gerassimova & Seravin, 1976; Lynn, 1976a, 1976c). One reason lies in the development of somatic and oral regions. The duplication of somatic kinetids in ciliates usually occurs closely adjacent to pre-existing kinetids, called cytotaxis or structural guidance (Frankel, 1991), and this may place severe constraints on the variability of the components. On the other hand, the organellar complexes of the oral region are not as intimately linked to pre-existing organelles and also, as more complex structures, there is a higher potential for change, at least in size and shape. Another reason that oral structures are more variable is that even slight structural alterations, if they resulted in increased capture and ingestion rates, would directly affect growth and reproductive rates, enhancing relative fitness and fixation of new variants. Thus, Lynn (1979b) concluded "somatic over oral", meaning that somatic structures have in general a "deeper" phylogenetic signal than oral ones.

The consistent application of these principles (i.e., structural conservatism and somatic over oral) resulted in the proposal of eight major classes by Small and Lynn (1981) (Table 1.3). During the Age of Ultrastructure, the classification was refined by Small and Lynn (1985) and Lynn and Small (1990), the latter revision beginning to consider the early results of molecular genetic research. Overall, somatic kinetids were used to identify monophyletic clades, called classes, and this approach often placed genera that had been assigned to different, older higher taxa together. The colpodeans provide a most dramatic example: *Sorogena* was a gymnostome; *Colpoda* was a vestibuliferan; *Cyrtolophosis* was a hymenostome; and *Bursaria* was a heterotrich (Fig. 1.4)!

Small and Lynn (1981, 1985) divided the phylum into three subphyla, based on ultrastructural features of the cortex: for the somatic cortex – the overlapping postciliary microtubular ribbons – for the Subphylum Postciliodesmatophora (Gerassimova & Seravin, 1976; Seravin & Gerassimova, 1978); and for the oral cortex – the presence of transverse microtubular ribbons supporting the cytopharynx in the Subphylum Rhabdophora and the presence of postciliary microtubular ribbons supporting the cytopharynx in the Subphylum Cyrtophora (Small, 1976) (Table 1.3). However, Huttenlauch and Bardele (1987) demonstrated in an ultrastructural study of oral development that the supposed oral transverse ribbons of the prostomate rhabdophoran *Coleps* were in fact postciliary microtubules that

TABLE 1.3. Classifications systems proposed by Small and Lynn (1981, 1985).[a]

Small & Lynn (1981)	Small & Lynn (1985)
Phylum Ciliophora	Phylum Ciliophora
Postciliodesmatophora	Postciliodesmatophora
KARYORELICTEA	**KARYORELICTEA**
SPIROTRICHEA	**SPIROTRICHEA**
Rhabdophora	*Heterotrichia*
PROSTOMEA	*Stichotrichia*
LITOSTOMEA	*Choreotrichia*
Haptoria	Rhabdophora
Vestibuliferia	**PROSTOMATEA**
Cyrtophora	**LITOSTOMATEA**
PHYLLOPHARYNGEA	*Haptoria*
Phyllopharyngia	*Trichostomatia*
Chonotrichia	Cyrtophora
Suctoria	**PHYLLOPHARYNGEA**
NASSOPHOREA	*Phyllopharyngia*
Hypostomia	*Chonotrichia*
Polyhymenophoria	*Suctoria*
COLPODEA	**NASSOPHOREA**
OLIGOHYMENOPHOREA	*Nassophoria*
Hymenostomia	*Hypotrichia*
Peritrichia	**COLPODEA**
Astomia	**OLIGOHYMENOPHOREA**
Apostomia	*Hymenostomatia*
	Peritrichia
	Astomatia
	Apostomatia
	Plagiopylia

[a]Classes are indicated in bold capital letters; subclasses, in italics.

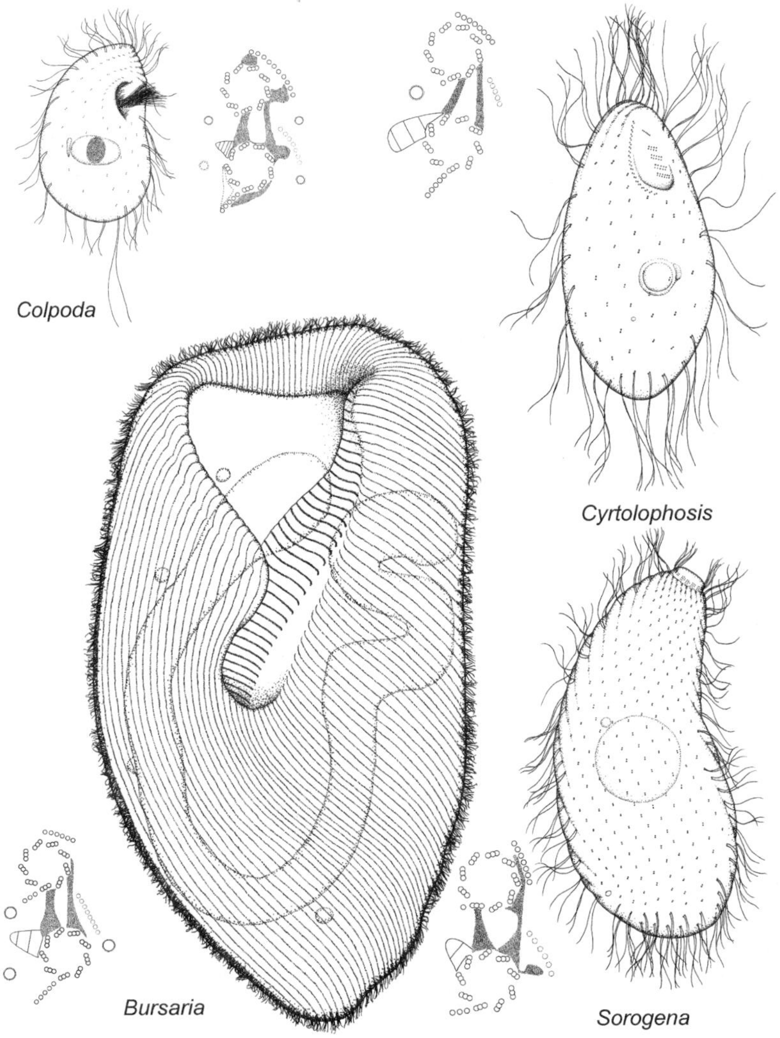

FIG. 1.4. Colpodeans and their somatic kinetids as a demonstration of the more conservative nature of the somatic kinetid and its "deeper" phylogenetic signal over the oral structures and general morphology of a group of ciliates. *Sorogena* was a gymnostome; *Colpoda* was a vestibuliferan; *Cyrtolophosis* was a hymenostome; and *Bursaria* was a spirotrich

became twisted during division morphogenesis, making them appear to be transverse microtubules. So, this rhabdophoran was really a cyrtophoran! This undercut our confidence that these characters had deep phylogenetic significance, and led Lynn and Corliss (1991) to abandon the subphyla, retaining only the eight classes of Small and Lynn. Later, de Puytorac et al. (1993) suggested three different subphyla, also based on significant cortical ultrastructural features proposed by Fleury, Delgado, Iftode, and Adoutte (1992): the Subphylum Tubulicorticata – a microtubular cortex; the Subphylum Filicorticata – a microfibrillar cortex; and the Subphylum Epiplasmata – an epiplasmic cortex (Table 1.4). Fleury et al. (1992) had used molecular phylogenies derived from large subunit rRNA gene sequences to support these morphology-based subdivisions. Nevertheless, Lynn and Small (1997) argued that given the variability of cortical ultrastructures in ciliates it was extremely difficult to circumscribe the limits of these subphyla. For example, virtually

TABLE 1.4. A comparison of the macrosystems of the Phylum Ciliophora of de Puytorac (1994a) and the system proposed herein. Authorships for names will be found in Chapter 17.[a]

de Puytorac (1994a)	Proposed system
Phylum Ciliophora	***Phylum Ciliophora***
Tubulicorticata	Postciliodesmatophora
POSTCILIODESMATOPHORA	**KARYORELICTEA**
KARYORELICTEA	**HETEROTRICHEA**
Trachelocercia	Intramacronucleata
Loxodia	**SPIROTRICHEA**
Protocruziidia	*Protocruziidia*
Protoheterotrichia	*Phacodiniidia*
HETEROTRICHEA	*Hypotrichia*
Heterotrichia	*Oligotrichia*
Clevelandellidia	*Choreotrichia*
SPIROTRICHA	*Stichotrichia*
HYPOTRICHEA	*Licnophoria*
Euplotia	**ARMOPHOREA**
Oxytrichia	**LITOSTOMATEA**
OLIGOTRICHEA	*Haptoria*
Oligotrichia	*Trichostomatia*
Strobilia	**PHYLLOPHARYNGEA**
TRANSVERSALA	*Cyrtophoria*
COLPODEA	*Rhynchodia*
Colpodia	*Chonotrichia*
Bryometopia	*Suctoria*
PLAGIOPYLEA	**NASSOPHOREA**
Filicorticata	**COLPODEA**
LITOSTOMATEA	**PROSTOMATEA**
VESTIBULIFEREA	**PLAGIOPYLEA**
Epiplasmata	**OLIGOHYMENOPHOREA**
CILIOSTOMATOPHORA	*Peniculia*
PHYLLOPHARYNGEA	*Scuticociliatia*
Cyrtophoria	*Hymenostomatia*
Chonotrichia	*Apostomatia*
Rhynchodia	*Peritrichia*
Suctoria	*Astomatia*
MEMBRANELLOPHORA	
NASSOPHOREA	
Prostomatia	
Nassulia	
OLIGOHYMENOPHOREA	
Peniculia	
Scuticociliatia	
Peritrichia	
Hysterocinetia	
Astomatia	
Hymenostomatia	
Apostomatia	

[a]Superclasses are indicated in capital letters; classes, in bold capital letters; subclasses, in italics.

all ciliates could be described as having a "cortical cytoskeleton of superficial microtubules associ ated, or not, with cortical kinetosomes" – the major feature distinguishing ciliates in the Subphylum Tubulicorticata (de Puytorac et al., 1993).

This reduced emphasis on oral structures as being of great phylogenetic significance extended to the homology of "membranelles" or oral polykinetids, used to establish the Classes Kinetofragminophora, Oligohymenophora, and Polyhymenophora. De Puytorac and Grain (1976) and Grain (1984) had demonstrated the variety of "membranelles" or oral polykinetids in their reviews of the diversity of cortical ultrastructures of ciliates. This variety lead to a proliferation of names to capture some of these differences. Oral polykinetids in kinetofragminophorans could be pseudomembranelles, in oligohymenophorans could be membranoids or membranelles, and in polyhymenophorans could be paramembranelles or heteromembranelles (see definitions in **Chapter 2**). This diversity suggested that these different complex oral structures were probably not homologues. In fact, what they undoubtedly illustrate are diverse solutions to the "problem" of filter feeding that had arisen through convergent evolution in a much larger number than three independent lineages or classes. Small and Lynn (1981, 1985) recognized these lineages as eight classes, established primarily on the basis of the ultrastructure of the somatic cortex, applying the principles of "structural conservatism" and "somatic over oral" (Fig. 1.4).

1.4 The Age of Refinement (1990–Present)

Greenwood et al. (1991a) suggested that 1990 might be designated as the beginning of the next age in ciliate systematics, the Age of Refinement, for it is in this period that tremendous advances have been made in confirming our basic notions derived from research on ciliate ultrastructure. As with the other ages, the technological roots of this age precede its formal beginning, and are based in the molecular phylogenetic work of Sogin's lab on small subunit rRNA gene sequences (Elwood, Olsen, & Sogin, 1985; Lynn & Sogin, 1988) and Adoutte's lab on large subunit rRNA gene sequences (Adoutte, Baroin, & Perasso, 1989; Baroin, Perasso, Qu, Brugerolle, Bachellerie, & Adoutte, 1988). Thus, it might also be called the Age of Genetic Diversity, since the sequences of these highly conserved genes (see **Chapter 16**), enabled us to test the structural conservatism of the ciliate somatic cortex, using the "molecular skeletons" of the ribosomal subunits – the small and large subunit rRNAs.

These early papers demonstrated tremendous genetic diversity within the phylum, a level of genetic diversity similar to differences among the "kingdoms" of multicellular organisms, like the plants, animals, and fungi. Further, the major clades established on the basis of ultrastructural research were generally confirmed, indicating that the somatic kinetid was a generally reliable feature to establish common descent (Lynn, 1991, 1996a; Lynn & Small, 1997). However, the molecular data suggested the need for further separation of clades, both at the "class" level and higher (Lynn, 1996b; Lynn & Small). De Puytorac (1994a) had presaged this by elevating to class rank two groups that molecular genetic data confirmed to be distinct – the Class PLAGIOPYLEA and the Class HETEROTRICHEA, removing heterotrichs from the spirotrich assemblage (cf. Table 1.3, 1.4). However, de Puytorac (1994a) elevated several groups to class rank (e.g., HYPOTRICHEA, OLIGOTRICHEA, VESTIBULIFEREA) for which there is as yet no strong molecular genetic evidence (see **Chapter 16**). Two new clades differentiated by small subunit rRNA gene sequences and now recognized as classes are the Class ARMOPHOREA (see Affa'a, Hickey, Strüder-Kypke, & Lynn, 2004; van Hoek, Akhmanova, Huynen, & Hackstein, 2000a) and the Class PLAGIOPYLEA (see Embley & Finlay, 1994; Lynn & Strüder-Kypke, 2002) (Fig. 1.5, Table 1.4). Lynn (2004) highlighted a difficulty with each of these so-called "riboclasses": the Class ARMOPHOREA associated genera, such as *Metopus* and *Nyctotherus*, whose somatic kinetids were dissimilar, while the Class PLAGIOPYLEA separated some genera, such as *Plagiopyla* and *Trimyema*, whose kinetids were quite similar to those of the Class OLIGOHYMENOPHOREA to which the plagiopyleans had been transferred as a subclass by Small and Lynn (1985) (Table 1.3). Thus, somatic kinetid structure seems *not* to be highly conserved in armophoreans and to be more highly conserved in some plagiopyleans! We have apparently reached the limits of structural conservatism of the somatic cortex as a principle, and we can only say that these are the exceptions that prove the rule!

By the mid-1990s there was ample evidence from a variety of independent phylogenetic analyses of both small subunit and large subunit rRNA gene sequences to demonstrate a fundamental bifurcation in the phylum (Baroin-Tourancheau, Tsao, Klobutcher, Pearlman, & Adoutte, 1995; Hammerschmidt, Schlegel, Lynn, Leipe, Sogin, & Raikov, 1996; Hirt et al., 1995) (Fig. 1.5). One branch, which separates the ciliates with postciliodesmata *sensu stricto*, corresponds to

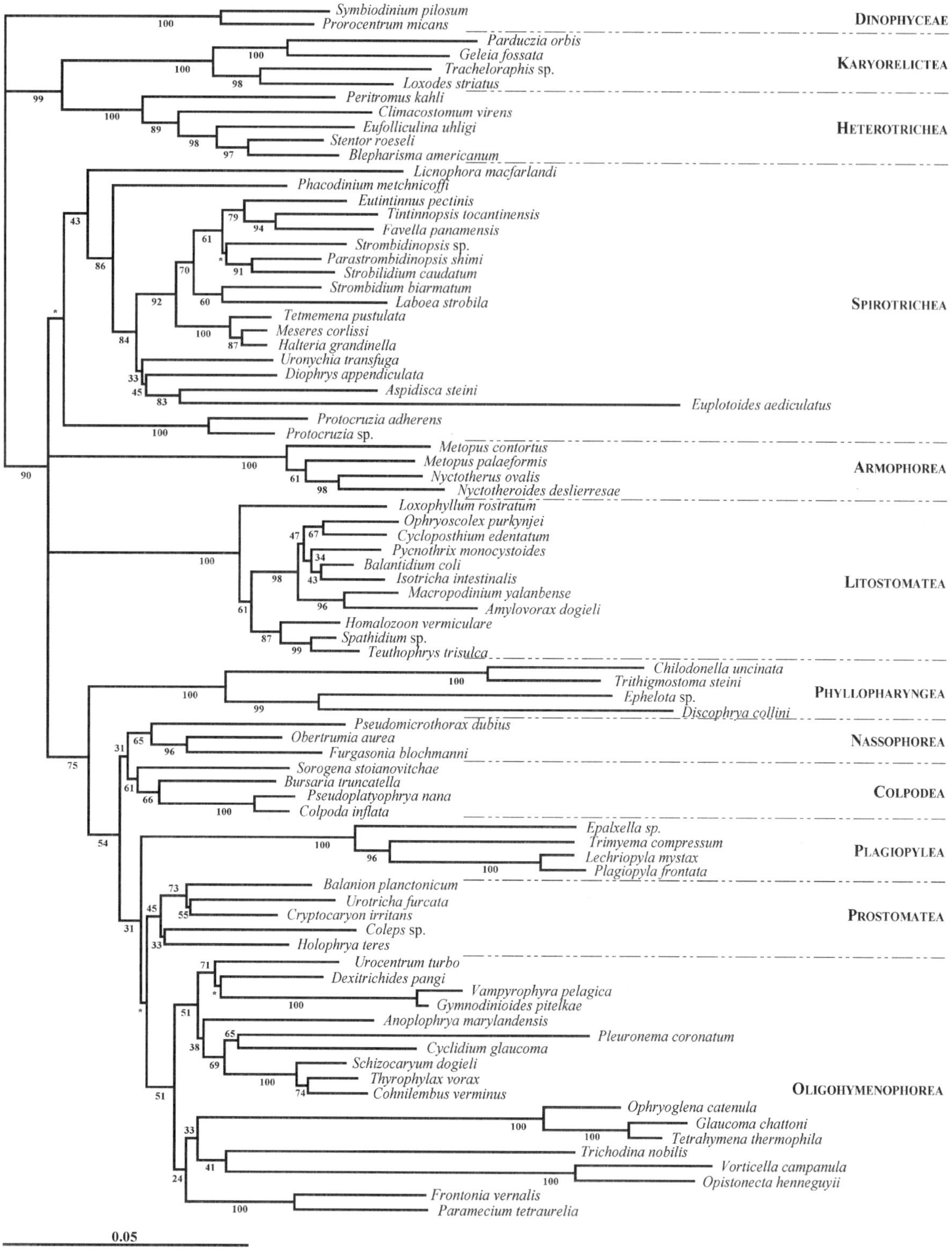

FIG. 1.5. A molecular phylogeny of the Phylum Ciliophora based on small subunit rRNA gene sequences. Several representatives of each class have been chosen to demonstrate the genetic diversity within the phylum and the distinctness of the different clades that are considered to be of class rank in the classification proposed herein (see Table 1.4) (see **Chapter 16** for further discussion of molecular phylogenetics)

the Subphylum POSTCILIODESMATOPHORA, a concept proposed by Gerassimova and Seravin (1976). This subphylum now includes only the Classes KARYORELICTEA and HETEROTRICHEA; it excludes the spirotrich clade, which was included by Small and Lynn (1985) (cf. Tables 1.3, 1.4). While karyorelicteans do not have dividing macronuclei, the heterotrichs do, apparently relying primarily on extramacronuclear microtubules for this process (Diener, Burchill, & Burton, 1983; Jenkins, 1973). Lynn (1996a) named the other branch, the Subphylum INTRAMACRONUCLEATA, because all ciliates in this clade have a dividing macronucleus that relies predominantly on intramacronuclear microtubules for completion of division. The suggestion that macronuclear division has arisen separately twice during the evolution of ciliates is not unreasonable, considering that at least two kinds of nuclear division, using both extranuclear and intranuclear microtubules also occur in the dinoflagellates (Perret, Albert, Bordes, & Soyer-Gobillard, 1991), the sister clade to the ciliates (Leander & Keeling, 2003; Van de Peer, Van der Auwera, & De Wachter, 1996).

1.5 Major Differences in the New Scheme

Corliss (1979) noted in his discussion of the major differences of schemes that an obvious trend has been the inflation of taxa as our discovery and understanding of diversity have changed from the 1880s until the present. As discussed above, approaches have been influenced both by technological advances – light microscopy, cytological staining, electron microscopy, molecular biology – and by new conceptual views. With respect to the latter, the emphasis on the somatic cortex by Small and Lynn (1981) caused a major revision in our understanding of relationships between the mid-1970s and the mid-1980s. Currently, there are two recent classification systems of ciliates seeking adherents; one proposed by de Puytorac (1994a) and his colleagues in the second volume of the Traité de Zoologie and the other proposed most recently by Lynn (2004) and presented in a slightly revised version herein (Table 1.4). Since various differences between these views have been discussed above, this section will serve to summarize these.

1. The subphyletic divisions in the two systems are different: three by de Puytorac (1994a) and two here (Table 1.4). Data on genetic diversity support a major division into two subphyla, the Subphylum POSTCILIODESMATOPHORA and the Subphylum INTRAMACRONUCLEATA.
2. De Puytorac (1994a) recognizes five superclasses, one essentially equivalent to our Subphylum POSTCILIODESMATOPHORA, while we provide no such subdivisions (Table 1.4). It is the case in molecular phylogenies that there is substructure within the Subphylum INTRAMACRONUCLEATA. For example, six classes (i.e., PHYLLOPHARYNGEA, NASSOPHOREA, COLPODEA, PLAGIOPYLEA, PROSTOMATEA, and OLIGOHYMENOPHOREA) are often consistently supported as a clade (Fig. 1.5). This grouping may represent a natural assemblage, and therefore represent a superclass assemblage. However, there is no obvious shared derived morphological feature uniting these taxa, and at this time we do not recognize it as a taxonomic category.
3. De Puytorac (1994a) recognizes 11 classes as does the system proposed here (Table 1.4). However, the classes are different. De Puytorac (1994a) includes the prostomates in the Class NASSOPHOREA. Differences in the somatic kinetid (Eisler, 1989; Lynn, 1991), stomatogenesis (Eisler; Huttenlauch & Bardele, 1987), and small subunit rRNA gene sequences (Stechmann, Schlegel, & Lynn, 1998) between nassophoreans and prostomateans argue against uniting them in the same class (Fig. 1.5). While both systems recognize the spirotrichs as a larger assemblage, the elevation of the oligotrichs to class rank, equivalent to hypotrichs, is not justified by the molecular data, which suggest at least seven separate lineages in the Class SPIROTRICHEA, here recognized as subclasses (Strüder-Kypke & Lynn, 2003).

Finally, we cannot agree with de Puytorac (1994a) that elevation of the vestibuliferians to class rank is warranted. We prefer to refer to this clade by the Bütschlian moniker, Trichostomatia (Table 1.4). The trichostomatians in this sense and the haptorians share virtually identical somatic kinetid patterns (Lynn, 1981, 1991). This varies only in the entodiniomorphids where Lynn (1991) has interpreted the appearance of a transient microtubule during kinetid replication (see Furness & Butler, 1986) to be the homologue of the T2 transverse microtubular ribbon of litostomes. Moreover, extensive analyses of litostome small subunit rRNA gene sequences consistently group the haptorians and trichostomes (Wright & Lynn, 1997b; Strüder-Kypke, Wright, Foissner, Chatzinotas, & Lynn, 2006).

De Puytorac (1994a) elevated a considerable number of taxa to subclass and ordinal ranks, totalling 25 subclasses and 70 orders. Comparison with the scheme presented here will demonstrate considerable agreement in the basic groups or clades, despite possible differences in rank (Table 1.4 and the original references). While Small and Lynn (1981, 1985) established 15 subclasses and 48 orders, our revised scheme has 19 subclasses and 59 orders. Many of these changes have been influenced by genetic data obtained in the last few years, and these are discussed both by Lynn (1996b, 2004) and in the chapters devoted to each class (see **Chapters 5–15**).

1.6 Guide to Remaining Chapters

This book takes its basic form from the 3rd edition of "The Ciliated Protozoa" by Corliss (1979). Following this Introduction, we have revised Chapter 2, but used Corliss as the solid grounding for the glossary of terms. Whenever appropriate, cross-reference has been made to terms, and the plural of non-English words has been included. Figures are explicitly referred to by number so that it should be easy to find illustrative support for many of the definitions.

Chapter 3 provides a discussion of the approach to constructing our macrosystem. The important characters used to establish different ranks in the hierarchy are described and justification is provided for their use. Some of this is a repetition of the material in this chapter, but in a different context.

Chapters 4 through 15 are structured along the lines of the Traité de Zoologie edited by de Puytorac (1994a). The phylum (Chapter 4) and each class (Chapters 5–15) are treated under the following topics: overview of the group; taxonomic structure of the group and its diversity; life history and ecology, including symbioses; somatic structures, cortical and cytoplasmic; oral structures; division and morphogenesis; sexuality and life cycle, including nuclear features; and other, a final section that may include aspects of the applied relevance of a group.

Chapter 16, a preamble to Chapter 17, deals particularly with important research papers on the genetic diversity of the phylum, especially as these results impact on refining the relationships of taxa. There is also some discussion of character evolution within the phylum, particularly as it relates to the classes and subphyla, and as revealed by the topologies of gene trees.

Chapter 17 is *the* taxonomic chapter, again relying heavily on Corliss (1979) for the basic characterization of groups from the family level and higher. As in Corliss, genera are assigned to families, but there is rarely any discussion of these assignments. Valid genera primarily follow the recommendations of Aescht (2001), whose important work should be referred to for the detailed nomenclatural background to problematic names. While not considered valid by the International Code of Zoological Nomenclature, nomina nuda have been included and are clearly indicated as such.

The References section includes an extensive literature cited section. In this, we have been conscious of including reference to the classic literature as both Corliss (1961) and Corliss (1979) are now out of print. However, we have also included, as appropriate, citation to important works bearing on the topics of Chapters 4 through Chapters 16. We do regret that we have often been unable to include all relevant literature on a topic, and trust that expert readers will understand and agree with our selection of references.

Chapter 2
Glossary of Terms and Concepts Useful in Ciliate Systematics

With the increasing contribution of data from other fields to the systematics of ciliates and the growing interest of biologists of all persuasions in these microorganisms, we ought today to be familiar with a far greater range of terms than was required in the past. The information offered below not only provides, in the aggregate, a "thumbnail sketch" of most aspects in the overall biology of ciliates, but also permits use of the terminology in succeeding chapters without the need to reexplain it there. Our treatment is not exhaustive – and many commonly known anatomical, cytological, and ecological words are purposely omitted – but this compilation is longer than those presented for ciliates by Corliss (1959, 1961, 1979). It is principally based on those of Corliss (1979) and Lee (2002).

An attempt has been made to keep the definitions or explanations succinct. However, brief descriptive comments have been added when terms are particularly complicated or important. Almost always we have included information identifying the taxonomic group or groups of ciliates that possess or show the character or trait being described. Cross-referencing is frequently employed, either directly or indirectly by italicizing selected words in the definition. We have tried to point out controversies, present alternative meanings, and give synonyms. Furthermore, we have indicated our own preferences. For many words derived from Greek or Latin, we have provided the suggested plural. If a plural is not provided, it is generally acceptable to add an "s" or "es" to the singular, as appropriate. A number of morphological features are illustrated in the figures at the end of this chapter, and are indicated by a direct reference to the labelled structure in the figure and its part (e.g., Kd, Fig. 2.1A).

Glossary

A

Aboral: away from the *oral region*; in ciliatology almost always used in the most extreme sense, meaning at the opposite (usually antapical) pole from the (other) end (usually apical) of the body bearing the mouth; but – as in the case of peritrichs – the aboral pole is not necessarily the posterior pole of the organism, functionally and/or morphologically and/or evolutionarily (Ab, Fig. 2.8Aj).

Abyssal: pertaining to the great depths of the ocean well beyond the continental shelf.

Accessory Comb: a conspicuous ridge lying between the oral polykinetids, especially in tintinnids.

Acellularity Concept: once popular hypothesis that protozoa, being individually complete (yet organized without tissues) cannot be cellular and thus must be considered noncellular or acellular organisms; it is now widely recognized that setting such eukaryotic cells as ciliates – notwithstanding their complex subcellular organization and extracellular specializations – apart from those of the "higher" metazoan groups is a decision basically neither defensible on (ultra)structural grounds nor judicious from a comparative evolutionary point of view; "cell" and "organism" need not be thought of as mutually exclusive terms.

Acetabulum (pl. **Acetabula**): term rarely used in ciliatology; see **Sucker**.

Acidosome: *vesicle* filled with acid that fuses with the *food vacuole* to promote digestion; particularly used in *Paramecium*.

Aciliferous: see **Nonciliferous**.

Acmocyst: *extrusome* of rhynchodid phyllopharyngeans.

Acontobolocyst: see **Trichocyst**.

Actinophore: structure bearing several, a bundle or a *fascicle* of *suctorial tentacles*; characteristic of certain suctorians.

Adhesive Disc: thigmotactic cup-shaped organelle at the *aboral* pole of mobiline peritrichs and some other ciliate (e.g., some spirotrichs) used for attachment to the substratum, usually the surface of another organism that serves as host (Fig. 2.9Bg, 2.9Bh).

Adhesive Organelle: often restricted to a secretory structure responsible for or structurally involved in production of a substance, other than cilia or some ciliary derivative, allowing the organism possessing it to adhere or attach to the substratum (e.g., the *podite* of dysteriid cyrtophorines); but also see **Holdfast Organelle** (AO, Fig. 2.9Bf).

Adoral Ciliary Fringe: see **Frange**.

Adoral Ciliary Spiral: often applies to the spiralling pattern of the oral *haplokinety* and *polykinety* of peritrichs; see **Adoral Zone of Oral Polykinetids** (Fig. 2.6b; ACS, Fig. 2.11B).

Adoral Zone of Membranelles: see **Adoral Zone of Oral Polykinetids**.

Adoral Zone of Oral Polykinetids: orderly arrangement of three or more oral polykinetids serially arranged typically along the left side of the *oral region*; this arrangement has evolved by convergence in different groups of ciliates (e.g. heterotrichs, spirotrichs, colpodeans) (Opk, Fig. 2.5Aa).

Adult Form: generalized term; depending on the situation, the mature form, the *parental form*, the interfissional form, the sessile or sedentary form, the *trophont*.

Afferent Canal: one of usually several cytoplasmic channels transporting excreted fluid from the *spongioplasm* to a *contractile vacuole*; also called a pulsating, nephridial, collecting, or radial canal; see **Secretory Ampulla**.

Aire Sécante: see **Secant System**.

Akontobolocyst: a synonym of spindle trichocyst; see **Trichocyst**.

Algivorous: feeding on algae; see also **Macrophagous** and **Microphagous**.

Alpha Membranoid: see **Membranoid**.

Alveolate(s): protists whose *pellicle* is characterized by pellicular *alveoli*; includes at least apicomplexans, ciliates, dinoflagellates, and colpodellids.

Alveolocyst: an extension of the *pellicular alveolus* into the underlying cortex; particularly characteristic of nassulids.

Alveolus (pl. **Alveoli**), **Pellicular**: flattened vesicle or sac, bounded by a *unit membrane*, lying just beneath the surface or *plasma membrane* of the cell (organism) and above the *epiplasm*; may occur in pairs in a given *kinetosomal territory*; also known as a cortical vesicle (PA, Fig. 2.1B).

Amacronucleate: without a *macronucleus*; rare, unnatural, unstable condition, realizable only experimentally (e.g., in *Paramecium*).

Ambihymenium (pl. **Ambihymenia**): condition of having oral ciliary "membranes" completely surrounding the mouth-area; claimed by some workers to be the situation obtaining in a number of the cyrtophorine phyllopharyngeans.

Amicronucleate: without a *micronucleus*, whether the loss has been brought about naturally or experimentally (e.g., in *Tetrahymena*).

Amitosis: nuclear division that results in the irregular distribution of chromosomes; a pathological kind of mitosis; however, unequal distribution of "chromosomal" elements in the polyploid *macronucleus* of ciliates might be considered a kind of amitosis.

Amphiparakinetal: *parakinetal stomatogenesis* in which the curved oral *anlage* intersects many *postoral somatic kineties* at two sites, enclosing few to many, short, non-proliferating *kinetofragments*; found in some heterotrichs.

Ampliploid: the nuclear condition in which there are numerous replicates of fragments of chromosomes, sometimes as small gene-sized pieces, produced by an amplification process and not by mitosis; characteristic of the ciliate *macronucleus*; see **Polyploid**.

Ampule: small, ovoid *extrusome* in the dorsal cortex of hypotrich spirotrichs (e.g. *Euplotes*), and associated in clusters with *kinetids* of the dorsal *bristle kinety*; may be involved in sexual interactions.

Ampulla (pl. **Ampullae**), **Secretory**: glandular organelle, generally multiple in number, which produces a thigmotactic substance or structure in some cyrtophorines (see **Adhesive Organelle**); also used for the *collecting canal* (or its enlarged distal end) that connects to the contractile vacuole of a number of ciliates (e.g., certain peniculine hymenostomes); see **Afferent Canal** (AS, Fig. 2.9Bf).

Ampullocyst: kind of *mucocyst* found in certain karyorelicteans; see **Mucocyst**.

Analogous: characters, traits, structures or organelles that have a similar function and are exhibited by organisms that do not share a recent common ancestry; see **Homologous**.

Anarchic Field: group of *barren* or *nonciliferous kinetosomes*, in an apparently dis- or unorganized array, giving rise to the infraciliary bases of the *oral ciliature* (e.g., in some hymenostomes); a transient primordial field or ciliary *anlage* in an early stage of *stomatogenesis* (AF, Fig. 2.11Dd, 2.11De).

Anisogamont: see **Gamont**.

Anisotomic: literally "unequal parts"; see **Fission**.

Anlage (pl. **Anlagen**): primordium; a developing, differentiating, or even presumptive structure or organelle; used with numerous modifiers, such as nuclear-, cytoplasmic-, cortical-, oral-, somatic-, ciliary-; in ciliate morphogenesis, often a group of kinetosomes; see **Anarchic Field**, **Germinal Row**, and **Scutica**.

Annulus (pl. **Annuli**): used variously; the non-living portion, also called the sheath, of the contractile stalk of some peritrichs, which surrounds the central membrane-bound myofibrillar bundle or *spasmoneme*; band(s) of fibrous, filamentous or dense amorphic material encircling at various levels, the *cytopharyngeal apparatus* of certain litostomes, phyllopharyngeans, and nassophoreans; various ring-like structures or markings in general, including the *pellicular striae* on the zooid of certain peritrichs.

Antecorpy, Rule of: new somatic kinetosomes always arise anterior to old ones.

Apical Funnel: distally drawn-out anterior end of many chonotrichs, sometimes lined with *atrial ciliature* and leading posteriorly to the cytostome; ontogenetically and phylogenetically, the body's ventral surface (ApF, Fig. 2.11Ca).

Apokinetal: type of *stomatogenesis* in which the kinetosomes of the *anlage* have no apparent pre-association with either somatic kineties or the parental oral apparatus; found in some entodiniomorphids and many spirotrichs; formerly known as the de novo kinetosomal mode of stomatogenesis; see **Epiapokinetal** and **Hypoapokinetal** (Fig. 2.11Dh, 2.11Di).

Apomorphic (adj.): see **Apomorphy**.

Apomorphy: any derived and homologous character; a character or attribute occurring or arising at a branching point and carried through one descending group in a phyletic lineage; a derived character that is less like or has less resemblance to the ancestral condition of the attribute in question or to that of the homologous character in the phylogenetically most closely related group; see **Plesiomorphy**.

Arboroid Colony: *zooids* disposed in a dichotomous branching or tree-like manner, interconnected by either stalks or loricae (e.g., as shown by many sessiline peritrichs); dendroid and dendritic are synonyms of arboroid (Fig. 2.8Ak, 2.8Al, 2.8Bc).

Argentophilic: literally silver-loving, referring to structures or associated elements that react positively to silver compounds; see **Silver-impregnation Techniques** (Figs. 2.3, 2.4).

Argyrome: whole system of pellicular or cortical argentophilic structures or markings revealed by *silver impregnation techniques*, particularly wet and dry silver nitrate methods; often may be indication of cortical filaments or of points or lines of contact of alveolar membranes; is *not* identical with the *infraciliature*, although superficially it shows overlapping in some components (e.g., the all-important *kinetosomes* are part of both systems); highly useful in taxonomy, comparative morphology, and morphogenetic studies; principal synonyms are silverline system, Silberliniensytem, and neuroformative system; see **Silver-impregnation Techniques**.

Asexual: typically a kind of *reproduction* (i.e., binary *fission*), which does not involve *sexual phenomena*.

Asexual Reproduction: see **Fission**.

Astomy (adj. **Astomatous**): condition of being mouthless, without a cytostome, whether naturally or experimentally derived; one entire subclass, the Astomatia, exhibits this naturally.

Atrial Ciliature: type of cilia or ciliary organelles associated with the *atrium* and limited to occurrence in certain cyrtophorine phyllopharyngeans; relatively simple in organization, with infraciliary bases of kinetofragmental origin; the only cilia in some chonotrichs, where it may line part of the *apical funnel* (AtC, Fig. 2.11Ca).

Atrium (pl. **Atria**): a non-ciliated *oral cavity* or depression around or in close proximity to the *cytostome* of certain phyllopharyngeans often bordered by *atrial ciliature* (At, Fig. 2.5Cd).

Attachment Disc: see **Adhesive Disc**.

Attachment Knob: enlarged distal end of a *suctorial tentacle* or of the *sucking tube* of rhynchodine phyllopharyngeans; adheres to or embeds itself in the body of a prey or host cell; when the knob is particularly prominent, the tentacle is said to be a capitate tentacle (AK, Fig. 2.9Cd).

Attachment Organelle: nonspecific name for all sorts of *adhesive discs*, crochets, *filaments*, hooks, *mucrons*, *spine*, *stalk*, *suckers*, *tentacles*, or even *thigmotactic cilia* used in a temporary or permanent manner to attach an individual cell or a colony to some living or inanimate substratum; see **Holdfast Organelle** (Fig. 2.9B).

Aufwuchs Community: a non-preferred term; see **Biofilm**.

Autogamy: self-fertilization type of *sexual phenomenon*; ultimately results in extreme inbreeding, since only single organisms are involved; believed to increase the longevity of a *clone*; when autogamy occurs in each member of a paired set of temporarily fused organisms, the phenomenon is sometimes termed cytogamy or autogamy in pairs; the process is always followed by *fission* of the organism(s).

Autogamy in Pairs: see **Autogamy**.

Autonomous: now discarded term, along with semi-autonomous, for a mode of *stomatogenesis*; see **Buccokinetal**.

Autotrophic: capable of synthesizing its own organic molecules, principally using photosynthesis in eukaryotes (i.e., photoautotrophic); see **Heterotrophic** and **Mixotrophic**.

Auxomorphy: morphological-evolutionary relationship between two forms, in a postulated ancestor-descendant relationship in which is shown a sameness of certain structures possessed by both but with an apparent increase in the size or number of component parts comprising those structures by the process of *polymerization* in the presumed descendant (e.g., compare *Dexiotricha* and *Loxocephalus*).

Axenic Culture: literally, "without strangers"; laboratory growth of organisms in a "pure" medium, although not necessarily chemically defined, in which no other living organisms of any kind can be present.

Axoneme: see **Cilium** (Axn, Fig. 2.1B).

Axosome: see **Cilium** (Axs, Fig. 2.1B).

AZM: see **Adoral Zone of Membranelles**.

B

Bactivorous: see **Bacterivorous**.

Bacterivorous: feeding on bacteria; the preferred term; see **Microphagous**.

Bacteriovorous: see **Bacterivorous**.

Barren Kinetosome: basal body not associated, always *or* at a given time, with a cilium; exhibition of a nonciliferous (or aciliferous) state; a particularly common condition of certain kinetosomes during some stages of *stomatogenesis*.

Basal Body: *kinetosome*; blepharoplast of flagellates; in a popular usage, a synonym of the kinetosome of ciliated protozoa.

Basal Disc: see **Adhesive Disc**.

Basal Fibers: see **Basal Microtubules**.

Basal Granule: now discarded name for *kinetosome* or *basal body*.

Basalkörper: see **Cilium**.

Basal Microtubules: set, group, ribbon or bundle of *microtubules*, typically very few in number, extending along the side of somatic kineties at or below the level of the proximal end of the *kinetosomes*; found in a number of oligohymenophoreans; sometimes confounded with *subkinetal microtubules*.

Basal Plate: see **Cilium**.

Bell: body proper, minus the *stalk*, of many sessiline peritrichs: see **Zooid**.

Benthic: pertaining to the bottom or near-bottom of an ocean, sea or lake; often implied by the term are the bottom sediments at the greatest depths of the body of water, but bottom-dwelling forms of marine life are also described by the term.

Beta Membranoid: see **Membranoid**.

Binary Fission: see **Fission**.

Biofilm: in the broadest sense, a loose association of organisms living on/and/or attached to various submerged substrata, often plant material or inanimate objects; the predominantly sessile forms, including many ciliates, which comprise this community may be found in marine, fresh-water or brackish habitats; synonyms Aufwuchs and periphyton.

Biogenetic Law: ancestral resemblance during *ontogeny*; recognition, in a broad sense, that some characters or structures seen during the development of an organism may be generally reminiscent of some structure or character known to be possessed by members of an alleged ancestral group, in either ontogenetic or adult stages of that predecessor, and often in a more or less modified form (e.g., as proposed in Sewertzoff's principles of phylembryogenesis); *to this highly qualified extent and with extreme caution*, Haeckel's old aphorism – "ontogeny recapitulates phylogeny" – may be applied to some phylogenetic-systematic problems in ciliatology; a synonym is the Law of Recapitulation; see also **Morphogenesis** and **Ontogeny**.

Biological Species: an assemblage of populations of organisms that are able to actually or potentially interbreed; see **Morphological Species**.

Biparakinetal: amphiparakinetal stomatogenesis in which two *oral anlagen* are formed independently; found in folliculinid heterotrichs.

Bipolar Kinety: somatic kinety running from pole to pole of the organism without interruption, without circling the body transversely, without going over the top and down the other side; fundamentally, kineties are assumed to be bipolar; in fact, there are many exceptions, but they may generally be considered secondary modifications of the basic original plan of cortical organization in ciliates – and it is noteworthy that the *Rule of Desmodexy* is never violated; see **Kinety**.

Birth Pore: opening or site of emergence of an internally budded *larval form* during the budding process of *reproduction* in some suctorians (BPr, Fig. 2.11Cb).

Blepharismin: photosensitive cytoplasmic "pink" pigment found in *pigmentocysts* in various species of the heterotrich *Blepharisma*; formerly called zoopurpurin (not to be confused with *blepharismone* or *blepharmone*: **see below**).

Blepharismone: the *conjugation*-inducing compound 3-(2′-formylamino-5′-hydroxy-benzoyl) lactate or Gamone 2, isolated from the heterotrich *Blepharisma*; probably a derivative of tryptophan; see **Blepharmone**.

Blepharmone: *conjugation*-inducing glycoprotein isolated from the heterotrich *Blepharisma*.

Border Membrane: finely striated circumferential band with fibers and proteinaceous *radial pins*, which are associated with and reinforce the *adhesive disc* of mobiline peritrichs; sometimes called a corona (BM, Fig. 2.9Bg, 2.9Bh).

Boring Apparatus: see mention under **Rostrum**.

Bristle: common name for long or short, generally single, rather stiff, resilient cilia of several kinds; perform a variety of presumed functions, including sensory, tactile, thigmotactic, locomotor, and attachment; occurring on one or more parts of the body of various ciliates; see **Brosse, Caudal Cilium, Saltatorial Cilia, Sensory Bristle**.

Bristle Kinety (pl. **Kineties**): somatic kineties restricted to one side, typically "dorsal", of the body (e.g. the karyorelictean, *Loxodes*; the hypotrich

Euplotes); characterized by kinetids whose cilia are straight or bristle-like.

Brood Chamber: see **Brood Pouch**.

Brood Pouch: temporary in some suctorians or permanent in some chonotrichs internal chamber or cavity formed by invagination of the pellicle and within which *budding* occurs, producing the *larval form* in the life cycle; preferably called a *crypt* in chonotrichs; also known as an embryo sac or marsupium (BPch, Fig. 2.11Cb).

Brosse: distinctive "brush" of *clavate cilia* arising from specialized short kineties or *kinetal segments*, often oriented obliquely to the body axis on the anterodorsal surface of the nondividing organism; characteristically found in haptorian litostomes and prorodontine prostomes.

Brow Kinety (pl. **Kineties**): see **Ophryokinety**.

Brush: see **Brosse**, the preferred term.

Buccal Apparatus: whole complement of oral *polykinetids* or compound *ciliary organelles* whose bases are located in or associated with the *buccal cavity* or *peristome* (e.g., in oligohymenophoreans, heterotrichs, spirotrichs, and armophoreans); includes *paroral* (*membranes*) and *membranelles sensu lato* (plus homologues and possible nonhomologues of these structures) and their infraciliary bases (= buccal infraciliature) and peristomial ciliature; the whole apparatus functions primarily in food-getting, sometimes in locomotion.

Buccal Area: region around the *cytostome* in ciliates that possess a *buccal apparatus*; strictly speaking, *not*, therefore, a synonym of the much broader and more generalized term *oral region*.

Buccal Cavity: typically a quite deep *oral cavity*, though sometimes secondarily flattened out or everted; often at or near the apical end of the body and/or on the ventral surface; contains the bases of the oral *polykinetids* or compound ciliary organelles and inwardly leads ultimately to the organism's *cytostome-cytopharyngeal complex*, sometimes via a specialized portion of itself known as the *infundibulum*; often applied to the oral cavity of oligohymenophoreans, but it is considered to be the structural equivalent of the *peristome* of heterotrichs and spirotrichs.

Buccal Ciliature: see **Buccal Apparatus**.

Buccal Infraciliature: see **Buccal Apparatus**.

Buccal Membranelles: see **Membranelle**.

Buccal Organelles: see **Buccal Apparatus**.

Buccal Overture: outer or distal opening or aperture of the *buccal cavity*, though essentially unrecognizable (i.e., disappears) when the buccal cavity is everted or flattened out; typical of and easily visible in some oligohymerophoreans, such as *Paramecium*, *Tetrahymena*, and *Ophryoglena*.

Bucco-anal Stria (pl. **Bucco-anal Striae**): see **Director-meridian**.

Buccokinetal: type of *stomatogenesis* in which at least some of the fields of kinetosomes involved – as the ultimate *anlage* – have an apparent origin from the organelles of the parental *oral apparatus sensu lato*; characteristic of many oligohymenophoreans; formerly known as the autonomous and/or semi-autonomous mode(s) of stomatogenesis; see **Ophryobuccokinetal** and **Scuticobuccokinetal** (Fig. 2.11Df, 2.11Dg).

Bud: filial product of a single or multiple fission, characteristically much smaller than the *parental form* and typically quite unlike it in both form and function; generally ciliated, playing a dispersal role in the organism's life cycle; results from a variety of methods of *budding*; it is a form occurring universally among suctorians and chonotrichs, not uncommonly in rhynchodians, some peritrichs, some apostomes, and occasionally in other groups, including even a species of hypotrich (Bud, Fig. 2.11C).

Budding: binary (though typically *anisotomic*) or multiple method of asexual *reproduction*, producing a single (*monogemmic*) or two or more (*polygemmic*) *filial products*, simultaneously or in succession; the phenomenon (also known as gemmation) is classifiable into several types: modified transverse fission, *strobilation, endogemmy* (*endogenous budding*: a subtype is cryptogemmy), *exogemmy* (*exogenous budding*), *evaginogemmy* (evaginogenous or *evaginative budding*), plus additional refinements of some of these (*viz.*, circumvaginative, inva-circumvaginative, invaginative, pseudo-transverse, semi-circumvaginative, and semi-invaginative); typical of suctorians and chonotrichs, but characteristic also of some other taxa (see **Bud**); here it is **not** considered to embrace *palintomy* (where perhaps *strobilation* also belongs) or *catenoid colony*-formation, although distinctions are not always clear-cut (Fig. 2.11C).

Bulge Microtubules: see **Rhabdos**.

C

Cannibalistic: see **Carnivorous**.

Capitate Tentacle: *suctorial tentacle* enlarged at its distal end; see **Attachment Knob** (Fig. 2.9Cd).

Capitulum (pl. **Capitula**): amorphous material capping the proximal end of the *nematodesma* in some dysteriid cyrtophorines; this maxillary armature or tooth, sometimes quite prominent in appearance, may enclose kinetosomes that were involved in the development of the nematodesma (Cap, Fig. 2.5D).

Capsules: see mention under **Tentaculoid**.

Capsules Torquées: see mention under **Tentaculoid**.

Carnivorous: literally "meat-devourèr"; eater of or feeder on some other or the same (= cannibalistic) species of ciliate, zooflagellate, or metazoan; generally refers to a *holozoic* and predatory, not a *parasitic* or even *histophagous*, mode of life.

Case: synonym of *lorica*, which is the preferred term.

Catenoid Colony: see **Catenulation**.

Catenulation: temporary line or chain of individuals brought about by repeated (and generally *anisotomic*) *binary fissions* without separation of the resulting *filial products*; found in some astomes, in certain apostomes, and rarely in species of other groups; see **Strobilation** (Fig. 2.8Ba).

Cathetodesma (pl. **Cathetodesmata**): periodically striated, subpellicular fiber, transversely oriented, arising from or near the anterior right region of the posterior somatic kinetosome of a somatic dikinetid, literally, "cutting" to the left toward the next kinety; found only in certain clevelandelline armophoreans; a short *kinetodesma*, arising from nearly the same location, is also present in the same ciliates (Cd, Fig. 2.1Ed).

Cathrobic: see **Kathrobic**.

Caudal Cilium (pl. **Cilia**): distinctly longer *somatic* cilium (occasionally more than one) at or near the posterior or antapical pole, sometimes used in temporary attachment to the substratum; arises from a *polar basal body-complex* (CC, Fig. 2.5Ac).

Caudalia: ciliary tufts (of *syncilia*), on short non-retractable stalks, at the posterior or antapical pole of some entodiniomorphids.

Cavernicolous: cave-dwelling; ciliates speleologically inclined!

Cell Anus (pl. **Anuses**): see **Cytoproct**.

Cell Division: see **Fission**.

Cell Envelope: see **Cortex**.

Cell Mouth: see **Cytostome**; but for usage in a very broad and general way, see **Oral Region**.

Chain Formation: see **Catenulation**.

Chondriome: total *mitochondrial* complex of a cell (or ciliate).

Ciliary Corpuscle: see **Kinetid**.

Ciliary Girdle: in a general way, the term is restricted to peritrichs, yet it is also used for any encircling band of somatic ciliature (e.g., as seen in *Didinium*-like haptorians); see **Locomotor Fringe** (CG, Fig. 2.5Ad).

Ciliary Meridian: the *argentophilic* line (= primary meridian) coursing *above* the kinetosomes, with recognition of secondary, and even tertiary, meridians located interkinetally; historically notable in the tetrahymenine hymenostomes, where it is especially visible with dry and wet *silver impregnation techniques*; see **Kinety** (CM, Fig. 2.7e).

Ciliary Organellar Complex: any specific structure, *oral* or *somatic*, compounded of cilia or cilia-derivatives; see **Ciliature** and **Compound Ciliature**.

Ciliary Rootlet: generally rare in ciliates; sometimes formerly used to include various structures, fibrillar or microtubular in nature, arising from or associated with *kinetosomes*, particularly the *nematodesma*; a special case is represented by the striated fibers extending centripetally from the vicinity of certain kinetosomes and plunging deep into the cytoplasm, for example, in the adhesive disc of mobiline peritrichs.

Ciliary Row: longitudinal line or file of *somatic cilia*; see **Kinety**.

Ciliary Territory: see **Kinetosomal Territory**.

Ciliary Wreath: see **Ciliary Girdle**.

Ciliatology: the study of ciliates; the investigators are therefore ciliatologists.

Ciliature: general term referring to assemblages of *cilia*; see definitions under terms denoting specific kinds of ciliature: *atrial, buccal, circumoral, cirral, compound, coronal, oral, perioral, peristomial, perizonal, prebuccal, scopulary, simple, somatic, synciliary, thigmotactic,* and *vestibular*.

Ciliferous: literally "cilium-bearing"; used in reference to *kinetosomes* that regularly produce cilia; used for the cilium-bearing stage in those ciliates that have cilia only at some stage(s) in the life cycle of the organism and not at others (at which time they are non-ciliferous, naked, or barren).

Ciliophore: an anterior protuberance that bears ciliature in some entodiniomorphids.

Ciliospores: now outmoded word for certain *tomites* arising by *palintomy* (e.g., in the hymenostome *Ichthyophthirius*).

Cilium (pl. **Cilia**): cylindrical organelle (diameter *ca.* 0.26 µm; length variable, often 5–10 µm) arising from a kinetosomal base and projecting from the body surface of an organism though covered with the common *plasma membrane*; internally complex, with an axoneme comprised ultrastructurally of microtubular structures in a "9 + 2" arrangement; the nine peripheral doublets continuous with the kinetosomal microtubules and the central pair arising from this axosome; typically disposed over the body in longitudinal rows or files, though with many exceptions; in "*compound*" forms – see appropriate terms under **Ciliature** – occurring both on the body (e.g., as *cirri*) and in oral regions (e.g., as *oral ciliature*); may function in locomotion, especially, with diversity of beating patterns, and in feeding, attachment, and sensing; various kinds of specialized cilia are recognized (e.g., *caudal, clavate, marginal*) (Figs. 2.1B, 2.1C, 2.3k, 2.4E, 2.5Aa, 2.5Ac).

Circumoral Ciliature: line, circle or band of essentially simple *ciliature* encircling (i.e. periorally) all or part of the apical end, including the cytostome, of the body of a number of litostomes and phyllopharyngeans; basically organized as dikinetids (but not *dyads*), only one *kinetosome* of which is typically *ciliferous*; often comprised of the anterior extremities of the more or less regularly arranged somatic kineties or occasionally, of extra, interpolated kinetal segments; variations exist, particularly at the level of the infraciliature, and these are of taxonomic value.

Circumoral Connective: fibril-like line seeming to encircle the *buccal overture*; a silverline structure, but of some value, along with other parts of the *argyrome* in general, in the comparative taxonomy of such forms as the tetrahymenine hymenostomes (CoC, Fig. 2.7e).

Circumoral Kinety (pl. **Kineties**): name sometimes used for the two posteriormost of the three oral *kinetofragments* found in cyrtophorine phyllopharyngeans anteriad to the complex *cyrtos* (in the interfissional stage); see **Preoral kinety** (Fig. 2.4O).

Cirral Ciliature: see **Cirrus**.

Cirromembranelle: name used by some workers for parts of the highly organized *oral ciliature* of certain colpodeans.

Cirrus (pl. **Cirri**): kind of *compound somatic ciliature* typical of hypotrich and stichotrich spirotricheans, though not exclusively found there; a composite tuft of cilia, few to >100, functioning as a unit, though typically with no special enveloping membrane, and tapering distally or else fimbriate; its infraciliary kinetosomes are also interlinked and joined to other cirral bases by connecting fibers or tracts of microtubules; characteristically, a cirrus is a thick conical locomotor organelle, but some are also occasionally of aid in food-gathering; occurring in lines or in groups in definite patterns on the ventral surface, with subtypes identifiable by their location (*viz.*, buccal, frontal, frontoventral, midventral, transverse (anal), caudal, and marginal) (Fig. 2.4M, 2.4N; Cir, Figs. 2.3Aj, 2.3Al, 2.5Ab, 2.7k).

Cisterna (pl. **Cisternae**): see mention under **Endoplasmic Reticulum** and **Golgi Apparatus**.

Clade: a *monophyletic* lineage.

Cladistics: the branch of systematics devoted to identifying *clades*.

Clathrocyst: cytoplasmic *extrusome* in *Didinium* involved in production of the elaborate middle or mesocystic layer of the *cyst* wall (Fig. 2.9Cc).

Clathrum: total abandonment of the term "clathrum" (Latin for "lattice") is here proposed; see **Rhabdos**.

Clavate Cilium (pl. **Cilia**): short immobile *cilia* lacking the central pair of microtubules in their axoneme; allegedly sensory in function, occurring in a number of ciliates, typically in the *brosse* of haptorian litostomes (e.g., *Didinium*) and prostomes, but also present in the *scopula* of many peritrichs; also called *stereocilium*; see **Condylocilium**.

Clonal Life Cycle: see **Life Cycle, Clonal**.

Clone: population of organisms established by culturing the descendants of a single individual.

Cnidocyst: special kind of *extrusome* generally found in certain dinoflagellates but now (also) reported from several species of karyorelicteans; in karyorelicteans, it appears to be a pear-shaped, laterally-flattened *extrusome* that contains a multiply coiled filament.

Cnidotrichocyst: while a synonym for *toxicyst*, use of this term is not recommended.

Code of Nomenclature: see **International Code of Zoological Nomenclature**.

Collar: term used variously; the neck area between the often flared apical end and the body proper in chonotrichs; variously differentiated region beneath the opening in the *lorica* and used as a diagnostic characteristic in taxonomy (e.g. loricate peritrichs, tintinnid spirotrichs); an open or closed zone of oral polykinetids encircling the anterior end of choreotrichs (Figs. 2.6B; Col, 2.8Ae, 2.8Af, 2.8Ah, 2.11Ca).

Collarette: apical, peristomial lip that circumscribes the retractable *epistomial disc* in many sessiline peritrichs; equipped with a sphincter "muscle" (or **Myoneme**) (Colt, Fig. 2.11B).

Collecting Canal: see **Afferent Canal**.

Colony: assemblage of cells derived by *fission* from a founder individual; see definitions under major types (i.e., **Arboroid, Catenoid,** and **Spherical**) (Fig. 2.8A, 2.8B).

Commensal: see **Commensalism**.

Commensalism: a *symbiosis* in which the *symbiont* benefits by the association but the host does not particularly suffer; endo- and ectocommensals exist in many groups; category *sensu lato* could include epibionts and *symphorionts*, typical forms in the life cycle of many attached or sessile ciliates, which live on the outside of their host; an *inquiline* may be considered a special kind of endocommensal.

Compound Ciliary Organelle: see **Compound Ciliature**.

Compound Ciliature: general term for all *ciliature* not comprised of single, separate or isolated individual cilia or of dikinetids of somatic cilia; various kinds, both somatic and oral in origin, occur throughout the phylum; particularly well developed in the spirotrichs (e.g. *cirrus*, *membranelle*, *polykinetid*).

Compound Oral Ciliature: *compound ciliature* found in the *oral region*.

Compound Somatic Ciliature: *compound ciliature* found in the *somatic region*.

Concrement Vacuole: curious and quite complex subpellicular cytoplasmic inclusion, one to an organism, containing refractile, probably mineral grains, having no opening, and sometimes strengthened by surrounding microtubules; characteristic of certain trichostome endocommensals; function unknown, but considered by some workers as a kind of statocyst.

Condylocilium (pl. **Condylocilia**): kind of *clavate cilium* found in hypotrich spirotrichs.

Condylopallium: ovoid, bulb-like extension of the ventral right anteriormost part of the hypotrich spirotrich *Certesia*, containing a *vesicle* with dense granules; function unknown, but perhaps excretory or statocyst-like.

Conjugation: reciprocal-fertilization type of *sexual phenomenon* in which a meiotic/mitotic product of the micronucleus is typically reciprocally exchanged between the two individuals, except in *total conjugation*; presumably occurs only between members of differing mating types; allegedly significant to both vitality of the clone and survival of the species, although the phenomenon is unknown (unobserved) in many ciliates from various groups; may involve *iso-* or *macro-* and *microconjugants*, with *temporary* (most widespread) or *total* (as in *all* peritrichs and chonotrichs) fusion of members of the pair; pre- and exconjugant stages are recognized; the process is always followed by *fission(s)* of the exconjugants.

Conocyst: a small, cone-shaped *extrusome*, probably a *toxicyst*, in the cortex of haptorians (e.g., *Homalozoon* and *Loxophyllum*).

Constellation of Characters Principle: use of multiple characters from diverse approaches or fields in assessment of differences or similarities between or among groups of organisms under comparative taxonomic study; its application helps overcome biases and prevents the extreme splitting likely when only very few characters – or data from but a single field – are used to draw conclusions concerning phylogenetic and taxonomic relationships.

Contractile Vacuole (CV): liquid-filled organelle (sometimes multiple), serving as an osmoregulator in the cytoplasm of nearly all ciliates; generally pulsates with regular frequency under natural conditions: grows (diastole) to a certain size and then "contracts" (systole), typically emptying its contents, which may include dissolved "waste materials", to the exterior via one or more pores; the CV is more widespread in ciliates than the (*cytoproct,* CYP); synonymous, but non-preferred terms, are water expulsion vesicle and nephridial apparatus (CV, Fig. 2.9Bf).

Contractile Vacuole Pore (CVP): minute permanent opening in the *pellicle*, with *argentophilic* rim and a canal reinforced by microtubules through which contents of the *contractile vacuole* are expelled to the outside milieu; CVP's are *corticotype* structures, characteristically stable in number and location and thus of diagnostic value in comparative taxonomy; also known by non-preferred term – expulsion vesicle pore (CVP, Figs. 2.3Ad, 2.3Af, 2.3Ai, 2.4A, 2.4F, 2.4K, 2.7a, 2.7f, 2.7i, 2.7k, 2.11Aa).

Convergent Evolution: development of similar characters separately in two or more groups that do not share a close common ancestry; such characters, preserved by natural selection, arising through adaptation to similar ecological pressures or habitats.

Corona (pl. **Coronae**): apical, cytostome-bearing extremity of the body of certain haptorians; often distinguished by longer *coronal ciliature* from a posteriorly adjacent neck; also used for the *border membrane* of mobiline peritrichs.

Coronal Ciliature: term used in different ways, but most commonly referring to the circlet of relatively long cilia at the ends of somatic kineties or isolated from them and surrounding the apical cytostomal area, the *corona*, of various haptorians and located anterior to the neck region of the body.

Cortex (pl. **Cortices**): in the broadest sense, the outer portion or "layer" of the ciliate body, sometimes termed the cell envelope; includes the *pellicle* and the *infraciliature sensu lato* and bears the *cilia*; its various openings, *pellicular ridges*, *alveoli*, *ciliferous kinetosomes*, and their fibrous and microtubular associates comprise the *corticotype*; mitochondria are in the cortex of many ciliates.

Cortical Vesicle: see **Alveolus, Pellicular**.

Corticotype: specific pattern of cortical structures or organelles found to be characteristic of a given organism or population of organisms within a species; cortical pattern especially as made visible following application of *silver-impregnation techniques* (Figs. 2.3, 2.4).

Cosmopolitan: capable of population growth in many different places worldwide.

Crista (pl. **Cristae**): see mention under **Mitochondrion**.

Crochet: see **Attachment Organelle**.

Crypt: see **Brood Pouch**; but this is the preferred term for use with chonotrichs; also occasionally employed in a more general sense for any cleft or depression in the body or elsewhere (Crp, Fig. 2.11Ca).

Cryptogemmy: see **Budding** and **Endogenous Budding**.

Cryptotelokinetal: *telokinetal stomatogenesis* in which the *oral anlage* originates as subequatorial *kinetofragments* derived from non-ciliated *kinetosomes* residing in subcortical pouches; found in entodiniomorphid trichostomes.

Crystallocyst: minute body, quite numerous in the cortex of the scuticociliate *Conchophthirus*; possibly a kind of *extrusome*.

Cursorial: adapted to or specialized for "running"; true of some hypotrich and stichotrich *cirri*.

Cuticular Pore: see **Pellicular Pore**.

Cyrtocyst: a short, curved *extrusome* found subpellicularly in the haptorian *Didinium*.

Cyrtos: tubular *cytopharyngeal apparatus*, often curved, the walls of which are strengthened by longitudinally arranged *nematodesmata* derived from circumoral kinetosomes and lined with extensions of *postciliary microtubules*; the nematodesmata may be interconnected and/or wrapped circumferentially by annular sheaths of diffuse fibrous material and an amorphous dense substance that may form *capitula* proximally; contains no toxicysts; septa and specialized *phagoplasm* may be present; typical of members of the Classes PHYLLOPHARYNGEA and NASSOPHOREA, including especially the "cyrtophorine gymnostomes" of older classifications; principal synonyms include nasse and pharyngeal basket; of Greek derivation, the word literally means "curved" but also may be extended to imply "basket" or "cage", all three descriptively appropriate to its usage here; the cyrtos is to be compared with the *rhabdos*, the other major type of cytopharyngeal apparatus characteristic of litostome ciliates, allegedly less complex, non-curved, often containing toxicysts, and lined with *transverse* and *bulge microtubules* (Figs. 2.5D, 2.7j).

Cyst: non-motile inactive stage in the life cycle of many ciliates, generally thought to serve the roles of protection and dispersal; when considered a protective stage, organism typically rounded up, mouthless, and surrounded by three or more secreted layers or "membranes" (i.e. pericyst, ectocyst, mesocyst, endocyst, metacyst); sculptured on the outside, and with or without an emergence pore, which may have an *operculum*; several types have been described, indicating the diverse functions of cysts – (1) digestive, (2) division, multiplicative, propagative or reproductive, (3) infective, (4) invasion (cuticular), (5) phoretic, (6) protective, (7) reorganization or reconstructive, (8) resting, and (9) temporary (Fig. 2.9A).

Cystation: processes involved in formation of and departure from a *cyst*.

Cystic Membrane: see **Cyst**.

Cyst Wall: see **Cyst**.

Cytobrain: see **Neuromotorium**.

Cytogamy: *autogamy* in pairs; non-preferred term.

Cytokinesis: strictly applied meaning division of the cytoplasm; often used in a general sense as a synonym of cell division (see **Fission**).

Cytopharyngeal Apparatus (pl. **Apparati**): the complex of microtubular and microfilamentous components that support the *cytopharynx*; most conspicuously developed in ciliates that have the *cyrtos* or the *rhabdos*; see **Cytopharynx**.

Cytopharyngeal Armature: refers to the nematodesmal elements of a *cytopharynx*, especially of the *cyrtos*.

Cytopharyngeal Basket: see **Cyrtos** and **Rhabdos**.

Cytopharyngeal Pouch: reservoir-like enlargement or receiving vacuole of the cytopharynx of a few ciliates (e.g. the carnivorous macrostome stage of *Tetrahymena vorax*); when food-filled, it pinches off as a regular membrane-bound *food vacuole*.

Cytopharyngeal Rod: synonym for a *nematodesma* that supports the *cytopharynx*; see **Cyrtos** and **Rhabdos**.

Cytopharynx (pl. **Cytopharynges**): non-ciliated tubular passageway of varying length in different ciliates, leading from the *cytostome* proper into the inner cytoplasm of the organism; typically, *food vacuoles* are formed at its inner or distal end, when it retains its own integrity during the feeding process; when its walls are particularly strengthened, the *cytopharyngeal apparatus* may be known by such specialized names as *cyrtos* or *rhabdos* (Cph, Figs. 2.5Ca, 2.6Ac).

Cytoproct (**Cyp**): cell anus; when present, generally permanent, slit-like opening in the *pellicle*, near the posterior end of the body, through which egesta may be discharged; its edges, resembling a kind of pellicular ridge and reinforced with microtubules, are *argentophilic*; in some species, the **Cyp** is a cortical landmark of taxonomic significance, located in or just to the left of the posterior portion of *kinety number 1* (Cyp, Figs. 2.3Ai, 2.4K, 2.7f, 2.7i, 2.11Aa).

Cytopyge: a non-preferred term; see **Cytoproct**.

Cytoskeleton: the complex of microtubular and microfilamentous components in the cytoplasm that provide structure and form to the cell body; generalized term referring to any secreted inorganic or proteinaceous material within or below or on the surface of a ciliate, covering or involving all or some specific part of the body and lending considerable rigidity to the shape of the organism; see **Skeletogenous Structure**, a term with which it is broadly synonymous, although the emphasis in meaning may vary in usage by different workers.

Cytospindle: sets of longitudinally-oriented cortical *microtubular ribbons* that appear transiently during *cytokinesis* of some nassophoreans and peniculines.

Cytostome: literally "cell mouth"; the "true" mouth or oral opening; simply a two-dimensional aperture through which food materials pass into the endoplasm of the organism via a more or less distinct *cytopharynx*; may open directly to the exterior or be sunken into a depression or *oral cavity* of some kind, such as an *atrium*, *vestibulum*, or *buccal cavity*; often definable as the level at which pellicular alveolar sacs are no longer present, it may occur as an angled or tipped elliptical opening with a long axis of considerable length (Cs, Figs. 2.5Ca, 2.6Aa).

Cytostome-cytopharyngeal Complex: convenient generalized term to refer to the inseparable complement of the *cytostome* and the adjacent and really continuous *cytopharynx*.

Cytotaxis: broad concept that considers the ordering of cortical structure(s) to be determined by the preexisting organization of the cytoplasm in the particular site concerned; the ordering and arranging of new cell structure under the influence of pre-existing cell structure; see also **Structural Guidance Principle**.

D

***Dactylophrya* – Stage**: see **Dactylozoite**.

Dactylozoite: infective stage in the most unusual life cycle of the suctorian *Tachyblaston*; dactylozoites arise by several rapid fissions of the loricate parental organism, and then, though non-ciliated and with but one tentacle, reach the body of another suctorian, *Ephelota*, "burrow" in, and develop into forms producing a ciliated *swarmer* stage; swarmers settle down, often on the stalk of the same *Ephelota*, produce their own stalk and lorica, and repeat the cycle; a synonym is *Dactylophrya*-stage.

Dargyrome: *argyrome* on the dorsal surface of hypotrich spirotrichs.

Daughter Organisms: see **Filial Products**.

Deme: population within a species; the concept is particularly used by parasitologists to indicate populations of different types: (e.g. monodeme, nosodeme, serodeme, topodeme, and xenodeme); the organisms comprising different demes may possess distinctive morphological and/or physiological characters and, in some cases, may be incipient subspecies or even unrecognized biological species.

Dendritic Colony: synonym of *arboroid colony*.

Dendroid Colony: synonym of *arboroid colony*.

De Novo Cytoplasmic: now discarded term for a mode of *stomatogenesis*; see **Telokinetal** for its modified replacement.

De Novo Kinetosomal: now discarded term for a mode of *stomatogenesis*; see **Apokinetal**.

Denticle: one of many similar structures or interlocked component parts of a supporting ring underlying the *adhesive disc* of mobiline peritrichs; a proteinaceous subpellicular skeletal element composed of a conical centrum and typically an inwardly directed spine or ray and outwardly directed blade (Dent, Fig. 2.9Bg, 2.9Bh).

Denticulate Ring: skeletal organelle, made up of *denticles*, found in mobiline peritrichs; see **Denticle** (Fig. 2.4L).

Derived Character: see **Apomorphy**.

Desmodexy, Rule of: true *kinetodesmata*, when present, lie to or extend anteriad and/or to the organism's right of the *kinety* with which they are associated; thus, polarity of the cell can be deduced (see **Bipolar Kinety**) (Figs. 2.1, 2.2).

Detritivorous: feeding on organic particles.

Deuterostomisation: an evolutionary process in which oral kinetosomes are formed anew from somatic kinetosomes after the loss of the original oral ciliature; supposed to have occurred in litostomes and phyllopharyngeans.

Diastole: see **Contractile Vacuole**.

Dictyosome: see **Golgi Apparatus**.

Dikinetid: a *kinetid* composed of two *kinetosomes* and their fibrillar associates; see **Dyad**, **Monokinetid**, and **Polykinetid** (Figs. 2.1E, 2.2c, 2.2d).

Diploid: 2N number of chromosomes; characteristic of ciliate micronuclei whose chromosomes may be visible and enumerated at mitosis, although the true ploidy of many ciliates is unknown; see **Haploid**.

Diplokinety (pl. **Diplokineties**): often, applied to a *kinety* with its kinetosomes doubled in some specific fashion; see **Diplostichomonad**.

Diplostichomonad: type of double *paroral* whose infraciliature is composed of two parallel rows or files of kinetosomes – the inner or endoral membrane and the outer or *paroral membrane*; the kinetosomes are never in *dyads* nor do they form a zigzag pattern, and all are *ciliferous*; a type of *diplokinety*; characteristic condition of clevelandellid armophoreans, many hypotrichs, and many stichotrichs.

Director-meridian: *argentophilic* non-kinetosomal line on the midventral surface, coursing from the posterior margin of the *buccal cavity* to the *cytoproct* near the posterior end of the body; occasionally with non-ciliferous basal bodies near its anterior end, its locale is part of the site of formation of new buccal organelles during *stomatogenesis*; characteristic of oligohymenophoreans, particularly scuticociliates; a little-used synonym is bucco-anal stria (Fig. 2.4D; DM, Fig. 2.7i).

Discoidal Vesicle: *vesicle* abundant in the *phagoplasm* and involved in building the *food vacuole* membrane.

Dorsal Bristle: see **Sensory Bristle**.

Dorsal Brush: see **Brosse**.

Dorsal Zone of Membranelles: an older non-preferred term referring to tufts of *syncilia*, which are not *membranelles*, located anteriorly and dorsally on the bodies of many entodiniomorphids.

Dyad: a paired set or couplet of *kinetosomes* in which the kinetosomal axes are at right angles (perpendicular) to the axis of the line or file of the kinetal structure of which they are a part; in the case of many *paroral* structures, only the outermost kinetosome of the dyad is *ciliferous*; the term is not used for the differently arranged pairs of kinetosomes comprising the *circumoral ciliature*, for which the preferred term is *oral dikinetid* (Fig. 2.1C).

E

Ectocommensal: see **Commensal**.

Ectocyst: see **Cyst**.

Ectoplasmic Flange: see **Flange**.

Ectosymbiont (of Ciliates): microorganism, such as bacteria or other ciliates, attached to the outside of the host ciliate; common examples of ciliates as hosts include many psammophilic species (e.g., the karyorelictean *Kentrophoros*) and commensalistic scuticociliates (e.g., *Echinocyclidium*), and sessiline peritrichs in which both bell and stalk may be involved (e.g., *Zoothamnium niveum*).

Edaphic: in a broad sense, pertaining to all kinds of soil, forest litter, and other types of terrestrial habitats, including mosses, lichens, and trunks and leaves of trees; synonym of terrestrial.

Elineation: process by which a kinety "divides" or separates longitudinally to produce a file of kinetosomes parallel to itself, increasing the number of kineties by one.

Embryo: see **Larval Form**.

Embryo Sac: see **Brood Pouch**.

Enantiotropic: a kind of *fission* typical of oligotrichous spirotrichs, said to involve a condition of inverse *homothetogenic fission* and shifting body axes via pronounced morphogenetic movements.

Encystment: the process of *cyst* formation; see **Cystation**, **Excystment**.

Endemic: regularly or only found in a certain geographic region; see **Cosmopolitan** and **Ubiquitous**.

Endocommensal: see **Commensal**.

Endogemmy: see **Budding** and **Endogenous Budding**.

Endocyst: see **Cyst**.

Endogenous Budding: type of single or multiple *fission* taking place within a *brood pouch*, with the embryo or *larval form(s)* completely free of the *parental form* before emergence through the *birth pore*; characteristic mode of reproduction of certain chonotrichs, where the process is called cryptogemmy, and, especially, of certain suctorians (Fig. 2.11Ca, 2.11Cb).

Endoplasmic Reticulum (ER): system of internal membranes in the form of flattened cisternae, and/or vesicles that are related to or derived from the *Golgi apparatus*; surfaces of the ER membranes are sites of ribosomal activity involved in synthesis of secretory proteins, which are then processed by the Golgi apparatus.

Endoral Membrane: see **Diplostichomonad**, **Paroral**.

Endoskeletal System: the term preferred by many students of the astome oligohymenophoreans (see **Cytoskeleton**).

Endosome: somewhat transient, brightly Feulgen-positive body in the *paramere* of the *heteromerous macronucleus* of cyrtophorine and chonotrich phyllopharyngeans; the term is sometimes misused for the Feulgen-negative, RNA-containing *nucleolus* so commonly found, generally as numerous small bodies, in the nucleoplasm of the *homomerous macronucleus* and the *orthomere* of the heteromerous macronucleus (End, Fig. 2.12r, 2.12bb).

Endosprit: old term for the very short *suctorial tentacle* of the curious suctorian *Cyathodinium*.

Endosymbiont (of Ciliates): *symbiont*, generally bacterial or algal in nature, living within the cytoplasm or nucleoplasm of a ciliate cell; ranging from *kappa particles* and other "Greek-letter parasites" of *Paramecium*, bacteria often intimately involved genetically and metabolically with their host, to the common *zoochlorellae* and *zooxanthellae* found widespread among ciliate groups; see **Xenosome**.

Envelope: used variously; for example, **Cortex**, **Lorica**, and **Nuclear Envelope**.

Epiapokinetal: *apokinetal stomatogenesis* in which the *oral anlage* develops on the cell surface; found in some spirotrichs.

Epibionts: see **Commensal**.

Epilorica (pl. **Epiloricae**): additions to *protolorica* or *paralorica* of tintinnids; often appearing as supernumerary collars or annuli whose form, structure, height, and number are variable; see **Lorica**, **Paralorica**, **Protolorica**.

Epiplasm: fibrillar or filamentous pellicular layer directly underlying *alveoli* and/or *plasma membrane*; see **Lamina corticalis**.

Epistomial Disc: retractable, non-ciliated, vaulted center of the *peristomial field* characteristic of many sessiline peritrichs; in some of the Operculariidae, the form of a prominent stalked *operculum*.

Epistomial Lip: a cortical ridge that overlies oral structures, such as the *paroral* and its *cilia* (e.g. in some oligotrich spirotrichs and peritrich oligohymenophoreans).

Ergastoplasm: a generally discarded term, formerly variously used for the so-called lifeless cell inclusions (i.e., stored fats, starches), for cytoplasmic components with affinity for basic dyes, and for a form of the *endoplasmic reticulum*.

Erratic Kinetosomes: a single or a few kinetosomes (*ciliferous* or *non-ciliferous*) that appear to have "wandered off" from some larger, more stable, infraciliary structure or organelle; may appear, or be revealable, only at certain stages in the life cycle of the organism, becoming involved (in effect as an *anlage* or as a *vestige*-turned-*anlage*) in some morphogenetic process, such as *stomatogenesis* or *budding* (e.g., in suctorians).

Esophagus: outmoded term not needed and misleading in description of any structure or cavity in the oral region.

Eukaryotic: literally "having a true nucleus", organisms having a *unit membrane*-bound *nucleus* containing chromosomes or chromosome-derived gene sequences.

Eupelagic: see **Pelagic**.

Eutrophic: pertaining to an aquatic habitat with high primary productivity; high rate of anabolism; referring to a habitat rich in minerals and dissolved organic nutrients, but often with low oxygen content; similar to *polysaprobic* in some ways, but generally *not* used with regard to pollution; see **Oligotrophic**.

Evaginative Budding: type of *fission* involving formation of a temporary *brood pouch* but in which the larval from is not freed within the *parental form*; in emergence, the entire wall of the pouch evaginates and *cytokinesis* takes place on the outside of the parental form; this is the characteristic mode of reproduction of members of an entire order of suctorians; in contrast see **Endogenous Budding** (Fig. 2.11Cd).

Evaginogemmy: see **Budding** and **Evaginative Budding**.

Evolutionary Series: arrangement of groups of organisms in a supposed phylogenetic sequence, using some major character or *constellation of* *characters*, as a basis for indicating a graded series from, for example, ancestral to more derived forms; see **Orthogenetic Lines** for the danger of possibly arbitrary and non-phylogenetic sequencing.

Excystment: the process of leaving a *cyst*; see **Cystation**, **Encystment**.

Exogemmy: see **Exogenous Budding**.

Exogenous Budding: type of single or multiple *fission* taking place essentially on the surface of the *parental form*; larvae are pinched off singly or multiply, and if multiply, either synchronously or consecutively; the characteristic mode of reproduction in certain chonotrichs and in one large group of suctorians (Fig. 2.11Cc).

Explosive Radiation: rapid diversification of forms (e.g., into many different taxa at a given level) brought about evolutionarily by invasion of a vast new and quite different biotope by some "stem" group, with subsequent adaptation to the variety of specialized habitats and niches thus made available to the (presumably unopposed) invader; the chonotrichs, largely *symphorionts* on certain crustaceans, may well serve as an example of a group that has taken advantage of such a situation.

Explosive Trichocyst: see **Trichocyst**.

Expulsion Vesicle: see **Contractile Vacuole**.

Expulsion Vesicle Pore: see **Contractile Vacuole Pore**.

Extensor Membrane: largely disused term referring to the ciliature of the anterior part of the *paroral* of certain ciliates when it is in an immobile state; the paroral cilia, forming a coalesced stiff membrane, aid in guiding food particles into the oral cavity; called Lachmann's bristle in *peritrichs*; at least partially synonymous with another seldom used term, semi-membrane.

External Budding: see **Exogenous Budding**.

Extramacronuclear Microtubules: *microtubules* that assemble and elongate outside the macro*nuclear envelope* and are used in its division; characteristic of the Class HETEROTRICHEA; see **Intramacronuclear Microtubules**.

Extrusive Organelle: see **Extrusome**.

Extrusome: *unit membrane*-bound extrusible body located in the cortex and assembled by the *Golgi apparatus*; a generalized term useful in referring to various types of probably non-homologous structures (e.g., *clathrocyst*, *cnidocyst*, *conocyst*, *crystallocyst*, *cyrtocyst*, *fibrocyst*, *haptocyst*, *lepidosome*, *mucocyst*, *pexicyst*, *rhabdocyst*, *toxicyst*, and *trichocyst*); extrusion occurs under conditions of appropriate chemical or mechanical stimulation.

Exuviotrophic: feeding on tissues or exuvial fluids of dead or molted hosts; particularly characteristic of one group of apostome ciliates.

F

Fascicle: generalized term, but used specifically with reference to a group or bundle of *suctorial tentacles* on the body of a suctorian, sometimes (but not necessarily) born on an *actinophore*.

Fibers, Postciliary: see **Postciliary Microtubules**.

Fibers, Transverse: see **Transverse Microtubules**.

Fibrils: see **Microfibrils**.

Fibrocyst: unique *trichocyst* characteristic of the microthoracine nassophoreans; fusiform, explosive, and revealing a conspicuous parachute- or umbrella-like tip after discharge; also called a compound trichocyst.

Fibrous Trichocyst: see **Fibrocyst** and **Trichocyst**.

Filamentous Annulus (pl. **Annuli**): "elastic," expansible binding of fine microfilaments surrounding and considered a part of the *rhabdos* near its proximal (outer) end; allows for the great expansion required by these carnivorous ciliates when feeding; said to be continuous with the *lamina corticalis* in the vicinity of the *corona* of various haptorians.

Filamentous Reticulum: three-dimensional lattice of kinetosome-associated microfibrils present in the wall of the *oral cavity* or *infundibulum* of certain ciliates; often united at condensation nodes, giving a striking hexagonal pattern at the ultrastructural level (e.g., in some peritrichs) (FR, Fig. 2.6Ac).

Filaments: any fine fibrous components of the *cytoskeleton*; see **Attachment Organelle**; also an extracellular secreted structure used for attachment (e.g., *Strobilidium*); see **Microfilaments**.

Filial Products: generalized term for (daughter) organisms resulting from any mode of ciliate *fission*; includes *tomites* and *buds*, as well as the usual *proter* and *opisthe*.

Fission: cell division; *asexual* reproduction; the sole mode of reproduction (nuclear mitosis and excluding meiosis) in ciliates; many kinds or types – *iso-* or *anisotomy* (*filial products* of equal or unequal size), *palintomy, strobilation*, and *budding*; a cystic stage is sometimes regularly involved; in the usual binary fission, the anterior filial product is called the *proter* and the posterior filial product, the *opisthe*; see also **Homothetogenic Fission**, **Interkinetal**, **Perkinetal**, and **Symmetrogenic Fission**.

Fixation Organelle: see **Holdfast Organelle**.

Flange: literally, "a projecting rim", used variously in ciliatology (e.g., an ectoplasmic flange underlies part of the *paroral* in the hymenostome *Glaucoma*).

Food Vacuole: intracellular vacuoles formed (usually) at the distal end of the *cytopharynx* and containing food materials in either a particulate or dissolved state; the food vacuolar membrane, supplied in the region of the *cytopharynx*, may originate in *discoidal vesicles* or packets delivered with the aid of certain *microtubules* in the vicinity; digestion takes place within the food vacuoles after fusion of *acidosomes* and *lysosomes*; solid egesta are often discharged through a *cytoproct*; also called phagosomes, phagocytic ("cell engulfing") vacuoles or gastrioles.

Forma: see mention under **Variety**.

Fragmon: see **Kinetofragment**.

Frange: band of *perioral ciliature* characteristic of certain nassophoreans; varying in composition from an extensive line of specialized ciliature winding around much of the anterior end of the organism to a short linear group of as few as three *pseudomembranelles* or *pavés* adjacent to the *cytostome* proper; sometimes called an adoral ciliary fringe, but more often the hypostomial frange (ACF, Fig. 2.5Ae; HF, Fig. 2.3i).

Fringe: on occasion, used alone or in other combinations, with different meanings; sometimes incorrectly used when *frange* is meant; see **Locomotor Fringe**.

Fusiform Trichocyst: see **Trichocyst**.

G

Gamma Membranoid: see **Membranoid**.

Gametic Nucleus: the *haploid* nucleus derived by meiosis from the diploid *micronucleus* prior to *conjugation*; meiosis may be followed by mitosis to produce the gametic nuclei.

Gamone: soluble substances active in inducing conjugation (e.g., see **Blepharmone**).

Gamont: members of a conjugating pair (iso- or anisogamonts, equal or unequal in size, with the latter kind including micro- and macrogamonts or more commonly *micro-* and *macroconjugants*); usage of this terminology is not widespread in ciliatology; see **Conjugation**.

Gastriole: see **Food Vacuole**.

Gemmation: see **Budding**.

Generative Nucleus: see **Micronucleus**.

Germinal Field: line of *non-ciliferous kinetosomes* associated with the terminal portion of the infraciliary base of the *paroral* or *haplokinety* of peritrichs; serves as an *anlage* in *stomatogenesis* in peritrichs, and may be homologous with the *scutico-vestige* of scuticociliates (GF, Fig. 2.6Ac).

Germinal Kinety: see **Germinal Field**.

Germinal Row: see **Germinal Field**.

Glandule, Secretory: see **Ampulla, Secretory**.

Golgi Apparatus (pl. **Apparati**): intracytoplasmic membranous structure consisting of flattened saccules (cisternae), often stacked in parallel arrays, and *vesicles*; involved in elaboration or storage of secretory products, such as *lysosomes* and *extrusomes*; the Golgi apparatus is not prominent in ciliates, in contrast to the condition in many other protozoa; often called a dictyosome.

Golgi Body: see **Golgi Apparatus**.

Grain Convention: see **Numbering Conventions**.

Gullet: non-preferred term used for the *buccal cavity* of ciliates, such as *Paramecium*.

Gymnostome: literally meaning "naked mouth", and not really appropriate since ciliates with a cytostome have some kind of oral ciliature, the suctorians being a notable exception.

H

Haploid: N set of chromosomes; in ciliate life cycles, haploidy is characteristic of the meiotically-reduced gametic nuclei.

Haplokinety (pl. **Haplokineties**): once-popular term for the infraciliary base of a generalized *paroral*, especially in ciliates belonging to the class Oligohymenophorea; typically a double row of kinetosomes (paired tangentially as *stichodyads*), joined in a zigzag pattern, generally with only the outermost *kinetosomes ciliferous*; also used to mean the entire *paroral*, the ciliated portion plus its infraciliary base; in scuticociliates, the haplokinety or zeta membranoid has been described as comprised of one, two, or three kinetosomal segments (see remarks under **Membranoid**), depending on the species under consideration (Hk, Figs. 2.6Cc, 2.7a, 2.7b, 2.7e, 2.7i).

Haptocyst: minute *extrusome* in the *suctorial tentacles* of suctorians; presumed to contain lytic enzymes useful in the capture of prey organisms; sometimes still referred to as a microtoxicyst, a missile-like body, or a phialocyst (Fig. 2.9Cd).

Haptotrichocyst: rod-shaped *extrusome* of rhynchodid phyllopharyngeans; synonym for *acmocyst.*

Head: generalized term, variously used in ciliatology, but usually in a nonspecific way.

Heterokaryotic: possessing more than one kind of nucleus; characteristic of the great majority of ciliates, with their *micro-* and *macronucleus*; see **Nuclear Dualism** (Fig. 2.12).

Heteromembranelle: specialized term for each of the several to many adoral *polykinetids* of the clevelandellid armophoreans; the infraciliary bases of the anterior or third row of *kinetosomes* are joined to the posterior row by a different (hence "hetero") set of interkinetosomal connectives; see **Paramembranelle**.

Heteromerous Macronucleus (pl. **Macronuclei**): nucleus partitioned into karyomeres (*orthomere* and *paramere*) with strikingly different DNA and RNA contents, and therefore, with differential staining capacities; found especially in cyrtophorian and chonotrich phyllopharyngeans; see also **Homomerous Macronucleus** (Fig. 2.12r, 2.12bb).

Heterotrophic: requiring organic molecules, typically derived from other organisms, to provide nutrients; see **Autotrophic** and **Mixotrophic**.

Histophagous: literally "tissue-eating"; the feeding habit of ciliates living on or in the usually unhealthy (i.e., wounded, moribund, or decaying) bodies of aquatic or edaphic metazoa, including vertebrates (generally larval forms), as well as many kinds of invertebrates of all sizes; blood is one of the preferred tissues for certain ciliates; examples of histophagous forms include species of the hymenostomes *Ophryoglena* and *Tetrahymena* and such scuticociliates as *Anophryoides*, *Mesanophrys*, and *Porpostoma*; often misspelled as "*histiophagous*".

Holdfast Organelle: any structure by which a ciliate can affix or attach, temporarily or permanently, to a living or inanimate substratum (e.g., by use of cilia, hooks, uncini, crochets, tails, loricae, mucous filaments, spines, stalks, suckers, tentacles, and the like); in the usual, more restricted sense, a specialized organelle, such as *stalks* of various kinds, the *adhesive disc* of mobiline peritrichs, the *sucker* of some astomes or clevelandellid armophoreans, or the localized *thigmotactic ciliature* of many thigmotrichine scuticociliates; see **Attachment Organelle** (Fig. 2.9B).

Holotelokinetal: *telokinetal stomatogenesis* in which the *oral anlage* is derived by proliferation of *kinetofragments* from all *somatic kineties*; found in haptorians (Fig. 2.11Da).

Holotrichous: having somatic *cilia* evenly distributed over the body surface; see **Oligotrichous**.

Holotype: the individual organism to which is attached the species-group name; often with ciliates, it is accompanied on a type slide by a number of other individuals, which can be considered lectotypes.

Holozoic: mode of nutrition or feeding in which particles or whole prey are ingested; for a contrasting mode; see **Saprozoic**.

Homokaryotic: fundamentally possessing but one kind of nucleus, neither a micronucleus nor a macronucleus; among the ciliates, this characteristic was limited to *Stephanopogon* species, which are no longer considered to be ciliates; not to be confused with the *amicronucleate* condition.

Homologous: characters, traits, structures or organelles that resemble one another due solely to inheritance from a common ancestor; if such ancestry is unknown or unknowable, it may be inferred, keeping in mind the possibility of *convergent evolution* and other such confounding factors; see **Analogous**.

Homomerous Macronucleus (pl. **Macronuclei**): nucleus with no differentiation into zones containing differing DNA and RNA contents; essentially uniform staining capacity exhibited, except for heterochromatin granules and nucleoli; this is the type of macronucleus found in the great majority of ciliates; see **Heteromerous Macronucleus** (Fig. 2.12).

Homonym: one of two or more names identical in orthography (spelling) applied to different organisms or taxa; the earlier published name of two is the senior homonym and must prevail; the other is the junior homonym and must be replaced, unless it is a junior synonym; see **Synonym**.

Homopolar Doublet: individual ciliate with two sets of mouthparts, separated by 180° at the anterior end of the body; typically, a teratological and unstable condition.

Homothetogenic Fission: type of division of a *parental form* in such a manner that there is a point-to-point correspondence (i.e., exhibition of the condition of homothety) between structures or "landmarks" in both *filial products*, the *proter* and the *opisthe*; generally transverse or *perkinetal fission*; to be contrasted with the *symmetrogenic* or mirror-image *fission* of flagellates (*interkinetal* division) (Fig. 2.11A).

Homothety: see **Homothetogenic fission**.

Hook: see **Attachment Organelle**.

Host: independent or so-called dominant member of a symbiotic pair, unless the relationship is a *mutualism*; the dependent partner, the *symbiont*, lives in or on the host.

Hydrogenosome: a cytoplasmic *organelle* derived from the ciliate *mitochondrion*; it produces hydrogen and ATP, and may or may not have remnants of the mitochondrial cristae.

Hyperparasitism: see **Parasitism**.

Hypoapokinetal: *apokinetal stomatogenesis* in which the *oral anlage* develops or begins development in a subsurface pouch or an "intracellular" tube; found in some spirotrichs (Fig. 2.11h).

Hypostomial Frange: see **Frange**.

I

Indicator Organism: an organism whose presence reflects a certain set of ecological or environmental conditions; used with respect to "pollution ecology" studies; see **Saprobity System**.

Infraciliary Lattice: branching filamentous tract or mat at the boundary of ectoplasm and endoplasm, running parallel to and not far from the surface of the organism's body, but at a deeper level than that of either the *epiplasm* or the layer of *striated bands* (as known in *Paramecium*).

Infraciliature: assembly of all *kinetosomes* and associated microfibrillar or microfilamentous and microtubular structures, both somatic and oral in location; lying below the *pellicle*; the *argentophilic* nature of most such organelles and structures and their universality and stability make the infraciliature an ideal system for study to gain significant information in areas of morphogenesis, evolution, and phylogeny, as well as in comparative systematics at all taxonomic levels (Fig. 2.4).

Infundibulum (pl. **Infundibula**): lower or inner or posterior part or section of the *buccal cavity* in certain ciliates, particularly peritrichs; an often long, funnel-shaped tube or canal; may contain some of the *oral ciliature* and its *infraciliature* (e.g., the *oral polykinetids* of peritrichs) (Inf, Fig. 2.6Aa).

Ingestatory Apparatus (pl. **Apparati**): see **Oral Apparatus**.

Inquiline: term often used for *commensal* organisms living in a body cavity of some *host* and not obtaining nourishment directly from or at the expense of it; employed primarily for echinophilic ciliates essentially endemic in the digestive tract of their hosts (e.g., certain scuticociliates).

Interkinetal: between the kineties; as a kind of division in protozoa; see **Symmetrogenic Fission**.

Intermeridional Connectives: apically located concentric silverlines encircling the anterior part of the body of certain ciliates (e.g., *Tetrahymena*); probably *argentophilic* artifacts of some sort, but their constancy and invariate nature, make them of value taxonomically; see **Circumoral Connective** (IC, Fig. 2.7e).

Internal Budding: see **Endogenous Budding**.

International Code of Zoological Nomenclature: authoritative dicta regarding all nomenclatural matters for animals essentially to the level of familial taxa; assumed to include the protozoa, provisions of "the Code" affect systematics and classification to the obvious yet significant extent that nearly all taxonomic decisions ultimately require use of scientific names.

Interstitial: living between, among, or in the interstices of sand grains or similar sediments; see **Psammophilic**.

Intertelokinetal: *telokinetal stomatogenesis* in which the *oral anlagen* of *kinetofragments* are produced by proliferation of *kinetosomes* both at the ends and beside the ends of all *somatic kineties*; thus, the number of *oral kinetofragments* exceeds the number of somatic kineties; found in some vestibuliferian trichostomes.

Intrabuccal Kinety (pl. **Kineties**): a *kinety* extending deep into the *oral cavity* of loxodid karyorelicteans.

Intraclonal Conjugation: *conjugation* within a *clone*, apparently pure for one mating type; sometimes referred to as selfing.

Intracytoplasmic Pouch: temporary depression, cavity or "vacuole" in which the *oral anlage* appears during *stomatogenesis*; found especially in entodiniomorphids, oligotrichs, and some hypotrichs; see **Hypoapokinetal** (IcP, Fig. 2.11h, 2.11i).

Intramacronuclear Microtubules: *microtubules* that assemble inside the *macronuclear envelope* during division by *amitosis* of the *macronucleus*; considered to be the derived character for the Subphylum Intramacronucleata.

Isoconjugants: conjugants of the same size; see **Conjugation**.

Isogamonts: gamonts of the same size; see **Gamonts**.

Isotomic: literally "equal parts"; see **Fission**.

K

Kappa Particles: members of the bacterial groups, alpha-proteobacteria and gamma-proteobacteria, endosymbionts of *Paramecium* species; see **Endosymbiont**.

Karyoklepty: literally "nuclear stealing"; a phenomenon in which a host ciliate captures the nuclei of its symbionts and uses these to maintain symbiont cytoplasm within the ciliate host's cytoplasm; only demonstrated so far in the litostome *Myrionecta rubra*.

Karyokinesis: synonym of *mitosis* (i.e. the division of the nucleus).

Karyological Relict: organism that is presumed to be a remnant or direct and little-changed descendant of an early or phylogenetically ancient group of ciliates, at least with respect to its nuclear condition or properties; particularly assumed to be the case for karyorelicteans with diploid, non-dividing *macronuclei*; see also **Macronuclear Evolution Hypothesis**.

Karyomere: see **Heteromerous Macronucleus**.

Karyonide: *clone* bearing the descendants of one of the new *macronuclei* produced during nuclear differentiation following *conjugation*.

Karyophore: strands or sheets of specialized and generally conspicuous fibers emanating from subpellicular locations and surrounding and suspending the *macronucleus*; found in some clevelandellid armophoreans (Kph, Fig. 2.12dd).

Kathrobic: preferring cold environments.

Kinetal Segment: in a broad sense, used for a section of any row or file of *kinetosomes*; see **Kinetofragment**.

Kinetal Suture System: see **Secant System**.

Kinetid: elementary repeating *organellar complex* of the typical ciliate *cortex*, consisting of a *kinetosome* (or two more kinetosomes) and its fibrillar associates, which include *cilium, unit membranes, alveoli, kinetodesma*, and various ribbons, bands, or bundles of *microtubules*, including some *nematodesmata*, and sometimes also *microfibrils, myonemes, parasomal sacs,* and *extrusomes*; synonyms are kinetosomal territory and ciliary corpuscle (Figs. 2.1E, 2.2).

Kineties: see **Kinety**.

Kinetodesma (pl. **Kinetodesmata**): typically periodically striated, subpellicular fiber arising close to the base of a somatic kinetosome, near Triplets Numbers 5–8 (see *Numbering Conventions*), and extending right or anteriad and toward or parallel to the organism's pellicular surface and on the right side of the kinety involved (*Rule of Desmodexy*); due to fixation artifacts (?), some fibres, like the *retrodesmal fiber*, in the same triplet position, may be positional homologues; when kinetodesmata are of a length greater than the interkinetosomal distance along the kinety, they overlap, producing a bundle of fibers (Kd, Figs. 2.1, 2.2, 2.4J).

Kinetodesmal Fiber: see **Kinetodesma**.

Kinetodesmal Fibril: see **Kinetodesma**.

Kinetodesmata: see **Kinetodesma**.

Kinetofragment: segment, patch or short file of basically *somatic kinetids* in the general vicinity of the *oral region*, originating from the nearby anterior terminations of the somatic kineties converging onto the general oral region; the *frange* and *pseudomembranelle* may be considered to be kinetofragments.

Kinetofragmon: the assembly of *kinetofragments* around the *oral region*; found in nassophoreans.

Kinetome: an *organellar system* composed of all *kinetids* (i.e., the *kineties*) covering the body of a given ciliate; the total mosaic of an organism's kinetids.

Kinetorhiza: a little-used synonym of *ciliary rootlet*.

Kinetosomal Territory: see **Kinetid**.

Kinetosome: homologue of centriole; cortical tubular cylinder of nine longitudinally oriented, equally spaced, skewed, peripheral triplets, each composed of three *microtubules*; when viewed from deeper in the cytoplasm of the organism looking outward, the nine triplets of microtubules are skewed inwardly, clockwise; typical size, *ca.* 1.0 μm long × 0.25–0.3 μm diameter; when *ciliferous*, produces a *cilium* at its distal end (Ks, Figs. 2.1, 2.2).

Kinetosome Triplet Numeration: see **Numbering Conventions**.

Kinety (pl. **Kineties**): single structurally and functionally integrated somatic file or row of *kinetids*, typically oriented longitudinally; may be composed of *monokinetids*, *dikinetids* or *polykinetids*; ancestral condition presumed to be a *bipolar kinety*, with derived states as fragmented, intercalated, partial, and shortened; asymmetry of kinetids allowing recognition of anterior and posterior poles of the organism itself (see **Rule of Desmodexy**); not to be used in reference to oral infraciliary structures (Fig. 2.2).

Kinety Number 1: the somatic *kinety* to the immediate right of the *oral region* or terminated anteriorly by the posterior margin of the *oral region* and/or identified as the rightmost *postoral meridian*; in tetrahymenine hymenostomes, has two unique features or properties – (1) it is the so-called stomatogenic kinety or stomatogenous meridian (see **Parakinetal Stomatogenesis**), and (2) it bears, or is topologically associated with, the *cytoproct* at its extreme posterior end; see **Numbering Conventions** (K1, Fig. 2.7a, 2.7b, 2.7e, 2.7f, 2.7g).

Kinety n: the last kinety obtained by numbering clockwise around the ciliate beginning with *Kinety 1*; see *Numbering Conventions* (Kn, Fig. 2.7a, 2.7b, 2.7e, 2.7f, 2.7g).

Kinety Numeration: see **Numbering Conventions**.

Kinoplasm: see mention under **Spasmoneme**.

Km Fiber: synonym for *postciliodesma*; see **LKm Fiber**.

Knob: see **Attachment Organelle**.

L

Lachmann's Bristle: see **Extensor Membrane**.

Lamina Corticalis (pl. **Laminae Corticalis**): dense fibrillar or filamentous layer beneath the pellicle, marking the ecto-endoplasmic boundary; in certain groups, seems to be indistinguishable from the *epiplasm*; appears to be continuous with the *filamentous annulus* of the *rhabdos* in certain haptorians; a synonym of tela corticalis.

Larval Form: a motile migratory form or dispersive form in the life cycle of free-living sessile or sedentary ciliates; includes the *bud* of suctorians and chonotrichs, but also the *telotroch* of peritrichs and the migratory stage of loricate heterotrichs; sometimes called *swarmers*; usually morphologically dissimilar to their *parental forms* or even the other *filial product* of the *fission*; *tomites* or *phoronts* of *histophagous* and *parasitic* species, which may well serve the same purpose, are traditionally not referred to as larval forms (Fig. 2.11B, 2.11C).

Lasiosome: literally "woolly-body"; dense linear array of granules in the axoneme of the *cilia* of some hypotrich spirotrichs.

Lepidosome: epicortical structures that often appear scale-like and that cover the body surface of a ciliate; a kind of *extrusome*.

Lieberkühn, Organelle of: lenticular refractile structure invariably and exclusively found beneath the pellicle close to the left side of, or in the left wall of, the *buccal cavity* of ophryoglenine hymenostomes; may function in phototaxis.

Life Cycle, Clonal: the physiological and genetic changes undergone by a clone of cells beginning with *conjugation* as the start of the life cycle "clock"; a series of physiological and genetic states described as immaturity, adolescence, maturity, and senescence; genetic research has demonstrated that senescent cells (e.g. *Paramecium*) can be rejuvenated partially by *autogamy* and completely by *conjugation*, which essentially restarts the life cycle "clock".

Lips: generalized term, variously used; experts on different groups of ciliates may employ it for specific yet non-homologous structures (e.g., the lips of the tintinnine lorica, the lips of the lorica of peritrichs).

Lithosome: *vesicle* containing some inorganic material, often laid down in concentric layers.

Littoral: pertaining to the zone of the shore between high- and low-water marks; this intertidal zone is the biotope of many marine *psammophilic* ciliates.

LKm Fiber: structure composed of an assemblage of overlapping *transverse microtubules* originating near Triplet Numbers 4,5 (see **Numbering Conventions**) of the posterior *kinetosome* of a somatic *dikinetid*; first described in the colpodean *Woodruffia*, the assemblage of *microtubular ribbons* runs on the left side of the associated *kinety*; totally different from a *kinetodesma*, with which some workers have confused it; see **Postciliodesma** (LKm, Fig. 2.2).

Locomotor Fringe: ring of specialized "compound" ciliature (sometimes called *pectinelles*) around the posterior part of the body of the *telotroch* of a sessiline peritrich and around the *adhesive disc* of a mature mobiline peritrich; used in swimming by the migratory *larval form*, and generally resorbed in the *adult form*; also known as a *trochal band* (LF, Fig. 2.11B).

Locus of Stripe Contrast: a ventral region of the body showing the greatest contrast with respect to width of contiguous granular, pigmented stripes; in *Stentor*, the site of *oral anlage* formation and also the region of stripe proliferation; see also **Secant System**.

Longitudinal Microtubule(s): a single microtubule or ribbon or band in the *pellicle* subjacent to the *plasma membrane*, running longitudinally down the body between the kineties; found in some oligohymenophorean ciliates.

Lorica (pl. **Loricae**): secreted and/or assembled test, envelope, case, shell, or theca; may be calcareous, composed of some proteinaceous or mucopolysaccharide secretion, including chitin, pseudochitin, or tectin, or made up of foreign matter (e.g., sand grains, diatom frustules, coccoliths, debris); found most commonly in peritrichs, folliculinids, and tintinnines, with the important properties of fitting the body loosely, opening at one (anterior) end (or occasionally both ends), and being either attached to the substratum or carried about by the freely-swimming organism (e.g., by tintinnines); may occur in a multiple (arboroid-tree) state; such a "house" or "tube" may be occupied only temporarily (e.g., as is true in the case of some stichotrichs) (Figs. 2.4S, 2.8A).

Loricastome: specialized opening or aperture, surrounded by thickened but movable lips, in the rigid *lorica* of lagenophryid peritrichs; the *buccal ciliature* may be extended through the aperture when it is open, the migratory *larval form* exits through it, and *microconjugants* can enter through it.

Lysosome: cytoplasmic organelle bounded by a *unit membrane* and containing hydrolytic enzymes; see **Golgi Apparatus**.

M

Macroconjugant: larger member of a pair in *conjugation*, and the only surviving conjugant in cases of *total conjugation*, such as in peritrichs; a little-used synonym is *macrogamont*.

Macrogamont: see **Gamont**, **Macroconjuant**.

Macronuclear Anlage (pl. **Anlagen**): nucleus that begins development from one of the diploid division products of the *synkaryon* and finishes development as the typically highly polyploid *macronucleus*.

Macronuclear Evolution Hypothesis: origin of the complex, nearly autonomous, polyploid *macronucleus* typical of the great majority of contemporary ciliates from a preceding diploid and non-dividing form (stage 2), which, in its turn, supposedly arose – concomitant with the first micronucleus – from a single nucleus (i.e., *homokaryotic* stage 1), before the differentiation that led to *nuclear dualism*; application of this idea to ciliate systematics is enhanced by recognition of the actual existence today of supposed *karyological relicts* assignable to the first two postulated stages as well as to the last, stage 3, which is today the predominant condition.

Macronucleus (pl. **Macronuclei**): so-called vegetative, trophic or transcriptionally active nucleus; controls the organism's phenotype; may be multiple, but even then is typically much larger than the *micronucleus*; most often compact, spherical or ellipsoidal, but sometimes of diverse other shapes (e.g., reniform, moniliform, filiform, dendritic, halteriform, C- or E-shaped); typically *ampliploid* or *polyploid*, but *diploid* or *paradiploid* in the karyorelicteans, with respect to its genomic content; commonly contains numerous small *nucleoli*; may be *homomerous* or *heteromerous*; divides by *amitosis*, though totally incapable of division in the karyorelicteans; has regenerative powers, but normally is resorbed during *sexual phenomena* and replaced by products of a *synkaryon*, itself derived from fusion of *gametic nuclei* (Ma, Fig. 2.9Aa, 2.9Af, 2.12).

Macrophagous: feeding on relatively large particles of food; see **Algivorous**, **Carnivorous**, and **Microphagous**.

Macrostome: a stage in the polymorphic life cycle in which the oral apparatus undergoes morphogenesis to become enlarged and capable of ingesting larger prey items, typically other ciliates, and sometimes conspecifics; see **Microstome** (Fig. 2.4B).

Macrozooid: see **Zooid**.

Marginal Cilia: circumferential band of long, stout cilia located above the aboral *locomotor fringe* of many mobiline peritrichs; sometimes called *cirri* because of their stoutness.

Marsupium: see **Brood Pouch**.

Maternal Form: see **Parental Form**.

Mating Type: a physiological state of the mature stage in the clonal *life cycle* of a ciliate enabling it to engage in *conjugation* with other individuals of different or so-called complementary mating type; can be developmentally determined either genetically or epigenetically (e.g., by cytoplasmic factors or environmental factors).

Maxillary Armature: see **Capitulum**.

M-band: see **Myoneme**.

Meganucleus: see **Macronucleus**, the preferred word.

Membrana Quadripartita: see **Quadrulus**.

Membrane: generalized term with a variety of particular meanings depending on its specific modifier: (e.g., see Cystic Membrane, Nuclear Membrane, Paroral Membrane, **Plasma Membrane**, **Undulating Membrane**); in ciliate systematics, often understood, to mean a ciliary membrane, such as the *paroral*.

Membranelle: one of the several serially arranged oral *polykinetids*, often known as the *adoral zone of membranelles* (AZM) or adoral zone of oral polykinetids (AZOPk), typically found on the left side of the *buccal cavity* or *peristomial field*; its *cilia*, sometimes seemingly fused or partially coalesced, if only hydrodynamically, may be used in food-getting or locomotion; the generally rectangular infraciliary base is commonly composed of two, three or more rows of densely set *kinetosomes*, which may or may not be associated with *parasomal sacs* and may be linked by microtubular or microfibrillar structures in specific patterns not necessarily identical for each row of the base or for the similar-appearing "membranelle" in a different taxonomic group; membranelle *sensu stricto* may be used to refer to the left-hand oral ciliary organelles of the tetrahymenine hymenostomes; membranelle *sensu* *lato* includes organelles very likely not homologous (e.g., see **Heteromembranelle, Membranoid, Paramembranelle, Peniculus, Polykinety,** and **Quadrulus**) (M1, M2, M3, Fig. 2.7e).

Membranoid: an oral *polykinetid*, including either the definitive membranellar fields or some stage in their development, in species belonging to the oligohymenophorean scuticociliates; at one time, alpha, beta, gamma and zeta types were differentially defined, the first three referring to the oral *polykinetids* on the left side of the buccal area, from anterior to posterior, and the fourth to the *paroral* on the right; the zeta membranoid or paroral in some scuticociliate species may consist of three, more or less separable, segments ("a", "b" and "c"), with terminal fragmentation (probably of "b", in this case) into a dozen additional pieces in the genus *Schizocalyptra* and with "a" most anteriad and "c" (= *scutico-vestige*) at the posterior end, sometimes far to the left (e.g., in *Pleuronema*).

Meridian: see **Ciliary Meridian**.

Merotelokinetal: *telokinetal stomatogenesis* in which the *oral anlage* is derived by proliferation of *kinetofragments* from a limited number of *somatic kineties*; found in colpodeans, cyrtophorians, and prostomes (Fig. 2.11Db, 2.11Dc).

Mesocyst: see **Cyst**.

Mesosaprobic: see mention under **Polysaprobic**.

Metacyst: a "granular layer" between the ciliate cell surface and the *endocyst*; see **Cyst**.

M Fibers: see **Myoneme**.

Microbiocenosis: restricted natural community of interacting microorganisms, including ciliates, with a stability of limited duration, but temporary equilibrium may be repeatedly regained.

Microconjugant: smaller member of a pair in *conjugation*; completely absorbed by the *macroconjugant* in cases of *total conjugation* (e.g., in chonotrichs and peritrichs); a little-used synonym is microgamont.

Microfibril: generalized term, perhaps better considered without the prefix; many structures and organelles that are (micro)fibrillar in composition are composed of non-hollow filaments 4–10 nm in diameter; the term is most frequently used in its adjectival form; fibrillar or microfibrillar constituents may include such prominent and organized structures as the *kinetodesma*, *myoneme*, and *spasmoneme*, and perhaps the *karyophore* and the *filamentous annulus*; "filamentous" may be used to describe the very same organelles; see **Microfilament**.

Microfilament: generalized term, perhaps better considered without the prefix; the finer or finest composition (*ca.* 5 nm in diameter) of a number of important organelles appears to be microfilamentous in nature, often densely so and with or without nodes; microfilamentous structures may include the *epiplasm*, the *filamentous reticulum*, and the *infraciliary lattice*; if this is the ultimate or lowest macromolecular level of organization, then there should be a distinction, even in very generalized usage, between this term and *microfibrils*, but this has not always been the case in the literature – the terms have been used interchangeably by some workers; see **Microfibril**.

Microgament: see **Gamont**.

Micronucleus (pl. **Micronuclei**): so-called generative nucleus, typically much smaller than the *macronucleus*; may be multiple, generally spherical or ovoid in shape, and typically *diploid* in its genomic content; without *nucleoli* and typically showing no transcriptional activity; its *nuclear envelope* with pores in some species, without them in others; divides mitotically or meiotically, playing a major role in sexual phenomena, such as *autogamy* and *conjugation*; absent in *amicronucleate* strains or races (Mi, Figs. 2.9Af, 2.12).

Microphagous: feeding on small or very small particles of food; a generalized term embracing especially *bacterivorous* and sometimes *algivorous* feeding; to be contrasted with *carnivorous, histophagous, saprozoic,* and especially *macrophagous*.

Micropyle: differentiated pore in the wall of a resting *cyst* through which the ciliate emerges on excystment; the pore canal is sealed by cyst wall material; found in some spirotrichs and colpodeans (Mpy, Fig. 2.9Ag).

Microstome: a stage in the polymorphic life cycle in which the oral apparatus undergoes morphogenesis to become reduced in size and capable of ingesting only small prey items, typically bacteria; see **Macrostome** (Fig. 2.4A).

Microstome-Macrostome Transformation: see **Stomatogenesis**.

Microtoxicyst: used as a synonym of *haptocyst*, but might also refer to some other minute *toxicyst*.

Microtubular Ribbons: a set of *microtubules* aligned laterally to form a flat "ribbon-like" structure; the most striking microtubular ribbons include the *transverse* and *postciliary microtubules*, and the microtubular arrays in the *suctorial tentacle* (Figs. 2.1, 2.2, 2.10).

Microtubule: hollow, cylindrical structure of indeterminate length, *ca*. 20–25 nm in diameter, composed of subunits of *tubulin*; rigid, often cross-linked with others to form a *microtubular ribbon* or *nematodesma*; microtubules in the cytoplasm are typically associated with the *kinetosome* (Figs. 2.1, 2.2, 2.10).

Microzooid: see **Zooid**.

Migratory Form: see **Larval Form**.

Missile-like Body: see **Haptocyst**.

Mitochondrion (pl. **Mitochondria**): generally conspicuous organelles in the cytoplasm, composed of a complex membrane system with the inner membrane appearing to form cristae of several types, usually tubular in ciliates, and indispensably functioning as the "powerhouse" of the cell; in some ciliates, arranged in specific (often linear) patterns or formations; in scuticociliates, apparently fused (?) in a single interconnected "compound" mitochondrion, a giant *chondriome* located immediately under the pellicular alveoli; in some ciliates, independently transformed to a *hydrogenosome* (e.g., armophoreans, litostomes, plagiopyleans).

Mixokinetal: *stomatogenesis* in which both parental *somatic kineties* and parental *oral structures* are simultaneously involved in development of the *opisthe's oral anlage*; found in nassophoreans, apostomes, and the spirotrich *Protocruzia*.

Mixotrophic: capable of using two or more modes of nutrition (e.g., *autotrophic* and *heterotrophic*).

Monogemmic: production of a single *bud* (at a time); a mode of *fission*.

Monokinetid: a *kinetid* composed of one kinetosome and its fibrillar associates; see **Dikinetid**, **Dyad**, and **Polykinetid** (Figs. 2.1E, 2.2).

Monoparakinetal: *parakinetal stomatogenesis* in which only one *somatic kinety* is involved in formation of the *oral anlage*; found in tetrahymenids (Fig. 2.11Dd).

Monophyletic: condition of a taxon being comprised of a common ancestor and descendants all presumed to be derived from this common ancestor; established by the *cladistic* approach through the sharing of *apomorphic* or derived *characters*; see **Clade**, **Paraphyletic**, and **Polyphyletic**.

Monostomy: condition of having but one *cytostome*.

Monotelokinetal: *telokinetal stomatogenesis* in which the *oral anlage* is derived by proliferation of *kinetosomes* in the *somatic* portion of *oral kineties*; found in pleurostome haptorians.

Monotomic: division of a single individual into but two *filial products*; the mode of *fission* typical of most ciliates.

Monotypic: a taxonomic group having only one included nominal taxon; for example, a monotypic genus includes only one species.

Monoxenic Culture: literally "one stranger" culture; laboratory growth of two kinds or species of living organisms with no others present; for example, a ciliate plus one "stranger" – a bacterium, an alga, a yeast, or another ciliate species; the second organism is typically present in the medium to serve as food for the ciliate of interest, which is usually being studied biochemically or ecologically.

Morphogenesis: coming-into-being of characteristic and specific form; the transformation involved in growth and differentiation or *ontogeny*, resulting in reproduction of the preexisting form, with the same patterned array of cytoarchitectural substructures; morphogenetic movements are involved in the process of *fission*, but also in *cystation*, *conjugation*, *regeneration*, and particularly in *stomatogenesis*; the consistent patterns of such dynamic ontogenetic phenomena may be of considerable value in both phylogenetic and comparative taxonomic work; see also **Biogenetic Law**.

Morphological Species: an assemblage of populations of organisms that share a strong and stable morphological similarity; often assumed by taxonomists to represent a *biological species*, but likely to represent a number of different biological entities.

Morphospecies: see **Morphological Species**.

Motorium: see **Neuromotorium**.

Mouth: of value only as a very general term, used in reference to the *oral region* of any mouth-bearing ciliate; the "true" mouth of a ciliate should be called the *cytostome*.

Mucigenic Body: see **Mucocyst**.

Mucocyst: cortical, membrane-bound, saccular or rod-shaped *extrusome* with a paracrystalline structure; dischargeable as an amorphous, mucus-like mass through an opening in the pellicle; probably involved in cyst formation, among other possible functions; occurs in regular, longitudinal, interkinetal rows in many ciliates; formerly known as a protrichocyst (especially), a mucous trichocyst, or a mucigenic body; an *ampullocyst* has been considered a special type of mucocyst (Fig. 2.9Cb).

Mucous Trichocyst: not a *trichocyst*; see **Mucocyst**.

Müller's Vesicle: small vacuole containing mineral concretions, and functioning as a gravity receptor; found in karyorelicteans such as *Loxodes*.

Mutualism: kind of *symbiotic* relationship in which both partners benefit from the association, the host as well as the *ecto-* or *endosymbiont*; verifiable cases rare in which ciliates are the symbiont, but there are several in which the ciliate is the host (e.g., *Omikron* in *Euplotes*; *zoochlorellae* of *Paramecium*).

Myoneme: fibrillar, ultimately filamentous, organelle with a known or presumed contractile function; in the broadest sense, may include the *spasmoneme* found in the stalk of many peritrichs, the M-bands or M fibers coursing beneath or beside the *kineties* in the bodies of certain contractile heterotrichs, retractors and sphincters in various other groups, and still additional (micro)filamentous strands, bands, sheets or bundles active in contraction or retraction of all or part of a ciliate's body; the fibrils, sometimes running deep in the cytoplasm, may be interconnected to one another, the pellicle, and/or certain *kinetosomes*.

N

Naked: see **Barren Kinetosome**.

Nasse: see **Cyrtos**.

Nebenkörper: literally "neighboring body" and used variously in protozoology; in the case of ciliates, it has been applied, formerly but now inappropriately, to the *parasomal sac*, which is so often found in the near vicinity of kinetosomes.

Neck: term used variously; the often highly extensible region of the body that is immediately posterior to the apical *cytostome* and *corona* in some haptorians; the non-extensible, sometimes quite elongate, part beneath the flared apical end of chonotrichs, but better called a *collar*.

Nematocyst: not a preferred term; see **Orthonematocyst**.

Nematodesma (pl. **Nematodesmata**): birefringent bundle of parallel *microtubules*, often showing a hexagonal, paracrystalline arrangement in cross-section; typically, *kinetosome*-associated; plunging into the cytoplasm at right angles to the pellicle, forming with others the major reinforcements of the *cytopharyngeal apparatus* (*rhabdos* and *cyrtos*) of haptorians, nassophoreans, and cyrtophorines, but also found in other groups (e.g., in frontoniids); formerly identified with light microscopy as trichites, cytopharyngeal rods, or the cytopharyngeal basket (Nd, Figs. 2.5B, 2.7j).

Neoformation Organelle: a permanent tube-like invagination of the cell surface in which the oral structures of some oligotrichs (e.g., *Pelagostrombidium*) develop; see **Intracytoplasmic Pouch**.

Neoteny: retention of major larval characters in the mature or *adult form*; the *trophont* of mobiline peritrichs is sometimes considered a matured or permanently arrested *telotroch* (i.e., the *larval form* in sessiline peritrichs).

Neotype: single specimen designated as the name-bearing type of a species; established when it is believed that no *holotype*, lectotype or syntype(s) exist.

Nephridial Apparatus: see **Contractile Vacuole**.

Nephridial Canal: see **Afferent Canal**.

Nephridioplasm: see **Spongioplasm**.

Neritic: pertaining to the region of shallow water along a seacoast; the biotope near the shoreline edge of an ocean; to be contrasted with *pelagic*.

Nesselkapseltrichocyste: see **Toxicyst**.

Neuroformative System: see **Neuromotorium**.

Neuromotor Apparatus: see **Neuromotorium**.

Neuromotor Concept: see **Neuromotorium**.

Neuromotorium: presumed center or cytobrain or motorium of a ciliate's entire neuromotor apparatus (associated with the now discarded but once very popular Neuromotor Concept); a chromophilic fibrillar bundle formerly thought to play a conductive or active coordinating role in locomotion, feeding, avoidance, and other behaviours; identified as the rest of the "neural" apparatus were various parts of the *argyrome* and/or structures today known to be microtubular or microfibrillar organelles of diverse sorts.

Nomen Conservandum: name to be conserved; with appropriate permission, a name preserved as an exception to some provision of the *International Code of Zoological Nomenclature*.

Nomen Dubium: dubious or doubtful name; a name of uncertain application through lack of sufficient information about it or the organism or taxon with which it might be associated.

Nomen Novum: new name; a name expressly proposed and published as replacement of another name, usually a *junior homonym* requiring such action; often abbreviated to "nom. nov."

Nomen Nudum: "naked" name; a name published without description of its associated taxon; a diagnosis is necessary to validate both the name and taxon involved.

Nomen Oblitum: forgotten name; a name unused as a senior *synonym* for more than 50 years; a long unused invalid name often literally "forgotten" by taxonomists of the group and generally best left in that condition.

Nonciliferous Kinetosome: see **Barren Kinetosome**.

Nonhomologous: the exact opposite of a *homologous character*; however, either character may have an *analogous* function with some other structure.

Nuclear Dualism: presence or existence of two different kinds of nuclei; for example, the *micro-* and *macronucleus* so characteristic of the great majority of ciliates; exhibition of the *heterokaryotic* condition (Fig. 2.12).

Nuclear Envelope: system of membranes or coverings of a nucleus; composed of two *unit membranes*, typically continuous with the *endoplasmic reticulum* and often replete with minute pores.

Nuclear Membrane: older term for *nuclear envelope*.

Nucleolus (pl. **Nucleoli**): typically visible region of the nucleus where assembly of ribosomes is organized around the ribosomal RNA genes; see **Endosome** (Nuc, Fig. 2.12g, 2.12i).

Nucleus (pl. **Nuclei**): see **Macronucleus** and **Micronucleus**.

Numbering Conventions: (1) *Kineties* are numbered – following the method of Chatton and Lwoff – around the body clockwise when viewed from the apical pole, with *Kinety Number 1*, for example, being the rightmost *postoral meridian*, which, in certain hymenostomes, also bears the *cytoproct* posteriorly and is normally the *stomatogenic kinety*; no matter the total number, the last one, immediately to the viewer's right of *Kinety Number 1*, is conventionally labelled as "n" (Fig. 2.7a, 2.7b, 2.7e). For counting kineties, the method of von Gelei, but subsequently generally ignored, gives results exactly the opposite from those of the Chatton and Lwoff system: Number 1 is the same, but the suggested direction of counting is counter-clockwise, and thus the Number "n" meridian is on the right rather than the left side of the first kinety.

(2) Microtubular triplets of a *kinetosome* are numbered – following the convention of Grain (1969) – clockwise around the proximal end of the basal body, viewed as in cross-section from the inside of the organism looking out, with Number 1 being the triplet lying in the axis of the kinety, but it is often less ambiguous to make use of the location of the *postciliary microtubules*, which are assumed to be associated with Number 9, the last triplet (Fig. 2.1D). For counting kinetosomal triplets, the convention of Pitelka (1969) considers Numbers 1–3 to be on the right anterior margin of the kinetosome, looking at a cross-section from the base outward, and associated with the *kinetodesma* (of *Paramecium*); the counting similarly proceeds clockwise around the base. In the Pitelka convention, the triplet associated with the postciliary microtubular ribbon is always number 5. This

equals Number 9 of the Grain convention, the system adopted in this book (Fig. 2.1D).

O

Occam's Razor: equivalent to the principle of parsimony, *viz.*, when faced with two or more hypotheses of equally explanatory value, choose the simplest.

Ogival Field: transitory group of kinetosomes, bearing *thigmotactic cilia*, which appear anterior to the *rosette* during *tomitogenesis* in many apostomes; this pointed, arch-shaped patch of specialized cilia facilitates attachment of the *tomite* to a new substratum, generally a crustacean integument.

Oligomerization: postulated evolutionary process of reduction or diminution, but not necessarily simplification, in the usual numbers of some organelle (e.g., in numbers of *kineties* over a ciliate's body); see **Polymerization**.

Oligoploidy: see **Polyploid**.

Oligosaprobic: see mention under **Polysaprobic**.

Oligotrichous: having sparse somatic *cilia*; typically of ciliates in the spirotrich Subclasses Oligotrichia and Choreotrichia, but also found in some stichotrichs (e.g., *Halteria*) and haptorians (e.g., *Didinium*); see **Holotrichous**.

Oligotrophic: see **Eutrophic**.

Omikron: Gram-negative bacterial *endosymbiont* in the cytoplasm of the hypotrich *Euplotes*; often indispensable to their hosts' life; see **Xenosome**.

Omnivorous: eats everything(!); such ciliates are not at all "fussy" in their feeding habits.

Ontogeny (pl. **Ontogenies**; adj. **Ontogenetic**): history of an individual, from egg to adult; by analogy, in the case of a ciliate, it is the growth and development from a *filial product*, the *tomite*, to the mature *trophont* or *tomont*, ready for another fission, in the full life cycle of the organism; comparative study of the patterns revealed in the *morphogenesis* associated with such ontogenetic development may throw light on the *phylogeny* of the group concerned (phylembryogenesis); see also **Biogenetic Law**.

Operculum (pl. **Opercula**): literally, lid or covering flap; used variously (e.g., as the cover of the emergence pore of some cysts), but mostly for two quite different structures both in sessiline peritrichs: (1) the stalked *epistomial disc* present in many of the operculariids; and (2) the organelle attached to the anterior end of the body, as a stalked "cap" at an oblique angle to the epistomial disc, which may wholly or partially cover the opening of the *lorica* on retraction of the organism into its case in some of the loricate vaginicolids (e.g., *Pyxicola*) (Opr, Figs. 2.8A, 2.9Ad).

Ophryobuccokinetal: *buccokinetal stomatogenesis* in which the *opisthe's oral anlage* derives from one to several *ophyrokineties* and the *paroral*; found in some peniculians.

Ophryokinety (pl. **Ophyrokineties**): literally "brow" *kinety*; one of three or more somatic kineties, often with *dikinetids* and single associated parasomal sac forming a triangular group as revealed in silver-impregnated material; on the ventral surface near the anterior end of the body and located immediately to the right of the buccal cavity proper (e.g., in the peniculine *Frontonia*); generally, but inappropriately called vestibular kineties, may represent a legitimate part of the buccal ciliature *sensu lato* in the organisms bearing them, and hence be considered *perioral ciliature* (OK, Fig. 2.7h).

Opisthe: posterior *filial product* of a regular binary *fission* of the *parental form*; the anterior ciliate resulting from such a division is the *proter* (Fig. 2.11Aa).

Oral Anlage: see **Anlage** and **Oral**.

Oral Apparatus (pl. **Apparati**): the entire complex of structures and organelles involved in or directly related to the *cytosome* and functionally integrated for the acquisition and ingestion of food; multiple in suctorians and absent in astomatous ciliates (OA, Fig. 2.11Aa).

Oral Area: see **Oral Region**.

Oral Atrium (pl. **Atria**): see **Atrium**.

Oral Cavity: an indentation or depression that contains part or all of the *oral apparatus*; see **Buccal Cavity** (OC, Fig. 2.5c).

Oral Ciliature: simple or compound cilia that are directly associated with the *oral apparatus*; associated with it would be the bases of all such structures, the oral *infraciliature* (as opposed to kinds of *somatic ciliature*).

Oral Disc: specialized name for the apically located *oral region* of a ciliate when it is conspicuously separated from the rest of the body (e.g. in the hourglass-shaped spirotrich *Licnophora*).

Oral Groove: generalized term for a depression leading to a *buccal cavity* or a *cytostome*; widely used in the past for *Paramecium* to indicate what was more recently termed *vestibulum* and now considered to be a kind of *prebuccal area* in that organism; see **Vestibulum**.

Oral Infraciliature: see **Oral Ciliature**.

Oral Polykinetid: general term for *organellar complexes* in the *oral region* that are composed of many, usually ciliferous kinetosomes; see **Polykinetid** (OPk, Figs. 2.5Aa, 2.5Ab, 2.7a, 2.7b, 2.7i; 2.4G, 2.4R, 2.7k, 2.7l).

Oral Primordium: synonym of *oral anlage*.

Oral Region: that part of the ciliate's body bearing the *oral apparatus*; convenient to use in a non-specific way; to be contrasted with the *somatic region* (the rest or bulk of the body); *buccal area*, a more restrictive term, is not to be considered a synonym.

Oral Replacement: see **Stomatogenesis**.

Oral Ribs: *argentophilic* pellicular ridges of a non-naked *ribbed wall*; appearing, under light microscopy, to represent lines coursing inwardly in a one-to-one ratio from the kinetosomal bases of the right-hand *paroral*; found in many oligohymenophoreans (OR, Fig. 2.7e).

Organellar Complex: consistently recognizable subcellular structures responsible for subsidiary cell functions and composed of a specific association of *unit organelles*; see **Organellar System**.

Organellar System: an organization of *organellar complexes* integrated to perform a major (i.e., systemic) cellular function (e.g., locomotion, osmoregulation, feeding and ingestion, digestion).

Organelle of Fixation: see **Attachment Organelle**.

Organic Pollution: see **Polysaprobic** and **Saprobity System**.

Orthogenetic Line: supposed *evolutionary series* that has allegedly followed a predetermined pathway and has not invoked nor been subject to the laws of natural selection; such proposed phylogenetic lines are rejected by modern evolutionary theory.

Orthography: correct or conventional spelling.

Orthomere: DNA-rich karyomere of a *heteromerous macronucleus*; to be contrasted with the *paramere*, the other kind of karyomere in that type of nucleus (Om, Fig. 2.12r, 2.12bb).

Orthonematocyst: *extrusome* in which the material to be extruded appears as a capped, straight tubular filament embedded in a matrix whose outer portion appears to be composed of myelin-like sheets; the ciliate organelle is unlikely homologous to the nematocyst of the cnidarians; found in the karyorelictean *Remanella*.

Osmotrophic: see **Saprozoic**

P

Palintomic (adj.): see **Palintomy**.

Palintomy (adj. **Palintomic**): rapid sequence of binary *fissions*, typically within a *cyst* and essentially without intervening growth, resulting in production of numerous, small-sized *filial products* or *tomites*; characteristic of various parasitic ciliates, including some apostomes, the hymenostome *Ichthyophthirius*, and a few others; the net result is similar to that of *polytomic divison* (Fig. 2.9Af, 2.9Ah).

Palp: variously used, often for a protuberance of the body with an alleged sensory function.

Papilla (pl. **Papillae**): variously used; often referring to the pellicular or extrapellicular wart-like bumps or small protuberances on the surface of an organism (e.g., on the *bell* of some sessiline peritrichs and on the body of certain chonotrichs); in a broad sense, tubercle may be considered a synonym.

Paradiploid: condition of ploidy of the macronucleus of karyorelicteans; very close to the diploid DNA amount, hence "para"-diploid; see **Diploid** and **Polyploid**.

Parakinetal: type of *stomatogenesis* in which the *anarchic field* of *kinetosomes* involved in the developing *opisthe* appears to derive directly from or appears alongside one or more of the *postoral somatic kineties* (i.e. stomatogenic kinety) of the *parental form* and at a level destined to be slightly posterior to the eventual fission furrow; the primordial field (*anlage*) for the opisthe's oral apparatus thus appears subequatorially on the ventral surface at a location far removed from the parental oral apparatus; partial or full replacement or restructuring of the parental (now *proter*) oral organelles, involving oral kinetosomes and kinetosomes from the anterior termination of the stomatogenic kinety, may occur simultaneously; characteristic of some hymenostomes and some spirotrichs; it was formerly known as somatic-meridional stomatogenesis; see **Amphiparakinetal**, **Biparakinetal**, **Monoparakinetal**, **Polyparakinetal**, and **Teloparakinetal** (Fig. 2.11Dd, 2.11De).

Paralabial Organ: enigmatic structure in a crypt near one of the adoral *syncilia* in certain entodiniomorphids; composed of pellicular folds and cilia; considered a kind of *sensory organelle*, but this function not proven.

Paralorica (pl. **Paraloricae**): complete *lorica* reconstructed by a tintinnid during the interphase period because the ciliate has abandoned or lost its *protolorica*; constructed more slowly than the protolorica and therefore often having a very different form (e.g., coxliella-form lorica of *Favella* species); see **Epilorica**.

Paramembranelle: specialized term for each of the several adoral *polykinetids* characteristic of free-living heterotrichs and spirotrichs; all its kinetosomes are linked by similar-appearing connectives and its *transverse microtubules* are limited to the kinetosomes of the left (outermost, distal) row of its infraciliary base.

Paramere: DNA-poor karyomere of a *heteromerous macronucleus*; to be contrasted with the *orthomere*, the other kind of karyomere in that type of nucleus (Pm, Fig. 2.12r, 2.12bb).

Paraphyletic: condition of a taxon being comprised of a common ancestor but only some of its presumed descendants; see **Monophyletic** and **Polyphyletic**.

Parasitic (adj.): see **Parasitism**.

Parasitism: *symbiosis* in which one member, the parasite, lives to various degrees at the expense of the other member, the *host*; from the point of view of the parasite, the association may be facultative or obligate; many ciliate species loosely called "parasitic" are more likely just exhibiting *commensalism*; in a general way, often used (e.g., "parasitism," "parasite," and "parasitic" *sensu lato*) as an admittedly imprecise synonym of *symbiosis*, *symbiont*, and *symbiotic;* hyperparasitism, relatively even rarer among ciliates, is the parasitic association of a form with a host (protozoan or metazoan) that is itself a parasite on or in still another host (e.g., the several apostome species that have stages on or in other apostomes, which themselves are parasitic on crustaceans; or chonotrichs on "whale-lice" on whales).

Parasomal Sac: small, *unit membrane*-lined, pit-like invagination or diverticulum in the *pellicle*, characteristically alongside, usually to the right of, a *ciliferous kinetosome*; a site of *pinocytosis* and exocytosis; perhaps the *pellicular pore* of peritrichs is a kind of parasomal sac (PS, Fig. 2.1B).

Paratene: see **Parateny**.

Parateny: condition or presence of recognizable repeating *kinetid* patterns at right angles to the longitudinal axis of the ciliate's body, thus parallel to the equator or eventual fission furrow; paratenes superficially give the impression that the organism's kineties run circumferentially rather than longitudinally in the part of the body affected (e.g., the anterior end of *Dexiotricha*, around the oral region on *Paramecium* and *Disematostoma*) (Par, Figs. 2.3c, 2.3d, 2.3h, 2.4J, 2.4K).

Parental Form: generalized term to denote the mature or about-to-divide stage (e.g., *trophont-tomont*) in the life cycle; the form capable of producing offspring – one or more – depending on the mode of *fission* invoked; generally, this form is itself lost in the process, typically by becoming one of the individuals of the reproduced generation; considerable *morphogenesis* occurs when the parental form persists (e.g., living to produce subsequent generations of *filial products*, as is true of budding in many suctorians and chonotrichs) and/or even dies a natural death itself in due time.

Paroral Kinety (pl. **Kineties**): see **Paroral**; kinety should be used only for *somatic* structures, though it is sometimes used in connection with various oral organelles.

Paroral: preferred term, used in a broad sense, for the ciliary organelle(s) lying along the right side or border of the *oral region*; its cilia may be undulatory or membrane-like, behaving as a single unit because of their fully or partially coalescent nature (see **Undulating Membrane**); different types – some very likely nonhomologous – are recognized by their variation in the pattern and organization of their infraciliature (e.g., *haplokinety* or *stichodyad* of oligohymenophoreans, the *stichomonad* and *diplostichomonad* arrangements in spirotrichs); other kinds of parorals may show additional, if minor, ultrastructural differences, but in all cases the topological position and the probable function are at least analogous; analogues, or possible homologues, include *endoral membrane* (e.g., in *Paramecium*), *undulating membrane* (e.g., in tetrahymenines), and zeta *membranoid* (e.g., in scuticociliates) (Pa, Figs. 2.5Aa, 2.5Ac, 2.7a, 2.7b, 2.7e, 2.7i, 2.7k).

Paroral Membrane: see **Paroral**.

Pavés: "blocks" of ciliary organelles or *kinetofragments*, also called *pseudomembranelles*, characteristic of the *frange* of certain nassophoreans; their infraciliary bases are particularly clearly revealed by methods of silver impregnation (Pav, Fig. 2.3Ai).

PBB-complex: see **Polar Basal Body-complex**.

Pecilokont: seldom used word once proposed to include both "cilium" and "flagellum".

Pectinelle: one of a circumferential band of short rows of closely apposed cilia oriented at an oblique angle to the long axis of the body; sometimes used to describe the composition of both the *locomotor fringe* of peritrichs and the *ciliary girdle* of didiniid haptorians.

Pedicel: term used variously in ciliatology, but generally with reference to a very short attachment *stalk*, such as in certain chonotrichs.

Peduncle: a synonym of a short *stalk*; often reserved for long, highly visible stalks, such as those, not necessarily homologous organelles, found in many peritrichs and suctorians; the adjectival form, "peduncular," is also often used with reference to stalk structures (Pdc, Fig. 2.11B, 2.11C).

Pelagic: pertaining to the open ocean beyond the continental slope or the "high seas" as an ecological habitat, in contrast to the near-shore or *neritic* biotope; eupelagic, for our purposes, is essentially a synonym.

Pellicle: outer "living" zone of the cortex, lying beneath any non-living secreted materials; composed of the typical cell or *plasma membrane* plus the *unit membrane*-lined *alveoli* and, often, the closely apposed underlying fibrous *epiplasm*; sometimes loosely used as synonymous with *cortex*, but the majority of the infraciliary cortical structures and organelles are mostly subpellicular in location.

Pellicular Alveolus (pl. **Alveoli**): see **Alveolus, Pellicular**.

Pellicular Crest: see **Pellicular Ridge**.

Pellicular Pore: self-explanatory term, but particularly used in reference to the numerous minute openings in the pellicle on the *bell* (and perhaps in the area of the *scopula*) of sessiline peritrichs through which are secreted substances involved in mucus-coatings, *lorica*-formation, and *stalk*-production; in many cases, these pores may be, in effect, some kind of *parasomal sac*, even kinetosome-less in the case of the bell of peritrichs and the *scopuloid* of suctorians; they have also been called cuticular pores.

Pellicular Ridge: in a general way, any ridge or crest formed on the surface of the body by the underlying pellicle; often revealed as an *argentophilic* line of contact or juncture of the (membranes of the) adjacent, contiguous *pellicular alveoli*; when, in the *buccal cavity* of various oligohymenophorans, the ridges are underpinned by *postciliary microtubules* and identified as *oral ribs* of a so-called non-naked *ribbed wall*.

Pellicular Stria (pl. **Striae**): ridges or markings in or on the *pellicle*; particularly applied to the circumferential *annuli* on the *zooid* of many sessiline peritrichs, rings that may be comprised of argentophilic *pellicular pores* and/or *pellicular ridges* (PelStr, Fig. 2.11B).

Peniculus (pl. **Peniculi**): kind of *oral polykinetid* in the form of a long band of often short, seemingly fused cilia; its infraciliary base, typically coursing along the left wall of a *buccal cavity*, may be as many as 11 kinetosomes in width but is usually only 3–7, with a tapering to still lower numbers at either end; known classically in peniculines like *Paramecium*, where there is one dorsal and one ventral peniculus; also used for the oral polykinetids in the *infundibulum* of peritrichs (although see **Polykinety**) and various other oligohymenophorans (P1, P2, P3, Figs. 2.6Ac, 2.7h).

Perforatorium: see mention under **Rostrum**.

Pericyst: a layer of material produced and deposited prior to the ectocyst layer and so lying upon it; often a more or less voluminous coat of mucus that may adhere the *cyst* to the substrate or may increase its bouyancy to enable dispersal.

Pericytostomial Ciliature: *cilia* adjacent to and/or surrounding the *cytosome*; see **Oral Ciliature**.

Perilemma: additional outermost "*unit membrane*-like" covering the *pellicle*, especially in various spirotrichs.

Perioral Ciliature: used to include any ciliature, properly *somatic*, even buccal, which is, in effect, around and/or adjacent to the *oral region*; see **Circumoral Ciliature**.

Periphyton Community: see **Biofilm Community**.

Peristome: in a broad sense, a synonym of *oral region*, is well entrenched in the literature to mean the entire expansive *oral region* or *peristomial field* of peritrichs, heterotrichs, and various spirotrichs, in which the *oral ciliature* has often emerged from an *oral cavity* to encircle, though usually only partially, much of the anterior end or pole of the organism's body (Pst, Figs. 2.5Cg, 2.6A).

Peristomial Area: see **Peristome**.

Peristomial Cavity: see **Buccal Cavity**.

Peristomial Ciliature: see **Buccal Apparatus**.

Peristomial Field: in the strict sense, a part of the *oral region* delimited by the *oral polykinetids* or *AZM* of spirotrichs and heterotrichs; this field may be barren (e.g. the spirotrich *Licnophora*) or have *kineties* coursing across it (e.g. the heterotrich *Stentor*).

Perizonal Ciliature: *somatic ciliature*, usually to the right of the *oral region*, the rows of which appear to run transversely (see **Parateny**); the often closely packed cilia are said to function in intensification of the food-carrying water currents that are being directed toward the *oral region*; found particularly in armophorids and odontostomatids.

Perkinetal: across or through the *kineties*; the common mode of *homothetogenic fission* in ciliates in which the division furrow cuts across the body at essentially right angles to the somatic kineties (Fig. 2.11A).

Pexicyst: type of small toxicyst-like *extrusome* in certain haptorians (e.g. *Didinium*), which, on discharge, adheres to the pellicle of the prey without subsequent penetration.

Phagocytic Vacuole: see **Food Vacuole**.

Phagocytotic Vacuole: see **Food Vacuole**.

Phagoplasm: specialized cytoplasm, rich in *discoidal vesicles*, found in or around the *cytopharyngeal apparatus*.

Phagosome: see **Food Vacuole**.

Phagotrophic: a kind of *heterotrophic* nutrition in which particulate food is engulfed in a *food vacuole*.

Pharyngeal Basket: see **Cyrtos** and **Rhabdos**.

Pharynx: see **Cytopharynx**.

Phialocyst: see **Haptocyst**.

Phoront: stage in a polymorphic life cycle during which the organism is carried about on or in the integument of another organism, generally a metazoan; used in a much more restrictive sense to indicate the condition exhibited primarily by certain polymorphic apostomes where it is a stage that is typically preceded by a *tomite* and followed by a *trophont*; see **Symphoriont** (Phor, Fig. 2.9Aj).

Phylembryogenesis: see mention under **Biogenetic Law** and **Ontogeny**.

Phylla (pl. **Phyllae**): *microtubular ribbons* arrayed in a somewhat radial fashion in the *oral apparatus* of phyllopharyngeans; see **Sucking Tube** and **Suctorial Tentacle** (Fig. 2.10a–2.10f).

Phylogeny: history of the race; lines of evolution involving groups of organisms through time and space; a continuum of ontogenies.

Phytoplankton: see **Plankton**.

Pigment Granules: see **Pigmentocyst**.

Pigmentocyst: vesicles providing endogenous pigmentation, of various colors, either in the cortex near the pellicle or deeper in the cytoplasm; many "colored" ciliates derive their hues exogenously (e.g., from endosymbiotic *zoochlorellae* or from pigments of ingested food materials) and not from pigmentocysts; see **Blepharismin** and **Stentorin**.

Pinocytosis: literally "cell drinking"; formation of a small *vesicle* by endocytosis; see **Parasomal Sac** and **Saprozoic**.

Pinocytotic Vesicle: see **Pinocytosis**.

Pitelka Convention: see **Numbering Conventions**.

Plankton: community of predominantly passively floating or weakly motile organisms (including various stages in their life cycles) on or near the surface of a body of water, fresh, brackish, or marine; if the plankton is largely plant-like (e.g., algae), it is called phytoplankton; if largely animal-like (e.g., eggs, larval stages of microcrustaceans), it is called *zooplankton*; classification into net-, micro-, or nanoplankton is based on body size (diameter), >200 μm, 20–200 μm, and 2–20 μm, respectively.

Plasmalemma: synonym for *plasma membrane*; sometimes used as a synonym of *pellicle*; not to be confused with *perilemma*.

Plasma Membrane: the *unit membrane* bounding the surface of the cell; see **Unit Membrane**.

Plesiomorphic (adj.): see **Plesiomorphy**.

Plesiomorphy: an ancestral character in a phyletic lineage; primitive, as in the ancestral condition; see **Apomorph**.

Pleurotelokinetal: *telokinetal stomatogenesis* in which the *oral anlage* is derived by subequatorial proliferation of *kinetosomes* within several right lateral *somatic kineties*; found in some colpodeans.

Podite: the often conical-shaped projection from the ventral surface, near the posterior pole, of certain dysteriid cyrtophorines; a foot-like appendage, rigid though usually slightly rotatable; the structure through which or from which a glutinous, mucus-like filament may be extruded to attach or anchor the ciliate, usually only temporarily, to or over a desirable substratum; also know as a stylet, stylus or style; may be homologous to the basal, secretory part of the *stalk* of the related chonotrichs (Pod, Fig. 2.9Bf).

Polar Basal Body-complex: grouping of *kinetosomes* and sometimes *parasomal sacs* at the posterior end of the body of a number of forms, especially scuticociliates; the kinetosome(s) may bear a long and often stiff *caudal cilium* (PBB, Figs. 2.3c, 2.3d, 2.4F, 2.7f, 2.7i).

Polybrachykinety: a band-like patch of kinetids arranged in multiple short kineties perpendicular or oblique to the longitudinal axis of the band; especially applied to the oral ciliature of ophryoscolecids; see **Syncilium**.

Polyenergid: state of having either multiple nuclei and/or multiple ploidy in a nucleus within a single cell or protistan body; all *heterokaryotic* ciliates exhibit this condition, generally to a high degree.

Polygemmic: production of multiple *buds*, synchronously or consecutively; a mode of *fission* exhibited by some suctorians and chonotrichs (Fig. 2.11Cc).

Polygenomic: synonym for *polyploid*, with respect to ciliate macronuclei; also used to mean the presence of many non-homologous genomes in the same eukaryotic cell (e.g., nuclear, mitochondrial, chloroplast, bacterial).

Polyhymenium: little-used term for denoting the multiple *membranelles* in the *oral region* of heterotrichs and spirotrichs.

Polykinetid: a *kinetid* composed of three or more *kinetosomes* and their fibrillar associates; see **Dikinetid**, **Dyad**, and **Monokinetid**.

Polykinety (pl. **Polykineties**): non-prefered term for the *oral polykinetids* or buccal *membranelles* of certain groups of ciliates; in peritrichs, the polykinety is essentially an extension of oral polykinetid 1 or peniculus 1 onto the peristome (Pk, Fig. 2.6Ac).

Polymerization: postulated evolutionary process of multiplication or increase in usual numbers of some organelle (e.g., in numbers of *membranelles* comprising an *AZM*); may lead to hypertelic development of certain organelles or structures and may be involved in *somatization* as well as in *auxomorphy*; see **Oligomerization**.

Polyparakinetal: *parakinetal stomatogenesis* in which two or more *postoral somatic kineties* are involved in formation of the *oral anlage*; found in many heterotrichs (Fig. 2.11De).

Polyphyletic: condition of a taxon being comprised of some members that are descended from or presumed to have been descended from a common ancestor that is quite different from other members of that taxon; an undesirable situation to be avoided when building a "natural" classification; see **Monophyletic** and **Paraphyletic**.

Polyploid: multiple sets of the *haploid* chromosome number within a single nucleus; characteristic of the ciliate *macronucleus*; a low polyploid condition may be called oligoploid; see **Ampliploid**.

Polysaprobic: pertains to an aquatic habitat poor in dissolved oxygen and rich in decomposition products, generally including high production of ammonia and hydrogen sulfide; exhibiting a high degree of organic pollution; physicochemically similar to *eutrophic*; to be contrasted with mesosaprobic and oligosaprobic habitats that show, respectively, either a medium degree or a low degree of organic pollution; broadly synonymous of polysaprobic are terms such as sapropelic and sapropelebiotic; see **Saprobity System**.

Polystichomonad: type of multiple *paroral* whose infraciliature is composed of more than two parallel rows or files of *kinetosomes*; found in a few spirotrichs; see **Diplostichomonad**.

Polystomy: having many or multiple mouths (e.g., suctorians with their typically numerous *suctorial tentacles*).

Polytomic: division of a single individual into numerous *filial products*, presumably at one time or in quick succession; generally rare in ciliates, but this type of *fission* may occur in certain kinds of *budding*; see **Palintomy**.

Pore: generalized term for variety of holes or generally small openings into or through the "cell" surface (e.g., plasma membrane, envelopes, pellicle, loricae, cysts, brood pouch); the *contractile vacuole pore* may serve as an example of a pore of considerable taxonomic value.

Postciliary Fiber: see **Postciliary Microtubule**.

Postciliary Microtubule(s): singlet, ribbon or band of *microtubules* associate with Triplet Number 9 of the *kinetosome* (see **Numbering Conventions**), first extending diagonally to the right upward into a *pellicular ridge* and then – if well developed – continuing posteriorly, parallel to and between the *kinety* containing its kinetosome and the next kinety to the right, with the ribbon either perpendicular to or parallel to the pellicle (e.g., postciliodesma of karyorelicteans and heterotrichs); in the buccal cavity of many oligohymenophoreans, postciliary microtubules are implicated in formation of the *ribbed wall* (Pc, Figs. 2.1, 2.2).

Postciliodesma (pl. **Postciliodesmata**): the conspicuous fiber, running posteriorly on the right side of the associated kinety and composed of stacked ribbons of overlapping *postciliary microtubules*, and involved in extension of the body following contraction by the *myonemes*; a shared-derived character or *apomorphy* for the classes KARYORELICTEA and HETEROTRICHEA in the subphylum Postciliodesmatophora (Pcd, Fig. 2.2b).

Posterior Microtubule: see **Postciliary Microtubule**.

Postoral Meridian: see **Postoral Somatic Kinety**.

Postoral Somatic Kinety (pl. **Kineties**): ventral kinety terminating anteriorly at the posterior border of the *buccal overture* or of the general *oral region*; in a number of ciliates, the postoral kinety (**POK**) #1, the rightmost POK if there is more than one, is the "stomatogenic kinety" in *parakinetal* stomatogenesis and bears the *cytoproct* in its left posterior extremity as well; particularly characteristic of tetrahymenine hymenostomes (POK, Fig. 2.7f).

Postoral Suture: typically, a midventral *secant system* or line coursing from the *oral region* toward the posterior pole of the organism and onto which the posterior extremities from both sides converge or run roughly parallel to it; see **Preoral Suture** (POS, Figs. 2.3h, 2.7h)

Prebuccal Area: the depression or *oral groove* leading to the *buccal cavity* and lined with somatic or slightly modified somatic ciliature (e.g., in the peniculine *Paramecium*); see **Vestibulum**.

Prebuccal Ciliature: the *somatic ciliature*, more or less modified, lining the *oral groove* or the *prebuccal area*; formerly termed vestibular ciliature; see **Vestibulum**.

Prehensile Tentacle: non-ingestatory cell extension with pointed rather than knobbed end, found in a few suctorians, such as *Ephelota*; allegedly used to capture or hold a prey organism in such a manner as to bring it into contact with the more common *suctorial tentacle*; its microtubules are arrayed in a complex arrangement of interconnected ribbons (Fig. 2.10g).

Preoral Ciliary Apparatus (pl. **Apparati**): all-inclusive term to indicate all categories of *oral ciliature* (from *atrial* to *buccal*), differentiating them from *somatic* ciliature.

Preoral Kinety (pl. **Kineties**): used for the anteriormost of the three oral *kinetofragments* found in certain cyrtophorine phyllopharyngeans; see **Circumoral Kinety**.

Preoral Suture: typically, a short, midventral line or secant system extending, often to the left, from the *oral region* to the apical pole of the organism and onto which the anterior ends of a number of *somatic kineties* from either side may converge (PrS, Figs. 2.3Aa, 2.3Ag, 2.3Ai, 2.4H, 2.4I, 2.7e, 2.7h).

Primary Meridian: see **Ciliary Meridian**.

Primary Ribbed Wall: see **Ribbed Wall**.

Primordium (pl. **Primordia**): see **Anlage**.

Primordial Field: see **Anlage**.

Priority, Principle of: that the valid name of a taxonomic group is the oldest available name, provided that the name is not invalidated by other provisions of the *International Code of Zoological Nomenclature*.

Proboscis: trunk-like extension of the anterior end of certain ciliates (e.g., *Dileptus*); differs from a neck in that the *oral region* is situated at its base rather than at its distal extremity; heavily armed with *toxicysts*, and – though ciliated and active – it is not capable of effecting extreme changes in its length; the non-homologous "proboscis" of *Didinium* is that organism's everted *cytopharyngeal apparatus*.

Protargol: see **Silver Impregnation Techniques**.

Proter: anterior *filial product* of a regular binary *fission* of the *parental form*; the posterior daughter is the *opisthe* (Fig. 2.11Aa).

Protolorica: (pl. **Protoloricae**): *lorica* constructed by the *proter* of tintinnid spirotrichs after *cytokinesis*; see **Epilorica** and **Paralorica**.

Protomite: relatively brief stage in the polymorphic life cycle of a few ciliates (e.g., some apostomes), recognizable by features of its *kinetome* as a separate form between the *tomont* and the *tomite*.

Protomont: relatively brief stage in the polymorphic life cycle of a few ciliates (e.g., some apostomes), recognizable by features of its *kinetome* as a separate form between the feeding *trophont* and the often encysted true *tomont*.

Protrichocyst: older, once popular term for *mucocyst*; a stage in the development of a trichocyst.

Psammophilic: literally "sand-loving"; descriptive term for *interstitial* forms found in, on or at least temporarily associated with, the sands of intertidal zones in marine *littoral* biotopes or in fresh-water beaches and the like.

Pseudobuccal Kinety (pl. **Kineties**): an oral kinety that may have been derived from somatic kinety 1; it is an inverted kinety based on the inverted orientation of its fibrillar associates; found in the Class KARYORELICTEA (e.g., *Loxodes*).

Pseudolorica: an enclosure derived by allometric growth of the external sheath of the stalk, caused by very rapid secretion of outer stalk material; found in peritrichs, such as *Opercularia*.

Pseudomembranelle: rather imprecise term used variously in the literature to describe oral or somatic ciliary complexes that seem to defy classification, but do appear to resemble some kind of *membranelle sensu lato* or complex *kinetofragment*; see **Frange** and **Pavés**.

Pseudonasse: see **Rhabdos**.

Pseudoperistome: term formerly used for the *vestibulum* of trichostomes and colpodeans.

Pulsating Canal: see **Afferent Canal**.

Pulsating Vacuole: see **Contractile Vacuole**.

Q

Quadrulus (pl. **Quadruli**): buccal *polykinetid* with long cilia and an infraciliary base, typically four kinetosomes in width and many in length; the lengthy rows are more loosely associated than is the case in *peniculi* and *membranelles sensu stricto* (e.g., in *Paramecium*); synonyms include membrana quadripartita, Vierermembran, and vierteilige Membran.

R

Radial Canal: see **Afferent Canal**.

Radial Fibers: see **Postciliary Microtubules**.

Radial Pins: see mention under **Border Membrane**.

Reactive Budding: *budding* in response to stressful environmental conditions.

Recapitulation, Law of: see **Biogenetic Law**.

Receiving Vacuole: see **Cytopharyngeal Pouch**.

Regeneration: a process in which parts of the body are developed anew after loss by either natural accidents or experimental manipulations. See **Morphogenesis**.

Reorganization Band: see **Replication Band**.

Replication Band: lightly staining, though with a narrow Fuelgen-positive leading edge, cross-band of a *macronucleus* that migrates or sweeps along the length of the nucleus of spirotrichs; in short macronuclei, one band and in longer macronuclei (e.g., in some hypotrichs) with a similar band traversing the other half either from midpoint out to the ends or from ends into the center; the replication band is involved in DNA replication and histone synthesis, the amounts of these substances doubling just behind the moving bands; in the *homomerous macronuclei* of spirotrichs, preceding macronuclear fission and cytokinesis of the organism itself; two zones may be recognized – the reticular ("forward zone" or "solution plane," as formerly known) and the diffuse ("rear zone" or "reconstruction plane", which is the locus of the DNA synthesis); an analogous (?) structure may also occur in the *heteromerous macronucleus* of certain phyllopharyngeans where there is only one band, moving across the *orthomere* (RB, Fig. 2.12s, 2.12t, 2.12x).

Reproduction: note that, though there are a number of types of *fission*, the only kind of reproduction in ciliated protozoa is *asexual*, textbook statements notwithstanding (i.e., *conjugation*, for example, is a *sexual phenomenon* but not sexual reproduction); see **Fission**.

Reticulated Fiber: see **Filamentous Reticulum**.

Retractor Fibers: generalized term for bundles of *myonemes*, used to draw back some extended part of the body or a protruding oral region (RF, Fig. 2.5Af).

Retrodesma (pl. **Retrodesmata**): see **Retrodesmal Fiber**.

Retrodesmal Fiber: rarely occurring non-striated fiber arising close to the base of a somatic *kinetosome* near its microtubular Triplets Numbers 5–7 and, unlike the *kinetodesma*, extending posteriad and parallel to the pellicle (e.g., in certain clevelandellid armophoreans); may be a homologue by positional similiarity of the kinetodesma; could be called a retrodesma (pl. retrodesmata) (Rd, Fig. 2.1Ec, 2.1Ed).

Rhabdocyst: rod-like *extrusome* composed of a shaft topped by a conical cap; on extrusion, the cap and distal part of the shaft remain unchanged, but are anchored(?) in the cell by a bulbous expansion of the basal portion of the organelle; found in certain karyorelicteans (Fig. 2.9Ca).

Rhabdos: the tubular *cytopharyngeal apparatus* whose walls are strengthened on the outside by bundles of *nematodesmata* and often lined longitudinally both by *transverse microtubules* derived from *circumoral kinetosomes* and by *bulge microtubules* whose origin is undetermined; contains specialized *phagoplasm*, sometimes with included *toxicysts*; may be bound, near its proximal (outer) end, by an expansible *filamentous annulus*; showing a range of complexity in its own composition from a loose organization in some vestibuliferans, which lack toxicysts, to the more elaborate structure in prorodontids and haptorians; principal synonyms of rhabdos include pseudonasse, and the recently used clathrum; clathrum is here considered totally inappropriate in view of its clear implication of a lattice work, whereas the rhabdos is actually both overall, and in its principal separate parts, highly reminiscent of a rod or rods, thus its name, arranged in a straight, non-curved, encircling palisade formation, with perhaps a suggestion of fluting; see **Cyrtos** (Fig. 2.7j).

Ribbed Wall: non-ciliated lining or surface of the right side of the *buccal cavity* of many oligohymenophoreans, ultrastructurally, appearing ribbed due to the presence there of *microtubular ribbons*, presumed to be *postciliary microtubules* that arise in association with the *kinetosomes* of the nearby *paroral*; considered to be naked, typically when no *pellicular alveoli* are involved posterior to the *cytostome*, thus in the *cytopharynx*; considered to be non-naked when *oral ribs* are present; the ribbed wall on the right side is sometimes called the primary ribbed wall, while the much rarer secondary ribbed wall is said to occur on the left side of the buccal cavity.

Rod, Cytopharyngeal: see **Nematodesma**.

Rod, Pharyngeal: see **Nematodesma**.

Rootlet: any fibrillar or microtubular structure originating from or near a *kinetosome* and extending into the cytoplasm away from the *pellicle*; see **Nematodesma**.

Rosette: unique septate structure near the *cytostome* of many apostomes; also used to describe the result of several rapid preconjugation divisions of certain peritrich *zooids* in production of free-swimming microconjugants (e.g., in *Carchesium*).

Rostellum: small *rostrum*; see **Rostrum**.

Rostrum: usually employed in a generalized way, with reference to the apical end of an organism's body when it has the appearance of a beak or shows a distinctive protuberance of some kind; may bear the *cytostome*, as in the haptorian *Chaenea* or a *sucking tube* as in the rhynchodians; the apically located perforatorium or boring apparatus might better be referred to by this less specific term.

Rule of Desmodexy: see **Desmodexy, Rule of**.

Rules of Nomenclature: see **International Code of Zoological Nomenclature**.

S

Saltatorial Cilia: long cilia distributed sparsely around the body (e.g., in *Halteria*), often stiff or heavy when not in motion and used in a quick, jerky sort of jumping locomotion.

Sanguicolous: living in the circulatory system or blood of the host.

Saprobity System: method of classification of aqueous habitats and their contained communities of microorganisms by recognizing both that distinct zones exist with respect to degrees of pollution and that these zones provide certain protists as *indicator organisms* with optimal conditions for their own growth; see **Polysaprobic**.

Sapropelebiotic: see **Polysaprobic**.

Sapropelic: see **Polysaprobic**.

Saprozoic: type of nutrition in which the organism feeds on, takes in, or absorbs food substances in the dissolved state from the surrounding medium, either by active transport or pinocytosis; this osmotrophic mode is to be contrasted with the *carnivorous, histophagous*, *holozoic*, *macrophagous*, *microphagous*, or other feeding or nutritional habits that essentially involve the ingestion of sizable particulate materials, often including whole prey organisms.

Scale: typically a small, sometimes complex structure, organic or mineralized, and of a shape characteristic for a group; origin, when known, by secretion from the *Golgi apparatus*.

Scopula (pl. **Scopulae**): compound organelle, structure or area, at the *aboral* pole of sessiline peritrichs especially; often cup-shaped with a thickened peripheral border or lip comprised of scopulary organelles, such as a plaque or field of *kinetosomes*, typically equipped with very short and immobile cilia, and *pellicular pores*; may function directly as a *holdfast organelle* or, more commonly, may be involved in secretion or elaboration of a *peduncle* or *stalk*; see **Scopuloid** (Sa, Fig. 2.11B).

Scopulary Ciliature: see **Scopulary Organelles**.

Scopulary Kinetosomes: see **Scopulary Organelles**.

Scopulary Organelles: basically the *kinetosomes* of the *scopula*, although their *clavate cilia*, when present, may be included in the definition as well as the associated *pellicular pores*; various additional fibrillar and microtubular structures are associated with these scopulary kinetosomes, and presumably they are also involved in assembly of *stalk* components, when one is present.

Scopuloid: organelle found at the posterior pole of the body of most suctorians; comprised mainly of some kind of *pellicular pores*, which are presumably involved in assembly of the sometimes lengthy, complex, non-living, never contractile *stalk* characteristic of suctorians (Sd, Fig. 2.11Cb, 2.11Cc).

Scutica (pl. **Scuticae**): transient "compound" *kinetosomal* structure or organelle of scuticociliates; identifiable by its shape, location, and presence at a late ontogenetic stage during *stomatogenesis*; quite conspicuous but generally non-ciliated at the time of its existence, the scutica represents the remainder of an often much larger stomatogenic field of *kinetosomes* located near and slightly to the right of the posterior termination of the presumptive infraciliary base of the *paroral* in both the proter and opisthe; typically, manifests a hook-like or whiplash configuration (giving it its name), recurving back to the right; presumably its kinetosomes have arisen from parts of the buccal *infraciliature* of the *parental form*; its typical ultimate fate, if it does not disappear altogether or become entirely incorporated into the paroral, is to persist as a ciliferous or non-ciliferous *scutico-vestige* of varying size and shape, in close juxtaposition to the base of the *paroral* and/or at the anterior end of the *director-meridian*; the scutica is thought to be limited to members of its namesake, the scuticociliates, but its homologue may be present in species of other taxa; erroneously spelled scuticus (Sc, Figs. 2.4P, 2.7i, 2.11Df).

Scuticobuccokinetal: *buccokinetal stomatogenesis* in which the *opisthe*'s *oral anlage* derives either from the *paroral* and the *scutica* or solely from the paroral; found in scuticociliates (Fig. 2.11Df).

Scutico-field: often used with reference to the slightly earlier multi-kinetosomal anlage stage of the *scutica*.

Scutico-hook: term emphasizing what is the most typical appearance of the *scutica*, its hook-like configuration.

Scutico-kinetosomes: kinetosomes comprising the *scutica*.

Scutico-vestige: structure visibly remaining in the *proter* and *opisthe* after the identifiable stage of the dynamic *scutica* has passed; residual field of recognizable *scutico-kinetosomes*.

Scuticus: a misspelling of *scutica*.

Secant System: various lines of convergence of *kineties* in the *somatic region*; *pre-* and *postoral sutures* and the convergence at the antapical pole are typical representatives of such systems; *suture lines* may also occur consistently elsewhere, especially in heavily ciliated organisms that do not have simple *bipolar kineties*; particularly striking in thigmotrichs, astomes, and clevelandellid armphoreans where such stabilized boundary lines or aires sécantes are of considerable taxonomic utility; non-preferred synonym is kinetal suture system (SS, Figs. 2.3, 2.7d).

Sedentary: permanently attached to the substrate, which can be sediment, alga, another organism, or even the inside of a *lorica*; see **Sessile**.

Seizing Organ: a special, structured, discrete organelle associated with the *proboscis* of *Didinium*; now known to be a bundle of discharged *toxicysts* and *pexicysts* used by *Didinium* in feeding.

Secondary Meridian: see **Ciliary Meridian** (2CM, Fig. 2.7e).

Secretory Organelle: used in a broader, more generalized way to refer to any vesicles, glands, pores, adhesive structures, and the like if they are involved in some form of secretion; see **Ampulla, Secretory**.

Selfing: see **Intraclonal Conjugation**.

SEM: scanning electron microscopy.

Seme: unit of phylogenetic information; a unit character, either ancestral or derived, of high information content, usable in reference to any structural part or function of an organism, from the molecular level up to large and *complex unit organelles* or *organellar systems*.

Semi-autonomous: now discarded term for a mode of *stomatogenesis*; see **Buccokinetal**.

Semi-membrane: formerly used as a synonym of the *undulating membrane*; see **Extensor Membrane**.

Sensory Bristle: rather widely applied term to many *bristles* or setae, even when the exact function is unknown; particularly used to describe both (1) the several short rows of *clavate cilia* in such haptorians as *Didinium* and (2) the non-homologous, very short, non-motile cilia occurring in several longitudinal rows of pits on the dorsal surface of many hypotrichs and stichotrichs; also called *dorsal bristles*; the Tastcilien of older literature (SB, Figs. 2.3Ae, 2.7l).

Sensory Organelle: generalized term probably often improperly or imprecisely applied to a variety of structures found in ciliates that may or may not actually possess a sensory function; frequently implicated organelles include diverse *bristles* and *setae*, the *brosse*, other specialized cilia (e.g., *caudal cilia*, *clavate cilia*, tactile cilia, *thigmotactic cilia*), the *concrement vacuole* (and *Müller's vesicle*), *organelle of Lieberkühn*, *palps*, and the *paralabial organ*.

Sessile: attached to substrate either by *lorica*, *stalk*, *holdfast*, *peduncle*, or other cell process; see **Sedentary**.

Seta (pl. **Setae**): see **Bristle**.

Sexual Phenomenon (pl. **Phenomena**): meiosis, haploid *gametic nuclei*, and a diploid *synkaryon* are involved; any *reproduction* that takes place occurs at the end of the process and is purely by *asexual* fission; see **Autogamy** and **Conjugation**.

Sheath: the outer portion of a peritrich stalk; see **Annulus**.

Shell: preferred term is *lorica*.

Sibling Species: one or more *biological species* that are difficult, if not impossible, to distinguish based on morphological critera; the *syngens* of the *Paramecium "aurelia"* complex and the *Tetrahymena "pyriformis"* complex are groups of sibling species of ciliates.

Silberliniensystem: see **Argyrome**.

Silver-impregnation Techniques: cytological staining methods that permit deposition of silver ions onto *argentophilic* sites where they are reduced, under UV light or appropriate chemicals, blackening the coated structures or areas affected and thus rendering them beautifully visible under subsequent light microscopic examination; the *argyrome* or silverline system so revealed in ciliates has proven of immeasurable value in comparative taxonomy and morphogenesis; the Klein "dry" method and the "wet" methods of von Gelei and especially of Chatton and Lwoff show up the argyrome *sensu stricto*, the more or less superficial "cortical" structures, such as the silverline *meridians*, the *contractile vacuole pores*, the *cytoproct*, and – most importantly – the (general sites of the) *kinetosomes*, both somatic (comprising the *kineties* proper) and oral (e.g., the infraciliary bases of the *oral or buccal organelles*); other methods, especially Bodian's Protargol (activated silver albumose) technique and the Rio-Hortega method, additionally blacken many truly cortical organelles of the *infraciliature sensu lato*, deeper in the organism, such as the *kinetodesmata*, the *nematodesmata*, *extrusomes*, *microtubular ribbons*, and *myonemes*, and even the *nuclei*, *mitochondria*, and *contractile vacuoles*, as well as the *cilia* themselves, thus allowing distinction between *ciliferous* and *barren kinetosomes* (Figs. 2.3, 2.4).

Silverline System: once popular synonyms are Silberliniensystem and neuroformative System; see **Argyrome**.

Simple Ciliature: general term restricted to meaning individual ciliated *monokinetids* (e.g., those comprising a *somatic kinety*) or single isolated ciliated monokinetids, such as most *bristles*, or ciliated *dikinetids*; excluded are formations or arrangements of cilia that are closely apposed in special groups or packets or blocks with some sort of interconnection, such as found in the case of *cirri*, somatic *polykinetids*, *syncilia*, some *atrial* and *vestibular ciliature*, and all oral "compound" ciliature, such as the *paroral*, *undulating membranes*, *polykinetids*, and *membranelles sensu lato*.

Skeletal Plaques: term recently applied to the numerous polysaccharide granules assembled in the unique *sucker* of certain clevelandellid armophoreans.

Skeletal Plates: term usually reserved for the long recognized and generally highly conspicuous subpellicular structures composed of polysaccharide reserves (i.e. amylopectin) within a fibrillar lattice; found in the entodiniomorphid vestibuliferans (i.e., ophryoscolecids and relatives); also used in reference to the uniquely calcified cuirass of the prorodontid *Coleps* (SP, Fig. 2.5Af).

Skeletal Ring: see **Denticulate Ring**.

Skeletogenous Structure: non-specific term usable for any organelle or system (e.g., various *microtubular ribbons*, *kinetodesmata*, *nematodesmata*, various proteinaceous rods, and polysaccharide formations, which may lend a certain firmness or rigidity to the *cortex* or to all or part of the body of an organism); see **Cytoskeleton**.

Solenocyst: dense vesicles found in the tentacles and subjacent cytoplasm of the suctorian cell body; fusing with the food vacuole membrane as it forms at the tentacle tip, they provide membrane and presumably contain lytic enzymes that aid in the preliminary digestion of the prey as it is ingested.

Solitary Form: an individual ciliate; used principally in reference to noncolonial forms in a contrasting sense; for example, there are colonial and solitary peritrich species, sometimes within a single family.

Somatic Area: see **Somatic Region**.

Somatic Ciliature: all-inclusive term for any *cilia* or compound ciliary organelles found anywhere on the body outside the *oral region*; associated with it would be the bases of all such structures, the somatic *infraciliature*; compare to **Oral Ciliature**.

Somatic Infraciliature: see **Somatic Ciliature**.

Somatic Kinety (pl. **Kineties**): *kinety* confined to the *somatic region*.

Somatic-meridional: now discarded term for a mode of *stomatogenesis*; see **Parakinetal**.

Somatic Region: general term for all of a ciliate's body except the *oral region*; may be functionally subdivided (e.g., *thigmotactic area*); its primary functions are locomotion, attachment to the substratum, and maintenance of form.

Somatization: evolutionary process of increasing the separation of "generative" from "somatic" functions in protozoa, demonstrated in ciliates in the development of *nuclear dualism*, in the complications of *sexual phenomena*, in the manifestation of epigenetic *morphogenesis*, and in the general diversification and differentiation, often involving *polymerization*, of more complex structures and functions that approach almost a metazoan level of organization.

Somatogenesis: the replication and development of all somatic components of the cell, usually occurring during the interfission period, but also at fission (e.g., contractile vacuole pores, cytoproct); see **Stomatogenesis**.

Sorocarp: "fruiting body" developing atop an aerial stalk after aggregation of cells; found only in the colpodean *Sorogena*, in which the individual cells encyst as a sorocyst, a component of the sorocarp.

Sorocyst: see **Sorocarp**.

Sorogenesis: production of a *sorocarp*.

Spasmin: see **Spasmoneme**.

Spasmoneme: used to describe the membrane-bound bundle of contractile protein, predominantly spasmin, found in the *stalks* of various sessiline peritrichs; arises from and maintains continuity with microfilaments in the *bell* or *zooid* proper; its former structural subdivision into thecoplasm and kinoplasm has not been confirmed by electron microscopy; see **Myoneme** (Sn, Fig. 2.9Bi, 2.9Bj).

Spherical Colony: *zooids* dispersed throughout a rounded, usually gelatinous but firm, colonial mass, with body axes perpendicular to the colony surface; because of attachment to a flat substratum, the overall shape may more often be hemispherical; the framework of such a globular colony may be basically arboroid, as shown by some species of the peritrich *Ophrydium* whose zooids are interconnected by long and slender "penduncular fibers" produced by their *scopulae* (Fig. 2.8Bb).

Spica: a *secant system* on the right anterior surface of amphileptid pleurostomatids.

Spindle Trichocyst: see **Trichocyst**.

Spines: variously used, though especially for certain apically located *holdfast organelles* (e.g., in some astomes); also applied to quite elaborately developed structures on the outside of the body of a number of chonotrichs; see **Attachment Organelle** (Sp, Fig. 2.9Bc–2.9Be).

Spongiome: see **Spongioplasm**.

Spongioplasm: specialized secretory cytoplasm, of spongy appearance (i.e., the spongiome), found in the vicinity of the *contractile vacuole*, functioning to collect fluid for elimination.

Springborsten: see **Saltatorial Cilia**.

Stalk: term broadly used for any kind of cylindrical and generally tubular supporting structure, either totally non-living or with a non-living sheath or *annulus*, running from the posterior end of a ciliate's body to a point of fixation on the underlying substratum; typically found in attached, sedentary or sessile forms or stages, with or without involvement of a lorica, serving as an *attachment* or *holdfast organelle*; may be of varying length, composition, and origin; produced with involvement of *secretory ampullae*, *kinetosomes*, and/or *pellicular pores*; may be contractile with a *spasmoneme* or non-contractile; ramified in some groups, associated with colonial organization of the supported *zooids*; most commonly known in chonotrichs, suctorians, and peritrichs, but undoubtedly is a non-homologous structure in these diverse taxonomic groups; see **Peduncle** (St, Fig. 2.11B, 2.11C).

Statocyst: see **Concrement Vacuole** and **Müller's Vesicle**.

Statolith: see **Concrement Vacuole** and **Müller's Vesicle**.

Stentorin: "blue" cytoplasmic pigment distributed in *pigmentocysts* and appearing in longitudinal rows in certain species of the heterotrich *Stentor*.

Stereocilum (pl. **Stereocilia**): see **Clavate Cilium**.

Stichodyad: type of *paroral* whose infraciliature is composed of *dikinetids* so oriented that each is perpendicular to the anteroposterior axis of the *paroral*; these *kinetosomes* are often arranged in a zigzag pattern with only the outer kinetosomes being *ciliferous*; common condition in oligohymenophoreans; see **Haplokinety**, **Stichomonad**.

Stichomonad: type of *paroral* whose infraciliature is composed of a single file or line of identically oriented *kinetosomes*; the characteristic condition found in many spirotrichs; see **Diplostichomonad**, **Stichodyad**.

Stock: now seldom-used term referring to any named or numbered *clone* that is maintained in culture separately from other such isolates.

Stomatogenesis: literally "mouth-formation"; in the broadest sense, this dynamic phenomenon embraces neoformation or replacement of all oral structures and infrastructures and any associated openings, depressions or cavities in both the *proter* and *opisthe*, typically prior to and during binary *fission*; major kinds or modes are now recognized (see **Apokinetal**, **Buccokinetal**, **Mixokinetal**, **Parakinetal**, and **Telokinetal**), under which the names of older descriptive categories are generally listed, as synonyms; *oral replacement*, a stomatogenic phenomenon that may occur periodically, refers to the *in situ* remodeling of the parental oral structures or their total substitution by new organelles, because of the partial or complete dedifferentiation or resportion of the former ones; reversible *microstome-macrostome transformation*, a special kind of oral replacement, involves the growth and/or replacement of a small oral apparatus with a greatly enlarged one or *vice versa* (e.g., in some species of *Tetrahymena*); see **Somatogenesis** (Fig. 2.11D).

Stomatogenic Field: general term for the group of *kinetosomes*, non-ciliferous throughout most of the process, actively involved – as the *anlage* – in the production of new *oral ciliature* by any of the described modes of *stomatogenesis*; see **Anarchic Field**, **Germinal Row**, and **Scutica** (SF, Fig. 2.11Dg).

Stomatogenic Kinety (pl. **Kineties**): see **Kinety Number 1** and **Parakinetal** (*stomatogenesis*) (SK, Fig. 2.11Dd).

Stomatogenous Meridian: an older term for *stomatogenic kinety*.

Strain: a named *clone* of a particular species of ciliate that usually differs in minor ways (e.g. genetic, phenotypic, physiological) from other strains.

Stria (pl. **Striae**): a beaded, longitudinal cytoplasmic strand found beneath the *perilemma* in the oral polykinetids of some spirotrichs, especially tintinnids.

Striated Bands: name for a (micro)fibrillar system discovered to lie just below the *epiplasm* in some ciliates (e.g., *Paramecium*).

Strobilation: kind of multiple *fission* in which successive *tomites* or *buds* are fully or partially separated or pinched off, sometimes within the confines of a cyst and usually resulting in a temporary linear chain of small individuals requiring subsequent metamorphosis to regain the form typical of the normal *trophont* stage of the life cycle; see **Catenoid Colony** (Fig. 2.8Ba).

Structural Conservatism Hypothesis: maintenance (i.e., conservation) of a structure through time is inversely related to the level of its biological organization; in the evolution of ciliates, then, among the most stable and most conservative taxonomic characters would be those involving *unit organelles*, ultrastructures at a relatively low organizational level (e.g., the various *microtubular ribbons* and *kinetodesma* associated with monokinetids and dikinetids).

Structural Guidance, Principle of: positioning and orientation of newly arising organelles (e.g., kinetosomes on the surface) under the localized influence of nearby pre-existing structures; see **Cytotaxis**.

Style: see **Stylet**.

Stylet: variously used, but generally as a synonym of *podite*.

Stylus: see **Stylet**.

Subkinetal Microtubules: one or more *microtubular ribbons* running under the proximal ends of *kinetosomes* in phyllopharyngeans (e.g., *Brooklynella*); compare to **Basal Microtubules**.

Subpellicular Microtubules: *microtubules*, single or in ribbons, in or just under the *pellicle* and coursing along parallel to the outer surface of the cell; originate either independently from kinetosomes (e.g., *longitudinal microtubules*) or may be extensions of *transverse* or *postciliary microtubules*.

Sucker: variously used, but generally for the cup-shaped concavity forming the often non-homologous thigmotactic area or *adhesive organelle* in scattered species belonging to quite different taxa (e.g., astomes, clevelandellids, licnophorids, peritrichs, thigmotrichs); also used to denote the *sucking tube* of rhynchodids; may be rich in fibrils, polysaccharide plaques or *thigmotactic cilia*; a rare synonym, for protozoa, is acetabulum; see **Attachment Organelle** (S, Fig. 2.9B).

Sucking Tentacle: see **Suctorial Tentacle**.

Sucking Tube: apically located, complex septate structure composed of *microtubular ribbons*, the *phyllae*, and serving as the ingestatory apparatus of rhynchodids and grossglockneriid colpodeans (Fig. 2.10f).

Suctorial Organelle: see **Suctorial Tentacle**.

Suctorial Tentacle: extensible and retractable tubular extension of the body of suctorians, containing a complex array (set or sets) of longitudinally arranged *microtubular ribbons* or *phyllae* and equipped both with *haptocysts* at its often *capitate* tip and vesicles functioning in formation of food vacuolar membrane; the tentacle serves for prey capture but is also the organism's ingestatory apparatus, analogous, if not homologous, with the *cytostome-cytopharyngeal complex* of other phyllopharyngeans; ranging from one in number to many, their grouping in *fascicles* on the body, whether on *actinophores* or not, is often of taxonomic significance; the larval or *bud* stages in suctorian life cycles possess none, as a rule; a type of extremely short, non-extensible tentacle, the *endosprit*, is known in several families (e.g., *Cyathodinium*) (SuT, Figs. 2.8i, 2.11Cd; 2.10a–2.10e).

Supernumerary Kinetosomes: apparently "extra" *ciliferous* or *barren kinetosomes* observed in various instances, in more than one stage in the life cycle, in differing but specific locations on the body or in the *oral region*; "overproduction" during *stomatogenesis* appears to be a source of some of these supernumeraries (e.g. in the *parakinetal* stomatogenesis of *Tetrahymena*); in some cases, they may be *erratic kinetosomes* and/or parts of an *anlage* and/or parts of a *vestige*; might also include cases of intercalated *kineties*, whole or partial, which involve a line or file of several or even many additional somatic kinetosomes.

Supraepiplasmic Microtubules: microtubules, single or in ribbons, lying above the *epiplasm* and coursing along parallel to the outer surface of the cell.

Suture Lines: simply folds or creases in the *pellicle*; preferably associated with the important concept of the *secant system*, the converging of kineties from different areas of the surface of the ciliate onto suture lines forming a pattern consistent within a given taxonomic group; see also **Postoral Suture** and **Preoral Suture**.

Swarmer: dispersive form in the life cycle of a number of ciliates; see **Larval Form**.

Symbiont: so-called dependent member or partner, except in cases of *mutualism*, of a pair of organisms exhibiting *symbiosis*, the other being the *host*; see **Commensalism**, **Mutualism**, and **Parasitism**.

Symbiosis: the living together, more or less intimately and contiguously, of two organisms, the *host* and the *symbiont*; see **Commensalism**, **Mutualism**, and **Parasitism**.

Symbiotic (adj.): see **Symbiosis**.

Symmetrogenic Fission: type of *fission*, generally longitudinal, of a *parental form* in such a manner that the two *filial products* are, in effect, mirror images of one another with respect to principal structures; typical of non-ciliate protozoa; compare to **Homothetogenic Fission** (Fig. 2.11Ab).

Symphoriont: *symbiont* exhibiting a kind of *commensalism* in which the *host*, usually via its integument, appears to serve solely as a convenient substratum for attachment of the typically stalked sessile ciliate; not always clearly distinguishable from other degrees of intimacy between hosts and their associated ectocommensals, but it represents a convenient term with reference to many peritrich and suctorian species; not generally used, by convention, for parasitic ciliates exhibiting a *phoront* stage, which, however, can really be functionally very similar; see **Phoront**.

Symplesiomorphic: shared ancestral *homologous character*; compare with **Synapomorphic**.

Synapomorphic: shared derived *homologous character* or shared *apomorphies*; used to unambiguously define a *clade* or *monophyletic* group.

Synciliary Ciliature: see **Syncilium**.

Syncilium (pl. **Syncilia**): a group of closely packed somatic cilia forming a special tuft exhibiting considerable internal coherence and arising from a packet of kinetosomes that are interconnected at their proximal ends with other syncilia; characteristic of entodiniomorphid vestibuliferans, syncilia were formerly called *membranelles*, with recognition of *adoral* and *dorsal zones*; they also occur as part of the *caudalia* present at the posterior end of the body of certain cycloposthiid entodiniomorphids; see **Polybrachykinety** (Syn, Fig. 2.5Af).

Syndesmogamy: a *conjugation* found only in apostomes during which two *trophonts* encyst together and undergo preconjugation division by linear *palintomy*, after which the *filial products* fuse and conjugate; also called zygopalintomy.

Syngamy: fusion of the *gametic nuclei* during *conjugation*; also called fertilization or karyogamy.

Syngen: complex of two or more sexually compatible mating types (e.g., in *Paramecium* or *Tetrahymena*), formerly known as "*varieties*"; long recognized as reproductively isolated biological units, hence *biological species*; see **Variety**.

Synhymenium: single apparent "membrane" resulting from an uninterrupted joining of the pavés of the *frange* of some nassophoreans.

Synkaryon: nucleus formed by fusion of two haploid *gametic nuclei* or pronuclei in the *sexual phenomena* of *conjugation* or *autogamy*; its division products differentiate into the new diploid *micronuclei* and the typically *polyploid macronuclei*; see **Zygotic Nucleus**.

Synonym: one or two or more names applied to the same organism or taxon; the oldest or earliest published is the senior synonym, which name usually must prevail; the later or younger of two is the junior synonym; an objective synonym is one based on study of the same material as the original describer, whereas a subjective synonym is based on material that is different but alleged to represent the original organism or taxon; see **Homonym**, **Rule of Priority**.

Système Sécant: see **Secant System**.

Systole: see **Contractile Vacuole**.

T

Tactile Cilium (pl. **Cilia**): see **Sensory Bristle** and **Thigmotactic Cilia**.

Tail: generalized term, variously used in nonspecific ways; a caudal appendage, ranging from specialized cilia or mucous filaments to narrow and lengthy extensions of the body proper.

Tangential Fibers: see preferred term, **Transverse Microtubules**.

Tastcilie (pl. **Tastcilien**): see **Sensory Bristle(s)**.

Tectin Granules: small, subpellicularly located bodies involved in secretion of a substance, probably mucopolysaccharide in nature, used to construct the *lorica* in a number of ciliates, especially among the sessiline peritrichs.

Teeth: nonspecific term; but also, perhaps unwisely, used to describe the nematodesmal *capitula* characteristic of some dysteriid cyrtophorines; see **Capitulum**.

Tela Corticalis: synonym of *lamina corticalis*.

Telokinetal: type of *stomatogenesis* in which formation of the new oral structures occurs by direct involvement either of *kinetosomes* at the anterior extremities of all or some of the encircling somatic *kineties* or of kinetosomes comprising the short *kinetofragments* available in the vicinity; see **Cryptotelokinetal, Holotelokinetal, Intertelokinetal, Merotelokinetal, Monotelokinetal**, and **Pleurotelokinetal** (Fig. 2.11Da, 2.11Db).

Teloparakinetal: *parakinetal stomatogenesis* in which the *oral anlage* originates by proliferation of *kinetosomes* at the anterior ends of *postoral kinetofragments* and at the "broken ends" of *bipolar somatic kineties*; found in ophryoglenids.

Telotroch: migratory free-swimming *larval form*, especially in the life cycle of sessiline peritrichs; *trophonts* of mobiline peritrichs are sometimes thought of as permanent telotrochs exhibiting *neoteny* (Fig. 2.11B).

TEM: transmission electron microscopy.

Temporary Conjugation: fusion with subsequent separation of the members of the conjugating pair; the mode of *conjugation* shown by most ciliates except peritrichs, chonotrichs, and some suctorians, which show *total conjugation*; see **Conjugation**.

Tentacle: tubular extension of or projection from the surface, of several different and probably non-homologous kinds, typically supported by *microtubular ribbons*: (1) *suctorial tentacle*, the (only) *ingestatory apparatus* in suctorians; (2) *prehensile tentacle*, a non-ingestatory structure present in some suctorians, solely for prey capture; (3) the short, non-extensible, apical *sucking tube* of rhynchodids and grossglockneriid colpodeans; (4) the non-suctorial but highly extensible and retractable prey-capturing organelle of such unusual haptorians as *Actinobolina*, composed of microtubular arrays often enclosing a prominent *toxicyst* and found in abundance, over the surface of the body; and (5) scattered other projections, lobes, or palps, and the like, which are or have been occasionally referred to by the term "tentacle", but properly excluding any *proboscis* (Fig. 2.10).

Tentaculoid: small finger-like extensions of the cytoplasm, possibly contractile, found between the oral polykinetids of some tintinnid spirotrichs, containing curious, little understood extrusomes called *capsules* or *capsules torquées*, which are subspheroid, 200–600 nm in length.

Tertiary Meridian: see **Ciliary Meridian**.

Test: see **Lorica**.

Tetrahymenal Buccal Apparatus: "tetrahymenal" refers to the "four-membraned" nature of the ciliary organelles found in the oral region of many oligohymenophoreans (e.g., members of the family Tetrahymenidae), which have a *paroral* or *undulating membrane* on the right of the *buccal cavity* and three *membranelles* on the left; see **Buccal Apparatus** (Fig. 2.7e).

Tetrahymenium: see **Tetrahymenal Buccal Apparatus**.

Theca: used in reference to the unusual envelope supporting certain operculariid peritrich species in particular; see **Lorica**.

Thecoplasm: see mention under **Spasmoneme**.

Theront: literally "hunter"; the dispersal stage in the polymorphic life cycle of a number of *parasitic* or *histophagous* ciliates (e.g., ophryoglenine hymenostomes); essentially a more or less transformed *tomite* searching for a new *host* or for a fresh source of food; on finding food, the theront transforms to a *trophont*.

Thigmotactic Area: see **Somatic Region** and **Thigmotactic Cilia**.

Thigmotactic Cilia: generally used to denote a patch, area, tuft, field or zone of more or less specialized *somatic ciliature* functionally modified to serve a presumed sensory-tactile or an adhering function, often localized (e.g., as a group of contiguous portions or segments of *kineties* occurring on the anterodorsal surface of the body in many thigmotrich scuticociliates); in certain astomes, and some other ciliates, the surface covered by the cilia may be concave and known as a *sucker*; the *scopula* of many sessiline peritrichs may, in a broad sense, be considered to possess thigmotactic cilia; see **Bristle** and **Holdfast Organelle** (TC, Fig. 2.7c).

Thigmotactic Zone: see **Thigmotactic Cilia**.

Tissue-eating: see **Histophagous**.

Tomite: a small, free-swimming, and non-feeding form derived by one or more fissions of a *tomont* (or sometimes of a *protomite*); a stage in the polymorphic life cycle of a number of *parasitic* or *histophagous* ciliates; usually emerges with numerous others from a cyst within which the divisions of the *tomont* have typically taken place; the next stage is the *theront* or *phoront* or *trophont*, depending on the species; the *filial products* of any binary or multiple *fission* could be called tomites, but generally – by convention – they are not (Fig. 2.9Ah).

Tomitogenesis: production of tomites; see also **Palintomy** (Fig. 2.9Af, 2.9Ah).

Tomont: prefission or dividing stage in the polymorphic life cycle of a number of *parasitic* or *histophagous* ciliates (e.g., apostomes and ophryoglenine hymenostomes); a large form, typically encysted, which may divide a number of times in quick succession; see **Tomitogenesis**.

Tooth: see **Capitulum**.

Total Conjugation: complete fusion of the *micro-* and *macroconjugant*; a phenomenon exhibited by all peritrichs and chonotrichs, a number of suctorians, and a scattered few other ciliates; incorrectly considered to be *syngamy*; see **Conjugation**.

Toxicyst: slender tubular *extrusome*, located in the cytoplasm of predaceous, carnivorous haptorians (e.g., *Didinium* and *Dileptus*); often concentrated in great numbers at or near the apical end of the organism and in the oral cytoplasm; also found in the nonsuctorial *tentacle* of *Actinobolina*; everting on discharge and apparently containing both paralytic and proteolytic enzymes, it penetrates, immobilizes, and commences to cytolyze the prey; the Nesselkapseltrichocyste and, less familiarly, the cnidotrichocyst and the tubular trichocyst of the older literature (Fig. 2.9Ce).

Transpodial Kinety (pl. **Kineties**): *somatic kineties* posterior to the *podite* or adhesive organelle; found in cyrtophorine phyllopharyngeans (TR, Fig. 2.4O).

Transverse Fibers: see **Transverse Microtubules**.

Transverse Fibrous Spur: dense (micro)fibrillar material associated with the proximal end of the kinetosome, arising near Triplet Number 3 (see **Numbering Conventions**) and extending a short distance to the left and upward into the nearby *pellicular ridge* (TFS, Fig. 2.1Ee).

Transverse Fission: see **Fission** and **Homothetogenic Fission**.

Transverse Microtubules: *microtubular ribbon* arising at the left anterior side of the kinetosome close to Triplet Numbers 3, 4, and sometimes 5 (see **Numbering Conventions**); the ribbon, which may be composed of 4–6 (occasionally more?) cross-linked microtubules, may originate tangentially or radially to the kinetosomal perimeter, first extends upward toward the *pellicle* and then continue to the left; in the *oral region* of litostomes, extensions of transverse microtubules are involved in the composition of the *cytopharyngeal apparatus* and/or *rhabdos* (T, Figs. 2.1, 2.2).

Transversodesma (pl. **Transversodesmata**): a complex set of overlapping transverse microtubular ribbons; characteristic of the Class COLPODEA; also called the *LKm fiber* (Td, Fig. 2.2).

Trichite: term used in at least two senses: (1) as an older and once highly popular, but now preferably discarded, name for a *nematodesma,* so prominent especially in the *cyrtos* and *rhabdos*; and (2) to describe the unique, rod-like, proteinaceous *extrusome* found in abundance in certain oligotrichs (e.g. *Strombidium*), usually radially arranged beneath the pellicle.

Trichocyst: in the past, term used to embrace nearly all *extrusomes* found in ciliates; now properly limited to the rather prominent, spindle-shaped, non-toxic, explosive extrusome of peniculines, like *Paramecium*; in the mature stage, consisting of an apical tip, shaped like an inverted golf tee, and a long, fusiform, fibrous shaft; on ejection, following an appropriate stimulus, acquiring a characteristic periodic structure; their function is often defensive; the *fibrocyst* of microthoracid nassophoreans is considered a special case; see **Extrusome** (Fig. 2.9Cf).

Trochal Band: synonym for *locomotor fringe* (TBd, Fig. 2.11B).

Trochal Girdle: see **Trochal Band**.

Trophic Nucleus: see **Macronucleus**.

Trophont: mature, vegetative, *adult form* as an interfissional or feeding or growing stage in the life cycle of any ciliate; the term is most often used, however, in reference to the specific stage between *tomite* (or *theront*) and *tomont* in the polymorphic life cycle of *parasitic* or *histophagous* species (e.g., as found among apostomes and hymenostomes); a term with identical meaning is *trophozoite* (but see remarks under that word, below).

Trophozoite: typically used for the "feeding" stage of truly parasitic species of protozoa, such as apicomplexans; see **Trophont**, the preferred synonym for ciliates.

Tubercle: variously used; see **Papilla**.

Tubicolous: tube-dwelling; used now and then with reference to loricate species (e.g. some folliculinids and a few spirotrichs), which may only temporarily occupy their loosely fitting, tube-shaped, often gelatinous housing, in a manner reminiscent of some of the true tube-dwellers among the polychaete annelids (Fig. 2.8Ak, 2.8Al).

Tubular Trichocyst: see preferred term, **Toxicyst**.

Tubulin: specific class of globular protein serving as the principal macromolecular constituent of all *microtubules*.

Type-genus: nominal genus designated as the type of a family-group taxon and not to be removed from that taxon; the familial name must be formed from the stem of this generic name plus the appropriate suffix (-idae for family, -inae for subfamily).

Type-species: nominal species designated as the type of a genus-group taxon; it cannot be removed from that genus.

Type-specimen: single specimen (perhaps whole slide of cloned organisms will be acceptable for protozoological materials in the future) known as the type or type-material of a taxon in the species-group; major kinds include the *holotype* (first specimen and more important), lectotype (named later from the type series if no holotype exists), and *neotype* (if all other type material lost); paratypes and syntypes are the extra specimens in a series from which a holotype or lectotype has been chosen (refer to the *International Code of Zoological Nomenclature* for detailed definitions).

U

Ubiquitous: the worldwide dispersal of individuals of a species.

Undulating Membrane: an older term for the *oral ciliature* on the right side of the *oral cavity*, implying a function that is not always realized, whereas *paroral* refers solely to the structure's location; also widely used for a different structure in certain flagellate protozoa; made popular as the UM of the tetrahymenines in which its presence is so neatly revealed by *silver-impregnation techniques*, its base – on the right side of the buccal cavity – standing in bold contrast to the three *membranelles* on the left side; see **Paroral** and **Tetrahymenal Buccal Apparatus**.

Undulipodium (pl. **Undulipodia**): see **Cilium**.

Unit Membrane: the phospholipid-protein layer that appears as a trilaminar structure in transmission electron microscopy; it delimits the boundaries of the cell (i.e. *plasma membrane*) and of many organelles (e.g. *mitochondria,* nucleus, *vesicles*).

Unit Organelle: subcellular structure (e.g., cilium, kinetosome, food vacuole) that is directly involved in subsidiary cell functions and that is composed of specific aggregations of macromolecules, often themselves morphologically recognizable as suborganelles (e.g., microtubules).

V

Vacuole: generalized term used for all sorts of sizable, fluid-filled, *unit membrane*-bound cavities or sacs in the cytoplasm (e.g., *concrement vacuole*, *contractile vacuole*, *food vacuole*); compare with **Vesicles**.

Variety: used as a synonym for the preferred term *syngen*; also, in a second and inaccurate meaning, considered as a formal taxonomic rank at an infraspecific level; but, along with forma, which is used by specialists of some groups, such as the entodiniomorphids, should not be so used in ciliatology; culture, *deme*, ecophenotype, population, *stock*, *strain*, and race are preferable terms.

Vegetative Nucleus: see **Macronucleus**.

Vegetative Reproduction: synonym for *asexual* reproduction or binary *fission*.

Veloid: see **Velum**.

Velum: variously used for the *paroral* membrane of some scuticociliates, sometimes termed a veloid; the *extensor membrane* of peritrichs; the *flange*; the skirt-like pellicular fold covering the *marginal cilia* in mobiline peritrichs.

Ventralization: presumed evolutionary process whereby the *oral region* comes to be located on the anterior ventral surface rather than in a wholly apical or near-apical position; position of *cytostome* fundamentally ventral, with no shift during fission.

Vesicle: *unit membrane*-bound cavity or sac in the cytoplasm of a size usually much less than that of a typical *vacuole* (e.g., *discoidal vesicles*, parts of the *Golgi apparatus* and *endoplasmic reticulum*, the *pellicular alveolus*, and pinocytotic vesicle); water expulsion "vesicle" is a misnomer for the *contractile vacuole*.

Vestibular Ciliature: rows or files of oral *cilia* only slightly modified from and sometimes only extensions of *somatic kineties;* characteristically located in a cavity or depression, the *vestibulum*; vestibular kinetids may have cilia longer or closer together, and may show minor differences in the organization of the *kinetids* compared to kinetids of the *somatic region*; compare to **Ophryokinety** and **Prebuccal Ciliature**.

Vestibular Kinety (pl. **Kineties):** see **Vestibular Ciliature**.

Vestibule: see **Vestibulum**.

Vestibulum (pl.**Vestibula**): a ciliated depression or invaginated *oral region*, at either pole, leading directly or indirectly to the *oral cavity* and adorned with vestibular ciliature.

Vestige: a visible trace or part of a structure or organelle persisting, usually as a nonfunctional remnant, in one stage (typically the mature *adult form*) as a carry-over from an earlier ontogenetic stage in which it was fully developed and functional; curiously enough, the remnant or vestige of an ontogenetic structure may serve as the *anlage* for production of that same structure in the next (repeated) stage in the life cycle (e.g., the *scuticovestige* in the mature form of some scuticociliates is apparently a major source of the kinetosomes forming the *anlage* for the *scutica* proper); see also **Erratic Kinetosomes**.

Vierermembran: see **Quadrulus**.

Vierteilige Membran: see **Quadrulus**.

W

Watchglass Organelle: see **Lieberkühn, Organelle of**.

Water Expulsion Vesicle: see **Contractile Vacuole**.

X

Xenodeme: see mention under **Deme**.

Xenosomes: literally "alien bodies"; a bacterial *endosymbiont* in the cytoplasm of certain marine scuticociliates (e.g., *Parauronema*); see also **Endosymbiont (of Ciliates)**.

Xeric: pertaining to a terrestrial habitat having a very low content of water (e.g., desert sands).

Xylophagous: literally "wood-eating"; capable of digesting cellulose, such as certain trichostomes and entodiniomorphids in the digestive tract of various mammalian herbivores.

Z

Zeta Membranoid: see **Membranoid**, **Haplokinety**, and **Paroral**.

Zone of Stripe Contrast: see **Locus of Stripe Contract**

Zoochlorella (pl. **Zoochlorellae**): endosymbiotic green algae, typically chlorophytes, found widely in the cytoplasm of ciliates belonging to nearly all major taxa; see **Endosymbiont (of Ciliates)**.

Zooid: generally restricted to mean only the body proper of an attached sessile form (e.g., the *bell* of many peritrichs), minus the *stalk*; the individual members of a free or attached colony, but usually only of the *arboroid colony* so typical of peritrichs; macrozooids and microzooids are distinguishable by size and exhibition of certain functional differences (e.g., only macrozooids of *Zoothamnium* are capable of starting new colonies) (Z, Fig. 2.8Bb, 2.8Bc).

Zooplankton: see **Plankton**.

Zoopurpurin: see **Blepharismin**.

Zooxanthella (pl. **Zooxanthellae**): endosymbiotic "non-green" algae, typically *mutualistic* dinoflagellates, cryptomonads or chrysomonads, found in the cytoplasm of ciliates; see **Endosymbiont (of Ciliates)**.

Zweigliedrige Kultur: see **Monoxenic Culture**.

Zygopalintomy: see **Syndesmogamy**.

Zygotic Nucleus (pl. **Nuclei**): fusion product of two *gametic nuclei*; see **Synkaryon**.

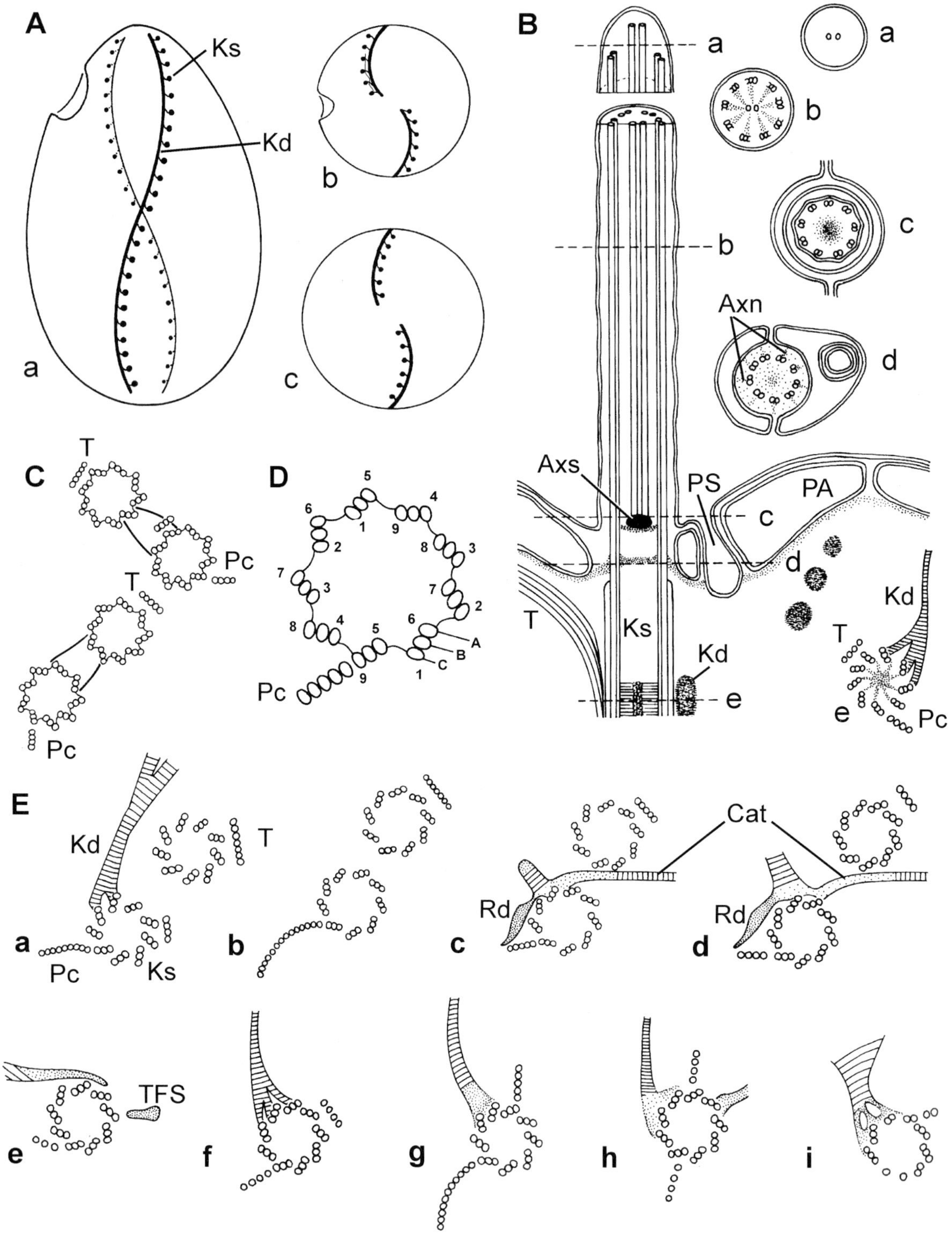

FIG. 2.1. Kineties and kinetosomes of the somatic cortex. **A** Structure of somatic kineties. **a**. Somatic kineties are files of kinetosomes (Ks) linked by kinetodesmata (Kd), which appear on the left side of the kinety, if viewed from the **outside** (**a**, bold) or top (**b**) and bottom (**c**), and on the right side of the kinety, if viewed from the inside (**a**, not bold). **B** Detailed structure of a single kinetosome (Ks) and its cilium at five different levels (**a**, **b**, **c**, **d**, **e**). The axoneme (Axn) is composed of 9 peripheral doublets in the cilium (**a–d**) that transform to triplets in the kinetosome (**e**). The central pair of ciliary microtubules arise from the axosome (Axs). A parasomal sac (PS) is adjacent to the cilium, which is surrounded by pellicular alveoli (PA) underlying the plasma membrane. The kinetosome, viewed from the **inside**

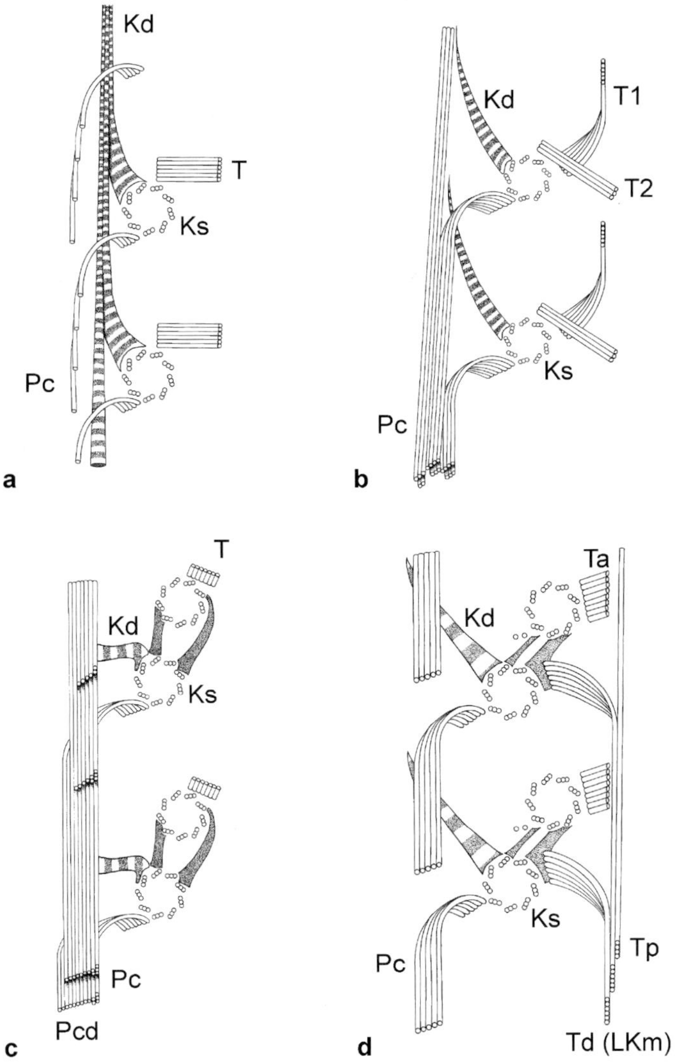

FIG. 2.2. Schematic drawings of two somatic kinetids from four classes of ciliates, two with somatic monokinetids (**a**, **b**) and two with somatic dikinetids (**c**, **d**). (**a**) The somatic kinetids of the Class OLIGOHYMENOPHOREA. Note that the transverse ribbons (T) are radial to the perimeter of the kinetosome (Ks). The kinetodesmata (Kd) from adjacent kinetids overlap but the postciliary ribbons (Pc) do not. (**b**) The somatic kinetids of the Class LITOSTOMATEA. Ciliates in this class typically have two sets of transverse ribbons (T1, T2) and the postciliary ribbons often lie side-by-side in the cortex. (**c**) The somatic kinetids of the Class HETEROTRICHEA in which the postciliary ribbons (Pc) overlap laterally to form the Km fiber or postciliodesma (Pcd). (**d**) The somatic kinetids of the Class COLPODEA in which the transverse ribbons (Tp) of the posterior kinetosome of the dikinetid overlap to form the transversodesma (Td) or LKm fiber

FIG. 2.1. (continued) (**e**) has a kinetodesma (Kd) and postciliary ribbon (Pc) on its right and a transverse ribbon (T) on its left (cf. Fig. 2.1E). **C** A pair of kinetosomes (upper) and a dyad (lower) in relation to the body axis (anterior is towards the top of the page). **D** Cross-section of a kinetosome as viewed from the **outside** of the cell showing the numbering system of Grain (1969) on the outside of the triplets and the numbering system of Pitelka (1969) on the inside of the triplets. The postciliary ribbon (Pc) is numbered as 9 or 5, respectively. **E** Examples of somatic kinetids of ciliates from different classes showing the diversity of patterns with dikinetids (**a–d**) and monokinetids (**e–i**). Note the kinetodesma (Kd), postciliary ribbon (Pc), and transverse ribbon (T) associated with kinetosomes (Ks). A retrodesmal fibril (Rd) may extend posteriorly to support the postciliary ribbon and a cathetodesmal fibril (Cat) may extend towards the left into the pellicle. Occasionally a transverse fibrous spur (TFS) replaces the transverse microtubules. (**a**) The karyorelictean *Loxodes*. (**b**) The heterotrichean *Spirostomum*. (**c**) The clevelandellid *Sicuophora*. (**d**) The clevelandellid *Nyctotherus*. (**e**) The rhynchodid *Ignotocoma*. (**f**) The peniculid *Paramecium*. (**g**) The scuticociliate *Porpostoma*. (**h**) The scuticociliate *Conchophthirus*. (**i**) The astome *Coelophrya*

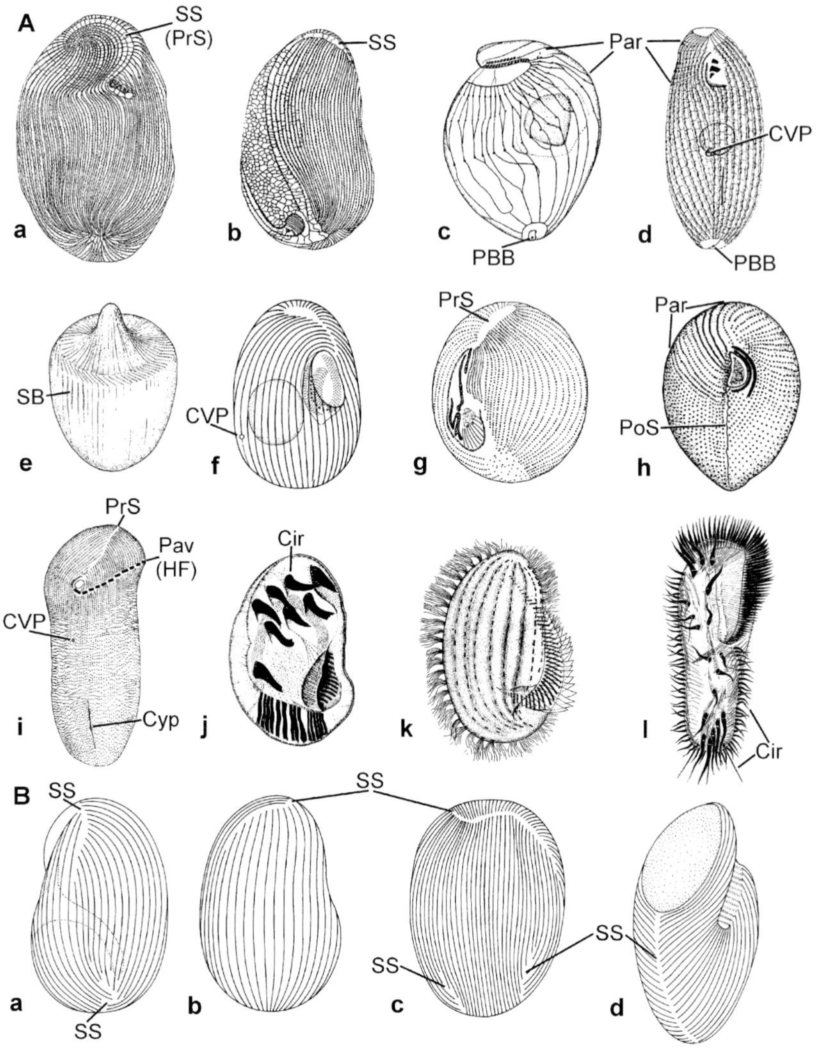

FIG. 2.3. Drawings of specimens after they have been stained by various silver impregnation techniques. (**a–c**) The dry silver technique of Klein showing the secant system (SS) or preoral suture (PrS) of *Colpidium* (**a**, Klein) and *Ancistrum* (**b**, Raabe) and the paratenes (Par) and polar basal body (PBB) of *Trimyema* (**c**, Jankowski). (**d**) *Dexiotricha* (Jankowski) stained by the von Gelei-Horváth technique to reveal the paratenes (Par), the contractile vacuole pore (CVP) and the polar basal body (PBB). (**e–i**) The Chatton-Lwoff wet silver technique, showing the sensory bristles (SB) of *Monodinium* (**e**, Dragesco), the contractile vacuole pore (CVP) of *Glaucoma* (**f**, Corliss), the preoral suture (PrS) of *Pleurocoptes* (**g**, Fauré-Fremiet), paratenes (Par) and postoral suture (PoS) of *Disematostoma* (**h**, Dragesco), and the preoral suture (PrS), contractile vacuole pore (CVP), cytoproct (Cyp), and pavés (Pav) of the hypostomial frange (HF) of *Obertrumia* (**i**, Fauré-Fremiet). (**j–l**) Protargol or silver proteinate impregnation, showing the cirri (Cir) of *Aspidisca* (**j**, Tuffrau) and *Stylonychia* (**l**, Dragesco), and the cilia of *Phacodinium* (**k**, Dragesco). **B** Secant systems (SS) where somatic kineties converge on the left ventral (**a**) and right dorsal (**b**) cortex of the clevelandellid *Nyctotheroides*, (**c**) the astome *Paracoelophrya*, and (**d**) the clevelandellid *Sicuophora*

FIG. 2.4. (continued) showing the complex pattern of denticles in the aboral sucker. **M**, **N** *Euplotes* sp. (Tuffrau) showing the complex pattern of the argyrome (**M**) after wet-silver staining and the complex subpellicular rootlets (**N**) after protargol staining. **O** *Brooklynella hostilis* (Lom) showing two circumoral kineties just anterior to the oral region and the transpodial kineties (TR) encircling the podite at the posterior end. **P** The scuticociliate *Pleuronema* (Small) in early stomatogenesis, demonstrating the scutica (Sc). **Q**, **R** Ventral view (**Q**) of *Philaster* sp. and a detail of the structure of its oral polykinetid 2 (OPk, **R**). **S** The tintinnid *Tintinnopsis* (Brownlee) with its two macronuclear nodules, residing in its lorica (**L**)

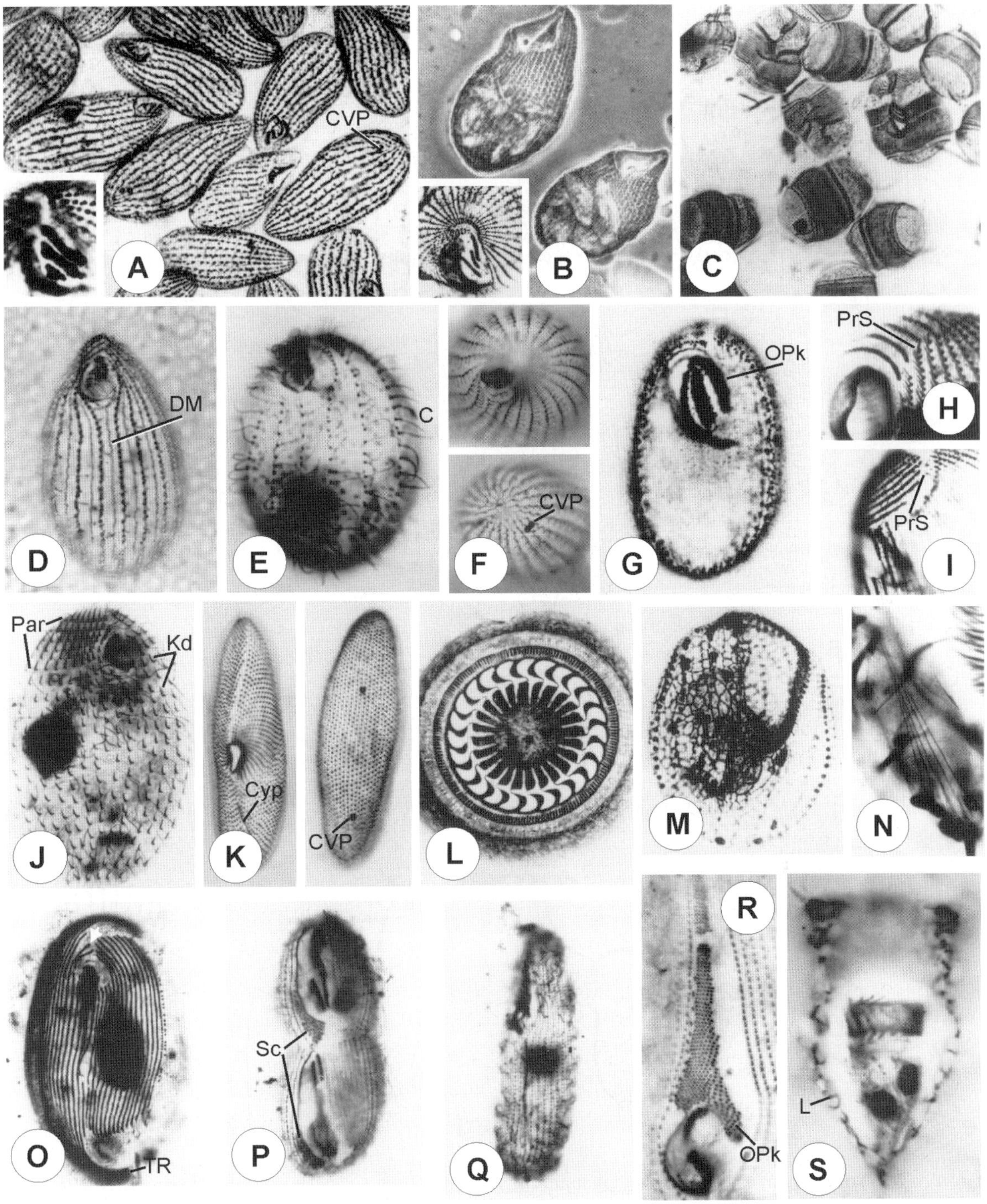

FIG. 2.4. Photomicrographs of specimens treated by various techniques of silver impregnation. **A–D**, **F**, **G**, **K–M** – Chatton-Lwoff technique. **J** Rio-Hortega method. **E**, **H**, **I**, **N–S** – Protargol or silver proteinate impregnation. **A** *Tetrahymena pyriformis* showing the microstome-type oral apparatus with a paroral and three membranelles (inset). Note the contractile vacuole pores (CVP). **B** Macrostome form of *Tetrahymena patula* adapted to ingesting smaller ciliates with view of the transformed oral apparatus (inset). **C** *Urocentrum turbo*. **D**, **E** *Tetrahymena* sp. showing the director meridian (DM) and the cilia (C). **F** Apical (upper) and antapical (lower) poles of *Tetrahymena setosa*. Note the contractile vacuole pores (CVP). **G**, **H** Ventral view of *Glaucoma scintillans*, showing its oral polykinetids (OPk, **G**) and preoral suture (**H**). **I** Preoral suture of *Colpidium* sp. **J** *Dexiotricha* (Fernández-Galiano) showing paratenes to the anterior right of the cell and demonstrating short kinetodesmata. **K** *Paramecium* sp. (Dippell) ventral view (left) showing the cytoproct (Cyp) and a dorsal view with the two densely staining contractile vacuole pores. **L** Ventral view of *Trichodina* sp. (Lom)

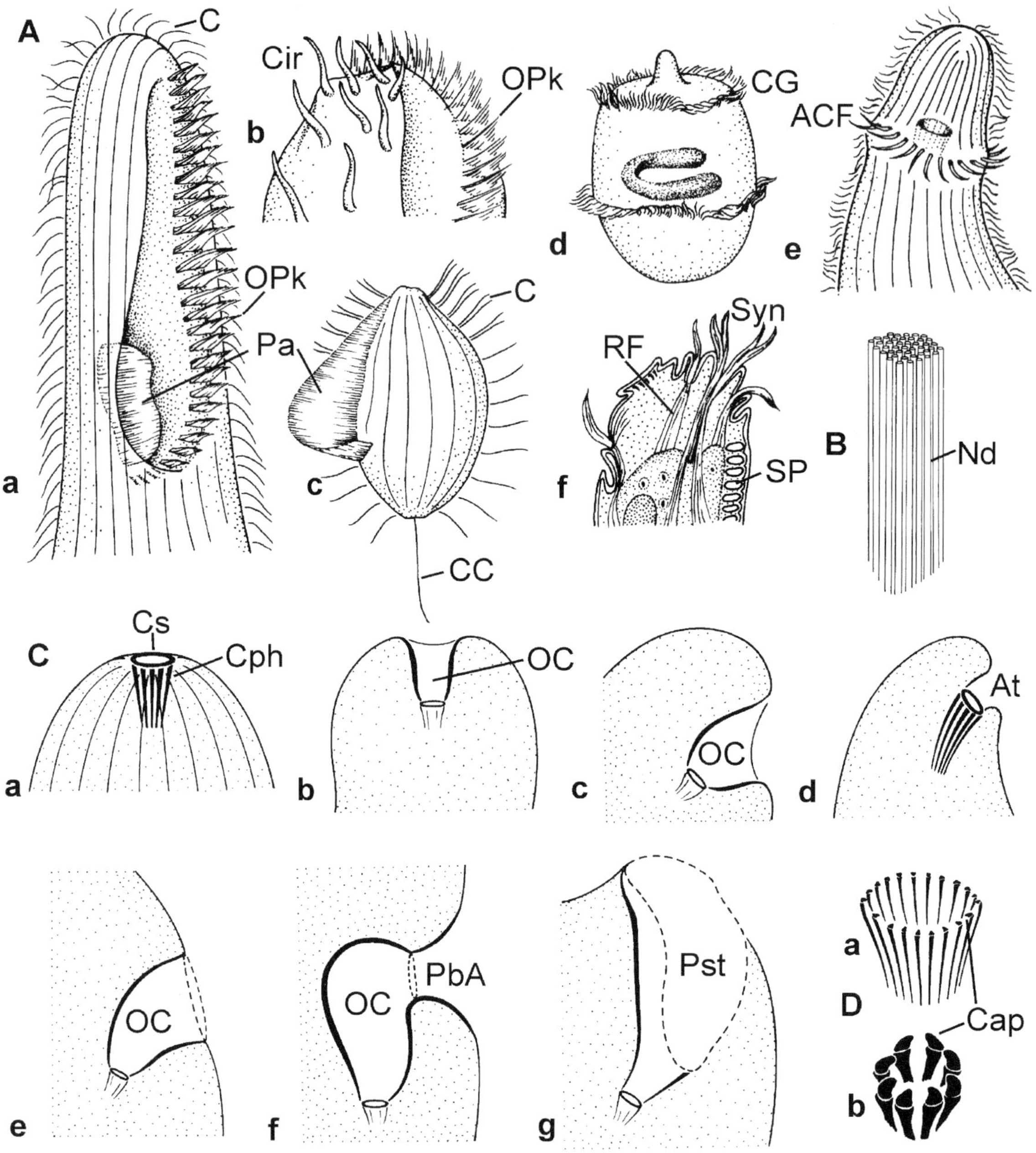

FIG. 2.5. Oral structures of ciliates. **A** Oral ciliature. (**a**) The heterotrich *Gruberia* covered by somatic cilia (**C**) and with an oral region bordered by an adoral zone of oral polykinetids (OPk) on its left and a paroral (P) on the right. (**b**) The hypotrich *Euplotes* showing its complex cirri (Cir) and an adoral zone of oral polykinetids (OPk). (**c**) The scuticociliate *Cyclidium* covered by somatic cilia (C) with a specialized caudal cilium (CC) extending to the posterior and the cilia of the paroral (P) raised in a curtain-like velum. (**d**) The haptorian *Didinium* with its anterior feeding protuberance surrounded by a ciliary girdle (CG). (**e**) The nassophorean *Nassulopsis* showing its adoral ciliary fringe (ACF) of pavés. (**f**) A longitudinal section through the anterior end of the entodiniomorphid *Epidinium*, showing the retractor fibres (RF), skeletal plates (SP) supporting the cortex, and the compound ciliary organellar complexes, called syncilia (Syn) surrounding the oral region. **B** Three-dimensional representation of the complex bundle of microtubules that makes up a typical nematodesma (Nd). **C** Schematic representations of oral regions. (**a**) Apical cytostome (Cs) and cytopharynx (Cph) of a prostomial form. Note the cytostome appears as a ring in (**b–g**). (**b**) Cytostome at

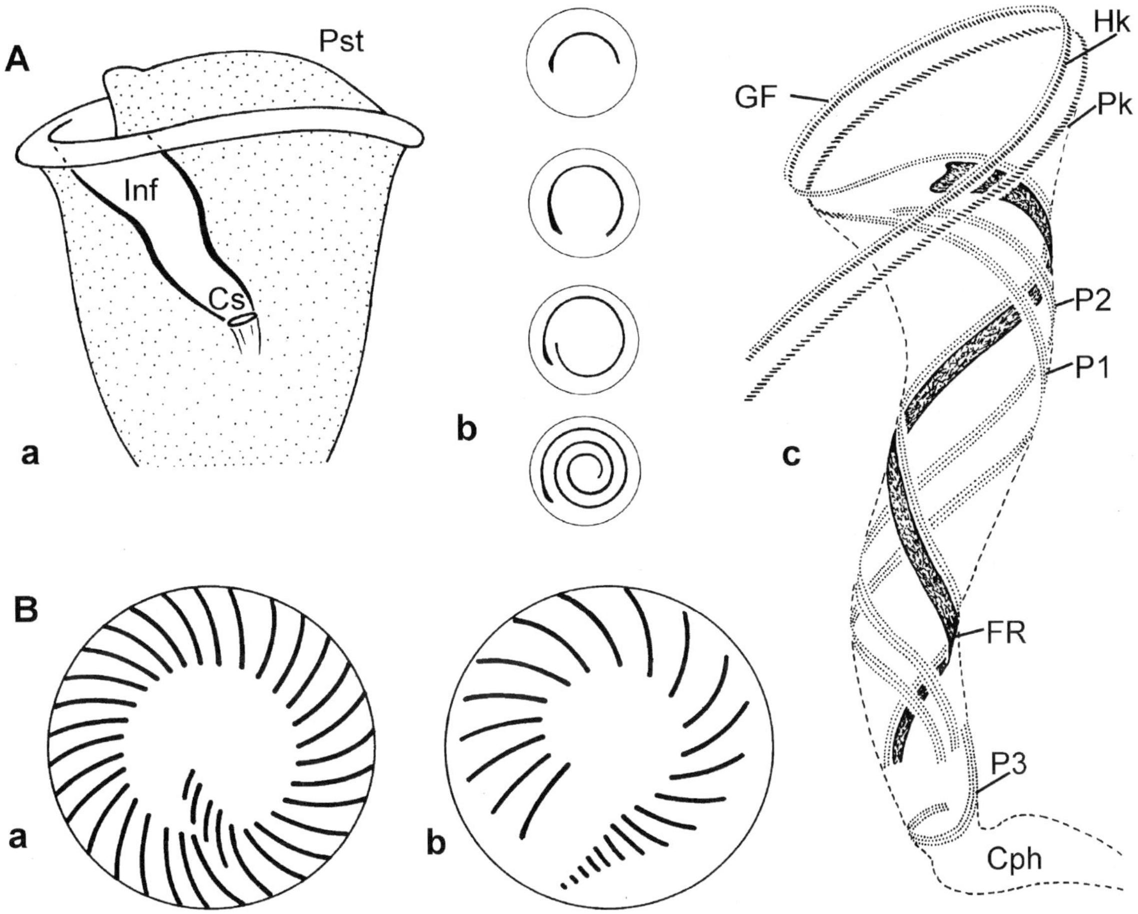

FIG. 2.6. Spiralling oral structures. **A** Oral structures of peritrichs. (**a**) The general arrangement of the peritrich oral region with the cytostome (Cs) at the base of a deep infundibulum (Inf), which leads out to the peristome (Pst) on which the oral ciliature spiral. (**b**) Varying degrees of complexity in the oral spiral of the mobiline peritrichs (from top to bottom) – *Semitrichodina*, *Trichodinella* or *Tripartiella*, *Trichodina* or *Urceolaria*, *Vauchomia*. (**c**) Detail of the oral infraciliature and related structures in the infundibulum of a peritrich. The haplokinety (Hk) and polykinety (Pk), actually peniculus 1 (P1) encircle the peristome, accompanied along part of their length by the germinal field (GF). As the Hk and Pk enter the infundibulum they are joined by peniculus 2 (P2) supported along the length by the filamentous reticulum (FR). Peniculus 3 (P3) and the cytopharynx (Cph) are at the base of the infundibulum. **B** Patterns of oral polykinetids in spirotrich ciliates. (**a**) The "closed" pattern of oral polykinetids in choreotrich ciliates, such as *Tintinnopsis* and *Strobilidium*. (**b**) The "open" pattern of an outer "collar" and ventral "lapel" of oral polykinetids in genera such as the stichotrich *Halteria* and the oligotrich *Strombidium*

FIG. 2.5. (continued) the base of an anterior oral cavity. (**c**) Cytostome at the base of a ventral oral cavity with an ill-defined opening. (**d**) Cytostome at the base of a subapical atrium (At), which is not lined with cilia. (**e**) Cytostome at the base of a ventral oral cavity with a well-defined opening (dashed line). (**f**) Prebuccal area (PbA) preceding a well-defined oral cavity. (**g**) Oral ciliature emerging onto the cell surface in a prominent peristomial area (Pst). **D** Schematic arrangement of the nematodesmata in the cyrtos of two cyrtophorians, *Aegyriana* (**a**) and *Brooklynella* (**b**). Each nematodesma is topped by a tooth-like capitulum (Cap) used in ingestion

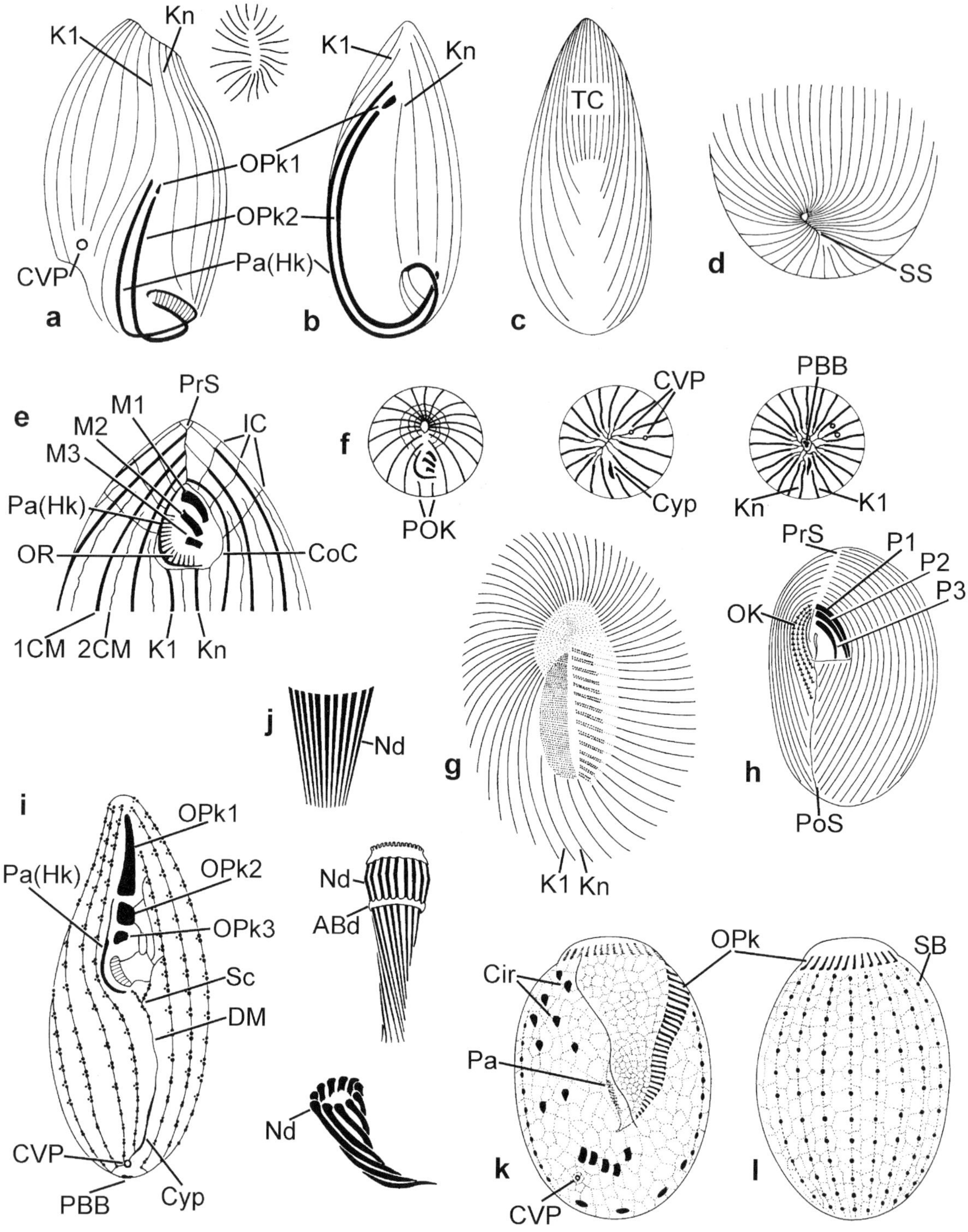

FIG. 2.7. Somatic and oral infraciliary patterns, as revealed particularly by Chatton-Lwoff silver impregnation. (**a**) The thigmotrich *Proboveria* showing the positions of the contractile vacuole pore (CVP) and the placement of Kinety 1 (K1) and Kinety n (Kn). Oral structures include two oral polykinetids (OPk1, OPk2) and the paroral (Pa) or haplokinety (HK). An apical view is to the top right of the cell. (**b**, **c**) Ventral (**b**) and dorsal (**c**) views of the thigmotrich *Ancistrum*. Note similar somatic and oral features to *Proboveria*. The dorsal anterior has a zone of densely packed thigmotactic ciliature (TC). (**d**) Posterior region of the hymenostome *Curimostoma*, showing a secant system (SS). (**e**) Anterior ventral surface of *Tetrahymena*, showing primary ciliary meridians (1CM) and secondary ciliary meridians (2CM) of the silver-line system, as well as intermeridianal connectives (IC) and circumoral connective (CoC). Two postoral kineties (K1, Kn) abut against the oral region, which is composed of three membranelles (M1, M2, M3) and a paroral (Pa) or haplokinety (HK) from which the oral ribs (OR) extend towards the cytostome. Somatic kineties abut on a preoral suture (PrS). (**f**) Apical (left) and antapical (middle) views of

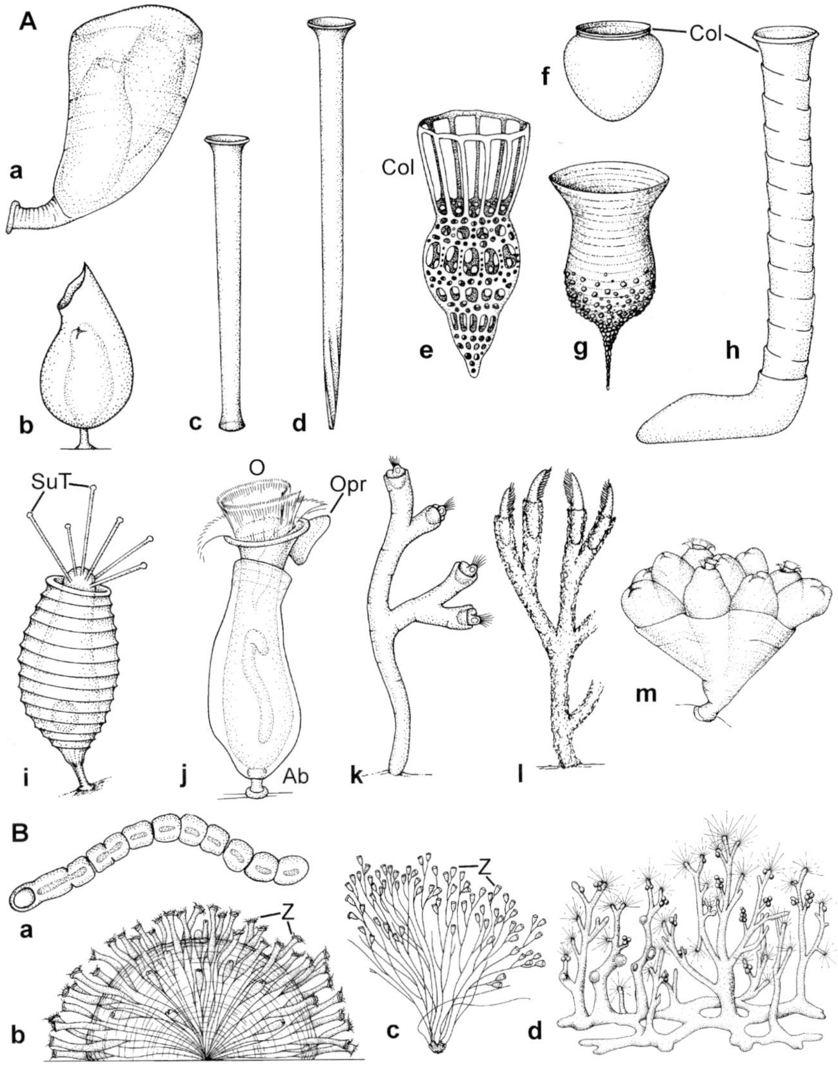

FIG. 2.8. Variations in form. **A** A variety of lorica types. (**a**) The peritrich *Cyclodonta*. (**b**) The peritrich *Cothurnia*. (**c–g**) Tintinnid loricae, including *Eutintinnus* (**c**), *Salpingella* (**d**), *Dictyocysta* with its perforated collar (Col) (**e**), *Metacylis* (**f**), and *Tintinnopsis* (**g**). (**h**) The folliculinid *Metafolliculina*. (**i**) The suctorian *Thecacineta* showing its sucking tentacles (SuT). (**j**) The peritrich *Pyxicola* with its ciliated oral (O) end, protected by the operculum (Opr) when it withdraws into the lorica. Ab, aboral. (**k, l**) The tube-like loricae of the colpodean *Maryna* (**k**) and the stichotrich *Stichotricha* (**l**). (**m**) The lorica or theca of *Orbopercularia*, which contains several zooids. **B** Colonial organizations. (**a**) The catenoid colony of the astome *Cepedietta*. (**b**) The spherical and dendritic colony of the peritrich *Ophrydium* with its zooids (Z) embedded in the matrix. (**c**) The dendritic colony of the peritrich *Epistylis*. (**d**) The arboroid or dendritic colony of the suctorian *Dendrosoma*

←

FIG. 2.7. (continued) *Tetrahymena pyriformis*, showing placement of the postoral kineties (POK), contractile vacuole pores (CVP), and cytoproct (Cyp). Antapical view of *Tetrahymena setosa* showing the placement of the polar basal body complex (PBB). (**g**) An apical view of *Colpoda magna* in a late stage of stomatogenesis. Kinety 1 (K1) is the rightmost postoral kinety. (**h**) Ventral view of the peniculine *Frontonia* showing somatic kineties converging on preoral (PrS) and postoral (PoS) sutures. The oral region is bounded on the right by the densely packed ophryokineties (OK) and contains on its left the three peniculi (P1, P2, P3). (**i**) Ventral view of the scuticociliate *Paranophrys*, showing features described previously (CVP, Cyp, HK, OPk1, OPk2, OPk3, Pa, PBB). The director meridian (DM) is a silver-line that extends posteriorly from the scutica (Sc). (**j**) Cytopharyngeal baskets of three ciliates: the rhabdos of a prorodontid (upper); the nasse or cyrtos of the nassophorean *Nassula*, bound in the middle by an annular band (ABd); and the cyrtos of the cyrtophorian *Chilodonella*. (**k, l**) Ventral (**k**) and dorsal (**l**) views of the hypotrich *Euplotes* showing the silver-line system of both surfaces. The large dark spots on the ventral surfaces are the bases of cirri (Cir) while the smaller dots in the dorsal kineties are sensory (SB) or dorsal bristles

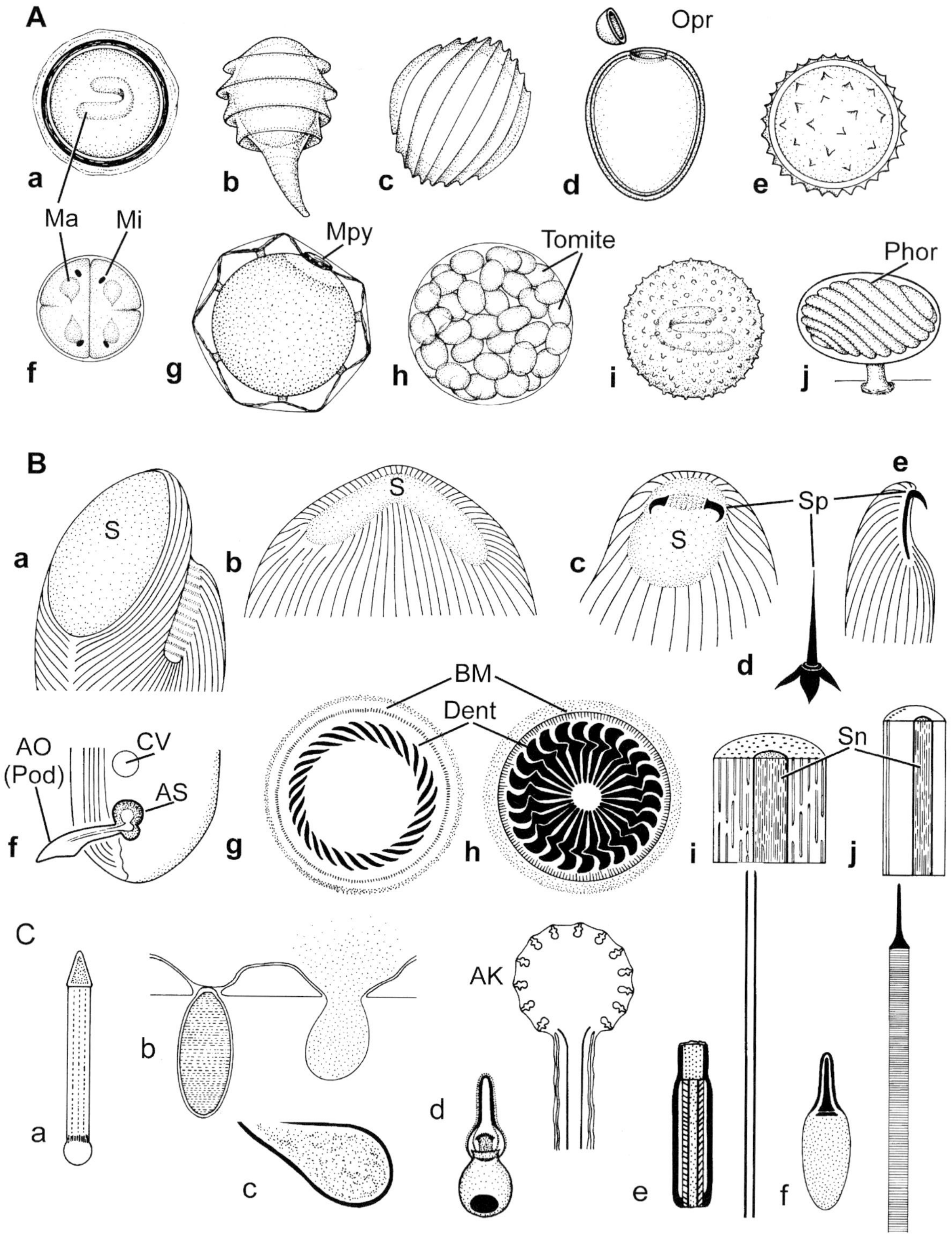

FIG. 2.9. **A** Cysts. (**a–e**, **g**, **i**) Resting cysts of the haptorian *Didinium* (**a**), the suctorian *Podophrya* (**b**), the hypotrich *Euplotes* (**c**), the clevelandellid *Nyctotherus* with its operculum (Opr) (**d**), the stichotrich *Oxytricha* (**e**), the colpodean *Bursaria* with its micropyle (Mpy) (**g**), and the peritrich *Vorticella* (**i**). (**f**, **h**) Division cyst of the colpodean *Colpoda* (**f**; Note the macronuclei (Ma) and micronuclei (Mi)) and the ophryoglenid *Ophryoglena* with its many tomites (**h**).

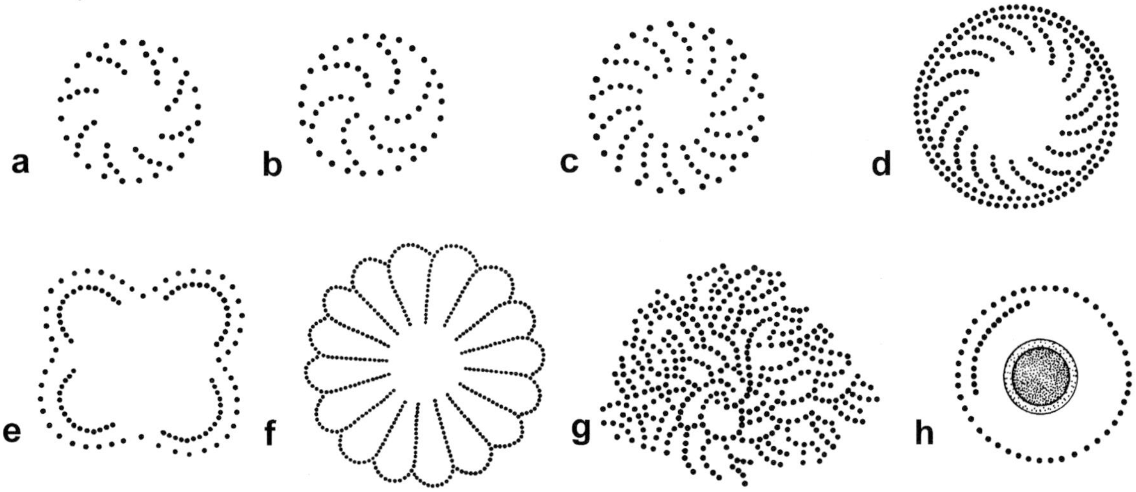

FIG. 2.10. Patterns of microtubules in cross-sections of various tentacle-like structures. (**a–e**) Sucking tentacles of the suctorians *Sphaerophrya* (**a**), *Acineta* (**b**), *Loricodendron* (**c**), *Dendrocometes* (**d**), and *Cyathodinium* (**e**). Note that there is an outer ring(s) enclosing the ribbon-like phyllae. (**f**) The sucker of the rhynchodid *Ignotocoma*. (**g**) The prehensile tentacle of the suctorian *Ephelota*. (**h**) The toxicyst-bearing, non-sucking tentacle of the haptorian *Actinobolina*

FIG. 2.9. (continued) (**j**) Resting cyst of the apostome *Spirophrya*, which is attached to the crustacean host cuticle and encloses the phoront (Phor) stage. **B** Attachment structures and holdfast organelles. (**a–c**) The attachment suckers (S) of the clevelandellid *Prosicuophora* (**a**), the scuticociliate *Proptychostomum* (**b**), and the astome *Steinella* (**c**). (**c–e**) Spines (Sp) may aid attachment in the astomes *Steinella* (**c**), *Maupasella* (**d**), and *Metaradiophrya* (**e**). (**f**) Posterior end of a dysteriid phyllopharyngean showing its attachment organelle (AO) or podite (Pod) at the base of which is a secretory ampulla (AS). CV, contractile vacuole. (**g, h**) Denticles (Dent) and border membrane (BM) are organized in the holdfast disk of the mobiline peritrichs *Trichodinopsis* (**g**) and *Trichodina* (**h**). (**i, j**) Longitudinal section of the attachment stalks of a peritrich with a central spasmoneme (Sn) (**i**) and an eccentric spasmoneme (**j**). **C** Extrusomes. (**a**) The rhabdocyst of the karyorelictean *Tracheloraphis*. (**b**) A mucocyst, resting (left) and discharging (right). (**c**) The clathrocyst of the haptorian *Didinium*. (**d**) Resting haptocyst of the suctoria (left) and their distribution at the tip of the attachment knob (AK) of the sucking tentacle. (**e**) Toxicyst, resting (left) and ejected (right). Not to the same scale. (**f**) Trichocyst of *Paramecium*, resting (left) and ejected (right). Not to the same scale

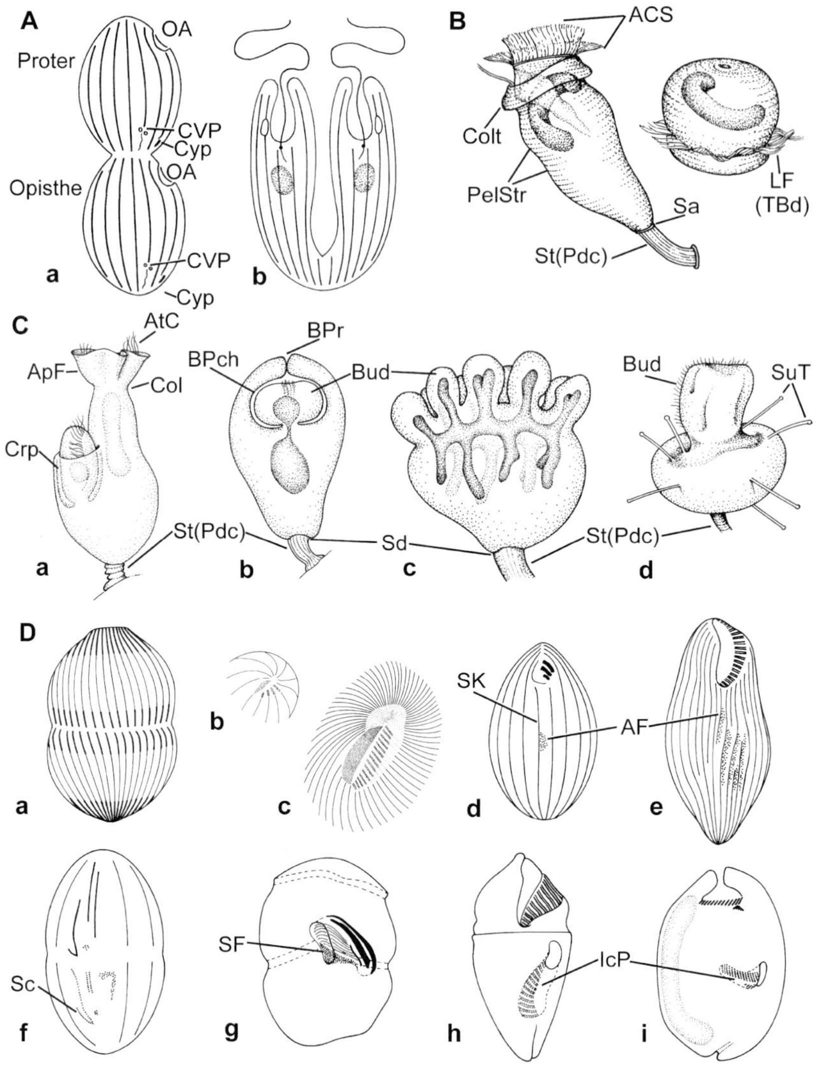

FIG. 2.11. Various kinds of fission processes. **A** A comparison of homothetogenic fission (**a**) in the ciliate *Tetrahymena* with symmetrogenic fission (**b**) in an idealized flagellate. The proter is the anterior cell and the opisthe the posterior cell, both of which are replicating cortical structures, such as the oral apparatus (OA), contractile vacuole pores (CVP), and cytoproct (Cyp). **B** An adult of the peritrich *Epistylis* (left) and its telotroch or bud (right). The adoral ciliary spiral (ACS) encircles the anterior end above the collarette (Colt). Pellicular striae (PelStr) adorn the body of the zooid, which has attached to the substratum by secreting a stalk (St) or peduncle (Pdc) using the scopula (Sa). The telotroch swims using the cilia of the locomotor fringe (LF) or telotroch band (TBd). **C** Kinds of budding. (**a**) Cryptogemmous budding in the chonotrich *Cristichona*. The atrial ciliature (AtC) of the adult lines the apical funnel (ApF), separated from the body by the collar (Col). The bud forms in the crypt (Crp). (**b–d**) Budding in suctoria. (**b**) Endogenous budding in *Tokophrya* occurs in a brood pouch (BPch) and the bud exits through a birth pore (BPr). (**c**) Multiple exogenous budding of *Ephelota*. (**d**) Evaginative budding of *Discophrya* with its sucking tentacles (SuT). These four forms are attached to the substratum by a stalk (St) or peduncle (Pdc). In the suctorians, the stalk is secreted by the scopuloid (Sd). **D** Major modes of stomatogenesis. (**a–c**) Telokinetal. Holotelokinetal in the litostome *Alloiozona* (**a**) and merotelokinetal in a small and larger colpodean *Colpoda* spp. (**b**, **c**). (**d**, **e**) Parakinetal. The anarchic field (AF) develops along the stomatogenic kinety (SK) in the monoparakinetal mode in the hymenostome *Tetrahymena* and along several somatic kineties in the polyparakinetal mode in the heterotrich *Condylostoma*. (**f**, **g**) Buccokinetal. (**f**) Scuticobuccokinetal with involvement of the scutica (Sc) in the scuticociliate *Pseudocohnilembus*. (**g**) In the peniculine *Urocentrum*, a stomatogenic field (SF) forms adjacent to the parental oral structures. (**h**) Apokinetal. Kinetosomal proliferation may occur in an intracytoplasmic pouch (IcP) in the oligotrich *Strombidium*. (**i**) Cryptotelokinetal. Kinetosomal replication may occur in an intracytoplasmic pouch (IcP), arising from non-ciliated cortical kinetosomes as in the entodiniomorphid *Entodinium*

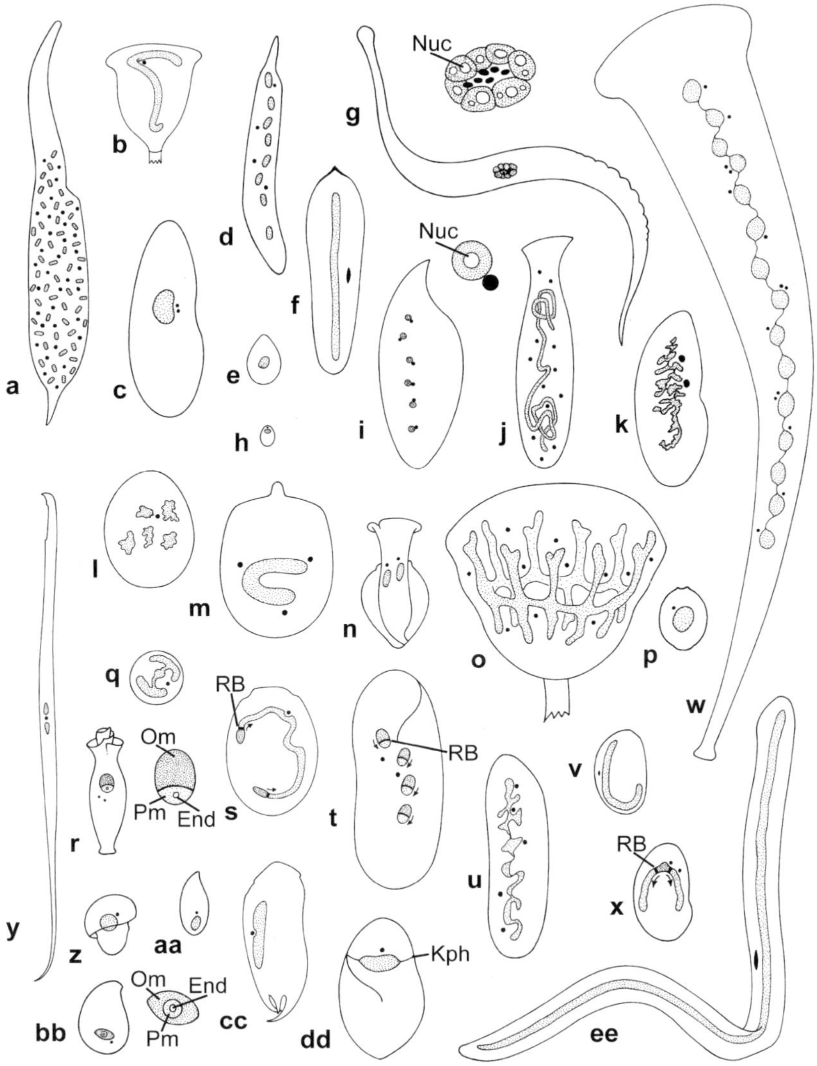

FIG. 2.12. Macronuclei (stippled) and micronuclei (solid) of diverse ciliates. Nuclei in general are not distinctive of different major groups of ciliates. The outlines of the bodies are shown roughly to scale, with the exception of *Loxodes* (**i**) and *Stentor* (**w**), which are reduced a further 50%. (**a**) The haptorian *Dileptus*. (**b**) The peritrich *Vorticella*. (**c**) The peniculine *Paramecium*. (**d**) A stichotrich. (**e**) An amicronucleate *Tetrahymena*. (**f**) The astome *Durchoniella*. (**g**) The karyorelictean *Tracheloraphis*, partly contracted with its aggregrate of nuclei above, bearing nucleoli (Nuc). (**h**) The scuticociliate *Cyclidium*. (**i**) The karyorelictean *Loxodes* with its paired macronucleus with a nucleolus (Nuc) and micronucleus. (**j**) The haptorian *Spathidium*. (**k**) The stichotrich *Plagiotoma*. (**l**) The scuticociliate *Schizocaryum*. (**m**) The haptorian *Didinium*. (**n**) The tintinnid *Tintinnopsis*. (**o**) The suctorian *Ephelota*. (**p**) The prostome *Urotricha*. (**q**) The mobiline peritrich *Leiotrocha*. (**r**) The chonotrich *Spirochona* whose heteromerous macronucleus (right) has an orthomere (Om) and a paramere (Pm) with an endosome (End). (**s**) The hypotrich *Euplotes* with two replication bands (RB). (**t**) The stichotrich *Parastylonychia* with replication bands (RB) in each nodule. (**u**) The hymenostome *Deltopylum*. (**v**) The rhynchodine *Parahypocoma*. (**w**) The heterotrich *Stentor*. (**x**) The hypotrich *Aspidisca*. (**y**) The karyorelictean *Remanella*. (**z**) The armophorian *Brachonella*. (**aa**) The rhynchodine *Insignicoma*. (**bb**) The cyrtophorine *Chilodonella* with its heteromerous macronucleus (right) showing the paramere (Pm) with its endosome (End) embedded in the orthomere (Om). (**cc**) The entodiniomorphid *Epidinium*. (**dd**) The clevelandellid *Nyctotherus* whose macronucleus is anchored by a karyophore (Kph). (**ee**) The astome *Protanoplophrya*

Chapter 3
Characters and the Rationale Behind the New Classification

Abstract Classification of a group of organisms typically begins "at the bottom" with an examination of the variation in characters of species – the genus-species level. A variety of methods have been used to determine species of ciliates from the interbreeding criterion of the biological species to a variety of features related to the morphology and ecology of a species, including life history traits, behavior, and size and shape of a variety of structures revealed by observation of living or stained cells. Genetic approaches are becoming increasingly more popular, especially molecular genetic ones. These have included the use of isoenzymes, randomly amplified polymorphic DNA (RAPD), and restriction fragment length polymorphisms (RFLP). The current cutting edge approaches are the sequencing of genes, such as small and large subunit rRNA, histone, actin, heat shock proteins, tubulins, and translation factors. Most recently, the mitochondrial cytochrome c oxidase 1 (*cox1*) gene has been chosen by some as a species "barcode".

Above the genus-level, establishing groups is more problematic, but should always rely on the establishment of monophyly using synapomorphies or shared-derived characters. These characters can be ultrastructural features of the somatic and oral kinetids, patterns of morphogenesis, and gene sequences. Taxonomy ultimately uses nomenclature and its rules to establish priority, ensure consistency, and maintain stability.

Keywords Biological species, morphological species, holotype, priority, synonym

As noted in Chapter 1, Simpson (1961) defined systematics as "the scientific study of the kinds and diversity of organisms and of any and all relationships among them" (p. 7). Systematists uncover patterns of variation in natural populations, discover the mechanisms by which species originate, and determine the phylogenetic history of organisms. These activities often lead systematists to establish a classification system to reflect the evolutionary history of a group. Ciliate systematists are no less concerned with these issues, and a variety of approaches have been discussed (e.g., Berger, 1978; Corliss, 1974a, 1976, 1979; Gates, 1978a; Lynn, 1996b). This diversity of approaches in philosophy and methodology, coupled with the technological advances of the discipline, from cytology to electron microscopy to molecular phylogenetics, has deepened our understanding of the processes of ciliate evolution (see **Chapter 1**).

While the main approach of this book is to discuss variation at a suprafamilial level, our systematic approach must begin "at the bottom", so to speak, at the level of species and genera. By first understanding the breadth of variation in characters at these levels, we can begin to assemble groups into larger and larger sets, ultimately establishing a hierarchical classification of the phylum. Since there are fairly clear differences in approach as one proceeds up the hierarchy (Corliss, 1980), it is appropriate to discuss ciliate systematics at two different "levels" – at the genus-species level and above the genus-species level.

3.1 At the Genus-Species Level

The perennial question for biologists is "What is a species?" The criterion of interbreeding as an operational definition of species – the biological species concept (see Mayr, 1970) – has been one of the most popular definitions of a species. We have known for many years that, in principle, it likely applies to ciliates, ever since the discovery of conjugation by Maupas (1889). To complicate matters, Sonneborn (1937, 1938) and Jennings (1938) recognized in the genus *Paramecium* that there were also cryptic or sibling species complexes (i.e., species that are morphologically extremely similar but yet separated into genetically distinct reproductive units). Gruchy (1955) discovered a similar species complex in the genus *Tetrahymena*. Thus, if a species is known to be sexual, one could in principle apply this methodological approach. Of course, one must first maintain in culture or in some preserved state ready to be cultured, representatives of all known isolates. Then, each time a new isolate is discovered, one must conduct all possible mating tests with known isolates to determine whether or not it is indeed new (see 3.1.3 GENETICS).

This genetic approach is undoubtedly the most definitive. Nevertheless, conjugation is rarely observed in natural populations of ciliates, many of which might be permanently asexual (Lucchesi & Santangelo, 2004). Moreover, it is practically impossible to maintain in culture all isolates of species as a set of "standards" against which to assess new isolates. Furthermore, it is often not easy to ensure that isolates have been treated perfectly so that one can confidently conclude that they are mating-incompatible. Thus, most taxonomists use morphology to describe new species, and assume that if the morphology of the new isolate is different in significant ways from the morphology of all previously described species, it is certain that the new isolate is a valid new species. Increasingly, as discussed below, molecular genetic approaches are being used as reliable substitutes for the interbreeding criterion to establish the genetic and taxonomic distinctness of new isolates.

An approach to the description of a new genus-species level taxon should begin with a search of the recent literature, which will provide direction as to the characters considered to be important for the taxonomy of the group. Below, we discuss in general terms some broad categories of descriptive characters: life history, ecology, and cultivation; morphology and multivariate morphometrics; genetics; isoenzymes and biochemistry; and gene sequences.

3.1.1 Life History, Ecology, and Cultivation

Corliss (1976, 1979) argued that a multiplicity of characters is essential for a complete description of a new species – his "constellation of characters" principle. Some of the first and possibly least problematic characters are those associated with life history and ecology. What habitat is the ciliate found in – freshwater, brackish, marine or terrestrial? What is its biogeography in these habitats – endemic or ubiquitous? What habitat variables appear to be important – temperature, pH, oxygen concentration, saprobic index (see Bick, 1972; Foissner, Berger, & Kohmann, 1994; Foissner, Blatterer, Berger, & Kohmann, 1991)?

If possible, features of its life cycle should be determined (Fig. 3.1). Has conjugation been observed? If so, how do the partners pair – anteriorly, laterally, ventrally? A common polymorphism is encystment. Have cysts been observed and are there any distinguishing characteristics of the cysts – wall ornamentation, wall layers, pigmentation? If the ciliate is free-living, is the life cycle also polymorphic in terms of ciliated forms – macrostome or cannibal forms, microstome forms? On what do these stages feed – bacteria, algae and other protists, other ciliates? If the ciliate is parasitic, is the life cycle polymorphic in other ways – trophont, protomont, tomont, tomite, theront (Fig. 3.1)?

If possible, it is very helpful to establish the ciliate in culture. This enables more detailed morphological observations to be made, including the analysis of biochemical and genetic features (**see below**). Culture methods will vary with the group. Various approaches are described in *Protocols in Protozoology* (Lee & Soldo, 1992), and an older paper published by the Society of Protozoologists has helpful directions focused on different taxa (Committee on Cultures, 1958). Reference to literature on the cultivation of taxa can also be found in the chapters in this book devoted to different classes. While an attempt is made to culture some individuals, other cells

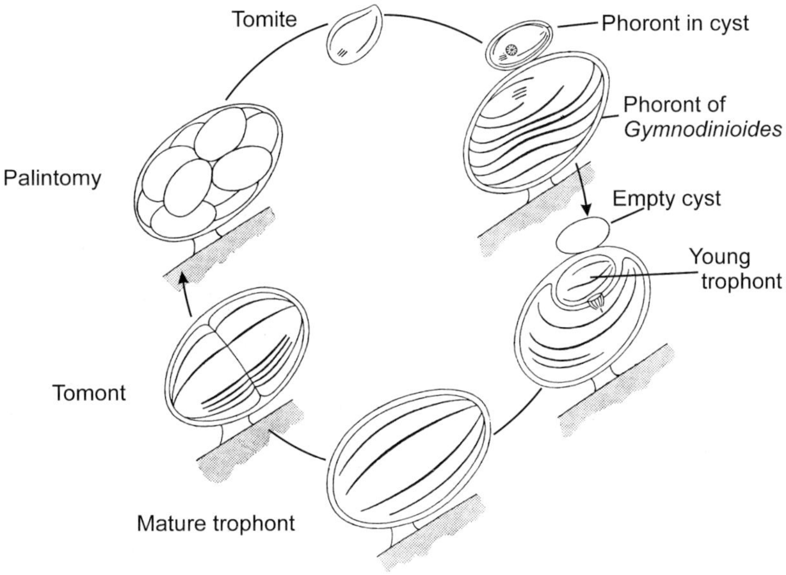

FIG. 3.1. Life history of the predatory apostome ciliate *Phtorophrya* as an example of the richness of characters that can be derived from a study of the life cycle. *Phtorophrya* is a "hyperparasite" feeding on the exuviotrophic apostome ciliate *Gymnodinioides*, which itself feeds on the exuvial fluids of its crustacean host. After *Gymnodinioides* encysts as a phoront on the crustacean host's cuticle (stippled area), the tomite of *Phtorophrya* encysts as a phoront on *Gymnodinioides*! *Phtorophrya* then penetrates the *Gymnodinioides* phoront wall and transforms to a young trophont that grows to a mature trophont by feeding upon the cytoplasm of *Gymnodinioides*. The mature trophont of *Phtorophrya* then becomes a tomont, dividing many times in palintomy to form multiple tomites, which excyst to find a new host. (Modified from Chatton & Lwoff, 1935a.)

should be more carefully observed cytologically and by molecular techniques.

3.1.2 Morphology and Multivariate Morphometrics

Cytological observations should begin with living cells, if possible. However, this may not be possible if only fixed environmental samples are available. For living cells, various aspects of their behaviour can be observed – the nature and speed of swimming. Other initial observations can include gross morphological features: body shape; body size (e.g., length, width); the kind and extent of ciliation; the general placement of the oral area (e.g., prostomial, ventrostomial); and details of the oral ciliature (Fig. 3.2). Other features that might be observed in living cells include: kinds of prey items in food vacuoles; types of food reserves; pigmentation; kinds of endosymbionts; kind, number, and distribution of contractile vacuoles; and kind, number, and distribution of extrusomes (e.g., mucocysts, toxicysts) (Fig. 3.2). While a great deal can be learned by careful observation of living cells, as demonstrated by the detailed observations of earlier microscopists (e.g., Bütschli, 1887–1889; Kahl, 1930–1935; Kent, 1880–1882; Stein, 1854, 1859), and especially today with differential interference contrast microscopy, it is essential for modern descriptions to also use staining techniques.

Descriptions of the four common methods of silver staining – the dry silver nitrate method, the wet silver nitrate method, protargol or silver-proteinate staining, and silver carbonate staining – can be found in *Protocols in Protozoology* (Lee & Soldo, 1992) and in Foissner (1991). A modified method of protargol staining, which uses cellulose acetate filters, is particularly useful for describing and quantitatively enumerating samples from the plankton (Montagnes & Lynn, 1993) and from soils (Acosta-Mercado & Lynn, 2003). Electron microscopical examination of silver-stained ciliates has demonstrated silver deposits in morphologically

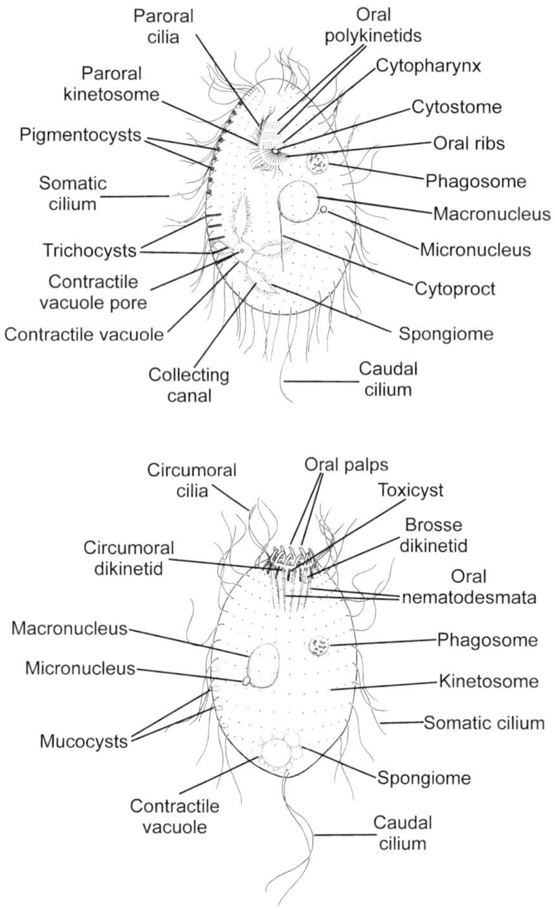

FIG. 3.2. A ventrostomatous and a prostomatous ciliate with morphological features labelled that are significant in the taxonomic description of morphological species. Reference should be made to the Glossary (Chapter 2) for definitions of these structures and for a more complete list of significant features

significant structures, such as the boundaries between cortical alveoli, microtubular organelles of the cortex and deeper cytoplasm, nuclear structures, and extrusomes (Foissner, 1975, 1977; Tellez, Small, Corliss, & Maugel, 1982; Zagon, 1970).

In general, the body shape and size of the fixed and stained ciliates should also be measured as fixation and staining can introduce artifacts (e.g. see Lynn & Berger, 1972, 1973). The gross morphological features observed for the living cells – such as, the kind and extent of ciliation and the general placement of the oral area (e.g., prostomial, ventrostomial) and details of the oral ciliature – can be confirmed. In addition, these silver methods reveal characters of the argyrome that can be taxonomically useful (Fig. 3.3). Even more generally useful, are features of the kinetome – the organellar system composed of all kinetids covering the body of a given ciliate. Here, a large number of characters can be measured, depending upon the ciliate. These could include qualitative characters, such as, whether and where there are monokinetids, dikinetids, or polykinetids and what is the nature of their ciliation (Fig. 3.2). As well, quantitative characters can be measured, such as the total number of somatic kineties, the number of kinetids in a particular kinety, the number of postoral kineties, and the number of somatic kineties on the left side and right side of the body. How the kineties converge to form suture lines or secant systems can also be important (Fig. 3.3). Ultimately, reference to the pertinent recent literature that properly describes new species will provide an exhaustive set of characters. It is important to obtain measurements on at least a statistical minimum number of cells, ideally at least 30 (Berger, 1978).

It is generally essential to use stained preparations to discover the details of the oral ciliature. Prostomial ciliates often have simple oral structures, like monokinetids and dikinetids, which are not easily visible in living specimens (Fig. 3.2). Ventrostomial ciliates may have significantly larger oral structures and many more of them. Although their detailed structure may be concealed from view in the living specimen, well-stained specimens can provide a wealth of information. Again, in addition to qualitative features of the shape of each unit and the pattern of their organization, measurements can be made, for example, on the number of oral structures, the numbers of rows and numbers of kinetosomes in each structure, and the length and width of each structure. If an actively growing population was discovered or cultivation has been possible, dividing individuals may be discovered and the detailed characterization of division morphogenesis may reveal features that could distinguish the species, but these features are more typically used at the genus level and above (see 3.2.2 MORPHOGENETIC PATTERNS).

Finally, as Corliss (1979) emphasized, nuclear features can be extremely important. Nuclear cytology can be revealed by protargol staining and also by the Feulgen nucleal stain (Lee & Soldo, 1992). The shape, size, number, and placement of both

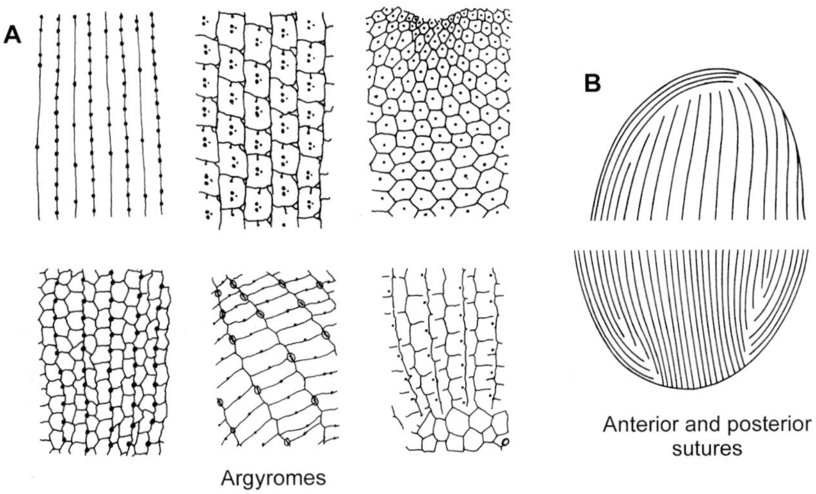

FIG. 3.3. **A** Argyromes of six types, demonstrating the diversity of patterns that can provide significant taxonomic character information, particularly at the species level. Top row: the hymenostome *Colpidium*, the peniculine *Frontonia*, the prostome *Bursellopsis*; Bottom row: the prostome *Pelagothrix*, the colpodean *Pseudoplatyophrya*, and the prostome *Urotricha*. Note that the three prostomes have quite different patterns (redrawn from various sources). **B** Examples of an anterior suture or secant system (top) and two posterior suture or secant systems (bottom)

micronuclei and macronuclei should be recorded. As well, the distribution of chromatin can be important and whether the macronucleus can be classified as homomerous or heteromerous (Raikov, 1982). While large chromatin aggregates in the macronucleus may indeed be nucleoli, they cannot be concluded to be so without objective evidence that they are the actual site(s) for the production of ribosomes (e.g., Postberg, Alexandrova, & Lipps, 2006). Without this evidence, they must only be referred to as chromatin bodies.

If there is a sufficient database on related species, some assessment, often using statistical approaches, can determine whether the new isolate is significantly different morphologically. For example, Berger (1965) and Lynn and Berger (1972, 1973) used a univariate statistical approach to demonstrate that demes of the echinoid endocommensal ciliates *Plagiopyliella* and *Thyrophylax* were significantly different on a number of morphological characters. Biometric characterization has now become standard practice, enabling future researchers to compare new isolates. Multivariate morphometric techniques have been applied with some success to distinguish the cryptic species of *Paramecium* and *Tetrahymena* (Gates & Berger, 1974; Gates, Powelson, & Berger, 1975; Powelson, Gates, & Berger, 1975). Others have applied these approaches to resolving morphological species within genera, such as *Colpoda* (Foissner & Schubert, 1983; Lynn & Malcolm, 1983) and *Ancistrum* (Berger & Hatzidimitriou, 1978). Gates (1977, 1978b, 1979) assessed the pattern of variation on the ventral surface of the hypotrich *Euplotes* by measuring all possible distances between cirri. By converting these intercirral distances to a relative frequency distribution of scaled intercirral distances, he was able to show that these distributions corresponded with relationships determined by mating tests.

3.1.3 Genetics

Interbreeding, in which two populations are considered to be members of different species if mating tests fail to produce fertile offspring, is the essential criterion for recognition of a biological species. Ciliates are usually stimulated to conjugate in the laboratory by starvation, the analogue of the natural stimulus, which is depletion of the food resource. Once starved, known species and mating types can be used as testers to identify unknowns. Studies recognizing new species of *Tetrahymena* (Nyberg, 1981a; Simon, Meyer, & Preparata, 1985) and *Paramecium* (Aufderheide, Daggett, & Nerad,

1983) have used this approach. Genetic techniques have been used to explore the biology of the sibling species complexes of *Euplotes* (e.g., Dini & Gianni, 1985; Génermont, Machelon, & Demar, 1985; Luporini & Dini, 1977) and the biogeography of *Paramecium* (Komala & Przybos, 1990; Przybos & Fokin, 1997).

Many morphological species may, in fact, be sibling or cryptic species groups (Curds, 1985). However, the genetic approach, while most rigorous, is difficult in practice since both a complete set of viable reference strains must be maintained and the taxonomist must have competence with genetic techniques. Fortunately, biochemical and genetic correlates have now been found for several sibling species complexes (see 3.1.4 ISOENZYMES AND BIOCHEMISTRY and 3.1.5 GENE SEQUENCES), and these studies provide metrics to discover how common cryptic species of ciliates are.

3.1.4 Isoenzymes and Biochemistry

Isoenzymes are enzymatic proteins that share the same biochemical function, but they are coded by structurally different alleles. This structural difference is revealed by their differential movement in an electrophoretic gel. Based on earlier work on isoenzymes (e.g., Allen, Byrne, & Cronkite, 1971; Tait, 1970), Sonneborn (1975) established Linnean names for the sibling species or syngens of the *Paramecium aurelia* sibling species complex. Allen et al. (1983a, 1983b) have applied this approach to other species of *Paramecium*. Nanney and McCoy (1976) likewise established Linnean names for the 12 syngens of the *Tetrahymena pyriformis* sibling species complex, following earlier isoenzyme studies (e.g., see Allen & Weremuik, 1971; Borden, Whitt, & Nanney, 1973a, 1973b). Species of the hypotrich *Euplotes* (Machelon & Demar, 1984; Schlegel, Kramer, & Hahn, 1988; Valbonesi, Ortenzi, & Luporini, 1985) and the stichotrich *Stylonychia* (Ammermann, Schlegel, & Hellmer, 1989) have also demonstrated different isoenzyme patterns.

Genetic diversity between ciliate species is exceedingly great, indicating a considerable evolutionary age of species or extremely rapid molecular evolution at these isoenzyme loci. Thus, these techniques are generally robust and reliable for distinguishing and identifying species (but see 3.1.5 GENE SEQUENCES). However, there are two major disadvantages to using isoenzymes. First, there is the need to have an efficient cultivation technique for the species of interest, one that yields significant protein biomass to enable resolution of these molecules. Second, since isoenzyme patterns are often complex, there is a strong possibility that "homologues" are not being identified unambiguously. Because of this, and because DNA techniques can now be carried out on much smaller numbers of cells, even single cells, isoenzymes have been displaced as systematic molecules of choice.

3.1.5 Gene Sequences

Allen and Li (1974) began sequence diversity studies on ciliates with their analysis of DNA-DNA hybridization of *Tetrahymena* species against syngen 1 (i.e., *Tetrahymena thermophila*) as the reference standard. They showed considerable genetic diversity within the genus. However, as with isoenzyme techniques, this approach required substantial amounts of DNA and therefore was best used for ciliates that could be easily cultivated. Moreover, the invention of DNA sequencing technologies (e.g., Sanger, Nicklen, & Coulson, 1977) enabled a direct comparison of primary DNA sequence similarity between species. This provides greater resolution than the single percentage that DNA hybridization could provide, and also obviates the need to maintain cultures to continually obtain the DNA of the reference standard(s). Ultimately the invention of the polymerase chain reaction (PCR) (Mullis & Faloona, 1987) has enabled us to amplify DNA from small numbers of cells, even single cells, so that cultivation is no longer an absolute necessity for the application of molecular genetic techniques.

Genetic diversity among *Tetrahymena* species was first compared using small subunit rRNA (SSUrRNA) gene sequences by Sogin, Ingold, Karlok, Nielsen, & Engberg (1986a), who cloned the SSUrRNA genes and demonstrated sequence identity in some species pairs and up to 33 differences between others. The histone H3II/H4II regions of the *Tetrahymena* genome were amplified by PCR and sequence analyses demonstrated relationships among species similar to those derived from SSUrRNA comparisons, and furthermore differentiated all species uniquely (Brunk, Kahn, & Sadler, 1990; Sadler & Brunk, 1992). A similar approach using a portion of the large subunit rRNA (LSUrRNA) gene showed considerable genetic diversity among *Tetrahymena* species, and generally corroborated groupings based on other molecular methods (Nanney, Meyer, Simon, & Preparata, 1989; Nanney, Park, Preparata, & Simon, 1998; Preparata et al., 1989). *Paramecium* species can be distinguished using SSUrRNA gene sequences (Strüder-Kypke, Wright, Fokin, & Lynn, 2000a) and heat shock protein 70 (Hori, Tomikawa, Przybos, & Fujishima, 2006).

In addition to direct sequence comparisons, PCR has also been used to generate randomly amplified polymorphic DNA (RAPD) and has been used in conjunction with restriction enzymes to digest SSUrRNA. Both approaches generate fragments of varying length that provide patterns diagnostic for species. Jerome and Lynn (1996) showed that different *Tetrahymena* species could be identified by discrete restriction fragment length patterns or riboprints. The application of RAPD fingerprinting has been used to assess differences among *Paramecium* (Fokin, Stoeck, & Schmidt, 1999; Skotarczak, Przybos, Wodecka, & Maciejewska, 2004; Stoeck & Schmidt, 1998; Stoeck, Przybos, Kusch, & Schmidt, 2000a, Stoeck, Welter, Seitz-Bender, Kusch, & Schmidt, 2000b) and *Euplotes* species (Chen, Song, & Warren, 2001; Kusch, Welter, Stremmel, & Schmidt, 2000; Mollenbeck, 1999). Since RAPD fingerprinting depends upon PCR, large numbers of cells are, in principle, not required. However, the technique does have significant problems, including variation introduced due to inefficiencies in the PCR and due to variations in band intensity. For these reasons, more predictable approaches are to be preferred.

The techniques discussed so far have all assessed variation based on nuclear genetic variation, which may be more constrained both within and between species. A promising new approach is the "barcode" gene, mitochondrial cytochrome c oxidase 1 (*cox*1), which has been successfully applied to a variety of animal groups (Hajibabaei, Janzen, Burns, Hallwachs, & Hebert, 2006; Hebert, Cywinska, Ball, & DeWaard, 2003; Hebert, Stoeckle, Zemlak, & Francis, 2004). Barth, Krenek, Fokin, and Berendonk (2006) demonstrated that *cox*1 could be effectively used to separate out several *Paramecium* species, with interspecific divergences ranging from 12–27%, while Lynn and Strüder-Kypke (2006) and Chantangsi, Lynn, Brandl, Cole, Hetrick, and Ikonomi (2007) have demonstrated similar levels of divergence in *cox* 1 between species of *Tetrahymena* that are identical based on the SSUrRNAgene sequence. Barth et al. (2006) showed significant intrahaplogroup variation within *paramecium caudatum* and *Paramecium multimicronucleatum*, suggesting that these species may, in fact, be sibling species complexes, while Chantangsi et al. (2007) have demonstrated that isolates of *Tetrahymena* identified to species on the basis of isozyme patterns have apparently been misclassified.

3.1.6 Summary

The approaches presented above provide different methods of assessing variation within species and between species within genera. We cannot recommend one of these approaches over another. Rather, a modern description of a new species of ciliate should, where possible, include data provided by observation of living organisms, stained organisms, and gene sequence data (e.g. see Agatha, Strüder-Kypke, Beran, & Lynn, 2005; Modeo, Petroni, Rosati, & Montagnes, 2003; Rosati, Modeo, Melai, Petroni, & Verni, 2004). Comparison of these datasets with previous descriptions should then enable one to conclude whether an isolate is indeed new. As our databases of gene sequences increase, it has been demonstrated that fluorescence in situ hybridization can be used to identify species (Fried, Ludwig, Psenner, & Schleifer, 2002), and environmental gene sequences can be linked to morphology using both light and scanning electron microscopy (Stoeck, Fowle, & Epstein, 2003).

While body size is important, body size on its own is seldom sufficient to distinguish a species. Indeed, there are many other quantitative traits not correlated with size that may ultimately be discriminatory. Just as there are no hard and fast rules for determining whether an isolate is a new species, it is also difficult to provide any for the genus level. In general, one can say that genera should be differentiated on the basis of significant qualitative characters. And one may reasonably ask – what is a significant qualitative character? Again, there are no hard and fast rules, and what characters are considered important may depend upon whether the taxonomist is a "lumper" or a "splitter" – what is a significant qualitative character for a "splitter" may not be so for a "lumper" (Corliss, 1976). In general, it is our view that "significant" at the generic level should at least included qualitative differences in body shape, pattern of the somatic kineties, and organization of the oral structures. As noted in Chapter 1, oral variations are likely to directly affect growth and reproductive rates, enhancing the relative fitness and fixation of new oral variants (Lynn, 1979b). Thus, it is often the case that new genera are distinguished on the basis of variations in oral features, as well as qualitative variations in somatic features.

3.2 Above the Genus-Species Level

Above the level of genus and species, it is even more difficult to provide guidance on what features can be used to generally distinguish a family, an order, a class, or a subphylum. Corliss (1976, 1979) discussed the "gap size of distinctness" as a conceptual way to identify the discontinuities that separate these higher taxa. As he noted, "one should be able to recognize a gap of 'sufficient' (how defined?!?) magnitude between any two groups of species before proposing their formal separation into different higher taxa" (p. 59, Corliss, 1979). Indeed, it is often the case that higher taxa show these discontinuities with respect to each other, and they often exhibit what Corliss (1979) termed a shared "constellation of characters", which further supports their separation. While a 'sufficient' gap size of distinctness and a shared constellation of characters often characterize higher taxa, there must be at least one synapomorphic or shared derived character that can be used to establish the monophyly of the group.

Thus, to identify major monophyletic clades, we must ultimately search for characters that are highly conserved over time. As Lynn (1976a, 1981) has argued, conservation of biological structure, especially in regard to the ciliate cortex, becomes more conserved as we investigate lower levels of biological organization (i.e., organellar complexes, organelles), which we discuss in more detail below (see 3.2.1 ULTRASTRUCTURE, ESPECIALLY OF THE CORTEX). These highly conserved 'characters' may also be morphogenetic sequences or developmental patterns, which appear as structural similarities, especially in the division ontogeny of ciliates, uniting different major taxa into higher assemblages (see 3.2.2 MORPHOGENETIC PATTERNS). In the present day, the ultimate signals of common descent are the primary and secondary structures of gene and amino acid sequences (see 3.2.3 GENE AND PROTEIN SEQUENCES).

3.2.1 Ultrastructure, Especially of the Cortex

Since the late 1960s and early 1970s, electron microscopic investigations of ciliates have provided a substantial increase in the number of characters available to determine relationships. As argued in Chapter 1 and elsewhere (Lynn, 1976a, 1981), there are good reasons to believe that similarities at this level of biological organization reveal much more ancient common ancestry. The diversity of somatic and oral kinetids of ciliates has been described (Grain, 1969, 1984; Lynn, 1981, 1991; de Puytorac & Grain, 1976). Lynn (1976a, 1979a, 1981) has argued that somatic kinetid features are more strongly conserved than oral features (Fig. 3.4). Application of these criteria – lower levels of biological organization more conserved *and* "somatic over oral" – has enabled us to establish a number of the major classes of ciliates (Lynn & Small, 1997, 2002; Small & Lynn, 1981, 1985).

While cortical characters have been of primary importance, the fine structure of other features has also been helpful: variations in the particle distributions on ciliary membranes (Bardele, 1981) and in the substructure of extrusomes, like toxicysts and trichocysts (Hausmann, 1978; Rosati & Modeo, 2003).

The multitude of ultrastructural characters has meant that several studies have used both phenetic and cladistic approaches assisted by computer to assess relationships among ciliates. These studies have ranged from a broad assessment at the phylum level (Lynn, 1979a; de Puytorac, Grain, & Legendre, 1994; de Puytorac, Grain, Legendre, & Devaux, 1984) to focussed treatments of classes and orders (Lipscomb & Riordan, 1990, 1992).

Nevertheless, there are clear signs that morphostatic structures, even at the ultrastructural level, have found their limits. Lynn (1991) noted that genera such as *Transitella*, *Phacodinium*, *Plagiopyla*, *Lechriopyla*, and *Schizocaryum* exhibit cortical features that cannot be used to confidently

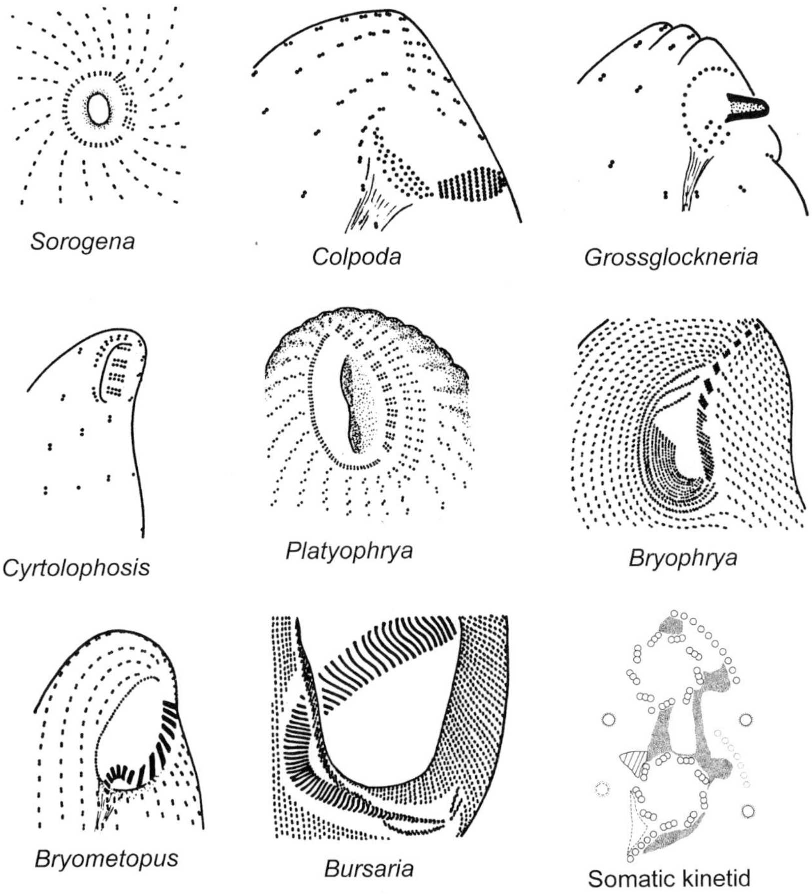

FIG. 3.4. A demonstration of the structural diversity of the oral region among ciliate genera from the class COLPODEA. When examined, all of these genera have the same basic somatic kinetid pattern (bottom right). However, prior to the Age of Ultrastructure, they were placed in different higher taxa: *Colpoda* and *Bryophrya* were trichostomes; *Cyrtolophosis* was a hymenostome; *Platyophrya* was a gymnostome; *Bryometopus* and *Bursaria* were heterotrichs (see Corliss, 1961). *Sorogena* and *Grossglockneria* were described during the Age of Ultrastructure. (Modified from Foissner, 1993a.)

assign them to a major class. In fact, the latter four genera have now been assigned to a class based only on gene sequences!

3.2.2 Morphogenetic Patterns

The analysis of developmental patterns in revealing phylogenetic relationships has its roots in the 19th century in the work of Haeckel and von Baer (see Gould, 1977). While this approach must be applied with caution, Corliss (1968, 1979) noted that application of the Biogenetic Law to ciliate development – similarities in the ontogenies of different major groups of ciliates "recapitulating" their phylogeny – had provided important insights. Sewertzoff (1931) discussed a number of other principles, such as oligomerization and polymerization, which have been applied to the evolution of protozoan lineages (von Gelei; 1950; Poljansky & Raikov, 1976; Raabe, 1971a). Oligomerization and polymerization, and the related principle of auxomorphy (Fauré-Fremiet, 1968), have been used to explain the origin of new types, primarily at the genus and species levels, but have not provided insights relating major groups of ciliates.

Ontogenetic patterns that have been the most illuminating in relating higher taxa are those exhibited by ciliates at asexual reproduction during which new mouthparts are formed and the somatic region of the parent is partitioned for the progeny. These two related processes are termed cortical stomatogenesis and cortical somatogenesis, respectively (Lynn & Corliss, 1991). As noted in Chapter 1, similarities in oral structures and division morphogenesis of hymenostomes, thigmotrichs, and peritrichs lead Fauré-Fremiet (1950a) to conclude these groups shared a common ancestry, while Guilcher (1951) used developmental patterns to infer the common ancestry of cyrtophorians, suctorians, and chonotrichs (Fig. 1.1). In addition to uniting major groups, stomatogenetic patterns can be used to separate major groups. For example, Small (1967) recognized a pattern of stomatogenesis among the hymenostomes in which the scutica or scutico-hook was formed, now called scuticobuccokinetal stomatogenesis, and this formed the basis for his distinguishing the Order Scuticociliatida within the oligohymenophorean clade.

Bardele (1989) has described a research program devoted to the careful analysis of morphogenesis using electron microscopy, arguing that this will enable us to identify with certainty the true developmental origin of the microtubular associates of kinetosomes and so determine homologies. In fact, Huttenlauch and Bardele (1987) proved the importance of this approach when they demonstrated that the microtubular ribbons supporting the cytopharynx of the prostome *Coleps* were in fact postciliary microtubular ribbons. These postciliary ribbons change their orientation during stomatogenesis to appear like transverse microtubular ribbons, an interpretation previously given to them based on morphostatic ultrastructure (e.g., see Lynn, 1985; de Puytorac & Grain, 1972). Furness and Butler (1986) examined somatic kinetosomal replication in entodiniomorphids and revealed the transient appearance of a second transverse microtubular ribbon microtubule, providing support for alignment of entodiniomorphids to the litostome clade (Lynn, 1991; Lynn & Nicholls, 1985). Patterns of oral kinetid assembly can involve the joining and rotation of kinetosomes and the appearance and disappearance of fibrillar structures in predictable sequences (Eisler, 1989; Jerka-Dziadosz, 1980, 1981a, 1981b, 1982), and these features may ultimately prove useful to systematists when the database becomes sufficiently large.

Finally, brief mention should be made of the ultrastructure of the nuclear apparatus. Nuclear dimorphism – the micronucleus and macronucleus – is a synapomorphy for the Phylum Ciliophora. The inability of the macronucleus to divide has long been considered a "karyological relict" character (Corliss, 1979; Raikov, 1969, 1982, 1996). Related to this, Orias (1991a) has suggested that differences in the use of extra-macronuclear and intra-macronuclear microtubules during macronuclear division argue that macronuclear division evolved twice in the phylum. Lynn (1996a) used this feature to establish the Subphylum Intramacronucleata, uniting all ciliates that primarily use intra-macronuclear microtubules during macronuclear division.

Thus, morphogenetic patterns at a variety of levels of biological organization and of a variety of structures have proved useful in revealing common descent among major groups of ciliates. These will continue to be useful at these higher levels, in addition to providing subtle differences that may distinguish species and genera.

3.2.3 Gene and Protein Sequences

The distinctive nature of the major groups of ciliates revealed both by light and electron microscopy means that relatively few characters can be used to assemble them into larger groups. As noted above and in Chapter 1, features of the somatic kinetid, similar patterns of division morphogenesis, and nuclear features have been most useful. Sequences of genes and proteins provide a large number of characters, either as nucleotides or amino acids, and many of these may evolve independently of each other. This structural conservation with variation, at yet a lower level of biological organization, is presumed to signal deep common ancestry, and therefore permits us to test relationships established using morphological criteria, such as the ultrastructure of the somatic kinetid.

The earliest studies used sequences of the 5S and 5.8S rRNAs, but their shorter lengths and conserved nature were not helpful in resolving deep relationships (Van Bell, 1985). Initial studies to sequence cloned small subunit ribosomal RNA (SSUrRNA) genes (Elwood, Olsen, & Sogin, 1985; Sogin, Swanton, Gunderson, & Elwood, 1986b) and reverse transcripts of both SSUrRNA and large subunit rRNA (LSUrRNA) (Baroin et al., 1988; Lynn & Sogin, 1988; Nanney et al., 1989; Preparata et al., 1989) provided clear evidence that these data would prove useful in elucidating deep phylogeny. Analyses were enabled by powerful sets of phylogenetic inference packages. Recent updates are found in Felsenstein (2004) and Swofford (2002).

The application of PCR using universal eukaryote primers provided both a rapid alternative to cloning genes and the potential to obtain sequence information from small numbers of cells, even a single cell (Medlin, Elwood, Stickel, & Sogin, 1988). The PCR sequencing approach has been extended to other genes, including histones (Bernhard & Schlegel, 1998), tubulins (Baroin-Tourancheau, Villalobo, Tsao, Torres, & Pearlman, 1998; Israel, Pond, Muse, & Katz, 2002), actins (Hogan, Hewitt, Orr, Prescott, & Müller, 2001), and translation factors (Moreira, Kervestin, Jean-Jean, & Philippe, 2002). Some of this research will be discussed in more detail in Chapter 16. Nevertheless, it appears for the moment, as judged by its agreement with morphology, that the rRNA genes provide the most reliable signals for deep phylogenetic relationships, in contrast to actins (Philippe, Chenuil, & Adoutte, 1994) and tubulins (Israel et al., 2002), which do not recover the same major clades as morphology, and may not even recover ciliates as a clade!

The SSUrRNA and LSUrRNA gene sequences have confirmed the major clades established using the somatic kinetid (Lynn, 1996b), and have enabled placement of enigmatic genera, such as *Phacodinium* (Shin et al., 2000) and *Schizocaryum* and *Licnophora* (Lynn & Strüder-Kypke, 2002). On the other hand, there is strong indication of several lineages that can only be termed "riboclasses" (Lynn, 2004), since there are no obvious morphological synapomorphies for these groups. These include the genera *Plagiopyla*, *Lechriopyla*, and *Trimyema* now assigned to the "ribo"-Class PLAGIOPYLEA, along with the odontostomatids, exemplified by *Epalxella* (see **Chapter 14**) and the genera *Nyctotherus* and *Metopus* now assigned to the "ribo"- Class ARMOPHOREA (see **Chapter 8**). Perhaps we will discover morphological traits that will corroborate these gene sequence data or perhaps we will discover additional supporting molecular signals.

3.2.4 Summary

The three approaches discussed above outline the major avenues to determining relationships and providing criteria to establish groups above the genus-species level. Again, there are no easy directions that enable one to identify higher taxa. In general, somatic kinetids have proved very diagnostic as they exhibit universality, constancy, and consistency within clades. We have argued in Chapter 1 and elsewhere (Lynn, 1981) that similarities in pattern at this level can be inferred to be homologous and are therefore strong indicators of common ancestry.

Gene and protein sequences continue to be the characters of choice in resolving the deeper relationships within the phylum. As discussed in Chapter 16, representatives of most of the major groups have now been sequenced. Thus, it is unlikely that our views on the major lines of diversification within the phylum will be radically changed in the immediate future.

3.3 Taxonomy and Nomenclature

We use scientific names to label taxa and provide a vocabulary for communicating about these organisms. Taxonomy is the discipline devoted to discovering, describing, and improving the characterization of taxa or groups of organisms. However, nomenclature is the discipline devoted to naming these taxa in ways that satisfy the criteria of availability, which are set out in the International Code of Zoological Nomenclature, herein called the Code. The most recent, 4th edition, of the Code was published by the International Commission on Zoological Nomenclature (1999a). Since the names of organisms are essential for scientific communication and are important as a key to the literature, taxonomy and nomenclature deserve a central place in biology. It is important to remember the difference between these two disciplines. An old example from the ciliate literature is illustrative. Hill (1752), as a taxonomic author, first described the genus *Paramecium* and provided a name. However, since "year zero" for the Code began on 1 January 1758, O. F. Müller (1773) is credited as the nomenclatural author of the formal genus name – *Paramecium* O. F. Müller, 1773 – since he was the first to publish the name in a way that satisfied the criteria of availability of the Code. Names can change for a variety of reasons: for legalistic reasons related to the rules of nomenclature; for philosophical reasons related to whether a given taxonomist might be a "lumper" or "splitter"; or because novel characters have been discovered often after the application of new methods to the study of the taxa (Aescht, 2001).

In a number of papers, Corliss (1962a, 1962b, 1972a, 1976, 1980, 1995) has dealt extensively with nomenclature and taxonomy as they relate to protozoa and protists. More recently, Aescht (2001) has provided a summary of important principles that should be observed, especially noting matters raised in the new edition of the Code. Both authors direct readers to the original rules of the Code as the primary authority. Corliss (1962a) noted that errors or bad habits can arise from "ignorance of rules, carelessness in their application, lack of clarity in the rules themselves, and total lack of a pertinent directive anywhere in the rules" (p. 307). In this section, only brief mention will be made of some important issues in nomenclature.

The Code was established to provide rules to establish priority, to ensure consistency in naming of organisms, and to maintain stability or universality in names. As Corliss (1972a) emphasized, common sense and courtesy should be used and deference should be paid to the stability of the names while always being mindful of the provisions of the Code. While the Code does not apply to taxa above the family level, Corliss (1962a) argued that it is common sense to apply these principles at the suprafamilial level, and we have followed this recommendation. When a suprafamilial taxon has been simply transferred within a higher taxon or between higher taxa, even if it has changed its rank, we have retained the priority date from the original publication along with the original authorship. This promotes stability and recognizes priority. On the other hand, if in our view the proposed change involves a new taxonomic concept, then we have recognized a new authorship and date (see Corliss, 1972a). While not required by the Code, we have adopted a uniformity for the endings of the higher taxa of ciliates: for class – "-ea"; for subclass – "-ia"; for order – "-ida"; and for suborder – "-ina". The principle of typification is a primary principle in the Code. The fixation of the name-bearing type of a nominal taxon provides an objective standard of reference: the concept of species is linked to a concrete specimen, the holotype; the concept of genus to a definite species, the type-species; and the concept of family to a definite genus, the type-genus. While this principle could be applied to the higher suprafamilial taxa of ciliates, we have not done so nor have we indicated the type genera of the families, a task that will need to be undetaken by a future revision.

While this monograph only treats ciliate taxa to the level of genus, nevertheless the principles and rules of the Code apply to the genus and family ranks treated herein. Aescht (2001) has served as our principle resource for the valid generic names of ciliates, and her excellent monograph should be the first source for all literature prior to about 13th March 2000 when her revisions on this monograph stopped. We have aimed to include all the literature subsequent to that date and up to 31st December 2006. A brief review of some important nomenclatural matters follows.

3.3.1 The Matter of Types

Species names are linked to concrete specimens, ideally designated by the original author of the name as holotypes and lectotypes. Corliss (1962a) discussed the difficulty of preserving types, as individual specimens, for the protozoa. However, much has changed since this time. With the development of more reliable mounting media and the refinement of silver-staining techniques, type specimens of ciliates on type slides can be deposited in a variety of museum collections (see Corliss, 1972b). These type specimens must be recognized as "the property of science" and should be kept safely, labeled clearly and completely, and provided with minimal difficulty and cost to any competent researcher who wishes to study them. After 1999, there must be an explicit fixation of a holotype or syntypes and indication of where these specimens are depositied.

The type for the genus is a species, which is often established by original designation, by some indication, or by subsequent designation. Since many ciliate genera were monotypic when first established, there is no ambiguity regarding the type-species. At the family level, the type-genus is usually easily recognized as the family name is typically based on it, although there are exceptions (Corliss, 1962a, 1962b, 1977).

3.3.2 Important Dates

As noted above, "year zero" for the Code is 1st January 1758, dating from the year in which Linnaeus (1758) first published his *Systema Naturae* (Art. 3.2; Code). Prior to 1900, names published as vernacular names and generally accepted by the specialist community are available (e.g., Bursariens of Dujardin as Bursariidae Dujardin, 1841; see Corliss, 1962a). Beginning 1st January 1931, it was necessary to designate a type species to establish a valid genus name (Art. 13, 68; Code). If this was not done, the name technically becomes a *nomen nudum* or "naked name." Prior to 1931, the name may be acceptable provided it was at least accompanied by a description, definition or other indication. After 1999, there must be an explicit fixation of a holotype and indication of where this specimen is deposited, and the taxon must be explicitly indicated as new by using "n. sp.", "n. gen.", and "n. fam." or equivalent designation. Foissner and Berger (1999) provide an excellent treatment of the problems arising from *nomina nuda* that arose during molecular biological investigations of the oxytrichid stichotrichs.

If a name has not been in practical use for at least 50 years – the "50-year" rule, it can be considered a *nomen oblitum* or forgotten name. Thus, an unused senior synonym (i.e., older name) cannot replace a junior synonym (i.e., younger name) that has been in general use.

3.3.3 About Names

Once fixed by a nomenclatural author, a name-bearing type cannot be changed. Names are considered available when they are published in a work that is in hard copy, publicly available, and produced in sufficient copies. Names published in theses and abstracts are generally considered not to be available.

The principle of homonymy states that no two scientific zoological names can be spelled identically. Thus, all other things being equal, the principle of priority will dictate which name shall remain valid and which name must be replaced. It is recommended that differences in one letter should be avoided. Whether names have been used before can be discovered by consulting indexes in the Zoological Record and of S. A. Neave. Aescht (2001), noting that these are not perfect records, emphasized that there is no substitute for a thorough personal knowledge of the relevant literature.

If two different names refer to the same name-bearing type, they are called objective synonyms or nomenclatural synonyms. The nomenclatural decision here is therefore unambiguous based on the rule of priority: the junior objective synonym, that is the more recent name, must be taken out of use. Sometimes, however, there is ambiguity in regard to the name-bearing type, especially in the protozoological literature in which written descriptions and/or figures may be the only means of understanding the features of the name-bearing type. In this case, a later worker may decide from the evidence that two different names, in their opinion, refer to the same species. These names would be considered subjective synonyms because they are based on the subjective judgement of that particular taxonomist. Subjective decisions are never definitive since they are a matter of opinion. Nevertheless, the reviser may invoke the rules of priority and recommend that the junior subjective synonym be taken out of use.

There are numerous rules and recommendations with regard to the technical formation of names. Simply, scientific names of organisms should be Latin or latinized, regardless of their etymological origin. The genus name begins with a capital letter and is a substantive or noun or adjective of a substantive or noun. When publishing the name, it is advisable to state its derivation or etymology. In addition, the gender of the genus name should be indicated. The gender can be determined by referring to standard Greek and Latin dictionaries. If the genus name is a compound word, it should take its gender from the last component (Art. 30; Code). Refer to Corliss (1962a, 1962b), Aescht (2001), and the Code for more detailed information and advice.

3.3.4 Summary

The above discussion is meant to provide a brief introduction to the rules of nomenclature. Nothing substitutes for a reading of the most recent edition of the Code. While the Code was established to promote stability, circumstances arise from time-to-time when the preservation of names that contravene the Code is seen to be in the best interests of the scientific community: stability is preserved in these cases even though the rules of the Code would be violated. In this event, a petition may be submitted to the International Commission on Zoological Nomenclature, arguing the case for the conservation of a name. For example, an appeal for conservation of the genus *Tetrahymena* Furgason, 1940 was made by Corliss and Dougherty (1967) while a more recent case, for example, was made by Corliss and Foissner (1997) for conservation of authorship for *Trachelocerca* Ehrenberg, 1840. Supportive rulings in relation to these petitions were respectively made by the International Commission on Zoological Nomenclature (1970, 1999b).

Chapter 4
Phylum CILIOPHORA – Conjugating, Ciliated Protists with Nuclear Dualism

Abstract The ciliated protozoa are a distinct group of protists characterized by (1) the presence of cilia derived from kinetosomes with three fibrillar associates; (2) nuclear dimorphism; and (3) conjugation as a sexual process. They are exceedingly diverse in shape and size, and may have evolved over 2 billion years ago. The ciliate body form likely evolved from a flagellate that proliferated kinetids near its oral apparatus, and these kinetids became first the ciliate paroral, which itself replicated to provide somatic kineties and finally the adoral kinetids. Ciliates are now divided into two subphyla and 11 classes based on features of nuclear division and the pattern of fibrillar associates in their somatic kinetids. They are found in a diversity of microhabitats, with the majority of species likely cosmopolitan. However, endemism appears to be characteristic of perhaps as many as 30% of the species. Ciliates are important components of the microbial loop, often responsible for consuming the majority of primary production and bacterial production in certain habitats.

Their development is complex since the kinetid patterns of the somatic and oral cortex are themselves complex. This complexity has made them useful models for developmental biologists while systematists have used these patterns to relate different taxa to each other. The life cycles of ciliates are also complex and clearly separate sexual processes from asexual reproduction. Sexual process, conjugation, occurs in the context of a breeding strategy, ranging from an inbreeding one to an outbreeding one. A new macronucleus typically develops at each conjugation cycle through a complicated set of processes of DNA fragmentation, diminution, and replication. However, this process likely occurs at each cell cycle in karyorelictean ciliates whose macronuclei do not divide.

Keywords Cytotaxis, structural guidance, extrusome, metachrony, alveoli

The ciliates, without doubt, are a most homogeneous group, which have long been separated from other protists. They are briefly distinguished by three major features: (1) by their cilia, variable in number and arrangement, distributed over the body surface and derived from kinetosomes with three typical fibrillar associates – the kinetodesmal fibril, the postciliary microtubular ribbon, and the transverse microtubular ribbon; (2) by their two kinds of nuclei – a macronucleus and a micronucleus, the former controlling the physiological and biochemical functions of the cell and the latter as a germ-line reserve; and (3) by conjugation, a sexual process in which partners typically fuse temporarily to exchange gametic nuclei. All ciliates are heterotrophic and most possess a mouth, although some are mouthless (e.g., the astomes) and others could be called polystomic (e.g., suctorians with their tentacles).

Ciliates vary in shape and size. There are stalked and colonial species that have unusual forms, but generally the shapes are simple geometric ones – spheres, cones, oblate spheroids, and cylinders, which may be flattened dorsoventrally in substrate-oriented species. Body form is relatively permanent since the cortex is supported by a complex microtubular and microfilamentous cytoskeleton. Ciliates range in size from 10 μm in very small spheroid forms to 4,500 μm in highly elongate and contractile ones. Biovolume ranges give another impression of size: in the Class COLPODEA, cell volume ranges from 10^2 µm^3 for *Nivaliella* to 10^8 µm^3 for *Bursaria*; and in the Class OLIGOHYMENOPHOREA, cell volumes range from 10^3 µm^3 for some *Cyclidium* species to $>10^9$ µm^3 for the trophonts of *Ichthyophthirius* (Lynn & Corliss, 1991).

Ciliates were first observed microscopically by van Leeuwenhoek (1674), the founder of protozoology (Corliss, 1975b). Ciliates were visible as blooms or colored waters in both marine and freshwater habitats, probably thousands of years before van Leeuwenhoek's discoveries. Their formal nomenclatural history begins in 1767 with the establishment of *Vorticella* by Linnaeus (Aescht, 2001). They were named the INFUSORIA throughout the 19th century, a name that was replaced in the early 20th century by the present name of the phylum – CILIOPHORA. The ciliates were an isolated group until 1991 when it was proposed that the membrane-bound sacs underlying their plasma membrane, the alveoli, were a synapomorphy or shared-derived character for a major clade of protists, the ALVEOLATA, which initially included dinoflagellates, apicomplexans, and ciliates (Cavalier-Smith, 1991; Wolters, 1991). This clade was reaffirmed in the revised classification of the protists proposed by Adl et al. (2005). Based on strong molecular evidence, this clade now includes these three major groups, along with the aberrant "dinoflagellate" *Oxyrrhis* and the perkinsids associated with the dinoflagellate clade and several flagellates, such as *Colpodella*, associated with the base of the apicomplexan clade. Dinoflagellates and apicomplexans are strongly supported as sister taxa, leaving the ciliates as a separate lineage (Leander & Keeling, 2003). When and how the ciliates diverged from the alveolate common ancestor are still open questions, but we have some ideas.

With regard to when, ciliates probably arose in the Precambrian era. While there are no fossils from this time, Wright and Lynn (1997c) argued that the phylum could be over 2 billion years old, based on the rate of evolution of the small subunit rRNA molecular clock (Fig. 4.1). While this estimate can be challenged as it makes some strong assumptions on the constancy of the molecular clock, there is no dispute over fossils of the loricate tintinnines, which stretch from the Ordovician period of the Paleozoic era, some 400–450 million years before present (myBP) into the Pleistocene, about 1 myBP. Tintinnine fossils are most abundant in the sediments of the Jurassic and Cretaceous periods of the Mesozoic era (see Bonet, 1956; Colom, 1965; Tappan & Loeblich, 1968, 1973). Fossils of other loricate forms have also been found: Deflandre and Deunff (1957) reported a fossil folliculinid heterotrich while Weitschat and Guhl (1994) described several fossil loricate peritrichs from the Lower Triassic period. Even more recently, extremely well-preserved ciliates, a *Paramecium* species and several colpodeans, have been described from amber derived from southern Germany in amber-bearing sandstone deposits (Schönborn, Dörfelt, Foissner, Krientiz, & Schäfer, 1999), which have now been assigned to the Late Cretaceous, some 90 million years ago (Schmidt, von Eynatten, & Wagreich, 2001). However, these fossil forms can generally be placed in contemporary families and even genera, so they provide little insight into the origin of the ciliates. Nevertheless, this has not deterred some speculation on how ciliates diverged. Given the molecular phylogenies, there is no doubt that the ciliates evolved from a flagellate ancestor. Two major features of ciliates – their complex cortex and their nuclear dualism – have preoccupied those who have speculated on how a flagellate ancestor might have evolved into the ancestral ciliate.

The complex ciliate cortex undoubtedly evolved from a simpler flagellate dikinetid. Earlier evolutionary schemes of Orias (1976) and Small (1984) argued that the dikinetids of the flagellate ancestor replicated to form first a "torsionless chain", similar to the polynergid dinoflagellate *Polykrikos*, and then a ribbon-like form with multiple somatic kineties. Orias (1976) imagined that the ribbon-like from rolled up to ultimately form an ancestral prostomial form, while Small (1984) imagined that the ribbon-like form might have been very similar to the contemporary karyorelictean *Kentrophoros*, with an elongate, flattened "ventral" surface, which it used for ingestion. Both these schema suggested that oral ciliature derived from somatic ciliature.

Eisler (1992) proposed a contrary model in which the paroral dikinetids of ciliates represent the first ciliature of the ancestral ciliate and that the somatic kinetids derived from this. This model, most recently summarized by Schlegel and Eisler (1996),

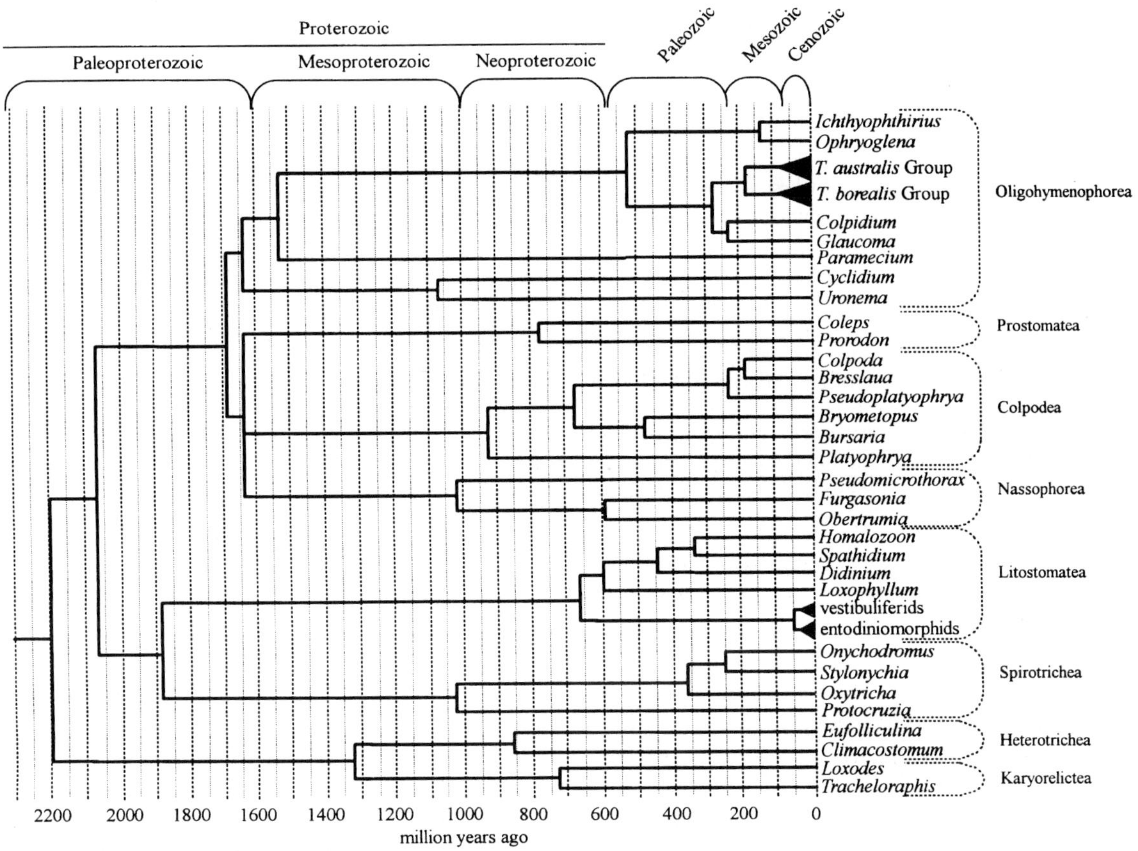

FIG. 4.1. A phylogeny of the ciliates demonstrating the estimated time of divergence of some major lineages as estimated by the divergence rate of small subunit rRNA gene sequences. The upper limit of 1% divergence per 80 million years was used to determine the lengths of the branches on the tree. (from Wright & Lynn, 1997c.)

begins with a similar polyenergid stage as above. However, this "first" kinety is interpreted to be a paroral, lying adjacent to a tube-like cytopharyngeal apparatus supported by microtubules, as is found in some dinoflagellates (Fig. 4.2). At this stage, the paroral dikinetids are considered to lie orthogonal to the longitudinal axis of the cell and their postciliary microtubular ribbons extended to support the cytopharynx. In the second stage, somatic Kinety 1 (K1, Fig. 4.2) was derived from these paroral dikinetids by separation and rightward migration of the anterior or rightmost kinetosome of each paroral dikinetid (Step a, Fig. 4.2). Replication of these kinetosomes followed to reconstitute the ancestral dikinetid state (Lynn & Small, 1981). Eisler (1992) suggested that multiple repetitions of this process could give rise to multiple somatic kineties (Step b, Fig. 4.2). Alternatively, one could invoke the processes of either torsion and fragmentation (see Small, 1984) or elineation to increase the number of somatic kineties. In a final stage, the adoral or "lefthand" oral structures are imagined to derive from the differentiation of somatic kinetids to the left of the oral region (Step c, Fig. 4.2). To support this, Schlegel and Eisler (1996) noted that the adoral ciliature is derived from a somatically-derived anlage in many contemporary ciliates. The paroral model has the advantage that, from the beginning, the ciliate ancestor, like many contemporary dinoflagellates, has a cytopharynx supported by microtubules derived from the "paroral" dikinetids: there is no need to invoke an independent evolution

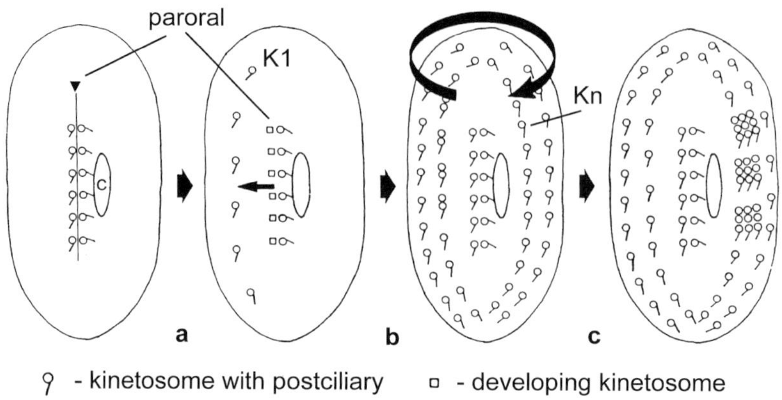

FIG. 4.2. Scheme of evolution of the ancestral ciliate oral and somatic cortex as proposed by Eisler (1992). Step **a** – an ancestral flagellate with a cytostome (c) and paroral of dikinetids separates the rightmost kinetosome of each dikinetid (arrowhead) to form somatic Kinety 1 (K1). Step **b** – this process is repeated (arrow) a number of times until the entire somatic cortex is covered by somatic kineties (Kn). Step **c** – adoral structures derive from the differentiation of somatic kinetids to the left of the cytostome. (Modified from Schlegel & Eisler, 1996.)

of the oral apparatus as in the competing models (see Orias, 1976; Small, 1984). Stomatogenesis and morphogenesis of the progeny cell or opisthe would have been of a buccokinetal type.

Speculations on the evolution of another major feature distinguishing ciliates – nuclear dualism – are intriguing. Orias (1991b) used the life cycle of heterokaryotic foraminifera as an analogue to provide a rationale for how a heterophasic life cycle of a ciliate, ancestral to the karyorelicteans, might have evolved. This heterophasic life cycle had an alternation of haploid and diploid generations. From this, Orias (1991b) developed the possible steps to the evolution of a karyorelicteancell cycle in which nuclear dualism occurs but with divisionless macronuclei differentiated at every cell cycle. While we now know that ciliates and foraminiferans are not closely related (Nikolaev et al., 2004), it is still possible that a heterophasic kind of life cycle might have been an intermediate stage in the evolution of ciliate nuclear dualism.

Orias (1991a) argued further that macronuclear division must have originated de novo *within* the ciliates. He noted that macronuclei in most ciliates fail to divide during the first postzygotic cell cycle following conjugation: more macronuclei are differentiated than is typical and these "excess" macronuclei are segregated *without* macronuclear division during postconjugation fissions, reminiscent of the phenomenon that occurs *at every cell cycle* in karyorelicteans. Macronuclear "dividers" may have had selective advantages: the division process is much faster than the nuclear differentiation process and so "dividers" would outcompete "non-dividers"; and dividing macronuclei may have increased the capability to assort intraclonally the genetic diversity of the parents and so increase the probability of generating more fit variants (Orias, 1991a). Why do karyorelicteans still persist? Orias (1991a) suggested that they may still be more fit in the "refugial" relict environments in which they are often found, and that many karyorelicteans may have compensated for macronuclear polyploidy by increasing the number of macronuclei, so supporting their increased cell size.

As additional evidence that macronuclear division originated *within* the ciliates, Orias (1991a) noted that diversity in the modes of karyokinesis suggested at least two independent origins of macronuclear division – heterotrichs use *extramacronuclear* microtubules and other ciliates use *intramacronuclear* microtubules. Herrick (1994) argued that macronuclear division may have evolved independently perhaps three or more times, given the diversity of molecular mechanisms underlying macronuclear differentiation. However, Katz (2001) has used current molecular phylogenies of ciliates to argue for a single origin of a differentiating mechanism relying on *trans*-acting factors that emanate from the parental macronucleus to influence the differentiation of the developing macronuclei. This epigenetic mechanism is conceived to be plastic enough to have generated the molecular genomic diversity that we see in contemporary ciliates.

Nevertheless, the number of origins of macronuclear division is still unsettled. Two models have been proposed. For Model 1, the ancestral ciliate had a dividing macronucleus that lost its capacity to divide in the karyorelicteans. For Model 2, there were two independent origins of macronuclear division from a "non-dividing" ancestor – one with *extramacronuclear* microtubules in the heterotrich lineage and one with *intramacronuclear* microtubules in the intramacronucleate lineage (Hammerschmidt et al., 1996; Lynn, 1996a). We currently favor Model 2, which is consistent with the evidence presented by Orias (1991a, 1991b). For it, the macronucleus of intramacronucleate ciliates would "regain" the capacity to use intranuclear microtubules that continued to be used by the micronucleus during its mitoses while the heterotrich lineage would have "re-invented" the use of extranuclear microtubules to divide their macronuclei, an invention used also by some dinoflagellates (Perret, Albert, Bordes, & Soyer-Gobillard, 1991).

In summary, what would our ancestral ciliate look like? It would have had a pellicle with alveoli underlying the plasma membrane. If the paroral model is used for cortical evolution, the ancestral ciliate would have possessed a paroral, would have had a ventral oral region, would have had a cytopharynx supported by postciliary microtubular ribbons, and would have undergone buccokinetal stomatogenesis (Schlegel & Eisler, 1996). Its micronucleus would have divided by using an intranuclear spindle of microtubules and kinetochores and its macronucleus would have been non-dividing and paradiploid.

4.1 Taxonomic Structure

The ciliates are among the top five groups of protists in terms of numbers of species (Corliss, 2004). There are likely at least 8,000 species; this includes about 200 fossil forms and close to 3,000 symbiotic species, but there is some dispute over these numbers (**see below: Life History and Ecology**). As noted in **Chapter 1**, the ciliates are now regarded as a phylum divided into two major subphyla and eleven classes. Among the classes, the classes SPIROTRICHEA, PHYLLOPHARYNGEA, and OLIGOHYMENOPHOREA have over two-thirds of the described species.

Lynn (1996a) argued for the establishment of two subphyla based primarily on data emerging from molecular phylogenetic studies, which showed a strongly supported bifurcation in the phylogenies of ciliates based on the small subunit rRNA gene (e.g., Hammerschmidt et al., 1996; Hirt et al., 1995). This bifurcation separated postciliodesmatophoran ciliates, which are placed in the Subphylum Postciliodesmatophora and are distinguished from all other ciliates by their somatic kinetids with postciliodesmata (Table 4.1). While the other clade is strongly supported by a variety of genes (see **Chapter 16**), the only morphological feature uniting this clade appears to be division of the macronucleus by intramacronuclear microtubules, hence the Subphylum Intramacronucleata (Table 4.1) (Lynn, 1996a). The history of the macrosystem presented in the following chapters was developed in **Chapter 1** and the distribution of significant characters on molecular phylogenies will be presented in **Chapter 16**. In the remainder of this section, we briefly characterize the 11 classes that are now recognized.

To begin with, the primary characters used to distinguish ciliate taxa reside in the cortex, although some non-cortical characters, such as nuclear features, are also important. The cortex, which is the main interface between the organism and its environment, can be divided into a somatic region and an oral region. The somatic region functions in locomotion, provides protective coverings and defensive responses, and enables attachment to the substrate. The oral region functions in the acquisition and ingestion of nutrients. Features of both regions, and particularly the kinds and arrangements of ciliary structures, are important in the characterization of the major groups of ciliates.

The subphylum Postciliodesmatophora is divided into the Classes KARYORELICTEA and HETEROTRICHEA (Table 4.1). Karyorelicteans, like *Loxodes* (Fig. 4.3), have non-dividing macronuclei and somatic kinetids with postciliodesmata (Fig. 4.7) in which the postciliary microtubular

TABLE 4.1. Classification of the Phylum Ciliophora.

Phylum CILIOPHORA Doflein, 1901
Subphylum *POSTCILIODESMATOPHORA* Gerassimova & Seravin, 1976
- Class **KARYORELICTEA** Corliss, 1974
  - Order Protostomatida Small & Lynn, 1985
  - Order Loxodida Jankowski, 1980
  - Order Protoheterotrichida Nouzarède, 1977
- Class **HETEROTRICHEA** Stein, 1859
  - Order Heterotrichida Stein, 1859

Subphylum *INTRAMACRONUCLEATA* Lynn, 1996
- Class **SPIROTRICHEA** Bütschli, 1889
- Subclass Protocruziidia de Puytorac, Grain, & Mignot, 1987
  - Order Protocruziida Jankowski, 1980
- Subclass Phacodiniidia Small & Lynn, 1985
  - Order Phacodiniida Small & Lynn, 1985
- Subclass Licnophoria Corliss, 1957
  - Order Licnophorida Corliss, 1957
- Subclass Hypotrichia Stein, 1859
  - Order Kiitrichida Nozawa, 1941
  - Order Euplotida Small & Lynn, 1985
    - Suborder Discocephalina Wicklow, 1982
    - Suborder Euplotina Small & Lynn, 1985
- Subclass Choreotrichia Small & Lynn, 1985
  - Order Tintinnida Kofoid & Campbell, 1929
  - Order Choreotrichida Small & Lynn, 1985
    - Suborder Leegaardiellina Laval-Peuto, Grain, & Deroux, 1994
    - Suborder Lohmanniellina Laval-Peuto, Grain, & Deroux, 1994
    - Suborder Strobilidiina Small & Lynn, 1985
    - Suborder Strombidinopsina Small & Lynn, 1985
- Subclass Stichotrichia Small & Lynn, 1985
  - Order Stichotrichida Fauré-Fremiet, 1961
  - Order Sporadotrichida Fauré-Fremiet, 1961
  - Order Urostylida Jankowski, 1979
- Subclass Oligotrichia Bütschli, 1887/1889
  - Order Strombidiida Petz & Foissner, 1992
- Class **ARMOPHOREA** Lynn, 2004[a]
  - Order Armophorida Jankowski, 1964
  - Order Clevelandellida de Puytorac & Grain, 1976
- Class **LITOSTOMATEA** Small & Lynn, 1981
- Subclass Haptoria Corliss, 1974
  - Order Haptorida Corliss, 1974
  - Order Pleurostomatida Schewiakoff, 1896
  - Order Cyclotrichiida Jankowski, 1980 *incertae sedis*
- Subclass Trichostomatia Bütschli, 1889
  - Order Vestibuliferida de Puytorac et al., 1974
  - Order Entodiniomorphida Reichenow in Doflein & Reichenow, 1929
    - Suborder Archistomatina de Puytorac et al., 1974
    - Suborder Blepharocorythina Wolska, 1971
    - Suborder Entodiniomorphina Reichenow in Doflein & Reichenow, 1929
  - Order Macropodiniida order nov.[a]
- Class **PHYLLOPHARYNGEA** de Puytorac et al., 1974
- Subclass Cyrtophoria Fauré-Fremiet in Corliss, 1956
  - Order Chlamydodontida Deroux, 1976
  - Order Dysteriida Deroux, 1976
- Subclass Chonotrichia Wallengren, 1895
  - Order Exogemmida Jankowski, 1972
  - Order Cryptogemmida Jankowski, 1975
- Subclass Rhynchodia Chatton & Lwoff, 1939
  - Order Hypocomatida Deroux, 1976
  - Order Rhynchodida Chatton & Lwoff, 1939
- Subclass Suctoria Claparède & Lachmann, 1858
  - Order Exogenida Collin, 1912
  - Order Endogenida Collin, 1912
  - Order Evaginogenida Jankowski, 1978
- Class **NASSOPHOREA** Small & Lynn, 1981
  - Order Synhymeniida de Puytorac et al., 1974
  - Order Nassulida Jankowski, 1967
  - Order Microthoracida Jankowski, 1967
  - Order Colpodidiida Foissner, Agatha & Berger, 2002 *incertae sedis*
- Class **COLPODEA** Small & Lynn, 1981
  - Order Bryometopida Foissner, 1985
  - Order Bryophryida de Puytorac, Perez-Paniagua, & Perez-Silva, 1979
  - Order Bursariomorphida Fernández-Galiano, 1978
  - Order Colpodida de Puytorac et al., 1974
  - Order Cyrtolophosidida Foissner, 1978
  - Order Sorogenida Foissner, 1985
- Class **PROSTOMATEA** Schewiakoff, 1896
  - Order Prostomatida Schewiakoff, 1896
  - Order Prorodontida Corliss, 1974
- Class **PLAGIOPYLEA** Small & Lynn, 1985[a]
  - Order Plagiopylida Jankowski, 1978
  - Order Odontostomatida Sawaya, 1940 *incertae sedis*
- Class **OLIGOHYMENOPHOREA** de Puytorac et al., 1974
- Subclass Peniculia Fauré-Fremiet in Corliss, 1956
  - Order Peniculida Fauré-Fremiet in Corliss, 1956
  - Order Urocentrida Jankowski, 1980
- Subclass Scuticociliatia Small, 1967
  - Order Philasterida Small, 1967
  - Order Pleuronematida Fauré-Fremiet in Corliss, 1956
  - Order Thigmotrichida Chatton & Lwoff, 1922
- Subclass Hymenostomatia Delage & Hérouard, 1896
  - Order Tetrahymenida Fauré-Fremiet in Corliss, 1956
  - Order Ophryoglenida Canella, 1964
- Subclass Apostomatia Chatton & Lwoff, 1928
  - Order Apostomatida Chatton & Lwoff, 1928
  - Order Astomatophorida Jankowski, 1966
  - Order Pilisuctorida Jankowski, 1966
- Subclass Peritrichia Stein, 1859
  - Order Sessilida Kahl, 1933
  - Order Mobilida Kahl, 1933
- Subclass Astomatia Schewiakoff, 1896
  - Order Astomatida Schewiakoff, 1896

[a] A taxon based on molecular phylogenetics, but still lacking a morphological synapomorphy.

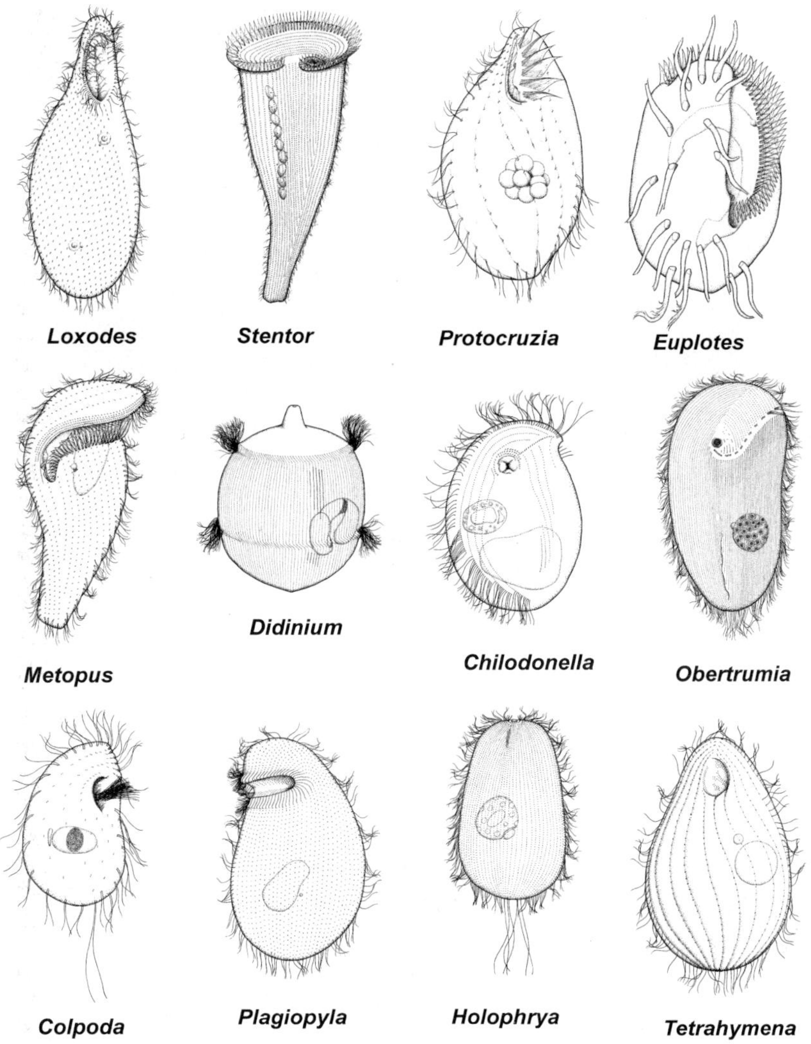

FIG. 4.3. Stylized drawings of genera representative of each class in the Phylum Ciliophora: *Loxodes* – Class KARYORELICTEA; *Stentor* – Class HETEROTRICHEA; *Protocruzia*, *Euplotes* – Class SPIROTRICHEA; *Metopus* – Class ARMOPHOREA; *Didinium* – Class LITOSTOMATEA; *Chilodonella* – Class PHYLLOPHARYNGEA; *Obertrumia* – Class NASSOPHOREA; *Colpoda* – Class COLPODEA; *Plagiopyla* – Class PLAGIOPYLEA; *Holophrya* – Class PROSTOMATEA; and *Tetrahymena* – Class OLIGOHYMENOPHOREA

ribbons are separated by 1 + 2 microtubules (see **Chapter 5**). Heterotricheans, like *Stentor*, *Blepharisma*, and *Fabrea* (Figs. 4.3, 4.4), have macronuclei that divide by *extramacronuclear* microtubules and somatic kinetids with postciliodesmata (Fig. 4.7) in which the postciliary microtubular ribbons are separated by only a single microtubule (see **Chapter 6**). Oral structures in the karyorelicteans are quite variable, ranging from prostomial with simple circumoral ciliature to ventrostomial with developed paroral and adoral ciliature. Heterotricheans, like *Stentor*, are virtually all bearers of a paroral and an elaborately developed adoral zone of polykinetids (Figs. 4.3, 4.4).

The Subphylum Intramacronucleata includes the remaining nine classes, each of which will

FIG. 4.4. Scanning electron micrographs of ciliate diversity. **A–C** Class HETEROTRICHEA. *Blepharisma* (**A**), *Fabrea* (**B**), and *Stentor* (**C**). **D–I** Class SPIROTRICHEA. The oligotrich *Strombidium* (**D**), the tintinnids *Dictyocysta* (**E**) and *Tintinnopsis* (**F**), the stichotrich *Stylonychia* (**G**), and the hypotrichs *Euplotes* (**H**) and *Uronychia* (**I**). (Micrographs courtesy of E. B. Small and M. Schlegel.)

be briefly characterized here (Table 4.1). Lynn (2004) has noted that four of the classes – the LITOSTOMATEA, PHYLLOPHARYNGEA, NASSOPHOREA, and COLPODEA – are strongly supported by **both** molecular and morphological characteristics. The remaining five classes – the SPIROTRICHEA, ARMOPHOREA, PLAGIOPYLEA, PROSTOMATEA, and OLIGOHYMENOPHOREA – have less strong support from molecules and morphology. Two of these latter five, the ARMOPHOREA and PLAGIOPYLEA, are only supported by molecules, and hence called "riboclasses" (Lynn, 2004).

Spirotricheans, like *Protocruzia*, *Euplotes*, *Strombidium*, *Dictyocysta*, *Tintinnopsis*, *Stylonychia*, and *Uronychia* (Figs. 4.3, 4.4, Table 4.1), are a diverse group, typically having a paroral and a well-developed adoral zone of polykinetids. The class is rarely strongly supported as a clade by molecular phylogenetics. Most of the taxa exhibit somatic dikinetids with a poorly developed kinetodesmal fibril (Fig. 4.7) and two of the included subclasses, the Hypotrichia and Stichotrichia, have compound ciliary organellar complexes called cirri (Figs. 4.3, 4.4, 4.7). The strongest morphological synapomorphy for the class is the replication band that occurs during macronuclear DNA S-phase (see **Chapter 7**). The replication band has been confirmed in members of all subclasses except the two monotypic Subclasses Protocruziidia and Phacodiniidia. The phacodiniids are undoubtedly spirotrichs by their placement well within the spirotrichean clade using the small subunit rRNA (SSUrRNA) molecule (Shin et al., 2000). The situation for *Protocruzia*, also the only member of its subclass, is more uncertain as it is typically placed as the basal lineage in the spirotrich branch. We have placed protocruziids in this class because they are tenuously associated with it by SSUrRNA gene sequences (see **Chapter 16**). However, they may warrant separate class status in the future because their histone sequences are divergent from other spirotrichs (Bernhard & Schlegel, 1998), and they exhibit a unique mode of macronuclear division (Ammermann, 1968; Ruthmann & Hauser, 1974).

Armophoreans, like the armophorid *Metopus* and the related clevelandellid *Nyctotherus* (Figs. 4.3, 4.5, Table 4.1), are strongly supported only by SSUrRNA gene sequences (van Hoek et al., 2000b). The somatic kinetids of the two subclasses within this class are quite different (Fig. 4.7; and see **Chapter 8**). However, Villeneuve-Brachon (1940) speculated that these ciliates might be related to each other, a view later endorsed by Jankowski (1968b) and Albaret (1975). Both free-living and endosymbiotic armophoreans are found in anoxic or close to anoxic environments and all are presumed to have hydrogenosomes. Future research on armophorean hydrogenosomes may reveal synapomorphies in their anaerobic metabolism that may more strongly confirm this class.

Litostomateans, like the haptorians *Didinium* and *Dileptus* and the trichostomes *Isotricha*, *Entodinium*, and *Ophryoscolex* (Figs. 4.3, 4.5, Table 4.1) are strongly supported by both SSUrRNA gene sequences and by features of the somatic kinetid, which is a monokinetid with two transverse ribbons (Fig. 4.7; and see **Chapter 9**). The two included subclasses may not be monophyletic: it now appears the haptorians may be a paraphyletic group (Strüder-Kypke, Wright, Foissner, Chatzinotas, & Lynn, 2006).

Phyllopharyngeans, like the cyrtophorians *Chilodonella* and *Trithigmostoma* and the suctorian *Podophrya* (Figs. 4.3, 4.6, Table 4.1), form a diverse group, strongly supported by both SSUrRNA gene sequences and by features of the somatic kinetid, which is a somatic monokinetid that has a laterally-directed kinetodesmal fibril and whose kinetosomes are underlain by subkinetal microtubules (Fig. 4.7; and see **Chapter 10**). A significant feature of the phyllopharyngean oral apparatus is a set of radial ribbons of microtubules that support the cytopharynx, the phyllae.

Nassophoreans, like *Obertrumia* (Fig. 4.3, Table 4.1), are also strongly supported by both SSUrRNA gene sequences and by features of the somatic kinetids, which are monokinetids that can be linked as dikinetids by filaments near the base of the transverse microtubular ribbon (Fig. 4.7; and see **Chapter 11**). In addition, the cytopharyngeal basket or nasse of these ciliates exhibits a suite of characters not found together in any other class.

Colpodeans, like *Colpoda* (Figs. 4.3, 4.5, Table 4.1), are typically well supported by both SSUrRNA gene sequences and by features of the somatic kinetids, which are dikinetids whose posterior kinetosomes have well-developed transverse microtubular ribbons extending posteriorly along the kinety, forming the transversodesma or LKm fibre (Fig. 4.7; and see **Chapter 12**).

Prostomateans, like *Holophrya* and *Coleps* (Figs. 4.3, 4.5, Table 4.1), are not strongly supported by molecular signals, but this may in part be due to the low taxon sampling for this group. Their somatic kinetids show similarities to those of the next three classes (Fig. 4.7), and it is only in the details of their oral structures and stomatogenesis that the group may be distinguished (see **Chapter 13**).

Plagiopyleans, like *Plagiopyla* (Fig. 4.3, Table 4.1), are strongly supported by SSUrRNA gene sequences even though the taxon sampling is low. Nevertheless, this is the second "riboclass" within the phylum because there is no strong synapomorphy for the group. The somatic kinetids are similar to those of the Classes PROSTOMATEA and OLIGOHYMENOPHOREA, showing a well-developed anteriorly-directed kinetodesmal fibril and a radially-oriented transverse microtubular ribbon (Fig. 4.7). A remarkable recent discovery

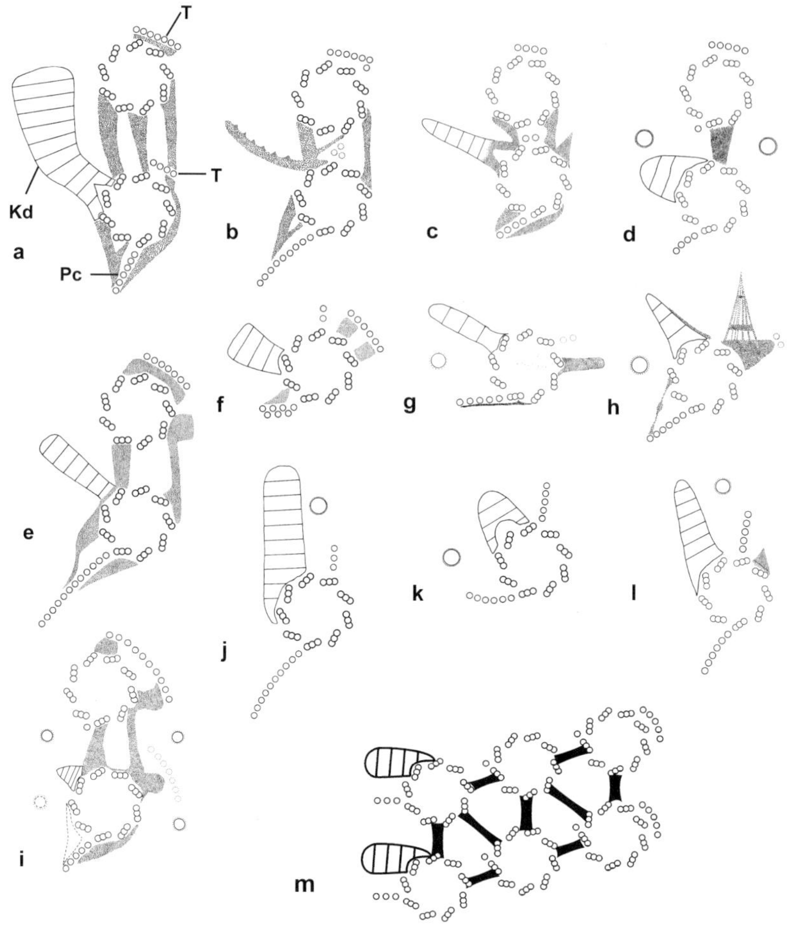

FIG. 4.7. Schematics of somatic kinetids of genera representative of each class in the Phylum Ciliophora. (**a**) *Loxodes* – Class KARYORELICTEA; (**b**) *Blepharisma* – Class HETEROTRICHEA; (**c**, **d**) *Protocruzia* (**c**), *Euplotes* (**d**) – Class SPIROTRICHEA; (**e**) *Metopus* – Class ARMOPHOREA; (**f**) *Balantidium* – Class LITOSTOMATEA; (**g**) *Chilodonella* – Class PHYLLOPHARYNGEA; (**h**) *Obertrumia* – Class NASSOPHOREA; (**i**) *Colpoda* – Class COLPODEA; (**j**) *Plagiopyla* – Class PLAGIOPYLEA; (**k**) *Holophrya* – Class PROSTOMATEA; (**l**) *Tetrahymena* – Class OLIGOHYMENOPHOREA; (**m**) *Plagiotoma* – Class SPIROTRICHEA. Kd – kinetodesmal fibril; Pc – postciliary microtubular ribbon; T – transverse microtubular ribbon (from Lynn, 1981, 1991)

and, related to this, differences in the survivability of the non-encysted stages (Jackson & Berger, 1985a, 1985b). Often, the starving trophont transforms into a highly motile form, the theront, which may be adapted both for dispersal and very long survival (Fig. 4.8) (Fenchel, 1990; Nelsen & DeBault, 1978).

A common adaptive variation is the differentiation of macrostome cannibal forms – ciliates that differentiate a new oral apparatus large enough to ingest their microstomatous siblings (de Puytorac, 1984b) (Fig. 4.8). This transformation is often induced by starvation, like the theront transformation mentioned above. More dramatic examples of adaptation are found in symbiotic forms, especially parasitic ones (Bradbury, 1996). *Ichthyophthirius*, the parasite of fish gills and epithelium, apparently lacks a typical resting cyst stage. Instead, it grows to a considerable size as a trophont on the fish host, then drops off the host and becomes a tomont in a reproductive cyst. The tomont divides to produce thousands of tomites, which may reside for some time within the cyst before breaking out to find the next host. Finally, hyperparasites or hyperpredators can be

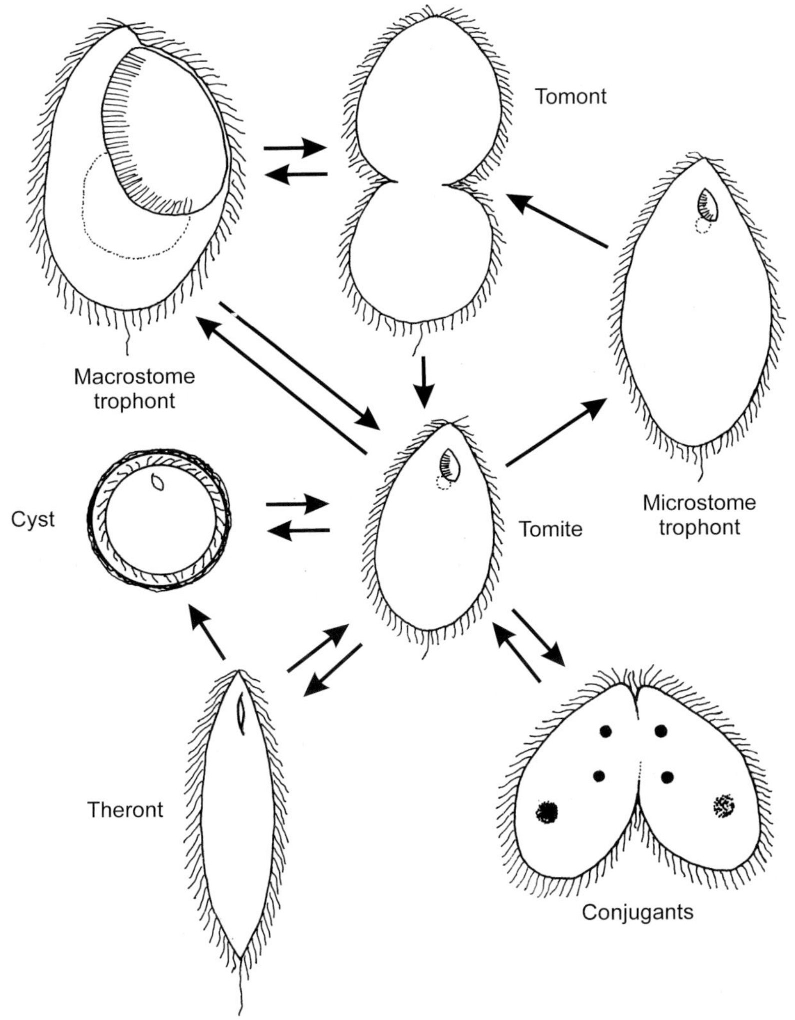

FIG. 4.8. Life cycle stages of ciliates. A microstome trophont, typically feeding on bacteria, grows from the tomite stage until it roughly doubles in size to become a dividing tomont. This vegetative or asexual cycle can repeat itself as long as food is present. If food becomes limiting the ciliate may transform to a macrostome trophont, which is a cannibal form that can eat tomites and smaller microstome trophonts or other ciliates. If food is limiting or other stressful environmental circumstances prevail, the ciliate may form a cyst or may transform into a theront, a rapidly swimming dispersal stage. If the theront does not find food, it too may encyst. In unusual circumstances, when food is depleted and a complementary mating type is present, the ciliates may fuse together as conjugants and undergo the sexual process of conjugation

found among the apostome oligohymenophoreans: *Phtorophrya insidiosa* is an apostome that attacks other apostomes, which are themselves symbionts on the cuticle of crustaceans (see Fig. 3.1).

Ciliates are heterotrophic, exhibiting a wide range of feeding behaviours, and occupying a diversity of ecological niches (Dragesco, 1984b; Finlay & Fenchel, 1996). As noted above, some can transform to feed on their siblings, which in the vast majority of cases are suspension feeders (Fenchel, 1980a, 1980b). The "particles" removed from suspension can be very small, like viruses and bacteria, moderately-sized, like various kinds of unicellular algae, and relatively large, like other ciliates. The varieties of specific prey chosen by ciliates in the different classes are detailed in the following chapters. Bacterivorous ciliates are particularly important in maintaining the "quality" of effluent from sewage treatment plants as they can reduce bacterial densities ten-fold by their feeding (Curds & Cockburn, 1970a, 1970b; Foissner, 1988a; Madoni, 2003) and may even consume viruses (Pinheiro et al., 2007).

In addition to being symbionts in other organisms (Bradbury, 1996; Fernández-Leborans & Tato-Porto, 2000a, 2000c; Levine, 1972; Song, 2003), a variety of other organisms can use ciliates as their host (Ball, 1969). These endosymbionts of ciliates can range from bacteria living in the micronucleus (Görtz, 1983, 1996; Hovasse, 1984b) to various species of algae (Hovasse, 1984a; Lobban et al., 2002; Reisser, 1986).

Ciliates are distributed globally in a diversity of habitats where they function as important trophic links in a variety of food webs (Adl, 2003; Finlay & Fenchel, 1996; Foissner, 1987; Pierce & Turner, 1992; Sanders & Wickham, 1993). They are found in the world's oceans, in the plankton (Edwards & Burkhill, 1995; Lynn & Montagnes, 1991; Pierce & Turner, 1993; Strom, Postel, & Booth, 1993), on ocean shores (Agamaliev, 1971; Al-Rasheid, 1999d; Dragesco, 1965; Kovaleva & Golemansky, 1979), in ocean depths (Fenchel et al., 1995; Hausmann, Hülsmann, Polianski, Schade, & Weitere, 2002; Silver, Gowing, Brownlee, & Corliss, 1984), and associated with sea ice (Lee & Fenchel, 1972; Song & Wilbert, 2000b). They are found in a variety of "land-locked" waters, including freshwater bodies, such as lakes (Beaver & Crisman, 1982, 1989a; Esteban, Finlay, Olmo, & Tyler, 2000; Taylor & Heynen, 1987), freshwater ponds (Finlay & Maberly, 2000; Taylor & Berger, 1976), rivers (Balazi & Matis, 2002; El Serehy & Sleigh, 1993; Foissner, 1997b), and streams (Madoni & Ghetti, 1980; Taylor, 1983a), and hypersaline lagoons and lakes (García & Niell, 1993; Post, Borowitzka, Borowitzka, Mackay, & Moulton, 1983; Yasindi, Lynn, & Taylor, 2002). Ciliates are also recorded from terrestrial habitats,primarily soils and mosses (Buitkamp, 1977; Foissner, 1998a; Ryan et al., 1989). Along with their association with mosses, ciliates can also be found in the liquid in pitcher plants leaves (Addicott, 1974; Rojo-Herguedas & Olmo, 1999) and in the axils of tropical plants, such as bromeliads (Foissner, Strüder-Kypke, van der Staay, Moon-van der Staay, & Hackstein, 2003). The species composition and diversity of ciliates have been used as bioindicators of the state of ecosystems (e.g., Foissner, 1988a, 1997b, 1997e).

How have ciliates come to be distributed as we now see them? Bamforth (1981) reviewed the factors that, in his view, explained the biogeography of both free-living and symbiotic species. For free-living species, these included characteristics of the autecology of the species and environmental conditions, such as wind patterns and ocean currents. For example, a variety of species are distributed by wind currents (Maguire, 1963b). The distribution of tintinnids in the Adriatic Sea, for example, is strongly influenced by ocean currents: still certain tintinnid species, despite the absence of vertical barriers to migration, can be characterized as surface, mesopelagic, or deep-sea forms (Krsinic & Grbec, 2006). Symbiotic ciliates have a biogeography that is influenced by the historical biogeography of their hosts. However, even species that we do not normally imagine as symbiotic, such as *Paramecium*, can be transported in tropical snails from flower to flower (Maguire & Belk, 1967)! Humans may have also played a role in dispersing species as a variety of taxa has been observed in the ballast tanks of ocean-going vessels (Galil & Hülsmann, 1997).

Nevertheless, the opinions on how the diversity of free-living ciliates is geographically distributed have become polarized into two major views. On one hand, ciliates are considered ubiquitous and cosmopolitan, and on the other, many ciliates are considered moderately endemic. Some of the controversy centers around semantics. Finlay, Esteban, and Fenchel (2004) have offered the following definitions to focus debate. They suggested that ubiquitous refer to the *process* of continuous, worldwide dispersal of organisms while cosmopolitan should refer to species that thrive *wherever* their habitat is found worldwide. Endemic refers to organisms of low dispersal ability and restricted distribution. Many years ago, Beijerinck (1913) made the argument for bacterial species that "everything is everywhere, the environment selects". Finlay and Clarke (1999) and Finlay and Fenchel (1999) have taken up this argument for protists, emphasizing that the typically small size and extremely high abundances of protist species, including most ciliates, should permit them to defy barriers to migration, making allopatric speciation almost impossible. While it is undoubtedly impossible that everything be everywhere, cosmopolitan species, as defined above, have been observed. For example, similar freshwater species assemblages have been found in the northern and southern hemispheres (Esteban et al., 2000); marine ciliates have been recorded in inland saline environments (Esteban & Finlay, 2004); and allegedly endemic "flagship" ciliates may be more broadly distributed than previously thought (Esteban, Finlay, Charubhun, & Charubhun, 2001). Moreover, there is now genetic evidence to suggest that the effective population sizes of ciliates might be quite large (Snoke, Berendon, Barth, & Lynch, 2006), although there is debate on how large (Katz, Snoeyenbos-West, & Doerder, 2006).

On the other side, Foissner (1999c) takes the view that many species show limited geographical distributions and low dispersal abilities. For example, the large tropical peniculine *Neobursaridium gigas*, a "flagship" tropical freshwater species, was described over 60 years ago in Africa, and yet it has only been recorded in the Southern Hemisphere despite intensive sampling of Northern Hemisphere habitats. Foissner (2005a) has described two large, scaled trachelophyllid haptorians that he describes as new "flagship" species from the Southern Hemisphere, to which can be added large-bodied species of the nassophorean *Frontonia* and the stichotrichian *Gigantothrix* (Foissner, 2006). Thus, he argued that endemism and a biogeography may be properties of a much larger subset of species than currently reported, perhaps up to one-third. This proportion has been supported by a more extensive analysis of over 300 soil samples from five continents (Chao, Li, Agatha, & Foissner, 2006), but a contrary view was provided by Finlay, Esteban, Clarke, and Olmo (2001) who found no evidence for geographic restriction of species across local and global scales.

The debate has important implications, as pointed out by Mitchell and Meisterfeld (2005). If species have global distributions, then overall diversity will be low; if species have more restricted distributions, not just due to narrow niche breadths, then overall diversity will be high. For ciliates, Finlay, Corliss, Esteban, and Fenchel (1996) concluded that there may only be 3,000 morphospecies of free-living ciliates. On the other hand, Foissner (1999c) argued that the number could be considerably higher, perhaps two or three times as many, since up to 80% of the morphospecies at some sites were new in his global studies of soil ciliate species diversity. New species are being discovered even in regions of Central Europe, which have been intensively investigated (Foissner, Berger, Xu, & Zechmeister-Boltenstern, 2005b). Of crucial importance to this debate is one's conception of a species: "splitters" might conclude that there are high rates of endemism while "lumpers" might conclude just the opposite (Mitchell & Meisterfeld, 2005).

Finlay et al. (1996) concluded that a pragmatic approach to ciliate biodiversity should be to recognize the "morphospecies" as the operational unit for analyses of biodiversity. They defined a morphospecies as "a collection of forms that all fit into a defined range of morphological variation – forms that, so far as we can tell, occupy the same ecological niche" (p. 232, Finlay et al.; see also Esteban & Finlay, 2004). Given the broad physicochemical tolerances of many ciliates species, they suggested that niche breadths are probably broad, and so morphospecies provide us a reasonable understanding of the functional role of ciliate biodiversity in ecosystems. There are certainly a number of studies that suggest that subunits of morphospecies, such as sibling species and particular genotypes, are not geographically restricted (e.g., Ammerman, Schlegel, & Hellmer, 1989; Bowers, Kroll, & Pratt, 1998; Przybos & Fokin, 2000; Stoeck, Przybos, & Schmidt, 1998; Stoeck, Przybos, Kusch, & Schmidt, 2000a).

In contrast, however, there is preliminary evidence that some genotypes may have restricted ranges (Stoeck et al., 1998) or appear at particular seasons of the year (Doerder, Gates, Eberhardt, & Arslanyolu, 1995; Doerder et al., 1996). Katz et al. (2005) have presented convincing evidence that gene flow was high and diversity was low in planktonic spirotrichs that inhabit open coastal waters (e.g., *Laboea*), while gene flow was high and diversity was also high in oligotrichs that inhabit ephemeral habitats (e.g., *Halteria*, *Meseres*). Furthermore, there are ecologically significant differences in growth rates and responses to temperature between geographically distant isolates of species of *Uronema* (Pérez-Uz, 1995) and *Urotricha* (Weisse & Montagnes, 1998; Weisse et al., 2001), and even among clones of planktonic *Coleps* and *Rimostrombidium* species (Weisse & Rammer, 2006). Dini and Nyberg (1999) have shown that ecological differentiation of genotypes occurs at all levels among species of *Euplotes* – at the morphospecies level, breeding system level, breeding group level, and stock level. Thus, we must put pragmatism aside if we are to advance our understanding of the interactions between the ecological factors and the evolutionary forces shaping ciliate diversity, and we must also move beyond the concept of morphospecies. A major first step would be to consider models for speciation other than the allopatric one, which is clearly inappropriate in its classical interpretation.

4.3 Somatic Structures

The surface of ciliates is covered by a plasma membrane underlain by cortical alveoli (Figs. 4.9B, 4.10A, 4.10B). In some nassophoreans, the alveoli can extend into the cortex as the so-called alveolocysts (Fig. 4.10G). The plasma membrane is characterized by a variety of intramembranous particles (Allen, 1978; Bardele, 1983; Hufnagel, 1992). The surface membranes are sites of ion channels that enable ciliates to sense mechanical, chemical, and temperature stimuli (Machemer & Teunis, 1996). The alveoli can be the sites of Ca^{2+} ion storage in some ciliates, thus playing a role in modulating locomotion (Mohamed et al., 2003; Plattner, Diehl, Husser, & Hentschel, 2006; Stelly, Halpern, Nicolas, Fragu, & Adoutte, 1995). All these inputs to the cell are "translated" into complex behavioral sequences that Ricci (1990, 1996) has described as an ethogram – a quantitative description of the behavioral repertoire of a species.

At various points on the cell surface, typically associated with the emergence of cilia, parasomal sacs or coated pits extend into the cytoplasm (Fig. 4.10E). These sacs can be the sites of pinocytosis (Nilsson & Van Deurs, 1983). The membranous junctions of neighboring alveoli or fibrous components associated with these boundaries (Fig. 4.10E) form characteristic patterned networks that are revealed upon silver-staining – the so-called argyrome (Foissner & Simonsberger, 1975a, 1975b). Underlying the alveoli is a fibrous or filamentous layer called the epiplasm, which is constructed, in part, by specific proteins called epiplasmins and articulins (Figs. 4.9B, 4.9C, 4.10B) (Coffe, Le Caer, Lima, & Adoutte, 1996; Huttenlauch & Stick, 2003; Huttenlauch, Peck, Plessmann, Weber, & Stick, 1998b; de Puytorac, 1984a). Genetic interference with some of these "cortical" genes can influence cell shape (Williams, 2004).

The most prominent features of the somatic surface of the vast majority of ciliates are the cilia. Membranous particles are also distributed in ciliary membranes, and undoubtedly function in the movement of Ca^{2+} ions influencing the ciliary beat pattern (Machemer & Teunis, 1996; Plattner, 1975; Plattner et al., 2006), and these patterns of intramembranous particles on the cilia may characterize different groups of ciliates (Bardele, 1981). The cilia beat with a straight effective stroke and a curved recovery stroke, typically moving the ciliate through the medium, whether it be the water of an ocean or pond or the digestive contents of the intestine of a sea urchin (Sleigh & Barlow, 1982). The often thousands of cilia on the cell surface are coordinated by a hydrodynamic coupling that is manifested in the metachronal waves observed passing along the cell's surface (Guirao & Joanny, 2007; Sleigh, 1984, 1989).

The ciliary axoneme with its 9 + 2 arrangement of microtubules underlies the ciliary membrane. The major force for the ciliary beat derives from active sliding of the nine peripheral doublet microtubules driven by dynein motors and ATP (Satir & Barkalow, 1996). The central pair of microtubules may rotate in a counterclockwise direction, viewed from the outside of the cell, making a complete rotation with every beat cycle (Omoto & Kung, 1980). Furthermore, this defined directional rotation, if true, means that when a patch of cortex is rotated, as sometimes happens following conjugation of *Paramecium* when the two cells separate, the cilia on the reversed patch beat in the opposite direction to the surrounding cortex that has a normal polarity (Tamm, Sonneborn, & Dippell, 1975). The central-pair microtubules are anchored in the axosome, which lies in a region of extreme complexity – the transition zone – between the ciliary axoneme and the basal body or kinetosome (Dute & Kung, 1978). In reviews, Fokin (1994, 1995) has demonstrated a considerable diversity in transition zone structures in ciliates, and suggested that transition-zone types may characterize some of the major clades of ciliates.

The axonemal microtubules arise out of the kinetosome, which is composed in most ciliates of nine sets of microtubular triplets. Associated with the ciliate kinetosome are three fibrillar associates

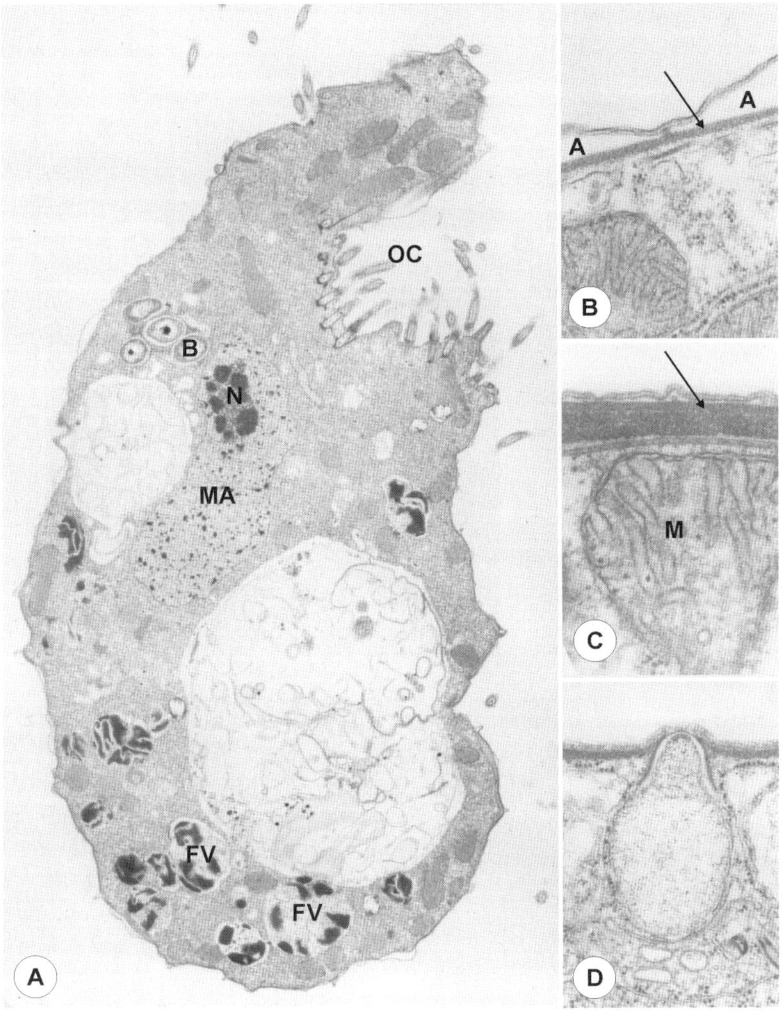

FIG. 4.9. Ultrastructural features of ciliates. **A** Longitudinal section of the colpodean *Colpoda steinii*. Note the anterior oral cavity (**OC**), macronucleus (**MA**) with its large nucleolus (**N**), and food vacuoles (**FV**) filled with bacteria (**B**). **B** Section through two cortical alveoli (**A**) of the colpodean *Colpoda cavicola*. Note the thin epiplasmic layer (arrow) in this small ciliate. **C** Section through the pellicle of the colpodean *Colpoda magna*. Note the much thicker epiplasmic layer (arrow) in this large colpodid and the mitochondrion (**M**) with tubular cristae. **D** A mucocyst in the cortex of the colpodean *Bresslaua insidiatrix*

– the striated kinetodesmal fibril, the transverse microtubular ribbon, and the postciliary microtubular ribbon (Fig. 4.10C, 4.10F, 4.10H–K). All these elements together – cilium, kinetosome, fibrillar associates – form the kinetid. Theoretical calculations support the notion that these fibrous structures function as anchors for the kinetid (Sleigh & Silvester, 1983). These fibrillar systems have diversified in form and pattern as ciliate lineages have evolved, providing a variety of patterns that have proved useful in characterizing major clades (Fig. 4.7; see **Taxonomic Structure** above). The fibrillar associates extend in a various directions, depending upon the ciliate, and form an elaborate cortical cytoskeleton (Figs. 4.10D, 4.11). This cytoskeleton functions both to "passively" support the cortex, since disassembly of the microtubules changes the form of the cell (Lynn & Zimmerman, 1981), and to "actively" change cell shape, since active sliding of postciliary microtubular ribbons in the heterotrich *Stentor* extends the body after contraction (Huang & Pitelka, 1973). Electron micro-

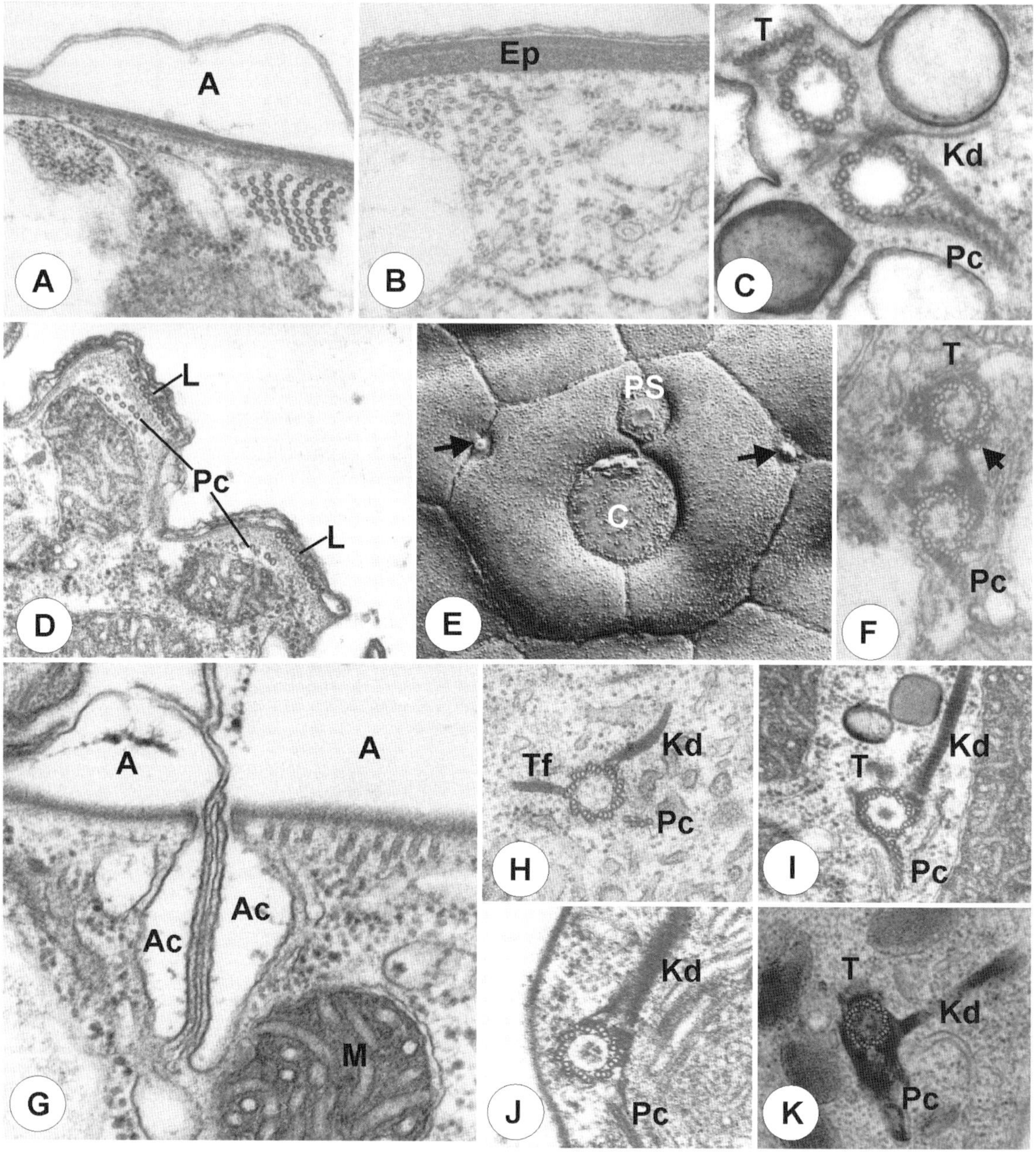

FIG. 4.10. Ultrastructural features of the somatic cortex of ciliates. **A** Section through a cortical alveolus (A) of the colpodean *Colpoda cavicola*. Note the epiplasm underlain by overlapping ribbons of cortical microtubules. **B** Section through the pellicle of the colpodean *Colpoda magna* showing microtubules underlying the thicker epiplasm (**Ep**). **C** Cross-section of the somatic dikinetid of the heterotrichean *Climacostomum virens*, showing the transverse microtubular ribbon (**T**), kinetodesmal fibril (**Kd**), and postciliary microtubular ribbon (**Pc**) (from Peck, Pelvat, Bolivar, & Haller, 1975). **D** Section through two cortical ridges of the oligohymenophorean *Colpidium campylum*. Note the longitudinal microtubules (**L**) above the epiplasm and the postciliary microtubules (**Pc**) underlying the epiplasm (from Lynn & Didier, 1978). **E** Freeze-fracture replica of the external faces of the inner alveolar membranes of the nassophorean *Nassula citrea*. Note the cilium (**C**) emerging between two alveoli, the parasomal sac (**PS**) anterior to the cilium, and in-pocketings of the alveolocysts (arrows) (see Fig. 4–10G) (from Eisler & Bardele, 1983). **F** Cross-section of the somatic dikinetid of the colpodean *Colpoda magna*. Note the single postciliary microtubule (arrow) associated with the anterior kinetosome. **G** Section through two adjacent alveoli (**A**) in the cortex of the nassophorean *Furgasonia blochmanni*. Note that the alveoli extend into the cell in the form of alveolocysts (**Ac**). **M** – mitochondrion (from Eisler & Bardele, 1983). **H** Cross-section of the somatic monokinetid of the phyllopharyngean *Trithigmostoma steini* (from Hofmann & Bardele, 1987). **I** Cross-section of the somatic monokinetid of the oligohymenophorean *Colpidium campylum* (from Lynn & Didier, 1978). **J** Cross-section of the somatic monokinetid of the prostomatean *Coleps bicuspis*. **K** Cross-section of the somatic kinetid of the litostomatean *Lepidotrachelophyllum fornicis*

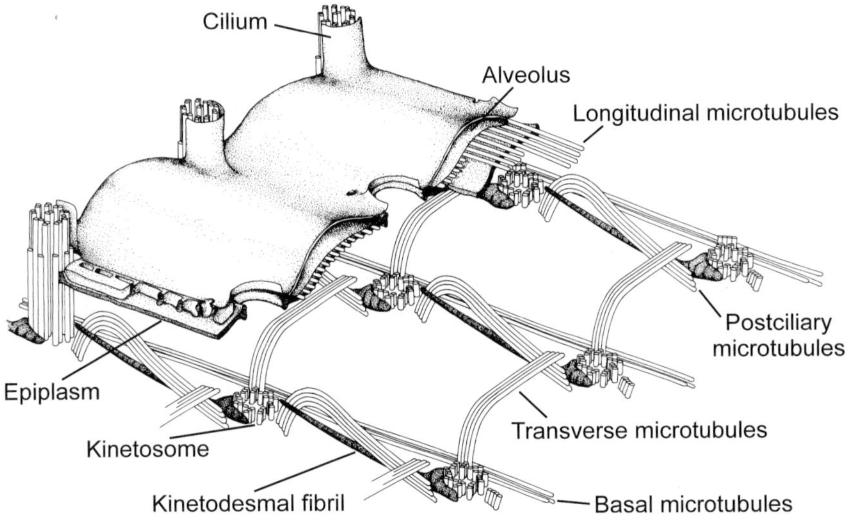

FIG. 4.11. Schematic drawing of the somatic cortex of a ciliate illustrating the interrelationships of the various structures

scopy has demonstrated that several silver-staining methods are highly specific for these fibrillar components (Foissner & Simonsberger, 1975a; Tellez, Small, Corliss, & Maugel, 1982; Zagon, 1970). Thus, the patterns observed after such staining procedures are grounded in the cytoskeletal structures of the cells, further confirming their essential usefulness as tools for systematists.

How this 9 + 2 structure evolved is still open to speculation. Hartman (1993) imagined its gradual evolution from a 3 + 0 structure, still found in some parasitic gregarines, like *Diplauxis*, by additions of "nucleating three's" – from a 6 + 0 to a 9 + 0 structure also found in some gregarines, like *Stylocephalus* (Kuriyama, Besse, Gèze, Omoto, & Schrével, 2005; but see Mitchell, 2004).

In addition to the kinetosomal fibrillar associates, there is a variety of other fibrous and filamentous components in the cortex, which also function to maintain or change cell shape (Adoutte & Fleury, 1996; Allen, 1971; Garreau de Loubresse, Keryer, Viguès, & Beisson, 1988; de Haller, 1984a, 1984b; Huang & Pitelka, 1973).

The somatic cortex is also differentiated in many species to provide a means of attachment to the substrate. These differentiations range from special thigmotactic cilia to complex attachment structures, like stalks and hooks (Fauré-Fremiet, 1984). Undoubtedly the most complex attachment structure exhibited by any ciliate is the attachment disc of the mobiline peritrichs, underlain by a complex set of fibres and denticles to form a "suction cup-like" structure (Fig. 4.6E) (Favard, Carasso, & Fauré-Fremict, 1963; Hausmann & Hausmann, 1981b).

More details on the somatic cortex can be found in later chapters and in reviews by Adoutte and Fleury (1996), Grain (1984), Lynn (1981), Lynn and Corliss (1991), Paulin (1996), and de Puytorac (1984a).

4.4 Oral Structures

The oral region shows great diversity among ciliates, a reflection of the ecological diversity within the phylum (Figs. 4.3–4.6, 4.12). If we use Eisler's model (Eisler, 1992), we assume that the simplest and earliest oral ciliature was a set of dikinetids extending along the righthand side of the oral region (Fig. 4.2). The oral dikinetids typically bear a single postciliary microtubular ribbon (Fig. 4.13). These microtubules often extend towards the cytopharynx, directing the movement of precursor or disc-shaped vesicles to the food vacuole-forming region where they fuse with the plasma membrane to provide membrane for the forming food vacuole (Allen, 1984). A typical oral organization has a set of adoral polykinetids or membranelles on the lefthand side of the oral region (Figs. 4.2, 4.3). These oral polykinetids are often initially constructed of dikinetids that assemble side-by-side into the organellar complexes to form

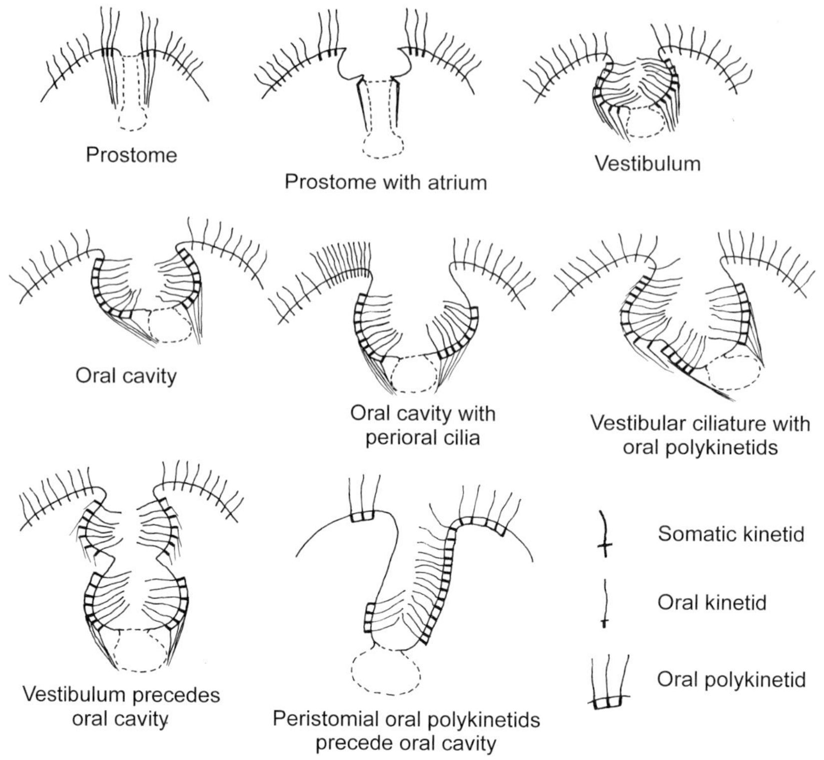

FIG. 4.12. Schematic drawings illustrating the diversity of kinds of oral regions in the Phylum Ciliophora

two rows to which a third and fourth rows may be added by additional kinetosomal replication (Fig. 4.14) (Frankel, 1989; Jerka-Dziadosz, 1981a). The complexity and diversity of these adoral polykinetids has given rise to a proliferation of terms that help to classify this diversity – cirromembranelle, membranelle, membranoid, heteromembranelle, paramembranelle, peniculus, polykinety, and quadrulus (see **Chapter 2. Glossary** for details). Further details of each of these structures and references to the primary literature are provided in the following chapters describing the features of each class.

Oral dikinetids are also found in prostomial forms (Fig. 4.12). In the Class PROSTOMATEA, ultrastructural evidence suggests that these develop from a paroral primordium that migrates and encircles the cytostomial region (Huttenlauch & Bardele, 1987). However, in the Class LITOSTOMATEA, Foissner and Foissner (1985, 1988) have proposed that the "original" oral ciliature has been lost and the oral dikinetids that we now see have been derived secondarily from the "oralization" of somatic kinetids. This is certainly consistent with the orientation of these oral dikinetids, which are not rotated and/or inverted like those of the paroral. Instead, the transverse or "anterior" microtubular ribbons of litostome oral dikinetids extend directly to support the cytopharynx (Fig. 4.13). Another novel hypothesis has been proposed for the oral structures of the Class PHYLLOPHARYNGEA. Bardele and Kurth (2001) proposed that the ancestral phyllopharyngean, now extinct, had also lost its primary oral ciliature, and had instead a suctorial oral apparatus, possibly similar to present–day rhynchodines. Therefore, the complex oral ciliature of cyrtophorine phyllopharyngeans was derived later from "oralization" of somatic kinetids during stomatogenesis. This formation of oralized somatic kinetosomes in litostomes and phyllopharyngeans has been called deuterostomisation (Bardele & Kurth, 2001).

Acquiring food can be a simple process of encountering edible food particles and then ingesting them, a behavior that is typical of prostomial forms and those with simpler arrangements of oral

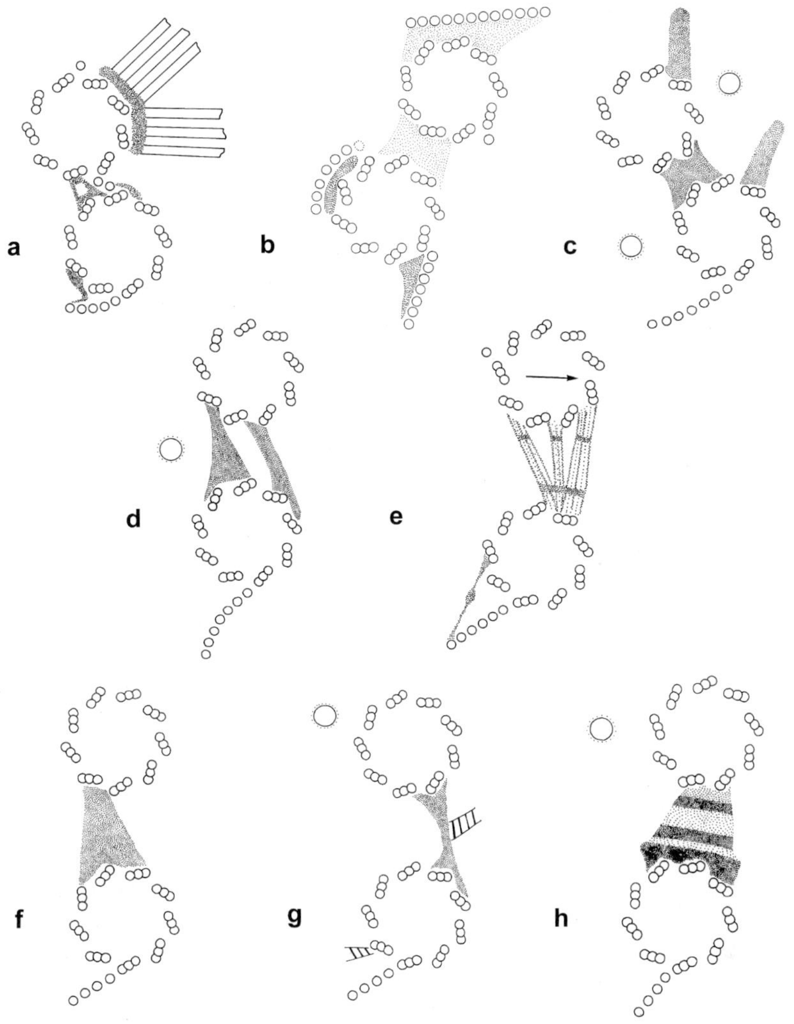

FIG. 4.13. Cross-sections of the paroral dikinetids of genera representative of classes in the Phylum Ciliophora. (**a**) *Eufolliculina* – Class HETEROTRICHEA. (**b**) *Lepidotrachelophyllum* – Class LITOSTOMATEA. (**c**) *Chilodonella* – Class PHYLLOPHARYNGEA. (**d**) *Woodruffia* – Class COLPODEA. (**e**) *Furgasonia* – Class NASSOPHOREA. (**f**) *Paramecium* – Class OLIGOHYMENOPHOREA. (**g**) *Cyclidium* – Class OLIGOHYMENOPHOREA. (**h**) *Colpidium* – Class OLIGOHYMENOPHOREA (from Lynn, 1981, 1991)

ciliature (Fig. 4.12) (Peck, 1985; Tucker, 1968; Wessenberg & Antipa, 1970). Ciliates with a paroral and adoral polykinetids are characterized as suspension feeders. The polykinetidal cilia can be used to both create the current and filter particles out of the suspension – the so-called upstream filter feeders – or the current can be created by these cilia and the particles filtered by the cilia of the paroral – the so-called downstream filter feeders (Fig. 4.15) (Fenchel, 1980a). Suspension feeding ciliates are typically considered to be non-selective feeders, "discriminating" among particles primarily on the basis of size (Fenchel, 1980b, 1980c). However, ciliates with filter-feeding oral apparati do demonstrate some selectivity, so feeding and ingestion may be more complicated than a simple mechanical process (Sanders, 1988; Stoecker, 1988; Stoecker, Gallager, Langdon, & Davis, 1995).

Food or food particles are sequestered in a food vacuole or phagosome. The food vacuole membrane is constructed when hundreds of disc-shaped vesicles fuse with the cytopharyngeal plasma membrane. Digestion occurs by processes typical of most eukaryotes, although in *Paramecium* an unusual set

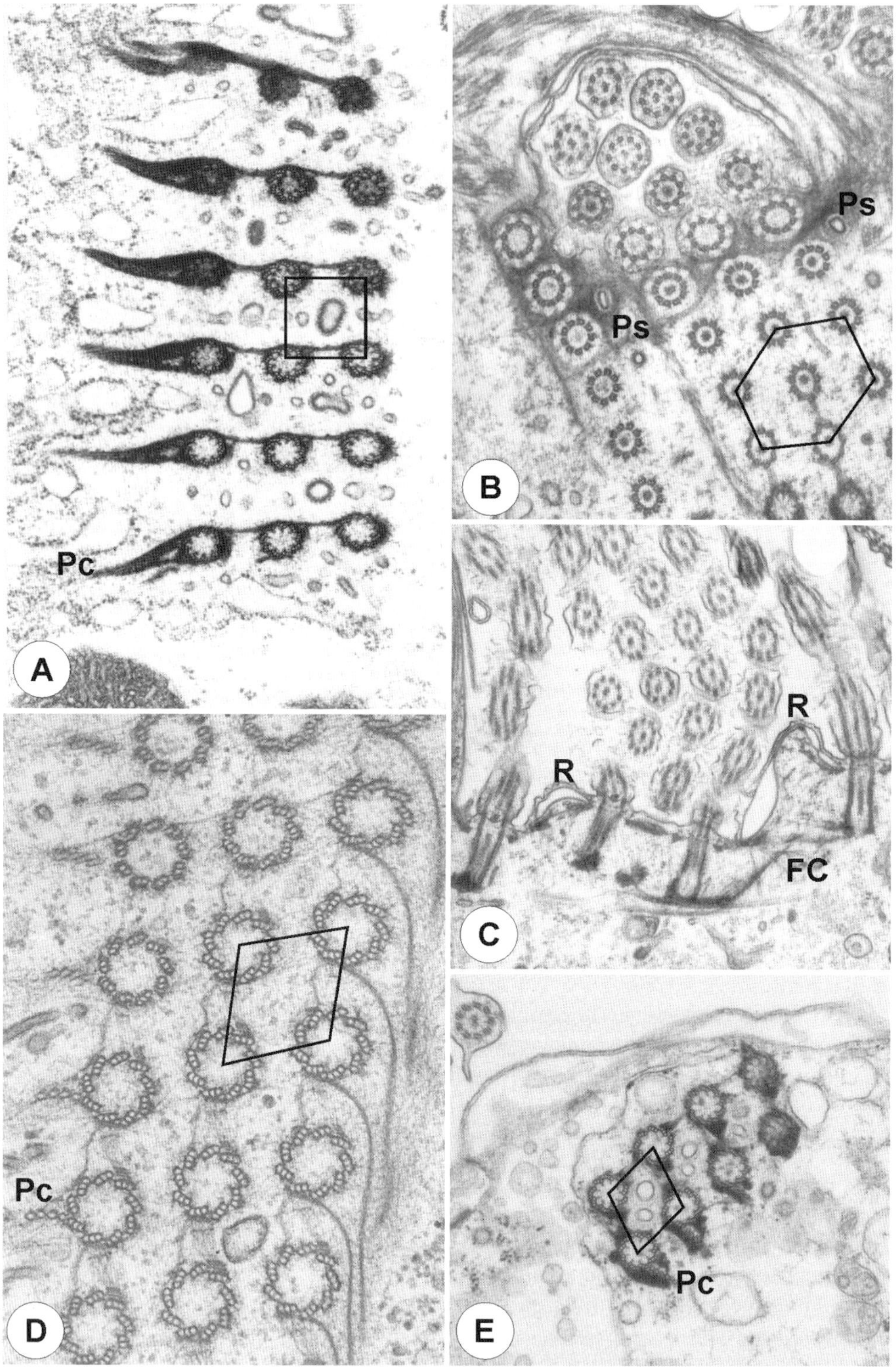

FIG. 4.14. Ultrastructure of the oral polykinetids of ciliates. **A** A square-packed oral polykinetid of the nassophorean *Nassula citrea* with the posterior row of kinetosomes bearing postciliary microtubular ribbons (**Pc**) (from Eisler & Bardele, 1986). **B** A hexagonally-packed oral polykinetid of the oligohymenophorean *Colpidium campylum*. Note the parasomal sacs (**Ps**) lying on either side of the three rows of kinetosomes (from Lynn & Didier, 1978). **C** Cross-section through the oral cavity of *C. campylum* shows the three oral polykinetids separated by two cortical ridges (**R**) underlain by alveoli. The polykinetids are connected by filamentous connectives (**FC**) (from Lynn & Didier, 1978). **D** A rhomboid-packed oral polykinetid of the oligohymenophorean *Thuricola folliculata* (from Eperon & Grain, 1983). **E** A slightly off square-packed oral polykinetid of the colpodean *Woodruffia metabolica*

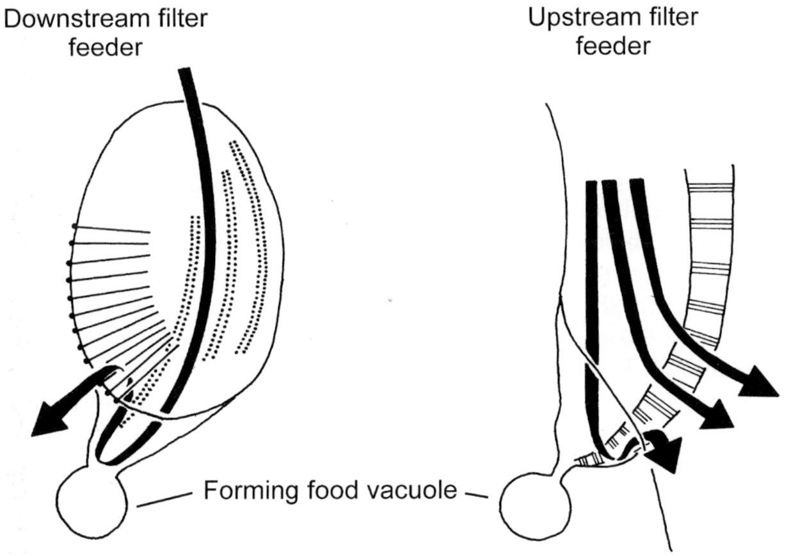

FIG. 4.15. Filter feeding ciliates can use their oral structures to function as a downstream filter feeder, which creates a current with the cilia of the oral polykinetids and captures particles in the cilia of the paroral, or as an upstream filter feeder, which both creates the current and captures the particles using the cilia of the oral polykinetids. (Redrawn after Fenchel, 1980a.)

of vesicles, called acidosomes, fuse with the phagosome to first acidify the phagosomal compartment prior to fusion of lysosomes (Allen, 1984; Allen & Fok, 2000). The old food vacuoles ultimately arrive in the region of the cytoproct where their contents are expelled to the outside. Excess food vacuole membrane is then recycled to the cytopharyngeal region as disc-shaped vesicles (Allen, 1994; Allen & Fok, 2000; Allen & Wolf, 1974).

More details on the oral region and its function can be found in later chapters and in reviews by Grain (1984), de Haller (1984c), Lynn (1981), Lynn and Corliss (1991), Paulin (1996), de Puytorac (1984a), de Puytorac and Grain (1976), and Radek and Hausmann (1996).

4.5 Division and Morphogenesis

Ciliates can be studied as cells, and like all cells during the interphase period of the cell cycle, they can be expected to faithfully duplicate all their component parts (Berger, 2001; Méténier, 1984a). This is what is called balanced growth. Following this duplication, ciliates as unicellular organisms reproduce by cell division. Unlike animals, this reproductive process in ciliates is separate from sexual processes (**see below, Nuclei, Sexuality, and Life Cycle**) so that months to years of asexual reproduction can take place between sexual events.

Ciliates typically divide by binary fission, in which the parental cell divides into two filial products, offspring, or progeny (Fig. 4.8). The anterior "daughter" cell is termed the proter and the posterior "daughter" cell is called the opisthe (Chatton & Lwoff, 1935b). This binary fission is usually equal or isotomic, that is both filial products are the same size, but it can be unequal or anisotomic. Budding is a common type of anisotomy, which is found especially in sessile taxa, such as suctorians and chonotrichs (see **Chapter 10**). Ciliate fission is also termed homothetogenic in the vast majority of cases, meaning that the cell axes of proter and opisthe have the same orientation or polarity: typically the posterior end of the proter is in contact with the anterior end of the opisthe. This is modified in two main ways. Some spirotrichs, especially oligotrichs and choreotrichs, undergo a modified division mode called enantiotropic division: the axes of proter and opisthe of these planktonic ciliates shift during cell division so that they have an almost opposite polarity. The second modification is found in peritrichs whose sessile life style is accompanied by a seemingly parallel type of cell division: the proter and opisthe appear to develop "alongside" each other with the fission furrow separating them "longitudinally". However, Lom (1994) has argued that this may just be a highly modified form of homothetogenic fission, easily re-interpreted by assuming that the stalk of peritrichs arises out of the dorsal surface, and is not the posterior end of the cell.

In some ciliates, binary fission may not occur when the ciliate doubles all its components. For example, the parasitic ciliate *Ichthyophthirius* may grow several orders of magnitude as a parasite in the epithelium of its fish host before dropping off, encysting, and dividing up to eight times sequentially to yield over 1,000 offspring. Even free-living ciliates, which may undergo a period of starvation as they disperse from one food patch to another, may undergo a period of "unbalanced" growth, presumably as they replenish and "balance" cell constituents that were differentially more exhausted during the starvation period. Upon refeeding, these free-living ciliates, like *Tetrahymena*, may grow larger than the typical size during balanced growth and then, undergo several sequential cell divisions without intervening growth (Lynn, 1975; Lynn & Tucker, 1976; Lynn, Montagnes, & Riggs, 1987). The process of multiple divisions without intervening growth is termed palintomy. It can occur sequentially in a cyst, as it does in *Ichthyophthirius* and some colpodean ciliates, or it may occur in a linear fashion in highly elongate ciliates, as it does in some astomes. In the latter case, it can also be called catenulation or strobilation.

Cell division can be thought of as being composed of two processes: division of the cytoplasm or cytokinesis and division of the nucleus or karyokinesis, often called mitosis. Cytokinesis is most obvious in its last stages where a fission furrow appears near the equator in ciliates undergoing isotomy. The furrow develops in some ciliates by assembly and then contraction of special kinds of microfilaments (Yasuda, Numata, Ohnishi, & Watanabe, 1980). Prior to furrow formation, special microtubules may appear in the cortical ridges, above the epiplasm, the so-called cytospindle of *Paramecium* (Sundararaman & Hanson, 1976). As the isthmus between the cells narrows further, the twisting and pulling movements of the progeny achieve the final separation.

Karyokinesis is more complicated in ciliates, since they have two nuclei. The typically globular or ellipsoid micronucleus undergoes a eukaryotic cell mitosis except that the nuclear membrane does not break down. Spindle microtubules assemble within the nuclear envelope and are used to separate the sister chromatids (LaFountain & Davidson, 1980). Raikov (1982) categorized the ciliate micronuclear mitosis as a closed intranuclear orthomitosis. The macronuclei of ciliates may take a variety of shapes and may be subdivided into apparently disconnected nodules. Prior to division, these macronuclear nodules often condense so that the many nodes, for example, may ultimately comprise a single ellipsoid body. The macronucleus then divides in two phases – an elongation phase and a constriction phase (Raikov, 1982). The elongation phase is likely driven by both the assembly and sliding of microtubules, which may assemble *inside* the macronuclear envelope (e.g., Tucker, Beisson, Roche, & Cohen, 1980; Williams & Williams, 1976) or *outside* the macronuclear envelope (e.g., Diener, Burchill, & Burton, 1983).

While duplication of all cell constituents occurs during the cell cycle, developmental biologists and systematists have been particularly fascinated by the duplication of the cortical components. Lynn and Corliss (1991) have separated this development into cortical somatogenesis and cortical stomatogenesis: the replication of the components of the somatic cortex and the oral cortex, respectively, which are often highly co-ordinated processes. Frankel (1989) has provided a detailed review of these processes from the perspective of a developmental biologist. In particular, ciliate systematists have long been fascinated with the ontogeny of the oral apparatus (see **Chapter 1**; Corliss, 1968; Fauré-Fremiet, 1950a, 1950b). Foissner (1996b) has provided a detailed discussion of the comparative stomatogenesis of ciliates, but see also Tuffrau (1984).

Briefly, the conspicuous elements of cortical somatogenesis that have attracted attention are the kinetosomes, contractile vacuole pores, and cytoproct. Ciliates were one of the first groups of organisms to be investigated for replication of kinetosomes, demonstrating that the "daughter" kinetosome developed in close proximity to and in a well-defined relationship with the parental kinetosome (Allen, 1969; Dippell, 1968). This process, now called cytotaxis or structural guidance, is responsible for the precise positioning of new cortical units (Aufderheide, Frankel, & Williams, 1980; Frankel, 1989). Kinetosomal replication can occur throughout the cell cycle or be confined to a period close to the time of cytokinesis and be highly correlated with cortical stomatogenesis. Initiation of kinetosomal replication undoubtedly involves participation of gene products that diffuse through the cytoplasm: for example, the product of one such gene, *sm19*$^{+}$, appears to be involved in kinetosomal replication in *Paramecium* (Ruiz, Garreau de Loubresse, & Beisson, 1987). New contractile vacuole pores (CVPs) are typically replicated at cell division, although in some ciliates with large numbers of contractile vacuoles the replication process may be uncoupled from cell division. In *Tetrahymena*, the proter develops new CVPs adjacent to somatic kineties in a predictable location in its posterior right quadrant, defined by the "central angle". This angle is a manifestation of a mechanism that places the new pores in a roughly proportional fashion in relation to the total number of somatic kineties (Frankel; Nanney, 1980; Nanney, Nyberg, Chen, & Meyer, 1980b). In *Chilodonella* species, a proportioning mechanism may also exist, but in this case the many contractile vacuole pores, which are distributed over the ventral surface, are newly placed in both proter and opisthe, apparently in relation to major features of the cortex, such as somatic kineties, the oral region, and the boundaries of the ventral surface. During this somatogenesis in *Chilodonella*, the old contractile vacuoles and their pores dedifferentiate and disappear (Kaczanowska, 1981; Kaczanowska, Wychowaniec, & Ostrowski, 1982). The old cytoproct, since it is typically in the posterior end of the cell, is inherited by the opisthe, and a new cytoproct develops in the appropriate position in the proter, presumably positioned by mechanisms similar to those specifying the position of CVPs.

In addition to these conspicuous cortical elements, we should remember that all other organelles are typically duplicated during each interfission period – mitochondria, extrusomes, Golgi apparati, ribosomes, lysosomes, and all the smaller molecular constituents not visible as discrete entities by the microscopist.

Cortical stomatogenesis is literally the formation of a mouth. This process is usually the most conspicuous cortical ontogenetic event, since the oral region is generally the most obvious cortical differentiation. Since the oral apparatus was historically considered highly significant as a taxonomic feature, its development in different taxa has preoccupied ciliate systematists. In the chapters that follow, stomatogenesis of each of the classes is briefly characterized, based on the primary literature and the comprehensive review of Foissner (1996b). Stomatogenic patterns are now divided into five major types with subtypes – apokinetal, parakinetal, buccokinetal, telokinetal, and mixokinetal (Corliss, 1979; Foissner, 1996b). However, all subtypes within a pattern of stomatogenesis should not be regarded as diversifying from an ancestral type: they should not be considered as homologous. Rather, the several kinds of telokinetal stomatogenesis probably have evolved independently in different classes as the morphology of these ciliates diversified. For example, cyrtophorids, prostomateans, colpodeans, and litostomateans all exhibit different kinds of telokinetal stomatogenesis (Foissner, 1996b), but molecular phylogenetic analyses clearly demonstrate that these classes are not closely related. Thus, typifying stomatogenesis using this classification system should be viewed only as a descriptive approach, enabling a systematic characterization of the existing diversity. It may be of phylogenetic significance in relating groups **within** the classes. More complete definitions of these kinds of stomatogenesis can be found in the **Glossary** (see **Chapter 2**) and in Foissner (1996b).

The conspicuousness of stomatogenesis has also attracted the attention of developmental biologists who have investigated a variety of its aspects. The primordium or anlage for the new oral apparatus may be positioned by mechanisms that are influenced by the global properties of the cell, ensuring that the new oral apparatus is placed in some proportional manner in relation to the whole (Frankel, 1989; Lynn, 1977b). However, the assembly of the oral apparatus adds a level complexity to cortical developmental processes as it encompasses at least three levels of biological organization – organelles (e.g., kinetosomes), organellar complexes (e.g., membranelles, polykinetids), and organellar systems (e.g., the entire apparatus itself). There is a complex interplay of controls at these levels and different processes that coordinate assembly at each level (Frankel, 1989). Furthermore, this development takes place in the context of the cell, so that the entire apparatus, both in terms of the size of each oral polykinetid, for example, and sometimes the numbers of oral polykinetids are strongly related, for example, to cell size (Bakowska & Jerka-Dziadosz, 1980; Bakowska, Nelsen, & Frankel, 1982a; Jacobson & Lynn, 1992). Thus, systematists must be aware of all of these potential constraints on oral development when they consider which aspects of the process and which features of the differentiated oral apparatus are significant from a systematic perspective. For example, are the differences in number and size of oral polykinetids in two isolates of a genus evidence of different species or of the phenotypic plasticity of these components in a single species as it varies in cell size? Even somatic structures, such as numbers of somatic kineties, can be strongly correlated with cell size (Lynn & Berger, 1972, 1973).

Finally, once the cells have separated, there are often significant morphogenetic processes subsequent to cell division that are necessary to complete differentiation. For example, in many sessile forms, like folliculinid heterotrichs, suctorians, or chonotrichs, the offspring are quite different from the parents. These so-called buds or swarmers must undergo considerable development once they themselves have found a suitable place to settle. These morphogenetic processes can be complex, and include, for example, the development of the characteristic oral arms in folliculinids, the development of attachment stalks in suctorians and chonotrichs, and the development of oral structures, such as tentacles, in suctorians.

4.6 Nuclei, Sexuality and Life Cycle

As noted in the characterization of the phylum, ciliates are typified by having two nuclei – the macronucleus is typically "polyploid" or ampliploid, and the micronucleus is presumed to be diploid, but is likely polyploid in some taxa (Figs. 4.9A, 4.19A) (Aury et al., 2006; Génermont, 1984; Raikov, 1996). Prescott (1994) categorized macronuclei into two types: (1) those with gene-sized DNA molecules, roughly 0.4–20 kb in size, each with telomeres and typically including one gene; and (2) those with subchromosome-sized DNA molecules, roughly 100–2,000 kb pairs, also with telomeres. During development of the macronucleus from the micronucleus, the micronuclear genome size can be considerably reduced before amplification, especially in the gene-sized macronuclei, hence the term ampliploid was introduced, since the entire genome is not duplicated as it would be in a true polyploid (Raikov, 1982, 1996; Schwartz, 1978). Regardless of the type of macronucleus, chromosome-like elements are difficult to observe in macronuclei, and in contrast to the micronucleus, there also are no centromeres and so no means of attachment for spindle microtubules during karyokinesis.

There is a huge range of variation in size and shape of macronuclei, ranging from 1.4 pg of DNA in *Uronema* to over 38,000 pg of DNA in *Bursaria* (Raikov, 1995). However, DNA amount can vary depending upon the stage in the cell cycle, the age of the cell, and the nutritional state of the cell (Berger, 2001; Raikov, 1995). While macronuclei are typically single, the tintinnid choreotrichs, for example, generally always have two nodules, and other spirotrichs can have dozens. Macronuclear nucleoli are also variable in size and number, but can only be unambiguously discriminated from larger chromatin aggregates when either ribosomal precursors or a nucleolar organizing center can be demonstrated (Figs. 4.9A, 4.19A). Thus, it is a mistake to describe nucleoli unless at least one of these features has been definitively demonstrated.

Raikov (1982, 1994a, 1996) has characterized in detail the range of variation in the macronuclei of the Class KARYORELICTEA, which have near diploid to paradiploid DNA amounts. Measurements of DNA amounts in the karyorelicteans indicate that *Loxodes*, for example, can have macronuclei with up to 6C DNA (Bobyleva, Kudriavtsev, & Raikov, 1980). Karyorelictean macronuclei do not divide, and their number is maintained by division of micronuclei: the karyorelictean micronucleus divides twice at each cell division, once to reproduce itself and once to provide a new macronucleus. After division, the micronucleus differentiates, a process that might include some sequence elimination followed by amplification (Kovaleva & Raikov, 1978).

This differentiation process occurs in all other classes of ciliates when macronuclei differentiate following conjugation (**see below**).

As noted above, ciliates spend most of their life cycle reproducing asexually by binary fission. Late in the 19th century, E. Maupas (1889) discovered that *Paramecium* had a clonal cycle superimposed on these eukaryotic cell cycles: cells could be classified as immature, adolescent, mature, and senescent (Fig. 4.16) (Hiwatashi, 1981; Miyake, 1996; Sonneborn, 1957). These periods are operationally defined by the ability of cells to mate or undertake conjugation: in the immature period, cells are unable to conjugate; in the adolescent period, there is some unpredictability in the ability to conjugate; in the mature period, cells are completely sexually competent; and finally in the senescent period, the ability to conjugate becomes initially unpredictable and then is lost (Fig. 4.16). Conjugation will rejuvenate the clonal life cycle, "turning the clock back", so to speak to the immature period. If cells are not able to find partners to conjugate, some species can undergo autogamy, a kind of self–fertilization, to "restart the clock."

Conjugation is often stimulated in the laboratory setting by starvation (i.e., depriving the ciliates of food), and this is likely a stimulant in natural settings as well. Other stimulants to conjugation have been observed, for example, temperature and light (Rapport, Rapport, Berger, & Kupers, 1976; Vivier, 1984). There need to be cells of complementary mating type present to ensure success. Prior to fusion of the cells, cell-to-cell communication needs to take place, either by direct contact between cells or through indirect means. Direct contact occurs when individuals of *Tetrahymena* and *Paramecium* touch each other over a period of time prior to forming successful pairs (Watanabe, 1978, 1983; Wolfe & Grimes, 1979). Indirect "contact" occurs when, for example, individuals of *Blepharisma* and *Euplotes* secrete soluble substances called gamones, which prepare potential partners for mating (Miyake, 1981, 1996; Miyake & Beyer, 1974; Heckmann & Kuhlmann, 1986;

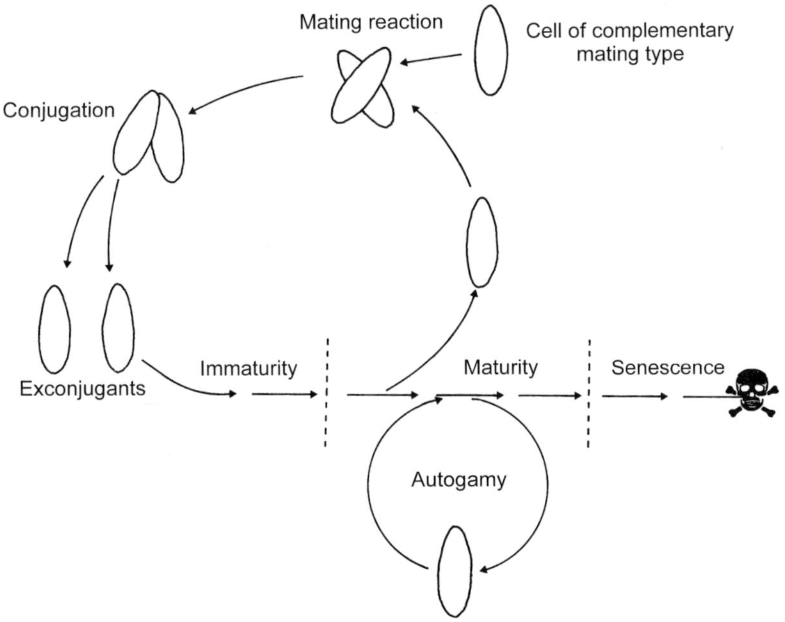

FIG. 4.16. The clonal life cycle of a ciliate, modeled after *Paramecium*. After conjugation, the exconjugants separate and undergo growth and binary fissions transiting through an immaturity stage during which conjugation is not possible. In maturity, the ciliates can conjugate with cells of complementary mating type. If cells in the clone are unable to conjugate they undergo a period of senescence with death temporarily delayed by autogamy or self-fertilization. (Redrawn after Hiwatashi, 1981.)

Luporini, Miceli, & Ortenzi, 1983; Luporini, Vallesi, Miceli, & Bradshaw, 1995; Vivier, 1984).

Once stimulated, cells will fuse in a variety of ways: side-to-side, anterior-to-anterior, among others (Fig. 4.17). During this fusion process, the region of fusion becomes differentiated in preparation for the exchange of the gametic nuclei, which derive by meiosis from the micronuclei of each partner. This conjugation bridge or conjugation basket is often supported by microtubules and microfilaments, which are believed to be involved in the transfer of the migratory gametic nucleus from partner to partner (Geyer & Kloetzel, 1987a, 1987b; Lanners & Rudzinska, 1986; Orias, Hamilton, & Orias, 1983). The migratory gametic nucleus then fuses with the stationary gametic nucleus in karyogamy, forming the synkaryon or zygotic nucleus. The synkaryon may divide twice to form four products, two of which develop into macronuclei and two of which develop into micronuclei (Fig. 4.18), but there is much variation in postkaryogamic development (Raikov, 1972). During this postkaryogamic phase, the restoration of the original nuclear condition occurs. This involves the programmed death of the parental macronucleus (Ejercito & Wolfe, 2003; Endoh & Kobayashi, 2006; Kobayashi & Endoh, 2003) and the simultaneous differentiation of the new macronucleus, with the elimination of sequences and the amplification of the genome (Jahn & Klobutcher, 2002; Prescott, 1994; Raikov, 1995).

There is a great range of variation in the features of conjugation among and even within the different

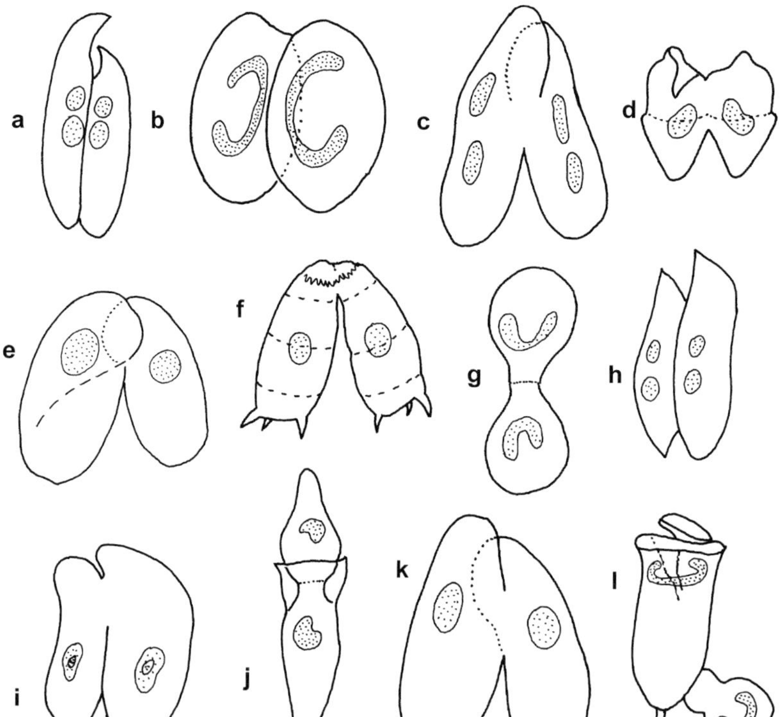

FIG. 4.17. Conjugation involves fusion of the two cells of complementary mating type. This fusion can occur in different body regions depending upon the group of ciliates. (**a**) *Loxodes* – Class KARYORELICTEA. (**b**) *Euplotes* – Class SPIROTRICHEA. (**c**) *Stylonychia* – Class SPIROTRICHEA. (**d**) *Strombidium* – Class SPIROTRICHEA. (**e**) *Metopus* – Class ARMOPHOREA. (**f**) *Coleps* – Class PROSTOMATEA. (**g**) *Actinobolina* – Class LITOSTOMATEA. (**h**) *Litonotus* – Class LITOSTOMATEA. (**i**) *Chilodonella* – Class PHYLLOPHARYNGEA. (**j**) *Spirochona* – Class PHYLLOPHARYNGEA. (**k**) *Paramecium* – Class OLIGOHYMENOPHOREA. (**l**) Vorticellid peritrich – Class OLIGOHYMENOPHOREA. (Redrawn from Kahl, 1930.)

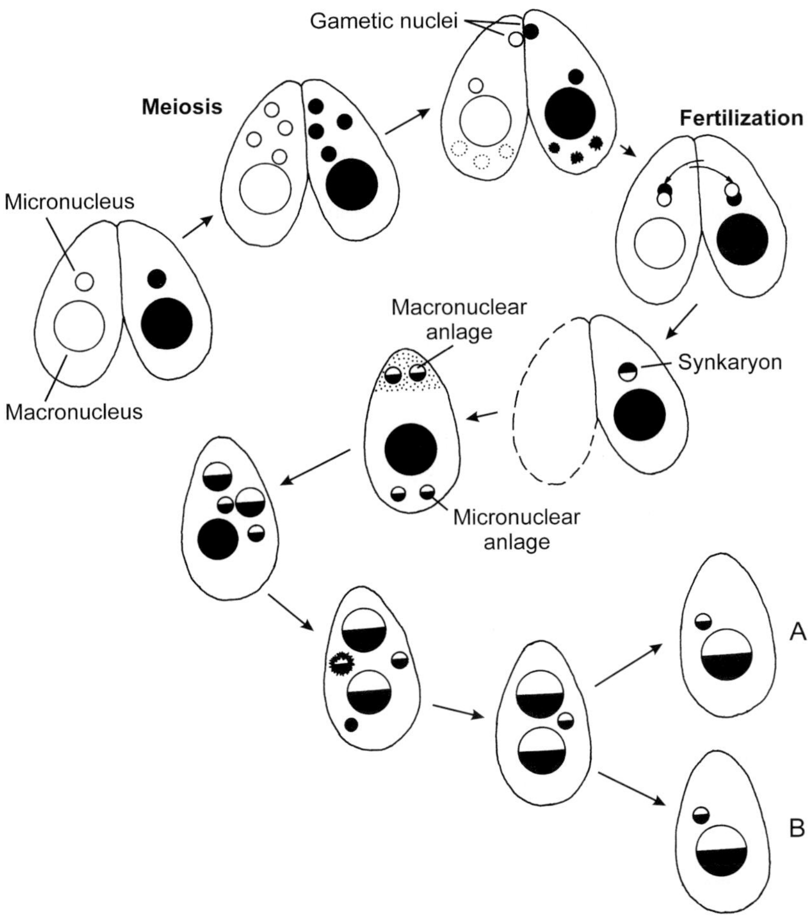

FIG. 4.18. The nuclear events of conjugation, modeled after *Tetrahymena*. Two ciliates of complementary mating type fuse (on the left) and their micronuclei undergo meiosis. One of the meiotic products survives and divides mitotically, giving rise to two gametic nuclei – one stationary and one migratory. Fertilization occurs after the migratory gametic nuclei cross the conjugation bridge. The synkaryon divides twice, in this case, and two products differentiate as macronuclei and two differentiate as micronuclei. The old macronucleus becomes pycnotic and is resorbed. (Redrawn after Nanney, 1980.)

classes of ciliates. Some of this variation is touched on in the section **Nuclei, Sexuality, and Life Cycle** in each chapter, but see reviews by Raikov (1972), Vivier (1984), and Miyake (1996). Briefly, conjugating cells are typically the same size, hence isomorphous conjugation, but cells can differ in size, hence anisomorphous conjugation. In anisomorphous conjugation, which occurs often in sessile forms, a migratory microconjugant disperses and may totally fuse with a stationary macroconjugant resulting in only one exconjugant cell (Fig. 4.17j, 4.17l). Mating type systems are either bipolar or multipolar. In bipolar systems, there are only two mating types: for example, the "odd" and "even" mating types of *Paramecium* (Sonneborn, 1957). In multipolar systems, there are many more than two mating types: in the stichotrich *Stylonychia*, there may be over 50 mating types (Ammermann, 1982). A further variation occurs in the length of the period of immaturity; when this period is short, the species is classified as a relative inbreeder and when it is long, the species is classified as a relative outbreeder (Bleyman, 1996; Landis, 1986; Sonneborn, 1957; Stoeck et al.,

2000a). Even though extreme "inbreeding", identified as selfing or intraclonal conjugation, has been identified in some *Tetrahymena* species, it does not always lead to clonal death, although viability is typically much reduced (Simon & Meyer, 1992; Simon & Orias, 1987).

A discussion of the nuclei of ciliates would not be complete without brief mention of the recent successful genome projects on *Tetrahymena* (Eisen et al., 2006) and *Paramecium* (Aury et al., 2006) and the earlier discovery of genetic code deviations among ciliates. In reference to the latter phenomenon, detailed investigation of protein-coding genes in ciliates demonstrated that the universal stop codons UAA and UAG coded glutamine in the oligohymenophoreans *Tetrahymena* and *Paramecium*, which used only UGA as the stop (Caron & Meyer, 1985; Horowitz & Gorovsky, 1985; Preer, Preer, Rudman, & Barnett, 1985). Subsequently, genetic deviations were discovered in the spirotrich *Euplotes* (Harper & Jahn, 1989), the heterotrich *Blepharisma* (Liang & Heckmann, 1993), and representatives of several other classes (Baroin-Tourancheau, Tsao, Klobutcher, Pearlman, & Adoutte, 1995; Kim, Yura, Go, & Harumoto, 2004; Sánchez-Silva, Villalobo, Morin, & Torres, 2003). Baroin-Tourancheau et al. (1995) concluded that evolution of these genetic code deviations must have occurred independently during the evolutionary diversification of the phylum. These variations are mechanistically explained by altered tRNAs (Caron, 1990; Grimm, Brunen-Nieweler, Junker, Heckmann, & Beier, 1998; Hanyu, Kuchino, & Nishimura, 1986; Sánchez-Silva et al., 2003) and by changes in the specificity of eukaryotic release factor 1 (Caron, 1990; Lozupone, Knight, & Landweber, 2001; Moreira, Kervestin, Jean-Jean, & Philippe, 2002).

4.7 Other Conspicuous Structures

Three other prominent kinds of structures are briefly mentioned below. More details on each of these can be found in the subsequent chapters relating to each class.

The osmoregulatory system of ciliates is centered on the contractile vacuole and its complex of vesicles and canals, which have long been known as responsive to ionic changes in the environment (Allen, 2000; Estève, 1984a; Kitching, 1967). As noted by Patterson (1980), the ciliate contractile vacuole complex is one of the most elaborately organized of those exhibited by protists. The cytoplasm surrounding the contractile vacuole is termed the spongioplasm, the region of cytoplasm responsible for the sequestration of water and ions, in part through the action of proton-translocating V-type ATPases (Allen; Stock, Gronlien, Allen, & Naitoh, 2002). The spongioplasm tubules may connect directly to the contractile vacuole or, as is often the case in larger ciliates, indirectly by collecting canals that radiate out from the contractile vacuole – Types A and B of Patterson (1980). This organelle received its name because of the rapid expulsion of its contents, inferred to be caused by a contractile mechanism. However, it now appears that cytosolic pressure is sufficient to explain the expulsion dynamics (Naitoh et al., 1997). The fluid is expelled through one or more pores that are typically permanent features of the somatic cortex. The pores are supported by a thickened epiplasm, a special set of helically-disposed microtubules, and a set of radial microtubules that contact the contractile vacuole itself (Fig. 4.19F) (McKanna, 1973a; Patterson, 1980). In addition to its activity being related to the external environment, the contractile vacuole is also influenced by ambient temperature and the size of the cell (Lynn, 1982; Nematbakhsh & Bergquist, 1993).

Mitochondria are also prominent organelles, typically several microns long and about 1 μm wide, distributed in the cortex of ciliates, underneath the cortical ridges and often in close association with the epiplasm (Figs. 4.9C, 4.10G, 4.19A) (Aufderheide, 1983). In all ciliates so far examined, the mitochondria have tubular cristae (Fokin, 1993a).

Two variations in mitochondria merit brief discussion. First, scuticociliates are typified by having perhaps a single mitochondrion, at least extending from the anterior to the posterior of the cell beneath each cortical ridge. These adjacent long mitochondria may extend laterally to join with their neighbors, forming one giant mitochondrion underlying the entire cortex – a structure truly worthy of the term chondriome (Antipa, 1972; Beams & Kessel, 1973). Second, mitochondrial variation occurs amongst anaerobic ciliates from different classes (i.e., Classes ARMOPHOREA, LITOSTOMATEA, OLIGOHYMENOPHOREA), which have mitochondria in which the tubules are

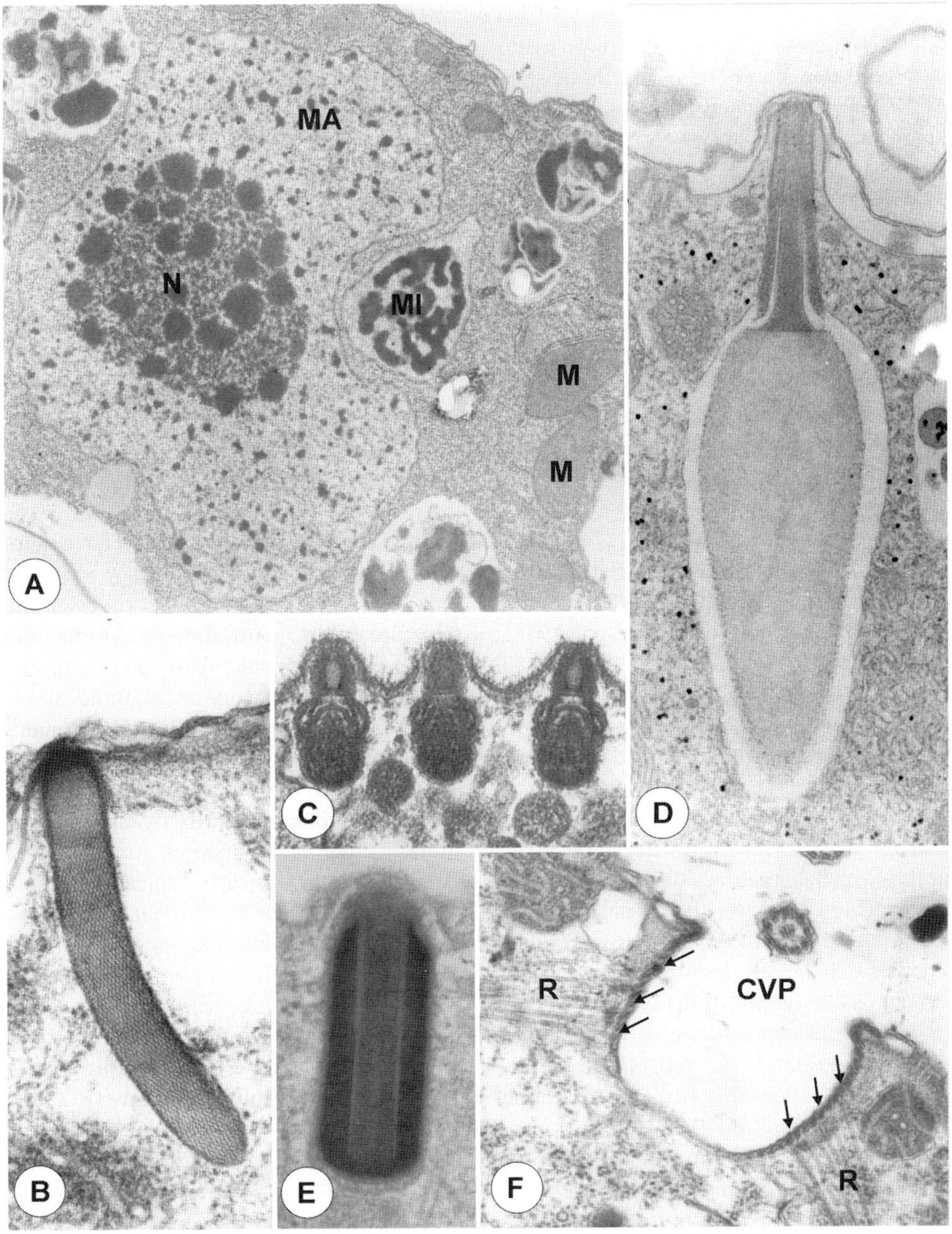

FIG. 4.19. Ultrastructural features of conspicuous organelles of ciliates. **A** The macronucleus (**MA**) and its nucleolus (**N**) of the colpodean *Colpoda steinii*. Note the closely adjacent micronucleus (**MI**) with its condensed chromosomes and several mitochondria (**M**). **B–E**. Extrusomes of ciliates. **B** A rod–shaped mucocyst of the oligohymenophorean *Colpidium campylum* (from Lynn & Didier, 1978). **C** Three haptocysts at the tip of the tentacle of the suctorian *Ephelota gemmipara* (from Grell & Benwitz, 1984). **D** The trichocyst of the oligohymenophorean *Paramecium tetraurelia* (from Kersken et al., 1984). **E** A short toxicyst from the litostomatean *Enchelydium polynucleatum* (from Foissner & Foissner, 1985). **F** A longitudinal section through the contractile vacuole pore (**CVP**) of the oligohymenophorean *Colpidium campylum*. Note that there is a set of helically disposed microtubules (arrows) supporting the pore canal and a set of radially disposed microtubules (**R**) that position the contractile vacuole. (from Lynn & Didier, 1978.)

reduced or absent (André & Fauré-Fremiet, 1984). These mitochondria-like organelles, which cannot accomplish oxidative phosphorylation, have independently evolved in these several ciliate classes to ferment pyruvate into acetate and H_2, and hence are referred to as hydrogenosomes (Fenchel & Finlay, 1991a). With the isolation of a genome from a ciliate hydrogenosome, there is now no doubt that these organelles are derived from mitochondria (Boxma et al., 2005; van Hoek, Akhmanova, Huynen, & Hackstein, 2000a). These anaerobic ciliates often have endosymbiotic and ectosymbiotic bacteria, typically methanogens, associated with the hydrogenosomes. This relationship, at least in the case of the endosymbiotic methanogens, provides the ciliate with increased efficiencies in growth (Fenchel & Finlay, 1991b).

Finally, a variety of extrusomes are prominent features of the somatic cortex. The different orders and classes of ciliates have different types of extrusomes (see reviews by Dragesco, 1984a; Hausmann, 1978; Rosati & Modeo, 2003). All these organelles are membrane-bound, likely synthesized in the endoplasmic reticulum-Golgi system, transported to the cell cortex, and stimulated to fuse with the plasma membrane by ionic changes (Hausmann, 1978). Mucocysts, broadly distributed throughout the classes, function to provide a surface coat for the cell, sometimes during the process of encystment (Figs. 4.9D, 4.19B) (Lynn & Corliss, 1991). Upon ejection, both their length and diameter become much larger than those dimensions in the resting state (Hausmann, 1978). Possible modifications of the mucocysts are the scale-like structures or lepidosomes secreted on the surface of some haptorians (Foissner, Müller, & Weisse, 2005a; Nicholls & Lynn, 1984). Clathrocysts and lepidosomes may also be used to construct the cyst wall of the haptorian *Didinium* (Holt & Chapman, 1971) and the spirotrich *Meseres* (Foissner et al., 2005a). Trichocysts, restricted primarily to some nassophoreans and some peniculine oligohymenophoreans, are extrusomes that maintain the diameter of the resting state but extend as thread-like filaments many times the resting length (Fig. 4.19D) (Hausmann, 1978). Trichocysts of the peniculine *Paramecium* appear to function to protect the ciliate from predators, such as *Climacostomum*, *Monodinium*, and *Dileptus* (Harumoto, 1994; Miyake & Harumoto, 1996; Sugibayashi & Harumoto, 2000).

While trichocysts may protect their ciliate bearer from predators, the last two common categories of extrusomes – toxicysts and haptocysts – are typically oral and enable the ciliate to switch roles and become the predator. Toxicysts are typical of the Subclass Haptoria (Class LITOSTOMATEA), and as the name suggests, are extrusomes with toxic potential. Upon extrusion, their tube-within-a-tube structure everts, maintaining the same width as in the resting state, but rapidly increasing in length to deliver the poisonous material now at the tip to the prey (Fig. 4.19E) (Hausmann, 1978). The compounds within the toxicyst can enable attachment of the predator to its prey and also immobilize the prey, partly by causing lysis of the somatic cilia (Wessenberg & Antipa, 1969, 1970). Finally, haptocysts are typically found at the tips of tentacles of the Subclass Suctoria (Class PHYLLOPHARYNGEA), and are small bottle-like organelles with a complex internal structure (Fig. 4.19C) (Hausmann, 1978). When prey contacts the suctorian tentacle, the haptocyst everts, cementing the two cells together and rapidly causing the prey to become immobile (Benwitz, 1982, 1984). Other extrusome types have been described as restricted to a particular group, and will be treated briefly in the appropriate chapter.

Chapter 5
Subphylum 1. POSTCILIODESMATOPHORA: Class 1. KARYORELICTEA – The "Dawn" or Eociliates

Abstract The ciliates in this class are thought to represent the nature of the ancestral ciliate lineage. Their non-dividing macronuclei make them "karyological relicts". They are a strongly supported clade, characterized by postciliodesmata arising from the somatic kinetids, their non-dividing macronuclei, and by robust phylogenetic support based on small subunit rRNA gene sequences. The class is divided into three orders, based primarily on oral features. These ciliate are conspicuous inhabitants of benthic marine habitats. Their elongated worm-like bodies can be seen crawling between sand grains and detrital particles. Thus, they are quite contractile and flexible, and also capable of regeneration. The extrusomes of this class are also unique with cnidocysts and orthonematocysts being found nowhere else in the phylum. Oral structures are quite variable, ranging from simple circumoral dikinetids to somewhat complex adoral ciliature. Stomatogenesis can be either parakinetal or buccokinetal, although much remains to be done on this aspect of their biology. Their non-dividing macronuclei, which arise at each cell division from division of a micronucleus, are often numerous and typically clustered around a micronucleus. Two unusual features of taxa in the group are the harvesting of epibiontic bacteria by *Kentrophoros* and the use of mineral crystals in the Müller's vesicle to sense gravity by *Loxodes*.

Keywords Postciliodesma, paradiploid, interstitial

The ciliates assigned to this class are considered by some to represent the nature of the "dawn" or eociliates that first diverged from the alveolate lineage. They have been labeled "karyological relicts", a term introduced by Grell (1962) and publicized by Raikov (1969, 1982, 1985), because they exhibit a simple form of nuclear dualism: the macronucleus is paradiploid but non-dividing. They have also been labeled "cortical relicts" because the cortex in some forms is thought to represent the ancestral condition: *Kentrophoros* does not have differentiated oral ciliature, but it does have somatic dikinetids, which are presumed to be the ancestral condition for the phylum (Lynn & Small, 1981; Small, 1984). There are over 130 species of these primarily interstitial ciliates, commonly found in the sands and sediments of marine littoral environments (Foissner, 1998b). Intertidal sands are the habitat for "relict" forms of various groups of small invertebrates, leading one to believe that the psammophilic karyorelicteans are also of ancient vintage (Corliss, 1974b, 1975b; Raikov, 1969). Finlay, Span, and Harman (1983) have suggested, based on their research on *Loxodes*, which is the only freshwater representative of the class, that these ciliates might also be "biochemical relicts" because of the odd mitochondrial potential of nitrate respiration under low oxygen conditions, which are common in interstitial environments.

The karyorelicteans are united by two major features: the presence of a non-dividing paradiploid macronucleus or macronuclei; and by postciliodesmata in which the microtubules are arranged as 2 + ribbon + 1 in a repeating fashion (see **Somatic Structures**). The class is supported robustly by small subunit rRNA gene sequences (Hammerschmidt et al., 1996; Hirt et al., 1995). The actin of *Loxodes* is quite divergent from other ciliates (Kim, Yura, Go,

& Harumoto, 2004). It is a matter of opinion whether this supports the ancestral nature of karyorelicteans or demonstrates again the unreliability of actin as a phylogenetic marker for ciliate evolution (see Philippe & Adoutte, 1998).

They form a diverse assemblage when one considers their oral structures. Some genera are ventrostomous (e.g., *Geleia*, *Loxodes*); some genera are prostomous (e.g., *Trachelocerca*, *Trachelolophos*); and some genera have apparently no differentiated oral ciliature (e.g., *Kentrophoros*). Bardele and Klindworth (1996) have observed that this parallels the evolution of oral structures in other groups. They argued that *Kentrophoros* may, in fact, have secondarily lost its oral apparatus when it acquired the obligatory symbiosis with thiotrophic or sulfur bacteria, an interpretation consistent with the observations of Foissner (1995a).

The distribution of these obligatorily psammobiotic species is global, though they are "endemic" with respect to their biotope. Means of dispersal remain unknown: Corliss and Hartwig (1977) supposed that continental drift may have been partially responsible.

5.1 Taxonomic Structure

We recognize three orders in this class: Order Protostomatida; Order Loxodida; and Order Protoheterotrichida. Alternative classifications have been proposed: Foissner (1998b) has argued that the bristle kinety, which frames the glabrous stripe or non-ciliated somatic cortex of protostomatids is homologous to that of loxodids, and so he supports uniting these in the Subclass Trachelocercia de Puytorac, Grain, and Mignot, 1987. We remain sceptical of this homology until ultrastructural investigation has demonstrated clear similarities in the kinetid structures or more extensive gene sequence data resolves the phylogeny of this class.

The Order Protostomatida includes the prostomous Family Trachelocercidae and the "astomous" Family Kentrophoridae (Fig. 5.1). Oral structures are simple and ingestion may be either at the anterior end or along the glabrous stripe (see **Oral Structures**). The Order Loxodida includes the ventrostomous Families Loxodidae and Cryptopharyngidae. These ciliates typically swim on the right surface of their flattened bodies. The oral cavity has a simplified ciliature of dikinetids (Fig. 5.1). The Order Protoheterotrichida, which includes the ventrostomous Family Geleiidae, are holotrichously ciliated and contractile, resembling their namesakes the heterotrichs (see **Chapter 6**). Their non-dividing macronuclei relate them to the other karyorelicteans even though their oral structures are more complex with simple adoral polykinetids and unusual paroral polykinetids on the right side of the oral region (Fig. 5.1).

A number of recent works have provided details of the morphology of these taxa: Trachelocercidae (Foissner, 1996c, 1997g; Foissner & Dragesco, 1996a, 1996b), Kentrophoridae (Foissner, 1995a, 1998b), Loxodidae (Foissner, 1995/1996, 1996b, 1998b), and Geleiidae (Dragesco, 1999), but refer to **Chapter 17** for detailed descriptions.

5.2 Life History and Ecology

These typically elongate and highly contractile ciliates are conspicuous consituents of interstitial habitats, especially sands and sediments of the marine littoral or brackish estuaries. Karyorelicteans have been recorded from interstitial habitats, often sandy ones in the marine sublittoral, in Africa (Dragesco, 1965), western and eastern Europe (Agamaliev, 1971; Azovsky & Mazei, 2003; Dragesco, 1963, 2002; Fernández-Leborans & Fernández-Fernández, 1999; Kovaleva & Golemansky, 1979; Mazei & Burkovsky, 2003), North America (Borror, 1963), and the Arabian Gulf (Al-Rasheid & Foissner, 1999). The only recorded exception is *Loxodes*, which is found in freshwater sediments (Finlay, 1982; Finlay & Berninger, 1984). Most species are classified as microaerophilic, restricted to sediments because these regions contain reduced oxygen concentrations, often becoming anoxic within a few centimeters of the sediment-water interface. However, *Loxodes* can move into the water column if the interstitial waters of the sediments become anoxic (Goulder, 1980). Finlay, Fenchel, and Gardener (1986) suggested that cytochrome oxidase is the oxygen receptor for *Loxodes* whose response to oxygen concentrations is modified by light (Fenchel & Finlay, 1986b).

Karyorelicteans are predaceous macrophages, using their filiform or vermiform bodies to crawl between the grains in the sediments in search of food. They have been recorded to ingest bacteria,

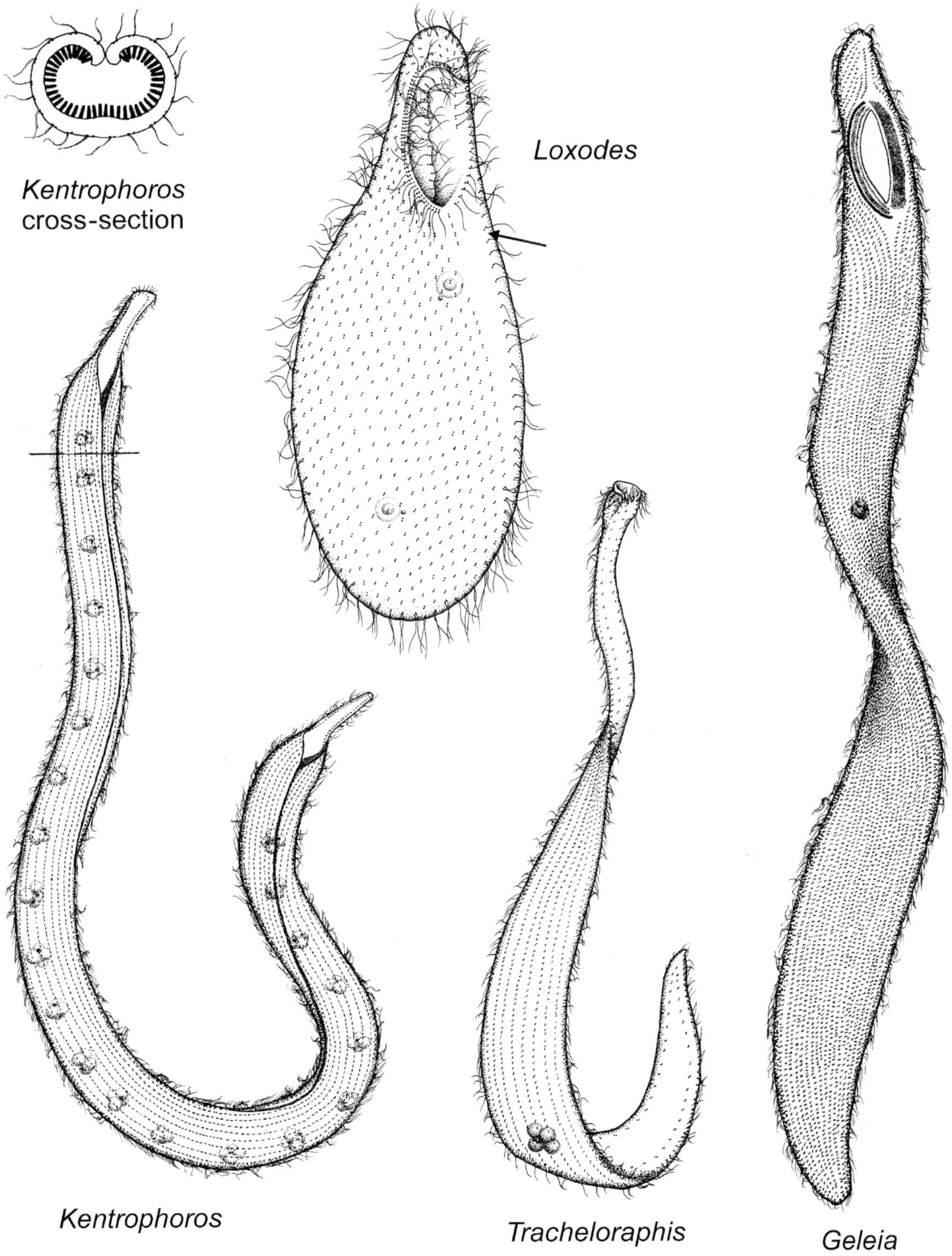

FIG. 5.1. Representative genera of the Class KARYORELICTEA. The protostomatid *Kentrophoros* whose body in cross-section is ciliated on one surface (the right?) and harbors a "kitchen garden" of epibiotic bacteria on its glabrous zone (after Foissner, 1995a). The loxodid *Loxodes* whose ventral oral region has a paroral along its right border and an intrabuccal kinety extending posteriorly into the tube-like oral cavity. Note the bristle kinety along the ventral left surface of the cell (arrow) (after Bardele & Klindworth, 1996). The protostomatid *Tracheloraphis* showing its prostomial oral region and the glabrous zone bordered by the bristle kinety (After Foissner & Dragesco, 1996b). The protoheterotrichid *Geleia*, which is holotrichous and shows a complex oral region of dikinetid files and simple polykinetids. (after Dragesco, 1999.)

diatoms (e.g., *Coscinodiscus*, *Phaeodactylum*), both autotrophic and heterotrophic flagellates (e.g., *Euglena*, dinoflagellates), other ciliates (e.g., *Euplotes*, *Strombidium*, and smaller karyorelicteans), and even micrometazoans, such as rotifers and copepods (Foissner, 1998b). The many species sharing an interstitial habitat probably coexist in part by partitioning food resources: different-sized *Loxodes* species coexist in the same lake as the larger species consumes the larger food particles (Finlay & Berninger, 1984).

Conjugation is rarely observed (see **Sexuality and Life Cycle** below). Since it does occur in some taxa, we presume it to be an ancestral feature of the group. Cysts are not known. Thus, explaining the presumed global distribution of some of these ciliates is problematic as it is with any group that does not form resistant phases in the life cycle.

5.3 Somatic Structures

The karyorelictean cell body is typically long, sometimes >5,000 µm, and frequently flattened to about 5–10 µm in thickness. In several genera, the cell surface on which the organism "crawls" is more densely ciliated (e.g., *Loxodes*, *Kentrophoros*). The body is often pigmented, brown or yellowish, possibly due to pigmentocysts or extrusomes. The pigmentocysts apparently have a defensive function, at least in *Loxodes* (Buonanno, Saltalamacchia, & Miyake, 2005). The cell surface may have a conspicuous glycocalyx, but is not underlain by a regular layer of cortical alveoli. When present, the alveoli are irregular and small. Parasomal sacs have not been observed.

The somatic dikinetids of these ciliates are composed of two kinetosomes joined by desmoses (Fig. 5.2), and oriented at 20–40° to the kinety axis. Both kinetosomes may be ciliated or only the anterior one. The postciliary microtubular ribbon of the posterior kinetosome is divergent, extending up to the cortex and posteriorly to overlap the ribbons of 10 or more anterior kinetids, and so forming the postciliodesma. The number of overlapping ribbons will vary depending upon the contractile state of the ciliate, as these microtubules are assumed to play the same role in cell elongation as those of *Stentor* (Huang & Mazia, 1975; Huang & Pitelka, 1973). It is not clear how the organization of the postciliary ribbons changes from their origin as a ribbon to the modified structure at the cell surface. There are two microtubules closest to the kinetosome followed by a ribbon of up to 20 microtubules perpendicular to the cell surface, and then a single microtubule. This 2 + ribbon + 1 pattern can be repeated for each overlapping set (Fig. 5.2) (Klindworth & Bardele, 1996; Raikov, 1994b; Raikov & Kovaleva, 1995; Raikov, Gerassimova-Matvejeva, & de Puytorac, 1976). The postciliary microtubules are accompanied by dense material on either side near their base. The posterior kinetosome may also have a tangential transverse ribbon associated with triplets 3–5 (Fig. 5.2). The kinetodesmal fibril originates near triplets 5, 6, and 7 and is variable in form. It is striated and elongate in *Remanella* (Raikov, 1994b), striated and shovel-shaped in *Loxodes* (Bardele & Klindworth, 1996), and short and hooked with only a faint periodicity in *Tracheloraphis* (Raikov & Kovaleva, 1995) and *Geleia* (de Puytorac, Raikov, & Nouzarède, 1973a). The kinetodesmal fibril structure in the latter two genera is very reminiscent of *Stentor*'s as described by Huang and Pitelka (1973). In *Loxodes,* the shovel-shaped kinetodesmal fibril becomes branched, one branch of which extends to contact the postciliary ribbon of the next anterior kinetid. The anterior kinetosome has a tangential transverse ribbon associated with triplets 3–5. There may be ribbons of subkinetal microtubules that originate from the bases of the somatic kinetosomes and extend posteriorly beneath the kinety (Raikov & Kovaleva, 1995) or towards the left (Klindworth & Bardele, 1996).

There are two kineties on the left side of *Loxodes* that have been interpreted to be one continuous kinety. Klindworth and Bardele (1996) have disproved this by showing that the kinetodesmal fibrils are oriented in the manner expected for two kineties: these kineties just happen to abut near the anterior end and so appear to be continuous at the level of the light microscope. Until it is demonstrated otherwise by electron microscopy, we assume that the bristle kineties bordering the non-ciliated stripe in *Kentrophoros* are bipolar, contrary to the interpretations of Foissner (1995a, 1998b).

Myonemes are arranged longitudinally and parallel to the somatic kineties in most karyorelicteans: to the right of the kinety in *Tracheloraphis*, to the left of the kinety in *Remanella*, and on both sides in *Geleia*. Since these ciliates are often not evenly ciliated around the body, contraction may cause

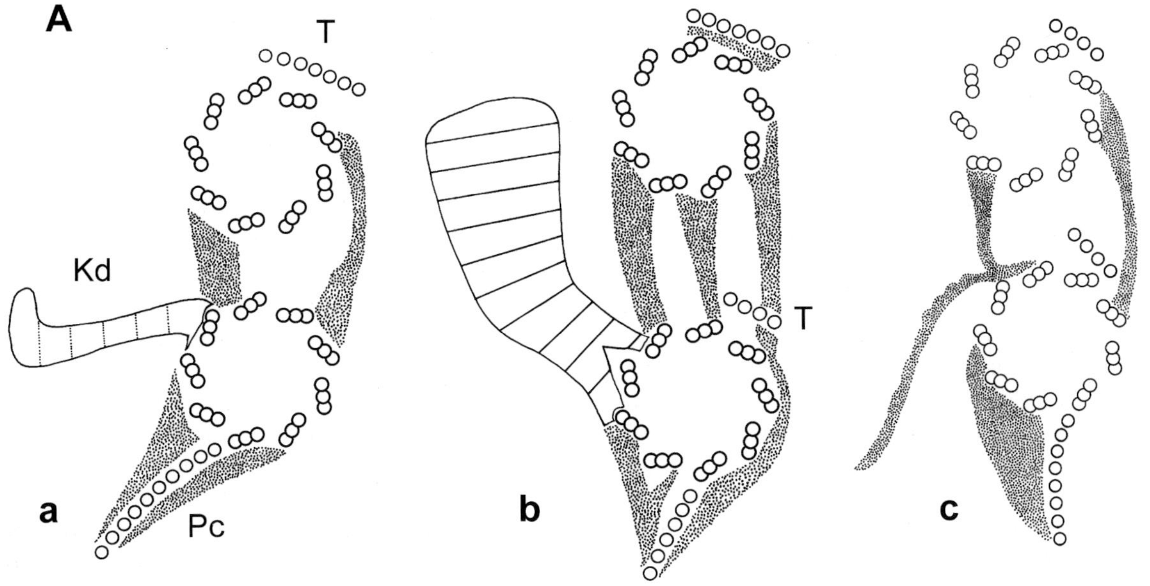

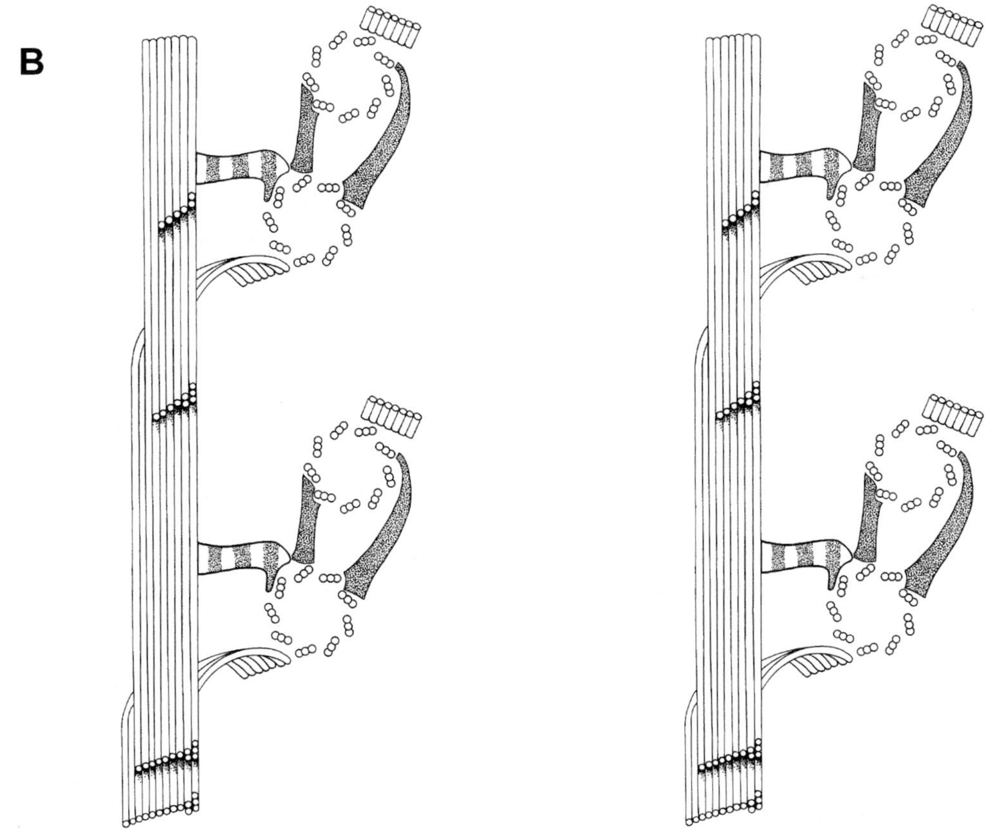

FIG. 5.2. Ultrastructure of the cortex of the Class KARYORELICTEA. **A** Somatic dikinetids. (a) The protostomatid *Tracheloraphis* (after Raikov & Kovaleva, 1995). (b) The loxodid *Loxodes*. (after Klindworth & Bardele, 1996.) (c) The protoheterotrichid *Geleia* (after de Puytorac, Raikov, & Nouzarède, 1973a) kinetodesmal fibril (Kd), postciliary ribbon (Pc), transverse ribbon (T). **B** Somatic cortex of the protostome *Tracheloraphis* with postciliodesmata composed of overlapping ribbons in the 2 + ribbon + 1 arrangement. (Redrawn after Raikov et al., 1976.)

the cell to become banana-shaped or roll up. In some species, transverse myonemes occur, possibly ensuring an even longitudinal contraction.

The contractile vacuole system is not well-developed, except in freshwater *Loxodes* species, and is often absent.

Extrusomes in the karyorelicteans are very diverse. Rhabdocysts have been recorded in *Tracheloraphis* and *Kentrophoros* (Raikov, 1974b), ampullocysts in *Kentrophoros* (Raikov), and cnidocysts and orthonematocysts in *Remanella* (Foissner, 1996a; Raikov, 1978, 1992, 1993). The aberrant character of karyorelictean extrusomes in relation to those of other ciliates and the apparent similarities of the cnidocysts of some karyorelicteans to the extrusomes of dinoflagellates, another alveolate group, have been used as another feature to indicate the ancestral nature of the karyorelicteans (Raikov, 1992).

5.4 Oral Structures

The taxa in this class are distinguished from each other primarily on the basis of oral structures, which, as we learn more about the detailed cyto-anatomy of this group, are quite diverse. Oral kinetosomes bear cilia that are usually slightly longer than the somatic cilia, and may have simple nematodesmata, which reinforce the cytopharyngeal walls.

Protostomatids have a dome-like oral region surrounded by circumoral dikinetids, which may form an uninterrupted ring around the cytostome or which may be interrupted by brosse kinetofragments (Fig. 5.1). Until electron microscopy demonstrates otherwise, we assume that the glabrous stripe is delimited by two kineties, one on its left and one on its right (Fig. 5.1). Foissner and Dragesco (1996b) interpreted these as the bristle kinety, which they assumed to be continuous around the glabrous stripe. Protostomatids may ingest food through the anterior end (Al-Rasheid & Foissner, 1999) or along the glabrous stripe (Lenk, Small, & Gunderson, 1984; Lenk, Hollander, & Small, 1989). We place *Kentrophoros* in this group because it has two kineties bordering the glabrous stripe and it ingests the symbiotic bacteria from its glabrous stripe (Raikov, 1974b) in a fashion similar to that reported for ingestion by *Tracheloraphis* (Lenk et al., 1984, 1989). Foissner and Dragesco (1996b) argued that protostomatids are derived with respect to their oral region, which they regarded as having become apicalized from that of a ventrostomous ancestor, based on the arguments of Eisler (1992).

Loxodids have a slit-like ventral oral region that is bordered by files of dikinetids (Fig. 5.1). There is a paroral of dikinetids bordering the right side of the oral region. Slightly inside of this is a file of dikinetids that extends into a posterior extension of the oral cavity. This file has been called an intravestibular kinety (Bardele & Klindworth, 1996) or an intrabuccal kinety (Foissner, 1995/1996). Since these oral structures are derived from buccal structures (**see below**), this oral cavity is not a vestibulum (see **Glossary**); therefore, we prefer the term intrabuccal kinety. The cytostome may be placed between the paroral and this kinety (Klindworth & Bardele, 1996). The left side of the oral region is bordered by the left pseudobuccal kinety, which may have been derived from somatic kinety 1 (Foissner, 1995/1996), since it is an inverted kinety based on the inverted orientation of its fibrillar associates (Bardele & Klindworth, 1996). Just interior to this, is a file of several anterior left oral dikinetids. A ventral kinetofragment of several dikinetids extends posterior from the ventral slit; it behaves like a scutica during stomatogenesis (Bardele & Klindworth, 1996).

Protoheterotrichids have been recently described in detail by Dragesco (1999). Their oral region varies from slit-like to almost rounded, and is bordered by more complex oral structures than found in the previous two orders (Fig. 5.1). The cytostome is bounded on its right side by a paroral of dikinetids to the right of which are right paroral polykinetids. The structure of these polykinetids appears variable at the light microscopic level: files of closely spaced monokinetids lie perpendicular to the paroral (e.g., *Geleia*, *Avelia*, *Parduczia*); and files of dikinetids lie parallel to the paroral (e.g., *Gellertia*). Adoral polykinetids are arrayed perpendicular to the longitudinal axis of the cell to the left of cytostome in all genera but *Parduczia*, which has oralized somatic kineties on this side of the oral region. When present, the adoral polykinetids appear to be composed of files of dikinetids (e.g., *Geleia*, *Gellertia*) or monokinetids (e.g., *Avelia*). Confirmation of Dragesco's interpretation from these protargol-stained specimens must await electron microscopic examination.

5.5 Division and Morphogenesis

Karyorelicteans divide while swimming freely. They also have considerable powers of regeneration, like the heterotrichs. Thus, they may increase in numbers by fragment regeneration when they become severed by sediment action in their natural habitat.

Foissner and Al-Rasheid (1999) have classified stomatogenesis in the protostomatid *Sultanophrys* as parakinetal. The proter oral apparatus does not reorganize. The oral primordium of the opisthe is apparently derived from proliferation of kinetosomes in the somatic kinety to the right of the glabrous stripe (Fig. 5.3). The subequatorial primordium differentiates from an anarchic field of kinetosomes into kinetofragments of dikinetids. These kinetofragments assemble as circumoral dikinetids and three small brosse kinetofragments.

Bardele and Klindworth (1996) concluded that stomatogenesis in *Loxodes* is buccokinetal. The parental oral apparatus is only slightly reorganized. Stomatogenesis begins with proliferation of kinetosomes from the ventral kinetofragment, just posterior to the oral region, and this is followed by proliferation of kinetosomes from the paroral and the left pseudobuccal kinety. The ventral kinetofragment gives rise to the opisthe paroral while the proliferation of the proter left pseudobuccal kinety gives rise to that of the opisthe. The intrabuccal kinety and the anterior left oral dikinetids, whose origin is not yet resolved, appear later in stomatogenesis at the anterior end of the opisthe oral region, as the oral anlagen reach the cell equator.

Thus, stomatogenesis among the karyorelictean orders demonstrates different modes: parakinetal in protostomatids and buccokinetal in loxodids. What mode the protoheterotrichs demonstrate awaits the results of further investigations. However, it is clear that there is as much diversity in stomatogenesis in this class as can be found in other classes of ciliates (e.g., Class OLIGOHYMENOPHOREA, Foissner & Al-Rasheid, 1999). Stomatogenesis is therefore not likely to be an indicator of deep phylogenetic relationships.

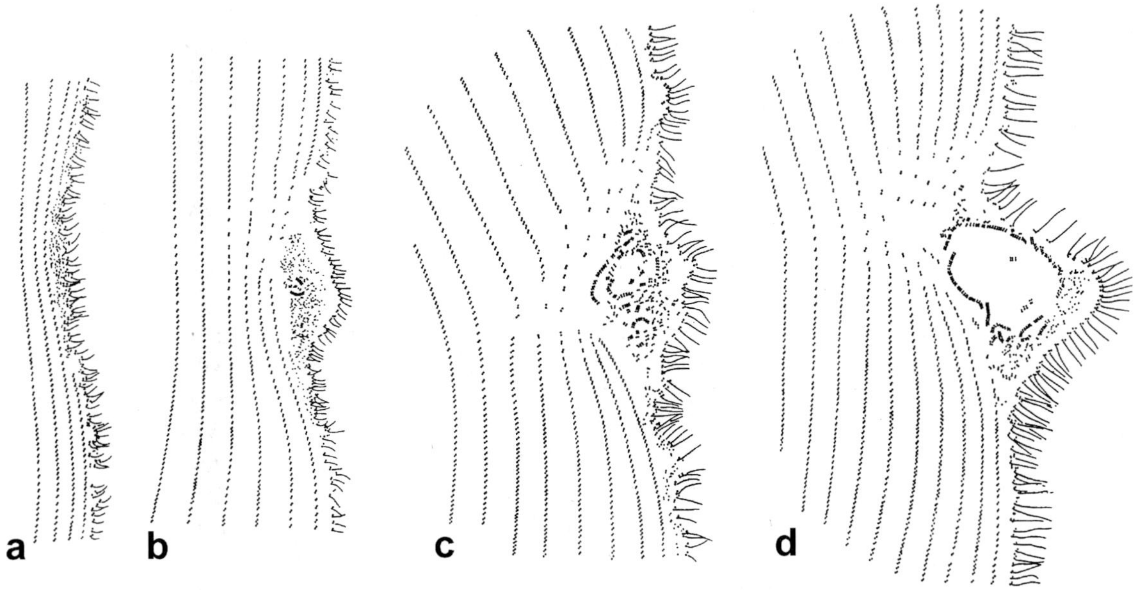

FIG. 5.3. Stomatogenesis of the protostomatid *Sultanophrys*. (**a**) The process begins in the mid-region of the body as kinetosomes proliferate, forming an anarchic field to the right of the ventral (?) or right bristle kinety. (**b–d**) The process continues until a ring of circumoral dikinetids forms accompanied by 3 minute brosse kinetids. (from Foissner & Al-Rasheid, 1999.)

5.6 Nuclei, Sexuality and Life Cycle

As discussed above, the karyorelicteans have the simplest form of nuclear dimorphism in the phylum: the macronucleus is diploid or paradiploid and non-dividing. It is rich in RNA with a conspicuous nucleolus and may have proteinaceous inclusions (Raikov, 1982). Since it does not divide, the macronucleus(ei) must be replaced at each cell division by differentiation of the products of micronuclear division. The micronucleus is presumed to be diploid and is capable of mitosis and meiosis. Both nuclei are small in size, typically globular or ellipsoid. They may be found in a variety of close associations: a single micronucleus and one macronucleus (e.g., *Loxodes*), a single micronucleus and two associated macronuclei (e.g., some protostomatids), or present in clusters or complexes of several micronuclei and several macronuclei with or without a surrounding envelope.

It is not clear what factors stimulate conjugation in karyorelicteans. Raikov (1972) reviewed the general features of conjugation in karyorelicteans, based on observations of *Loxodes* and *Tracheloraphis*, and summarized it thus. Preconjugants differ from vegetative cells by the incomplete differentiation of the nuclei, especially the macronuclei. Meiosis is typical but it may lead to the formation of multiple pronuclei and, ultimately after exchange of migratory gametic nuclei, multiple synkarya in *Tracheloraphis* species. Thus, genetic identity of the two separating exconjugants is not assured. After separation, the "parental" macronuclei do not degenerate. Thus, the exconjugant is presumed to be a genetic chimaera with a "parental" phenotype expressed by the old macronuclei and a developing phenotype expressed by the genome residing in the "new" macronucleus. Replacement of the "parental" macronuclei occurs over several cell divisions following conjugation, and the new phenotype presumably becomes established as the "parental" macronuclei are diluted out by cell division cycles. Since renewal of the macronuclei is not directly connected with conjugation, Raikov supposes this to be an ancestral feature of the conjugation process (see also Orias, 1991a, 1991b).

5.7 Other Features

Gram-negative bacteria are commonly associated with karyorelicteans. *Geleia* species may have perhaps 10,000 bacteria as epibionts on their cell surface (Epstein, Bazylinski, & Fowle, 1998). Other Gram-negative bacteria are found in the cytoplasm. Since they are often not bounded by a ciliate vacuolar membrane, it is assumed that they are endosymbionts. Their functional relationship to their hosts is unknown.

A novel feature restricted to the loxodid karyorelicteans is the organelle known as Müller's vesicle. The vesicle, about 7 μm in diameter, encloses the Müller's body, which itself is bounded by a cell membrane that encloses barium salt-dominated crystals in the freshwater *Loxodes* and strontium salt-dominated crystals in the marine *Remanella* (Rieder, Ott, Pfundstein, & Schoch, 1982). Movements of the Müller's body, in response to gravity and the orientation of the ciliate, may deform ion channels on the cell surface and thereby modulate cell movement. Movement in *Loxodes* is also dependent on external oxygen concentration in the water: *Loxodes* swim faster upward when the water is anoxic and faster downward when the water is oxygen-saturated (Fenchel & Finlay, 1984, 1986b).

Müller's body is suspended in the vesicle by a stalk that is supported by postciliary microtubules from an adjacent dorsal left somatic dikinetid (Fenchel & Finlay, 1986a). These dikinetids comprise the loxodid dorsolateral kinety that Foissner (1998b) has presumed to be homologous to a somatic kinety of kentrophorids that is in a similar position but does not "bear" Müller's vesicles. Foissner (1998b) thus argues that loxodids and kentrophorids are sister taxa. We remain skeptical until ultrastructural homologies of the kinetids are proved or until gene sequence data confirm these two groups as sister taxa.

Chapter 6
Subphylum 1. POSTCILIODESMATOPHORA: Class 2. HETEROTRICHEA – Once Close to the Top

Abstract Ciliates in this class were thought to represent the pinnacle of ciliate evolution, along with the spirotrichs. However, small subunit rRNA gene sequences and the presence of postciliodesmata in the somatic cortex strongly relate members of this class to the Class KARYORELICTEA. The heterotrichs are typically majestic ciliates of large cell size and with a conspicuous adoral zone of polykinetids or membranelles (AZM) that extend out over the peristomial surface. The ciliates in this class are not subdivided, and so there is one order – Order Heterotrichida. Heterotrichs are found in a diversity of habitats, from the marine benthos and hydrothermal vents to the plankton of high altitude oligotrophic lakes. They feed on a diversity of prey, ranging from bacteria up to small metazoa, like rotifers, and sometimes are conspicuous by carrying symbiotic zoochlorellae. Their body is highly contractile, elongated by postciliodesmal microtubules and shortened by contractile myonemes. The oral structures have a paroral and multiple paramembranelles. Stomatogenesis is parakinetal. Macronuclei can be nodular, and are divided by extramacronuclear microtubules. Conjugation has not been studied in any breadth in the class with the gamone-receptor system of *Blepharisma* being the only model. The heterotrich *Spirostomum* has been developed as a bioassay model for heavy metals.

Keywords Ampliploid, microbiotest

Heterotrichs are common and large ciliates, some *Spirostomum* species achieving body lengths of up to 4,000 μm. They include some of the best-known and most common ciliates in the phylum. *Stentor*, a typical representative, has long attracted attention from protozoologists and cell biologists because of its size, ubiquity, ease of general laboratory culture, contractility, and regenerative powers (Fig. 6.1). Typically heterotrichs are free-swimming and holotrichously ciliated, although members of the Family Folliculinidae secrete attached loricas in which they live. The group is at least 100–200 million years old as demonstrated by the fossilized lorica of *Priscofolliculina* (Deflandre & Deunff, 1957).

Heterotrichs were so-named because of the marked difference between their holotrichous somatic ciliation and the conspicuous, typically spiralling adoral zone of membranelles or oral polykinetids. They were conceived as the pivotal group in the evolution of the "higher" or polyhymenophorean ciliates (Corliss, 1961, 1979). Doubt about this vision began to emerge in the late 1970s. Ultrastructural data indicated dramatic differences in their somatic kinetids compared to other polyhymenophoreans; and similarities in the heterotrich somatic cortex to that of the karyorelicteans suggested a closer relationship between the presumably most derived and the presumably most ancestral groups (Gerassimova & Seravin, 1976; Lynn, 1981, 1991). The nature of membrane particle arrays in different ciliate groups also suggested a stronger relationship between heterotrichs and karyorelicteans (Bardele, 1981). Finally, sequences of nuclear ribosomal RNA genes from heterotrichs and karyorelicteans supported their sister group status in an early diverging lineage (Baroin-Tourancheau, Delgado, Perasso, & Adoutte, 1992; Greenwood, Schlegel, Sogin, & Lynn, 1991b; Hirt et al., 1995).

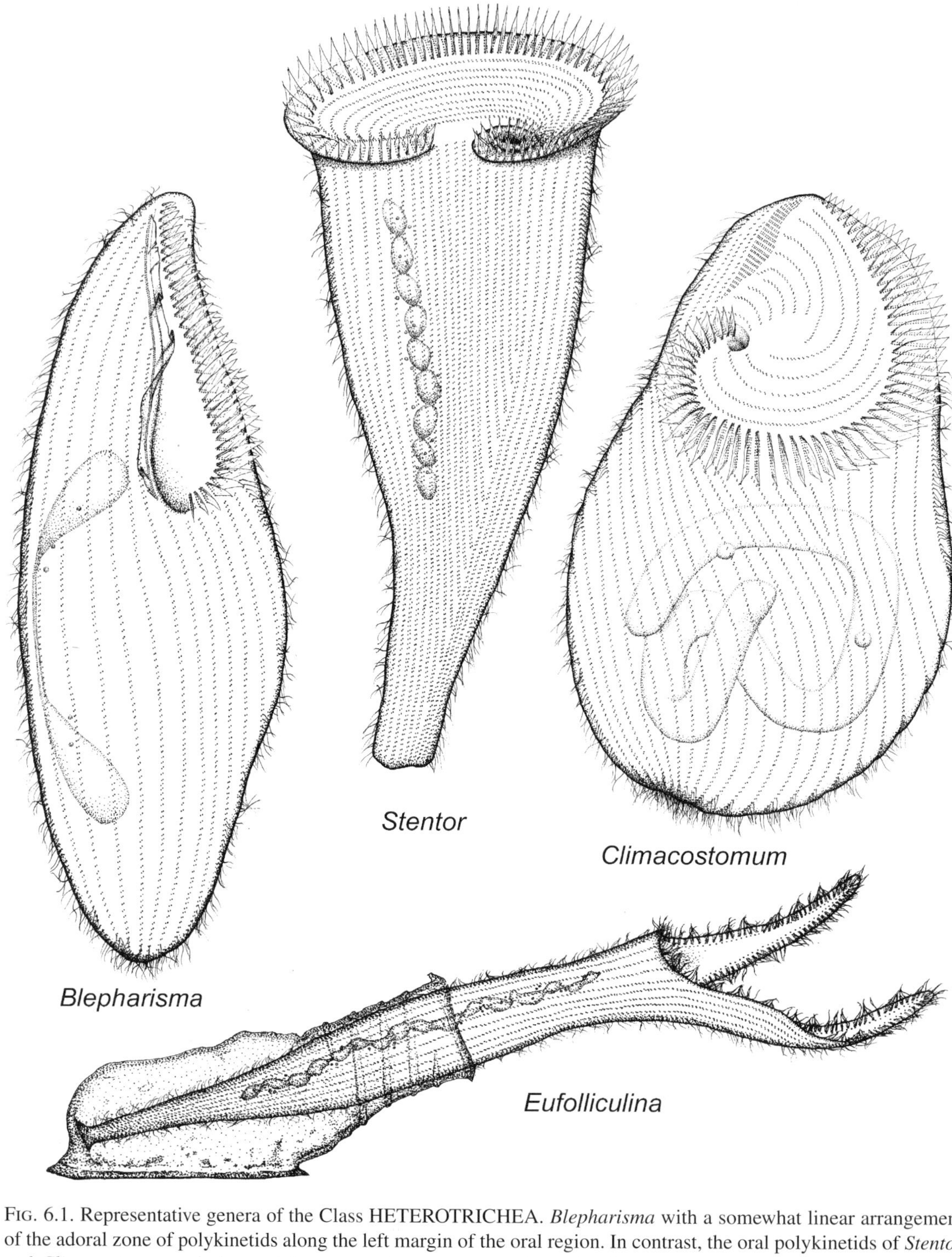

FIG. 6.1. Representative genera of the Class HETEROTRICHEA. *Blepharisma* with a somewhat linear arrangement of the adoral zone of polykinetids along the left margin of the oral region. In contrast, the oral polykinetids of *Stentor* and *Climacostomum* spiral out of the oral cavity in a counter-clockwise direction, bounding a peristomial field that is covered by kineties. The folliculinid *Eufolliculina* exemplifies this unique family of heterotrichs in being anchored in a lorica and in having its oral region drawn out into two extensive peristomial "wings"

Thus, we are now certain that the postciliodesmata, shared by both karyorelicteans and heterotrichs, demonstrate their shared common ancestry.

The ciliates now assigned to this class are united by four major features. First, they have highly ampliploid and dividing macronuclei. Macronuclear karyokinesis is accomplished primarily by extramacronuclear microtubules (Diener, Burchill, & Burton, 1983; Jenkins, 1973; Lynn & Small, 1997) and probably evolved independently of macronuclear karyokinesis in the Subphylum Intramacronucleata (Lynn, 1996a; Orias, 1991a; and see **Chapter 4**). Second, the microtubular components of their postciliodesmata are more simply organized than those of the karyorelicteans: they appear as ribbons oriented perpendicular to the cell surface, only separated by a single microtubule. Third, the oral polykinetids on the left side of the oral region are characterized as paramembranelles (de Puytorac & Grain, 1976), which form a conspicuous adoral zone often extending out onto the anterior cell surface. Fourth, the differentiation of these oral polykinetids during stomatogenesis occurs from the center of the oral primordium towards the anterior and posterior, a unique pattern within the phylum (Aescht & Foissner, 1998).

6.1 Taxonomic Structure

There has been considerable change in the composition of this taxon since Corliss (1961, 1979). Corliss (1979) recognized six suborders within the Order Heterotrichida: (1) Suborder Heterotrichina; (2) Suborder Clevelandellina; (3) Suborder Armophorina; (4) Suborder Coliphorina; (5) Suborder Plagiotomina; and (6) Suborder Licnophorina. Results from the study of ultrastructure and molecular sequences now suggest the following. The somatic kinetids of clevelandellids (Affa'a, unpublished, 2007; de Puytorac & Grain, 1969), armophorids (Schrenk & Bardele, 1991), plagiotomids (Albaret & Grain, 1973), and licnophorids (Da Silva Neto, 1994a) do not exhibit postciliodesmata and also have different patterns of fibrillar associates (Lynn, 1981, 1991). Furthermore, nuclear small subunit rRNA (SSUrRNA) gene sequences separate the clevelandellids (van Hoek, van Alen, Sprakel, Hackstein, & Vogels, 1998, 2000b) and armophorids (Embley, Finlay, Thomas, & Dyal, 1992; van Hoek et al., 1998, 2000b) to a new class, the Class ARMOPHOREA (see **Chapter 8**), while plagiotomids (Affa'a, Hickey, Strüder-Kypke, & Lynn, 2004) and licnophorids (Lynn & Strüder-Kypke, 2002) are transferred to the Class SPIROTRICHEA (see **Chapter 7** for details).

The Suborder Coliphorina only included the Family Folliculinidae. However, the somatic kinetids of *Eufolliculina* are extremely similar to other heterotrichs in their fibrillar associates and the character of the postciliodesmata (Mulisch, Barthlott, & Hausmann, 1981), while SSUrRNA sequences indicate this genus falls within the heterotrich radiation and is the sister taxon to *Maristentor* (Miao, Simpson, Fu, & Lobban, 2005). Thus, this group should not be separated at such high rank, and we do not now recognize this suborder.

Within the Suborder Heterotrichina, Corliss (1979) included the following: Family Bursariidae, Family Chattonidiidae, Family Climacostomidae, Family Condylostomatidae, Family Metopidae, Family Peritromidae, Family Phacodiniidae, Family Reichenowellidae, Family Spirostomidae, and Family Stentoridae. The somatic kinetids of *Phacodinium* (Didier & Dragesco, 1979; Da Silva Neto, 1993a), *Transitella*, a reichenowellid (Foissner, Adam, & Foissner, 1982; Iftode, Fryd-Versavel, Wicklow, & Tuffrau, 1983), metopids (Schrenk & Bardele, 1991), and bursariids (Gerassimova, Sergejeva, & Seravin, 1979; Lynn, 1980) do not form postciliodesmata, while redescriptions of the reichenowellid *Balantidioides* suggest that it has affinities to the spirotrichs (Foissner et al., 1982). While SSUrRNA gene sequences support placement of *Phacodinium* among the spirotrichs (Shin et al., 2000), these same gene sequences confirm the heterotrich affinities of *Peritromus* (Rosati, Modeo, Melai, Petroni, & Verni, 2004), *Chattonidium*, (Modeo et al., 2006), and *Condylostomides* (Schmidt, Foissner, Schlegel, & Bernhard, 2007).

In conclusion, we now recognize one order within the class, the Order Heterotrichida with characters of the class, and eight families: Family Blepharismidae [but see Aescht & Foissner, 1998], Family Chattonidiidae, Family Climacostomidae, Family Condylostomatidae, Family Maristentoridae, Family Peritromidae, Family Spirostomidae, and Family Stentoridae (see **Chapter 17. Ciliate Taxa**).

They are distinguished primarily by features of the oral region and variations in their overall body form.

A number of works have treated different genera in detail: *Spirostomum* (Repak & Isquith, 1974); *Blepharisma* (Repak, Isquith, & Nabel, 1977); *Stentor* (Foissner & Wölfl, 1994); and a recent report on the rare genus *Copemetopus* (Al-Rasheid, 2001). We should not forget the classic works on *Stentor*, the majestic "king of the ciliates", by Tartar (1961) and on *Blepharisma*, the light-sensitive protozoon by Giese and collaborators (1973). Hadži (1951) is the classic work on the folliculinids.

6.2 Life History and Ecology

Because of their typically large size, heterotrichs can be conspicuous members of microbial foodwebs and have a widespread distribution. Heterotrichs have been recorded from freshwater lakes in subtropical Florida (Beaver & Crisman, 1989b), Antarctica (Kepner, Wharton, & Coats, 1999), Europe (Finlay, 1982), and high altitude lakes in South America (Woelfl & Geller, 2002), and streams in Europe (Madoni & Ghetti, 1980). They are found in a variety of marine habitats, including anaerobic sediments in Europe (Fenchel & Finlay, 1990a), the marine sublittoral in Europe (Agamaliev, 1971; Azovsky & Mazei, 2003; Kovaleva & Golemansky, 1979; Mazei & Burkovsky, 2003) and even deep marine habitats (Fenchel et al., 1995) and hydrothermal vents (Small & Gross, 1985; Bergquist et al., 2007). Heterotrichs are often dominant members of the low diversity ciliate communities of hypersaline habitats across the globe – in Europe (Esteban & Finlay, 2004), Africa (Yasindi, Lynn, & Taylor, 2002), Arabia (Al-Rasheid, Nilsson, & Larsen, 2001; Elloumi et al., 2006), and Australia (Post, Borowitzka, Borowitzka, Mackay, & Moulton, 1983). They are occasionally found in soils (Buitkamp, 1977; Foissner, 1998a; Griffiths, 2002).

Most species are free-swimming, but some, such as *Stentor*, have the ability to use a holdfast to temporarily attach to the substrate (Fauré-Fremiet, 1984). A few species of *Stentor* secrete a mucoid sheath and all species of folliculinids secrete a lorica in which they can retract to avoid predation. The substances for these external coverings originate from extrusomes (Bussers, 1984; Mulisch & Hausmann, 1983), and in the folliculinids may contain chitin fibrils (Mulisch, Herth, Zugenmaier, & Hausmann, 1983). Substrates to which heterotrichs attach include inorganic substrates and macrophytes. Folliculinids attach to the integument of various invertebrates (Matthews, 1968; Fernández-Leborans & Córdoba, 1997), and may cause the skeletal eroding band or brown band diseases of scleractinian corals (Antonius, 1999; Cróquer et al., 2006). *Maristentor* is found on corals, but does not appear to cause disease (Lobban et al., 2002).

Some genera, like *Fabrea*, are strictly marine or brackish water forms, which can attain abundances of $10^5 l^{-1}$ (Elloumi et al., 2006; García & Niell, 1993). *Stentor* species can reach more than $10^3 l^{-1}$ in some lakes in the southern hemisphere, perhaps due to the absence of larger microcrustacean predators (James, Burns, & Forsyth, 1995; Laybourn-Parry, Perriss, Seaton, & Rohozinski, 1997). Dispersal generally occurs by swimming, but cysts may also be involved (**see below**). Kusch (1998) has demonstrated clear evidence of relatively high gene flow among populations of *Stentor* separated by as much as 400 km. Genera in the Family Folliculinidae are typically marine although the fresh-water species *Folliculina boltoni* has been recorded from Europe (Penard, 1919), North America (Hamilton, 1952), and South America (Dioni, 1972) while *Ascobius lentus* has been recorded recently in European freshwaters (Mulisch, Heep, Sturm, & Borcherding, 1998). Folliculinids are dispersed in part by the movements of their host, but the proter or anterior daughter differentiates at cell division as a "mouthless swarmer" stage that is adapted for dispersal.

Heterotrichs are omnivorous, upstream filter feeders (Fenchel, 1980a), showing little preference for prey species. Bacteria, autotrophic and heterotrophic flagellates, and ciliates are ingested, with some prey species proving more nutritious than others (Rapport, Berger, & Reid, 1972; Repak, 1983, 1986). Heterotrichs may change the shape of the oral region (Liebsch, 1976) and the spacing between the cilia of the oral polykinetids (Rickards & Lynn, 1985) in response to physiological states and prey types. When smaller food items become scarce, heterotrichs can become cannibalistic (Foissner & Wölfl, 1994; Giese, 1973; Pierce, Isquith, & Repak, 1978) and have also been known to ingest smaller metazoans

(Foissner & Wölfl; Tartar, 1961). In an unusual turn of the tables, it appears that *Mirofolliculina limnoriae*, an epibiont on the wood-boring isopods of the genus *Limnoria*, may outcompete its host for food and hinder host dispersal, suggesting it can be considered an ectoparasite (Delgery, Cragg, Busch, & Morgan, 2006).

Heterotrichs harbor a variety of endosymbionts: bacteria can be found in the cytoplasm and in the macronucleus (Fokin, Schweikert, Brummer, & Görtz, 2005; Görtz, 1983; Görtz & Wiemann, 1987). The bacterial endosymbionts do not appear to be harmful; in fact, some bacteria may be essential symbionts (Hufschmid, 1984). A variety of *Chlorella* species provide their *Stentor* and *Climacostomum* hosts with the "by-products" of photosynthesis (Fernández-Leborans & Zaldumbide, 1983; Kawakami, 1984; Reisser, 1984; Woelfl & Geller, 2002), and may compete with bacterial endosymbionts for the host cytoplasmic niche (Hufschmid, 1984). Laybourn-Parry et al. (1997) determined that *Stentor amethystinus* could contribute almost 70% of the total plankton photosynthesis in some Australian lakes.

Heterotrichs themselves are prey for other ciliates and metazoans. *Stentor* has mechanoreceptors distributed on its cell surface that may enable response to predator contact (Wood, 1989). When contact is made with toxicyst-bearing litostome ciliates, like *Dileptus* (see **Chapter 9**), *Blepharisma* (Harumoto et al., 1998; Miyake, Harumoto, Salvi, & Rivola, 1990), *Climacostomum* (Masaki et al., 1999), and *Stentor* (Miyake, Harumoto, & Iio, 2001) induce a massive release of their pigmentocysts, respectively containing the pigments blepharismin, climacostol, and stentorin, which have proved lethal to this predator. However, the pigment does not inhibit predation by the heterotrich *Climacostomum* on its heterotrich relative *Blepharisma* (Terazima & Harumoto, 2004).

Pigmented heterotrichs also exhibit light-sensitive behavior (Giese, 1973). Their photophobic response appears as a ciliary reversal when the intensity of incident light suddenly increases. The mechanism is likely due to a release of H^+ by the pigment. These ions are then translocated to the cytoplasm, causing electrical potential changes in the plasma membrane and subsequent ciliary reversal and reorientation of the cell (Fabczak, Fabczak, & Song, 1993a; Fabczak et al., 1993b; Matsuoka & Kotsuki, 2001; Menzies, Das, & Wood, 2004; Sobierajska, Fabczak, & Fabczak, 2006), perhaps involving a G-protein-mediated signalling pathway (Fabczak, Sobierajska, & Fabczak, 2004). Certainly, a photophobic response might be a selective advantage, keeping the ciliate hidden from potential predators. However, one cannot help but wonder which trait is under selection: the photophobic response mediated by the pigment chemicals or the toxic nature of the pigment chemicals themselves (Lobban, Hallam, Mukherjee, & Petrich, 2007).

Heterotrichs can survive several weeks without food (Jackson & Berger, 1985a, 1985b). Nevertheless, encystment is a common feature of this class, stimulated by a variety of factors, such as absence of food or excess metabolites (Giese, 1973; Repak, 1968). The cyst wall is formed of several layers, which may contain chitin (Mulisch & Hausmann, 1989). Excystment occurs through a cyst pore plug or micropyle. It may be induced by freshly bacterized medium (Giese, 1973; Repak, 1968), and is perhaps promoted by a substance liberated from excysting conspecifics (Demar-Gervais & Génermont, 1971).

Conjugating *Stentor* have been observed rarely in nature (Burchill, George, Lindberg, & Sims, 1974; Tartar, 1961). *Stentor coeruleus* mates with some frequency in the laboratory, perhaps induced by elevated temperatures (Rapport, Rapport, Berger, & Kupers, 1976), while *Blepharisma* has served as a model for understanding heterotrich sexual processes (Miyake, 1996). The marine heterotrich *Fabrea* requires a complex organic medium to complete conjugation (Demar–Gervais, 1971).

6.3 Somatic Structures

The heterotrich cell body is quite variable in shape (Fig. 6.1) depending upon whether the ciliate is a benthic or substrate oriented species or a more planktonic species. The body is covered by numerous bipolar somatic kineties composed of dikinetids (Tuffrau, 1968). There is a fine glycocalyx on top of the plasma membrane, which is underlain by an alveolar layer that is often not well-developed and appears to be discontinuous. The epiplasm is very thin and inconspicuous. Mucocysts and pigmentocysts are found beneath the alveolar layer, giving the range of "living colors" in this group – black, blue, blue-green, brown, rose, and yellow. *Chlorella* symbionts may impart a grass-green color to those species harboring them.

The somatic kinetids are typically dikinetids that lie 20–40° to the long axis of the kinety (Grain, 1984; Lynn, 1981, 1991; *Fabrea* – Da Silva Neto & Grolière, 1993; *Condylostomides* – Da Silva Neto, 1994b). The anterior kinetosome is often the only one ciliated and it bears a tangential transverse ribbon of about 6 microtubules near triplets 3, 4, and 5. This ribbon is usually doubled by a single microtubule on the right inside edge (Fig. 6.2). The posterior kinetosome is less often ciliated and may have a transverse ribbon associated with it, oriented in a variety of ways. There is a kinetodesmal fibril homologue originating near triplets 5, 6, which extends laterally, usually associating with the postciliary ribbon originating from the next anterior dikinetid. It is called a homologue because it usually does not have the obvious periodic striation found in other ciliates, although it arises from the same triplet region. The postciliary ribbon of this kinetosome is divergent and extremely well-developed, numbering 12 or more microtubules, which extend towards the cortex as ribbons oriented perpendicular to the cell surface and separated by a single microtubule (Fig. 6.2). These postciliary ribbons are accompanied by dense material called a retrodesmal fibril (Grain, 1984) or a postciliary accessory fibre (Peck, Pelvat, Bolivar, & Haller, 1975). Yogosawa-Ohara, Suzaki, and Shigenaka (1985) suggest that this fibre may induce the twisting of the body during contractions of *Spirostomum*. A number of postciliary ribbons are integrated together to form the conspicuous postciliodesma or Km fiber in the cortex of these ciliates.

Myonemes are typically present and always so in contractile forms. They are predominantly longitudinally arranged around the entire body, and are either in direct contact with the somatic kinetosomes or indirectly contact them via intermediate fibres. Transverse myonemes may integrate the longitudinal bundles, either locally or throughout the cortex. These cortical contractile systems created considerable interest among cell biologists who sought to explain their role in changing cell shape. Huang and Pitelka (1973) first experimentally demonstrated the antagonistic relationship between the myonemes and postciliodesmata in *Stentor*: the myonemes are responsible for contraction of the body while microtubule-on-microtubule sliding achieves the slow elongation back to the "relaxed" form. This system has now been demonstrated in normal contractions of *Spirostomum* (Yogosawa-Ohara & Shigenaka, 1985; Yogosawa-Ohara et al., 1985) and in the light-induced contractions of the posterior portion of the body in *Blepharisma* (Ishida, Suzaki, & Shigenaka, 1991a; Ishida Suzaki, Shigenaka, & Sugiyama, 1992). Contraction and elongation involve calcium (Ishida et al., 1992; Legrand, 1971), which may be stored as hydroxyapatite in crystalline intracellular deposits (Takagui & Silveira, 1999) or in cortical alveoli (Ishida, Shigenaka, Suzaki, & Sugiyama, 1991b). However, there are conflicting reports regarding the inhibitory action of cytochalasin, an actin antagonist, on contraction (Ettienne & Selitsky, 1974; Yogosawa-Ohara & Shigenaka). Although we cannot yet conclude that this is an actin-based system, a caltractin-like protein has been localized to the myonemes of *Stentor* (Maloney, McDaniel, Locknar, & Torlina, 2005).

The contractile vacuole system is well-developed in heterotrichs, especially the fresh-water forms, which may have conspicuous collecting canals (Patterson, 1976, 1980).

Mucocysts are probably widespread among heterotrichs, since they often encyst. However, there are relatively few studies on these. Mulisch and Hausmann (1983) documented their role in lorica construction in folliculinids. Pigmentocysts are considered to be a special type of mucocyst (Hausmann, 1978).

6.4 Oral Structures

Tuffrau (1968) presented a detailed description of the heterotrich oral region and its fibrillar supports, but this diversity has been considerably reduced with the removal of a number of groups from this class (see **Taxonomic Structure** above). The oral region is characterized by an adoral zone of polykinetids, which typically form a small dextral spiral around the cytostomal region deeper in the oral cavity. The oral polykinetids then extend out of this deeper cavity onto the cell surface of the oral region where variations in the pattern become more conspicuous: they may be organized as a linear file along the left border; may form a more or less complete circle around the anterior end, or may extend out on two wing-like projections in the folliculinids (Fig. 6.1). The peristomial region circumscribed by

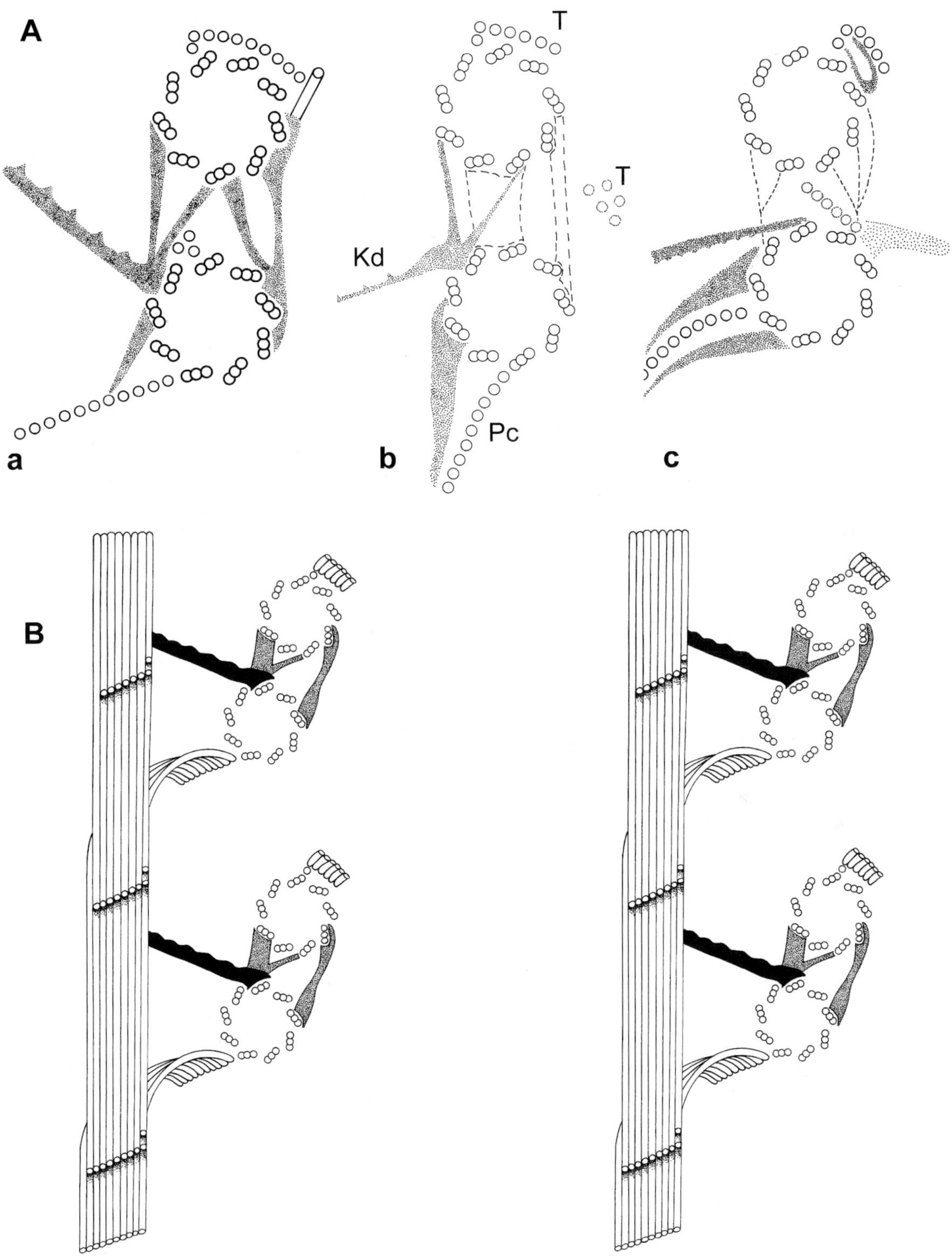

FIG. 6.2. Ultrastructure of the cortex of the Class HETEROTRICHEA. **A** Somatic dikinetids. (**a**) *Blepharisma* (after Ishida et al., 1991a). (**b**) *Climacostomum* (after Peck et al., 1975). (**c**) *Eufolliculina* (after Mulisch et al., 1981). Note the single transverse microtubule at the right end of the transverse ribbon (T) of the anterior kinetosome and the variable arrangement of transverse microtubules (T) associated with the posterior kinetosome. Kd, kinetodesmal fibril homologue; Pc, postciliary microtubular ribbon. **B** Somatic cortex of *Blepharisma* with postciliodesmata composed of overlapping ribbons in the ribbon + 1 arrangement. (Redrawn after Ishida et al., 1992.)

the oral polykinetids may be covered by somatic kineties in some forms (e.g., *Stentor*) or not in others (e.g., *Condylostoma*).

The oral polykinetids, termed paramembranelles (de Puytorac & Grain, 1976), are minimally composed of a row of dikinetids, whose anterior kinetosomes bear a transverse ribbon and posterior kinetosomes bear a divergent postciliary ribbon. Additional complete or incomplete rows of kinetosomes may be added to these first two "rows", depending upon the species and upon the position in the adoral zone at which the oral polykinetid lies. Kinetosomes on the right border of the second and third rows may also have postciliary microtubules. For example, in *Stentor*, those polykinetids furthest from the cytostome have only two to four kinetosomes in the third row while closer to the cytostome this number increases to twenty, close to the number of kinetosomes in the first two rows (Jacobson & Lynn, 1992). This is also the case for other genera (Mulisch & Hausmann, 1984; de Puytorac & Grain, 1976; Da Silva Neto, 1993b, 1994b). However, the number of kinetosomes in each oral polykinetid may be reduced as the cell size of the heterotrich decreases (Jacobson & Lynn, 1992).

It is in the structure of the paroral that most variation is seen. This is presumably because the conservation of the paroral structure is less critical to feeding function in upstream filter feeders like the heterotrichs. De Puytorac and Grain (1976) have provided definitions for a variety of terms applied to describe the structure of the heterotrich paroral (see also the **Glossary**; Mulisch & Hausmann, 1984; Da Silva Neto, 1994b). These include paroral in pairs (e.g., *Climacostomum*), stichodyad (e.g., *Blepharisma, Stentor*), stichomonad (e.g., *Spirostomum*), and polystichomonad (e.g., *Condylostoma*). Our knowledge of this diversity is increased with each new description (e.g., *Fabrea*, Da Silva Neto, 1993b; *Condylostomides*, Da Silva Neto, 1994b). It is further complicated by variations within the same species along the course of the paroral from near the cytostome to out onto the body surface (e.g., *Stentor*, Bernard & Bohatier, 1981; *Climacostomum*, Fischer-Defoy & Hausmann, 1981; *Condylostomides*, Da Silva Neto, 1994b). We conclude that it is futile to capture this diversity by establishing new terms for each new paroral structure described, and we recommend that descriptions preface the character of the paroral with the taxonomic name of the group (e.g., *Stentor* paroral).

Tuffrau (1968) diagrammed the variety of fibrillar structures that support the cell surface of the oral region and the cytopharynx. Ultrastructural studies have demonstrated these to be kinetosomal postciliary ribbons and nematodesmata that extend from the bases of the oral polykinetids and paroral kinetosomes (Grain, 1984). The nematodesmata from adjacent polykinetids join to form overlapping microtubular rootlets, which may serve a cytoskeletal function for the oral region and may also be involved in re-extension of the oral region, especially in folliculinids with their elongated peristomial wings (Mulisch & Hausmann, 1984). The cytopharynx may be supported by paroral postciliary microtubules (*Climacostomum*, Fischer-Defoy & Hausmann, 1981) or oral polykinetid postciliary microtubules (*Eufolliculina*, Mulisch & Hausmann, 1984).

Filamentous structures are often observed in the cytostomal region. As for other ciliates, Mulisch and Hausmann (1984) have proposed that these function to facilitate the pinching off of the forming food vacuole. This hypothesis has received support in experiments using cytochalasins, "anti-actin" drugs, which inhibit phagocytosis in *Spirostomum* (Zackroff & Hufnagel, 1998). Once food is ingested, food vacuoles and primary lysosomes fuse, and eventually the food vacuole fragments into smaller vesicles, which may distribute digesting materials (Fischer-Defoy & Hausmann, 1982). The defecation vacuole ultimately fuses with the plasma membrane at the cytoproct and its membrane is presumably recycled by the cell (Fischer-Defoy & Hausmann, 1992).

6.5 Division and Morphogenesis

Heterotrichs typically divide while swimming freely, and are characterized as having parakinetal stomatogenesis (Foissner, 1996b). The early studies of Fauré-Fremiet (1932) and Villeneuve-Brachon (1940) demonstrated that kinetosomal replication in one to several subequatorial kineties produced the initial anarchic field of the oral primordium. This often occurs in the zone of stripe contrast, a region where there is a marked contrast in the width of the pigment stripes or interkinetal separation (Frankel, 1989; Tartar, 1961). This anarchic field may be in the ventral region in the long axis of the body

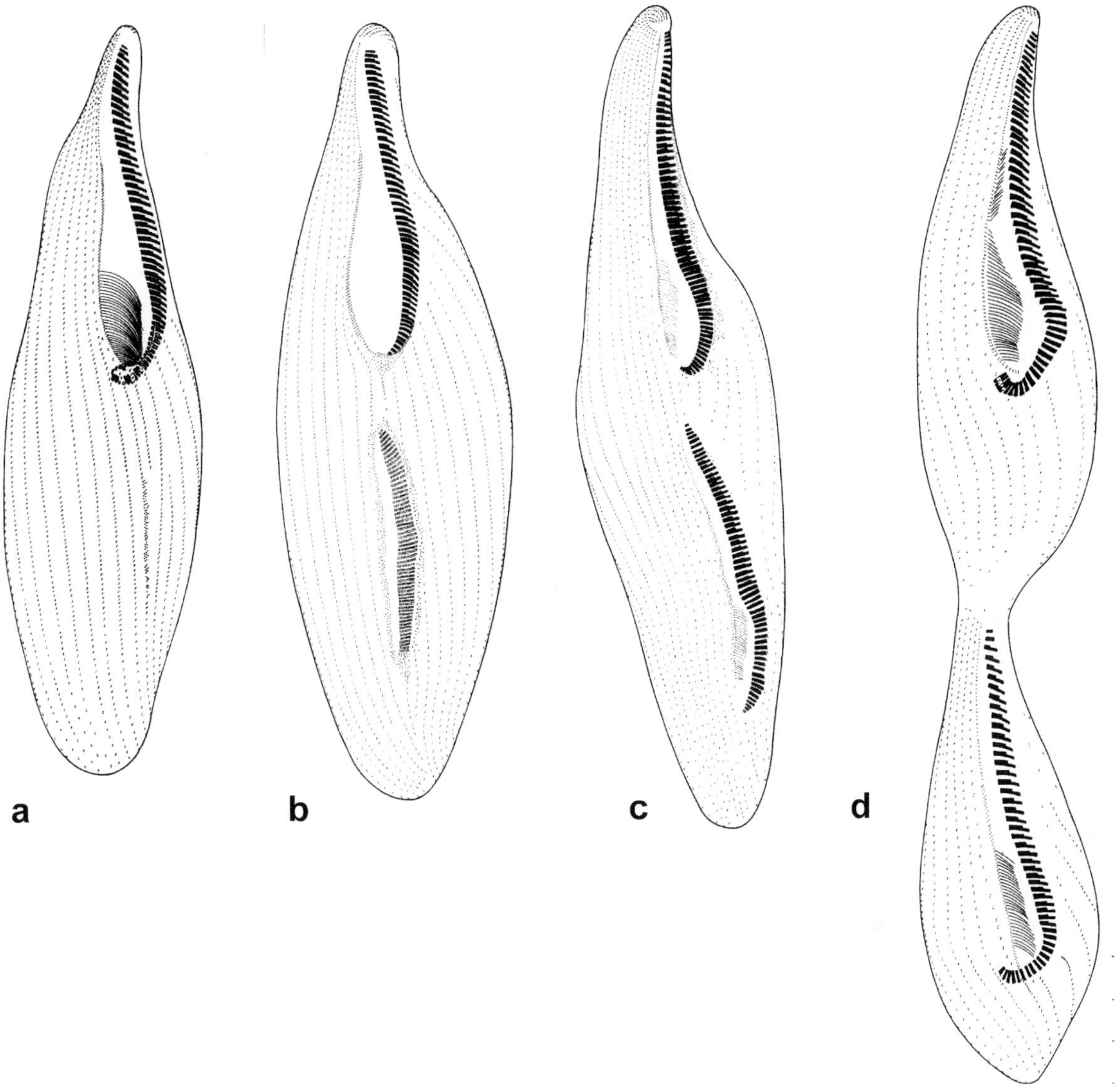

FIG. 6.3. Stomatogenesis of the heterotrich *Blepharisma*. (**a**) The process begins with proliferation of kinetosomes and dikinetids along a ventral postoral kinety. (**b**) The oral polykinetids begin to differentiate as dikinetids align beginning in the middle of the anlage and extending towards each end. (**c**) Differentiation continues towards the ends, visible at this stage by the addition of a third row to oral polykinetids initially in the middle of the anlage. The paroral begins to differentiate in the posterior right region (**d**) The paroral continues its differentiation as the adoral zone begins to curve towards and right in preparation for invagination of the opisthe's oral cavity. Note that there is some dedifferentiation and redifferentiation of the oral structures of the proter. (from Aescht & Foissner.)

posterior to the proter oral region (e.g., *Blepharisma*) (Fig. 6.3) or might even occur on the dorsal surface (e.g., *Fabrea*). Initially, the anarchic field is composed of unoriented kinetosomal pairs (Bernard & Bohatier, 1981; Mulisch & Hausmann, 1988). Eventually, the anarchic field divides longitudinally and paroral structures differentiate on its right side while adoral structures differentiate on its left side. Oral polykinetids initially differentiate in the centre of the oral primordium as dikinetids assemble from the right towards the left. Additional oral polykinetids join these central ones as the adoral field develops from the center towards the two ends of the primordium (Fig. 6.3). This latter feature is considered a strong apomorphy for the class (Aescht & Foissner, 1998; see also Shao et al., 2006).

Foissner (1996b) has described and defined the variations in heterotrich stomatogenesis: it is parakinetal when one kinety is involved, polyparakinetal when more than one kinety is involved, and amphiparakinetal when the oral primordium curves to intersect somatic kineties at both its anterior and posterior ends. These intersected somatic kineties become the peristomial field kineties in *Stentor* and *Fabrea*, but are resorbed in folliculinids.

The parental or proter oral apparatus may be only slightly dedifferentiated during cell division (e.g., *Blepharisma*, *Condylostoma*, *Stentor*) or entirely dedifferentiated and regenerated (e.g., *Spirostomum*). In folliculinids, the entire proter oral apparatus is dedifferentiated as the proter becomes the dispersal "swarmer" stage, which has a very reduced spiral of membranelles and does not form food vacuoles. Differentiation of the proter oral apparatus occurs upon settling and shows a similar pattern to the development of the opisthe oral apparatus prior to cell division (Mulisch & Hausmann, 1988).

Heterotrichs, like the karyorelicteans, have remarkable regenerative abilities that have long been exploited to probe how the development of cell pattern is regulated at the morphological (e.g., De Terra, 1985; Fahrni, 1985; Tartar, 1961; Uhlig, 1960) and biochemical levels (Bohatier, 1981, 1995), a subject area that is well beyond the scope of our treatment here, but see Frankel (1989) for a comprehensive review.

6.6 Nuclei, Sexuality and Life Cycle

The macronuclei of heterotrichs range from compact ellipsoid to ribbon-like and finally to moniliform or beaded. Correlated with their large cell size, heterotrich macronuclei are large and highly amplified. Nucleoli are often prominent. In *Stentor*, the nuclear envelopes of both macronucleus and micronuclei are surrounded by an additional membrane system that integrates them structurally to the endoplasmic reticulum (Mulisch, 1988). Since Mulisch (1988) used special fixation techniques to obtain this result, it may be a more general property of heterotrichs than is presently realized. Heterotrichs typically have many micronuclei distributed along the length of filiform macronuclei or associated singly or in groups with each bead of moniliform macronuclei.

During cell division, filiform and moniliform macronuclei become compact and nucleoli dedifferentiate. Macronuclear division is accomplished primarily by extramacronuclear microtubules (Diener et al., 1983; Jenkins, 1973). These extramacronuclear microtubules may be intimately associated with the cortex since De Terra (1983) has demonstrated cortical control over the direction of macronuclear elongation in *Stentor*. Since the majority of ciliates use intramacronuclear microtubules in macronuclear karyokinesis, Orias (1991a) has argued that extramacronuclear microtubules represent an independent evolution of the capacity to divide the macronucleus (see **Chapter 4**).

Relatively little research has been done on the molecular biology of heterotrich nuclei. The macronuclear DNA molecules or "chromosomes" are typically long, some being up to 20 μm (Hufschmid, 1983; Pelvat & De Haller, 1976). Thus, very little fragmentation of micronuclear chromosomes appears to occur, in contrast to other classes (Riley & Katz, 2001; Steinbrück, 1990). The heterotrich *Blepharisma* at least shows a deviation from the use of the universal stop codons – UAA, UGA, and UAG. Of the three, it uses at least UAA, like the spirotrich *Euplotes* (Liang & Heckmann, 1993) (see **Chapter 7**).

As noted above (see **Life History and Ecology**), conjugation may be initially stimulated by starvation conditions or changes in temperature. These conditions stimulate transcription of a gamone gene (Sugiura, Kawahara, Iio, & Harumoto, 2005), which is followed by excretion of diffusible mating type substances, called gamones (Miyake, 1996) or mating pheromones (Luporini & Miceli, 1986). All species of *Blepharisma* excrete two gamones: blepharismone, a tryptophan derivative resembling serotonin (Kubota, Tokoroyama,Tsukuda, Koyama, & Miyake, 1973; Miyake & Bleyman, 1976) common to all species; and blepharmone, a 20-kDa glycoprotein that is species specific (Braun & Miyake, 1975). Typically, the diffusion of these substances is sufficient to stimulate conjugation in receptive individuals of fresh-water species of *Blepharisma* but is apparently not sufficient in marine species (Ricci & Esposito, 1981).

There is considerable debate about the precise mechanisms that stimulate conjugation in heterotrichs at the cellular and molecular levels. Miyake (1996) favors his gamone-receptor hypothesis while Luporini and Miceli (1986) reinterpret the results from *Blepharisma* in the context of their self-recognition hypothesis. Whichever interpretation is true, the heterotrichs have not evolved a stable mating type system, as stable lines are quite rare (Demar-Gervais, 1971; Miyake & Harumoto, 1990). Luporini and Miceli (1986) argued that *Blepharisma* (and therefore heterotrichs in general?) has not yet evolved a mating type system and that it is only the formation of blepharmone-blepharismone complexes that stimulate conjugation.

Once conjugation is stimulated and micronuclear meiosis has occurred, a variable number of micronuclei enter the third or pregametic division: one in *Spirostomum* and *Blepharisma americanum*, but two or three in *Fabrea* and *Blepharisma japonicum* (Raikov, 1972). Ultimately, the products of only one of these micronuclei form the stationary and migratory gametic nuclei, which fuse to form the synkaryon. Thus, the exconjugants are genetically identical to each other although different from their parents. The synkaryon typically divides three times to produce eight products, several of which develop as macronuclear anlagen that may fuse to form the macronucleus, following separation of the conjugants (Raikov, 1972).

6.7 Other Features

Given the widespread distribution of ciliates, and heterotrichs in particular, and their large cell size and ease of cultivation, several laboratories have developed low-cost bioassays or microbiotests using *Spirostomum* species. *Spirostomum* turns out to be quite a sensitive indicator species, especially to some heavy metal contaminants (Madoni, 2000; Nalecz-Jawecki, 2004; Nalecz-Jawecki & Sawicki, 2002; Twagilimana, Bohatier, Grolière, Bonnemoy, & Sargos, 1998).

Chapter 7
Subphylum 2. INTRAMACRONUCLEATA: Class 1. SPIROTRICHEA – Ubiquitous and Morphologically Complex

Abstract The Class SPIROTRICHEA is an extremely diverse assemblage that, except for protocruziids and phacodiniids, is characterized by having replication bands pass through the macronucleus during the DNA synthesis phase. The name of the group refers to the prominent adoral zone of paramembranelles (AZM) that spirals out over the anterior end, sometimes completely enclosing it. The class is divided into seven subclasses. These ciliates are found in almost every microhabitat that ciliates can be encountered, and there are even symbiotic forms. Spirotrichs feed on a diversity of prey, from bacteria to other ciliates. They locomote typically using somatic polykinetids, called cirri, although several included groups have simpler somatic kinetids. Division morphogenesis is as diverse as the included subclasses, ranging from mixokinetal to hypoapokinetal. Spirotrichs typically have multipolar mating systems, and hold the phylum record for the number of mating types - over 100! Molecular biological research on the developing macronuclei of spirotrichs demonstrated the presence of polytene chromosomes and the ultimate differentiation of macronuclear DNA characterized by gene-sized pieces.

Keywords Protocruziidia, Phacodiniidia, Licnophoria, Hypotrichia, Oligotrichia, Choreotrichia, Stichotrichia

The SPIROTRICHEA is a diverse assemblage of ciliates that are cosmopolitan, being found in almost any habitat where ciliates are encountered even from hydrothermal vent sites at over 2,000 m deep (Small & Gross, 1985). The class name arises from the characteristically spiralling nature of the adoral zone of membranelles or adoral zone of oral polykinetids, which emerge from the oral cavity and spiral in a counter-clockwise direction as they wrap around the anterior body surface. The vast majority of species have sparse somatic ciliation, which may take the form of compound ciliary structures like cirri and bristles. Those forms with a holotrichous ciliation and somatic dikinetids (e.g., *Protocruzia*) are considered closer to the ancestral archetype, and recent molecular phylogenies bear this out (i.e., Bernhard et al., 2001; Hammerschmidt et al., 1996; Shin et al., 2000). Spirotrichs are typically medium-sized ciliates (i.e., 100-200 μm in length), although exceptionally small oligotrichs can be 5 μm in length while exceptionally large, cannibalistic stichotrichs, up to 900 μm long, may differentiate in some species (e.g., *Onychodromus*, Wicklow, 1988). Spirotrichs are behaviorally either planktonic (i.e., oligotrichs, choreotrichs) or substrate-oriented and benthic (i.e., hypotrichs, stichotrichs). Spirotrichs are predominantly free-swimming although the most distinctive group in the class, the tintinnid choreotrichs, might be considered sessile as tintinnids secrete a lorica in which they reside attached while using their oral cilia for locomotion. Some tintinnids can attach to the substrate with the lorica (e.g., Foissner & Wilbert, 1979). The loricae of more than 30 genera have been fossilized, in most cases genera with no known contemporary species. Fossils date from the Ordovician period of the Paleozoic Era, 400-500 million years ago, up to the Pleistocene Period about 1 million years ago, but the greatest diversity of tintinnid fossils occurred in the Mesozoic Era (Loeblich & Tappan, 1968; Tappan & Loeblich, 1968, 1973).

Since Corliss's observation (1979) that the literature on this group was select, it has exploded, driven by interests at two ends of the biological hierarchy – the molecular and the ecological. Stimulated both by the description using molecular techniques of DNA processing in the macronucleus of the stichotrich *Stylonychia mytilus* by Ammermann, Steinbrück, von Berger, and Hennig (1974) and by mass culturing methods in both stichotrichs (Laughlin, Henry, Phares, Long, & Olins, 1983) and hypotrichs (Roth, Lin, & Prescott, 1985; Schmidt, 1982), the developmental genetics of these ciliates, particularly *Oxytricha*, *Stylonychia*, and *Euplotes*, has become a fertile area of research (e.g., for reviews see Gall, 1986; Klobutcher & Herrick, 1997; Prescott, 1994, 1998, 2000). At the other extreme, Kofoid and Campbell (1929, 1939) had demonstrated the abundance and diversity of tintinnids in the oceanic plankton, highlighting their possible ecological significance. Pomeroy's (1974) vision of the importance of microbial organisms in ocean food webs coupled with the conception of the "microbial loop" by Azam et al. (1983) led to an explosion of interest in the ecology of planktonic ciliates, which are dominated particularly by oligotrichs and choreotrichs among which the tintinnids are placed. Our understanding of the autecology of these latter two groups, which often represent the bulk of abundance and biomass of ciliates in the plankton (e.g., for reviews see Lynn & Montagnes, 1991; Pierce & Turner, 1993; Sherr & Sherr, 1988), has also benefited from the development of laboratory culture methods (Gifford, 1985; Gold, 1973).

Finally, stichotrichs and hypotrichs have been of particular interest to biologists interested in the development of pattern in cells. Stichotrichs, like heterotrichs, have considerable regenerative powers, which have also made them useful models for developmental biologists. The literature is rich and has demonstrated among other things that pattern is controlled at both a global or cellular level to position organelles and at the organellar-organellar complex levels to construct individual "pattern units" (Frankel, 1989; Grimes, 1982; Grimes, McKenna, Goldsmith-Spoegler, & Knaupp, 1980). Gates (1978b, 1988) has shown by quantitative analysis of ventral cirral and dorsal kinetid patterns that there is an underlying morphometric theme within the genus *Euplotes*. Genetic variations of pattern have been demonstrated in both the ventral (Génermont, Demar, Fryd-Versavel, Tuffrau, & Tuffrau, 1992) and dorsal (Heckmann & Frankel, 1968) ciliature of *Euplotes* species. In the stichotrich *Paraurostyla weissei*, Jerka-Dziadosz and Dubielecka (1985) have demonstrated the genetic basis for a pattern mutant that develops multi-left marginal cirral files. Moreover, Jerka-Dziadosz (1976) has shown that numbers of oral polykinetids and marginal cirri are proportional to cell size in *P. weissei*. This research on pattern formation and variation in spirotrichs should give taxonomists cause to reflect on the significance of the characters chosen to distinguish species. Dramatic differences in the number of cirral files may be due only to a single gene mutation (Jerka-Dziadosz & Dubielecka, 1985), while differences in the absolute number of elements may merely be a function of cell size (Jerka-Dziadosz, 1976). Ultimately, the strong test for a new species will be demonstration of mating segregation. Demonstration of mating incompatibility involves considerable work, especially for the geneticists of stichotrichs and hypotrichs, whose species typically have multipolar mating-type systems (Ammermann, 1982; Caprette & Gates, 1994; Dini & Luporini, 1979; Heckmann, 1963), possibly including over 100 mating types in *Stylonychia mytilus* (Ammermann, 1982) and hiding cryptic species (Dini, Bracchi, & Gianní, 1987)!

Corliss (1979) included the heterotrichs (see **Chapter 6**) and armophorids – clevelandellids – odontostomatids (see **Chapter 8**) in the Subclass Spirotricha of the Class POLYHYMENOPHOREA. These groups are now excluded from the Class SPIROTRICHEA (**see below Taxonomic Structure**), a position supported by ultrastructural features of the somatic cortex (Grain, 1984; Lynn, 1981, 1991) and by recent molecular phylogenetic analyses (Baroin-Tourancheau, Delgado, Perasso, & Adoutte, 1992; Bernhard et al., 2001; Hirt et al., 1995; Shin et al., 2000), which have built on earlier work (Elwood, Olsen, & Sogin, 1985). Spirotrichs appear to have diverged early in the evolution of the Subphylum Intramacronucleata, possibly from a *Protocruzia*- or *Phacodinium*-like ancestor, which had multiple adoral polykinetids along the left side of the oral cavity and somatic dikinetids or simple linear somatic polykinetids (Da Silva Neto, 1993a; Didier & Dragesco, 1979; Grolière, de Puytorac, &

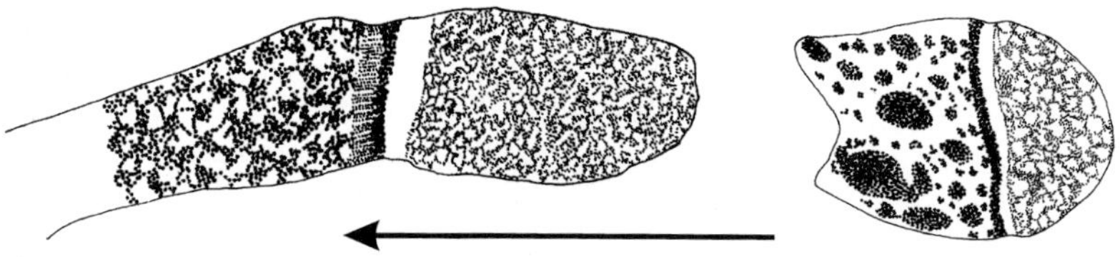

FIG. 7.1. Replication bands move from one end of the macronucleus (arrow) and are the structures responsible for macronuclear DNA synthesis in spirotrichs. These bands are characteristic of the majority of spirotrichs, such as the hypotrich *Euplotes* (left) and the oligotrich *Strombidium* (right). (Redrawn from Salvano, 1975.)

Detcheva, 1980a). It is thought that polymerization of the adoral polykinetids and their extension over the anterior body surface, accompanied by a reduction in the somatic ciliature, gave rise to the body forms exemplified by the more speciose subclasses in the class, such as the stichotrichs and tintinnid choreotrichs.

There is no strong synapomorphy uniting the ciliates assigned to this class, although they repeatedly form a robust cluster based on SSUrRNA sequences (Bernhard et al., 2001; Hammerschmidt et al., 1996; Shin et al., 2000). Three features typify the group. First, the adoral zone of membranelles is a prominent feature of the oral region, typically extending out onto the anterior end in a counter-clockwise spiral. Yet, this is also a feature of members of the Class HETEROTRICHEA (see **Chapter 6**) and of some colpodeans (see **Chapter 12**). Second, macronuclear DNA is replicated in a single replication band that passes from one end of smaller macronuclei to the other end of the nucleus or by two replication bands proceeding from the ends to the middle in more elongate macronuclei (Fig. 7.1) (Raikov, 1982). Nevertheless, members of the Class PHYLLOPHARYNGEA have been reported to have a type of "replication band" although DNA synthesis has not yet been demonstrated in it and its morphological substructure is not similar to that of the spirotrichs (Raikov, 1982). On the other hand, replication bands have not been observed in *Protocruzia* (Ammermann, 1968; Ruthmann & Hauser, 1974) and *Phacodinium* (Fernández-Galiano & Calvo, 1992; Da Silva Neto, 1993a), which are both members of the "molecular spirotrich clade" (Fig. 7.2). Thirdly, somatic ciliation tends to be reduced in all species but those assumed to represent an ancestral type, which include, as examples, the holotrichously ciliated *Protocruzia* (Subclass Protocruziidia), *Phacodinium* (Subclass Phacodiniidia), *Plagiotoma* (Subclass Stichotrichia), *Kiitricha* (Subclass Hypotrichia), and *Strombidinopsis* (Subclass Choreotrichia). A molecular or cell biological trait may ultimately be found as a synapomorphy for this clade, but we currently still seek a strong synapomorphy for the class.

7.1 Taxonomic Structure

Corliss (1979) placed the spirotrichs as the Subclass Spirotricha in the Class POLYHYMENOPHORA because of their adoral zone of multiple membranelles. He recognized four orders within the class: (1) Order Heterotrichida; (2) Order Odontostomatida; (3) Order Hypotrichida; and (4) Order Oligotrichida. We have discussed above (see **Chapter 6**) the reasons for removal of the heterotrichs from this assemblage, based on the structure of the somatic kinetid and its postciliodesma (Lynn, 1981, 1991), the absence of replication bands in the macronuclei, and the use of extramacronuclear microtubules during macronuclear division (Lynn, 1996a). Schrenk and Bardele (1991) compared the somatic kinetid of the odontostomatid *Saprodinium* to that of the armophorid *Metopus*. Although the kinetid similarities between these two major groups are not strong, their kinetids differ from those of the hypotrichs, stichotrichs, oligotrichs, *Protocruzia*, and *Phacodinium*. We now know that at least one odontostomatid, *Epalxella*, has affinities at the molecular level with plagiopylids (Stoeck, Foissner, & Lynn, 2007), and so we transfer the odontostomatids out of the spirotrichs and place them with the plagiopylids (see **Chapter 14**). Lynn and Strüder-Kypke (2002) have confirmed that *Licnophora* is a spirotrich, and this genus now establishes the type of a new spirotrich subclass.

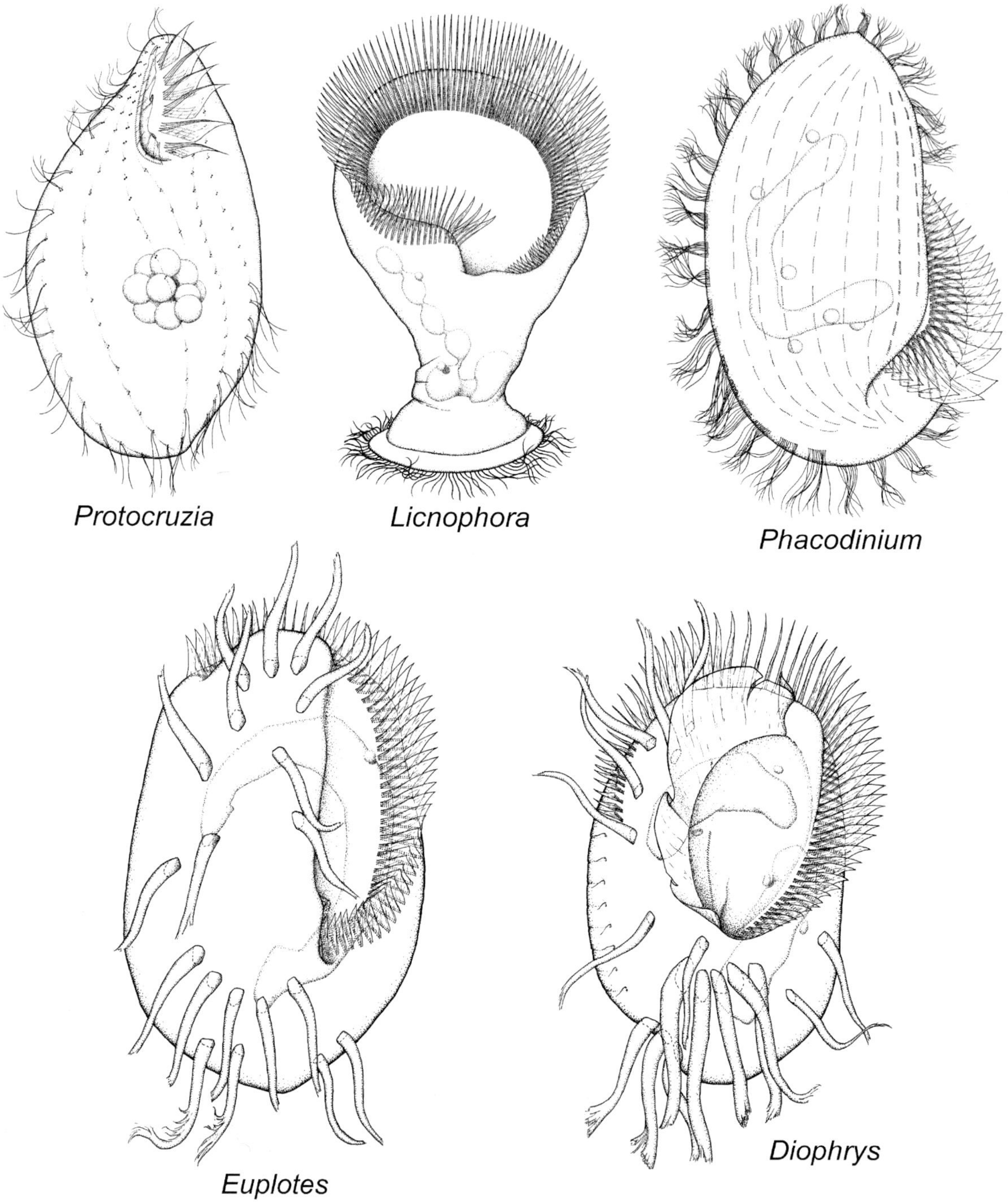

FIG. 7.2. Stylized drawings of representative genera from subclasses in the Class SPIROTRICHEA. Subclass Protocruziidia: *Protocruzia*. Subclass Licnophoria: *Licnophora*. Subclass Phacodiniidia, *Phacodinium*. Subclass Hypotrichia: *Euplotes*; *Diophrys*

We have elevated the Orders Oligotrichida and Hypotrichida to Subclass rank, following Lynn and Small (1997) and now recognize seven subclasses: (1) Subclass Protocruziidia; (2) Subclass Phacodiniidia; (3) Subclass Licnophoria: (4) Subclass Hypotrichia; (5) Subclass Oligotrichia;

(6) Subclass Choreotrichia; and (7) Subclass Stichotrichia. Corliss (1979) removed *Protocruzia* from the Family Spirostomidae and placed it incertae sedis in the Suborder Philasterina (Fig. 7.2). Small and Lynn (1985) recognized the Family Protocruziidae Jankowski, 1980 and established this as the type family for the Order Protocruziida Jankowski, 1980. De Puytorac, Grain, and Mignot (1987) established the Subclass Protocruziidia, which we recognize herein (see **Chapter 17**). As noted above, SSUrRNA gene sequeness clearly relate *Protocruzia* to the spirotrichs, although the levels of statistical support are often not strong (Hammerschmidt et al., 1996; Shin et al., 2000). Should it be recognized as type for a new monotypic class?

Corliss (1979) established the Family Phacodiniidae, placing the genera *Phacodinium* and *Transitella* in it (Fig. 7.2). As noted above, SSUrRNA gene sequences clearly relate *Phacodinium* to the spirotrichs (Bernhard et al., 2001; Shin et al., 2000). Fernández-Galiano and Calvo (1992) noted that *Phacodinium* can be related to the hypotrichs by the following: its dorsoventral differentiation; some somatic polykinetids with two rows of kinetosomes, which they called pseudocirri; and the polykinetid-like nature of its paroral. Small and Lynn (1985) established the Order Phacodiniida with the type family Phacodiniidae. Following the recommendation of Shin et al. (2000), we elevate the order to subclass rank, attributing authorship to Small and Lynn.

As noted above, Corliss (1979) also placed *Transitella* in the Family Phacodiniidae. However, Iftode, Fryd-Versavel, Wicklow, and Tuffrau (1983) drew attention to the structural differences in the somatic polykinetids, the development of oral nematodesmata forming a basket-like cytopharyngeal structure, and the presence of a paroral composed of two parallel files of kinetosomes to substantiate their establishment of the Family Transitellidae (Fryd-Versavel & Tuffrau, 1978). However, here we follow Foissner, Adam, and Foissner (1982) who argued that the genus *Transitella* did not differ substantially from the genus *Balantidioides*. They placed *Transitella* as a junior synonym to *Balantidioides*, and placed the Family Transitellidae as a junior synonym of the Family Reichenowellidae, in which they included *Reichenowella* and *Balantidioides*. We place the Family Reichenowellidae incertae sedis in the Subclass Hypotrichia and await molecular evidence to refine its taxonomic position.

Corliss (1979) placed the dorsoventrally flattened spirotrichs with prominent ventral cirri and less conspicuous dorsal bristle cilia in the Order Hypotrichida. Characters that support the monophyly of this group include the pattern of development of the ventral cirri and the overall organization of the body plan, which are claimed to be too similar to have evolved convergently (Fleury, 1988; Martin, 1982; Tuffrau, 1987). At the same time, there is a recognition of considerable diversity within the group, both in respect to morphology, morphogenesis, and genetics. Fleury and coworkers (Fleury, 1988; Fleury, Iftode, Deroux, & Fryd-Versavel 1985a; Fleury, Iftode, Deroux, Fryd-Versavel, & Génermont, 1985b; Fleury, Iftode, Deroux, & Fryd-Versavel, 1986) have used these morphological and morphogenetic differences to establish the suborder Euhypotrichina (Fleury et al., 1985a) to include stichotrich-like forms and the suborder Pseudohypotrichina (Fleury et al., 1985b) to include euplotid-like forms. Euhypotrichs were so-named because they manifested truly derived "hypotrich" characters: (1) the complete turn-over or replacement of the somatic ciliature, both ventral and dorsal, during all morphogenetic processes; (2) cysts that typically resorb all infraciliary constituents (i.e., kinetosome-resorbing cysts); and (3) dorsal bristles that lose kinetodesmal fibrils at the completion of development. In contrast, pseudohypotrichs were so-named because they appeared "hypotrich"-like but lacked the derived characters noted above. Instead, they exhibit: (1) turn-over or replacement of **only** the ventral somatic infraciliature during morphogenetic processes; (2) encystment typically **not** accompanied by resorption of all kinetosomes; and (3) mature dorsal bristles that **retain** kinetodesmal fibrils. Small and Lynn (1981, 1985) had used these kinetid differences, along with differences in the organization of the somatic cortex, to separate the hypotrichs into two clades: the Subclass Stichotrichia Small and Lynn, 1985, which is equivalent to the euhypotrichs (Fleury et al., 1985b) and the Oxytrichia (Tuffrau & Fleury, 1994) of others; and the Subclass Hypotrichia Stein, 1859, which is equivalent to the pseudohypotrichs (Fleury et al., 1985b) and the Euplotia (Tuffrau & Fleury, 1994). Small and Lynn (1985) placed the Subclass Hypotrichia within the Class NASSOPHOREA, based on similarities in the dikinetid of members of these two groups. However, molecular evidence (Lynn & Sogin, 1988) clearly refuted this relationship, demonstrating instead that stichotrichs and hypotrichs belong to the same major clade, a fact that has led Lynn and Corliss (1991) and Lynn and Small (1997) to place both groups in the Class SPIROTRICHEA.

The stichotrichs and hypotrichs are separated deeply on small subunit ribosomal RNA (SSUrRNA) gene trees with oligotrich and choreotrich genera diverging **after** the separation of the hypotrich lineage (Bernhard et al., 2001; Chen & Song, 2001; Petroni, Dini, Verni, & Rosati, 2002; Snoeyenbos-West, Salcedo, McManus, & Katz, 2002; Strüder-Kypke et al., 2003). Moreover, there are substantial and deep differences in the allele frequencies of the isoenzymes of hypotrichs and stichotrichs (Schlegel & Steinbrück, 1986). We have favored retaining separate subclasses for these four groups notwithstanding the fact that the apparently fast molecular clock of the hypotrich ribosomal RNA genes may be creating a "treeing artifact" for these molecular sequences due to the long branch attraction artifact (Felsenstein, 1978; Morin, 2000; Philippe, Chenuil, & Adoutte, 1994).

Molecular phylogenetic studies on a number of hypotrich genera, such as *Euplotes* and *Diophrys*, consistently demonstrate them as basal lineages in the spirotrich radiation (Fig. 7.2). Depending upon taxon sampling and variations in sequence alignment, the hypotrichs appear to be monophyletic (Li & Song, 2006; Petroni et al., 2002) or paraphyletic (Chen & Song, 2001, 2002; Song, Wilbert, Chen, & Shi, 2004). Subdivision of the genus *Euplotes* is supported by some molecular analyses (Bernhard et al., 2001; Borror & Hill, 1995). We now recognize two orders in the Subclass Hypotrichia: (1) Order Kiitrichida to include spirotrichs with uniformly small and multiple cirri arranged in curving files; and (2) Order Euplotida to include those with well-developed and sparse ventral cirri arranged in frontal, ventral, and transverse groups (see **Chapter 17**).

Features of division morphogenesis have also been used to reevaluate systematic relationships among oligotrich ciliates and these have now been tested by sequences of the small subunit ribosomal RNA genes. Oligotrich and choreotrich SSUrRNA gene sequences support retention and separation of genera assigned to these two subclasses (Snoeyenbos-West et al., 2002; Strüder-Kypke et al., 2003). Corliss (1979) placed the Order Oligotrichida in the Class POLYHYMENOPHOREA, noting that its two suborders, the Oligotrichina and Tintinnina, were united by a reduced somatic ciliature and an anterior adoral zone of membranelles, which are primarily used in locomotion and feeding. Fauré-Fremiet (1970) had already drawn attention to the diversity in the oral structures of the oligotrichs, recognizing strombidiids, like *Limnostrombidium* and *Laboea* (Fig. 7.3) and halteriids, like *Halteria* (Fig. 7.4), to have an "open" adoral zone with what has been called a "collar" and "lapel" arrangement of the oral polykinetids. In contrast, there were the strombidinopsids, like *Strombidinopsis*, strobilidiids, like *Strobilidium*, and tintinnids, like *Codonella* and *Tintinnopsis*, with a "closed circle" of oral polykinetids (Fig. 7.3).

Fauré-Fremiet (1970) retained the classical separation of the tintinnids, based on their being loricate, even while noting the similarity in the "closed circle" oral structures of strobilidiids and tintinnids. Small and Lynn (1985) established the Subclass Choreotrichia Small and Lynn, 1985 to include these latter two groups, placing the aloricate choreotrichs in the Order Choreotrichida Small and Lynn, 1985 and retaining the Order Tintinnida for loricate choreotrichs. Division morphogenesis of oligotrichs was initially described by Fauré-Fremiet (1953), who described their division as enantiotropic and noted that the oral primordium of *Strombidium* developed in an invagination. Deroux (1974) confirmed that oral development occurred in a pocket in *Strombidium sulcatum*, provided a new description for stomatogenesis in *Strobilidium gyrans* (= *Strobilidium caudatum*?), and drew attention to the similarity with the early stages of division morphogenesis in *Euplotes*, which also occurs in a cortical invagination (Wise, 1965). Petz and Foissner (1992) confirmed this "pocket" oral development in *S. caudatum* and also observed it in a *Tintinnidium* species while Dale and Lynn (1998) observed the same pattern of "pocket" oral development in the aloricate choreotrich *Strombidinopsis*. Petz and Foissner (1992) used this as an additional character to support the monophyly of the choreotrichs. *Strombidium*-like species may begin development of

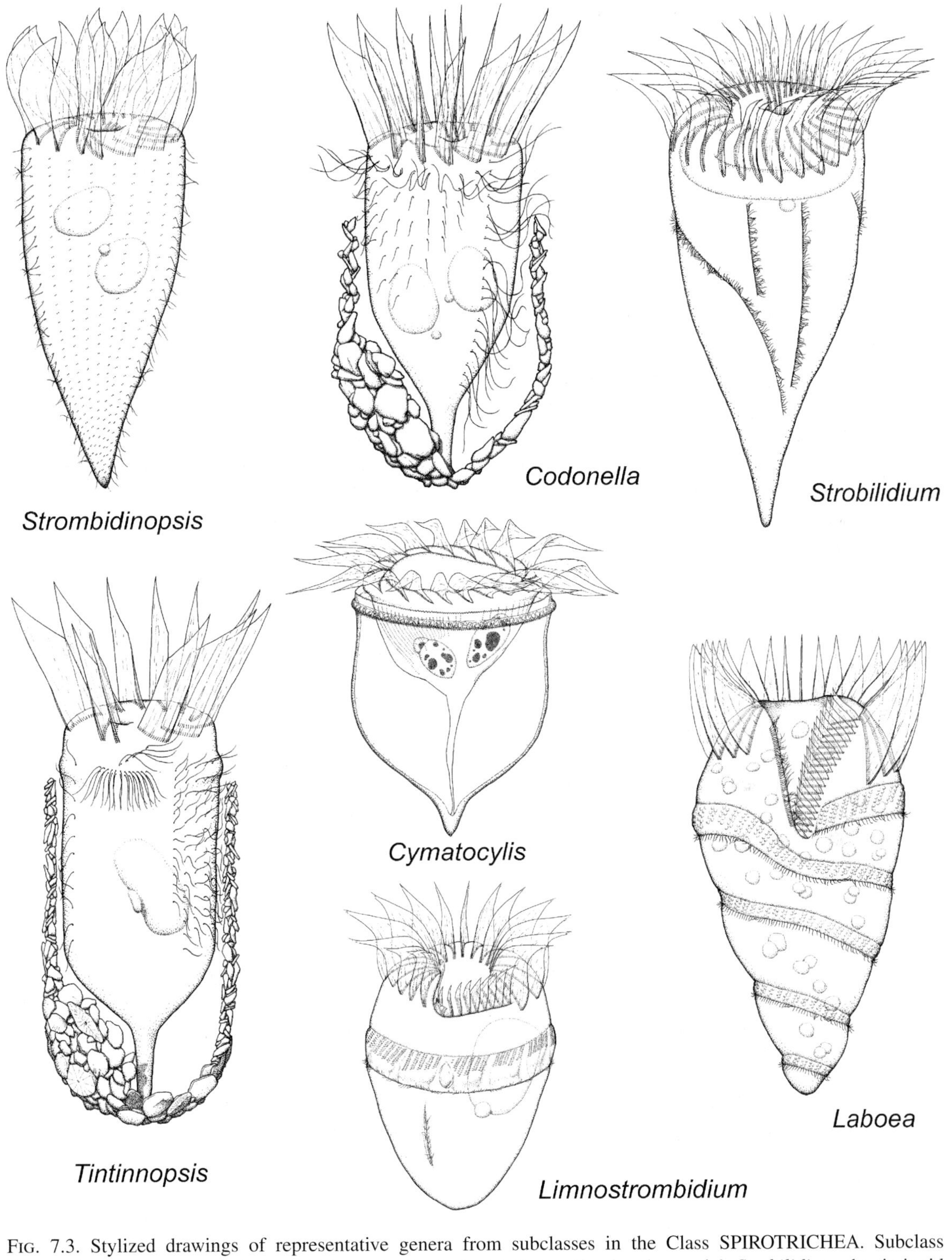

FIG. 7.3. Stylized drawings of representative genera from subclasses in the Class SPIROTRICHEA. Subclass Choreotrichia: the choreotrich *Strombidinopsis*; the tintinnid *Codonella*; the choreotrich *Strobilidium*; the tintinnid *Tintinnopsis*; the tintinnid *Cymatocylis*. Subclass Oligotrichia: *Limnostrombidium*; *Laboea*

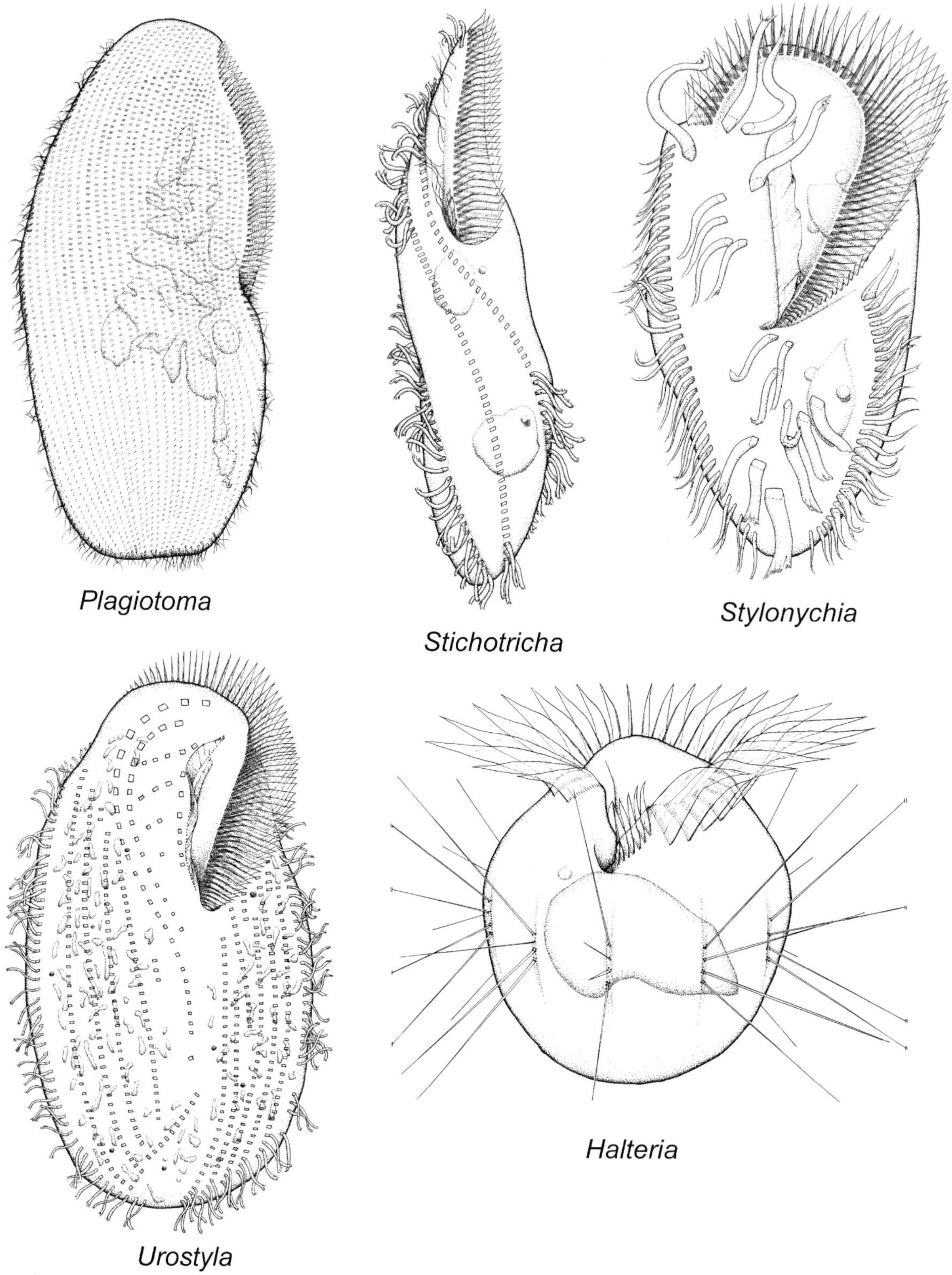

FIG. 7.4. Stylized drawings of representative genera from subclasses in the Class SPIROTRICHEA. Subclass Stichotrichia: *Plagiotoma*, *Stichotricha*, *Stylonychia*, *Urostyla*, and *Halteria*, formerly an oligotrich (compare to *Strombidium* and *Laboea*). Note that the bristles of *Halteria* have been shortened to accommodate the space on the page

the oral primordium in an intracellular tube or neoformation organelle (Petz & Foissner, 1992) or on the cell surface (Song & Wang, 1996). Oligotrichs, such as *Cyrtostrombidium*, *Strombidium*, *Laboea*, and *Tontonia*, demonstrate a considerable diversity of somatic ciliary patterns. Agatha (2004a) has provided a detailed analysis of these patterns and used this to justify establishment of several new families and genera (see **Chapter 17**).

Oral anlagen development is epiapokinetal, that is it occurs on the cell surface of the halteriids *Halteria* (Petz & Foissner, 1992; Song, 1993) and *Meseres* (Petz & Foissner, 1992), and these two "classical oligotrich" genera undergo a complete turn-over or replacement of somatic ciliature during division (Agatha, 2004b; Petz & Foissner, 1992; Song, 1993). Consistently, the "classical oligotrich" *Halteria* falls among the stichotrich clade, based on the SSUrRNA gene (Foissner et al., 2004; Snoeyenbos-West et al., 2002; Strüder-Kypke et al., 2002), ITS1 and ITS2, 5.8S rRNA, and the large subunit rRNA gene (Hewitt et al., 2003), DNA polymerase α (Hoffman & Prescott, 1997), and actin (Croft et al., 2003). These morphogenetic and molecular features support their affinities to the stichotrichs. Agatha (2004b) argued that halteriids are related to strombidiids (= oligotrichs) *sensu lato* on two main synapomorphies: (1) the enantiotropic division mode; and (2) the *de novo* origin of the paroral. We suggest that both these features are strongly convergent: the enantiotropic division mode is found in a form in the distantly-related prostome *Balanion* (Foissner, Oleksiv, & Müller, 1990), while many taxa may differentiate the paroral *de novo*. Instead, we would emphasize the importance of epiapokinetal stomatogenesis, the complete turnover of the somatic ciliature, and the molecular affinities, all of which relate halteriids to the stichotrichs. Thus, the proposed oligotrich taxonomies (e.g., Agatha, 2004b; Laval-Peuto, Grain, & Deroux, 1994; Lynn & Small, 1997; Petz & Foissner) are refuted: the Family Halteriidae must be transferred to the Subclass Stichotrichia (see **Chapter 17**). Halteriids can be considered as stichotrichs, highly adapted to the planktonic habitat. We only recognize the Order Strombidiida in the Subclass Oligotrichia (see **Chapter 17**).

The tintinnids deserve special mention because of their long-standing independent status as a group, their conspicuousness in the marine plankton, and the exceedingly large number of described species, well over 1,200. Stomatogenetic (Dale & Lynn, 1998; Petz & Foissner, 1992) and molecular features (Strüder-Kypke et al., 2002; Snoeyenbos-West et al., 2002) place them in the Subclass Choreotrichia alongside their aloricate relatives. In addition to the lorica, they are distinguished by a number of peculiar features: tentaculoids, a contractile body, a lateral cytoplasmic lobe apparently used in lorica construction, short somatic cilia arranged in characteristic patterns dependent upon species, and a perilemma surrounding the entire body and ciliature. Doubts have been cast on the taxonomic utility of lorica morphology, both since the description by Laval-Peuto (1977) of a *Favella* species making a *Coxliella*-like lorica and by quantitative analyses of lorica variation that conclude it is not feasible to objectively distinguish many species based on lorica form (Davis, 1981). Despite the recent success in automatic categorization by lorica form of five species of *Cymatocylis* by an artificial neural network (Culverhouse et al., 1994) and linear discriminant analysis (Williams, McCall, Pierce, & Turner, 1994), we support Laval-Peuto and Brownlee (1986) who recommended a systematic approach based on cytology as revealed by protargol staining. Only a handful of species have been stained so far, but clear somatic kinetid patterns are emerging, such as the presence of longer dorsal and ventral kineties that separate left and right fields of shorter kineties (e.g., Agatha & Riedel-Lorjé, 2006; Agatha & Strüder-Kypke, 2007; Choi, Coats, Brownlee, & Small, 1992; Foissner & Wilbert, 1979; Laval-Peuto & Brownlee, 1986). Nevertheless, lorica form has been the major diagnostic feature for the only new family of tintinnids established since 1979 (Sniezek, Capriulo, Small, & Russo, 1991; Snyder & Brownlee, 1991)! Because of the lack of comparative data on kinetid patterns, we believe it is premature to redistribute tintinnid genera among families. In the Subclass Choreotrichia, we therefore continue to recognize the loricate Order Tintinnida with its classical "loricate" families while genera in the aloricate Order Choreotrichida are divided into four monotypic orders based primarily on variations in somatic kinetid patterning (see **Chapter 17**).

Finally, researchers continue to explore relationships within the stichotrichs. The exceedingly complex ventral cirral patterns of stichotrichs have lead taxonomists to use features of division morphogenesis as a means to resolve relationships among taxa. This approach is premised on the conservative nature of developmental patterns as argued by Fauré-Fremiet (1948a) and Corliss (1961, 1967, 1968, 1979). There is a rich literature using these morphogenetic patterns to resolve relationships within the stichotrichs (e.g., Berger & Foissner, 1997; Borror, 1979; Borror & Hill, 1995; Eigner, 1997, 1999, 2001; Fleury et al., 1985a, 1985b, 1986; Martin, 1982; Wicklow, 1982). Other researchers have primarily used the SSUrRNA genes. While the Subclass Stichotrichia appears to be monophyletic and now includes the halteriids (**see above**), it is still quite difficult to reconcile morphological and molecular approaches at the family and genus levels, although current molecular evidence at least supports a clade of *Stylonychia*-related species (Bernhard et al., 2001; Berger & Foissner, 1997; Foissner et al., 2004; Hewitt et al., 2003; Schmidt, Bernhard, Schlegel, & Foissner, 2007), suggested already in a cladistic analysis of morphological traits (Berger & Foissner, 1997). We have done our best to reconcile these data, but the effort is obviously unfinished as there is a considerable amount of convergence in morphological traits (see Wiackowski, 1988). We have remained conservative in our taxonomic treatment of this subclass and recognize three orders: (1) Order Stichotrichida, including genera such as *Plagiotoma* and *Stichotricha*, whose cirri are arranged, often in many, linear files; (2) Order Sporadotrichida, such as *Stylonychia*, in which cirri are distributed "sporadically" in conspicuous frontal, ventral, and transverse groupings; and (3) Order Urostylida, such as *Urostyla*, in which the frontoventral cirri are arranged in two or more zig-zag files on the ventral surface (Fig. 7.4) (see **Chapter 17**).

In conclusion, we recognize seven subclasses in the Class SPIROTRICHEA: (1) Subclass Protocruziidia; (2) Subclass Phacodiniidia; (3) Subclass Licnophoria; (4) Subclass Hypotrichia; (5) Subclass Oligotrichia; (6) Subclass Choreotrichia; and (7) Subclass Stichotrichia. Molecular phylogenetics suggests that the taxa are ordered in this manner with *Protocruzia* at the base of the spirotrich lineage and various stichotrichs, like *Stylonychia* and *Sterkiella*, at the tip (Bernhard et al., 2001; Snoeyenbos-West et al., 2002; Strüder-Kypke et al., 2002). The possession of macronuclear replication bands unites the latter five subclasses while a dorsoventral differentiation unites phacodiniids with hypotrichs and stichotrichs. Since it is likely that halteriids have evolved a planktonic life style from benthic stichotrich ancestors like *Oxytricha*, it is very likely that oligotrichs and choreotrichs are also derived from benthic dorsoventrally-differentiated ancestors. Perhaps future phylogenetic analyses based on genes other than the SSUrRNA gene will shed light on these unresolved phylogenetic relationships.

Applications of molecular methods have enabled the identification of a number of spirotrichs species. Ammermann, Schlegel, and Hellmer (1989) distinguished the sibling species *Stylonychia mytilus* and *Stylonychia lemnae* using isozymes of isocitrate dehydrogenase. Haentzsch et al. (2006) have now used a PCR method based on this gene to clearly identify these sibling species, while Schmidt, Bernhard, Schlegel, and Fried (2006b) have used fluorescence *in situ* hybridization (FISH) of SSUrDNA probes for the same purpose. Others have employed either random amplified polymorphic DNA (RAPD) fingerprinting or restriction fragment length polymorphisms (RFLPs) to distinguish species and populations of hypotrichs, such as *Euplotes* (Chen, Song, & Warren, 2001; Kusch & Heckmann, 1996; Shang, Chen, & Song, 2002) and *Uronychia* (Chen, Song, & Warren, 2003).

Finally, monographic works have recently appeared on the spirotrichs. These include monographs on the hypotrichs and stichotrichs by Berger (1999, 2001, 2006a, 2006b) and tintinnids (Alder, 1999).

7.2 Life History and Ecology

As noted above, spirotrichs are broadly distributed, both as symbionts and free-living forms. There is not a large number of symbiotic spirotrichs, but what species there are colonize a diverse array of hosts. For example, *Licnophora* is found in holothurian echinoderms (Stevens, 1901), on the eyes of scallops (Harry, 1980), on cyanobacterial filaments (Fauré-Fremiet, 1937), and in the mantle cavity of limpets (Van As, Van As, & Basson, 1999); *Kerona polyporum* is found on the body of *Hydra* (Hemberger & Wilbert, 1982); *Trachelostyla tani* is found in the mantle cavity of the scallop *Chlamys* (Hu & Song, 2002); *Strombidium* is found associated with echinoid echinoderms (Jankowski, 1974b; Xu, Song, Sun, & Chen, 2006); *Euplotes tuffraui* is also associated with echinoid echinoderms (Berger, 1965); *Euplotaspis* associates with the chambers of ascidian tunicates (Chatton & Séguéla, 1936); and *Plagiotoma* is an obligate endosymbiont in the digestive tract of oligochaete annelids (Albaret, 1973; Mandal & Nair, 1975).

Free-living hypotrichs and stichotrichs occupy a wide range of habitats from freshwater to brackish to marine, in sands and soils, and edaphic habitats. They are typically substrate-oriented and benthic, and are particularly prominent in soil ecosystems from Europe to Antarctica (Berger & Foissner, 1987, 1989a; Blatterer & Foissner, 1988; Foissner, 1980a, 1982; Petz & Foissner, 1997) where they may represent over 30% of the species richness and may attain abundances of up to $1{,}000\,g^{-1}$ of dry soil. Hypotrichs and stichotrichs have also been recorded in aquatic habitats from arctic and antarctic ice (Agatha, Spindler, & Wilbert, 1993; Agatha, Wilbert, Spindler, & Elbrächter, 1990; Garrison et al., 2005), in benthic communities in the deep Mediterranean Sea (Hausmann, Hülsmann, Polianski, Schade, & Weitere, 2002), in freshwater ponds in Great Britain (Finlay & Maberly, 2000), in streams and rivers in Europe (Cleven, 2004; Foissner, 1997b), in tropical littoral habitats (Al-Rasheid, 1999a; Dragesco & Dragesco-Kernéïs, 1986), and associated with leaf litter in mangrove forests in India (Dorothy, Satyanarayana, Kalavati, Raman, & Dehairs, 2003). Isolates from different regions have demonstrated adaptations to those environments: Antarctic *Euplotes* species show growth rate optima at lower temperatures than temperate and tropical isolates (Lee & Fenchel, 1972) as do temperate isolates of the halteriid *Meseres* compared to tropical and subtropical isolates (Gachter & Weisse, 2006). These cold adaptations arise in some *Euplotes* species through gene mutations in protein-coding genes that increase the molecular flexibility of these proteins at low temperatures (Pucciarelli, Marziale, Di Giuseppe, Barchetta, & Miceli, 2005).

Oligotrichs and choreotrichs have been extensively studied in recent years in both freshwater (e.g., Beaver & Crisman, 1989a; Foissner, Berger, & Schaumburg, 1999; Obolkina, 2006; Sonntag, Posch, Klammer, Teubner, & Psenner, 2006; Zingel, Huitu, Makela, & Arvola, 2002) and brackish water lakes (Pfister, Auer, & Arndt, 2002) and marine (e.g., Lynn & Montagnes, 1991; Pierce & Turner, 1992; Sanders & Wickham, 1993) ecosystems, where they can be found in both neritic and eupelagic regions. For marine species, the vast majority of studies are not unexpectedly localized near sites of marine laboratories (see Lynn & Montagnes, 1991). Tintinnid choreotrichs have been quite extensively sampled over all the world's oceans, ranging from the arctic to the antarctic (Pierce & Turner, 1993). Some taxa show unusual biogeographic distributions, but these must be interpreted with some caution since tintinnids can survive in ballast water and be distributed globally by humans (Pierce, Carlton, Carlton, & Geller, 1997). Marine oligotrichs are equally broadly distributed with recent reports from northern (Dale & Dahl, 1987a; Stelfox-Widdicombe, Archer, Burkill, & Stefels, 2004) and southern oceans (Pettigrosso, Barria de Cao, & Popovich, 1997) to the tropics (Al-Rasheid, 1999; Lynn, Roff, & Hopcroft, 1991a) and Antarctica (Garrison et al., 2005; Leakey, Fenton, & Clarke, 1994; Petz, 1994).

Oligotrichs and choreotrichs are now considered to play an important role in microbial food webs (e.g., Beaver & Crisman, 1989b; Pierce & Turner, 1993; Sanders & Wickham, 1993; Weisse, 2006). Aloricate oligotrichs (e.g., *Strombidium*, *Tontonia*, *Laboea*) and choreotrichs (e.g., *Strobilidium*, *Strombidinopsis*) tend to dominate the abundance, ranging from tens to thousands per liter. Mean annual abundances are typically $1{,}000\,l^{-1}$ (Lynn, Roff, & Hopcroft, 1991a; Montagnes, Lynn, Roff, & Taylor, 1988; Moritz, Montagnes, Carleton, Wilson, & McKinnon, 2006), but exceptionally high abundances of over $10^6\ l^{-1}$ have been recorded (e.g., Dale & Dahl, 1987a). Tintinnid choreotrichs typically comprise a smaller fraction of the total ciliate abundance, ranging up to 25% of the abundance of their aloricate relatives (e.g., Abboud-Abi Saab, 1989; Gilron, Lynn, & Roff, 1991; Lynn et al., 1991a). Because of their loricae, which are easily preserved, there is an extensive literature on the ecology of tintinnids. Tintinnid abundances average in the 10s l^{-1} in places as geographically distant as the Mediterranean (Abboud-Abi Saab; Cariou, Dolan, & Dallot, 1999; Krsinic & Grbec, 2006), Greenland Sea (Boltovsky, Vivequin, & Swanberg, 1995), Barents Sea (Boltovsky, Vivequin, & Swanberg, 1991), and off the cost of Japan (Dohi, 1982).

However, abundances of $500 l^{-1}$ (Cariou et al., 1999) may occur seasonally, with particularly high abundances, to over $10{,}000 l^{-1}$, being associated with estuarine and near shore locations, for examples, in North America (Capriulo & Carpenter, 1983; Verity, 1987), South America (Barría de Cao, 1992), and China (Zhang & Wang, 2000). Variations have been reported in lorica dimensions of tintinnids, correlated with the season (e.g., Gold & Morales, 1975) and water temperature (e.g., Boltovsky et al., 1995). Variation in community lorica oral diameter has been correlated with the average size of available prey, suggesting some kind of trophic specialization of tintinnids on their prey (Middlebrook, Emerson, Roff, & Lynn, 1987; Verity). However, Dolan (2000) concluded that trophic specialization, at least, is not the dominant factor determining tintinnid diversity in the Mediterranean Sea. Tintinnid diversity has been used to indicate different water masses in the North Pacific Ocean (Kato & Taniguchi, 1993).

Spirotrichs are considered to be upstream filter feeders that "select" particles primarily on the basis of the structural nature of the oral apparatus (Fenchel, 1980a, 1980b; Jonsson, 1986; Wilks & Sleigh, 1998). These theoretical predictions using various beads have been corroborated by feeding experiments on natural prey, which show a positive relation between size of the ciliate and the average size of the prey that it can efficiently ingest (Bernard & Rassoulzadegan, 1990; Kamiyama & Arima, 2001). Benthic or substrate-oriented spirotrichs, like the hypotrich *Euplotes* and some substrate-oriented choreotrichs like *Strobilidium*, can likely survive on bacterivory in the wild (Lawrence & Snyder, 1998; Sime-Ngando, Demers, & Juniper, 1999). It is unlikely that pelagic oligotrichs and choreotrichs will achieve maximal growth rates by bacterivory since bacterial abundances are often less than the critical concentration of 10^6 ml^{-1} necessary to support growth (Bernard & Rassoulzadegan; Fenchel, 1980c; Macek, Šimek, Pernthaler, Vyhnálek, & Psenner, 1996; Šimek, Macek, Pernthaler, Straškrabová, & Psenner, 1996). Nevertheless, behavioral modification of the swimming pattern may enable some oligotrichs to exploit food patches of bacteria exceeding these minimum abundances (Fenchel & Jonsson, 1988) and enable some tintinnids to remain in patches of dinoflagellate prey (Buskey & Stoecker, 1989). Furthermore, chemosensory responses to prey particles may enable truly selective feeding by some tintinnids and oligotrichs (Burkill, Mantoura, Llewellyn, & Owens, 1987; Stoecker, 1988; Stoecker, Gallager, Langdon, & Davis, 1995; Taniguchi & Takeda, 1988; Verity, 1991). Selective food capture may also be enhanced by a lectin-type binding of prey to food-capturing cell membranes (Wilks & Sleigh, 2004).

The puzzle of average bacterial abundances being too low to maintain the growth of planktonic spirotrichs has had at least three solutions. First, these ciliates may be omnivorous in nature. Second, they may browse on particles where bacterial abundances are higher. Third, these ciliates may have behavioral mechanisms that maintain them in small-scale patches (Tiselius, Jonsson, & Verity, 1993; Montagnes, 1996). Patches of oligotrichs and choreotrichs ranged from <13 m to <77 m in size in a tropical coast lagoon under non-turbulent conditions (Bulit, Díaz-Avalos, Signoret, & Montagnes, 2003). Turbulence will destroy this patch structure, and it may also reduce the feeding efficiency of these ciliates, by changing either the rate or pattern of their locomotion (Dolan, Sall, Metcalfe, & Gasser, 2003).

In recent years there has been considerable interest in the feeding biology of oligotrichs and choreotrichs (Sanders & Wickham, 1993), particularly as they are dominant grazers in planktonic food webs. They can consume significant proportions of the primary production, up to 25% (Capriulo & Carpenter, 1983; Verity, 1985) and over 30% of the bacterial standing stock (Lavrentyev, McCarthy, Klarer, Jochem, & Gardner 2004; Rassoulzadegan, Laval-Peuto, & Sheldon, 1988; Sime-Ngando et al., 1999). Furthermore, their excretion of phosphorus and ammonia may fuel over 15% of the net primary production (Dolan, 1997; Stoecker, 1984; Taylor, 1984; Verity, 1985). Field and laboratory studies have demonstrated that tintinnids consume cyanobacteria, picoflagellates, chlorophytes, prymnesiophytes, dinoflagellates, diatoms, euglenophytes, prasinophytes, and raphidophytes (Aelion & Chisholm, 1985; Bernard & Rassoulzadegan, 1993; Christaki, Jacquet, Dolan, Vaulot, & Rassoulzadegan, 1999; Dolan, 1991; Kamiyama & Arima, 2001; Rassoulzadegan & Etienne, 1981; Stoecker, 1984; Verity; Verity & Villareal, 1986). Toxic dinoflagellates, like *Alexandrium*,

Gyrodinium, and *Pfiesteria* may increase tintinnid and oligotrich mortality, possibly due to secretion or ingestion of the toxins (Hansen, 1995; Hansen, Cembella, & Moestrup, 1992; Stoecker, Parrow, Burkholder, & Glasgow, 2002) as do thread-bearing diatoms (Verity & Villareal, 1986). However, the toxic effect is at least species-specific as some tintinnids and oligotrichs thrive on these toxic dinoflagellates (Kamiyama, Suzuki, & Okumura, 2006; Stoecker et al., 2002). Aloricate choreotrichs, such as *Strobilidium*, *Lohmanniella*, and *Strombidinopsis*, consume bacteria, prymnesiophytes, cryptophytes, dinoflagellates, chlorophytes, and prasinophytes (Burkill et al., 1987; Christaki, Dolan, Pelegri, & Rassoulzadegan, 1998; Jeong et al., 2004; Jonsson, 1986; Kamiyama & Matsuyama, 2005; Montagnes, 1996, 1999; Sime-Ngando et al., 1999). Diatoms, kinetoplastids, and eustigmatophytes can be added to this list for aloricate oligotrichs, such as *Strombidium* species (Bernard & Rassoulzadegan, 1990; Burkill et al., 1987; Christaki et al., 1998, 1999; Fenchel & Jonsson, 1988; Jonsson; Montagnes, 1996; Ohman & Snyder, 1991). Tintinnids and oligotrichs generally consume food particles that are less than 20 μm in diameter (Rassoulzadegan, 1982; Rassoulzadegan et al., 1988).

Stichotrichs, like *Halteria*, *Oxytricha*, and *Stylonychia*, are probably omnivorous. They have been shown to feed on bacteria, diatoms, dinoflagellates, chrysophytes, cryptophytes, and chlorophytes (Balczon & Pratt, 1996; Kaul & Sapra, 1983; Skogstad, Granskog, & Klaveness, 1987). *Oxytricha* and *Onychodromus* can feed on other ciliates, including members of their own species for which they may undergo a developmental polymorphism to become cannibal giants (Foissner, Schlegel, & Prescott, 1987; Riggio, Ricci, Banchetti, & Seyfert, 1987; Wicklow, 1988). *Euplotes* species are the only hypotrichs that have recently been examined for feeding preferences. They can be omnivorous, ingesting bodonids and a variety of heterotrophic flagellates, in addition to those prey mentioned for stichotrichs, and including other ciliates (Dolan & Coats, 1991a, 1991b; Gast & Horstmann, 1983; Lawrence & Snyder, 1998; Premke & Arndt, 2000; Wilks & Sleigh, 1998). Dini and Nyberg (1999) have convincingly demonstrated that ecologically important differences in feeding responses among nine reproductively isolated groups of *Euplotes* are genetically determined. They concluded that morphospecies inadequately represent the ecological diversity of ciliates.

In addition to heterotrophy, some spirotrichs exhibit varying degrees of mixotrophy, either by sequestering chloroplasts from their prey or by harboring symbiotic *Chlorella* species (Sanders, 1991). Retention of prey chloroplasts is common in oligotrichs, with a significant fraction of the species exhibiting this trait in some oligotrophic lakes (Macek, Callieri, Simek, & Vazquez, 2006), in the Mediterranean (Bernard & Rassoulzadegan, 1994; Dolan & Pérez, 2000; Laval-Peuto & Rassoulzadegan, 1988), and in temperate oceans (Stoecker, Taniguchi, & Michaels, 1989). The chloroplasts appear to originate from a variety of groups of chromophytic protists, many of the groups noted above that can serve as food for these ciliates (Laval-Peuto & Febvre, 1986; Stoecker, Silver, Michaels, & Davis, 1988/1989). These chloroplasts remain functional for several days in *Laboea* and *Strombidium*, enabling the ciliate to fix carbon during this time (Perriss, Laybourn-Parry, & Jones, 1994; Stoecker, Michaels, & Davis, 1987a; Stoecker, Silver, Michaels, & Davis, 1988). Mixotrophy may serve a variety functions for the ciliate, including providing fixed carbon during periods when prey are not abundant (Dolan & Pérez, 2000; Stoecker et al., 1987a) and oxygen in those species that prefer to live at the oxic-anoxic boundary in freshwater ponds (Berninger, Finlay, & Canter, 1986).

As abundant components of aquatic ecosystems, spirotrichs can also be consumed by other organisms, serving as links to higher trophic levels (Sanders & Wickham, 1993). Choreotrichs and tintinnids are consumed by rotifers (Gilbert & Jack, 1993), copepods (Burns & Gilbert, 1993; Gismervik, 2006; Pérez, Dolan, & Fukai, 1997; Stoecker & Sanders, 1985; Turner, Levinsen, Nielsen, & Hansen, 2001; Wiackowski, Brett, & Goldman, 1994; Wickham, 1995), cladocerans (Jack & Gilbert, 1993; Wickham & Gilbert, 1991; Wickham, Gilbert, & Berninger, 1993), barnacle nauplii (Turner et al., 2001), euphausiids (Nakagawa, Ota, Endo, Taki, & Sugisaki, 2004), scyphozoan jellyfish (Stoecker, Michaels, & Davis, 1987b), larval and post-larval ctenophores (Stoecker, Verity, Michaels, & Davis, 1987c), nematode worms

(Hamels, Moens, Muylaert, & Vyverman, 2001), freshwater oligochaetes (Archbold & Berger, 1985), oysters (Loret et al., 2000), and larval fish (Nagano, Iwatsuki, Kamiyama, & Nakata, 2000; Olivotto et al., 2005; Stoecker & Govoni, 1984). They are even consumed by other protists, both ciliates and dinoflagellates (Bockstahler & Coats, 1993; Dolan, 1991; Smalley & Coats, 2002; Smalley, Coats, & Adam, 1999).

Planktonic ciliates, in their turn, have independently evolved jumping movements to escape predation (Tamar, 1979). In the case of the suspension feeding crustaceans, the choreotrichs, like *Strobilidium*, and the stichotrich *Halteria* have independently evolved a jumping behavior that significantly reduces encounters with predators (Gilbert, 1994), even relative to loricate tintinnids (Broglio, Johansson, & Jonsson, 2001). The ciliates are apparently responding to a hydromechanical signal, very likely deformations of fluid flows caused by the suction-feeding currents of the predators (Jakobsen, 2000, 2001). *Euplotes* undergoes an "escape response" in the presence of the turbellarian predator *Stenostomum* (Kuhlmann, 1994).

For "softer"-bodied predators, like other ciliates, oligochaetes, and turbellarians, *Euplotes* species develop an enlarged circular, 'winged' form that is much less easily consumed (Fyda, Warren, & Wolinska, 2005; Kuhlmann & Heckmann, 1985, 1994; Kusch, 1995). This 'winged' form develops under the influence of a diffusible protein morphogen or kairomone when the ciliate predator is *Lembadion* (Kuhlmann & Heckmann; Peters-Regehr, Kusch, & Heckmann, 1997). The kairomone induces the assembly of new microtubular structures to support the 'wings' (Jerka-Dziadosz, Dosche, Kuhlmann, & Heckmann, 1987). *Onychodromus quadricornutus* develops large dorsal spines in response to the predator *Lembadion* (Wicklow, 1988), although the ultrastructural basis of this development has not yet been determined. These morphological defenses are genetically variable among clones, suggesting that natural selection can act on them to favor the more fit variants (Duquette, Altwegg, & Anholt, 2005; Wiackowski, Fyda, Pajdak-Stós, & Adamus, 2003).

Marine *Euplotes* species also secrete a variety of terpenoids, which are strain and species specific. These have been shown to have biological activity, killing both conspecifics that could be potential competitors (Dini, Guella, Giubbilini, Mancini, & Pietra, 1993) and predators, such as the haptorian *Litonotus* (Guella, Dini, & Pietra, 1995; Guella, Dini, & Pietra, 1996). The most bizarre predator defense, probably among all the ciliates, is that exhibited by the hypotrich *Euplotidium*: it carries bacterial ectosymbionts that explode on contact with the predatory ciliate *Litonotus*, defending their host against predation (Verni & Rosati, 1990; Petroni, Spring, Schleifer, Verni, & Rosati, 2000)!

Symbionts of spirotrichs can be categorized as both beneficial (i.e., mutualistic) and harmful (i.e., parasitic). Bacteria, as in *Euplotidium* above, have been observed on the outside of tintinnid loricae, and also as intracellular forms (Laval-Peuto, 1994). Their role in tintinnids is not known. However, both freshwater and marine species of the hypotrich *Euplotes* depend upon bacterial symbionts for cell growth. Aposymbiotic *Euplotes* fail to divide and ultimately die (Heckmann, 1975, 1983; Heckmann & Schmidt, 1987; Heckmann, Ten Hagen, & Görtz, 1983; Vannini, Schena, Verni, & Rosati, 2004). However, Fujishima and Heckmann (1984) rescued aposymbiotic freshwater *Euplotes* by transfer of bacterial symbionts even among species of *Euplotes*, suggesting that the common ancestor of this group of *Euplotes* species inherited a hereditary defect in cell division that was rescued by its bacterial endosymbiont. At the other extreme, some bacterial endosymbionts carried by *Euplotes* species kill sensitive cells of the same species, either by liberating toxins into the medium or by transfer of toxins and bacteria during mating, a phenomenon also reported in *Paramecium* species (Verni, Rosati, & Nobili, 1977; see review by Heckmann, 1983). Fenchel and Bernard (1993) have described an unusual association between the obligately anaerobic *Strombidium purpureum* and its photosynthetic purple non-sulphur bacteria, an analogy to the symbiosis that is believed to have lead to the evolution of mitochondria.

Eukaryotic symbionts of spirotrichs have also been observed. *Chlorella* species are often found in hypotrichs and stichotrichs, whose photobehavior may be influenced by the presence of the symbiont (Reisser, 1984; Reisser & Häder, 1984). These *Chlorella* species are typical photosynthetic symbionts that presumably at least provide organic sugars to their hosts. Other protists are not as friendly. Species of the kinetoplastid *Leptomonas* infect the macronucleus of the stichotrich *Paraholosticha sterkii* and the hypotrich *Euplotes*, causing death in some strains of these species (Görtz & Dieckmann, 1987; Wille, Weidner, & Steffens, 1981). One of the most unusual flagellate infections of spirotrichs is that of the parasitic dinoflagellate *Duboscquella* (Coats, 1988; Coats, Bockstahler, Berg, & Sniezek, 1994). Species of *Duboscquella* that kill the tintinnid host may even regulate host abundances in natural populations (Coats & Heisler, 1989). Foissner and Foissner (1986) have described a zoosporic fungus, *Ciliomyces spectabilis*, that invades and kills the cysts of the stichotrich *Kahliella simplex*.

Spirotrichs have complex behaviors that have been extensively investigated. Ricci and his research group have introduced the ethogram as a quantitative approach to characterizing behavior, especially of hypotrichs and stichotrichs (e.g., Banchetti, Erra, Ricci, & Dini, 2003; Leonildi, Erra, Banchetti, & Ricci, 1998; Ricci, 1990; Ricci, Cionini, Banchetti, & Erra, 1999). In addition to thigmotactic responses by benthic forms to the texture of substrate surfaces (Ricci, 1989) and water current (Ricci et al., 1999), planktonic forms demonstrate a negative geotaxis that is influenced by temperature and light and that explains their vertical distribution in turbulent water columns (Jonsson, 1989). Negative geotaxis may also enable these ciliates to move towards light, whose presence may be correlated with more abundant prey. Light may also enhance the digestion and growth rates of tintinnids, possibly by photooxidative breakdown of ingested food (Strom, 2001). In more stable aquatic habitats, where oxyclines become established, spirotrichs may distribute vertically both in the water column (Bernard & Fenchel, 1994; Fenchel et al., 1995) and in benthic microbial mats (Fenchel & Bernard, 1996), seeking a preferred oxygen pressure. The chemosensory behavior of benthic forms, such as *Euplotes*, has been interpreted as a form of the classical random walk, which enables populations to accumulate very quickly at point sources of attractants (Fenchel, 2004).

A behaviour unique to the tintinnid choreotrichs is the construction of the lorica in which the cell resides and which may have some selective advantage for avoiding predation (Capriulo, Gold, & Okubo, 1982). The progeny of *Favella* constructs its hyaline lorica within 10 min of leaving the parental lorica (Laval-Peuto, 1977), while species with agglutinated loricae, like *Tintinnopsis* species, may take several hours to complete lorica construction (Gold, 1979). The materials agglutinated to the lorica secretory material may vary with the environment in which the tintinnid is found and with the season (Bernatzky, Foissner, & Schubert, 1981; Gold, 1979). However, in field samples, there also appears to be some selectivity, both with respect to the size of the particles on the lorica relative to those in the environment and with respect to the quality of the particles (Gold & Morales, 1976; Rassoulzadegan, 1980). The basis of this selectivity is not understood, and cannot yet be duplicated in the laboratory (Gold, 1979).

Spirotrichs, like other ciliates, survive from days to weeks without food (Jackson & Berger, 1985b; Montagnes, 1996). If they are able to encyst, survival can be extended to years, and this is especially true of species found in soils and on mosses (Corliss & Esser, 1974). Both freshwater and marine choreotrichs and oligotrichs encyst; tintinnids form their cyst within the lorica (Jonsson, 1994; Müller & Wunsch, 1999; Paranjape, 1980; Reid & John, 1978). Encystment has been correlated with tidal rhythms (Fauré-Fremiet, 1948b; Jonsson, 1994), and reductions in prey density or increases in predator density (Müller & Wunsch, 1999; Reid, 1987). Excystment may be stimulated by changes in temperature (Kim & Taniguchi, 1995; Kim, Suzuki, & Taniguchi, 2002; Müller, 2002; Paranjape), light (Kim & Taniguchi, 1995), algal exudates (Kamiyama, 1994), and soil extracts (Müller, Foissner, & Weisse, 2006). Cysts of planktonic species settle through the water column during the productive season. These cysts may ultimately reach the sediments from which they are resuspended by turbulent events and turnover currents (Kim & Taniguchi, 1997; Müller & Wunsch, 1999; Reid, 1987). Excystment may be facilitated by an expanding excystment vacuole (Rawlinson & Gates, 1985), often through a specialized region of the cyst wall, such as the papula in oligotrich and choreotrich cysts (Kim & Taniguchi, 1995; Kim et al., 2002; Müller & Wunsch, 1999; Reid, 1987). The "cystment cycle" can be complex, vary seasonally, and is likely maintained in all cases by natural selection of correctly phased individuals. This may be especially important for the maintenance of the unusual tidal rhythmicity of the cystment cycle in the tide-pool oligotrich *Strombidium*

Licnophora is one of two genera in the Subclass Licnophoria (Fig. 7.2). Its hourglass shape has the adoral zone at the 'anterior' and several rings of cilia surrounding the 'posterior' attachment disk (Van As et al., 1999). The ultrastructure of these somatic structures is highly complex and not similar to any of the dikinetid structures so far described (Da Silva Neto, 1994a).

The infraciliature of Subclasses Oligotrichia and Choreotrichia has only recently been the subject of detailed studies, and there is considerable variation in somatic kinetid pattern (Fig. 7.3). Strombidiid oligotrichs typically have a girdle kinety that may be composed of monokinetids in *Strombidium* species (Agatha, 2004a, 2004b; Lynn, Montagnes, & Small, 1988) or dikinetids (Lynn & Gilron, 1993; Petz & Foissner, 1992). The ventral kinety has at least a posterior portion that is always composed of dikinetids (Agatha, 2004a, 2004b; Lynn & Gilron, 1993; Lynn et al., 1988). There has been no ultrastructural study of these somatic kinetids. Choreotrichs are also variable in somatic kinetid structure. Although *Strombidinopsis* species typically have somatic dikinetids distributed in bipolar somatic kineties (Fig. 7.3) (Lynn et al., 1991c), tintinnids can have both monokinetids and dikinetids in their somatic kineties (Fig. 7.3) (Agatha & Strüder-Kypke, 2007; Choi et al., 1992; Foissner & Wilbert, 1979; Laval-Peuto & Brownlee, 1986). Among other aloricate choreotrichs, *Strobilidium* species have somatic monokinetids while *Leegaardiella* and *Lohmanniella* have somatic dikinetids (Agatha & Strüder-Kypke, 2007; Lynn & Montagnes, 1988; Montagnes & Lynn, 1991; Petz & Foissner, 1992). *Strobilidium* species have ciliated somatic monokinetids whose cilia extend out under a cortical ridge, parallel to the cell surface (Fig. 7.3) (Grim & Halcrow, 1979); The kinetid appears to have a bilayered transverse ribbon composed of an anterior and posterior row of microtubules (Grim, 1987).

Corliss (1979) classified members of our Subclasses Hypotrichia and Stichotrichia in the Order Hypotrichida while Tuffrau and Fleury (1994) placed them both in their Class HYPOTRICHEA. This view is based on the strong resemblances in the global patterning of the infraciliatures of both groups: a ventral infraciliature of scattered cirri and a dorsal infraciliature of widely spaced files of dikinetids. Marginal cirri may be distributed along the body edges, delimiting the dorsal and ventral surfaces. Ventral cirri can be arranged in anterior-posterior files (e.g., *Urostyla, Kahliella, Plagiotoma*) or more asymmetrically scattered (e.g., *Diophrys, Euplotes, Oxytricha, Stylonychia*) (Figs. 7.2, 7.4). Small and Lynn (1985) emphasized that the subtle differences in kinetid structure meant that these two groups were not closely related, and they separated them into the Subclasses Hypotrichia and Stichotrichia, even placing them in separate classes (**see above**) while Fleury et al. (1985b) had placed them in separate suborders. Division morphogenesis in hypotrichs and stichotrichs is also different (**see below**) while SSUrRNA gene sequences separate them at some distance from each other (e.g., Bernhard et al., 2001). The dorsal dikinetids of hypotrichs are characterized as follows: a ciliated anterior kinetosome with a tangential transverse ribbon, probably near triplets 3-5 and sometimes a single postciliary microtubule; a posterior kinetosome with a short condylocilium, a divergent postciliary ribbon, and a laterally-directed, striated kinetodesmal fibril at triplets 6, 7 (Fig. 7.5) (Lenzi & Rosati, 1993; Lynn, 1991; Rosati, Verni, Bracchi, & Dini, 1987; Wicklow, 1983). Lasiosomes, whose function is not known, may be associated with the axonemal base of the anterior cilium while ampules may surround the kinetid in some euplotids (e.g., Ruffolo, 1976a; Görtz, 1982a). The stichotrich dorsal dikinetid is much more variable but typically is characterized as: the ciliated anterior kinetosome bears a tangential transverse ribbon; the non-ciliated posterior kinetosome, if present, bears a small divergent postciliary ribbon and typically loses its kinetodesmal fibril during development (Fig. 7.5) (Fleury et al., 1986; Görtz, 1982a; Grimes & Adler, 1976; Lynn, 1991).

The ventral somatic kinetids of hypotrichs and stichotrichs are typically polykinetids, called cirri. These complex ciliary structures enable the complex movements of hypotrichs and stichotrichs, allowing them to rapidly dart a short distance forward, quickly withdraw, and change directions rapidly (e.g., Erra, Iervasi, Ricci, & Banchetti, 2001; Ricci, 1990). However, cirri undoubtedly develop from the assembly of dikinetid units (see **Division and Morphogenesis**; Jerka-Dziadosz, 1980). In fact, the ventral kinetids of some stichotrichs, such as *Engelmanniella*, may complete development as dikinetids, which are characterized as follows: an anterior ciliated kinetosome with a single postciliary microtubule and a tangential transverse ribbon at triplets 4, 5; and a posterior ciliated kinetosome with a divergent postciliary ribbon and a kinetodesmal fibril near triplets 6-8 (Fig. 7.5) (Wirnsberger-Aescht, Foissner, & Foissner, 1989). The ultrastructure of hypotrich cirri is quite variable, but typically kinetosomes are hexagonally packed and joined by a basal plate and a distal plate of dense or filamentous material. Microtubules originate from the lateral edges of the distal plate and extend out into the cortex. These microtubules are joined by those of the transverse and postciliary ribbons, which originate from the basal plate adjacent to the kinetosomal bases. Kinetodesmal fibrils may be associated with kinetosomes along the right edge of the polykinetid (Lynn, 1991; Tuffrau & Fleury, 1994). The ultrastructure of stichotrich cirri is quite consistent: kinetosomes are typically hexagonally packed and joined at the basal level and at mid-height. Microtubules originate from the mid-height connective material and extend out into the cortex. As in the hypotrichs, these microtubules are joined by the microtubules of transverse and postciliary ribbons, which arise from the basal plate adjacent to kinetosome bases. Kinetodesmal fibrils may be associated with kinetosomes along the right edge of the polykinetid (Fig. 7.6) (Fleury et al., 1985a; Grain, 1984; Lynn; Matsusaka et al., 1984; Tuffrau & Fleury, 1994). However, kinetodesmal fibrils or ciliary rootlets may be resorbed in a domain-specific fashion in some stichotrichs (Jerka-Dziadosz, 1990). Therefore, it is difficult to generalize about the presence or absence of these structures unless a thorough analysis has been done of the entire infraciliature. Even the somatic polykinetids of *Plagiotoma*, which was assigned by Corliss (1979) to the heterotrichs, demonstrate features of the stichotrich cirrus (Fig. 7.6) (Albaret & Grain, 1973). This supports transfer of *Plagiotoma* to the Subclass Stichotrichia (see also Tuffrau & Fleury), a fact that is also corroborated by SSUrRNA gene sequences (Affa'a et al., 2004).

Members of the Family Halteriidae are now placed in the Subclass Stichotrichia (**see above** and **Chapter 17**) (Fig. 7.4). *Meseres* has somatic dikinetids while *Halteria* and *Pelagohalteria* have fused, bristle-like cilia arising from dikinetids (Petz & Foissner, 1992; Song, 1993). The somatic dikinetids of *Halteria* are highly unusual. They apparently lack the classical fibrillar associates of the somatic kinetid, but are surrounded by dense material from which cortical microtubules originate, a feature shared with the cirri of other stichotrichs (Grain, 1972).

We have placed the Family Reichenowellidae incertae sedis in the Subclass Hypotrichia. The ultrastructural study of *Balantidioides* (= *Transitella*?)

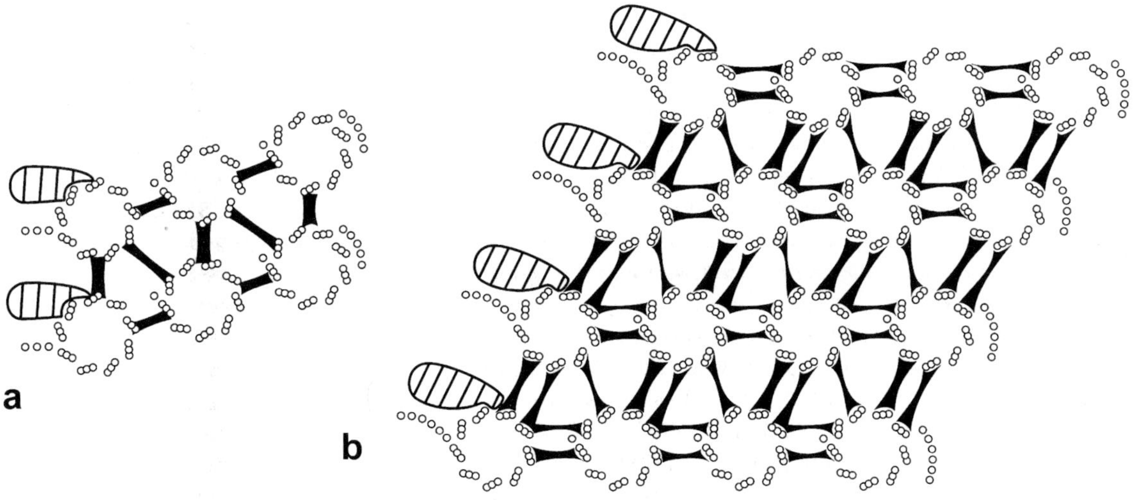

FIG. 7.6. Schematics of the somatic polykinetids or cirri of representatives of the Class SPIROTRICHEA. (**a**) Somatic polykinetid or cirrus of the stichotrich *Plagiotoma* (based on an electron micrograph of Albaret & Grain, 1973). (**b**) Somatic polykinetid or cirrus of the stichotrich *Histriculus*. (Based on an electron micrograph of Matsusaka et al., 1984.)

shows its somatic kinetids to be assembled into groups of typically 2-6 dikinetids whose structure is characterized as follows: a ciliated anterior kinetosome with a tangential transverse ribbon near triplets 3-5, a single postciliary microtubule, and possibly two microtubules near triplet 1, which could be transverse microtubules for the posterior kinetosome; and a ciliated posterior kinetosome with a large divergent postciliary ribbon and a posteriorly-directed kinetodesmal fibril. The postciliary microtubules originate in a large dense structure that is the base of an interiorly directed nematodesma (Fig. 7.5) (Iftode et al., 1983; Lynn, 1991).

The structural variation among spirotrich somatic structures contradicts the hypothesis of "structural conservatism" of the somatic cortex (Lynn, 1976a, 1981) (see **Chapter 4**), and begs the question "Why?". In his discussion of the unusual nature of the somatic kinetids of choreotrichs, Grim (1987) suggested one explanation: when somatic kinetids are no longer used in locomotion, selection may be relaxed on the structure since it no longer performs a critical locomotor function. Although this is helpful in explaining variations in somatic kinetids of oligotrichs and choreotrichs and the dorsal dikinetids of hypotrichs and stichotrichs, it is not helpful in explaining the structural diversity of locomotory somatic kinetids among spirotrichs (Fig. 7.5). We currently have no explanation for this deviation from structural conservatism, except to suggest that these lineages could be extremely ancient, as suggested by the branch lengths in molecular phylogenetic analyses (e.g., Bernhard et al., 2001; Snoeyenbos-West et al., 2002; Strüder-Kypke et al., 2002).

The cirri of hypotrichs and stichotrichs can perform complex "walking" or "running" movements (Erra et al., 2001; Sleigh, 1989). These movements in *Euplotes* and *Stylonychia* are controlled by membrane hyperpolarizations that produce forward movement (i.e., rearward beating) and membrane depolarizations that produce rearward movement (i.e., forward beating) (Epstein & Eckert, 1973; Deitmer, Machemer, & Martinac, 1984). Rhythmic depolarizations determine the rhythm of walking in *Euplotes* (Lueken, Ricci, & Krüppel, 1996). Similar to the model for *Paramecium*, Ca^{2+} ions probably interact with ciliary axonemal components, serving as the intracellular messengers for membrane potential changes. The cirri of *Stylonychia* also respond to slow changes in membrane potential by "inclining" or bending at the base without beating, which adds a further degree of sophistication to their movement (Machemer & Sugino, 1989).

Spirotrichs can be quite flexible or contractile. However, in choreotrichs, this contractility apparently does not rely on filamentous structures but on highly unusual membranous elements, for example in the tail of *Tontonia* (Greuet, Gayol, Salvano, & Laval-Peuto, 1986) and the posterior end of tintinnids (Laval-Peuto, 1994). We do not know how these structures work.

Contractile vacuoles are common throughout the group, especially in freshwater forms. Marine tintinnids do not have contractile vacuoles while their freshwater relatives do.

A wide variety of extrusomes has been described in the group. Mucocyst-like extrusomes are found in this class (*Phacodinium*, Didier & Dragesco, 1979; *Stylonychia*, Görtz, 1982b; *Transitella*, Iftode et al., 1983). Cortical ampules found around the dorsal dikinetids of hypotrichs may be a special type of mucocyst (*Aspidisca*, Rosati et al., 1987; *Certesia*, Wicklow, 1983; *Euplotes*, Görtz). Other extrusomes with a distinctly lamellar nature appear to be involved in cyst formation (e.g., Grim & Monganaro, 1985; Walker et al., 1980; Verni, Rosati, & Ricci, 1990), while the highly unusual lepidosomes, mentioned above, form a surface coat on the cysts of *Meseres* (Foissner et al., 2005a). The trichites of oligotrichs have been demonstrated to be extrusomes, but their function remains unknown (Modeo, Petroni, Bonaldi, & Rosati, 2001).

Finally, to round out this brief treatment of other somatic structures, lithosomes or calculi have been observed in both hypotrichs (Lenzi & Rosati, 1993; Rosati et al., 1987; Ruffolo, 1978) and stichotrichs (Wirnsberger-Aescht et al., 1989). Of unkown function, they can be composed of calcium salts (Hausmann & Walz, 1979).

7.4 Oral Structures

Spirotrichs, like the heterotrichs, are characterized by a prominent adoral zone of polykinetids or membranelles, which are typically composed of three or four rows of kinetosomes. Because of the tremendous diversity in form of members of this class, there are few general statements that can be made about the pattern of the oral structures. Benthic forms with dorsoventrally flattened bodies typically have an adoral zone of polykinetids that extend along the left side of the oral region and may extend over the anterior end and a short distance down the right side of the oral region (Figs. 7.2, 7.4). Planktonic forms, which are generally spherical or conical, have an adoral zone of polykinetids that wraps incompletely or completely around the anterior end of the cell (Figs. 7.3, 7.4). De Puytorac and Grain (1976) suggested the term paramembranelle to describe the oral polykinetids of heterotrichs and some spirotrichs. As noted for the heterotrichs (see **Chapter 6** and also below), conspicuous variations in ultrastructure argue against inventing a new term for the organellar complexes of each taxon. We now briefly characterize the oral structures of the seven subclasses, referring to the structures in a taxonomic fashion (e.g., protocruziid paroral).

The sole representative of the Subclass Protocruziidia (i.e., *Protocruzia*) has six adoral polykinetids and a paroral of dikinetids (Fig. 7.2) (Grolière et al., 1980a; Song & Wilbert, 1997). There are typically four rows of square-packed kinetosomes in the oral polykinetids with the first two rows more widely separated from each other than the last three. The posterior kinetosomes bear postciliary ribbons while the anterior kinetosomes may have radially oriented transverse ribbons. The paroral is composed of dikinetids, having postciliary ribbons extending from the left-most kinetosome. The paroral dikinetids form a file along the right and posterior sides of the oral region (Grolière et al., 1980a).

Phacodinium, the sole representative of the Subclass Phacodiniidia, has an extensive adoral zone of polykinetids along the left side of the oral region and a short paroral on the right side of the oral cavity (Fig. 7.2). The oral polykinetids are composed of four rows of hexagonally-packed kinetosomes (Fernández-Galiano & Calvo, 1992). The kinetosomes on the right side of the oral polykinetids bear divergent postciliary ribbons while those of the anterior row apparently bear convergent postciliary ribbons (Da Silva Neto, 1993a). The paroral is a unique structure, a polymerized stichomonad termed a polybrachystichomonad by Fernández-Galiano and Calvo (1992). It consists of a series of oblique rows of 6-7 kinetosomes, lying at the base of a ridge. The left-most file of kinetosomes, which bear a postciliary ribbon, extends as a single file deeper into the oral cavity adjacent to the first oral polykinetids (Da Silva Neto, 1993a).

Licnophora is a representative of the Subclass Licnophoria. The anterior portion of its hour-glass shaped body is encircled by an adoral zone of oral polykinetids (Fig. 7.2) that have the structure of paramembranelles (Da Silva Neto, 1994a). The paroral is a single file of monokinetids (Song, Warren, Ji, Wang, & Al-Rasheid, 2003).

Members of the Subclass Oligotrichia, typified by *Strombidium*, have an adoral zone of polykinetids divided into the "lapel" or oral cavity polykinetids and the "collar" or anterior oral polykinetids, which encircle the anterior end of the cell (Fig. 7.3). These oral polykinetids have three rows of kinetosomes while the paroral extends along the right side of the oral cavity as a single file of kinetosomes (Agatha, 2004a; Lynn et al., 1988; Petz & Foissner, 1992; Song, Wang, & Warren, 2000). There has been no ultrastructural study of oligotrich oral organelles.

The Subclass Choreotrichia includes such genera as *Strobilidium*, *Pelagostrobilidium*, *Lohmanniella*, *Leegaardiella*, *Strombidinopsis*, and the tintinnids, in which an adoral zone of polykinetids completely encircles the anterior end of the cell (Fig. 7.3). These oral polykinetids are typically composed of three rows of kinetosomes, but they may be divided into two parts as in *Leegaardiella* (Dale & Lynn, 1998; Lynn & Montagnes, 1988; Petz & Foissner, 1992; Song & Bradbury, 1998). Some of these polykinetids may extend into the oral cavity (e.g., *Strobilidium*, tintinnids; Foissner & Wilbert, 1979; Petz & Foissner, 1992) or there may be separate, smaller polykinetids that line the oral cavity (e.g., *Lohmanniella*, *Leegaardiella*, *Strombidinopsis*; Dale & Lynn, 1992; Lynn & Montagnes, 1988). When appropriately stained, the paroral appears to be composed of a file of monokinetids (Petz & Foissner, 1992). This is confirmed by study of choreotrich oral ultrastructure: the paroral is a file of monokinetids bearing transverse (?) microtubules (Grim, 1987). The kinetosomes of the oral polykinetids appear to be square-packed. Kinetosomes of the morphologically "anterior" row may bear a transverse (?) ribbon while those of the posterior row bear a postciliary (?) microtubular ribbon (Grim,

1987). Clearly, a detailed ultrastructural study of choreotrich oral structures is needed. The cell surface between the oral polykinetids of tintinnids bears evaginations called tentaculoids, which contain the "capsules torquées" or twisted capsules. Along the oral polykinetid cilia of tintinnids are the striae or streaks, bulge-like evaginations that may also contain the twisted capsules, which may function as extrusomes in prey capture (Laval-Peuto, 1994; Laval-Peuto, Gold, & Storm, 1979).

The adoral polykinetids of members of the Subclasses Hypotrichia and Stichotrichia were characterized as paramembranelles (de Puytorac & Grain, 1976). Their polykinetids are composed of 3-4 rows of kinetosomes, depending upon the position along the adoral zone. The kinetosomes are square-packed with the anterior row kinetosomes bearing transverse ribbons and the posterior row kinetosomes bearing postciliary ribbons, in such genera as *Euplotidium*, *Halteria*, *Kahliella*, *Parastrongylidium*, *Paraurostyla*, *Stylonychia* (Figs. 7.2, 7.4) (Bakowska & Jerka-Dziadosz, 1980; Fleury et al., 1985a, 1985b, 1986; Grain, 1972; Lenzi & Rosati, 1993; de Puytorac, Grain, & Rodriguez de Santa Rosa, 1976; Tuffrau & Fleury, 1994). Foissner and Al-Rasheid (2006) have provided a detailed description of the stichotrich oral cavity and revealed an unusual structure, the buccal seal, which can apparently cover the entire oral opening like a sheet. They also identify lateral membranellar cilia, which derive from the fourth row of membranellar kinetosomes, extend rightward across the oral cavity, and may be used in prey selection and feeding. Paroral structures are typically a polykinetid-like structure in hypotrichs (Curds, 1975a; Grim, 1982; Tuffrau, 1960; Wicklow, 1983). In stichotrichs, de Puytorac and Grain (1976) apply the term diplostichomonad to the paroral and endoral files, which are typically composed of single kinetosomes with associated microtubular rootlets (Albaret & Grain, 1973; de Puytorac & Grain, 1976). The halteriids are an exception among the stichotrichs as they have apparently lost either the endoral or paroral and bear only a single file of paroral kinetosomes (Grain, 1972).

The physiology of oral cilia in the stichotrich *Stylonychia* appears to be different from that of the somatic cilia. Oral cilia are continually active, enabling the organism to continuously probe the environment for food while it moves forwards or backwards or remains stationary on it somatic cirri (Deitmer et al., 1984). The oral polykinetids and paroral of spirotrichs can be underlain by microtubules and/or a nodal filamentous reticulum, which may confer on the region a highly contractile nature (**see references above**). For example, tintinnids and other choreotrichs are able to not only contract the body but also the oral region, possibly due to these filamentous elements (Grim, 1987; Laval-Peuto, 1994). The microtubular elements likely provide structural support for the oral region (Grain, 1984; Tuffrau & Fleury, 1994). Large accumulations of pharyngeal disks in the oral region may enable spirotrichs, like *Euplotes*, to rapidly form as food vacuole membranes the rough equivalent of the entire surface area of the cell and so exploit a periodically abundant food source (Kloetzel, 1974).

7.5 Division and Morphogenesis

Spirotrichs typically divide while free-swimming with few notable exceptions, such as the stichotrich *Paraholosticha* (Dieckmann, 1988; Tuffrau & Fryd-Versavel, 1977). Foissner (1996b) presents a comprehensive review of the types of stomatogenesis and division morphogenesis in the ciliates. The literature on stomatogenesis of stichotrichs is particularly rich as the features of division morphogenesis have been instrumental in establishing phylogenetic hypotheses about relationships among families and genera (e.g., see Berger & Foissner, 1997; Eigner, 1997, 1999, 2001; Foissner, 1996b; Petz & Foissner, 1992). These, in turn, are now being tested by molecular phylogenetics (Bernhard et al., 2001; Foissner et al., 2004; Schmidt et al., 2007). Our review below will briefly characterize the nature of division morphogenesis in the various spirotrich subclasses. Because of Foissner's comprehensive review, few references are made to earlier literature.

Stomatogenesis of the Subclass Protocruziidia (i.e., *Protocruzia*) has been characterized as mixokinetal since elements of both somatic and parental infraciliature are involved in oral primordium formation (Foissner, 1996b). However, Grolière et al. (1980a) have provided the only published description, which does not demonstrate involvement of parental oral structures, typing it as parakinetal. The oral polykinetids assemble from anterior to posterior and from right to left in the primordial field while the paroral differentiates as a file of dikinetids along the right side of the primordial field (Fig. 7.7).

Division morphogenesis has not been described for *Phacodinium* or *Licnophora* (Foissner, 1996b). This presents an opportunity for future research.

Deroux (1974) provided the first detailed description of stomatogenesis in a choreotrich, *Strobilidium*. Foissner (1996b) classifies choreotrich stomatogenesis as hypoapokinetal since it occurs for most of the process in a subsurface cortical pouch. Initial kinetosomal proliferation appears to occur on the cell surface, and the region invaginates as oral development proceeds (Dale & Lynn, 1998) (Fig. 7.8). The developing oral polykinetids form a particularly characteristic "barrel stave-like" formation across the outer or surface end of which the developing paroral extends (Fig. 7.8). Kinetosomal proliferation occurs within the somatic kineties, which lengthen and are subdivided at cytokinesis. This has been observed in *Strombidinopsis*, *Strobilidium*, and tintinnids (Agatha, 2003a; Dale & Lynn, 1988; Deroux; Petz & Foissner, 1992, 1993).

Hypotrichs also share the feature of oral primordium development within a subsurface pouch, and have been characterized as hypoapokinetal (Foissner, 1996b) (Fig. 7.7). In hypotrichs, the earliest kinetosomes of the oral primordium appear in a subsurface pouch (e.g., *Aspidisca*, Song, 2003; *Certesia*, Wicklow, 1983; *Diophrys*, Hill, 1981; Song & Wilbert, 1994; *Euplotes*, Wise, 1965; *Uronychia*, Hill, 1990). However, the primordium develops on the cell surface of the related discocephalines (Wicklow, 1982). The somatic ciliature may be renewed differently on dorsal and ventral surfaces. On the ventral surface to the posterior right of the proter's oral region, typically five, but up to ten, primordial streaks form by kinetosomal replication (Fig. 7.7). These kinetosomes appear not to be associated with parental somatic kinetids, but may acquire kinetosomes from parental structures as the parental structures dedifferentiate. These ventral streaks elongate and eventually split into two sets, one giving rise to the proter's and the other to the opisthe's ventral ciliature (e.g., *Certesia*, Wicklow, 1983; *Diophrys*, Hill, 1981; Song & Wilbert, 1994; *Euplotes*, Wise, 1965; *Uronychia*, Hill, 1990; discocephalines, Wicklow, 1982). Movements of cirral primordia both posteriorly and anteriorly are driven by the assembly of microtubular structures associated with them (Fleury, 1991a; Ruffolo, 1976b). Wallengren (1900) devised a method of numbering these ciliary streaks and the subsequently differentiating cirri in *Euplotes* and this has served as an important means of comparing the development of hypotrichs and stichotrichs (see also Martin, 1982; Wise, 1965). Dorsal kinety streaks may appear in both proter and opisthe alongside parental kineties, and with kinetosomal replication ultimately replacing the parental structures (e.g., *Diophrys*, Song & Wilbert, 1994). In contrast, dorsal kinetosomal replication in euplotids occurs within each dorsal kinety (Frankel, 1975; Song, 2003). The pattern of intensity of replication is guided by global positional systems (Frankel, 1989).

Oligotrich stomatogenesis appears to fall into two types, possibly related to the extent of development of the cortical polysaccharide plates. Foissner (1996b) typed it as epiapokinetal because the oral primordium forms on the cell surface independent of parental infraciliature in some *Strombidium* species (Agatha, 2003b; Agatha, Strüder-Kypke, & Beran, 2004; Song & Wang, 1996). However, Fauré-Fremiet (1953) described it to occur in a long inpocketing beneath the polysaccharide plates of *Strombidium oculatum*, and this neoformation organelle was confirmed in *Pelagostrombidium fallax* by Petz and Foissner (1992). In the early stages of oral primordium formation, proliferation may begin on the cell surface, followed by invagination as development of the oral structures proceeds (Petz, 1994) (Fig. 7.8). This is similar to the process in the choreotrichs, and later dividers in both groups may be characterized as showing an enantiotropic kind of cell division (Fauré-Fremiet, 1953; Petz & Foissner, 1993).

There is a rich literature on the patterns of division morphogenesis in stichotrichs with the perspectives of different investigators leading to very different sets of relationships (e.g., see Berger & Foissner, 1997; Eigner, 1997, 1999; Martin, 1982; Wicklow, 1981). The pattern of development has been interpreted using the system of Wallengren (1900), which has been modified to accommodate a larger diversity of patterns (e.g., Martin, 1982). Pattern development in ciliates is controlled at both global and local levels, and although we have some ideas of the properties of the developmental processes,

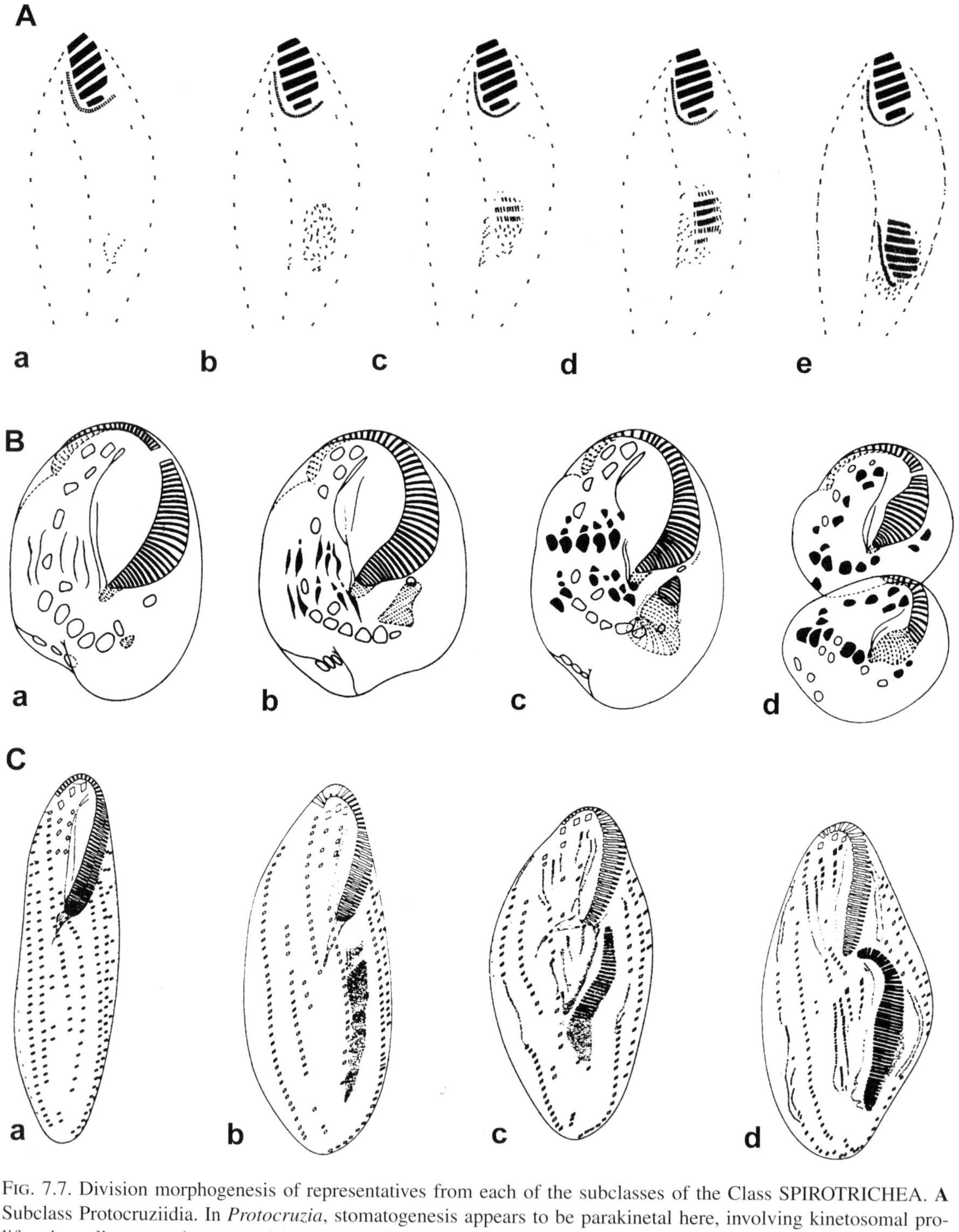

FIG. 7.7. Division morphogenesis of representatives from each of the subclasses of the Class SPIROTRICHEA. **A** Subclass Protocruziidia. In *Protocruzia*, stomatogenesis appears to be parakinetal here, involving kinetosomal proliferation adjacent to the equatorial region of Kinety 1 (**a**, **b**) and then differentiation of the adoral polykinetids and paroral (**c-e**) (from Grolière et al., 1980a). However, Foissner (1996b) has evidence that it is mixokinetal, involving elements from the parental oral region. **B** Subclass Hypotrichia. In *Diophrys*, the oral primordium begins development in a subsurface pouch while five ventral streaks appear at the cell equator (**a**). The ventral streaks divide into an anterior or proter and posterior or opisthe group (**b**, **c**). Cirral differentiation and migration occur as the oral ciliature develops (**c**, **d**). (from Hill, 1981.) **C** Subclass Stichotrichia. In *Parakahliella*, the oral primordium develops by kinetosomal proliferation on the ventral surface (**a**, **b**). Two sets of ventral streaks - an anterior proter set and a posterior opisthe set develop and cirri differentiate and migrate as the oral primordium continues to develop (**c**, **d**). (from Berger et al., 1985.)

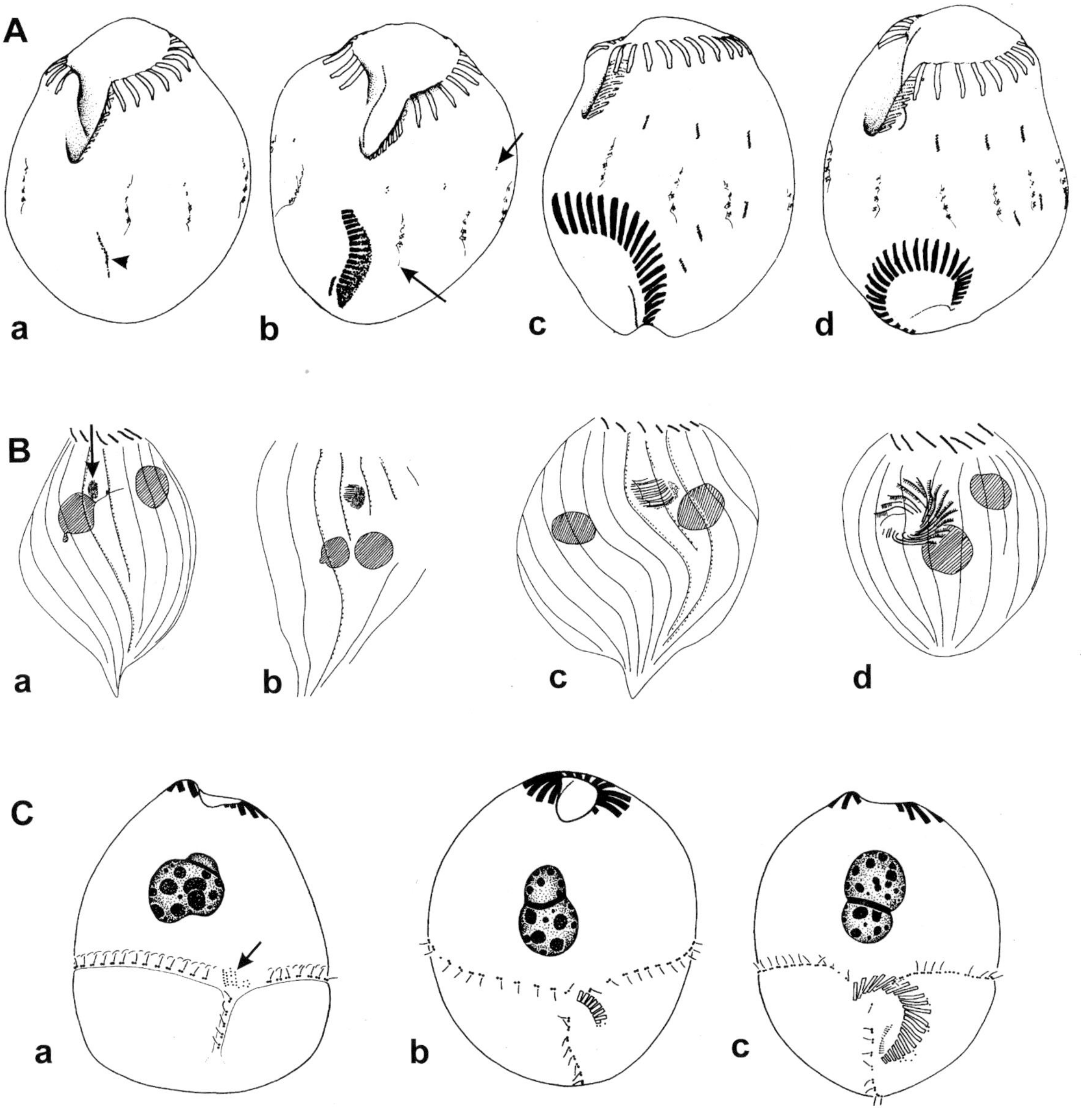

FIG. 7.8. Division morphogenesis of representatives from each of the subclasses of the Class SPIROTRICHEA. **A** Subclass Stichotrichia. In *Halteria*, formerly an oligotrich, the oral primordium (arrowhead) develops on the cell surface (**a**). New sets of bristle kinetosomes appear anterior and posterior (arrows) to parental bristle kinetosomes, which eventually dedifferentiate as development proceeds (**b–d**). (from Song, 1993.) **B** Subclass Choreotrichia. In *Strombidinopsis*, the oral primordium begins development in a subsurface pouch (arrow) (**a**). Oral development proceeds to a "barrel stave-like" formation (**b, c**), and then the opisthe's oral structures expand out onto the cell surface (**d**). Kinetosomal replication of somatic kinetids occurs within the kineties. (from Dale & Lynn, 1998.) **C** Subclass Oligotrichia. In *Strombidium*, the oral primordium (arrow) begins development on the cell surface in the region of junction between the ventral kinety and the girdle kinety (**a**). As development of the oral primordium proceeds, there is kinetosomal replication in the girdle and ventral kineties and a complex series of morphogenetic movements (**b, c**). (from Petz, 1994.)

the actual cellular mechanisms remain to be determined (Frankel, 1989). Martin (1982), for example, assumes homology between hypotrichs and stichotrichs using a Wallengren-like numbering system. We believe it premature to assume that these patterns are homologous until we have a firmer understanding of the underlying cellular mechanisms determining pattern development. Thus, we present only a general description of stichotrich division morphogenesis, touching on some major differences in the development of pattern (Fig. 7.7). It may be that the application of molecular approaches will not only provide refutation of competing schemes derived by morphologists, but application of molecular phylogenetics may inform the evolution of division morphogenesis in stichotrichs. A general assumption is made that the ancestral stichotrich was a ciliate with many files of small cirri (e.g., *Phacodinium*, Fig. 7.2; *Plagiotoma*, Fig. 7.4), and that as evolution proceeded this number was reduced to a few scattered cirri (e.g., *Stylonychia*, Fig. 7.4), although it now seems that *Plagiotoma* is a derived form (Foissner et al., 2004; Schmidt et al., 2007). Foissner (1996b) tentatively characterized stomatogenesis in *Plagiotoma* as parakinetal. Somatic kineties are completely renewed during division morphogenesis by proliferation of streaks within the parental kineties in both the proter and opisthe (Albaret, 1973; Fleury, 1983). These features, along with macronuclear replication bands (Dworakowska, 1966), support its placement within the Subclass Stichotrichia.

Among stichotrichs, there is a bewildering array of patterns, placed into parakinetal and epiapokinetal types by Foissner (1996b), who admits that all stichotrichs may be epiapokinetal since electron microscopy does not clearly implicate parental structures in kinetosomal replication (Grimes, 1972, 1973). Whether or not the oral primordium proliferates in relation to the parental infraciliature, the ventral cirral primordia may arise in at least two ways in the opisthe. The proter ventral cirral primordia almost always arise separately from those of the opisthe, so two sets of somatic cirral streak primordia are present in stichotrichs at the outset, rather than one set that divides as in the hypotrichs (Fig. 7.7). To simplify the diversity of patterns considerably, in the vast majority of genera, a series of streaks, from 3 to more than 20, arise separately from the oral primoridium often in association with the dedifferentiation of parental cirri (Fig. 7.7) (e.g., *Amphisiellides*, Eigner & Foissner, 1994; *Bakuella*, Eigner & Foissner, 1992; *Coniculostomum*, Kamra & Sapra, 1990; *Deviata*, Eigner, 1995; *Gastrostyla*, Hu & Song, 2000; *Hemigastrostyla*, Song & Hu, 1999; *Holosticha*, Hu & Song, 2001a; *Histriculus*, Berger, Foissner, & Adam, 1985; *Kahliella*, Berger & Foissner, 1988; *Lamtostyla*, Petz & Foissner, 1996; *Laurentiella*, Martin, Fedriani, & Perez-Silva, 1983; *Parakahliella*, Berger & Foissner, 1989b; *Paraurostyla*, Wirnsberger, Foissner, & Adam, 1985; *Steinia*, Voss & Foissner, 1996; *Stylonychia* (= *Tetmemena*), Wirnsberger, Foissner, & Adam, 1986; *Thigmokeronopsis*, Hu, Song, & Warren, 2004; Wicklow, 1981; *Urosomoida*, Ganner, Foissner, & Adam, 1986/1987). The other way is for a series of streaks, from five to more than ten, to appear to derive from the opisthe oral primordium (e.g., *Amphisiella*, Voss, 1992; *Circinella*, Foissner, 1994a; *Gonostomum*, Song, 1990a; *Metaurostylopsis*, Song, Petz, & Warren, 2001; *Pseudokeronopsis*, Hu & Song, 2001b; *Urosoma*, Foissner, 1983a). In both cases, the relation of cirral structures to the oral primordium may be more a function of the density of ciliation on the ventral surface. Where cirri are dense, cirral primordia appear to arise separately from the oral primordium; and where cirri are sparse, cirral primordia appear to emerge by kinetosomal proliferation from the oral primordium. Until we have more concrete understanding of the mechanisms underlying primordium formation, we should not put too much weight on subtle differences in these spatial patterns.

Primordia for marginal cirri and for dorsal kineties typically develop *within* the parental files and as proliferation proceeds, the parental kinetosomes are resorbed. However, the primordia may also appear *beside* the parental files and subsequent migration may make it appear that proliferation has occurred *within* the parental cirral file (see Wirnsberger et al., 1985). Finally, Eigner (1995, 1997, 2001) has defined neokinetal proliferation, especially in forms with longitudinal cirral files, in which additional new primordia or anlagen derive from the primary anlagen and migrate anteriorly or posteriorly from it to provide new structures.

Two unusual groups bear special mention. First, *Paraholosticha* divides within a cyst, dedifferentiating all parental structures first and then developing new structures in an epiapokinetal fashion (Dieckmann, 1988). Foissner (1996b) speculated that this may have evolved as an adaptation to the highly variable terrestrial and semiterrestrial habitats in which *Paraholosticha* is found, demonstrating a parallel evolution with division morphogenesis in some colpodeans (see **Chapter 12**). Second, the halteriids, such as *Halteria* and *Meseres*, are now placed by molecular sequences *within* the Subclass Stichotrichia (Bernhard et al., 2001), so the planktonic "oligotrich" body form has evolved convergently in stichotrichs. Stomatogenesis in halteriids is typed as epiapokinetal like that of other stichotrichs (Foissner, 1996b). The somatic infraciliature is replaced completely, also similar to many stichotrichs, from primordia that develop beside or in between the parental infraciliature (Fig. 7.8) (Song, 1993; Petz & Foissner, 1992).

Finally, brief mention should be made of division morphogenesis in the Family Reichenowellidae. This has been described for *Balantidioides* (= *Transitella*?) (Iftode et al., 1983) and characterized as pleurotelokinetal by Foissner (1996b) since the oral primordium arises by kinetosomal replication in several right lateral kineties. Replication in somatic kineties apparently occurs throughout the length of the kineties and without replacement of the parental kinetids. These features are more akin to the heterotrichs and some colpodeans (Foissner, 1996b; Iftode et al., 1983), thus justifying our current placement of this family as incertae sedis in the Subclass Hypotrichia.

Hypotrichs and stichotrichs have been model organisms for developmental biologists who are interested in probing the underlying mechanisms of pattern formation (e.g., see Frankel, 1989, 1991; Nanney, 1980). The pattern formed in each species is undoubtedly under genic control, as has been demonstrated for the ventral ciliature of both groups (Génermont et al., 1992; Jerka-Dziadosz & Czupryn, 1995; Jerka-Dziadosz & Dubielecka, 1985) and for the dorsal ciliature of hypotrichs (Heckmann & Frankel, 1968). These genetic traits in patterns of the infraciliature can be conserved across groups of species, as has been carefully demonstrated for *Euplotes* species (Gates, 1977, 1978b, 1988; Machelon & Génermont, 1984), suggesting that the positional information systems determining pattern can be conserved while other genetic traits, such as enzyme polymorphisms evolve. Nevertheless, there can be considerable phenotypic variation in the numbers of cirri on the ventral surface of hypotrichs (Walker & Grim, 1973). As occurs in the heterotrichs, this variation can be related to the cell size: smaller cells have proportionately fewer oral polykinetids in the adoral zone (Jerka-Dziadosz, 1976) and proportionately fewer kinetosomes in each polykinetid (Bakowska & Jerka-Dziadosz, 1980). These are clearly issues that practising taxonomists need to consider as they decide whether to establish a new morphospecies based on quantitative differences in cirral patterns. Qualitative differences occurring **within** hypotrich species further confound the taxonomist's job: Génermont et al. (1992) reported on a mutant of *Euplotes* that had disturbed positioning of the ventral cirral pattern compared to that of the wild type!

In a now classical series of papers on the ultrastructure of morphogenesis in spirotrichs, Jerka-Dziadosz (1980, 1981a, 1981b, 1982) provided not only a model for future studies but also revealed the basic details of the assembly of the somatic and oral infraciliature during the division morphogenesis of the stichotrich *Paraurostyla* (Fig. 7.9). The model described below is consistent with the light microscopic observations on the division morphogenesis within the spirotrich subclasses. Specifically, the anarchically arranged dikinetids on the left side of the oral primordium are assembled into rows of dikinetids, and subsequent kinetosomal replication adjacent to the anterior kinetosome of each dikinetid adds additional kinetosomal rows (Fig. 7.9a) (Jerka-Dziadosz, 1981a). Dikinetids on the right side of the oral primordium rotate so that the dikinetid axis is transverse to the cell anterior-posterior axis, forming an anterior-posterior file of kinetosomes as the primordium for the paroral structures, the paroral and endoral "membranes", develop (Fig. 7.9b) (Jerka-Dziadosz, 1981b). Within the somatic ventral streak primordia, dikinetids align in an anterior-posterior file. Subsequently, these become partitioned in groups by intrastreak microtubules while kinetosomal replication at the structurally "anterior" end of each dikinetid adds kinetosomes to complete cirral development (Fig. 7.9c) (Jerka-Dziadosz, 1980). Fleury, Le Guyader, Iftode, Laurent, and Bornens (1993) demonstrated using immunocytochemistry that a protein scaf-

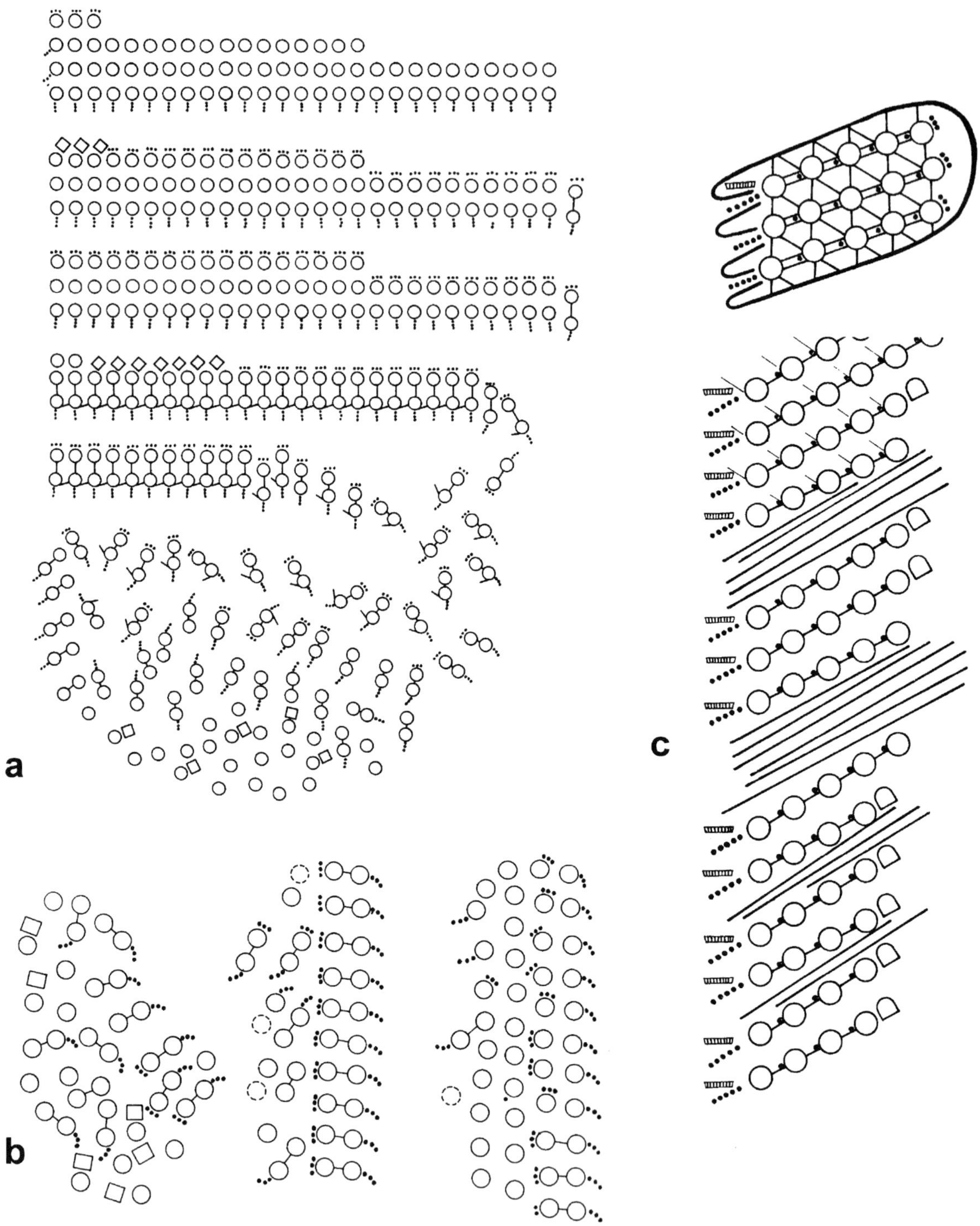

FIG. 7.9. Schematic of the development of cortical structures in the stichotrich *Paraurostyla weissei*. (**a**) Development on the left side of the oral primordium field showing the sequential formation of five oral polykinetids by assembly of kinetosomes and dikinetids in the anarchic region. (from Jerka-Dziadosz, 1981a.) (**b**) Development on the right side of the oral primordium field showing the alignment and then dissociation of dikinetids to form the endoral and paroral structures. (from Jerka-Dziadosz, 1981b.) (**c**) Development of the somatic polykinetids in the marginal cirral files from a linear file of kinetids (bottom) to separate hexagonally-packed polykinetids (top), separated during the process by groups of intrastreak microtubules (oblique lines). (from Jerka-Dziadosz, 1980.)

fold appears transiently and cirral kinetosomes are distributed along this. Replication and development of dorsal dikinetids begins similarly to that in ventral streaks, but differences exist in the fibrillar associates, kinetosomal connectives, and the ultimate arrest of development at the dikinetid stage (Jerka-Dziadosz, 1982). This is similar to the pattern already described by Grimes and Adler (1976) and markedly different from that of *Euplotes* (Ruffolo, 1976b).

Following cell division, there are relatively few species that elaborate extracellular structures: the loricae of tintinnids and the tubes of *Stichotricha* being exceptional. In tintinnids, it is the proter that constructs the new lorica while the opisthe is left with the old one (Laval-Peuto, 1994). Typically, construction of the protolorica takes only a few minutes following division and is constructed from the posterior to the anterior. Lorica construction involves the somatic cilia, which manipulate the secreted contents of the formative vesicles. Lorica length can be quite variable while lorica diameter is less so. If the tintinnid becomes separated from the protolorica, it may secrete a new lorica, called the paralorica. The paralorica can appear quite different, possibly due to its formation, which takes place over several hours, much more slowly than the few minutes taken for protolorica construction (Laval-Peuto, 1994).

7.6 Nuclei, Sexuality and Life Cycle

There is a tremendous diversity in macronuclear shapes among the spirotrichs. Generally, macronuclei are globular to ellipsoid and often in pairs, which are joined together by a membranous isthmus that may only be visible by electron microscopy (e.g., Walker & Goode, 1976). In some instances, there can be dozens of smaller macronuclei (e.g., *Urostyla*, Fig. 7.4) while in other instances, the macronucleus may be an elongate "C"- or "E"-shape as in the hypotrichs where macronuclear shape may even be diagnostic of a species (Fig. 7.2) (Curds, 1975a; Tuffrau, 1960). Micronuclei can be globular to ellipsoid and range from one to many more than ten.

A characteristic feature of the macronuclear cycle of spirotrichs is the emergence during the time of cell division of a replication band, formerly called a reorganization band (Fig. 7.1). Stein (1859) was probably the first to observe these in *Euplotes*, whose replication bands have been the subject of cell biological studies (Bonifaz & Plaut, 1974; Gall, 1959; Olins & Olins, 1994; Olins et al., 1988). Replication bands in *Euplotes* begin at each end of the elongate macronucleus and proceed towards the middle. The bands themselves are composed of a forward zone in which the chromatin is decondensed and a posterior zone in which the DNA is replicated (Lin & Prescott, 1985; Olins & Olins; Raikov, 1982, 1996). A similar substructure has been observed in the oligotrich *Strombidium* (Salvano, 1975) and in other oligotrichs and choreotrichs (Laval-Peuto, 1994; Laval-Peuto et al., 1994). Our recent assignments to the spirotrichs also demonstrate replication bands: certainly in *Licnophora* (Da Silva Neto, 1994a; Villeneuve-Brachon, 1940) and probably in *Plagiotoma* (Dworakowska, 1966). Thus, replication bands are characteristic of what one might call the "higher" subclasses of spirotrichs, but to our knowledge, they have not been observed in the macronuclei of *Phacodinium* (Subclass Phacodiniidia) and *Protocruzia* (Subclass Protocruziidia). Spirotrich macronuclear division very likely involves participation of intramacronuclear microtubules, as observed for the stichotrichs *Gastrostyla* and *Stylonychia* (Walker & Goode, 1976). However, we do not yet have any details on the ultrastructure of the unusual "chromosomal" structures during "mitosis-like" macronuclear division of *Protocruzia* (Ammermann, 1968; Ruthmann & Hauser, 1974). Micronuclear division involves participation of intramicronuclear microtubules (Walker, 1976b).

The chromosome-like structures of *Protocruzia*, which may be the basal taxon of the spirotrich clade (Bernhard et al., 2001; Shin et al., 2000), are reminiscent of the polytene chromosomes that appear during the development of the "higher" spirotrich macronuclear anlage following conjugation (Alonso, 1978; Ammermann, 1971; Ammermann et al., 1974; Jareño, 1976; Kuhlmann & Heckmann, 1991; Prescott, 1994). Following amplification of the DNA in the polytene chromosomes, DNA is eliminated and then followed by a further amplification to complete development of the mature macronucleus (Ammermann et al., 1974; Jahn &

Klobutcher, 2002; Prescott, Murti, & Bostock, 1973). The mature macronucleus is composed of gene-sized pieces, 0.5-25 kb in size, a fact that has been confirmed in several stichotrich genera, *Halteria*, and the hypotrich *Euplotes* (Lawn, Heumann, Herrick, & Prescott, 1978; Prescott et al., 1973; Riley & Katz, 2001; Steinbrück, 1990). Typically, these macronuclear "chromosomes" contain a single gene, although two-gene chromosomes have been described (McFarland, Chang, Kuo, & Landweber, 2006). Different gene-sized pieces may be differentially amplified and their copy number may be controlled through the vegetative cell cycles of the hypotrichs and stichotrichs (Baird & Klobutcher, 1991; Steinbrück, 1983). Creation of these gene-sized pieces by processing the macronuclear chromosomes generates literally thousands of chromosome ends or telomeres to which the telomeric sequence CCCCAAAA is added (Klobutcher, Swanton, Donini, & Prescott, 1981). Telomerases are the enzymes responsible for addition of these sequences (Blackburn, 1992; Greider & Blackburn, 1987), and telomerase transcripts are tightly regulated during macronuclear development (Price, Adams, & Vermeesch, 1994; Shippen-Lentz & Blackburn, 1989). The telomeric sequences of hypotrichs and stichotrichs have different numbers of GT/CA repeats and show some differences in primary sequence (Prescott, 1994; Steinbrück, 1990). Since much of the amplified micronuclear DNA in the polytene chromosomes is eliminated during anlage development, this raised the question of what molecular signals were used to recognize the non-genic eliminated sequences. It is now clear that transposon-like elements distributed throughout the genome are the first sets of sequences to be eliminated in hypotrichs and stichotrichs (Baird, Fino, Tausta, & Klobutcher, 1989; Herrick et al., 1985; Jahn, Kirkau, & Shyman, 1989; Prescott, 1994). These may have originated from the invasion of the hypotrich/stichotrich micronuclear genome by transposons that originally populated the micronuclear genome but that now are excised by host-directed mechanisms (Klobutcher & Herrick, 1997). The chromosome fragmentation process appears to use unique sites in *Euplotes* but multiple, closely spaced sites in *Oxytricha* (Baird & Klobutcher, 1989). Not only are sequences eliminated **between** the ends of micronuclear genes, but sequences are also eliminated **within** micronuclear genes leading to what are called internally eliminated sequences or IESs and macronuclear destined sequences or MDSs (Klobutcher, Jahn, & Prescott, 1984; Prescott, 1994, 1998). The macronuclear destined sequences are then ligated to construct the functional genes. The story becomes even more bizarre: the macronuclear destined sequences in some genes are actually scrambled so that their linear order in the micronuclear genome would not yield a functional gene if ligation occurred simply by joining the cut ends (Greslin, Prescott, Oka, Loukin, & Chappell, 1989). Genes have now been discovered with up to 48 scrambled, macronuclear destined sequences, stimulating intriguing explanations as to how the spirotrichs have solved the complex computational problem of assembling a functional gene (Landweber & Kari, 1999; Landweber, Kuo, & Curtis, 2000; Prescott, 1999). Internally eliminated sequences were apparently added successively into genes, first without scrambling. Scrambling occurred later likely by recombination pathways that gave rise to divergent arrangements in the descendant lineages (Hogan, Hewitt, Orr, Prescott, & Müller, 2001; Wong & Landweber, 2006).

A final unusual aspect to the stichotrich genome is the change in the universal genetic code with deviations from the universal stop codons - UAA, UGA, and UAG. Helftenbein (1985) demonstrated that tubulin genes of *Stylonychia* use the universal UAA stop codon to code for the amino acid glutamine. There are UAA and UAG internal codons in a putative *Oxytricha* gene (Herrick, Hunter, Williams, & Kotter, 1987), a feature that presumably evolved independently in *Tetrahymena* and *Paramecium* (see **Chapter 15**). In contrast, the hypotrich *Euplotes* continues to use UAA as the stop codon (Harper & Jahn, 1989; Miceli, La Terza, & Melli, 1989) and codes cysteine using UGA (Meyer et al., 1991). We do not know how often codon deviations have occurred during the evolution of the spirotrichs. Do the oligotrichs and choreotrichs have a codon useage similar to hypotrichs or stichotrichs? The most plausible explanation for these deviations in the spirotrichs, and other ciliates, is the evolution of translational release factors with a higher specificity for one or other of the universal stop codons (Caron, 1990). There is now evidence, in both hypotrichs and stichotrichs, of coevolution between these genetic code changes and the recognition site of eukaryotic release factor 1, which is the protein that recognizes stop codons and terminates translation (Inagaki & Doolittle, 2001; Lozupone, Knight, & Landweber, 2001).

Micronuclear genome organization has been the subject of several studies. The micronuclear genome shows considerably more sequence complexity than the macronuclear genome that is derived from it. In both hypotrichs and stichotrichs, "macronuclear" genes are typically clustered together along the micronuclear chromosomes, and long stretches of eliminated unique sequences are uninterrupted by repetitive sequences (Jahn, Nilles, & Krikau, 1988a; Jahn, Prescott, & Waggener, 1988b; Klobutcher, 1987).

Conjugation in spirotrichs is characterized as temporary. In rare instances, total conjugation, that is the fusion of both conjugants, has been recorded in tintinnids (Gold, 1971) and stichotrichs (e.g., *Urostyla*, *Pseudourostyla*, Heckmann, 1965; Raikov, 1972, Takahashi, 1974). Cells are typically of similar size although cells of different size have been shown to mate preferentially under some conditions (Gold & Pollingher, 1971). Laval-Peuto (1983) observed cultured forms of *Favella* to mate even when the loricae of cells were of different types. Fusion of cells typically occurs near the oral regions often accompanied by disassembly of cortical alveoli and some oral structures and assembly of microtubules and microfilaments (Dallai & Luporini, 1989; Geyer & Kloetzel, 1987a, 1987b; Laval-Peuto, 1983; Rosati, Verni, & Dini, 1998). As in the Class OLIGOHYMENOPHOREA (see **Chapter 15**), microtubules and microfilaments are presumably involved in the positioning and movement of the gametic nuclei.Completion of macronuclear maturation and post-conjugation morphogenesis (**see below**) in hypotrichs and stichotrichs may take several days during which feeding does not occur. During this time, energy may be supplied by programmed autophagocytosis of cell constituents (Sapra & Kloetzel, 1975). As in other ciliates, spirotrichs are induced to conjugate by a variety of factors, with reduction in food resources being a dominant one (e.g., Adl & Berger, 2000; Gates & Ramphal, 1991). Their sexual life cycle is characterized by a period of immaturity in which conjugation cannot effectively occur, a period of maturity or conjugation competence, and a period of senescence (Miyake, 1996; Smith-Sonneborn, 1981). The development to maturity is not an instantaneous switch in all cells (Dini & Nyberg, 1994), while the length of the immaturity period can be determined by cytoplasmic factors (Dini, Bleyman, & Giubbilin, 1990). The intensity of mating reactivity may show daily rhythms in some *Euplotes* species (Gates & Ramphal, 1991; Miyake & Nobili, 1974).

The breeding systems of spirotrichs are generally multipolar, that is with many more than two mating types: many more than 100 mating types have been recorded in the stichotrich *Stylonychia* (Ammermann, 1982; Ammermann & Schlegel, 1983); and up to 38 have been recorded in *Euplotes* species (Dini & Luporini, 1979, 1985; Heckmann, 1964; Kimball, 1942; Nobili, 1966). On the other hand, bipolar breeding systems have been observed in other *Euplotes* species (Katashima, 1959) and its hypotrich relative *Aspidisca* (Dini et al., 1987). Strains isolated from geographically distant localities may show varying intensities of mating reactivity, but this has not resulted in genetic partitioning as there is still clear evidence of gene flow (Ammermann et al., 1989; Kusch, Welter, Stremmel, & Schmidt, 2000; Mollenbeck, 1999; Valbonesi, Ortenzi, & Luporini, 1992).

Preparation for conjugation in stichotrichs and hypotrichs may be mediated by chemical signals, called pheromones or gamones, which diffuse through the medium and induce cells of complementary mating type to prepare for conjugation (Esposito, Ricci, & Nobili, 1976; Heckmann & Kuhlmann, 1986; Luporini & Miceli, 1986; Luporini, Vallesi, Miceli, & Bradshaw, 1995). Secretion of multiple pheromones appears to occur sequentially as the cells mature sexually (Kuhlmann & Heckmann, 1989). The pheromones are proteins whose genes have been sequenced (Miceli, La Terza, Bradshaw, & Luporini, 1992; Raffioni, Luporini, & Bradshaw, 1989). Several models now exist to explain how pheromones from different mating types may interact with each other to induce mating reactivity in cells of complementary mating types (Luporini & Miceli; Miyake, 1996; Ortenzi et al., 2000). Pheromones may also function in an autocrine fashion, stimulating mitosis in some *Euplotes* species (Luporini, Alimenti, Ortenzi, & Vallesi, 2005). There is suggestive evidence that pheromones may also attract cells of the complementary mating type (Kosaka, 1991a) and influence the locomotory behavior of cells by eliminating rhythmic, spontaneous membrane depolarizations (Stock, Kruppel, Key, & Lueken, 1999). While diffusible pheromones appear to be the rule among hypotrichs and stichotrichs, we do not know how widespread they are within the Class SPIROTRICHEA. Indeed, there are some species of *Euplotes* in which the signal substances appear to be firmly bound to the cell surface (Heckmann & Siegel, 1964).

There are typically three prezygotic divisions of the micronucleus of spirotrichs, but as few as two and as many as four have been observed (Raikov, 1972). In some *Euplotes* species, the gametic nuclei may not be sister nuclei. Thus, the two exconjugants will not be isogenic (Baird & Klobutcher, 1988; Katashima, 1960; Kuhlmann & Heckmann, 1991) nor will autogamous forms be homozygous (Dini, Raikov, & Bracchi, 1999).

Once fertilization has occurred, cells separate when the fusion zone is resorbed and/or contractile processes operate (Geyer & Kloetzel, 1987a). During conjugation and the long postconjugation period, hypotrichs typically undergo two rounds of cortical reorganization while stichotrichs undergo three rounds of cortical reorganization (Ng, 1990; Tuffrau, Fryd-Versavel, & Tuffrau, 1981; Tuffrau, Tuffrau, & Genermont, 1976). The first reorganization is correlated with separation of the conjugants. Since the cytostome is non-functional, this necessitates additional rounds of reorganization. Ng (1990) has analyzed these processes in terms of developmental heterochrony in which he believes the sexual cycle overlaps the preceding asexual cycle. The first reorganization is similar to asexual development in that it bears similarities to asexual cortical development and the micronucleus is dispensable to this process (Ng, 1990; Zou & Ng, 1991). In the hypotrich *Euplotes*, the second cortical reorganization requires the presence of maternal macronuclear fragments (Fidler, Jayaraman, & Kloetzel, 1985), while in stichotrichs a macronuclear anlage and micronuclei are necessary to proceed through both the second and third cortical reorganizations (Lu, Shi, & Ng, 1991; Ng, 1990). In *Paraurostyla weissei*, which undergoes total conjugation, an exconjugant zygocyst is formed as the entire infraciliatures of both donor and recipient cells are resorbed. Nevertheless, there are still three rounds of cortical reorganization (Frontczak-Baniewicz & Jerka-Dziadosz, 1992). Retention of the interphase or vegetative pattern of superficial cortical microtubules indicates that the "cell pattern" is retained throughout this complex conjugation process (Fleury & Laurent, 1994). Whether these differences in numbers of cortical reorganizations during and following conjugation are phylogenetically significant is an open question. Ng (1990) noted that the stichotrich *Kahliella* apparently undergoes only two cortical reorganizations (Fleury & Fryd-Versavel, 1982). Could this reflect it's presumed basal position in the adaptive radiation of this subclass? And we do not yet know how many cortical reorganizations oligotrichs and choreotrichs might undergo.

Senescence follows the maturity period of the life cycle, and is characterized by reduced growth rate and increased mortality (Smith-Sonneborn, 1981). Clones can be rejuvenated by undergoing conjugation. Hypotrichs and stichotrichs typically have long immaturity and maturity periods, which means that they are typically outbreeding organisms: the probability of encountering cells of like type being decreased the longer the cell has a chance to disperse (Nanney, 1980; Sonneborn, 1957). In circumstances where complementary mating types are not present, some hypotrichs may undergo intraclonal conjugation or selfing (Akada, 1986; Kosaka, 1990; Machelon, 1986). This has the advantage of resetting the life cycle clock, but the disadvantage that it can only be a short term strategy as over several generations it can lead to lethal inbreeding depression (Kosaka, 1982).

Another route to resetting the life cycle clock is autogamy, a process of self-fertilization undertaken by a single cell. Some autogamous strains of *Euplotes* species are determined by a dominant allele at a single locus (Heckmann & Frankel, 1968; Dini & Luporini, 1980). Although autogamy is an extreme form of inbreeding, heterozygosity is maintained for longer periods in these species because the meiotic products of non-sister nuclei may form the zygotic nucleus or synkaryon (Dini et al., 1999; Luporini & Dini, 1977). Nevertheless, autogamous strainsare less tolerant to stresses, for example, mercury toxicity, than non-autogamous or outbreeding strains (Dini, 1981). Further, changes in body proportions of autogamous strains relative to non-autogamous strains may inhibit effective cell pairing (Gates, 1990).

Sibling or cryptic species are found among spirotrichs (Valbonesi, Ortenzi, & Luporini, 1988; Valbonesi, Ortenzi, & Luporini, 1992) as they are among other groups of ciliates (Nanney & McCoy, 1976; Sonneborn, 1957, 1975). Most genetic work on the species problem in spirotrichs has focused on *Euplotes* species where there are competing conclusions on whether one or another "species" of *Euplotes* is reproductively isolated. For example, using mating tests, Valbonesi et al. (1988, 1992) claimed that *Euplotes crassus* is not a sibling species complex, but it is a species separate from *Euplotes vannus*. Caprette and Gates (1994) claimed that these two "species" were not reproductively isolated. Nevertheless, they cautioned that until the extent of interbreeding is known in nature, results of laboratory experiments must be interpreted with caution.

Valbonesi et al. (1988) have also used characteristics of isoenzymes to distinguish *E. crassus*, *E. vannus*, and *Euplotes minuta*, all of which demonstrated discretely different patterns in five isoenzymes, differences that are as great as those used to separate species of the *Tetrahymena* and *Paramecium* sibling species complexes (Nanney & McCoy, 1976; Sonneborn, 1975). Isoenzyme differences clearly distinguish morphologically different species of the hypotrich *Euplotes* (Machelon & Demar, 1984; Schlegel, Kramer, & Hahn, 1988) and the stichotrich *Stylonychia*, even when isolated from separate continents (Ammermann et al., 1989). Schmidt, Ammermann, Schlegel, & Bernhard (2006a) have identified a single nucleotide difference in the SSUrRNA genes of *Stylonychia lemnae* from Eurasia and North America. This tentatively suggests a biogeography, a conclusion that was also tentatively reached in a study of strains of the soil stichotrich *Gonostomum affine* from Europe, Africa, and Asia (Foissner, Stoeck, Schmidt, & Berger, 2001).

More recently, random amplified polymorphic DNA or RAPD fingerprinting has been used to demonstrate genetic diversity **within** *Euplotes aediculatus* (Kusch et al., 2000) and *Euplotes octocarinatus* (Mollenbeck, 1999) and also **between** morphospecies of *Euplotes* (Chen, Song, & Warren, 2000). The intraspecific analyses concluded that there was no geographic subdivision of species despite continental separation of some strains, confirming the results of isoenzyme studies on stichotrichs (Ammermann et al., 1989). This indicates that conjugation must be frequent enough across intercontinental geographic distances to essentially maintain a single gene pool, even though it is rarely observed in natural populations (Lucchesi & Santangelo, 2004). The rarity of conjugation in *Euplotes* was supported by RAPD analysis of a population of *Euplotes daidaleos* in Germany: the genetic diversity was very low, indicating a clonal population structure rarely undergoing conjugation (Kusch & Heckmann, 1996).

7.7 Other Features

As with heterotrichs (see **Chapter 6**), the widespread distribution of hypotrichs and stichotrichs coupled with the ease of culturing them has led to their use in monitoring environmental quality. Hypotrichs and stichotrichs can be found in extremely acidic environments (Packroff & Wöfl, 2000) although some oligotrichs may be quite sensitive (Pedersen & Hansen, 2003). They are also very abundant in the biofilms of water treatment facilities (Curds, 1969; Martin-Cereceda, Serrano, & Guinea, 2001a; Perez-Uz et al., 1998), presumably playing a role by feeding upon bacteria in the biofilms (Lawrence & Snyder, 1998). Hypotrichs and stichotrichs have been used to bioassay copper, nickel, cadmium, and organic compounds (Albergoni et al., 2000; Madoni, 2000; Piccinni, Irato, Cavallini, & Ammermann, 1992; Stebbing, Soria, Burt, & Cleary, 1990). They showed broad variations in sensitivities to different toxicants: *Euplotes* species can be highly tolerant of nickel (Madoni, 2000) or highly sensitive to nickel (Madoni & Romeo, 2006) and to copper (Albergoni et al.); *Halteria* can be highly sensitive to cadmium (Madoni & Romeo, 2006). Resistance to heavy metals may be conferred on stichotrichs by the presence of unique metal-binding proteins, very different from metallothioneins and chelatins isolated from other protozoa (Piccinni et al., 1992).

As noted above (see **Life History**), spirotrichs can be important predators in microbial food webs, ingesting a variety of prey organisms from bacteria to other ciliates and metazoa. This can have important consequences for humans. For example, Tso and Taghon (1999) demonstrated that *Euplotes* did not show selectivity for contaminant-degrading bacteria, which may have important implications for bioremediation initiatives. On the other hand, feeding by hypotrichs and stichotrichs might remove *Cryptosporidium* oocysts from wastewaters, helping to decrease the incidence of waterborne outbreaks of cryptosporidiosis (Stott, May, Matsushita, & Warren, 2001).

Chapter 8
Subphylum 2. INTRAMACRONUCLEATA: Class 2. ARMOPHOREA – Sapropelebionts that Once Were Heterotrichs

Abstract The Class ARMOPHOREA represents a new assemblage of ciliates, and one of the two "riboclasses" as their establishment is completely dependent upon small subunit rRNA gene sequences that showed affinities of the two included orders – Armophorida and Clevelandellida. The ciliates in this class occupy anoxic habitats. Armophorids are typical of sapropelic habitats, but can be benthic or planktonic, while clevelandellids are endosymbionts in the digestive systems of a wide variety of invertebrates, particularly insects, and some vertebrates, particularly amphibians. While their somatic dikinetids are quite different, armophoreans are all characterized by having their mitochondria transformed to hydrogenosomes, organelles that provide hydrogen to the methanogenic bacterial symbionts of these ciliates. The oral structures of the two orders are also divergent: membranelle-like in armophorids and heteromembranelles in clevelandellids. Stomatogenesis is pleurotelokinetal. The macronucleus is of simple form, but polytene chromosomes develop after conjugation and the macronuclear DNA ultimately differentiates into gene-sized pieces. Armophorids, because of their habitat preferences, are particularly good bioindicators of anoxic aquatic environments.

Keywords Endosymbiont, cathetodesmal fibril, sulfureta, secant system

The ciliates included in this class are typically small to medium-sized cells. Armophoreans are free-swimming and typically holotrichously ciliated. However, their body ciliation can vary from many, densely ciliated kineties in some clevelandellids to only anterior and posterior cirrus-like tufts in some armophorids. All species have multiple adoral polykinetids, ranging from around a dozen in some armophorids to several dozens in some clevelandellids. These ciliates are very restricted in their distribution. Although world-wide, they are confined to sediments, both aquatic (Fenchel, 1993) and terrestrial (Foissner, 1987), and the water column (Fenchel et al., 1995), where oxygen tensions are extremely reduced to absent. They are also found as endocommensal symbionts in the digestive tracts of a variety of metazoans, ranging from selected invertebrates (Albaret, 1970b; Hackstein & Stumm, 1994) through to amphibians (Affa'a, Ndongo, & Granosik, 1995). Interest has increased in the group recently because they harbor endosymbiotic methanogenic bacteria, which can themselves produce the greenhouse gas, methane. There can be thousands of methanogenic bacteria per ciliate (van Bruggen, Stumm, & Vogels, 1983), producing significant quantities of methane, which is then liberated into the environment (Fenchel & Finlay, 1992; Hackstein & Stumm, 1994).

The name of the class, ARMOPHOREA, is derived from the subordinal name originally proposed by Jankowski (1964a, 1964b) to include only the caenomorphid heterotrichs, which he argued derived from a *Metopus*-like ancestor. It derives from the Latin *arma*, meaning weapons (or it derives from the Latin *armus* meaning shoulder), and refers to the fact that caenomorphids have the appearance of military helmets (or the *caenomorphid* body is twisted to give the appearance of a shoulder). Although not highly similar, a number of clevelandellids have conspicuous polysaccharide "skeletal" elements in their cortex, an "armor" of a different sort (see Albaret, 1970a).

Like the Class SPIROTRICHEA, there is no conspicuous synapomorphy for members of this class. They are united by the following three features. First, they are restricted to anaerobic habitats and are typically dependent upon methanogenic symbionts. Although this is not a unique feature for the Class ARMOPHOREA (see particularly **Chapter 12. Class PLAGIOPYLEA**), we predict that the metabolic dependence on hydrogenases in this class will be shown to have a common phylogenetic origin. Second, clevelandellids and armophorids share pleurotelokinetal stomatogenesis of the adoral polykinetids, a feature shown by members of other classes (Foissner & Agatha, 1999). Finally, they show strong similarities in the sequences of their small subunit rRNA (SSUrRNA) genes (Embley et al., 1995; Hackstein, Van Hoek, Leunissen, & Huynen, 2001; van Hoek, van Alen, Sprakel, Hackstein, & Vogels, 1998). This class could be called the first "riboclass" of ciliates, since its monophyly is predicted by sequence analyses of the SSUrRNA genes. However, we do not yet have a signature sequence that would characterize the class.

8.1 Taxonomic Structure

The two major groups – the clevelandellids and armophorids – included in this class have long been considered heterotrichs because of their possession of multiple adoral polykinetids (Fig. 8.1). Corliss (1979) considered them to be suborders within the Order Heterotrichida. However, early ultrastructural analysis demonstrated clear differences between the somatic and oral structures of clevelandellids and their presumed "heterotrich" relatives. The somatic dikinetids do not give rise to postciliodesmata, their kinetodesmal fibril is differently shaped, and there is a prominent left-directed striated cathetodesmal fibril arising adjacent to the anterior kinetosome (Paulin, 1967; de Puytorac & Grain, 1969, 1976). Although there is still no published account devoted solely to the ultrastructure of armophorids, Schrenk and Bardele (1991) have indicated differences between the somatic kinetid of the armophorid *Metopus* and those of clevelandellids. It does appear that *Metopus* may have cathetodesmal-like fibrils, which do not appear striated. Little research has been done on members of this class, outside the recent interest in their symbiotic methanogens (**see below Life History and Ecology**).

We place armophorids and clevelandellids in the Class ARMOPHOREA primarily based on their strong association derived from sequence similarities of the SSUrRNA gene: the clevelandellids *Nyctotherus* and *Nyctotheroides* strongly group with the armophorids *Metopus* and *Caenomorpha* (Embley et al., 1995; van Hoek et al., 1998). Both Jankowski (1968b) and Albaret (1975) have suggested that clevelandellids may have derived from metopids through transformation of the cortical patterning, following a suggestion by Villeneuve-Brachon (1940). Therefore, we place these two groups together and elevate them to ordinal status, as others have done (Lynn & Small, 1997, 2002; de Puytorac, 1994a; Small & Lynn, 1985). Following Jankowski (1964a, 1964b, 1968b) and Albaret (1975), we assume that the free-living armophorids represent the descendants of the ancestral group from which the endosymbiotic clevelandellids evolved.

The Order Armophorida includes two families: the Family Metopidae and the Family Caenomorphidae (Fig. 8.1). In most forms, there is a slight twist left to the anterior end of the body, which is covered by up to five perizonal or epistomial kineties (e.g., Fernández-Galiano & Fernández-Leborans, 1980; Jankowski, 1968b). This twist becomes pronounced in derived forms and in all caenomorphids (Fig. 8.1). Caenomorphids are not typically holotrichous, but rather may have the somatic ciliation restricted to anterior and posterior cirrus-like tufts.

The Order Clevelandellida has not changed in composition since Corliss (1979). It contains five families: the Family Nyctotheridae, the Family Sicuophoridae, the Family Clevelandellidae, the Family Inferostomatidae, and the Family Nathellidae. The latter two families are monotypic. Clevelandellids are densely ciliated, often laterally compressed ciliates with many left serial oral polykinetids that are hidden in a groove-like peristome and deep oral cavity or infundibulum (Fig. 8.1). These obligate endosymbionts are commensal in a wide range of hosts: *Nyctotherus* is

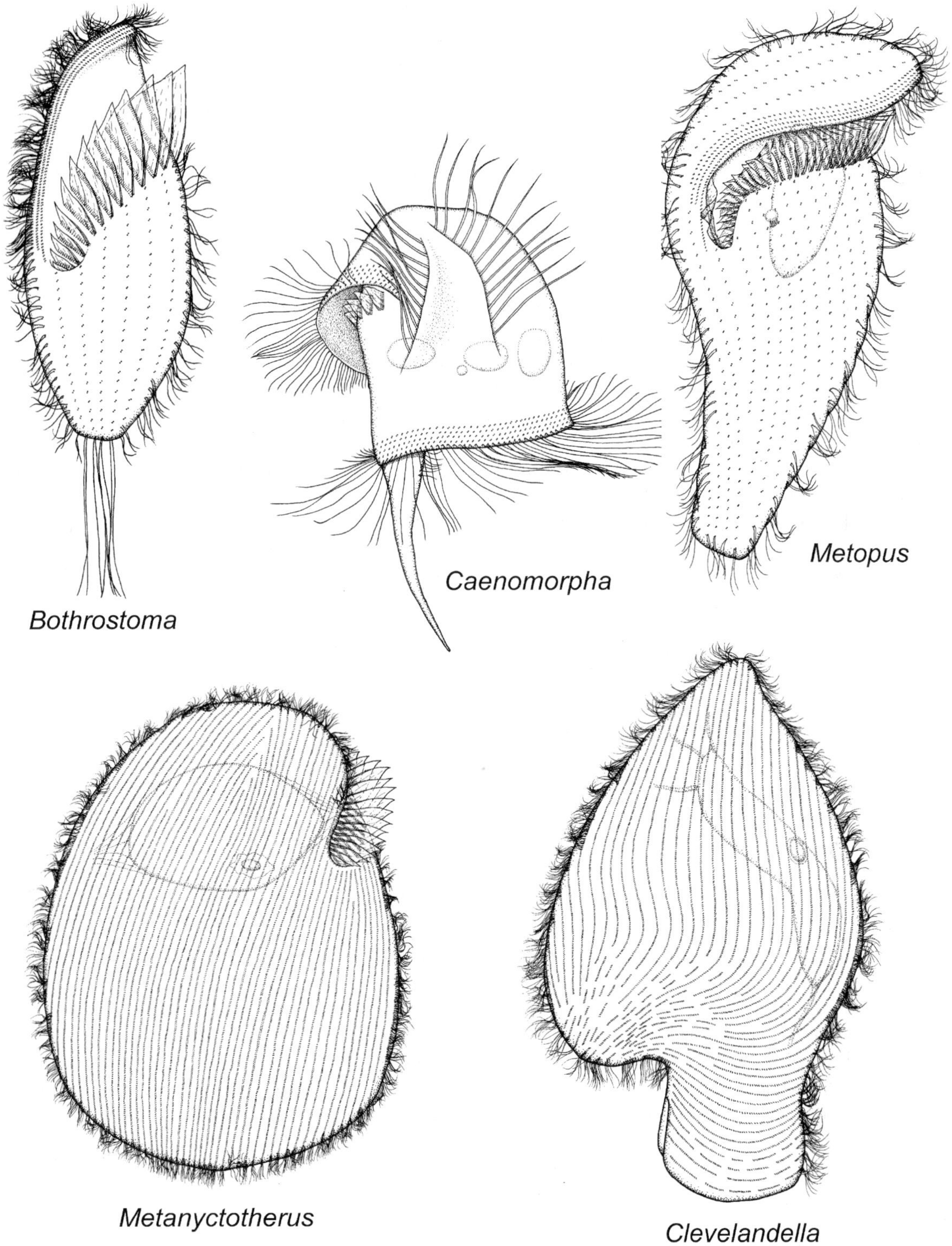

FIG. 8.1. Stylized drawings of representative genera from the two orders in the Class ARMOPHOREA. Order Armophorida: the metopids *Bothrostoma* and *Metopus*, and the caenomorphid *Caenomorpha*. Order Clevelandellida: *Nyctotherus* and *Clevelandella*

found in oligochaetes, insects, and myriapods; *Nyctotheroides* is found in frogs and toads; and *Clevelandella* is found in wood-feeding roaches and termites (see **Life History and Ecology**).

Systematic research on members of this group has been done by literally a handful of investigators, following monographic work on the armophorids and caenomorphids by Jankowski (1964a, 1964b) and on the clevelandellids by Albaret and coworkers (Albaret, 1975; Albaret & Njiné, 1976). Exploration of the biodiversity of clevelandellid symbionts of African anurans has been expanded considerably by Affa'a (1980, 1983, 1988b) while Affa'a (1989) and Grim (1992) have described new genera symbiotic in fishes (see also Earl, 1991).

Esteban, Fenchel, and Finlay (1995) have taken a conservative approach in their revision of *Metopus*, reducing 76 nominal species to 22 morphospecies. It will be up to molecular systematists to determine if these morphospecies are really as phenotypically variable as presumed by Esteban et al. (1995).

8.2 Life History and Ecology

Armophoreans, like most ciliates, are globally distributed. A novel technical approach to their study used electromigration to extract these often sediment-dwelling ciliates from their habitats (Wagener, Stumm, & Vogels, 1986). Free-living armophorids have been found in freshwater and marine habitats in Eurasia (e.g., Agamaliev, 1974; Finlay & Maberly, 2000; Grolière & Njiné, 1973; Guhl, Finlay, & Schink, 1996; Madoni & Sartore, 2003) and North America (Bamforth, 1963; Borror, 1963), and chloride lakes (Madoni, 1990). In these habitats, they are part of the sulfureta community, which may also include ciliates from the Classes HETEROTRICHEA, PLAGIOPYLEA, and OLIGOHYMENOPHOREA (Dyer, 1989; Fenchel, 1987). Foissner (1987, 1995b) recorded metopids from temperate and tropical soils in which they survive by encystment. Encystment is crucial to the transmission between hosts of the clevelandellids, all of which are endosymbionts in both terrestrial and aquatic metazoans. These ciliates have been recorded from diverse hosts: insects (Hackstein & Stumm, 1994; Lalpotu, 1980a, 1980b; Zeliff, 1933), millipedes (Albaret, 1970b; Hackstein & Stumm; Lalpotu, 1980c), molluscs (Laval & Tuffrau, 1973), sea urchins (Biggar & Wenrich, 1932; Grolière, de Puytorac, & Grain, 1980b), fishes (Grim, 1998; Grim, Clements, & Byfield, 2002; Grim, Reed, & Fishelson, 1995/1996; Jankowski, 1974a), amphibians (Albaret, 1975; Affa'a et al., 1995; Wilbert & Schmeier, 1982), and reptiles (Geiman & Wichterman, 1937; Takahashi & Imai, 1989).

Free-living armophorids are restricted to anoxic or microaerobic habitats, such as the anoxic hypolimnion in lakes and bays or the anoxic layers in sediments. The armophorids *Caenomorpha* and *Metopus* can reach abundances of more than $5{,}000\,l^{-1}$ in the water column, but are typically much less abundant than this (Fenchel & Finlay, 1991a; Fenchel, Kristensen, & Rasmussen, 1990; Guhl & Finlay, 1993; Guhl et al., 1996). Armophorids increase their relative abundance in sediments during periods of anoxia, reaching more than $50\,ml^{-1}$ of sediment (Fenchel, 1993; Finlay, 1982). These ciliates survive best at low oxygen concentrations. They exhibit a chemosensory response to oxygen concentration: they increase their swimming speed at higher oxygen concentrations and show ciliary reversals when leaving anoxic conditions and entering an oxygen zone (Fenchel & Finlay, 1990a). The abundances of symbiotic clevelandellids depend partly on the host. Wilbert and Schmeier (1982) recorded hundreds of *Nyctotheroides* in some frog hosts while Gijzen and Barugahare (1992) recorded over 10^4 ml^{-1} *Nyctotherus* in the hindgut of the American cockroach *Periplaneta americana*.

Armophoreans typically feed on heterotrophic and phototrophic purple bacteria, and typically grow more slowly than comparably-sized aerobic ciliates with generation times in the order of days (Fenchel & Finlay, 1990b). *Metopus* requires bacterial abundances of more than 10^7 ml^{-1} for maximum growth (Massana, Stumm, & Pedrós-Alió, 1994). The abundance of *Caenomorpha* is correlated with the abundance of its photosynthetic bacterial prey, *Thiopedia*, suggesting that there is chemosensory tracking of prey by this ciliate predator (Guhl & Finlay, 1993). While Guhl and Finlay (1993) concluded that *Thiopedia* production is controlled by *Caenomorpha*, Massana and Pedrós-Alió (1994) concluded in another habitat that anaerobic ciliates do not likely control bacterial production. The growth efficiencies of anaerobic ciliates are quite low, less than 10%. Although these ciliates are not dependent upon their intracellular endosymbiotic methanogenic bacteria, their growth rates can, in some cases, be reduced if deprived of their bacteria. Although there is yet no direct evidence, the methanogens in these cases may be supplying the host ciliate with organic excretions to enhance the growth rate (Fenchel & Finlay, 1991b).

One of the first surveys of symbiotic bacteria was that of Fenchel, Perry, and Thane (1977) who reported both ectosymbiotic and endosymbiotic bacteria in the armophoreans *Caenomorpha* and *Metopus*. Endosymbiotic methanogenic bacteria have been reported in members of both orders of armophoreans (e.g., Fenchel & Finlay, 1991a; Gijzen & Barugahare, 1992). Many of these bacteria have been confirmed to be methanogens, which can number from hundreds to over 8,000 per ciliate (Fenchel, 1993). They can take various shapes from elongate rods, up to 7 μm in length, to coccoid forms, about 0.5 μm in diameter. Methanogens were identified first on the basis of a characteristic, fluorescent, deazaflavin coenzyme F_{420} (van Bruggen et al., 1983). Van Bruggen, Zwart, van Assema, Stumm, and Vogels (1984) and Van Bruggen et al. (1986) were first to isolate and characterize the methanogens to the genera *Methanobacterium* and *Methanoplanus*. Use of the polymerase chain reaction has increased the diversity of methanogens to include potentially other genera, such as *Methanolobus* and *Methanocorpusculum* (Embley & Finlay, 1994). In both free-living and symbiotic armophoreans, unrelated ciliates may contain the same methanogen species while the same ciliate species may at different times or in different hosts carry different methanogen species. This demonstrates that losses and acquisitions of methanogens are continually occurring and some may be quite recent acquisitions (Embley & Finlay, 1993; van Hoek et al., 2000b). We do not yet know how the association is established since the bacteria lie in the cytoplasm not surrounded by a cell membrane.

Methanogen symbiosis has attracted recent interest because methane is a greenhouse gas. Thus, ciliates could potentially contribute indirectly to greenhouse gases by "growing their own methane producers." Indeed, significant amounts of methane production have been attributed to these ciliate endosymbionts. Up to 95% of the methane production in certain marine habitats has been attributed to the ciliates (Fenchel, 1993), but in other habitats methanogenesis derived from ciliate endosymbionts is a transient and minor contribution (Schwarz & Frenzel, 2005). In contrast, over 80% of the methane produced by the American cockroach can be attributed to ciliates (Gijzen & Barugahare, 1992). In other anaerobic habitats, stimulation of bacterial production by ciliate grazing can enhance methane production, here not by endosymbiotic bacteria, but by the free-living methanogens. Organic acids, such as acetate and propionate, excreted by the ciliates may stimulate bacterial growth (Biagini, Finlay, & Lloyd, 1998).

Research on the endosymbiotic armophoreans, the clevelandellids, has primarily focussed on the symbionts of frogs and insects. The amphibians of Cameroon have provided a rich resource to probe the biology of the clevelandellids. Frog's eggs are not infected and frog's with a direct life cycle were never found to carry ciliates. The small frog *Phrynodon sandersoni* provides a "natural experiment" to confirm these facts. Its tadpoles develop **without** a digestive tract; of course, the tadpoles are uninfected and so are the adults (Amiet & Affa'a, 1985). Affa'a and coworkers (Affa'a, 1988a; Affa'a & Amiet, 1985, 1994; Amiet & Affa'a, 1985) have concluded that there are three general life histories of infaunation. First, the ciliates may be found only in the juvenile or tadpole stages of the host: this applies to such species as *Nyctotheroides brachystomus*, *Neonyctotherus reticulatus*, and *Parasicuophora aberrans*. Second, other species, such as *Nyctotheroides heterostomus* and *Prosicuophora basoglui*, infaunate only the adult stage. Finally, both tadpole and adult stages are infaunated by other species, such as *Nyctotheroides teochii*.

We do not know what factors control the disappearance of ciliates from the tadpole or the appearance of ciliates in the adults. Affa'a (1986b) has shown that gonadotropins induce encystment in *Prosicuophora* and *Nyctotheroides*. It may be that the changes at metamorphosis of the tadpoles induce encystment in those forms that occur only in the tadpole and induce excystment in those forms that occur only in the adult. Ingestion of cysts is probably the main mode of transmission, although infection by live ciliates may occur since the feces of adult frogs have an abundance of ciliates (Amiet & Affa'a, 1985). The prevalence of a ciliate species in a frog host varies from one locality to another, although it is not yet clear what factors determine this variability (Affa'a, 1986a, 1988a). Geographic variation has also been reported for *Nyctotherus* species that infect cockroaches: similar ciliate genotypes can occur in different insect genera at the same or distant localities (van Hoek et al., 1998). The ciliates apparently have no effect on the amphibian hosts. However, those resident in cockroaches may significantly increase the growth rate and body weight of their hosts (Gijzen & Barugahare, 1992).

Reid and John (1983) characterized the cysts of the clevelandellids as flask-shaped, noting similarities to those of the heterotrichs (see also Esteban et al., 1995; Takahashi & Imai, 1989). Cysts are crucial to the maintenance of the life histories of the endosymbiotic clevelandellids and must certainly be important for those armophorids, such as *Metopus*, which are found in soils. How widely cyst-forming is distributed in other members of the class remains to be determined.

8.3 Somatic Structures

Armophorean ciliates are quite variable in size. Clevelandellids are intermediate in size at around 100 µm; armophorids can range up to 300µm in in length. Shapes are also quite variable. Armophorids, especially caenomorphids, have a rigid, armor-like pellicle with processes and spines, but larger metopids can be quite flexible. The armophorid body is developed into an anterior lobe that can become quite twisted, and along which travel the perizonal or frontal kineties (Fig. 8.1). Smaller forms may have somatic ciliature reduced to anterior and posterior cirrus-like tufts.

On the other hand, clevelandellids are very densely ciliated with closely packed somatic kineties. These somatic kineties converge on each other forming what are called sutures or secant systems (Fig. 8.1). In clevelandellids, these are typically preoral, apical, caudal, and postoral; the length and precise positions of these secant systems is used in distinguishing genera (e.g., Affa'a, 1983; Albaret & Njiné, 1976; Earl, 1991; Grim, 1998).

The cell membrane is underlain by an alveolar layer that may be conspicuous in some caenomorphids (Fenchel et al., 1977), but it is apparently very compressed, or perhaps even absent, in metopids (Fenchel & Finlay, 1991a) and clevelandellids (de Puytorac & Grain, 1969).

Somatic kinetids are dikinetids throughout the class. However, as with the Class SPIROTRICHEA, there is considerable diversity in kinetid structure within the Class ARMOPHOREA. Unfortunately, much of this research remains to be published, appearing only in abstract form or as schematic drawings without micrographic support (Tuffrau & de Puytorac, 1994). We will rely on these but caution that detailed descriptions need to be published to corroborate the drawings (Fig. 8.2). The armophorid somatic dikinetid is characterized as follows: a ciliated anterior kinetosome with a tangential transverse ribbon at triplets 3, 4, 5 and a ciliated posterior kinetosome with a well-developed divergent postciliary ribbon and a laterally-directed kinetodesmal fibril at triplets 5, 6, 7 that may not be striated (Schrenk & Bardele, 1991). Other microtubules have been reported to accompany the anterior transverse ribbon near triplets 5 or 6 while a pair of presumably transverse microtubules is situated between the two kinetosomes opposite triplet 4 of the posterior kinetosome (Da Silva Neto in de Puytorac & Tuffrau, 1994; Esteban et al., 1995) (Fig. 8.2). Foissner and Agatha (1999) observed by silver-staining what might be well-developed cathetodesmal fibrils in several *Metopus* species. The postciliary microtubular ribbons extend alongside each other in the cortical ridges (Fig. 8.3).

Paulin (1967) and de Puytorac and Grain (1969) provided the first evidence of the clevelandellid somatic dikinetid of *Nyctotherus* and *Sicuophora*, respectively. Grim (1998) has provided some information on the dikinetid of the clevelandellid *Paracichlidotherus*. The clevelandellid dikinetid can now be characterized as follows: a ciliated anterior kinetosome that bears a tangential transverse ribbon at triplets 4, 5 and a striated cathetodesmal fibril extending to the lateral left from an origin near triplet 2; and a ciliated posterior kinetosome with a divergent postciliary ribbon and a kinetodesmal fibril homologue at triplets 5, 6 (Fig. 8.2). Grim (1998) reported two transverse microtubules associated with the posterior kinetosome of *Paracichlidotherus*. The striated cathetodesmal fibrils of clevelandellids may be bifurcated (Fernández-Galiano, 1986; de Puytorac & Grain; de Puytorac & Oktem, 1967). De Puytorac and Grain (1969)) illustrated the cathetodesmal fibril of *Sicuophora* as having two origins, one as indicated above on the anterior kinetosome and the other on the posterior kinetosome near the base of the kinetodesmal

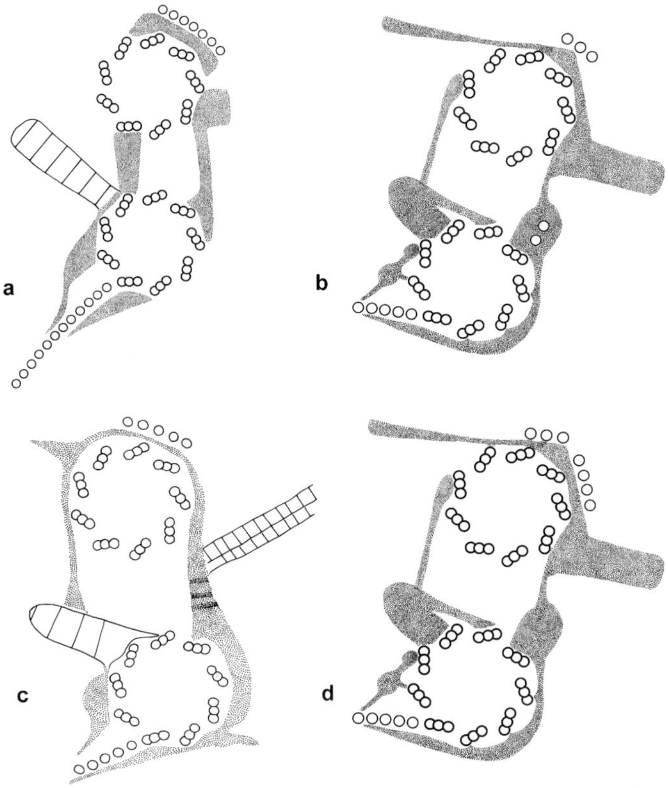

FIG. 8.2. Schematics of the somatic kinetids of representatives of the Class ARMOPHOREA. (**a**) Dikinetid of *Metopus*. (**b**) Dikinetid of *Paracichlidotherus*. (**c**) Dikinetid of *Nyctotherus*. (**d**) Dikinetid of *Sicuophora* (from Lynn, 1981, 1991)

fibril homologue. No micrographic evidence is presented for this interpretation so we have revised our drawing accordingly (Fig. 8.2).

We need to have some detailed reinvestigations of armophoreans before any generalizations can be made about their somatic dikinetids. A further intriguing physiological observation is that *Nyctotherus ovalis* switches swimming direction in response to voltage changes rather than showing a ciliary reversal. Moreover, this behavior appears to be influenced by host-dependent factors (van Hoek et al., 1999).

Contractile vacuoles are present in armophoreans. The cytoproct is often conspicuous, and in clevelandellids may open to the outside by a cilia-lined channel.

Mucocysts appear to be present in the cortex of clevelandellids (Paulin, 1967; de Puytorac & Grain, 1969) and armophorids (Esteban et al., 1995).

Finally, mention must be made of the apparent absence of mitochondria with tubular cristae in all armophoreans. The mitochondria in these ciliates have evolved into hydrogenosomes (van Hoek, Akhmanova, Huynen, & Hackstein, 2000a; Boxma et al., 2005). These hydrogenosomes have a hydrogenase that uses electrons derived from pyruvate oxidation to reduce protons and generate hydrogen (Fenchel & Finlay, 1991a; Müller, 1993; Voncken et al., 2002). The hydrogen is typically used in armophoreans by endosymbiotic methanogens (see **Life History and Ecology**).

8.4 Oral Structures

The armophoreans were placed until recently with the heterotrichs because of their holotrichous somatic ciliation and the presence of multiple oral polykinetids forming an adoral zone. The two or

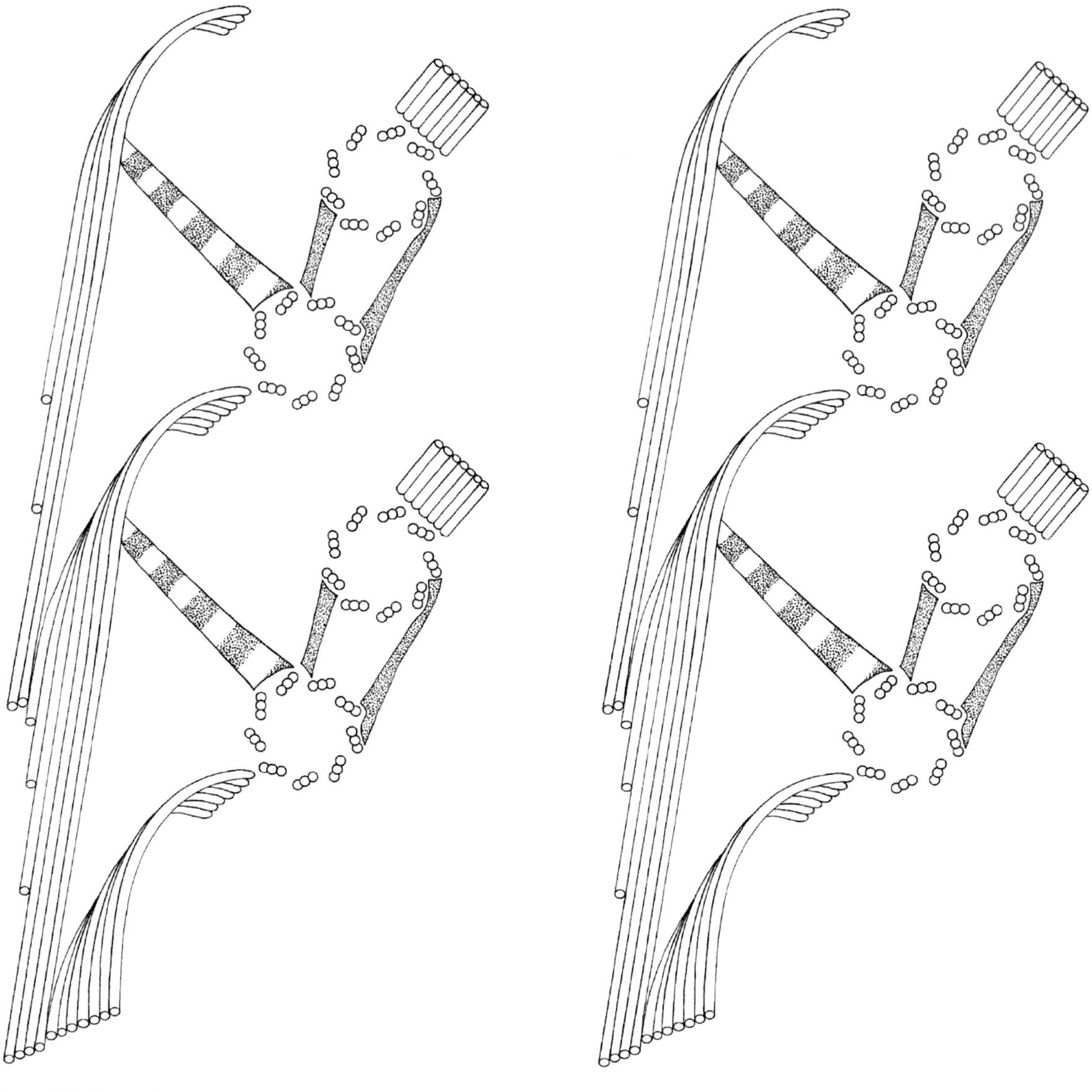

FIG. 8.3. Somatic cortex of *Metopus* whose postciliary ribbons extend alongside each other into the cortical ridges. This schema was constructed based on the brief descriptions provided in reports by Schrenk and Bardele (1991) and Esteban et al. (1995)

three rows of kinetosomes of the oral polykinetids are hexagonally packed. In armophorids, a third or fourth row of kinetosomes is added continuing the hexagonal packing (Esteban et al., 1995; Foissner & Agatha, 1999). Armophorid oral polykinetids have been called paramembranelles. Clevelandellids typically have three rows of kinetosomes hexagonally packed, but a fourth, shorter row lies directly opposite to, rather than hexagonally packed with, the kinetosomes of the third row, leading to their designation as heteromembranelles because of the different packing of these kinetosomes of the fourth row (de Puytorac & Grain, 1976). This different packing leads to a different orientation and beating of the cilia that was nicely revealed in some published micrographs (Paulin, 1967; Takahashi & Imai, 1989).

The adoral zones of armophorids and clevelandellids may be quite extensive, spiralling around the body one or more times in some armophorids

(Fig. 8.1). The clevelandellids have a deeper oral cavity called an infundibulum where the heteromembranelles typically occur (Tuffrau & de Puytorac, 1994). Postciliary ribbons are associated with the kinetosomes of the posterior row in both armophorids and clevelandellids (Tuffrau & de Puytorac, 1994).

Paroral structures are quite variable in the class. Armophorids appear to have a single file of cilia, which may be derived from linearly arranged oral dikinetids (Esteban et al., 1995; Foissner & Agatha, 1999; Sola, Serrano, Guinea, & Longás, 1992). Clevelandellids have a paroral with two sets of cilia deriving from two files of kinetosomes separated by a ridge (Grim, 1998; Paulin, 1967; de Puytorac & Grain, 1969; Takahashi & Imai, 1989), termed a diplostichomonad by de Puytorac and Grain (1976). The oral structures of armophoreans are underlain by complex fibrillar structures and microtubules. The filamentous components are implicated in the movement of vesicles to the food vacuole forming region (Eichenlaub-Ritter & Ruthmann, 1983).

8.5 Division and Morphogenesis

There have been only a few papers on cell division and division morphogenesis of armophoreans since Wichterman (1936) described division in *Nyctotheroides* (= *Nyctotherus*). He observed the oral primordium to develop subequatorially. Since silver-staining was not used, kinetosomal replication was not detailed. As far as we know, armophoreans divide while swimming freely. Foissner (1996b) has characterized stomatogenesis as pleurotelokinetal (i.e., occurring within or at the end of several somatic kineties).

Two studies on the armophorids, *Metopus* and *Caenomorpha*, demonstrated pleurotelokinetal stomatogenesis. Martín-González, Serrano, and Fernández-Galiano (1987) showed that the oral primordium in *Caenomorpha* develops by proliferation from the posterior ends of many perizonal somatic kineties. The primordial field splits later in development with an anterior portion developing into the paroral and the posterior portion developing into the oral polykinetids. In *Metopus*, a number of posterior dorsolateral somatic kineties begin to proliferate kinetosomes (Foissner & Agatha, 1999). These differentiate as the oral polykinetids (Fig. 8.4). The paroral differentiates later. Foissner and Agatha (1999) interpreted it to develop from kinetosomes derived from perizonal kineties. However, it is just as possible from the evidence presented that paroral dikinetids could derive from "anterior" or "right-side" kinetosomes in a fashion very similar to that reported for *Caenomorpha*. If this were the case, there would be strong similarities in

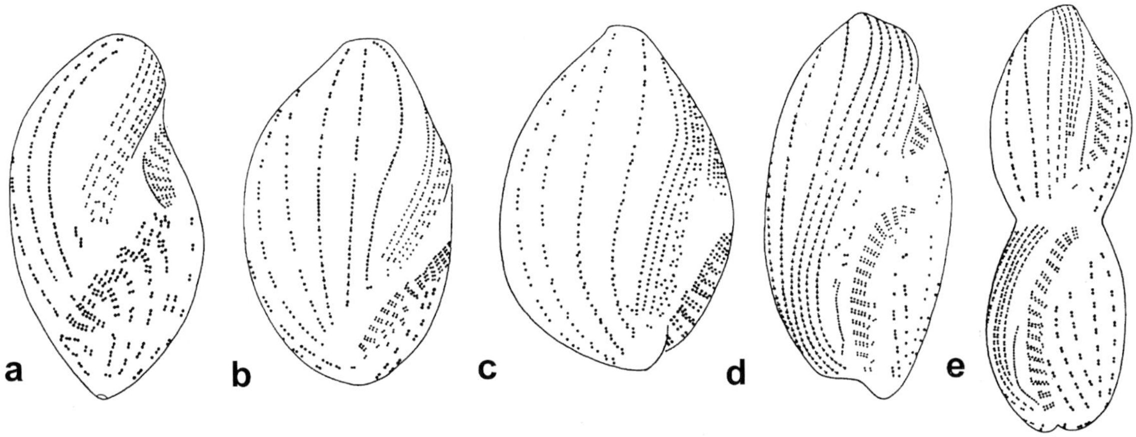

FIG. 8.4. Division morphogenesis of *Metopus*, a representative of the Class ARMOPHOREA. (**a**) Kinetosomal replication begins at the "equatorial ends" of a number of somatic kineties. (**b**) Oral polykinetids assemble through side-by-side alignment of dikinetids units. (**c**) The posterior ends of several somatic kineties adjacent to the developing oral region disassemble, and it may be that the paroral (**d, e**) is assembled from these as division proceeds. (from Foissner & Agatha, 1999.)

stomatogenesis between these two genera. *Caenomorpha* undergoes a complicated post-stomatogenesis morphogenesis, reminiscent of the enantiotropic division of some oligotrichous spirotrichs (Martín-González et al., 1987). Considering the current evidence, we are not convinced that the differences between metopids and caenomorphids are sufficient to justify ordinal status for these two groups, as suggested by Foissner and Agatha (1999).

Santos, Guinea, and Fernández-Galiano (1986) have provided a preliminary account of stomatogenesis in the clevelandellid *Nyctotherus*. Breaks occur in somatic kineties posterior to the oral region and kinetosomal proliferation occurs at the anterior ends of these breaks. A lateral groove develops as proliferation proceeds and primordium elements on the posterior wall of the groove differentiate as oral polykinetids while those on the anterior wall develop as paroral dikinetids, eventually forming the two files of the diplostichomonad (Santos et al., 1986). This is clearly a pleurotelokinetal stomatogenesis, showing significant similarities to that of the armophorids.

8.6 Nuclei, Sexuality and Life Cycle

Armophoreans have the typical complement of macronucleus and one or more micronuclei. The macronuclei can also be variable in number in caenomorphids, sometimes numbering more than four (Fig. 8.1). In smaller forms, the macronucleus is typically globular to ellipsoid, but in larger clevelandellids it can become elongated and quite irregular in shape. The macronucleus of some clevelandellids is "suspended" from the cortex by microfibrillar strands that collectively are called the karyophore (Fig. 8.1).

Eichenlaub-Ritter and collaborators have undertaken some detailed ultrastructural studies on micronuclear and macronuclear division in the clevelandellid *Nyctotherus cordiformis*. Macronuclei divide by intramacronuclear microtubules that are primarily responsible for the elongation of the macronucleus, which is also accompanied on its **outside** by scattered extramacronuclear microtubules (Eichenlaub-Ritter & Tucker, 1984; Hamelmann, Eichenlaub-Ritter, & Ruthmann, 1986). Micronuclear mitosis is an endomitosis, typical of ciliates (Raikov, 1982). There may be three "classes" of microtubules, identified by their differing responses to drugs and temperature, which function to accomplish micronuclear mitosis: (1) manchette microtubules underlying the nuclear envelope; (2) interpolar and kinetochore microtubules, which function during anaphase; and (3) stembody microtubules, which function during telophase to separate the putative micronuclei to each progeny cell (Eichenlaub-Ritter & Ruthmann, 1982a, 1982b). Microtubules in the dividing nuclei may have more than the canonical 13-protofilaments (Eichenlaub-Ritter, 1985; Eichenlaub-Ritter & Tucker).

Conjugation has been studied in only a few examples of armophoreans since the description of it in *Nyctotheroides* (= *Nyctotherus*) by Wichterman (1936). It is not established what factors stimulate conjugation in free-living forms. Wichterman (1936) observed it occurring only in transforming tadpoles of the frog *Hyla versicolor*. This lead to speculation that gonadotropins or some other physiological signal derived from the host may cue these ciliates to begin conjugation. However, Sandon (1941a) observed conjugation in *Paranyctotherus* isolated from the adult clawed frog *Xenopus laevis*, suggesting that other factors are involved. Affa'a and Amiet (1994) have confirmed that conjugation can occur in all stages of the frog life cycle – tadpoles, transforming individuals, and adults. Gonadotropin injections induced conjugation in *Prosicuophora*, even when immature stages were treated (Affa'a, 1986b). Thus, it is unlikely that one single factor stimulates conjugation.

Fusion of the conjugants occurs in the anterior region, and in some *Metopus* species total conjugation may occur (Noland, 1927). The micronuclei of each partner typically undergo three maturation divisions – two meiotic divisions followed by a mitosis of one of the four haploid products (Raikov, 1972; Martín-González et al., 1987). In the total conjugation of *Metopus*, the cytoplasm of one conjugant flows into the partner carrying the gametic nucleus or nuclei with it. However, the old macronucleus is left in the cortical shell of the disgarded partner (Noland, 1927). Following fusion of the gametic nuclei to form the synkaryon, armophoreans typically have one post-synkaryon division with one nucleus becoming the new micronucleus and the other becoming the new macronucleus. In species with more than one macronucleus, there may be additional post-synkaryon divisions (see Martín-González et al., 1987).

Development of the macronuclear anlage in armophoreans is an extremely long process: Golikova (1965) recorded it taking up to 2 weeks in *Nyctotheroides* (=*Nyctotherus*) while Noland (1927) observed a mininimum of 1 week in *Metopus*. In both these genera, it appears that polytene chromosomes are formed at one stage during anlage development. Golikova (1965) concluded that one giant polytene chromosome may form in *Nyctotheroides* by the end-to-end joining of the individual chromosomes. This giant chromosome later fragments both transversely and longitudinally to yield the macronuclear chromosomes (Vinnikova & Golikova, 1978). Ultimately, the macronuclear chromosomes fragment into gene-sized pieces as happens in the Class SPIROTRICHEA (see **Chapter 7**), a fact that Riley and Katz (2001) have confirmed by molecular analyses of the macronuclear DNA of both armophorids and clevelandellids.

8.7 Other Features

The free-living armophorids have been recognized for some time as strong indicators of anoxic aquatic environments (e.g., Bick, 1972; Foissner, 1988a; Sládeček, 1973). They are commonly found in soils (Foissner, 1987) and have been recorded from a variety of municipal landfill sites in the United Kingdom, where they undergo an encystment-excystment cycle in response to starvation and water loss (Finlay & Fenchel, 1991).

Chapter 9
Subphylum 2. INTRAMACRONUCLEATA: Class 3. LITOSTOMATEA – Simple Ciliates but Highly Derived

Abstract The ciliates in this class are divided into two major assemblages, represented by the Subclass Haptoria and Subclass Trichostomatia. The haptorians are predatory ciliates that are commonly found in a variety of habitats, feeding on flagellates and other ciliates, which they immobilize and kill with extrusomes called toxicysts. An exception to this rule is the marine planktonic haptorian *Myrionecta rubra*, which harbors a cryptophyte endosymbiont and which produces red tides that contribute up to 70% of the primary production. The trichostomes have lost toxicysts and are all endosymbionts in a variety of metazoans, ranging from fish to humans. In fact, the only ciliate known to be pathogenic to humankind is the trichostome *Balantidium*, which can be an intestinal parasite. The somatic monokinetid of litostomes is unique in possessing two transverse microtubular ribbons, T1 and T2 – a strong synapomorphy for the class. Oral structures are typically simple, hence the name *litos* (Gr.) for simple. Haptorians typically have either circumoral dikinetids or oralized monokinetids. Trichostomes show more diversity: some forms, like *Balantidium*, have a vestibulum with extensions of densely packed somatic kineties lining it, while the entodiniomorphids have polybrachykineties, more complex assemblages of short kinetofragments. Stomatogenesis is characterized as telokinetal, but there is a range of types from holotelokinetal to cryptotelokinetal.

Keywords Karyoklepty, pexicyst, bulge microtubules, conocyst

The ciliates included in this class are divided into two subclasses, the Subclass Haptoria and the Subclass Trichostomatia. The former includes free-living and the latter endosymbiotic forms. These ciliates are extremely variable in size and form. Small endosymbiotic trichostomes can be around 50 μm in length while some free-swimming haptorians, such as *Homalozoon* species, can be 2,500 μm long! Form is no less variable: small forms are typically ovoid while elongate forms are ribbon-like, flexible, and contractile. Endosymbiotic forms can have lobes, spines, and unusual cell processes, while some free-swimming forms have extensible "necks", flexible proboscises, and toxicyst-bearing tentacular processes. Body ciliation is also variable, ranging from isolated tufts and bands in entodiniomorphid trichostomes to holotrichous ciliation in most haptorians. Free-living haptorians are distributed world-wide in freshwater and marine habitats and are characterized as voracious predators of flagellates, other ciliates, and even small metazoans.

The endosymbiotic trichostomes are found in a variety of vertebrates, ranging from fish to reptiles and mammals where they typically consume bacteria and plant material. Trichostomes inhabit two major groups of mammals: (1) they are found in ruminants or foregut fermenters, such as cattle, sheep, hippopotamus, and kangaroos; and (2) they are found in various hindgut fermenters, such as horses, tapirs, and some anthropoid apes. The only ciliate known to be pathogenic to human beings is the trichostome *Balantidium coli*, which damages the intestinal mucosa of humans. *Balantidium* has also been found in some invertebrates and other vertebrates. In the vast majority of human infections, there has been a history of human contact with pigs

(Ferry et al., 2004; Zaman, 1978), although rare human cases occur apparently with no reported contact with pigs (Anargyrou et al., 2003).

Culture methods have been developed for some of the endosymbiotic forms. Entodiniomorphids can be cultured short-term or long-term on several different media that model the rumen microenvironment (Bonhomme, Fonty, & Senaud, 1982; Coleman, Laurie, & Bailey, 1977; Hillman, Williams, & Lloyd, 1991; Michalowski, Muszyński, & Landa, 1991). Essential lipids have extended cultivation of entodinia for more than 3 months (Hino, Kametaka, & Kandatsu, 1973). These cultivation methods have enabled exploration of the biochemistry and physiology of rumen ciliates. Methods for cultivation of *Balantidium* have been in use since the early 1900s, modeled on media that support the growth of *Entamoeba histolytica* (Zaman, 1978). Klaas (1974) has reported on media that enable cultivation of some isolates for well over 2 years. Cox (1963) has argued that initiation of cultures is the only reliable method for diagnosis of *Balantidium coli* infections.

The name of the Class LITOSTOMATEA is derived from the Greek *litos* meaning simple and the Greek *stoma* meaning mouth. Small and Lynn (1981) suggested it as a replacement name for the classical Gymnostomata (i.e., Greek *gymnos* meaning naked) to which a number of the included taxa were assigned by Corliss (1979) and others. Electron microscopic research on the litostomes had revealed that the mouth was not naked, but encircled by a specialized, but simple, circumoral infraciliature (Grain, de Puytorac, & Bohatier, 1973). The litostomes have long been regarded as little-modified descendants of the most primitive ciliates. However, ultrastructural research led several workers (Bardele, 1989; Small & Lynn, 1981) to argue that the simplified nature of litostomes was likely secondarily derived. This has been confirmed by rRNA gene sequences that show litostomes to be several branches removed from the base of the ciliate "tree" (Baroin-Tourancheau, Delgado, Perasso, & Adoutte, 1992; Baroin-Tourancheau, Villalobo, Tsao, Torres, & Pearlman, 1998; Leipe, Bernhard, Schlegel, & Sogin, 1994; Wright, Dehority, & Lynn, 1997). While α-tubulin sequences suggested that litostomes were indeed basal in the ciliate tree (Baroin-Tourancheau et al., 1998), it is thought that this may be due to the poor resolving power of this gene.

Small and Lynn (1981) established the monophyly of the class based on the ultrastructural pattern of the somatic kinetids. These are monokinetids that were ultimately shown to have a convergent postciliary ribbon, whose microtubules are arranged in a double-row configuration, and two transverse ribbons (Leipe & Hausmann, 1989; Williams, Williams, & Hogan, 1981). This kinetid pattern is the primary synapomorphy for the class. Three other features unite the ciliates in this class. First, the oral kinetids are either monokinetids or dikinetids whose transverse ribbons extend to support the cytopharyngeal apparatus, which is called a rhabdos. In all other ciliate classes, cytopharyngeal ribbons are derived from postciliary ribbons. Second, there are regions of at least several somatic kineties in holotrichous species whose ciliature is differentiated as clavate cilia, forming a clavate field or brush (Foissner, 1996b). Third, conjugation in litostomes is often preceded by a preconjugation cell division during which the first meiotic reduction division occurs leaving characteristically swollen nuclear division products. Although a preconjugation division has a scattered distribution among other classes (Raikov, 1972), we regard it as convergently evolved in each group since only litostomes show the micronuclear swelling, although there may be exceptions (see Xu & Foissner, 2004). Finally, McEwan et al. (2000) demonstrated a bias against the use of G in the third position of the codons for lysine, glutamine, and glutamic acid in entodiniomorphid ciliates (e.g., *Entodinium*, *Epidinium*, *Polyplastron*) and one haptorian (i.e., *Spathidium*). Could this be a molecular synapomorphy for the Class LITOSTOMATEA?

9.1 Taxonomic Structure

Corliss (1979) placed the major groups now included in the Class LITOSTOMATEA as orders in the Subclass Gymnostomata and Subclass Vestibulifera. Small and Lynn (1981) were the first to suggest unifying this assemblage of ciliates. They based this on the structure of the somatic kinetids, which at that time were characterized as having convergent postciliary ribbons, laterally to antero-laterally directed kinetodesmal fibrils, and a tangential transverse ribbon, and on the presence of the lamina corticalis or ecto-endoplasmic fibrillar layer. Lynn and Small (1997, 2002) have maintained these taxonomic assignments, which have been confirmed by sequence analyses of both ribosomal (Baroin-Tourancheau et al., 1998; Strüder-Kypke, Wright, Foissner, Chatzinotas, & Lynn, 2006; Wright & Lynn, 1997a, 1997b; Wright et al., 1997) and protein genes (Baroin-Tourancheau et al.). De Puytorac (1994a) united these same taxa in the Subphylum Filicorticata. Since the major feature of this subphylum is the presence of an ecto-endoplasmic fibrillar layer, already characterized for the Class LITOSTOMATEA by Small and Lynn, we do not believe this new name is warranted and retain the Class LITOSTOMATEA.

We recognize two subclasses within this class. The Subclass Haptoria is distinguished by the presence of toxicysts, typically in the oral region, and typically by a "ring" of circumoral dikinetids that surround the oral region. Members of the Subclass Trichostomatia have lost the toxicysts, and have only "oralized" somatic monokinetids in the oral region (Lipscomb & Riordan, 1992; Lynn & Small, 2002). The oral region of trichostomes may be invaginated as a vestibulum, a cavity lined by specialized extensions of somatic kineties. While some haptorians have been described to have only oral monokinetids (e.g., Foissner & Foissner, 1985), we believe gene sequence data will show that this is a convergent feature arising independently in haptorians and the endosymbiotic trichostomes (see also Lipscomb & Riordan, 1990, 1992). It is likely the case that the trichostomes evolved from a microaerophilic haptorian-like ancestor that had oral monokinetids and hydrogenosomes: *Balantidium* has both mitochondria and hydrogenosomes (Grain, 1994) while some haptorians (e.g., *Arcuospathidium*, *Chaenea*, *Lacrymaria*) appear to be adapted to anaerobic habitats by harboring endosymbiotic methanogens (Finlay & Maberly, 2000). Strüder-Kypke et al. (2006) found the free-living *Epispathidium papilliferum* grouped with the trichostomes using small subunit rRNA (SSUrRNA) gene sequences.

The Subclass Haptoria is divided into three orders: Order Haptorida, Order Pleurostomatida, and Order Cyclotrichiida. Grain et al. (1973) were among the first to suggest that details of oral kinetid structure could be used to distinguish clades of "gymnostome" ciliates. Foissner and Foissner (1988) described these three orders and added two more, the Order Spathidiida and the Order Pseudoholophryida. They also included in this subclass the Order Archistomatida, which we assign to the Order Entodiniomorphida (**see below**). Xu and Foissner (2005) argued that a *Dileptus*-like ancestor gave rise to the spathidiid diversity by several allometric models of kinety growth. The basal position of *Dileptus* in some SSUrRNA gene trees is consistent with this model (Strüder-Kypke et al., 2006).

Lipscomb and Riordan (1990) used cladistic analyses to assess relationships among haptorians and concluded that there should be two orders, the Order Haptorida and Order Pleurostomatida. They placed **within** the Order Haptorida trichostome vestibuliferids (e.g., *Balantidium*, *Isotricha*) and archistomatids (e.g., *Alloiozona*, *Didesmis*), which we assign to the Order Entodiniomorphida (**see below**). Lipscomb and Riordan (1992) affirmed these two major divisions using a successive weighting cladistic analysis, but concluded that a stable classification needed more data. They also concluded that the Subclass Ditransversalia proposed by Leipe and Hausmann (1989) was an unnecessary proposal. Using only an equal weighting analysis, Wright and Lynn (1997b) reanalyzed the Lipscomb and Riordan (1992) dataset, and affirmed the two major divisions of haptorians proposed here. Wright and Lynn (1997b) were able to separate the haptorians from the trichostomes and vestibuliferids. De Puytorac (1994a) essentially followed the system of Foissner and Foissner (1988). We are unconvinced of the phylogenetic significance of a number of the characters used to justify these ordinal taxa. Therefore, we conservatively recognize only three orders and no suborders within the Subclass Haptoria, until such time as molecular characters consistently confirm or refute these divisions (see Strüder-Kypke et al., 2006).

The Order Haptorida is characterized by an oral region that has somatic dikinetids or oralized somatic monokinetids whose transverse ribbons support the cytopharynx, which is lined by nematodesmata originating from the kinetids and supported also by an internal "pallisade" of bulge microtubules. We include the following families: Acropisthiidae, Actinobolinidae, Apertospathulidae, Didiniidae, Enchelyidae, Helicoprorodontidae, Homalozoonidae, Lacrymariidae, Pleuroplitidae, Pseudoholophryidae, Pseudotrachelocercidae, Spathidiidae,

Tracheliidae, and Trachelophyllidae. We have been conservative and retained the genus *Myriokaryon* as a spathidiid, rather than recognizing a new family as suggested by Foissner (2003).

The Order Pleurostomatida is characterized by a flattened elongated oral region along the ventral margin of a laterally compressed body. Somatic ciliation shows a right-left differentiation. We include here the Amphileptidae and Litonotidae. Foissner and Leipe (1995) established the new Family Loxophyllidae to include genera (i.e., *Loxophyllum* and *Siroloxophyllum*) that have dorsolateral kineties. They affirmed the two suborders recognized by Foissner and Foissner (1988). We again remain conservative here, preferring to await gene sequence data to support the separation of the loxophyllids and eschewing the establishment of monotypic suborders.

The Order Cyclotrichiida is monotypic, including the Family Mesodiniidae. These ciliates are distinguished by cirrus-like cilia that typically form two girdles around the equator of the cell. Although these ciliates (e.g., *Myrionecta*) do have toxicyts, Foissner and Foissner (1988) speculated that they may even belong to another subclass based on the details of the ciliature and infraciliature, which have been well described in the major genera and common species (Krainer & Foissner, 1990; Song, 1997; Tamar, 1992). Johnson, Tengs, Oldach, Delwiche, and Stoecker (2004) have sequenced the SSUrRNA genes of *Myrionecta rubra* and *Mesodinium pulex* and found them to be closely related, but widely separated from other litostomes, hence supporting the Order Cyclotrichiida. Strüder-Kypke et al. (2006) confirmed these results and demonstrated litostome features in the secondary structure of the SSUrRNA of cyclotrichiids. We have chosen to retain the order in the Class LITOSTOMATEA (see **Chapter 17**).

The Subclass Trichostomatia includes a diverse assemblage of ciliates predominantly endosymbiotic in vertebrates. These ciliates have haptorian somatic monokinetids and typically a conspicuous ecto-endoplasmic fibrillar layer. Their oral region is slightly more complex than that of the haptorians. It is typically surrounded by extensions of somatic kineties that have a higher kinetosomal density than the somatic portions and that may be invaginated into an oral cavity called a vestibulum. There are no oral toxicysts and most species have hydrogenosomes rather than mitochondria. We prefer the name Trichostomatia for the class rather than Vestibuliferia because some entodiniomorphids (i.e., Buetschliidae) do not have a vestibulum. The subclass is divided into three orders. The Order Vestibuliferida includes ciliates that are holotrichously ciliated and have a vestibulum, defined as an oral cavity or depression lined by densely ciliated kineties, typically as extensions of somatic kineties. Members of the Order Entodiniomorphida have somatic ciliation restricted as girdles, bands, and tufts, and may or may not have a deep oral cavity. A new order, Order Macropodiniida n. ord., form what might be called a "ribo-order" as there are no strong morphological synapomorphies for this group, and only share their habitat as endocommensals in the forestomach of macropodid and vombatid marsupials. Cameron and O'Donoghue (2004b) argued that it is premature to take this taxonomic step.

The Order Vestibuliferida includes six families: Balantidiidae, Isotrichidae, Paraisotrichidae, Protocaviellidae (= Hydrochoerellidae), Protohalliidae, and Pycnotrichidae. Grain (1966a, 1966b) provided details of the cytology and ultrastructure of vestibuliferids, demonstrating the nature of their oral structures as very slightly specialized extensions of somatic kineties whose kinetids bear transverse ribbons and nematodesmata that support the cytopharynx. Ito and Imai (2000a, 2000b) described several new genera and species from the cecum of the South American capybara, following the classic work of Da Cunha and Muniz (1925, 1927). Grim (1988, 1993a) has provided ultrastructural descriptions of *Balantidium* species from tropical fishes and the pycnotrichid *Vestibulongum*, confirming the haptorian nature of their somatic monokinetids. Strüder-Kypke et al. (2006) questioned the monophyly of the Order Vestibuliferida since *Balantidium* did not group with the other vestibuliferids based on SSUrRNA gene sequences. Again, we have remained conservative here until a larger sampling of vestibuliferids and *Balantidium* species justifies this conclusion.

There is a tremendous literature on the Order Entodiniomorphida, from the early work of the 20th century by Dogiel (1927, 1946), Noirot-Timothée's (1960) monograph on cytology and ultrastructure, to the study series of Latteur (1968/1969, 1970), Lubinsky (1957a, 1957b), and Wolska

(1971, 1981). The order, which includes three suborders, is consistently the sister clade to the vestibuliferids in gene sequence trees (Cameron & O'Donoghue, 2003a; Strüder-Kypke et al., 2006). The Suborder Archistomatina is monotypic: the Family Buetschliidae includes ciliates with a concrement vacuole overlain by a clavate field. There are typically girdles of somatic kineties at the anterior, middle, and posterior ends. The anterior "girdle" comprises the oral ciliature. The Suborder Blepharocorythina is also monotypic: members of the Family Blepharocorythidae have a vestibulum lined by several kinety fields. The concrement vacuole is absent. However, we accept Wolska's (1971) argument that the patch of somatic kineties remaining is homologous to the patch overlying the archistomatine concrement vacuole, an hypothesis that needs testing by gene sequence data.

The Suborder Entodiniomorphina is the largest of the three. We accept Wolska's (1971) hypothesis for the evolution of the entodiniomorphid oral ciliature (i.e., ophyroscolecid) from a blepharocorythine-like ancestor, and await gene sequence data that will test it. The suborder includes ten families: Cycloposthiidae, Gilchristidae, Ophryoscolecidae, Parentodiniidae, Polydiniellidae, Pseudoentodiniidae, Rhinozetidae, Spirodiniidae, Telamodiniidae, and Troglodytellidae. The Entodiniomorphina are characterized by somatic ciliature that is arranged in bands or tufts. The oral cavity of these ciliates is a vestibulum lined with densely ciliated, tightly spaced single rows of kinetosomes, sometimes termed polybrachykineties, which functionally behave like membranelles when they beat. This lead to the earlier placement of this group with the heterotrichs and hypotrichs (Corliss, 1961). Grain (1994) and others have recognized subfamilies while Bonhomme, Grain, and Collet (1989) have suggested the families might be grouped on the basis of the ultrastructure of the ecto-endoplasmic fibrillar layer. One group of families has only transverse strands in the ecto-endoplasmic layer (i.e., Cycloposthiidae, Ophryoscolecidae, Troglodytellidae) while the other group (i.e., Spirodiniidae, Tripalmariidae) has both transverse and longitudinal strands. This is yet another hypothesis that must await testing by gene sequence analysis. Lubinsky (1957a, 1957b) and Imai (1998) have presented phylogenetic analyses for the evolution of ophryoscolecid genera based on morphological features, primarily based on the increasing complexity of skeletal and ciliary structures. Wright and Lynn (1997a) have found some consistency between morphological and molecular phylogenies, but reserved judgement until a larger sampling of entodiniomorphid genera had been accomplished.

The Order Macropodiniida n. ord. is formally established here for a clade of entodiniomorophids that consistently groups as the sister clade of the previous two orders (Cameron, Wright, & O'Donoghue, 2003; Strüder-Kypke et al., 2006) (see **Chapter 17**). Dehority (1996) first recognized this group as a novel assemblage of ciliates, endosymbiotic in macropodid marsupials, and established the Family Macropodiniidae to represent that fact. Since then, Cameron and coworkers (Cameron, 2002; Cameron & O'Donoghue, 2002a, 2002b, 2003a, 2003b, 2004a, 2004b; Cameron, Adlard, & O'Donoghue, 2001a, 2001b) have established two new families, the Amylovoracidae and the Polycostidae, and considerably expanded the host range of macropodiniids in Australian marsupials. Except for their shared habit as marsupial endosymbionts, these ciliates are not strongly united by one feature, but their oral cavities, like some vestibuliferids, are lined by extensions of somatic kineties and are supported by oral nematodesmata.

At the species level, much research remains to be done, especially on entodiniomorphids. Species in this order have often been established based on the possession of novel spines or small shape differences. However, Poljansky and Strelkow (1938) recognized that alimentary tract conditions may stimulate the appearance of certain forms, and this has been confirmed by others (Chardez, 1983; Dehority, 1994). In vitro cultivation of clones of rumen ciliates has now confirmed this phenotypic plasticity (Dehority, 2006; Miltko, Michalowski, Pristas, Javorsky, & Hackstein, 2006). The latter study suggested that several morphospecies were in fact not genetically distinct, but could be derived from an *Ophryoscolex caudatus* progenitor. In the only study of genetic variability of populations of litostome endosymbionts, Wright (1999) demonstrated no genetic variability in the internally transcribed spacer regions from isolates of *Isotricha prostoma* derived from cattle on two continents.

9.2 Life History and Ecology

Litostomes cannot be discussed as a homogeneous assemblage since the life histories of the haptorians, as free-living predators, are much different from the life histories of trichostomes, which are endosymbionts of vertebrates. It is safe to say, as with other classes discussed so far, that the distributions of members of both subclasses are global. The limitation of the symbiotic forms being the presence of their hosts, none of which occur in Antarctica but are present on every other continent.

Haptorians are predators of smaller protists, both autotrophic and heterotrophic flagellates, other ciliates, and even small metazoans, such as rotifers. Common genera, such as *Didinium*, *Monodinium*, *Mesodinium*, *Dileptus*, and *Lagynophrya*, have been recorded from continental and coastal marine waters and sea ice of Antarctica (Garrison et al., 2005; Leakey, Fenton, & Clarke, 1994; Petz & Foissner, 1997), North America (Dolan, 1991) and Europe (Leakey, Burkhill, & Sleigh, 1993; Zingel, Huitu, Makela, & Arvola, 2002), deep waters of the Mediterranean Sea (Hausmann, Hülsmann, Polianski, Schade, & Weitere, 2002), temperate freshwater lakes in Europe (Carrias, Amblard, & Bourdier, 1994; Zingel & Ott, 2000), Asia (Obolkina, 2006), and North America (Hunt & Chein, 1983), reservoirs in South America (Barbieri & Orlandi, 1989), subtropical lakes in North America (Beaver & Crisman, 1982, 1989b), rivers (El Serehy & Sleigh, 1993; Foissner, 1997b), and ponds and streams (Domenech, Gaudes, Lopez-Doval, Salvado, & Munoz, 2006; Foissner, 1980b; Madoni & Sartore, 2003). They are found as part of the interstitial fauna of marine shores (Al-Rasheid, 1999) and are a conspicuous constituent of soils throughout the world (Berger, Foissner, & Adam, 1984; Buitkamp, 1977; Foissner, 1998a; Petz & Foissner). Haptorians are often recorded from anoxic sediments (Guhl, Finlay, & Schink, 1996; Madoni & Sartore, 2003). Environmental DNA analyses have even found *Spathidium* genes in a cryoconite hole in the Canada Glacier, Antarctica (Christner, Kvitko, & Reeve, 2003).

Their numbers vary more erratically than their prey and they are typically not as abundant as this prey. They can achieve high densities: *Didinium* can reach 2,800 l^{-1} (Dolan, 1991); *Mesodinium pulex* over 2,000 l^{-1} (Barbieri & Orlandi, 1989); *Askenasia stellaris* and *Lagynophrya* over 1,000 l^{-1} (Leakey et al., 1993). Haptorians represented almost 50% of the ciliate abundance at depths between 10 and 15 m in some lakes (Carrias et al., 1994). They generally range from 5–30% of the abundance across lakes of differing trophic status, while their abundance is positively correlated with the trophic status of the lake (Beaver & Crisman, 1982).

Special mention should be made of the autotrophic haptorian *Myrionecta rubra* (= *Mesodinium rubrum*). This ciliate is host to a cryptophycean endosymbiont (**see more below**). This ciliate has been recorded in all the oceans of the world (Crawford, 1989; Lindholm, 1985), and in antarctic brackish and saline lakes (Laybourn-Parry, Quayle, & Henshaw, 2002; Perriss, Laybourn-Parry, & Marchant, 1995). *Myrionecta* is often restricted to a stratum or layer and may migrate vertically at least 10 m on a daily basis (Dale, 1987; Owen, Gianesella-Galvão, & Kutner, 1992). Two discrete cell sizes have been reported for *Myrionecta*, a larger form at colder times of the year and a smaller form at warmer times (Modigh, 2001; Montagnes & Lynn, 1989). Its abundances can be very high, ranging to over 30,000 l^{-1} (Edwards & Burkhill, 1995; Sanders, 1995) so that it will cause red tides (Crawford, 1989; Lindholm, 1985; White, Sheath, & Hellebust, 1977). These abundances mean that *M. rubra* can make significant contributions to primary production, sometimes well over 20% (Leppänen & Bruun, 1986; Sanders, 1995; Smith & Barber, 1979).

Trichostomes are endosymbiotic primarily in vertebrates. The vestibuliferidan *Balantidium* has been reported from fish (Grim, 1989; Grim, Clements, & Byfield, 2002), frogs and toads (Affa'a, 1988a; Khan & Ip, 1986), turtles (Fenchel, 1980d; Geiman & Wichterman, 1937), ostriches and rheas (Gordo, Herrera, Castro, Buran, & Diaz, 2002), the caecum of a horse (Wolska, 1962), baboons and other primates, including humans (Müller-Graf, Collins, & Woolhouse, 1996; Zaman, 1978) and pigs (Zaman, 1978). A survey of recent reports of trichostomes provides the following brief synopsis. Some or all of buetschliids, isotrochids, paraisotrichids, blepharocorythids, ophryoscolecids, and cycloposthiids have been reported recently from ruminants such as cattle and sheep (Dehority, 1986; Gocmen, Dehority, Talu, & Rastgeldy, 2001; Imai, Han, Cheng, & Kudo, 1989, Towne & Nagaraja, 1990), the yak

(Guirong, Su, Hua, Zhu, & Imai, 2000), water buffalo (Dehority, 1979), bison (Towne, Nagaraja, & Kemp, 1988), musk oxen (Dehority, 1985), giraffe (Kleynhans & Van Hoven, 1976), and camel (Imai & Rung, 1990b). Ophryoscolecids, predominantly *Entodinium* species, and fewer dasytrichids and isotrichids are found in antelopes (Fernández-Galiano & Campos, 1992; Imai & Rung, 1990a; Kleynhans, 1982; Van Hoven, 1983; Van Hoven, Hamilton-Attwell, & Grobler, 1978). *Entodinium* species appear also to dominate the fauna in deer, elk, and pronghorn antelope (Dehority, 1990, 1995; Ito, Imai, & Ogimoto, 1993) although isotrichids have also been reported (Imai et al., 1995).

Overall, the non-ruminant mammals harbor a much higher diversity of trichostomes, although one particular host species may have a limited diversity of ciliate species. Most is known about the endosymbionts in the colon of horses (Bonhomme-Florentin, 1994; Grain, 1966a, 1994; Wolska, 1965). Buetschliids, hydrochoerellids, pycnotrichids, and cycloposthiids have been recorded in the rodents, such as the South African mole rat (Sandon, 1941b) and South American capybara (Ito & Imai, 2000a, 2000b). Ophyroscolecids were found in the collared peccary (Carl & Brown, 1983). Buetschliids, paraisotrichids, blepharocorythids, and entodiniomorphids were recorded in the stomach of the hippopotamus (Thurston & Grain, 1971; Thurston & Noirot-Timothée, 1973). Some unusual cycloposthiids, rhinozetids, buetschliids, paraisotrichids, blepharocorythids, and ditoxids have been recorded from the colon of rhinoceros (Gilchrist, Van Hoven, & Stenson, 1994; Van Hoven, Gilchrist, & Hamilton-Attwell, 1987; Van Hoven, Gilchrist, & Hamilton-Attwell, 1988). Buetschliids, paraisotrichids, ophryoscolecids, and cycloposthiids have been observed throughout the intestinal tract of elephants (Eloff & Van Hoven, 1979; Timoshenko & Imai, 1997). Several very recent reports have described novel isotrichids, cycloposthiids, and new families of macropodiniids from Australian macropodid marsupials (Cameron & O'Donoghue, 2003a; Cameron et al., 2000, 2001a, 2001b; Dehority, 1996). Finally, cycloposthiids and troglodytellids have been recorded in the feces of chimpanzees and gorillas (Freeman, Kinsella, Cipolletta, Deem, & Karesh, 2004; Goussard, Collet, Garin, Tutin, & Fernandez, 1983; Imai, Ikeda, Collet, & Bonhomme, 1991).

Except for the vestibuliferidan *Balantidium*, which can cause damage to the intestinal tract of pigs and humans (Zaman, 1978), most trichostomes are considered to be commensals. Indeed, non-pathogenic *Balantidium* species likely feed on a variety of bacteria and flagellates, which cohabit the gut (Grim, 2006). Nevertheless, there has been debate about the role of the rumen ciliates in the biology of their hosts since their first discovery by Gruby and Delafond (1843). The rumen ecosystem is composed of a variety of bacterial species, fungi, a few flagellates, and a considerable diversity and abundance of ciliates. Ciliate abundances can range from 10,000 ml^{-1} of rumen fluid in yak (Guirong et al., 2000) and zebu (Bonhomme-Florentin, Blancou, & Latteur, 1978), to over 100,000 ml^{-1} in cattle and sheep (Imai et al., 1989), antelopes (Imai & Rung, 1990a; Van Hoven et al., 1978), and in water buffalo (Dehority, 1979), to over 500,000 ml^{-1} in musk oxen (Dehority, 1985) and in some deer species (Dehority, 1990, 1995; Ito et al., 1993). Abundances of ciliates in non-ruminants have a similar range: cycloposthiids numbered up to 700 ml^{-1} in the cecum of the capybara, representing about 30% of the ciliate fauna (Ito & Imai, 2000a, 2000b); over 10,000 ml^{-1} throughout the intestinal tract of African elephant (Eloff & Van Hoven, 1979); typically over 100,000 ml^{-1} in the colon of rhinoceroses (Gilchrist et al., 1994) and caecum of horses (Bonhomme-Florentin, 1994); and abundances of ophryoscolecids of around 1,000,000 ml^{-1} in collared peccary (Carl & Brown, 1983).

The rumen ciliates have diverse interactions with the bacterial and fungal communities and with each other. Bacteria are the foundation of the rumen ecosystem, colonizing substrates minutes after ingestion and forming cellulolytic consortia that digest the plant tissues (McAllister, Bae, Jones, & Cheng, 1994). While most rumen ciliates ingest bacteria as a source of nitrogen, *Entodinium* species are particularly important predators of bacteria, consuming more than 10^5 bacteria per ciliate per hour (Coleman, 1989; Williams, Joblin, Butler, Fonty, & Bernalier, 1993). Entodiniomorphids are able to ingest plant fragments and digest these using their own cellulolytic enzymes (Akin & Amos, 1979; Bauchop, 1979; Benyahya, Senaud, & Bohatier, 1992; Bohatier, Senaud, & Benyahya, 1990; Grain & Senaud, 1985; Michalowski, Belzecki, Kwiatkowska, & Pajak, 2003; Stan, Belzecki, Kasperowicz,

Kwiatkowska, & Michalowski, 2006), and even chitin (Belzecki & Michalowski, 2006). Polysaccharides are stored as amylopectin, either as skeletal plates or as cytoplasmic particles. Eadie (1967) was one of the first to recognize that there were some ciliates that were consistently common in the rumen, namely *Entodinium* spp., *Isotricha* spp. and *Dasytricha ruminantium* while others varied. Eadie (1967) identified two assemblages, Type A and Type B. The Type A assemblage included *Polyplastron multivesiculatum*, *Diploplastron affine*, and *Ophryoscolex tricoronatus*, while the Type B assemblage included *Eudiplodinium maggii*, *Epidinium* spp., *Eremoplastron* spp., and *Ostracodinium* spp. The Type A assemblage typically displaces Type B as the former includes some predators of the latter (Eadie, 1967; Imai, Katsuno, & Ogimoto, 1979). *Eudiplodinium maggii* may be induced to develop to a larger size in the presence of its predator *Polyplastron multivesiculatum* (Eadie, 1979). *Polyplastron multivesiculatum*, *E. maggii*, and *Entodinium* sp. may also ingest fungal zoospores and rhizoids (Williams et al., 1993). Finally, *Isotricha* and *Dasytricha* species preferentially ingest starch grains. This feeding habit may provide a more stable rumen pH since it prevents the more rapid bacterial fermentation of starch to lactic acid, which may lead to lactic acid acidosis (Williams, 1986). Since most of the ciliates are retained in the rumen (Bonhomme-Florentin, 1994; Williams, 1986), ciliate biomass contributes little to host metabolism. Moreover, although both ciliates and bacteria are highly cellulolytic, bacterial activity can entirely replace the ciliate activity (Hidayat, Hillman, Newbold, & Stewart, 1993). Although there are contradictory reports, typically defaunation or removal of the ciliates has little or no impact on host growth (see reviews of Bonhomme, 1990; Jouany, 1994; Veira, 1986; Williams, 1986). Some recent research has demonstrated that defaunation improves ruminal nitrogen metabolism to the host (Ivan, Neill, & Entz, 2000; Koenig, Newbold, McIntosh, & Rode, 2000).

Bonhomme-Florentin (1994) noted that there is very little research on the importance of the ciliates of non-ruminant and primate hosts. Like their rumen relatives, *Cycloposthium* and *Didesmis* associate with plant fibres in the caecum of the horse, aiding in the digestion of these fibres (Bonhomme-Florentin, 1985). However, Moore and Dehority (1993) concluded that ciliates do not play an essential role in the equine hindgut: defaunation of the caecum and colon had no effect on levels of cellulose digestion.

One major impact of defaunation is reduction of methane production by the ruminant. Like the sapropelic armophoreans (see **Chapter 8**), methanogenic bacteria are associated as epibionts on entodiniomorphids (Krumholz, Forsberg, & Veira, 1983; Stumm, Gijzen, & Vogels, 1982) and as endosymbionts in vestibuliferidans (Finlay et al., 1994). These methanogens have been assigned to the genera *Methanobrevibacter*, *Methanosphaera*, and *Methanosarcina* (Hillman, Lloyd, & Williams, 1988; Tokura, Chagan, Ushida, & Kojima, 1999). Adult cattle can produce from 300–600 l of methane per day, translating to 80 million tonnes of methane worldwide (Jouany, 1994). This methane production may represent the greatest source of methane production in the European Union (Moss, Jouany, & Newbold, 2000) and over 50% of the methane emissions in Australia (Klieve & Hegarty, 1999). Defaunation to reduce methane production is thus a major priority in the context of global warming. Adding coconut oil to artificial rumen fermenters reduced methane formation by 40% (Dohme et al., 1999) while the common food preservative, nisin, reduced methanogen production by 36% (Klieve & Hegarty, 1999). There are other strategies for eliminating the protozoa, but as yet none have reached commercialization (Hegarty, 1999).

With the exception of *Balantidium*, trichostome ciliates do not form cysts. Thus, transmission from one host to the other must take place by various forms of contamination. Rumen ciliates and those in the forestomach of the host are transferred by salivary contamination from mother to offspring and contamination of drinking water (Bonhomme-Florentin, 1994; Van Hoven, 1978). *Entodinium* is typically the first rumen ciliate to appear (Crha, Stříž, Skřivánek, & Valach, 1991). Young horses become infected by actively eating the mothers' feces in the first week of life (Ike, Imai, & Ishii, 1985). It is not yet known how the macropodiniids are transmitted between marsupial hosts, although maternal grooming of the young may be a typical route (Cameron & O'Donoghue, 2003b).

The feeding preferences and strategies of litostomes are quite diverse. As noted above for the trichostomes, bacteria and plant material can be prey items in addition to other ciliates. The haptorians can be typified as fast-swimming, active and voracious predators, showing marked preferences for flagellates and other ciliates, even to becoming cannibalistic. Dragesco (1962) used high speed cinematography to provide some detailed descriptions of predation by *Enchelys*, *Litonotus*, *Chaenea*, *Didinium*, and *Dileptus* on other ciliates, like *Colpidium*. Feeding by *Dileptus* may even be entrained to a daily rhythm (Miller, 1968). This preference of haptorians for flagellates and other ciliates has been confirmed by others (e.g., Dolan & Coats, 1991b; Estève, 1982; Foissner & Leipe, 1995; Foissner, Berger, & Schaumburg, 1999; Johnson, Donaghay, Small, & Sieburth, 1995). In an ingenious series of experiments, Karpenko, Railkin, and Seravin (1977) used magnetic moving models to demonstrate that *Didinium* and *Dileptus* responded to prey movement, somehow sensing hydrodynamic disturbances of the medium. This sensitivity to hydromechanical signals has been confirmed for *Mesodinium pulex*, which probably uses its bristle girdle as the "detector" (Jakobsen, Everett, & Strom, 2006). However, some pleurostomatids show a marked preference for more sedentary prey, such as colonial peritrichs and rotifers (Canella, 1951, 1954).

Didinium nasutum and *Paramecium* have been the subjects of numerous studies exploring the relationship between a predator and its prey. *Didinium* can be "trained" to feed on a variety of *Paramecium* species (Berger, 1980; Hewett, 1980a) and even *Colpidium* (Berger, 1979). *Didinium* consumes *Paramecium bursaria* that have zoochlorellae less efficiently than apochlorotic cells, indicating that the zoochlorellae may have a protective function for their host (Berger, 1980). To support continued growth, the best *Paramecium* prey must be reared on a mixture of wild bacteria (Burbanck & Eisen, 1960). The size of *Didinium* varies throughout the growth cycle (Salt, 1975). Size of *Didinium* also is related to the prey size: *Didinium* feeding on *Colpidium* are much smaller than those feeding on the larger *Paramecium multimicronucleatum* (Berger, 1979; Hewett, 1980a). Although the situation is complex, *Didinium* tend to be most successful at capturing the size of prey on which they have been conditioned. However, large *Didinium* did have shorter handling times and encounter times regardless of prey size (Hewett, 1988), but capture rate is also influenced by predator density (Hewett, 1980b; Salt, 1974). The *Didinium-Paramecium* predator-prey system has also been used to model the stability of simple ecosystems. Luckinbill (1973) showed that populations of both species could be maintained for over 1 month if the Cerophyl culture medium was used at half-strength but thickened with methyl cellulose. Harrison (1995) reanalyzed the data from these experiments, concluding that the functional response curve might actually be of a more sigmoid, Type III form. Maly (1978) confirmed that spatial or temporal complexity is necessary to maintain the stable interaction between these two organisms.

Litostomes are also hosts for a variety of symbionts, both as epibionts and endobionts. Endosymbiotic methanogens have already been mentioned above in our discussion of rumen ciliates. Non-methanogens of the genera *Streptococcus* and *Ruminococcus* can be found attached to the cell surface of rumen ciliates, localizing in some cases around the cytostome (Imai & Ogimoto, 1978), while a *Balantidium* species infecting marine fish has epibiontic and endosymbiotic bacteria, some even residing in the macronucleus (Grim, 1993b). Haptorians, such as *Askenasia*, *Didinium*, *Dileptus*, *Homalozoon*, *Lacrymaria*, and *Monodinium*, harbor endosymbiotic *Chlorella* species (Foissner et al., 1999; Karpov, Goodkov, & Marinich, 1991). Undoubtedly the most famous "symbiotic" litostome is *Myrionecta* (= *Mesodinium*) *rubra*. Ultrastructural studies demonstrated that this ciliate was in fact a consortium of a cryptophyte living inside the ciliate so that three nuclei are characteristic of this ciliate (Grain, de Puytorac, & Grolière, 1982; Hibberd, 1977). The cryptophyte symbiont is necessary for growth of *M. rubra*, which must have continued access to these prey populations for continuous and optimal growth (Hansen & Fenchel, 2006; Johnson & Stoecker, 2005). Johnson, Oldach, Delwiche, and Stoecker (2007) have demonstrated that the cryptophyte nuclei are essential for continued function of cryptophyte chloroplasts. They have termed this relationship karyoklepty – nuclear stealing, since one cryptophyte nucleus may support the cytoplasm of several cryptophyte individuals whose nuclei have been lost from the consortium. They pose the question – is this truly a symbiosis?

Myrionecta does have oral extrusomes, presumably toxicysts, that are carried in oral tentacles.

These are presumably used to capture the symbiont and other prey organisms, such as bacteria (Lindholm, Lindroos, & Mörk, 1988; Myung, Yih, Kim, Park, & Cho, 2006). *Myrionecta* is an extremely active swimmer, achieving typical rates of 1–2 mm sec^{-1}, enabling it to migrate vertically to take advantage of changing light levels (Lindholm, 1985). It is functionally a member of the phytoplankton. When it is in bloom, numbering over 10^5 l^{-1}, it can contribute over 70% of the total primary production, but typically contributes much less than that (Crawford, 1989; Stoecker, Putt, Davis, & Michaels, 1991). In these abundances it can cause non-toxic red tides. These "photosynthetic" litostomes are technically mixotrophs. One of the most unusual mixotrophic litostomes is *Perispira ovum*. This ciliate has a preference for *Euglena proxima*, whose chloroplasts, mitochondria, and paramylon it sequesters. Although these intact organelles are surrounded by *Perispira* membranes and endoplasmic reticulum, we do not yet know how much they contribute functionally to the predator's physiology (Johnson et al., 1995).

Among the unicellular eukaryotes, litostomes, and particularly haptorians, are top predators in microbial food webs. Nevertheless, they are eaten by multicellular organisms, including copepods (Wickham, 1995), insect larvae (Addicott, 1974), and bivalve molluscs, such as mussels and scallops (Carver, Mallet, Warnock, & Douglas, 1996). Cyclotrichiids, as a group, are characterized by their ability to jump to avoid predators (Tamar, 1979). Escape speeds can exceed 100 body lengths per second, attaining speeds of up to 1.2 cm sec^{-1}, likely a speed record for ciliates (Fenchel & Hansen, 2006; Jakobsen, 2001). *Myrionecta* may be able to avoid predation because of its tremendous swimming ability, since it was hardly ingested by barnacle nauplii and some copepods (Jakobsen, 2001; Turner, Levinsen, Nielsen, & Hansen, 2001). However, ambush-feeding copepods, like *Acartia tonsa*, are able to consume *Mesodinium* species (Jakobsen, 2001). The endosymbiotic trichostomes can be preyed upon by the suctorian *Allantosoma*, which resides in the caecum and colon of the horse (Imai, 1979; Sundermann & Paulin, 1981). Ophryoscolecids can be parasitized by chytrid fungi (Lubinsky, 1955a, 1955b).

Like other ciliates, litostomes have chemosensory abilities. *Didinium* is attracted to a heat-stable chemoattractant isolated from bacterial cultures, a behavior that presumably gets it to where its typical prey, *Paramecium*, might be found (Antipa, Martin, & Rintz, 1983). *Litonotus lamella* feeds on *Euplotes crassus* and modifies its behavior when placed in the *Euplotes* cell-free fluid: the predator decreases its creeping speed and modifies its turning behavior so that it accumulates in regions of high prey density (Morelli, Ricci, & Verni, 1999). The swimming behavior of *Didinium* is apparently controlled in a fashion similar to other ciliates, with intracellular Ca^{2+} concentrations influencing ciliary beating mode (Pernberg & Machemer, 1995). *Didinium* also shows gravity-dependent swimming velocities, modulated by mechanically sensitive membrane channels, which keeps the cells stationary in the gravity field (Bräucker, Machemer-Röhnisch, & Machemer, 1994). *Myrionecta rubra* has been reported to avoid being flushed from an estuary by dispersing away from the top of the water column on the ebb tide. Crawford and Purdie (1992) speculated that the major cue to this may be the turbulence generated by the shearing surface currents. This is supported by Fenchel and Hansen (2006) who have demonstrated that *M. rubra* can detect fluid flows and orient appropriately.

Encystment is a common feature of most free-living litostomes, especially haptorians that are found in freshwater and soils. In contrast, the vast majority of endosymbiotic trichostomes of ruminants and non-ruminants have apparently lost this capacity to encyst. *Balantidium* is an exceptional trichostome, forming a cyst wall by synthesizing materials in mucocysts that are transported to the cell surface prior to encystment (Grain, 1968). *Didinium* has been a favored subject for encystment studies since the early work of Beers (1927). When *Didinium* starves, it modulates its swimming speed, apparently in response to both its own population density and the period of starvation. It swims fastest when starved over 1 h at the highest densities (Salt, 1979). Survival is very short-lived, in the order of several days if *Didinium* does not encyst (Jackson & Berger, 1985a). Survival rate is related to cell size: larger *Didinium* have greater survival rates (Hewett, 1987). Like *Balantidium*, *Didinium* synthesizes extrusomes prior to encystment. There is an ordered extrusion of these organelles to form the three layers of *Didinium*'s cyst wall. The ectocyst or outer layer is derived from mucocysts; the mesocyst or middle layer is derived from clathrocysts, special extrusomes whose internal matrix has a flattened honeycomb-like structure; and the endocyst, which is amorphous (Holt & Chapman, 1971; Rieder, 1971). Polysaccharides, proteins, and lipids are the major constituents of the cyst wall (Rieder, 1973). The cyst wall in *Dileptus* is also three-layered (Jones, 1951). There is considerable dedifferentiation of somatic and oral kinetosomal structures and the ecto-endoplasmic layer in encysted haptorians (Holt & Chapman, 1971; Kink, 1978). These structures gradually differentiate again as the excystment process proceeds (Holt, 1972).

9.3 Somatic Structures

The litostomes range is size from the small endosymbiotic buetschliids and blepharocorythids, typically less than 50 μm in length, to the larger free-living haptorians, such as *Dileptus* and *Homalozoon*, whose elongate and flexible bodies can sometimes exceed 1 mm in length (Fig. 9.1). Litostomes include some of the most bizarre forms in the phylum with some blepharocorythids and entodiniomorphids possessing elongate cortical processes and spines (Fig. 9.3). The rhinozetids, entodiniomorphids found in the digestive system of the rhinoceros, even have flexible processes that can be extended away from the body, which may decrease their transit time through the gut (Van Hoven et al., 1988). The cell processes of haptorians, on the other hand, tend to be related to food capture: *Dileptus* has an extensible proboscis preceding the cytostome (Fig. 9.1); *Lacrymaria* has an extremely active and very extensible neck at the end of which sits the cytostome (Fig. 9.1); and *Actinobolina* has distributed over the somatic surface extensible toxicyst-bearing tentacles that aid in prey capture (Holt, Lynn, & Corliss, 1974).

Somatic ciliation is variable (e.g., see Foissner et al., 1999; Grain, 1966a; Ito & Imai, 1998). In haptorians and vestibuliferids, the vast majority of the species have bipolar kineties and holotrichous ciliation (Figs. 9.1, 9.2). However, didiniids and cyclotrichiids, for example, have ciliation restricted to equatorial girdles. Somatic ciliation of entodiniomorphids is typically not holotrichous, but rather restricted to the anterior and posterior ends or to bands that may partially encircle the body in the midregions as well as anteriorly and posteriorly (Figs. 9.2, 9.3). The number of somatic kineties is variable, and at least in *Dileptus* is related to cell size: larger cells have more somatic kineties (Drzewińska & Golińska, 1987). Thus, taxonomists should be cautious about describing a new haptorian species based only on differences in numbers of somatic kineties.

There has been a tremendous amount of research on the cortical ultrastructure of litostomes, which has lead different research groups to varied conclusions about relationships among these forms (Foissner & Foissner, 1988; Grain, 1994; Lipscomb & Riordan, 1990, 1991, 1992). Our discussion will be limited to a general account of the major features and variations in these structures within the class.

The cell surface is covered by a glycocalyx of variable thickness, typically more conspicuous in endosymbiotic forms. In *Dileptus*, this surface coat can change in response to temperature, as measured by serotype changes, which have also been observed in oligohymenophoreans (see **Chapter 15**) (Uspenskaya & Yudin, 1992). Beneath the plasma membrane, alveoli are often very inconspicuous. In haptorians, alveoli are typically small and irregularly distributed beneath the plasma membrane in cortical ridges (Foissner & Foissner, 1985; Grain, 1970; Lipscomb & Riordan, 1990, 1991; Williams et al., 1981). In entodiniomorphids, alveoli may not be visible at all. Rather three cell membrane layers are observed, the inner two presumably being collapsed alveoli (Furness & Butler, 1983, 1985).

The epiplasm is variable in thickness: typically not conspicuous in haptorians and vestibuliferids but often quite thick in ophryoscolecids. Fauré-Fremiet and André (1968) noted that litostomes often had a conspicuous layer of filaments at the ecto-endoplasmic boundary, the so-called lamina corticalis of Bretschneider (1959). There is typically one layer of filaments at the ecto-endoplasmic boundary in haptorians (e.g., *Didinium* – Lipscomb & Riordan, 1992; *Homalozoon* – Leipe & Hausmann, 1989; *Lagynophrya* – Grain, 1970; *Litonotus* – Bohatier & Njiné, 1973; *Perispira* – Johnson et al., 1995) and one layer in some trichostomes (e.g., *Balantidium* – Grain, 1966a, Grim, 1993a; buetschliids – Grain, 1966a; entidiniomorphids – Furness & Butler, 1983,

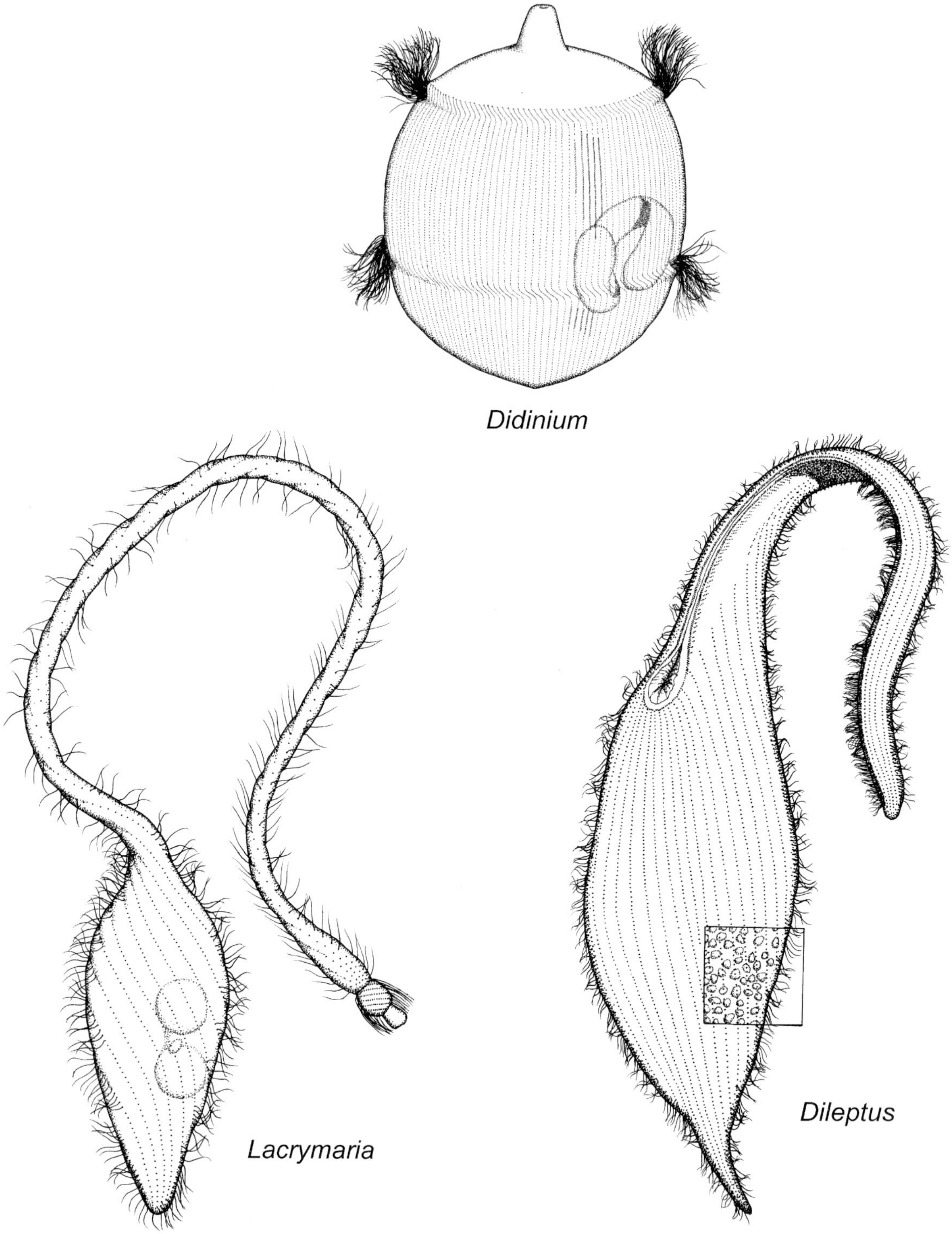

FIG. 9.1. Stylized drawings of representative genera from the Subclass Haptoria of the Class LITOSTOMATEA. The haptorids *Didinium*, *Lacrymaria*, and *Dileptus*. These are three classical encounter feeders: *Didinium* swims through the water bumping into prey; *Lacrymaria* probes the water above the substratum on which it crawls using its extremely extensible neck; and *Dileptus* swims through the water like a swordfish, sweeping it with its toxicyst-laden proboscis, whose extrusomes immobilize and kill prey that are then ingested. Inset shows many small macronuclei

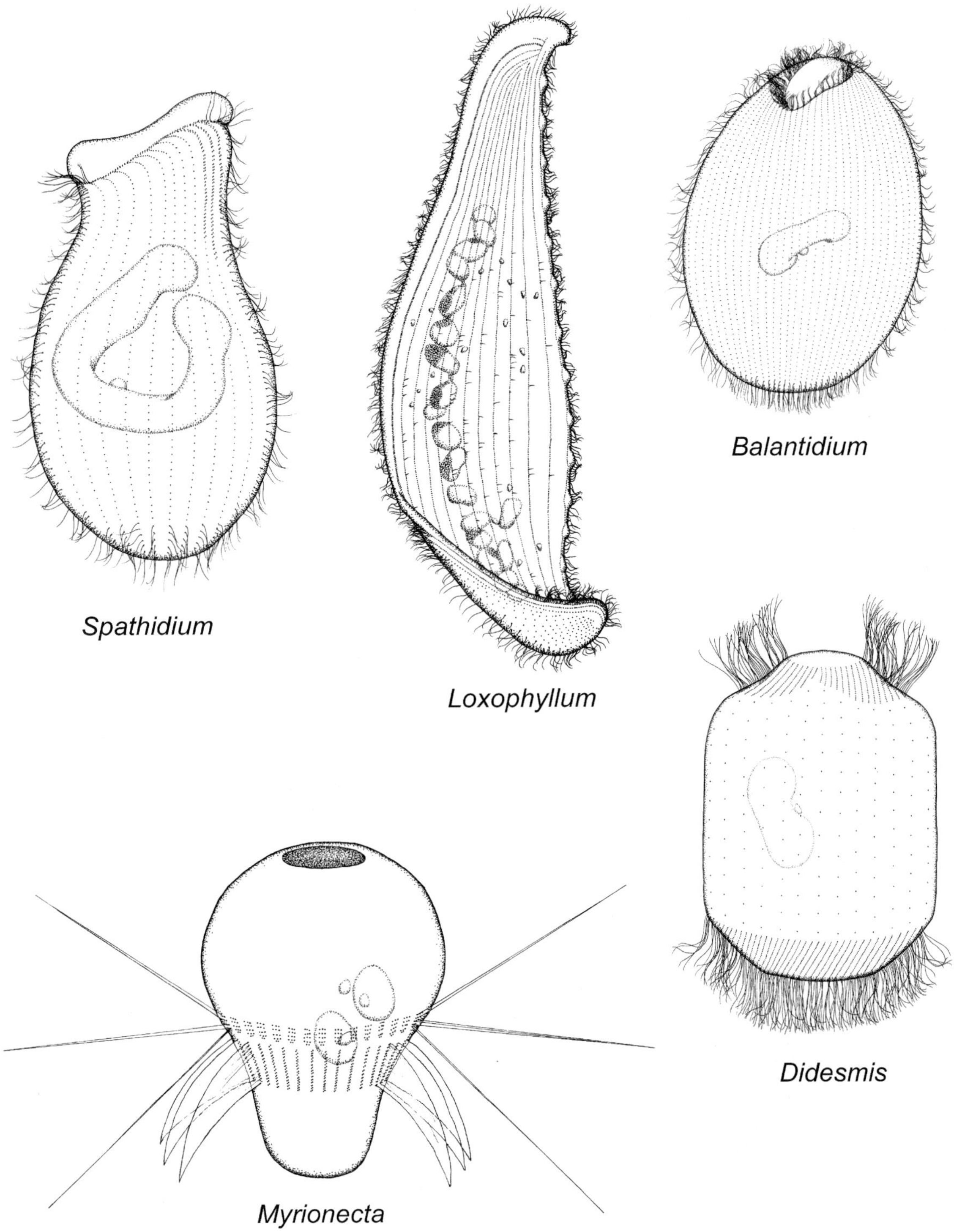

FIG. 9.2. Stylized drawings of representative genera from the Subclasses Haptoria and Trichostomatia of the Class LITOSTOMATEA. Subclass Haptoria: the haptorid *Spathidium*; the pleurostomatid *Loxophyllum*; and the cyclotrichiid *Myrionecta*. Subclass Trichostomatia: the vestibuliferid *Balantidium*; and the entodiniomorphid buetschliid *Didesmis*

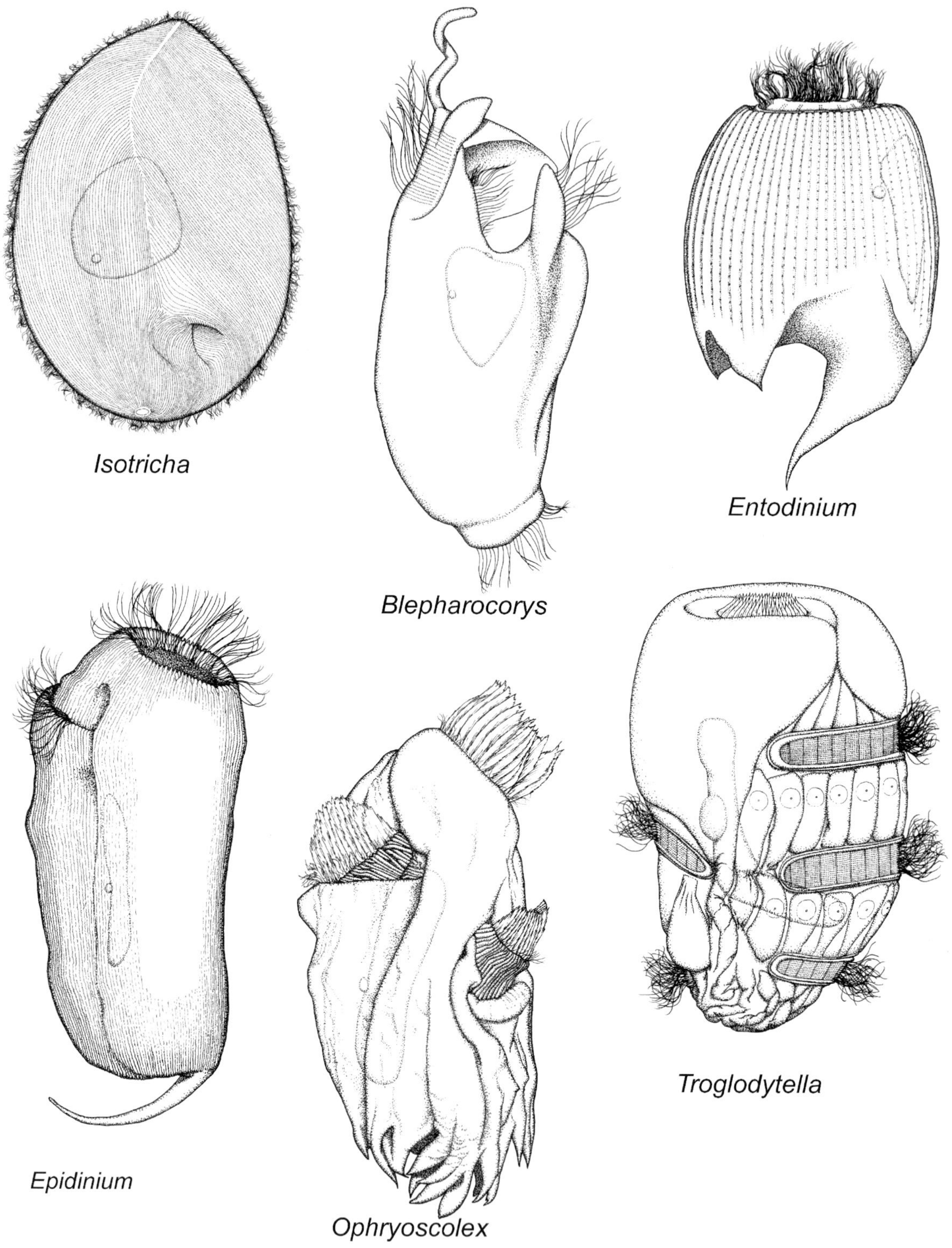

FIG. 9.3. Stylized drawings of representative genera from the Subclass Trichostomatia of the Class LITOSTOMATEA. The vestibuliferid *Isotricha*. The blepharocorythid *Blepharocorys*. The entodiniomorphids *Entodinium*, *Epidinium*, *Ophryoscolex*, and *Troglodytella*

1985). However, *Isotricha* and *Dasytricha* have an additional layer, internal to the layer commonly found (Grain, 1966a; Paul, Butler, & Williams, 1989). The proteins of the epiplasm and ecto-endoplasmic layer of trichostomes have been isolated and found to differ. Filamentous proteins, ranging in molecular weight between 58–96 kDa, are the major constituent of the epiplasm of *Entodinium* and *Polyplastron* (Vigues & David, 1989; Vigues, Méténier, & Sénaud, 1984b), while the 4 nm filaments of the ecto-endoplasmic layer of *Isotricha* are about 22 kDa in size (Vigues, Méténier, & Grolière, 1984a). These proteins are Ca^{2+}-binding proteins, which show antigenic similarity to those of the ecto-endoplasmic layer of *Polyplastron* (Vigues & Grolière, 1985). The filaments of the ecto-endoplasmic layer can be specialized to enable the contraction of the neck of *Lacrymaria olor* (Tatchell, 1980), retract the ciliophore, which bears the somatic cilia in entodiniomorphids (Grain, 1994), and can "carry" the nuclear apparatus in some trichostomes (Grain, 1966a).

Williams et al. (1981) were the first to observe that the somatic monokinetids of *Spathidium* had two transverse ribbons. The litostome kinetid can now be characterized as a monokinetid bearing a slightly convergent postciliary ribbon at triplet 9, a laterally directed kinetodesmal fibril at triplets 6 and 7, and two transverse ribbons of which a tangential one, T1, is associated with triplets 3 and 4 and a somewhat radial one, T2, is associated with triplet 5 at some time during kinetid development (Fig. 9.4) (Lynn, 1991 and references therein). More recent descriptions have confirmed this for additional genera of haptorians (Foissner & Leipe, 1995; Grim, 1993a; Johnson et al., 1995; Lipscomb & Riordan, 1991, 1992). Lipscomb and Riordan (1991) demonstrated that the tangential transverse microtubules might have their origin in the lamina corticalis of *Helicoprorodon*, a feature that might be shared by other haptorians. Furness and Butler (1986) observed the transient appearance of a single microtubule during somatic kinetid replication in the entodiniomorphid *Eudiplodinium* (Fig. 9.4). They concluded this to be homologous to T2 and used this to explain why other entodiniomorphids only exhibited the T1 (see also Lynn, 1991). The postciliary ribbons of haptorians may form a two-layered structure of "n + 1-over-n" microtubules as they reach the cortex; typically, this is a "4-over-3" assemblage (Fig. 9.5). Postciliary ribbons of trichostomes are also "bundled" near their origin and may be reduced in number (Fig. 9.4). Nematodesmal microtubules may extend from the somatic kinetosomes into the cytoplasm, and these may be responsible for determining asymmetries in cell shape, such as the tail and proboscis of *Dileptus* (Golińska, 1991).

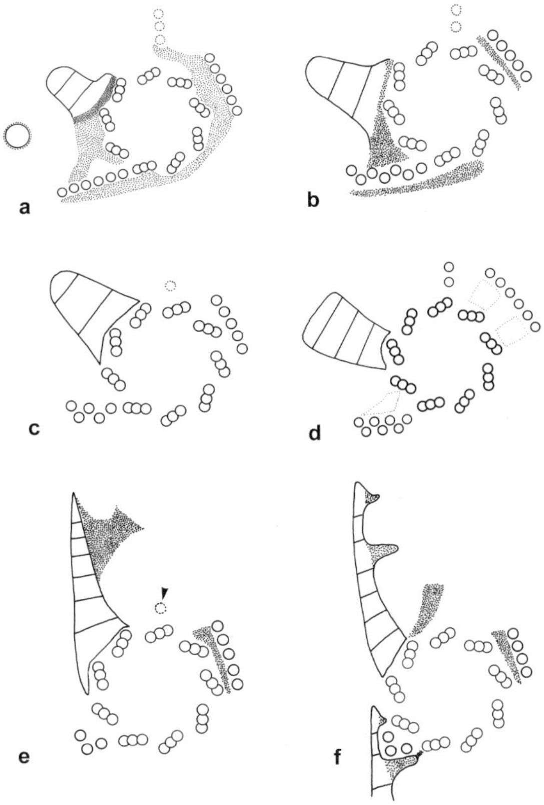

FIG. 9.4. Schematics of the somatic kinetids of the Class LITOSTOMATEA. (**a**) Monokinetid of *Homalozoon*. (**b**) Monokinetid of *Spathidium*. (**c**) Monokinetid of *Balantidium*. (**d**) Monokinetid of *Dasytricha*. (**e**) Monokinetid of *Eudiplodinium* showing transient appearance of T2 (arrowhead). (**f**) Monokinetid of *Entodinium*, showing interrelation of kinetodesmal fibrils between kinetids. Note how the postciliary microtubules appear to segregate into two rows (see Fig. 9.5) (from Lynn, 1981, 1991)

The girdle of somatic ciliature of the cyclotrichiid *Myrionecta* (= *Mesodinium*) is very unusual and unlike that of *Didinium* (Figs. 9.1, 9.2). The two ciliary elements of the girdle are a posterior file of kinetosomes in zig-zag arrangement but showing none of the typical litostome fibrillar associates, and a polykinetid of loosely arranged kinetosomes

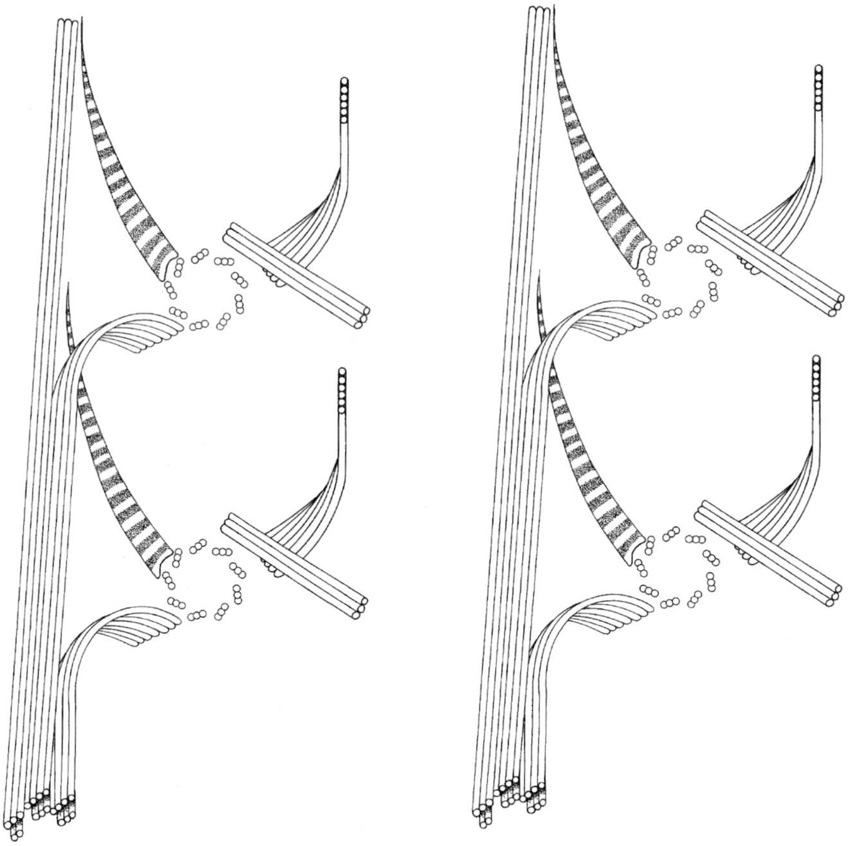

FIG. 9.5. Somatic cortex of a typical litostome whose postciliary ribbons, composed of two rows, extend alongside each other into the cortical ridges. Note that the tangential transverse ribbons extend anteriorly into the cortical ridge while the radial transverse ribbons extend somewhat posteriorly. (Modified after Leipe & Hausmann, 1989.)

is situated anterior to each of these files (Grain et al., 1982). This unusual kinetid structure is correlated with the great genetic divergence observed in the SSUrRNA genes of cyclotrichiids from other haptorians (Johnson et al., 2004; Strüder-Kypke et al., 2006).

Haptorians and some trichostomes have clavate cilia, typically localized as a longitudinal differentiation of two to four somatic kineties, depending upon the genus. These cilia tend to be shorter and swollen compared to other somatic cilia, primarily because the axonemal structure is aberrant. They are typically arranged in pairs (e.g., *Enchelydium* – Foissner & Foissner, 1985; *Fuscheria* – Foissner & Foissner, 1988; *Homalozoon* – Liepe & Hausmann, 1989; *Spathidium* – Bohatier, Iftode, Didier, & Fryd-Versavel, 1978), posterior to the oral region on what might be defined as the dorsal surface. However, in *Dileptus*, these clavate pairs are on the dorsal surface of the proboscis anterior to the cytostome (Grain & Golińska, 1969), while they may be distributed adjacent to each tentacle base in *Actinobolina* (Holt et al., 1974). Clavate cilia have been observed around the oral region in *Balantidium* (Paulin & Krascheninnikow, 1973), possibly making up the Villeneuve-Brachon field of kineties (see Guinea, Anadón, & Fernández-Galiano, 1992). Clavate cilia occur above the region of the concrement "vacuoles" of other trichostomes (**see below**), and on the paralabial organelle of entodiniomorphids (see **Oral Structure** below). Although there is no direct experimental evidence, clavate cilia are considered to be sensory structures, probably functioning like the dorsal bristles of some hypotrichs and stichotrichs (Görtz, 1982a). Locomotory cilia can differentiate to clavate cilia and the converse can occur during regeneration of *Dileptus* (Golińska, 1982a, 1982b), while high temperatures can increase the number of microtubules in the axonemes of clavate cilia (Golińska, 1987).

Mucocysts are a common feature of the cortex of the litostomes. In the trichostomes, if the cortex has a thickened epiplasm, mucocysts are not usually present. *Loxophyllum* is an unusual haptorian, which has "warts" along its body that bear clusters of extrusomes. An unusual small extrusome, called a conocyst, may also be found in these (Hausmann, 1977, 1978).

Mitochondria are common in free-living forms. Hydrogenosomes, typical of commensal litostomes, have been reported in vestibuliferids (Müller, 1993; Yarlett, Hann, Lloyd, & Williams, 1981; Williams, 1986) and entodiniomorphids (Grain, 1994; Müller; Paul, Williams, & Butler, 1990; Snyers, Hellings, Bovy-Kesler, & Thines-Sempoux, 1982; Yarlett, Coleman, Williams, & Lloyd, 1984). *Balantidium* may have both mitochondria and hydrogenosomes (Grain, 1994). Much remains to be learned about the enzymatic functioning of the hydrogenosomes of litostomes. However, it is clear that the hydrogenosomes from *Isotricha* and *Dasytricha* respond reversibly to oxygen tension levels in the rumen, producing hydrogen only when oxygen tensions are lower. This, in turn, influences production of methane by rumen methanogens (Lloyd, Hillman, Yarlett, & Williams, 1989).

Litostomes have a variety of different "storage" products in the cytoplasm. The buetschliid concrement "vacuoles" contain calcium carbonate and are surrounded by hydrogenosomes (Grain, 1994). Concrement "vacuoles" of paraisotrichids also contain calcium carbonate (Grain, 1994). Blepharocorythids have a "vacuole" that is considered homologous to the concrement "vacuoles" of other litostomes, although it does not contain calcium carbonate (Grain, 1994). Zinc granules have been reported in the cytoplasm of some entodiniomorphids (Bonhomme, Quintana, & Durand, 1980).The parapharyngeal mass of *Homalozoon* has granules containing high levels of magnesium, potassium, calcium, and phosphorus (Kuhlmann, Walz, & Hausmann, 1983).

Haptorians may store paraglycogen in large granules (e.g., *Homalozoon* – Kuhlmann et al., 1983). However, trichostomes apparently do not store glycogen. These endosymbionts store amylopectin, a starch-like polysaccharide, which appears either as isolated granules in the vestibuliferid endoplasm or as skeletal plates in the entodiniomorphids (Grain, 1994; Nakai & Imai, 1989; Wakita & Hoshino, 1980). Starch-like storage products have also been observed in the macropodiniids (Cameron & O'Donoghue, 2002a, 2003a).

Contractile vauoles are a typical feature of the litostomes with some endosymbionts having up to 20. The cytoproct is a typical feature of trichostomes, but often not reported in haptorians. Entodiniomorphids have a distinct cytoproct surrounded by a conspicuous filamentous sphincter.

9.4 Oral Structures

The litostome oral region in its simplest state is on the cell surface at the anterior end of the body as in haptorids and archistomatines (i.e., buetschliids). The cytostome is on a dome-like elevation surrounded by circumoral cilia, which are slightly longer than the adjacent somatic cilia. In pleurostome haptorians, the oral region has become slit-like, extending along the ventral edge of the laterally flattened body. In vestibuliferids, blepharocorythids, and entodiniomorphids, the oral region has become invaginated to varying degrees and the oral ciliature can become functionally differentiated, although still arising essentially from the anterior ends of somatic kineties. Prior to electron microscopy, the "membranelle-like" behavior of the oral cilia of entodiniomorphids suggested that they actually had an adoral zone of membranelles (Corliss, 1961). This functionality has now been redescribed by scanning electron microscopy of *Entodinium* and *Cycloposthium* (Imai & Yamazaki, 1988; Imai, Tashiro, & Ishii, 1983). However, there is no corresponding structural subdivision at the level of the kinetosomes where a "membranellar" ultrastructure as in the spirotrichs might be expected (Furness & Butler, 1983, 1985).

At the ultrastructural level, the circumoral kinetids of haptorians are typically dikinetids characterized as follows: a ciliated posterior or right kinetosome with a postciliary ribbon and sometimes a transverse ribbon; and a non-ciliated anterior or left kinetosome that bears a tangential transverse ribbon, sometimes a single postciliary microtubule, and a nematodesma (see Grain, 1994; Lynn, 1991). These oral dikinetids are often linked up by a filamentous annulus that is continuous with the filaments of the ecto-endoplasmic layer. The transverse ribbons of the anterior kinetosomes extend anteriorly and then bend posteriorly to support the cytopharnynx. The nematodesmata extend posteriorly to form the rhabdos or litostome cytopharyngeal apparatus. *Dileptus* species may have two rings of nematodesmata, the inner one of which is not associated with kinetosomes (Grain & Golińska, 1969). A set of bulge microtubules may be found interior to the nematodesmata and extending as a "pallisade" deep into the cytoplasm surrounding the cytopharynx from the point where the transverse ribbons bend interiorly (Foissner & Foissner, 1988; Grain; Kuhlmann, Patterson, & Hausmann, 1980; Lipscomb & Riordan, 1991; Lynn, 1991). The oral dikinetids can be rotated in some haptorians (e.g., *Didinium* – Lipscomb & Riordan, 1992; Foissner & Foissner, 1988) so that it may be preferable to refer to right and left kinetosomes rather than anterior and posterior ones (Grain, 1994). Finally, the slit-like oral region of pleurostomes is also bordered by oral dikinetids whose transverse ribbons are also associated with bulge microtubules (Foissner & Leipe, 1995). Oral monokinetids are only found in some haptorians (Foissner & Foissner, 1985). Lipscomb and Riordan (1990) used a parsimony analysis and cladistics to conclude that the haptorian oral monokinetids are homologous to the anterior or left kinetosome of the oral dikinetid (i.e., the posterior right kinetosome has not differentiated).

The trichostomes have what are termed "oralized" somatic kinetids since Lipscomb and Riordan (1990) concluded that these ciliates have completely lost the oral dikinetid. This is correlated with the loss of toxicysts in these ciliates. Transverse ribbons of several of the more anterior oral monokinetids of trichostomes may still extend to support the cytopharynx, while nematodesmata and bulge microtubules support the oral cavity or vestibulum and cytopharynx of some of these ciliates (Grain, 1994; Lynn, 1991). Typically, these "oralized" kinetosomes have kinetodesmal fibrils that are considerably reduced or absent (Grain, 1966a, 1994; Grim, 1993a; Guinea et al., 1992; Paul et al., 1989).

Feeding and ingestion by litostomes involve intimate contact with their prey since they are not suspension or filter-feeding ciliates. Clavate cilia may be involved in sensing prey or prey metabolites, while the girdle ciliature of the planktonic cyclotrichiids may detect prey by hydromechanical signals (Jakobsen et al., 2006). In the rumen ciliate *Ophryscolex*, the paralabial organelle is situated on the ventral side of the body adjacent to the oral region. This organelle, which has clavate cilia arising from files of dikinetids that border a central kinety of monokinetids, has been implicated as a sensory structure involved in feeding (Bretschneider, 1962; Schrenk & Bardele, 1987). Treatment of *Dileptus* with proteolytic enzymes has demonstrated that surface proteins are essential for ingestion, while ingestion itself is a calcium-dependant process (Estève, 1984b). Toxicysts are crucial to prey capture in haptorians and their discharge is also calcium-dependant (Iwadate, Katoh, Kikuyama, & Asai, 1999). Microtubules are also essential to the ingestive process as the anti-microtubule drug, colchicine, decreases food vacuole formation in *Dileptus* (Tołłoczko, 1980). The food vacuole is formed by the fusion of vesicles with the plasma membrane of the cytostome (Kuhlmann & Hausmann, 1983). The digestive process is similar to that in *Paramecium* (see **Chapter 15**) with the fusion of acid vesicles with the food vacuole membrane during the initial stages of digestion in *Litonotus* (Verni & Gualtieri, 1997). In the entodiniomorphids, the complexity of the cytopharyngeal apparatus is directly related to the diet of the ciliates: species that ingest large plant fragments have larger and more fibrous oral structures than those that ingest bacteria and starch grains (Furness & Butler, 1988).

Extrusomes with toxic capacities are typical of haptorians but have apparently been lost by trichostomes, possibly coincident with the loss of oral dikinetids. Several categories of toxicysts have been described and the number and their types are characteristic of certain genera. Toxicysts can be both proteolytic and paralytic. *Didinium* typically has three types: (1) the longer toxicysts that extend out tube-like to paralyze prey and begin proteolysis of the prey; (2) cyrtocysts that may be prototype toxicysts; and (3) pexicysts that are shorter and appear to fasten the predator to its prey (Hausmann, 1978; Wessenberg & Antipa, 1969, 1970). *Homalozoon* also has both long and short toxicysts. Even shorter extrusomes, called conocysts, are trichocyst-like in that their contents are rod-like, bearing a pin-like element at the tip when extruded (Hausmann, 1977; Kuhlmann & Hausmann, 1980). Toxicysts develop in the endoplasm and are probably assembled by the endoplasmic reticulum through the classical cellular secretory pathway (Dragesco, Auderset, & Baumann, 1965; Hausmann, 1978).

Finally, brief mention must be made of what we know about the regulation of oral structures in litostomes. Kink (1976) demonstrated a zone of kinetosome proliferation near the cytostomal region of *Dileptus* where kinetosomes are supplied to the enlarging circumoral kinety and to adjacent elongating somatic kineties. In contrast, Golińska (1984) examined the effects of reduced cell size on ultrastructural components of the cell: *Dileptus* reduced in size by microsurgery had oral nematodesmata and transverse ribbons that contained fewer microtubules. However, the reduction was not perfectly proportional to the reduction in cell size, suggesting that some lower limit may exist for a functional organelle (Golińska, 1984). In contrast, in overfed *Dileptus*, although the number of oral nematodesmata might increase, the number of microtubules in these nematodesmata and in transverse ribbons did not vary, suggesting an upper size-limit on these structures (Golińska, 1986). High temperatures can cause abnormalities in the oral development, probably by interfering with microtubule assembly (Golińska, 1988).

9.5 Division and Morphogenesis

Litostomes typically divide while swimming freely, although some pleurostomatids may divide within a cyst. Stomatogenesis is telokinetal and the parental oral structures are retained. Foissner (1996b) distinguished four subtypes in the class, primarily characterizing different ordinal or subordinal assemblages. The majority of taxa within the Subclass Haptoria are characterized as undergoing holotelokinetal stomatogenesis, with the "anterior" ends of all the opisthe somatic kineties undergoing kinetosomal differentiation and proliferation to produce the oral dikinetids (Fig. 9.6) (e.g., *Fuscheria*, *Spathidium* – Berger, Foissner, & Adam, 1983; *Homalozoon* – Leipe, Oppelt, Hausmann, & Foissner, 1992). *Dileptus* has also been characterized as holotelokinetal. However, the development of the proboscis and its ciliary fields means that kinetosomal proliferation at the equator gives rise to **both** the proboscis infraciliature **and** the circumoral dikinetids (Golińska, 1995). Pleurostomatid stomatogenesis is characterized as monotelokinetal as only the somatic portions of the oral kineties are involved (e.g., *Litonotus*, *Loxophyllum* – Fryd-Versavel, Iftode, & Dragesco, 1976). The endocommensal vestibuliferids are characterized as undergoing intertelokinetal stomatogenesis (Foissner, 1996b). Kinetosomal proliferation occurs at the ends and by elineation or laterally to produce kinetofragments **between** the anterior ends of the somatic kineties (e.g., *Balantidium* – Fauré-Fremiet, 1955; isotrichids, paraisotrichids, and buetschliids – Grain, 1966a). The buetschliid *Polymorphella* may be the only exception as it appears to have holotelokinetal stomatogenesis (Foissner 1996b; Grain 1966a). It might be more appropriate to classify the stomatogenesis of buetschliids as cryptotelokinetal (**see below**) since the region of oral kinetosomal proliferation usually has non-ciliated somatic kinetosomes.

The entodiniomorphids have long been considered to have apokinetal stomatogenesis since somatic kinetosomes were not visible by light microscopy. Furness and Butler (1986) have demonstrated that proliferation of oral kinetosomes occurs from non-ciliated somatic kinetosomes distributed in the cortex of *Eudiplodinium*. A similar process may explain the appearance of the new girdles in the haptorian *Didinium* as they appear "de novo" at some distance from the parental girdles (Small, Antipa, & Marszałek, 1972). This is presumably the case for all entodiniomorphids (Fernández-Galiano, Serrano, & Fernández-Galiano, 1985; Ito & Imai, 2005; Ito, Miyazaki, & Imai, 2001) and blepharocorythids (Wolska, 1966), and has lead to describing the process as cryptotelokinetal (Foissner, 1996b). Several primordial fields may develop, typically in cortical pockets, and then later fuse to form the differentiated oral structures (Fig. 9.6). Cameron and O'Donoghue (2001) have concluded that the stomatogenesis of *Macropodinium*, the genus endosymbiotic in kangaroos and wallabies, is very unusual, having features of epiapokinetal, endoapokinetal, and cryptotelokinetal stomatogenesis depending upon the infraciliary structures involved. Clearly, an ultrastructural study is needed to resolve these questions. Moreover, stomatogenesis has not been studied in representatives from all

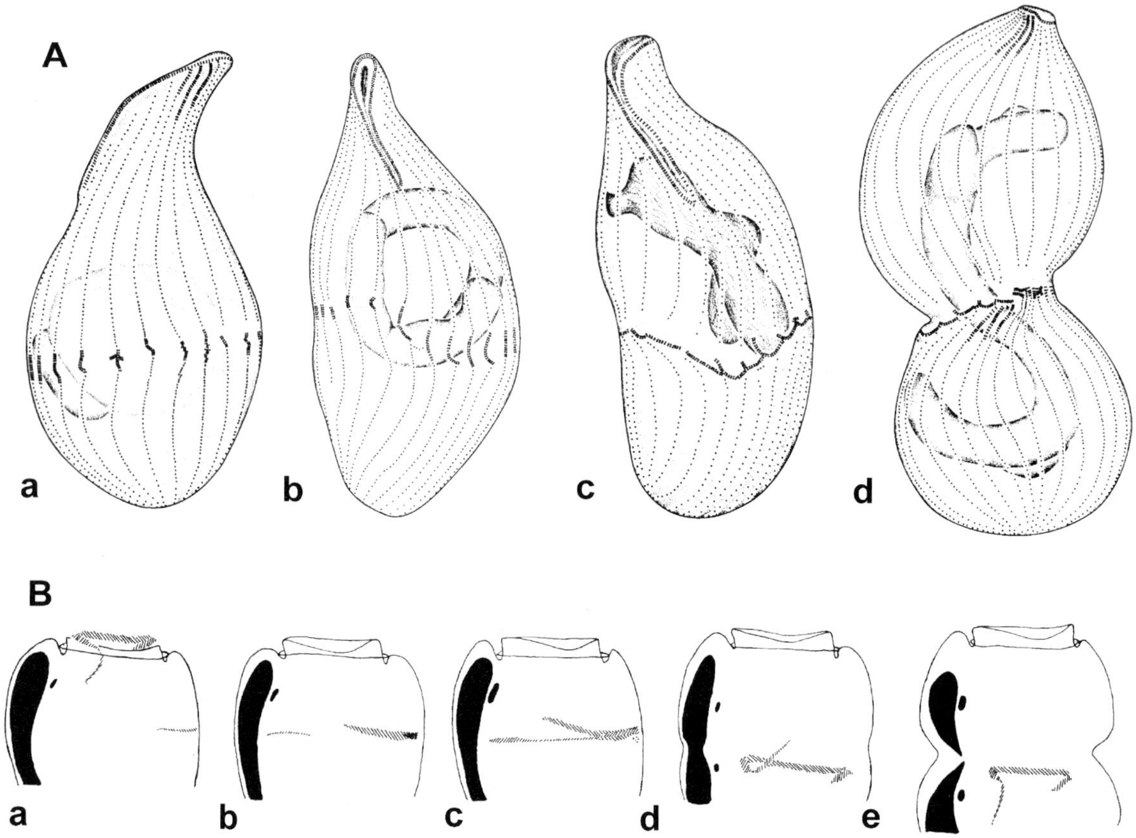

FIG. 9.6. Division morphogenesis of litostomes. **A** In the haptorian *Spathidium*, its holotelokinetal stomatogenesis begins with proliferation of circumoral dikinetids at the equator of all somatic kineties (**a**, **b**). These kinetofragments bend rightward to extend across the interkinetal space and so form the opisthe's circumoral dikinetid as division morphogenesis is completed (**c**, **d**) (from Berger et al., 1983). **B** In the entodiniomorphid *Entodinium*, what appears to be apokinetal is now considered cryptotelokinetal because there are somatic kinetosomes scattered throughout the cortex. Kinetosomes first appear on the right ventral side (**a**) and as these proliferate to form a polybrachykinety another field forms on the dorsal left side (**b**). These two primordia meet as replication continues (**c**). Ultimately, a portion invaginates to become the infundibular or vestibular portion (**d**, **e**) (from Fernández-Galiano et al., 1985.)

the families in this class, so there is clearly much comparative work to do.

There is an extensive literature on the morphogenesis and cell biology of regeneration in litostomes, particularly using the genera *Dileptus* and *Lacrymaria* (Frankel, 1989). Both ciliates can be sectioned by microsurgery and regeneration follows. If the proboscis or trunk of *Dileptus* is isolated, each part can proportionally regulate a smaller "body" over several days (Golińska, 1979; Golińska & Kink, 1977). Regeneration involves the proliferation of new kinetosomes in both genera (Bohatier, 1972; Golińska; Kink, 1972). Initially regeneration stomatogenesis apparently requires neither RNA nor protein synthesis. However, "stabilization" of the oral structures and development of toxicysts does require these two crucial cell processes (Bohatier & Kink, 1977; Golińska & Bohatier, 1975). A fuller discussion of this research can be found in Frankel (1989).

9.6 Nuclei, Sexuality and Life Cycle

Litostomes are typical ciliates, having at least one macronucleus and one micronucleus. Macronuclear DNA does not appear to be fragmented into gene-sized pieces (e.g., *Didinium* – Riley & Katz, 2001). Macronuclear shapes can be quite variable, ranging from small ovoid to very elongate and band-shaped (e.g., *Didinium*). Pleurostomatids often have the macronucleus in pairs or quartets. Some vestibuliferids (e.g., *Isotricha*) have filaments extending from the ecto-endoplasmic layer to form a karyophore (Grain, 1966a). The macronucleus and micronucleus are typically quite closely associated in trichostomes with the micronucleus often residing in a depression in the macronucleus (Grain, 1994). Although appearing multiple, the macronucleus may in fact be moniliform, like a string of beads, with the number of beads increasing as cell size increases (Leipe & Hausmann, 1992). These beads condense prior to macronuclear division, while the more elongate and band-shaped macronuclei become ellipsoid. Macronuclear andmicronuclear division are accomplished by intranuclear microtubules (e.g., *Isotricha* – Grain, 1966a; *Didinium* – Karadzhan & Raikov, 1977; *Homalozoon* – Leipe & Hausmann).

Relatively little research has been done on conjugation processes in litostomes, and the event is rarely observed in natural populations (Lucchesi & Santangelo, 2004). What we know derives mostly from research on conjugation in *Dileptus*, and so any generalizations from its features must be taken with caution. Three mating types, which show serial dominance, have been identified in this species (Miyake, 1996). Each mating type excretes a gamone that can attract cells of complementary mating type (Afon'kin & Yudin, 1987). The gamone is thought to be a small polypeptide (Parfenova, Afon'kin, Yudin, & Etingof, 1989). Homotypic pairs (i.e., pairs of cells of the same mating type) may form, but are usually not stable, possibly due to the cessation of the expression of adhesive molecules (Afon'kin, 1991). Intraclonal conjugation or selfing has been reported in *Spathidium* (Williams, 1980).

Litostomes are generally characterized as undergoing a preconjugation cell division, although this is not a universal trait (Xu & Foissner, 2004). The preconjugation division can be equal, but then yields cells that are smaller than trophic cells since no growth takes place prior to conjugation (Raikov, 1972). In entodiniomorphids, the preconjugation division can be unequal (e.g., *Opisthotrichum*) and so produce macroconjugants and microconjugants (Raikov). Tavrovskaja (1974 in Miyake, 1996) has demonstrated that the gamones of *Dileptus* stimulate the preconjugation cell division.

Cell fusion usually occurs in the oral region, often involving some dedifferentiation of oral structures in haptorians (Xu & Foissner, 2004). Cell membranes of the two partners fuse, often over a considerable area (e.g., *Didinium* – Karadzhan, 1979; *Dileptus* – Golińska & Afon'kin, 1993; *Homalozoon* – Leipe & Hausmann, 1993). Thus, cytoplasmic organelles, as well as the migratory pronuclei, might be exchanged during the process, which may take several days.

There are typically three maturation divisions during micronuclear meiosis in haptorians and trichostomes but only two in entodiniomorphids (Raikov, 1972; Xu & Foissner, 2004). In the vestibuliferids, the micronuclear mitotic spindle is much enlarged and the division products remain swollen and spindle-like during conjugation. The migratory gametic nucleus of entodiniomorphids has portions of the telophase spindle attached so that it appears "spermatozoon-like" (Raikov, 1972). Following fertilization, the synkaryon of haptorians typically undergoes two or three divisions (Raikov, 1972; Serrano, Martín-González, & Fernández-Galiano, 1990; Xu & Foissner, 2004). Trichostomes typically undergo one synkaryon division although one of these products may undergo a second division in *Paraisotricha* and *Balantidium* species, which is followed by fusion of these two division products (Raikov, 1972). We do not know whether this feature is a homologous or convergent one in these two genera.

Williams (1980) has observed clonal aging in the haptorian *Spathidium*. This aging process is observed as a reduction in daily fission rate. The time of greatest reduction is species specific and is reversed in part by intraclonal conjugation or encystment.

9.7 Other Features

The free-living litostomes have been recorded in a variety of habitats, including anaerobic ones (Foissner, 1988a; Madoni & Sartore, 2003), from which they have been collected by an electromigration apparatus (Wagener, Stumm, & Vogels, 1986). Some of the anaerobic species (e.g., *Lacrymaria*) may harbor endosymbiotic methanogens, while other genera (e.g., *Lagynophrya*) may harbor *Chlorella* species (Finlay & Maberly, 2000). *Didinium* and *Spathidium* have been recorded in waste treatment facilities where their prey is also abundant (e.g., Rivera et al., 1988). *Myrionecta rubra* appears to be sensitive to oil pollution in the marine environment when compared to other ciliates (Dale, 1988). *Didinium* appears to be more sensitive to copper stress than its prey *Paramecium*, a feature that has some impact on the stability of this predator-prey interaction (Doucet & Maly, 1990). If haptorians are generally more sensitive to toxicants than their prey, this may have significant impacts on the dynamics of microbial food webs. However, we know nothing yet of the generality of this susceptibility among haptorians or its impact on the stability of planktonic food webs, especially those dominated by protistan predators.

Chapter 10
Subphylum 2. INTRAMACRONUCLEATA: Class 4. PHYLLOPHARYNGEA – Diverse in Form, Related in Structure

Abstract Phyllopharyngeans are divided into four subclasses, only one of which is dominated by free-living forms. The Subclass Cyrtophoria includes common members of biofilm communities, from sea ice in Antarctica to waste water treatment plants. The Chonotrichia, Rhynchodia, and Suctoria are typically found as symbionts: chonotrichs are sessile ectocommensals on the appendages of crustaceans; rhynchodians are ectoparasites on the gills of bivalves; and suctorians are epibionts on metazoans, from crustaceans to turtles. The cyrtophorians and chonotrichs have heteromerous macronuclei, which suggests that they might be united into a larger taxon, and this is supported by small subunit rRNA gene sequences. The somatic kinetid is a monokinetid with a distinctly shaped and laterally directed kinetodesmal fibril, and is underlain by subkinetal microtubules. The cytopharyngeal apparatus is lined by microtubular ribbons, called phyllae, from which the class derives its name. Cyrtophorians and chonotrichs feed on bacteria and small algae. Rhynchodians use a single sucker to ingest cytoplasm of host cells, while suctorians typically have multiple sucking tentacles to catch primarily other ciliates. Division in cyrtophorians is characterized as merotelokinetal, but in chonotrichs and suctorians is characterized by specialized forms of budding. The macronuclear DNA is apparently highly fragmented, and as in the spirotrichs, differentiates after the formation of polytene chromosomes.

Keywords Internally eliminated sequences, polytene chromosome

The ciliates included in this class are very diverse in body form, influenced by the fact that three of the four subclasses are symbionts of metazoans. Members of the mainly free-living group, the cyrtophorians, range from small to medium-sized. Their oral region has a prominent cytopharyngeal basket or cyrtos, which gave rise to their name (Corliss, 1979). Typically dorsoventrally flattened, they are conspicuous members of the Aufwuchs or biofilm communities of marine and freshwater habitats to which they attach with their ventral ciliature and attachment organelles (Deroux, 1976a, 1976b, 1977). The chonotrichs are typically found on the appendages of a variety of crustaceans, and range from 40–350 µm in length (Fernández-Leborans, 2001; Jankowski, 1973b; Mohr, Matsudo, & Leung, 1970). As the name suggests, their body is cone-like, particularly at its apex where the ciliature is enclosed in a cone-like structure. The rhynchodids are very small to small (i.e., 20–50 µm in length), typically found on the gills and within the mantle cavity of molluscs and other invertebrates (Chatton & Lwoff, 1949, 1950; Raabe, 1967, 1970a, 1970b) where they attach to the tissues with a single tentacle-like oral structure. The suctorians, some, like *Dendrosoma radians*, as large as 5 mm, are epibionts on a wide range of metazoans from crustaceans (Fernández-Leborans & Tato-Porto, 2000a) to turtles (Matthes, 1988). Called tentaculiferans because of their feeding tentacles and "acinétiens" because they apparently lacked ciliated kineties, they have been related to heliozoans and hydrozoans at various times. Careful study of their complete life cycle showed that they were indeed heterokaryotic unicells or ciliates, had ciliated kineties in the dispersal swarmer stage, and underwent conjugation (Collin, 1912; Guilcher, 1951; Kormos & Kormos, 1957a, 1957b, 1958). Phyllopharyngeans are typically not dominant members in microbial communities. However, cyrtophorians, particularly in the genus *Chilodonella*, are important indicators of water quality (Foissner, 1988a) and can be troublesome ectoparasites on the skin and gills of fishes in aquaculture settings (Hoffman, 1988). Standardization of culture methods has meant that some *Chilodonella* species have become model organisms for studying the biology of this class (Radzikowski & Golembiewska, 1977).

Small and Lynn (1981) were the first to recognize this group as a class within the phylum, and united these four groups based on the structure of the somatic kinetid and the presence of *phyllae* or leaf-like ribbons of microtubules surrounding the pharyngeal tube. This gives rise to the class name PHYLLOPHARYNGEA, which is derived from *phyllos*, Greek, meaning leaf and *pharynx*, Greek, meaning mouth. Guilcher (1951) demonstrated the affinities of cyrtophorians and chonotrichs in her studies of division morphogenesis of these organisms. Ultrastructural data then prompted de Puytorac et al. (1974b) to include these two groups in their Superorder Phyllopharyngidea. De Puytorac, Grain, Legendre, and Devaux (1984) confirmed this using a numerical phenetic analysis and further supported the decision of Small and Lynn (1981) to unite these four groups into a class.

The Class PHYLLOPHARYNGEA must have originated at least in the Paleozoic Era. Mohr (1966) has argued that the chonotrichs themselves must have diverged not later than the Tertiary Period, some 225 million years ago. The class is established as monophyletic based on two strong synapomorphies (Small & Lynn, 1981). First, the structure of the somatic monokinetids is highly conserved with a distinctively shaped, laterally directed kinetodesmal fibril and subkinetal microtubules underlying the somatic kinetosomes. These monokinetids are observed in the somatic regions of the trophonts of cyrtophorians, chonotrichs, and rhynchodians, and in the kineties of the dispersive swarmers or buds of chonotrichs and suctorians. Second, the pharynx, or at least ingestatory structures in the case of the tentacles of suctorians, is lined by ribbons of arm-bearing microtubules called phyllae. These ribbons are often disposed in a somewhat radial fashion when the ciliate is not feeding. The phyllae lining the cytopharynx of cyrtophorians lie inside a palisade of rod-like nematodesmata, which together form the cyrtos or cytopharyngeal basket. Cyrtophorians and chonotrichs possess heteromerous macronuclei, a synapomorphy that unites them as a clade (Grell & Meister, 1982a), which might be recognized taxonomically in the future. The first molecular phylogeny of phyllopharyngeans supports these cytological affinities, strongly demonstrating that the chonotrich *Isochona* groups well within the cyrtophorine clade (Snoeyenbos-West, Cole, Campbell, Coats, & Katz, 2004).

10.1 Taxonomic Structure

Corliss (1979) placed the four subclasses that we now include in this class in two subclasses, the Subclass Hypostomata and Subclass Suctoria in the Class KINETOFRAGMINOPHORA. The Hypostomata, which included cyrtophorines, chonotrichs, and rhynchodines, also included what we now consider to be unrelated forms. Small and Lynn (1981) emphasized the features of the somatic kinetid (i.e., a distinctively shaped, laterally directed kinetodesmal fibril and subkinetal microtubules underlying the somatic monokinetids) and the similarities in the oral structures (i.e., phyllae lining cytopharynx) as shared derived characters to establish the Class PHYLLOPHARYNGEA, a name that had been proposed by de Puytorac et al. (1974b) to include only cyrtophorines and chonotrichs. De Puytorac et al. (1984) later supported the class as envisioned by Small and Lynn (1981). Lynn and Small (1997, 2002) have maintained these subclass assignments. Leipe, Bernhard, Schlegel, and Sogin (1994) demonstrated that the cyrtophorian *Trithigmostoma* and the suctorian *Discophrya* formed a strongly supported clade based on sequence analyses of the small subunit ribosomal RNA genes. Gene sequence data now affirm the affinities of cyrtophorians and chonotrichs (Snoeyenbos-West et al., 2004). We now need gene sequence data to complete testing the hypothesis of Grell and Meister (1982a) that the four groups divide into two major clades: (1) the cyrtophorians and chonotrichs with heteromerous macronuclei; and (2) the rhynchodians and suctorians with tentacle-like oral structures and toxic extrusomes. We currently recognize these four groups as subclasses.

The Subclass Cyrtophoria includes typically free-swimming forms, often dorsoventrally flattened with somatic ciliature restricted to the ventral surface. The oral ciliature is composed typically of three kinetofragments, a preoral and two circumoral kineties, which are moved into position during stomatogenesis by a complex counterclockwise rotation during division morphogenesis (see **Division and Morphogenesis** below). We have reverted to the older taxonomic name for this group because of its familiarity to many and here suppress the subclass name Phyllopharyngia. We recognize two orders within the subclass. The Order Chlamydodontida includes dorsoventrally-flattened forms with ventral thigmotactic ciliature and no adhesive organelle or podite (Fig. 10.1). "Classical" families included in this order are the Chilodonellidae, Chlamydodontidae, Gastronautidae, and Lynchellidae. Added to these four "classical" families are the monotypic Family Chitonellidae including an unusual loricate genus (Small & Lynn, 1981) and the monotypic Family Kryoprorodontidae, which includes the former enchelyid *Gymnozoum*, an unusual planktonic cyrtophorian (Alekperov & Mamajeva, 1992; Petz, Song, & Wilbert, 1995). The second order, Order Dysteriida, includes typically laterally compressed forms although the body morphology and oral structures can be quite bizarre in this group. Instead of thigmotactic ciliature, dysteriids typically use a posterior podite for attachment to the substrate (Fig. 10.1). Substrates may range from the substrate of marine and freshwater habitats for genera such as *Plesiotrichopus* and *Dysteria* (Deroux, 1976a) to the nasal passages of whales and dolphins for *Kyaroikeus* (Sniezek, Coats, & Small, 1995). We place four families in the order: Dysteriidae, Hartmannulidae, Plesiotrichopidae, and the newest addition, the Kyaroikeidae. Fauré-Fremiet (1965) provided one of the first modern analyses of dysteriid morphology. This was followed by the seminal works of Deroux and his collaborators, who have essentially set the modern groundwork for this subclass (Deroux, 1965, 1970, 1976a, 1976b, 1977; Deroux & Dragesco, 1968). Deroux (1994a) has revised the taxonomy of the group, introducing new orders, several

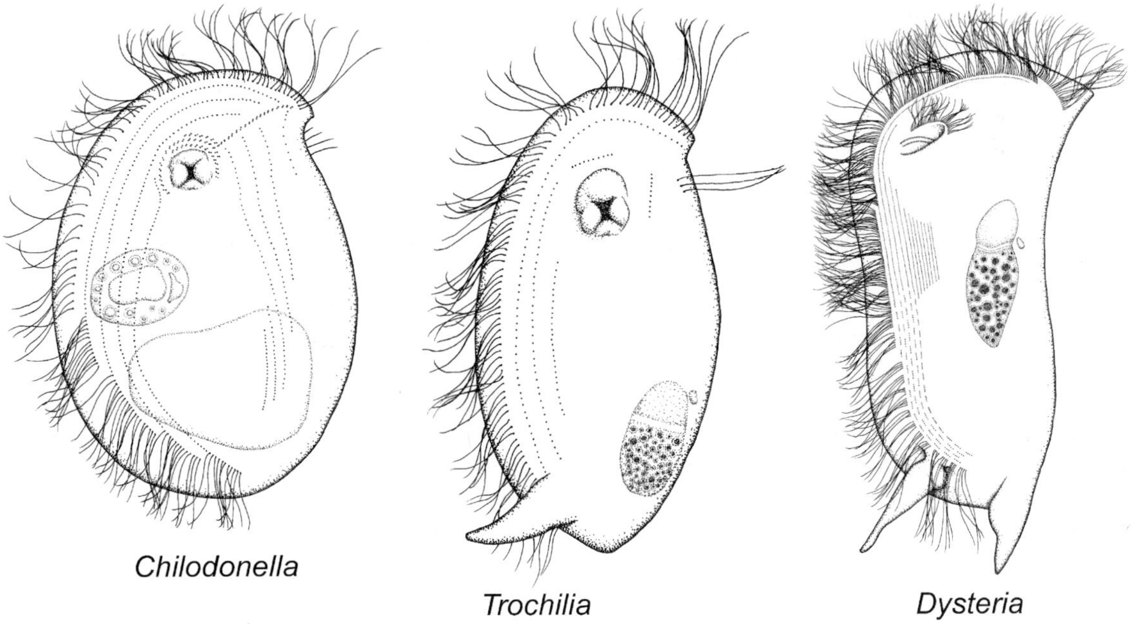

Fig. 10.1. Stylized drawings of representatives of the Subclass Cyrtophoria of the Class PHYLLOPHARYNGEA. The chlamydodontid *Chilodonella*. The dysteriids *Trochilia* and *Dysteria*. Note that the left side of *Dysteria* has not been drawn so that the somatic ciliation in the ventral groove can be revealed

new suborders, and several new families. He also reviewed the tremendous amount of morphological diversity within the dysteriids, suggesting several phylogenetic trends that bear testing by gene sequence data. We have followed Lynn and Small (2002), remaining conservative until molecular evidence can corroborate these further subdivisions proposed by Deroux (1994a) and others. Li and Song (2006b) have added the dysteriid *Hartmannula* to the phyllopharyngean small subunit rRNA (SSUrRNA) gene database and have demonstrated the monophyly of four of these cyrtophorian families.

The Subclass Chonotrichia includes ectosymbionts, typically on the mouthparts and body appendages of a variety of marine and freshwater crustaceans. First described on a gammarid amphipod by Stein (1851), their taxonomic history has had a checkered past (Mohr et al., 1970). Guilcher (1951) noticed their striking similarities to cyrtophorians in her study of chonotrich division morphogenesis (Fig. 10.2). Ultrastructural analysis confirmed this common ancestry (Grain & Batisse, 1974; Fahrni, 1982). These sessile ciliates have a vase- or funnel-shaped apex that forms an atrial cavity. This cavity is lined by ciliature that we now know is predominantly a somatic ciliature (Grain & Batisse, 1974; Fahrni). Jankowski (1972b, 1973a, 1973b, 1975) divided the group on the basis of the kinds of reproductive budding. The Order Exogemmida produces daughter cells or buds externally and includes six families: Chilodochonidae, Filichonidae, Heliochonidae, Lobochonidae, Phyllochonidae, and Spirochonidae. The Order Cryptogemmida produces buds, sometimes up to eight, within a crypt or brood pouch. It now also includes six families: Actinichonidae, Echinichonidae, Inversochonidae, Isochonidae, Stylochonidae, and the newest family Isochonopsidae, described by Batisse and Crumeyrolle (1988). Jankowski (1972b, 1973a, 1973b) and Mohr et al. (1970) have laid the modern groundwork for this group. Batisse (1994a) suggested placing the Family Chilodochonidae into a monotypic order, but we await a test of this hypothesis by gene sequence data.

The Subclass Rhynchodia has now been elevated from its ordinal level within the hypostomes. These predators of other ciliates and ectosymbionts of molluscs and other invertebrates have a single suctorial tube or "tentacle", appearing as a somewhat pointed protuberance at the anterior end of the cell (Fig. 10.3). Ultrastructural study by Lom and Kozloff (1968, 1970) clearly established affinities with other phyllopharyngeans: the somatic monokinetids were similar and the suctorial tube was lined by phyllae and enclosed elongate extrusomes called acmocysts or haptotrichocysts. Lynn and Small (2002) recognized two orders, Order Hypocomatida and Order Rhynchodida. Hypocomatids were transferred to this order based on the ultrastructural study of *Hypocoma* and the cladistic analysis of Grell and Meister (1982a). Hypocomatids, a monotypic order, are very similar to cyrtophorians with a dorsoventrally-flattened body, posterior adhesive region, and an external right kinety (Deroux, 1976b). Members of the Order Rhynchodida have considerably reduced somatic ciliature that may be divided into two fields leaving a large part of the ventral surface bare. We recognize two families within the order, the Ancistrocomidae and Sphenophryidae, which de Puytorac (1994b) and Lynn and Small (2002) recognized as types for two monotypic suborders. Raabe (1970b) has provided the modern synthesis of this group, following the pioneering research of Chatton and Lwoff (1939a, 1939b, 1950) and Kozloff (1946, 1955, 1961, 1965a).

The Subclass Suctoria has been one of the most puzzling groups in the phylum. Kahl (1930–1935) excluded it entirely from his treatment of the ciliates. These ciliates, like the chonotrichs, are dimorphic with a tentacled, non-ciliated feeding stage or trophont and a ciliated dispersal stage or swarmer. Toxic extrusomes are the peculiar haptocysts or phialocysts. Trophonts are not ciliated but have a field of kinetosomes typically near the contractile vacuole pore. Studying division morphogenesis, Guilcher (1951) confirmed the suggestion of Fauré-Fremiet (1950a) that these ciliates are specialized "holotrichs". Ultrastructural study by Batisse (1973) demonstrated that the somatic monokinetids of the swarmer of *Trematosoma* were similar to those of other phyllopharyngeans. With a bewildering diversity of forms, and possibly significant phenotypic plasticity, there is no consensus on subdivision within the subclass. Batisse (1975) described seven suborders within what was then the Order Suctorida. He has now supported three orders with eight included suborders in a Subclass Suctoria (Batisse, 1994b). These systems are modeled on

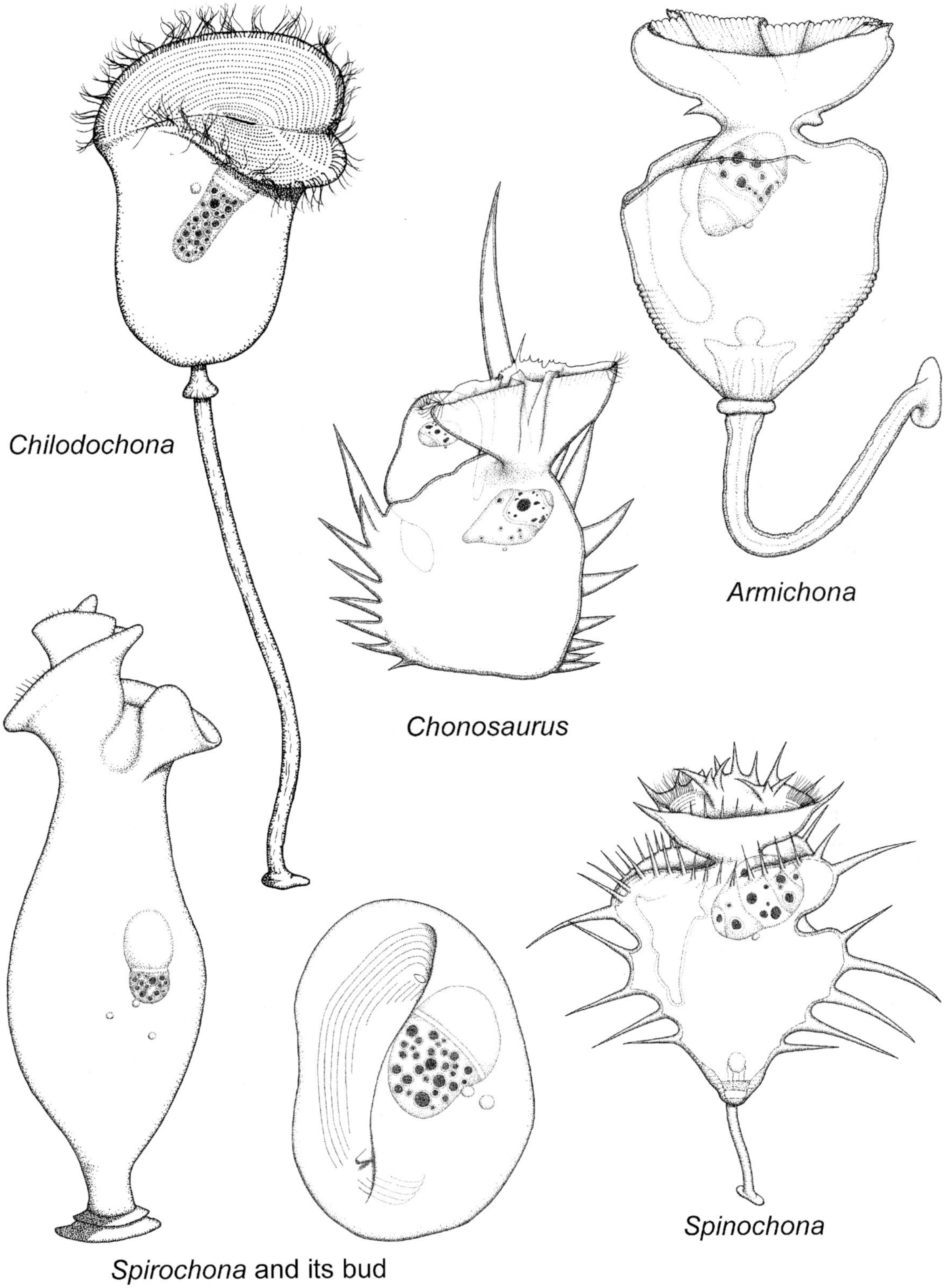

FIG. 10.2. Stylized drawings of representatives of the Subclass Chonotrichia of the Class PHYLLOPHARYNGEA. The exogemmid *Chilodochona*. The exogemmid *Spirochona* and its bud. Note the similarity of the bud's ciliary pattern to the cyrtophorians. The cryptogemmids *Chonosaurus*, *Armichona*, and *Spinichona*

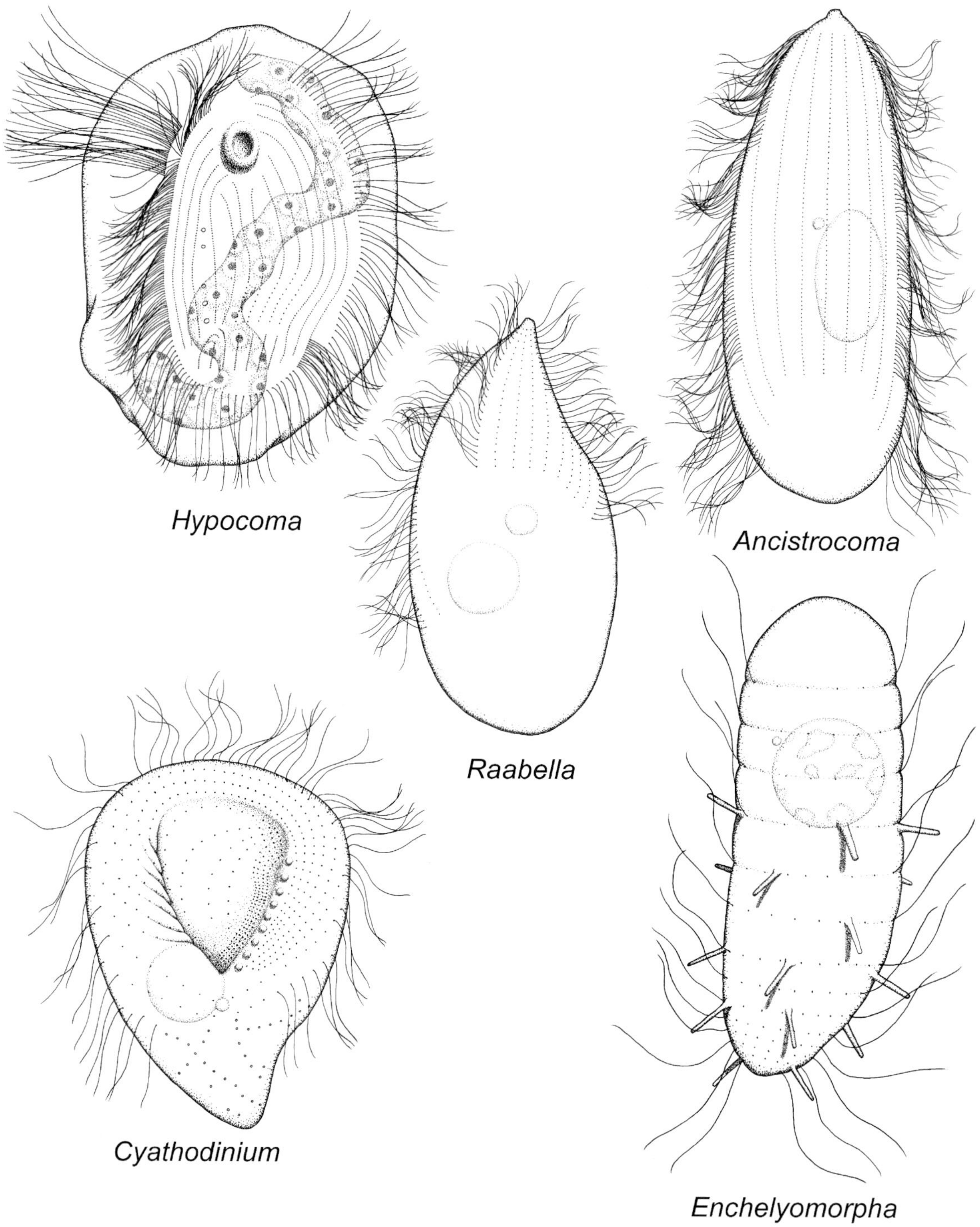

FIG. 10.3. Stylized drawings of representatives of the Subclasses Rhynchodia and Suctoria of the Class PHYLLOPHARYNGEA. Members of the Subclass Rhynchodia. The hypocomatid *Hypocoma*. The rhynchodids *Raabella* and *Ancistrocoma*. Members of the Subclass Suctoria. The highly unusual endoparasite of guinea pigs, the evaginogenid *Cyathodinium*. This ciliated suctorian has its tentacles reduced to small protuberances along the left border of a cortical depression. The bud of the endogenid *Enchelyomorpha*, exhibiting a rare condition in which the bud bears tentacles

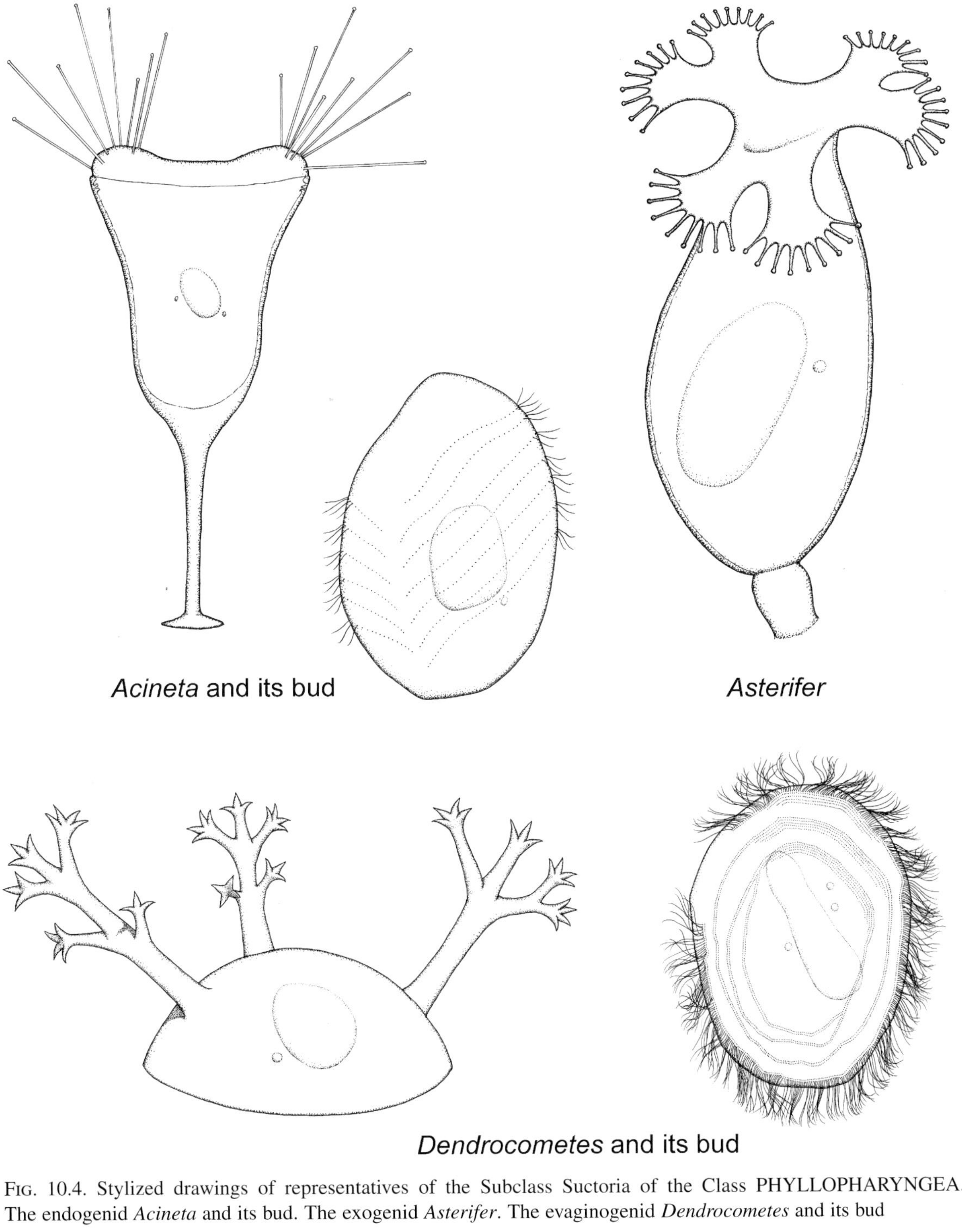

Fig. 10.4. Stylized drawings of representatives of the Subclass Suctoria of the Class PHYLLOPHARYNGEA. The endogenid *Acineta* and its bud. The exogenid *Asterifer*. The evaginogenid *Dendrocometes* and its bud

the morphogenetic analyses of Kormos and Kormos (1957a, 1957b, 1958). Jankowski (1980) listed five subclasses and 21 orders in a Class Suctoria. We conservatively follow Lynn and Small (2002) and Matthes et al. (1988) in recognizing three orders based on modes of asexual reproduction (Figs. 2.11Cb–d, 10.3–10.6). The Orders Exogenida and Endogenida were suggested by Collin (1912) an inorganic substrate, on aquatic plants, or on the body surface of animals. Thus, suctorians can be both free-living or symbiotic, and are probably the most widespread symbiotic group in the phylum. They are found in marine and freshwater habitats, in soils and mosses and have been recorded on all continents (e.g., Collin, 1912; López-Ochoterena, 1966; Wang, 1977). There are also some species that are found in the plankton. Symbiotic species are often found on crustaceans (Fernández-Leborans & Tato-Porto, 2000a; Morado & Small, 1995). *Acineta* and *Tokophrya* are two common genera that are found as free-living species or as facultative ectosymbionts, along with other suctorians, on copepods (Evans, Sell, & Beeton, 1981; Fernández-Leborans & Tato-Porto, 2000b; Grigorovich, Dovgal, MacIsaac, & Monchenko, 2001), isopods (Fernández-Leborans, Hanamura, & Nagasaki, 2002; Olafsdottir & Svavarsson, 2002), mysids (Fernàndez-Leborans & Tato-Porto, 2002), decapods (Fernàndez-Leborans & Gabilondo, 2006; Fernàndez-Leborans, Córdoba, & Gómez, 1997; Granados & Chinchilla, 1990; Vogelbein & Thune, 1988), and amphipods (Fernàndez-Leborans, Arndt, & Gabilondo, 2006; Walker & Roberts, 1982). The migratory swarmer, which is undescribed in many species, is the stage that infects new hosts or increases the colonization of the current host. Swarmers may be small and ciliated or large, worm-like and incapable of swimming. In the Laurentian Great Lakes, reinfestation of copepods by *Tokophrya quadripartita* occurs on a seasonal basis with peak prevalences of infestation found in the summer months (Evans, Sicko-Goad, & Omair, 1979). Suctorians can be broadly distributed on the appendages of their crustacean hosts (e.g., Batisse, 1986; Fernàndez-Leborans & Tato-Porto, 2002; Nicol, 1984; Walker & Roberts) while others can show high site specificity (e.g., Batisse, 1973; Hitchen & Butler, 1972). The intensity of the suctorian infestation has been correlated to the age of the host: mature hosts can have almost ten times the number of suctorians as juvenile hosts (Fernàndez-Leborans & Tato-Porto, 2002). Several unusual suctorians even colonize the **outside** of vertebrates, such as on the shell of turtles (Goodrich & Jahn, 1943) and the gills of Arctic char (Hofer, Salvenmoser, & Fried, 2005). At least two genera colonize the **inside** of vertebrates: *Allantosoma* is found in the cecum and colon of horses (Imai, 1979; Sundermann & Paulin, 1981); and *Cyathodinium* in the caecum of guinea pigs (Paulin & Corliss, 1964).

Suctorians feed primarily on other ciliates, typically showing little preference. However, haptorians and heterotrichs generally appear to be immune to suctorian predation (Batisse, 1994b). There is nevertheless some diversity in feeding habits. For example, *Trematosoma* may feed on bacteria (Batisse, 1973), *Choanophrya* feeds on particles of food debris derived from its copepod host (Hitchen & Butler, 1973a), and *Phalacrocleptes* feeds on the tissues of its polychaete host (Kozloff, 1966). Finally, there are some suctorians that are relatively small compared to their ciliate "prey" or host so that they have been described as parasites (Matthes, 1971). *Podophrya* species attach to the outside of their "host" ciliate, a *Nassula* species (Fauré-Fremiet, 1945) or *Paramecium* or *Urostyla* (Jankowski, 1963) or even become "internal" predators (Fig. 10.5). *Pseudogemma* species attack the suctorian *Acineta* (Batisse, 1968). *Podophrya* may reach abundances sufficient to sometimes control *Nassula* populations (Canter, Heaney, & Lund, 1990). *Pottsiocles* invades its folliculinid heterotrich host (Chatton & Lwoff, 1927) while *Endosphaera* is found in the cytoplasm of trichodinid peritrichs (Padnos & Nigrelli, 1947). Probably the most complex life cycle so far worked out for a suctorian is that of *Tachyblaston*, an unusual obligate parasite of the larger suctorian *Ephelota*. This parasite shows an alternation of generations – a generation parasitic on the trophont of *Ephelota* and then a generation epibiotic on the stalk of this host suctorian (Grell, 1950, 1973).

Most studies of feeding by phyllopharyngeans have been descriptive, and so we know very little of the quantitative aspects of their feeding ecology. Balczon and Pratt (1996) measured ingestion rates of the cyrtophorian *Trithigmostoma cucullulus* on the diatom *Navicula*. This ciliate can consume a diatom every 2 min and may consume almost 10% of the daily diatom production. In the marine environment, Epstein and Shiaris (1992) estimated that a *Chlamydodon* species consumed more than 125 bacteria per hour, which amounted to less than 1% of the bacterial standing stock per day. Thus, these ciliates have much less of an impact on their prey than the planktonic oligotrichs and choreotrichs appear to do (see **Chapter 7**).

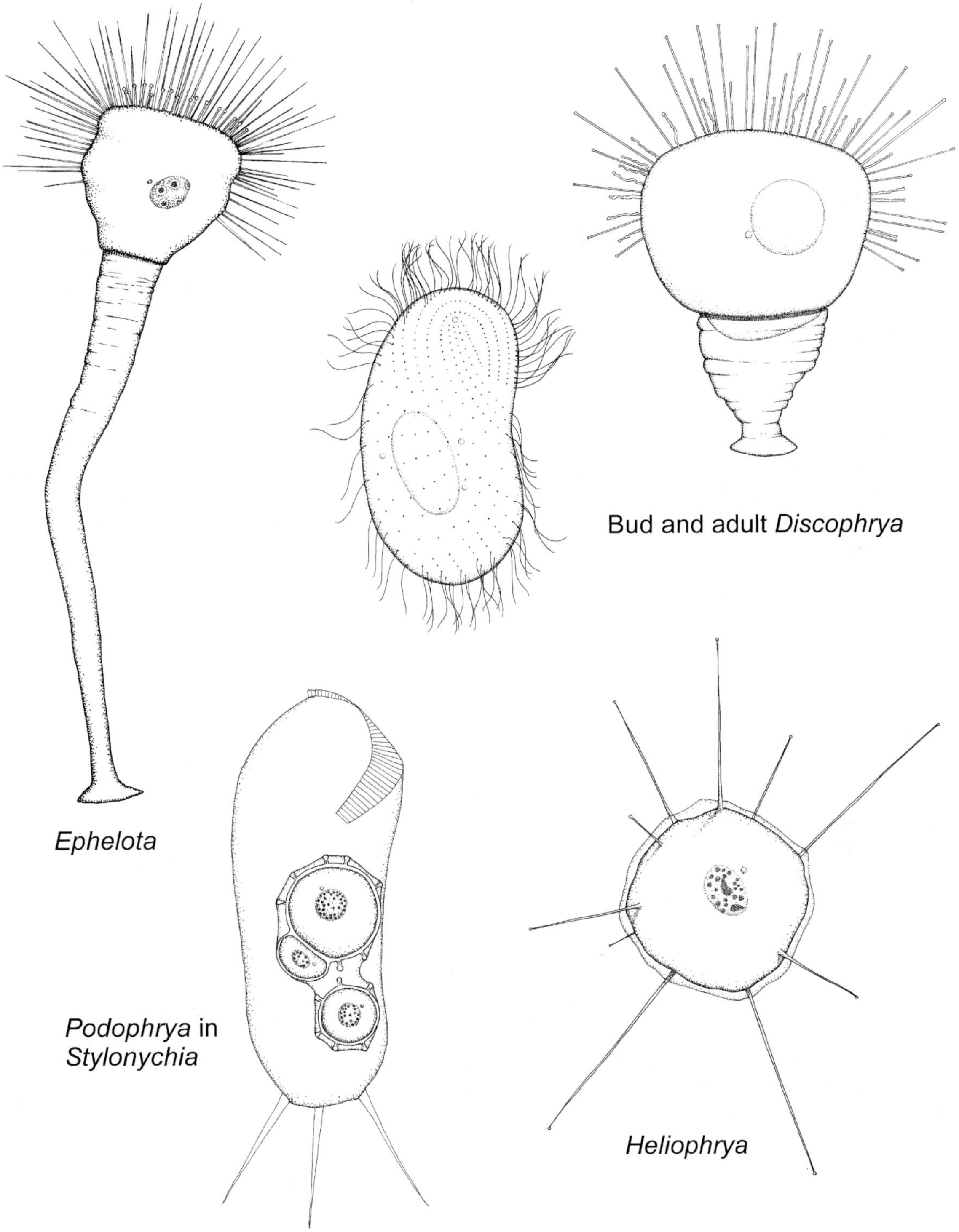

FIG. 10.5. Stylized drawings of representatives of the Subclass Suctoria of the Class PHYLLOPHARYNGEA. The exogenid *Ephelota*. The evaginogenid *Discophrya* and its bud. The exogenid *Podophrya* shown as three individuals parasitizing the stichotrich *Stylonychia*. A top view of the evaginogenid *Heliophrya*, attached to the substrate by secreted material

Suctorian species may occur in close proximity to each other and may be associated based on their tolerances of water conditions rather than competitive interactions (Bereczky, 1990). Kent (1980, 1981) explored the quantitative relationships between food levels and fecundity in *Tokophrya lemnarum*. She demonstrated that fecundity was directly related to food level while the number of non-attaching embryos was higher on a low level diet. This suggests that this suctorian tries to escape areas of low prey density by dispersing. Laybourn (1976) measured respiratory rates of *Podophrya* under different growth temperatures and conditions. She noted that this sessile ciliate, as with other sessile forms, had over an order of magnitude lower respiratory rate than similar-sized free-swimming species, like *Paramecium* or *Tetrahymena*. With so few studies, there clearly remains much to explore about the ecology of suctorians.

Given the few reports on the ultrastructure of cyrtophorians, it is not surprising that we know relatively little about the extent of symbioses of other organisms with or on them. Epibiotic bacteria have been observed in depressions on the surface of the cyrtophorian *Brooklynella* (Lom & Corliss, 1971), the chonotrich *Spirochona* (Fahrni, 1983), and the suctorian *Ephelota* (Grell & Benwitz, 1984). Matthes and Gräf (1967) identified *Flavobacterium buchneri* as the bacterial endosymbiont of *Discophrya*, a suctorian genus whose species often have bacterial endosymbionts (Matthes, 1973). Treatment with antibiotics reduced the bacterial population in *Discophrya* and led to gigantism, reduced reproduction, and a shortened lifespan of the suctorian, suggesting that the symbiosis was essential (Curry & Butler, 1975).

There are very few reports of predation on phyllopharyngeans, although we can imagine that all those predators previously reported to consume other ciliates will also consume phyllopharyngeans. Addicott (1974) notes that mosquito larvae in pitcher plants prey on the ciliate community, which included *Chilodonella* and *Dysteria* species. Batisse (1994b) noted that any predators that consume the hosts of symbiotic forms will indirectly be predators of the ciliates. He listed heliozoans, amoebae, and even other suctorians as predators of suctorians. For example, the suctorian *Acinetopsis rara* feeds exclusively on the suctorian *Ephelota gemmipara* (Grell, 1973). Phyllopharyngeans themselves may be parasitized by other organisms. Görtz and Maier (1991) described a bacterium that invades the macronucleus of the cyrtophorian *Trithigmostoma*, while Canter and Dick (1994) reviewed the literature on fungal parasites of suctorians and described a new genus of oomycete fungus, *Eurychasmopsis*, parasitizing the parasitic suctorian *Podophrya*.

Much of the behavior of phyllopharyngeans remains conjectural. Some cyrtophorians, like *Chlamydodon mnemosyne*, have a stigma apparatus that enables sensitivity to blue-light, suggesting flavins or flavin-like pigments are involved. Mildly starved cells are positively phototactic while well-fed cells are negatively phototactic (Kuhlmann, 1998; Kuhlmann & Hemmersbach-Krause, 1993a; Selbach & Kuhlmann, 1999; Selbach, Hader, & Kuhlmann, 1999). The swarmer stages of chonotrichs and suctorians are undoubtedly sensitive to chemical constituents in substrates as they often settle preferentially on particular body parts of their hosts or near to other members of their species. To our knowledge, there has been no experimentation to explore these sensitivities.

Encystment or resistant stages are common among phyllopharyngeans, especially freshwater and terrestrial cyrtophorians and suctorians (e.g., Canter et al., 1990; Fauré-Fremiet, 1945; Fernàndez-Leborans, Tato-Porto, & Sorbe, 1996; Foissner, 1979b). Encystment has not been reported for chonotrichs and rhynchodians. Deroux (1994a) reported that *Cyrtophoron*, a marine cyrtophorian that lives in the littoral splash zone, is unusual in that it encysts, and even divides in a "division cyst." Jackson and Berger (1985a) observed that *Tokophrya lemnarum* has a relatively long survival time even though it can encyst. They reported that food deprivation does not induce encystment in this species, although in the field study of Canter et al. (1990) there was a correlation between reducing prey densities and onset of encystment in *Podophrya*. Laybourn (1976) noted that encystment of *Podophrya* in culture correlated with lower temperature. Thus, it may be inappropriate to generalize on which factors are important in stimulating encystment. Kent (1981) reported that totally starved individuals had the longest mean life span. This is probably due to a reduced metabolic rate, recorded by Laybourn (1976) for starved *Podophrya*, and a reduced accumulation of toxic waste products. Rudzinska (1974) noted the appearance of autophagic vacuoles in starved *Tokophrya*, enabling survival for a period of time prior to encystment.

Suctorians are unusual among ciliates in lacking a cytoproct. Thus, wastes accumulate in the cytoplasm during the lifespan of a cell. Nevertheless, Batisse (1994b) noted that some suctorians, although lacking the cytoproct, can eliminate waste: *Thecacineta* may sequester wastes in the cell apex and sever this periodically, while *Dendrosoma* and *Ephelota* may eliminate wastes near the base of the tentacles. Rudzinska (1962) has demonstrated that overfeeding shortens the adult life span of *Tokophrya*, probably due to the accumulation of wastes. While lifespans of cells varied considerably, even within a clone, the average lifespan of cells decreased as the clonal life cycle progressed (see **Nuclei, Sexuality and Life Cycle** below) (Colgin-Bukovsan, 1979; Karakashian, Lanners, & Rudzinska, 1984).

10.3 Somatic Structures

It is difficult to generalize about the body shape and size of phyllopharyngeans. Some free-living cyrtophorians and most rhynchodians are very small ciliates, less than 50 µm in length, but typically ovoid in shape (Figs. 10.1, 10.3). Chonotrichs have a basic groundplan of basal portion and cone-shaped apex (Fig. 10.2). However, the cell body can reside on a stalk of considerable length, up to 600 µm in some *Oxychonina* species, while the neck can be extended and quite elongated in some *Filichona* species (Jankowski, 1973b). Suctorians, like chonotrichs, can be attached directly to the substrate or elevated off it on a stalk (Figs. 10.4–10.6). In those species attached to the substrate, body form can be quite variable. Batisse (1994b) recognized five major morphological types (i.e., monaxon – *Podophrya*; homaxon – *Sphaerophrya*; radial – *Cyclophrya*; bilateral – *Stylophryodendron*; and irregular – *Lernaeophrya*, *Dendrosoma*), but acknowledged it was not easy to classify a form unambiguously. This will be complicated further by any phenotypic plasticity exhibited by a species.

The basic pattern for the somatic ciliature of the class is considered to be that of a free-living cyrtophorian, such as *Chilodonella* or *Chlamydodon*

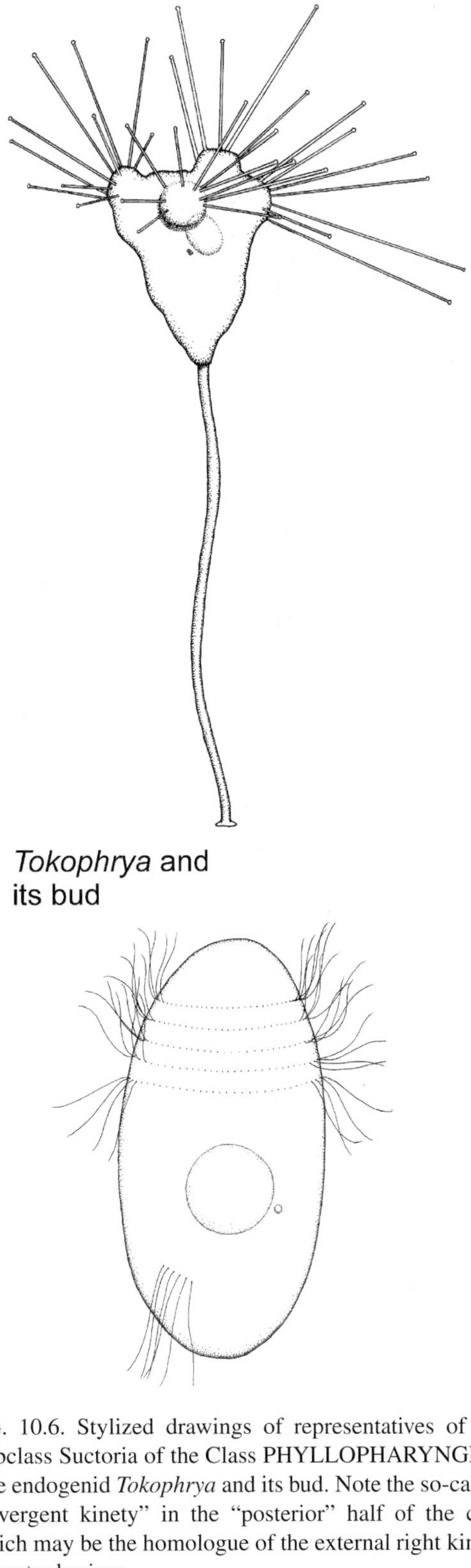

FIG. 10.6. Stylized drawings of representatives of the Subclass Suctoria of the Class PHYLLOPHARYNGEA. The endogenid *Tokophrya* and its bud. Note the so-called "divergent kinety" in the "posterior" half of the cell, which may be the homologue of the external right kinety of cyrtophorians

(Fig. 10.1). The somatic ciliature, which is highly thigmotactic, is ventral and divided into a right field and a left field. Typically, the right field is more developed than the left field, arching anteriorly in front of the oral region over onto the left ventral surface (Fig. 10.1). On the right edge of the right field are the remnants of a kinety, called the external right kinety; one kinetofragment is located at the cell equator and the other is often located on the dorsal left surface. (Deroux, 1970, 1976a, 1976b, 1977; Deroux & Dragesco, 1968). These external right kineties, and the overall pattern of the ciliature, are also present in hypocomatid rhynchodians, demonstrating their probable common ancestry with the cyrtophorians (Deroux, 1975). This general pattern of ciliature is also found in the buds of chonotrichs (Dobrzańska-Kaczanowska, 1963; Fahrni, 1984; Guilcher, 1951), suggesting an ontogenetic recapitulation of their common ancestry (cf. Figs. 10.1, 10.2). Finally, the somatic ciliature of the suctorian swarmer is believed to represent the right ventral ciliature of its cyrtophorian(?) ancestor, the left field having been lost. The remaining right field then extends in a horse-shoe around the anterior end in some species while the whole ciliature has "slipped" to an equatorial position in other species, forming a girdle of ciliature (Figs. 10.3–10.6). Foissner and Foissner (1995) demonstrated by ultrastructural study that the "polarity" of these girdle kineties is transverse rather than anterior-posterior in orientation. Suctorian swarmers swim with the tentacles in the "rear" and the scopuloid, used for attachment going first! Are they swimming backwards or forwards?

A glycocalyx covers the plasma membrane of phyllopharyngeans, reaching its full development in sessile forms (Fahrni, 1982; Henk, 1979; Sundermann & Paulin, 1985). This layer, which may function in prey capture, is very susceptible to fixation treatment and can be best demonstrated by either freeze-etching or ruthenium red staining, demonstrating its polysaccharide nature (Henk, 1979; Sundermann & Paulin, 1985). Underlying the plasma membrane is an alveolar layer that is typically conspicuous in members of this class, sometimes the alveoli contain material (Grain & Batisse, 1974; Lom & Kozloff, 1968). The epiplasm is a consistent feature of the phyllopharyngean pellicle, varying in thickness depending upon the region of the body. In some cyrtophorians, it can be somewhat thicker on the dorsal, non-ciliated surface, which is also underlain by triads of microtubules (Kurth & Bardele, 2001). The dorsal surface of rhynchodians can be underlain by many layers of microtubules (Lom & Kozloff, 1970). The epiplasm can be more than 1 μm thick in some chonotrichs (Fahrni, 1982; Karadzhan, 1976) and suctorians (Grell & Benwitz, 1984; Grell & Meister, 1982b). Pores penetrate through the non-ciliated pellicle of chonotrichs and suctorians, and there may be over 100,000 on an average *Spirochona* (Fahrni, 1982). These pores are sites of active pinocytosis in suctorians, providing the ciliate with macromolecules from the medium (Rudzinska, 1980).

It was the structure of the somatic kinetid that convinced Small and Lynn (1981) to unite these four major groups into the Class PHYLLOPHARYNGEA, based on the descriptions of several pioneering studies (e.g., Batisse, 1973; Grain & Batisse, 1974; Lom & Corliss, 1971; Lom & Kozloff, 1970; Sołtyńska, 1971). The characterization of the kinetid by Lynn (1981, 1991) is supported by recent descriptions (Foissner & Foissner, 1995; Kurth & Bardele, 2001). The phyllopharyngean monokinetid is as follows: a slightly convergent postciliary ribbon at triplet 9, a short, rapidly tapering and laterally directed kinetodesmal fibril at triplets 5 and 6, and a transverse fibre at triplet 3 (Fig. 10.7). Transverse microtubules may be associated with triplet 4 in some taxa (e.g., *Chlamydodon* – Kurth & Bardele, 2001; *Spirochona* – Fahrni, 1982; *Hypocoma* – Grell & Meister, 1983) (Fig. 10.7). The postciliary microtubules typically extend to overlap each other in a "triad" arrangement, accompanying other ribbons in the right cortical ridge, while the transverse microtubules, when present, extend slightly posteriorly and laterally to support the left cortical ridge (Fig. 10.8). Parasomal sacs typically occur on the right side of the kinetosome, but may also occur on the left (Fig. 10.7) (see Lynn, 1991). Finally, subkinetal microtubules originate as a flat ribbon from the base of the kinetosome and extend beneath those originating from more anterior or posterior kinetosomes (Kurth & Bardele, 2001; Lynn, 1991). Their orientation in some suctorians cannot be concluded with precision given the unusual orientations of their somatic kineties.

Small mucocysts are distributed throughout the ciliated cortex of cyrtophorians (Kurth & Bardele,

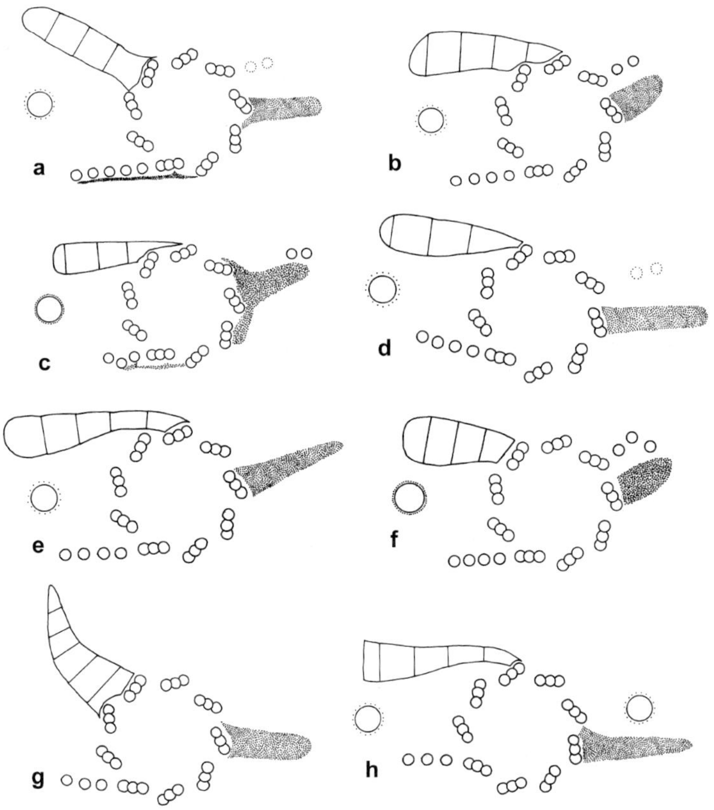

FIG. 10.7. Schematics of the somatic kinetids of the Class PHYLLOPHARYNGEA. (**a**) Monokinetid of the cyrtophorian *Chilodonella*. (**b**) Monokinetid of the cyrtophorian *Brooklynella*. (**c**) Monokinetid of the chonotrich *Chilodochona*. (**d**) Monokinetid of the chonotrich *Spirochona*. (**e**) Monokinetid of the rhynchodian *Hypocoma*. (**f**) Monokinetid of the rhynchodian *Ignotocoma*. (**g**) Monokinetid of the suctorian *Trematosoma*. (**h**) Monokinetid of the suctorian *Trichophrya* (from Lynn, 1981, 1991)

2001) and rhynchodians (Lom & Kozloff, 1970). Some dysteriid cyrtophorians have well-developed podites that are used for attachment to substrates (Deroux, 1975; Fauré-Fremiet, André, & Ganier, 1968a). Other cyrtophorians and hypocomatids have a posterior "attachment" region, a portion of the somatic cortex with some associated kineties with densely spaced kinetosomes and a secretory pellicular area (Deroux, 1975). Abundant secretory vesicles, likely containing a mucoprotein, are found in the holdfast organelle of cyrtophorians (Lom & Corliss, 1971; Fauré-Fremiet et al., 1968a) and chonotrichs (Fahrni, 1984) where they provide substances for the temporary attachment structures and for the basal disc and stalk, respectively. The scopuloid is likely the homologous structure in the suctorians. Secretions from vesicles in the scopuloid provide material for the basal disc, non-contractile stalk, and, in some species, the lorica. A thin dense outer layer surrounds a stalk matrix, which is highly variable in appearance: it can be composed of fibres that are periodically striated (e.g., *Acineta* – Batisse, 1967a; *Acinetopsis* – Grell & Meister, 1982b; *Thecacineta* – Batisse, 1969) or mostly not (e.g., *Tokophrya* – Batisse, 1970). The stalk is a highly resistant structure composed of proteins and sulfate groups, possibly also polysaccharides, and it can represent up to 15% of the total protein of the cell (Hascall, 1973). In loricate forms, the lorica can cover a portion or the entire cell body. Sometimes it is an extension of the outer covering of the stalk with loose fibrous material "gluing" it to the cell body (Batisse, 1967a; Bardele, 1968). The lorica or "shell" of *Metacineta* species has an exceedingly complex crystalline nodal substructure (Batisse, 1967b).

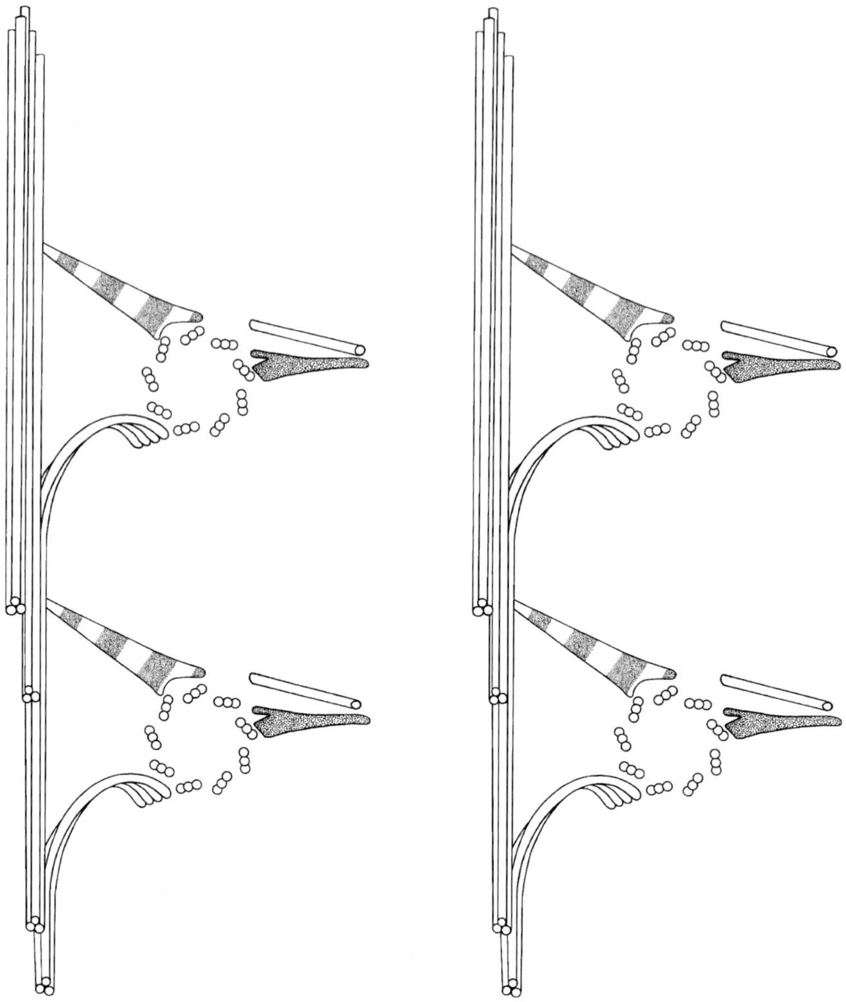

FIG. 10.8. Somatic cortex of a typical phyllopharyngean cyrtophorian whose postciliary microtubules extend as "triads" alongside each other into the right cortical ridges. Note that the transverse microtubules extend slightly posteriorly into the left cortical ridge. (Adapted from Sołtyňska, 1971.)

Mitochondria are the typical tubular mitochondria. Batisse (1994b) reported that *Allantosoma*, the endosymbiotic suctorian of the horse cecum, contains hydrogenosomes, while Foissner and Foissner (1995) described hydrogenosomes in *Enchelyomorpha*, an anaerobic suctorian collected from domestic sewage.

Contractile vacuoles are present throughout the class. Indeed, the sessile habit of suctorians made them fruitful models for our basic understanding of contractile vacuole function in the protozoa (Kitching, 1967). Patterson (1980) characterized them as having an irregular network of spongiome tubules, and this has been confirmed for the chonotrichs (Fahrni, 1983; Karadzhan, 1976). Contractions of the vacuole of *Heliophrya* were correlated with spontaneous depolarizing potentials of the plasma membrane (Eagles, Gregg, & Spoon, 1980).

The cytoproct is found in cyrtophorians, chonotrichs, and rhynchodians. In cyrtophorians it is not a defined area, but egestion typically occurs through the right posterior dorsal surface (Deroux, 1994a). In the chonotrich *Spirochona*, the cytoproct opens at the base of a 20 μm long "excretory" canal (Fahrni, 1983). The rhynchodian *Hypocoma* has a cytoproct canal about 1 μm long (Grell &

Meister, 1983). As noted above, suctorians do not have a cytoproct, but they may dispose of wastes by pinching off portions of the cytoplasm that are laden with wastes.

10.4 Oral Structures

The phyllopharyngeans represent a class where there is dramatic adaptive divergence in the structures of the oral region related to very different feeding functions. Nevertheless, their common ancestry is supported both by the strong similarities in the somatic kinetids (**see above Somatic Structures**) and by the presence of phyllae, or arm-bearing microtubular ribbons, lining the ingestatory apparatus, either as a true cytopharynx or a tentacle (see Lynn & Foissner, 1994). Cyrtophorians tend to be substrate-oriented, encounter feeders, ingesting single diatoms or several bacteria at once as they "browse" along the substrate (Epstein & Shiaris, 1992; Sawicka, Kaczanowski, & Kaczanowska, 1983). Chonotrichs are suspension feeders, using the entire ciliature, both somatic and oral, which lines the oral cone, to create feeding currents to bring particles to the cytostome. Rhynchodians and suctorians both have suctorial tube-like oral structures or tentacles. Rhynchodians are encounter feeders, attaching to the host, whether it be another ciliate or a metazoan. Their haptotrichocysts in their suctorial tube function in some fashion, not yet known, to aid ingestion of host cytoplasm. Suctorians can be characterized as passive encounter feeders – they are either sessile or float freely "waiting" for prey to contact the feeding tentacles. If appropriate, this contact will elicit extrusion of the toxic haptocysts that ensure fusion of predator-prey cells and paralysis of the prey (**see below**; Hausmann, 1978).

Cyrtophorians have an oral region that is typically bordered on its right by two circumoral kineties or kinetofragments and anteriorly by a preoral kinety or multiple preoral kinetofragments (Fig. 10.1). However, there is considerable variation on this "basic" plan: the dysteriid *Pithites* may have 5 or more small kinetofragments surrounding the cytostome while some *Lynchella* species may have oral kinetofragments that extend almost across the entire ventral surface of the cell (see Deroux, 1970, 1976a, 1977; Deroux & Dragesco, 1968). Early ultrastructural observations demonstrated the inverted nature of these oral kineties, predicted by their counter-clockwise migration occurring during cell division (Lom & Corliss, 1971; de Puytorac & Grain, 1976) (**see below Division and Morphogenesis**). Subsequent descriptions have confirmed this (Hofmann, 1987; Hofmann & Bardele, 1987). These oral kinetids are characterized as follows: a 'posterior' or right-most ciliated kinetosome with which are associated a slightly convergent postciliary ribbon and occasionally a transverse fibre; and an 'anterior' or left-most, non-ciliated kinetosome with which is associated a transverse fibre (Lynn, 1981; Grain, de Puytorac, & Bohatier, 1973). Parasomal sacs may occur on either side of these kinetosomes and additional dense fibres may be present (see Hofmann, 1987; Hofmann & Bardele, 1987). These oral kinetofragments are associated with the cyrtos, the complex cytopharyngeal "basket" of these ciliates, which shares strong similarities to the oral basket of nassophoreans (see **Chapter 11**). It is not certain that the cyrtos in both groups represents a demonstration of deep common ancestry or of convergent evolution. Although phylogenies based on small subunit ribosomal RNA genes clearly separate phyllopharyngeans and nassophoreans (see Strüder-Kypke, Wright, Fokin, & Lynn, 2000b), they are nevertheless topologically close on these trees suggesting the cyrtos could be homologous. In addition to the palisade of nematodesmata, three groups of microtubular ribbons (i.e., cytostomal or cytopharyngeal or Z lamellae; subcytostomal or Y lamellae; and nematodesmal or X lamellae, Eisler, 1988; Kurth & Bardele, 2001; Tucker, 1968) are associated with the cyrtos. Two of these, the Y and Z lamellae, are shared by nassophoreans and cyrtophorians. Ingestion is likely aided by dense differentiations at the outermost ends of the oral nematodesmata, variously called capitula, "teeth", "maxillae", or dens, while the arm-bearing microtubules of the cytopharyngeal lamellae propel food vacuole membrane enclosing food into the endoplasm (Tucker, 1972). A system of complex elongated tubules, found also in chonotrichs, is associated with the cyrtophorian cyrtos. These may function in food vacuole formation, although direct evidence for this hypothesis is still needed (*Chilodonella* – Pyne & Tuffrau, 1970; chonotrichs – Fahrni, 1982; Grain & Batisse, 1974).

As suspension feeders, chonotrichs engage the entire body ciliature in creating feeding currents. It is therefore likely that there have not been strong selective pressures to retain an organized and adaptive oral ciliature. Indeed, it was only through electron microscopy that the chonotrich *Chilodochona* was shown to have an oral ciliature. Grain and Batisse (1974) described a single inverted oral kinety composed of dikinetids patterned identically to those in cyrtophorians. This kinety is accompanied by two or three inverted somatic kineties, which were presumably carried into this orientation during division morphogenesis. Distinct oral kinetids have not been found in other chonotrichs, although a puzzling X-field has been described in *Spirochona*. Fahrni (1982) concluded that this was not likely "oral" in nature. Chonotrichs have lost the oral nematodesmata and retained only the phyllae with which are associated the elongated cytopharyngeal tubules (Fahrni; Grain & Batisse; Karadzhan, 1976). Clearly, this small sampling of chonotrichs gives us only partial insight into the structural diversity of the oral region of these ciliates.

Rhynchodians have also been little studied by modern techniques. There is only a handful of studies on electron microscopy. Lom and Kozloff (1968) first demonstrated that these ciliates have a "suctorial" tube composed of arm-bearing microtubules or phyllae radially disposed as in other phyllopharyngeans. This basic structure was also found in the hypocomid *Hypocoma* (Grell & Meister, 1982a, 1983). These studies provided the interesting revelation that the phyllae were surrounded by an outer "ring" of microtubules, making the hypocomid "suctorial" tube structurally identical to a suctorian tentacle (**see below**). Grell and Meister (1982a, 1983) also described elongated extrusomes called haptotrichocysts within the tube lumen. These presumably function like the haptocysts of suctoria (**see below**), although we have no direct evidence for this in hypocomids. Food vacuoles are observed in the suctorial tube and sometimes in the cytoplasm of rhynchodians. However, it is not yet clear how they feed.

It is safe to say that the most-studied body part of the suctorians has been their tentacles. Given the similarities in the ultrastructure of the tentacle to the cytopharyngeal components of other phyllopharyngeans, we can now conclude that suctorian tentacles are a very specialized cytostome-cytopharyngeal apparatus. However, suctorians are unusual among ciliates in that they can be considered polystomatous or many-mouthed. A few suctorians have only one or two tentacles (e.g., *Acinetopsis* – Grell & Meister, 1982b; *Rhyncheta* – Hitchen & Butler, 1972). Most have many tentacles regularly distributed over the body surface or clustered together in fascicles, sometimes borne on very elongate projections of the body called actinophores. The tentacle is an extension of the cell with a thinner glycocalyx layer on the plasma membrane and a much thinner epiplasm than the cell body. Intrinsic movements of the tentacle include bending, repeated short extensions and retractions, and complete retraction (e.g., see Grell & Meister, 1982b; Hitchen & Butler, 1973a). Tentacles retract when electrically stimulated and in elevated concentrations of external Ca^{2+} (Hackney & Butler, 1981a; Hackney, AL-Khazzar, & Butler, 1982). The microtubular axoneme of retracted tentacles is not changed, but glycerinated models suggest that actin-like filaments in the epiplasm may be the contractile elements (Hackney & Butler, 1981b). Elongation of tentacles may occur quite rapidly, and depending upon the suctorian may involve assembly of the axonemal microtubules, which may have been disassembled during retraction (Hauser & van Eys, 1976).

The tentacle tips of many suctoria are swollen or capitate and are loaded with haptocysts (e.g., Mogensen & Butler, 1984; Spoon, Chapman, Cheng, & Zane, 1976). Most suctorians can capture ciliates but a rare few cannot (e.g., *Choanophrya* – Hitchen & Butler, 1973a). Bardele and Grell (1967) and Rudzinska (1965, 1970) provided the first ultrastructural evidence of the feeding process in suctorians, implicating the haptocysts in attachment of predator to prey and the role of the axonemal microtubules in transport of food vacuoles into the cell's endoplasm. Later research confirmed the existence of arms on the inner microtubular lamellae or phyllae, confirming that the mechanism of food vacuole membrane transport was likely the same in suctorians as it was in other phyllopharyngeans (Bardele, 1974; Rudzinska, 1973; Tucker, 1974).

Suctorian phyllae are surrounded by an outer set of microtubules, which may form a complete ring or be separated into several ribbons, ranging from around 20 to over 100 microtubules (Batisse,

1994b; Lynn & Foissner, 1994). These two sets of microtubules are helically disposed. A complex set of movements at the time of contact between predator and prey, possibly involving contraction of the epiplasm and sliding of the microtubules, expands the tentacle tip to expose the haptocysts (Hauser & van Eys, 1976; Tucker & Mackie, 1975). Haptocysts enable the "gluing" of the predator to the prey, likely **without** fusion of the plasma membranes of the two ciliates (Benwitz, 1984). Haptocyst discharge probably makes the suctorian tentacle refractory to subsequent prey capture (MacKeen & Mitchell, 1977), a prediction that has been confirmed by a mathematical model (McNair, 1979). Haptocysts develop in association with the endoplasmic reticulum in the cell body. Unlike larger extrusomes (e.g., toxicysts, mucocysts), they differentiate synchronously in groups of over 20 within one vesicle (Benwitz, 1982). They are then transported on the outside of the tentacle axoneme up to the tentacle tip. Large numbers of osmiophilic granules, dense bodies, or solenocysts are also found within the lumen of non-feeding tentacles and are also transported upwards beneath the pellicle but outside of the axoneme of feeding tentacles (Bardele & Grell, 1967; Grell & Meister, 1982b). These dense bodies are thought to be primary lysosomes as they are positive for acid phosphatase (Rudzinska, 1974). They may also contain calcium deposits (Hackney & Butler, 1981c). The prehensile or capturing, but not ingestatory, tentacles of *Ephelota*, for example, have batteries of haptocysts along their length, reminding one of the tentacles of *Hydra* with its batteries of nematocysts (Grell & Benwitz, 1984). Tentacle morphogenesis may occur throughout the life of a suctorian as tentacles can be torn off by prey during unsuccessful captures, while new tentacles may be continually added as the cell body grows in size (e.g., Hull, 1954; Hitchen & Butler, 1973b). Tentacle morphogenesis has only been described in two suctorians. In both cases, a single non-ciliated kinetosome is associated with the early formation of a microtubule-organizing center around which the tentacle axoneme assembles (Curry & Butler, 1976; Hitchen & Butler, 1973b). From where do these non-ciliated kinetosomes originate and how are tentacle pattern and assembly determined?

A discussion of the suctoria would not be complete without mention of three unusual ciliates that are now recognized as members of this subclass, primarily based on ultrastructural studies. In historical order, *Phalacrocleptes*, recognized as a ciliate by its nuclear dimorphism, is a **non-ciliated** ciliate that feeds on the cilia of the pinnules of the sabellid polychaete *Schizobranchia* (Kozloff, 1966)! Lom and Kozloff (1967) described "tentacles" about 0.5 µm in length, each containing one haptocyst that is used to attach the ciliate to an annelid cilium, whose cytoplasm is presumably ingested! *Cyathodinium* is a puzzling ciliate found in the cecum of the guinea pig *Cavia* (Paulin & Corliss, 1964). Its endosprits turned out to be short tentacles containing haptocysts (Paulin & Corliss, 1969). This raises the question – is the permanently ciliated *Cyathodinium* a neotenous suctorian swarmer or a "living fossil" of the ancestral suctorian, preserved in the cecum of a vertebrate?! Finally, Foissner and Foissner (1995) conclusively demonstrated using electron microscopy that the strange tentacled "haptorian" *Enchelyomorpha* was, in fact, the swarmer of a small globular suctorian, based on the substructure of its tentacles and a complete study of its life cycle.

10.5 Division and Morphogenesis

Guilcher (1951) consolidated our current view that the subclasses in this class might be phylogenetically related with her descriptions of division morphogenesis, which Dobrzańska-Kaczanowska (1963) later confirmed. Foissner (1996b) characterized stomatogenesis of cyrtophorians as merotelokinetal because the opisthe anlagen form at the anterior ends of a small number of somatic kineties (Fig. 10.9). He does not characterize stomatogenesis for the other three subclasses because no oral ciliature has been described in these ciliates.

Stomatogenesis in cyrtophorians involves a counter-clockwise migration of the kinetofragments, when the ventral surface is viewed from outside the cell (Fig. 10.9). Thus, at completion of stomatogenesis, these three or more oral kinetofragments are inverted with respect to neighbouring somatic kineties (Deroux, 1970, 1976a, 1977; Sniezek & Coats, 1996). These movements have been confirmed and the interpretation of the inverted nature of the oral dikinetids in cyrtophorians has been verified by the

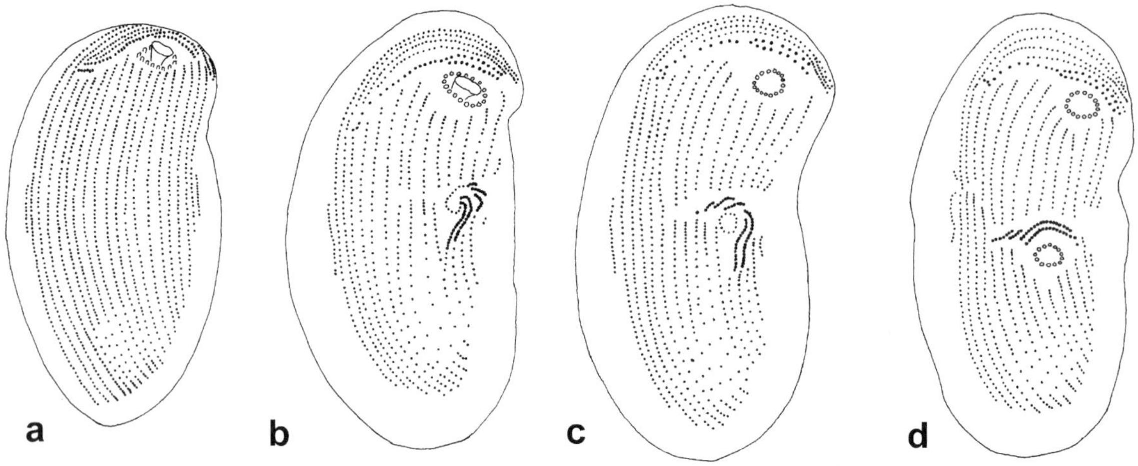

FIG. 10.9. Merotelokinetal division morphogenesis of the cyrtophorian *Chlamydonella pseudochilodon*. Note how the new oral structures appear in the equatorial region by kinetosomal replication of a few somatic kineties (**a**). These kinetosomes assemble as oral dikinetids (**b**) and undergo a counter-clockwise rotation as seen from outside the cell (**b–d**). (Redrawn from Deroux, 1970.)

detailed ultrastructural study of *Trithigmostoma* (Hofmann & Bardele, 1987) and *Chilodonella* (Hofmann, 1987). Kurth and Bardele (2001) verify the same pattern in *Chlamydodon* and put forward the intriguing hypothesis that the cyrtophorian oral apparatus is a secondary one based on the very divergent nature of these oral kinetids. Turning the traditional view of phylogeny within the Class PHYLLOPHARYNGEA upside-down, they claim that suctorians represent the basal branch with cyrtophorians and chonotrichs deriving "typical" cytostomes and body ciliature secondarily! Clearly, sampling of more phyllopharyngean genera followed by gene sequencing will help to resolve how basal the suctorians really are. Molecular phylogenetic analyses of SSUrRNA gene sequences do not resolve the question as suctorians are the sister clade to a cyrtophorine-chonotrich clade (Li & Song, 2006a; Snoeyenbos-West et al., 2004).

Chilodonella and *Trithigmostoma* have also been models in understanding the global parameters of pattern formation in ciliates (see Frankel, 1989). Contractile vacuole pore positioning at cell division suggests that new pores are positioned, in a probabilistic manner, with reference to the developing oral apparatus and the margins of the cell (Kaczanowska, 1974, 1981). Variability in the number of ventral kineties has been determined to arise in *Trithigmostoma* following cell division. This is primarily due to how many left-field kineties are separated by the fission furrow since kineties typically decrease in length towards the left margin of the cell (Fig. 10.1) (Radzikowski & Golembiewska-Skoczylas, 1999). At each cell division, the right-most "stomatogenetic" kinety releases an anterior fragment, which separates, moves to the right of this kinety, and elongates by replication. This at least compensates for the loss of one left-field kinety (Deroux, 1994a; Radzikowski & Golembiewska-Skoczylas, 1999). However, "stomatogenic" kineties can vary in position, leading to a phenomenon similar to cortical slippage in the oligohymenophorean *Tetrahymena* (Frankel, 1989; Radzikowski & Golembiewska-Skoczylas, 1999). This indicates that it is not the kinety per se that has the morphogenetic "potential" but rather some particular region of the cortex, specified in a probabilistic manner by a global patterning mechanism, like that for contractile vacuole positioning (see Frankel, 1989). Parental cytopharyngeal structures typically dedifferentiate and redifferentiate in synchrony with those of the opisthe.

Chonotrichs reproduce by two major kinds of budding, termed exogemmous and cryptogemmous (Jankowski, 1973b, 1975). The swarmer is produced probably by continuity with the ciliature of the parent in exogemmous forms. Cryptogemmous forms develop within a crypt, which may derive its kinetosomes by migration from the parental field (Gunderson, 1984). One bud is typically formed followed by regrowth of the parent. However, sequential reactive budding can occur at times when the host molts or dies (Batisse, 1994a; Jankowski, 1973b). Very few buds have been described from silver stained specimens. However, those that have been described remind one of a dysteriid-like cyrtophorian with a right ventral kinety field extending anteriorly over a smaller left ventral kinety field (Fig. 10.2). There is also an adhesive organelle in the posterior (see Dobrzańska-Kaczanowska, 1963; Fahrni, 1984; Guilcher, 1951; Jankowski, 1973b; Taylor, Lynn, & Gransden, 1995).

Rhynchodian cell division can be equal or unequal. Since there is no oral ciliature, it is an uncomplicated division of the somatic kineties. The parental cytopharyngeal apparatus dedifferentiates and redifferentiates in synchrony with that of the opisthe (de Puytorac, 1994b). Sphenophryids may have a division that is so unequal that it could be called budding (Chatton & Lwoff, 1950; Dobrzańska, 1961).

The suctorian bud or swarmer, like that of the chonotrichs, is a short-lived dispersal stage. It may be ciliated or it may be worm-like and non-ciliated. The bud "recapitulates" the phylogenetic origin of the group, under the hypothesis that suctorians are a derived group (but see Kurth & Bardele, 2001 and above). Budding can be simple or single or it can be multiple, either successive or simultaneous. Reactive budding, as in the chonotrichs, may occur under unfavourable conditions or when the host molts, possibly stimulated by ecdysone (Batisse, 1994b; Walker & Roberts, 1988). There are several schemes of classification for budding (Batisse, 1994b; Collin, 1912; Corliss, 1979; Kormos & Kormos, 1957a, 1957b). We follow Corliss (1979) until molecular evidence confirms the diversity suggested by Batisse (1994b) and Dovgal (2002). In exogenous budding, the bud infraciliature develops on the cell surface of the parent followed by an uncomplicated cell division, sometimes almost equal (e.g., *Podophyra* – Fauré-Fremiet, 1945). In evaginative budding, the bud infraciliature begins development in a pocket that erupts rapidly out of the parental cell surface (e.g., *Discophrya*– Henk & Paulin, 1977). In endogenous budding, the bud, develops and is completed within a brood pouch. The swarmer then exits through a "birth pore" (e.g., *Tokophrya* – Noble, 1932) (Fig. 2.11Cb–d).

There have been relatively few studies on the ultrastructural aspects of division morphogenesis in suctoria. Non-ciliated kinetosomes, often near the parental contractile vacuole pore, replicate to produce the infraciliature of the swarmer. The kinetal pattern of *Discophrya* is very reminiscent of a cyrtophorian with kineties curving around the "anterior" end (Fig. 10.5) (Canella, 1957; Plachter, 1979; Suárez, Guinea, & Fernández-Galiano, 1987a). However, these kineties curve around the scopuloid NOT the oral region, and the tentacle primordia are in the "posterior" of the cell, defined by its direction of swimming! Curry and Butler (1982) described the early proliferation of non-ciliated kinetosomes in the shallow embryonic cavity of *Discophrya* to form the kineties of the swarmer. Budding in *Tokophrya* has also been studied by electron microscopy. Its bud is distinguished by kineties that encircle the ovoid cell body. A so-called "divergent kinety" remains isolated in the "posterior" half of the cell (Fig. 10.6) (Guilcher, 1951; Noble, 1932; Suárez, Guinea, & Fernández-Galiano, 1987b). Could this "divergent kinety" in fact be the homologue of the external right kinety of cyrtophorians? The primordial field of kinetosomes also occurs near the contractile vacuole pore in *Acineta* and *Tokophrya*. As replication proceeds, the brood pouch enlarges by internal growth of the parental pellicle, which also forms the birth pore (Bardele, 1970; Millecchia & Rudzinska, 1970).

The life span of a swarmer lasts from minutes to hours prior to settling. It does not feed during this period and is presumably stimulated to settle in a particular place by chemical cues. Walker and Roberts (1988) noted that swarmers of *Dendrocometes* are probably triggered to settle near conspecifics on the gills of the amphipod *Gammarus*. Upon settling metamorphic changes are quite rapid: a basal disc is secreted; a stalk is formed; cilia are resorbed; and tentacles begin to grow and extend (Bardele, 1970; Fernàndez-Leborans & Tato-Porto, 2002; Hascall & Rudzinska, 1970; Henk & Paulin, 1977). After formation of the basal disc, the *Acineta* swarmer can produce a 70 μm long stalk in 5–10 min by secretion of material from scopuloid vesicles into a canal-like invagination of the body (Bardele, 1970). After the stalk reaches its mature length, the cell body of the adult *Acineta* is enclosed in a lorica, which is formed by the migration of the perimeter of the scopuloid up over the surface of the cell body, accompanied by continued secretion of material (Bardele, 1970).

10.6 Nuclei, Sexuality and Life Cycle

Phyllopharyngeans exhibit nuclear dimorphism, but the four subclasses are separated into two groups on the basis of a macronuclear feature. Like all other ciliates, rhynchodians and suctorians have homomerous macronuclei – a potential indicator of their ancestral nature? Cyrtophorians and chonotrichs have heteromerous macronuclei (Karadzhan, 1976; Lom & Corliss, 1971). Heteromerous macronuclei have two basic parts: the orthomere (orthos, Greek – right, proper; meros, Greek – part) is a "proper" part because it is DNA-rich and contains nucleoli; and the paramere (para, Greek – beside) is DNA-poor (Raikov, 1982). The typical arrangement is called juxtaposed in which the orthomere is beside the paramere (Figs. 10.1, 10.2). However, in some species, the arrangement is concentric with the orthomere surrounding the paramere (Fig. 10.1) (Fauré-Fremiet, 1957; Radzikowski, 1985). Macronuclear shape is most typically globular or ellipsoid (Figs. 10.1–10.6). However, some large suctorians, like some *Ephelota* species, can have complex ramified or ribbon-like macronuclei. There is usually one ellipsoid micronucleus. Both macronuclear and micronuclear division are accomplished by intranuclear microtubules (Millecchia & Rudzinska, 1971).

Reminiscent of the spirotrichs, chromosomes are fragmented in the macronuclei of phyllopharyngeans. This has been demonstrated at least in the Subclasses Cyrtophoria and Suctoria (Lahlafi & Metenier, 1991; Riley & Katz, 2001). The gene-sized pieces, ranging from 2–70 kb in size, including ribosomal DNA, are located in the orthomere (Radzikowski & Steinbrück, 1990; Steinbrück, Radzikowski, Golembiewska-Skoczylas, & Sapetto-Rebow, 1995). Raikov (1982) reported that cyrtophorians and chonotrichs also show a "replication-like band" in the macronucleus, reminiscent of that of the spirotrichs, but it has not yet been substantiated by autoradiography that DNA replication occurs in this region. As in the spirotrichs, development of the fragmented condition occurs as the macronuclear anlage differentiates after conjugation. A polytene chromosome stage has been observed in *Chilodonella* (= *Trithigmostoma*) *cucullulus* between 50–75 h after conjugation (Radzikowski, 1973). In a related *Chilodonella* species, polytene chromosomes have not been seen by light microscopy presumably because of a higher degree of chromosome despiralization that is only visible by electron microscopy (Pyne, 1978; Pyne, Ruch, Leeman, & Schneider, 1974). Although there is no apparent DNA-diminution stage during anlage development, as in the spirotrichs, over 30% of the macronuclear DNA is eliminated during the first cell division of the exconjugants (Radzikowski, 1979). In a further molecular similarity to the spirotrichs and oligotrichs, cyrtophorians have internally eliminated sequences (IESs) that are located in the coding region of micronuclear genes, but flanked by a different direct repeat – YGATTSWS (Katz, Lasek-Nesselquist, & Snoeyenbos-West, 2003).

Conjugation has been reported in members from all four subclasses of phyllopharyngeans. The micronucleus typically undergoes three maturation divisions and there is typically one division of the synkaryon following conjugation (Raikov, 1972). Conjugation can be between conjugants equal in size (i.e., isogamontic in *Chilodonella* – MacDougall, 1935; *Tokophrya* – Noble, 1932; Colgin-Bukovsan, 1977) or unequal in size (i.e., anisogamontic in *Spirochona* – Tuffrau, 1953). Anisogamontic conjugation occurs often in chonotrichs and suctorians, and is usually accompanied by total fusion of the conjugants rather than temporary fusion. Temporary, total isogamontic, and total anisogamontic conjugation can occur within one family of suctorians (i.e., Discophryidae) (Raikov, 1972). Fusion of cells typically occurs in the oral region, when it can be identified (e.g., see Dobrzańska, 1961; MacDougall, 1935; Tuffrau, 1953). However, some rhynchodians fuse in the posterior region (de Puytorac, 1994b).

As sessile organisms, suctorians prepare to conjugate by touching tentacles. Stalked species then approach each other by changes in the cell shape, often involving extension of a pseudopodium-like process. The two conjugants of *Tokophrya* only temporarily fuse (Noble, 1932), while the smaller conjugant or microgamont of *Ephelota* leaves its stalk and totally fuses with the macrogamont (Grell & Meister, 1984). Stalkless species can produce a special conjugation bridge into which the micronuclei migrate (e.g., *Heliophrya* – Lanners, 1973). The conjugation bridge between cells is formed by the fusion of the epiplasmic layers of the two cells

(Grell & Meister, 1984; Lanners, 1978). Similar to what has been described for *Tetrahymena* (Orias, Hamilton, & Orias, 1983), positioning and exchange of gametic micronuclei is facilitated by a cytoskeletal meshwork composed of microtubules and microfilaments (Lanners & Rudzinska, 1986; Hanke-Bucker, Lanners, & Hauser, 2000).

Relatively few studies have been done on the factors influencing conjugation in phyllopharyngeans and on the genetics of this process. It is likely that host-mediated factors may influence conjugation in symbiotic forms, but no definitive experiments have yet demonstrated this. Exhaustive searching has not revealed stable opposite mating types in *Chilodonella*: all isolates so far undergo intraclonal conjugation. Often this leads to abortive conjugation and retention of the old macronucleus in the "exconjugants" (Kaczanowski, Radzikowski, Malejczyk, & Polakowski, 1980). Nevertheless, interaction of cells requires participation of the surface glycocalyx as in other ciliates (Golembiewska & Radzikowski, 1980). Kaczanowski et al. (1980) speculated that the adaptive advantage of intraclonal conjugation or inbreeding to *Chilodonella steini* is that it is a rare ciliate and has a special feeding preference for living diatoms. Inbreeding species generally are adapted to narrow ecological niches (Nyberg, 1974).

The genetics of mating types in suctorians has been most thoroughly investigated in only one series of studies. Colgin-Bukovsan (1976) demonstrated that *Tokoprhya lemnarum* has two mating types, one being homozygous and the other heterozygous or hemizygous. Mating occurs between cells of complementary mating type. Although it occurs under all nutritive conditions, cells that are slightly starved showed peak reactivity (Colgin-Bukovsan, 1979). This suctorian shows a typical clonal life cycle with immaturity, maturity, and senescence stages. The long periods of immaturity and maturity (i.e., at least 800 fissions) characterize this species as a typical outbreeder (Colgin-Bukovsan, 1979). The life span of individual *Tokophrya* is quite variable, ranging from several days to over 1 month. However, individual lifespans are dramatically reduced as clones become senescent (Karakashian et al., 1984). Senescence is also accompanied by morphological abnormalities arising from incomplete budding (Batisse, 1994b). With so few studies, clearly much remains to be learned about sexual reproduction in this class.

10.7 Other Features

Cyrtophorians, especially chilodonellids, are conspicuous species in the biofilms in wastewater treatment facilities. Their presence has been used to assess efficiency of operation of these facilities where these ciliates were indicative of good water purification conditions, both within plants (Martin-Cereceda, Serrano, & Guinea, 2001a) and in the natural environment (Bick, 1972; Foissner, 1988a). They do show high sensitivities to heavy metals, such as cadmium and copper, and this may negatively impact their role in treatment facilities (Madoni, Davoli, Gorbi, & Vescovi, 1995). However, this high sensitivity to copper has been exploited to our advantage as a preventative measure to reduce their incidence as fish ectoparasites in aquaculture facilities (Horwath, Lang, & Tamas, 1978). Treatments with dilute sodium chloride and malachite green and formalin can also be effective (Hoffman, 1978; Lom, 1995; Rowland, Mifsud, Nixon, & Boyd, 2006). Finally, Henebry and Ridgeway (1979) suggested that the high prevalence of ectosymbiotic *Tokophrya* on planktonic microcrustaceans might be used as an indicator of eutrophic water conditions.

Chapter 11
Subphylum 2. INTRAMACRONUCLEATA: Class 5. NASSOPHOREA – Diverse, Yet Still Possibly Pivotal

Abstract Ciliates in the Class NASSOPHOREA have played a pivotal role in phylogenetic schemes of the evolution of diversity of ciliates. Their simplified oral structures were thought to represent the ancestral condition of the more well-developed oral polykinetids of oligohymenophoreans, heterotrichs, and spirotrichs. They are united by two ultrastructural features: alveolocysts are a presumed synapomorphy of all representatives, although they have not been observed yet in synhymeniids; and the nematodesmata of the nasse bear nematodesmal or X-lamellae, which are not found in the phyllopharyngean cytopharyngeal basket. The highly developed nasse is used to ingest various "algae", typically cyanobacteria such as *Anabaena* and *Oscillatoria*, whose natural populations in rare instances nassophoreans may control. The somatic cortex has a highly developed epiplasm. In addition to the nasse, there is a set of "oral" polykinetids that extends often around the body circumference as a linear assemblage called a frange or synhymenium. This is why stomatogenesis in these forms is considered mixokinetal because both somatic and oral kinetal elements are involved. The genetics of these ciliates is virtually unexplored so details of conjugation, mating type system, and nuclear development remain to be discovered.

Keywords Cyrtos, articulins, B-cartwheel, pavés, blue-green algae

The ancestors of *Pseudomicrothorax*, a ciliate now assigned to the Class NASSOPHOREA, were argued to have played a pivotal role in the evolution of the oligohymenophoreans (Corliss, 1958a, 1958b; Thompson & Corliss, 1958). This was based on both the revelation by silver staining of three adoral polykinetids, similar in position to those of the Class OLIGOHYMENOPHOREA, and in the mode of stomatogenesis. The "oral" ciliature of nassophoreans is typically arranged as a hypostomial "frange", an extensive ventral band of more complex kinetids that courses slightly posterior to the cytostome and may extend onto the dorsal surface (Fig. 11.1). Fauré-Fremiet (1967a, 1967b) analyzed this ciliary "frange" and the adoral structures of other nassulid-like ciliates, *Chilodontopsis*, *Nassulopsis*, *Nassula*, *Cyclogramma*, *Paranassula*, and *Pseudomicrothorax*, and argued that, despite their diversity, these oral structures could all be considered homologues, justifying the recognition of a clade of nassulid ciliates. De Puytorac, Grain, Legendre, and Devaux (1984) demonstrated that cortical ultrastructural features related peniculines (e.g., *Paramecium*, *Frontonia*) and nassulids, separating them from the hymenostomes (e.g., *Glaucoma*, *Tetrahymena*). This analysis expanded on the previous, more restricted analysis of Lynn (1979a) who had shown that nassulids, peniculines, and hymenostomes were all related using phyllopharyngeans as the outgroup taxon: nassulids were the basal clade of the three (Lynn, 1979a). Sequence analyses of the large and small subunit rRNA genes have confirmed a close relationship between nassulids, peniculines, and hymenostomes (Baroin-Tourancheau, Villalobo, Tsao, Torres, & Pearlman, 1998; Bernhard, Leipe, Sogin, & Schlegel, 1995; Strüder-Kypke, Wright, Fokin, & Lynn, 2000b). Histone gene sequence similarities

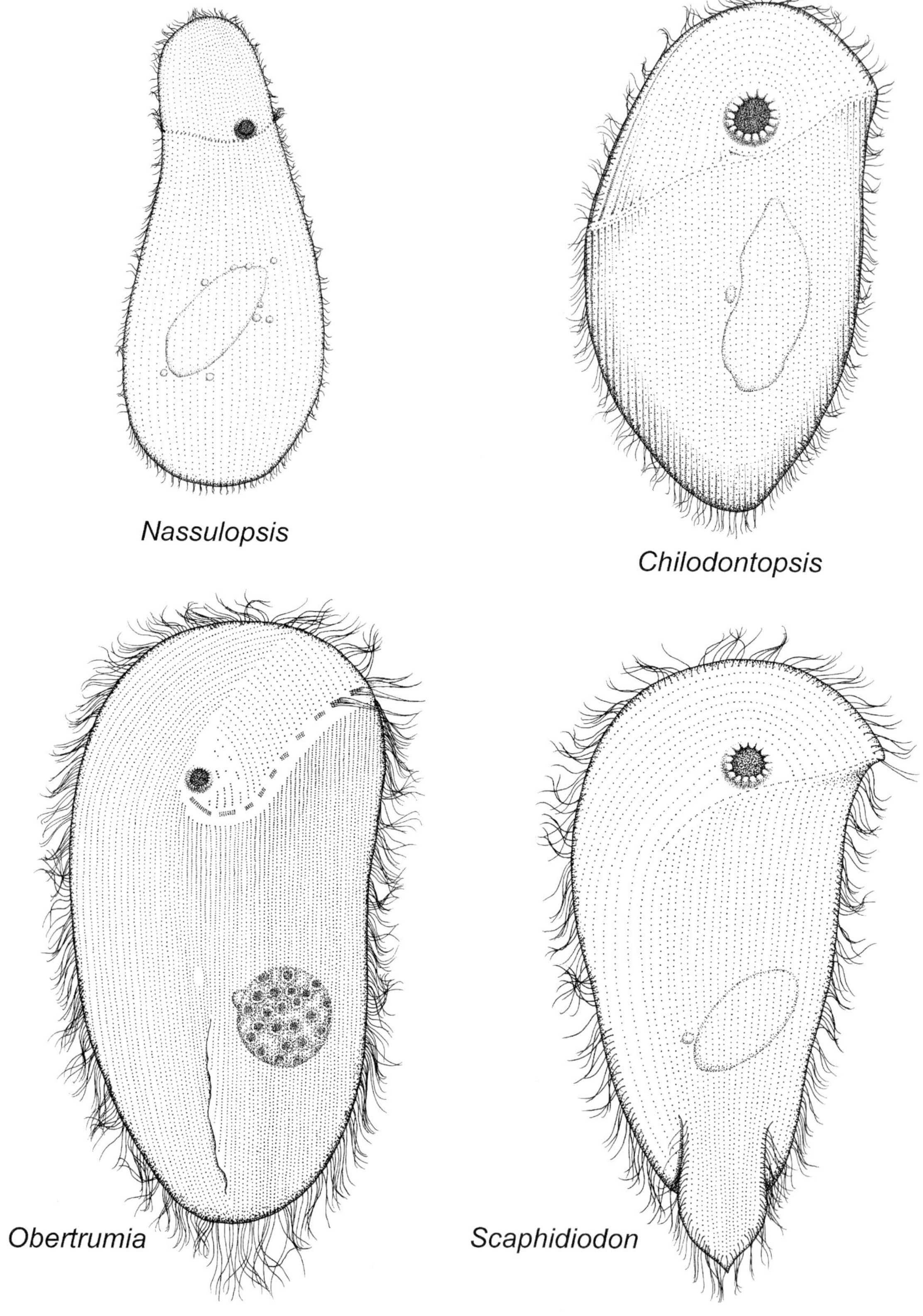

Fig. 11.1. Stylized drawings of representative genera from the orders in the Class NASSOPHOREA. The synhymeniids *Nassulopsis*, *Chilodontopsis*, and *Scaphidiodon*. The nassulid *Obertrumia*

related nassulids and hymenostomes (Bernhard & Schlegel, 1998) although the α-tubulin gene sequence of *Zosterodasys* does not support this relationship (Baroin-Tourancheau et al., 1998). Overall, the earlier conception that nassulid-like ciliates were ancestors for the oligohymenophoreans still seems a reasonable view (**see below Division and Morphogenesis**).

Ciliates in this class are typically holotrichous. Larger nassulids, which can be >200 μm in length, are densely ciliated. However, some of the smaller microthoracids, which may be about 10 μm in length, can exhibit regions of the cortex that are barren of cilia, including the dorsal surface in discotrichids. *Scaphidiodon* is tentatively placed in this class, although it has three features that relate it to the cyrtophorian phyllopharyngeans: (1) a non-ciliated dorsal surface; (2) right somatic kineties that arch over the anterior end onto the left ventral surface and terminate on the anterior suture; and (3) a podite-like appendage at the posterior end (Dragesco, 1965). The pattern of the somatic ciliation of other nassophoreans is also similar to that of cyrtophorians as the right somatic kineties may arch over the oral region onto the left ventral surface (Fig. 11.1) (Deroux, 1994b).

Small and Lynn (1981) were the first to elevate this group to the class level, establishing the Class NASSOPHOREA. The class derives its name from the French "nasse" meaning basket and the Greek *phoros* meaning to bear. This refers to the complex cytopharyngeal basket of nematodesmata that are used in feeding. Original descriptions of the ultrastructure of the nasse (Fauré-Fremiet, 1962a) stimulated later research on the structure, function, and development of this complex microtubular apparatus in *Nassula* (Tucker, 1968, 1970a, 1970b). Earlier demonstration of the thick epiplasm in *Pseudomicrothorax* (Fauré-Fremiet & André, 1967) has led to the discovery of a novel class of proteins, the articulins, which are found in ciliates and euglenoid flagellates (Huttenlauch & Stick, 2003; Huttenlauch, Peck, & Stick, 1998a). Cellular and biochemical research has been possible because these ciliates can be easily grown on filamentous cyanobacteria (Peck, 1977b; Tucker, 1968). Members of the class are united by two synapomorphies: (1) the presence of alveolocysts, extensions of the cortical alveoli into the cytoplasm; and (2) the presence of nematodesmal or X lamellae, accompanying the nematodesmata of the nasse (Eisler, 1989; Eisler & Bardele, 1983). These features, presumed present in synhymeniids, were not observed in *Zosterodasys*, the only synhymeniid examined so far (Kivimaki, Riordan, and Lipscomb, 1997).

11.1 Taxonomic Structure

Corliss (1979) placed nassophorean ciliates in the Subclass Hypostomata of the Class KINETOFRAGMINOPHORA based on the presence of a hypostomial "frange" that extends to varying degrees across the ventral surface of the cell and that may ultimately be restricted to the oral region. Small and Lynn (1981, 1985) were led by similarities in the somatic kinetids and extrusomes to include synhymeniids, nassulids, microthoracids, peniculines, and hypotrichs in their newly conceived Class NASSOPHOREA. Gene sequence data have now refuted a close relationship of hypotrichs with these taxa and demonstrated that peniculines are a basal clade in the oligohymenophorean radiation (e.g., Baroin-Tourancheau, Delgado, Perasso, & Adoutte, 1992; Lynn & Sogin, 1988; Strüder-Kypke et al., 2000b).

Fauré-Fremiet (1967a) set the conceptual perspective for phylogeny within this class by proposing a phylogenetic transformation series for the ciliary "frange", the French for fringe. Some synhymeniids are considered to represent its ancestral state: a transverse line of dikinetids, not well differentiated from the adjacent somatic monokinetids, extending completely across the ventral surface and onto the dorsal surface (Fig. 11.1) (e.g., *Zosterodasys*, formerly *Chilodontopsis*). It is imagined that these dikinetids became polymerized into the "pavés", French meaning paving-stone or tile, or small polykinetids (e.g., some *Nassulopsis* species). These polykinetids then gradually decreased in number as they became increasingly restricted to the left side of the ventral surface (e.g., some *Nassula* species) and then to the left side of the oral region. This ultimately resulted in hymenostome-like ciliates with three oral polykinetids (Fig. 11.2) (i.e., *Furgasonia*, *Pseudomicrothorax*) – a phylogenetic hypothesis that now requires more extensive testing by gene sequence data!

It is clear that there is a significant amount of diversity in the "oral" structures of these ciliates, and this has led to substantial high level splitting of the taxa. The French researchers have

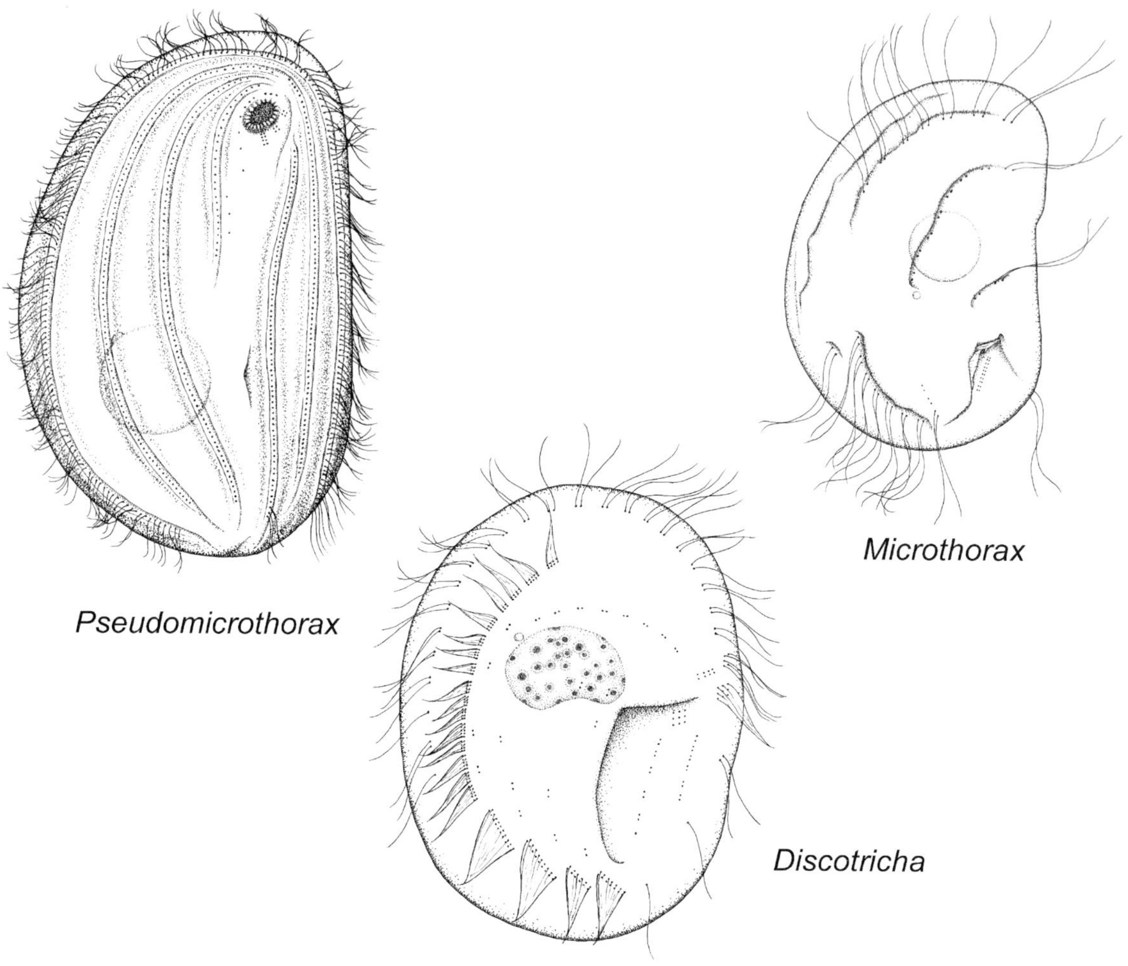

FIG. 11.2. Stylized drawings of representative genera from the orders in the Class NASSOPHOREA. The microthoracids *Pseudomicrothorax*, *Microthorax*, and *Discotricha*

recognized this by supporting six orders within a Subclass Nassulia (Deroux, 1994b; de Puytorac, 1994a). Jankowski (1968a) recognized two suborders within his Order Ambihymenida. Given that relatively little taxonomic research has focused on these ciliates while only two genera have received the bulk of research attention, we have remained conservative. Following Lynn and Small (2002), we include three orders in this class and anxiously await data derived from silver staining, electron microscopy, and gene sequences on the distinctiveness of the aberrant genera included in this class.

The Order Synhymeniida includes forms whose ciliary fringe or synhymenium is composed of dikinetids or small polykinetids, typically of 4–6 kinetosomes. The synhymenium extends from the right postoral body surface sometimes onto the left dorsal body surface. We include four families: Nassulopsidae, Orthodonellidae, Scaphidiodontidae, and Synhymeniidae. Deroux, Iftode, and Fryd (1974) and Deroux (1978) laid the modern groundwork for this group, based on Jankowski (1968a). Sola et al. (1990a) have speculated that *Nassulopsis* might be removed from this order and placed in the Order Nassulida. We await gene sequence data before making this transfer.

The Order Nassulida includes taxa whose synhymenium is composed of obvious polykinetids, restricted to the left ventral and sometimes dorsal surface. In some forms, these polykinetids have been reduced to three, which are restricted to the left side of the cytostome. Nevertheless, there is considerable variation from this "typical" tripartite left oral pattern: *Enneameron* (formerly *Nassula brunnea*; see Jankowski, 1968a) may have more than five rows of monokinetids in an oral atrium (Fauré-Fremiet, 1962a) while *Parafurgasonia* appears to have a paroral and a single oral polykinetid (Foissner & Adam, 1981). These variations have led some to elevate included families and genera to ordinal rank (e.g., Deroux, 1994b; Grain, Peck, Didier, & Rodrigues de Santa Rosa, 1976; de Puytorac, 1994a). We include conservatively three families: Furgasoniidae, Nassulidae, and Paranassulidae.

The third order, the Microthoracida, includes typically small ciliates with sparse somatic ciliation and a cyrtos that is reduced in size. Although three adoral polykinetids are typical, there is considerable variation among genera (e.g., Foissner, 1985a). Fibrous trichocysts with anchor-like tips are considered characteristic of the order. We include three families in the order: Leptopharyngidae, Microthoracidae, and Discotrichidae. Members of the latter family, which is monotypic, are highly aberrant: *Discotricha* has a non-ciliated dorsal surface, ventral somatic polykinetids that are cirrus-like, and extrusomes that do not have anchor-like tips (Foissner, 1997a; Tuffrau, 1954; Wicklow & Borror, 1977). Gene sequence data are clearly needed here!

We place one family incertae sedis in this class. We have removed the Colpodidiidae from the Order Nassulida, where it was placed by Lynn and Small (2002), as these species lack a cyrtos and have highly aberrant oral ciliature, and placed it incertae sedis in the Class NASSOPHOREA.

11.2 Life History and Ecology

Nassophoreans are only rarely observed in high abundances. Most species are found in freshwaters or soils with fewer in brackish and marine habitats. However, they have been found on all continents. Microthoracids are typical of soils in Europe (Foissner, 1981a, 1998a) and Africa (Buitkamp, 1977; Foissner, 1998a, 1999a). Nassulids and synhymeniids have been described from marine and freshwaters in Europe (Agamaliev, 1967; Alekperov, 1984; Burkovsky, 1970; Czapik & Jordan, 1976; Finlay & Maberly, 2000), Africa (Dragesco, 1965; Njiné, 1979), Asia (Ozaki & Yagiu, 1941; Song & Wei, 1998), North America (Borror, 1972; Bullington, 1940), and Antarctica (Thompson, 1972).

The larger nassulids and microthoracids appear to feed preferentially on cyanobacteria, such as *Anabaena*, *Aphanizomenon*, *Oscillatoria*, *Phormidium*, and *Synechococcus* (Canter, Heaney, & Lund, 1990; Peck, 1985; Tucker, 1978). They do show some feeding preferences: *Nassula aurea* was reported never to graze *Gomphosphaeria* and *Microcystis* (Canter et al., 1990) while *Pseudomicrothorax dubius* rarely ingested some *Anabaena* species (Peck, 1985). Both surface charges and phagocytosis-specific molecules on the cyanobacterial filaments are necessary to explain these feeding preferences (Kiersnowska, Peck, & de Haller, 1988). Feeding behavior of *Pseudomicrothorax* has been divided into two phases: (1) a contact swimming phase during which the ciliate guides itself along the cyanobacterial filament, typically finding an end to begin ingestion; and (2) a phagocytosis phase that involves first attachment and then ingestion. Ca^{2+} influx is probably essential for both the attachment phase of phagocytosis and for the exocytosis of lysosomes during the initial ingestion of the filaments (Peck & Duborgel, 1985). Some slightly starved *Nassula* species show a negative phototaxis to light when they also possess a conspicuous stigma-like structure. How this phototaxis is mediated has not been determined although its function is presumed to lead these ciliates towards slightly illuminated regions that are preferred by cyanobacteria (Kuhlmann & Hemmersbach-Krause, 1993b). Microthoracids are typically bacterivorous (Foissner, Berger, & Kohmann, 1994) and have been reported from the activated sludge biotope (Leitner & Foissner, 1997a).

Deroux (1994b) remarked that many nassophoreans harbor *Chlorella* symbionts. However, there has been little research on this relationship.

Nassophoreans are likely eaten by a variety of invertebrates, but records of this are scarce. Addicot (1974) implied that *Leptopharynx* might be eaten by mosquito larvae while Braband, Faafeng, Källqvist, and Nilssen (1983) observed fish fry and copepods to feed on *Nassula*. The suctorians, *Podophrya* (Canter et al., 1990; Fauré-Fremiet, 1945) and *Sphaerophrya* (Clément-Iftode, 1967), are repeatedly observed as predators of nassulids.

Encystment is typical of nassophoreans, which are stimulated to do so by the lack of food (Beers, 1966a; Canter et al., 1990; Mulisch & Hausmann, 1989). The cyst wall is composed of three layers with the mesocyst layer having chitin microfibrils, as has also been observed in heterotrichs (Mulisch & Hausmann, 1989).

11.3 Somatic Structures

Synhymeniids and nassulids are typically larger ciliates, holotrichously ciliated with cylindrical bodies. Microthoracids are smaller, often flattened, and with fewer somatic kineties whose kinetosomes may be more widely dispersed or even aggregated into polykinetid-like organellar complexes (e.g., *Discotricha*) (Figs. 11.1, 11.2).

The cell surface of these ciliates is undoubtedly covered by a glycocalyx, although it has only been clearly demonstrated in *Pseudomicrothorax* (Hausmann, 1979). Underlying the plasma membrane is a typical alveolar layer with the unusual feature that the alveoli may send invaginations through the epiplasm into the cortex of the ciliate. These alveolocysts are typically paired and on either side of the somatic kinetids (Eisler, 1989; Eisler & Bardele, 1983). We recognize these structures as a synapomorphy for the class NASSOPHOREA although they remain to be demonstrated in synhymeniids.

Some nassophoreans have a conspicuous epiplasm (e.g., *Pseudomicrothorax* – Peck, 1977b; *Furgasonia* – Eisler, 1988; *Nassula* – de Puytorac & Njiné, 1980; Tucker, 1971a). *Pseudomicrothorax* can be prepared as an "epiplasmic" ghost, retaining its cell shape without any of the cell membranes or cortical microtubular structures – a clear demonstration of the shape-maintaining function of the epiplasm (Peck, 1977b; Peck, Duborgel, Huttenlauch, & Haller, 1991). Immunocytochemistry has demonstrated that proteins from the ciliate epiplasm share common epitopes with those proteins from the pellicles of euglenoids and dinoflagellates (Vigues, Bricheux, Metivier, Brugerolle, & Peck, 1987). The epiplasm, especially adjacent to the inner alveolar membrane, has higher concentrations of glycoproteins (Curtenaz & Peck, 1992; Huttenlauch & Peck, 1991). The middle layer is composed of articulins, a novel kind of cytoskeletal protein found also in euglenoids, which has a characteristic valine-proline-valine (VPV) motif, presumed to provide stability to this layer (Huttenlauch, Geisler, Plessmann, Peck, Weber, & Stick, 1995; Huttenlauch, Peck, Plessmann, Weber, & Stick, 1998b). In addition, another class of proteins, the epiplasmins, are also found in the microthoracid epiplasm and related to epiplasmins in the peniculine epiplasm. Epiplasmins, although rich in valine and proline, do not show the VPV-motif of the articulins (Coffe, Le Caer, Lima, & Adoutte, 1996; Huttenlauch et al., 1998a).

The somatic kinetid of the nassophoreans has been resummarized by Eisler (1988). Monokinetids can now be characterized as follows: a divergent postciliary ribbon at triplet 9; an anterior and laterally-directed kinetodesmal fibril at triplets 5 and 6; and a small tangential transverse ribbon at triplets 3 and 4, arising from some dense material (Figs. 11.3, 11.4) (Lynn, 1991). Dikinetids can occur: a posterior ciliated kinetosome with the typical fibrillar pattern is connected to an anterior ciliated kinetosome with a single postciliary microtubule and sometimes a transverse ribbon (Fig. 11.3) (Eisler, 1988). The kinetosomes of nassulids have a distal B-cartwheel and may also have a proximal and standard A-cartwheel, while microthoracids may lack both cartwheels (Eisler, 1988; Njiné & Didier, 1980; Peck, 1977b; Tucker, 1971a).

The contractile vacuole system of nassophoreans is a Type A system (Patterson, 1980) with the contractile vacuole surrounded by a spongiome of irregularly arranged tubules, 20–80 nm in diameter (Hausmann, 1983; Prelle, 1966). Microthoracids may have an elongated contractile vacuole pore canal that extends into the cytoplasm.

Nassophoreans have rod-shaped extrusomes that have been called fibrocysts or fibrous trichocysts (Hausmann, 1978). Their structure and development have been particularly well studied in *Pseudomicrothorax*. Its trichocysts have anchor-like tips that splay out upon ejection. The 50-nm periodicity of the ejected shaft is very similar to that of the ejected trichocysts of *Paramecium* (Hausmann, 1978), which also show remarkable similarities in their constituent proteins (Eperon & Peck, 1993). Fibrocyst development occurs in Golgi vesicles and involves the unusual fusion of two types of vesicles, one containing shaft precursors and the other containing tip precursors (Peck, Swiderski, & Tourmel, 1993a, 1993b). Once developed, the trichocyst docks in the cortex by localized dissolution of the epiplasm and penetration

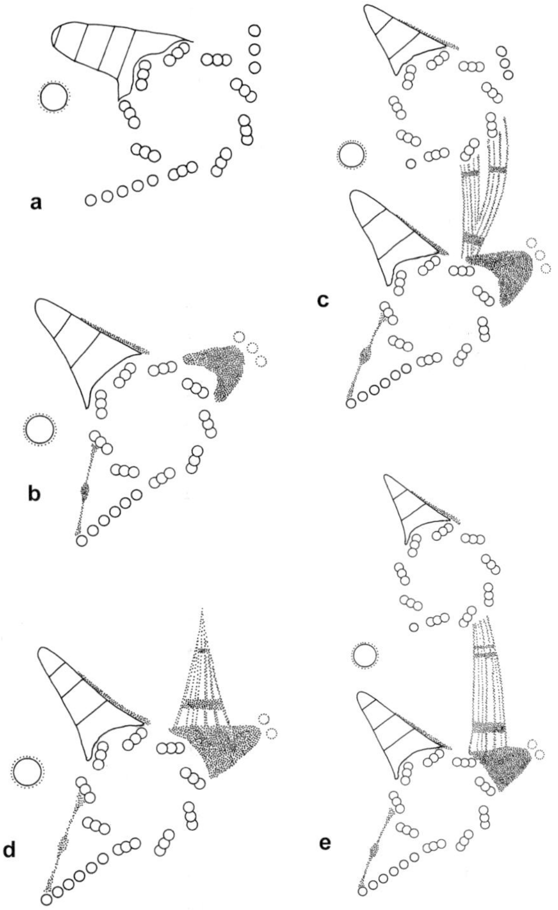

FIG. 11.3. Schematics of the somatic kinetids of the Class NASSOPHOREA. (**a**) Monokinetid of *Pseudomicrothorax*. (**b**) Monokinetid of *Furgasonia*. **c**. Dikinetid of *Furgasonia*. (**d**) Monokinetid of *Nassula*. (**e**) Dikinetid of *Nassula* (from Lynn, 1981, 1991)

of the alveolar layer before contacting the inner surface of the plasma membrane (Eisler & Peck, 1998). Although classified here as a microthoracid, *Discotricha* does not have anchor-like tips on its extrusomes (Wicklow & Borror, 1977). Does this mean that it is truly not a microthoracid although its oral structures suggest otherwise (**see below**)?

11.4 Oral Structures

Nassophoreans possess some kind of oral basket of nematodesmata – "nasse" or cyrtos, which can be quite conspicuous and well-developed. Ciliary structures may be associated with this basket in nassulids and microthoracids. The nassulid *Furgasonia* has a paroral of stichodyads and three adoral polykinetids (Figs. 11.1, 11.2) (Eisler, 1988). In *Pseudomicrothorax*, the paroral dikinetids dissociate during stomatogenesis so that "posterior" kinetosomes remain associated with the nematodesmata while a few "anterior" kinetosomes that are not resorbed remain as "residual kinetosomes" posterior to the cytostome (Peck, 1975; Thompson & Corliss, 1958). In most *Nassula* species, the "oral" polykinetids course on the left ventral surface, posterior to the cytostome, and may extend onto the dorsal surface.

"Oral" structures in the synhymeniids differ from that of nassulids in two ways. First, they extend across the entire ventral surface, even encircling the entire body as the so-called synhymenium (e.g., *Nassulopsis*). Second, they are composed of dikinetids or polykinetids of typically no more than six kinetosomes (Fig. 11.1). However, in scaphidiodontids and orthodonellids, the extension of the synhymenium into the anterior suture recalls the overall pattern of cyrtophorians (cf. Figs. 10.1, 11.1) (Deroux, 1994b). In the only ultrastructural study of a synhymeniid, Kivimaki, Riordan, and Lipscomb (1997) concluded *Zosterodasys* could either be primitively simple or secondarily reduced, based on the features of its organellar structures.

On the other hand, several studies have detailed nassulid and microthoracid oral ultrastructure. Eisler's (1988) detailed study has demonstrated that the kinetosomes of the paroral dikinetids of *Furgasonia* and probably *Nassula* are oriented perpendicular to each other: the right or "anterior" kinetosome is oriented in the long axis of the paroral while the left or "posterior" kinetosome is oriented at right angles to the paroral. The Z or cystostomal lamellae arise from the postciliary ribbons of the "posterior" kinetosomes (Eisler, 1988). The oral polykinetids of nassulids are square-packed organellar complexes of three rows. Kinetosomes of the posterior row bear postciliary ribbons and all kinetosomes bear presumably a single transverse microtubule at triplet 4. Parasomal sacs are distributed throughout the structure (Eisler, 1988; de Puytorac & Njiné, 1980).

The nassophorean cytopharyngeal basket or cyrtos has received the most detailed analysis by cell biologists who were attracted to it as perhaps the most complicated microtubular organellar complex

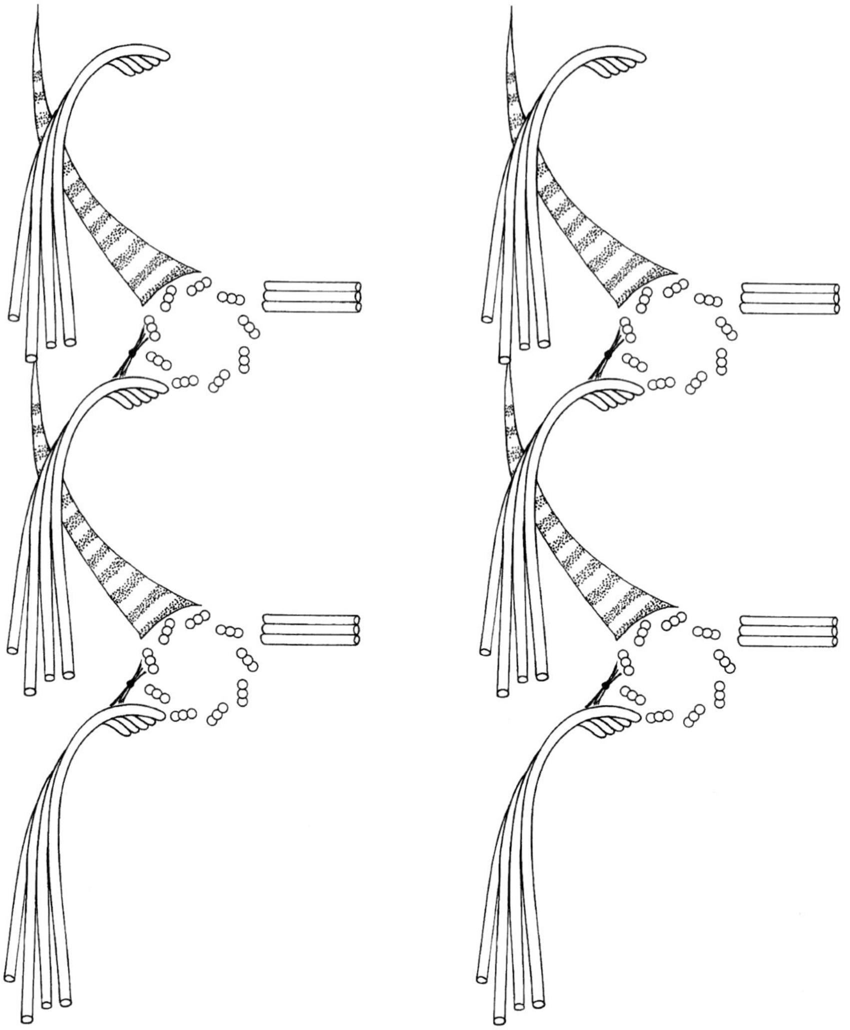

FIG. 11.4. Somatic cortex of a typical nassophorean interpreted based on the somatic cortex of *Pseudomicrothorax*. (Modified after Peck, 1977b.)

of any cell! Eisler (1988) has noted that nassulids and microthoracids have the X or nematodesmal lamellae, which are absent in phyllopharyngeans. These lamellae develop from the marginal microtubules of the nematodesmata, forming a ribbon and eventually gaining dynein-like arms. The cyrtos of nassulids has both Y or subcytostomal lamellae and, as noted above, Z or cytostomal lamellae, neither of which are found in the microthoracid cyrtos (Eisler, 1988).

The structure and function of the cytopharyngeal basket of nassophoreans has been described in detail for *Nassula* (Tucker, 1968) and *Pseudomicrothorax* (Hausmann & Peck, 1978). Microfilaments bind the nematodesmata at the oral or distal end and may extend along much of the length of the cyrtos while a denser annulus binds the nematodesmata of *Nassula* at a more proximal level. Displacement of the nematodesmata, possibly by contraction of the microfilamentous systems facilitates ingestion of the cyanobacterial filaments. The arm-bearing microtubules of the X or nematodesmal lamellae have been implicated in endocytosis of these filaments. Tucker (1978) argued that the arms in *Nassula* act **indirectly** on a highly gelated cytoplasm that is strongly associated with the food vacuole membrane. Hausmann and Peck (1979) argued that the arms in *Pseudomicrothorax* are associated with microfilaments that interact **directly** with the food vacuole membrane, transporting it inwards at up to 15 μm sec^{-1}. Subsequent research on *Pseudomicrothorax* has confirmed the presence of actin, α-actinin, and ATPase in the basket, implicating an actin-based motility system in feeding (Hauser & Hausmann, 1982; Hauser, Hausmann, & Jockusch, 1980).

Hundreds of square micrometers of food vacuole membrane must be formed in minutes during the ingestion of cyanobacterial filaments in these ciliates. Both Tucker (1978) and Hausmann and Peck (1979) have observed cytoplasm and vesicles entering the cyrtos between the nematodesmata at its oral or distal end. Many of these vesicles are probably primary lysosomes that serve a double function of providing membrane for the expanding food vacuole and hydrolases to begin the very rapid digestion of their food (Peck & Hausmann, 1980). Subsequent folding of the food vacuole membranes and vesiculation of the food vacuole may facilitate resorption of nutrients (Hausmann, 1980; Hausmann & Rüskens, 1984). Thus, we have now detailed knowledge of how oral structures function in both nassulids and microthoracids. How similar is the process in synhymeniids?

11.5 Division and Morphogenesis

Nassophoreans typically divide while swimming freely. The parental oral structures are almost completely dedifferentiated and then redifferentiated in synchrony with those of the opisthe (e.g., Eisler & Bardele, 1986; Tucker, 1970a). Foissner (1996b) established mixokinetal stomatogenesis to characterize division morphogenesis in these ciliates: both the parental oral apparatus and the somatic ciliature simultaneously participate in stomatogenesis – a **mix**ture of origins. Broadly, the parental paroral gives rise to the opisthe paroral while the synhymenium or hypostomial frange is derived from somatic kineties.

Eisler (1989) and Eisler and Bardele (1986) have provided the most detailed comparative analysis of stomatogenesis in the nassophoreans (Fig. 11.5). In nassulids, the parental paroral splits longitudinally to form a new Kinety 1' from its right kinetosomes and a new paroral from the left kinetosomes. The kinetosomes of the paroral serve as nucleation sites for the development of the oral nematodesmata, which subsequently close to form the circular palisade of the differentiated cyrtos (Eisler & Bardele, 1986; Tucker, 1970a). The microtubule nucleating template that develops in association with these oral kinetosomes probably controls the shape and pattern of the growing nematodesmata (Pearson & Tucker, 1977; Tucker, Dunn, & Pattisson, 1975).

Eisler and Bardele (1986) interpreted stomatogenesis in the microthoracids using their model for nassulid stomatogenesis. They concluded that the paroral and kinetal segments of the opisthe in *Pseudomicrothorax* and *Leptopharynx* originate from the parental paroral and are retained as the so-called "residual kinetosomes" at the next cell division. Peck (1975) and Njiné (1980) interpreted their origin to be from somatic Kinety 1. Regardless of this difference of opinion, the paroral kinetosomes play a key role in formation of the basket while the adoral polykinetids assume a highly similar relationship with the cytostome, strongly supporting the ultrastructural similarities in somatic and oral structures discussed above.

The stomatogenesis of the highly unusual microthoracid *Discotricha* may also be mixokinetal (Foissner, 1996b). Wicklow and Borror (1977) tentatively concluded that post-buccal Kinety 1 participated in stomatogenesis. This kinety itself may ultimately be an "oral" kinety, homologous to the "residual kinetosomes" of other microthoracids. Further study of the stomatogenesis of this highly unusual ciliated is warranted as is investigation of stomatogenesis in the synhymeniids.

Cytokinesis, at least in *Nassula*, coincides with the development of a contractile ring of microfilaments that presumably constrict against a girdle of several thousand longitudinally oriented microtubules, which are embedded in the epiplasm (Tucker, 1971b).

11.6 Nuclei, Sexuality and Life Cycle

There has been relatively little research on these aspects of the biology of nassophoreans. The single macronucleus is homomerous and typically globular to ellipsoid in shape (Figs. 11.1, 11.2). Species of smaller cell-size have one micronucleus while larger cells may have multiple micronuclei (e.g., *Nassulopsis* species – Sola

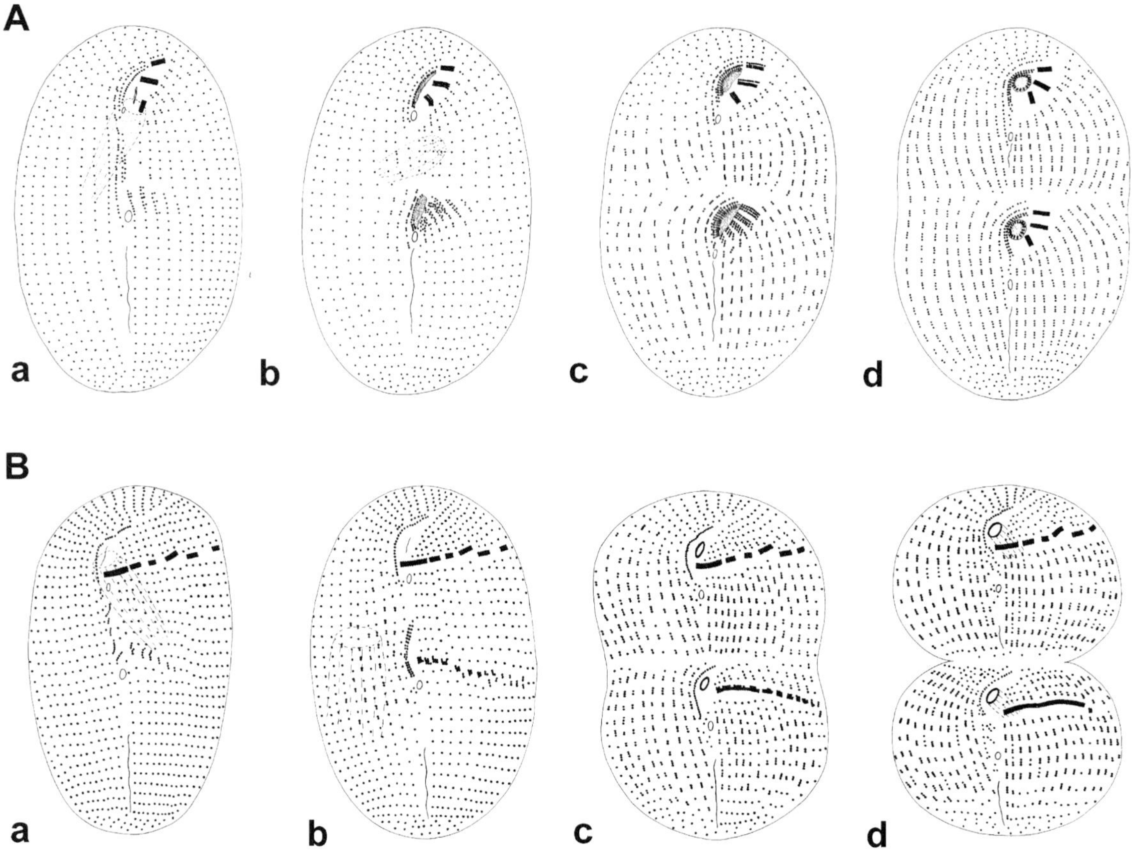

FIG. 11.5 Division morphogenesis of the nassulids **A** *Furgasonia* and **B** *Nassula*. Stomatogenesis in both these genera is mixokinetal, initially involving kinetosomal proliferation from both somatic and oral kinetosomes (**a**). In *Furgasonia*, assembly of the adoral structures involves proliferation from right to left (**b**), and as the developing oral polykinetids rotate (**c**), the differentiation is completed from anterior to posterior and right to left (**d**). In *Nassula*, proliferation (**b**) and assembly (**c, d**) of the polykinetids also occurs from right to left. (from Eisler & Bardele, 1986.)

et al., 1990a). Intranuclear microtubules are found during division of both the micronucleus and the macronucleus, and membrane bridges link micro- and macronuclei during late anaphase and early telophase, coordinating karyokinesis of the two nuclei (Tucker, 1967). Raikov (1982) characterized the nassulid macronucleus as a polyploid subnuclear type because the chromatin apparently aggregates as diploid subunits in both *Nassula* and *Nassulopsis* and whole genomes are believed to segregate at macronuclear division. These conclusions based on early work need to be verified by modern techniques.

To our knowledge, the detailed cytology of conjugation has not been described for any nassophorean except for four stages illustrated by Raikov (1972) who reported that conjugation in *Nassula* might be seasonal. The pattern of conjugation appears to be typical of the ciliates. As in the cyrtophorians, the cyrtos detaches from the cortex and is resorbed. During meiosis in *Nassula*, there are three maturation divisions, two meiotic and one mitotic. The micronucleus at zygotene assumes a "parachute stage", a stage homologous to the "crescent stage" in other ciliates. The conjugation "fusion zone" in *Nassula* appears as a region of homogeneous cytoplasm that encloses the four gametic nuclei. Fertilization occurs in this cytoplasmic region without apparent migration of the gametic nuclei (Fig. 34 in Raikov, 1972). In addition to details on the cytology, we can only assume that there is a life cycle and genetics of mating type determination as for other ciliates. But what it is and how it is determined remain among the many questions to be answered for this possibly pivotal group of ciliates.

Chapter 12
Subphylum 2. INTRAMACRONUCLEATA: Class 6. COLPODEA – Somatically Conserved but Orally Diverse

Abstract Ciliates now assigned to the Class COLPODEA were scattered throughout the phylum in unrelated classes and orders. However, it was the structural conservatism of the cortex and its somatic kinetids that enabled the identification of this natural assemblage based on cortical ultrastructure, a fact that has been confirmed by small subunit rRNA gene sequences. Colpodeans are the quintessential cyst-formers in the phylum, and are therefore typical in habitats that have a high probability of desiccation: mosses, soils, and leaf litter are typical habitats. However, colpodeans are found in ponds and lakes, although rarely in marine habitats. Their prey varies with their cell size: smaller colpodeans eat bacteria while the largest colpodean *Bursaria* can ingest *Paramecium*. The somatic kinetid is a dikinetid with well-developed overlapping transverse microtubular ribbons derived from the posterior kinetosome and forming what is called the LKm fiber or transversodesma – the strong synapomorphy for the group. Oral structures range from a paroral that almost encircles a prostomatous oral region to a huge deep oral cavity, almost spirotrich-like, adorned with many oral polykinetids. Stomatogenesis ranges from mero- to pleurotelokinetal, and in the colpodids typically occurs within a cyst after dedifferentiation of all parental oral structures. Conjugation has rarely been observed, and it is often assumed that colpodeans are completely asexual. This needs to be tested by molecular genetic approaches.

Keywords Palintomy, cryptobiosis

The revolution in our view of what morphological characters are significant to identifying clades of ciliates arose, in part, from investigations of this class of ciliates. Lynn (1976a) proposed the structural conservatism hypothesis based on his study of the cortical ultrastructure of colpodeans and a comparison of his results with those of the literature. This conclusion, independently arrived at by Gerassimova and Seravin (1976), established, as a general rule, the conservative nature of the somatic kinetid within clades of ciliates and provided a "rule" for establishing phylogenetic affinity.

In the Class COLPODEA, several examples illustrate the strength of this principle. Ultrastructural study of the cortex of *Woodruffia* by Golder and Lynn (1980) confirmed its relationship to colpodeans as suggested by Kahl (1930–1935), although von Gelei (1954) had originally considered it a heterotrich-like ciliate. *Bursaria*, for many years another "heterotrich" (Corliss, 1979), revealed colpodean-like stomatogenetic features on more careful study (Fernández-Galiano, 1979), and its colpodean "nature" was confirmed by ultrastructural studies (Gerassimova, Sergejeva, & Seravin, 1979; Lynn, 1980; de Puytorac & Perez-Paniagua, 1979). McCoy (1974a, 1977) drew attention to the colpodean affinities of the stomatogenesis of *Cyrtolophosis*, a genus that had been presumed to be a hymenostome (Corliss, 1961), but was confirmed to be a colpodean by ultrastructural features of the somatic kinetid (Didier, de Puytorac, Wilbert, & Detcheva, 1980). As a final example, the unusual sorocarp-forming ciliate, discovered by Olive (1978), was named *Sorogena* (Bradbury & Olive, 1980) and placed with the haptorians, primarily because of its oral structures. Lynn (1991) suggested that it might be regarded as a colpodean, a conclusion confirmed by the study of its ultrastrucure and stomatogenesis by Bardele, Foissner, and Blanton (1991). Thus, as beautifully illustrated by Foissner (1985b) (see Fig. 3.4), colpodean oral structures show such broad diversity that the class has really only received broad recognition in the past 20 years.

Colpodean ciliates are extremely common, especially in terrestrial habitats where the genus *Colpoda* is almost ubiquitous. Colpodeans range in size from about 10 μm in length for the genus *Nivaliella* to over 500 μm in length for *Bursaria truncatella*, an almost 200,000 times difference in cell volume (Foissner, 1993a). Colpodeans are generally holotrichously ciliated. The kineties are typically bipolar with a number of them ending on the perimeter of the oral region or coursing anteriorly to abut in a prominent anterior suture or keel in some *Colpoda* species. The body undergoes torsion to varying degrees, a feature that Stout (1960a) used to imagine a phylogeny for the group based on D'Arcy Thompson's Cartesian coordinate analysis. This torsion can be lost during cell division as was illustrated years ago by Tuffrau (1952).

Colpodeans have interested physiologists with their life cycle that typically includes a resistant stage or cyst. Early on, Burt (1940) carefully described species so physiologists would not misidentify forms. More recently, multivariate statistical approaches have been used to discriminate among morphospecies (Foissner & Schubert, 1983; Lynn & Malcolm, 1983). *Colpoda* species have now been grown in chemostat cultures on a variety of bacterial species, including *Escherichia coli* (Drake & Tsuchiya, 1977). Undoubtedly the most intriguing recent example of bizarre feeding strategies among ciliates has been the discovery of the mycophagous colpodeans in the Family Grossglockneriidae. Originally discovered by Foissner (1980d) in the European Alps, they have been described worldwide as predators of fungi, using their minute feeding tube to perforate the cell wall of fungi and remove the contents (e.g., Foissner, 1993a, 1999b; Foissner & Didier, 1983), and they have now been identified from fossilized amber dating from the Lower Cretaceous (Ascaso et al., 2005).

De Puytorac et al. (1974b) elevated the colpodeans to ordinal rank within the Subclass Vestibulifera. Small and Lynn (1981) elevated the group to class rank based on the structure of the somatic dikinetids, which really provide the only synapomorphy for the class. The special feature of the colpodean somatic dikinetid is the presence of a posteriorly-directed transverse ribbon of microtubules associated with the posterior kinetosome. These ribbons can extend for some distance posteriorly, overlapping each other in what has been called the LKm fibre (Golder, 1974; Golder & Lynn, 1980) or the transversodesma (Small & Lynn, 1985). The class name is derived from *Colpoda*, one of the most common genera of ciliates, and its name, in turn, is derived from the Greek *kolpos*, meaning breast, referring to the bulging shape of some *Colpoda* species.

12.1 Taxonomic Structure

As noted above, de Puytorac et al. (1974b) were the first to elevate the colpodids to ordinal rank within their Subclass Vestibulifera, a position that was maintained by Corliss (1979). Small and Lynn (1981, 1985) elevated the group to class rank, establishing the Class COLPODEA, based primarily on the structure of the somatic kinetid. This distinctness as a class, which has been maintained by Lynn and Small (1997, 2002), is supported both by phylogenetic analyses based on morphological features (de Puytorac, Grain, Legendre, & Devaux, 1984; de Puytorac, Grain, & Legendre, 1994) and by small subunit (SSU) rRNA gene sequences (Lynn, Wright, Schlegel, & Foissner, 1999). However, one recent molecular study suggests that the colpodeans may be paraphyletic (Lasek-Nesselquist & Katz, 2001).

The colpodeans have been related to the litostomes and the nassophoreans based on kinetid structures (Aescht, Foissner, & Mulisch, 1991). Phylogenetic trees derived from SSUrRNA gene sequences support an affinity with prostomateans (Lynn et al., 1999) or nassophoreans (Lasek-Nesselquist & Katz, 2001) while trees based on large subunit rRNA gene sequences place the colpodeans near to nassophoreans and oligohymenophoreans (Baroin-Tourancheau, Villalobo, Tsao, Torres, & Pearlman, 1998). Histone gene sequences (Bernhard & Schlegel, 1998) and α-tubulin nucleotide sequences (Baroin-Tourancheau et al., 1998) support this latter relationship.

Foissner (1993a, 1994b) recognized two subclasses, the Subclass Colpodia and Subclass Bryometopia, based primarily on features of the reticulate silverline system. There are three general types of silverline systems in the colpodeans: the colpodid, platyophryid, and kreyellid patterns. Foissner (1978) demonstrated the usefulness of the silverline system in identifying relationships among colpodeans. However, as shown many years ago by Taylor and Garnjobst (1939) and confirmed by Foissner (1993a, 1994b), the silverline system is a dynamic element of the cortex, even changing from one type to another within the same species! Thus, we are reluctant at this time to support the subclass taxa based only on this feature and will await gene sequence data to confirm this fundamental divison within the class. We currently recognize six orders in the class: Bryometopida, Bryophryida, Bursariomorphida, Colpodida, Cyrtolophosidida, and Sorogenida, based mainly on the monographic work of Foissner (1993a).

The Order Bryometopida is characterized as having an argyrome with a very highly reticulated and subdivided dense network (i.e., "kreyellid type", Foissner, 1993a). Oral structures include typically a paroral of dikinetids, sometimes modified, and several left oral polykinetids. There are four families in the order: Bryometopidae, Jaroschiidae, Kreyellidae, and Trihymenidae.

The Order Bryophryida, monotypic for the Family Bryophryidae, is characterized by right oral kineties that at least include a series of radially oriented kinetosomal rows, except for the genus *Notoxoma*, in which these are interpreted to have been reduced to dikinetids. There can be a single or multiple left adoral polykinetids composed of square-packed kinetosomes (Foissner, 1993a).

The Order Bursariomorphida, which includes the Families Bursariidae and Bursaridiidae, is characterized by an expansive oral cavity whose left side is lined by many, long and equidistantly spaced oral polykinetids (Fernández-Galiano, 1979; Foissner, 1993a). These oral polykinetids are typically of two to three rows of kinetosomes, arranged in a square-packed arrangement. The right oral ciliature is composed of many regularly or slightly irregularly and obliquely arranged "paroral" kineties.

The Order Colpodida includes colpodeans that undergo merotelokinetal stomatogenesis, developing a left oral polykinetid that is composed of several to many rows of regularly-spaced monokinetids. Foissner (1993a) recognized the monotypic Order Grossglockneriida as distinct from the Order Colpodida. However, the similarities in stomatogenesis and general morphology strongly suggest affinities between the two, whose genera are not greatly different based on SSUrRNA gene sequences (Lynn et al., 1999). In addition to the Family Grossglockneriidae, we also include the following families: Colpididae, Hausmanniellidae, Marynidae, Bardeliellidae, and Grandoriidae.

Of the remaining orders recognized by Foissner (1993a, 1994b), this leaves the Cyrtolophosidida and Sorogenida. Lasek-Nesselquist and Katz (2001) have noted the strong genetic similarities in SSUrRNA gene sequences between *Sorogena*, the only representative for the Order Sorogenida, and those of the cyrtolophosidid *Platyophrya*. They recommended suppression of the Order Sorogenida, which is considered distinct primarily on the basis of the unusual sorocarp development in the life cycle of its type genus (Foissner, 1993a). Furthermore, the stomatogenesis of *Sorogena* appears to be similar to that of *Platyophrya* and other cyrtolophosidids (Bardele et al., 1991). However, we have maintained this order, monotypic for the Family Sorogenidae, until a more complete analysis of molecular genetic diversity compels us to suppress it.

Finally, taxa in the Order Cyrtolophosidida show strong similarities in stomatogenesis to bryometopids, like *Bryometopus* (cf., Dragesco, Fryd-Versavel, Iftode, & Didier, 1977; McCoy, 1977; de Puytorac, Perez-Paniagua, & Perez-Silva, 1979b; Wirnsberger, Foissner, & Adam, 1985b). However, we have maintained the Order Cyrtolophosidida, which likely includes representatives similar to the common ancestor of the class. The order, characterized by a simple paroral of dikinetids and several to many left oral polykinetids, includes ciliates with the basic features of the class and is comprised of the following families: Cyrtolophosididae, Platyophryidae, Woodruffiidae, and Sagittariidae.

There is clearly a need for gene sequences from more representatives of this class to test the competing classifications (e.g., the one proposed here, Foissner, 1993a; de Puytorac et al., 1979b). Molecular studies would also resolve the placement of the Families Tectohymenidae and Pseudochlamydonellidae, which we consider incertae sedis in this class.

12.2 Life History and Ecology

Encystment is the life history feature that typifies the colpodeans. To our knowledge, all colpodeans can develop resistant resting cysts in response to desiccation. This has meant that more than any other class, their global distribution has been assured. Species of the genus *Colpoda* are notorious for this ability. Several studies have demonstrated the high vagility of *Colpoda* species by aerial dispersal (Maguire, 1963a, 1963b; Rivera et al., 1992). This high vagility is supported by studies that have not demonstrated any global biogeography in the genetic variation in ribosomal DNA within *Colpoda* species (Bowers & Pratt, 1995; Bowers, Kroll, & Pratt, 1998) nor in variations in physiological responses, such as growth rate and sensitivity to toxicants (Xu, Bowers, & Pratt, 1997). Nevertheless, Foissner (1994b) noted some restrictions in biogeography of other genera: *Orthokreyella* and *Tectohymena* have only been reported from Laurasia while *Puytoraciella*, *Apocolpoda*, and *Jaroschia* have only been reported from Gondwanaland.

Colpodeans are typically found in water associated with mosses, forest litters, and soils where the water may range from fresh to brackish and salty. These so-called "terrestrial" ciliates have been found in these habitats in Europe and Asia (Detcheva, 1973; Foissner, 1980d, 1993a, 1994b; Griffiths, 2002; Grolière, 1977; Hattori & Hattori, 1993; Tirjaková & Matis, 1985; Vargas & Hattori, 1990), the Americas (Bamforth, 1973, 1980; Foissner, 1997c; Rivera et al., 1992), Africa (Buitkamp, 1977; Foissner, 1988c), Australia (Foissner, 1988c, 1990, 1997c), Antarctica (Ryan et al., 1989), and Hawaii (Foissner, 1993b, 1994c). Several unusual habitats include pitcher plants (Addicott, 1974; Rojo-Herguedas & Olmo, 1999) and tree holes (Novotny, Lynn, & Evans, 1977).

Colpodeans have also been reported from streams, temporary ponds, and lakes in Europe (Foissner, 1979e, 1997b; Skogstad, Granskog, & Klaveness, 1987), the Americas (López-Ochoterena, 1966), and Africa (Dragesco, 1972; Njiné, 1979). However, relatively few species are truly limnetic, typically including species in the genera *Bursaridium*, *Cyrtolophosis*, and *Pseudochlamydonella* (Foissner, 1994b). Colpodeans are not typically marine although several species of *Platyophrya* and *Woodruffia* have been reported from marine environments (Kahl, 1930–1935) and soils of coastal dunes (Verhoeven, 2002).

Finally, there have been a few reports of colpodeans as symbionts, probably commensals, within other organisms. A report by Powers (1933) of a *Colpoda* from the sea urchin intestines has never been confirmed. Reynolds (1936) reported *Colpoda* as a facultative parasite of land slugs. *Colpoda* species have also been collected from the feces of deer (Bradbury & Outka, 1967) and amphibians and reptiles (Fernández-Galiano, Fernández-Galiano, & Madrigal-Sesma, 1986). Fernández-Galiano et al. (1986) concluded that this is likely due to ingestion by these animals of resting cysts attached to vegetation or in soil; the colpodids probably do not excyst in the intestinal environment.

The extreme body size range of colpodeans from 10 μm to over 500 μm is correlated with a similar breadth in the prey that can be consumed by different members of this class. Bacteria and picocyanobacteria, like *Synechococcus*, are the typical prey of the smaller species of *Colpoda* and *Cyrtolophosis* (Griffiths, 1986; Iriberri, Ayo, Santamaria, Barcina, & Egea, 1995; Šimek, Macek, Pernthaler, Straškrabová, & Psenner, 1996; Taylor & Berger, 1976). *Colpoda* may consume almost 1,000 bacteria per ciliate per hour (Hadas, Malinsky-Rushansky, Pinkas, & Cappenberg, 1998). Moderate-sized species can consume chlorophyte and cryptophyte algal cells (Skogstad et al., 1987; Wenzel & Winkler, 1984). Although some *Colpoda* species can ingest whole the cells of the yeast *Saccharomyces* (Wenzel & Winkler, 1984), the grossglockneriid colpodids apparently specialize on fungi with chitinous cells walls, which these ciliates penetrate with their tube-like cytopharynx (Petz, Foissner, & Adam, 1993). Finally, the larger colpodeans can ingest flagellates and other ciliates, including other colpodeans (Bradbury & Olive, 1980; Claff, Dewey, & Kidder, 1941; Foissner, 1990). *Woodruffia metabolica* appears to have a preference only for *Paramecium* species (Johnson & Larson, 1938; Salt, 1967).

The growth rates of colpodeans can be predicted by the amount of macronuclear DNA, and are generally related to cell size: the larger the cell size, the more macronuclear DNA, and the slower the growth rate (Wickham & Lynn, 1990). The smaller species of bacterivorous colpodeans have been characterized as *r*-strategists with doubling times often less than 6 h (Drake & Tsuchiya, 1977; Lüftenegger, Foissner, & Adam, 1985). These colpodids commonly divide twice in a division cyst and may compose the majority of the species in extreme habitats (Foissner, 1994c). At the same time, they are not good competitors against more *K*-selected species, like *Paramecium*, which apparently can exclude *Colpoda* species from favorable habitats (Maguire, 1963a). The larger colpodeans, such as *Bursaria* and *Woodruffia*, have much longer doubling times, in the order of 12 h (Salt, 1967), and might be typified as *K*-selected forms.

Colpodeans do not typically harbor symbionts. Endosymbiotic *Chlorella* have been described as mutuals in *Paracondylostoma*, *Platyophrya*, and *Thylakidium* species (Foissner, 1993a; Foissner & Kreutz, 1998; Kawakami, 1991). Parasites of colpodeans include the flagellate *Spiromonas* (Foissner & Foissner, 1984), *Ciliatosporidium*, a presumed microsporidian (Foissner & Foissner, 1995), and the proteomyxid-like *Endemosarca* (Erdos & Olive, 1971). Suctorians can also be parasites/predators of larger colpodeans, like *Bursaria* (Jankowski, 1973d) while insect larvae and cladocerans are among the metazoan predators of these ciliates (Addicott, 1974; Cochran-Stafira & von Ende, 1998; Jack & Gilbert, 1993).

A discussion of the life history of the Class COLPODEA must include some aspects of the resting or resistant cysts of these ciliates. The literature on this aspect of their biology, cryptobiosis, is significant and has been reviewed most recently by Gutiérrez, Izguierdo, Martín-González, & Callejas, (1998b). *Colpoda* species have survived from 5 years (Dawson & Hewitt, 1931) to 38 years (Goodey, 1915) in laboratory settings so it is likely that their survivability under natural conditions might be even more extensive. As with other ciliates, encystment may also be stimulated by reduced abundances of prey (Barker & Taylor, 1931; Johnson & Evans, 1941; Salt, 1967) and also increasing concentrations of ions, such as Ca^{+2}, Na^{+}, and K^{+}, possible signals for impending desiccation (Yamaoka, Watoh, & Matsuoka, 2004). Excystment is stimulated by the presence of prey and excystment rate and "efficiency" can be dependent, for example, upon the quality of the food preceding encystment (Wenzel & Meier-Tackmann, 1975). Earlier physiological research (e.g., Barker & Taylor, 1931; Beers, 1945, 1948; Johnson & Evans, 1939; Taylor & Strickland, 1939) was followed by research on the "cell biology" of encystment in the early days of electron microscopy. Kawakami and Yagiu (1964) completed their sixth paper on the changes in the fine structure of *Colpoda cucullus* during its life cycle, having described the formation of the cyst wall and the excystment of the ciliate. The cyst wall is composed of two layers, an outer ectocyst and an inner endocyst, probably created by exocystosis of the cortical mucocysts (Martín-González, Benitez, Palacios, & Gutiérrez, 1992b; Ruthmann & Kuck, 1985; Tibbs, 1968). Freeze-etching suggests a third outer mucous layer in *Colpoda* (Janisch, 1980). The resting cyst of *Bursaria* has a conspicuous emergence pore (Foissner, 1993a) and may have a fibrous middle or mesocyst layer connecting the ectocyst and endocyst (Sergejeva et al., 1995). The cyst wall components of *Colpoda* species are rich in glutamic acid while its glycoproteins have high mannose content (Izquierdo, Martín-González, Diaz, & Gutiérrez, 1999; Tibbs & Marshall, 1970). The profiles and distributions of glycoproteins may change in the cyst wall as the cysts age (Chessa et al., 2002). Since ciliature and infraciliature are partially, but not completely, resorbed during encystment, colpodean cysts are typed as partial kinetosome-resorbing cysts (Martín-González, Benitez, & Gutiérrez, 1992a).

Gutiérrez, Izguierdo, Martín-González, & Callejas, (1998b) have argued that colpodid cystment serves as an interesting model for exploration of the cell and molecular biology of cryptobiosis, a life history feature common to both prokaryotes and eukaryotes. Encystment requires RNA and protein synthesis (Ruthmann & Kuck, 1985). In the cryptobiotic state, there is protein turnover and evidence for encystment-specific mRNA (Benitez & Gutiérrez, 1997; Gutiérrez & Martín-González, 1990). In addition to cytoplasmic changes, the macronucleus in particular undergoes substantial changes during this process (Gutiérrez, Martín-González, & Callejas, 1998a). The macronuclear chromatin condenses into large bodies several times the size of those in interphase cells (Frenkel, 1992; Popenko, Cherny, Ivanova, & Yakovleva, 1998a). This process is accompanied by extrusion of macronuclear fragments, a process that appears to regulate the DNA amount in proportion to cell size (Morat, Chessa, & Crippa-Francheschi, 1981). Extrusion of condensed chromatin may also occur at excystment (Chessa, Gallus, Tiano, Trielli, & Corrado, 2001).

It would be remiss not to add to the discussion of cryptobiosis in colpodeans a description of the structure and development of the aerial sorocarps produced on a stalk by *Sorogena stoianovitchae*, which was originally found in tropical habitats (Olive, 1978), but has now been found in the canopy of European deciduous forests (Schnittler, Unterseher, & Tesmer, 2006). *Sorogena* is a predator of smaller colpodeans. Under an alternating photoperiod of light and dark and when prey abundance declines, *Sorogena* trophonts aggregate as a sorogen on plant fragments or other floating films or objects, typically just before sunrise (Sugimoto & Endoh, 2006). The sorogen secretes a stalk, elevating the secreting cells above the attachment point. These cells then encyst as sorocysts, which are discharged for dispersal as the sorus dries (Olive & Blanton, 1980). Secretory vesicles, probably homologues of the mucocysts of other colpodeans, are manufactured in the rough endoplasmic reticulum and their contents released by exocytosis during stalk formation (Blanton & Olive, 1983a). The stalk material expands to a hydrated, fibrillar matrix, pushing the sorogen upwards. The stalk material then solidifies to provide structural support for the sorus (Blanton & Olive, 1983b). Like the cyst wall materials of other colpodeans, the stalk is composed of polysaccharides, which probably include glucose and *N*-acetyl-d-glucosamine residues, and proteins, which have glycine as a predominant amino acid (Blanton, Warner, & Olive, 1983).

12.3 Somatic Structures

As noted above, colpodeans range in size from the very small *Nivaliella* to the almost macroscopic *Bursaria*. Somatic kineties of colpodeans are bipolar with a significant number terminating on the oral region, especially in species whose oral apparatus is at the anterior apex. The colpodean body can be slightly twisted, spiralling to the left in forms like *Platyophrya* and this twisting can become extreme in colpodids (Figs. 12.1, 12.2). Somatic ciliation can be characterized as holotrichous, although in smaller colpodids some regions of the cortex can have very reduced somatic ciliation (Fig. 12.1).

The cell surface of colpodeans is covered by the plasma membrane, which rarely demonstrates elements of a superficial glycocalyx (Bardele et al., 1991; Bradbury & Olive, 1980). The plasma membrane is underlain by conspicuous cortical alveoli (Golder & Lynn, 1980; Lynn, 1976a, 1977a, 1980). Fauré-Fremiet and André (1965a) reported an abundance of dense granules filling this space in *Colpoda* (formerly *Tillina*) *praestans*. The junctions and boundaries between these alveoli may give rise to the complicated silverline system or argyrome of these ciliates, although this is not always the case (Foissner, 1981b; Foissner & Foissner, 1994). As in other ciliates, the alveoli are underlain by an epiplasmic layer whose thickness appears to be correlated with the cell size of the ciliate, at least within the colpodids (Lynn, 1977a). *Bursaria* does not have a thick epiplasm, but like the larger colpodids, its somatic kinetids give rise to microtubular nematodesmata that insert upon a cortical microfilamentous network, which presumably supports the form of these larger cells (Lynn, 1977a, 1980).

Of all classes of ciliates, the somatic kinetids of colpodeans have caused considerable interest. Golder (1974) was the first to report in *Woodruffia* the complex overlapping microtubular ribbons, called LKm fibers, that extend along the left side of the somatic kineties. Lynn (1975) demonstrated that *Woodruffia* also had a typical, striated kinetodesmal fibril and thus was not an exception to the rule of desmodexy as applied to ciliates by Chatton and Lwoff (1935b). A few years later, several simultaneous reports demonstrated that the LKm microtubular ribbons derived from the transverse ribbon of the posterior kinetosome of the somatic dikinetids (Gerassimova, 1976; Lynn, 1976a). The colpodean somatic dikinetid is rotated about 10–20° clockwise to the kinety axis, viewed from outside the cell. The posterior ciliated kinetosome bears the following: a divergent postciliary ribbon that sometimes extends far enough to overlap those of more posterior kinetids; a short, laterally-directed kinetodesmal fibril that originates near triplets 5, 6, flattening to a fan-shape as it extends upwards in the cortical ridge; and a very well developed, tangentially-oriented transverse ribbon that originates from a medial desmose connecting the two kinetosomes and extends from triplets 3–5 posteriorly and often overlaps several other ribbons from more posterior kinetids (Figs. 12.3, 12.4). The anterior ciliated kinetosome bears the following: a single to occasionally several postciliary microtubules and a large tangentially-oriented

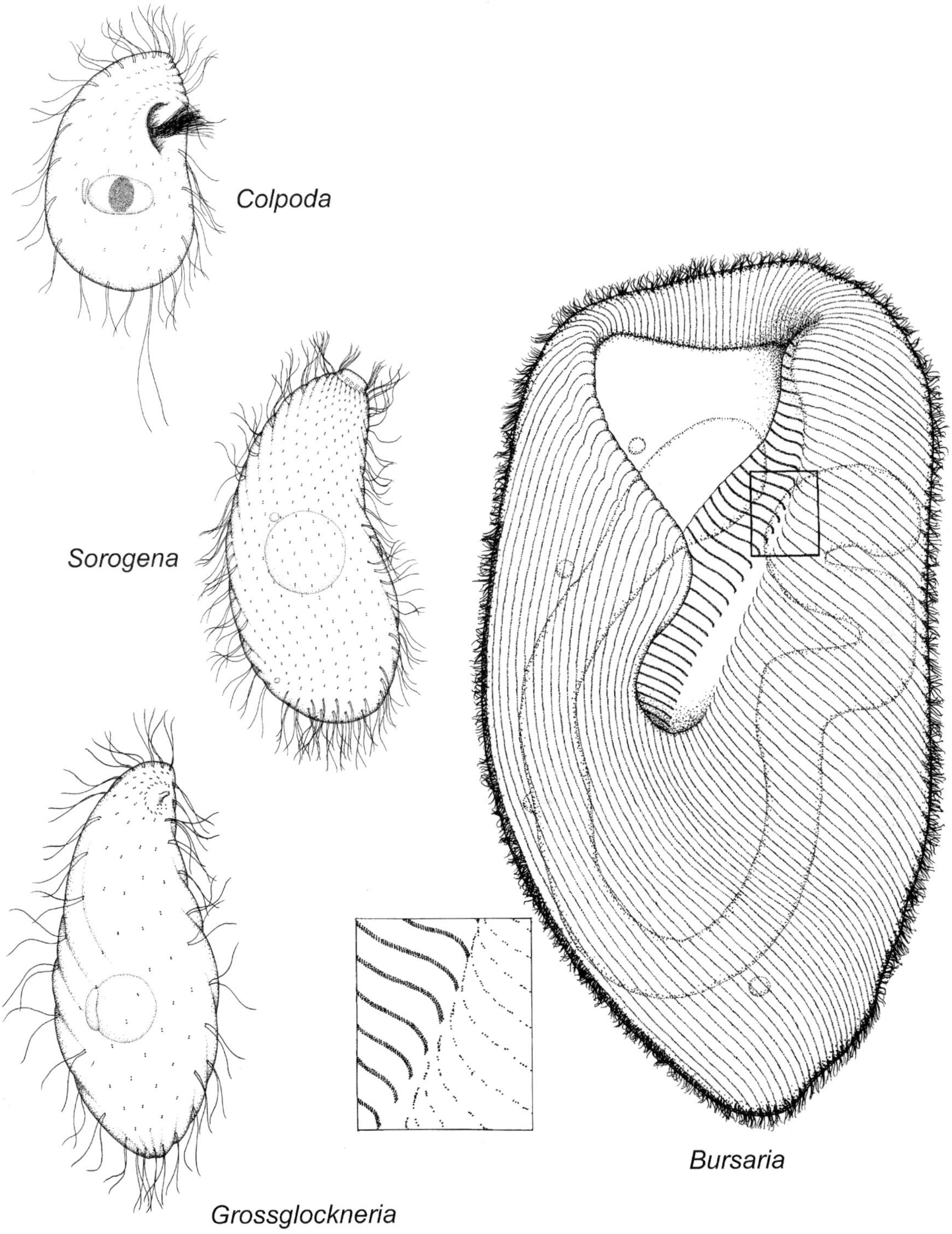

FIG. 12.1. Stylized drawings of representative genera from the orders in the Class COLPODEA. The colpodids *Colpoda* and *Grossglockneria*. The sorogenid *Sorogena*. The cyrtolophosidid *Cyrtolophosis*. The bursariomorphid *Bursaria*. Inset is a detail of the adoral polykinetids and adjacent somatic kineties

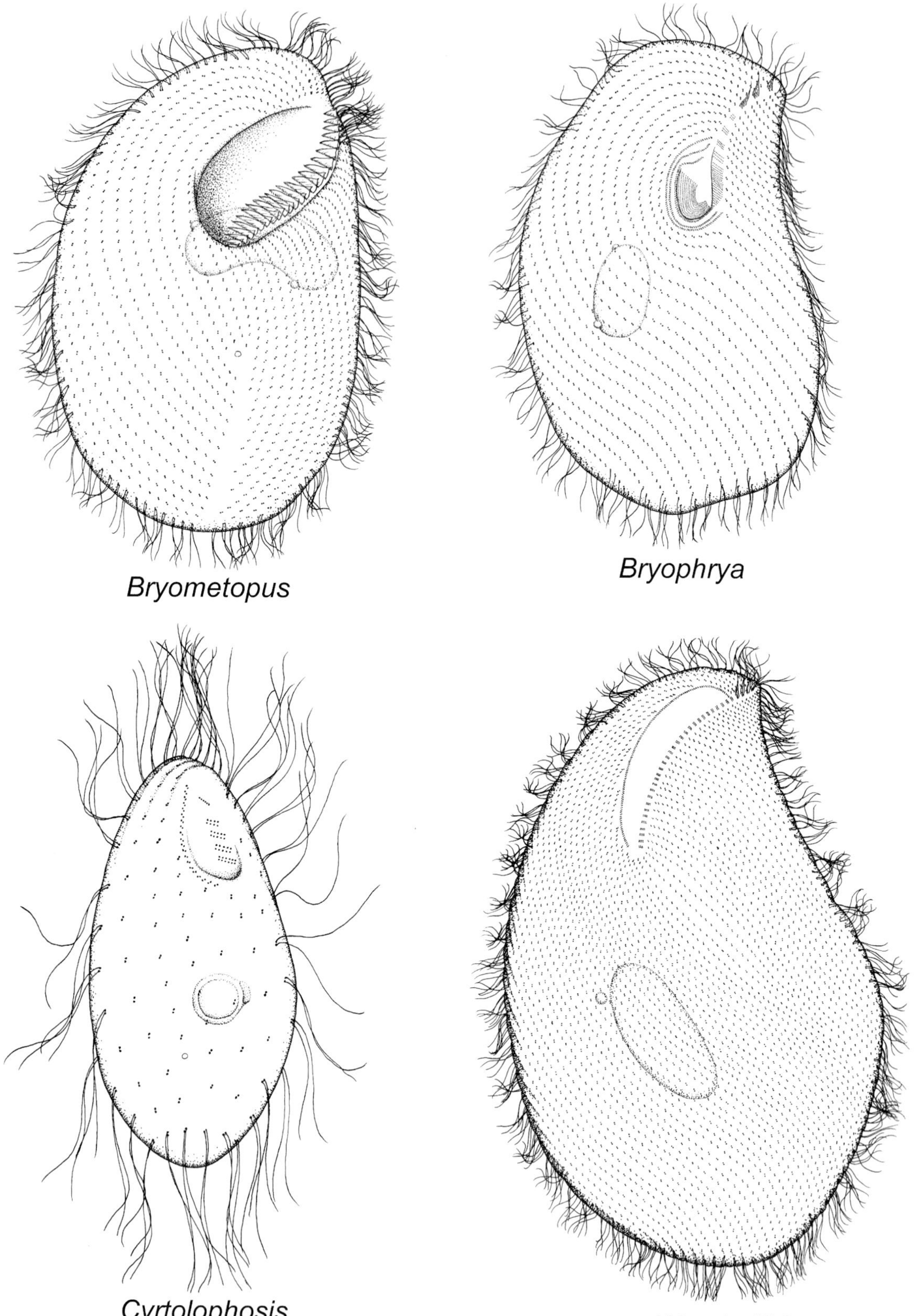

FIG. 12.2. Stylized drawings of representative genera from the orders in the Class COLPODEA. The bryometopid *Bryometopus*. The bryophryid *Bryophrya*. The cyrtolophosidids *Cyrtolophosis* and *Woodruffides*

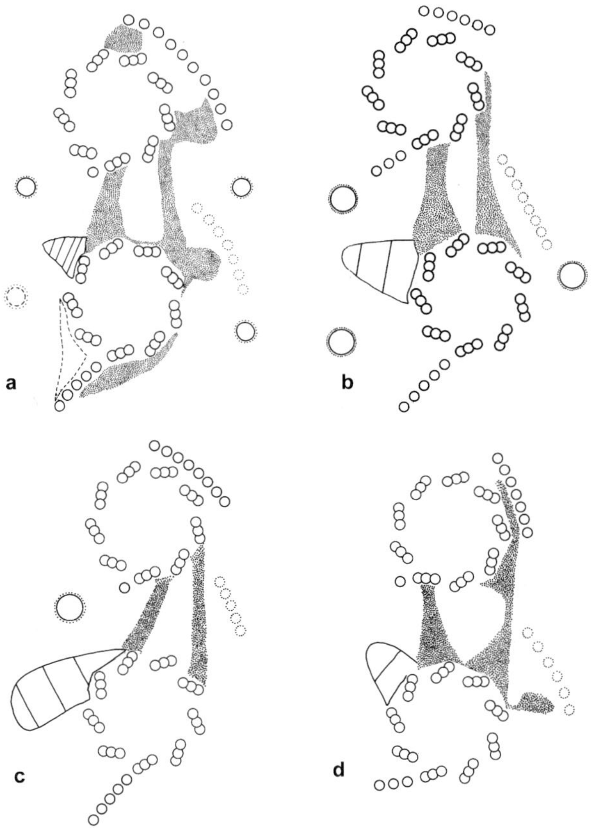

FIG. 12.3. Schematics of the somatic dikinetids of the Class COLPODEA. (**a**) The bryophryid *Bryophrya*. (**b**) The colpodid *Colpoda*. (**c**) The cyrtolophosidid *Cyrtolophosis*. (**d**) The bursariomorphid *Bursaria* (from Lynn, 1981, 1991)

transverse ribbon that arises near triplets 3–5 and extends laterally into the ridge to overlap the compound bundle of posterior kinetosomal transverse ribbons. There may be up to four parasomal sacs, two on each side of the kinetid (Fig. 12.3). Nematodesmata may also originate from the base of one or both kinetosomes (Lynn, 1981, 1991). More recent descriptions of colpodean kinetids have confirmed this structural pattern for the class (Aescht et al., 1991; Bardele et al., 1991; Foissner & Foissner, 1994; Platt & Hausmann, 1993). Like some oligohymenophoreans (Antipa, 1972), the cortex of colpodids can undergo structural differentiation. Larger colpodids have a somatic groove that forms the "hilum" in the kidney-bean shape. This somatic groove is supported longitudinally by strongly overlapping posterior transverse ribbons and underlain by orthogonally disposed nematodesmata that arise from the bases of the groove somatic dikinetids (Lynn, 1976c; Lynn & Zimmerman, 1981). The groove can be divided into an incurrent and excurrent component, which facilitates feeding in *Colpoda* species (Fenchel, 1980a; Lynn, 1976c, 1977a). The larger the *Colpoda* species, the greater the development of its somatic groove (Lynn, 1978).

Colpodeans typically have at least one contractile vacuole, located in the posterior end. Depending upon the species, they can be identified as showing Type A or Type B morphology (Patterson, 1980). Larger cells may have either multiple vacuoles (e.g., *Bursaria*, Foissner, 1993a) or may have long collecting canals radiating anteriorly in the cell's cortex (e.g., *Colpoda magna* (formerly *Tillina magna*, *Tillina canalifera*), Lynn, 1977a; Turner, 1937). As cell size increases in *Colpoda* species, the relative size of the contractile vacuole and its pore increases as does the output of the contracile vacuole itself (Lynn, 1977a, 1982).

Colpodeans have ovoid to rod-shaped mucocysts. These are particularly conspicuous in medium to large cells (Foissner, 1993a; Lynn, 1976c, 1977a),

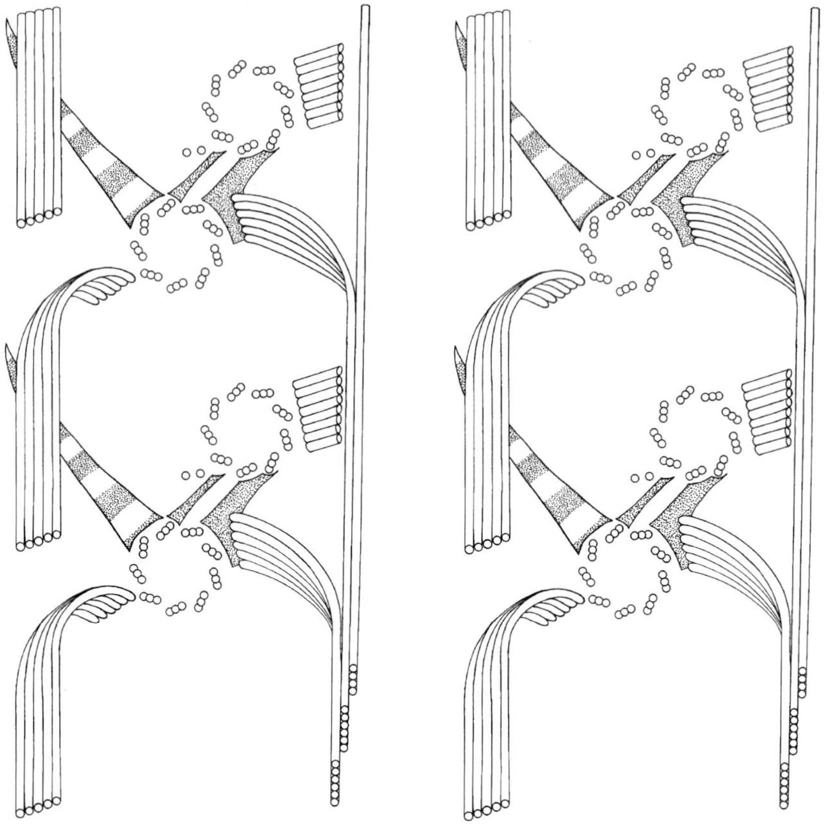

FIG. 12.4. Somatic cortex of a typical colpodean interpreted based on the somatic cortex of several colpodeans, such as *Colpoda* and *Bursaria*

but inconspicuous in smaller ones (Foissner, 1993a; Lynn, 1976a, 1976b). The matrix is granular (Lynn, 1977a) or often completely removed upon fixation (Delmonte Corrado, Chessa, & Pelli, 1996; Lynn, 1976c). The characteristic periodic substructure of mucocysts (Hausmann, 1978) is rarely observed (see Perez-Paniagua, Perez-Silva, & de Puytorac, 1979a, 1979b). Mucocyst abundance varies throughout the life cycle with depletions in their abundance correlated with encystment (Delmonte Corrado et al., 1996; Suhama, 1969).

Mitochondria are the typical tubular forms characteristic of ciliates. They may be distributed throughout the cortex or anchored in the cortex, possibly by filamentous elements associated with the microtubular ribbons of the somatic kinetids (Lynn, 1977a).

The cytoproct, when it has been described, is usually located near the contractile vacuole (Foissner, 1993a).

12.4 Oral Structures

Colpodeans have a permanent cytostome and an oral cavity that is typically in the anterior third of the cell (Figs. 12.1, 12.2). Marynids with their posterior oral cavity are an exception to this general description (Foissner, 1993a). The size of the oral apparatus essentially varies with body size: the larger ciliates, such as *Bursaria*, typically have a larger oral cavity (Fig. 12.1). However, this is not always the case. Some carnivorous colpodids, like *Bresslaua*, are smaller than some of the larger bacterivorous species of *Colpoda* (e.g., *Colpoda magna*) (Foissner, 1993a). As noted in **Chapter 1**, the colpodeans superficially have extremely variable oral structures (Fig. 3.4) (Foissner, 1993a). However, upon closer examination and especially through detailed analyses of stomatogenesis (**see below Division and Morphogenesis**), there is an underlying unity to the diversity of patterns.

The cyrtolophosidids, bursariomorphids, and bryophryids have several to many adoral polykinetids oriented along the left side of the oral cavity (Figs. 12.1, 12.2). These can be composed of a basic two rows of kinetosomes, constructed by the side to side assembly of dikinetids, joined by dense connectives in a square-packed or hexagonal arrangement. A third row can often appear during stomatogenesis (Bardele et al., 1991; Didier et al., 1980; Golder & Lynn, 1980). Colpodids have a large left oral polykinetid composed of regularly spaced rows, which may be linked by connectives both within and between rows (Garcia-Rodriguez, Perez-Paniagua, & Perez-Silva, 1981; Hofmann-Münz, 1991; Lynn, 1976a, 1976b, 1976c, 1977; Perez-Paniagua et al., 1979a, 1979b). The single, small adoral polykinetid of grossglockneriid colpodids also probably has an organized kinetosomal arrangement (Aescht et al., 1991; de Puytorac, Didier, Detcheva, & Foissner, 1983a). Postciliary ribbons extend from these kinetosomes to support the cytopharnynx.

The right oral kinetids in the cyrtolophosidids, bursariomorphids, and bryophryids are minimally organized as a paroral composed of dikinetids. Both kinetosomes are ciliated while the postciliary ribbon from the left kinetosome extends to support the cytopharynx (Bardele et al., 1991; Didier et al., 1980; Golder & Lynn, 1980).

Hofmann-Münz (1991) was the first to demonstrate that there was some order to the right oral field of colpodids, which had been described previously as unordered (Garcia-Rodriguez et al., 1981; Lynn, 1976c, 1977a; Perez-Paniagua et al., 1979a; but see also Perez-Paniagua & Perez-Silva, 1978). A dikinetid file may extend along the right border of the right oral polykinetid. This "paroral" is composed of a single file of kinetosomes whose orientation is similar to the left kinetosome of the paroral kinetids of the other orders, while the right kinetosome of the paroral dikinetids in colpodids is oriented at 180° as judged by the orientation of the postciliary microtubules. An unordered field of kinetosomes is arrayed to the right of this "paroral" (Hofmann-Münz, 1991). In the grossglockneriid colpodids, whose oral ciliature is quite reduced, the right oral ciliature is just a single file of kinetosomes bearing a postciliary ribbon that extends anteriorly to support the sucker (Aescht et al., 1991; de Puytorac et al., 1983a). As noted before, the grossglockneriids have a small feeding tube, supported by lamellae of microtubules reminiscent of the arrangement of the phyllae in the cytopharynx of the phyllopharyngeans and presumed to be derived during stomatogenesis from postciliary microtubular ribbons (Aescht et al.; de Puytorac et al., 1983a).

Except for the unusual mycophagous grossglockneriids (Petz, Foissner, & Adam, 1993), colpodeans can be characterized as suspension or filter feeders. Fenchel (1980a) has described *Bursaria* as an upstream filter feeder, using the spacing between its left oral polykinetids to trap prey ciliates larger than 8–10 µm, which is the average spacing between these structures. Fenchel (1980a) was unable to determine which oral structure of colpodids was responsible for retaining the particles since the species that he described fed while swimming. The smaller *Colpoda steinii* cleared larger particles, about 1 µm diameter, more efficiently than the larger *Colpoda cucullus*, which optimally cleared particles slightly less than 0.4 µm (Fenchel, 1980b). Since the left oral field of colpodids is ordered, one is tempted to speculate that this provides the filter in these species.

As in other ciliates, colpodeans form food vacuoles in which prey digestion occurs (Rudzinska, Jackson, & Tuffrau, 1966).

12.5 Division and Morphogenesis

Colpodeans divide either free-swimming or within a cyst. Parental oral structures may be partially or completely reorganized. The colpodids are the group classically considered to divide in reproductive cysts. However, under exceptional circumstances even colpodids may divide while swimming freely (Stuart, Kidder, & Griffin, 1939). Division within a cyst is typically palintomic, that is typically two divisions occur with no DNA S-phase occurring between them (Foissner, 1993a).

Tuffrau (1952) provided details of silver-stained dividing colpodids and demonstrated that all the parental oral structures dedifferentiated and the ciliates were essentially astomatous during palintomy. Following cytokinesis, kinetosomal proliferation occurred at the ends of some of these somatic kineties to provide kinetosomes for the new oral structures (Fig. 12.5). Hashimoto (1966) confirmed both that the same kind of stomatogenesis occurred in colpodids excysting from resting cysts and that the number of somatic

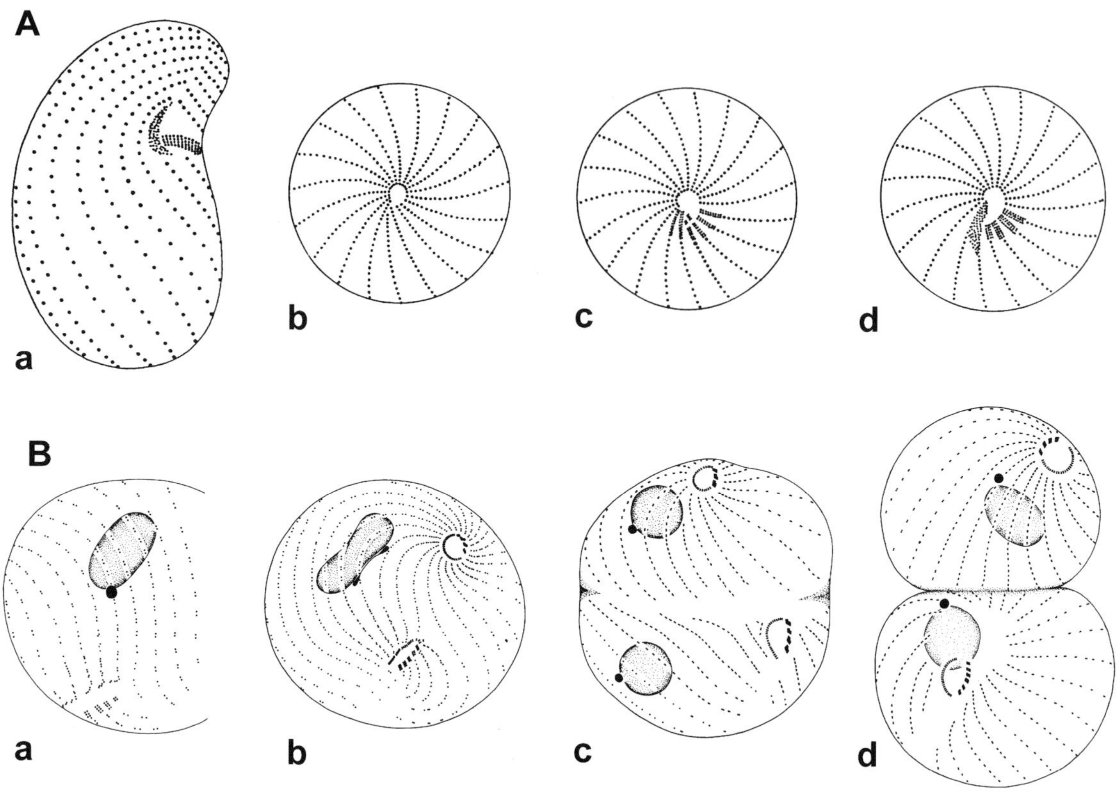

FIG. 12.5. Division morphogenesis in the Class COLPODEA. **A** The merotelokinetal stomatogenesis of *Colpoda* occurs within a division cyst and begins with complete dedifferentiation of the oral structures (**a**, **b**). Kinetosomal proliferation then occurs at the anterior ends of several somatic kineties (**c**) and the right and left oral structures differentiate from different subsets of these somatic kineties (**d**) (from Foissner, 1993a). **B** The pleurotelokinetal pattern is more widespread within the class, exemplified here by *Sorogena*. Kinetosomal proliferation begins on several kineties in the posterior right region of the body (**a**). A paroral and oral polykinetids begin to differentiate along the anterior and posterior borders of the primordial field, respectively (**b**). As the field rotates clockwise (**c**) and migrates anteriorly (**d**), these become the right and left oral structures respectively. (from Bardele et al., 1991.)

kineties involved increased as the size of the species increased. Foissner (1996b) has typed this as merotelokinetal stomatogenesis since the ends of only a limited number of somatic kineties are involved. Since Tuffrau (1952) and Hashimoto (1966), several studies have demonstrated that the left oral polykinetid is derived by the fusion of "subpolykinetids" composed of three rows of kinetosomes, which are initially derived from dikinetids. The right oral polykinetid is assembled by a file of dikinetids along the border of the primordial field and to its right there is a field of unordered kinetosomes derived by kinetosomal replication of somatic kinetosomes (Garcia-Rodriguez et al., 1981; Perez-Paniagua & Perez-Silva, 1978; Perez-Paniagua et al., 1979a, 1979b).

The pattern in colpodids is undoubtedly derived from the pattern demonstrated by representatives of the other colpodean orders, at least if we use the molecular phylogenies as the basis for polarizing stomatogenetic patterns in this class. Foissner (1996b) characterizes the alternate pattern as pleurotelokinetal since oral kinetosomal proliferation occurs within, and sometimes between, several kineties on the right side of the body and on both sides of the fission zone. Stomatogenesis in *Sorogena* is of this type, and occurs by proliferation within and between about six posterior, right ventral somatic kineties (Fig. 12.5) (Bardele et al., 1991). This is followed by a clockwise rotation of the field and differentiation of paroral dikinetids on the anterior and ultimately right side of the primordial field and several oral polykinetids on the posterior and ultimately left side of the primordial field (Fig. 12.5). This basic pattern has been observed in *Platyophrya* (Dragesco et al., 1977; Grolière, 1975b) and *Woodruffia* (= *Woodruffides*) (de Puytorac et al., 1979b; Prelle, 1963). A similar pattern has also been reported for *Bryometopus* (Wirnsberger et al., 1985b) and *Cyrtolophosis* (Díaz, Martín-González, Borniquel, & Gutiérrez, 2000).

There is as yet no study of bryophryid stomatogenesis while stomatogenesis of bursariomorphids has only been studied once. Perez-Paniagua, de Puytorac, and Savoie (1980) demonstrated that *Bursaria* undergoes a pleurotelokinetal stomatogenesis by kinetosomal replication within and between a number of right posterior somatic kineties. Similar to colpodids, the left oral polykinetids are composed of dikinetids aligned to make two rows to which a third is added to complete stomatogenesis. However, these never fuse, as in colpodids, but remain as a series of separated adoral polykinetids. The right oral structures derive from a more disordered kinetosomal proliferation that eventually develops into the multiple and oblique but parallel files of dikinetids of the "paroral" of this ciliate (Foissner, 1996b).

12.6 Nuclei, Sexuality and Life Cycle

Colpodeans typically have a single macronucleus. Depending upon cell size, larger colpodeans may have many micronuclei (e.g., *Colpoda magna*, *Bursaria truncatella*) (Foissner, 1994b). The macronucleus varies from globular to ellipsoid in smaller species to an elongate band-shape in larger forms (Figs. 12.1, 12.2). The micronucleus is globular to ellipsoid. Colpodids especially have a prominent nucleolus, which takes a variety of forms (Burt, Kidder, & Claff, 1941). The nucleolus is actually an aggregation of a number of nucleolar organizing regions that assume various, often species specific arrangements. Although giving the appearance of a heteromerous nucleus, distribution of nucleoli in other colpodeans confirms that the colpodean macronucleus is homomerous (Raikov, 1969). Another unusual feature of the colpodean nuclear apparatus is the fusion of macronuclear and micronuclear envelopes, especially in cyrtolophosidids (Detcheva, 1976; Dragesco et al., 1977; Golder, 1976). This has been used as a defining character for the Order Cyrtolophosidida (Foissner, 1994b; Lynn & Small, 2002; Small & Lynn, 1985). However, there are now reported exceptions within the genus *Cyrtolophosis* (Díaz et al., 2000) and in forms related by stomatogenesis, such as *Bryometopus* (Wirnsberger et al., 1985b). Thus, we have abandoned this feature as diagnostic of the Order Cyrtolophosidida.

Division of colpodean macronuclei occurs through participation of intramacronuclear microtubules (Kuck & Ruthmann, 1985). Micronuclear chromosomes are attached by kinetochores to single microtubules that apparently shorten as the interpolar microtubules elongate during micronuclear division (Kuck & Ruthmann, 1983). Micronuclei in multimicronucleate species probably all undergo fission but, due to the large size of the cells that they are found in, may not separate their products to different poles (e.g., see Beers, 1946a). Occasional amicronucleate clones may arise but they apparently have limited viability (Beers, 1946a, 1946b; Piekarski, 1939). The division of the macronucleus of small colpodids, like *Colpoda steinii*, involves the formation and separation of a small number of discrete chromatin aggregates (Burt et al., 1941; Piekarski, 1939; Frenkel, 1978). This was confirmed by electron microscopy in both small and larger species (Frenkel, 1980, 1982). Quantitation of the DNA in these nuclei led to the hypothesis that these were diploid subnuclei undergoing separation (Frenkel, 1978; Frenkel, Kudryavtsev, & Kudryavtseva, 1973). Although this may be the case in colpodeans with low macronuclear ploidy, such as *Colpoda steinii*, it seems less likely for colpodeans like *Bursaria* whose ploidy can exceed 5,000 n (Raikov, 1969). In *Bursaria*, composite chromosomes and polynemic structures have been observed, suggesting that there is a higher order organization to its DNA structure, even if it is not organized as diploid subnuclei (Raikov, 1969; Sergejeva & Bobyleva, 1995). Popenko et al. (1998b) demonstrated that the macronuclear DNA molecules of *Bursaria* range from 50–360 kbp, probably packed in chromatin aggregates that are themselves formed into higher-order structures. DNA from *Bursaria* is about 50% single copy sequence with much of the remainder being highly repetitive (Borchsenius & Sergejeva, 1979). Other colpodeans have macronuclear chromosomes that range from 90–2,000 kbp and show karyotypic variation among species in the chromosome size on which rDNA genes reside (Martín et al., 1997).

Conjugation has only been described in *Bursaria* and has never been reported in any colpodid (Raikov, 1972). Poljansky (1934) reported conjugation to occur in the fall in *Bursaria*, but the environmental cues are not known. Micronuclear meiosis is typical and demonstrates a "parachute stage" as is commonly observed in ciliates. Although many micronuclei undergo meiosis, only the one nearest the zone of contact and its denser cytoplasm undergoes a third division to produce the gametic nuclei. Raikov (1972) established a unique subtype for the development of macronuclear anlagen and micronuclei in *Bursaria*, which is characterized by the development of four macronuclear anlagen and four micronuclei. Poljansky and Sergejeva (1981) described oligotenic-like chromosomes during macronuclear anlagen development and likened the process to anlagen development in spirotrichs and phyllopharyngeans, but this observation has yet to be corroborated.

Sexual processes are known to "restart" the life cycle clock of ciliate populations that are becoming senescent (Sonneborn, 1954; Dini & Nyberg, 1993). Under this model, it is remarkable that some lines of colpodids, which never undergo conjugation (e.g., *Colpoda magna*), show no signs of senescence after over 500 generations (Beers, 1944), although senescence may begin at double that number of cell generations (Crippa Franceschi, Schieti Cavazza, & Boccardo Rinesi, 1967). Extrusion bodies have been consistently observed during division of the macronucleus in colpodids (e.g., Beers, 1946a; Burt et al., 1941). These extrusion bodies have been considered a method of eliminating defective DNA (Burt et al., 1941), but they may also serve to enable regulation of the nuclear-cytoplasmic ratio (Woodruff, 1941). Extrusion does not seem to regulate DNA content in vegetative cells, and so it may indeed be a "purification" mechanism (Frenkel, 1975). Extrusion also appears to maintain a strong nucleocytoplasmic ratio in cells preparing to encyst (Morat et al., 1981).

12.7 Other Features

Colpodeans have been recorded as dominant ciliate members of activated sludge plants (Aescht & Foissner, 1992) and, as noted above, they are common and dominant elements in soils and other terrestrial habitats all over the world (Foissner, 1993a, 1994b). This has lead to an interest in the impacts of toxicants, especially heavy metals on *Colpoda* species, which show reduced growth rates in response to heavy metals in laboratory and field situations (Forge, Berrow, Darbyshire, & Warren, 1993; Janssen, Oosterhoff, Heijmans, & Van der Voet, 1995), although some isolates appear to be quite resistant to metal toxicity (Díaz, Martín-González, & Gutiérrez, 2006). Nevertheless, strains of *Colpoda* from around the world appear not to differ significantly in their sensitivity to these toxicants (Xu et al., 1997).

Chapter 13
Subphylum 2. INTRAMACRONUCLEATA: Class 7. PROSTOMATEA – Once Considered Ancestral, Now Definitely Derived

Abstract Ciliates in this class were considered in the 19th and to the mid-20th centuries to be the ancestral kind of ciliate – almost perfectly radially symmetrical and prostomatous. They are now considered to be quite derived, and careful microscopic analyses have exposed some differentiated oral ciliature in one of the included orders. Prostomateans are commonly represented in both freshwater and marine habitats where they can achieve significant abundances under bloom conditions, sometimes exceeding 30,000 l^{-1}, and they are suspected therefore of being able to control prey populations, which sometimes may be toxic red tide algae. *Cryptocaryon irritans* is a prostome that causes a disease of the skin of marine fishes. The somatic monokinetids of prostomes do not distinguish them from several other classes, although their somatic kineties are often organized in paratenes. They are distinguished by their oral structures: a paroral that almost encircles the anterior end and a set of brush or brosse polykinetids that abut near the oral region. Stomatogenesis is merotelokinetal. Although these ciliates conjugate, we do not know the genetical basis of their mating system(s).

Keywords Cryptocaryoniasis, *Coleps*, histophagy

This class and the next, Class PLAGIOPYLEA, each include the smallest numbers of species and genera among classes of ciliates. Yet, their distinctive morphologies, their cortical ultrastructure, their genetic divergence, and their very different ecological roles differentiate them among ciliates. Members of the Class PROSTOMATEA are typically ovoid to cylindroid in body shape. Moreover, they are some of the smallest ciliates: <20 μm for some *Urotricha* species to moderate-sized for species of prorodontids, which typically do not exceed 500 μm in length. Body ciliation is typically holotrichous often with a single or multiple caudal cilia. The cytostome is apical to subapical in position. Most species are free-living microphages. However, a number of prorodontids are at least facultative histophages (Czapik, 1965), and one species, the fish ectoparasite *Cryptocaryon irritans*, has recently been transferred to this class based on the sequence of its small subunit (SSU) rRNA gene (Wright & Colorni, 2002).

The focus of attention on the role of ciliates in the microbial loop has revealed that small prostomateans can be extremely abundant. *Coleps hirtus* reached densities exceeding 60,000 cells l^{-1} in rice fields (Madoni, 1986) while prostomatids collectively can reach over 100,000 cells l^{-1} (Müller, 1989). The marine prostomatid *Tiarina fusus* has been observed in bloom conditions to exceed 30,000 l^{-1} (Dale & Dahl, 1987b). Ecologists have brought these ciliates into the laboratory and established long-term cultures of *Coleps* (Klaveness, 1984) and *Balanion* (Müller, 1991) on the flagellate *Rhodomonas*. Isolates of *Urotricha* grew well on the flagellate *Cryptomonas* (Weisse & Montagnes, 1998). Both species have been efficiently counted using flow cytometry and nucleic acid stains (Lindstrom, Weisse, & Stadler, 2002). Weisse (2006) has argued that *Urotricha* species and other planktonic ciliates are excellent ecophysiological models to explore the role of protists in microbial food webs. Parasitologists have established standardized methods for maintaining *C. irritans* in culture using experimental hosts (Burgess & Matthews, 1994a; Dan, Li, Lin, Teng, & Zhu, 2006), and initial attempts at in vitro culture show promise (Yambot & Song, 2004).

The prostomes have also occupied a pivotal position in our conceptions of evolutionary diversification within the phylum. The name of the class, PROSTOMATEA, is derived from the Greek *pro* meaning before and the Greek *stoma* meaning mouth. It refers therefore to the apical to subapical position of the cytostome, which is surrounded by a simple circumoral ring of ciliated dikinetids. Corliss (1979) argued that the apical to subapical cytostome and simple oral ciliature made prostomes the "most primitive" group of ciliates next to the karyorelicteans. While we are still convinced that the karyorelicteans represent descendants of an early diverging clade of ciliates, the prostomes are now conceived as a fairly derived group whose oral features have become secondarily simplified (Bardele, 1989; Hiller, 1993b). These conclusions based on ultrastructural research have been confirmed by rRNA gene sequences that show prostomes to be closely associated with the colpodeans, plagiopyleans, and oligohymenophoreans (Baroin-Tourancheau et al., 1992; Fleury, Delgado, Iftode, & Adoutte, 1992; Lynn, Wright, Schlegel, & Foissner, 1999; Stechmann, Schlegel, & Lynn, 1998).

Small and Lynn (1981) established the class based on both the unusual nature of the presumed transverse microtubular ribbons that arose from the oral dikinetids in prostomes (de Puytorac & Grain, 1972) and features of the somatic monokinetids, which were very similar to those of the oligohymenophoreans (Rodrigues de Santa Rosa, 1976). Subsequent ultrastructural study by Huttenlauch and Bardele (1987) demonstrated that these presumed transverse ribbons were actually postciliary microtubular ribbons, oriented into a "transverse"-like position during stomatogenesis. Thus, the cytopharynx of prostomes is lined by postciliary microtubules arising from oral dikinetids, a situation certainly found in most of the classes of this subphylum – the Class LITOSTOMATEA being the notable exception. Aside from these two features, stomatogenesis in prostomes appears to involve a clockwise migration of the oral dikinetids to form the circumoral crescent or ring characteristic of the class (Hiller, 1993b).

To close these introductory remarks, we must draw the reader's attention to a nomenclatural change that has complicated our subsequent discussion. Foissner, Berger, and Kohmann (1994) discovered that for years ciliates assigned to the genus *Prorodon* ought properly to have been assigned to the genus *Holophrya* based on the features of its type species, *Holophrya ovum* Ehrenberg, 1831, which exhibits brosse kineties. Furthermore, they concluded that *Prorodon* ought to include only species with a slit-like oral region and a brosse that extends the length of the body (Foissner et al., 1994), based on its type species, *Prorodon niveus* Ehrenberg, 1833. Foissner et al. (1994) established a new genus *Apsiktrata* based on *Urotricha gracilis*, a *Urotricha* species without a brosse, and a new family Apsiktratidae, to include those prostomes without a brosse. We have assumed throughout this chapter that ciliates identified as *Prorodon*, *Pseudoprorodon*, and *Holophrya* prior to Foissner et al. (1994) actually ought to have been identified as *Holophrya*, *Prorodon*, and *Apsiktrara*, respectively. Throughout the chapter, we have used the new names but followed these in parentheses by the genus name used in the literature cited.

13.1 Taxonomic Structure

Corliss (1979) assigned the prostome ciliates to a single order in the Subclass Gymnostomatia. He included three suborders, the Archistomatina, Prostomatina, and Prorodontina. Morphological and ultrastructural analysis of the archistomatines, which includes only the Family Buetschliidae, argues for a closer relationship with ciliates in the Order Vestibuliferida (see **Chapter 9. Class LITOSTOMATEA**; see also parsimony analysis of de Puytorac, Grain, & Legendre, 1994). Following earlier ultrastructural studies, Small and Lynn (1985) concluded that the cytopharyngeal structure of prostomes was a rhabdos, and united the Classes PROSTOMATEA and LITOSTOMATEA in the Subphylum Rhabdophora, respectively distinguishing ciliates in these two classes by the radial and tangential orientation of transverse microtubular ribbons on somatic monokinetids. Huttenlauch and Bardele (1987) undercut the entire conceptual unity of the Subphylum Rhabdophora by showing that the "cytopharyngeal" microtubules of prostomes were postciliary ribbons, and not transverse ribbons. De Puytorac (1994c) assigned the prostomes to the Subclass Prostomatia in the Class NASSOPHOREA, and recognized two orders, the Prorodontida and Prostomatida. However, there are few detailed similarities in morphology (**see below and Chapter 11. Class NASSOPHOREA**) and in molecules to support a close relationship between nassophoreans and prostomes (Baroin et al., 1992; Bernhard & Schlegel, 1998; Fleury et al., 1992; Lynn et al., 1999; Stechmann et al., 1998). Thus, we follow Lynn and Small (2002) in retaining a Class PROSTOMATEA in the Subphylum Intramacronucleata.

The Class PROSTOMATEA is based on the structure of the somatic kinetids, which have a divergent postciliary ribbon, anteriorly extending kinetodesmal fibril, and a somewhat radially oriented transverse ribbon, and on the structure of the circumoral dikinetids, which, at least in some prorodontids, have postciliary ribbons organized as ribbons of circumoral microtubules. We recognize two orders within this class, the Order Prorodontida and the Order Prostomatida.

The Order Prorodontida is characterized by having an oral region that is apical or slightly subapical. In some species (e.g., *Balanion*, *Bursellopsis*, *Urotricha*), the cytostome is surrounded by overlapping ribbons of circumoral microtubules derived from the postciliary ribbons of the circumoral dikinetids (Hiller, 1992, 1993b). Toxicysts are typical and may be carried in oral palps, which are internal to the cilia of these oral dikinetids (e.g., Bardele, 1999; Fauré-Fremiet & André, 1965b). Finally, typically outside the circumoral dikinetids, there are three or more "oral" polykinetids that form the so-called brosse. The brosse can be reduced to several dikinetids in *Balanion* (Bardele, 1999) or be composed of many units extending down the entire length of the body in *Pleurofragma* (see Lynn & Small, 2002). We include the following families: Balanionidae, Colepidae, Holophryidae, Lagynidae, Placidae, Plagiocampidae, Prorodontidae, and Urotrichidae.

The Order Prostomatida is characterized by having a truly apical oral region, by having perioral somatic kineties that form conspicuous paratenes (see Dragesco, Iftode, & Fryd-Versavel, 1974), and by two negative characters – lack of a brosse and lack of toxicysts. Bardele (1999) argued that this order is not justified. He imagined the taxa assigned to it, such as *Metacystis* and *Apsiktrata* (formerly *Holophrya*), as being at the end of an evolutionary process of adaptation to the planktonic habitat by ventrostomial ancestors of the prostomes: *Balanion* has remnants of the brosse while the almost perfectly radially symmetrical prostomatids have lost the brosse completely. While we sympathize with this view, there are yet no detailed ultrastructural studies on a "true" prostomatid and no molecular data. Since other higher taxa are also defined by the absence of characters (e.g., absence of toxicysts in the Subclass Trichostomatia) and have been subsequently supported by molecular data, we have retained the Order Prostomatida in which we include the following families: Apsiktratidae and Metacystidae.

Finally, we have placed the Family Malacophryidae incertae sedis in this class and await data from electron microscopy and gene sequencing to confirm its position.

There are no monographic works on this class. However, Foissner, Berger, and Schaumburg (1999) provide an excellent recent treatment in their monograph on the identification and ecology of limnetic plankton ciliates, while Foissner and Pfister (1997) have provided a key to some common *Urotricha* species.

13.2 Life History and Ecology

The life cycle of a typical prostome *Holophrya* (formerly *Prorodon*) has been characterized by Hiller and Bardele (1988) (Fig. 13.1). As noted above, there is a considerable recent literature on the distribution and abundance of prostome ciliates, especially those found in the plankton of a variety of aquatic environments. Prostomes have extremely broad feeding preferences, primarily due to the range in size of taxa in this class. Smaller prostome species are microphagous bacterivores while larger species can consume filamentous algae. Common genera, such as *Balanion* (now includes *Pseudobalanion*), *Bursellopsis*, *Coleps*, *Holophrya* (formerly *Prorodon*), and *Urotricha*, have been recorded from habitats around the world, as diverse as small temporary ponds (Foissner, 1984b; Madoni & Sartore, 2003) and soils in Europe (Berger, Foissner, & Adam, 1984) and globally (Foissner, 1998a). They have been found in saline lakes in Europe (Esteban, Finlay, & Embley,

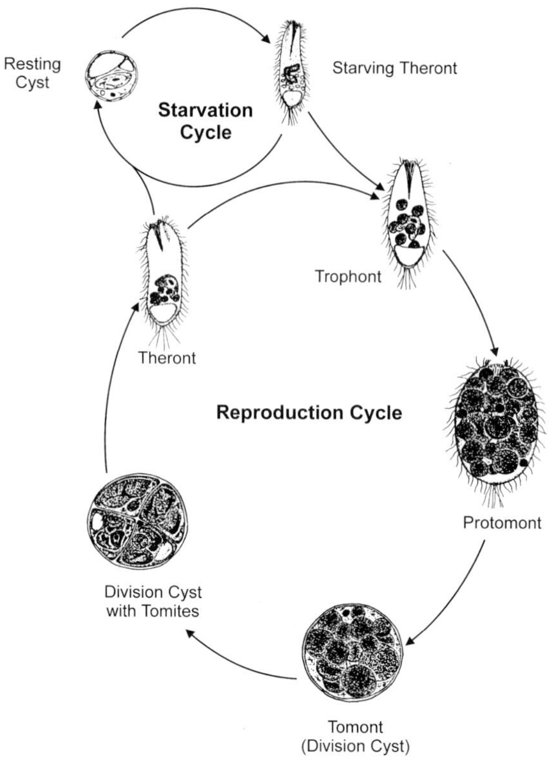

FIG. 13.1. Life cycle of a typical prostome, *Holophrya* (formerly *Prorodon*) showing the two typical phases in its life cycle. In the Starvation Cycle, the ciliate forms a resting cyst when deprived of food. In the Reproduction Cycle, the ciliate may feed for some time without division. The protomont then forms a division cyst in which the tomont undergoes palintomy to produce multiple tomites that either encyst in resting cysts if there is no food or begin feeding again. Perhaps lacking a resting cyst stage, the parasite of marine fishes, the prorodontid *Cryptocaryon*, has a life cycle similar to this, which shows remarkable convergence on the life cycle of the oligohymenophorean *Ichthyophthirius*, a parasite of freshwater fishes (see **Chapter 15**) (from Hiller & Bardele, 1988)

1993), Australia (Post, Borowitzka, Borowitzka, Mackay, & Moulton, 1983), and Chad (Pourriot, Iltis, & Leveque-Duwat, 1967). Prostomes have also been recorded in the interstitia of marine coastal habitats of Black and Caspian Seas (Agamaliev, 1971; Kovaleva & Golemansky, 1979) and in leaf litter associated with coastal mangroves (Dorothy, Satyanarayana, Kalavati, Raman, & Dehairs, 2003). They are conspicuous in the marine plankton off North America (Dolan, 1991; Martin & Montagnes, 1993; Montagnes, Lynn, Roff, & Taylor, 1988) and Europe (Dale & Dahl, 1987b; Edwards & Burkill, 1995; Leakey, Burkill, & Sleigh, 1993) where they show extreme tolerances to fluctuations in salinity (Ax & Ax, 1960). Prostomes have been recorded from rivers (Foissner, 1997b) and lakes worldwide, in South America (Barbieri & Orlandi, 1989), North America (Beaver & Crisman, 1989a; Lynn & Munawar, 1999), Europe (Carrias, Amblard, & Bourdier, 1994; Müller, 1991; Pfister, Auer, & Arndt, 2002; Stensdotter-Blomberg, 1998; Stenson, 1984; Zingel, Huitu, Makela, & Arvola, 2002), Israel (Hadas & Berman, 1998), Australia (Esteban, Finlay, Olmo, & Tyler, 2000; James, Burns, & Forsyth, 1995), and in Lake Baikal (Obolkina, 1995, 2006). In deeper lakes, prostomes are common in the epilimnion and metalimnion (Zingel & Ott, 2000), and even in the hypolimnion (Guhl, Finlay, & Schink, 1996). Other freshwater habitats include rice fields (Madoni, 1986, 1996) and small ponds (Finlay et al., 1988).

Since prostomes are found in soils, these ciliates must encyst. Cysts can be extremely cryptic, even in species found in freshwater (Hiller & Bardele, 1988; Leipe, 1989; de Puytorac, & Savoie, 1968). The cyst wall of the tomonts of *Cryptocaryon* is composed of several layers of fibrous material, presumably derived from the many cortical mucocysts (Matthews, Matthews, & Burgess, 1993). Tannreuther (1926) noted that the resting cysts of *Holophrya* (formerly *Prorodon*) have thicker cyst walls than the temporary cysts (Fig. 13.1).

Abundances vary due to a number of expected factors, such as food abundance (Finlay et al., 1988), but may also be influenced by ultraviolet irradiance (Barcelo & Calkins, 1980), pH (Weisse & Stadler, 2006), and temperature (Montagnes & Weisse, 2000; Weisse, 2006). Prostomes often tend to numerically dominate the plankton, especially in the early summer months in temperate lakes (Müller, 1989). As noted above, prostomes can reach exceedingly high densities. Nevertheless, typical peak abundances for *Urotricha* can be up to 2,500 cells l^{-1} in freshwater (Barbieri & Orlandi, 1989) and up to 4,000 cells l^{-1} in marine coastal habitats (Leakey et al., 1993). *Coleps* species can range from 1,000–8,000 cells l^{-1} (Barbieri & Orlandi). James et al. (1995) recorded a large *Holophrya* (formerly *Prorodon*) species contributing over 70% to the biomass during the fall overturn. However, prostomes, being typically small bodied, contribute less to the average biomass although they are often widespread and numerically dominant in lakes (Beaver & Crisman, 1989a).

Prostomes are encounter feeding or raptorial ciliates. They have been reported to feed on a variety of prey from bacteria to filamentous green algae. Bacteria feeders include planktonic and benthic species in the genera *Coleps* (Madoni, Berman, Hadas, & Pinkas, 1990; Šimek, Macek, Pernthaler, Straškrabová, & Psenner, 1996), *Urotricha* (Šimek et al., 1996), and *Holophrya* (formerly *Prorodon*) (Epstein & Shiaris, 1992). These ciliates graze from 60–160 bacteria h^{-1} and planktonic species may survive and grow exclusively on a bacterial diet (Epstein & Shiaris, 1992; Šimek et al., 1996). The majority of studies on prostome feeding have demonstrated that a variety of "algal" species support good growth. Algae consumed by both marine and freshwater prostomes of the genera *Balanion*, *Bursellopsis*, *Coleps*, *Holophrya* (formerly *Prorodon*), *Tiarina*, and *Urotricha* include the chlorophytes *Chlamydomonas* and *Chlorella* (Madoni et al., 1990), the cryptophytes *Rhodomonas* and *Cryptomonas* (Pedrós-Alió, Massana, & Latasa, 1995a; Klaveness, 1984; Müller, 1991; Weisse & Montagnes, 1998), the chrysophytes *Mallomonas* and *Synura* (Skogstad, Granskog, & Klaveness, 1987; Wilbert, 1986), and the dinoflagellates *Akashiwa*, *Heterocapsa*, *Scrippsiella*, and *Dinophysis* (Hansen, 1991; Jeong, Yoon, Kim, Yoo, & Seong, 2002; Nakamura & Hirata, 2006; Stoecker, Davis, & Provan, 1983). The marine prostome *Tiarina* has been implicated in controlling growth of some of these red-tide and toxic algae (Jeong et al., 2002). *Holophrya* (formerly *Prorodon*) species can also ingest filamentous green algae, like *Spirogyra* (Leipe, 1989), and other ciliates (Canella, 1951). Finally, some species of *Holophrya* (formerly *Prorodon*) are histophages, which can be cultured in the laboratory on tissues from a variety of organisms, including the annelid *Tubifex*, fish, and mammals (Czapik, 1965; Hiller & Bardele, 1988; de Puytorac & Savoie, 1968), which they likely encounter as moribund or dead in the natural environment.

Prostomes are rarely implicated in the deaths of animals: Székely and Bereczky (1992) reported an unusual case of *Coleps* killing the fry of three species of aquarium fishes. A real killer of marine fishes, especially in tropical and subtropical environments, is *Cryptocaryon irritans*, a ciliate ectoparasite of fishes (Dickerson & Dawe, 1995). Reports of *Cryptocaryon* are increasing as the marine aquaculture industry expands. Outbreaks and infections have been reported on a variety of fishes, including flounder *Paralichthys olivaceus* in Japan (Kaige & Miyazaki, 1985), sea perch *Lates calcarifer*, and grouper *Epinephelus coioides* (Yambot, Song, & Sung, 2003). With a life cycle convergent on that of *Ichthyophthirius multifiliis*, *Cryptocaryon* was placed in the Family Ichthyophthiriidae (Corliss, 1979). However, light microscopical (Diggles, 1997), ultrastructural (Colorni & Diamant, 1993), and molecular (Diggles & Adlard, 1995; Wright & Colorni, 2002) data now assign it unequivocally to the Class PROSTOMATEA. Trophonts of *C. irritans* tend to leave the host fish during the dark cycle of a photoperiod (Burgess & Matthews, 1994b). Trophonts and tomonts do not develop normally at 34°C and anoxic conditions killed tomonts (Yoshinaga, 2001). However, tomonts could survive hypoxia and were stimulated to develop and excyst when transferred to oxic conditions. Yoshinaga (2001) argued that this explains why outbreaks of cryptocaryoniasis in fish mariculture cages are often associated with the autumnal turnover of the water column. Several species of fish can acquire protective immunity to cryptocaryoniasis through controlled infections (Burgess & Matthews, 1995a; Yoshinaga & Nakazoe, 1997). It is interesting to note that this immunity does not protect these fish against ichthyophthiriasis (Burgess & Matthews, 1995a), supporting the distant phylogenetic relationship between the two species. Fish species vary in their susceptibility to cryptocaryoniasis. Susceptible fish can be successfully treated only with a rigorous regimen of repeated hyposaline baths (Rigos, Pavlidis, & Divanach, 2001) or by dietary supplements with the medium-chain fatty acid, caprylic acid (Hirazawa, Oshima, Hara, Mitsuboshi, & Hata, 2001). Based on rDNA sequence analysis, Yambot et al. (2003) provided evidence of marine and low-salinity variants of *C. irritans*. This suggests both that this may be a cryptic species complex and that treatments may need to be adapted to the genotype of the invading strain.

The distributions of prostomes and their prey are spatially correlated in both vertical and horizontal dimensions in aquatic habitats (Pedrós-Alió et al., 1995; Stoecker, Davis, & Anderson, 1984). In situ growth rates are typically about one doubling per day (Macek, Šimek, Pernthaler, Vyhnálek, & Psenner, 1996). Growth can be strongly influenced by the strains of prey organism: a strain of *Coleps* grew much better on some isolates of *Cryptomonas* than on others (Klaveness, 1984), while *Balanion* preferred cryptophytes as prey (Müller & Schlegel, 1999). These preferences are presumably due to chemical signals. *Coleps* will not eat living ciliates, such as *Blepharisma* and *Spirostomum*, but will readily ingest them when they are dead (Seravin & Orlovskaja, 1977).

While there are obvious preferences for prey, growth may equally be influenced by the strain of prostome predator doing the feeding. Different strains and species of *Urotricha* and *Balanion* grew significantly differently on the same strain of *Cryptomonas* (Weisse & Montagnes, 1998; Weisse et al., 2001). This niche separation extended also to physiological responses to temperature as different *Urotricha* species and strains showed different temperature optima for growth. At high temperatures, *Urotricha farcta* grew fastest while *Urotricha castalia* grew well at low temperatures (Weisse et al., 2001). *Balanion* species appear not to tolerate temperatures much in excess of 20°C (Müller, 1991; Weisse et al., 2001).

Mixotrophic prostomes have been observed to be dominant in the metalimnion of stratified temperate lakes (Zingel & Ott, 2000) and near the oxic-anoxic boundary in smaller bodies of water (Finlay & Maberly, 2000) where there is sufficient light for net photosynthesis. The symbionts can be a facultative association between a *Chlorella* and *Coleps* or *Holophrya* (formerly *Prorodon*) (Christopher & Patterson, 1983) or the *Holophrya* may actually sequester chloroplasts of prey (Blackbourn, Taylor, & Blackbourn, 1973). As with other mixotrophs, symbiont-bearing *Coleps* had growth rates higher than aposymbiotic forms when food resources were scarce (Stabell, Andersen, & Klaveness, 2002). Symbiotic bacteria have been reported in *Urotricha ovata* (de Puytorac & Grain, 1972).

As conspicuous components of microbial food webs, it is not surprising that prostomes are themselves eaten by other organisms. A variety of predatory ciliates, such as *Dileptus*, *Lagynophrya*, *Lacrymaria*, *Didinium*, and *Favella* (Jürgens, Skibbe, & Jeppesen, 1999; Seravin & Orlovskaja, 1977; Stoecker & Evans, 1985), prey upon prostomes, such as *Coleps* and *Balanion*. Even though *Tiarina* may feed upon the autotrophic dinoflagellate *Dinophysis*, when the dinoflagellate is heterotrophic, the table is literally turned and predator becomes prey (Hansen, 1991)! Prostomes are also eaten by the rotifers *Brachionus* (Mohr, Gerten, & Adrian, 2002) and *Keratella* (Jack & Gilbert, 1993). Microcrustacean zooplankton are also predators of prostomes: in fresh water by the cladocerans *Daphnia* and *Bosmina* (Jack & Gilbert, 1993; Wickham & Gilbert, 1991) and the copepod *Cyclops* (Wickham, 1995); and in marine environments by the copepod *Acartia* (Olsson, Granéli, Carlsson, & Abreu, 1992). Even the jellyfish *Aurelia* preferred the prostome *Urotricha* over dinoflagellates (Stoecker, Michaels, & Davis, 1987b). *Balanion* is able to avoid predation by marine copepods by escape jumping in response to fluid mechanical signals generated by the predator's feeding apparatus (Jakobsen, 2001), and some *Urotricha* species may have a similar ability (Tamar, 1979).

Finally, Hiller and Bardele (1988) make the important point that the stages in the life cycle of some prostomes, such as the histophagous *Holophrya* (formerly *Prorodon*), can be so different in size and shape that one might consider them to be different species. Thus, careful and continuous observation is necessary to confirm that this is in fact not the case.

13.3 Somatic Structures

Prostomes are generally ovoid to ovoid elongate in shape, but they can range from nanociliates (i.e., <20 µm in body length), as some *Urotricha* species do, to several hundreds of micrometers in length, as some *Holophrya* (formerly *Prorodon*) species do (Fig. 13.2). They are typically not highly contractile, although their bodies are flexible and, in histophagous and parasitic forms, able to penetrate the tissues of hosts. A few species, such as *Vasicola*, can form a lorica (Dragesco et al., 1974). Since the species diversity of this class is not great, there have also been relatively few studies of the ultrastructure on this group. Species in the genera *Balanion*, *Bursellopsis*, *Coleps*, *Cryptocaryon*, *Holophrya*

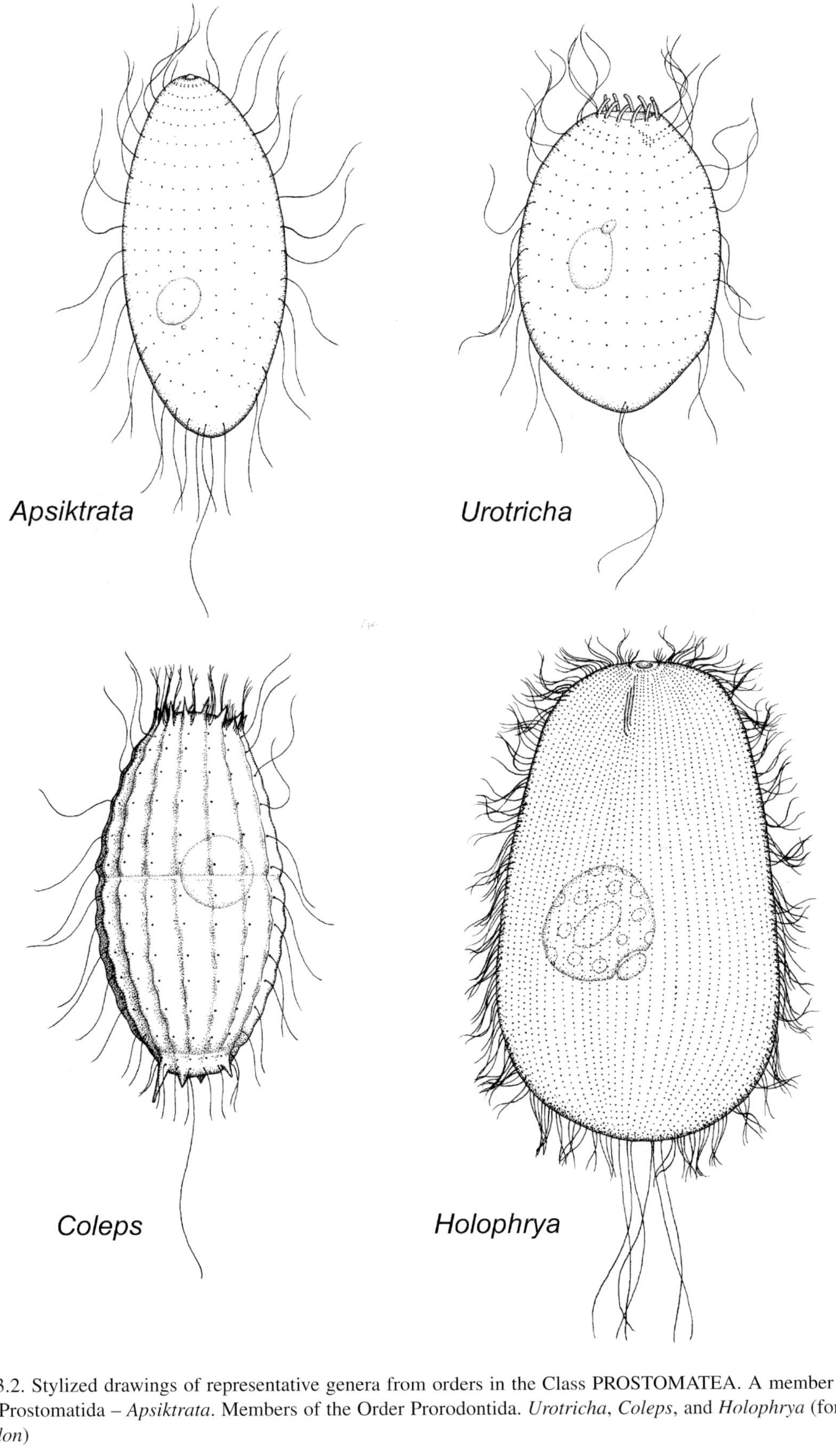

FIG. 13.2. Stylized drawings of representative genera from orders in the Class PROSTOMATEA. A member of the Order Prostomatida – *Apsiktrata*. Members of the Order Prorodontida. *Urotricha*, *Coleps*, and *Holophrya* (formerly *Prorodon*)

(formerly *Prorodon*), *Placus*, and *Urotricha* have been investigated.

The plasma membrane of prostomes rarely exhibits a glycocalyx, although a very thin and inconsistent layer can be observed in some micrographs of *Holophrya* (formerly *Prorodon*) (Hiller, 1993a). The underlying alveoli are often swollen and can be occupied partially with smaller particles in *Holophrya* (formerly *Prorodon*) (Hiller, 1993a), bacteria in *Placus* (Grain et al., 1979), completely with dense material in both *Holophrya* (de Puytorac, 1964) and *Cryptocaryon* (Colorni & Diamant, 1993; Matthews et al., 1993), and with calcium carbonate skeletal plates in *Coleps* (Fauré-Fremiet, André, & Ganier, 1968b). The fine structure of these skeletal plates differs in morphology among *Coleps* species (Huttenlauch, 1985). The epiplasm is conspicuous in larger forms with a fine layer observed in *Holophrya* (formerly *Prorodon*) (Hiller, 1993a) and a thicker layer recorded in *Cryptocaryon* (Colorni & Diamant, 1993; Matthews et al., 1993) and *Placus* (Grain et al., 1979). Longitudinal microtubules have been observed to lie above the epiplasm in the cortical ridges of *Holophrya* (formerly *Prorodon*) species, a feature demonstrated also in oligohymenophoreans (Hiller, 1993a).

The prostome somatic monokinetid can now be characterized definitively as bearing a slightly divergent postciliary ribbon at triplet 9, an anteriorly directed and typically short kinetodesmal fibril at triplets 5, 6, 7, and a radially oriented transverse ribbon at triplet 4 or 5 (Hiller, 1993a). The kinetodesmal fibril may be bifurcated at its beginning in some species (Fig. 13.3) (Hiller, 1993a; Lynn, 1991). The transverse ribbon microtubules typically extend upward and laterally into the cortical ridge and the postciliary microtubular ribbons may not overlap (Fig. 13.4). However, in *Placus*, the transverse ribbons extend downward and laterally

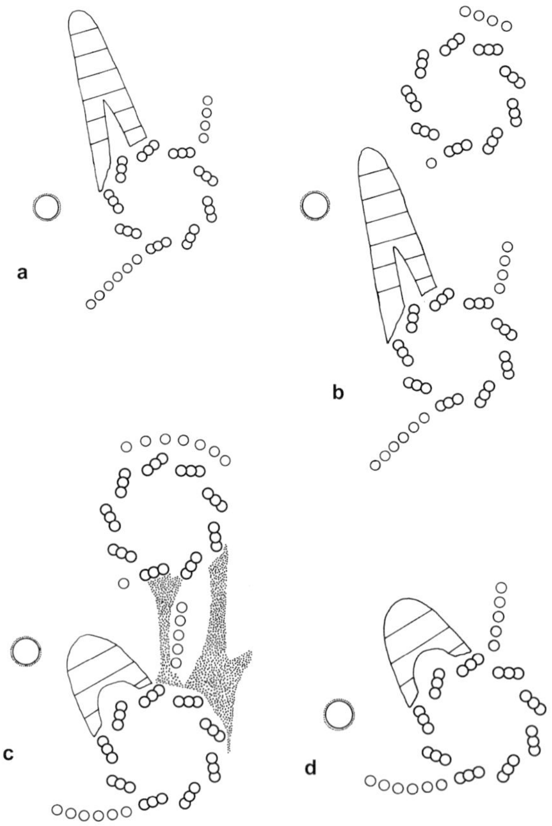

Fig. 13.3. Schematics of the somatic kinetids of the Class PROSTOMATEA. (**a**) Monokinetid of *Coleps*. (**b**) Dikinetid of *Coleps*. (**c**) Dikinetid of *Bursellopsis*. (**d**) Monokinetid of *Bursellopsis* (from Lynn, 1981, 1991)

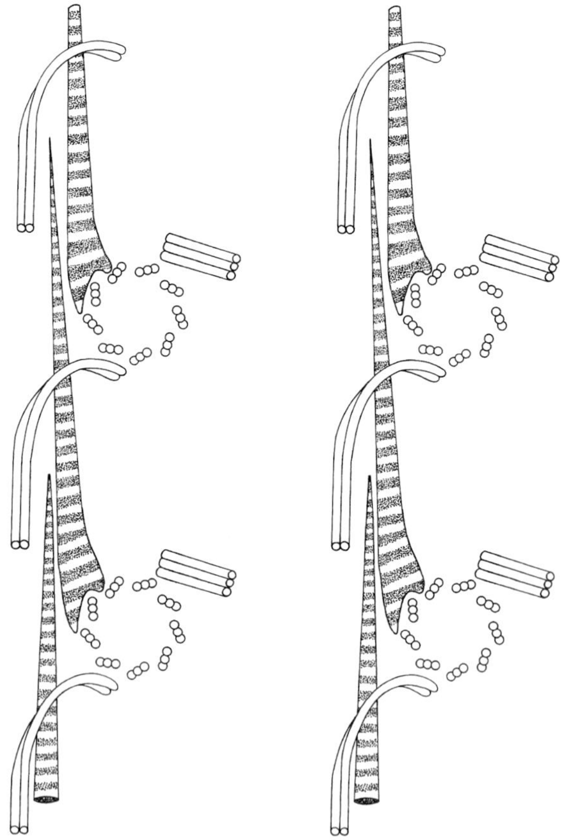

Fig. 13.4. Somatic cortex of a typical prostome interpreted based on the somatic cortex of *Bursellopsis*. (Modified after Hiller, 1993a.)

due to the position of the somatic kinetosomes near the tops of the cortical ridges in this genus (Grain et al., 1979). The kinetids at the anterior of somatic kineties may be dikinetids whose posterior kinetosome bears the same fibrillar associates in similar orientation to those of the somatic monokinetids. However, the anterior kinetosome bears a single postciliary microtubule and a tangentially oriented transverse ribbon adjacent to triplets 3, 4, 5 (Fig. 13.3) (Hiller, 1993a; Lynn, 1991). Hiller (1993a) has suggested a transformation series deriving prostome and colpodean dikinetids from a common ancestor with this somatic dikinetid structure. Microtubules have been observed underlying the kinetosomes of prostomes, originating from dense material at the kinetosome base and extending anteriorly in *H olophrya* (formerly *Prorodon*), the so-called prekinetosomal microtubules (Hiller, 1993a). Their association with the kinetosome has not been demonstrated in *Coleps* (Lynn, 1985). Rudberg and Sand (2000) have provided the only study of the electrophysiology of prostomes. They observed a unique biphasic membrane potential that correlates with the alternating periods of linear and circular swimming of *Coleps*.

Mucocysts, typically elongate ovoid, have been observed in the somatic cortex of prostomes (Hiller, 1991, 1993a; Huttenlauch, 1987; Matthews et al., 1993). Colorni and Diamant (1993) observed a bizarre phenomenon in *Cryptocaryon*: vesicles within the alveoli appear to fuse with the plasmalemma, presumably releasing material to the outside and leaving the cell's surface covered with numerous "pores". Mitochondria, often present in the cell cortex, are the typical forms with tubular cristae.

As with other ciliates, prostomes have mineral elements in the cytoplasm. Like those in the alveoli of *Coleps* species, the mineral concretions in the endoplasm have been demonstrated to be calcium carbonate (André & Fauré-Fremiet, 1967; Rieder, Ott, Pfundstein, & Schoch, 1982).

There is typically a contractile vacuole in the posterior of the cell, although some larger forms can have about 70 independently functioning vacuoles (Leipe, 1989). The contractile vacuole system exits to the exterior through a pore supported by helically disposed microtubules (Rodrigues de Santa Rosa, 1976). In some *Holophrya* (formerly *Prorodon*) species, there is a set of collecting canals extending anteriorly beneath the somatic cortex from the posterior contractile vacuole. These canals are associated with prekinetosomal microtubules, which may provide structural support (Hiller, 1993a).

13.4 Oral Structures

As the class name suggests, the prostomes are ciliates whose oral re gion is at the anterior end of the body, although it may be slightly subapical (Fig. 13.2). The oral opening is typically permanent and flanked by oral dikinetids that border one side of the oral region in *Plagiocampa* (Foissner, 1978) or surround the entire cytopharyngeal complex (Hiller, 1991, 1993b; Huttenlauch, 1987). Brosse polykinetids are situated on the opposite side to the oral dikinetids in *Plagiocampa* (Foissner, 1978), almost enclosed by them in some *Coleps* species (Wilbert & Schmall, 1976), or completely enclosed by them in *Balanion* species. The brosse kinetids of *Balanion* may be reduced to two kinetosomes with reduced ciliation (Bardele, 1999; Jakobsen & Montagnes, 1999). The brosse in other genera can be more extensive, ranging from one up to five rows of kinetosomes embedded in the somatic infraciliature. In *Holophrya* (formerly *Prorodon*) species, Hiller and Bardele (1988) have identified two major patterns: the *aklitoloph* pattern in which the somatic kineties do not terminate laterally on the brosse rows and the *enklitoloph* pattern in which the somatic kineties terminate on the right of the brosse rows (i.e., *dexiotrop*), on the left of the brosse rows (i.e., *aristerotrop*), and on both sides of the brosse (i.e., *syntrop*). They speculated that these patterns might indicate four genera of prorodontids. A brosse has also been identified in *Cryptocaryon* (Diggles, 1997).

Early ultrastructural research on prostomes suggested that it was transverse microtubules arising from the oral dikinetids that surrounded the oral region (de Puytorac & Grain, 1972). Huttenlauch and Bardele (1987) demonstrated that these microtubules were in fact postciliary microtubules and that the oral dikinetids rotated into place during stomatogenesis to give this unusual orientation (**see below Division and Morphogenesis**). Both anterior and posterior kinetosomes of the oral dikinetids bear a postciliary ribbon and may also have one or two transverse microtubules (Hiller,

1993b). The postciliary microtubules appear to be used in several different ways among the prostomes. In *Holophrya* (formerly *Prorodon*) species, the oral postciliaries may extend anteriorly supporting cortical ridges that cover the walls of the oral opening (Hiller, 1993b). In *Balanion*, *Coleps*, and *Plagiocampa*, oral palps, bearing a toxicyst, are placed internal to the oral dikinetids (e.g., Bardele, 1999; Fauré-Fremiet & André, 1965b). These palps are supported by microtubules, presumably derived from the postciliary ribbon of the anterior kinetosome of the oral dikinetids (Bardele, 1999). In a third group of species (e.g., *Balanion*, *Bursellopsis*, *Urotricha*), the postciliary microtubules of the posterior kinetosome of the oral dikinetids extend around the perimeter of the oral opening in a counter-clockwise direction as viewed from outside the cell. They may overlap as many as 12 other ribbons and appear to be joined together by intermicrotubule bridges, which may permit sliding to dilate or close the oral opening (Hiller, 1991, 1993b). The oral dikinetids typically sit atop two nematodesmata that are triangular to trapezoidal in cross-section. Each nematodesma extends into the cytoplasm and is joined by dense material deeper in the cytopharyngeal basket to the nematodesma arising from a neighboring oral dikinetid (Hiller, 1993b; Lynn, 1985). Since postciliary ribbons support the oral cavity and line the cytopharynx, the cytopharyngeal apparatus of prostomes must now be designated as a cyrtos – the rhabdos is now only found in the Class LITOSTOMATEA.

The oral ridges on the walls of the oral cavity have been observed in all prostomes, including *Cryptocaryon* (Colorni & Diamant, 1993; Diggles, 1997). These ridges are covered by cortical alveoli down to the level of the cytostome (Hiller, 1993b; Hiller & Bardele, 1988; Huttenlauch, 1987; de Puytorac, 1964). There are typically two sets of microtubules in these ridges, one set composed of more microtubules than the other (Hiller, 1993b; Lynn, 1985). However, these can be reduced to 3 + 2 in *Bursellopsis*, very reminiscent of the oral ridges in oligohymenophoreans (Hiller, 1991). These microtubules are presumably derived from the postciliary ribbons of the oral dikinetids (Huttenlauch & Bardele, 1987), and ultimately extend to line the cytopharynx where they may function to bring new food vacuole membrane to this region (Hiller, 1993b; Rodrigues de Santa Rosa, 1976).

The brosse or brush is considered here as an oral structure. As noted above, its kinetosomes can be within the circumoral dikinetids, opposite a paroral set of oral dikinetids or outside the oral perimeter entirely. In both colepids and prorodontids, the brosse is composed of rows of dikinetids, whose axes are typically oriented perpendicularly to the longitudinal axis of the brosse rows. The anterior (or left) kinetosome bears a typically shorter, clavate-like cilium and may or may not bear a tangential transverse ribbon while the posterior (or right) kinetosome is non-ciliated and bears a postciliary ribbon (Hiller, 1991, 1993b).

In species without oral palps, toxicysts are concentrated in the oral region, where they are presumably used to soften prey, like the filamentous alga *Spirogyra* (Leipe, 1989). Similar to the long toxicysts of *Dileptus*, the toxicysts of *Holophrya* (formerly *Prorodon*) demonstrate a combination of tubule evagination and telescoping (Hausmann, 1978). *Placus* exhibits an unusual fosette or pit at the posterior end of its single brosse row. The brosse row extends into this pit in which there is a dense aggregation of toxicysts (Fryd-Versavel, Iftode, & Dragesco, 1976; Grain et al., 1979).

13.5 Division and Morphogenesis

Prostomes typically divide while swimming freely. However, histophagous and parasitic prorodontids, like *Holophrya* and *Cryptocaryon*, can form reproduction or division cysts (Fig. 13.1) (Czapik, 1965; Diggles, 1997; Hiller & Bardele, 1988). Even microphagous forms can form temporary division cysts (Tannreuther, 1926). Hiller (1992) has reviewed the merotelokinetal stomatogenesis of prostomes and noted that since proliferation of kinetosomes occurs in a localized region of the cortex, this region should be denominated ventral. The oral region becomes prostomatous or apical following cytokinesis by allometric growth of somatic kineties that "push" the ventral oral region anteriorly (Fig. 13.5). The brush or brosse, therefore, cannot be considered a dorsal structure, although Hiller (1992) prefers not to consider it ventral.

As viewed from outside the cell, the circumoral ciliature forms by a clockwise migration of the circumoral dikinetids to form a partially or completely closed circle. In *Coleps*, the circumoral arises from

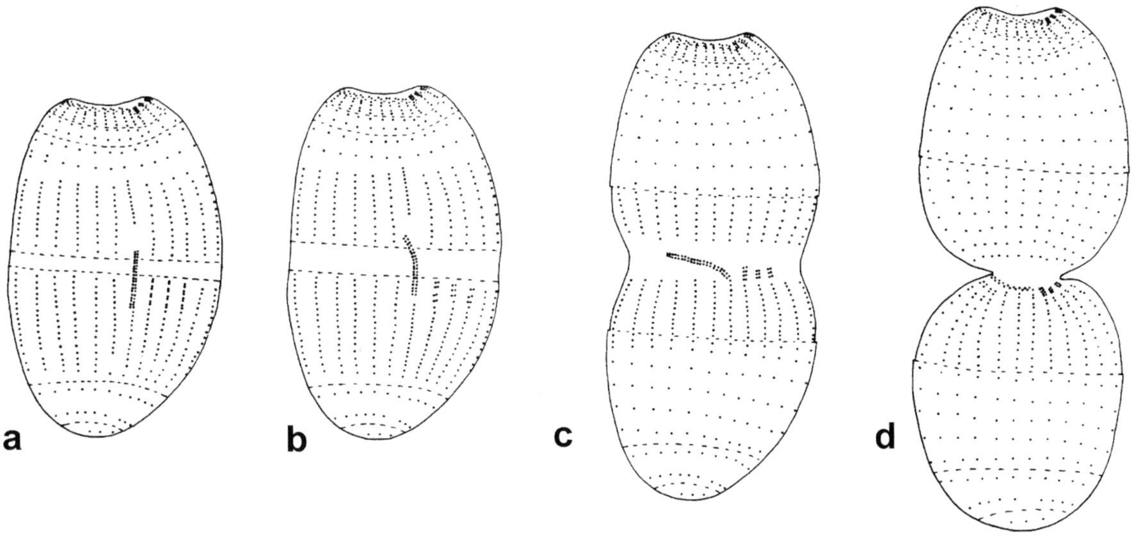

FIG. 13.5. Division morphogenesis of the prorodontid *Coleps*. The circumoral dikinetids and brosse kinetids begin to differentiate at the equatorial ends of four somatic kineties (**a**, **b**). As the brosse kinetids differentiate as small polykinetids, the circumoral dikinetids then begin a clockwise migration into the fission furrow to encircle the putative anterior end of the opisthe (**b–d**). (from Huttenlauch & Bardele, 1987.)

one somatic kinety (Fig. 13.5; Huttenlauch & Bardele, 1987). The number of kineties involved in circumoral dikinetid formation probably increases with increasing cell size: only two in *Urotricha* (Muñoz, Téllez, & Fernández-Galiano, 1989), three to five in *Bursellopsis* (Hiller, 1992), and up to six in *Holophrya* (formerly *Prorodon*) (Hiller, 1993b). As noted above (see **Oral Structures**), rotation of these circumoral dikinetids brings the postciliary ribbons into a position to support the oral opening and cytopharynx (Hiller, 1993b; Huttenlauch & Bardele, 1987). Using scanning electron microscopy, Bardele (1999) has described the formation of the circumoral in *Balanion*, in which a field, possibly of two rows of dikinetids, separates into an anterior and posterior portion. The space in the center becomes the cytostome as these two rows round up to form the circumoral. Bardele (1999) noted that this is a unique form of stomatogenesis in the prostomes. It should be studied by protargol staining and transmission electron microscopy to confirm his interpretation and resolve how the circumoral dikinetids rotate to form the circumoral supports.

The brush kinetids derive from kineties to the left of those involved in formation of the circumoral dikinetids (Hiller, 1992). In *Coleps*, the three brosse polykinetids originate one from the anterior end of each of three somatic kineties (Fig. 13.5) (Huttenlauch & Bardele, 1987). In the other prorodontids, variable numbers of kineties are involved and each kinety is involved in the formation of all brosse rows: two somatic kineties in *Urotricha*, up to eight in *Bursellopsis*, and up to 13 in *Holophrya* (formerly *Prorodon*) (Hiller, 1992, 1993b). This process occurs in several "waves", each of which includes proliferation of a kinetofragment, rotation of that kinetofragment, and then fusion of the kinetofragments to form the mature brosse row. Allometric growth of somatic kineties following cytokinesis "pushes" the brosse into its morphostatic position (Fig. 13.5). There is some dedifferentiation of the proter's oral nematodesmata during division and then regrowth as the opisthe's nematodesmata grow.

13.6 Nuclei, Sexuality and Life Cycle

Prostomes typically have a single, ellipsoid to elongate macronucleus of the homomerous type (Fig. 13.2) (Raikov, 1982). *Cryptocaryon* is a notable exception, having four linked ellipsoid to spheroid macronuclear segments in the phoront and young trophont stages. These segments fuse into one elongated and twisted macronucleus as cell growth proceeds to the protomont stage (Colorni & Diamant, 1993). The heterochromatin in the macronucleus is distributed throughout the macronucleus, but can be aggregated into one central condensation body or bodies in some *Holophrya* (formerly *Prorodon*) species (Hiller, 1993a; de Puytorac & Savoie, 1968). There is no report of the role of microtubules in macronuclear division in prostomes. We presume that intramacronuclear microtubules are involved.

There is typically a single micronucleus closely associated with the macronucleus. The micronucleus can be ellipsoid (e.g., *Coleps*, Rodrigues de Santa Rosa, 1976) or elongated and lenticular (Hiller, 1993a; Sola, Guinea, Longás, & Fernández-Galiano, 1990b). There can be up to five micronuclei in the theront and phoront stages of *Cryptocaryon* (Colorni & Diamant, 1993).

There has been no genetic research on prostomes to demonstrate their kind of mating type system, which undoubtedly exists as they have been reported to conjugate. Starvation appears to stimulate conjugation in *Coleps* (Serrano, Martín-Gonzàlez, & Fernández-Galiano, 1985) while *Holophrya* (formerly *Prorodon*) has been reported to conjugate immediately on emerging from resting cysts (Tannreuther, 1926). Prostomes typically first encounter each other and associate using the oral and anterior ciliature (Serrano et al., 1985; Sola et al., 1990b). Cell fusion takes place near the oral region, but in *Coleps* the ciliates can continue to feed (Serrano et al., 1985). The micronucleus typically undergoes two meiotic divisions during which the micronucleus may exhibit the "crescent stage" (Serrano et al., 1985). Three of these four products degenerate and the remaining nucleus undergoes mitosis to produce the stationary and migratory gametic nuclei (Raikov, 1972; Serrano et al., 1985). These gametic nuclei may remain connected by the telophase spindle as nuclear exchange proceeds (Serrano et al., 1985; Tannreuther, 1926). Raikov (1972) characterized the postconjugation process as involving a single division of the synkaryon to produce a new micronucleus and macronucleus. This was confirmed in *Coleps* by Serrano et al. (1985). However, Sola et al. (1990b) reported two or three postconjugation divisions in *Lagynus*.

Population genetic research on prostomes has been restricted to isolates of *Cryptocaryon*. Variation among isolates of *C. irritans* has been of interest to parasitologists since it may explain the outbreaks and incidence of cryptocaryoniasis in various parts of the world's oceans. Isolates of *C. irritans* show relatively low host specificity (Burgess & Matthews, 1995b; Diggles & Lester, 1996; Rigos et al., 2001). Rather, cryptocaryoniasis typically develops in susceptible fish at temperatures above 19°C. However, isolates have been reported that can grow faster at lower or at higher temperatures (Diggles & Adlard, 1997; Jee, Kim, Park, & Kim, 2000). Diggles and Adlard (1997) and Yambot et al. (2003) examined the genetic variation in the rDNA gene region of isolates of *C. irritans*, and recognized a variety of genetic strains within this species, some global in distribution, and perhaps two lineages, one adapted to marine and the other to low-salinity habitats. Are these in fact cryptic species? Wright (1999) noted that the alignment of Diggles and Adlard (1997) wrongly identified the 3'-end of the ITS-1 region. Nevertheless, there appears to be significant genetic variation among these isolates, an issue worthy of continued investigation for this commercially important parasite of fishes.

13.7 Other Features

There have been a few reports of the impacts of anthropogenic pollution on prostome ciliates. Dale (1988) reported that *Tiarina* and other aloricate heterotrophic ciliates were more susceptible to oil pollution in the marine environment than tintinnids. Rivera et al. (1988) reported that *Urotricha farcta* was the most abundant prostome in rotating biological contactor treatment plants in Mexico, along with five species of oligohymenophoreans and *Didinium nasutum*. Finally, Packroff (2000) noted that the community structure of acidic mining lakes in Germany appeared to favor prostomes over haptorians, scuticociliates, and oligotrichs. This, however, does not appear to be a global phenomenon as Beaver and Crisman (1981) noted oligotrichs dominated in acidic softwater lakes in Florida.

Chapter 14
Subphylum 2. INTRAMACRONUCLEATA: Class 8. PLAGIOPYLEA – A True Riboclass of Uncommon Companions

Abstract This class is truly a riboclass because it assembles three groups of ciliates that were never suspected of being phylogenetically related, and yet there is an extremely strong signal from the small subunit rRNA gene sequences that they are. The now "classic" plagiopyleans, the sonderiids and plagiopylids, are now united with the trimyemids and tentatively also the odontostomatids. These ciliates are all considered anaerobic to microaerophilic, and are often found in sapropelic habitats. Several species have conspicuous assemblages of hydrogenosomes and methanogens, which presumably enable these ciliates to survive in these anoxic habitats. There are really no unifying morphological features. The somatic kinetids are monokinetids in the sonderiids, plagiopylids, and trimyemids and highly unusual dikinetids in the odontostomatids. Oral structures in the plagiopylids and sonderiids are modified extensions of somatic kineties; trimyemids apparently have a kind of "circumoral" ciliature; and odontostomatids have several small oral polykinetids. Stomatogenesis is apparently holotelokinetal in all but the odontostomatids, and we are ignorant of how this latter group divides. There remains much to be learned about their life cycle, sexual processes, and nuclear features.

Keywords *Epalxella, Plagiopyla, Trimyema*

The Class PLAGIOPYLEA, like the Class ARMOPHOREA, is in essence a "riboclass" – a group whose monophyly is based only on the evidence of sequences of the small subunit (SSU) rRNA gene. Small and Lynn (1985) established the subclass Plagiopylia, including the sonderiids and plagiopylids, and transferred these ciliates to the Class OLIGOHYMENOPHOREA primarily on the basis of the ultrastructure of the somatic kinetids. De Puytorac et al. (1993) elevated the group to class status, a move supported by Lynn and Small (1997). Sequencing of the SSUrRNA genes of several species of *Trimyema* and several plagiopylid genera has now demonstrated these to be sister taxa (Baumgartner, Stetter, & Foissner, 2002; Embley & Finlay, 1994; Lynn & Strüder-Kypke, 2002). Stoeck, Foissner, and Lynn (2007) have evidence that the SSUrRNA gene sequence of the odontostomatid *Epalxella* clusters with strong support with these plagiopyleans, and so we have made the risky decision to assign the odontostomatids to this class as the second order, beside the Order Plagiopylida.

The plagiopyleans are anaerobic or microaerophilic ciliates that range in size from about 15 μm in length but rarely exceed 200 μm in length. They are typically ovoid or elongate in body shape and not contractile or flexible. In the larger genera, such as *Lechriopyla* and *Sonderia*, the ciliation is holotrichous. In smaller forms, such as trimyemids and odontostomatids, the number of somatic kinetids is reduced and much of the body surface is non-ciliated. In trimyemids, the kineties even appear to spiral, but both light and electron microscopic study of *Trimyema* refute this interpretation, and confirm the interpretation proposed by Fauré-Fremiet (1962b): the kinetosomes in each longitudinally oriented kinety are distributed in such a manner that they appear to be spiralling (Baumgartner et al., 2002; Detcheva, de Puytorac,

& Grolière, 1981). Plagiopyleans are typically found in anaerobic freshwater and marine habitats, ranging from hydrothermal vents to anoxic marine sediments to the intestines of sea urchins and to sewage treatment plants.

A notable feature of these ciliates is the presence of hydrogenosome-methanogen assemblages in their cytoplasm in which the methanogens are typically sandwiched between hydrogenosomes forming groups of up to a dozen units. These assemblages have been observed in representatives of the order Plagiopylida – in *Sonderia* (Fenchel, Perry, & Thane, 1977), *Plagiopyla* (Berger & Lynn, 1992), and *Trimyema* (Detcheva et al., 1981). However, the "sandwich" pattern can depend upon the particular species of methanogen involved: *Methanocorpusculum parvum* is polymorphic – ovoid when free in the cytoplasm of *Trimyema* and profusely dentate when associated with its hydrogenosomes (Finlay, Embley, & Fenchel, 1993).

Biochemical analyses supported the conclusion that the ciliate organelles are not mitochondria, but rather are hydrogenosomes: they exhibit hydrogenase activity (Zwart et al., 1988) and do not demonstrate cytochromes, cytochrome oxidase, and catalase activities (Goosen, Wagener, & Stumm, 1990). There are now techniques for culturing both *Trimyema* (Wagener & Pfennig, 1987) and *Plagiopyla* (Fenchel & Finlay, 1991c), using bacteria isolated from the environment or cultured bacterial strains. Furthermore, electromigration has been used to concentrate these ciliates from environmental sludge samples (Wagener, Stumm, & Vogels, 1986) and from mass cultures to enable biochemical research (Broers, Molhuizen, Stumm, & Vogels, 1992).

The name of the class PLAGIOPYLEA is derived from the Greek words, *plagios* meaning oblique and *pylon* meaning gate. This refers to the nature of the oral opening in plagiopylids, which is an oblique slit whose walls are covered by extensions of the somatic kineties. While the somatic kinetid of plagiopylids bears some resemblance to that of the oligohymenophoreans, the odontostomatid dikinetid is quite different (**see below Somatic Structures**). There is no morphological synapomorphy for the class, and so it was designated as one of the "riboclasses" of ciliates by Lynn (2004), since SSUrRNA gene sequences appear to be the only "strong" characters that support the clade.

14.1 Taxonomic Structure

Corliss (1979) retained both the plagiopylids and trimyemids in the Order Trichostomatida, following research by Fauré-Fremiet (1950a, 1962b, 1973) among others. Corliss (1979) did note that this order was a rather heterogeneous taxon with respect to the morphological diversity of the families placed in it. De Puytorac, Grain, Legendre, and Devaux (1984) used a phenetic analysis to place the trimyemids in an order Trimyemida in the subclass Gymnostomia, while de Puytorac, Grain, and Legendre (1994) used parsimony methods to provisionally place *Trimyema* adjacent to the phyllopharyngeans and vestibuliferians, noting that stomatogenetic characters might lead to reconsideration of this result. Berger and Lynn (1984) noted a peculiar microtubular ribbon associated with triplets 2, 3, tentatively unique for the plagiopylids. Partly based on this mistaken interpretation (see **Somatic Structures**), Small and Lynn (1985) established the subclass Plagiopylia within the Class OLIGOHYMENOPHOREA, based on features of the somatic kinetid, which had long anteriorly extending kinetodesmal fibrils and a divergent postciliary ribbon similar to that of other oligohymenophoreans. De Puytorac et al. (1993) elevated the subclass to the Class PLAGIOPYLEA. This position was maintained by de Puytorac (1994c) for the plagiopylids and sonderiids, although he placed the trimyemids in the subclass Prostomatia, based on the assumption that the oral dikinetids were homologues of the prostomatean brosse.

Lynn and Small (1997) also recognized the Class PLAGIOPYLEA, and included in it both plagiopylids, sonderiids, and trimyemids (Lynn & Small, 2002). This was rationalized by similarities in the somatic kinetids with their anteriorly directed kinetodesmal fibrils. To these features, we can now add the typical sandwich-like arrangement of the hydrogenosome-methanogen assemblages in plagiopylids. Finally, SSUrRNA gene sequences clearly confirmed plagiopylids and trimyemids as sister taxa (Baumgartner et al., 2002; Embley & Finlay, 1994; Lynn & Strüder-Kypke, 2002). To these, we can now add the odontostomatids, based on the SSUrRNA gene sequence of the odontostomatid *Epalxella* (Stoeck et al., 2007). We currently recognize two orders: the Order Plagiopylida and the Order Odontostomatida.

The Order Plagiopylida is characterized by the typical sandwich-like arrangement of the hydrogenosome-methanogen assemblages. This order includes the families, Plagiopylidae, Sonderiidae, and Trimyemidae. Genera in the former two families typically have a striated band structure (see **Somatic Structure**), which trimyemids lack.

The Order Odontostomatida was established as a group by Lauterborn (1908), and remains one of the smallest ordinal groups outside those that are monotypic (e.g., Licnophorida, Phacodiniida, Protocruziida). It includes three families: the Epalxellidae, the Mylestomatidae, and the Discomorphellidae. These are typically small ciliates with a prominent dorsal keel and often elongate, spine-like processes. The somatic ciliature is reduced to what are considered vestiges of the perizonal kineties of armophorid ciliates, to which odontostomatids were originally related (Jankowski, 1964b; Tuffrau, 1992; Tuffrau & de Puytorac, 1994).

14.2 Life History and Ecology

Plagiopylids and odontostomatids are key indicators of the ciliate sulfureta community, which also includes ciliates from the Classes HETEROTRICHEA and OLIGOHYMENOPHOREA (Dyer, 1989; Fenchel, 1987). They are consistently represented in surveys of these habitats, which are characterized primarily as being anoxic or at most with very low concentrations of oxygen. Plagiopyleans have been found in freshwater habitats in Europe (Finlay & Maberly, 2000; Madoni & Sartore, 2003; Sola, Guinea, Longás, & Fernández-Galiano, 1988), Africa (Dragesco, 1972), and North America (Bamforth, 1963; Beaver & Crisman, 1989b), and in chloride lakes (Madoni, 1990). They are typically restricted to the sediment layers, often in microhabitats with high concentrations of dissolved sulphide (Esteban, Finlay, & Embley, 1993) and of mesotrophic to hypereutrophic status (Beaver & Crisman, 1989b). Plagiopyleans have also been observed in the coastal sediments and sands of marine and estuarine habitats in Eurasia (Agamaliev, 1974; Dragesco, 1962; Fauré-Fremiet, 1973; Fauré-Fremiet & Tuffrau, 1955; Fenchel et al., 1977), North America (Borror, 1963; Dyer, 1989; Nerad, Schaffer, Small, & Mangold, 1995), the Gulf of Arabia (Al-Rasheid, 1999b), and the Sea of Japan (Ozaki & Yagiu, 1941). *Plagiopyla* may extend its distribution into the water column of marine habitats when the oxycline changes its vertical placement as the seasons progress (Fenchel, Kristensen, & Rasmussen, 1990). Where abundances have been recorded in the water column, *Plagiopyla* rarely exceeds $1\,ml^{-1}$ (Fenchel et al., 1990; Massana & Pedrós-Alió, 1994), while odontostomatids can increase their relative abundance in sediments during periods of anoxia, reaching more than $50\,ml^{-1}$ of sediment (Fenchel, 1993). Plagiopyleans have never been recorded from soils (Foissner, 1998a).

Plagiopyleans have also been conspicuous endosymbionts from hosts as diverse as sea urchins (Grolière, de Puytorac, & Grain, 1980b; Lynch, 1930; Poljansky & Golikova, 1959) and the hippopotamus (Thurston & Grain, 1971).

Our understanding of feeding and growth in plagiopyleans is primarily derived from research on *Trimyema* and *Plagiopyla*. Strains of *Trimyema compressum* have been fed over 50 strains of bacteria, including both Gram-positive and Gram-negative species, as well as a variety of methanogens. This ciliate indiscriminantly ingested all bacteria (Schulz, Wagener, & Pfennig, 1990). However, its growth was limited to a smaller subset of the strains, although both Gram-positive and Gram-negative strains and strains of Archaea supported some growth (Baumgartner et al., 2002; Schulz et al., 1990; Yamada, Kamagata, Nakamura, Inamori, & Nakamura, 1994). Bacterial carbohydrates are the most important energy source for these anaerobic ciliates (Holler, Gälle, & Pfennig, 1994). Yields of over 9,000 ciliates ml^{-1} were recorded when *T. compressum* ingested a strain of *Desulfovibrio vulgaris* (Yamada et al., 1994). Threshold concentrations of bacteria to support growth were in the range of $10^7\ ml^{-1}$ (Schulz et al., 1990). *Plagiopyla nasuta* has been grown on a mixed assemblage of natural sediment-derived bacteria and its feeding and growth dynamics studied by uptake of fluorescently labeled bacteria (FLB) (Massana, Stumm, & Pedrós-Alió, 1994). *Plagiopyla* could consume over 4,000 bacteria ciliate^{-1} h^{-1} but had growth rates that were very low compared to aerobic ciliates of similar size feeding at these uptake rates (Massana et al., 1994). This confirms the general view that gross growth efficiency of these anaerobic ciliates is about 25% that of aerobes, although it can be increased by the presence of symbiotic methanogens (Fenchel & Finlay, 1990b, 1991a, 1991b). In the field, the realized growth rates of *Plagiopyla* are much lower. Thus, the natural abundances of these ciliates are unlikely to exert control on natural bacterial populations (Massana & Pedrós-Alió, 1994).

Plagiopyleans have conspicuous and abundant symbiotic bacteria associated both as ectosymbionts and as endosymbionts (Fenchel et al., 1977; Berger & Lynn, 1984). It is now clear that these endosymbiotic associations have been established repeatedly, and therefore have also been lost repeatedly. Wagener, Bardele, and Pfennig (1990) demonstrated that *Methanobacterium formicicum* could be functionally integrated into *Trimyema* cells that were endosymbiont-free. In natural populations, repeated losses and functional integration are demonstrated by the sister species *P. nasuta* and *Plagiopyla frontata* that have endosymbionts related to the different methanogen genera *Methanocorpusculum* and *Methanolobus*, respectively (Embley & Finlay, 1994). As noted above, this symbiotic association increases the growth efficiency of the ciliate. Moreover, it is mutualistic as it also provides a refuge for the methanogens, which avoid competition with sulphate reducing bacteria, avoid the toxic effects of environmental oxygen, and have a ready supply of hydrogen (Fenchel & Finlay, 1992; Müller, 1993). *Plagiopyla frontata* may have over 3,000 methanogens per cell. The cell division of these bacteria, which is synchronous with their host ciliate's cell division, may be controlled somehow by the ciliate. Excess bacterial production is transferred to the ciliate host in a fashion similar to the endosymbiosis with *Chlorella* species found in other ciliates (Fenchel & Finlay, 1991c).

14.3 Somatic Structure

Plagiopylids are ovoid to elongate ovoid ciliates that may show some dorsoventral flattening (Fig. 14.1). Odontostomatids are laterally-compressed ciliates with a rigid and often ribbed, armor-like pellicle. Spiny processes are often present, both posteriorly and anteriorly (Fig. 14.2). Trimyemids and odontostomatids can be about 15 μm in length while some plagiopylids can exceed 200 μm in length (Nerad et al., 1995; Sola, Guinea, Longás, & Fernández-Galiano, 1989b). Trimyemids have an apparently helicoidally disposed and sparse somatic ciliature (Fig. 14.1). The helicoidal disposition is due to the patterning of kinetids in the up to 60 somatic kineties in some species. Trimyemids may also have a caudal cilium complex, which Baumgartner et al. (2002) interpreted to demonstrate affinities with the caudal cilium complex of oligohymenophoreans. The larger plagiopylids and sonderiids are typically holotrichous and densely ciliated (Fig. 14.1). *Paraplagiopyla*, if truly a plagiopylean, is an exception as its somatic kineties are restricted to a narrow furrow that extends around the edges of the flattened cell (Thurston & Grain, 1971). The somatic ciliature of odontostomatids is typically reduced to anterior and posterior cirrus-like tufts, although the infraciliature probably persists as non-ciliated kinetosomes (Fig. 14.2).

Only four studies have been published on which to base the description of the cortical ultrastructure of plagiopyleans (Berger & Lynn, 1984; Detcheva et al., 1981; de Puytorac et al., 1985; Schrenk & Bardele, 1991).

The plasma membrane is covered by a thin glycocalyx, which can appear somewhat granular (de Puytorac et al., 1985). The alveoli in plagiopylids are well developed and lie on a thin epiplasmic layer. Schrenk and Bardele (1991) claimed that the alveolar layer is absent in the odontostomatid *Saprodinium* in which the cell membrane is underlain only by a thick epiplasmic layer. The cortex is ridged with kinetosomes lying between the ridges in trimyemids and at the tops of the ridges in plagiopylids.

The kinetids of plagiopyleans can still only be tentatively characterized, and they differ dramatically between plagiopylids and odontostomatids (Fig. 14.3). However, it now appears that the characterization of the plagiopylid kinetid by Berger and Lynn (1984) was incorrect, and that the microtubular ribbon they interpreted as an unusual, anteriorly-directed transverse ribbon is probably a kind of basal microtubular system. The somatic kinetids are monokinetids with a divergent postciliary ribbon that extends into the cortical ridges. The well-developed, anteriorly-directed kinetodesmal fibril originates near triplets 5, 6, 7 (Fig. 14.3). The orientation of the transverse ribbon has not

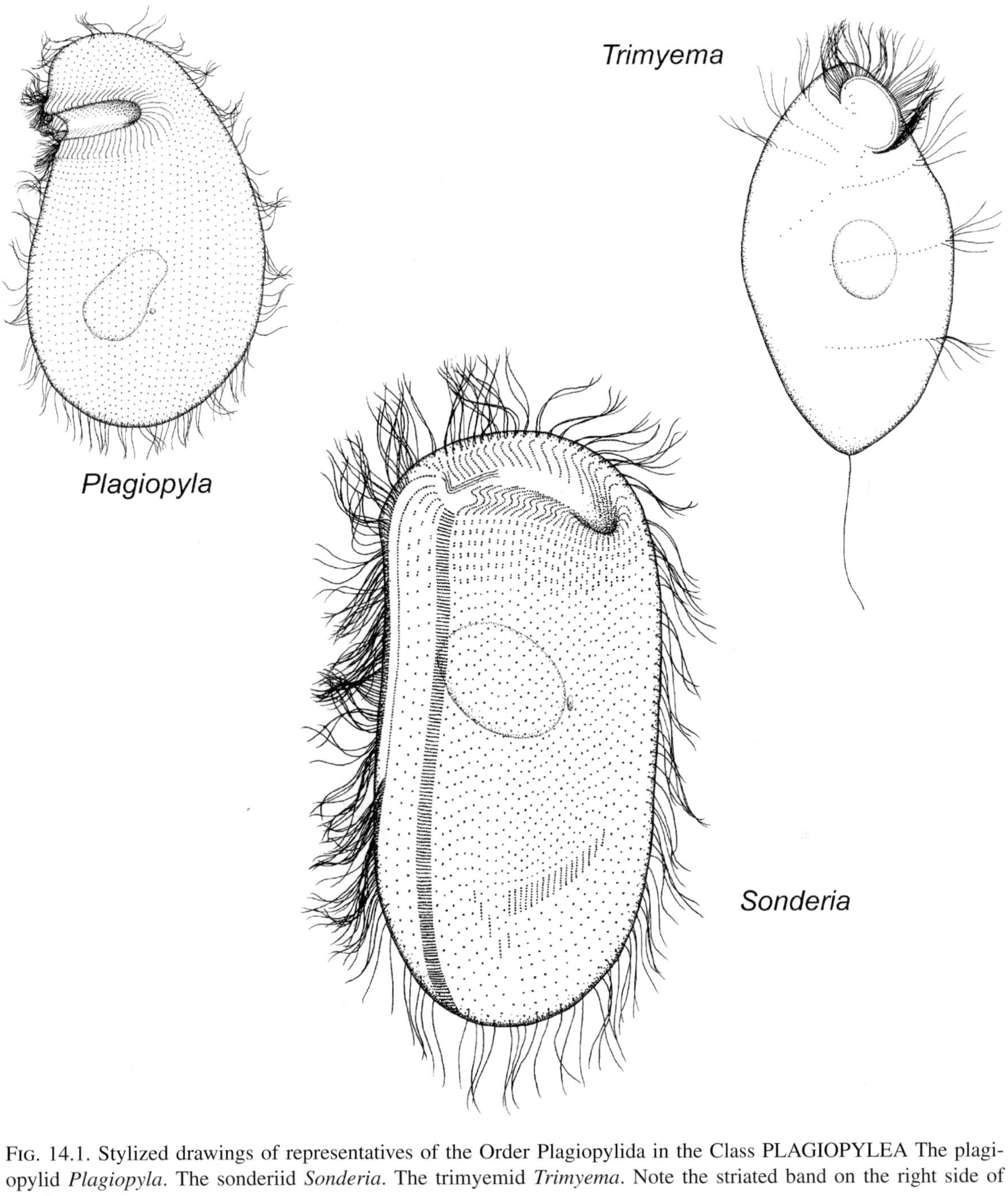

FIG. 14.1. Stylized drawings of representatives of the Order Plagiopylida in the Class PLAGIOPYLEA The plagiopylid *Plagiopyla*. The sonderiid *Sonderia*. The trimyemid *Trimyema*. Note the striated band on the right side of *Sonderia*

been definitively proven, although published and unpublished micrographs (C. Bardele, personal communication 2006; D. Lynn, 2006) suggest a radial orientation adjacent to triplet 4 and a very short trajectory, underlain by dense material, into the adjacent cortical ridge (Fig. 14.3). The overall pattern is very similar to that of the hymenostomes (see **Chapter 15**). Dense material adjacent to the base of the kinetosome near triplets 2, 3 provides the origin of several longitudinally orientated

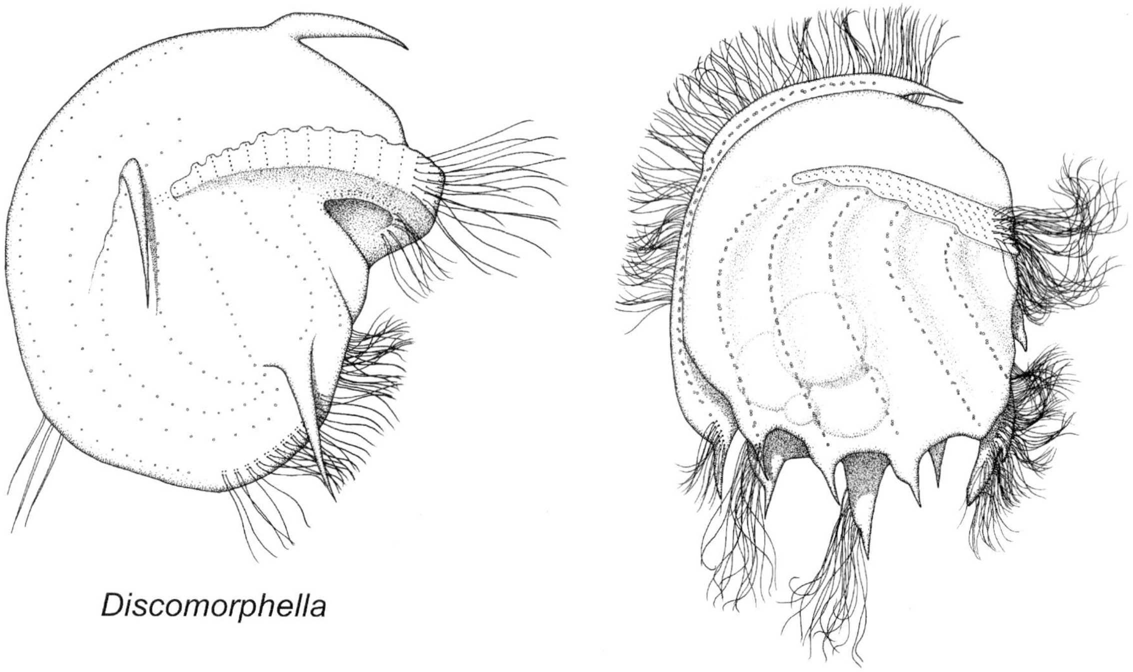

Discomorphella

Saprodinium

FIG. 14.2. Stylized drawings of representatives of the Order Odontostomatida in the Class PLAGIOPYLEA. The discomorphellid *Discomorphella*. The epalxellid *Saprodinium*

microtubules that extend along the left side of the kinety in *Lechriopyla*. These were originally interpreted incorrectly as transverse microtubules (Berger & Lynn, 1984). A parasomal sac is found anterior to the kinetosome.

The somatic kinetids of odontostomatids are also accompanied by parasomal sacs. However, odontostomatids have dikinetids, not monokinetids, throughout the cortex, although not all are ciliated (Fig. 14.3) (Schrenk & Bardele, 1991; Sola, Serrano, Guinea, & Longás, 1992). The odontostomatid somatic dikinetids can be characterized as follows: a ciliated anterior kinetosome that has a tangential transverse ribbon of microtubules associated with triplets 4, 5; and a ciliated posterior kinetosome with a divergent postciliary ribbon (Fig. 14.3). Cathetodesmal-like fibrils may originate near triplet 2 on the anterior kinetosome. Schrenk and Bardele (1991) concluded that there is no kinetodesmal fibril although there is a dense structure in the appropriate position near the posterior kinetosome and Sola et al. (1992) reported kinetodesmal fibrils adjacent to some anterior and posterior kinetosomes in the light microscopic description of *Saprodinium*. In the non-ciliated regions of the cortex, *Saprodinium* has its dikinetid kinetosomes without fibrillar associates and separated by extremely inflated parasomal sacs, which may be used for endocytosis (Schrenk & Bardele, 1991). An inverse kinety, whose origin is unclear, lies to the left of the oral region (Schrenk & Bardele, 1991; Sola et al., 1992).

A unique feature of the cortex of most plagiopylids and sonderiids, a feature that might suggest establishment of a subordinal category for members of these two families, is the striated band (Fig. 14.1) (Lynch, 1930). This band extends from the right side of the oral opening, in parallel with adjacent somatic kineties, sometimes turning anteriorly before turning posteriorly to extend almost to the posterior pole. It is composed of a series of thin ridge-like lamellae of cytoplasm, about 2 μm high, overlain by flattened cortical alveoli. The walls of the striated band appear to be supported by 8–9 macrotubules (Berger & Lynn, 1984). The function of the striated band is unknown.

Plagiopylids and sonderiids have conspicuous rod-shaped extrusomes, which may be up to 20 μm

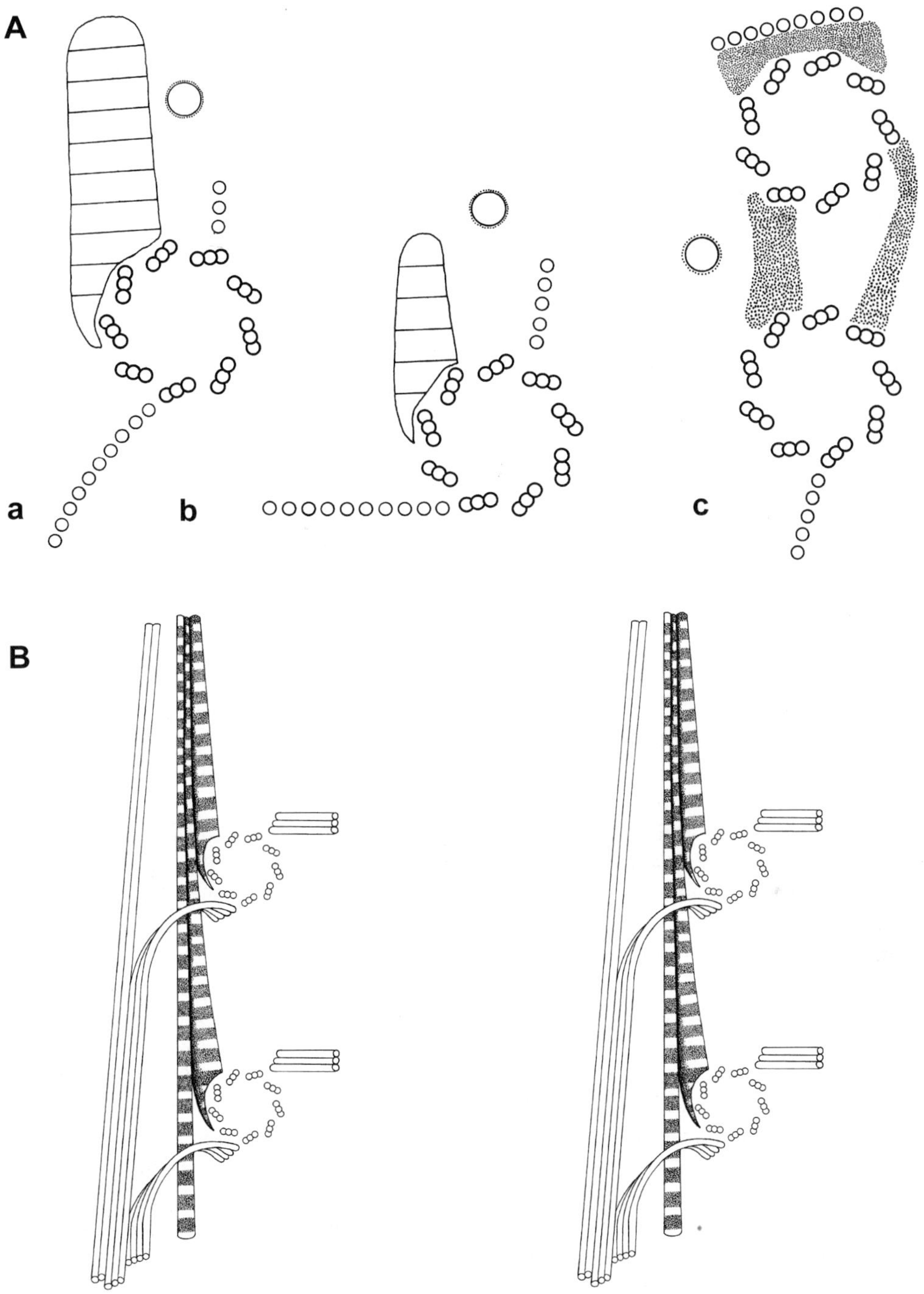

FIG. 14.3. **A** Schematics of the somatic kinetids of the Class PLAGIOPYLEA. (**a**) Monokinetid of *Plagiopyla*. (**b**) Monokinetid of *Trimyema*. (**c**) Dikinetid of *Saprodinium* (from Lynn, 1981, 1991). **B** Somatic cortex of a typical plagiopylid based on the somatic cortex of *Plagiopyla* and *Lechriopyla*

in length (Fauré-Fremiet & Tuffrau, 1955). On extrusion, the matrix extends as a striated rod from a retained cylindrical envelope (Berger & Lynn, 1984). Trimyemids have spheroidal mucocysts (Baumgartner et al., 2002; Detcheva et al., 1981) while *Plagiopyla* may also have some smaller extrusomes (de Puytorac et al., 1985). Mucocysts have not been observed in odontostomatids (Schrenk & Bardele, 1991).

Plagiopyleans do not have mitochondria, but rather "microbodies" without cristae that are now known to be hydrogenosomes (see **Life History and Ecology**) (Goosen et al., 1990; Zwart et al., 1988).

A contractile vacuole and a cytoproct are typically found in the posterior one-third of the cell.

14.4 Oral Structures

The oral structures divide the plagiopyleans into three groups – the odontostomatids, the trimyemids, and the sonderiids and plagiopylids (Figs. 14.1, 14.2). Whether detailed and careful ultrastructural investigations will eventually reveal homologies, at this stage we must treat them quite separately.

Odontostomatids have a small and complex oral cavity with a reduced number of oral polykinetids, typically less than a dozen (Schrenk & Bardele, 1991; Sola et al., 1992; Tuffrau, 1992). They are composed of three rows of kinetosomes, which are hexagonally packed, but only the oral polykinetid closest to the cytostome has fibrillar associates that are interpreted as postciliary ribbons (Schrenk & Bardele, 1991). These latter authors speculated that the oral region of *Saprodinium*, and perhaps other odontostomatids, is in an inverse orientation, but this will have to await morphogenetic studies. Odontostomatids may also have two files of paroral cilia (see Sola et al., 1992; Tuffrau, 1992), but this has not been confirmed by electron microscopy (Schrenk & Bardele, 1991).

The trimyemids have always been classified among ciliates with a simple oral ciliature and oral apparatus. It is now certain that they have at least an outer row of kinetosomes with kinetodesmal fibrils that borders the oral region on the anterior, left, and posterior portions (Detcheva et al., 1981; Serrano, Martín-González, & Fernández-Galiano, 1988). This row may be accompanied by a second row interior to it and perhaps even a third shorter fragment (Fig. 14.1) (Serrano et al., 1988). We could tentatively interpret Figure 6 of Detcheva et al. (1981) as indicating that the outer kinetosomes have postciliary ribbons and a kinetodesmal fibril while the inner kinetosomes have only a tangential transverse ribbon. There are no ultrastructural observations for the second set of oral structures in trimyemids: these oral dikinetids lie on the right side of the oral region, and can range from four independent dikinetids to two polykinetids, each composed of three dikinetids (Fig. 14.1) (cf. Baumgartner et al., 2002; Nerad et al., 1995; Serrano et al., 1988).

The sonderiids and plagiopylids share a basic plan to the oral ciliature, which lines a ventral transverse oral groove that becomes tubular as it extends inwards towards the cytostome. After a slight break, the ends of the somatic kineties that border the anterior (= dorsal) lip and posterior (= ventral) lip become much more densely packed with kinetosomes (Fig. 14.1) (Lynch, 1930; Serrano et al., 1988; Sola et al., 1988). In *Plagiopyla*, the density of the kinetosomes becomes thinner as these oral kineties extend to line the oral cavity (Sola et al., 1988). Small and Lynn (1985) distinguished genera based on the trajectory of the oral invagination: the oral cavity of *Plagiopyla* extends to the left while that of *Paraplagiopyla* extends directly dorsally. The oral kinetosomes of *Plagiopyla* and *Lechriopyla* lack "somatic" fibrillar associates but do have alveoli between them and parasomal sacs to the side. Microtubules of unknown origin have been observed between these oral kineties (de Puytorac et al., 1985). Two to three fibrous rootlets arise at the base of each of these kinetosomes and extend parallel to the cell surface. Rootlets from adjacent kinetosomes intertwine forming a complex cytoskeletal structure that departs from each kinety and assembles into an aggregate, which in *Lechriopyla* is fork-shaped and called the furcula (Berger & Lynn, 1984; Lynch, 1930). The cytopharynx is lined by ribbons of microtubules whose origin is undetermined.

14.5 Division and Morphogenesis

Plagiopyleans divide while swimming freely. There are only two recent studies of division morphogenesis in plagiopyleans, and no reports of stomatogenesis in odontostomatids.

Division morphogenesis in *Plagiopyla* has been redescribed by de Puytorac et al. (1985). It begins by kinetosomal replication occurring at the equator especially on the posterior side of the putative fission

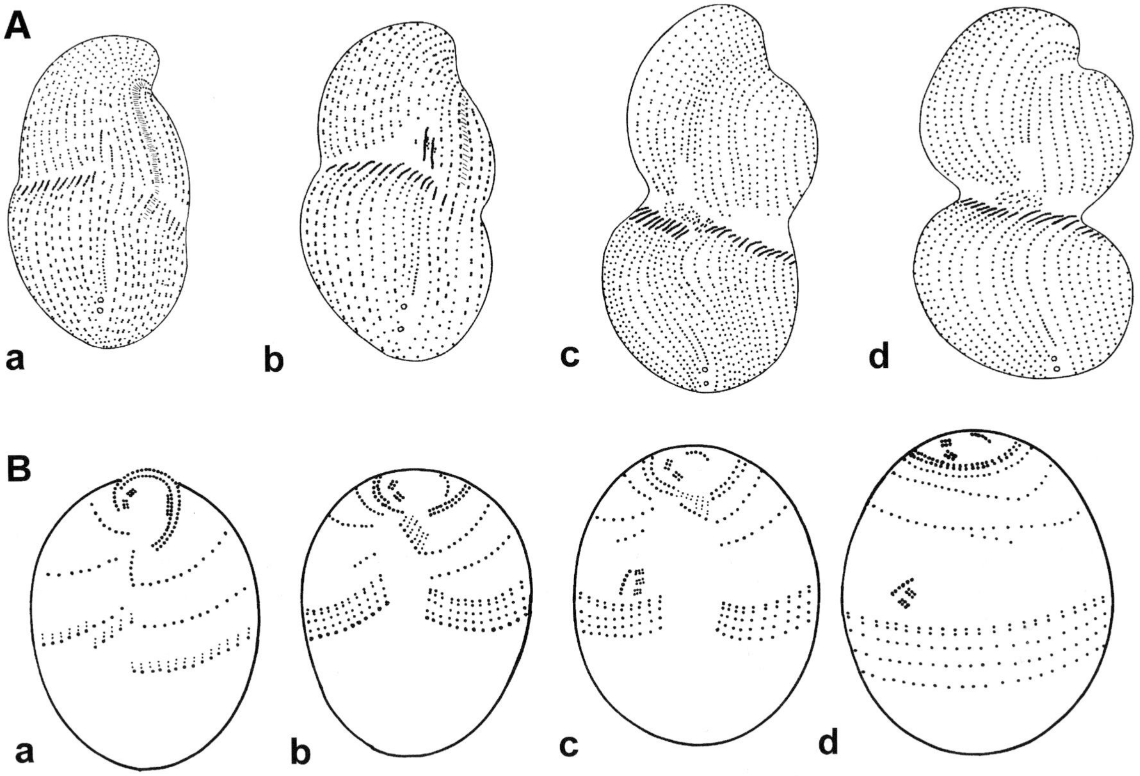

FIG. 14.4. Division morphogenesis of plagiopylids. **A** In the plagiopylid *Plagiopyla*, kinetosomal replication occurs at the anterior ends of all the somatic kineties (**a**–**d**). A set of kinetosomes appears in the fission furrow in the right dorsal area, and these may give rise to oral kinetosomes (**b**–**d**). (from de Puytorac et al., 1985.) **B** In the trimyemid *Trimyema*, a file of kinetosomes appears in the ventral anterior region (**a**) and this appears to organize into a file and two polykinetids of six kinetosomes. (from Serrano et al., 1988.)

furrow where the kinetosomes become dense as replication proceeds, approaching a density similar to that of the proter's oral kineties (Fig. 14.4). De Puytorac et al. (1985) remarked on the appearance in the right dorsal portion of the fission furrow of a field of irregularly arranged kinetosomes whose destiny remains to be determined (Fig. 14.4). Could these be the homologues of the dikinetids found in the oral region of trimyemids?

Serrano et al. (1988) demonstrated that *Trimyema* has a kind of holotelokinetal stomatogenesis. These authors claimed that kinetosomes from Kinety *n* also participate to form the dikinetids and the third inner row of kinetosomes (Fig. 14.4). However, this claim needs to be confirmed by demonstrating the intermediate stages to definitively justify this conclusion. What appears probable is that the outer two rows of the "circumoral" arise from the two anterior kinetosomes of each somatic kinety.

14.6 Nuclei, Sexuality and Life Cycle

The plagiopylean macronucleus is homomerous, ranging in shape from globular in the small odontostomatids and trimyemids to an elongate ellipsoid in larger sonderiids. Some odontostomatids can have multiple macronuclei (Figs. 14.1, 14.2). The macronucleus is typically accompanied by a single, globular micronucleus. Fauré-Fremiet (1973) noted that *Parasonderia kahli* had a highly unusual macronucleus: it was triangular in shape, flattened, and wrapped around the tubular oral cavity of this ciliate.

To our knowledge, there are no reports of conjugation in plagiopyleans. Thus, their genetics and details of nuclear development and differentiation remain to be determined.

Chapter 15
Subphylum 2. INTRAMACRONUCLEATA: Class 9. OLIGOHYMENOPHOREA – Once a Pivotal Group, Now a Terminal Radiation

Abstract This class includes the two model ciliates – *Tetrahymena* and *Paramecium* – laboratory genetic models of protists and the first two ciliates to have completed genome projects. Along with these representatives of the Subclasses Hymenostomatia and Peniculia, there is a diverse variety of species distributed in four other subclasses: Scuticociliatia, Peritrichia, Apostomatia, and Astomatia. While the archetype of the class has a paroral and three left oral polykinetids, there is really no strong morphological or ultrastructural synapomorphy for the class. Molecular phylogenetics generally recovers an oligohymenophorean clade with moderate bootstrap support. The somatic kinetid is similar to that of the last three classes: a generally robust, anteriorly-directed kinetodesmal fibril, a divergent postciliary ribbon, and a radial transverse ribbon, except in the peniculines where it is tangential. Oral structures depart from the archetype in the apostomes with their rosette and complex cytopharyngeal apparatus and are absent in the astomes. Oligohymenophoreans are as broadly distributed across habitats as any class, but they include obligate symbionts: the apostomes typically with crustaceans and the astomes with annelids. *Ichthyophthirius*, the parasite of freshwater fishes, is an hymenostome that can plague aquaculture operations. Division morphogenesis is characterized as buccokinetal and parakinetal. There is a rich literature on the genetics of these ciliates, particularly *Paramecium* and *Tetrahymena*, which can be easily induced to conjugate in the laboratory, enabling a deeper understanding of their molecular biology.

Keywords Autogamy

The Class OLIGOHYMENOPHOREA includes two ciliate genera, *Tetrahymena* and *Paramecium*, which one might describe respectively as the "white mouse" and the "white rat" of the Phylum Ciliophora, based both on their relative sizes and on their being the most "highly tamed" laboratory models in the phylum. Lwoff (1923) successfully cultured *Tetrahymena* axenically, that is in a sterile nutrient broth without bacteria, and so initiated the exploration of its physiology and biochemistry as a model organism. It was to be many years later that a partly defined medium was devised to axenically culture *Paramecium* (Lilly & Klosek, 1961). Now there are genome projects well underway for representative species of both these genera (Eisen et al., 2006; *Tetrahymena* – www.lifesci.ucsb.edu/~genome/Tetrahymena/; and Aury et al., 2006; *Paramecium* – carroll.vjf.cnrs.fr/pt/), bringing these two ciliates and molecular genetic approaches to understanding their biology into a competitive position with other model organisms.

The Class OLIGOHYMENOPHOREA is arguably the most diverse assemblage within the phylum. While there are fewer included subclasses in it compared to the Class SPIROTRICHEA (see **Chapter 7**), there have been many more species described. Oligohymenophoreans are commonly encountered in freshwater and marine habitats with peniculines being most conspicuous in the former habitats and scuticociliates being most conspicuous in the latter habitats. Some members of the class show a preference for moribund and dead metazoans (e.g., Corliss, 1972c; Fauré-Fremiet & Mugard, 1946), and this capacity has lead several times independently in the class to symbiotic and parasitic modes of life in diverse hosts: the apostomes in crustaceans (Bradbury, 1966a; Chatton & Lwoff, 1935a); the thigmotrich scuticociliates in bivalve molluscs (Chatton & Lwoff, 1949, 1950); hymenostomes as parasites of gastropods (Kozloff, 1956) and fishes (Hoffman, 1988; Hoffman et al., 1975); and peritrichs, both **on** invertebrates (Matthes, 1974) and vertebrates (Lom, 1995), and **in** invertebrates (Lom, 1994) and vertebrates (Lom, 1958, 1995). These infections on rare occasions can cause mass mortalities of their host, such as *Ichthyophthirius* and fishes (Wurtsbaugh & Tapia, 1988). In other instances, infections may significantly impact fecundity, such as the scuticociliate *Orchitophrya* infecting the testes of starfish (Stickle, Weidner, & Kozloff, 2001). The morphological polymorphism and complex life cycles often characteristic of these symbiotic ciliates are presaged by similar dramatic changes in form in free-living species, which may undergo form change in response to feeding conditions, such as starvation (Fauré-Fremiet & Mugard, 1949a; Nelsen & De Bault, 1978) or the presence of different kinds of foods (Small, Heisler, Sniezek, & Iliffe, 1986). The absence of bacteria may stimulate some species to transform into macrostome cannibals, consuming their conspecifics and other smaller species (Fig. 15.1) (Corliss, 1973; Njiné, 1972).

Despite these dramatic variations in body morphology, the ciliates in this class demonstrate their name, having few oral polykinetids or membranelles (*oligo*, Gr. – few; *hymen*, Gr. – membrane; *phoros*, Gr. – bearing). The typical arrangement is three, inconspicuous oral polykinetids on the left side of the oral cavity and a paroral of dikinetids or a stichodyad on the right side. Nevertheless, the oral structures of some pleuronematine scuticociliates, such as *Pleuronema*, and peritrichs can be quite prominent. Corliss (1979) argued that oligohymenophoreans were a linking assemblage between the "lower" kinetofragminophoran groups and the "higher" polyhymenophorans. Molecular phylogenetic analyses have now refuted this vision and demonstrate the oligohymenophoreans to be a "terminal" branch in the Subphylum Intramacronucleata (Greenwood, Sogin, & Lynn, 1991a; Strüder-Kypke et al., 2000b). With the notable exception of the peritrichs, the vast majority of oligohymenophoreans are holotrichously ciliated. They are medium-sized, typically 65–150 μm in length; some scuticociliate species can be less than 10 μm in length, the large carnivorous peniculine *Neobursaridium* can reach lengths of 600 μm (Dragesco & Tuffrau, 1967), and some astomes can reach 3 mm in length (de Puytorac, 1994g). As with the tintinnids, the loricate forms – fossilized peritrichs – can be placed in contemporary families even though they date from the Lower Triassic, some 200 million years ago (Weitschat & Guhl, 1994). Even more remarkably, a *Paramecium* species has been described from southern Germany in amber-bearing sandstone deposits (Schönborn, Dörfelt, Foissner, Krientiz, & Schäfer, 1999), which have been assigned to the Late Cretaceous, some 90 million years ago (Schmidt, von Eynatten, & Wagreich, 2001). Thus, it is very likely that the class originated over 250 million years ago, and even perhaps in the Paleoproterozoic, if the molecular clock can be trusted (Wright & Lynn, 1997c).

As noted above, the ease of cultivating species of *Tetrahymena* and *Paramecium* has made them two of the most widely studied genera in the phylum, and therefore in this class. These earlier approaches to their cultivation lead to the establishment of chemically defined media for both *Tetrahymena* (see Kidder & Dewey, 1951; Orias, Hamilton, & Orias, 2000) and *Paramecium* (Bihn, Lilly, & Napolitano, 1974; Fok & Allen, 1979; Soldo, Godoy, & van Wagtendonk, 1966). This tractability, together with tractable genetic techniques, has given rise to an enormous literature on these two genera, summarized in books, which can only be listed here: on *Tetrahymena* – Hill (1972), Elliott (1973a), Asai and Forney (2000); and on *Paramecium* – Beale (1954), Görtz (1988a), Jurand and Selman (1969), Sonneborn (1970), van Wagtendonk (1974), and Wichterman (1986). Both axenic (Finley & McLaughlin, 1965; Finley, McLaughlin, & Harrison, 1959; McLaughlin, Johnson & Bradley, 1974) and non-axenic (Finley et al., 1959; Vacchiano, Kut, Wyatt, & Buhse, 1991) techniques have been developed for cell biological research on peritrichs. Soldo and Merlin (1972, 1977) have successfully developed an axenic culture medium for marine scuticociliates, and this has enabled the cultivation of some species, reported as parasitic on shellfish and fish (Cawthorn et al., 1996; Crosbie & Munday, 1999; Iglesias et al., 2003; Messick & Small, 1996). There has not yet been a successful axenic cultiva-

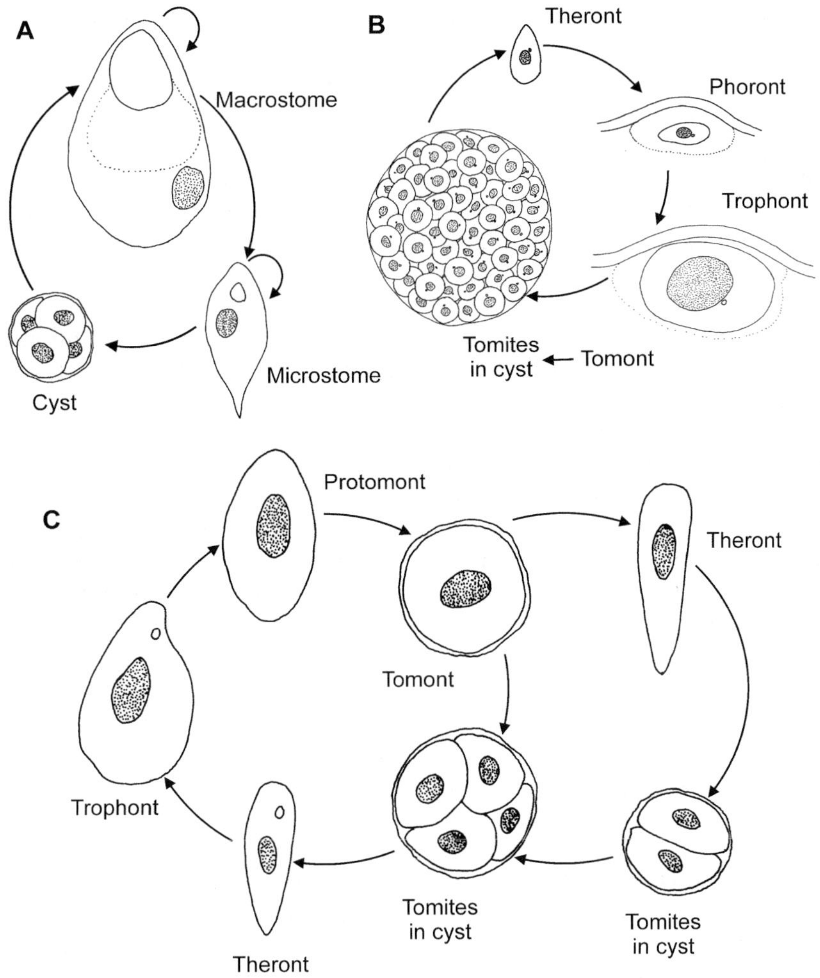

FIG. 15.1. Life cycles of oligohymenophoreans. **A** A hymenostome, like *Tetrahymena paravorax*, can transform between a microstome bacterivore and a macrostome carnivore depending upon food availability. (after Corliss, 1973.) **B** The theronts of the ophryoglenine *Ichthyophthirius multifiliis* seek out the epithelium of a freshwater fish host, burrow underneath as a phoront, and then begin to grow as a trophont. Trophonts later drop off the fish and undergo palintomy to produce sometimes more than 1,000 theronts. (after Lynn & Small, 2002.) **C** The ophryoglenine *Ophryoglena* typically feeds on dead or moribund invertebrates. After feeding, the trophont becomes a protomont, encysts as a tomont to undergo palintomy and produce theronts, the dispersal stage that seeks out other prey. (after Canella & Rocchi-Canella, 1976.)

tion of the fish parasite *Ichthyophthirius*, although Nielsen and Buchmann (2000) succeeded in inducing trophont transformation using a fish tissue culture cell line. Mizobuchi, Yokoigawa, Harumoto, Fujisawa, and Takagi (2003) have recently demonstrated that hydrogen peroxide is the toxic substance in the infusion of wheat grass powder used to grow *Paramecium*. This substance, which may also inhibit the growth of other ciliates, is detoxified by catalase excreted by bacteria, explaining why "conditioned" media may be more successful in cultivating some ciliate species.

Molecular phylogenies generally recover an oligohymenophorean clade with strong bootstrap support (Baroin-Tourancheau, Villalobo, Tsao, Torres, & Pearlman, 1998; Miao et al., 2001; Sánchez-Silva, Villalobo, Morin, & Torres, 2003; Strüder-Kypke et al., 2000b). Sánchez-Silva et al. (2003)

argued that contemporary oligohymenophoreans may all have arisen from an ancestor with a deviant nuclear genetic code for glutamine/glutamate. Nevertheless, there are no strong morphological synapomorphies for the class. The somatic kinetid has been typified as a monokinetid with a radial transverse microtubular ribbon and a typically well-developed kinetodesmal fibril (Lynn, 1981, 1991). However, inclusion of the peniculines in the class presents an exception to this rule. As noted above, the oral structures typically include three or four oral polykinetids on the left side of the oral cavity and a paroral of dikinetids on the right. Yet, representatives of the Subclass Astomatia entirely lack an oral apparatus while those in the Subclass Apostomatia have highly modified oral structures, albeit with presumed homologies to their less modified kin (Bradbury, 1989). We have relied on the molecular phylogenies to support the class and presume that modifications in morphology away from the type are manifestations of evolutionary divergence. The class includes six subclasses: Subclass Peniculia, Subclass Scuticociliatia, Subclass Hymenostomatia, Subclass Peritrichia, Subclass Apostomatia, and Subclass Astomatia.

15.1 Taxonomic Structure

Corliss (1979) divided the Class OLIGOHYMENOPHOREA into two subclasses – Subclass Hymenostomata and Subclass Peritrichia. He excluded the apostomes from this class, placing them as an order within the Class KINETOFRAGMINOPHORA. Bradbury (1989) has demonstrated a paroral-like kinetid arrangement in the apostomes. On this basis, and together with similarities in the somatic kinetid (**see below**), we have transferred the apostomes to the Class OLIGOHYMENOPHOREA. We now have sequences of thesmall subunit (SSU) rRNA gene of several genera of a postomes to confirm this transfer (Clamp, Bradbury, Strüder-Kypke, and Lynn, 2008). Our system include all of the subclasses that de Puytorac (1994a) recognized, except for the Subclass Hysterocinetia (**see below**), which we conservatively retain as a group within the Subclass Scuticociliatia, awaiting molecular genetic evidence to demonstrate that this group is clearly so different. Thus, we outline briefly the six subclasses: (1) Subclass Peniculia; (2) Subclass Scuticociliatia; (3) Subclass Astomatia; (4) Subclass Peritrichia; (5) Subclass Hymenostomatia; and (6) Subclass Apostomatia – and their major included groups.

Small and Lynn (1981, 1985) placed the peniculine ciliates, such as *Paramecium* and *Frontonia*, as an order in their Class NASSOPHOREA, based on similarities in the somatic kinetids, pellicular ultrastructure, and extrusomes. Molecular genetic studies on both SSUrRNA (Strüder-Kypke et al., 2000a, 2000b) and proteins (Sánchez-Silva et al., 2003) have refuted this association and confirmed the classical view that peniculines derived from the same ancestral lineage as the other oligohymenophorean groups. Nevertheless, these molecular data clearly confirm the distant relationship of the peniculines to other members of the class, a distance supported by their morphological features, seemingly homologous to those of the nassophoreans. The prime synapomorphies for peniculines are as follows: three oral polykinetids, called peniculi, that are aligned longitudinally in the oral cavity (Fauré-Fremiet, 1950a, 1950b); the typical, although not universal, presence of fibrous trichocysts (Didier, 1971; Jurand & Selman, 1969); and a stomatogenesis in which the parental paroral and its accompanying anarchic field produce the new oral structures (Beran, 1990; Foissner, 1996b; Yusa, 1957). We have included two orders within the subclass. The Order Peniculida, characterized by holotrichous somatic ciliation and the presence of fibrous trichocysts includes six families: the Frontoniidae, the Lembadionidae, the Maritujidae, the Neobursaridiidae, the Parameciidae, and the Stokesiidae. Since there is clear evidence, both from its distinctive girdle of somatic cilia, its lack of fibrous trichocysts (Didier; Didier & de Puytorac, 1969), and its divergent SSUrRNA gene sequence (Strüder-Kypke et al., 2000b) that *Urocentrum* is very divergent from other peniculines, we support the monotypic Order Urocentrida proposed by de Puytorac, Grain, and Mignot (1987) to include the Family Urocentridae.

Species diversity within the genus *Paramecium* continues to be exhaustively analyzed (Maciejewska, 2007). Corliss and Daggett (1983) particularly focused on taxonomic and nomenclatural issues surrounding those species, previously assigned to the *aurelia* species complex of *Paramecium*. They emphasized that there is no longer a species named *Paramecium aurelia* in the genus *Paramecium*. This taxonomic arrangement had been formalized by Sonneborn (1975) after research by Tait (1970) and Allen, Farrow, and Golembiewski (1973) had demonstrated categorical differences among isoenzymes, such as esterases and dehydrogenases, in this sibling species complex. The same year, morphologically-oriented systematists had used multivariate techniques to demonstrate the separability of several of these sibling species or syngens (Gates, Powelson, & Berger, 1975; Powelson, Gates, & Berger, 1975), but these techniques were unable to practically separate all species (Gates & Berger, 1976b). This morphological "similarity" in multivariate space has been confirmed for the *aurelia* species complex and also demonstrated for the *woodruffi* grouping of species (Fokin & Chivilev, 2000). Nevertheless, the only new species in the *aurelia* complex, *Paramecium sonneborni*, has been characterized as being distinct by using mating-type reactivity and isoenzyme patterns (Aufderheide, Daggett, & Nerad, 1983).

Biochemical differences and their genetic basis continue to be discovered for *Paramecium* species (Tait, 1978), and used to identify other sibling species within the genus (Allen, Nerad, & Rushford, 1983a; Allen, Rushford, Nerad, & Lau, 1983b; Usuki & Irie, 1983). While the SSUrRNA genes of *aurelia* complex species are very similar (Hoshino, Hayashi, & Imamura, 2006; Strüder-Kypke et al., 2000a), significant inter- and intra-specific genetic variation in the internal transcribed spacer regions has been discovered in *aurelia* complex species (Hoshino et al., 2006; Tarcz, Przybos, Prajer, & Greczek-Stachura, 2006). By modeling the secondary structure of internal transcribed spacer region 2 (ITS2), Coleman (2005) correlated the phylogeny of *Paramecium* species with their mating characterization. Stoeck and Schmidt (1998) have used fingerprints derived from randomly amplified polymorphic DNA (RAPD) to distinguish nine of thesespecies: these results confirmed genetic studies that demonstrated this complex to be widely distributed geographically (Przybos, 1993; Stoeck, Przybos, Kusch, & Schmidt, 2000a). RAPD - fingerprinting does not require the careful axenic cultivation of the ciliates, which is necessary for the proper isoenzyme characterization of the strains, nor does one need to maintain genetic stocks to do proper mating tests. It has been used to reject sibling species for *Paramecium caudatum* (Hori, Tomikawa, Przybos, & Fujishima, 2006; Stoeck, Welter, Seitz-Bender, Kusch, & Schmidt, 2000b) and support them for *Paramecium jenningsi* (Skotarczak, Przybos, Wodecka, & Maciejewska, 2004). Recently, Barth, Krenek, Fokin, and Berendonk (2006) showed significant intrahaplogroup variation within *P. caudatum* and *Paramecium multimicronucleatum* using the mitochondrial cytochrome c oxidase subunit 1 (cox-1) gene, suggesting that these "species" may, in fact, be sibling species complexes, contrary to what RAPD fingerprinting suggests.

Small (1967) was the first to formally recognize a monophyletic assemblage that he designated as an order at that time, but which has now been elevated to the subclass rank (Lynn & Small, 1997, 2002; de Puytorac, 1994a, 1994e). The Subclass Scuticociliatia is characterized by a paroral that is divided into three segments – anterior *a*, middle *b*, and posterior *c* or scutica. The scutica, named for its hook-like or "whiplash" configuration taken during stomatogenesis in some forms, is the major synapomorphy for the group. Although only a handful of the thousands of species has been examined by molecular techniques, the subclass has so far had strong support from small subunit rRNA sequences (Lynn & Strüder-Kypke, 2005; Miao, Fen, Yu, Zhang, & Shen, 2004b; Shang, Song, & Warren, 2003). The now "modern classic" research on this group (e.g., Evans & Corliss, 1964; Evans & Thompson, 1964; Raabe, 1967, 1972; Small, 1967; Thompson, 1963, 1964, 1965, 1966a, 1966b, 1967, 1968, 1969) is being revised with the use of protargol silver-staining, careful re-examination of living specimens, and biogeographic analyses. Global distributions of species have been confirmed using both morphological (e.g., Esteban & Olmo, 1997; Foissner & Wilbert, 1981) and molecular (Goggin & Murphy, 2000) criteria; new genera have been recognized (Olmo, Tellez, & Esteban, 1998; Song & Wilbert, 2002); and synonymies of species have been proposed (Esteban & Olmo, 1997; Song & Wilbert, 2000a). Furthermore, molecular genetic studies are now casting doubts on the assignment of genera to families, and even on the identity of genera (Lynn & Strüder-Kypke, 2005; Ma, Song, Gong, & Warren, 2004; Paramá, Arranz, Álvarez, Sanmartín, & Leiro, 2005; Shang et al.,

2003). Shang and Song (2005) have successfully used RAPD fingerprinting to identify and separate marine scuticociliate species.

We have maintained a conservative subdivision of the subclass, recognizing three included orders: Order Philasterida, Order Pleuronematida, and Order Thigmotrichida. The philasterids appear to be strongly supported as a group by molecular phylogenetics (Lynn & Strüder-Kypke, 2005; Shang et al., 2003). It is still too early to tell for the other two orders: there are gene sequences for only a few representative pleuronematids while no thigmotrich has yet been sequenced. In fact, thigmotrichs have received relatively little attention since the monographic works of Chatton and Lwoff (1949, 1950), Fenchel (1965a), and Raabe (1967, 1970a, 1970b, 1971b, 1972).

The Order Philasterida is characterized by having a paroral shorter than the other oral structures, typically by reduction of the paroral *a* and *c* segments and with the scutica separate and posterior to the paroral. We include 16 families in the order: the Cinetochilidae, the Cohnilembidae, the Cryptochilidae, the Entodiscidae, the Entorhipidiidae, the Loxocephalidae, the Orchitophryidae, the Paralembidae, the Parauronematidae, the Philasteridae, the Pseudocohnilembidae, the Schizocaryidae, the Thigmophryidae, the Thyrophylacidae, the Uronematidae, and the Urozonidae. The Family Schizocaryidae has been assigned to this order **only** on the basis of its SSUrRNA gene sequence (Lynn & Strüder-Kypke, 2005). The Order Pleuronematida is characterized by an expansive oral region along whose right border extend the prominent paroral cilia forming a curtain or velum as the organism filter feeds. The scutica is a permanent component of the paroral *c* segment. We include nine families in the order: the Calyptotrichidae, the Conchophthiridae, the Ctedoctematidae, the Cyclidiidae, the Dragescoidae, the Histiobalantidiidae, the Peniculistomatidae, the Pleuronematidae, and the Thigmocomidae.

Ngassam, de Puytorac, and Grain (1994) proposed the new Subclass Hysterocinetia to separate the hysterocinetid ciliates from the thigmotrichs. These ciliates, which are endosymbionts of oligochaetes and molluscs, are substantially different from other thigmotrichs (Ngassam & Grain, 2002; Njiné & Ngassam, 1993; de Puytorac, 1994f). However, given what we are already discovering about the differing views of relationships provided by morphological and molecular approaches on other scuticociliates (Lynn & Strüder-Kypke, 2005; Shang et al., 2003), we prefer to await molecular evidence of hysterocinetid distinctiveness before recognizing this group as a subclass within the Class OLIGOHYMENOPHOREA. Therefore, we characterize the Order Thigmotrichida to include ciliates having obviously differentiated thigmotactic somatic ciliature and a subequatorial oral region whose oral ciliature may spiral around the posterior end of the cell and whose oral polykinetids may quite often be reduced or even absent. We include four families in the order: the Ancistridae, the Hemispeiridae, the Hysterocinetidae, and the Paraptychostomidae.

The Subclass Hymenostomatia, which was much more broadly inclusive (see Corliss, 1979), includes only the two orders Tetrahymenida and Ophryoglenida. These "membrane-mouthed" ciliates are united by the position of a well-defined oral cavity with a paroral and three oral polykinetids, called membranelles, although some included taxa are astomatous (Kozloff, 1954). The ophryoglenids share the organelle of Lieberkühn as the synapomorphy for the group (Canella & Rocchi-Canella, 1964, 1976; Corliss; Lynn, Frombach, Ewing, & Kocan, 1991b). *Ichthyophthirioides* may have secondarily lost this organelle. We include the families Ichthyophthiriidae and Ophryoglenidae in this order. The tetrahymenids lack the organelle of Lieberkühn and do not demonstrate any synapomorphies at the morphological level. However, sequences of the SSUrRNA gene strongly separate the two orders and confirm their monophyly (Lynn & Strüder-Kypke, 2005; Miao et al., 2004b; Wright & Lynn, 1995). We confirm the placement of the Family Turaniellidae in the Order Tetrahymenida (Corliss, 1979) based on the similarities of its oral ultrastructure with that of the tetrahymenids (Lynn & Didier, 1978). The family now includes *Colpidium* because of similarities in division morphogenesis and ultrastructure (Iftode, Fryd-Versavel, & Lynn, 1984). In addition to this family, the Order Tetrahymenida includes the following five families: the Curimostomatidae, the Glaucomidae, the Spirozonidae, the Tetrahymenidae, and the Trichospiridae.

The tetrahymenids, and particularly the genus *Tetrahymena*, have been the focus of much systematic research. Corliss and Daggett (1983) reviewed the taxonomic and nomenclatural aspects of research on the *"Tetrahymena pyriformis"* complex, noting that there is still a species *Tetrahymena pyriformis*. Nanney and McCoy (1976) restricted this name to an amicronucleate form on the basis of isoenzyme features, and characterized 13 other species in the *pyriformis* complex, three of them amicronucleates and ten of them bona fide biological species, formerly called syngens. They relied heavily on isoenzyme variation among these taxa previously demonstrated as distinct by Allen and Weremiuk (1971), Borden, Whitt, and Nanney (1973a, 1973b), and Borden, Miller, Whitt, and Nanney (1977). This biochemical characterization of species had been necessary because, as with the *"Paramecium aurelia"* complex, there is strong conservation of morphological form among *Tetrahymena* species (Gates & Berger, 1976a; Nanney, Chen, & Meyer, 1978; Nanney, Cooper, Simon, & Whitt, 1980a). Multivariate techniques, however, have been successful at discriminating some strains (Gates & Berger, 1974). New species continue to be described based on combinations of mating-type reactivity and isoenzyme patterns (Nanney et al., 1980a; Nyberg, 1981a; Simon, Meyer, & Preparata, 1985). In addition to enzyme proteins, *Tetrahymena* species have been shown to vary in both ribosomal protein patterns (Cuny, Milet, & Hayes, 1979), surface proteins (Williams, Van Bell, & Newlon, 1980), and cytoskeletal proteins (Vaudaux, Williams, Frankel, & Vaudaux, 1977; Williams, Buhse, & Smith, 1984; Williams, Honts, & Dress, 1992). Williams et al. (1984) described a new species, *Tetrahymena leucophrys* in part based on the cytoskeletal protein pattern. Williams (1984) drew attention to the conspicuous disjunction between morphological and molecular variation among "*Tetrahymena pyriformis*" species: species are typically impossible to distinguish morphologically but demonstrate vast differences in cytoskeletal protein patterns. Meyer and Nanney (1987) concluded in their review of the isozyme approach to *Tetrahymena* that these molecules may be most useful in the future to analyze evolutionary processes, while systematic approaches will rely more heavily on nucleic acid sequences.

Allen and Li (1974) began the DNA approach to *Tetrahymena* taxonomy using DNA-DNA hybridizations, and demonstrated deep divergences among species. Van Bell (1985), using sequences of 5 S and 5.8 S rRNA genes, showed that the latter gene had one nucleotide change between two species. Sogin, Ingold, Karlok, Nielsen, & Engberg (1986a) used complete sequences of the SSUrRNA gene to demonstrate that most *Tetrahymena* species could be distinguished from each other while Morin and Cech (1988) demonstrated that mitochondrial large subunit (LSU) rRNA genes revealed similar phylogenetic relationships. Nanney, Meyer, Simon, and Preparata (1989) and Preparata et al. (1989) demonstrated congruence in the topologies of phylogenetic trees for evolution of *Tetrahymena* species derived from nuclear 5 S, 5.8 S, SSU-, and LSUrRNA genes, which also broadly confirmed the major clusters based on isozyme variation. These major clusters – the so-called *australis* and *borealis* clades – were also confirmed by sequences of the amino-terminal portion of the histone H4 gene (Sadler & Brunk, 1992) and telomerase RNA (Ye & Romero, 2002). What these phylogenetic trees thus clearly refuted was the assignment of the *Tetrahymena* species into three classical complexes, the *"pyriformis"* complex, *"patula"* complex of microstome-macrostome forms, and "*rostrata*" complex of histophages (Corliss, 1970, 1972c). Rather, it now appears that these three complexes represent similar life history strategies that have evolved by convergence (Nanney, Park, Preparata, & Simon, 1998; Strüder-Kypke, Wright, Jerome, & Lynn, 2001). Since true biological species of *Tetrahymena* are known to have identical SSUrRNA genes, taxonomists have used sequence differences to identify new species within the genus (Jerome, Simon, & Lynn, 1996; Lynn, Gransden, Wright, & Josephson, 2000). Jerome and Lynn (1996) provided a riboprinting strategy to identify those species whose sequences were not identical, but this leaves us unable to differentiate several species. Brunk, Lee, Tran, and Li (2003) have now embarked on a program to completely sequence the mitochondrial genomes of tetrahymenines, and have demonstrated homology of the entire organellar genomes of *T. thermophila* and *T. pyriformis*. Comparison of some mitochondrial genes suggested greater divergences among species than was found with nuclear genes. Barcode sequencing of the cytochrome *c* oxidase subunit 1 (cox-1) gene has not only differentiated those species identical based on nuclear rRNA genes (Lynn & Strüder-Kypke, 2006), but was shown to be a powerful identification tool (Chantangsi et al., 2007).

McCoy (1974b) has used the isoenzyme approach to examine "species" of *Colpidium*, and confirmed broadly earlier conclusions of morphologists (Jankowski, 1967b). Foissner and Schiffmann (1978) provided a slightly different vision of evolution within this genus, which remains to be tested by a molecular approach. Variation among strains of *Ichthyophthirius multifiliis* has also been demonstrated in the surface antigens (Dickerson, Clark, & Leff, 1993) and the amino-terminal third of histone H3 and H4 genes (Van Den Bussche, Hoofer, Drew, & Ewing, 2000). Whether this indicates that this fish parasite is also a species complex remains to be determined by future research.

The apostomes were placed by Corliss (1979) as an order in the Class KINETOFRAGMINOPHORA. However, Small and Lynn (1981, 1985) considered the similarities in the somatic kinetid of apostomes and the general features of their life cycle to demonstrate oligohymenophorean affinities, and so established them as the Subclass Apostomatia in the Class OLIGOHYMENOPHOREA. This subclass can be characterized by several synapomorphies, including the presence of a rosette opening and a polymorphic life cycle, often including palintomy within cysts. If the rosette is absent, we assume that it has been secondarily lost. Jankowski (1966a, 1966c, 1973c) suggested the division into three major groups, now recognized as orders: the Apostomatida, the Astomatophorida, and the Pilisuctorida. The Order Apostomatida is characterized by a highly modified "hymenostome" oral ciliature accompanied by a rosette and its associated *x*, *y*, and *z* kineties; there are three families – the Colliniidae, the Cyrtocaryidae, and the Foettingeriidae. Description of the complete life cycle of members of the Order Astomatophorida, monotypic for the Family Opalinopsidae, may confirm placement of these curious parasites of the internal organs of cephalopods, characterized by division by catenulation. The Order Pilisuctorida includes species that spend most of their life cycle attached to the cuticular setae of crustaceans (Bradbury, 1975; Mayén-Estrada & Aladro-Lubel, 2004). Bradbury (1982) has confirmed Jankowski's (1966a) hypothesis that the pilisuctorid *Conidophrys* has a rosette opening in its tomite stage, and so is legitimately an apostome. Bradbury (1989) has interpreted the features of the fine structure of the exuviotrophic apostome *Hyalophysa* as homologous to the paroral (i.e., two rows of staggered barren kinetosomes) of hymenostomes, an interpretation corroborated by gene sequence data (Clamp, Bradbury, Strüder-Kypke, and Lynn, 2008).

The peritrichs have long been presumed to have derived from a pleuronematine- or thigmotrich-like scuticociliate (Fauré-Fremiet, 1910, 1950a; Lom, Corliss, & Noirot-Timothée, 1968). The anterior thigmotactic region of these putative ancestors has been presumed to have given rise to the attachment structures – the scopula and adhesive disk – of the peritrichs, while the posterior oral ciliature was presumed to have evolved into the prominent peristomial ciliature of the Subclass Peritrichia. This oral ciliature is composed of a paroral, called the haplokinety, accompanied by oral polykinetid 1 – both encircle the apical end in a counterclockwise sense before entering into the oral cavity, called an infundibulum in this group. These oral organellar complexes are accompanied by two other oral polykinetids that appear peniculus-like in that they are oriented lengthwise in the oral cavity. Sequences of the SSUrRNA genes have confirmed the monophyly of the sessilid peritrichs. Moreover, SSUrRNA gene sequences place hymenostomes and peritrichs as sister lineages, refuting the "classical" hypothesis that peritrichs share a close common ancestry with the scuticociliates (Itabashi, Mikami, Fang, & Asai, 2002; Miao, Yu, & Shen, 2001; Miao et al., 2004b). However, these same gene sequences suggest that sessilids and mobilids may not be sister taxa (Gong, Yu, Villalobo, Zhu, & Miao, 2006)!

Monographic works on the peritrichs focus on the two major orders: (1) those that deal primarily with the sessiline forms, those peritrichs attached to substrates, both living and non-living, by the scopula or scopular products; and (2) those that deal with the mobiline forms, peritrichs that attach temporarily by means of an adhesive disk supported by a skeletal apparatus and surrounded by three ciliated girdles (Lom, 1994). Sessiline peritrichs can be solitary or colonial. Solitary species often aggregate, settling very close to each other to form so-called pseudocolonies. True colonial forms remain attached to the same stalk after cell division, and if zooids differentiate a monomorphic colony becomes polymorphic. The monographs on sessilines by Kahl (1935), Nenninger (1948), Stiller (1971), and Guhl (1979), for example, still remain useful. Foissner and Schiffmann (1975, 1976) have demonstrated that silver-staining can provide a rich set of characters to supplement those of cell size and shape, which were traditionally used to separate species. These surface structures can also be revealed by SEM when the peritrichs are relaxed by chlorbutol (Carey & Warren, 1983). This focus on surface features has lead to the revision of previous descriptions and the recognition of new genera (e.g., see Foissner & Schiffmann, 1976; Leitner & Foissner, 1997b; Warren, 1986, 1987, 1988). Roberts, Warren, and Curds (1983) have also demonstrated that multivariate and Fourier analyses of the outline shape of *Vorticella* species can resolve taxa to some degree. Sequencing of ITS regions has suggested that river-dwelling populations of *Carchesium polypinum* may show some vicariance biogeography (Miao et al., 2004a) while gene flow among lake-dwelling populations appears to be much higher when assessed using inter-sample sequence repeat (ISSR) fingerprinting (Zhang, Yang, Yu, Shu, & Shen, 2006). Clearly, these two studies only scratch the surface of the population genetics and biogeography of the peritrichs. Some sessiline groups secrete a lorica, which may be directly attached to the substrate or which may surround the zooid that is itself attached to the substrate by the scopula. Features of the lorica, such as its shape, character of the opening, and presence of an operculum, have proved useful in discriminating genera and species within genera (e.g., see Clamp, 1987, 1991; Finley & Bacon, 1965; Jankowski, 1985, 1986). The preliminary phylogenetic analyses based on SSUrRNA suggest that family assignments of sessilids based on morphology may not be correct. Nevertheless, we have remained conservative in our treatment and recognized the following 14 families in the Order Sessilida: the Astylozoidae, the Ellobiophryidae, the Epistylididae, the Lagenophryidae, the Operculariidae, the Ophrydiidae, the Opisthonectidae, the Rovinjellidae, the Scyphidiidae, the Termitophryidae, the Usconophryidae, the Vaginicolidae, the Vorticellidae, and the Zoothamniidae.

The Order Mobilida is characterized by a mobile zooid as a "permanent telotroch" or swarmer stage, which has a complex, ring-like, skeletal armature of denticles and fibres that support the adhesive disk on the aboral pole. We include the following five families: the Leiotrochidae, the Polycyclidae, the Trichodinidae, the Trichodinopsidae, and the Urceolariidae. Monographs on mobilines include the work of Wallengren (1897), Haider (1964), Raabe (1964), and Lom (1994). Others continue to record the morphological variability of trichodinids using numbers and sizes of the skeletal denticles, as well as other denticle characters (Kazubski, 1981, 1988, 1991; Van As & Basson, 1989). Denticle characters have been used to assess the phylogeny within the Family Trichodinidae and demonstrate that *Hemitrichodina* is a very divergent genus (Gong, Yu, Feng, & Shen, 2005).

The astomes, now as the Subclass Astomatia, have always presented a problem to ciliate systematists who have relied on oral characters to determine affinities. These endosymbionts, typically of annelids, are all mouthless, but have evolved elaborate holdfast structures in the form of hooks, spines, spicules, and suckers. The group may be polyphyletic as astomy has arisen independently within the hymenostomes (Kozloff, 1954), and astomatous mutants of *Tetrahymena* and *Glaucoma* have been isolated in the laboratory (Frankel, 1961; Orias & Pollock, 1975; Rasmussen & Orias, 1975). Nevertheless, the current phylogenetic hypothesis is that astomes arose from a thigmotrich-like ancestor with a reduced, posterior oral apparatus and an anterior thigmotactic zone (de Puytorac, 1954; de Puytorac, Grolière, & Grain, 1979). In our scheme, and different from de Puytorac (1994g), the subclass includes the single Order Astomatida with its nine families: the Anoplophryidae, the Buetschliellidae, the Clausilocolidae, the Contophryidae, the Haptophryidae, the Hoplitophryidae, the Intoshellinidae, the Maupasellidae, and the Radiophryidae. Affa'a, Hickey, Strüder-Kypke, and Lynn (2004) have presented the only molecular genetic evidence that confirms the placement of the astome *Anoplophrya* within the Class OLIGOHYMENOPHOREA. As with the thigmotrichs, monographic work on the group is now well over 30 years old (e.g., Corliss, de Puytorac, & Lom, 1965; de Puytorac,

1954, 1957, 1960, 1961, 1970, 1972), but see de Puytorac (1994g) for an update.

In conclusion, molecular phylogenies of SSUrRNA, LSUrRNA, and some proteins strongly support the monophyly of the Class OLIGOHYMENOPHOREA and the subclasses of predominantly free-living ciliates assigned to it. We await anxiously the results of more comprehensive analyses of gene sequences for representatives of the astomes and the apostomes, both to confirm preliminary results of their assignment to this class and to determine if each subclass is indeed monophyletic.

15.2 Life History and Ecology

It almost goes without saying that the ciliates in this class are broadly distributed throughout the world. For those species that are symbionts, both ectosymbionts and endosymbionts, their distribution is determined by that of their host or hosts. This would include all the species in the subclasses Astomatia and Apostomatia, a large number of species of peritrichs, including all mobilid peritrichs, a few species of scuticociliates and hymenostomes, and a rare species of peniculine (Maguire & Belk, 1967). Representatives of the subclasses Scuticociliatia, Peniculia, and Peritrichia are often reported as conspicuous members of free-living assemblages of ciliates, while hymenostomes are less often conspicuous. Oligohymenophoreans have some of the most complex life cycles among the ciliates (Fig. 15.1; see also Figs. 3.1, 4.8).

Scuticociliates, typically *Cyclidium* and *Pleuronema* species, have been found: in soils and mosses in Europe (Foissner, 1981a; Grolière, 1975c), South America (Steffens & Wilbert, 2002), Africa (Buitkamp, 1977; Steffens & Wilbert, 2002), and Antarctica (Ryan et al., 1989); in temporary ponds in Latin America (López-Ochoterena, 1966) and Antarctica (Thompson, 1972); in marine plankton, especially of coastal regions, in North America (Borror, 1963; Dolan, 1991), Latin America (Bulit, Díaz-Avalos, Signoret, & Montagnes, 2003; Silver, Gowing, Brownlee, & Corliss, 1984), in Europe (Edwards & Burkhill, 1995), sometimes as deep as 900 m (Hausmann, Hülsmann, Polianski, Schade, & Weitere, 2002), and in Antarctica (Song & Wilbert, 2000b); in marine sands as interstitial fauna from Western Europe (Dragesco, 1963; Fernández-Leborans & Fernández-Fernández, 1999; Fernández-Leborans, Valgañon, & Castro de Zaldumbide, 1999), from Eastern Europe (Agamaliev, 1971; Burkovsky, 1970; Kovaleva & Golemansky, 1979), and the Arabian Gulf (Al-Rasheid, 1999c); in hypersaline and solution lakes in Europe (Dyer, 1989; Esteban, Finlay, & Embley, 1993a), Africa (Yasindi, Lynn, & Taylor, 2002), and Australia (Post, Borowitzka, Borowitzka, Mackay, & Moulton, 1983); in leaf litter in mangrove forests in Asia (Dorothy, Satyanarayana, Kalavati, Raman, & Dehairs, 2003); in streams and rivers in Europe (Cleven, 2004; Domenech, Gaudes, Lopez-Doval, Salvado, & Munoz, 2006; Foissner, 1997b; Madoni & Ghetti, 1980); in freshwater ponds in Europe (Finlay et al., 1988; Madoni & Sartore, 2003) and North America (Wickham & Gilbert, 1993); and in freshwater lakes in Europe (Carrias, Amblard, & Bourdier, 1994; Finlay, Bannister, & Stewart, 1979; Schlott-Idl, 1984; Skogstad, Granskog, & Klaveness, 1987; Zingel & Ott, 2000; Zingel, Huitu, Makela, & Arvola, 2002), in North America (Beaver & Crisman, 1982, 1989b), and Asia (Obolkina, 2006; Song, 2000), often in the anoxic hypolimnion (Guhl, Finlay, & Schink, 1996). *Cyclidium* and other oligohymenophoreans have been recorded in the fluid from pitcher plants on three continents, Eurasia, North America, and Australia (Cochran-Stafira & von Ende, 1998; Rojo-Herguedas & Olmo, 1999) and bromeliads in Central and South America (Foissner, Strüder-Kypke, van der Staay, Moon-van der Staay, & Hackstein, 2003).

Scuticociliates are typically microphagous bacterivores, consequently they are often more abundant in eutrophic habitats (Beaver & Crisman, 1982), achieving abundances of almost $40{,}000\,l^{-1}$ in freshwater lakes (Song, 2000). In lakes and coastal marine habitats, they are often most common in the deeper waters (Zingel & Ott, 2000), often at or below the oxycline (Finlay & Maberly, 2000; Fenchel, Kristensen, & Rasmussen, 1990; Fenchel et al., 1995; Taylor & Heynen, 1987). These ciliates may first become associated with sinking detritus on which bacteria are growing (Silver et al., 1984) or they may be growing in situ in the water column where bacterial abundances may be higher (Fenchel et al., 1990).

Peritrichs, typically assigned to the genus *Vorticella*, are the second most abundant oligohymenophorean group recorded from a variety of habitats. They have been found: in soils and mosses in Europe (Foissner, 1981a) and Antarctica (Ryan et al., 1989); in permanent and temporary ponds in Europe and North America (Madoni & Sartore, 2003), Latin America (López-Ochoterena, 1966), and Antarctica (Thompson, 1972); in coastal marine habitats, primarily attached to substrates, in North America (Beech & Landers, 2002; Borror, 1963; Landers & Phipps, 2003), in Asia (Dorothy et al., 2003), and Antarctica (Song & Wilbert, 2002); in rivers and streams, attached to substrates, in Europe (Cleven, 2004; Foissner, 1997b; Harmsworth & Sleigh, 1993), North America (Small, 1973; Taylor, 1983a), and Asia (Kusuoka & Watanabe, 1987), and also in the plankton in Europe (Balazi & Matis, 2002) and North America (Clamp & Coats, 2000); in freshwater ponds in Europe (Finlay et al., 1988) and Arabia (Al-Rasheid, 1996); and in the pelagial of freshwater lakes in Europe (Foissner, 1979d; Packroff, 2000; Zingel & Ott, 2000), North America (Beaver & Crisman, 1989b; Kerr, 1983), South America (Barbieri & Orlandi, 1989), and Asia (Obolkina, 2006; Song, 2000).

Pelagic peritrichs may not be free-swimming, but rather are attached to filamentous algae (Davis, 1973; Kerr, 1983; Pratt & Rosen, 1983). Abundances of peritrichs are not often reported, and when planktonic, are difficult to interpret because they are often attached to algae in the plankton. They can attain abundances exceeding $2{,}000\,l^{-1}$ in freshwater habitats, growing on colonies of *Microcystis* and *Nostoc* (Barbieri & Orlandi, 1989; Kerr, 1983). Freshwater species of *Carchesium* and *Vorticella* have been recorded to exceed $1{,}000\,cm^{-2}$ and $100\,cm^{-2}$ respectively (Kusuoka & Watanabe, 1987), but total peritrich communities are typically below $50\,cm^{-2}$ on benthic substrates (Harmsworth & Sleigh, 1992, 1993). In marine habitats, colonization of artificial substrates by peritrichs can achieve densities exceeding $5{,}000\,cm^{-2}$, but typically abundances are well below $1{,}000\,cm^{-2}$ (Beech & Landers, 2002; Landers & Phipps, 2003).

Peniculines, typically assigned to the genera *Frontonia*, *Paramecium*, and more rarely *Stokesia*, have often been reported from a variety of habitats around the world, and are rarely more abundant than representatives from the previous two subclasses. Peniculines have been found – in soils from South America and Africa (Steffens & Wilbert, 2002); in temporary ponds from Latin America (López-Ochoterena, 1966); in marine habitats, from the water column in Europe (Fenchel et al., 1990) to coastal waters in North America (Borror, 1963) to Antarctica (Thompson, 1972); in the psammobiotic communities of Europe and Asia (Agamaliev, 1968; Burkovsky, 1970; Fernández-Leborans et al., 1999), including Arabia (Al-Rasheid, 1999d); in rivers and streams in Europe (Domenech et al., 2006; Foissner, 1997b; Komala & Przybos, 1990) and Africa (Dragesco, 1972); in freshwater ponds in Europe (Finlay et al., 1979, 1988; Komala & Pryzbos, 1994; Kosciuszko & Prajer, 1991; Madoni & Sartore, 2003; Przybos & Fokin, 1997), and in Asia (Przybos, Fokin, Stoeck, & Schmidt, 1999), including Arabia (Al-Rasheid, 1996); in hypersaline lakes in Africa (Yasindi et al., 2002); and in freshwater lakes in Europe (Schlott-Idl, 1984), in North America (Hunt & Chein, 1983), in Asia (James, Burns, & Forsyth, 1995; Obolkina, 2006), and in Africa (Dragesco & Dragesco-Kernéis, 1991). Peniculines are dominant in terms of abundance or biomass only in exceptional circumstances. For example, on occasion, Yasindi et al. (2002) noted *Frontonia* species dominating the community of the alkaline saline Lake Nakuru. The planktonic taxon, *Disematostoma*, has been recorded to exceed $10{,}000\,l^{-1}$ (Finlay et al., 1988) while its relative *Stokesia* rarely exceeds $100\,l^{-1}$ (Hunt & Chein, 1983).

Hymenostomes are rarely recorded in general surveys of habitats, suggesting that they are extremely patchy in their distributions and probably dependent upon high concentrations of bacteria associated with decaying organic matter. However, when appropriate sampling strategies are employed, *Tetrahymena* species have been recorded from the major continents (Elliott, 1973b; Simon et al., 1985). *Lambornella* species have been recorded from tree-hole habitats where they may parasitize mosquitoes (Washburn, Gross, Mercer, & Anderson, 1988), and a proposed new tetrahymenine genus, *Bromeliophrya*, has been found in water trapped in the leaf axils of bromeliads (Foissner et al., 2003).

While the above discussion has focused on free-living oligohymenophoreans, there is a much more abundant literature that deals with those species as posterior contractile vacuole and a somewhat pyriform body shape, suggests that it might have been a hymenostome.

Few hymenostomes have been reported as symbionts, with species in the three genera, *Tetrahymena*, *Ophryoglena*, and *Ichthyophthirius* being the most commonly reported. *Tetrahymena* species have been reported infecting natural populations of platyhelminthes (Wright, 1981), gastropod molluscs (Kazubski, 1964; Kozloff, 1956), and a variety of kinds of insects, such as black flies (Batson, 1983; Lynn, Molloy, & Lebrun, 1981), mosquitoes (Barros et al., 2006; Clark & Brandl, 1976; Corliss & Coats, 1976; Egerter & Anderson, 1985; Jerome et al., 1996), chironomids (Corliss, 1960b; Corliss, Berl, & Laird, 1979; Golini & Corliss, 1981), and megalopterans (Batson, 1985), and aquarium or aquacultured fishes (Astrovsky et al., 2002; Ferguson, Hicks, Lynn, Ostland, & Bailey, 1987; Hatai et al., 2001; Hoffman et al., 1975; Imai, Tsurimaki, Goto, Wakita, & Hatai, 2000). *Tetrahymena* has infected a broader array of organisms in experimental situations (Thompson, 1958), and it is likely that it is only a facultative parasite in nature (Corliss, 1960b, 1972c). Nevertheless, in some situations, these ciliates may cause significant mortality of their insect hosts (Barros et al., 2006; Grassmick & Rowley, 1973; Zaritsky, Ben-Dov, Zalkinder, & Barak, 1992). Could they be used as biocontrol agents for these vectors of important human diseases? The relationship between the predator-host mosquito and its ciliate symbiont is intimate enough to have evolved a predator-induced chemical signal that causes the free-living "symbionts" to transform as parasites, ultimately killing their host (Washburn et al., 1988). As a taxonomic aside, this tetrahymenid associated as a cuticular cyst on mosquitoes (Corliss & Coats, 1976; Egerter & Anderson, 1985) has been assigned to the genus *Lambornella*, although recent molecular evidence suggests that this genus rank is probably not justified (Strüder-Kypke et al., 2001).

Ophryoglena species are typically characterized as histophagous on moribund and dead aquatic organisms (Mugard, 1949), although they have been found infecting living bivalves (Karatayev, Burlakova, Molloy, Volkova, & Volosyuk, 2002; Molloy, Lynn, & Giamberini, 2005) and insects (Gaino & Rebora, 2000) (Fig. 15.1). There is now evidence that parasitic *Ophryoglena* have been very recently introduced to Ireland with their bivalve hosts, likely by humans from mainland Europe (Burlakova et al., 2006). Undoubtedly the most infamous ophryoglenine ciliate is *Ichthyophthirius multifiliis*, the causative agent of white spot disease of fishes. There is an extensive literature on this ciliate (see Dickerson & Dawe, 1995; Matthews, 2005). *"Ich"* is apparently attracted to host fish by serum factors in their mucus (Buchmann & Nielsen, 1999; Haas, Haberl, Hofmann, Kerschensteiner, & Ketzer, 1999), and can cause significant mortalities in aquaculture operations (e.g., Munderle, Sures, & Taraschewski, 2004). To our knowledge, only one, natural, mass mortality has ever been reported during which over 18 million killifish in Lake Titicaca were killed (Wurtsbaugh & Tapia, 1988). However, *"Ich"* infections can impact the swimming speed of eels and, while not killing them, may affect their ability to reach spawning sites in early spring (Münderle, Sures, & Taraschewski, 2004). The dispersal or theront stage of the life cycle arises by multiple palintomic divisions from an encysted tomont (Dickerson & Dawe, 1995; Ewing & Kocan, 1992) (Fig. 15.1). The theront contacts the host epithelium or gill tissue and penetrates between cells to enter the epidermis (Ewing, Kocan, & Ewing, 1985; Kozel, 1986). Fish hosts can be immunized against *"Ich"* using the surface immobilization antigen of the ciliate (Buchmann, Sigh, Nielsen, & Dalgaard, 2001; Wang & Dickerson, 2002; Xu, Klesius, & Panangala, 2006). While malachite green and formalin have long been effective treatments, both have carcinogenic properties. Malachite green has already been banned on fish farms in some countries. The search for other effective treatment compounds has included sodium percarbonate, garlic extract, triazinone, and crude extracts of plants (Buchmann, Jensen, & Kruse, 2003; Ekanem, Obiekezie, Kloas, & Knopf, 2004; Schmahl et al., 1989).

Apostomes have been reported as commensal symbionts, primarily from crustaceans (Bradbury, 1996; Chatton & Lwoff, 1935a). Their life cycles are complex and varied (Fig. 3.1). One group, represented by species of *Hyalophysa* and *Gymnodinioides*, is termed exuviotrophic because they excyst to feed on the exuvial fluids in the host moult, often increasing their body volume 60-fold before they encyst, divide, and disperse to find another host (Bradbury, 1966a; Grimes, 1976;

Landers, Confusione, & Defee, 1996). A second group, represented by *Terebrospira*, burrows through the endocuticle of the host shrimp and ingests the dissolved products (Bradbury, Clamp, & Lyon, 1974; Debaisieux, 1960). A third group, represented by *Vampyrophrya*, ingests tissues of the host calanoid copepod, either when it is injured or ingested (Grimes & Bradbury, 1992). A fourth group, represented by *Collinia*, lives endoparasitically in the body fluids of euphausiid crustaceans (Capriulo & Small, 1986; Lindley, 1978), and can cause mass mortalities of their hosts (Gómez-Gutiérrez, Peterson, De Robertis, & Brodeur, 2003; Gómez-Gutiérrez, Peterson, & Morado, 2006). Two other, highly unusual members of the subclass are the pilisuctorids *Conidophrys* and *Askoella*, which attach to the setae of the host crustacean (Bradbury, 1975; Mayén-Estrada & Aladro-Lubel, 2004), and the cyrtocarid *Cyrtocaryum*, which lives in the digestive caeca of polychaete annelids (Fauré-Fremiet & Mugard, 1949b). A final example are the chromidinid apostomes, which were studied by Chatton and Lwoff (1935a), and have been recently reported from the kidneys of Japanese cephalopods (Furuya, Ota, Kimura, & Tsuneki, 2004).

Astomes are obligate commensal symbionts, found typically in the digestive tract of annelids (de Puytorac, 1994g). Cépède (1910) and de Puytorac (1954) stand as the substantial 20th century monographic works on this group. Despite these intensive investigations with reports from Europe (Cépède; de Puytorac), North America (Bush, 1934; Powders, 1970), and Africa (de Puytorac & Dragesco, 1969a, 1969b; Ngassam, 1983), we still do not know how these ciliates are transmitted from one host to the next. They display a variety of cellular differentiations, such as hooks and suckers, to maintain their position in the intestine (de Puytorac, 1994g). The distribution of species of *Maupasella*, *Anoplophrya*, and *Metaradiophrya* along the digestive tract of their host worm *Allolobophrya savigni* is correlated with pH: each species apparently preferring a region characterized by a different pH (de Puytorac & Mauret, 1956). *Cepedietta* species are found in the intestine of salamanders, and their prevalence is inversely related to altitudes below 1,400 m, perhaps explained by temperature variations (Powders, 1970). There is yet no experimental evidence on how astomes feed. However, it is likely that they use receptor-mediated endocytosis, perhaps at the parasomal sacs, as has been demonstrated for *Tetrahymena* (Nilsson & van Deurs, 1983) and *Paramecium* (Allen, Schroeder, & Fok, 1992; Ramoino et al., 2001).

While we can only speculate at the moment on the feeding habits and preferences of astomes, there is no doubt that most free-living oligohymenophoreans are bacterivorous, down-stream filter feeders. The cilia of the paroral or undulating membrane typically are used to filter the water from the feeding current created by the oral polykinetids (Fenchel, 1980a, 1980b), although species without well-developed paroral cilia, such as *Glaucoma* species, may use the innermost oral polykinetid as the filter (Fenchel & Small, 1980). Hymenostomes, such as species in the genera *Colpidium*, *Glaucoma*, and *Tetrahymena*, can ingest a variety of bacterial species, which vary in how well they support growth (Dive, 1973; Taylor, 1979; Taylor & Berger, 1976; Taylor, Gates, & Berger, 1976). *Colpidium* (or *Dexiostoma*) and other hymenostomes may also supplement their diet with small detrital particles (Posch & Arndt, 1996). *Tetrahymena* may be a poor competitor in relation to *Colpidium* or *Paramecium* (Long & Karel, 2002). This may explain the selection for histophagous and endoparasitic feeding strategies in some *Tetrahymena* species (Corliss, 1972c; Roque, de Puytorac, & Savoie, 1971), although glaucomids, such as *Espejoia*, have also adopted histophagy, feeding on and in the gelatinous matrices of egg masses of aquatic insects and molluscs (Fryd-Versavel, Iftode, & Wilbert, 1975). Cannibalism has also evolved in *Tetrahymena* with species like *Tetrahymena vorax* and *Tetrahymena patula* able to respond to secretions from prey species and develop into large-mouthed or macrostome predators able to feed on smaller *Tetrahymena* species (Fig. 15.1) (Buhse, 1967; Corliss, 1973; Williams, 1960, 1961). Furthermore, macrostome forms of *T. vorax* appear to be highly selective feeders, preferring to ingest *T. thermophila* over latex beads and microstome forms of *T. vorax* (Grønlien, Berg, & Løvlie, 2002).

Peniculine feeding preferences range from bacterivory to mixotrophy. The peniculine *Paramecium* feeds on a variety of bacterial species, although some bacterial species may be toxic (Curds & Vandyke, 1966). *Paramecium* may also supplement its diet by ingesting detrital particles (Posch & Arndt, 1996). *Paramecium bursaria* typically hosts endosymbiotic *Chlorella* species (**see below**), and these influence the emphasis on bacterivory: in the dark *P. bursaria* relies on bacterivory, but in the light the predominant nutritional source derives from the photosynthetic products of its symbionts (Weis, 1974). However, Berk, Parks, and Ting (1991) observed that light itself may enhance ingestion rates since mixotrophic *P. bursaria* fed faster than aposymbiotic individuals. As an aside, *Chlorella*-bearing *P. bursaria* are not ingested by *Didinium* as rapidly as apochlorotic individuals, suggesting that metabolites from the consortium may discourage predation (Berger, 1980). On the other hand, species of the peniculine *Frontonia* are typically not bacterivorous, but flourish on chrysophytes, cryptophytes, chlorophytes, diatoms, and even testate amoebae (Dias & D'Agosto, 2006; Skogstad et al., 1987). Carnivorous peniculines include the giant *Neobursaridium*, first classified as a heterotrich because of its large size and convergently arranged somatic cilia that appeared like an adoral zone (Dragesco & Tuffrau, 1967; Nilsson, 1969), and *Lembadion*, which can adjust its size to the size of its prey, such as *Colpidium* and *Paramecium* (Fyda, 1998; Kopp & Tollrian, 2003).

Peritrichs, such as *Vorticella*, *Epistylis*, and *Zoothamnium*, are very efficient downstream filter feeders (Fenchel, 1980a, 1980b; Sleigh & Barlow, 1976). They can have significant grazing impacts on bacterial and picocyanobacterial communities, ingesting over 4,000 bacteria per cell per hour and sometimes over 500 picocyanobacteria per cell per hour (Callieri, Karjalainen, & Passoni, 2002; Šimek, Bobková, Macek, Nedoma, & Psenner, 1995). This demonstrates their importance as components of the community in sewage treatment facilities. Their influence in these communities, and in natural habitats, such as mangroves, extends to their creation of strong micro-currents, flowing sometimes at over 180 μm sec^{-1}. These currents can bring nutrient-rich waters to the bacterial biofilm on the surface of which the peritrichs are attached (Fried & Lemmer, 2003; Vopel, Reick, Arlt, Pohn, & Ott, 2002, Vopel, Thistle, Ott, Bright, & Roy, 2004). Even epibionts on turbellarians, such as *Urceolaria mitra*, feed on bacteria, chlorophytes, and chrysophytes (Reynoldson, 1955). Some peritrichs, particularly *Ophrydium* species, are known to harbor *Chlorella* symbionts. Non-colonial *Ophrydium* can be dominant ciliates in some lakes (Modenutti & Balseiro, 2002) while the colonies of *Ophrydium versatile* can be conspicuous components of the freshwater benthos (Duval & Margulis, 1995). Photosynthesis by the *Chlorella* symbionts of *O. versatile* is very efficient at low light levels and along with filter-feeding by the ciliates produces carbon sufficient to maintain the growth rate of the colony (Sand-Jensen, Pedersen, & Geertz-Hansen, 1997).

Scuticociliates, characterized as downstream filter feeders (Fenchel, 1980a, 1980b), are typically the dominant bacterivorous ciliates in the hypolimnion of freshwater (Amblard, Sime-Ngando, Rachiq, & Bourdier, 1993; Carrias, Amblard, & Bourdier, 1998; Taylor & Heynen, 1987) and brackish environments (Dolan, 1991). A variety of bacterial species has been observed to be ingested by *Cyclidium* (Šimek, Macek, Pernthaler, Straškrabová, & Psenner, 1996; Taylor, 1979) and *Uronema* species (Christaki, Dolan, Pelegri, & Rassoulzadegan, 1998, Christaki, Jacquet, Dolan, Vaulot, & Rassoulzadegan, 1999; Hamilton & Preslan, 1969; Iriberri, Ayo, Santamaria, Barcina, & Egea, 1995; Parker, 1976). Ingestion can exceed 500 bacteria per cell per hour for *Cyclidium* (Šimek et al., 1996). While *Cinetochilum* could ingest bacteria, its growth flourished on diatoms, dinophytes, chrysophytes, cryptophytes, and chlorophytes (Šimek et al., 1995; Skogstad et al., 1987). The larger planktonic scuticociliate *Histiobalantium* grows well on cryptophytes (Müller & Weiss, 1994). As predicted by bead experiments (Fenchel, 1980a, 1980b), *Cyclidium* does feed selectively on different-sized bacterial prey (Šimek, Vrba, & Hartman, 1994), and this can, in turn, influence size distribution of the bacterial community (Posch et al., 2001). Nevertheless, the situation is undoubtedly more complex. Sanders (1998) and Christaki et al. (1998, 1999) have demonstrated that the surface properties of particles can influence the ingestion rates of both freshwater and marine scuticociliates.

Oligohymenophoreans are some of the fastest growing ciliates, at least in the laboratory setting. *Tetrahymena* under appropriate axenic culture conditions can achieve doubling times of less than 1.5 h (Orias et al., 2000). However, more typically the doubling time is around 5.5 h in axenic proteose-peptone medium extending to 8.5 h in bacterized medium (Taylor et al., 1976). The doubling times of other hymenostomes, such as *Colpidium* and *Glaucoma* species, on bacterial prey achieve maxima around 2.5 h and 4 h, respectively (Taylor, 1978). Scuticociliates, such as *Uronema*, *Parauronema*, and *Cyclidium*, can achieve doubling times close to 2.5 h in the laboratory setting, but typically exceed 4 h (Pérez-Uz, 1995, 1996; Taylor). Laboratory growth rates can be significantly different among geographically diverse clones of *Uronema*, suggesting local physiological adaptation or genetic differentiation (Pérez-Uz, 1995). In contrast, growth rates of field populations of scuticociliates incubated over a 24-h period tend to be much slower, indicating that natural food supplies likely limit growth (Macek, Šimek, Pernthaler, Vyhnálek, & Psenner, 1996). This difference also applies to laboratory and field populations of peritrichs, such as *Vorticella*, *Epistylis*, and *Opercularia*: doubling times in the laboratory can be around 1–2 h (Curds & Vandyke, 1966) while estimates from field populations suggest 8 h to be typical (Taylor, 1983b). The peniculine *Paramecium* can also achieve doubling times approaching 2 h in the laboratory setting (Curds & Vandyke, 1966).

Oligohymenophoreans, in addition to being effective predators on organisms ranging from bacteria to fish, are themselves also prey. Hymenostomes, such as *Tetrahymena* and *Colpidium*, are consumed by copepods (Hartmann, Taleb, Aleya, & Lair, 1993; Kumar, 2003), cladocerans (Jack & Gilbert, 1993), rotifers (Gilbert & Jack, 1993), and mosquito larvae (Addicott, 1974). *Colpidium* species can escape predation by the macrostomatous peniculine *Lembadion* by significantly transforming their morphology, becoming more broad and almost spherical in shape (Fyda, 1998; Fyda, Kennaway, Adamus, & Warren, 2006). Peniculines, such as *Paramecium* and *Frontonia*, are consumed by rotifers (Maly, 1975), copepods (Hartmann et al., 1993), cladocerans (DeBiase, Sanders, & Porter, 1990; Jack & Gilbert, 1993), and mosquito larvae (Addicott, 1974). Peniculines are also the prey of other ciliates, such as the suctorian *Podophrya* (Jurand & Bomford, 1965), a variety of litostomes (Harumoto & Miyake, 1991; Miyake & Harumoto, 1996; Salt, 1974), and even larger colpodeans (Foissner, 1993a; Salt, 1967). While the trichocyst extrusomes of peniculines may not defend them against their metazoan predators, *Paramecium* species obtain some defensive function from these organelles against some litostome predators, an exception being *Didinium*, the infamous predator of *Paramecium* (Harumoto, 1994; Harumoto & Miyake, 1991; Miyake & Harumoto; Sugibayashi & Harumoto, 2000). Even though some peritrichs can contract their stalks at up to 60 cm sec^{-1} (Lom, 1994), they are still eaten. Peritrichs, such as *Epistylis* and *Opercularia*, show survivorships of about 50% per day (Taylor, 1983b), possibly due to predation by rotifers and oligochaetes (Kusuoka & Watanabe, 1989) and insects (Addicott, 1974). Their sessile nature makes them susceptible to other ciliates, such as the slow-moving pleurostome litostomes (Canella, 1951; Foissner, 1983b), and even hyphomycete "fungi" (Barron & Szijarto, 1982). Finally, scuticociliates are susceptible to predation by heliozoans (Pierce & Coats, 1999), copepods (Burns & Gilbert, 1993; Ederington, McManus, & Harvey, 1995), cladocerans (Wickham & Gilbert, 1993), and mosquito larvae (Addicott, 1974). While it has not yet been supported empirically, the jumps and darts of some planktonic scuticociliates, like *Cyclidium*, may reduce their susceptibility to predation (Tamar, 1979), as has been demonstrated for prostomes and oligotrichs (Jakobsen, 2001).

Oligohymenophoreans can also be infected or colonized by a variety of smaller organisms, ranging from bacteria to other protozoa. The literature on these bacterial groups is extensive and is only briefly introduced here. Probably the best-known symbionts are the so-called "killer particles" of *Paramecium*, originally reported by Sonneborn (1938). They have now been assigned to the alpha-proteobacterium genus *Caedibacter* (Pond, Gibson, Lalucat, & Quackenbush, 1989; Preer, Preer, & Jurand, 1974; Quackenbush, 1988; Fokin & Görtz, 1993) or to the gamma-proteobacteria (Beier et al., 2002). It is the so-called R body, constructed by these obligate endosymbionts, which kills susceptible host cells (Preer et al., 1974; Quackenbush, 1988). The existence of "killer" *Paramecium* in nature is probably maintained by natural selection as "killers" are rarely found with "sensitives" in natural collections (Landis, 1981, 1986). Another obligately symbiotic bacterial species group is an assemblage belonging to the genus *Holospora* (Görtz, 1988b, 1996). The nuclei of several *Paramecium* species are infected: the macronucleus by some *Holospora* species (e.g., Fokin, Brigger, Brenner, & Görtz, 1996; Fujishima, Sawabe, & Iwatsuki, 1990) and the micronucleus by others (e.g., Görtz & Dieckmann, 1980; Ossipov, Borchsenius, & Podlipaev, 1980). These latter species can essentially genetically castrate the *Paramecium* by destroying the micronucleus (Görtz, 1988b). On the other hand, conjugation may be a way that some *Paramecium* can rid themselves of the macronuclear endosymbionts, although symbiont strategies can avoid this by migrating to the anlage or by causing the fusion of infected pycnotic fragments with the anlage (Fokin, 1998). Some clones of *Paramecium* are resistant to infection by *Holospora*, and this resistance can evolve (Lohse, Gutierrez, & Kaltz, 2006). Microsurgical transfers of nuclei suggest that the lytic abilities of the host are mediated by macronuclear activity (Fokin & Skovorodkin, 1997). A variety of other bacterial endosymbionts has been described in *Paramecium* species (Fokin, Sabaneyeva, Borkhsenius, Schweikert, & Görtz, 2000; Görtz, 1996), in the scuticociliates *Uronema* (Soldo, Brickson, & Vazquez, 1992; Soldo, Godoy, & Brickson, 1974), *Schizocaryum* (Lynn & Frombach, 1987), *Cyclidium* (Esteban et al., 1993b), and *Conchophthirus* (Fokin, Giamberini, Molloy, & de Vaate, 2003), and in the hymenostome *Ophryoglena* (Fokin et al., 2003). *Holospora* may enhance the success of entry of these other symbionts into the nuclei of *Paramecium* (Fokin, Skovorodkin, Schweikert, & Görtz, 2004). While the host-symbiont relationship of most of these symbionts is unknown, scuticociliates from anaerobic habitats are known to harbor methanogens, which make use of hydrogen generated by the host (Esteban & Finlay, 1994; Esteban et al., 1993b).

From a human perspective, possibly the most insidious examples of "endosymbiotic" bacteria carried by ciliates are species of the "pneumonia-causing" genus *Legionella*, which have been found infecting and confirmed to proliferate in *Tetrahymena* species (Barbaree, Fields, Feeley, Gorman, & Martin, 1986; Steele & McLennan, 1996). This led to *Tetrahymena* being called "Trojan Horses of the microbial world" (Barker & Brown, 1994). Bacteria are occasionally observed also as epibionts on the cell surface of oligohymenophoreans (e.g., Bauer-Nebelsick, Bardele, & Ott, 1996; Beams & Kessel, 1973; Esteban & Finlay, 1994; Lynn & Frombach, 1987).

The diversity of eukaryotic endosymbionts of oligohymenophoreans pales in comparison to the prokaryotes. Gillies and Hanson (1963) described a *Leptomonas* species that infected the macronucleus of *Paramecium* species. Suctorians have been reported as "parasites" of peniculines and peritrichs (Jankowski, 1963; Padnos & Nigrelli, 1947; Pérez Reyes & López-Ochoterena, 1963), and rhynchodids can "parasitize" peritrichs (Chatton & Lwoff, 1939b). The astome *Spirobuetschliella* was reported to be parasitized by the microsporidian *Gurleya* (Hovasse, 1950).

The vast majority of research on eukaryotic symbionts has focused on the endosymbiotic *Chlorella* species of *Paramecium bursaria* (Görtz, 1996; Reisser, 1986). These *Chlorella* symbionts enhance the growth rate, maximum population density, and survival of their host *Paramecium* (Karakashian, 1975). Several strains and species of *Chlorella*, which typically release several times more sugar by cell dry weight than non-infective isolates, have been isolated from different strains of *P. bursaria* world-wide (Reisser, Vietze, & Widowski, 1988; Weis, 1979). Karakashian and Rudzinska (1981) demonstrated that vacuoles containing infective *Chlorella* inhibited lysosomal fusion, and speculated that this was due to alteration of the vacuolar membrane, a prediction confirmed by Meier, Lefort-Tran, Pouphile, Reisser, and Reisser (1984). While "infection-capable" *Chlorella* species may influence food vacuolar membrane properties, these species are also distinguished by the presence of glucosamine in the cell wall (Takeda, Sekiguchi, Nunokawa, & Usuki, 1998). Nevertheless, *Chlorella* cells can be digested by the host ciliate, and this is particularly enhanced in the dark (Gu, Chen, Ni, & Zhang, 2002). Perhaps darkness increases the mortality of the *Chlorella*, which cannot then "control" their vacuolar environment. *Chlorella*-type symbionts have also been observed in another peniculine *Frontonia* (Finlay & Maberly, 2000) and in the peritrichs *Vorticella* (Graham & Graham, 1980) and *Ophrydium* (Woelfl & Geller, 2002). The abundances of these ciliates, coupled with the photosynthetic activity of their symbionts, can at times make them significant contributors to the primary production of some waters (Sand-Jensen et al., 1997; Woelfl & Geller, 2002).

Oligohymenophoreans demonstrate behavioral responses to a variety of environmental parameters. The repertoire of these responses, in turn, can help to explain their ecology. Symbiont-bearing *P. bursaria* show positive photokinesis and photoaccumulation (Cronkite & Van den Brink, 1981; Nakaoka, Kinugawa, & Kurotani, 1987). This response occurs with different *Chlorella* species and requires algal photosynthesis. At least 50 individual *Chlorella* cells must be present in an individual *Paramecium* to induce this behavior (Niess, Reisser, & Wiessner, 1982). Hymenostomes, such as the ophryoglenid *Ophryoglena*, and scuticociliates, such as *Porpostoma*, have cup-like organelles in the oral region. These ciliates also demonstrate complex life cycles with theront, trophont, protomont, tomont, and tomite stages (Fig. 15.1). Photobehavior is related to the life-cycle stage: the tomont or dividing stage typically exhibits negative phototaxis while the dispersing theront stage exhibits little preference (Kuhlmann, 1993; Kuhlmann, Bräucker, & Schepers, 1997). Theronts of *Ophryoglena* do exhibit chemotaxis (Kuhlmann, 1993), a behavior that has been thoroughly investigated in species of *Paramecium* and *Tetrahymena*. Species in these latter genera are attracted to inorganic compounds and organic compounds, including amino acids (Almagor, Ron, & Bar-Tana, 1981; Levandowsky et al., 1984; Hellung-Larsen, Leick, & Tommerup, 1986; Van Houten, 1975, 1982). *Tetrahymena* is particularly sensitive, at 3×10^{-8} M, to some proteins, such as platelet-derived growth factor (Hellung-Larsen et al., 1986), and this may explain the facultative histophagy exhibited by a number of species in this genus (**see above**). Finally, *Paramecium* shows behavioral hypothermia as these organisms seek lower temperatures apparently to survive hypoxic conditions, a behavior also exhibited by a variety of animal species (Malvin & Wood, 1992).

As noted above, oligohymenophoreans can exhibit complex polymorphic life cycles. Stages in these life cycles exhibit different behaviors, undoubtedly of adaptive significance (Fenchel, 1990). A prime stress in the life cycle of any protist is the disappearance of food. There are two typical responses to starvation or the absence of appropriate food. The first is transformation to the theront or "hunter" phenotype (Fig. 15.1; also Fig. 4.8). This phenotype is typically characterized by a more elongate cell shape than the trophont or "feeding" phenotype and by more rapid swimming with fewer tumbles and turns (Fenchel, 1990; Nelsen & Debault, 1978). Depending upon the species, after a certain period of starvation, during which autophagy occurs (Nilsson, 1984), the second typical response to starvation – encystment – occurs. Encystment has been reported in all major groups of oligohymenophoreans, except the astomes. The cyst wall, derived in part from extrusomes, is often multilayered, composed of an ectocyst, mesocyst, and endocyst. The cyst wall is composed of chitin, other complex carbohydrates, and some proteins. Cysts have been analyzed in the peniculines *Furgasonia* (= *Cyclogramma*) and *Pseudomicrothorax* (Bussers, 1976), the scuticociliate *Pseudocohnilembus* (Olendzenski, 1999), the hymenostomes *Tetrahymena* (McArdle, Bergquist, & Ehret, 1980) and *Ichthyophthirius* (Ewing, Kocan, & Ewing, 1983), the peritrichs *Telotrochidium* (Walker, Edwards, & Suchard, 1989) and *Opisthonecta* (Calvo, Fernandez-Aliseda, Garrido, & Torres, 2003), and the apostome *Hyalophysa* (Bradbury, 1974; Landers, 1991a, 1991b). Excystment is stimulated by a number of factors (for review see Corliss & Esser, 1974). Ultimately and in all probability, a small set of "signal" molecules may be responsible: for example, glycogen is very effective in inducing excystment of the phoront stage of apostomes (Bradbury & Trager, 1967). *Paramecium* species, which are not known to encyst, can survive for more than a month without food (Jackson & Berger, 1985b).

15.3 Somatic Structures

While oligohymenophoreans are an assemblage more speciose than the spirotrichs, the range of variation in body form is less dramatic. The "typical" oligohymenophorean is ovoid in shape, ranging in size from about 10 μm in small scuticociliates, to almost 1 mm in the infective stage of the hymenostome *Ichthyophthirius*, and up to 3 mm in some astomes (Figs. 15.2–15.5). As a group, the peritrichs demonstrate considerable diversity of body form: zooids can be borne on stalks, attach directly to the substrate on a modified scopulary

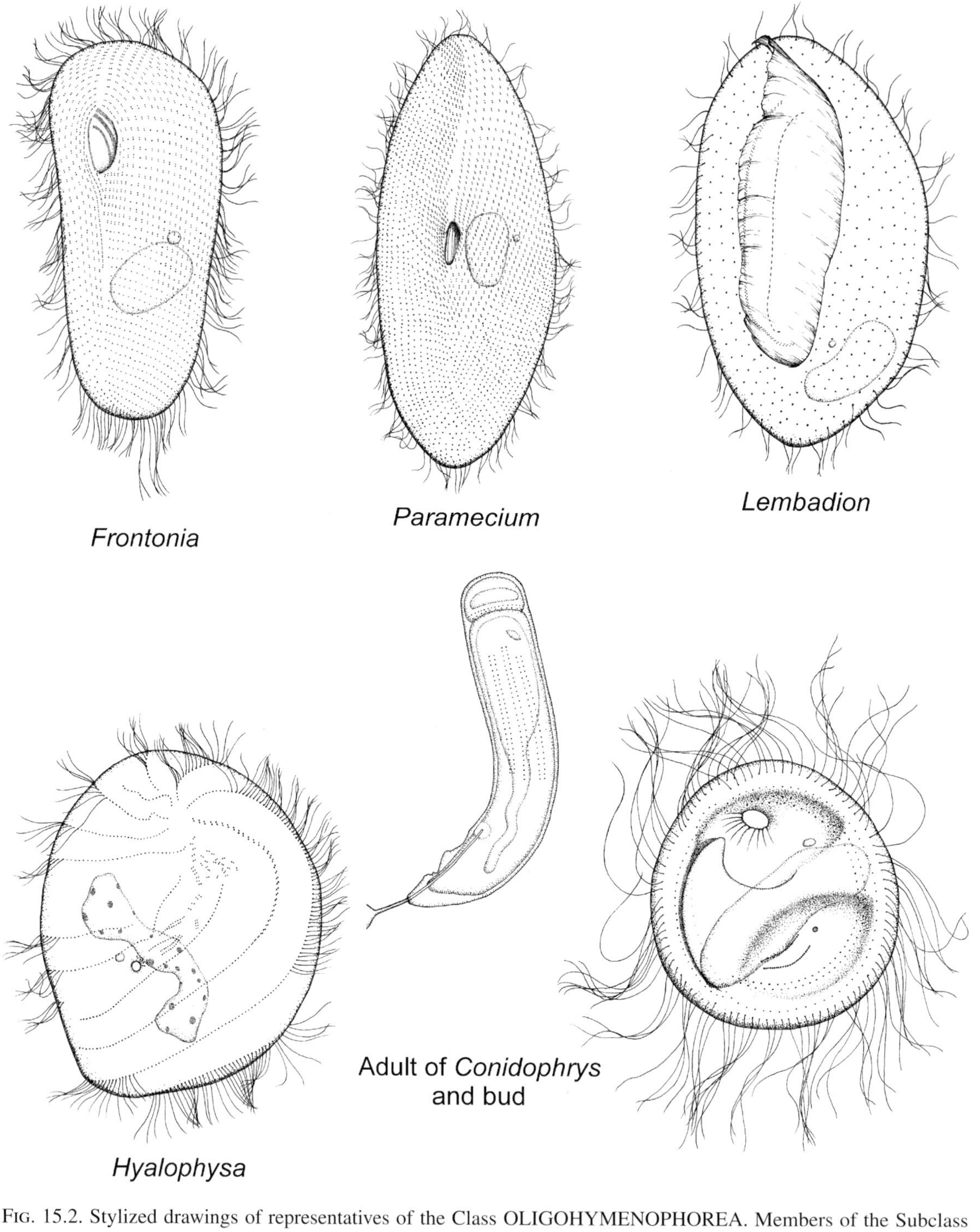

Fig. 15.2. Stylized drawings of representatives of the Class OLIGOHYMENOPHOREA. Members of the Subclass Peniculia – *Frontonia*, *Paramecium*, and *Lembadion*. Members of the Subclass Apostomatia – *Hyalophysa* and the adult of *Conidophrys* "impaled" on the seta of a crustacean and its ciliated dispersive bud

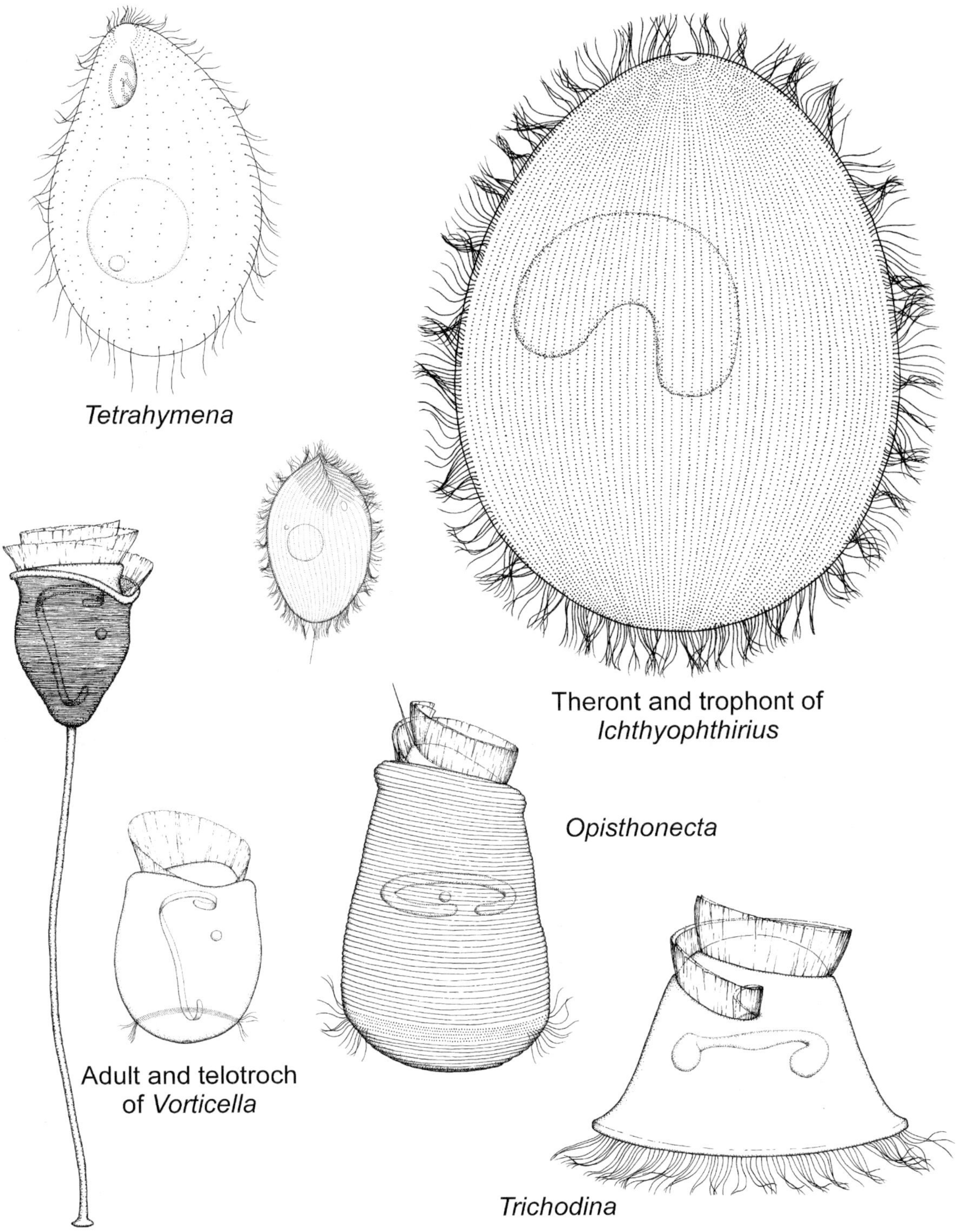

FIG. 15.3. Stylized drawings of representatives of the Class OLIGOHYMENOPHOREA. Members of the Subclass Hymenostomatia – the tetrahymenid *Tetrahymena* and the ophryoglenid *Ichthyophthirius* with its small theront and gigantic trophont, which causes "*Ich*". Members of the Subclass Peritrichia – two sessilids, the stalked and sessile *Vorticella* and its telotroch or swarmer and the permanently mobile and stalkless *Opisthonecta*; and the mobilid *Trichodina*, which causes trichodinosis

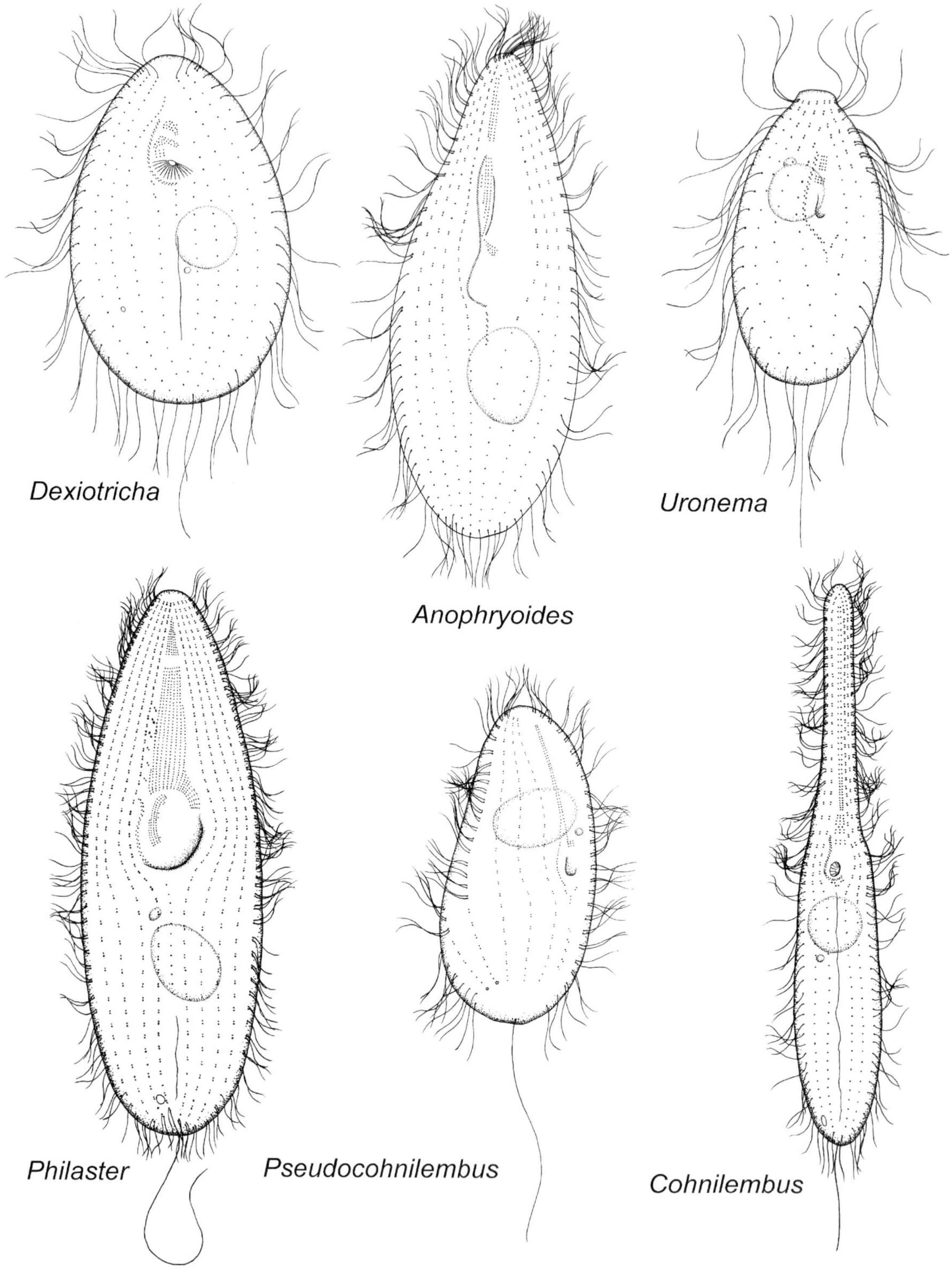

FIG. 15.4. Stylized drawings of representatives of the Class OLIGOHYMENOPHOREA. Members of the Subclass Scuticociliatia – the philasterids *Dexiotricha*, *Anophryoides*, *Uronema*, *Philaster*, *Pseudocohnilembus*, and *Cohnilembus*

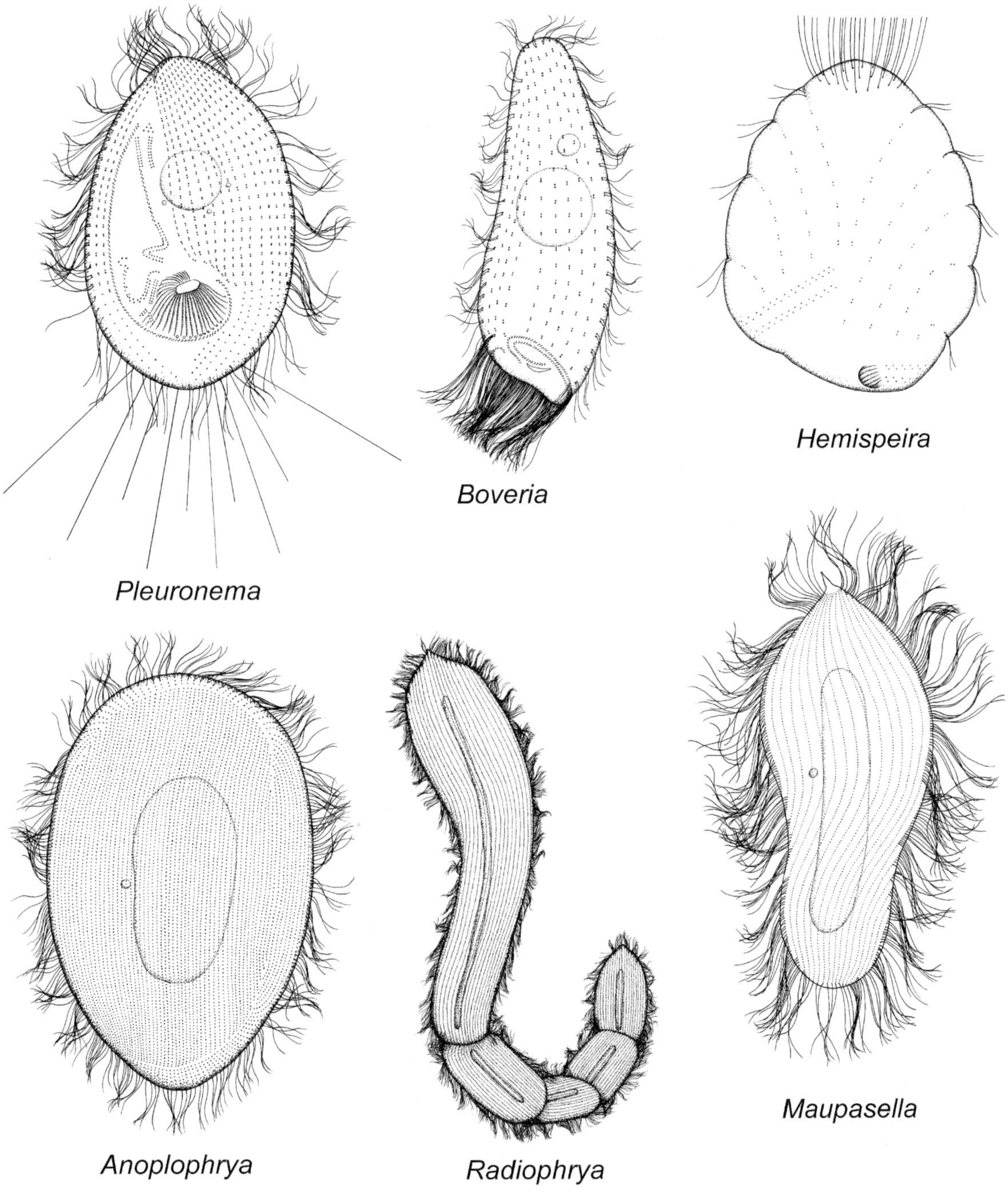

FIG. 15.5 Stylized drawings of representatives of the Class OLIGOHYMENOPHOREA. Members of the Subclass Scuticociliatia – the pleuronematid *Pleuronema* and the thigmotrichids *Boveria* and *Hemispeira*. Members of the Subclass Astomatia – *Anoplophrya*, *Radiophrya*, and *Maupasella*

region or attach and migrate along substrates with a complex adhesive disk, as does the mobiline peritrich *Trichodina* (Fig. 15.3).

Oligohymenophoreans are typically holotrichous with somatic kinetids as either monokinetids or dikinetids (Figs. 15.2–15.5). Dikinetids are often more common in the anterior half of the body and can be organized as paratenes, which are seemingly transverse rows of kinetids (Ehret, 1967). The somatic kineties can converge forming a preoral suture or anterior secant system at the anterior end and a postoral suture or posterior secant system at the posterior end. The shape, kind, and arrangement of these convergence regions is systematically important, especially in the astomes. The posterior end of oligohymenophoreans, especially scuticociliates, often bears kinetids whose cilia are significantly longer than the general body ciliature (Fig. 15.4). These so-called caudal cilia are typically not active, but can be used for attachment to the substrate.

The oligohymenophorean cortex, especially those of *Tetrahymena* and *Paramecium*, has been and continues to be exhaustively studied by cell and molecular biologists. The literature on this is extensive, and we provide here a few selected references as an introduction to the literature and to demonstrate the structural attributes of the ciliates in this class. The oligohymenophorean cell is covered by a plasmalemma, which is underlain by cortical alveoli (Allen, 1967, 1971, 1978). Intramembranous particles are distributed over the surface of the plasmalemma, joining this surface membrane with the subsurface alveoli and trichocysts and integrating them with the cilia (Allen, 1978; Plattner, Miller, & Bachmann, 1973). The lipid content of these surface membranes can be varied so that the ciliate can maintain a particular membrane fluidity over large ranges of environmental temperatures: at lower temperatures, membranes have more unsaturated fats and at higher temperatures, membranes have more saturated fats (Kitajima & Thompson, 1977; Nozawa, Iida, Fukushima, Ohki, & Ohnishi, 1974; Wunderlich, Speth, Batz, & Kleinig, 1973). Variations in thermotolerance among species of *Paramecium* may be due to genetic variation in the fatty acid profiles of their cell membranes (Sasaki et al., 2006).

The plasma membrane is covered by a surface coat, which is about 20 nm thick, and composed of glyco- and other proteins (Allen, 1978). Prominent among the proteins are the immobilization antigens or i-antigens, so named because antibodies to them bind the cilia together and prevent ciliary locomotion (Beale, 1954). These proteins are GPI-anchored proteins (Capdeville, Cardoso de Almeida, & Deregnaucourt, 1987; Clark, Gao, Gaertig, Wang, & Cheng, 2001; Ko & Thompson, 1992). They show considerable variation in both laboratory and natural populations (Lin et al., 2002; Saad & Doerder, 1995). There has long been speculation that these antigens protect the plasma membrane from the environment. Clear seasonal differences in the appearance of particular antigens suggests an important, but undetermined, ecological role for them (Doerder et al., 1996; Gerber, Lopez, Shook, & Doerder, 2002; Saad & Doerder, 1995).

The plasma membrane is underlain, as in other ciliates, by cortical alveoli, which are conspicuous in the oligohymenophoreans. The alveoli were likened to the membrane cisternae of muscle cells (Allen, 1971; Satir & Wissig, 1982), a speculation that predicted they would sequester calcium. Calcium, which was later directly visualized in the alveoli of *Paramecium* using secondary ion mass spectrometry, was shown to decrease on extrusion of trichocysts – a calcium-induced process. Calcium refilled the alveoli over several hours following depletion (Mohamed et al., 2003; Stelly, Halpern, Nicolas, Fragu, & Adoutte, 1995). In addition to calcium, alveoli of some oligohymenophoreans may also contain phosphatases (Lobo da Cunha & Azevedo, 1990).

As in other ciliates, the alveoli are underlain by an epiplasmic layer (Allen, 1967, 1971). The epiplasm of hymenostomes includes some of the same proteins as found in other protists – the multigene families of articulins and epiplasmins (Huttenlauch & Stick, 2003; Huttenlauch, Peck, & Stick, 1998; Pomel et al., 2006). Differential extraction techniques demonstrated that this layer, isolated as a ghost cell, provides a cortical scaffold for the cell (Collins, Baker, Wilhelm, & Olmsted, 1980; Keryer et al., 1990). On one hand, the varieties of cortical proteins in related *Tetrahymena* species are vastly different, and yet the form of these cells is very similar (Williams, 1984). On the other hand, the microstome and macrostome phenotypes of *T. vorax*, which are morphologically dramatically different, show very similar profiles of cortical proteins (Buhse & Williams, 1982). Nevertheless, Keryer et al. (1990) demonstrated the over-abundance of a particular band in one cortical mutant of *Paramecium*. Thus, a molecular change in a single component can have dramatic morphological effects. This has now been confirmed also for *Tetrahymena*: knockout constructs of the cortical protein Epc1p have altered cell shape (Williams, 2004). How these very different molecules can assemble similarly- or differently-shaped cells remains to be explained.

In peritrichs, the epiplasm is often quite thick on the non-ciliated body surface, and it is penetrated by pores that are presumably homologues of the parasomal sacs of other oligohymenophoreans (Lom, 1994). In scuticociliates and hymenostomes, there are one to many supraepiplasmic microtubules, called longitudinal microtubules, which extend the length of the cell (Allen, 1967; Antipa, 1972; Peck, 1977a). Presumed homologues of these microtubules appear transiently during division of peniculines (Sundararaman & Hanson, 1976).

The most prominent features of the ciliate cortex, the cilia associated with the somatic kinetids, have been extensively studied in *Tetrahymena* and *Paramecium*. Freeze-fracture analyses demonstrated that the somatic cilia have few randomly distributed particles over most of their length. However, at the base, distal to the ciliary necklace, were nine plaques of three longitudinal rows of particles at the ciliary base (Plattner, 1975; Sattler & Staehelin, 1974). These plaques are associated with Ca^{2+}–binding sites and are linked via an internal plaque complex to the peripheral doublets of the ciliary axoneme (Dute & Kung, 1978; Plattner, 1975). Ca^{2+} and cyclic nucleotides affect ciliary movement, and so influence the behavioral responses of oligohymenophoreans (Machemer & Sugino, 1989; Noguchi, Kurahashi, Kamachi, & Inoue, 2004). Parasomal sacs are associated with the base of the cilium. These sacs are regions of pinocytosis as cationized ferritin is internalized by them (Nilsson & Van Deurs, 1983). Moreover, there is suggestive evidence that they may also be a route for the exocytosis of certain enzymes (Allen & Fok, 2000; Nielsen & Villadsen, 1985).

The somatic kinetid of oligohymenophoreans has been characterized as a monokinetid as follows: a divergent, well-developed postciliary ribbon that extends usually to the next kinetid in the kinety but not to overlap its postciliary ribbon; a well-developed, anteriorly-directed kinetodesmal fibril that originates near triplets 5–7 and tapers as it overlaps fibrils from other kinetids; a reduced to well-developed, radially-oriented transverse ribbon that extends typically from triplet 4 laterally towards the adjacent kinety; and, in some cases, a transverse fibre that originates near triplet 3 and extends laterally in association with the transverse ribbon (Figs. 15.6–15.9) (Lynn, 1981, 1991). Dikinetids, often in the anterior part of the cell, have similar fibrillar associates to the monokinetid, but the anterior kinetosome sometimes bears a tangential transverse ribbon (Figs 15.6–15.8) (Lynn, 1981). While the monokinetid description applies very well to the kinetids of scuticociliates, hymenostomes, apostomes, and some astomes, the kinetids of peniculines, predominantly dikinetids, differ in that **both** sets of transverse ribbons are tangential to the perimeter of the kinetosome (Fig. 15.6) (Lynn, 1981, 1991). Somatic kinetids of peritrichs are so highly modified that there are really no obvious similarities (**see below**). Extending the length of the kineties, near the base of the kinetosomes, are several basal microtubules that may supply additional structural support to the cortex (Allen, 1967; Antipa, 1972). In peniculines, somatic kinetids can be connected by striated bands at mid-kinetosome level while their bases are surrounded by a complex network of filaments, called the infraciliary lattice (Allen, 1971). This lattice does have contractile properties and demonstrates cross-reaction to antibodies that recognize the filamentous layer of the litostomes (Garreau de Loubresse, Keryer, Viguès, & Beisson, 1988). Antipa (1972) carefully described structural differentiation of somatic kinetids in the cortex of the scuticociliate *Conchophthirius*: the kinetids in the thigmotactic region of the cortex of this ciliate were modified compared to those of the locomotory cortex.

Since the reviews of Lynn (1981, 1991) and Grain (1984), there have been relatively few reports of the ultrastructure of oligohymenophorean kinetids. Those that have appeared have confirmed these basic patterns. Some selected older and recent references are: for peniculines – *Paramecium* (Allen, 1971); *Frontonia*, *Urocentrum* (Didier, 1971); for the scuticociliates – *Cinetochilum* (de Puytorac, Didier, Detcheva, & Grolière, 1974a), *Conchophthirus* (Antipa, 1972), *Dexiotricha* (Peck, 1977a), *Myxophthirus* (Da Silva Neto, 1992), *Paranophrys* (Didier & Wilbert, 1976), *Proboveria* (de Puytorac, Grain, Grolière, & Lopez-Ochoterena, 1978); for hymenostomes – *Colpidium* (Lynn & Didier,

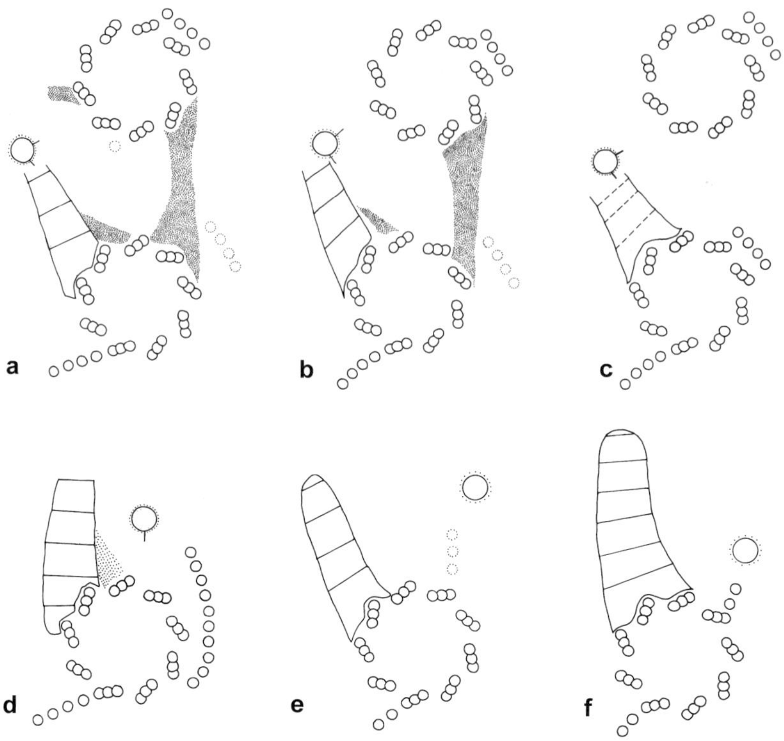

FIG. 15.6. Schematics of the somatic kinetids of the Class OLIGOHYMENOPHOREA. (**a–d**) Kinetids of the Subclass Peniculia. (**a–c**) Dikinetids of the Order Peniculida – *Paramecium* (**a**), *Disematostoma* (**b**), *Frontonia* (**c**). (**d**) Monokinetid of the Subclass Peniculia and Order Urocentrida – *Urocentrum*. (**e**, **f**) Monokinetids of the Subclass Apostomatia – *Hyalophysa* (**e**) and *Collinia* (**f**) (from Lynn, 1981, 1991)

1978), *Glaucoma* (Peck, 1978), *Ichthyophthirius* (Chapman & Kern, 1983), *Turaniella* (Iftode et al., 1984); for apostomes – *Hyalophysa* (Bradbury, 1966b), *Collinia* (de Puytorac & Grain, 1975); for astomes – *Coelophrya*, *Dicoelophyra* (Grain & de Puytorac, 1974). Undoubtedly the most unusual oligohymenophorean somatic kinetid is that of the scuticociliate *Schizocaryum*, whose somatic cortex is covered by cirrus-like polykinetids "organized" adjacent to a typical oligohymenophorean monokinetid (Fig. 15.9) (Lynn & Frombach, 1987).

The vast majority of sessiline peritrich species only display somatic ciliature at the time of cell division or when stimulated to leave their stalk by adverse environmental circumstances (Barlow & Finley, 1976; Rose & Finley, 1976). At this time, the daughter zooid or telotroch differentiates a band of cilia, called the trochal band, at the pole opposite the oral region, composed of from one row in *Lagenophrys* up to eight rows in *Ophrydium* (Fig. 15.3) (Lom, 1994). The trochal band of sessiline peritrichs, such as *Opisthonecta*, which are permanently motile, can be a complex arrangement of ciliated kinetosomes. The fibrillar associates and arrangement provide no evidence of homology with the somatic kinetids of other oligohymenophoreans (Bradbury, 1965). The trochal band surrounds the scopula, a structure at the extreme aboral pole, which includes kinetosomes with modified and reduced cilia, microtubular rootlets extending into the cytoplasm, and secretory granules (Fauré-Fremiet, 1984; Lom & Corliss, 1968; Willey & Walkosz, 1975). The transformation of the telotroch to a stalked zooid has been studied in the colonial *Zoothamnium* and solitary *Vorticella*, under controlled culture conditions (Suchard & Goode, 1982; Vacchiano et al., 1992). Stalk formation requires microtubules to transport secretory vesicles to the sites of exocytosis as the stalk forms and elongates (Suchard & Goode, 1982).

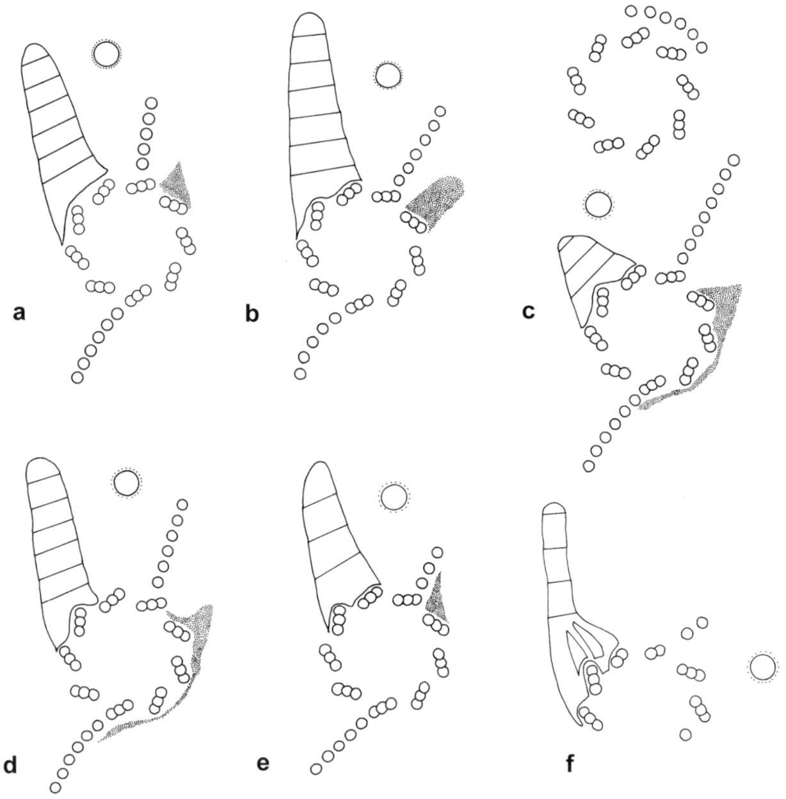

FIG. 15.7. Schematics of the somatic kinetids of the Class OLIGOHYMENOPHOREA. (**a–e**) Kinetids of the Subclass Hymenostomatia – Monokinetids of *Tetrahymena* (**a**), *Glaucoma* (**b**), *Colpidium* (**d**), and *Ichthyophthirius* (**e**). Dikinetid of *Colpidium* (**c**). (**f**) Monokinetid of the Subclass Astomatia – *Coelophrya* (from Lynn, 1981, 1991)

The stalk of sessiline peritrichs is composed of the secreted outer sheath that surrounds, in some species, an extension of the zooid's cytoplasm. The bulk of this extension is filled with a contractile myoneme, called the spasmoneme (Allen, 1973a, 1973b). The spasmoneme is continuous with the filamentous myonemes in the zooid itself, and these, in turn, are attached by a linkage complex to adjacent cisternae of endoplasmic reticulum, which may be the sites for Ca^{2+} necessary to induce contraction (Allen, 1973a, 1973b). The contractile proteins of the spasmoneme are not related to actin or myosin, and have therefore been termed spasmins (Amos, Routledge, & Yew, 1975; Routledge, 1978). Spasmins, which are related to centrin and calmodulin, may not be solely responsible for contraction (Asai, Ninomiya, Kono, & Moriyama, 1998). In addition to these filamentous myonemes of the zooid, there are other contractile filaments that retract the epistomial disk and there is a collar sphincter that closes the apical pole (Lom, 1994).

The other major group of peritrichs, the mobilines, are distinguished by an aboral adhesive disk (Fig. 15.3). On its oral side, the adhesive disk is bounded by a locomotor fringe, organized in three components in trichodinids: an "oral" or superior girdle of solitary or paired lateral cilia; a locomotor wreath or middle girdle, composed of oblique rows of 3–8 kinetosomes, reminiscent of the telotroch girdle of *Opisthonecta*; and an "aboral" or inferior girdle of groups, typically pairs of kinetosomes and cilia (Hausmann & Hausmann, 1981a; Maslin-Leny & Bohatier, 1984). Fibrillar rootlets attach the kinetosomes of the inferior and middle girdles respectively to the outer peripheral pin and middle radial pin of the skeletal components of the adhesive disk (Hausmann & Hausmann, 1981b). The radial pins articulate with the innermost skeletal element, the

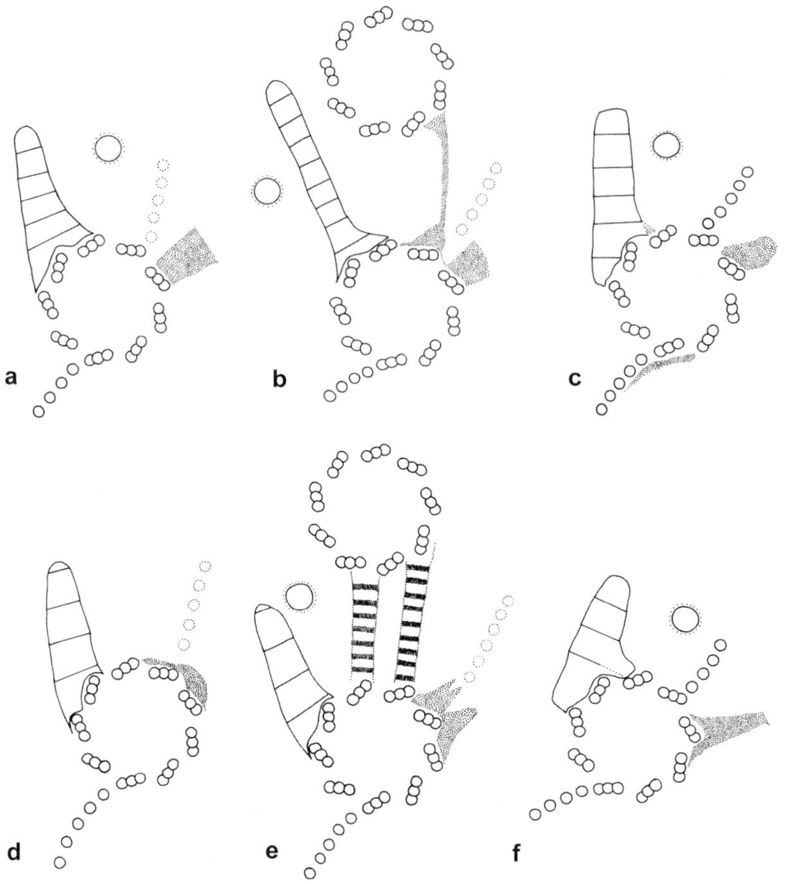

FIG. 15.8. Schematics of the somatic kinetids of the Subclass Scuticociliatia of the Class OLIGOHYMENOPHOREA. (**a**, **b**) Monokinetid and dikinetid of *Cinetochilum*. (**c**) Monokinetid of *Dexiotricha*. (**d**, **e**) Monokinetid and dikinetid of *Cohnilembus*. (**f**) Monokinetid of *Conchophthirus* (from Lynn, 1981, 1991)

denticles, which have been traditionally used after silver staining, to discriminate genera and species. The denticles have a complex structure whose detailed anatomy has only recently been revealed by scanning electron microscopy preceded by either nitric acid dilution (Kruger, Basson, & Van As, 1993) or sonication (Gaze & Wootten, 1999). To further complicate the interpretation of these denticles in a taxonomic context, their relative morphology can change as the diameter of the mobiline body increases (Kazubski, 1967). Given these insights, molecular genetic studies are now required to test the robustness of a taxonomy that is based primarily on denticle morphology.

Oligohymenophoreans are rarely found within loricas or other secreted structures. Among the free-living forms, the scuticociliate *Calyptotricha* is an exception, building a tube-like lorica in which it lives (Wilbert & Foissner, 1980). However, the peritrichs include several families characterized by the form and diversity of their loricas, which can have a smooth or more architectured surface (e.g., Couch, 1973; Clamp, 1987, 1990, 1991, 1992; Finley & Bacon, 1965; Warren & Carey, 1983). The lorica is a structure, primarily proteinaceous, which is likely secreted by the scopula in some species (González, 1979). The lorica of *Platycola* apparently sequesters heavy metals, such as manganese, preferentially concentrating them above their environmental concentrations (Warren & Carey, 1983). Undoubtedly the most conspicuous secreted structure among peritrichs, and perhaps ciliates as a phylum, is the gelatinous matrix secreted by *Ophrydium* species, which can range from 2 cm in diameter up to 30 cm in diameter (Duval & Margulis, 1995; Winkler & Corliss, 1965).

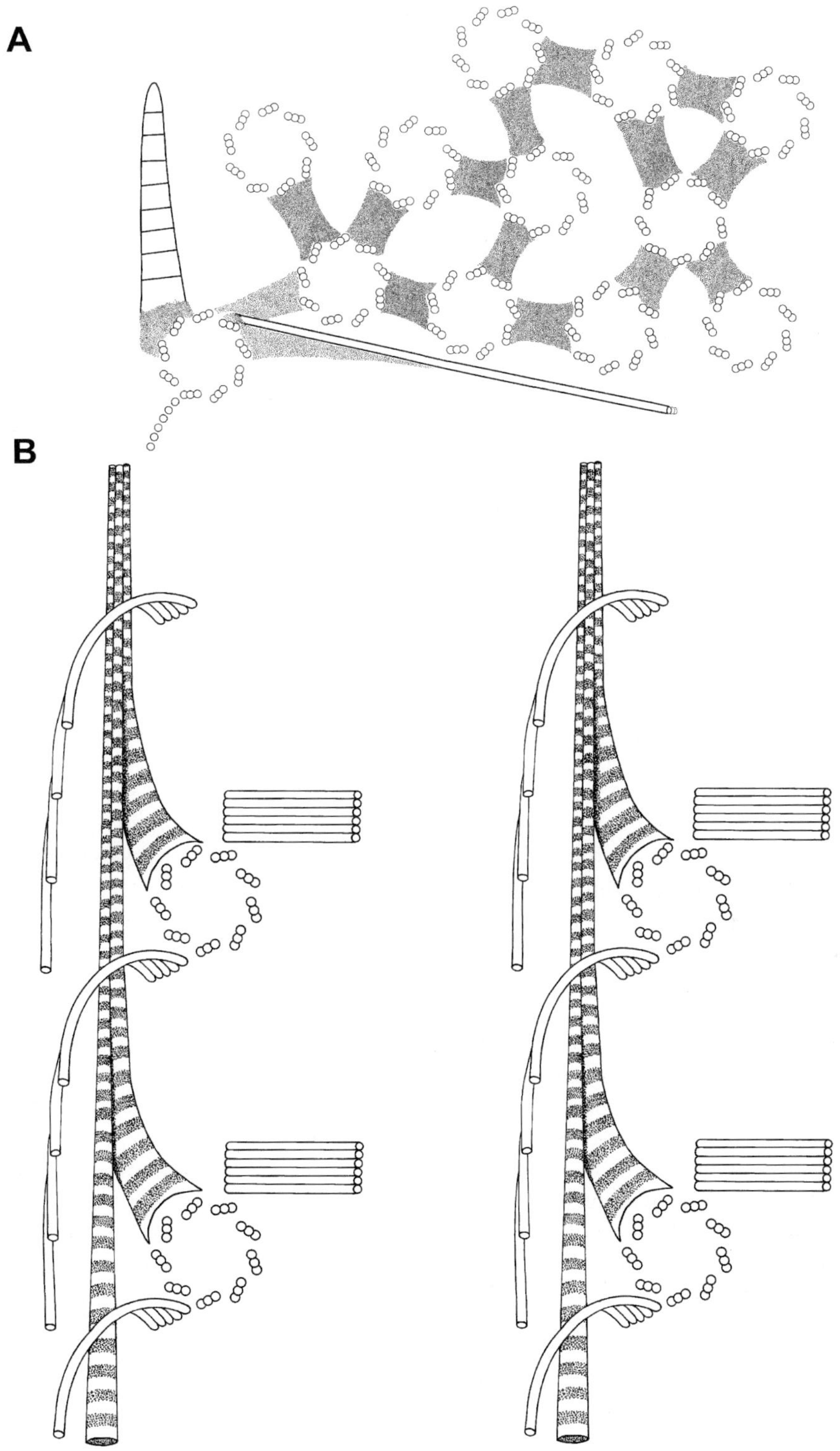

FIG. 15.9. **A** Schematics of the somatic polykinetid of the scuticociliate *Schizocaryum* of the Class OLIGOHYMENOPHOREA (from Lynn, 1981, 1991). **B** Somatic cortex of a typical oligohymenophorean based on the somatic cortex of *Tetrahymena* and *Colpidium*

Contractile vacuoles are a conspicuous organelle in many oligohymenophoreans. They are typically solitary and restricted to the posterior half of the cell. However, peniculines often have two, one in each half of the cell while astomes can have dozens, whose distribution patterns have been used to characterize species (de Puytorac, 1994g). The contractile vacuole of peritrichs is cryptic as it opens into the infundibulum at the anterior end of the cell. New contractile vacuoles are positioned at each cell division by a global positioning system that is able to proportionally "measure" cell size (Frankel, 1989; Nanney, 1980; Nanney, Nyberg, Chen, & Meyer, 1980b). The contractile vacuole complex of oligohymenophoreans, such as *Paramecium*, *Tetrahymena*, and *Vorticella*, is composed of two major membranous compartments, the spongiome and the contractile vacuole itself (Patterson, 1980; Allen & Naitoh, 2002). Much remains to be learned of the details of the mechanism of water sequestration. It is clear that the spongiome membrane has vacuolar-ATPases that pump ions, probably K^+ and Cl^-, into this compartment to create an osmotic gradient drawing water and metabolic wastes from the cytosol (Allen & Naitoh; Allen, Ueno, Pollard, & Fok, 1990; Stock, Gronlien, Allen, & Naitoh, 2002). Fluid is then moved to the contractile vacuole itself, which periodically rounds prior to expulsion. The fluid is excreted through the contractile vacuole pore, which is supported by helically coiled microtubules, and which serves as the origin for ribbons of microtubules that radiate out from the pore to support the membranous components of the spongiome (Chapman & Kern, 1983; McKanna, 1973a). Similar to natural populations of marine ciliates, *Paramecium* species continue to adapt and maintain contractile vacuole function, even in environments with high osmotic strength. They do this, in part, by regulating the amounts of the free amino acids proline and alanine in the cytoplasm (Cronkite & Pierce, 1989). Stock, Allen, and Naitoh (2001) argued that the maintenance of contractile vacuole function, even at these high environmental osmolarities, indicates involvement of this organellar complex in both the elimination of excess water and in the excretion of metabolic wastes.

The cytoproct of the oligohymenophoreans *Paramecium* and *Tetrahymena* is a "somatic" cortical structure that is the "terminal" component of the "digestive system" of these cells. Like the contractile vacuole complex, microtubules extend into the cytoplasm from dense material supporting the cytoproct. These microtubules guide food vacuoles to the cell cortex where their contents can be egested, and where the membranes can be recycled back to the oral region to form new food vacuoles (see **Oral Structures**) (Allen & Wolf, 1974, 1979).

Extrusomes of oligohymenophoreans are either mucocysts, which can be quite rod-like in some scuticociliates, or spindle trichocysts (Hausmann, 1978). Their similarity to secretory granules in other eukaryotes has made *Tetrahymena* and *Paramecium* model systems to deepen our understanding of cellular secretion processes in general. Models for this process, involving a variety of molecules, such as rosette particles, Ca^{2+}-ATPase, parafusin, and annexins, have been presented for both *Tetrahymena* (e.g., Satir, 1989; Satir, Schooley, & Satir, 1973; Turkewitz, 2004) and *Paramecium* (e.g., Froissard et al., 2002; Gilligan & Satir, 1983; Knochel et al., 1996; Plattner et al., 1980; Satir, 1989). These extrusomes develop from Golgi-like membranous systems in the endoplasm and are then transported out to the somatic cortex (Ehret & de Haller, 1963; Hausmann, 1978). Trichocyts are composed of as many as 100 polypeptides that are processed and arranged as elementary units into a crystal lattice (Hausmann, 1978; Vayssie, Garreau de Loubresse, & Sperling, 2001). While the shape changes of wild-type trichocysts depend on Ca^{2+} (Adoutte, 1988; Adoutte, Garreau de Loubresse, & Beisson, 1984; Sperling, Tardieu, & Gulik-Krzywicki, 1987), several non-discharge trichocyst mutations have now been described in *Paramecium* (Beisson, Cohen, Lefort-Tran, Pouphile, & Rossignol, 1980; Pollack, 1974).

The function of trichocysts has long been debated (Haacke-Bell, Hohenberger-Bregger, & Plattner, 1990), and trichocyst mutants have permitted the first test of the defensive function hypothesis of these organelles (see **Life History and Ecology**). Trichocyst non-discharge mutants of *Paramecium* are up to 45 X more susceptible to predation by the litostomes *Dileptus* and *Monodinium*, and by the heterotrich *Climacostomum* than wild-type cells (Harumoto & Miyake, 1991; Miyake & Harumoto, 1996; Sugibayashi & Harumoto, 2000). Backward swimming, which often accompanies an attack by these predators, does not enable a more effective escape than forward swimming, as mutants unable to swim backwards are caught as frequently as wild-type cells (Harumoto, 1994; Sugibayashi & Harumoto, 2000). Intriguingly, trichocysts do not protect *Paramecium* against predation by *Didinium*, suggesting that this predator is currently ahead in the arms race between predator and prey (Miyake & Harumoto, 1996).

Mucocysts are the other major extrusome type in oligohymenophoreans. Mucocysts provide for a variety of cell functions: they are involved in the formation of cyst walls (e.g., Ewing et al., 1983; McArdle et al., 1980) and loricas (e.g., González, 1979; Wilbert & Foissner, 1980). In the apostome *Hyalophysa*, Landers (1991a) has observed that the rod-shaped mucocysts of this ciliate are digested in autophagic vesicles during the phoretic stage, perhaps serving as a nutrient source.

Mitochondria in the Class OLIGOHYMENOPHOREA are typical of those of the phylum – primarily cortical organelles with tubular cristae. They are anchored to the somatic cortex through fibrous connections between the outer mitochondrial membrane and cortical microtubules and the epiplasm (Aufderheide, 1983). The mitochondria grow primarily by elongation and divide when their length is doubled. This growth and division maintains the population of mitochondria in the cytoplasm, but it is not tightly coupled to the cell cyle in *Paramecium* (Perasso & Beisson, 1978). In scuticociliates, perhaps all taxa have exceedingly large mitochondria, often extending the entire length of the ciliate beneath the cortical ridges, and perhaps are even connected between kineties (Antipa, 1972; Kaneshiro & Holz, 1976; Peck, 1977a; de Puytorac et al., 1974a). In rare instances, the mitochondria have transformed into hydrogenosomes in anaerobic species, such as the scuticociliates *Cristigera* and *Cyclidium* (Clarke, Finlay, Esteban, Guhl, & Embley, 1993; Fenchel & Finlay, 1991a).

A variety of other organelles typical of eukaryotes have been described in oligohymenophoreans. Golgi complexes, composed of a few flattened cisternae, have been reported in representatives of all the major subclasses (Estève, 1972; Kurz & Tiedtke, 1993; Lobo-da-Cunha & Azevedo, 1994). In *Tetrahymena*, they are often localized in the cortex adjacent to mitochondria (Kurz & Tiedtke, 1993). Peroxisomes have also been reported in hymenostomes (Fok & Allen, 1975; Lobo-da-Cunha & Azevedo, 1993) and peniculines (Stelly, Balmefrezol, & Adoutte, 1975). Finally, there may be crystals, excretory in function, whose abundance depends on the physiological state of the cell, and which may contain calcium (Nilsson & Coleman, 1977) and/or the purines guanine and hypoxanthine (Creutz, Mohanty, Defalco, & Kretsinger, 2002; Soldo, Godoy, & Larin, 1978).

15.4 Oral Structures

The oral region of the oligohymenophoreans, quite similar in four of its six included subclasses, typically includes, on the right side of the oral region, a ciliated paroral and, on the left side, three oral polykinetids of from 3–8 rows of kinetosomes (Figs. 15.2–15.5). This general pattern applies well to the peniculines, scuticociliates, hymenostomes, and peritrichs, but it does not to the apostomes and astomes. The latter two groups are undoubtedly derived from within this radiation, based on SSUrRNA gene sequences (Affa'a et al., 2004; Clamp et al., 2008): astomes lack an oral region altogether while apostomes have a highly modified oral region (**see below**).

The oral structures of oligohymenophoreans are also influenced by the polymorphic life histories typical of many of the included species, especially scuticociliates, hymenostomes, and apostomes. As the ciliate transforms from one life history stage to the next, its morphology, both somatic and oral, changes as an adaptation to the new mode of living. A typical change is in the size and shape of the oral organelles, which are adapted to feed on different prey species: *Ichthyophthirius* has a diminutive oral cavity as the dispersive theront and a larger, seemingly undifferentiated cavity as the feeding trophont (Fig. 15.3) (Canella & Rocchi-Canella, 1976). Hymenostomes, such as some species of *Tetrahymena* and *Glaucoma*, and scuticociliates may have microstome forms that feed on bacteria and macrostome forms, sometimes cannibals, which feed on their smaller conspecifics (Fig. 15.1) (Corliss, 1973; Njiné, 1972; de Puytorac, Savoie, & Roque, 1973b; Small et al., 1986; Williams, 1960, 1961). The macrostome-microstome transformation in some *Tetrahymena* species is induced by a "stomatin" preparation derived from the prey (Buhse, 1967; Méténier, 1977).

The cell biology of ingestion, digestion, and egestion of ciliates has relied heavily on research on *Tetrahymena* (Nilsson, 1979) and *Paramecium* (Allen, 1984; Allen & Fok, 2000; Plattner & Kissmehl, 2003), both ciliates serving as model systems for phagotrophy by other eukaryotic cells, like macrophages and leukocytes. Briefly, the model, which can probably be applied generally to all ciliates, involves formation at the cytopharynx of the nascent food vacuole by fusion of membrane vesicles. The food vacuole or phagosome separates from the oral region. In *Paramecium*, some of these vesicles are called acidosomes because of the acidic nature of their contents. As it circulates through the cytoplasm it fuses with lysosomes, the ingested prey are killed and digested, and nutrients are absorbed. The spent vacuoles make their way to the cytoproct where the contents are egested. Much of the vacuolar membrane is recycled to the cytopharyngeal region via microtubular tracts and in readiness for formation of the next food vacuole. While this model has not been demonstrated completely in any oligohymenophorean except *Paramecium* (Allen, 1984; Allen & Fok, 2000), components of the model have been observed in representatives of most subclasses in this class, and indeed even in other classes of ciliates.

The now classic work of Didier (1971) provided working definitions for the oral polykinetids of the oligohymenophorean subclasses. These definitions have stood the test of time (Grain, 1984; de Puytorac & Grain, 1976), although they may only have a loose application (Peck, 1977a, 1978). Among the ciliates obviously exhibiting oral polykinetids, four major kinds have been identified, primarily identified with a subclass: the peniculus and quadrulus of the peniculines; the membranoid of the scuticociliates; the membranelle of the hymenostomes; and the polykinety of the peritrichs. These will be briefly described below with some reference to the primary literature. This section will close with a discussion of the apostome oral region.

Peniculines typically have three longitudinally oriented oral polykinetids on the left side of the oral cavity. The subclass is named for the peniculus, a term originally proposed by von Gelei (1934a) for the two, left-most oral polykinetids of *Paramecium* whose kinetosomes are closely packed. Lund (1941) provided the term quadrulus for the peniculine polykinetid whose kinetosomes are more loosely packed. It is now agreed that the quadrulus is only a developmentally differentiated peniculus, peculiar to some, but not all, peniculines (Didier, 1971; Roque, 1961). The peniculus is typically longitudinally oriented in the oral cavity, has postciliary microtubular ribbons associated with the right-most row of kinetosomes, has alveoli between the rows, and has parasomal sacs restricted to the outsides of the kinetosomal grouping. The kinetosomes may be linked by distal and proximal connectives and each polykinetid may be linked to its neighbor by a deeper network of fibrils (Grain, 1984; Lynn, 1981; de Puytorac & Grain, 1976; but see Peck, 1977a, 1978). This structure has been recorded, for example, in *Paramecium* (Didier, 1971), *Urocentrum* (Didier, 1971; Guinea, Gil, & Fernández-Galiano, 1987), and *Frontonia* (Didier, 1971; Gil, 1984). Minor components of these filament macromolecules include actin (Cohen, Garreau de Loubresse, & Beisson, 1984) and tetrin-related elements (Clerot et al., 2000). *Paramecium* is a filter feeder, using its oral cilia to not only concentrate particles from the environment, but also to propel them into the nascent digestive vacuole (Fenchel, 1980a; Ishida, Allen, & Fok, 2001). Actin, as in *Tetrahymena* (**see below**), is involved in food vacuole formation (Cohen et al., 1984; Zackroff & Hufnagel, 1998). Acid hydrolases, at least, are found in lysosomal vesicles that fuse with the digestive vacuole (Allen, 1984; Estève, 1970), along with other vesicles that may derive from endocytosis at the parasomal sacs (Ramoino et al., 2001). Actin paralogs may also function in vesicle fusion and phagosome movement through the cytoplasm (Sehring, Reiner, Mansfeld, Plattner, & Kissmehl, 2007). In the frontoniid peniculines, the elongated cytostomal region is bounded by robust nematodesmata (Didier, 1971; Gil, 1984), which probably aid these ciliates in the ingestion of large diatoms and filamentous cyanobacteria. While Small and Lynn (1981, 1985) argued that this was another feature relating nassophorean and peniculine ciliates, SSUrRNA gene sequences now strongly suggest that peniculines are related to other oligohymenophoreans (Strüder-Kypke et al., 2000b). Thus, the frontoniid nematodesmata very probably evolved by convergent evolution as an adaptation for their feeding on these larger prey particles.

The scuticociliates, like the peniculines, typically have three oral polykinetids on the left side of the oral cavity. These function to provide a filter-feeding current that directs particles towards the cilia of the paroral, which then filters out these particles (Figs. 15.4, 15.5) (Fenchel, 1980a, 1980b). Paraxonemal bodies bounded by the ciliary membrane of some oral polykinetid cilia may provide strength and resilience to these current-creating cilia (Didier, 1976). However, the oral polykinetids can become highly modified, fragmenting during stomatogenesis into structurally independent components numbering more than three: oral polykinetid 1 of *Porpostoma* can have up to 20 "parts" while oral polykinetid 2 of *Pleuronema* is typically divided into two widely separated parts (Fig. 15.5) (Small, 1967; Song, 2000). At the ultrastructural level, these oral polykinetids demonstrate some diversity – some being called membranoids and others membranelles. Membranoids might be considered the archetypical scuticociliate oral polykinetid. Membranoid kinetosomes are irregularly arranged and linked at distal and proximal levels, and only irregularly do the kinetosomes on the right side bear a postciliary ribbon (e.g., *Cohnilembus* – Didier & Detcheva, 1974; *Paranophrys* – Didier & Wilbert, 1976). Nevertheless, the scuticociliates *Cinetochilum* and *Dexiotricha* (Peck, 1977a; de Puytorac et al., 1974a) have oral polykinetids with the basic features of a membranelle (**see below**). Alveoli have been observed between the cilia of oral polykinetids of thigmotrich scuticociliates (Da Silva Neto, 1992; de Puytorac et al., 1978), and this has suggested to de Puytorac et al. (1978) their closer affinities to the peniculines. Peck (1977a, 1978) carefully reviewed this early literature and came to the conclusion, which still is justified, that it is very difficult to make broad categorical characterizations of peniculi, membranoids, and membranelles. At most, the terms can only be applied loosely to general configurations of oral kinetosomes.

The hymenostomes have served as the archetype of the class, exhibiting an oral structure with a paroral on the right and three oral polykinetids on the left (Fig. 15.3) (Corliss, 1979; Lynn & Small, 2002). The ophryoglenines are further distinguished by the presence of the organelle of Lieberkühn, a dense "watch-glass" shaped structure placed between oral polykinetids 2 and 3. In *Ophryoglena*, oral polykinetid 2 has an enlarged posterior portion whose cilia form the conspicuous "brush" or "flamme" (Canella & Rocchi-Canella, 1976; Lynn, Frombach, Ewing, & Kocan, 1991b). Filter-feeding on bacteria and smaller particles is typical of most hymenostomes, which can use either the cilia of the paroral or the cilia of the innermost or third oral polykinetid to filter out particles (Fenchel, 1980a, 1980b; Fenchel & Small, 1980). The oral cilia of some hymenostomes have differing intramembranous particle distributions, have bristles on the outside of the ciliary membrane, and paraxonemal bodies extending their length (Didier, 1976; Sattler & Staehelin, 1974). Didier (1976) drew particular attention to the development of paraxonemal bodies in *Tetrahymena paravorax* and *Glaucoma ferox*, two ciliates that are predatory, and sometimes cannibalistic. Perhaps, the development of these paraxonemal bodies increases the capture efficiency of these predators.

The oral cavity of hymenostomes, and particularly its organization and development in *Tetrahymena*, has been the subject of both extensive and intensive investigations (Forer, Nilsson, & Zeuthen, 1970; Frankel, 1991; Frankel, Jenkins, Bakowska, & Nelsen, 1984a; Frankel, Nelsen, Bakowska, & Jenkins, 1984b; Nilsson, 1976; Williams & Bakowska, 1982). The oral polykinetids of hymenostomes are characterized as membranelles. The membranelle is typically oriented transversely in the oral cavity, has postciliary microtubular ribbons associated with the right-most row of kinetosomes, has no alveoli between the rows, and has parasomal sacs distributed irregularly between the kinetosomes. Membranelles have been characterized as being linked within by distal and proximal filamentous systems and between by a proximal filamentous system (Grain, 1984; Grain & de Puytorac, 1976). This structure has been observed in *Tetrahymena* (Nilsson, 1976), *Glaucoma* (Peck, 1978), and *Colpidium* (Lynn & Didier, 1978). Nevertheless, Peck (1977a, 1978) makes a reasonable argument against this view, suggesting that there is considerable variation. A case in point is the hymenostome *Turaniella*, a predator on other ciliates. Its oral cavity is much-expanded and its oral polykinetids and underlying filamentous systems are extremely well-developed, bearing some resemblance to the peniculines with which it was formerly associated (Iftode & Grain, 1975; Iftode, Versavel, & Didier, 1970). However, the ultrastructure of its somatic cortex and its stomatogenesis demonstrated clear affinities to the hymenostomes with which it is now classified (Didier, Iftode, & Versavel, 1970; Iftode et al., 1984). It is very likely, therefore, that the complicated fibrillar systems of the oral structures of *Turaniella* have converged on the peniculine model, correlated with the predatory feeding preference of this macrostomatous hymenostome.

Functioning of the hymenostome oral apparatus has been elucidated by studies on other macrostomatous hymenostomes, *Tetrahymena vorax* and *Tetrahymena paravorax*. The macrostome forms of these species have an expanded oral region with a large cytopharyngeal pouch in which to capture ciliate prey. The oral apparatus is modified, through morphogenesis, from that of the microstome form by increasing the number and arrangement of kinetosomes in the paroral and oral polykinetids (Smith, 1982). Food vacuoles appear to be formed by a contractile mechanism that involves the microtubules of the ribbed wall, which extends from near the kinetosomes of the paroral, and contractile proteins around the cytostome (McLaughlin & Buhse, 2004; Méténier, 1984b; Smith-Somerville & Buhse, 1984). The disruption of food vacuole formation by actin antagonists, such as cytochalasin and latrunculin B, implicates this filamentous protein in the process (Grønlien et al., 2002; Zackroff & Hufnagel, 2002). Exploitation of genetic constructs in *Tetrahymena* has now corroborated the important role of actin in food vacuole formation (Williams et al., 2006) and, in association with myosin, in the movement of food vacuoles through the cytoplasm (Hosein, Williams, & Gavin, 2005). The ribbed wall microtubules of the microstomatous *Tetrahymena* species have also been implicated in feeding (Sattler & Staehelin, 1979). Once the phagosome is formed, digestion occurs in a process very similar to that of *Paramecium*, except that acidosomes are not involved (Nilsson, 1976, 1979, 1987). As has been reported from *Paramecium*, membrane retrieval and recycling likely occurs from both the early phagosome during its condensation stage and after its fusion with the cytoproct (Mislan & Smith-Somerville, 1986).

The peritrichs, as their name suggests, are characterized by having ciliary structures around the perimeter of the peristome (Fig. 15.3). Two oral structures are involved – the paroral, traditionally called the haplokinety, and oral polykinetid 1, traditionally called a polykinety. These two structures circle the peristome in a counter-clockwise direction, if viewed from the top, up to five times in some *Campanella* species. They then plunge into the oral cavity, traditionally called the infundibulum. The peritrich oral polykinetid 1 is composed of three rows, parallels the paroral in its counter-clockwise course into the infundibulum, and terminates near oral polykinetids 2 and 3, which lie deeper in the infundibulum. Similar to other oligohymenophorean oral polykinetids, there are postciliary ribbons associated with the kinetosomes of the rightmost row, sometimes only visible during stomatogenesis (Bradbury, 1965; Eperon & Grain, 1983; Maslin-Leny & Bohatier, 1984). Alveoli are absent between the polykinetidal cilia, parasomal sacs may be distributed between the kinetosomes, and a complex set of fibres and filaments links the kinetosomes to each other and to a filamentous reticulum bordering the leftmost row. These features have been observed in *Opisthonecta* (Bradbury, 1965), *Trichodina* (Hausmann & Hausmann, 1981a; Maslin-Leny & Bohatier, 1984), *Thuricola* (Eperon & Grain, 1983), *Tripartiella* (Maslin-Leny & Bohatier, 1984), and *Astylozoon* (Guinea, Gil, Serrano, & Sola, 1990). There has been much speculation about these divergent filamentous structures compared to those of the oral polykinetids of other oligohymenophoreans. It is most likely that they are correlated with the highly contractile ability of peritrichs, which can bring all their oral ciliature "inside" the peristome as they withdraw from irritating stimuli.

The peritrichs create filtering-feeding currents by metachronal beating of the cilia of the paroral and oral polykinetid 1. This creates a "peristaltic" flow between the cilia that traps particles and forces them into the infundibulum where the particles are essentially trapped on the deeper paroral cilia before being directed to the food vacuole (Fenchel, 1980a; Sleigh & Barlow, 1976). Oral ribs direct particles, on the outside, to the cytostome, while, on the inside, the ribbed wall microtubules direct diskoidal vesicles to the cytostome where they fuse to form the nascent phagosome (Allen, 1984; McKanna, 1973b). As in other ciliates, excess membrane, as cup-shaped vesicles, is removed from the early phagosome and recycled to the food vacuole forming region (Goff & Stein, 1981; McKanna, 1973b).

The paroral, stichodyad or haplokinety is a typical feature of the oral apparatus of the four preceding classes. Stichodyad refers to the dikinetid nature of this paroral, with the pairs of kinetosomes so oriented after stomatogenesis that they are almost perpendicular to the long axis of the paroral so that the postciliary ribbons of the more oral or inner kinetosome are "on the left" (Grain, 1969, 1984). Haplokinety refers to there being only one, the outer, of the two kinetosomes ciliated (Grain, 1984; de Puytorac & Grain, 1976). While there are variations in the nature of the links both that connect the kinetosomes of each dikinetid and that link dikinetids together in the paroral, this basic structure is typical of the oligohymenophorean paroral. It has been reported, for example, in the following: the peniculines *Paramecium*, *Frontonia*, and *Urocentrum* (Didier, 1971); the scuticociliates *Cinetochilum*, *Myxophthirus*, and *Paranophrys* (Didier & Wilbert, 1976; de Puytorac et al., 1974a; Da Silva Neto, 1992); the hymenostomes *Colpidium*, *Glaucoma*, *Tetrahymena*, and *Turaniella* (Iftode et al., 1984; Lynn & Didier, 1978; Nilsson, 1976; Peck, 1978; Williams & Bakowska, 1982); and the peritrichs *Termitophrya*, *Trichodina*, and *Thuricola* (Eperon & Grain, 1983; Maslin-Leny & Bohatier, 1984; Noirot-Timothée & Lom, 1965). In turaniellid hymenostomes, paroral kinetosomes may not be ciliated, along only part of or the whole length of the paroral (Iftode et al., 1984; Lynn & Didier, 1978).

For some time, it was conjectured that the oral rib microtubules arose from the postciliary microtubules of the paroral dikinetids. If so, these microtubules must break during stomatogenesis, because there is firm evidence now that the oral rib microtubules arise in dense material, often taking the form of a spur or papilla, which may extend over top of the nonciliated kinetosomes of the paroral (Hausmann & Hausmann, 1981; Iftode et al., 1984; Lynn & Didier, 1978). This origin is reminiscent of the origin of the cytopharyngeal microtubules of the prostomes, such as *Coleps* (Lynn, 1985). The oral ribs are separated by alveoli and are typically supported by microtubules arranged as 4 + 2 in hymenostomes and scuticociliates (Lynn & Didier, 1978; Nilsson, 1976; Peck, 1978; de Puytorac et al., 1974a; Sattler & Staehelin, 1979), and 3 + 1, 3 + 2, and 4 + 2 in peritrichs (Hausmann & Hausmann, 1981; McKanna, 1973b).

Bradbury (1989) finally clinched the phylogenetic affinities of the apostomes by discovering a paroral during the development of the free-swimming tomite of *Hyalophysa*. Molecular genetic evidence now supports this conclusion (Clamp et al., 2008; Lynn, Strüder-Kypke, & Bradbury, 2005). The apostomes show a bewilderingly bizarre set of cortical kinetosomal assemblages, and all have had various authors suggest that they are homologues of the oral structures of other ciliates. The rosette, although not always present, can also clinch membership of a ciliate to this subclass of oligohymenophoreans. The rosette is a tube-like invagination, lined by ridges or septa that are covered by cortical alveoli and provided with some cilia and an aggregation of dense vesicles. Its function is not known, but its ultrastructure has been described in typical apostomes like *Hyalophysa* (Bradbury, 1966b) and atypical ones like *Collinia* (de Puytorac & Grain, 1975) and *Conidophrys* (Bradbury & Tyson, 1982). The *x*, *y*, and *z* kineties of foettingeriid apostomes have been characterized as perioral (Bradbury, 1966a) or as oral (Lynn & Small, 2002). Given that kinetodesmal fibrils are associated with the kinetosomes of the falciform kinetosomal fields of apostomes (Bradbury, 1966b, 1989; de Puytorac & Grain, 1975), it is more likely that these are highly modified arrays of somatic kinetosomes. Apostomes feed on the exuvial fluids of the moulted exoskeleton of their crustacean host (Bradbury, 1973, 1975b), on the blood of their crustacean host (de Puytorac & Grain, 1975), by dissolving the host's exoskeleton (Bradbury, 1975b; Bradbury & Goyal, 1976; Bradbury, Deroux, & Campillo, 1987), and by penetrating setae on the crustacean exoskeleton to presumably feed on tissue fluid (Fig. 15.2) (Bradbury & Tyson, 1982). The ingestatory region in exuviotrophic forms extends between somatic kinety 1 and the *x*, *y*, and *z* kineties. Food may be ingested via a cytopharynx lined with microtubular ribbons and to which diskoidal vesicles are directed for phagosome formation, reminiscent of other ciliates (Bradbury, 1973, 1975b; Bradbury et al., 1987). Deviants from this typical pattern occur in *Terebrospira*, which ingests the solubilized components of the exoskeleton by pinocytosis over its entire cell surface (Bradbury, 1975b; Bradbury & Goyal, 1976) and by *Conidophrys*, which ingests, presumably setal tissue fluids, through a broad cytostome characterized by delicate tubules (Bradbury & Tyson, 1982). Digestive enzymes, such as phosphatases, may be secreted outside the cell in some apostomes (Bradbury & Goyal, 1976). Phagosomes are acidified during the digestive process and lysosomes are likely involved (Bradbury & Goyal, 1976; Landers, Treadaway, Johnson, & Luckie, 2001).

In comparison to the above, we are essentially ignorant about the feeding biology of the astomes. A reasonable conjecture is that they may feed like the astomatous mutants of *Tetrahymena* and apostomes, presumably by pinocytosis via the parasomal sacs, but in the case of the astomes, over the entire cell surface (Nilsson & Van Deurs, 1983; Rasmussen & Orias, 1975).

15.5 Division and Morphogenesis

Division by oligohymenophoreans is typically as free-swimming cells, which divide equally or isotomically. In rare cases, division in peritrichs, thigmotrichs, and some astomes can be anisotomic or unequal. Multiple division, catenulation, or linear palintomy can occur in which chains of cells are formed in apostomes, like *Polyspira*, and astomes, like *Hoplitophyra* and *Radiophrya* (Fig. 15.5) (de Puytorac, 1994g, 1994h). The isolation of cell division arrest mutants in *Tetrahymena* suggests that chain formation may require a modification of perhaps only one gene product (Frankel, 1989; Frankel, Nelsen, & Jenkins, 1977). Palintomy can also occur in a reproductive cyst: the tomont stages of the ophryoglenines *Ophryoglena* and *Ichthyophthirius* divide within a reproductive cyst, the latter yielding over 4,000 tomites; and the apostome tomonts and in some special cases, like *T. vorax* and *Porpostoma*, divide within a cyst (Fig. 15.1).

Foissner (1996b) remarked that the oligohymenophoreans include "the pets of the ciliatologists" – the genera *Paramecium* and *Tetrahymena*. Species in these two genera have provided models for advancing our understanding at two cellular levels: (1) the replication of organelles and organellar complexes; and (2) the replication of the pattern of the entire cell. This developmental biological literature has been extensively treated by Frankel (1989), who provides a thorough review of the literature and a thoughtful treatment of a variety of issues. As we have noted previously, we still have much to learn about the cellular and molecular processes underlying division morphogenesis. Much of the systematic literature attributes a particular role to a particular kinetosomal structure, and yet there are unambiguous refutations of this role attribution. For example, somatic kinety 1 of *Tetrahymena* is the rightmost postoral kinety, and it has been called the "stomatogenic kinety" because the oral primordium typically develops in association with it (Fig. 15.11). However, Nanney (1967) demonstrated that the primordium can develop along other kineties, a phenomenon that he called "cortical slippage". If there were a consistent directional bias in this process, ultimately every somatic kinety would become a stomatogenic kinety. The inevitable conclusion is that the relative position on the somatic cortex is the important causal determinant and **not** an association with a **particular** kinety or kinetosomal structure (Frankel, 1989; Nanney, 1967). It is worth remembering this important point in the subsequent discussion.

Our reference to the literature on developmental biology will be very selective, highlighting a few papers that have relevance to the systematics. From this perspective, Foissner (1996b) has thoroughly reviewed the systematic literature on ontogenesis of oligohymenophoreans. Our citations to this literature prior to Foissner will mainly highlight some exemplary studies as we proceed below to characterize each subclass. Contrary to Small and Lynn (1985), the molecular genetic evidence now strongly links the peniculines to the other oligohymenophoreans (Strüder-Kypke et al., 2000a, 2000b). Moreover, Beran (1990) has made the most recent compelling argument for ontogenetic homologies between peniculines and other oligohymenophoreans in his investigations of the early development of the oral anlage. Beran (1990) considered as homologues the following structures involved in oligohymenophorean stomatogenesis: the ophryokineties of the peniculine *Frontonia*, the anarchic field of the peniculine *Paramecium*, and the scutica of the scuticociliates, and this may be extended to the germinal row of peritrichs (Figs. 15.10, 15.11). With this assumption of homology, these ontogenetic features corroborate the monophyly of this class.

Peniculine stomatogenesis has been characterized as ophryobuccokinetal (Foissner, 1996b). While there is no doubt on molecular and morphological grounds that the frontoniids and parameciids are sister taxa, there are differences in their division

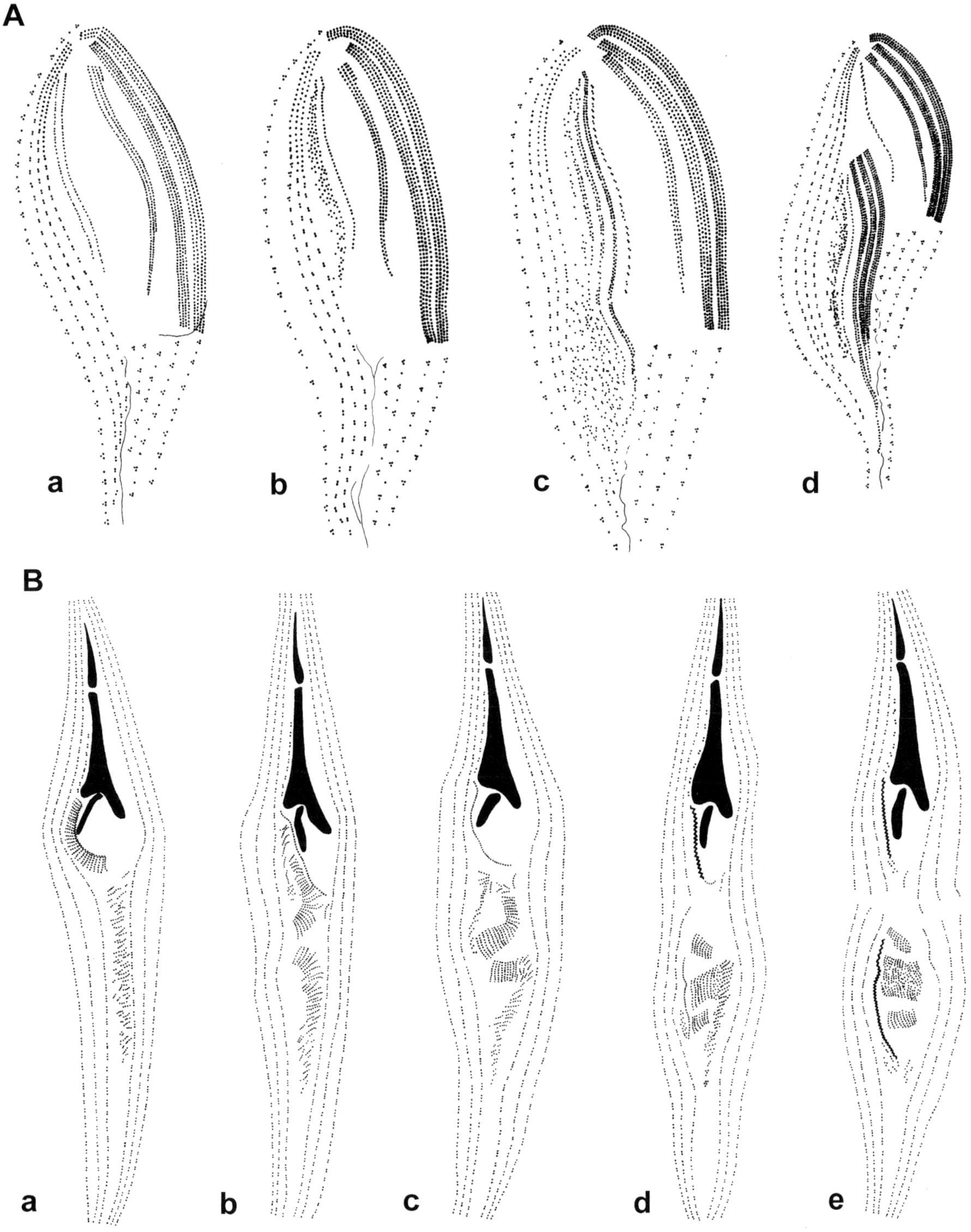

FIG. 15.10. Division morphogenesis of representatives from subclasses of the Class OLIGOHYMENOPHOREA. **A** In the Subclass Peniculia, represented by *Frontonia*, stomatogenesis is considered ophryobuccokinetal because it involves proliferation of kinetosomes from the parental paroral and from several "ophryokineties" to the right of the oral region (**a**, **b**). As stomatogenesis proceeds, a new paroral differentiates on the left of the field for the proter while the opisthe's oral apparatus differentiates into the three peniculi and a paroral as it migrates posteriorly (**c**, **d**). (from Song, 1995.) **B** In the Subclass Scuticociliatia, *Philaster* represents the Order Philasterida. Stomatogenesis begins by proliferation of kinetosomes from the parental paroral and the scutica, which resides in the director meridian between Kineties 1 and n (**a**, **b**). Kinetosomes from the paroral migrate posteriorly along the right to form the opisthe's paroral and part of the oral polykinetids (**c**). As the proter's paroral reconstitutes itself (**d**, **e**), the opisthe's oral structures take shape with the scutica appearing as a "hook-like" attachment at the posterior end of the developing paroral. (from Coats & Small, 1976.)

until a clearly rationalized alternative arrangement is proposed (see **Chapter 17**).

There has been only a handful of recent studies on pleuronematine and thigmotrich scuticociliates using protargol silver impregnation, the staining procedure that best reveals kinetosomal structures. Thus, generalizations for these groups cannot easily be made. Dolan and Antipa (1985) suggested two stomatogenetic patterns for these non-philasterine scuticociliates: (1) oral primordia derived from the paroral and scuticovestige, observed in *Ancistrum* (Hatzidimitriou & Berger, 1977), *Cyclidium* (Grolière, 1980), and *Conchophthirus* (Antipa & Hatzidimitriou, 1981); and (2) oral primordia derived only from the paroral, accompanied by considerable dedifferentiation of the parental oral apparatus, observed in *Histiobalantium* (Dragesco & Iftode, 1972), *Mytilophilus* and *Peniculistoma* (Dolan & Antipa, 1985), and *Pleuronema* (Grolière & Detcheva, 1974; Ma, Gong, & Song, 2003a; Small, 1967) (Fig. 15.11). Njiné and Ngassam (1993) have shown that stomatogenesis in the thigmotrich hysterocinetid *Ptychostomum* begins as a parakinetal kinetosomal proliferation, anterior to the parental oral apparatus. While this distances the hysterocinetids from other thigmotrichs, we do not believe it yet warrants placing them in a separate subclass, as suggested by some (de Puytorac, 1994f; Ngassam & Grain, 2002). It is important to remember that position is developmentally "prior" to pattern. Thus, we conclude that hysterocinetid stomatogenesis may be one of those instances where our interpretation of the underlying morphogenetic "pattern" is confused by the surface kinetosomal structures. We await molecular evidence to refute or corroborate their position as thigmotrichs within the Subclass Scuticociliatia.

The hymenostomes have been reduced to a smaller subset of oligohymenophorean families, which Foissner (1996b) categorizes as showing parakinetal stomatogenesis. *Tetrahymena* is the archetypical hymenostome, showing monoparakinetal stomatogenesis (Fig. 15.11) (Frankel, 1989; Grolière, 1975a). The oral primordium typically forms at the equator by kinetosomal proliferation at what is defined as Kinety 1, the "stomatogenic kinety", but which is reported to occur at other kineties also (Frankel, 1966, 1989; Nanney, 1967). This typical hymenostome pattern has been observed in *Glaucoma* (Frankel, 1960; Peck, 1975), *Tetrahymena* (Bakowska, Nelsen, & Frankel, 1982a; Grolière, 1975a), and *Turaniella* (Iftode et al., 1970). The polykinetids assemble through a process reminiscent of that described in the spirotrichs by Jerka-Dziadosz (1981a). Dikinetids are formed that later align to form the two primary rows of each oral polykinetid, followed by replication of a third and sometimes fourth row (Bakowska et al., 1982b; Frankel et al., 1984a, 1984b). The gradient of differentiation proceeds from the upper left of the primordial field to the lower right (Peck, 1974). The paroral finally develops along the right border of the primordial field by assembly of a ciliated file of single kinetosomes that replicate an external kinetosome to form the paroral dikinetids. These external kinetosomes become ciliated as the internal kinetosomes lose their cilia (Bakowska et al., 1982a, 1982b; Nelsen, 1981). What is intriguing is the pattern of paroral assembly and disassembly in the proter: the external kinetosomal file separates from the internal and a new external kinetosome is replicated prior to the dedifferentiation of the "old" kinetosomes (Bakowska et al., 1982a, 1982b; Nelsen, 1981). Bakowska et al. (1982b) make the intriguing proposal that this is a phylogenetic signal of the common ancestry of hymenostomes and scuticociliates, and by extension even to the peniculines – in the hymenostomes, these "old" kinetosomes normally dedifferentiate rather than remain to participate in the next fission.

This archetypical hymenostome pattern is modified in two circumstances. First, during oral replacement, the oral apparatus is dedifferentiated and replaced without cell division in microstome hymenostomes when proliferation of kinetosomes from the "old" paroral and a region posterior to the parental oral apparatus provides the source of kinetosomes for the new oral apparatus (Frankel, 1989; Williams & Frankel, 1973). This pattern of stomatogenesis is also characteristic of those *Tetrahymena* species that transform into macrostomes, developing the larger oral apparatus by a process of replacement similar to that in the microstomatous species described above (Buhse, 1966; Méténier & Grolière, 1979; Njiné, 1972). The second modification of this pattern occurs in the ophryoglenines whose pattern is characterized as teloparakinetal (Foissner, 1996b). In this group, adapted for histophagy and parasitism, cell division of the encysted tomont is preceded by complete dedifferentiation of the parental oral structures. Only at the end of the series of palintomic divisions does the oral apparatus differentiate from an oral primordium derived by replication at the anterior ends of a number of somatic kineties. The paroral in *Ophryoglena* and *Ichthyophthirius* is completely dedifferentiated during the final stages of stomatogenesis, a feature these ophryoglenines share along with the differentiation of the organelle of Lieberkühn (Foissner, 1996b; de Puytorac et al., 1983b).

Division morphogenesis of peritrichs can be relatively simple in solitary forms, and can become increasingly more complex in colonial forms and in symbiotic forms. In colonial forms, such as *Zoothamnium* species, cell division and subsequent development can explain the structure of the colony itself and the differentiation of several types of zooids (Fauré-Fremiet, 1930; Summers, 1938). In symbiotic epibionts, division and formation of telotrochs appear to be correlated with the molt cycle of their crustacean hosts (Clamp, 1973; Walker, Roberts, & Usher, 1986). Lom (1964) provided the modern schema for peritrich stomatogenesis by studying protargol impregnations of dividing *Telotrochidium*. A germinal kinety or field, adjacent to the parental paroral, proliferates to provide kinetosomes for the opisthe's paroral and oral polykinetids 2 and 3 (Fig. 2.6Ac). The parental paroral provides kinetosomes for the opisthe's oral polykinetid 1, the so-called peristomial polykinety, and the proter's paroral or haplokinety. A new germinal kinety proliferates from the paroral of both filial cells prior to completion of stomatogenesis. This pattern has been confirmed for *Astylozoon* (Guinea, Sola, Rueda, & Fernández-Galiano, 1988), *Carchesium* (Esteban & Fernández-Galiano, 1989), *Opercularia* (Fernández-Galiano, Esteban, & Munoz, 1988), *Opisthonecta* (Sola, Guinea, & Fernández-Galiano, 1985), and *Thuricola* (Eperon, 1980). Foissner (1996b) characterizes this as an ophryobuccokinetal stomatogenesis since the opisthe's oral apparatus derives from an ophryo– or germinal kinety, suggesting homologies to the process in peniculines, but also to that of the scuticociliates. Indeed, it is to the latter group, and particularly the thigmotrichs, to which the common ancestry of the peritrichs has been linked (Fauré-Fremiet, 1950a; Lom, 1964). We currently need some gene sequences from thigmotrichs to explicitly test this hypothesis. However, the gene sequence database currently does not support it: peritrichs are consistently a strongly supported sister clade to the hymenostomes and **not** to the scuticociliates (Affa'a et al., 2004; Miao et al., 2004b).

The last two groups of oligohymenophoreans, the apostomes and astomes, are problematic because they are so divergent. Astomes, of course, have no stomatogenesis, since by definition they have no mouth. They divide transversely, equally or unequally (Fig. 15.5). In the latter case, they may remain attached as chains of cells or catenoid "colonies" (Beers, 1938; de Puytorac, 1954, 1994g). Subsequent cell growth and division may involve only the anterior cell (e.g., *Hoplitophrya*) or each filial cell may grow and divide (e.g., *Cepedietta*) but not separate (de Puytorac, 1994g).

Apostome division morphogenesis demonstrates no clear homologies with other oligohymenophoreans, presumably a result of the highly unusual life cycle of these ciliates. "Stomatogenesis" and morphogenesis during the life cycle have been studied in *Hyalophysa* using protargol staining (Fig. 15.11) (Bradbury, Song, & Zhang, 1997; Landers, 1986). The anterior kinety, kinety *a*, plays a central role in the replication of cortical structures. It elongates by replication of a small, anterior fragment in the trophont, and apparently differentiates into three bipolar kineties, named *a*, *b*, and *c*. The latter kinety dedifferentiates completely, *b* may differentiate as the paroral, and *a* provides continuity as the kinety *a* for the next round of fission (Bradbury et al., 1997). Bradbury et al. (1997) noted that kinety *a* in *Foettingeria* derives from Kinety 1. Thus, this is a kind of monoparakinetal stomatogenesis, like that of *Tetrahymena*, since kinety *a* provides an oral structure, the paroral homologue. This is also consistent with preliminary gene sequence data that place apostomes within the oligohymenophorean clade, although not close to the hymenostomes (Clamp, Bradbury, Strüder-Kypke, and Lynn, 2008).

A discussion of division morphogenesis of oligohymenophoreans would not be complete without some reference to the extensive literature on the cell and developmental biology of the process, most recently reviewed by Frankel (1989, 1991). Simply, the process can be viewed as a duplication of structure controlled at two major levels – at the level of the organelles and organellar complexes and at the level of the cell as a whole. At the organellar complex level, the working hypothesis has long been that the local environment and pre-existing structure play determining roles, so-called structural guidance or cytotaxis (Frankel, 1989; Sonneborn, 1964; Williams, 1986). There is now convincing evidence for this in the propagation, over many cell cycles, of a patch of inverted somatic kineties (Beisson & Sonneborn, 1965; Ng & Williams, 1977). Furthermore, successful replication is dependent upon the presence of specific, kinetosome-associated structures (Iftode & Fleury-Aubusson, 2003; Kaczanowska et al., 1996). The inversion of these kinetids also causes the beat cycle of their cilia to be opposite to those of adjacent, normally-oriented kinetids (Tamm, Sonneborn, & Dippell, 1975). For the *Paramecium* cell as a whole, there is suggestive evidence that morphogenetic waves, originating from the oral apparatus and fission furrow, induce duplication and reorganization processes (Iftode et al., 1989). Migration of the new oral structures, essential to the completion of normal division in all oligohymenophoreans but hymenostomes, depends upon proper disassembly and reassembly of cortical structures (Kaczanowska et al., 1995). When the oral development in the two cells is almost complete, cytokinesis occurs, accompanied by the appearance of a contractile ring of microfilaments at the fission furrow (Eperon, 1985; Jerka-Dziadosz, 1981c; Yasuda, Numata, Ohnishi, & Watanabe, 1980). Assembly of a functional contractile ring depends upon Ca^{2+} and several proteins, including calmodulin and actin (Gonda & Numata, 2002; Williams et al., 2006).

Oligohymenophoreans have limited powers of regeneration. Nevertheless, as in other classes, regeneration after microsurgery has been demonstrated in some peniculines (Chen-Shan, 1979) and hymenostomes (Mugard & Lorsignol, 1956).

15.6 Nuclei, Sexuality and Life Cycle

The oligohymenophoreans present a broad diversity of forms in the macronucleus. Typically, the macronucleus is single and globular to ellipsoid (Figs. 15.2–15.5). Variations do exist: peritrichs are typified by the horseshoe- or band-shaped macronucleus (Fig. 15.3) (Lom, 1994); astomes may have a macronucleus extending along the entire length of the body, sometimes with irregular extensions (Fig. 15.5) (de Putyorac, 1994g); a rare scuticociliate can have multiple fragments of the macronucleus (Lynn & Frombach, 1987); and apostomes demonstrate a variety of macronuclear forms with one form showing a complex network (Fig. 15.2) (de Puytorac, 1994h).

The micronucleus is typically solitary, although some species are typified by having two micronuclei. In rare exceptions, over 40 micronuclei have been observed in particularly large-bodied forms (Lynch, 1929; Lynn & Berger, 1973). The micronucleus of oligohymenophoreans can have from five chromosomes in *Tetrahymena* (Ray, 1956) to several hundreds in *Paramecium* species, and some *Paramecium* species may be polyploid (Aury et al., 2006; Raikov, 1982). Micronuclear morphology can vary both between and within genera. For example, four different types of micronuclei have been identified by Fokin (1997) among ten different *Paramecium* species: these are vesicular, endosomal, chromosomal, and compact types. Macronuclear ploidy varies typically with the sizes of the cell and the macronucleus: the larger macrostome species of *Tetrahymena* may be 450 × n; larger *Paramecium* species over 850 × n; and the large trophonts of *Ichthyophthirius* set an oligohymenophorean record of 6,300 × n (Raikov, 1982).

Both kinds of nuclei in oligohymenophoreans divide with the aid of microtubules. Intramacronuclear microtubules have been observed in dividing *Paramecium* and *Tetrahymena* (Nilsson, 1976; Tucker, Beisson, Roche, & Cohen, 1980) and myosin has been implicated by immunofluorescence studies (Hauser, Beinbrech, Gröschel-Stewart, & Jockusch, 1975). Analysis of mutant phenotypes of *Paramecium* and drug and heat treatments of *Tetrahymena* provided support for the model that microtubular sliding elongates the macronucleus (Cohen, Beisson, & Tucker, 1980; Nilsson, 1976). Nevertheless, microtubule-deficient macronuclei can divide as the somatic cortex appears to play a crucial role in positioning and even elongating the macronucleus (Jaeckel-Williams, 1978; Tucker et al., 1980). Microtubules and microfilaments are also implicated in micronuclear division, although the relative importance of each to chromosomal movement has not been resolved (LaFountain & Davidson, 1980; Lewis, Witkus, & Vernon, 1976). Spindle microtubules in the dividing micronucleus of *Paramecium* typically have 15 protofilaments compared to the "normal" 13 for other organelles (Eichenlaub-Ritter & Tucker, 1984).

The macronucleus in oligohymenophoreans, as for ciliates in general, varies in size, and therefore DNA amount, with cell size (e.g., Kazubski, 1963; Lynn & Berger, 1972, 1973; Morat, 1982). Division of the macronucleus is often slightly unequal (Berger, 2001; Morat, 1982). Yet, ciliates maintain a roughly proportional nuclear:cytoplasmic ratio over the course of many cell cycles. The mechanisms responsible for this regulation have been extensively explored, especially in *Paramecium* (Berger, 2001). Nilsson (2000) has argued that the minimal units of segregation during macronuclear division in *Tetrahymena* represent full genomes, although how this is accomplished remains to be explained.

The macronucleus in oligohymenophoreans develops from a product of the zygotic nucleus, which typically divides two or three times to provide nuclei for differentiation (Raikov, 1972). The macronuclear anlage undergoes a series of changes in its fine structure until it has developed nucleoli and begins transcription (Weiske-Benner & Eckert, 1985). As in the spirotrichs (see **Chapter 7**), the development of the oligohymenophorean anlage involves DNA amplification, chromosome fragmentation, sequence elimination, addition of telomeres, and amplification of some genes, particularly ribosomal genes (Prescott, 1994; Schmidt, 1996). Moreover, this development may be epigenetically regulated by the parental or maternal macronucleus, even to the level of the precise excision of eliminated sequences (Meyer & Duharcourt, 1996; Preer, 2000).

Much of the research on oligohymenophoreans has focused on *Paramecium* and *Tetrahymena* in which chromosome fragmentation and sequence elimination occur by different mechanisms. In *Paramecium*, a terminal inverted repeat unit flanks the internally eliminated sequences (IESs) and bears some similarity to the transposable elements found in spirotrichs (Klobutcher & Herrick, 1995, 1997). The quality of this flanking sequence is critical as single base pair changes in it can prevent IES elimination (Mayer, Mikami, & Forney, 1998; Matsuda, Mayer, & Forney, 2004). In *Tetrahymena*, a consensus sequence has not been identified, apparently leading to less precision in the elimination of sequences (Austerberry, Snyder, & Yao, 1989; Yao, Duharcourt, & Chalker, 2002). However, there is an internal 10-bp core to the chromosome breakage sequence – AAACCAACC?C – that is completely conserved and possibly represents a regulatory protein binding site (Hamilton et al., 2006). Sequence–specific information may also be provided by molecules, for example by small RNAs, that derive from the parental macronucleus and that can control DNA rearrangements and processing in the developing macronucleus through homology-dependent mechanisms (Kowalczyk, Anderson, Arce-Larreta, & Chalker, 2006; Le Mouel, Butler, Caron, & Meyer, 2003; Meyer, Butler, Dubrana, Duharcourt, & Caron, 1997).

In both cases, the processes of fragmentation of and excision from micronuclear chromosomes result in macronuclear "chromosomes" that are shorter than the micronuclear chromosomes, although not as short as the gene-sized pieces of spirotrichs (see **Chapter 7**). Oligohymenophorean macronuclear chromosomes range in size from from 20–2,500 kb for *Paramecium* species (Rautian & Potekhin, 2002; Steele, Barkovy-Gallagher, Preer, & Preer, 1994), from 21–1,500 kb in *Tetrahymena* (Altschuler & Yao, 1985; Conover & Brunk, 1986), and from 2–300 kb in *Glaucoma* (Katzen, Cann, & Blackburn, 1981). Breakage of micronuclear chromosomes forms many new chromosome "ends", and this has provided ciliate molecular biologists with a useful model to investigate the structure, formation, and maintenance of telomeres (Blackburn, 1986). The CCCCAA oligonucleotide repeat characterizes the telomeres of the macronuclear chromosomes of *Tetrahymena* (Blackburn & Gall, 1978; Yao & Yao, 1981), *Glaucoma* (Katzen et al., 1981), and *Paramecium* (Yao & Yao, 1981). The abundance of ends and the need for their reconstruction and maintenance lead to the discovery of the ribonucleoprotein enzyme complex responsible for these processes, now called telomerase (Blackburn, 1992; Greider & Blackburn, 1987).

Some years prior to these discoveries, it had been possible to separate micronuclei and macronuclei, so that at least pure macronuclear preparations could be analyzed for sequence complexity (Gorovsky, Yao, Keevert, & Pleger, 1975; Soldo & Godoy, 1972). With renaturation kinetic analyses, most macronuclear DNA sequences behaved as unique sequences, with very little highly repeated sequences and up to 20% moderately repetitive (McTavish & Sommerville, 1980; Soldo & Godoy; Yao & Gorovsky, 1974). The highly and moderately repetitive fractions are apparently eliminated during development of the macronuclear anlage (Yao & Gall, 1979).

As for other classes of ciliates, examples are now accumulating of genetic code deviations among oligohymenophoreans. Of the three universal stop codons – UAA, UGA, and UAG, oligohymenophoreans, such as the hymenostome *Tetrahymena* (Horowitz & Gorovsky, 1985), the peniculine *Paramecium* (Caron & Meyer, 1985; Preer, Preer, Rudman, & Barnett, 1985), and the peritrichs *Vorticella* and *Opisthonecta* (Sánchez-Silva et al., 2003) use only UGA. The codons UAA and UAG are now used by *Paramecium* and *Tetrahymena* as sense codons for glutamine and glutamic acid (Caron & Meyer, 1985; Preer et al., 1985), while UAA has been confirmed as the "glut" codon in peritrichs (Sánchez-Silva et al., 2003). There are two explanations for these deviations. First, tRNAs have been described that decode for these novel amino acid assignments (Hanyu, Kuchino, & Nishimura, 1986; Sánchez-Silva et al., 2003). As with genetic code deviations in other ciliates, another explanation is the evolution of translational release factors with a higher specificity for one or other of the universal stop codons (Caron, 1990). This prediction was confirmed by analyses of the spirotrich eukaryotic release factor 1, which is the protein that recognizes stop codons and terminates translation (Inagaki & Doolittle, 2001; Lozupone, Knight, & Landweber, 2001). This higher specificity then allowed the evolution of tRNA anticodons to use the now-available and unused stop codons.

Conjugation of oligohymenophoreans, typed as temporary and equal or isogamontic in most groups, was first described by Hertwig (1889) in *Paramecium* (Raikov, 1972). However, the peritrichs conjugate by total cell fusion, typically of unequal-sized conjugants; they are therefore anisogamontic (Raikov, 1972). The macroconjugant is large and sessile while the microconjugant is small and free-swimming. Microconjugants may arise by unequal cell division, rapid successive divisions without intervening growth or direct transformation of microzooids in *Zoothamnium* species that show colonial polymorphism (Raikov, 1972). The endosymbiotic apostome *Collinia*, described by Collin (1909a) as an *Anoplophrya* species, appears to undertake a typical conjugation, although modified by a bizarre mutual exchange of portions of macronuclei. Other apostomes engage in what has been called syndesmogamy or zygopalintomy: two trophonts encyst together, undergo synchronized palintomic divisions, and ultimately conjugate as tomites (Chatton & Lwoff, 1935a; Minkiewicz, 1912; Raikov, 1972). Conjugating endosymbionts may often be smaller than the average trophont, suggesting that they might undergo preconjugation divisions (Kazubski, 1963; Raikov, 1972).

Paramecium and *Tetrahymena* served as the early model organisms for exploration of the genetics of the ciliates, and these efforts have been complemented and amplified by research on spirotrich genetics (see **Chapter 7**). There is a very large literature on these two genera, representing the Subclasses Peniculia and Hymenostomatia, which we have used very selectively to touch on the characteristics of these sexual processes for the Class OLIGOHYMENOPHOREA. Late in the 19th century, Maupas (1889) discovered the clonal life cycle of ciliates in his investigations of nuclear phenomena in *Paramecium*. During this clonal life cycle, cells of a clone pass through stages characterized as immature, adolescent, mature, and senescent, dependent upon the number of fissions since the last conjugation and upon their ability to engage in conjugation at the moment (Hiwatashi, 1981; Sonneborn, 1957; Takagi, 1999). As with other ciliates, oligohymenophoreans can be induced to conjugate after moderate starvation and if they are sexually mature (Bruns, 1986; Fujishima, 1988). Relatively little research, other than that of a descriptive nature, has been focused on members of the other oligohymenophorean subclasses. Finley (1936) was able to predictably induce conjugation in *Vorticella* species, but only after they had encysted and excysted. Conjugation can also be induced in *Paramecium* by chemical means (Cronkite, 1974) and by mixing a single mating type with detached cilia from mature cells of the complementary mating type (Miyake, 1964). In *Paramecium*, the natural mating-type substances aiding agglutination are very likely proteins, and in complementary mating types they may even have a precursor-derivative relationship (Xu et al., 2000).

Sonneborn (1937) discovered complementary mating types of *Paramecium "aurelia"* species while Elliott and Nanney (1952) reported a similar phenomenon in *Tetrahymena*. These discoveries opened the way to exploration of the genetics of ciliates (Nanney, 1980; Sonneborn, 1947). The binary mating-type system in *Paramecium* with its two complementary mating types is probably the ancestral state in the oligohymenophoreans, and from it the multiple mating-type systems described for *Paramecium bursaria* and *Tetrahymena thermophila* have likely evolved (Bleyman, 1996; Miyake, 1996). Exhaustive sampling of species in these two genera has not expanded the number of mating types beyond a maximum of 8 and 7 respectively, in dramatic contrast to the potentially hundreds of mating types recorded for some stichotrichs (Doerder et al., 1996; see **Chapter 7**). Jankowski (1972b) has demonstrated that *Paramecium putrinum* also has a multiple mating-type system. Mating type in binary systems has been further classified into three categories. In the A or caryonidal mating-type system, each developing macronucleus following conjugation is independently determined to express either mating type. In the B or clonal mating-type system, the parental macronucleus epigenetically determines the expression of the mating type by the new macronucleus. In the C or genotypic mating-type system, the mating type expressed is under genotypic control (Bleyman, 1996; Simon & Orias, 1987; Sonneborn, 1977). A variety of environmental factors can influence the expression of the ultimate mating type in *Tetrahymena* with temperature and nutrition being important variables (Arslanyolu & Doerder, 2000; Doerder et al., 1996). One of the most interesting environmental influences is the circadian rhythm of mating-type expression in *Paramecium multimicronucleatum* (Barnett, 1966). While the complete analyses of the genomes of *Paramecium* and *Tetrahymena* may eventually resolve the matter, the mating-type alleles of *Tetrahymena* exhibit serial dominance at a single locus, explained by a model for somatic DNA rearrangements during macronuclear development (Orias, 1963, 1981). Those of *Paramecium bursaria* are explained by combinations of alleles at two or three different loci (Siegel, 1963).

In the oligohymenophoreans, the molecules signalling the readiness for mating are firmly bound to the cell surface of the ciliate, either on the ciliary membranes or the plasma membrane; Miyake (1996) calls this type, gamone-carrying. Mature *Tetrahymena* prepare for cell fusion in two stages, called initiation (Bruns & Brussard, 1974) and costimulation (Finley & Bruns, 1979). In contrast, mature *Paramecium* agglutinate upon first contact after which the cells prepare for conjugation (Fujishima, 1988; Nanney, 1980). Other morphological differentiations occur in cells prior to fusion. In *Tetrahymena*, a region of the cortex anterior to the oral region becomes smooth and flattened. This tip transformation prepares the cells for fusion (Wolfe & Grimes, 1979). In *Paramecium*, a broad region of cortex anterior and posterior to the oral region becomes deciliated prior to cell fusion (Watanabe, 1978). Cell membranes ultimately fuse to enable transfer of the pronuclei, at the anterior end in *Tetrahymena* and at the paroral cone region in *Paramecium* (Fujishima; Inaba, Imamoto, & Suganuma, 1966; Wolfe, 1985). In *Tetrahymena*, microtubules have been implicated in the movements of micronuclei prior to formation of the gametic nuclei (Nakajima, Ishida, & Mikami, 2002), and in the formation of the complex, microtubular-microfilamentous transfer basket that envelopes the migratory pronucleus in each partner, enabling transfer (Orias, Hamilton, & Orias, 1983).

As with other ciliates, the micronuclei of oligohymenophoreans undergo typically three maturation divisions with the exception that some peritrich micronuclei may only undergo two maturation divisions (Raikov, 1972). These patterns are confirmed in more recent reports for scuticociliates (Coppellotti, 1990), hymenostomes (Martín-Gonzalez, Serrano, & Fernández-Galiano, 1984), and peritrichs (Sola, Guinea, & Fernández-Galiano, 1989a). Raikov (1972) noted that the "crescent stage" is probably typical of micronuclear meiosis in peniculines, scuticociliates, hymenostomes, and some peritrichs, but a "parachute stage", found in nassophoreans, has been observed as examples, in one astome, one apostome, and in *Paramecium putrinum* (Jankowski, 1972b, Raikov, 1972). The "crescent stage" micronucleus is probably elongated by microtubular growth (Suganuma & Yamamoto, 1992). The chromosomes appear to be arranged in parallel with their telomeres aggregated near one end of the developing "crescent", which may facilitate the pairing of homologues (Loidl & Scherthan, 2004). There is either mutual exchange of gametic nuclei, or in the case of total conjugation in peritrichs, the fusion of the single "migratory" gametic nucleus of the microconjugant with the single "stationary" gametic nucleus of the macroconjugant. This makes the exconjugants isogenic. The zygotic nucleus or synkaryon then undergoes typically two or three divisions, although in some *Frontonia* species it may divide only once while in some *Paramecium* species it may divide four times (Raikov, 1972). This means the macronuclear anlagen developing from these division products of the synkaryon can range from one up to 15. This is complicated by the fact that variable numbers of these division products may degenerate without development (Raikov, 1972). There is really insufficient breadth of analysis to draw any firm conclusions of patterns, if indeed there are any, in relation to the subclasses of oligohymenophoreans.

Autogamy results from the fusion in one cell of the haploid meiotic products of the maturation divison of the micronucleus (Corliss, 1952; Diller, 1936). The progeny are thus homozygous, and this has been advantageous for the genetic exploration of *Paramecium* in that mutations can be brought to full expression by inducing autogamy. *Tetrahymena thermophila* cannot be induced to autogamy. However, geneticists can now achieve homozygosity in this species by matings with so-called star strains, for example, strain A^*. During this process, called genomic exclusion, the star strain loses its micronucleus during meiosis. A migratory gametic nucleus is transferred to this star strain partner, after which both partners, now isogenic, become diploid by endoreduplication (Allen, 1967; Bruns, 1986).

Conjugation and/or autogamy are now considered crucial to the continued existence of strains of ciliates. Sonneborn (1954) originally showed their importance, demonstrating that periodic bouts of autogamy in a so-called "Methuselah" strain of *Paramecium biaurelia* extended its clonal life. Later the same phenomenon was demonstrated for *Tetrahymena* species (Corliss, 1965). Without these sexual processes, senescence sets in at from 200–350 cell divisions in members of the *Paramecium "aurelia"* complex and up to 1,500 cell divisions in *Tetrahymena* (Takagi, 1988, 1999). A single known exception is the amicronucleate *Tetrahymena pyriformis*, which has remained in culture for over 60 years: while it is "genetically dead", it is so-far physiologically immortal (Nanney, 1974). A variety of features indicates cells have entered senescence, among others: unequal distribution of macronuclear DNA at cytokinesis, a decreased viability of progeny after conjugation, a decreased ability to form food vacuoles, and a decreased fission rate (Smith-Sonneborn, 1981; Takagi, 1988). Clonal life span is undoubtedly under genetic control as mutants with variations in the clonal life cycle have been discovered (Komori, Sato, Harumoto, & Takagi, 2005; Takagi, Suzuki, & Shimada, 1987). Environmental factors can influence longevity, including UV and other forms of ionizing radiation (Smith-Sonneborn, 1981).

Conjugation is rarely observed in natural populations of oligohymenophoreans (Lucchesi & Santangelo, 2004). However, populations of *Paramecium* and *Tetrahymena* can be dominated by immature individuals, suggesting that sex may be quite frequent in nature (Doerder, Gates, Eberhardt, & Arslanyolu, 1995; Kosaka, 1991b), although a population dominated by senile individuals has also been discovered (Kosaka, 1994).

Sonneborn (1957) also related breeding systems of *Paramecium* to characteristics of the life history of the species. He proposed an inbreeding-outbreeding continuum: extreme inbreeders would have two mating types, a short period of immaturity, high fission rates, and local distributions, while extreme outbreeders would have the opposite set of characters (reviewed by Landis, 1986; Nyberg, 1988). At that time, some *Paramecium "aurelia"* species represented the extreme inbreeders while *Paramecium "bursaria"* species represented extreme outbreeders. *Tetrahymena* species with their multiple mating types would be considered relative outbreeders. While this has been an attractive thesis, Nyberg (1988) concluded that there are contradictory data to refute it. *Paramecium "bursaria"* species, supposed extreme outbreeders, appear to be restricted in their geographic distributions while some *Paramecium "aurelia"* species, typical inbreeders, are globally distributed. Furthermore, Nyberg (1981b) demonstrated that continental geographic distances did not reduce the fertility of several *Tetrahymena* species while Przybos (1995) has demonstrated that North American and European isolates of "inbreeding" *Paramecium triaurelia* are not genetically isolated.

Nevertheless, our ideas may be refined in the future as more molecular data accumulate. Stoeck, Przybos, and Schmidt (1998) have shown, using RAPD fingerprinting, that European populations of *Paramecium sexaurelia*, an extreme inbreeder, are more genetically isolated than populations of *Paramecium triaurelia*, a moderate inbreeder, consistent with Sonneborn's predictions. Stoeck et al. (2000a) have also used this approach to characterize *P. novaurelia* as a moderate inbreeder and *P. pentaurelia* as a weak inbreeder.

Sonneborn (1957) provided evidence that the "genetic species" of the *Paramecium "aurelia"* complex were identical to the sibling species of the fruit fly *Drosophila*. Nevertheless, because of the relatively onerous task of operationally identifying a species of *Paramecium "aurelia"*, Sonneborn (1957) was reluctant to name them as taxonomic species and instead chose to place them in syngens (*syn*, Gr = same, *gens*, Gr = kind). A similar situation was soon discovered for the *Tetrahymena "pyriformis"* species complex (Elliott, 1973b; Gruchy, 1955; Nanney, 1980). While analyses of cortical patterns suggested that some species of tetrahymenine hymenostomes might be separated morphologically (Cho, 1971; Nanney, 1966, 1968), multivariate morphometric analyses finally demonstrated that four species of the *P. "aurelia"* complex could be separated but others could not (Gates & Berger, 1976b; Powelson et al., 1975). The discoveries of isozyme variation among species of *Paramecium* by Tait (1970) and Allen, Byrne, and Cronkite (1971) and *Tetrahymena* (Allen & Weremiuk, 1971; Borden et al., 1973a, 1973b) were to provide an easy operational method to distinguish "genetic species". These results lead Sonneborn (1975) and Nanney and McCoy (1976) to establish nominate species for the syngens of *Paramecium "aurelia"* and *Tetrahymena "pyriformis"*. DNA fingerprinting is now being used to distinguish species of *Paramecium* (Skotarczak et al., 2004; Stoeck et al., 1998), and to demonstrate that other morphological species of *Paramecium*, such as *Paramecium duboscqui*, are probably also species complexes (Fokin et al., 1999).

While it had been difficult to morphologically resolve free-living species of *Paramecium* and *Tetrahymena*, morphological variability even among populations of symbiotic species has been well established. For example, mobiline peritrichs on breeding carp showed statistically significant seasonable variability (Kazubski & Migala, 1968); scuticociliate endosymbionts of sea urchins (Lynn & Berger, 1972, 1973) and bivalves (Berger & Hatzidimitriou, 1978) showed statistically significant variation on a number of traits between host populations; and apostome symbionts showed significant variation among host crustaceans (Landers, Zimlich, & Coate, 1999). These variations are likely due to a combination of factors, including invasion of the host by one to a few founders and adaptive responses to differing host environments (Berger & Hatzidimitriou, 1978). This dramatic morphological variation is contrasted with genetic uniformity in some symbionts from around the world: isolates of *Orchitophrya stellarum* have apparently identical nuclear genotypes in different starfish hosts from around the world (Goggin & Murphy, 2000). Whether this holds for cytoplasmic genes, such as those from mitochondria, awaits future research.

15.7 Other Features

Oligohymenophoreans are prominent members of the ciliate communities in water treatment plants (Curds, 1969, 1975b). The peritrichs are a particularly important group, responsible for clarification of the water by their bacterivory (Fried et al., 2000; Lee, Basu, Tyler, & Wei, 2004; Martin-Cereceda et al., 2001a, 2001b; Rivera et al., 1988). The widespread distribution of scuticociliates, hymenostomes, and peritrichs, and the ease with which they can be cultivated has also established them as model organisms in the assessment of toxicants. Hymenostomes, in particular *Tetrahymena* and *Colpidium*, have often been used in a wide variety of applications testing a diversity of toxicants from heavy metals to detergents to hydrocarbons. Recent reviews discuss the various end-points used (Gilron & Lynn, 1996; Sauvant, Pepin, & Piccinni, 1999). These include growth inhibition (Dive et al., 1991; Miyoshi et al., 2003; Schultz, 1997; Schultz & Dumont, 1977; Zilberg & Sinai, 2006), survival (Komala, 1993; Madoni & Romeo, 2006; Sartory & Lloyd, 1976; Schlenk & Moore, 1994), respiration (Slabbert, Smith, & Morgan, 1983), chemosensory behavior (Berk, Gunderson, & Derk, 1985; Gilron et al., 1999; Roberts & Berk, 1990), mutagenicity (Smith-Sonneborn, 1981), and ingestion rate (Juchelka & Snell, 1995). Recently, viability has been assessed using fluorescent dyes indicative of the integrity of particular cell functions (Dayeh et al., 2004; Dias et al., 2003; Wang, Zhang, & Wang, 2000).

Toxicants, such as heavy metals and oils, are likely to impact communities, both in natural habitats (Caron & Sieburth, 1981) and in waste-water treatment plants (Madoni et al., 1996). Nevertheless, the scuticociliate *Uronema* can acquire tolerance to heavy metals (Berk et al., 1978), and this is likely due to the induction of metallothioneins, cysteine-rich metal-binding proteins, which have been identified in *Tetrahymena* (Fu & Miao, 2006; Santovito et al., 2001). Nyberg and Bishop (1983) concluded that selective forces have operated to generate the variation in tolerance to heavy metals, particularly copper and mercury, among stocks of *Paramecium primaurelia*.

Finally, *Tetrahymena* may be engineered to play a new role in the manufacturing of proteins for human use. Gaertig et al. (1999) were able to express the surface antigen of *Ichthyophthirius multifiliis* on the cell surface of *Tetrahymena*, demonstrating that *Tetrahymena* might be used as the "vehicle" to establish immunity to "*Ich*" in fishes. Of more relevance to humans, Peterson et al. (2002) targeted a protein from the malaria parasite *Plasmodium falciparum* to the cell surface of *Tetrahymena*, demonstrating that this ciliate might be used in vaccine development against malaria. As a last example, Weide et al. (2006) have used *Tetrahymena* as an expression system for human enzymes and shown that a functional enzyme results. Is it too much of a stretch to imagine oligohymenophoreans, like *Tetrahymena*, playing crucial roles in future human health applications?

Chapter 16
Deep Phylogeny, Gene Sequences, and Character State Evolution – Mapping the Course of Ciliate Evolution

Abstract Our understanding of the evolutionary diversification of ciliates in the past two decades particularly has depended upon the interaction between conceptual views and technological advances. Transmission electron microscopy precipitated a revolution in our views of what characters might be significant in inferring deep phylogenetic relationships. The fibrillar patterns of somatic kinetids were considered crucial, based on the notion of the structural conservation of these cortical components. Molecular phylogenetic analyses have been used to test the conclusions based on electron microscopy. In the main, phylogenetic relationships inferred from sequences of the small subunit and large subunit rRNA genes have confirmed the major classes, and suggested several new ones (i.e., Classes ARMOPHOREA and PLAGIOPYLEA). In addition, the rRNA genes demonstrated a fundamental subphyletic division – now named the Subphyla Postciliodesmatophora and Intramacronucleata. Protein gene sequences (e.g., elongation factor 1α, α-tubulin, and histone H3 and H4) provide confirmation for some clades. Using the rRNA phylogeny, the evolution of some major character states, particularly nuclear ones, can be assessed.

Keywords Phosphoglycerate kinase, intramembranous particles, ciliary necklace

The progress in our understanding of the evolutionary diversification of ciliates has depended upon an interaction between conceptual views and technological advances. On the conceptual side, our views of which characters or features of ciliates were most important in revealing common ancestry have changed (see **Chapter 1**). Briefly, in the 18th and 19th centuries, overall ciliation patterns and the dominance of the "spirotrich" oral region divided the ciliates into "holotrichs" and "spirotrichs". In the first half of the 20th century, ontogenetic patterns, particularly revealed by silver-staining organisms at cell division, received greater weight and aligned taxa that had previously been distantly separated (e.g., chonotrichs and suctoria were related to the cyrtophorines). In the latter half of the 20th century, transmission electron microscopy revealed a whole new set of cytoskeletal characters, particularly the somatic kinetid patterns. The diversity of these somatic kinetid patterns initially suggested eight major clades or classes (Small & Lynn, 1981, 1985).

In the 1970s, microbiologists studying prokaryotes had been successfully using small subunit (SSU) rRNA genes to resolve relationships among this group whose members were not rich in morphological features (Stackebrandt & Woese, 1981). By the mid-1980s, several research groups began sequencing SSUrRNA genes of ciliates (Elwood, Olsen, & Sogin, 1985; Sogin & Elwood, 1986; Sogin, Swanton, Gunderson, & Elwood, 1986a), demonstrating that ciliates, even with this small sampling of species, appeared to be monophyletic and yet showed very deep divergences, equivalent to the genetic distances between the classical plant and animal "kingdoms". The first denser samplings of species, using both the SSUrRNA (Lynn & Sogin, 1988; Sogin & Elwood, 1986) and the large subunit (LSU) rRNA (Baroin et al., 1988), provided enough taxon density to demonstrate utility in testing the deeper relationships predicted by ultrastructural research.

The molecular phylogenetic approach is now a recognized method for testing and establishing phylogenetic relationships among organisms, and has been particularly fruitful in revealing the broad lines of evolutionary descent among eukaryotes. However, it rests on the basic assumption that phylogenetic trees based on genes truly represent the phylogeny of the organisms. Ultimately, our confidence in so-called "gene trees" increases when multiple and unlinked genes show patterns congruent with each other and with organismal phylogenies constructed on other features, such as morphology. It is the purpose of this chapter to briefly review the deep phylogeny of ciliates as inferred from features of cortical ultrastructure, primarily, and then to examine how this topology is congruent with gene tree topologies derived from rRNA genes and several protein coding genes. This will provide a consensus phylogenetic tree of the currently recognized classes of ciliates, which will provide the basis for a final discussion of the evolution of character states in the phylum. It is this distribution of character states that, in part, forms the rationale for the higher classification presented in **Chapter 17**.

16.1 Deep Phylogeny and Ultrastructure

The transmission electron microscope provided a technical approach that opened up literally a vast array of detailed character information with which to investigate the cellular morphology of protists. Initially, there was a preoccupation with cortical fibrillar systems, an approach pioneered by Pitelka (1969). Later, comparative analyses of these cortical patterns, especially of somatic kinetids, suggested eight major clades or classes of ciliates: (1) Class KARYORELICTEA; (2) Class SPIROTRICHEA; (3) Class LITOSTOMATEA; (4) Class PHYLLOPHARYNGEA; (5) Class COLPODEA; (6) Class NASSOPHOREA; (7) Class PROSTOMATEA; and (8) Class OLIGOHYMENOPHOREA (Lynn, 1981; Small & Lynn, 1981, 1985). As discussed in **Chapter 1**, arrangement of these classes into subphyla based on morphology has not been supported by molecular analyses (**see below**). While divided into subphyla by Small and Lynn (1985), the classes emerged "bush-like" from the common ancestor (Fig. 16.1).

Bardele (1981) analyzed the arrays of intramembranous particles of cilia in 68 genera, representing a broad diversity of ciliates. These particle array patterns were classified into a ciliary necklace that ringed the base of the cilium, ciliary plaques, ciliary rosettes, single- and double-stranded longitudinal rows, and orthogonal arrays covering most of the cilium. His analysis suggested six major assemblages: (1) SPIROTRICHA, corresponding to the Class SPIROTRICHEA; (2) GYMNOSTOMATA, which included representatives of the Classes LITOSTOMATEA and PROSTOMATEA; (3) TRICHOSTOMATA, which included representatives from the Classes LITOSTOMATEA and COLPODEA; (4) ENTODINIOMORPHA, which included representatives from the Class LITOSTOMATEA; (5) HYPOSTOMATA+SUCTORIA, corresponding to the Class PHYLLOPHARYNGEA; and (6) HYMENOSTOMATA+PERITRICHA+ASTOMATA, corresponding to the Class OLIGOHYMENOPHOREA. Bardele's "ciliate bush" was anchored in a gymnostome-like form and radiated out from there. While there was some broad agreement with the clades based on cortical ultrastructure, the particle array character set was not rich enough to tease out the details of this diversification (Fig. 16.2).

Bardele (1987, 1989) turned his "ciliate bush" upside down as he reviewed the data arising from his laboratory on the ultrastructure of ontogeny, and particularly stomatogenesis, in ciliates. These observations, coupled with the conception that the ciliate ciliature arose by proliferation from the paroral (Eisler, 1989, 1992), suggested that gymnostomy – a simple, anterior oral region – may have arisen repeatedly as a derived and secondary feature of oral apparatus evolution and not as a primary feature. Bardele (1989) concluded by doubting that many of the major groups suggested by Small and Lynn (1981, 1985) would be confirmed to be monophyletic, and he strongly argued that a research program in ontogeny would reveal this view to be true.

By the early 1990s, there was general agreement among morphologists that the ciliates could by arranged into from 8 to 11 major clades or classes, although there was some disagreement on

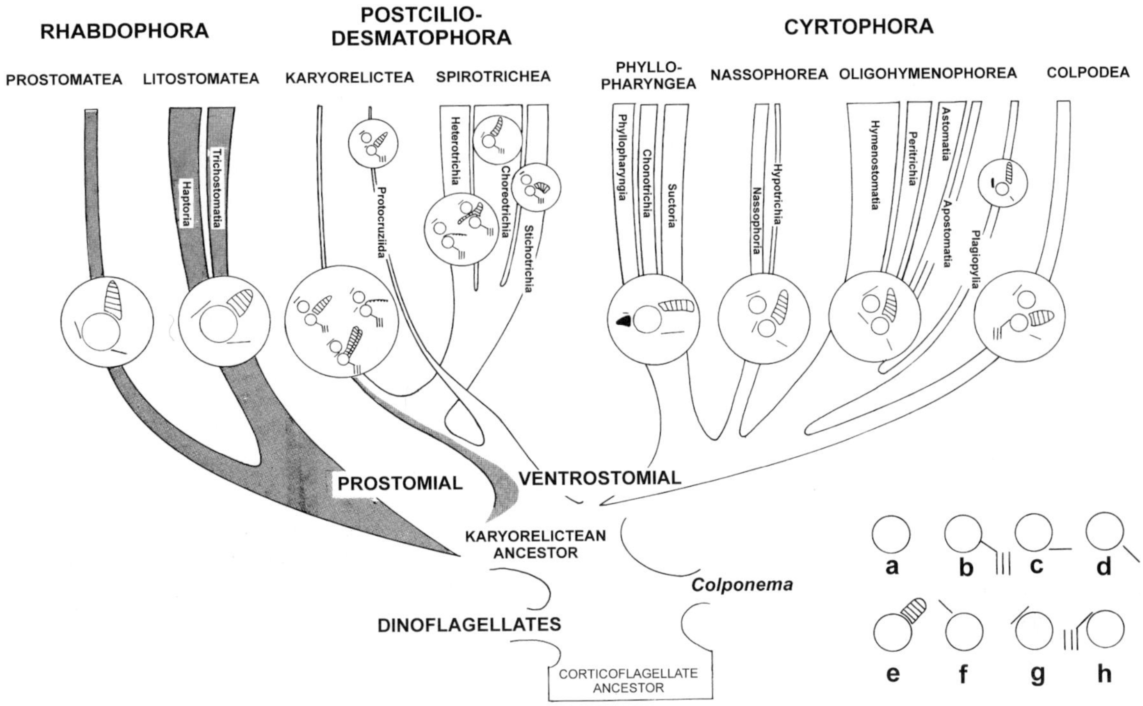

FIG. 16.1. Phylogeny of the Phylum Ciliophora as presented by Small and Lynn (1985). Eight major monophyletic lineages (= classes) are thought to have diversified from a karyorelictean ancestor, one that exhibited the ancestral state of nuclear dimorphism. The thickness of each clade represents generic diversity. Each clade is characterized by a schematic of its kinetid, which is diagrammed as if viewed from the **inside** of the cell. The key to the kinetid structures is as follows: (**a**) kinetosome; (**b**) overlapping postciliary microtubular ribbons forming postciliodesma; (**c**) convergent postciliary microtubular ribbon; (**d**) divergent postciliary microtubular ribbon; (**e**) striated kinetodesmal fibril; (**f**) radial transverse microtubular ribbon; (**g**) tangential transverse microtubular ribbon; (**h**) overlapping transverse microtubular ribbons, the so-called transversodesma. (Redrawn from Small & Lynn, 1985.)

how these might be related at deeper levels (Lynn & Corliss, 1991; de Puytorac, 1994a; de Puytorac et al., 1993). The early researches into rRNA gene sequences suggested that molecular phylogenetics would be a productive approach to test the robustness of these morphology-based phylogenies and classifications.

16.2 Deep Phylogeny and Gene Sequences

It is not our intention in this section to present an exhaustive review of molecular phylogenetic studies on ciliates. Instead, studies will be cited that have tested the monophyly of the major classes, as suggested by morphological analysis, and that also provide some evidence of the deeper structure to the relationships among classes. Often, these deeper relationships have not been strongly supported by "statistical" approaches, like bootstrap analysis or likelihood probabilities. However, if a consensus emerges based on different genes, both rRNA and proteins, we will use this to construct a tree with which to examine the broad evolution of character states within the phylum (**16.3 Character State Evolution**).

The basic approach for gene sequencing remains the same, but has developed to be much more efficient since the days of cloning genes into vectors in the 1980s. In brief, conserved regions of genes are used to design polymerase chain reaction (PCR) primers, which enable amplification of the gene of interest (e.g., Bernhard

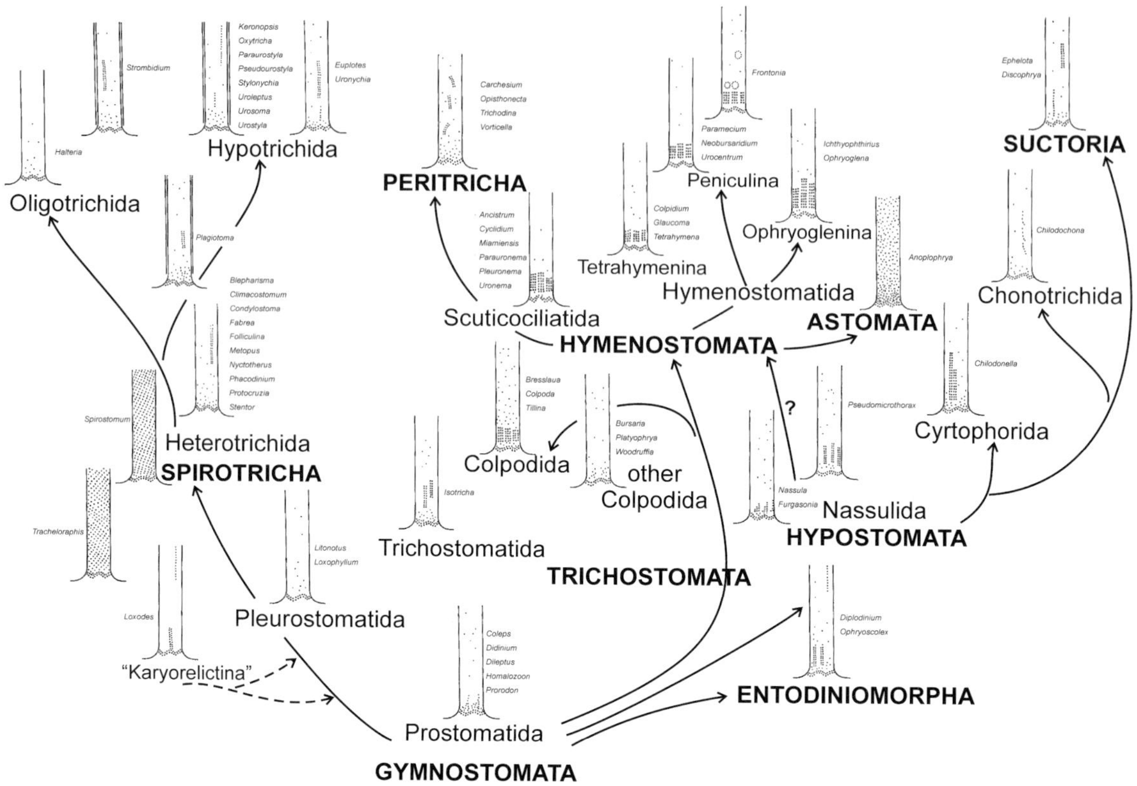

FIG. 16.2. Schematic view of the phylogeny of ciliates based on characterization of the particle arrays in ciliary membranes, revealed by the freeze fracture technique. The particle array patterns can be classified into a ciliary necklace that ringed the base of the cilium (virtually all groups), ciliary plaques (see Hymenostomatida), ciliary rosettes (see *Frontonia*), single- (see Hypotrichida, "Karyorelictina", and SUCTORIA) and double-stranded (see SPIROTRICHA, PERITRICHA, and HYPOSTOMATA) longitudinal rows, and orthogonal arrays (see *Tracheloraphis* and *Spirostomum*). (Redrawn from Bardele, 1981.)

& Schlegel, 1998; Medlin, Elwood, Stickel, & Sogin, 1988). The PCR-amplified genes may then be cloned into a plasmid vector, amplified in bacteria, purified, and then sequenced (e.g., Baroin-Tourancheau, Villalobo, Tsao, Torres, & Pearlman, 1998; Greenwood, Schlegel, Sogin, & Lynn, 1991b; Hirt et al., 1995). As is often the case now, the PCR-amplified genes are directly sequenced (e.g., Lynn & Strüder-Kypke, 2005). In either case, both strands of the DNA should be sequenced to corroborate the sequence reads.

16.2.1 Ribosomal RNA Sequences

The initial studies on rRNA gene sequences, using both SSUrRNA (Lynn & Sogin, 1988) and LSUrRNA (Baroin et al., 1988), confirmed the ciliates as a monophyletic group. Later studies have served to solidify this confirmation and provide substantial support for the ciliates as the sister taxon to the dinoflagellates and apicomplexans in the alveolate clade (Leander & Keeling, 2003; Van de Peer, Van der Auwera, & De Wachter, 1996). Thus, the classical view of ciliates long being regarded as monophyletic is strongly supported by rRNA gene sequences.

In the intervening years, species sampling has increased with the aim of determining how robust the monophyly of the major classes has been. Based on partial LSUrRNA gene sequences, Baroin-Tourancheau, Delgado, Perasso, and Adoutte (1992) provided evidence of the deep genetic divergences among five of the major classes (i.e. Classes KARYORELICTEA, SPIROTRICHEA, LITOSTOMATEA, COLPODEA, and NASSOPHOREA), and their results united the Classes PROSTOMATEA and

OLIGOHYMENOPHOREA. They did not sample the Class PHYLLOPHARYNGEA.

Numerous studies on the SSUrRNA have now confirmed the major classes, but also suggested the recognition of new ones. Greenwood et al. (1991b) demonstrated the basal branching of the heterotrichs, separating them from the other spirotrichs, a result confirmed by subsequent studies (Hirt et al., 1995; Rosati, Modeo, Melai, Petroni, & Verni, 2004), and justifying their elevation to class rank (de Puytorac, 1994a). This added a ninth class to the Small and Lynn (1981, 1985) system. Greenwood, Sogin, and Lynn (1991a) added sequences of oligohymenophoreans to demonstrate the integrity of this group, which has been confirmed by later studies (Strüder-Kypke, Wright, Fokin, & Lynn, 2000b). Phyllopharyngeans were shown to be genetically distinct by Leipe, Bernhard, Schlegel, and Sogin (1994), and this has been subsequently confirmed (Riley & Katz, 2001; Snoeyenbos-West, Cole, Campbell, Coats, & Katz, 2004). Leipe et al. (1994) first demonstrated the genetic distinctness of the Class LITOSTOMATEA, and this has been subsequently confirmed (Cameron, Adlard, & O'Donoghue, 2001; Wright & Lynn, 1997b). Hirt et al. (1995) added members of the Classes KARYORELICTEA and HETEROTRICHEA to confirm the sister group relationship of these two taxa, and also demonstrated their genetic distinctness. In their study of the evolution of ciliate hydrogenosomes, Embley et al. (1995) demonstrated the genetic distinctness of the plagiopyleans, intriguingly including *Plagiopyla* and *Trimyema*, two genera not suspected to be closely related on the basis of morphology – a so-called "riboclass" (Lynn, 2004). This has been subsequently confirmed (Lynn & Strüder-Kypke, 2002), supporting the elevation of plagiopylids as the tenth class (de Puytorac, 1994a). Bernhard, Leipe, Sogin, and Schlegel (1995) provided evidence of the genetic distinctness of nassulid ciliates, now placed in the Class NASSOPHOREA. Throughout these intervening years, the Class SPIROTRICHEA with the heterotrichs removed, was confirmed as a monophyletic group to which *Protocruzia* was attached (Hammerschmidt et al., 1996) as well as the morphologically distinct genera – *Phacodinium* (Shin et al., 2000) and *Licnophora* (Lynn & Strüder-Kypke, 2002). Stechmann, Schlegel, and Lynn (1998) provided evidence of the distinctness of the Classes PROSTOMATEA and COLPODEA, while Lynn, Wright, Schlegel, and Foissner (1999) added species density to solidify the genetic distinctness of the COLPODEA.

Embley et al. (1995) had demonstrated that the armophorid *Metopus* spp. were not closely related to the heterotrichs, disproving this classical relationship. The independence of this lineage was clinched by the addition of a substantial number of additional armophorid sequences, demonstrating them to form a sister taxon with several species of the clevelandellid nyctotherids (van Hoek et al., 2000b). Lynn (2004) elevated this group to class rank as the Class ARMOPHOREA, establishing the eleventh class in our macrosystem.

The deeper relationships among these clades have not been strongly resolved. Cameron et al. (2001) performed statistical analyses and concluded that there was good statistical support for the Classes KARYORELICTEA, HETEROTRICHEA, SPIROTRICHEA, LITOSTOMATEA, PHYLLOPHARYNGEA, PROSTOMATEA, and PLAGI OPYLEA. The Classes COLPODEA and NASSOPHOREA were often associated in their analyses, while the Class OLIGOHYMENOPHOREA often did not form a well supported clade.

Review of the deeper topology demonstrated in the studies cited above provides no doubt of a deep bifurcation in the phylum, providing confirmation for the Subphylum Postciliodesmatophora to include the Classes KARYORELICTEA and HETEROTRICHEA, and providing support for the Subphylum Intramacronucleata (Lynn, 1996a, 2004). There is no consistent deep topology within the intramacronucleates, although the following assemblages receive some support: SPIROTRICHEA + ARMOPHOREA; NASSOPHOREA + COLPODEA; PROSTOMATEA + PLAGIOPYLEA; and PHYLLOPHARNYGEA + (NASSOPHOREA + COLPODEA) + (PROSTOMATEA + PLAGIOPYLEA) + OLIGOHYMENOPHOREA. Based on an analysis of our SSU rRNA database, a summary tree provides support for some of these groupings (Fig. 16.3).

16.2.2 Protein Gene Sequences

There is a handful of studies that examine protein sequences, both as nucleotides and as amino acids, to provide further tests of the robustness of our understanding of relationships among ciliates. An underlying problem with using protein genes to

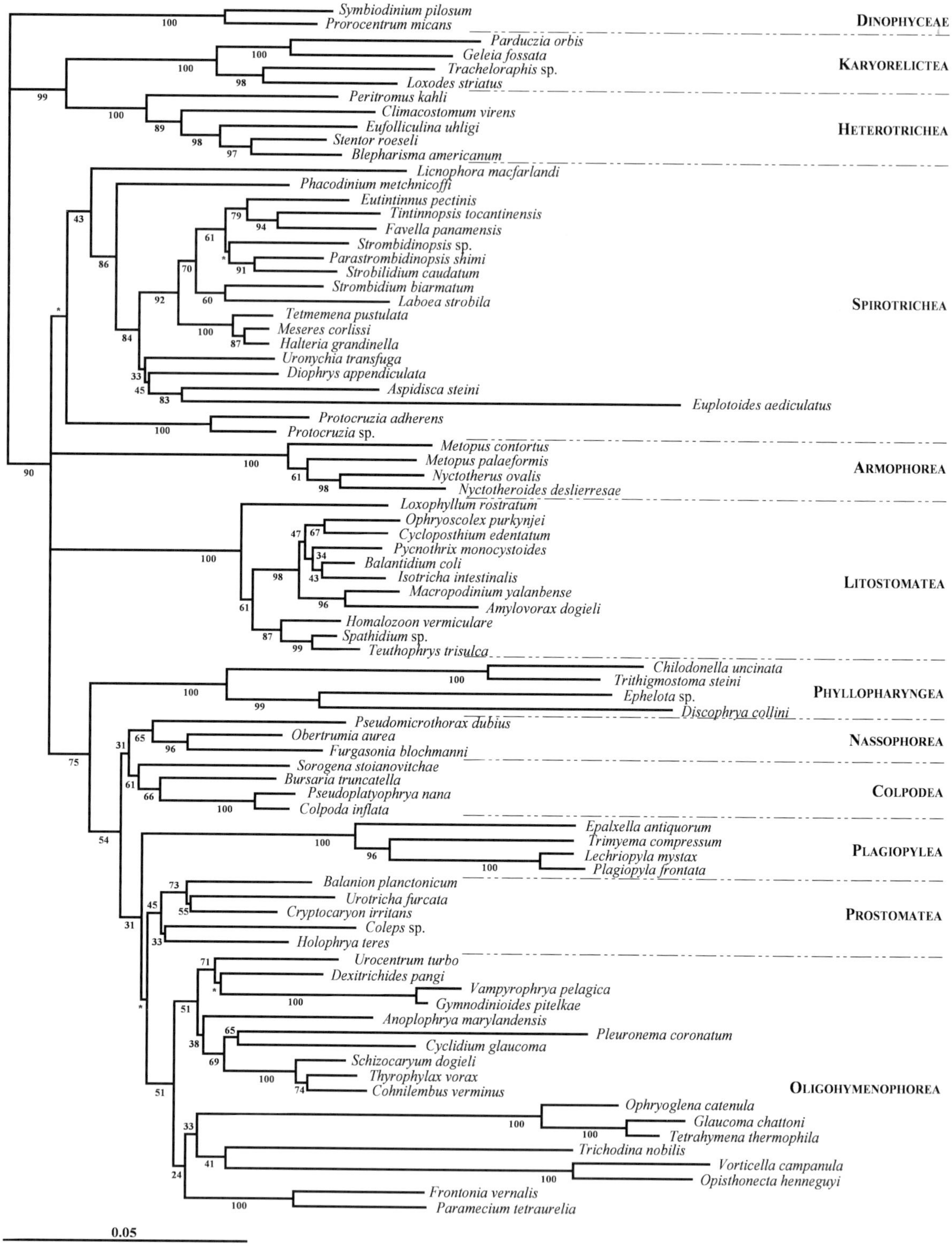

FIG. 16.3. A phylogenetic tree based on sequences of the small subunit rRNA gene and using the profile-neighbor-joining method implemented in Profdist ver. 0.9.6.1 (Friedrich et al., 2005). Note that the two subphyla – Postciliodesmatophora and Intramacronucleata - are strongly supported at >90%. Some classes are strongly supported (e.g., KARYORELICTEA, HETEROTRICHEA, ARMOPHOREA, LITOSTOMATEA, PHYLLOPHARYNGEA, PLAGIOPYLEA). Six "terminal" clades consistently cluster: the Classes PHYLLOPHARYNGEA, COLPODEA, NASSOPHOREA, PLAGIOPYLEA, PROSTOMATEA, and OLIGOHYMENOPHOREA) (cf. Fig. 16.5). We still have no rationalization outside of sequence data for this grouping. *Indicates support <20%

reconstruct the phylogeny of ciliates is the relatively high rate of protein diversification in the phylum, and especially in ciliate clades whose macronuclear genomes are extensively fragmented (Zufall, McGrath, Muse, & Katz, 2006). Nevertheless, protein phylogenetic studies can be divided into two groups – those that have sequenced a small number of representative genera from across the phylum and those that have provided a larger sampling of species.

Initial studies of the actin genes of ciliates indicated that the phylum was not recovered as a monophyletic group due to the high relative evolutionary rate of this gene in ciliates (Philippe & Adoutte, 1998). Kim, Yura, Go, and Harumoto (2004) have extended the sampling to about 20 genera of ciliates from five classes. Again, the ciliates are not recovered as a monophyletic group, although several classes appear to be: the Class LITOSTOMATEA and Class OLIGOHYMENOPHOREA.

Elongation factor 1α (EF-1α) is a protein that, in addition to its role in protein synthesis, probably interacts with actin in the cytoskeleton of ciliates. It also shows unusually high rates of evolution, and again ciliates are not recovered as a monophyletic assemblage (Moreira, Le Guyader, & Philippe, 1999). In an update of this research, Moreira, Kervestin, Jean-Jean, and Philippe (2002) provided sequences of eukaryotic release factor 1 (eRF1) and factor 3 (eRF3) in addition to sequences of EF-1α and elongation factor 2 (EF-2). The genus sampling of eRF3 was too low to draw any definitive conclusions, but ciliates again were not recovered as monophyletic using either EF-1α or eRF1. With seven genera representing five classes, the ciliates were recovered as monophyletic with EF-2 (Moreira et al., 2002). Moreira et al. (2002) speculated that these accelerated rates of evolution in the ciliates may be due to loss of interaction of these proteins with cytoskeletal elements or may be a co-evolutionary phenomenon linked with the extremely fast-evolving actins of ciliates. The 70 kDa heat shock proteins (Hsp70) comprise a multigene family that has been divided into three major subfamilies: (1) prokaryotic, mitochondrial, and chloroplast proteins; (2) eukaryotic cytosolic and nuclear proteins; and (3) eukaryotic proteins localized in the endoplasmic reticulum (Budin & Philippe, 1998). Budin and Philippe (1998) demonstrated that Hsp70 subfamily sequences from *Euplotes* and *Paramecium* confirmed the ciliates as a monophyletic group.

Baroin et al. (1998) provided sequences of phosphoglycerate kinase (PGK) for seven species representing three classes – Classes HETEROTRICHEA, SPIROTRICHEA, and OLIGOHYMENOPHOREA – and showed that the phylum was monophyletic, although these data could be compared to only a limited sampling of other eukaryotes. Thus far, only three protein genes – EF-2, Hsp70, and PGK – have confirmed the monophyly of the ciliates. The last two proteins that have been studied – the tubulins and histones – also comprise multigene families, but they have been much more extensively sampled across the phylum.

Baroin et al. (1998) provided nucleotide and amino acid sequences for α-tubulins from representatives of seven classes – Classes KARYORELICTEA, HETEROTRICHEA, SPIROTRICHEA, LITOSTOMATEA, COLPODEA, NASSOPHOREA, and OLIGOHYMENOPHOREA. Israel, Pond, Muse, and Katz (2002) have added sequence data for the Classes ARMOPHOREA and PHYLLOPHARYNGEA. Although both studies only compared the ciliate sequences to alveolate sister taxa, the ciliates were monophyletic. Overall, although taxon sampling was low, most classes appeared to be monophyletic, excepting the Classes HETEROTRICHEA and SPIROTRICHEA. While the classes were generally supported, there was no consistently recoverable deep topology (Fig. 16.4) (Israel et al., 2002). The ciliates were also recovered as a monophyletic group based on β-tubulin sequences (Philippe & Adoutte, 1998).

Bernhard and Schlegel (1998) provided the first analyses of variation among the histone genes H3 and H4 in six classes – Classes HETEROTRICHEA, SPIROTRICHEA, COLPODEA, NASSOPHOREA, PROSTOMATEA, and OLIGOHYMENOPHOREA. Katz, Bornstein, Lasek-Nesselquist, and Muse (2004) have expanded the database, adding sequences from representatives of the Classes ARMOPHOREA, COLPODEA, PHYLLOPHARYNGEA, and OLIGOHYMENOPHOREA. Thus, only representatives of the Class PLAGIOPYLEA are missing. In unconstrained analyses of H4 nucleotides, the ciliates were not monophyletic, but they were monophyletic based on amino acid sequences (Katz et al., 2004). Based on amino acids, classes were generally monophyletic (Fig. 16.5). The deep

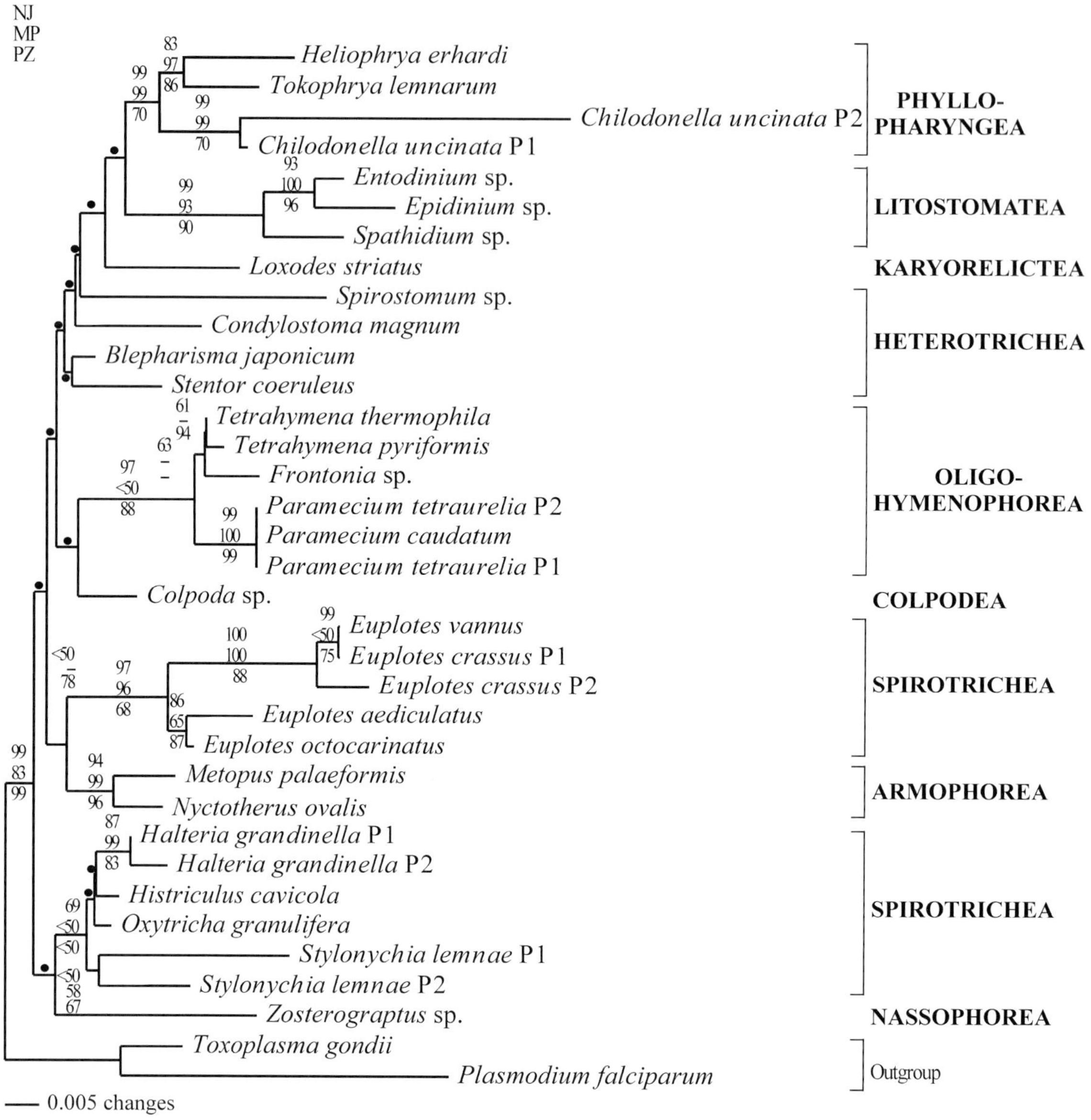

FIG. 16.4. A phylogenetic tree derived from a neighbor-joining analysis of the amino acid sequences of the α-tubulin gene. The numbers on the branches represent bootstrap percentages for neighbor-joining (NJ) and maximum parsimony (MP) while support estimates are provided for puzzle quartet analysis (PZ). The dots indicate branches with very low support values or inconsistent topology; P1 and P2 refer to paralogs of the α-tubulin gene. (Redrawn from Israel et al., 2002.)

topology was generally unresolved, although four classes were often associated – Classes COLPODEA, NASSOPHOREA, PROSTOMATEA, and OLIGOHYMENOPHOREA (CONP, Fig. 16.5) (Bernhard & Schlegel, 1998; Katz et al., 2004). The unusual ciliate *Protocruzia*, which we place in the Class SPIROTRICHEA (see **Chapter 17**), is associated with karyorelicteans (Bernhard & Schlegel, 1998) or the four-class assemblage (Katz et al., 2004), based on H4 **nucleotide** sequences. However, this genus is at the base of the intramacronucleate clade (Bernhard & Schlegel, 1998) or associated with the spirotrichs (Katz et al., 2004), based on **amino acid** sequences (Fig. 16.5).

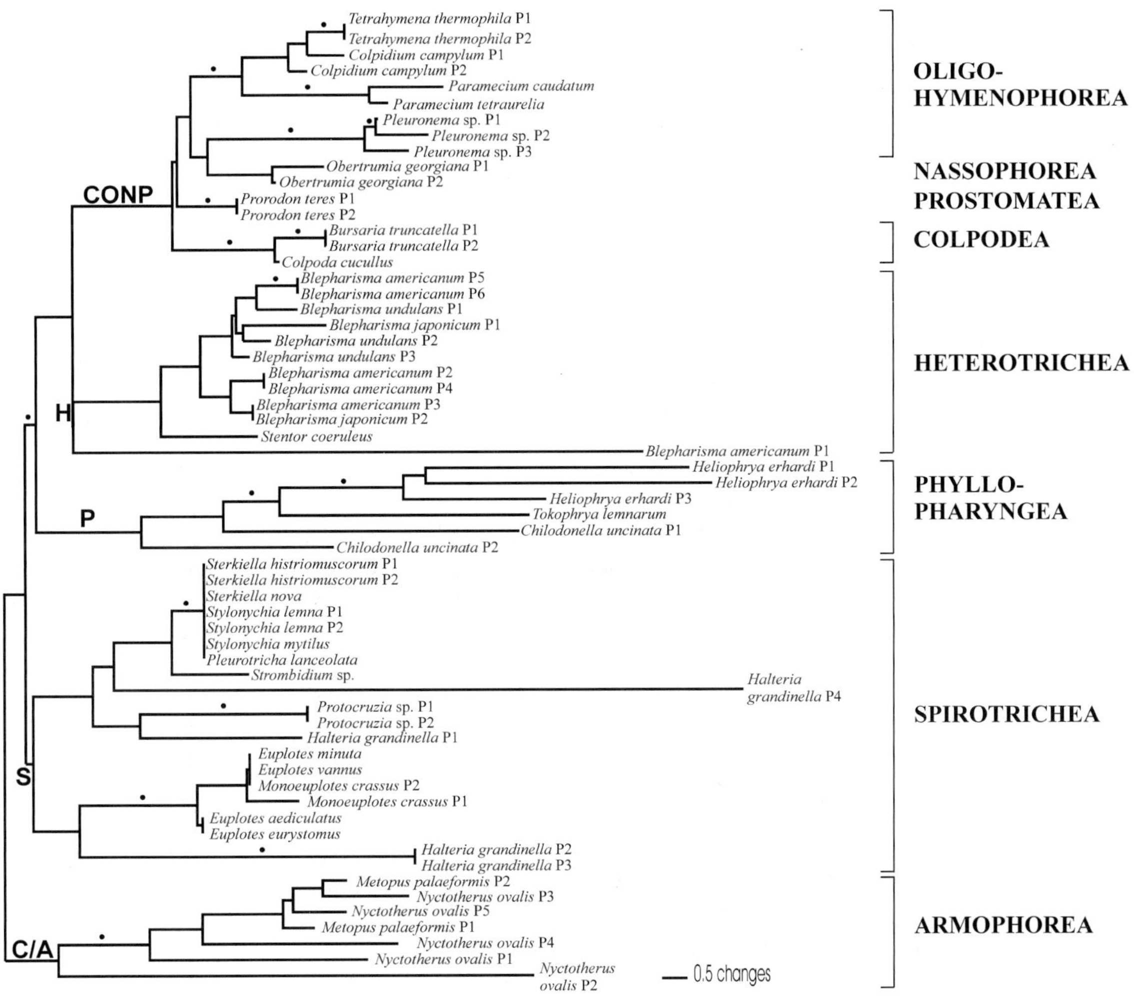

FIG. 16.5. A phylogenetic tree derived from a neighbor-joining analysis of the amino acid sequences of the histone H4 gene. The dots indicate bootstrap percentages >70%. Clades indicated by capital letters correspond to the respective classes. Note that only the Classes COLPODEA and PROSTOMATEA are supported >70%, but species sampling in these is very low. P1, P2, etc. indicate paralogs. (Redrawn from Katz et al., 2004.)

Overall, the protein sequence database provides us with little confidence in the deep phylogeny of the ciliates. Proteins refute or confirm the monophyly of the phylum. Since there is no doubt from a morphological perspective that the ciliates are monophyletic, reinforced strongly by the rRNA sequence databases, we must consider those protein molecules refuting this monophyly to be aberrant in some way, perhaps due to very high relative rates of evolution (Katz et al., 2004; Moreira et al., 2002; Zufall et al., 2006). The major assemblages suggested by the SSUrRNA database, including the Classes COLPODEA, NASSOPHOREA, PROSTOMATEA, and OLIGOHYMENOPHOREA, are supported at least by H4 amino acid sequences (cf. Figs. 16.3, 16.5).

16.3 Character State Evolution

The review of gene sequence data for rRNA and protein genes, excluding those proteins with unusually high relative rates of evolution (i.e., actins, elongation factors), leaves us to conclude that the Phylum Ciliophora is monophyletic, supporting the classical view based on morphology. The sampling density of sequence information across the phylum is really only significant for the SSUrRNA gene, for which we now have representatives sequenced for all major classes and most major subclasses or orders. Based on this gene, a simplified topology has been constructed to use in our evaluation of the evolution of character states within the phylum (Figs. 16.6, 16.7). This analysis will provide some of the evidential basis for the higher classification presented in **Chapter 17**.

The ciliate tree is deeply divided into two major lineages. Mapping the presence of postciliodesmata on the tree demonstrates that this character is restricted to one of these two major lineages, which is now recognized as the Subphylum Postciliodesmatophora (Fig. 16.6A) (Lynn, 1996a).

The next five characters are all related to nuclear features. The other major lineage of ciliates has the major unifying feature of dividing the macronucleus primarily by using intramacronuclear microtubules. Distribution of this character on the tree supports recognition of the Subphylum Intramacronucleata (Fig. 16.6B) (Lynn, 1996a). The other major lineage with dividing macronuclei uses extramacronuclear microtubules in the division process. Distribution of this character on the tree supports recognition of the Class HETEROTRICHEA, which is also characterized by postciliodesmata whose ribbons are separated by a single microtubule (Fig. 16.6C) (see **Chapter 6**). The third nuclear character is the presence of non-dividing macronuclei. Distribution of this character on the tree supports recognition of the Class KARYORELICTEA, which is also characterized by postciliodesmata whose ribbons are separated by the 2+ribbon+1 microtubular arrangement (Fig. 16.6D) (see **Chapter 5**). As noted earlier, the topology of the tree does not permit us to unambiguously conclude how dividing macronuclei evolved within the phylum. One view is that macronuclei gained the ability to divide using both intra- and extramacronuclear microtubules. This was followed by a loss of division in the karyorelicteans, an emphasis on extramacronuclear microtubules in heterotrichs, and an emphasis on intramacronuclear microtubules in all other ciliates (Hammerschmidt et al., 1996). The other view is that dividing macronuclei evolved twice independently from non-dividing macronuclei (Katz, 2001; Orias, 1991a).

The next two nuclear characters are related to the molecular processing of macronuclear DNA. Following conjugation, the formation of polytene chromosomes and extensive chromosomal fragmentation can occur as the new macronucleus differentiates (Jahn & Klobutcher, 2002; Prescott, 1994; Raikov, 1996). The distribution of this combined feature is restricted to three classes – SPIROTRICHEA, ARMOPHOREA, and

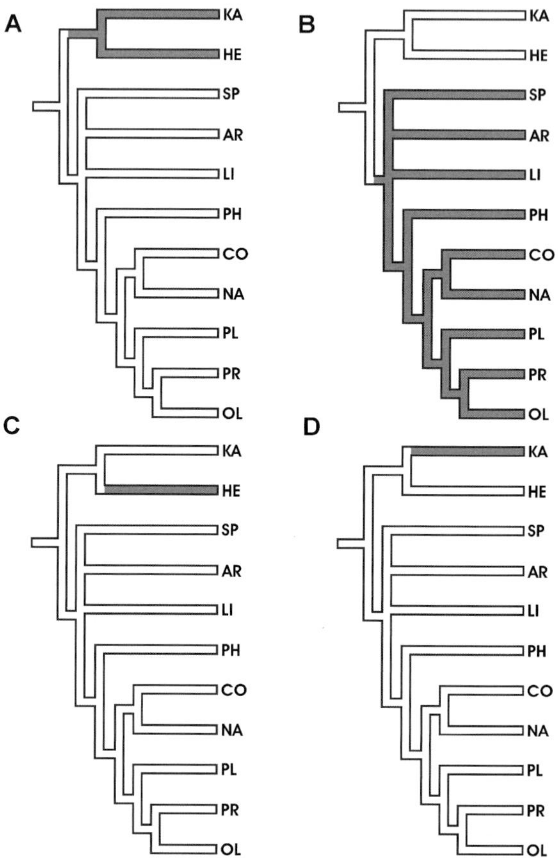

Fig. 16.6. Character evolution in the ciliates using a phylogenetic tree whose deep topology is based on the consensus of gene sequences, primarily from the small subunit rRNA and histone H4 genes (cf. Figs. 16.3, 16.5). **A** Presence of postciliodesmata. **B** Presence of intramacronuclear microtubules to divide macronucleus. **C** Presence of extramacronuclear microtubules to divide macronucleus. **D** Presence of non-dividing macronuclei. **KA**, Class KARYORELICTEA; **HE**, Class HETEROTRICHEA; **SP**, Class SPIROTRICHEA; **AR**, Class ARMOPHOREA; **LI**, Class LITOSTOMATEA; **PH**, Class PHYLLOPHARYNGEA; **CO**, Class COLPODEA; **NA**, Class NASSOPHOREA; **PL**, Class PLAGIOPYLEA; **PR**, Class PROSTOMATEA; **OL**, Class OLIGOHYMENOPHOREA

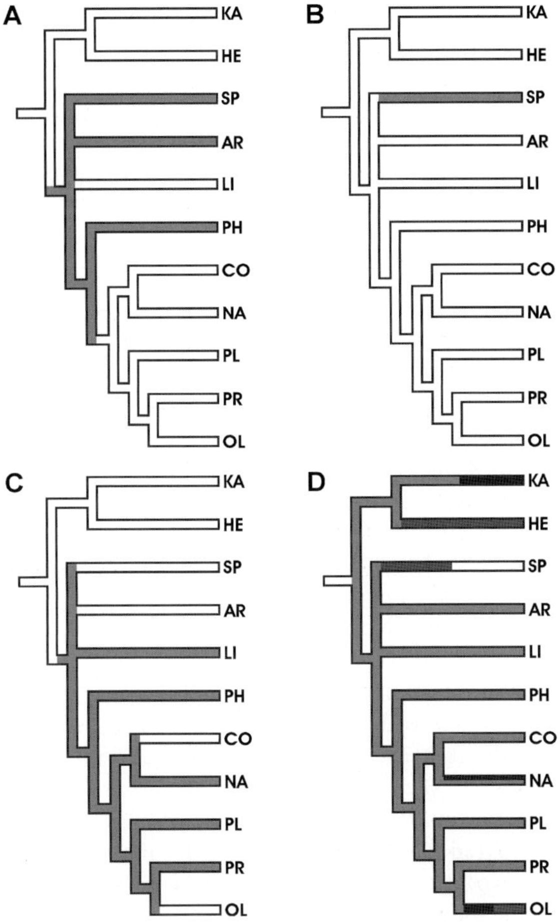

FIG. 16.7. Character evolution in the ciliates using a phylogenetic tree whose deep topology is based on the consensus of gene sequences, primarily from the small subunit rRNA and histone H4 genes (cf. Figs. 16.3, 16.5). **A** Presence of polytene chromosomes and chromosal fragmentation during macronuclear development. **B** Presence of replication bands during S phase of macronuclear DNA synthesis. Note that the genus *Protocruzia* does not have this feature although it clusters with the Class SPIROTRICHEA (cf. Figs. 16.3, 16.5). **C** Presence of somatic monokinetids. **D** Presence of buccokinetal (black), parakinetal (dark grey), telokinetal (grey), apokinetal (white), and mixokinetal (half black: half grey) modes of stomatogenesis. **KA**, Class KARYORELICTEA; **HE**, Class HETEROTRICHEA; **SP**, Class SPIROTRICHEA; **AR**, Class ARMOPHOREA; **LI**, Class LITOSTOMATEA; **PH**, Class PHYLLOPHARYNGEA; **CO**, Class COLPODEA; **NA**, Class NASSOPHOREA; **PL**, Class PLAGIOPYLEA; **PR**, Class PROSTOMATEA; **OL**, Class OLIGOHYMENOPHOREA

PHYLLOPHARYNGEA (Fig. 16.7A). Riley and Katz (2001) argued that chromosomal fragmentation may have had multiple origins. However, these three lineages often find their place at the "base" of the intramacronucleate radiation in gene sequence trees, sometimes separated by the Class LITOSTOMATEA (Fig. 16.7A). Thus, a common molecular mechanism of polytenization and genome fragmentation possibly underlies the explosive diversification of intramacronucleates. This mechanism has been refined or lost secondarily, at least twice, as this radiation diverged: it may have been lost in the common ancestor to the Class LITOSTOMATEA and in the common ancestor of the NASSOPHOREA-OLIGOHYMENOPHOREA clade (Fig. 16.7A).

The final nuclear feature is the presence of replication bands, which pass through the macronuclear karyoplasm during the S phase of DNA synthesis. Distribution of this character is restricted to lineages in the Class SPIROTRICHEA, and with the exception of *Protocruzia*, provides a rationale for the monophyly of this group (Fig. 16.7B) (see **Chapter 7**).

Finally, two features that have been considered important in systematic discussions are the presence of somatic monokinetids or somatic dikinetids and the kinds of stomatogenesis. Lynn and Small (1981) argued that the dikinetid state was likely the ancestral state for the ciliates, considering that the majority of flagellate taxa believed to be sister taxa to the ciliates had dikinetids. Distribution of the monokinetid character state on the ciliate tree is consistent with this view as four of the "early" emerging classes – KARYORELICTEA, HETEROTRICHEA, SPIROTRICHEA, and ARMOPHOREA – are characterized by somatic dikinetids (Fig. 16.7C). In fact, the character state distribution of monokinetids suggests a "gain" of this character as the common ancestor of the litostomes, phyllopharyngeans, and their sister taxa arose, with an independent secondary evolution of the somatic dikinetid character in the Class COLPODEA and within the Class OLIGOHYMENOPHOREA (Fig. 16.7C).

Ontogenetic features have assumed a central place in ciliate systematics since the early researches of Fauré-Fremiet and his group (Fauré-Fremiet, 1948a, 1950a, 1950b). Corliss (1968)

affirmed this view, and presented the basis of the current classification of stomatogenetic types (Corliss, 1979). Foissner (1996b) has updated and refined the classification of types, and provided a phylogenetic scenario for the evolution of these stomatogenetic types, assuming that the buccokinetal mode was ancestral or plesiomorphous. Foissner (1996b) noted that evidence for this assumption is weak, but he used as support the model proposed by Eisler (1992) for the evolution of the ciliate cortex. Distribution of all buccokinetal modes on the tree is not consistent with this view (Fig. 16.7D). Instead, the most broadly distributed mode is the telokinetal mode (Fig. 16.7D). Thus, Eisler's model (Eisler, 1992; Schlegel & Eisler, 1996) may be incorrect. Alternatively, soon after the ancestral cortex evolved by this "paroral model" of evolution (Eisler, 1992), a telokinetal mode of stomatogenesis may have evolved as the cell division process. As we have argued elsewhere, and is confirmed by this analysis, modes of stomatogenesis should be used only as descriptive features at this deep level. The usefulness of stomatogenetic characters is highest when characterizing and comparing genera and species. It is also useful in broadly associating ciliates into different clades based on the details of the stomatogenetic process rather than the mode itself (e.g. phyllopharyngean merotelokinetal vs. colpodean merotelokinetal; see Foissner, 1996b).

A final feature that we have not mapped on the tree, but which has been discussed by several research groups, is the evolution of hydrogenosomes from mitochondria (Embley et al., 1995; van Hoek et al., 2000b). Hydrogenosomes have been found in all species so far examined of the Classes ARMOPHOREA and PLAGIOPYLEA, which are not closely related (Figs. 16.6, 16.7), and in select members of the Classes LITOSTOMATEA and OLIGOHYMENOPHOREA. The latter evidence – origin within a class – demonstrates unambiguously the adaptive nature of the hydrogenosome (Fenchel & Finlay, 1990b, 1991a).

16.4 Summary

We have provided this discussion as an approach to demonstrating how to rationalize morphological and molecular features of the ciliates. This approach can also serve as the basis for providing evidence of the robustness of a classification or suggesting deeper subdivisions, which may not be inspired immediately by morphology (e.g., Subphylum Intramacronucleata; see Lynn, 1996a). As the species sampling for our gene sequence database expands, this approach may be productively extended "higher" in the tree, testing relationships among subclasses within classes and orders within subclasses. For example, the increased species sampling of SSUrRNA genes of suctorians provided very preliminary genetic evidence that the Orders Exogenida, Endogenida, and Evaginogenida may capture the evolutionary diversification of the suctorians (Snoeyenbos et al., 2004). Extensive sampling within the Class OLIGOHYMENOPHOREA has confirmed the monophyly of the major subclasses classically based on morphology (Affa'a, Hickey, Strüder-Kypke, & Lynn, 2004; J.C. Clamp et al., 2008; Greenwood et al., 1991a; Lynn & Strüder-Kypke, 2005; Strüder-Kypke et al., 2000b). Yet, clearly, much work remains to be done!

Chapter 17
The Ciliate Taxa Including Families and Genera

This chapter, in large measure, is independent of the others. Here, we diagnose or characterize in a succinct manner, all suprageneric taxa assignable to the Phylum Ciliophora. In each characterization, we have highlighted text in bold that we believe refers to synapomorphies or shared derived traits for these taxa. These characterizations are revisions of those provided in Corliss (1979), Lynn and Small (2002), de Puytorac and collaborators (1994a), and other references subsequent to these publications. Kahl's (1930–1935) great series has invaluable species descriptions, but omitted entirely certain major higher groups, and there is no comparison between the size of the total assemblage then and now. All genera are listed, supplying author and date. Synonyms are occasionally indicated where genera are considered problematic, and homonyms are noted. We have also indicated those genera designated as *nomen nudum* by Aescht (2001) as these highlight generic concepts for which additional research may confirm the conclusions of the original describer. Nevertheless, our coverage does not go to the nomenclatural and taxonomic depth of Aescht (2001), whose scholarly work should be consulted for a comprehensive treatment of the literature in relation to genera published up to 13 March 2000. Nor do we provide diagnoses and characterizations of genera or subgenera, many of which can be found in the second volume of the Traité edited by de Puytorac and collaborators (1994a), Jankowski (2007), and in other specialist works: the chapters of the Traité volume have been separately cited in the previous chapters devoted to treatment of the classes. We have included newly described genera published since Aescht (2001) and Berger (1999, 2001), and have indicated these genera by marking them with an asterisk (*).

Details, discussions of controversial matters, revelation of the rationale behind the new classification, evolutionary interrelationships, definition of terminology employed, and rich citation of pertinent literature sources are still to be found only in the preceding chapters.

17.1 Style and Format

The classes are arranged, in supposed phylogenetic order, based on the data provided by molecular phylogenetic analyses and ultrastructural studies (see **Chapter 16**). However, other taxa are presented alphabetically, each name followed by its author(s) and date of first valid description. Throughout, we have generally followed Aescht (2001) as the nomenclatural authority on genera. Remarks and comments are kept to a minimum but are included when considered indispensable.

Characterizations of taxa are as complete as possible at the levels of phylum, subphylum, class, and family. Other taxa are provided with a minimum characterization. Diagnostic or shared-derived characters for each taxon are highlighted in bold typeface. Whenever reference is made to size, the following criteria were used: small means < 80 μm in body length; medium means 80–200 μm in body length; and large means >200 μm in body length. Taxa are considered "free-swimming" if they are typically not attached to a substrate by a stalk or lorica or some other attachment organelle. In relation to life history, we make the broad distinction between "free-living" (i.e., NOT associated with other organisms) and symbiotic. We use the following classification of symbioses: commensalism – the symbiont benefits but the host does not; mutualism – both symbiont and host benefit; and parasitism – the symbiont benefits and the host is obviously harmed. Ciliates have often been called parasitic, but they are most often commensalistic. In the characterizations of families, it is common that a general feature might be followed by "(?)" (e.g., feeding (?)). This is meant to indicate that we have not been able to discover information about that character from the literature. We would be grateful to receive correspondence that would remove these areas of ignorance, either through personal observations by the correspondent or direction to the literature on the group.

We have reserved the use of the category *incertae sedis* for families and genera that seem to demand restudy to determine their appropriate assignment in our overall scheme.

17.2 Nomenclatural Notes, Abbreviations, and Figure References

Names of all authors of a taxonomic name are written out in full, even when the authorship is multiple, with a single exception. Whenever "de Puytorac et al., 1974" appears after a name, it is to be understood to stand for the following complete list of co-authors: de Puytorac, Batisse, Bohatier, Corliss, Deroux, Didier, Dragesco, Fryd-Versavel, Grain, Grolière, Hovasse, Iftode, Laval, Roque, Savoie, and Tuffrau.

Abbreviations (*not* italicized if directly following a *generic* name) include: hom. = homonym; *n. n.* = *nomen novum*; *non* = not; *p.p.* = *pro parte*; *s.l.* = *sensu lato*; *s.s.* = *sensu stricto*; subj. syn. = subjective synonym; syn(s). = synonym(s). In the case of subjective synonyms, we have noted all those that were listed by Aescht (2001) and for which we currently have reserved judgement on their validity. When comments are used, they are preceded by "**NOTE**" or are in brackets.

Synonymies have been established either by reference to the original literature or, when that was unavailable, by using the information provided in the *Zoological Record*. For many of the genera established, for example, by Jankowski (1978, 1979, 1980, 1981, and sometimes later) and others, and for which authors have not provided assignment to family, we have assumed that the genus remains within the family to which its type species was assigned at the time the author proposed the new genus. For example, *Pelagotrichidium* Jankowski, 1978 has *Hypotrichidium faurei* as the type species, but Jankowski (1978) did not assign the new genus to a family. Therefore, we have assumed that the new genus *Pelagotrichidium* remains in the Family SPIROFILIDAE to which *Hypotrichidium* was assigned.

Since Corliss (1979), numerous monotypic families have been established. In the vast majority of cases, we have not affirmed these families, but rather placed them in synonymy with the family from which the type genus of the new monotypic family originated. Future research, especially using molecular genetic approaches, may demonstrate the genetic distinctiveness of these proposed monotypic families. This would provide the additional support necessary to remove them from synonymy. For example, Corliss (1979) synonymized the monotypic Family SCHIZOCARYIDAE with the Family PLAGIOPYLIDAE and placed its type genus *Schizocaryum* as *incertae sedis* in the latter family. Examination of *Schizocaryum* by electron microscopy (Lynn & Frombach, 1987) and gene sequencing (Lynn & Strüder-Kypke, 2002) demonstrated in the former case that its morphology was very different from the plagiopylids, justifying the establishment of a monotypic family, and demonstrated in the latter case that the family should be placed within the Order Philasterida (see p. 414, 417).

17.3 The Ciliate Taxa to Genus

Phylum CILIOPHORA Doflein, 1901
(syns. Ciliae, Ciliozoa, Cytoidea, Eozoa, Heterocaryota, Heterokaryota, Infusoria, also Ciliata [Ciliatea, Ciliaside, Euciliara] + Suctoria [Suctorea], Gymnostomea + Ciliostomea + Tentaculifera, Kinetodesmatophora + Postciliodesmatophora, Rhabdophora + Cyrtophora, Kinetofragminophora + Oligohymenophorea + Polyhymenophorea, Tubulicorticata + Filicorticata + Epiplasmata + Membranellophora)

Eukaryotic, unicellular, protists; size, 10–4,500 µm; free-swimming or sessile; pellicular alveoli; **with simple or compound cilia in at least one stage of the life cycle**; complex cortical infraciliature, divided into somatic and oral regions; **cortical microtubular or microfibrillar structures associated with the kinetosome including a laterally- or anteriorly-directed kinetodesmal fibril at kinetosomal triplets 5–7, a tangential or radial transverse microtubular ribbon at triplets 3–5, and a postciliary microtubular ribbon at triplet 9**; parasomal sac often adjacent to the base of the somatic cilium; extrusomes, common, with somatic extrusomes as mucocysts and oral extrusomes as toxicysts; oral region, generally monostomic, but some groups mouthless or polystomic; stomatogenesis apokinetal, parakinetal, mixokinetal, buccokinetal or telokinetal; **fission homothetogenic** and often perkinetal, isotomic or anisotomic, and occasionally multiple; **nuclear dualism with one or more presumed diploid micronuclei and one to several ampliploid (rarely diploid or oligoploid) macronuclei**, with acentric mitosis; **sexual process by conjugation, which may be temporary or total, with gametic nuclei formed by meiotic division of micronucleus**; contractile vacuole, typically present; cytoproct, often present; typically heterotrophic, feeding modes ranging from osmotrophy to phagotrophy, and some mixotrophy; broadly distributed in diverse aquatic and terrestrial habitats; ecto- or endosymbionts common; two subphyla.

Subphylum POSTCILIODESMATOPHORA Gerassimova & Seravin, 1976
(syns. Ciliostomatophora *p.p.*, Heterotricha + Kinetofragminophora *p.p.*, Homotricha *p.p.*, Polyhymenophora *p.p.*, Tubulicorticata *p.p.*)

Size, generally large; shape, typically elongate, highly contractile; sessile and free-swimming; cortical alveoli, poorly developed; **somatic ciliation with dikinetids that have postciliodesmata, a special arrangement of overlapping postciliary microtubular ribbons**; parasomal sacs, absent; somatic extrusomes as mucocysts, rhabdocysts, and/or pigmentocysts; oral structures, highly variable, from prostomatous to a highly differentiated adoral zone of polykinetids; during stomatogenesis, the oral apparatus of the proter usually undergoing regression and redevelopment prior to and/or during cytokinesis; fission almost always isotomic; conjugation, temporary and isogamontic; diverse feeding habits; species widely distributed; rarely symbiotic; two classes.

Class **KARYORELICTEA** Corliss, 1974
(syns. Antostomatina *p.p.*, Epitrichina, Inferotrichina, Karyorelictida, Karyorelictina, Loxodina, Primociliatida-Karyorelictina, Rhynchophorina)

Size, generally, large; shape, long, vermiform, fragile, flattened, often contractile; free-swimming and highly thigmotactic; many genera having a more-or-less conspicuous non-ciliated or sparsely ciliated surface region (= glabrous zone or stripe), and all taxa having a bristle-kinety surrounding the glabrous zone; **somatic kinetids with postciliodesmata whose major microtubular ribbons are typically separated by one microtubule**; stomatogenesis, parakinetal or buccokinetal; **macronuclei, non-dividing, two to many, containing approximately, sometimes slightly more than, the diploid amount of DNA; macronuclei form only by division of micronuclei, then differentiate**; conjugation, temporary; contractile vacuole, often apparently absent; in marine and freshwater habitats, typically interstitial forms of marine sands; three orders.

NOTE: Foissner (1995a), and see also Foissner and Dragesco (1996b), placed the Family KENTROPHORIDAE in the Order Loxodida, whose members he defined as having a specialized dorsolateral 'bristle' kinety and an epipellicular mucus and/or scale layer on the left body surface. However, members of the Family LOXODIDAE do not have the scale layer, while some members of the Family TRACHELOCERCIDAE show a 'bristle-like' kinety. Molecular genetic data should demonstrate the phylogenetic affinity of the kentrophorids and help to demonstrate which morphological features are ancestral and which derived.

Order Protostomatida Small & Lynn, 1985
(syns. Kentrophorida *p.p.*, Symbiophagina *p.p.*, Thysanophorida *p.p.*, Trachelocercida, Trachelocercina)

Shape, long to very long, highly contractile; **ventral (= right?) somatic surface densely ciliated, dorsal (= left?) somatic surface glabrous (= non-ciliated) and of varying width; border between somatic surfaces ringed by a 'bristle' or 'bristle-like' kinety that may extend from the left margin around the posterior end and anteriorly along the right margin, possibly with circumoral kinetids inserted in the anterior right region**; oral region, apical or ventral; in marine or estuarine interstitial habitats; two families.

Family KENTROPHORIDAE Jankowski, 1980

Size, large; shape, flattened, vermiform, often C-shaped in cross-section; free-swimming; ventral (= right?) surface densely ciliated, covered by kineties of ciliated dikinetids; **dorsal (= left?) surface only bordered by 'bristle' kinety and covered by mucous material colonized by symbiotic sulphur bacteria**; extrusomes as somatic secretory ampullae (= mucocysts?); oral region, apical, and oral ciliature, if present, highly reduced to remnants of kinetids; cytostome, not permanent; nuclei in clusters, typically more than 10, commonly with four macronuclei surrounding one micronucleus; contractile vacuole (?); cytoproct (?); feeding primarily by ingesting 'its' symbiotic bacteria through the naked dorsal surface; in marine interstitial habitats; one genus.

- *Kentrophoros* Sauerbrey, 1928

Family TRACHELOCERCIDAE Kent, 1881
(syn. Prototrachelocercidae, Sultanophryidae)

Size, large; shape, elongate, contractile, often with distinct 'head and neck'; free-swimming; somatic ciliature as files of dikinetids, which may cover the body completely or leave a glabrous longitudinal zone of varying width on the dorsal (= left?) side; glabrous zone, if present, bordered by a 'bristle-like' kinety; extrusomes as small somatic mucocysts; **oral area apical, surrounded by circumoral ciliature of dikinetids and accompanied by a brosse of either short ciliated rows or an unstructured ciliated tuft; cytostome, inconspicuous, not permanent**; nuclei in one or more clusters, with several macronuclei surrounding one or two micronuclei; contractile vacuole, absent; cytoproct (?); feeding on bacteria and smaller protists; in marine interstitial habitats; six genera.

- *Kovalevaia* Foissner, 1997
- *Prototrachelocerca* Foissner, 1996
- *Sultanophrys* Foissner & Al-Rasheid, 1999
- *Trachelocerca* Ehrenberg, 1840
- *Trachelolophos* Foissner & Dragesco, 1996
- *Tracheloraphis* Dragesco, 1960

Order Loxodida Jankowski, 1980
(syns. Cryptopharyngina, Loxodina)

Size, medium to large; **shape, laterally flattened, non-contractile**; free-swimming; somatic cilia as files of dikinetids mainly on the right surface with the left surface barren except for single marginal (= 'bristle'?) kinety; extrusomes as somatic cnidocyst-like organelles in some genera; **oral region, subapical on the narrow ventral surface; oral kinetids as dikinetids surrounding the oral area as two perioral kineties and one intraoral (= intrabuccal) kinety**; stomatogenesis, monoparakinetal or buccokinetal; nuclei in clusters, typically of two macronuclei and one micronucleus; in marine and freshwater (only *Loxodes*) habitats, typically in anoxic sediments and anoxic water columns; two families.

Family CRYPTOPHARYNGIDAE Jankowski, 1980

Size, small to medium; shape, ovoid, flat, margin often serrate; **epipellicular ornamental scales embedded in mucous layer on left body surface**; free-swimming; somatic ciliature as files of dikinetids on the right (= ventral?) surface and a single left lateral kinety bordering the left (= dorsal?) surface; extrusomes, not reported; **oral region, circular to ovoid, ringed by inner and outer dikinetid files with**

an intrabuccal kinety, which may be short or long; nuclei in clusters, typically two macronuclei and one micronucleus; contractile vacuole, absent; cytoproct (?); feeding on bacteria and smaller protists; in marine interstitial habitats; two genera.

- *Apocryptopharynx* Foissner, 1996
- *Cryptopharynx* Kahl, 1928

Family LOXODIDAE Bütschli, 1889
(syns. Ciliofaureidae, Drepanostom(at)idea)

Size, medium to very large; **shape, long, flat, with beak-like anterior rostrum, which interrupts perioral kineties at the anterior end**; free-swimming; somatic ciliature as files of dikinetids on the right (= ventral?) surface and a single left lateral kinety bordering the left (= dorsal?) surface; extrusomes as somatic cnidocyst-like organelles in *Remanella*; **Müllerian vesicles in the endoplasm, containing strontium (*Remanella*) and barium (*Loxodes*) salts**; **oral area in long ventral groove behind rostrum, with inner and outer dikinetid files and a long, rectilinear intrabuccal kinety**; nuclei in clusters with one or two macronuclei and a single micronucleus; contractile vacuole, in freshwater species; cytoproct (?); feeding on bacteria and other protists, such as microalgae; in freshwater (*Loxodes*) and marine (*Remanella*) habitats, typically in sediments but ranging into the water column when oxygen levels decline; two genera.

- *Loxodes* Ehrenberg, 1830
- *Remanella* Foissner, 1996

Order Protoheterotrichida Nouzarède, 1977

Size, large; shape, elongate, highly contractile; often pigmented, appearing black in transmitted microscopic light; free-swimming; somatic ciliation, holotrichous; **oral region, ventral, with an elaborate oral ciliature including a series of transverse rows of monokinetids, conspicuous either on the left or the right side of the oral region, depending upon the genus**; stomatogenesis, not described; nuclei in clusters, typically two macronuclei and one micronucleus; contractile vacuole, absent; cytoproct (?); feeding on microalgae and other protists; in marine or brackish coastal sands and gravels; one family.

Family GELEIIDAE Kahl, 1933
(syn. Aveliidae)

With characteristics of the order; four genera.

- *Avelia* Nouzarède, 1977
- *Geleia* Kahl in Foissner, 1998
- *Gellertia* Dragesco, 1999
- *Parduczia* Dragesco, 1999

Incertae sedis in Class KARYORELICTEA

- *Ciliofaurea* Dragesco, 1960
- *Corlissia* Dragesco, 1954

Class **HETEROTRICHEA** Stein, 1859
(syns. Heterotricha, Heterotrichorina, Membranellata *p.p.*, Membranellophora *p.p.*, Spirotricha *p.p*, Spirotrichophora *p.p.*)

Size, medium to large; shape, variable, from compressed to conical, often elongate and contractile; free-swimming, but some species are temporarily or permanently sessile; some species are pigmented, often brightly, with pigment in specialized pigmentocysts; somatic ciliation, holotrichous; **somatic kinetids as dikinetids with postciliodesmata whose major microtubular ribbons are typically separated by 1 + 2 microtubules**; extrusomes, sometimes as mucocysts, but pigmentocysts can also be secreted; **left serial oral polykinetids conspicuous, typically paramembranelles, encircling the anterior end clockwise before plunging into the oral cavity**; one or more "parorals" on the right side of the oral cavity; stomatogenesis, parakinetal; **macronucleus, highly polyploid, divided by extra-macronuclear microtubules**; micronuclei, typically multiple; conjugation, temporary; contractile vacuole, often conspicuous, with long collecting canals; cytoproct, present; distributed widely in marine, freshwater, and terrestrial habitats; one order.

Order Heterotrichida Stein, 1859
(syns. Blepharismina *p.p.*, Coliphorida *p.p.*, Coliphorina *p.p.*, Condylostomatina *p.p.*, Peritromida *p.p.*, Stentorina *p.p.*)

With characteristics of the class; nine families.

Family BLEPHARISMIDAE Jankowski in Small & Lynn, 1985

Size, medium to large; shape, pyriform or ellipsoid, somewhat narrowed anteriorly, laterally compressed; free-swimming; pigmentocysts, common, filled with the pigment blepharismin; somatic ciliation, holotrichous; extrusomes as sacculate mucocysts; oral polykinetids extending along the left margin of the oral region and circling counter-clockwise into a shallow posterior oral cavity; **paroral dikinetids anterior of cytostome (precytostomal)**; macronucleus, ovoid to sometimes nodular; micronucleus, may be multiple; contractile vacuole, present; cytoproct, present; feeding on bacteria, microalgae, and other protists, including ciliates, with some species becoming cannibalistic; in marine, freshwater, and terrestrial habitats; four genera.

- *Anigsteinia* Isquith, 1968
- *Blepharisma* Perty, 1849
- *Parablepharisma* Kahl, 1932 [nomen nudum]
- *Pseudoblepharisma* Kahl, 1927

Family CHATTONIDIIDAE Villeneuve-Brachon, 1940

Size, large; shape, somewhat rotund, contractile, with gently pointed posterior end; free-swimming; somatic ciliation, holotrichous; **a unique somatic "posteroaxial cavity," containing 6–7 ciliary organelles, opening at posterior pole**; oral cavity opening apically and bearing full circle of strong oral polykinetids; macronucleus, very long, with loops; micronucleus, present; contractile vacuole (?); cytoproct (?); feeding on bacteria, microalgae, and other protists, including ciliates; in hypereutrophic brackish waters; one genus.

- *Chattonidium* Villeneuve, 1937

Family CLIMACOSTOMIDAE Repak, 1972
(syn. Pediostomatidae)

Size, medium to large; shape, ovoid, often anteriorly pointed; free-swimming; somatic ciliation, holotrichous; extrusomes as sacculate somatic mucocysts; **oral region anterior, with very prominent, ciliated peristomial field occupying much of the anterior part of the body, bordered at least on the posterior and right margins by the serial adoral polykinetids**; paroral ciliature, inconspicuous; macronucleus, usually as a thick ribbon, often coiled and lengthy; micronucleus, present; contractile vacuole, at least in freshwater forms; cytoproct (?); symbiotic zoochlorellae in some species; feeding on bacteria, microalgae, and other protists, including ciliates; in marine and freshwater habitats; three genera.

- *Climacostomum* Stein, 1859
- *Fabrea* Henneguy, 1890
- *Pediostomum* Kahl, 1932

Family CONDYLOSTOMATIDAE Kahl in Doflein & Reichenow, 1929
(syns. Condylostomidae, Condylostentoridae)

Size, medium to large; shape, very elongate in some forms, nearly ellipsoidal in others; free-swimming; somatic ciliation, holotrichous; extrusomes as sacculate somatic mucocysts in some species; **oral region expansive; oral ciliature, including adoral polykinetids and a prominent paroral membrane**; **peristomial field absent**; macronucleus, typically long and moniliform; micronucleus, typically multiple; contractile vacuole often with long collecting canal; cytoproct, present; feeding on bacteria, microalgae, and other protists; in marine, freshwater, and terrestrial habitats; seven genera and one genus *incertae sedis*.

- *Condylostoma* Bory de St. Vincent, 1824
- *Condylostomides* Da Silva Neto, 1994
- *Copemetopus* Villeneuve-Brachon, 1940
- *Electostoma* Jankowski, 1979
- *Linostomella* Aescht in Foissner, Berger, & Schaumberg, 1999
- *Predurostyla* Jankowski, 1978
- *Procondylostoma* Jankowski, 1979

Incertae sedis in Family Condylostomatidae

- *Dellochus* Corliss, 1960

Family FOLLICULINIDAE Dons, 1914
(syn. Coliphorida, Coliphorina)

Size, medium to large; shape, typically elongate, contractile; often pigmented; **mature forms, sessile and sedentary, always residing in a lorica; body, especially in neck region in some species, with conspicuous pair of "peristomial wings" bearing the prominent oral ciliature**; **at division, complex morphogenesis with vermiform migratory larval stage that "recapitulates" a typical heterotrich**

with reduced spiralling of the adoral zone; macronucleus, variable, as single ellipsoid to multiple and moniliform; micronucleus, single to multiple; contractile vacuole, present at least in freshwater forms; cytoproct (?); feeding on bacteria, microalgae, and other protists; widely distributed in marine habitats, but a few species in freshwater, attached to algae, higher aquatic plants, or integument or shells of invertebrates (e.g., molluscs, various crustaceans, bryozoa, coelenterates); 30 genera including some fossil forms.

- *Ampullofolliculina* Hadzi, 1951
- *Ascobius* Henneguy, 1884
- *Atriofolliculina* Hadzi, 1951 [nomen nudum]
- *Aulofolliculina* Hadzi, 1951
- *Botticula* Dioni, 1972
- *Claustrofolliculina* Hadzi, 1951
- *Diafolliculina* Hadzi, 1951 [nomen nudum]
- *Echinofolliculina* Dons, 1934
- *Epifolliculina* Hadzi, 1951
- *Eufolliculina* Hadzi, 1951
- *Folliculina* Lamarck, 1816
- *Folliculinopsis* Fauré-Fremiet, 1936 [nomen nudum]
- *Halofolliculina* Hadzi, 1951 [nomen nudum]
- *Lagotia* Wright, 1857
- *Latifolliculina* Hadzi, 1951
- *Magnifolliculina* Uhlig, 1964 [nomen nudum]
- *Metafolliculina* Dons, 1924
- *Mirofolliculina* Dons, 1928
- *Pachyfolliculina* Hadzi, 1951
- *Parafolliculina* Dons, 1914
- *Pebrilla* Giard, 1888
- *Perifolliculina* Hadzi, 1951
- *Planifolliculina* Hadzi, 1951
- *Platyfolliculina* Hadzi, 1938
- *Priscofolliculina* Deflandre & Deunff, 1957 (fossil)
- *Pseudofolliculina* Dons, 1914
- *Pseudoparafolliculina* Andrews & Nelson, 1942
- *Splitofolliculina* Hadzi, 1951 [nomen nudum]
- *Stentofolliculina* Hadzi, 1938
- *Valletofolliculina* Andrews, 1953

Family MARISTENTORIDAE Miao, Simpson, Fu, & Lobban, 2005

Size, large, majestic when fully extended; **shape, trumpet-shaped with apical area expanded into a bilobed "cap" divided by a ventral indentation, very contractile**; free-swimming, but typically temporarily attached to substrate; pigmented, due both to blood-red pigmentocysts and to the presence of symbiotic zooxanthellae; somatic ciliation, holotrichous; **oral ciliature of over 300 oral polykinetids, spirals around flared-out anterior end, encircling an anterior peristomial field with scattered ciliate dikinetids, not arranged in kineties**; paroral, very reduced; macronucleus, ellipsoid; micronuclei, multiple; contractile vacuole, absent; cytoproct, not reported; feeding on bacteria, microalgae, and other protists; in marine habitats, at least associated with coral reefs; one genus.

- *Maristentor* Lobban, Schefter, Simpson, Pochon, Pawlowski, & Foissner, 2002*

Family PERITROMIDAE Stein, 1867

Size, medium; shape, ellipsoidal, dorsoventrally flattened, contractile; free-swimming; somatic ciliation primarily on ventral (= right?) surface; **single somatic kinety on slightly convex dorsal (= left?) surface; spine-like cilia of dorsal kinetids emerging from wart-like papillae**; extrusomes, not reported; **oral region expansive with left serial oral polykinetids extending anteriorly from equatorial oral cavity along anterior border of cell and paroral extending parallel to serial oral polykinetids**; macronucleus, typically in two lobes; micronucleus, present; contractile vacuole, present; cytoproct, present; feeding on bacteria, microalgae, and other protists; generally in marine habitats, including salt marshes; one genus.

- *Peritromus* Stein, 1863

Family SPIROSTOMIDAE Stein, 1867
(syn. Spirostomatidae)

Size, medium to large; shape, often elongate, cylindrical and very contractile; some pigmented forms; free-swimming; somatic ciliation, holotrichous; extrusomes, not reported; oral region in anterior half; **peristomial field, long, narrow, not ciliated; oral ciliature sometimes relatively inconspicuous, but still with many serial oral polykinetids, extending 1/3–1/2 body length**; paroral may extend almost entire length of oral region, paralleling the adoral zone; macronucleus, moniliform; micronucleus, present; contractile vacuole posterior, frequently large, and may have lengthy collecting canal; cytoproct, present;

feeding on bacteria, microalgae, and other protists; predominantly in freshwater habitats; two genera and two genera *incertae sedis*.

- *Gruberia* Kahl, 1932
- *Spirostomum* Ehrenberg, 1834

Incertae sedis in Family Spirostomidae

- *Diplogmus* Mansfeld, 1923
- *Propygocirrus* Mansfeld, 1923

Family STENTORIDAE Carus, 1863

Size, medium to large, becoming majestic in size and movement; **shape, trumpet-shaped with apical area not bilobed, very contractile**; often pigmented due to pigmentocysts with the pigment stentorin and/or with symbiotic zoochlorellae; free-swimming, but typically temporarily attached to the substrate, with a few species residing in mucilaginous loricae; somatic ciliation, holotrichous, with posterior end having thigmotactic cilia permitting temporary attachment to substrate; **oral ciliature spirals nearly 360° around flared-out anterior end, encircling an anterior peristomial field that is covered by ciliated kinetids arranged in ordered kineties; paroral accompanying entire length of adoral zone of polykinetids**; macronucleus, ellipsoid to ribbon-like and moniliform; micronucleus, one to many; contractile vacuole, may be multiple; cytoproct, present; feeding on bacteria, microalgae, and other protists, including other ciliates; typically in freshwater habitats only; one genus and three genera *incertae sedis*.

- *Stentor* Oken, 1815

Incertae sedis in Family Stentoridae

- *Heterostentor* Song & Wilbert, 2002*
- *Parastentor* Vuxanovici, 1961
- *Stentoropsis* Dogiel & Bychowsky, 1934

Subphylum INTRAMACRONUCLEATA Lynn, 1996 (Ciliostomatophora *p.p.*, Homoiotricha *p.p.*, Homotricha *p.p.*, Kinetodesmatophora *p.p.*, Postciliodesmatophora *p.p.*, Tubulicorticata + Filicorticata + Epiplasmata + Membranellophora *p.p.*, Transversonematophora *p.p.*, Ventrata *p.p.*)

Size, small to large; shape, variable, from globular to ellipsoid to elongate; free-swimming or sessile; cortical alveolar system typically well-developed; somatic ciliation, holotrichous, but forms with girdles and strips, and even non-ciliated taxa are known; parasomal sacs, present; extrusomes as somatic mucocysts and trichocysts and oral and somatic toxicysts; oral structures, variable, minimally with oral dikinetids, either encircling the cytostome or on the right side as a paroral, but some forms also with several to many oral polykinetids in an adoral zone and other forms astomatous or with oralized somatic kinetids; stomatogenesis, variable from telokinetal to buccokinetal; fission, typically isotomic, rarely anisotomic and multiple; **macronuclear genome typically differentiated by fragmentation of micronuclear chromosomes during anlage development; ampliploid macronucleus dividing by intramacronuclear microtubules**; micronucleus, present; conjugation, typically temporary and isogamontic, but some forms showing complete conjugation with anisogamonty; feeding habits, diverse, including several major classes and subclasses as obligate symbionts, sometimes parasitic; widely distributed in marine, freshwater, and terrestrial habitats; nine classes.

NOTE: Lynn (1996a) suggested that the rapid radiation within this subphylum arose from a fundamentally different property of ciliate cellular organization, perhaps related to fragmentation of the micronuclear chromosomes and processing of the genomic DNA during macronuclear development (see also Riley & Katz, 2001).

Class **SPIROTRICHEA** Bütschli, 1889 (syns. Halteriia *p.p.*, Halteriida *p.p.*, Membranellata *p.p.*, Membranellophora *p.p.*, Polytrichidea + Oligotrichidea, Postciliodesmatophora *p.p.*, Spirotricha *p.p.*, Spirotrichia *p.p.*)

Size, small to large; shape, variable, from spheroid to cone-shape to dorsoventrally flattened; free-swimming or sessile, with some loricate forms that may be attached to substrates and/or sedentary attached within lorica; somatic ciliation, holotrichous in some groups, but nearly devoid of cilia in others; somatic dikinetids usually with the anterior or both kinetosomes ciliated or somatic polykinetids, called cirri; extrusomes as mucocysts or trichocyst-like trichites; **oral ciliature conspicuous, with adoral zone of oral polykinetids, typically as paramembranelles, especially prominent and often encircling oral region clockwise before entering the oral cavity**; one or more "parorals" on the right, and if two, as paroral (= outer) and endoral (= inner) membranes; stomatogenesis, typically parakinetal or apokinetal, but mixokinetal in *Protocruzia*; fission almost always isotomic; **macronuclear DNA replication by replication bands (except in Subclasses Protocruziidia and probably Phacodiniidia** (but latter needs confirmation)); conjugation, typically temporary and isogamontic, but at least one case of total conjugation; feeding habits, diverse, ranging from microphagous bacterivores to predators on other ciliates and even small metazoans; widely distributed in marine, freshwater, and terrestrial habitats; seven subclasses.

NOTE: Molecular evolution studies suggest that the Subclasses Protocruziidia, Phacodiniidia, Licnophoria, Hypotrichia, Choreotrichia, Oligotrichia, and Stichotrichia belong to a monophyletic group here called the Class SPIROTRICHEA (see Lynn & Strüder-Kypke, 2002). Macronuclear replication bands have not been demonstrated in protocruziids and phacodiniids. Protocruziids demonstrate a form of macronuclear division unique within the phylum (Ammermann, 1968; Ruthmann & Hauser, 1974) and may deserve separate status as a monotypic class. Phacodiniids need careful restudy to determine whether or not they have replication bands or something akin to them.

Subclass Protocruziidia de Puytorac, Grain, & Mignot, 1987

Size, small; shape, ovoid; free-swimming; alveoli not well-developed; body with monokinetid or dikinetid field on right side and dorsal surface; somatid dikinetids with short kinetodesmal fibrils and overlapping postciliary microtubular ribbon, which do not form postciliodesmata; extrusomes, trichocyst-like; adoral zone of 6 (5–8) oral polykinetids on left of oral region; paroral to posterior and right of oral region, composed of dikinetids; stomatogenesis, mixokinetal; **nuclear apparatus a cluster of similar-sized nuclei with paradiploid macronuclei surrounding one or more micronuclei; macronuclear division apparently as separation of two composite (?) chromosomes per macronucleus**; conjugation, not reported; contractile vacuole (?); cytoproct (?); microphagous, on bacteria, microalgae, and smaller protists; in marine and brackish water habitats; one order.

Order Protocruziida Jankowski, 1980

With characters of subclass; one family.

Family PROTOCRUZIIDAE Jankowski, 1980

With characters of order; one genus.

– *Protocruzia* de Faria, da Cunha, & Pinto, 1922

Subclass Phacodiniidia Small & Lynn, 1985

Size, medium; shape, ovoid, compressed laterally; free-swimming; cortex with rigid and ribbed pellicle (= cuirass) and well-developed alveoli; **somatic ciliature, in widely spaced rows, of linear polykinetids of 6–8 kinetosomes with delicate cilia**; only a few cirrus-like somatic polykinetids, composed of two rows of kinetosomes; **oral region, long, with an adoral zone of conspicuous oral polykinetids, terminating at the cytostome, very near posterior pole of organism; paroral, series of obliquely-oriented, short files of kinetosomes**; macronucleus horseshoe-shaped; micronuclei, multiple; contractile vacuole, present; cytoproct, present; feeding on bacteria, microalgae, and smaller protists; mainly in freshwater and terrestrial habitats (e.g., moss on trees); one order.

Order Phacodiniida Small & Lynn, 1985
(syn. Protohypotrichina *p.p.*)

With characters of the subclass; one family.

FAMILY PHACODINIIDAE Corliss, 1979

With characters of the order; one genus.

– *Phacodinium* Prowazek, 1900

Subclass Licnophoria Corliss, 1957
(syn. Scaiotricha *p.p.*)

Size, medium; **shape, in form of hour-glass, with prominent oral disc apically and conspicuous aboral attachment disc at posterior pole**; free-swimming, but typically attached to substrate; **somatic cilia, essentially absent, except for posterior ciliary rings encircling attachment disc**; adoral zone of oral polykinetids encircling oral region; paroral as single file of kinetosomes; stomatogenesis, apokinetal; macronucleus, typically moniliform; micronucleus, one to several; conjugation, temporary; contractile vacuole, absent; cytoproct, absent; microphagous on bacteria, microalgae, and perhaps organic detrital particles derived from host's feeding activities; in marine habitats as "ectocommensals" on organisms ranging from an alga (substratum for one species) to a variety of invertebrates (e.g., tunicates, coelenterates, annelids, molluscs, and the respiratory trees of sea cucumbers); one order.

Order Licnophorida Corliss, 1957
With characteristics of subclass; one family.

Family LICNOPHORIDAE Bütschli, 1887
With characteristics of order; two genera.

- *Licnophora* Claparède, 1867
- *Prolicnophora* Jankowksi, 1978

Subclass Hypotrichia Stein, 1859
(syns. Euplotia, Hypotricha *p.p.*, Hypotrichea *p.p.*, Hypotrichida *p.p.*, Hypotrichina *p.p.*, Hypotrichorida *p.p.*, Pseudohypotrichina)

Size, small to medium; shape, dorsoventrally flattened, typically rigid, oval to rectangular; free-swimming; **alveoli well-developed and, at least in euplotids, filled with a protein, called platein**; somatic ciliature commonly represented by rows or localized groups of polykinetids, called cirri, conspicuous on the ventral surface; **dorsally, files of widely spaced dikinetids with short cilia ("sensory bristles") and retention of a laterally-directed kinetodesmal fibril**; files of marginal cirri, incomplete or absent; **somatic infraciliature typically retained during encystment**; prominent adoral zone of generally numerous oral polykinetids, as paramembranelles, on left-anterior portion of the ventral surface, bordering a broad, non-ciliated peristomial field and sometimes continuing over apical end of body onto the dorsal surface; paroral as paroral and/or endoral in diplo- or polystichomonad condition; stomatogenesis, generally apokinetal, beginning in a cortical pocket in some forms, but sometimes parakinetal; macronucleus, ellipsoid to band-shaped or in fragments, with replication bands moving from ends to middle when the nucleus is elongated; micronucleus, one to several; conjugation, temporary; contractile vacuole, at least present in freshwater forms; microphagous and macrophagous; in marine, freshwater, and terrestrial habitats, widely distributed as free-living forms, but a few species as ectocommensals on various invertebrates and one one inquilinic in an echinoid; two orders.

NOTE: Sequences of small subunit rRNA genes of species assigned to this subclass generally show rapid sequence evolution. This may explain why the representative genera that have been sequenced do not form a strong monophyletic grouping (Chen & Song, 2001). We conservatively maintain this subclass until sequence data from additional genes provide evidence that the class should be subdivided.

Order Kiitrichida Nozawa, 1941
(syn. Protohypotrichina *p.p.*)

Size, medium; shape, small, rounded-elliptical in outline; **frontoventral cirri, relatively small polykinetids, uniform in size, in 7–10 curving files along right side of ventral surface**; oral region, broad area, on right; adoral zone of polykinetids bordering left margin of body and extending from near posterior end to near anterior end; paroral, bordering almost entire length of right margin of oral region; macronucleus, ovoid; micronucleus, present; contractile vacuole, absent; cytoproct (?); feeding on smaller protists; in marine habitats; one family.

Family KIITRICHIDAE Nozawa, 1941
With characteristics of order; three genera.

- *Caryotricha* Kahl, 1932
- *Kiitricha* Nozawa, 1941
- *Musajevella* Alekperov, 1984

Order Euplotida Small & Lynn, 1985

Size, small to medium; shape, ovoid to rectangular; free-swimming; frontoventral cirri, sporadically scattered over ventral surface, but never forming more than one conspicuous file on ventral surface, except in Gastrocirrhidae; oral structures, as for subclass; **during cell division, only the ventral somatic infraciliature is replaced while replication of the dorsal ciliature typically occurs within an equatorial band and within the parental kineties (i.e., intrakinetally)**; caudal cirri, when present, derived from dorsal kinety anlagen; two suborders.

Suborder Discocephalina Wicklow, 1982

Size, small to medium; **shape, elongate ovoid with anterior "head-like" part bearing oral region made distinct from main body by more or less obvious neck-like constriction**; free-swimming, but quite thigmotactic; right marginal cirri usually present, but not on *Discocephalus*; **file of left marginal cirri typically divided into anterior and posterior-lateral**

parts; transverse cirri, conspicuous with well-developed microtubular rootlets; caudal cirri, present; oral structures, as for subclass; macronucleus, often in many fragments; micronucleus, present; contractile vacuole, absent; cytoproct (?); feeding on microalgae and smaller protists; in marine habitats, especially sands; one family.

NOTE: The Family Erionellidae was placed in this suborder by Lynn and Small (2002). However, its sole genus *Erionella* Jankowski, 1978 is likely a synonym of *Holosticha* (Aescht, 2001).

Family DISCOCEPHALIDAE Jankowski, 1979
(syns. Discocephalinae, Discocephaloidea, Marginotrichinae)

With characteristics of the suborder; three genera.

NOTE: *Psammocephalus* Wicklow, 1982 has been included in this family. Lin, Song, and Warren (2004) view it as a junior synonym of *Prodiscocephalus* Jankowski, 1979.

- *Discocephalus* Ehrenberg in Hemprich & Ehrenberg, 1831
- *Marginotricha* Jankowski, 1978
- *Prodiscocephalus* Jankowski, 1979

Suborder Euplotina Jankowski, 1979
(syn. Euplotia, Euplotiidea, Gastrocirrhida *p.p.*, Uronychiida *p.p.*)

Size, small to medium; shape, ovoid to ovorectangular; free-swimming; **right marginal cirri, absent**; **left marginal cirri, when present, not as two distinct groups**; oral structures, as for subclass; **contractile vacuole, typically in right posterior of body**; in marine, freshwater, and terrestrial habitats; five families.

Family ASPIDISCIDAE Ehrenberg, 1830
(syn. Aspidiscina, Aspidiscoidea, Euplotaspinae, Paraeuplotidae)

Size, small; shape, flattened and disc-like; free-swimming, highly thigmotactic; dorsal surface may be ridged; **no left marginal cirri**; transverse cirri, conspicuous; **caudal cirri, absent; reduced number of oral membranelles, located centrally and inconspicuously on ventral surface**; paroral, reduced or absent; macronucleus usually C-shaped; micronucleus, present; contractile vacuole, present; cytoproct (?); feeding on bacteria, microalgae, and smaller protists; in marine and freshwater habitats, widely distributed, often benthic but including commensals in the echinoid gut and ascidian branchial cavity; two genera and one genus *incertae sedis*.

- *Aspidisca* Ehrenberg, 1830
- *Euplotaspis* Chatton & Séguéla, 1936

Incertae sedis in Family Aspidiscidae

- *Paraeuplotes* Wichterman, 1942

Family CERTESIIDAE Borror & Hill, 1995
(syn. Certesiina, Certesiinae)

Size, small; shape, broadly ovoid; free-swimming; **unique condylopallium in anterior end of cell; left marginal cirri more than three**; transverse cirri, large, well-developed; caudal cirri, absent; adoral zone of polykinetids, well-developed and continuous; paroral polykinetid conspicuous, extending along two thirds length of the oral cavity; macronucleus, ellipsoid, two to four in number; micronuclei, several; contractile vacuole, present; cytoproct (?); feeding on diatoms and other smaller protists; in marine habitats; one genus.

- *Certesia* Fabre-Domergue, 1885

Family EUPLOTIDAE Ehrenberg, 1838
(syn. Euplotiidea, Euplotidiidae, Euplotidiinae, Euplotina, Euplotinae, Ploesconiidae)

Size, small to medium; shape, ovoid, ventrally-flattened; body, rigid; free-swimming; extrusomes as small vesicles (i.e., ampules) associated with dorsal bristle dikinetids; frontoventral and transverse cirri, dispersed in conspicuous groups; **left marginal cirri, reduced typically to fewer than three**; caudal cirri, ventral; **ventrally-oriented oral cavity with distinct, contiguous, adoral zone of oral polykinetids forming a "collar" and "lapel"**; **paroral as polykinetid on right of oral area accompanied by single endoral file of kinetosomes**; macronucleus, more or less C-shaped; micronucleus, present; contractile vacuole, present; cytoproct (?); feeding on bacteria, microalgae, and smaller protists; in marine, freshwater, and terrestrial habitats, widely distributed but predominantly marine, with one *Euplotes* species found in the intestines of sea urchins; four genera.

- *Euplotes* Ehrenberg in Hemprich & Ehrenberg, 1831
- *Euplotoides* Borror & Hill, 1995
- *Euplotopsis* Borror & Hill, 1995
- *Moneuplotes* Jankowski, 1978

Family GASTROCIRRHIDAE Fauré-Fremiet, 1961 (syns. Cytharoidinae, Gastrocirrhida, Gastrocirrhina, Gastrocirrhinae, Gastrocirrhoidea)

Size, medium; shape, conoid, nearly round in cross-section; free-swimming; **frontoventral cirri apparently in two files; left marginal cirri, inconspicuous or absent**; **transverse cirri, many, conspicuous, in U-shape; expansive, anteriorly-opened oral cavity with anterior end of body remarkably truncate**; macronucleus, ellipsoid, typically in two fragments; micronucleus, present; contractile vacuole, absent; cytoproct (?); feeding on bacteria, microalgae, and smaller protists; in marine habitats, either planktonic or psammophilic; three genera.

NOTE: *Paraeuplotidium* Lei, Choi, and Xu, 2002 is considered a junior synonym of *Euplotidium* since the single left marginal cirrus proposed by Lei, Choi, and Xu (2002) as a character of generic distinctiveness must be corroborated first by gene sequence data to confirm its significance.

- *Cytharoides* Tuffrau, 1975
- *Euplotidium* Noland, 1937
- *Gastrocirrhus* Lepsi, 1928

Family URONYCHIIDAE Jankowski, 1975 (syns. Diophryidae, Swedmarkiidae, Uronychiida)

Size, small to medium; shape, blunt ovoid, nearly circular in cross-section; free-swimming; cirri, generally conspicuous with frontoventral cirri reduced to groupings on right side; transverse cirri, well-developed; **right caudal cirri, dorsal, well-developed**; left marginal cirri, may be conspicuous; **oral region, expansive, with oral polykinetids of "lapel" and "collar" separated, the latter anteriodorsal; paroral, prominent, as polystichomonad, encircling right border of the oral region from its right rear to its anterior left**; macronucleus, ellipsoid, as two or more separated nodules or moniliform; micronucleus, present; contractile vacuole, may be present; cytoproct (?); feeding on bacteria, microalgae, and smaller protists; in marine habitats, free-living and in the mantle cavity of molluscs; five genera.

- *Diophryopsis* Hill & Borror, 1992
- *Diophrys* Dujardin, 1841
- *Paradiophrys* Jankowski, 1978
- *Swedmarkia* Dragesco, 1954
- *Uronychia* Stein, 1859

Incertae sedis in Subclass Hypotrichia

Family REICHENOWELLIDAE Kahl, 1932 (syn. Transitellidae)

Size, medium; shape, ellipsoidal; free-swimming; somatic ciliation, holotrichous, with kineties occasionally slightly spiraled; **ventral somatic kinetids as groupings of 2–6 dikinetids forming delicate "cirri" and dorsal somatic dikinetids as bristles**; extrusomes, not reported; **oral region, a narrowed peristomial field with an oral cavity supported by a basket of nematodesmata, originating from kinetosomes of the oral polykinetids and paroral**; adoral zone of polykinetids on left side of oral region; paroral and/or endoral on right side of oral region; macronucleus, elongate; micronucleus, present; contractile vacuole, present; cytoproct (?); feeding on microalgae and smaller protists; in freshwater or terrestrial habitats; two genera.

- *Balantidioides* Penard in Kahl, 1930
- *Reichenowella* Kahl, 1932

Subclass Choreotrichia Small & Lynn, 1985 (syns. Oligotricha *p.p.*, Oligotrichorida *p.p.*, Strobilia *p.p.*)

Size, small to large; shape, typically conical or bell-shaped, sometimes tailed; free-swimming (even when loricate, as in Order Tintinnida); a perilemma, often present external to the cell (plasma) membrane; extrusomes, restricted to the oral region as capsules torqueés, at least in tintinnids; somatic ciliature, as dikinetids or monokinetids, poorly developed, ranging from weakly holotrichous (e.g. *Strombidinopsis*) to extremely reduced (e.g. *Lohmanniella*); **adoral zone of oral polykinetids, used in locomotion and feeding, forming closed, outer circle around broader anterior end, but slightly open at least in the living forms of *Parastrombidinopsis***; inner ends of some of these outer polykinetids may extend into the oral cavity where they may accompany a smaller number of inner oral polykinetids restricted to the inner oral cavity; paroral, typically composed of single file of kinetosomes (monostichomad); stomatogenesis, apokinetal, ultimately developing in a below-surface pouch; macronucleus, typically as two ellipsoid fragments, but various other shapes possible; micronucleus, present; contractile vacuole, present at least in freshwater forms; cytoproct, possibly absent; feeding on bacteria, microalgae, and other protists; mainly in marine habitats, but some freshwater forms, typically planktonic; two orders.

Order Tintinnida Kofoid & Campbell, 1929
(syns. Archaetintinnoinea, Eutintinnina *p.p.*, Tintinnina, Tintinnoida, Tintinnoidea, Tintinniona, Tintinnoinea)

Size, small to large; shape, cylindrical or cone-shaped, highly contractile, often with elongate posterior end; **attached to inside of lorica, and sedentary within lorica**; loricae, 100–200 (24–1,000) μm in length, but up to 3,000 μm if certain aberrant questionable fossil material is included; loricae, typically rigid, but gelatinous in Family Tintinnidiidae, hyaline or agglomerated with mineral or organic particles; free-swimming or sessile; tentaculoids containing extrusomes ("capsules torquées") interspersed between oral polykinetids in some taxa; macronuclei, typically two; micronucleus, present; in marine and freshwater habitats, typically marine, widespread in pelagic and neritic plankton (with fossil evidence for past aeons); 15 families, excluding several families based on fossil taxa (see **NOTE**).

NOTE: If phylogenies derived from small subunit rRNA gene sequences represent the true phylogenetic relationships, the taxonomy of this order, based on lorica morphology, is very probably incorrect (Strüder-Kypke & Lynn, 2003). Agatha and Riedel-Lorjé (2006) noted that fewer than 20 species have the kinetome described in sufficient detail from silver impregnation to permit a rigorous comparative morphological analysis. Until more complete morphological and gene sequence data are available, we have conservatively retained the classification based on loricae.

We do not accept the all-fossil Families Calpionellidae Bonet, 1956, Calpionellopsidae Makarieva, 1982, Colomiellidae Bonet, 1956, Chitinoidellidae Grün and Blau, 1997, Crassicolariidae Makarieva, 1982, Remaniellidae Makarieva, 1982, and Semichitinoidellidae Nowak, 1978, distributing their genera as *incertae sedis* among other families (see earlier review by Loeblich & Tappan, 1968). Several recently established families for fossil genera are listed at the end of this section with their included genera, which cannot be easily placed within the families that include contemporary taxa. A thorough review of these fossil data by a taxonomist familiar with the contemporary diversity of the tintinnids would provide a useful perspective on the "historical" diversity of this group.

Family ASCAMPBELLIELLIDAE Corliss, 1960
(for Craterellidae)

Size, small; lorica, cup-shaped, not elongate, with smooth to denticulate oral rim and trilaminar wall; **lorica rim of collar as inner collar and outer flared rim, with gutter or trough between**; lorica of some species with agglutinated coccoliths; in marine habitats, mainly eupelagic; no fossil species; two genera and two genera *incertae sedis*.

- *Acanthostomella* Jörgensen, 1927
- *Ascampbelliella* Corliss, 1960

Incertae sedis in Family Ascampbelliellidae

- *Luxiella* Lecal, 1953
- *Niemarshallia* Corliss, 1960

Family CODONELLIDAE Kent, 1881
(syns. Calpionellidae *p.p.*, Chitinoidellidae, Crassicollariidae, Remaniellidae)

Size, small to medium; lorica flask-, bowl- or chalice-shaped, with aboral end sometimes pointed and closed; lorica collar not clear, but if present may or may not have a nuchal constriction; **lorica wall, unilaminar, commonly reticulate and agglommerated**; predominantly in marine habitats, neritic and eupelagic forms, but a few species (e.g., of *Codonella* and *Tintinnopsis*) abundant in the plankton of freshwater lakes, rivers, and ponds; numerous fossil as well as widespread contemporary forms; 28 genera and 16 fossil genera *incertae sedis*.

- *Amphorellina* Colom, 1948 (fossil)
- *Bignotella* Willems, 1975 (fossil)
- *Chitinoidella* Doben, 1963 (fossil)
- *Claretinella* Keij, 1974 (fossil)
- *Codonaria* Kofoid & Campbell, 1939
- *Codonella* Haeckel, 1873 (some fossils)

- *Codonopsis* Kofoid & Campbell, 1939
- *Coxliellina* Colom, 1948 (fossil)
- *Crassicollaria* Remane, 1962 (fossil)
- *Dicloeopella* Eicher, 1965 (fossil)
- *Durandella* Dragastan, 1972 (fossil)
- *Lorenziella* Knauer & Nagy, 1964 (fossil)
- *Parachitinoidella* Trejo, 1972 (fossil)
- *Poroecus* Cleve, 1902
- *Praetintinnopsella* Borza, 1969 (fossil)
- *Pseudarcella* Spandel, 1909 (fossil)
- *Remanellina* Tappan & Loeblich, 1968 (fossil)
- *Remaniella* Catalano, 1965 (fossil)
- *Salpingellina* Colom, 1948 (fossil)
- *Savroniella* Belokrys, 1995 (fossil)
- *Spinarcella* Keij, 1969 (fossil)
- *Spinophenia* Szczechura, 1969 (fossil)
- *Tintinnopsella* Colom, 1948 (fossil)
- *Tintinnopsis* Stein, 1867 (some fossils)
- *Tytthocorys* Tappan & Loeblich, 1968 (fossil)
- *Urnulella* Szczechura, 1969 [not listed in Aescht] (fossil)
- *Vautrinella* Cuvillier & Sacal, 1963 (fossil)
- *Yvonniellina* Tappan & Loeblich, 1968 (fossil)

Incertae sedis in Family Codonellidae

- *Bacculinella* Konenkova, 2000 (fossil)
- *Biconvexellina* Konenkova, 1999 (fossil)
- *Bicornella* Bugrova, 2003* (fossil) [junior homonym of ostracoda]
- *Borzaiella* Makarieva, 1982 (fossil)
- *Borziella* Pop, 1987 (fossil)
- *Carpathella* Pop, 1998 (fossil)
- *Cubanites* Aescht, 2001 (fossil)
- *Cylindriconella* Aescht, 2001 (fossil)
- *Daciella* Pop, 1998 (fossil)
- *Dobeniella* Pop, 1997 (fossil)
- *Foliacella* Makarieva, 1979 (fossil)
- *Longicollaria* Pop, 1997 (fossil)
- *Popiella* Rehakova, 2002*(fossil)
- *Rossielella* Aescht, 2001 (fossil)
- *Scalpratella* Makarieva, 1979 (fossil)
- *Tianella* Bugrova, 2003* (fossil)

Family CODONELLOPSIDAE Kofoid & Campbell, 1929
(syns. Calpionellidae *p.p.*, Calpionellopsidae, Colomiellidae, Semichitinoidellidae)

Size, small to large; lorica, top-shaped with aboral end rounded to apiculate; **lorica collar, hyaline, cylindrical, delicate but sometimes ridged**; lorica wall, thick, but wall of bowl thicker than collar and agglomerated with mineral particles; in marine habitats, neritic and eupelagic forms; many fossil as well as extant forms; 12 genera and 12 fossil genera *incertae sedis*.

- *Calpionella* Lorenz, 1902 (fossil)
- *Calpionellites* Colom, 1948 (fossil)
- *Calpionellopsella* Trejo, 1975 (fossil)
- *Calpionellopsis* Colom, 1948 (fossil) (subj. syn. *Remaniella*)
- *Codonellopsis* Jörgensen, 1924
- *Colomiella* Bonet, 1956 (fossil)
- *Deflandronella* Trejo, 1975 (fossil)
- *Laackmanniella* Kofoid & Campbell, 1929
- *Luminella* Kofoid & Campbell, 1939
- *Praecalpionellopsis* Borza, 1971 (fossil)
- *Stenosemella* Jörgensen, 1924
- *Stenosemellopsis* Colom, 1948 (fossil)

Incertae sedis in Family Codonellopsidae

- *Baranella* Nagy, 1989 (fossil)
- *Borzaites* Aescht, 2001 (fossil)
- *Calpionelloides* Colom, 1948 (fossil)
- *Calpionellopsites* Nagy, 1986 (fossil)
- *Crassicalpionella* Nagy, 1989 (fossil)
- *Furssenkoiella* Makarieva, 1979 (fossil)
- *Lorenziellites* Nagy, 1986 (fossil)
- *Lorenziellopsis* Nagy, 1989 (fossil)
- *Praecalpionellites* Pop, 1986 (fossil)
- *Semichitinoidella* Nowak, 1978 (fossil)
- *Sopianella* Nagy, 1989 (fossil)
- *Sturiella* Borza, 1981 (fossil)

Family CYTTAROCYLIDIDAE Kofoid & Campbell, 1939
(syns. Cyttarocylidae)

Size, small to medium; lorica, bell- or kettle-shaped, sometimes elongate; **lorica wall with very conspicuous meshwork between wall layers and conspicuous wall material in broad bars separating polygons**; lorica collar slightly flared, with inner suboral shelf; in marine habitats, predominantly eupelagic; no fossil species; one genus.

- *Cyttarocylis* Fol, 1881

Family DICTYOCYSTIDAE Haeckel, 1873

Size, small to medium; lorica, ovoid to conical with dense bowl; **lorica collar with one or**

two rows of open-arched frames, windows or fenestrae (latter with or without panes); lorica wall of bowl reticulate and, in some species, agglomerated with coccoliths or mineral particles; in marine habitats, eupelagic; no fossil species; two genera.

- *Dictyocysta* Ehrenberg, 1854
- *Wangiella* Nie, 1934

Family EPIPLOCYLIDIDAE Kofoid & Campbell, 1939
(syns. Epiplocyclididae, Epiplocylidae)

Size, small to medium; **lorica, short, acorn-shaped with aboral end blunt acuminate or with horn**; lorica oral rim, smooth, with collarette or suboral shelf; **lorica wall, thin, hyaline, with its surface partially or entirely ornamented with polygons bounded by raised edges**; in marine habitats, eupelagic; no fossil species; three genera.

- *Epicancella* Kofoid & Campbell, 1929
- *Epiplocylis* Jörgensen, 1924
- *Epiplocyloides* Hada, 1938

Family METACYLIDIDAE Kofoid & Campbell, 1929
(syns. Calpionellidae *p.p.*, Coxliellidae *p.p.*, Metacycoidinae, Metacylin[e]ae)

Size, medium to large; **lorica, tubular or goblet-shaped, delicate and with rings or spirals as wound lamina, at least in oral half**; lorica aboral end sometimes with horn; **lorica wall, three-layered, usually with distinct and delicate alveoli, sometimes with agglomerated coccoliths**; paralorica, coxlielliform; sometimes an epilorica; in marine habitats, eupelagic and neritic, though occasionally one found in fresh water; a number of fossil species; eight genera and two genera *incertae sedis*.

NOTE: The genus *Coxliella* via its type-species, at least, may actually have no reality. Laval-Peuto (1977, 1994a) has demonstrated that there is a "coxlielliform" stage in the life cycle of many tintinnines, including species from a number of families. Thus, Corliss (1979) placed the generic name in a "questionable" status (**see below**) and replaced the former familial name (Coxliellidae) with the next name available for the group, *viz.*, Metacylididae (originally the second included subfamily).

- *Climacocylis* Jörgensen, 1924
- *Favelloides* Thalmann, 1942 (fossil)
- *Helicostomella* Jörgensen, 1924
- *Metacylis* Jörgensen, 1924
- *Pseudometacylis* Balech, 1968
- *Rhabdonelloides* Colom, 1939 (fossil)
- *Spiroxystonellites* Knauer, 1969 (fossil)

Nomen inquirendum: *Coxliella* Brandt, 1906 (fossil)

Incertae sedis in Family Metacylididae

- *Rhizodomus* Strelkov & Wirketis, 1950
- *Stylicauda* Balech, 1951

Family NOLACLUSILIIDAE Sniezek, Capriulo, Small, & Russo, 1991

Size, small; **lorica, hyaline, bell-shaped, with hinged, oral flaps that close lorica opening when ciliate retracts**; macronucleus, bilobed; in marine habitats, particularly estuarine and coastal plankton; no fossil species; one genus.

- *Nolaclusilis* Snyder & Brownlee, 1991

Family PETALOTRICHIDAE Kofoid & Campbell, 1929

Size, medium; lorica, bell-shaped with flared lip on top of vertical cone; **lorica lip and bowl with small, suboral fenestrae;** in marine habitats, eupelagic; no fossil species; two genera.

- *Parapetalotricha* Hada, 1970
- *Petalotricha* Kent, 1881

Family PTYCHOCYLIDIDAE Kofoid & Campbell, 1929
(syns. Ptychocyclidae, Ptychacyclididae, Ptychocylidae)

Size, small to large; **lorica conical or chalice-shaped with several annular bulges and pointed or blunt pedicel**; lorica oral rim, often denticulate, with no collar; **lorica wall, apparently trilaminate, alveolar in midsection**; paralorica, coxlielliform; epilorica, if present, possibly spiralled; in marine habitats, mostly eupelagic; no fossil species; five genera.

- *Cymatocylis* Laackmann, 1910
- *Favella* Jörgensen, 1924
- *Protocymatocylis* Kofoid & Campbell, 1929
- *Ptychocylis* Brandt, 1896
- *Wailesia* Kofoid & Campbell, 1939

Family RHABDONELLIDAE Kofoid & Campbell, 1929

Size, medium to large; lorica, acorn to chalice-shaped, often very long with aboral horn in some species; lorica oral rim, smooth; **lorica wall, trilaminate, hyaline with longitudinal, low ridges that may be simple, branched or anastomosing, sometimes with pores**; in marine habitats, mostly eupelagic; no fossil species; four genera.

- *Epirhabdonella* Kofoid & Campbell, 1939
- *Protorhabdonella* Jörgensen, 1924
- *Rhabdonella* Brandt, 1906
- *Rhabdonellopsis* Kofoid & Campbell, 1929

Family TINTINNIDAE Claparède & Lachmann, 1858
(syns. Salpingellinae *p.p.*, Stelidiellinae *p.p.*, Tintinninae *p.p.*)

Size, medium to large; lorica elongate, typically tubular, and in some species, with *both* ends of lorica open; lorica oral end often flared, smooth or denticulate; **lorica wall, thin, hyaline, probably unilaminar, appearing homogeneous with clear inner and outer layers**; lorica surface ornamentation frequently as ridges and crests, which may be spiralled; primarily in marine habitats, eupelagic, but a few found in brackish habitats; no fossil species; 24 genera.

- *Albatrossiella* Kofoid & Campbell, 1929
- *Amphorellopsis* Kofoid & Campbell, 1929
- *Amphorides* Strand, 1928
- *Brandtiella* Kofoid & Campbell, 1929
- *Bursaopsis* Kofoid & Campbell, 1929
- *Buschiella* Corliss, 1960
- *Canthariella* Kofoid & Campbell, 1929
- *Clevea* Balech, 1948
- *Dadayiella* Kofoid & Campbell, 1929
- *Daturella* Kofoid & Campbell, 1929 [nomen dubium]
- *Epicranella* Kofoid & Campbell, 1929
- *Eutintinnus* Kofoid & Campbell, 1939
- *Funnela* Li & Zhang, 2006* (fossil)
- *Odontophorella* Kofoid & Campbell, 1929
- *Ormosella* Kofoid & Campbell, 1929
- *Proamphorella* Kofoid & Campbell, 1939
- *Prostelidiella* Kofoid & Campbell, 1939
- *Rhabdosella* Kofoid & Campbell, 1929
- *Salpingacantha* Kofoid & Campbell, 1929
- *Salpingella* Jörgensen, 1924
- *Salpingelloides* Campbell, 1942
- *Steenstrupiella* Kofoid & Campbell, 1929
- *Stelidiella* Kofoid & Campbell, 1929
- *Tintinnus* Schrank, 1803

Family TINTINNIDIIDAE Kofoid & Campbell, 1929
(syn. Tintinnididae)

Size, small to large; lorica, tubular or flaring, with aboral end open or closed, may be attached to the substrate; lorica collar, rarely visible; **lorica wall soft, gelatinous, with agglomerated particles**; in marine, brackish, and (occasionally) freshwater habitats; no fossil species; three genera.

- *Leprotintinnus* Jörgensen, 1900
- *Membranicola* Foissner, Berger, & Schaumburg, 1999
- *Tintinnidium* Kent, 1881

Family UNDELLIDAE Kofoid & Campbell, 1929

Size, small to medium; loricae goblet- or urn-shaped, occasionally elongate; lorica oral rim sometimes with suboral ledge and perhaps an inner collar; **lorica wall, hyaline, conspicuously thick, obviously trilaminate, with obvious inner and outer layers**; lorica surface, smooth, and sometimes with annuli; in marine habitats, eupelagic; no fossil species; seven genera and one genus *incertae sedis*.

- *Amplectella* Kofoid & Campbell, 1929
- *Amplectellopsis* Kofoid & Campbell, 1929
- *Cricundella* Kofoid & Campbell, 1929
- *Micrundella* Loeblich & Tappan, 1968
- *Proplectella* Kofoid & Campbell, 1929
- *Undella* Daday, 1887
- *Undellopsis* Kofoid & Campbell, 1929

Incertae sedis in Family Undellidae

- *Rotundocylis* Kufferath, 1952 [*nomen nudum*]

Family XYSTONELLIDAE Kofoid & Campbell, 1929
(syn. Xistonellidae)

Size, medium to large; lorica elongate, chalice-shaped with aboral end long and narrow, sometimes with pedicel; lorica oral rim, usually denticulate; **lorica wall, hyaline, trilaminate, reticulate, showing an irregular and conspicuous meshwork between lorica wall layers, with inconspicuous, thin wall material separating polygons**; in marine habitats, mainly eupelagic; a few fossil forms known; five genera and one genus *incertae sedis*.

- *Parafavella* Kofoid & Campbell, 1929 (fossil)
- *Parundella* Jörgensen, 1924
- *Spiroxystonella* Kofoid & Campbell, 1929
- *Xystonella* Brandt, 1906
- *Xystonellopsis* Jörgensen, 1924

Incertae sedis in Family Xystonellidae

- *Parafavelloides* Deflandre & Deflandre, 1949

ADDENDUM: The following families and included genera of fossil forms are listed here as belonging to the Order Tintinnida. Assignment to families that include contemporary genera has not been possible at this time.

Incertae sedis in Order Tintinnida: unassigned genera – *Aubertianella* Szczechura, 1969, *Daturellina* Radoičić, 1959, *Praecolomiella* Borza, 1979, *Spirocystomellites* Colom, 1988; *Syringella* Paulmier, 1997 [not listed in Aescht; junior homonym of anthozoan/poriferan], *Tintinnoidella* Elicki, 1994; unassigned Family Berounkellidae Koshevoj, 1987 with included genera *Batiola* Koshevoj, 1987, *Berounkella* Koshevoj, 1987, *Chervurtskella* Koshevoj, 1987, *Kejvia* Koshevoj, 1987, *Olgella* Koshevoj, 1987, *Ollella* Aescht, 2001, *Tundrella* Koshevoj, 1987, and *Velavella* Koshevoj, 1987; unassigned Family Cadosinidae Wanner, 1940 with included genera *Cadosina* Wanner, 1940, *Cadosinopsis* Scheiber, 1967, and *Crustocadosina* Rehanek, 1985; unassigned Family Calcisphaerulidae Bonet, 1956 with included genera *Bonetcardiella* Dufour, 1968 and *Calcisphaerula* Bonet, 1956; unassigned Family Causiidae Koshevoj, 1987 with included genus *Causella* Aescht, 2001; and unassigned Family Stomiosphaeridae Wanner, 1940 with included genera *Carpistomiosphaera* Nowak, 1968, *Colomisphaera* Nowak, 1968, *Committosphaera* Rehanek, 1985, *Inocardion* Masters & Scott, 1978, and *Stomiosphaera* Wanner, 1940.

Order Choreotrichida Small & Lynn, 1985

Size, small to large; shape, globular to ellipsoid; free-swimming or not sessile, but may be temporarily attached to substratum; **no lorica**; in marine and freshwater habitats, widespread, planktonic with some benthic and symbiotic species; four suborders.

Suborder Leegaardiellina Laval-Peuto, Grain, & Deroux, 1994

Size, small to medium; shape, spheroid; free-swimming; somatic kineties composed of dikinetids, restricted to aboral half; **serial oral polykinetids of the outer circle divided into an inner and outer part, with inner oral polykinetids, apparently clearly separated from outer oral polykinetids**; oral cavity with a set of internal oral polykinetids on the outer wall and paroral along the margin of the inner wall; macronuclei, two ellipsoid nodules; micronucleus, not observed; contractile vacuole, absent; cytoproct (?); feeding on microalgae and smaller protists; in marine habitats, planktonic; one family.

Family LEEGAARDIELLIDAE Lynn & Montagnes, 1988

With characters of the suborder; one genus.

- *Leegaardiella* Lynn & Montagnes, 1988

Suborder Lohmanniellina Laval-Peuto, Grain, & Deroux, 1994

Size, small; shape, spheroid; **somatic kineties, short, posterior, possibly composed of monokinetids**; serial oral polykinetids of outer circle undivided; inner oral polykinetids, apparently as extensions of several outer oral polykinetids, but with a break between the outer and inner parts; macronucleus, curved, ellipsoid; micronucleus, not observed; contractile vacuole, absent; cytoproct (?); feeding on microalgae and smaller protists; marine plankton; one family.

Family LOHMANNIELLIDAE Montagnes & Lynn, 1991

With characters of the suborder; one genus.

- *Lohmanniella* Leegaard, 1915

Suborder Strobilidiina Small & Lynn, 1985

Size, small to medium; shape, spheroid to conoid; free-swimming, but some forms attached to substrates by mucous thread; **somatic kineties, one to several, usually not as long as the body, composed of monokinetids with a cortical flap covering the bases of cilia and directing them to the cell's left and underlain (when protargol-stained) by densely-stained material**; somatic cilia, relatively short, typically <5 µm long; outer and inner oral polykinetids typical of the subclass; macronucleus, typically single, variably shaped from ellipsoid to an elongate band; micronucleus, if observed, typically single; contractile vacuole, at least present in freshwater forms; cytoproct (?); feeding on bacteria, microalgae, and smaller protists; in marine and freshwater habitats, with some species commensals of echinoids; one family.

Family STROBILIDIIDAE Kahl in Doflein & Reichenow, 1929
(syn. Torquatellidae)

With characters of the suborder; three genera and two *incertae sedis*.

- *Pelagostrobilidium* Petz, Song, & Wilbert, 1995
- *Rimostrombidium* Jankowski, 1978
- *Strobilidium* Schewiakoff, 1893

Incertae sedis in Family Strobilidiidae

- *Ciliospina* Leegaard, 1915 [*nomen dubium*]
- *Patronella* Corliss, 1979 (subj. syn. *Strombidium*)

Suborder Strombidinopsina Small & Lynn, 1985

Size, small to medium; shape, spheroid to conoid; free-swimming; **somatic kineties of ciliated dikinetids, equally distributed around body, extending as simple files from the outer ring of serial oral polykinetids towards cell's posterior**; outer and inner oral polykinetids typical of the subclass, although *Parastrombidinopsis* may not have completely closed outer circle when alive; macronuclei, ellipsoid, typically two; micronucleus, when observed, typically single; contractile vacuole, at least present in freshwater forms; cytoproct (?); feeding on bacteria and microalgae; typically in marine habitats, planktonic; one family.

Family STROMBIDINOPSIDAE Small & Lynn, 1985

With characters of the suborder; two genera.

- *Parastrombidinopsis* Kim, Jeong, Strüder-Kypke, Lynn, Kim, Kim, & Lee, 2005*
- *Strombidinopsis* Kent, 1881

Subclass Stichotrichia Small & Lynn, 1985
(syns. Euhypotrichina, Hypotricha *p.p.*, Hypotrichea *p.p.*, Hypotrichida *p.p.*, Hypotrichina *p.p.*, Hypotrichorida *p.p.*)

Size, small to large; shape, often elongate, sometimes very drawn out posteriorly, in cross-section round to dorsoventrally compressed; free-swimming with a few loricate forms; perilemma in some groups; **pellicular alveoli weakly developed; somatic ventral ciliature as ventral cirri ranging from small and quite inconspicuous, occasionally as few as 2–3 cilia per cirrus, arranged in longitudinal, sometimes spiraled, files to a few, larger cirri in scattered groups, with in the latter case marginal files of cirri differentiated**; transverse cirri, may or may not be present; caudal cirri, may or may not be present; dorsal somatic ciliature as one to many kineties – typically three – of dikinetids without kinetodesmal fibril, but with short, bristle cilium on anterior kinetosome; adoral zone of oral polykinetids as paramembranelles in "collar" and "lapel", each typically of four rows of kinetosomes, with the first two rows equally long and the fourth row quite short; right oral cilia variable, but usually as a paroral and endoral; stomatogenesis, parakinetal in those with conspicuous kineties to apokinetal in those with scattered cirri; **division morphogenesis may involve replacement of all somatic ciliature of both proter and opisthe**; macronucleus, typically two nodules, but often multiple, each component typically with one replication band; micronucleus, one to many; conjugation, typically temporary, but sometimes total; contractile vacuole, at least present in freshwater and terrestrial forms, and typically on the middle left of the body; cytoproct, very likely present; feeding strategies ranging from bacterivorous to cannibalistic; **encysted forms typically dedifferentiate all kinetosomes**; in marine, freshwater, and terrestrial habitats, free-living with some symbiotic forms as endo- and ectocommensals; three orders.

NOTE: The taxonomy of stichotrichians is one of the most confused in the phylum. The revision below relies heavily on the morphology of the differentiated individual. There is, however, a trend in recent years to rely more heavily on the similarities in the pattern of division morphogenesis (e.g., Berger, 1999, 2006b; Eigner & Foissner, 1994). Stability may only be achieved when complete division morphogenetic patterns and molecular genetic information for several genes are available on the majority of genera (see Foissner et al., 2004).

Order Stichotrichida Fauré-Fremiet, 1961
(syns. Chaetospirina *p.p.*, Oxytrichida *p.p.*, Oxytrichina *p.p.*, Plagiotomida *p.p.*, Plagiotomoidea *p.p.*)

Size, small to large; shape, often elongate, sometimes very drawn out posteriorly; free-swimming with a few loricate forms; **ventral cirri as one or more longitudinal files of varied lengths, linear (not zig-zag as in Urostylida)**; dorsal ciliature, typically regularly distributed in longitudinal files; oral structures as for subclass; stomatogenesis, parakinetal or apokinetal; six families.

Family AMPHISIELLIDAE Jankowski, 1979
(syns. Gastrostylidae *p.p.*, Gastrostylina *p.p.*, Orthoamphisiellidae)

Size, small to large; shape, elongate ovoid; free-swimming; **ventral cirral file, single with anterior segment of this file formed by cirri from right-most ventral anlage and posterior segment from the second ventral anlage from right**; marginal files of cirri, typically extending from anterior to posterior on left and right sides; transverse cirri, may or may not be present; caudal cirri, may or may not be present; dorsal kineties, typically fewer than ten, composed of dikinetids; oral structures as for order with paroral and endoral; macronucleus, from two to many globular to ellipsoid nodules; micronuclei, one often accompanying each macronuclear nodule; contractile vacuole, present; cytoproct, likely present; feeding on bacteria and smaller protists; in marine, freshwater, and terrestrial habitats; 11 genera and two genera *incertae sedis*.

- *Afroamphisiella* Foissner, Agatha, & Berger, 2002*
- *Amphisiella* Gourret & Roeser, 1888
- *Amphisiellides* Foissner, 1988
- *Hemiamphisiella* Foissner, 1988
- *Nudiamphisiella* Foissner, Agatha, & Berger, 2002*
- *Lamtostyla* Buitkamp, 1977
- *Orthoamphisiella* Eigner & Foissner, 1991
- *Paramphisiella* Foissner, 1988
- *Pescozoon* Jankowski, 1978
- *Uroleptoides* Wenzel, 1953
- *Urospinula* Corliss, 1960 (subj. syn. *Psilotricha*)

Incertae sedis in Family Amphisiellidae

- *Balladyna* Kowalewski, 1882 (subj. syn. *Cyrtohymena*)
- *Circinella* Foissner, 1994

Family KAHLIELLIDAE Tuffrau, 1979
(syns. Banyulsellidae, Cladotrichidae, Lacazeidae, Parakahliellidae)

Size, medium to large; shape, elongate ovoid; free-swimming; somatic ventral ciliature with at least two, typically more than two, ventral cirral files, often not distinctly different from right and left marginal cirral files; **ventral cirral files may be preserved through a variable number of cell divisions (= cell generations) before being resorbed and replaced through additional new (= neokinetal) anlagen**; transverse cirri, typically absent; caudal cirri, typically absent; dorsal ciliature as several files of dikinetids; oral ciliature as for order with paroral and endoral; macronucleus, two to many nodules; micronuclei, several to many; contractile vacuole, present; cytoproct, likely present; feeding on bacteria, microalgae, and smaller protists; in marine, freshwater, and terrestrial habitats; ten genera and four genera *incertae sedis*.

- *Cladotricha* Gajewskaja, 1926
- *Deviata* Eigner, 1995
- *Engelmanniella* Foissner, 1982
- *Kahliella* Corliss, 1960
- *Neogeneia* Eigner, 1995
- *Parakahliella* Berger, Foissner, & Adam, 1985
- *Plesiotricha* Dragesco, 1970 (subj. syn. *Kahliella*)
- *Pseudokahliella* Berger, Foissner, & Adam, 1985
- *Trachelochaeta* Šrámek-Husek, 1954
- *Wallackia* Foissner, 1976

Incertae sedis in Family Kahliellidae

- *Banyulsella* Dragesco, 1954
- *Fragmocirrus* Foissner, 2000

- *Lacazea* Dragesco, 1960
- *Pseudouroleptus* Hemberger, 1985

Family KERONIDAE Dujardin, 1840
(syns. Keronina, Keronopsidae, Keronopsina)

Size, medium; shape, from broad to elongate, even tailed in some species; free-swimming; somatic ventral ciliature as frontoventral cirri generally in several oblique rows across the ventral surface, coursing between right and left marginal cirral files; **first ventral row, so-called "frontal cirri", as curved row along ventral anterior border of left serial oral polykinetids and differentiating from one to several anlagen**; transverse cirri, present; caudal cirri, present; dorsal somatic ciliature as several files of bristle dikinetids; oral ciliature as for order with paroral and endoral; cell division in cyst, except for *Kerona*; macronucleus, two to many globular to ellipsoid nodules; micronucleus, present; contractile vacuole, present; cytoproct (?); feeding on bacteria, microalgae, and smaller protists; in marine and freshwater habitats, including sand, and one species well-known ectocommensal on *Hydra*; four genera.

- *Kerona* Müller, 1786
- *Keronopsis* Penard, 1922
- *Paraholosticha* Wenzel, 1953
- *Parakeronopsis* Shi, Song, & Shi, 1999 [nomen nudum]

Family PLAGIOTOMIDAE Bütschli, 1887

Size, medium; shape, laterally flattened, elongate-ovoid, with right side slightly concave; free-swimming; **somatic ciliation, holotrichous, dense, of small polykinetids or cirri with no differentiation into cirral groups on either body surface**; oral ciliature of extensive adoral zone on its left side, coursing from apical end to subequatorial position and then entering a deep oral cavity and with paroral and endoral along right wall of oral cavity; stomatogenesis, parakinetal; macronucleus, as an irregular bunch of nodules; micronuclei, several, relatively large; contractile vacuole, present; cytoproct (?); feeding on bacteria and organic matter in host's digestive system (?); **in terrestrial habitats as an endocommensals solely in certain species of lumbricid oligochaete annelids**; one genus.

- *Plagiotoma* Dujardin, 1841

Family PSILOTRICHIDAE Bütschli, 1889
(syn. Psilotrichinae)

Size, small to medium; shape, oval to elliptical in outline, with posterior spiny extensions in some species, and sometimes with zoochlorellae in cytoplasm; free-swimming; **somatic ventral ciliature as long and sparse cirri in seven slightly curved cirral files with the postoral oblique cirral file developing from the anlage file IV**; frontal cirri; marginal cirri, strongly reduced; transverse cirri, present; caudal cirri, present; dorsal somatic ciliature as several files of monokinetids (?); peristomial areal limited to anterior third of organism with oral ciliature typical of order, including paroral and endoral; macronucleus, two globular to ellipsoid nodules; micronucleus, present; contractile vacuole, present; cytoproct (?); feeding (?); in freshwater and terrestrial habitats; two genera.

- *Psilotricha* Stein, 1859
- *Hemiholosticha* von Gelei, 1954

Family SPIROFILIDAE von Gelei, 1929
(syns. Atractidae, Chaetospiridae, Chaetospirina, Chaetospirinae, Chaetospiroidea, Hypotrichidiidae, Microspirettidae, Spiretellidae, Spirofilopsidae, Stichotrichinae, Strongylidae, Strongylidiidae, Strongylidiinae, Strongylidioidea)

Size, small to medium; shape, varied, with some forms tailed and others more elongate at the anterior end; free-swimming, but lorica produced by some species; **somatic ventral ciliature as inconspicuous ventral cirri in files curved or spiralling obliquely around body, some ending on dorsal surface**; transverse cirri, present or absent; caudal cirri, present or absent; **somatic dorsal ciliature as several dorsal files of bristle dikinetids in "dorsal strip" helically winding around the body**; adoral zone not highly prominent with oral ciliature typical of order and with paroral and endoral; macronucleus, globular to ellipsoid, single to several nodules; micronucleus, present; contractile vacuole, present; cytoproct, likely present; feeding on bacteria, algae, and smaller protists; in marine and freshwater habitats, both benthic and planktonic; 12 genera and one genus *incertae sedis*.

- *Atractos* Vörösváry, 1950
- *Chaetospira* Lachmann, 1856
- *Hypotrichidium* Ilowaisky, 1921

- *Microspiretta* Jankowski, 1975
- *Mucotrichidium* Foissner, Oleksiv, & Müller, 1990
- *Parastrongylidium* Fleury & Fryd-Versavel, 1985
- *Pelagotrichidium* Jankowski, 1978
- *Planitrichidium* Jankowski, 1979 [nomen nudum]
- *Spirofilopsis* Corliss, 1960
- *Stichotricha* Perty, 1849
- *Strongylidium* Sterki, 1878
- *Urostrongylum* Kahl, 1932

Incertae sedis in Family Spirofilidae

- *Kahliela* Tucolesco, 1962

Incertae sedis in Order Stichotrichida

- *Balladinopsis* Ghosh, 1921 [nomen dubium]
- *Klonostricha* Jankowski, 1979
- *Psilotrix* Gourret & Roeser, 1888
- *Stylonethes* Sterki, 1878 [nomen dubium]

Order Sporadotrichida Fauré-Fremiet, 1961
(syn. Halteriina *p.p.*, Oxytrichina *p.p.*, Pleurotrichina *p.p.*, Sporadotrichorina)

Size, small to large; shape, sometimes elongate, even tailed, but often oval to elliptical in outline; free-swimming; **somatic ventral ciliature as frontoventral cirri, typically heavy and conspicuous, arranged in specific, localized frontal and ventral groups, except in a few taxa (e.g., Family Halteriidae, *Laurentiella*, *Onychodromus*, and *Styxophrya*)**; marginal cirri, typically present; transverse cirri, may or may not be present; caudal cirri, may or may not be present; dorsal somatic ciliature, typically as files of dikinetids with a single bristle cilium; oral ciliature as for subclass; **stomatogenesis, apokinetal, usually with five or six anlagen streaks in two groups for differentiation of ventral somatic ciliature**; in marine, freshwater, and terrestrial habitats, widely distributed, primarily benthic with some forms planktonic, others interstitial, and a few species symbiotic, either as ectocommensals on the integument or in the branchial cavity of several invertebrates or as intestinal inquilines of echinoids; three families.

NOTE: Gene sequences of both actins and small subunit rRNA unambiguously place several genera of the Family Halteriidae (e.g., *Halteria*, *Meseres*) close to the oxytrichid clade (Croft et al., 2003; Foissner et al., 2004; Hewitt et al., 2003). We have therefore transferred this family to this order from the Subclass Oligotrichia. We consider halteriids as planktonic descendants of this primarily benthic lineage.

Family HALTERIIDAE Claparède & Lachmann, 1858
(syn. Meseridae [for Lieberkuehnidae])

Size, small; shape, spheroid to subovoid and conical; free-swimming, often darting through the water; **somatic ciliature as somatic kinetids with long cilia (i.e., >10 µm long), often as cirrus-like "bristles" or stiff cilia**; oral ciliature, with the "collar" as an "open" circle of apical oral polykinetids, the "lapel" on the left side of the oral cavity, and only with the paroral (or endoral?) on the right side of the oral cavity; **macronucleus, single, ellipsoid**; micronucleus, present; contractile vacuole, present; cytoproct (?); feeding on bacteria and microalgae; predominantly in freshwater habitats, typically planktonic but some in terrestrial habitats; five genera and two *incertae sedis*.

- *Cephalotrichium* Meunier, 1910 (subj. syn. *Strobilidium*)
- *Halteria* Dujardin, 1841
- *Halterioforma* Horváth, 1956 [*nomen oblitum*]
- *Meseres* Schewiakoff, 1893
- *Pelagohalteria* Foissner, Skogstad, & Pratt, 1988

Incertae sedis in Family Halteriidae

- *Jeannellia* Tucolesco, 1962 (subj. syn. *Halterioforma*)
- *Octocirrus* Madhava Rao, 1929 [*nomen dubium*]

Family OXYTRICHIDAE Ehrenberg, 1830
(syns. Ancystropodiinae, Gastrostylidae *p.p.*, Gastrostylina *p.p.*, Onychodromusidae, Oxytrichinae, Oxytrichoidea, Pattersoniellidae, Pleurotrichidae, Rigidotrichidae(?), Stylonychinae)

Size, small to large; shape, relatively elongate; free-swimming; **somatic ventral ciliature as frontoventral and transverse cirri, typically 18 in number, large and distinctive, scattered over mid-area of ventral surface, between right and left marginal cirral files, usually with three frontoventral cirri posterior to posterior vertex of the oral region, but several genera with conspicuous files of cirri (e.g., *Laurentiella*, *Onychodromus*)**;

right and left marginal files of cirri, obvious; caudal cirri, may or may not be present; somatic dorsal ciliature as several files of bristle dikinetids, often showing fragmentation; adoral zone, typical of order, generally restricted to anterior half or quarter of body; **division morphogenesis with six fronto-ventral anlagen streaks in two groups that make a longitudinal fan-like pattern;** macronuclei, typically as two ellipsoid nodules; micronucleus, present; contractile vacuole, at least in freshwater and terrestrial forms; cytoproct, dorsolateral left; feeding on bacteria, microalgae, and smaller protists, but several of the included genera are macrophagous carnivores on other ciliates and even smaller metazoa; in a variety of marine, freshwater, and terrestrial habitats, widely distributed; 42 genera and three genera *incertae sedis*.

NOTE: This family, the most well-characterized among the stichotrichs, has been subdivided into several subfamilies (e.g., Oxytrichinae, Stylonychinae) based primarily on body flexibility and involvement of postoral cirrus V/3 in anlagen formation. The monophyly of the Subfamily Stylonychinae, but not the Subfamily Oxytrichinae, has been confirmed by gene sequences (Foissner et al., 2004). Foissner and Stoeck (2006) established the Rigidotrichidae to include *Afrophrya*, *Rigidothrix*, *Territricha*, and *Uroleptus*. Since these genera clustered among the oxytrichids based on gene sequences, we have tentatively placed them *incertae sedis* herein.

- *Actinotricha* Cohn, 1866 (subj. syn. *Oxytricha*)
- *Allotricha* Sterki, 1878
- *Ancystropodium* Fauré-Fremiet, 1907
- *Apoamphisiella* Foissner, 1997
- *Apourosomoida* Foissner, Agatha, & Berger, 2002*
- *Architricha* Gupta, Kamra, & Sapra, 2006*
- *Australocirrus* Blatterer & Foissner, 1988
- *Coniculostomum* Njiné, 1979
- *Cyrtohymena* Foissner, 1989
- *Erimophrya* Foissner, Agatha, & Berger, 2002*
- *Gastrostyla* Engelmann, 1862
- *Gigantothrix* Foissner, 1999
- *Hemigastrostyla* Song & Wilbert, 1997
- *Hemiurosoma* Foissner, Berger, & Agatha, 2002*
- *Histriculus* Corliss, 1960
- *Laurentiella* Dragesco & Njiné, 1971
- *Neokeronopsis* Warren, Fyda, & Song, 2002*
- *Notohymena* Blatterer & Foissner, 1988
- *Onychodromopsis* Stokes, 1887
- *Onychodromus* Stein, 1859
- *Oxytricha* Bory de St. Vincent in Lamouroux, Bory de St. Vincent & Deslongchamps, 1824
- *Parahistriculus* Grolière, 1976 (subj. syn. *Histriculus*)
- *Parastylonychia* Dragesco, 1963
- *Parentocirrus* Voss, 1997
- *Parurosoma* von Gelei, 1954
- *Paraurostyla* Borror, 1972
- *Pattersoniella* Foissner, 1987
- *Pleurotricha* Stein, 1859
- *Ponturostyla* Jankowski, 1989
- *Pseudostrombidium* Horváth, 1933
- *Rigidicortex* Berger, 1999
- *Rubrioxytricha* Berger, 1999
- *Steinia* Diesing, 1866
- *Sterkiella* Foissner, Blatterer, Berger, & Kohmann, 1991
- *Stylonychia* Ehrenberg, 1830
- *Styxophrya* Foissner, Moon-van der Staay, van der Staay, Hackstein, Krautgartner, & Berger, 2004*
- *Tachysoma* Stokes, 1887
- *Territricha* Berger & Foissner, 1988
- *Tetmemena* Eigner, 1999
- *Urosoma* Kowalewski, 1882
- *Urosomoida* Hemberger in Foissner, 1982
- *Vermioxytricha* Foissner, Agatha, & Berger, 2002*

Incertae sedis in Family Oxytrichidae

- *Afrophrya* Foissner & Stoeck, 2006*
- *Anatoliocirrus* Özbek & Foissner in Foissner, Agatha, & Berger, 2002*
- *Rigidothrix* Foissner & Stoeck, 2006*

Family TRACHELOSTYLIDAE Small & Lynn, 1985
(syns. Gonostomatidae, Gonostomidae)

Size, small to medium; shape, elongate and sometimes tailed; free-swimming; **somatic ventral ciliature as frontoventral cirri scattered on anterior near peristomal region, sometimes in posterior, but never on mid-area of ventral surface, which is bordered by right and left marginal cirral files**; transverse cirri, present or absent; caudal cirri, present or absent; somatic dorsal ciliature as several files of bristle dikinetids; oral ciliature as for order with paroral and endoral, but in some genera the adoral zone of oral polykinetids is divided into an anterior and posterior part; macronucleus, two to many ellipsoid nodules; micronuclei, several; contractile vacuole, present at least in freshwater forms; cytoproct, likely present; feeding on bacteria, algae, and smaller protists; in marine, freshwater, and terrestrial habitats; seven genera.

- *Cossothigma* Jankowski, 1978
- *Gonostomum* Sterki, 1878
- *Hemisincirra* Hemberger, 1985
- *Paragonostomum* Foissner, Agatha, & Berger, 2002*
- *Spirotrachelostyla* Gong, Song, Li, Shao, & Chen, 2006*
- *Terricirra* Berger & Foissner, 1989
- *Trachelostyla* Borror, 1972

Incertae sedis in Order Sporadotrichida

- *Cinetoconia* Renault & Roche, 1898 [*nomen dubium*]
- *Etoschothrix* Foissner, Agatha, & Berger, 2002*
- *Gruberella* Corliss, 1960

Order Urostylida Jankowski, 1979
(syns. Pseudokeronopsina *p.p.*, Urostylina)

Size, small to large, up to 800 μm; shape, elongate-elliptical in outline, sometimes quite broad; free-swimming; **somatic ventral ciliature as fronto-ventral cirri in zig-zag files, running almost the full length of ventral surface between right and left files of marginal cirri and ranging from a "single" file of zig-zag or offset cirri to multiple and short files of cirri whose anterior and sometimes posterior ends are offset (= developed zig-zag) (e.g., *Eschaneustyla*)**; transverse cirri, present or absent; caudal cirri, present or absent; somatic dorsal ciliature as three or more kineties of bristle dikinetids; **during division morphogenesis, zig-zag cirri differentiating from anlagen of many short oblique kinetofragments**; four families.

Family EPICLINTIDAE Wicklow & Borror, 1990

Size, medium to large; shape, very flexible, elongate with distinct, elongate tail; free-swimming; **somatic ventral ciliature as many oblique files of ventral cirri, arising by elineation from five to seven anlagen**; marginal cirral files, present; **transverse cirri as prominent oblique file, parallel to left marginal file**; caudal cirri, present; dorsal somatic ciliature, several files of dikinetids; oral structures as for order; macronucleus, several to many nodules; micronucleus, present; contractile vacuole, present; cytoproct (?); feeding on bacteria and diatoms; in marine habitats; two genera.

- *Epiclintes* Stein, 1863
- *Eschaneustyla* Stokes, 1886

Family PSEUDOKERONOPSIDAE Borror & Wicklow, 1983
(syns. Pseudokeronopsinae, Thigmokeronopsinae)

Size, medium to large; shape, elongate; free-swimming; **somatic ventral ciliature as frontal cirri forming a conspicuous, arc-like file that parallels the anterior left serial oral polykinetids, which may be doubled by an arc-like extension of the frontoventral zig-zag file; left and right marginal cirri as 1 (rarely 2) file(s)**; transverse cirri, present or absent; caudal cirri, present or absent; dorsal somatic ciliature as three or more files of bristle dikinetids; oral ciliature as for order with paroral and endoral; macronucleus, ellipsoid, typically many more than two; micronucleus, present; contractile vacuole, present at least in freshwater forms; cytoproct, likely present; feeding on bacteria, algae, and smaller protists; in marine, freshwater, and terrestrial habitats; three genera and three genera *incertae sedis*.

- *Pseudokeronopsis* Borror & Wicklow, 1983
- *Thigmokeronopsis* Wicklow, 1981
- *Uroleptopsis* Kahl, 1932

Incertae sedis in the Family Pseudokeronopsidae

- *Bicoronella* Foissner, 1995
- *Keronella* Wiackowski, 1985
- *Tricoronella* Blatterer & Foissner, 1988

Family PSEUDOUROSTYLIDAE Jankowski, 1979
(syn. Pseudourostyloidea)

Size, medium to large; shape, elongate ovoid; free-swimming; **somatic ventral ciliature as frontal cirri forming a conspicuous, arc-like file that parallels the anterior left serial oral polykinetids and a frontoventral cirral zig-zag with**

1–6 files to the right of left marginal cirral file and also with several cirral files to the left of the right marginal cirri; multiple "marginal files" derive from unique anlagen during morphogenesis; transverse cirri, present or absent; caudal cirri, absent; dorsal somatic ciliature as several files of bristle dikinetids; oral ciliature as for order with paroral and endoral; macronucleus, ellipsoid, multiple; micronucleus, present; contractile vacuole, present; cytoproct, likely present; feeding on bacteria, algae, smaller protists, including testate amoebae and ciliates; in freshwater and terrestrial habitats; three genera.

- *Hemicycliostyla* Stokes, 1886
- *Pseudourostyla* Borror, 1972
- *Trichotaxis* Stokes, 1891

Family UROSTYLIDAE Bütschli, 1889
(syns. Bakuellidae, Bakuellinae, Erionellidae, Holostichidae, Holostichina, Holostichinae, Holostichoidea, Psammomitrinae Pseudoamphisiellidae, Urostylinae, Urostyloidea)

Size, small to large; shape, elongate ovoid with some tailed forms; free-swimming; **somatic ventral ciliature with several frontal cirri somewhat larger than other frontoventral cirri, and with frontoventral cirri as a single zig-zag file of paired cirri or a series of shorter files offset at their anterior and posterior ends (e.g., *Bakuella*, *Eschaneustyla*) and typically not with additional "marginal files" on both sides of this zig-zag (cf. Pseudourostylidae)**; transverse cirri, may be numerous; caudal cirri, present or absent; dorsal somatic ciliature as three to many files of bristle dikinetids; oral ciliature as for order with paroral and endoral; during division morphogenesis, frontoventral cirri differentiate from a longitudinal field of more than five oblique ciliary streaks; macronucleus, ellipsoid, two to many nodules; micronucleus, present; contractile vacuole, present; cytoproct, likely present; feeding on bacteria, algae, and smaller protists, including ciliates; in marine, freshwater, and terrestrial habitats; 25 genera.

- *Afrothrix* Foissner, 1999
- *Anteholosticha* Berger, 2003*
- *Australothrix* Blatterer & Foissner, 1988
- *Bakuella* Agamaliev & Alekperov, 1976
- *Biholosticha* Berger, 2003*
- *Birojimia* Berger & Foissner, 1989
- *Caudiholosticha* Berger, 2003*
- *Diaxonella* Jankowski, 1979
- *Holosticha* Wrzesniowski, 1877
- *Holostichides* Foissner, 1987
- *Metabakuella* Alekperov, 1989
- *Metaurostylopsis* Song, Petz, & Warren, 2001*
- *Notocephalus* Petz, Song, & Wilbert, 1995
- *Parabirojimia* Hu, Song, & Warren, 2002*
- *Paragastrostyla* Hemberger, 1985
- *Paramitrella* Jankowski, 1978
- *Paruroleptus* Wenzel, 1953 (subj. syn. *Uroleptus*)
- *Periholosticha* Hemberger, 1985
- *Perisincirra* Jankowski, 1978
- *Psammomitra* Borror, 1972 (subj. syn. *Uroleptus*)
- *Pseudoamphisiella* Song, 1996
- *Pseudobakuella* Alekperov, 1992
- *Tunicothrix* Xu, Lei & Choi, 2006*
- *Uroleptus* Ehrenberg, 1831
- *Urostyla* Ehrenberg, 1830

Incertae sedis in Subclass Stichotrichia

- *Erniella* Foissner, 1987
- *Gastrosticha* Kahl, 1932 [nomen dubium]
- *Saudithrix* Berger, Al-Rasheid, & Foissner, 2006*
- *Stenotricha* Jankowski, 1978
- *Uncinata* Bullington, 1940

Subclass Oligotrichia Bütschli, 1887/1889
(syns. Oligotricha, Oligotrichorida)

Size, small to large, shape, typically rounded or gently pointed posteriorly, sometimes tailed; free-swimming; a perilemma present in many species; internal polysaccharide plates in some species; **somatic kineties, reduced in number and variable in pattern, forming girdles and spirals, typically derived from an "equatorial" girdle kinety and a "ventral" kinety of dikinetids that originates near the posterior pole**; extrusomes often prominent rod-like "trichites"; **oral region on anterior half with oral polykinetids extensive and conspicuous in two sections – a "collar" out on the body surface encircling anterior pole of organism and a "lapel" inside the oral cavity proper**; paroral, a single file of kinetosomes (i.e., monostichomonad); stomatogenesis, apokinetal, often in a below-surface pouch; division morphogenesis, enantiotropic-like; macronucleus, globular to ellipsoid to band-like, often multiple; micronucleus, present; contractile vacuole, at least present in freshwater forms; cytoproct, possibly absent; feeding on bacteria, microalgae, and smaller protists, but mixotrophic species common; in marine and freshwater habitats, free-living as plankton but several species endocommensals in echinoids; one order.

NOTE: Our classification reflects the molecular genetic analyses that suggest the halteriids, classically considered oligotrichs, arose *within* the stichotrichs while the strombidiids are a separate lineage probably related to the choreotrichs (Snoeyenbos-West, Salcedo, McManus, & Katz, 2002; Strüder-Kypke & Lynn, 2003).

Order Strombidiida Petz & Foissner, 1992
(syn. Strombidiina)

With characterististics of the class; two families.

Family STROMBIDIIDAE Fauré-Fremiet, 1970
(syns. Cyrtostrombidiidae, Pelagostrombidiidae)

With characteristics of the order; **without an elongate and conspicuous contractile tail, but tail is often lost in fixed specimens** (see Tontoniidae below); 17 genera.

NOTE: We have listed all the names of strombidiid genera considered valid by Aescht (2001) and Agatha (2004). However, it is highly doubtful given recent research on the molecular evolution of strombidiids that all these genera represent monophyletic clades (e.g., see Agatha, Strüder-Kypke, Beran, & Lynn, 2005; Snoeyenbos-West et al., 2002; Strüder-Kypke & Lynn, 2003).

- *Buehringa* Busch, 1921 (subj. syn. *Strombidium*)
- *Cyrtostrombidium* Lynn & Gilron, 1993
- *Echinostrombidium* Jankowski, 1978
- *Laboea* Lohmann, 1908
- *Limnostrombidium* Krainer, 1995
- *Lissostrombidium* Jankowski, 1978
- *Metastrombidium* Fauré-Fremiet, 1924
- *Novistrombidium* Song & Bradbury, 1998
- *Omegastrombidium* Agatha, 2004*
- *Parallelostrombidium* Agatha, 2004*
- *Pelagostrombidium* Krainer, 1991
- *Peristrombidium* Jankowski, 1978
- *Pseudostrombidium* Horváth, 1933
- *Seravinella* Alekperov & Mamajeva, 1992
- *Spirostrombidium* Jankowski, 1978
- *Strombidium* Claparède & Lachmann, 1859
- *Thigmostrombidium* Jankowski, 1978

Family TONTONIIDAE Agatha, 2004

With characteristics of the order; **with elongate and conspicuous contractile tail, but tail is often lost in fixed specimens**; four genera.

- *Paratontonia* Jankowski, 1978 (subj. syn. *Tontonia*)
- *Pseudotontonia* Agatha, 2004*
- *Spirotontonia* Agatha, 2004*
- *Tontonia* Fauré-Fremiet, 1914

Class **ARMOPHOREA** Lynn, 2004

Size, small to large; shape, varied from top-shaped with spines to laterally flattened and leaf-like; alveoli, conspicuous to absent; somatic dikinetids with both kinetosomes bearing cilia, typically distributed in kineties covering entire body, but in smaller forms ciliature reduced to tufts or cirri; **hydrogenosomes, with remnant genome retained in some forms, replacing mitochondria;** oral polykinetids on left side of oral cavity, few and inconspicuous to many, forming an adoral zone; stomatogenesis, typically pleurokinetal; macronucleus, single, large, and ellipsoid to elongate, but sometimes multiple; conjugation, typically temporary, but may be total in some armophorids; contractile vacuole, present, sometimes with collecting canals; cytoproct, may be present; bacterivorous, but also with endosymbiotic methanogens; in marine, freshwater, and rarely terrestrial anaerobic habitats (i.e. polysaprobic), typically in sediments and the intestinal tracts of diverse hosts, such as echinoids, arthropods, and some vertebrates; two orders.

NOTE: Lynn (2004) established this as one of the two "riboclasses" within the phylum as representatives from the Orders Armophorida and Clevelandellida are strongly associated based on small subunit rRNA gene sequences (Affa'a, Hickey, Strüder-Kypke, & Lynn, 2004; van Hoek, van Alen, Sprakel, Hackstein, & Vogels, 1998). The odontostomatids, which have been historically associated with the armophorids (see Jankowski, 1964) are now tentatively removed to the Class PLAGIOPYLEA (Stoeck, Foissner, & Lynn, 2007).

Order Armophorida Jankowksi, 1964
(syn. Metopina)

Size, generally small to medium; shape, top-like, usually twisted to left, often much so; free-swimming; somatic ciliature, holotrichous, but may be absent except for caudal tuft and several anteriorly located cirri; **oral region spiralled, with series of 3–5 perioral or perizonal somatic kineties along its anterior edge**; oral polykinetids as paramembranelles, extending into an oral cavity with cytostome near the antapical pole in some forms; paroral as diplostichomonad; typically in marine and freshwater benthic anaerobic habitats (i.e., sapropel), but some are endosymbionts of echinoids; two families.

Family CAENOMORPHIDAE Poche, 1913
(syns. Gyrocoridae, Gyrocorycidae, Gyrocorythidae, Ludiidae, Ludioidae)

Size, small; shape, round or conical, rigid, twisted left less prominently than Metopidae (**see below**); free-swimming; **somatic cilia as small kineties or cirrus-like tufts**; oral polykinetids, several, in a small oral cavity in the posterior half of the cell; paroral, not described; macronucleus, globular to ellipsoid; micronucleus, present; contractile vacuole, present; cytoproct (?); feeding on bacteria; in brackish and freshwater anaerobic habitats; four genera.

- *Caenomorpha* Perty, 1852
- *Cirranter* Jankowski, 1964
- *Ludio* Penard, 1922
- *Sulfonecta* Jankowski, 1978

Family METOPIDAE Kahl, 1927

Size, small to medium; **shape, contorted with anterior part of body twisted to left, and posterior part sometimes tailed and/or bearing a tuft of longer caudal cilia**; free-swimming; somatic ciliation, holotrichous; oral polykinetids, multiple, sometimes extending out of a more posterior oral cavity onto a broader peristomial region; macronucleus, globular to ellipsoid; micronucleus, present; contractile vacuole, present; cytoproct (?); feeding on bacteria; in marine and freshwater anaerobic habitats; nine genera.

- *Bothrostoma* Stokes, 1887
- *Brachonella* Jankowski, 1964
- *Eometopus* Small & Lynn, 1985
- *Metopus* Claparède & Lachmann, 1858
- *Palmarella* Jankowski, 1975
- *Parametopidium* Aescht, 2001
- *Spirorhynchus* da Cunha, 1915 (subj. syn. *Metopus*)
- *Tesnospira* Jankowski, 1964 (subj. syn. *Metopus*)
- *Tropidoatractus* Levander, 1894

Order Clevelandellida de Puytorac & Grain, 1976
(syns. Clevelandellidia, Nyctotherina *p.p.*, Paranyctotherina *p.p.*)

Size, medium to large, often > 150 μm; shape, typically flattened; free-swimming; somatic ciliature, holotrichous, dense, with somatic kineties forming a variety of sutures or complex secant systems, which are used, in part, to distinguish families and genera; **somatic dikinetids with non-microtubular retrodesmal and cathetodesmal fibrils**; sometimes conspicuous dorsoanterior sucker region; **oral structures as many left serial oral polykinetids or heteromembranelles, not usually conspicuous, arranged in a long peristomial groove that precedes a well-developed infundibulum**; paroral as diplostichomonad; macronucleus anchored in a karyophore in many species; conjugation often synchronized with reproductive life cycle of the host; contractile vacuole, present; cytoproct in several forms lined with cilia; feeding on bacteria and organic detritus; in marine, freshwater, and terrestrial habitats as endocommensals in the digestive tracts of oligochaetes, insects, centipedes, millipedes, molluscs, and some vertebrates; five families.

Family CLEVELANDELLIDAE Kidder, 1938
(for Clevelandiidae)

Size, small to medium; **shape, basically ovoid or elongate-ovoid, flattened, with posterior pole as oddly shaped projection, which bears the inconspicuous opening of the oral cavity**

and its infundibular opening; free-swimming; somatic ciliation, holotrichous, often with very developed preoral secant system; oral cavity, extending from the posterior opening anteriorly into the body, with oral polykinetids and paroral and endoral; macronucleus, globular to ellipsoid, supported by karyophore; micronucleus, present; contractile vacuole, may be present; cytoproct (?); feeding on bacteria and organic detritus; in terrestrial habitats in the digestive tracts of termites and wood-feeding roaches only; three genera.

- *Clevelandella* Kidder, 1938
- *Metaclevelandella* Uttangi & Desai, 1963
- *Paraclevelandia* Kidder, 1937

Family INFEROSTOMATIDAE Ky, 1971
(syns. Nathellidae, Nathelliidae)

Size, small to medium; shape, roughly ovoid, flattened, but distorted somewhat by **huge sucker on right side at anterior end**; free-swimming; **somatic ciliation, holotrichous, with one right caudal secant system, one left caudal secant system, and one right transverse secant system**; **oral region as extensive peristome bearing oral polykinetids that extend anteriorly out onto the body surface from the infundibular opening at the truncate posterior pole**; macronucleus, ellipsoid, may be supported by a karyophore; micronucleus, present; contractile vacuole (?); cytoproct (?); feeding (?); in freshwater habitats in the intestine of certain fishes; three genera.

- *Ichthyonyctus* Jankowski, 1974
- *Inferostoma* Ky, 1971
- *Nathella* Singh, 1953 [nomen nudum]

Family NEONYCTOTHERIDAE Affa'a, 1987

Size, small to medium; shape, ovoid, flattened; **polysaccharide elements forming a reticulated subpellicular system under the entire cortex**; free-swimming; **somatic ciliation, holotrichous, with one preoral secant system and one apical right secant system**; oral cavity, inconspicuous, but with oral polykinetids and paroral and endoral; macronucleus, globular to ellipsoid; micronucleus (?); contractile vacuole, present; cytoproct, quite long, opening near posterior end; feeding (?); in freshwater and terrestrial habitats in the digestive tract of amphibians; one genus.

- *Neonyctotherus* Affa'a, 1983

Family NYCTOTHERIDAE Amaro, 1972
(syn. Paranyctotherida *p.p.*)

Size, small to large; shape, ovoid to slightly reniform, plump; free-swimming; somatic ciliation, holotrichous, with secant systems, varying significantly with included genera, but **never a transverse secant system; no skeletal apparatus beneath concave surface and "sucker" not obvious**; oral ciliature running from near-apical to sub-equatorial position, in a sigmoid-like curve as it enters conspicuous infundibulum; macronucleus, ellipsoid, large, compact, in anterior half of body, supported by more or less well-developed karyophore; micronucleus, present; contractile vacuole (?); cytoproct (?); feeding (?); in marine, freshwater, and terrestrial habitats as endocommensals in wide variety of hosts from oligochaetes, insects (cockroach), and myriapods (centipede, millipede) to molluscs (shipworm), fish, amphibians (frog, toad), and reptiles; 15 genera.

- *Cameronyctus* Jankowski, 1986 [nomen nudum]
- *Cichlidotherus* Affa'a, 1989
- *Cryptonyctus* Jankowski, 1978
- *Falconyctus* Jankowski, 1978
- *Indonyctus* Jankowski, 1978
- *Metanyctotherus* Albaret, 1970
- *Micronyctus* Jankowski, 1978
- *Nyctositum* Affa'a, 1979
- *Nyctotheroides* Grassé, 1928
- *Nyctotherus* Leidy, 1849
- *Paracichlidotherus* Grim, 1992
- *Paranyctotherus* Sandon, 1941 (subj. syn. *Balantidium*)
- *Pronyctotherus* Albaret & Njiné, 1976
- *Pygmotheroides* Affa'a, 1980
- *Vesonyctus* Jankowski, 1978

Family SICUOPHORIDAE Amaro, 1972

Size, medium to large; shape, plump-ovoid to ellipsoid, occasionally tailed; **inferior (= right) concave surface in part or whole as "sucker" with**

supporting polysaccharide skeletal elements, which may also extend to support other parts of the body; somatic ciliation, holotrichous, with one apical right secant system and one caudal right secant system; oral ciliature running from near-apical to sub-equatorial position, in a sigmoid-like curve as it enters conspicuous infundibulum; macronucleus, ellipsoid, large, compact, in anterior half of body, supported by more or less well-developed karyophore; micronucleus, present; contractile vacuole (?); cytoproct (?); feeding (?); in freshwater and terrestrial habitats as endosymbionts in the digestive tracts of vertebrate hosts only, such as amphibians and reptiles; seven genera.

- *Albaretia* Affa'a in Aescht, 2001
- *Geimania* Albaret, 1975
- *Metasicuophora* Albaret, 1973
- *Parasicuophora* Albaret, 1968
- *Prosicuophora* de Puytorac & Oktem, 1967
- *Sicuophora* de Puytorac & Grain, 1969
- *Spiroperistomatus* Amaro & Sena, 1967

Class **LITOSTOMATEA** Small & Lynn, 1981
(syns. Apicostomata *p.p.*, Homotricha *p.p.*, Rhabdophora *p.p.*, Transversonematophora *p.p.*)

Size, small to large; shape, varied; free-swimming; alveoli, poorly to well-developed; somatic ciliation, holotrichous to sparse in pleurostomes and some endosymbionts; **somatic monokinetids, typical, with laterally directed kinetodesmal fibril that does not overlap those of adjacent kineties, slightly convergent postciliary ribbon, and two transverse ribbons, one of which is tangential to the kinetosome perimeter and extends anteriorly into the somatic ridge to the left of the kinetid while the other transverse ribbon is radial to the kinetosome perimeter and extends transversely into the adjacent somatic ridge; one to several dorsal somatic kineties differentiated as a brosse or brush kinetids with specialized dikinetids bearing clavate cilia**; lamina corticalis or ecto-endoplasmic fibrillar layer often present and well-developed; oral ciliature as simple kinetids from which nematodesmata arise to support the cytopharynx, but nematodesmata may also arise from so-called "oralized" somatic kinetids adjacent to the oral region, and in some symbionts, oral ciliature is organized into polykinetid-like structures called syncilia; stomatogenesis, telokinetal; macronucleus, typically single, variously shaped from globular to band-shaped or moniliform; micronucleus, present; conjugation, temporary; contractile vacuole, present; cytoproct, present; feeding, extremely diverse, on bacteria and plant debris in some symbionts to carnivorous in others; in marine, freshwater, and terrestrial habitats, free-living and as endosymbionts in wide variety of vertebrates, especially; two subclasses.

Subclass Haptoria Corliss, 1974
(syns. Acrostomatina, Haptorida, Paramastigina *p.p.*, Prionostomatina, Raptorida, Rhynchostomata *p.p.*, Rhynchostomatida *p.p.*, Sciadophorida [-ina] *p.p.*, Telostomata *p.p.*, Toxistomia *p.p.*)

Size, small to large; shape, variable, some species equipped with proboscis and a few species with non-suctorial tentacles; free-swimming; poorly developed alveoli; somatic ciliation, holotrichous, but reduced to girdles in some forms, and sparse in pleurostomes; somatic kinetid as for the class, but postciliary microtubules overlapping longitudinally; extrusomes as somatic mucocysts, clathrocysts, and lepidosomes, and oral and/or somatic toxicysts; oral region, typically anterior, with cytostome, apical or subapical, oval or slit-like, rarely permanently open, so that the cytopharynx becomes eversible in some species; **oral dikinetids, rarely monokinetids, on border of cytostome-cytopharynx, typically with outer or posterior kinetosome bearing a slightly longer cilium and inner or anterior non-ciliated kinetosome with a transverse microtubular ribbon that extends anteriorly and then reflects posteriorly to support the cytopharynx; cytopharynx, supported by the rhabdos, which is formed by bulge microtubules and transverse microtubular ribbons and nematodesmata arising from oral dikinetids; toxicysts localized in or near the oral area, typically between the oral transverse ribbons and bulge microtubules of the rhabdos**; stomatogenesis, telokinetal; conjugation, temporary; rapacious carnivores of flagellates, ciliates, and other protists; two orders and one order *incertae sedis*.

NOTE: Small and Lynn (1985), Foissner and Foissner (1988), Lipscomb and Riordan (1992), and Grain (1994) have suggested different ordinal, subordinal, and familial classifications for these ciliates. Recent molecular phylogenetic analysis does not provide unambiguous support for any proposed taxonomy of haptorians (Strüder-Kypke, Wright, Foissner, Chatzinotas, & Lynn, 2006). Thus, until unambiguous, high-weight morphological synapomorphies are supported by molecular genetic evidence, we cannot support substantial subdivision and have remained conservative.

Order Haptorida Corliss, 1974
(syns. Acropisthiina *p.p.*, Belonophryina *p.p.*, Didiniina *p.p.*, Dileptida *p.p.*, Enchelyina *p.p.*, Helicoprorodontida *p.p.*, Helicoprorodontina *p.p.*, Inferotrichida *p.p.*, Lacrymariina *p.p.*, Pseudoholophryida *p.p.*, Pseudoholophryina *p.p.*, Spathidiida *p.p.*, Spathidiina *p.p.*, Trachelophyllina *p.p.*)

Somatic ciliation, holotrichous, but restricted to girdles in didiniids; **oral region, typically circular or elliptical, surrounded by circumoral dikinetids whose microtubules extend to support the cytostome-cytopharynx, but where circumoral dikinetids are absent, oralized somatic monokinetids bear nematodesmata for the rhabdos**; 14 families.

Family ACROPISTHIIDAE Foissner & Foissner, 1988
(syn. Fuscheriidae)

Size, small to medium; shape, ovoid to elongate; free-swimming; somatic ciliation, holotrichous, often more dense in the anterior half; brosse kineties, 2–4; extrusomes as somatic mucocysts and oral toxicysts; **oral region, apical, with oral dikinetids evenly surrounding cytostome, accompanied by some oralized somatic monokinetids**; oral nematodesmata arising from oral dikinetids *and* adjacent oralized somatic monokinetids; macronucleus, globular to ellipsoid, band-like or in many nodules; micronucleus, present; contractile vacuole, present; cytoproct (?); feeding on flagellates; in freshwater and terrestrial habitats; nine genera.

- *Acropisthium* Perty, 1852
- *Actinorhabdos* Foissner, 1984
- *Chaenea* Quennerstedt, 1867
- *Clavoplites* Foissner, Agatha, & Berger, 2002*
- *Coriplites* Foissner, 1988
- *Dioplitophrya* Foissner, Agatha, & Berger, 2002*
- *Diplites* Foissner, 1998
- *Fuscheria* Foissner, 1983
- *Sikorops* Foissner, 1999

Family ACTINOBOLINIDAE Kahl, 1930
(for Actinobolidae; syn. Legendreidae)

Size, small to medium; shape, ovoid; free-swimming; somatic ciliation, holotrichous, with kineties more or less spiralling; brosse kineties, at least two; extrusomes as somatic mucocysts and somatic toxicysts; **retractable, non-suctorial tentacle-like processes, widely distributed over body, containing toxicysts, and associated with somatic monokinetids**; oral region, apical, with oral dikinetids surrounding apical cytostome; macronucleus, globular to ellipsoid to extremely elongate and ribbon-like; micronucleus, present; contractile vacuole, present; cytoproct (?); feeding on flagellates and other ciliates; in brackish and freshwater habitats; four genera.

- *Actinobolina* Strand, 1928
- *Belonophrya* André, 1914
- *Dactylochlamys* Lauterborn, 1901
- *Legendrea* Fauré-Fremiet, 1908

Family APERTOSPATHULIDAE Foissner, Xu, & Kreutz, 2005

Size, small to medium; shape, elongate ovoid; free-swimming; somatic ciliation, holotrichous; brosse kineties, three; extrusomes as somatic mucocysts and oral toxicysts; **oral region, apical to subapical, forming a lasso-shaped bulge surrounded by an unclosed ring of circumoral dikinetids that extends further posteriorly on the right side of the oral region than on the left side**; macronucleus, ellipsoid to elongate ellipsoid; micronucleus, present; contractile vacuole, present; cytoproct (?); feeding on flagellates, amoebae, and other ciliates; in freshwater and terrestrial habitats; three genera.

- *Apertospathula* Foissner, Agatha, & Berger, 2002*
- *Longispatha* Foissner, Xu, & Kreutz, 2005*
- *Rhinothrix* Foissner, Xu, & Kreutz, 2005*

Family DIDINIIDAE Poche, 1913
(for Cyclodinidae; syns. Cyclotrichiidae, Didiniina, Liliimorphidae)

Size, small to medium; shape, ovoid to ellipsoid, often with flattened anterior end, and some taxa with an anterior protuberance; free-swimming; **somatic cilia as series of apparently short kinetofragments in one or more girdles around body, but in non-ciliated regions, non-ciliated kinetosomes are arranged in meridional kineties; brosse, typically a field of clavate cilia or "sensory bristles" usually clearly detectable in 3–5 kineties**; extrusomes as somatic mucocysts and multiple kinds of oral toxicysts; oral region, apical, often in a conical form with eversible cytopharynx in some species; macronucleus, ellipsoid to extremely elongate and ribbon-like; micronucleus, present; contractile vacuole, present; cytoproct (?); feeding on flagellates, microalgae, and other ciliates; in marine and freshwater habitats, typically planktonic; eight genera.

- *Choanostoma* Wang, 1931
- *Cyclotrichium* Meunier, 1910
- *Dicyclotrichium* Xu, Song, & Hu, 2005*
- *Didinium* Stein, 1859
- *Liliimorpha* Gajewskaja, 1928
- *Monodinium* Fabre-Domergue, 1888
- *Pelagovasicola* Jankowski, 1980
- *Zonotrichium* Meunier, 1910

Family ENCHELYIDAE Ehrenberg, 1838
(syns. Enchelidae, Enchelyina *p.p.*, Enchelynidae, Enchelyodontidae *p.p.*)

Size, small to medium; shape, ovoid to flask-like; free-swimming; somatic ciliation, holotrichous, with its kineties abutting on oral region without arching; brosse kineties, 2–4; **dikinetids or oralized somatic monokinetids surrounding cytostome, and all as anterior extremities of somatic kineties**; extrusomes as somatic mucocysts and oral toxicyts; **oral region, apical, typically flat with cytostome in number of species located at distal end of flexible neck**; macronucleus, single or multiple, ellipsoid to elongate; micronucleus, present; contractile vacuole, present; cytoproct (?); feeding on flagellates and other ciliates; typically in freshwater habitats, often somewhat anaerobic; 20 genera and two *incertae sedis*.

NOTE: Some of the genera included in this family have oralized somatic monokinetids, while others have the typical circumoral dikinetids. It may be that oralized somatic monokinetids should be used as a synapomorphy for a Family Enchelyodontidae Foissner, Agatha, and Berger, 2002, but we await confirmation of this by gene sequences.

- *Apoenchelys* Foissner, Agatha, & Berger, 2002*
- *Balantidion* Eberhard, 1862
- *Crobylura* André, 1914
- *Enchelydium* Kahl, 1930
- *Enchelys* O. F. Müller, 1773
- *Haematophagus* Woodcock & Lodge, 1921
- *Ileonema* Stokes, 1884
- *Lagynurus* Mansfeld, 1923
- *Microregma* Kahl, 1930
- *Nannophrya* Kahl, 1933
- *Obliquostoma* Foissner, Agatha, & Berger, 2002* [junior homonym of bryozoan]
- *Papillorhabdos* Foissner, 1984
- *Pithothorax* Kahl, 1926
- *Quasillagilis* Busch, 1920
- *Rhopalophrya* Kahl, 1926
- *Schewiakoffia* Corliss, 1960 [*nomen dubium*]
- *Spasmostoma* Kahl, 1927
- *Sphaerobactrum* Schmidt, 1920
- *Thalassiomastix* Busch, 1923
- *Urochaenia* Savi, 1913

Incertae sedis in Family Enchelyidae

- *Microcardiosoma* Vuxanovici, 1963
- *Microchoanostoma* Vuxanovici, 1963

Family HELICOPRORODONTIDAE Small & Lynn, 1985

Size, large, may be >1,000 μm; shape, elongate, vermiform, contractile; free-swimming; somatic ciliation, holotrichous; brosse kineties, 2–5; extrusomes as somatic mucocysts and up to several types of oral toxicysts; **toxicysts, distributed along perioral ridge that makes from one turn in** *Trachelotractus* **to a spiral of perioral kineties in** *Helicoprorodon*; oral region, apical, with oral dikinetids at ends of perioral kineties, and "oralized" somatic monokinetids also giving rise to nematodesmata; macronucleus, many, as isolated nodules or in moniliform grouping; micronucleus, not observed; contractile vacuole, present; cytoproct (?); feeding on flagellates and smaller protists; in marine sands; two genera.

- *Helicoprorodon* Fauré-Fremiet, 1950
- *Trachelotractus* Foissner, 1997

Family HOMALOZOONIDAE Jankowski, 1980 (syn. Homalozo[on]idea)

Size, medium to large; **shape, worm-like, laterally compressed, contractile**; free-swimming, typically gliding on the substrate; **somatic ciliation, holotrichous, with kineties packed much more densely on the right (?) side**; brosse kineties, three; extrusomes as somatic mucocyts and oral toxicysts; oral region, apical, with temporary cytostome; macronucleus, a single elongate ribbon to multiple ellipsoid nodules, which may be in a moniliform grouping; micronucleus, present; contractile vacuole, present; cytoproct (?); feeding on flagellates and smaller protists; in freshwater habitats, especially ponds; one genus.

- *Homalozoon* Stokes, 1890

Family LACRYMARIIDAE de Fromentel, 1876 (syn. Lacrymariina *p.p.*)

Size, small to medium; shape, elongate, often flask-shaped, with some species having an extremely extensible neck-like anterior region; free-swimming, rarely sedentary; somatic ciliation, holotrichous; brosse kineties, two or more; extrusomes as somatic mucocysts and oral toxicysts; **anterior region of the body (= head), bulb-like, covered by short oblique kineties with densely packed kinetids that abut the circumoral dikinetids**; oral region, apical, with oral dikinetids and cytostome at the anterior end of the bulb-like swelling; macronucleus, single or multiple, typically ellipsoid; micronucleus, present; contractile vacuole, present; cytoproct (?); feeding on flagellates and ciliates; in marine, freshwater, and terrestrial habitats, with some species planktonic; four genera.

- *Lacrymaria* Bory de St. Vincent, 1824
- *Pelagolacrymaria* Foissner, Berger, & Schaumberg, 1999
- *Phialina* Bory de St. Vincent, 1824
- *Phialinides* Foissner, 1988

Family PLEUROPLITIDAE Foissner, 1996

Size, small; shape, elongate; free-swimming; somatic ciliation, holotrichous; brosse kineties, several; **extrusomes as somatic toxicysts in an extracytostomal bundle on the ventral side**; oral region, apical; macronucleus, ellipsoid; micronucleus, present; contractile vacuole, present; cytoproct (?); feeding on flagellates and small ciliates; in terrestrial habitats; two genera.

- *Pleuroplites* Foissner, 1988
- *Pleuroplitoides* Foissner, 1996

Family PSEUDOHOLOPHRYIDAE Berger, Foissner, & Adam, 1984

Size, small; shape, ovoid to elongate; free-swimming; somatic ciliation, holotrichous; **somatic kineties having a slight right spiral; brosse kineties of many rows in which clavate dikinetids alternate with typical somatic monokinetids**; extrusomes as oral toxicysts; oral region, apical, round or elliptical; macronucleus, globular to elongate ellipsoid; micronucleus, present; contractile vacuole (?); cytoproct (?); feeding on various heterotrophic protists; in freshwater and terrestrial habitats; three genera and one genus *incertae sedis*.

- *Ovalorhabdos* Foissner, 1984
- *Paraenchelys* Foissner, 1983
- *Pseudoholophrya* Berger, Foissner, & Adam, 1984

Incertae sedis in Family Pseudoholophryidae

- *Songophrya* Foissner, 2003*

Family PSEUDOTRACHELOCERCIDAE Song, 1990

Size, medium; shape, cylindrical or bottle-shaped; free-swimming; somatic ciliation, holotrichous, having bipolar kineties with anterior kinetids more densely packed; **kinetids of brosse kinety irregularly arranged and continuous with only one somatic kinety**; extrusomes as oral toxicysts; oral region, apical; macronucleus, elongate band; micronucleus (?); contractile vacuole, present; cytoproct (?); feeding possibly on bacteria; in marine habitats; one genus.

- *Pseudotrachelocerca* Song, 1990

Family SPATHIDIIDAE Kahl in Doflein & Reichenow, 1929

(syns. Arcuospathidiidae, Bryophyllidae, Myriokaryonidae, Paraspathidiidae, Perispiridae, Protospathidiidae, Spathidiina *p.p.*, Teuthophryidae)

- *Apoamphileptus* Lin & Song, 2004*
- *Epiphyllum* Lin, Song, & Warren, 2005* [junior homonym of anthozoan]
- *Kentrophyllum* Petz, Song, & Wilbert, 1995
- *Opisthodon* Stein, 1859
- *Pseudoamphileptus* Foissner, 1983

Family LITONOTIDAE Kent, 1882
(syn. Loxophyllidae)

Size, typically medium to large; shape, flattened ovoid with narrowing at the anterior and posterior ends; free-swimming, typically gliding on the substrate; **somatic ciliation, holotrichous, with right somatic kineties gradually terminating along rightmost perioral kinety, thus spica absent, and with one or two dorsolateral kineties in some forms**; extrusomes as somatic mucocysts and somatic and/or oral toxicysts, but toxicysts in some forms distributed on the perimeter of the flattened body in protuberances (e.g., *Loxophyllum*); **oral region along the ventral edge, with two right perioral kineties and one left perioral kinety, with rightmost perioral kinety of monokinetids and other kineties of dikinetids**; macronucleus, typically two ellipsoid nodules; micronucleus in between macronuclear nodules; contractile vacuole, present; cytoproct (?); feeding on flagellates and smaller protists; in marine, freshwater, and terrestrial habitats, with some species planktonic; five genera.

- *Acineria* Dujardin, 1841
- *Heminotus* Kahl, 1933
- *Litonotus* Wresniowski, 1870
- *Loxophyllum* Dujardin, 1841
- *Siroloxophyllum* Foissner & Leipe, 1995

Incertae sedis in Subclass Haptoria

- *Baznosanuia* Tucolesco, 1962
- *Celerita* Tucolesco, 1962
- *Racovitzaiella* Aescht, 2001

Incertae sedis in Subclass Haptoria

Order Cyclotrichiida Jankowski, 1980
(syn. Mesodiniida)

Size, small to medium; shape, globular to subspheroid; free-swimming; **somatic cilia, bristle-like, of at least two types, arranged in girdles around the body**; brosse kineties absent; extrusomes as oral toxicysts; **oral region, apical, domed, circular, and delimited by circumoral dikinetids, but apparently without nematodesmata and bulge microtubules of rhabdos**; macronucleus, ellipsoid to band-shaped; micronucleus, present; contractile vacuole, present; cytoproct (?); feeding on flagellates and microalgae, and sometimes containing symbiotic algae; planktonic in marine and freshwater habitats; one family.

NOTE: Johnson, Tengs, Oldach, Delwiche, and Stoecker (2004) have demonstrated that the small subunit rRNA sequences of two *Mesodinium* species are highly divergent to other haptorian sequences, placing them at a basal position in the phylum. Based on the presence of toxicysts and features of the secondary structure of the small subunit rRNA molecule, which suggest homology to the litostomes (see Strüder-Kypke et al., 2006), we are maintaining these ciliates in the Subclass Haptoria until other genes suggest otherwise.

Family MESODINIIDAE Jankowski, 1980

With characteristics of the order; four genera.

- *Askenasia* Blochmann, 1895
- *Mesodinium* Stein, 1863
- *Myrionecta* Jankowski, 2007
- *Rhabdoaskenasia* Krainer & Foissner, 1990

Subclass Trichostomatia Bütschli, 1889
(syn. Synciliostoma *p.p.*)

Size, small to large; shape, ovoid to elongate, sometimes with bizarre processes and cell appendages; free-swimming; alveoli, typically well-developed, often filled with "skeletal" material; somatic ciliation, variable, from holotrichous to reduced to girdles, bands, and tufts, but with somatic kinetids as for the class; concrement vacuole(s) present in a few forms; extrusomes as somatic mucocysts **and oral toxicysts absent; oral region or cavity, typically a densely ciliated vestibulum, with oral cilia sometimes as "polykinetids" or syncilia**; cytostome, and therefore vestibulum, sometimes antapical; stomatogenesis, telokinetal, but cryptotelokinetal in entodiniomorphids; macronucleus, typically elongate ellipsoid; micronucleus, present; contractile vacuole, present; cytoproct, often conspicuous; **hydrogenosomes, typically replace mitochondria**; feeding on bacteria, detritus, plant material ingested by the host, and other ciliates; majority of species endocommensals in vertebrate hosts; three orders.

Order Vestibuliferida de Puytorac et al., 1974
(syns. Balantidiida *p.p.*, Balantidiina *p.p.*, Isotrichida *p.p.*, Isotrichina *p.p.*, Infundibuloriina *p.p.*, Paraisotrichida *p.p.*, Rimostomata *p.p.*, Synciliophora *s.l.*, Vestibulifera *s.l.*)

Somatic ciliation, holotrichous and dense; cortex, often with thick microfilamentous layer between ecto- and endoplasm; **oral region a depression or vestibulum, densely ciliated by extensions of somatic kineties, whose cilia do not appear organized as "polykinetids"**; cytostome at base of vestibular cavity; **endocommensals in herbivorous placental mammals, but not in marsupials, with balantidiids endocommensals in selected vertebrates from fish to great apes, but parasitic in humans at least**; six families.

Family BALANTIDIIDAE Reichenow in Doflein & Reichenow, 1929
(syn. Paranyctotheridae)

Size, small to medium; shape, ovoid to elongate; free-swimming; somatic ciliation, holotrichous, with clavate cilia in several kineties, as possible homologues of the haptorian brosse; no concrement vacuole; extrusomes as somatic mucocysts; **oral cavity apico-ventral, as a vestibular groove less than one-half body length and lined by extensions of somatic kineties, which are accompanied by supernumerary kineties**; macronucleus, elongate ellipsoid; micronucleus, present; contractile vacuole, present; cytoproct, likely absent; encystment probable; feeding on bacteria and organic detritus, but also may feed on host intestinal epithelial cells; in marine, freshwater, and terrestrial habitats as endocommensals in the digestive tracts of diverse hosts, such as insects, fish, frogs, snakes, guinea pig, pig, monkey, chimpanzee, gorilla, and orang-utang, and parasitic at least in man; three genera.

- *Balantidium* Claparède & Lachmann, 1858
- *Dilleria* Earl, 1973
- *Metacollinia* Jankowski, 1980

Family ISOTRICHIDAE Bütschli, 1889
(syn. Dasytrichidae)

Size, small to medium; shape, ovoid, flattened; free-swimming; somatic ciliation, holotrichous, dense, with sometimes up to 200 kineties; no concrement vacuole, but endoplasmic polysaccharide reserves; extrusomes as somatic mucocysts; **oral cavity at or near antapical pole, lined by extensions of somatic kineties, with parental vestibulum migrating anteriorly during stomatogenesis to become the vestibulum of the proter**; macronucleus, ellipsoid, may be anchored by a karyophore; micronucleus, present; contractile vacuoles, present, may be multiple; cytoproct in posterior; feeding on bacteria and organic detritus; in terrestrial habitats, widely found as endocommensals in ungulate ruminants, but *Protoisotricha* is from rodents and an *Isotricha* was once reported from the cockroach; four genera.

- *Dasytricha* Schuberg, 1888
- *Isotricha* Stein, 1859
- *Oligoisotricha* Imai, 1981
- *Protoisotricha* Kopperi, 1937

Family PARAISOTRICHIDAE da Cunha, 1917
(syn. Enterophryidae *p.p.*, Helicozosteridae *p.p.*)

Size, small to medium; shape, ovoid to pyriform; free-swimming; somatic ciliation, holotrichous, dense, sometimes slightly spiralled, often with anterior tuft of cilia, but ciliation of *Latteuria* confined more to posterior half; extrusomes as somatic mucocysts; **concrement vacuole(s) at apical pole**; oral region, anterior, with vestibulum ciliated on ventral wall and with cytostome at its base; macronucleus, ellipsoid; micronucleus, present; contractile vacuole, present; cytoproct (?); feeding on bacteria and organic detritus; in terrestrial habitats, common as endocommensals in horses, but also in capybaras, guinea pigs, and elephants; three genera and one genus *incertae sedis*.

- *Latteuria* Timoshenko & Imai, 1997
- *Paraisotricha* Fiorentini, 1890
- *Rhizotricha* Wolska, 1964

Incertae sedis in Family Paraisotrichidae

- *Helicozoster* Latteur, 1967

Family PROTOCAVIELLIDAE Grain in Corliss, 1979
(syn. Enterophryidae *p.p.*, Hydrochoerellidae)

Size, small to medium; shape, ovoid to elongate; free-swimming; somatic ciliation, from sparse, somewhat holotrichous to dense holotrichous, sometimes kineties slightly spiralling, with longer cilia as a tuft or band anterior to the oral region; extrusomes (?); no concrement vacuole; oral region, subapical to subequatorial; **oral cavity, funnel- or trumpet-shaped with adoral polybrachykinety running along the anterior and right edge of the oral cavity opening and vestibular polybrachykinety running along the left wall of the oral cavity, but adoral ciliature absent in *Enterophrya* and *Ogimotopsis***; macronucleus, globular to elongate ellipsoid; micronucleus, present; contractile vacuole, present; cytoproct, present; feeding on bacteria and organic detritus (?); in terrestrial habitats as endocommensals in the hindgut of capybaras, guinea pigs, and lagomorphs; nine genera.

- *Anacharon* Ito & Imai, 2000
- *Cunhamunizia* Ito & Imai, 2000
- *Enterophrya* Hasselmann, 1918
- *Hydrochoerella* da Cunha & Muniz, 1925
- *Ogimotoa* Ito & Imai, 2000
- *Ogimotopsis* Ito & Imai, 2000
- *Paracunhamunizia* Ito & Imai, 2000
- *Protocaviella* Kopperi, 1937
- *Uropogon* Ito & Imai, 2000

Family PROTOHALLIIDAE da Cunha & Muniz, 1927
(for Halliidae [for Rhipidostom(at)idae])

Size, small; **shape, ovoid, with apical disc and posterior anal papilla**; free-swimming; somatic ciliation, holotrichous; extrusomes (?); oral region, apical, with oral ciliature short, inconspicuous; **adoral ciliature divided into three regions, one densely ciliated, crown-shaped on the left anterior of the oral region and the other two, inconspicuous, on the posterior of the oral region; cytostome apical, with cytopharynx supported by prominent, basket-like cytopharyngeal apparatus (homologous to the rhabdos of the haptorian litostomes?)**; macronucleus, ellipsoid; micronucleus, present; contractile vacuole, present; cytoproct, in posterior; feeding on bacteria, smaller protists, and organic detritus; in terrestrial habitats as endocommensals in the cecum of capybaras; one genus.

NOTE: Ito and Imai (2000a) suggested transferring this family to the Class PHYLLOPHARYNGEA, Subclass Cyrtophoria, because it is characterized by a prominent, basket-like "cyrtos". Until there is evidence that this basket is truly a cyrtos or until gene sequence demonstrate a different affinity, we maintain the family here and consider the cytopharyngeal basket as a rhabdos.

- *Protohallia* da Cunha & Muniz, 1927

Family PYCNOTRICHIDAE Poche, 1913
(syns. Infundibuloriidae, Miniziellidae, Muniziellidae, Nicollellidae, Pycnothricidae, Pycnotrichida)

Size, small to large; shape, ovoid; free-swimming; somatic ciliation, holotrichous, dense; cortex, thick, apparently of up to three layers; no concrement vacuole; extrusomes (?); **oral cavity as a long vestibular groove at least one-half body length, lined on its edges by extensions of somatic kineties**; cytostome can be subequatorial, posterior or dorsal, depending on the length and orientation of the vestibulum; macronucleus, globular to ellipsoid; micronucleus, present; contractile vacuole, present, sometimes with collecting canals; cytoproct, present; feeding (?); in terrestrial habitats as endocommensals in various herbivorous vertebrates, including fishes, gundis, hyraxes, capybaras, camels, cattle, water buffaloes, and the red spider monkey; seven genera and one genus *incertae sedis*.

- *Collinina* Chatton & Pérard, 1924
- *Infundibulorium* Bozhenko, 1925
- *Muniziella* da Fonseca, 1939
- *Nicollella* Chatton & Pérard, 1919
- *Pycnothrix* Schubotz, 1909
- *Taliaferria* Hegner & Rees, 1933
- *Vestibulongum* Grim, 1988

Incertae sedis in Family Pycnotrichidae

- *Buxtonella* Jameson, 1926

Incertae sedis in the Order Vestibuliferida

- *Microcetella* Aescht, 2001

Order Entodiniomorphida Reichenow in Doflein & Reichenow, 1929
(syns. Entodiniomorpha, Entodiniomorphina, Entodiniorida, Syntricha *p.p.*)

Size, small to large; shape, ovoid, often flattened; pellicle firm and thickened, often drawn out into posterior spines; cortex with thick microfilamentous layer between ecto- and endoplasm;

somatic ciliature, typically greatly reduced, appearing only in bands, zones or tufts, often as polybrachykineties, and functioning as syncilia; concrement vacuole may be present; oral area as only a slight depression to a deep one, often with well-differentiated "polykinetids"; cytoproct distinct, sometimes at the base of a ciliated tube; in terrestrial habitats, widely found as commensals in mammalian hosts, mainly artiodactyls and perissodactyls, with species of the Family Troglodytellidae restricted to anthropoid apes; three suborders.

Suborder Archistomatina de Puytorac et al., 1974 (syns. Archiciliatida, Cyclotrichina *p.p.*, Didesmida, Didesmina)

Size, often small; shape, ovoid to pyriform; free-swimming; somatic ciliation, holotrichous or limited to girdles, tufts or bands; extrusomes as somatic mucocysts; **concrement vacuole present, overlain by 4–5 somatic kineties, presumed to be homologous to the brosse of haptorians**; oral region, apical, with permanent cytostome surrounded by circumoral monokinetids, closely packed and derived from oralized somatic kinetids, as an adoral polybrachykinety; macronucleus, ellipsoid to elongate ellipsoid; micronucleus, present; contractile vacuole, present, sometimes multiple; cytoproct in posterior; feeding (?); in terrestrial habitats as endocommensals of vertebrates, typically in horses and camels, but also in rodents, hippopotami, and sometimes ruminants; one family.

Family BUETSCHLIIDAE Poche, 1913 (syns. Blepharoconinae, Didesminae, Paraisotrichopsidae, Polymorphellinae, Sulcoarcidae)

With characteristics of suborder; 33 genera.

- *Alloiozona* Hsiung, 1930
- *Ampullacula* Hsiung, 1930
- *Amylophorus* Pereira & Almeida, 1942
- *Blepharocodon* Bundle, 1895
- *Blepharoconus* Gassovsky, 1919
- *Blepharomonas* Kopperi, 1937
- *Blepharoplanum* Kopperi, 1937
- *Blepharoposthium* Bundle, 1895
- *Blepharosphaera* Bundle, 1895
- *Blepharozoum* Gassovsky, 1919
- *Buetschlia* Schuberg, 1888
- *Buissonella* de Cunha & Muniz, 1925
- *Bundleia* da Cunha & Muniz, 1928
- *Cucurbella* Thurston & Grain, 1971
- *Didesmis* Fiorentini, 1890
- *Hemiprorodon* Strelkow, 1939
- *Holophryoides* Gassovsky, 1919
- *Holophryozoon* Jirovec, 1933
- *Hsiungella* Imai in Aescht, 2001
- *Kopperia* Corliss, 1960
- *Levanderella* Kopperi, 1937
- *Meiostoma* Sandon, 1941
- *Parabundleia* Imai & Ogimoto, 1983
- *Paraisotrichopsis* Gassovsky, 1919
- *Pingius* Hsiung, 1932
- *Plexobundleia* Kornilova, 2005
- *Polymorphella* Corliss, 1960
- *Prorodonopsis* Gassovsky, 1919
- *Protolutzia* da Cunha & Muniz, 1925
- *Pseudobuetschlia* Jirovec, 1933
- *Sciurula* Corliss, 1960
- *Sulcoarcus* Hsiung, 1935
- *Wolskana* Ito, Imai, Ogimoto, & Nakahara, 1996

Suborder Blepharocorythina Wolska, 1971 (syns. Apotrichina, Blepharocorythida, Pharyngotrichina)

Size, small; shape, ovoid, laterally flattened, with a prominent frontal lobe in some species, and distinctive corkscrew-like process in others; free-swimming; somatic ciliation markedly reduced, as tufts and bands; **presumed remnant of concrement vacuole present only as its overlying somatic kinetids**; oral area, apical or subapical, with non-retractable cilia; oral cilia inconspicuous, non-retractable, as extensions of somatic ciliature, forming two groups with presumed homologies to those of entodiniomorphines (i.e., the ventral vestibular kineties or vestibular polybrachykinety and the adoral polybrachykinety); macronucleus, globular to ellipsoid; micronucleus, present; contractile vacuole, present; cytoproct, likely in posterior; feeding (?); in terrestrial habitats, as endocommensals principally in the hindgut of horses, with a few species in elephants and cattle; one family.

Family BLEPHAROCORYTHIDAE Hsiung, 1929

With characteristics of the suborder; eight genera.

- *Blepharocorys* Bundle, 1895
- *Charonina* Strand, 1928

- *Charonnautes* Strelkow, 1939
- *Circodinium* Wolska, 1971
- *Ochoterenaia* Chavarria, 1933
- *Pararaabena* Wolska, 1968
- *Raabena* Wolska, 1967
- *Spirocorys* Wolska, 1969

Suborder Entodiniomorphina Reichenow in Doflein & Reichenow, 1929
(syns. Entodiniida *p.p.*, Entodiniina *p.p.*, Entodiniorida, Spirodiniina, *p.p.*, Syntricha *p.p.*)

Size, small to medium, rarely large; shape, typically laterally flattened; somatic ciliature greatly reduced, appearing in tufts, sometimes elongated as spiraled bands, often arranged as polybrachykineties, and functioning as syncilia; pellicle firm and thickened, often drawn out into spines; **prominent skeletal plates characteristic of many species, composed of polysaccharide reserves, such as - amylopectin granules or plaques**; oral area, apical to subapical, often retractable; **oral cilia often functioning as syncilia, of two parts, a prevestibular band in the peristomial region and a vestibular part(s)** sensu stricto; cytoproct distinct; in terrestrial habitats, widely found as endocommensals in mammal hosts, mainly in artiodactyls and perissodactyls, with species of the Family Troglodytellidae in anthropoid apes; ten families.

Family CYCLOPOSTHIIDAE Poche, 1913
(syns. Cycloposthiinae *p.p.*, Monoposthiinae *p.p.*, Prototapirellidae, Tripalmariidae)

Size, small to large; shape, elongated, often ovoid, with several genera having bizarre finger-like projections (e.g., *Arachnodiniella*, *Phalodinium*); free-swimming; **somatic cilia, essentially non-retractable, with from none to four caudal tufts or caudalia; skeletal plates, at least one large one, and up to four**; oral ciliature in adoral zone, retractable, with adoral polybrachykinety and dorsal perivestibular polybrachykinety; macronucleus, ellipsoid to elongate ellipsoid; micronucleus, present; contractile vacuole, one to several; cytoproct, present; feeding on bacteria and plant fibres; in terrestrial habitats as endocommensals in the cecum and colon of horses, zebras, rhinoceroses, and tapirs, and occasionally in elephants, capybaras, and hippopotami; 17 genera.

NOTE: Grain (1994) and others subdivide this family into three families and subfamilies, based primarily on the numbers of caudalia. We await confirmation of the significance of these traits using molecular genetic data.

- *Arachnodinella* van Hoven, Gilchrist & Hamilton-Attwell in Aescht, 2001
- *Bertolinella* Carpano, 1941
- *Bozasella* Buisson, 1923
- *Carinoposthium* Jankowski, 1980
- *Cycloposthium* Bundle, 1895
- *Dicycloposthium* Strelkow, 1939
- *Lavierella* Buisson, 1923
- *Monoposthium* Thurston & Noirot-Timothée, 1973
- *Paracycloposthium* Grain, 1994
- *Phalodinium* van Hoven, Gilchrist, & Hamilton-Attwell, 1987
- *Prototapirella* da Cunha, 1918
- *Rhabdothoracella* Aescht, 2001
- *Toxodinium* da Cunha, 1938
- *Tricaudalia* Buisson, 1923
- *Trifascicularia* Strelkow, 1931
- *Tripalmaria* Gassovsky, 1919
- *Triplumaria* Hoare, 1937

Family GILCHRISTIDAE Ito, Van Hoven, Miyazaki, & Imai, 2006

Size, medium to large; shape, ellipsoid, laterally flattened; free-swimming; **somatic cilia, non-retractable, as several equatorial and/or posterior bands**; skeletal plates, one or two; **oral ciliature of retractable adoral polybrachykinety, accompanied by paralabial kineties, and with vestibular polybrachykinety extending longitudinally into the oral cavity**; macronucleus, elongate; micronucleus, present; contractile vacuoles, dorsal, multiple; cytoproct in posterior; feeding (?); in colon of rhinoceros; two genera.

- *Digilchristia* Ito, Van Hoven, Miyazaki, & Imai, 2006*
- *Gilchristia* Ito, Van Hoven, Miyazaki, & Imai, 2006*

Family OPHRYOSCOLECIDAE Stein, 1859
(syn. Caloscolecinae *p.p.*, Cunhaiidae, Diplodiniinae *p.p.*, Entodiniidae, Entodiniina. Epidiniinae

p.p., Ophryoscolecinae *p.p.*, Opisthotrichinae *p.p.*)

Size, small to large; shape, ovoid to fusiform, more or less flattened; free-swimming; **retractable dorsal ciliary tuft, absent in *Entodinium*, may cover at least 1/3 of body perimeter**; skeletal plates commonly present; **oral ciliature of retractable adoral polybrachykinety, accompanied by paralabial kineties, with vestibular polybrachykinety extending longitudinally into the oral cavity**; macronucleus, typically elongate; micronucleus, present; contractile vacuole, present, variable in number; cytoproct, present; feeding on bacteria, plant detritus, and other ciliates; in terrestrial habitats as endocommensal in the rumen of artiodactylan ruminants, such as cattle, sheep, goats, deer *s.l.*, antelope, caribou, bison, buffalo, ox, and close relatives, and camels, and intestine of the guinea pig; 20 genera.

NOTE: Grain (1994) and others subdivide this family into subfamilies. We await confirmation of the significance of the distinguishing traits using molecular genetic data.

- *Anoplodinium* Dogiel, 1927
- *Caloscolex* Dogiel, 1926
- *Campylodinium* Jankowski, 1975
- *Cunhaia* Hasselmann, 1924
- *Diplodinium* Schuberg, 1888
- *Diploplastron* Kofoid & MacLennan, 1932
- *Elytroplastron* Kofoid & MacLennan, 1932
- *Endoralium* Eloff & van Hoven, 1980
- *Enoploplastron* Kofoid & MacLennan, 1932
- *Entodinium* Stein, 1859
- *Eodinium* Kofoid & MacLennan, 1932
- *Epidinium* Crawley, 1923
- *Epiplastron* Kofoid & MacLennan, 1933
- *Eremoplastron* Kofoid & MacLennan, 1932
- *Eudiplodinium* Dogiel, 1927
- *Metadinium* Awerinzew & Mutafowa, 1914
- *Ophryoscolex* Stein, 1859
- *Opisthotrichum* Buisson, 1923
- *Ostracodinium* Dogiel, 1927
- *Polyplastron* Dogiel, 1927

Family PARENTODINIIDAE Ito, Miyazaki & Imai, 2002

Size, small; shape, ovoid, laterally compressed, with longitudinal surface striations; free-swimming; somatic ciliature, non-retractable, and no caudalia; no skeleton; **oral ciliature as a retractable adoral polybrachykinety, completely encircling vestibular opening, with several vestibular kineties and a set of paralabial kineties to the right of the adoral synciliary group**; macronucleus, globular to ellipsoid; micronucleus, present; contractile vacuole, present; cytoproct, with synciliary tuft; feeding (?); in terrestrial habitats as endocommensals in the stomach of hippopotamus and the rumen of cattle; one genus.

- *Parentodinium* Thurston & Noirot-Timothée, 1973

Family POLYDINIELLIDAE Corliss, 1960 (for Polydiniidae)

Size, large; shape, fusiform, slightly laterally flattened; free-swimming; **somatic ciliature, non-retractable, as 4–12 "accessory ribbons" partially encircling body**; skeletal plates present, variable in number and size; **oral ciliature as an adoral zone of non-retractable synciliary tufts; vacuole with granular contents in a caudal lobe that bears cilia**; macronucleus, ellipsoid to elongate and twisted; micronucleus, present; contractile vacuoles, small, very numerous, arranged in transverse rows; cytoproct, present; feeding on bacteria, flagellates, plant debris, and other organic particles; in terrestrial habitats as endocommensals in the cecum and colon of elephants only; four genera.

- *Elephantophilus* Kofoid, 1935
- *Polydiniella* Corliss, 1960
- *Pterodiniella* Aescht, 2001
- *Thoracodinium* Latteur, 1958

Family PSEUDOENTODINIIDAE Wolska, 1986

Size, small; shape, ovoid, slightly flattened; free-swimming; skeletal plates, anterior, slat-like; somatic ciliature, absent; **oral ciliature, a retractable adoral zone of a single, broad adoral polybrachykinety with dorsal part divided into an anterior and posterior fragment and longitudinal files extending along the vestibular wall**; macronucleus, ellipsoid; micronucleus, present; contractile vacuole, present; cytoproct, present; feeding (?); in terrestrial habitats as endocommensals in the digestive tract of elephants; one genus.

- *Pseudoentodinium* Wolska, 1986

Family RHINOZETIDAE Van Hoven, Gilchrist, & Hamilton-Attwell, 1988

Size, small; shape, ovoid, flattened; free-swimming; **somatic synciliary tufts, non-retractable, occurring in three to five short bands on left and right body surfaces**; skeletal plates in varying numbers and sizes; oral ciliature as an adoral zone on a retractable cone; macronucleus, elongate band-form; micronucleus, present; contractile vacuoles, 2–6, between macronucleus and cell surface; cytoproct, present; feeding (?); **in terrestrial habitats in the cecum and colon of rhinoceros**; one genus.

NOTE: Information about the detailed structure of the adoral ciliature is needed for this family.

–*Rhinozeta* van Hoven, Gilchrist, & Hamilton-Attwell, 1988

Family SPIRODINIIDAE Strelkow, 1939
(syns. Ditoxidae, Triadiniidae)

Size, medium; shape, elongated to globose, often markedly laterally flattened; free-swimming; **somatic cilia as 2–4 non-retractable ribbons or bands, spiralling around body at different levels**; **no skeletal plates**; oral ciliature, non-retractable, in two bands, with adoral polybrachykinety and dorsal perivestibular polybrachykinety; macronucleus, elongate and band-form; micronucleus, present; contractile vacuole, one or two; cytoproct, present; feeding (?); in terrestrial habitats as endocommensals predominantly in the colon and cecum of horses; six genera.

- *Cochliatoxum* Gassovsky, 1919
- *Ditoxum* Gassovsky, 1919
- *Gassovskiella* Grain, 1994
- *Spirodinium* Fiorentini, 1890
- *Tetratoxum* Gassovsky, 1919
- *Triadinium* Fiorentini, 1890

Family TELAMODINIIDAE Latteur & Dufey, 1967
(for Telamodidae)

Size, medium; shape, elongate; free-swimming; **somatic cilia as five, non-retractable, "accessory ciliary ribbons" partially encircling body**; two or three skeletal plates present; oral ciliature of adoral zone, retractable; macronucleus, elongate; micronucleus, present; contractile vacuole, present, may be multiple; cytoproct, present; feeding (?); **in terrestrial habitats as endocommensals in the colon of the desert wart-hog**; three genera.

- *Megadinium* Latteur & Dufey, 1967
- *Telamodinium* Latteur & Dufey, 1967
- *Teratodinium* Latteur & Dufey, 1967

Family TROGLODYTELLIDAE Corliss, 1979

Size, medium to large; shape, ovoid, fusiform, laterally flattened; free-swimming; **somatic cilia as 3–5 non-retractable bands encircling body, essentially perpendicular to longitudinal axis**; skeletal plates, large, both dorsal and ventral, fusing to envelope anterior half of organism; **cell surface between ciliary bands divided by deep cortical grooves into elongated rectangles**; oral ciliature of adoral zone, retractable, beating as syncilia; macronucleus, L-shaped; micronucleus, present; contractile vacuole, multiple, in transverse rows; cytoproct, present; feeding on bacteria and detritus; **in terrestrial habitats as endocommensals in the colon of anthropoid apes only**; two genera.

- *Gorillophilus* Imai, Ikeda, Collet, & Bonhomme, 1991
- *Troglodytella* Brumpt & Joyeux, 1912

Order Macropodiniida order nov.
(syn. Reikostomatida *p.p.*)

Size, small to medium; shape, ovoid to elongate; free-swimming; somatic ciliation, holotrichous, but may be reduced to a dorsoventral groove in macropodiniids; extrusomes as somatic mucocysts; **oral cavity, anterior, a shallow to deep vestibulum lined by extensions of somatic kineties, and supported by nematodesmata arising from these kinetids**; stomatogenesis, telokinetal or cryptotelokinetal, possibly apokinetal; macronucleus, spheroid to elongate band-form; micronucleus, present; contractile vacuole, present; cytoproct, present; feeding on bacteria, starch, and in larger forms, other ciliates; **in terrestrial habitats as endocommensals in the forestomach of macropodid and vombatid marsupials;** three families.

NOTE: Cameron and O'Donoghue (2004a) have noted the strongly supported monophyly of representatives of these three families using small subunit rRNA gene sequences. Aside from their restriction to marsupial hosts, there are no strong and obvious morphological synapomorphies for the group. Thus, this could be called a "ribo-order".

Family AMYLOVORACIDAE Cameron & O'Donoghue, 2002

Size, small to medium; shape, ovoid, slightly flattened; free-swimming; **somatic ciliation, holotrichous, sometimes spiralling and/or separated into longitudinal bands of closely adjacent kineties by broad interkinetal ridges**; extrusomes as somatic mucocysts; oral ciliature as extensions of somatic kineties or in isolated vestibular fields; macronucleus, ellipsoid to elongate bandform; micronucleus, present; contractile vacuole, present; cytoproct, may be present; feeding mainly on starch and bacteria; in terrestrial habitats as endocommensals in the forestomach of macropodid and vombatid marsupials; three genera.

- *Amylovorax* Cameron & O'Donoghue, 2002*
- *Bandia* Cameron & O'Donoghue, 2002*
- *Bitricha* Cameron, O'Donoghue, & Adlard, 2000

Family MACROPODINIIDAE Dehority, 1996

Size, small to medium; shape, ovoid, flattened; free-swimming; **cortical alveoli divided by transverse grooves, elaborated into strikingly trapezoidal and parallelogram shapes, filled with dense material**; somatic cilia, holotrichous or restricted to kineties lying in a dorsoventral groove that encircles the body; extrusomes as somatic mucocysts; oral cavity, conical, bordered or lined by extensions of somatic kineties; macronucleus, globular to ellipsoid; micronucleus, present; contractile vacuole, present; cytoproct, present; feeding on bacteria, starch grains, and smaller ciliates; in terrestrial habitats as endocommensals in the stomach of macropodid marsupials; one genus and one genus *incertae sedis*.

- *Macropodinium* Dehority, 1996

Incertae sedis in Family Macropodiniidae

- *Megavestibulum* Cameron & O'Donoghue, 2003*

Family POLYCOSTIDAE Cameron & O'Donoghue, 2003

Size, small to medium; shape, stout ovoid; free-swimming; **somatic ciliation, holotrichous, with meridional kineties that can be separated by broad interkinetal ridges filled with many, small dense bodies (extrusomes?); oral cavity, a conical, round or flattened vestibulum in cross-section, lined by extensions of some somatic kineties only along right side**; macronucleus, globular to ellipsoid; micronucleus, present; contractile vacuole, one or more; cytoproct, present; feeding on bacteria and starch; in terrestrial habitats as endocommensals in the forestomach of macropodid marsupials; one genus.

- *Polycosta* Cameron & O'Donoghue, 2003*

Class **PHYLLOPHARYNGEA** de Puytorac et al., 1974
(syn. Homotricha *p.p.*)

Size, small to large; shape, extremely variable, globular to ellipsoid to bizarre spinous and tentacled forms; free-swimming and sessile or both, depending on the life cycle stages (e.g. free-swimming dispersal larval forms of sessile chonotrichs and suctorians); alveoli, generally well-developed, especially on non-ciliated surfaces; somatic kineties, typically arranged in two fields, which may be continuous over the body surface of ciliated stages; **somatic kinetids as monokinetids that each have a lateral kinetodesmal fibril, a reduced or absent transverse microtubular ribbon, usually accompanied by a left-directed transverse fiber, and a somewhat convergent postciliary ribbon extended posteriorly to accompany ribbons of more anterior monokinetids; ribbon-like subkinetal nematodesmata arise from somatic monokinetids, extending beneath kineties as subkinetal ribbons**, which in cyrtophorids and chonotrichs probably extend anteriorly and in rhynchodids and suctorians probably extend posteriorly; extrusomes vary with subclass (e.g., acmocysts in rhynchodians; haptocysts in suctorian tentacles); **oral region with radially arranged microtubular ribbons, the phyllae, supporting the cytopharynx**; "cytopharynx" may be restricted to a tentacle in the Suctoria or a tentacle-like tube in the Rhynchodia; stomatogenesis, mixokinetal or merotelokinetal; macronucleus, homomerous in rhynchodians and suctorians, and heteromerous in cyrtophorians and chonotrichs; conjugation, temporary or total, the latter may involve micro- and macroconjugants; micronucleus, single to many; contractile vacuoles, common; cytoproct, apparently absent in suctoria and many chonotrichs; feeding strategies, diverse, from algivorous and bacterivorous in cyrtophorians to carnivorous on other ciliates in suctorians; encystment, common; in marine, freshwater, and terrestrial habitats, distributed widely, with many suctorian species as epibionts on a wide diversity of aquatic invertebrates and some vertebrates, chonotrichs primarily restricted to the appendages of crustaceans, and rhynchodians typically as ectoparasites on invertebrates; four subclasses.

NOTE: Grell and Meister (1982a) argued for two lineages within this clade that has phyllae lining the cytopharynx. One clade included the Subclasses Cyrtophoria and Chonotrichia, which have a heteromerous macronucleus and subkinetal microtubules that extend anteriorly beneath the somatic kineties. The other clade included the Subclasses Rhynchodia and Suctoria, which have toxic "oral" extrusomes, acmocysts and haptocysts respectively, enclosed within an ingestatory tentacle(s), and have subkinetal microtubules that extend posteriorly beneath the somatic kineties. We have not recognized these two lineages in our classification, although preliminary molecular evidence suggests that chonotrichs arose from *within* the cyrtophorine clade with suctorians as a separate branch (Snoeyenbos-West, Cole, Campbell, Coats, & Katz, 2004).

Subclass Cyrtophoria Fauré-Fremiet in Corliss, 1956
(syns. Cyrtophorina *p.p.*, Cyrtohymenostomata *p.p.*, Hypostomata, Hypostomatida, Phyllopharyngia, Phyllopharyngidea *p.p.*)

Size, small to large; shape, frequently dorsoventrally flattened; free-swimming, may be sessile but usually not sedentary, often thigmotactic, sometimes using an adhesive organelle at the posterior end; alveoli, well-developed, revealed as a complex argyrome on the dorsal surface; **somatic ciliature predominantly restricted to ventral surface with preoral suture skewed far to left and with rightmost somatic kinety often divided into a dorsal kinetofragment and a midventral kinetofragment; oral ciliature typically composed of one preoral kinety and two circumoral kineties as several short double files of kinetosomes located anteriad to the cytostome; cytopharyngeal apparatus a complex cyrtos with phyllae surrounded by rod-shaped nematodesmata**; stomatogenesis, merotelokinetal, but involving extensive morphogenetic movements of preoral and circumoral kineties; macronucleus heteromerous; conjugation, temporary; feeding on bacteria and algae, with some parasitic species possibly ingesting epithelial tissues of host, such as fish; in marine and freshwater habitats, broadly distributed, mostly marine, with numerous free-living forms and many epibionts of which a few species are parasites on fish; two orders.

NOTE: The genus *Cyrtohymenostomata* Das, Chatterjee, and Ray, 1969, which may be placed in the Subclass Cyrtophoria, is an unavailable name according to Aescht (2001) and therefore the Family Cyrtohymenostomatidae Jankowski, 1980 would also be unavailable.

Order Chlamydodontida Deroux, 1976
(syns. Chilodonellida, Chilodonellina, Chlamydodontina, Gymnozoida *p.p.*)

Size, small to large; shape, typically dorsoventrally flattened, broad; free-swimming, but may attach to substrate by thigmotactic ventral somatic cilia; **somatic kineties typically ventrally disposed in two roughly equal fields, which may be separated midventrally (except in Family Kryoprorodontidae); without non-ciliated adhesive region or flexible podite**; six families.

Family CHILODONELLIDAE Deroux, 1970
(syn. Chilodontidae [for Odontohypotrichidae], Phascolodontinae *p.p.*)

Size, small to large; shape, width < 2/3 length, usually with pronounced anterior projection of body or "beak" extending to left; free-swimming (i.e., without lorica); **somatic ciliation with right ventral somatic kineties arching preorally to left into "beak" so that the anterior preoral kinetal arcs of all right ventral somatic kineties are continuous with the more posterior parts of those kineties**; oral ciliature, typically as one preoral and two circumoral kineties, but some variations; macronucleus, centric heteromerous, globular to ellipsoid; micronucleus, present; contractile vacuole, present, may be multiple; cytoproct (?); feeding on bacteria and microalgae, but parasitic forms may feed on host tissues; in marine, freshwater, and terrestrial habitats, free-living but some *Chilodonella* species as facultative parasites of fishes; seven genera and two genera *incertae sedis*.

- *Chilodonatella* Dragesco, 1966
- *Chilodonella* Strand, 1928
- *Phascolodon* Stein, 1859
- *Pseudochilodonopsis* Foissner, 1979
- *Talitrochilodon* Jankowski, 1980
- *Thigmogaster* Deroux, 1976
- *Trithigmostoma* Jankowski, 1967

Incertae sedis in Family Chilodonellidae

- *Odontochlamys* Certes, 1891
- *Phyllotrichum* Engelmann in Bütschli, 1889

Family CHITONELLIDAE Small & Lynn, 1985

Size, small; shape, ovoid to spheroid; **sedentary (?) in stalkless lorica, attached to substrate; somatic kineties as only two right kineties and about four left kineties; oral ciliature as only one circumoral kinety**; macronucleus, centric heteromerous, globular; micronucleus, present; contractile vacuole, present; cytoproct (?); feeding (?); in marine habitats, recorded only once from a salt marsh, kathrobic; one genus.

- *Chitonella* Small & Lynn, 1985

Family CHLAMYDODONTIDAE Stein, 1859
(syn. Chlamidodontidae)

Size, small to medium; shape, nearly ellipsoidal, with width >2/3 length; free-swimming; **dorsal and ventral surfaces separated by the "railroad track" groove, which is supported by regular cytoskeletal elements; ventral somatic kineties running from the right ventral body surface to dorsal right and anterior left surfaces; local region of thigmotactic cilia at posterior of ventral surface, but not developed as a non-ciliated adhesive region**; oral ciliature as preoral and two circumoral kineties; macronucleus, juxtaposed heteromeric, globular to ellipsoid; micronucleus, present; contractile vacuole, may be multiple; cytoproct (?); feeding typically on diatoms and filamentous algae; in marine habitats; three genera.

- *Chlamydodon* Ehrenberg, 1835
- *Cyrtophoron* Deroux, 1975
- *Lynchellodon* Jankowski, 1980

Family GASTRONAUTIDAE Deroux, 1994

Size, small; shape, ovoid, flattened; free-swimming; **oral opening, a large, elongate, and transverse groove, oriented across body axis so that some somatic kineties on the right side are broken into preoral and postoral fragments; oral ciliature, apparently as one kinety that encircles the perimeter of the large oral opening**; macronucleus, centric heteromerous, ellipsoid; micronucleus, present; contractile vacuole, present; cytoproct (?); feeding (?); in freshwater and terrestrial habitats, even as a commensal (?) in small freshwater mussels; two genera.

- *Gastronauta* Engelmann in Bütschli, 1889
- *Paragastronauta* Foissner, 2001

Family KRYOPRORODONTIDAE Alekperov & Mamajeva, 1992
(syn. Gymnozoonidae)

Size, small to medium; shape, circular in cross-section; free-swimming; **somatic kineties evenly disposed around the body, with several somatic kineties extending to encircle the apical cytostome; oral kineties as a series of small fragments accompanied by dikinetids**; macronucleus, juxtaposed heteromerous, ellipsoid; micronucleus, present; contractile vacuole (?); cytoproct (?); feeding (?); in marine habitats, planktonic and near sea ice; one genus.

- *Gymnozoum* Meunier, 1910

Family LYNCHELLIDAE Jankowski, 1968

Size, small to medium; shape, elongate to discoid, somewhat rounded anteriorly; free-swimming; dorsal and ventral surfaces may be separated by "railroad track" groove (see Family

Chlamydodontidae); **typically anterior preoral arcs of some-to-all right ventral somatic kineties not continuous with more posterior parts of those kineties, and left somatic kineties reduced to fewer than 7**; thigmotactic zone broad, posterior, with curious structureless protrusions in several species; **oral kineties, anterior to cytostome, typically more than three, of variable pattern, but often as flattened "Y", opened to the left**; tips of oral nematodesmata often toothed; macronucleus, juxtaposed heteromeric, ellipsoid; micronucleus, present; contractile vacuole, may be multiple; cytoproct (?); feeding (?); in marine and freshwater benthic habitats, often in sands; six genera and one genus *incertae sedis*.

- *Atopochilodon* Kahl, 1933
- *Chlamydonella* Petz, Song, & Wilbert, 1995
- *Chlamydonellopsis* Blatterer & Foissner, 1990
- *Coeloperix* Gong & Song, 2004*
- *Lynchella* Kahl in Jankowski, 1968
- *Wilbertella* Gong & Song, 2006*

Incertae sedis in Family Lynchellidae

- *Lophophorina* Penard, 1922

Order Dysteriida Deroux, 1976
(syns. Dysteriina, Hartmannulina)

Size, small to large; shape, ovoid to elongate ovoid; free-swimming, but often temporarily attached; **ventral cilia not thigmotactic, but ciliate attached to substrate by non-ciliated adhesive region or by flexible podite (except** *Atelepithites*); macronucleus, juxtaposed heteromerous; widespread and numerous, mainly marine, but some ectosymbiotic forms with members of the Family Kyaroikeidae exclusively on cetaceans; four families.

Family DYSTERIIDAE Claparède & Lachmann, 1858
(syns. Erviliidae, Trochiliidae)

Size small; shape, ovoid to almost rectangular, may be conspicuously laterally compressed; free-swimming; **somatic ciliature, typically reduced, with left ventral somatic kineties as midventral postoral field, typically separated from an anterior preoral field; flexible podite used for attachment**; oral ciliature as two or more small kinetofragments disposed around the cytostome; **nematodesmata of cyrtos reduced to six or fewer, with cytopharyngeal capitula or "teeth" often prominent**; macronucleus, juxtaposed heteromerous, globular to ellipsoid; micronucleus, present; contractile vacuole, present, may be multiple; cytoproct (?); feeding on bacteria and microalgae; in marine and freshwater habitats, widely distributed, mainly in marine habitats, and frequently as symphorionts; seven genera.

- *Agnathodysteria* Deroux, 1977
- *Dysteria* Huxley, 1857
- *Hartmannulopsis* Deroux & Dragesco, 1968
- *Mirodysteria* Kahl, 1933
- *Schedotrochilia* Deroux, 1977
- *Orthotrochilia* Song, 2003*
- *Trochilia* Dujardin, 1841

Family HARTMANNULIDAE Poche, 1913
(for Onychodactylidae; syns. Aegyrianidae, Aegyrianinae, Allosphaeriidae, Trichopodiellidae, Trochilioididae, Trochilioidinae)

Size, small to medium; shape, ovoid, flattened; free-swimming, but may attach to substrate, sometimes making a "byssal" filament; **somatic ciliature with left ventral somatic kineties, which may be quite short, as continuous field (i.e., not fragmented); ventral kineties behind podite (i.e. transpodial kineties) with more closely packed kinetosomes**; oral ciliature, variable, ranging from a single circumoral kinetofragment to the typical preoral and two circumoral kinetofragments; nematodesmata of cyrtos, ranging from thin and inconspicuous to prominent, typically many; macronucleus, juxtaposed heteromerous, ellipsoid; micronucleus, present; contractile vacuole, may be multiple; cytoproct (?); feeding on bacteria, diatoms, and other microalgae; in marine habitats, free-living but *Brooklynella* harmful as gill parasite of marine fishes; eleven genera.

- *Aegyriana* Song & Wilbert, 2002*
- *Allosphaerium* Kidder & Summers, 1935
- *Brooklynella* Lom & Nigrelli, 1970
- *Chlamydonyx* Deroux, 1977
- *Hartmannula* Poche, 1913
- *Horocontus* Deroux, 1977
- *Microxysma* Deroux, 1977
- *Paratrochilia* Kahl, 1933

- *Sigmocineta* Jankowski, 1967
- *Trichopodiella* Corliss, 1960
- *Trochilioides* Kahl, 1931 (*nomen nudum*)

Incertae sedis in Subclass Cyrtophoria

- *Dysterioides* Matthes, 1950

Family KYAROIKEIDAE Sniezek & Coats, 1996

Size, medium; shape, elongate, ovoid, circular in cross-section; free-swimming; **somatic ciliation, essentially holotrichous, but with left-ventral non-ciliated strip at whose anterior end is the left somatic field as four kinetal fragments anteriorly and midventrally; adhesive region at posterior tip of cell; oral ciliature as one preoral and two circumoral kinetofragments in a deep oral cavity**; macronucleus, juxtaposed heteromerous, elongate ellipsoid; micronucleus, present; contractile vacuole (?); cytoproct, present; feeding on organic detritus and sometimes epithelial cells of host; **in marine habitats, known so far only as ectosymbionts in the nasal cavities of cetaceans, and collected from the mucus discharged from the blowholes of living or dead hosts**; two genera.

- *Kyaroikeus* Sniezek, Coats, & Small, 1995
- *Planilamina* Ma, Overstreet, Sniezek, Solangi, & Coats, 2006*

Family PLESIOTRICHOPIDAE Deroux, 1976

Size, small to medium; shape, ovoid to elongate, somewhat dorsoventrally compressed; free-swimming, but may temporarily attach by podial filament; somatic ciliation with right ventral kineties arcing preorally to left, and a field of left ventral kineties abutting at a preoral suture system; oral ciliature varying from a preoral and two circumoral kinetofragments to multiple circumoral kinetofragments; **ventral adhesive region, glandular, non-ciliated (except in *Atelepithites*), but may secrete podial filament**; macronucleus, juxtaposed heteromerous, ellipsoid; micronucleus, present; contractile vacuole, may be multiple; cytoproct (?); feeding on bacteria and microalgae; in marine habitats; five genera.

- *Atelepithites* Deroux, 1976
- *Parachilodonella* Dragesco, 1966
- *Pithites* Deroux & Dragesco, 1968
- *Plesiotrichopus* Fauré-Fremiet, 1965
- *Trochochilodon* Deroux, 1976

Subclass Chonotrichia Wallengren, 1895
(syns. Phyllopharyngidea *p.p.*, Scaiotricha *p.p.*)

Size, small to medium; shape, often vase-shaped, with pellicle quite rigid and frequently adorned with collar, lobes, and/or spines; sessile and sedentary, except as dispersive larval forms; posterior adhesive organelle or podite produces stalk or peduncle, always non-contractile; **somatic kineties only on walls of perioral funnel or cone-shaped region, which may be flared, or compressed**; **somatic kineties in two fields – a right field whose kineties are typically arrayed parallel to the margin of the cone and a left field whose kineties are typically arranged obliquely to the margin of the cone**; oral cilia, apparently absent or only as several inverted kineties next to the cytostome; **cytopharyngeal apparatus with phyllae, but no nematodesmata**; reproduction solely by unequal division or budding, within a crypt or "marsupium" in one order; polygemmy may occur on death of host; **migratory larval forms or tomites bear an adhesive gland posteriorly and two fields of cilia on deep concave ventral surface or gutter, reminiscent of their putative ancestry among the cyrtophorine-like forms**; conjugants, of unequal size, invariably undergoing total fusion; macronucleus, heteromerous; contractile vacuoles, when present, connected to body surface by an excretory canal; cytoproct, may be present; microphagous, typically feeding on bacteria and food particles derived from host; in marine, brackish, and freshwater habitats as ectosymbionts on the gills, mouthparts, and/or other appendages of crustaceans, principally amphipods, isopods, and copepods, but also decapods, nebaliids, and others, with one species attaching to a marine alga; two orders.

NOTE: The monograph of Jankowski (1973b) still stands as the major modern treatise on this group. Batisse (1994a) has provided a revision, including a new order (i.e. Order Chilodochonida), which we do not recognize. Molecular genetic evidence may justify its distinctness. We have maintained the divisions based on the kind of budding (e.g., external budding and internal budding) as proposed by Jankowski (1973b).

Order Exogemmida Jankowski, 1972
(syn. Chilodochonida *p.p.*, Lobochonina *p.p.*, Spirochonina *p.p.*)

Shape, typically long and cylindrical, and typically with a well-developed collar (except Family Chilodochonidae); spines absent or poorly developed; usual attachment by undistinguished peduncle (rather than "true" stalk, except in Family Chilodochonidae); **a few to several tomites or buds produced by external budding; macronucleus, heteromerous, with orthomere directed apically towards funnel**; six families.

Family CHILODOCHONIDAE Wallengren, 1895

Size, medium; **shape, ovoid or pyriform, massive, not flattened; apical end large, cylindrical, flaring slightly; collar indistinct or absent**; no spines; sessile; cortex, markedly thickened; **long, wide, solid stalk, rather than peduncle typical of order**; somatic ciliature as two subparallel ciliary fields; oral ciliature as a circumoral kinety bordering the left side of the cytostome; macronucleus, heteromerous, elongate; micronucleus, present; contractile vacuole, absent; cytoproct, absent; feeding (?); in marine habitats on the mouthparts of decapods in littoral and sublittoral habitats; two genera.

- *Chilodochona* Wallengren, 1895
- *Vasichona* Jankowski, 1972

Family FILICHONIDAE Jankowski, 1973

Size, medium; shape, cylindrical or bottle-shaped, not flattened, with simple apical end, conical and unadorned; sessile; **collar distinct and markedly elongate, with spine-like processes on hypocollar between collar and body**; peduncle, low and broad; macronucleus, heteromerous, ellipsoid; micronucleus, present; contractile vacuole, absent; cytoproct, absent; feeding (?); in marine habitats on isopods; two genera.

- *Aurichona* Jankowsi, 1973
- *Filichona* Jankowski, 1973

Family HELIOCHONIDAE Jankowski, 1972

Size, small to medium; shape, bottle-like, elongate to sac-like, not flattened; sessile; **cone a simple funnel that may have spines of different structure on each side of cone**; collar, distinct, usually short; **somatic ciliature as a left field and larger right field, which is subdivided into an upper horizontal component of usually <6 kineties and a lower oblique component**; peduncle, short; macronucleus, heteromerous, ellipsoid; micronucleus, present; contractile vacuole, may be present; cytoproct, absent; feeding (?); in marine or brackish habitats, particularly on gammarid amphipods; two genera.

- *Heliochona* Plate, 1889
- *Heterochona* Jankowski, 1972

Family LOBOCHONIDAE Jankowski, 1967

Size, medium; shape, elongate, bottle-like, not flattened; sessile; **apical end simple, conical, slightly flared, often with two dorsal lobes; cone a simple funnel**; collar, distinct; **ciliation as a left field and a larger right field, which is divided, having >6 kineties in each part**; peduncle, short, distinct; macronucleus, heteromerous, ellipsoid; micronucleus, present; contractile vacuole, absent; cytoproct, absent; feeding (?); only in marine or brackish habitats – but with very wide distribution – on isopods and amphipods, with one unique species on an alga; five genera.

- *Lobochona* Dons, 1941
- *Oenophorachona* Matsudo & Mohr, 1968
- *Physochona* Batisse & Crumeyrolle, 1988
- *Segmentochona* Jankowski, 1989
- *Toxochona* Jankowski, 1972

Family PHYLLOCHONIDAE Jankowski, 1972

Size, small; **shape, leaf-like, not elongate, flattened dorsoventrally, contorted; cone with leaf-shaped preoral outgrowths**; collar absent; ciliation as two fields; sessile; **peduncle in form of wide disc; macronucleus, heteromerous, massive**; micronucleus, present; contractile vacuole, absent; cytoproct, absent; feeding (?); in marine habitats on isopods; one genus.

- *Phyllochona* Jankowski, 1972

Family SPIROCHONIDAE Stein, 1854

Size, small to medium; shape, vase-like, elongate, typically not flattened; **apical end flared,**

with dorsal wall multiply-spiraled in characteristic helical coils so that the margin of the cone spirals at least twice around the central column, but may spiral up to a half a dozen full turns in some species; no spines; collar, short, broad; ciliation with posterior part of the right field covering the spiralling cone and the left field at the base of the cone; sessile; peduncle, low and broad; macronucleus, heteromerous, ellipsoid; micronucleus, present; contractile vacuole, possibly present; cytoproct, present; feeding (?); typically in freshwater habitats, usually on the gills of gammarid amphipods; three genera.

- *Cavichona* Jankowski, 1973
- *Serpentichona* Jankowski, 1973
- *Spirochona* Stein, 1852

Order Cryptogemmida Jankowski, 1975
(syns. Endogemmina, Dorsofragmina + Ventrofragmina)

Size, small; shape, often flattened, leaf-like, and angular; spines common and of several types; collar, reduced; stalk, typically present, of varying length; **internal budding, with up to eight tomites produced in a crypt or marsupium; macronucleus, heteromerous, with orthomere directed antapically away from funnel**; in marine habitats, occurring solely on littoral and open ocean crustaceans (i.e. amphipods, copepods, cyamids, nebaliids), including crustacean epibionts of whales; six families.

Family ACTINICHONIDAE Jankowski, 1973

Size, small to medium; shape, sac-like, usually flattened; cortex, often thickened; **apical end conical, not flattened, sometimes with a fold, and with conspicuous spines in some species; cone rotated 90° to right, relative to body and point of attachment**; collar may be elongate; ciliation with left field considerably reduced; sessile; peduncle, present, rather than stalk, with broad part of body often closely applied to substrate; crypt of varying size; macronucleus, heteromerous, ellipsoid; micronucleus, present; contractile vacuole, absent; cytoproct, absent; feeding (?); in marine habitats, exclusively on nebaliids; six genera.

- *Actinichona* Jankowski, 1973
- *Carinichona* Jankowski, 1973
- *Crassichona* Jankowski, 1973
- *Cristichona* Jankowski, 1973
- *Kentrochonopsis* Doflein, 1897
- *Rhizochona* Jankowski, 1973

Family ECHINICHONIDAE Jankowski, 1973

Size, small to medium; shape, rhombic or spindle-like, markedly flattened dorsoventrally; cone flattened, not rotated; **cone with smooth wall and small teeth on its margins**; collar distinct, narrow, low; **ciliation with long and narrow right field and very reduced left field**; sessile; stalk, quite long in some species; **crypt, very deep and broad**; macronucleus, heteromerous, ellipsoid; micronucleus, present; contractile vacuole, absent; cytoproct, absent; feeding (?); in marine habitats, on nebaliids; three genera.

- *Coronochona* Jankowski, 1973
- *Echinichona* Jankowski, 1973
- *Eurychona* Jankowski, 1973

Family INVERSOCHONIDAE Jankowski, 1973
(syn. Pleochonidae)

Size, small to medium; shape, sometimes elongate, flattened dorsoventrally; very heavy, well-developed body spines in some species; **apical end very broad, flattened, usually simple, but occasionally with a few spines**; cone flattened, not rotated; collar, distinct; **ciliation with left field larger than right field, which may be subdivided into two components**; sessile; peduncle, exceedingly short; **crypt relatively shallow**; macronucleus, heteromerous, elongate; micronucleus, present; contractile vacuole, absent; cytoproct, absent; feeding (?); in marine habitats, on nebaliids; five genera.

- *Ceratochona* Jankowski, 1973
- *Chonosaurus* Jankowski, 1973
- *Inversochona* Jankowski, 1973
- *Kentrochona* Rompel, 1894
- *Pleochona* Jankowski, 1973

Family ISOCHONIDAE Jankowski, 1973

Size, medium; shape, cylindrical, elongate, not flattened; **cone, rounded, simple, funnel-shaped, rather small and undistinguished, and in line**

with main axis of long body; collar, short; ciliation with right field not subdivided and left field relatively large; sessile; stalk, sometimes long; crypt of moderate size; macronucleus, heteromerous, ellipsoid; micronucleus, present; contractile vacuole, absent; cytoproct, absent; feeding (?); in marine habitats, with very wide distribution on appendages or shell of amphipods, nebaliids, and cyamids, including the "whale-lice" found on species of several genera of whales from various oceans; five genera

- *Cyamichona* Jankowski, 1971
- *Inermichona* Jankowski, 1971
- *Isochona* Jankowski, 1973
- *Thalassochona* Jankowski, 1971
- *Trichochona* Mohr, 1948

Family ISOCHONOPSIDAE Batisse & Crumeyrolle, 1988

Size, small to medium; shape, cylindrical, elongate; cone, rounded, funnel-shaped, in line with main axis of long body; **cone margin indented by flexible folds that are able to close the opening to the oral region**; collar, short; ciliation with left field and a right field, which is subdivided into an horizontal upper band and an oblique lower band; sessile; peduncle, short; macronucleus, heteromerous, ellipsoid; micronucleus, present; contractile vacuole, absent; cytoproct, connected by tube to the cell surface; feeding (?); in marine habitats, on the periopods of copepods; one genus.

- *Isochonopsis* Batisse & Crumeyrolle, 1988

Family STYLOCHONIDAE Mohr, 1948

Size, small to medium; shape, triangular or rhomboid, leaf-like, markedly flattened dorsoventrally, spines, may be large, elongate or may be as rows of papillae; cone, flattened, not rotated; **conal margin often spiny, with pockets and folds in conal wall**; collar, very short; ciliation with larger right field and a left field that may be reduced to an almost vertical band; sessile; stalk, of varying length, sometimes unusually long; **crypt, often very deep**; macronucleus, heteromerous, ellipsoid; micronucleus, present; contractile vacuole, absent; cytoproct, absent; feeding (?); in marine habitats, on nebaliids; ten genera.

- *Armichona* Jankowski, 1973
- *Ctenochona* Jankowski, 1973
- *Dentichona* Jankowski, 1973
- *Eriochona* Jankowski, 1973
- *Flectichona* Jankowski, 1973
- *Oxychonina* Corliss, 1979
- *Paraoxychona* Jankowski, 1973
- *Pterochona* Jankowski, 1973
- *Spinichona* Jankowski, 1973
- *Stylochona* Kent, 1881

Subclass Rhynchodia Chatton & Lwoff, 1939
(syns. Rhynchodea *p.p.*, Toxistomia *p.p.*)

Size, small to rarely medium; shape, ovoid, somewhat flattened, typically with pointed anterior end; free-swimming, but if parasitic, typically attached to host tissue; adult forms, either devoid of somatic ciliature or with it mostly restricted to an anteroventral thigmotactic field; **oral region not bounded by oral kinetal structures; oral apparatus a suctorial tube supported only by phyllae; oral extrusomes, as toxic (?) acmocysts or haptotrichocysts**; reproduction, isotomic fission or often by budding; larval forms typically with two ciliated fields; macronucleus, homomerous; micronucleus, often large, sometimes multiple; predators of other ciliates, especially suctorians and peritrichs, or parasitic (?) on gills or mouthparts of diverse invertebrates in marine and freshwater habitats, but most often on gills of marine bivalve molluscs; two orders.

NOTE: The classic monographs on this group are by Chatton and Lwoff (1949, 1950). Raabe (1970b) provided the last major taxonomic treatment.

Order Hypocomatida Deroux, 1976
(syns. Hypocomatina, Hypocomida, Hypocomina + Macrostomatina)

Size, small; shape, dorsoventrally flattened; **somatic kineties, essentially restricted to the ventral surface with a short anterio-lateral left kinety, a presumed homologue of the dorsal right kinetofragment of cyrtophorines; posterior adhesive region bounded by somatic kineties in right-ventral pit or fosette**; oral ciliature, absent or reduced to a few pericytostomal kinetosomes; **macronucleus, homomerous, band-like**;

micronucleus, present; contractile vacuole, present; cytoproct (?); predators of peritrichs and suctorians or on tissues (?) of host; in marine habitats as endosymbionts in wide range of hosts, such as ascidians, barnacles, brittle stars, and tunicates; one family.

Family HYPOCOMIDAE Bütschli, 1889
(syn. Crateristomatidae)

With characteristics of order; five genera.

- *Crateristoma* Jankowski, 1967
- *Harmocoma* Jankowski, 1980
- *Hypocoma* Gruber, 1884
- *Parahypocoma* Chatton & Lwoff, 1939
- *Rhynchocoma* Jankowski, 1975

Order Rhynchodida Chatton & Lwoff, 1939
(syns. Ancistrocomina, Rhynchodina, Sphenophryina)

Size, small to medium; shape, variable; free-swimming, but typically attached to the host by the oral region; **somatic kineties, sometimes with non-ciliated kinetosomes, typically organized in a thigmotactic field, which may extend to cover the entire body or which may be divided in two, leaving a large part of the cell surface bare**; **no posterior adhesive region**; macronucleus, variably shaped, typically not in a band-form; parasites of the gills of invertebrates, commonly bivalve molluscs; two families.

Family ANCISTROCOMIDAE Chatton & Lwoff, 1939
(syns. Ancistrocominae, Cepedellidae, Hypocomellinae, Hypocomidinae)

Size, small to medium; shape, typically pear- or banana-like with a pointed anterior end; free-swimming, but typically attached to host; **somatic kineties, with thigmotactic cilia more or less developed, at least near the anterior end, tending to reduction to a small anterior thigmotactic ventral field; with apical sucker**; oral ciliation, absent; macronucleus, globular to ellipsoid; micronucleus, present; contractile vacuole, may be present; cytoproct (?); feeding on cell contents of host tissues; in marine and freshwater habitats as parasites of invertebrates, such as polychaetes, a phoronid, and others, but found principally in the mantle cavity of molluscs; 21 genera and one genus *incertae sedis*.

- *Ancistrocoma* Chatton & Lwoff, 1926
- *Anisocomides* Chatton & Lwoff, 1950
- *Colligocineta* Kozloff, 1965
- *Crebricoma* Kozloff, 1946
- *Enerthecoma* Jarocki, 1935
- *Goniocoma* Chatton & Lwoff, 1950
- *Heterocinetopsis* Jarocki, 1935
- *Holocoma* Chatton & Lwoff, 1950
- *Hypocomagalma* Jarocki & Raabe, 1932
- *Hypocomatidium* Jarocki & Raabe, 1932
- *Hypocomella* Chatton & Lwoff, 1924
- *Hypocomides* Chatton & Lwoff, 1922
- *Hypocomidium* Raabe, 1938
- *Hypocomina* Chatton & Lwoff, 1924
- *Ignotocoma* Kozloff, 1961
- *Insignicoma* Kozloff, 1946
- *Isocomides* Chatton & Lwoff, 1950
- *Kozloffiella* Raabe, 1970
- *Raabella* Chatton & Lwoff, 1950
- *Stegotricha* Bower & Meyer, 1993
- *Syringopharynx* Collin, 1915

Incertae sedis in Family Ancistrocomidae

- *Cepedella* Poyarkoff, 1909

Family SPHENOPHRYIDAE Chatton & Lwoff, 1921
(syns. Gargariidae, Lwoffidae, Pelecyophyridae)

Size, small to medium; shape, ovoid to elongate, flattened; except in one species, **adult form or trophont unciliated, but with an infraciliature in two fields that may diverge from a central (= posterior) apex towards the ends of the body; attached to host by adhesive "sole" and short tentacle**; reproduction, isotomic fission or by budding; larval form, typically ciliated with kinetal pattern reminiscent of ancistrocomids; macronucleus, globular to elongate band-form; micronucleus, present; conjugation, often occurring epidemically; contractile vacuole, present; cytoproct (?); feeding on cell contents of host tissues; in marine and freshwater habitats as parasites in the mantle cavities of bivalve molluscs; three genera.

- *Gargarius* Chatton & Lwoff, 1934
- *Pelecyophrya* Chatton & Lwoff, 1922

- *Sphenophrya* Chatton & Lwoff, 1921

Incertae sedis in Order Rhynchodida

- *Lwoffia* Kozloff, 1955

Subclass Suctoria Claparède & Lachmann, 1858 (syns. Acinet[e], Acinetaria, Acinetina, Acinet[o]idea, Actinifera, Actinosuctorifera, Atricha, Dystricha, Suctorasina, Suctorea, Suctoriae, Suctorifera, Suctoriorida, Tentaculifer[id]a, Tentaculiferiae, Toxistomia *p.p.*)

Size, small to large; shape, variable, from simple spheroid to flattened discs to complex branching forms; polymorphic, with free-swimming, typically ciliated larval form and typically sessile, adult form, usually non-ciliated, although with an infraciliature; alveoli, well-developed, underlain by a thick epiplasm; extrusomes as toxic "oral" haptocysts in tips of suctorial tentacles or arrayed along the length of prehensile tentacles; **oral structures as one to many multiple, rarely none, ingestatory suctorial tentacles, short (e.g., *Cyathodinium*, *Phalacrocleptes*) or long and extensible (e.g., *Rhyncheta*, *Rhynchophrya*), usually supported by an outer ring of microtubules and an inner set of microtubular ribbons (= the presumed phyllae) with extrusomes as haptocysts at the tips**; stalk, often present, always non-contractile, of varying length and produced by the scopuloid; migratory motile ciliated larval form or swarmer, produced by some mode of budding, but typically bearing neither tentacles nor stalk; macronucleus, homomerous; conjugation of different kinds, but frequently total with unequal conjugants; contractile vacuole, present; cytoproct, absent; cyst, often present; feeding primarily on other ciliates, but some species parasites of other eukaryotes; in marine, freshwater, and rarely terrestrial habitats, widespread, predominantly as ectosymbionts on diverse invertebrates, but some as endocommensals in hosts ranging from other ciliates to vertebrates; three orders.

NOTE: There is as yet no strong consensus on the evolutionary diversification of the suctorians. We have remained conservative, and tried to assign taxa to the included orders based on the modes of budding proposed by Collin (1912). Kormos and Kormos (1957a) have proposed a more complex classification of budding, which Batisse (1994) has partly followed. Revisionary monographs have also been published by Jankowski (1981), Curds (1985a, 1985b, 1985c, 1986, 1987), and Matthes (1988). Dovgal (2002) has undertaken an extensive cladistic analysis and discussed the status of taxa at the generic level and above. In the main, we have followed Dovgal (2002) for taxonomy within the group, and strongly recommend this monograph as a starting point for future taxonomic investigations. However, we do not agree with Dovgal (2002) on two major points. First, we do not recognize the class status of the suctorians. Second, we have not recognized the vermigenids as a separate group, and instead placed these families within the Order Exogenida, especially considering that Dovgal (2002) suggested that vermigenids may have been derived from exogenid ancestors. While a comprehensive analysis using gene sequences may prove one of these schemes most appropriate, the first small subunit rRNA gene sequences suggest that Collin's (1912) system may have validity (Snoeyenbos-West et al., 2004).

Order Exogenida Collin, 1912 (syns. Allantosomatida *p.p.*, Asteriferina, Dendrosomidida *p.p.*, Ephelophagina, Ephelotida, Ephelotina, Exogenea, Exotropida, Metacinetida *p.p.*, Nemertodendrina *p.p.*, Oligostomatida *p.p.*, Ophryocephalida *p.p.*, Ophryodendrida, Ophryodendrina, Paracinetida, Paracinetina, Phalacrocleptida *p.p.*, Podophryida, Podophryina, Spelaeophryida, Spelaeophryina, Stylostomatina *p.p.*, Thecacinetina, Tomogenea, Urnulida *p.p.*, Vermigemmida, Vermigenea, Vermigenia)

Size, small to large; shape, diverse; often stalked and loricate; tentacles borne on actinophores in some species, and others with prehensile as well as suctorial tentacles; **exogenous budding, most often monogemmic, but polygemmic in some species, or by binary fission with no appreciable invagination of parental cortex**; small permanent field of non-ciliferous kinetosomes in vicinity of contractile vacuole; migratory larval form typically large or long, the former with complex ventral ciliature, derived from the parental kinetosomal field, but some of the longer larvae practically devoid of cilia, vermiform, and incapable of swimming; majority marine, typically solitary forms, and free-living or ectocommensal; 17 families.

Family ALLANTOSOMATIDAE Jankowski, 1967

Size, small to medium; **trophont, elongated, cylindroid; tentacles, capitate or rod-like, in fascicles or rows, at the poles of the body or on actinophores, or evenly distributed**; reproduction by binary fission; macronucleus, globular to ellipsoid; micronucleus, present; contractile vacuole, present; **in terrestrial habitats as endosymbionts in the digestive tracts of mammals, such as horses, elephants, and rhinoceros**; five genera.

- *Allantosoma* Gassovsky, 1919
- *Allantoxena* Jankowski, 1978
- *Arcosoma* Jankowski, 1967
- *Strelkowella* Kornilova, 2004*
- *Vanhovenia* Dovgal, 2002*

Family DENTACINETIDAE Batisse, 1992

Size, small to medium; **trophonts, elongate ovoid to pyramidal, with longitudinal cortical ribs**; with stalk, that sometimes extends over body as a pseudolorica; **tentacles, clavate and agile, in single, centroapical fascicle, conspicuously folded on retraction; swarmers, vermiform with a long terminal neck, bearing a lozenge-like, apical adhesive organelle for exploration prior to attachment by the "posterior" scopuloid**; macronucleus, ellipsoid; micronucleus, 1–3; contractile vacuole, present; in marine habitats as ectocommensals on harpacticoid copepods; two genera.

- *Dentacineta* Jankowski, 1978
- *Pleurophryodendron* Jankowski, 1978

Family DENDROSOMIDIDAE Jankowski, 1978

Size, medium; trophonts with ramified body; **tentacles, capitate, in fascicles or rows on well-developed actinophores or branches of the body**; swarmers, vermiform; macronucleus, globular to ribbon-like and ramified; micronuclei, numerous; contractile vacuole, present; in marine habitats as ectocommensals on crustaceans; four genera.

- *Asterifer* Jankowski, 1967 (subj. syn. *Ophryodendron*)
- *Dendrosomides* Collin, 1905
- *Leboransia* Dovgal, 2002*
- *Rondosomides* Jankowski, 1981

Family EPHELOTIDAE Kent, 1882
(syn. Hemiophryidae, Ophryocephalidae, Tunicophryidae)

Size, large; trophonts, truncate-spherical; some species loricate and stalked; **tentacles, of two kinds – shorter, extensible, feeding tentacles with flat tips and longer, pointed, non-feeding, prehensile ones, both bearing haptocysts**; swarmers, ellipsoidal and flattened, with ciliary field horseshoe-shaped; swarmers, produced synchronously and multiply by polyexogemmy; macronucleus, usually ramified, crown-like; micronuclei, numerous; contractile vacuole, present; in marine habitats as ectocommensals on various marine invertebrates; eight genera.

- *Ephelota* Wright, 1858
- *Metephelota* Willis, 1945
- *Ophiurephelota* Jankowski, 1981
- *Ophryocephalus* Wailes, 1925
- *Podocyathus* Kent, 1882
- *Shellephelota* Jankowski, 1981
- *Thaumatophrya* Collin, 1912
- *Tunicophrya* Jankowski, 1973

Family LECANOPHRYIDAE Jankowski, 1973

Size, small to medium; trophont, goblet-shaped or laterally flattened; pellicle, thick, girdle-like; stalked; **tentacles, capitate, apical, arranged in rows or in fascicles on actinophores**; **swarmer with invaginated gutter**; macronucleus, ellipsoid; micronucleus, present; contractile vacuole, present; in brackish and freshwater habitats as ectocommensals on the antennules of harpacticoid copepods; two genera.

- *Lecanophrya* Jankowski, 1994
- *Lecanophryella* Dovgal, 1985

Family METACINETIDAE Bütschli, 1889
(syns. Beckmaniidae, Urnulidae)

Size, small; **trophont, spheroid, not basally attached to lorica, which has several radial slits in the distal half, splitting it into triangular valves**; possibly stalked; tentacles, capitate, single or numerous, arranged in fascicles or rows, and extending out through slits in the lorica; reproduction by semi-circumvaginative budding with a lateral protomite; swarmer, ovoid, with spiral kineties; macronucleus, ellipsoid; micronucleus, present; contractile vacuole, present; in marine and freshwater habitats, free-living and sometimes as ectocommensals on aquatic invertebrates or as parasites on peritrichs and other suctorians; two genera.

- *Metacineta* Bütschli, 1889
- *Urnula* Claparède & Lachmann, 1857

Family MANUELOPHRYIDAE Dovgal, 2002

Size, small; **trophont, spherical or sac-like, attached to host by a single rod-like tentacle or by a basal protuberance of the stylotheca**; some forms loricate; reproduction by a lateral semi-circumvaginative budding; macronucleus, globular; micronucleus, present; contractile vacuole, present; **in marine and freshwater habitats as ectoparasites of sessile ciliates**; three genera.

- *Manuelophrya* Matthes in Jankowski, 1997
- *Mistarcon* Jankowski, 1997
- *Pseudogemmides* Kormos, 1935

Family OPHRYODENDRIDAE Stein, 1867
(syns. Corethriidae, Crevicometidae, Loricodendridae, Nemertodendr(on)idea, Stylostom(at)idae)

Size, large; **trophonts of somewhat baggy or irregular shape, broadly attached to substratum**; with or without definitive stalk; some loricate forms; **tentacles, rod-like or ramified, in fascicles on one or more prominent, extensible branches, often called trunks or actinophores**; swarmers, vermiform; macronucleus, extensively ramified; micronuclei, often numerous; contractile vacuole, present; having a complex endoplasmic canal network containing endosymbiotic bacteria; in marine habitats as ectocommensals on hydrozoans and crustaceans; seven genera.

- *Corethria* Wright, 1859 (subj. syn. *Ophryodendron*)
- *Crevicometes* Jankowski, 1981
- *Loricodendron* Jankowski, 1973 (subj. syn. *Ophryodendron*)
- *Ophryodendron* Claparède & Lachmann, 1859
- *Schizactinia* Jankowski, 1967
- *Spongiarcon* Jankowski, 1980
- *Syllarcon* Jankowski, 1981

Family PARACINETIDAE Jankowski, 1978
(syns. Loricophryidae, Luxophryidae)

Size, small to medium; **trophont, spherical or sac-like, basally attached to conical lorica on long stalk**; lorica, without slits or notches; tentacles, capitate, grouped apically in a single fascicle or row; swarmers, ovoid, formed apically, with somatic kineties disposed in "U" around body; macronucleus, globular to ellipsoid; micronucleus, present; contractile vacuole, may be multiple; resting cysts; in marine, brackish, freshwater, and terrestrial habitats on inanimate substrates and as ectosymbionts on aquatic plants and animals; seven genera.

- *Actinocyathula* Corliss, 1960
- *Distarcon* Jankowski, 1987 [not listed in Aescht]
- *Limnoricus* Jankowski, 1981
- *Loricophrya* Matthes, 1956
- *Luxophrya* Jankowski, 1978
- *Nipponarcon* Jankowski, 1981
- *Paracineta* Collin, 1911

Family PHALACROCLEPTIDAE Kozloff, 1966

Size, small; **shape, flattened hemispherical, with neither cilia nor infraciliature at any stage of the life cycle; tentacles, very short, serving for attachment to host**; reproduction by binary fission; macronucleus, ellipsoid; micronucleus, present; contractile vacuole, absent; in marine habitats as a parasite on the oral structures of polychaete annelids; one genus.

- *Phalacrocleptes* Kozloff, 1966

Family PODOPHRYIDAE Haeckel, 1866
(syns. Parapodophryidae, Sphaerophryidae)

Size, small to medium; **trophonts, pyriform or spherical, typically without lorica (except *Podophrya* life cycle)**; usually stalked; **tentacles, capitate, apical or evenly distributed; swarmers, commonly apically produced one at a time**

and sometimes as large as adult; swarmers, frequently cylindrical, with broad equatorial band of cilia; macronucleus, globular; micronucleus, present; contractile vacuole, present; **encystment common, on a stalk, with cyst wall having transverse circular ribs**; generally in freshwater habitats, often attached to other ciliates as parasites; three genera.

- *Parapodophrya* Kahl, 1931
- *Podophrya* Ehrenberg, 1834
- *Sphaerophrya* Claparède & Lachmann, 1859

Family PRAETHECACINETIDAE Dovgal, 1996

Size, medium; **trophont, pyriform or sac-like, attached to the bottom of a lorica**; stalked; **tentacles, capitate, arranged in a single, apical fascicle**; swarmer, elongate, ciliated, formed laterally; macronucleus, globular to ellipsoid; micronucleus, present; contractile vacuole, present; in marine habitats as ectocommensals on invertebrates; one genus.

- *Praethecacineta* Matthes, 1956

Family RHABDOPHRYIDAE Jankowski, 1970 (syn. Trophogemmidae)

Size, small to medium; **trophont, laterally flattened, ribbon-like or sac-like**; stalk, short, broad; **tentacles, rod-like, short, evenly distributed or in transverse groupings along the longitudinal axis of the body processes, sometimes borne on actinophores**; swarmers, vermiform, ciliated, cylindrical; macronucleus, ellipsoid; micronucleus, present; contractile vacuoles, two to three; in marine habitats as ectocommensals on crustaceans, such as shrimp; five genera.

- *Hastarcon* Jankowski, 1981
- *Rhabdophrya* Chatton & Collin, 1910
- *Spinarcon* Jankowski, 1981
- *Trophogemma* Jankowski, 1970
- *Vostonica* Jankowski, 1994

Family SEVERONIDAE Jankowski, 1981

Size, small; trophont, globular to ellipsoid, attached to the substrate by a body protuberance or basal "button"; **tentacles, capitate, evenly distributed on apical surface; swarmer with narrow equatorial ciliated girdle**; macronucleus, ellipsoid; micronucleus (?); contractile vacuole (?); in marine habitats as an ectocommensal of sponges; one genus.

- *Severonis* Jankowski, 1981

Family SPELAEOPHRYIDAE Jankowski in Batisse, 1975

Size, medium to large; trophont, cylindrical, conical or trumpet-shaped; stalk, short; **tentacles, capitate, in an apical corona or in groups along the body, possibly prehensile as well as suctorial types**; swarmer, vermiform, cylindrical, non-ciliated; macronucleus, ellipsoid or ribbon-like; micronucleus, present; contractile vacuoles, multiple; in marine and freshwater habitats as ectocommensals on crustaceans, such as decapod shrimp; two genera.

- *Cucumophrya* Kunz, 1936
- *Spelaeophrya* Stammer, 1935

Family TACHYBLASTONIDAE Grell, 1950

Size, small to medium; **with two alternating generations – one, loricate, often attached to various marine hydrozoans and even to the stalk of *Ephelota*, producing up to 16 small non-ciliated, unitentaculate forms that pierce the pellicle of the *Ephelota* body and become the second generation, which lives parasitically within the cytoplasm of *Ephelota* and produces large ciliated larvae that, in turn, attach to the host stalk, become loricate, and repeat the cycle**; reproduction by lateral, sequential semi-circumvaginative budding; macronucleus, globular to ellipsoid; micronucleus, present; contractile vacuole (?); in marine habitats; one genus.

- *Tachyblaston* Martin, 1909

Family THECACINETIDAE Matthes, 1956

Size, small to medium; **trophont, sac-like, attached to bottom of lorica near stalk**; stalked; **tentacles, clavate, grouped on rounded, distal, narrow end of body**; swarmers, ellipsoidal, flattened, or vermiform, ciliated on one margin; macronucleus, ellipsoid; micronucleus, present; contractile vacuole, present; predominantly in marine habitats as ectosymbionts on algae, crustaceans, and nematodes; one genus.

- *Thecacineta* Collin, 1909

Order Endogenida Collin, 1912
(syns. Acinetida, Acinetina, Astrosomatida, Dendrosomatida, Dendrosomatina, Endogenea, Endogenia, Endosphaeriida, Endosphaeriina, Entotropida *p.p.*, Heliophryida, Marinectida, Oligostomatida *p.p.*, Pseudogemmida *p.p.*, Solenophryina *p.p.*, Stylophryina, Tokophryina, Trichophryida *p.p.*)

Size, small to large; trophonts, ovoid to spheroid, but ramified and of enormous size in some groups; often loricate; tentacles, frequently in fascicles; **endogenous budding occurring in a pouch, monogemmic or polygemmic, with swarmers produced completely internally and becoming free-swimming in brood pouch** ***before*** **emergence through birth pore**; small permanent field of non-ciliferous kinetosomes near contractile vacuole responsible for larval ciliature; **swarmer, small, ciliated**; in marine and freshwater habitats, with ectosymbiotic forms common and some endocommensals; 13 families.

Family ACINETIDAE Stein, 1859
(syn. Cryptophryidae)

Size, small to medium; **trophonts, laterally flattened, trapezium-like, triangular or rarely disc-like; in lorica, which is often triangular in shape; stalked, with stalk persisting in some but not all endosymbiotic forms**; tentacles, in two or three rows or fascicles, typically on actinophores; **swarmers small, ovoid, with oblique, longitudinal somatic kineties**; macronucleus, ellipsoid or ribbon-like; micronucleus, present; contractile vacuole, present; in marine and freshwater habitats as free-living forms or if attached, never intracellular; 12 genera.

- *Acineta* Ehrenberg, 1834
- *Acinetides* Swarczewsky, 1928
- *Anthacineta* Jankowski, 1978
- *Cryptacineta* Jankowski, 1978
- *Cryptophrya* Jankowski, 1973
- *Phyllacineta* Jankowski, 1978
- *Rondacineta* Jankowski, 1978 (subj. syn. *Tokophrya*)
- *Soracineta* Jankowski, 1978 (subj. syn. *Pelagacineta*)
- *Squalorophrya* Goodrich & Jahn, 1943 (subj. syn. *Tokophrya*)
- *Trematosoma* Batisse, 1973
- *Vasacineta* Jankowski, 1981 (subj. syn. *Metacineta*)
- *Veracineta* Jankowski, 1978

Family ACINETOPSIDAE Jankowski, 1978

Size, small to medium; trophonts, trapezium-like, laterally flattened; loricate; stalked; **tentacles of two types – hypertrophied, agile prehensile ones and regular feeding ones**; macronucleus, globular to ellipsoid; micronucleus, present; contractile vacuole, present; in marine and freshwater habitats as ectosymbionts on plants and invertebrates; one genus.

- *Acinetopsis* Robin, 1879

Family CHOANOPHRYIDAE Dovgal, 2002

Size, small; trophont, globular to ellipsoid; stalked; **tentacles, funnel-like, lacking the inner microtubular phyllae**; macronucleus, globular to ellipsoid; micronucleus, present; contractile vacuole, present; feeding on liquid remains of host's food; in freshwater habitats as ectocommensals on cyclopoid crustaceans; one genus.

- *Choanophrya* Hartog, 1902

Family CORYNOPHRYIDAE Jankowski, 1981

Size, large; trophont, spheroid to cylindroid; in lorica, but not basally attached; stalk, well-developed; **tentacles, capitate, extensible and contractile, but not flexible, arranged in a single apical fascicle or evenly distributed; swarmer, club-shaped with large marginal ciliated field**; macronucleus, ellipsoid to ribbon-like; micronucleus, present; contractile vacuole, present; in marine habitats as ectosymbionts on algae, hydroids, molluscs, and crustaceans; two genera.

- *Andrusoviella* Dovgal, 2005*
- *Corynophrya* Kahl in Curds, 1987

Family DACTYLOSTOMATIDAE Jankowski, 1978

Size, small to medium; trophont, sac-like, not flattened; **stalk, massive with apical widening; tentacles, bottle-like, arranged in two apical rows**; macronucleus, ellipsoid; micronucleus, present; contractile vacuole, present; in freshwater habitats as ectocommensals on crustaceans from Lake Baikal; one genus.

- *Dactylostoma* Jankowski, 1967

Family DENDROSOMATIDAE Fraipont, 1878
(syns. Dendrosomidae, Stylophryidae)

Size, medium to large; trophont, pyriform to truncate to branching; stalkless, with rare exception, but rather attached to the substratum by broad part of body or protuberance; **indeterminate growth**; aloricate, occasionally planktonic; **tentacles, capitate, evenly distributed or arranged in fascicles at ends of conspicuous finger-like processes, sometimes highly specialized or greatly reduced in number**; budding, often multiple; swarmers, small, with transverse band of kineties; macronucleus, globular to ramified; micronucleus, present; contractile vacuole, multiple throughout body; in brackish and freshwater habitats, free-living in the periphyton, some as ectosymbionts on turtles, others endosymbiotic, and still others as parasites on crustacean gills; four genera.

- *Astrophrya* Awerintzew, 1904 (subj. syn. *Dendrosoma*)
- *Dendrosoma* Ehrenberg, 1837
- *Gorgonosoma* Swarczewsky, 1928 (subj. syn. *Dendrosoma*)
- *Stylophrya* Swarczewsky, 1928

Family ENDOSPHAERIDAE Jankowski in Corliss, 1979
(syn. Endosphaeriidae)

Size, small to medium; trophonts, ovoid to spheroid; without stalk; **tentacles, not present**; swarmers, spheroid to ellipsoid, with several 'transverse' kineties; swarmers, produced by monogemmy or polygemmy; macronucleus, globular to ellipsoid; micronucleus, present; contractile vacuole, present; **in marine and freshwater habitats solely as endoparasites of cells and tissues of other organisms, such as ciliates (e.g., folliculinids, peritrichs, and even other suctoria), turbellarians, and bivalve molluscs**; two genera.

- *Acoelophthirius* Jankowski in Dovgal, 2002*
- *Endosphaera* Engelmann, 1876

Family ERASTOPHRYIDAE Jankowski, 1978

Size, small to medium; **trophonts, ovoid to irregular, attaching to peritrich host by arm-like appendages called the cinctum or hemicinctum**; tentacles, capitate, evenly distributed on body surface or arranged in fascicles on short actinophores; macronucleus, ellipsoid to ribbon-like; micronucleus, present; contractile vacuole, present; in freshwater habitats as hypercommensals on peritrich ectosymbionts of fishes; two genera.

- *Chenophrya* Dovgal, 2002*
- *Erastophrya* Fauré-Fremiet, 1943

Family PSEUDOGEMMIDAE Jankowski, 1978

Size, small; trophonts, globular to ellipsoid; loricate; **tentacles, rod-like, one to several, serving both for feeding and attachment**; macronucleus, globular to ellipsoid; micronucleus, present; contractile vacuole, present; in marine and freshwater habitats as parasites of other ciliates, such as folliculinids and suctorians; two genera.

- *Pottsiocles* Corliss, 1960
- *Pseudogemma* Collin, 1909

Family RHYNCHETIDAE Jankowski, 1978
(syn. Riftidae)

Size, small; trophont, pyriform to ovoid, attaching to substrate by basal body surface or protuberance; **tentacles, agile, very flexible**; macronucleus, globular to ellipsoid; micronucleus (?); contractile vacuole, present; in freshwater habitats as symphorionts on crustaceans; two genera.

- *Rhyncheta* Zenker, 1866
- *Riftus* Jankowski, 1981 (subj. syn. *Tokophrya*)

Family SOLENOPHRYIDAE Jankowski, 1981

Size, small; trophont, spheroid to ovoid, attaching to substrate by basal surface of the lorica;

stalkless; **tentacles, capitate**; macronucleus, ellipsoid; micronucleus, present; contractile vacuole, present; in brackish and freshwater habitats in periphyton or in plankton; two genera.

- *Solenophrya* Claparède & Lachmann, 1859
- *Sphaeracineta* Jankowski, 1987 [not listed in Aescht]

Family TOKOPHRYIDAE Jankowski in Small & Lynn, 1985

Size, small to medium; **trophonts, ovoid, cylindrical or triangular-shaped, often laterally flattened; without lorica**; stalked, of varying length; tentacles, capitate, typically in two or rarely more fascicles; swarmers, ovoid with oblique somatic kineties; macronucleus, globular to ribbon-like; micronucleus, present; contractile vacuole, present; in marine and freshwater habitats, both free-living and as ectocommensals on copepods, amphipods, and even other ciliates (e.g., stalk of peritrichs); seven genera.

- *Lecanodiscus* Jankowski, 1973
- *Listarcon* Jankowski, 1982
- *Parastylophrya* Jankowski, 1978
- *Pelagacineta* Jankowski, 1978
- *Talizona* Jankowski, 1981
- *Tokophrya* Bütschli, 1889
- *Tokophryopsis* Swarczewsky, 1928

Family TRICHOPHRYIDAE Fraipont, 1878
(syns. Actinobranchiidae, Caprinianidae [for Capriniidae], Marinectidae, Mucophryidae, Peltacinetidae, Staurophryidae)

Size, small; **trophont, flattened, attached to substratum by broad part of body or a body protuberance**; stalkless; some loricate forms, mainly with a mucous lorica; tentacles, capitate or rod-like, may be in rows or fascicles, rarely on poorly developed actinophores; **determinate growth; swarmer, discoid, flattened, with equatorial kineties**; macronucleus, ellipsoid, ribbon-like or ramified; micronucleus, present; contractile vacuole, may be multiple; in marine and freshwater habitats as ectocommensals on aquatic invertebrates and vertebrates, with some found on gills of fishes; eleven genera.

- *Anarma* Jankowski, 1981 (subj. syn. *Discophrya*)
- *Brachyosoma* Batisse, 1975
- *Capriniana* Strand, 1928 (subj. syn. *Trichophrya*)
- *Marinecta* Jankowski, 1973
- *Mucophrya* Gajewskaja, 1928
- *Paramucophrya* Chen, Song, & Hu, 2005*
- *Peltacineta* Jankowski, 1978 (subj. syn. *Trichophrya*)
- *Rhizobranchium* Jankowski, 1981
- *Staurophrya* Zacharias, 1893
- *Tetraedrophrya* Zykoff in Dovgal, 2002*
- *Trichophrya* Claparède & Lachmann, 1859

Order Evaginogenida Jankowski, 1978
(syns. Cyathodiniida, Cyathomorphida, Cyathomorphina, Dendrocometida, Dendrocometina, Discophryida, Discophryina, Heliophryida, Evaginogenea, Evaginogenia, Inversogenea, Neotenea *p.p.*, Stylocometina *p.p.*, Tripanococcina *p.p.*)

Size, small to large; trophonts, sessile; with or without stalk, occasionally in lorica; tentacles either scattered singly or in fascicles at the ends of sometimes massive arms or trunks; **kinetosomes of larval kineties first develop on "parental" surface of a brood pouch, but cytokinesis of a single swarmer completed exogenously** ***after*** **full emergence of the "everted" bud (i.e., evaginative budding)**; swarmer, often ellipsoidal, flattened; in marine and freshwater habitats, widespread, especially as symphorionts, with species of one endosymbiotic genus showing a strikingly aberrant life cycle; 11 families.

Family COMETODENDRIDAE Jankowski, 1978

Size, small to large; **trophont, vase-like, branched, lifted off substrate but attached to it by a basal protuberance**; **tentacles, ramified**; macronucleus, globular; micronucleus, present; contractile vacuole, present; in freshwater habitats as ectocommensals on gammarid amphipods; one genus.

- *Cometodendron* Swarczewsky, 1928

Family CYATHODINIIDAE da Cuhna, 1914
(syn. Enterophryidae *p.p.*)

Size, small; trophont, pyriform to ovoid; stalkless; adult stage fleeting, but typically produces two ciliated buds simultaneously; larval form as trophont, pyriform in shape, persisting as the dominant stage in the life cycle; **swarmers retaining extensive ciliature and having tentacles, called endosprits, which are reduced to a series of short protuberances along the left side of the anterior ciliated cavity**; macronucleus, globular to ellipsoid; micronucleus, present; contractile vacuole, one to several; **in terrestrial habitats as endocommensals in the digestive tract of domestic and wild guinea pigs**; one genus.

- *Cyathodinium* da Cunha, 1914

Family DENDROCOMETIDAE Haeckel, 1866

Size, small to medium; **trophont, hemispherical or disc-shaped; tentacles, with conical or tapered tips, ramified, and borne on arms or trunks**; swarmers, lenticular; macronucleus, globular to ellipsoid; micronucleus, present; contractile vacuole, present; in freshwater habitats as ectocommensals on gammarid amphipods; two genera.

- *Dendrocometes* Stein, 1852
- *Niscometes* Jankowski, 1987 [not listed in Aescht]

Family DISCOPHRYIDAE Collin, 1912
(syns. Coronodiscophryidae, Cyathodiscophryidae, Multifasciculatidae)

Size, small to medium; trophont, with flattened disc-like body, rarely sac-like; with or without stalk; without lorica; **tentacles, capitate, in fascicles or evenly distributed; swarmer, typically large, flattened, or elongate ovoid, with somatic kineties typically marginal, in two fields**; macronucleus, ellipsoid, ribbon-like or ramified; micronucleus, present; contractile vacuole, multiple as a rule; in freshwater habitats, in the periphyton, but many species ectocommensal on adult crustaceans and the larval forms of aquatic insects; four genera.

- *Discophrya* Lachmann, 1859
- *Misacineta* Jankowski, 1978 (subj. syn. *Discophrya*)
- *Multifasciculatum* Goodrich & Jahn, 1943
- *Setodiscophrya* Jankowski, 1981 (subj. syn. *Discophrya*)

Family ENCHELYOMORPHIDAE Augustin & Foissner, 1992

Size, small; **trophont, ovoid to spheroid**; without stalk; without lorica; tentacles, rod-like, randomly distributed on one side of body; **budding occurring evaginatively, usually in pairs; swarmer, spindle-shaped, with several rod-like tentacles and "transverse" kineties**; macronucleus, globular; micronucleus, present; contractile vacuole, present; **hydrogenosomes, present**; in brackish, freshwater, and terrestrial habitats, especially anaerobic ones, such as activated sludge; one genus.

- *Enchelyomorpha* Kahl, 1930

Family HELIOPHRYIDAE Corliss, 1979

Size, small to medium; **trophonts, discoid, often with flattened body, attached directly to the substrate by tectinous adhesive disc**; tentacles, knobbed, extensible, solitary or arranged in several fascicles; macronucleus, ellipsoid or ramified; micronucleus, present; contractile vacuole, multiple, around periphery of cell; in freshwater habitats, free-living in the periphyton or as ectocommensals on invertebrates; two genera.

- *Cyclophrya* Gönnert, 1935
- *Heliophrya* Saedeleer & Tellier, 1930

Family PERIACINETIDAE Jankowski, 1978
(syns. Caracatharinidae, Catharinidae)

Size, small to medium; **trophont, laterally flattened or rarely sac-like**; **lorica or stylotheca, tectinous**; tentacles, clavate, arranged in fascicles; macronucleus, ellipsoid, ribbon-like or ramified; micronucleus, present; contractile vacuoles, typically several; in freshwater habitats, free-living in the periphyton or as ectocommensals on invertebrates; four genera.

- *Elatodiscophrya* Jankowski, 1978 (subj. syn. *Discophrya*)
- *Kormosia* Dovgal, 2002*
- *Periacineta* Collin, 1909 (subj. syn. *Discophrya*)
- *Peridiscophrya* Nozawa, 1938

Family PRODISCOPHRYIDAE Jankowski, 1978

Size, small; trophont, spheroid; stalked; tentacles, capitate, evenly distributed over the body surface; macronucleus, globular; micronucleus, present; **conjugation, anisogamous, with ciliated microconjugant similar to a swarmer**; contractile vacuole, present; in freshwater habitats in periphtyon; one genus.

– *Prodiscophrya* Kormos, 1935

Family RHYNCHOPHRYIDAE Jankowski, 1978

Size, small; trophont, laterally flattened, elongate; stalked; **tentacles, several, agile and contractile**; macronucleus, ribbon-like; micronucleus, present; contractile vacuole, multiple; **in freshwater habitats as ectoparasites of discophryid suctorians**; one genus.

– *Rhynchophrya* Collin, 1909

Family STYLOCOMETIDAE Jankowski, 1978
(syn. Discosomatellidae)

Size, small to medium; trophont, ovoid, sac-like or disc-like, spread over the substrate; some stalked forms; **tentacles, rod-like, unramified, evenly distributed or arranged in rows**; macronucleus, elongate ellipsoid; micronucleus, several; contractile vacuole, present; in freshwater habitats as ectocommensals on isopod and amphipod crustaceans; three genera.

– *Discosomatella* Corliss, 1960
– *Echinophrya* Swarczewsky, 1928
– *Stylocometes* Stein, 1867

Family TRYPANOCOCCIDAE Dovgal, 2002
(syn. Tripanococcidae)

Size, small; trophont, sac-like; without stalk; **tentacles, absent**; **swarmer, ellipsoid, laterally flattened, with several longitudinal kineties; swarmer, produced by sequential polyinversogemmy**; macronucleus, globular; micronucleus, present; contractile vacuole, present; **in freshwater habitats as parasites of the tissues of rotifers**; one genus.

– *Trypanococcus* Stein in Zacharias, 1885

Incertae sedis in Class PHYLLOPHARYNGEA

– *Silenella* Fenchel, 1965

Class **NASSOPHOREA** Small & Lynn, 1981
(syns. Clinostomata *p.p.*, Cyrtostomata *p.p.*, Gymnostomatida-Cyrtophorina, Gymnostomorida *p.p.*, Homotricha *p.p.*, Hypostomatida, Hypostomea *p.p.*, Hypostomina, Hypostomata *p.p.*, Parahymenostomata *p.p.*)

Size, small to large; shape, flattened dorsoventrally or cylindrical; free-swimming; **somatic alveoli well-developed with paired alveolocysts present in at least two orders – the Nassulida and Microthoracida**; somatic ciliation, very dense to often reduced in smaller forms; somatic cilia as monokinetids, dikinetids, or polykinetids; monokinetid with anterior, tangential transverse ribbon, a divergent postciliary ribbon, and anteriorly directed kinetodesmal fibril; for dikinetids, only the anterior kinetosome has a transverse ribbon while the posterior kinetosome has a postciliary ribbon and kinetodesmal fibril; polykinetids are cirrus-like in one family, the Discotrichidae; somatic extrusomes as fibrocysts, fibrous trichocysts or rod-shaped mucocysts; cytostome ventral; **cytopharyngeal apparatus typically of the cyrtos type, well-developed in several groups**; oral area may be sunk into an atrium, with more or less organized atrial ciliature; **oral polykinetids with alveoli between kinetosomal rows, may be confined to oral area or extend around body as hypostomial frange or synhymenium**; stomatogenesis, mixokinetal, and morphogenesis of fission may be complex; macronucleus, typically homomerous; micronuclei, one to several; conjugation, temporary; contractile vacuoles, often multiple; cytoproct, typically midventral; microphagous to algivorous; in marine, brackish, freshwater, and terrestrial habitats, with ecto- and endocommensals also common, usually with invertebrate hosts; three orders and one order *incertae sedis*.

Order Synhymeniida de Puytorac et al. in Deroux, 1978
(for Synhymenida; syns. Nassulopsida, Nassulopsina, Scaphidiodontida, Scaphidiodontina, Synhymen[i]ina)

Size, small to medium; shape, cylindrical; somatic ciliation, typically holotrichous with bipolar kineties; **hypostomial frange or synhymenium of dikinetids or small polykinetids (i.e.,**

usually of 4 kinetosomes), extending from right postoral body surface to left dorsal body surface, almost encircling the body in some forms; no atrium; cyrtos, conspicuous; free-living, predominantly freshwater forms, though some marine and a number interstitial species; four families.

Family NASSULOPSIDAE Deroux in Corliss, 1979

Size, medium to large; shape, elongate and radially symmetrical; free-swimming; somatic ciliation, holotrichous with polar anterior suture separated from oral region and synhymenium; **synhymenium, extending almost completely around body circumference just below level of the cytostome-cytopharyngeal apparatus**; cytostome in anterior 1/4 of body; macronucleus, globular to ellipsoid; micronucleus, present; contractile vacuole, medially located on ventral surface; feeding on diatoms, other microalgae, and cyanobacteria; in marine, freshwater, and occasionally terrestrial habitats; two genera.

- *Beersena* Jankowski, 1989 (for preoccupied *Phasmatopsis* Deroux, 1978)
- *Nassulopsis* Foissner, Berger, & Kohmann, 1994

Family ORTHODONELLIDAE Jankowski, 1968

Size, small to medium; shape, roughly ovoid, sometimes with asymmetrical lobe or beak to the left; free-swimming; somatic ciliation, holotrichous; **synhymenium, thickly ciliated, extending from right postoral body surface, just below level of the cytostome-cytopharyngeal apparatus, to left preoral body surface into preoral suture, which is formed by right somatic kineties extending around anterior end onto the left side**; cytostome in anterior 1/4 of body; macronucleus, ellipsoid to elongate band-form; micronucleus, present; contractile vacuole, present; feeding on diatoms, other microalgae, and cyanobacteria; predominantly in marine and brackish habitats; two genera and one genus *incertae sedis*.

- *Orthodonella* Bhatia, 1936
- *Zosterodasys* Deroux, 1978

Incertae sedis in Family Orthodonellidae

- *Eucamptocerca* da Cunha, 1914 [nomen dubium]

Family SCAPHIDIODONTIDAE Deroux in Corliss, 1979

Size, small; shape, somewhat dorsoventrally flattened with slightly broader anterior end and gently tapered posterior; free-swimming; somatic ciliation, holotrichous, but reduced on much of the dorsal surface of *Scaphidiodon*; **synhymenium or hypostomial frange, sparsely ciliated, extending from only slightly right postoral region onto left preoral body surface into preoral suture**; cytostome in anterior 1/4 of body; macronucleus, globular to ellipsoid; micronucleus, present; contractile vacuole, present; feeding on diatoms, other microalgae, and cyanobacteria; in marine and freshwater habitats; two genera.

- *Chilodontopsis* Blochmann, 1895
- *Scaphidiodon* Stein, 1859

Family SYNHYMENIIDAE Jankowski in Small & Lynn, 1985

Size, medium; shape, elongate; free-swimming; somatic ciliation, holotrichous; **synhymenium or hypostomial frange, extending into a left preoral suture and also with a second preoral suture to right of the first, into which synhymenium does not extend**; cytostome in anterior 1/4 of body; macronucleus, ellipsoid; micronucleus (?); contractile vacuole, present; feeding (?); in freshwater habitats; one genus.

NOTE: This family and genus need careful redescription.

- *Synhymenia* Jankowski, 1968

Order Nassulida Jankowski, 1967
(syns. Ambihymenida, Cyrtohymenostomatida *p.p.*, Parahymenostomatida, Paranassulida, Pronassulida *p.p.*)

Size, small to large; shape, elongate, ovoid; alveolocysts, present; somatic ciliation, holotrichous, usually dense with kineties closely adjacent; distinct preoral suture; **somatic kinetosomes with a proximal and distal cartwheel**; somatic extrusomes, rod-like, when present; **synhymenium or hypostomial frange, beginning in postoral region, always to right of the stomatogenic kinety, and extending to lateral left onto dorsal surface, but sometimes reduced to 3–4 polykinetids**

restricted to a shallow oral cavity; cyrtos, typically large, with complete palisade of nematodesmata; cystment, common; three families.

Family FURGASONIIDAE Corliss, 1979
(for Cyclogrammidae)

Size, small to medium; shape, ovoid to ellipsoid; free-swimming; somatic ciliation, holotrichous; somatic extrusomes as fusiform trichocysts; **oral region in anterior 1/3 of the body with one to three left oral polykinetids and one short, right paroral in the oral area, but not confined to an oral cavity**; cyrtos, conspicuous, surrounding cytostome-cytopharynx; macronucleus, globular to ellipsoid; micronucleus, present; contractile vacuole, midventral; feeding on bacteria and smaller protists, including microalgae; typically in freshwater, but sometimes terrestrial, habitats; four genera.

- *Furgasonia* Jankowski, 1964
- *Parafurgasonia* Foissner & Adam, 1981
- *Urliella* Foissner, 1989
- *Wolfkosia* Foissner, Agatha, & Berger, 2002*

Family NASSULIDAE de Fromentel, 1874
(for Odontoholotrichidae; syns. Cyrtohymenostomatidae, Enigmostomatidae, Liosiphonidae)

Size, small to large; shape, roughly ellipsoid; free-swimming; somatic ciliation, holotrichous, dense; **synhymenium hypostomial frange, with few to many polykinetids composed of at least four kinetosomes extending from postoral region to left, sometimes onto dorsal surface**; paroral dikinetids sometimes conspicuous; cyrtos, prominent, surrounding cytostome-cytopharynx in anterior 1/3 of body; macronucleus, globular to ellipsoid; micronucleus, present; contractile vacuole, typically single, midventral; algivorous; in marine, occasionally brackish, and freshwater habitats, widespread; five genera and three genera *incertae sedis*.

- *Nassula* Ehrenberg, 1834
- *Nassulides* Foissner, Agatha, & Berger, 2002*
- *Naxella* Fryd-Versavel, Iftode, & Deroux, 1980
- *Obertrumia* Foissner & Adam, 1981
- *Rhinakis* Iftode, Fryd & Deroux in Deroux, 1994

Incertae sedis in the Family Nassulidae

- *Archinassula* Kahl, 1935
- *Chilodina* Srämek-Husek, 1957
- *Stomatophrya* Kahl, 1933

Family PARANASSULIDAE Fauré-Fremiet, 1962
(syns. Enneameronidae, Gullmarellidae)

Size, medium, typically >100 µm long; shape, ovoid to elongate-ovoid; free-swimming; somatic ciliation, holotrichous; **oral region in anterior 1/4 of cell with oral structures as 3–4 polykinetids to the left of the cytostome in a shallow oral pit**; cyrtos, conspicuous; macronucleus, ellipsoid to ribbon-like; micronucleus, present; contractile vacuole, on dorsal surface; feeding (?); in marine habitats; two genera and one genus *incertae sedis*.

- *Enneameron* Jankowski, 1964 (subj. syn. *Nassulopsis*)
- *Paranassula* Kahl, 1931

Incertae sedis in Family Paranassulidae

- *Gullmarella* Fenchel, 1964

Order Microthoracida Jankowski, 1967
(syns. Cyrtopharyngina, Microthoracina, Propeniculida)

Size, small to medium, usually <100 µm long; shape, frequently broadly ellipsoidal, with right side more rounded, occasionally crescentic, and often laterally flattened; alveolocysts, present; pellicle, firm and rigid, with thickened epiplasm in some forms; **somatic ciliation, holotrichous, but typically with a few somatic kineties, separated by wide interkinetal spaces, composed of monokinetids, dikinetids, but polykinetids in the Family Discotrichidae; somatic extrusomes as fibrous trichocysts with anchor-like tip (fibrocysts), except in Family Discotrichidae**; oral cavity, usually three left oral polykinetids, with oral cavity sometimes displaced to posterior due to differential growth of cortex; **right paroral dikinetid, variably developed, but its vestige always appears in stomatogenesis**; cyrtos, small, with complete palisade of nematodesmata; often cyst-forming; microphagous and algivorous; typically in freshwater and terrestrial habitats, but *Discotricha* is marine; three families.

Family DISCOTRICHIDAE Jankowski, 1967

Size, small; shape, slightly reniform; free-swimming; **cortex, forming papillae distributed between somatic kineties; somatic kineties composed of cirrus-like polykinetids, most of more than four kinetosomes, distributed over right lateral and ventral surfaces**; somatic extrusomes as rod-shaped mucocysts, not fibrocysts; oral cavity, anterior, with three small oral polykinetids and a paroral of reduced size; macronucleus, globular; micronucleus, present; contractile vacuole, present; microphagous; in marine habitats, benthic; two genera.

NOTE: This family has been placed here for some time, a placement that requires confirmation by molecular genetic analysis since the ultrastructure of this ciliate shows no clear affinities to other microthoracids.

- *Discotricha* Tuffrau, 1954
- *Lopezoterenia* Foissner, 1997

Family LEPTOPHARYNGIDAE Kahl, 1926
(syns. Pseudomicrothoracidae, Trichoderidae [for Trichopelm(at)idae])

Size, small; shape, ovoid, slightly flattened; pellicle thrown into obvious cortical ridges, underlain by well-developed epiplasm; free-swimming; somatic ciliation, holotrichous, but sparse; **left border of oral region has perioral somatic kinety of dikinetids and right border has three oral polykinetids**; **cyrtos, long, tubular, in anterior 1/3 of body**; macronucleus, globular to ellipsoid; micronucleus, present; contractile vacuole, present; microphagous, on bacteria and microalgae, including cyanobacteria, even filamentous forms; in freshwater and terrestrial habitats, such as mosses; two genera.

- *Leptopharynx* Mermod, 1914
- *Pseudomicrothorax* Mermod, 1914

Family MICROTHORACIDAE Wrzesniowski, 1870
(syns. Conchophryidae, Drepanomonadidae, Trochiliopsidae)

Size, typically very small; shape, crescentic and flattened; free-swimming; **somatic ciliation, very sparse, typically with three right preoral somatic kineties, usually of dikinetids, anterior to oral region, and left somatic kineties very reduced**; oral region, variable in position, anterior in some genera (e.g. *Stammeridium*) and displaced to the posterior in others (e.g. *Microthorax*); typically a reduced paroral and three small oral polykinetids confined to the oral cavity; macronucleus, globular; micronucleus, present; contractile vacuole, present; microphagous; in terrestrial habitats, especially mosses; four genera and two genera *incertae sedis*.

- *Drepanomonas* Fresenius, 1858
- *Microthorax* Engelmann, 1862
- *Stammeridium* Wenzel, 1969
- *Trochiliopsis* Penard, 1922

Incertae sedis in Family Microthoracidae

- *Conchophrys* Chatton, 1911
- *Hexotricha* Conn in Conn & Edmondson, 1918

Incertae sedis in Class NASSOPHOREA

Order Colpodidiida Foissner, Agatha, & Berger, 2002

Size, small; shape, elongate, ovoid; free-swimming; somatic ciliation, holotrichous, with slight twist to kineties in anterior end; **oral region in middle 1/3 of cell, with a paroral and three oral polykinetids that can be reduced in size to only one or two kinetosomes; cytostome-cytopharynx, supported by a delicate cyrtos (?), which extends anteriorly, then dorsally and posteriorly**; stomatogenesis, mixokinetal; macronucleus, globular to ellipsoid, may be as two nodules; micronucleus, present; contractile vacuole, present; cytoproct, present; feeding on bacteria and flagellates; in terrestrial habitats, particularly saline soils; one family.

NOTE: Foissner, Agatha, and Berger (2002) report that their cytological stained preparations suggest the presence of alveolocysts, a diagnostic feature of two orders in the class, and they provide clear evidence for nassophorean features in the stomatogenesis of these forms. The placement of this family requires confirmation by ultrastructural and molecular genetic analyses.

Family COLPODIDIIDAE Foissner, 1995

With characteristics of the order; five genera.

- *Apocolpodidium* Foissner, Agatha, & Berger, 2002*
- *Colpodidium* Wilbert, 1982

- *Pedohymena* Foissner, 1995
- *Phagoon* Foissner, Agatha, & Berger, 2002
- *Pseudocolpodidium* Foissner, Agatha, & Berger, 2002

Class **COLPODEA** Small & Lynn, 1981
(syns. Bryometopia *p.p.*, Colpodia, Rimostomata *p.p.*, Transversala *p.p.*, Stichofragmina *p.p.*)

Size, small to large; shape, variable, but often somewhat twisted; free-swimming with a few species building gelatinous loricae; **alveoli, typically well-developed and revealed as a prominent argyrome, typically reticulated**; somatic kineties of dikinetids with a cilium on each kinetosome; **somatic dikinetids having one transverse microtubular ribbon and at least one postciliary microtubule associated with the anterior kinetosome, and one transverse ribbon, one postciliary ribbon, and one short kinetodesmal fibril, directed towards the right, associated with the posterior kinetosome; posterior transverse ribbons of microtubules extending posteriorly and overlapping one another as the LKm fibre or transversodesma**; parasomal sacs, from two to four, on both sides of kinetid; somatic extrusomes as saccular or rod-shaped mucocysts; oral structures based on a paroral of dikinetids on the right and one to many square-packed polykinetids on the left, but variations from this pattern characterize included taxa; stomatogenesis, mero- or pleurokinetal; fission, often palintomic in a reproductive cyst; macronucleus, homomerous, sometimes with single large nucleolus; conjugation, reported only for bursariomorphids; contractile vacuole, typically posterior, with collecting canals in some larger cells; resting cysts common; feeding, highly variable, with small-sized cells as bacterivores and larger cells as algivores and carnivores; in freshwater and terrestrial habitats, rarely marine; six orders.

NOTE: Foissner (1993a) has written an outstanding and authoritative monograph on this group. Foissner (1993a, 1994b) has divided this class into the subclasses Colpodia and Bryometopia based primarily on characteristics of the argyrome. The argyrome can be a variable feature of cells, changing with their physiological state. Thus, before we accept this subdivision, we await corroboration of the taxonomic significance of this trait by other kinds of data, such as gene sequences.

Order Bryometopida Foissner, 1985

Size, small to large; shape, ovoid; **argyrome, "kreyellid type", a very highly reticulated, subdivided dense network**; somatic kineties, slightly spiralled; oral region subapical to equatorial with paroral of dikinetids or multiple kinetosomes on right extending to right posterior region of the oral cavity and usually several left oral polykinetids; stomatogenesis, pleurotelokinetal; fission may be in reproductive cyst; four families and one family *incertae sedis*.

NOTE: Foissner (1993a, 1994b) established this taxon based on the character of the argyrome of the included families. We have united these families in this order despite our concerns noted above about the significance of this character to establish a subclass. The order needs corroboration using other characters.

Family BRYOMETOPIDAE Jankowski, 1980
(syn. Thylakidiidae)

Size, small to large; shape, ovoid; free-swimming; **somatic ciliation, holotrichous, forming a conspicuous postoral suture**; oral region, large relative to body size; **paroral typically extending along right border of oral cavity; left serial oral polykinetids conspicuous, typically more than 10, extending along left side of oral cavity**; macronucleus, globular to ellipsoid; micronucleus, may be multiple; contractile vacuole, present; cytoproct (?); feeding on bacteria and algae; in freshwater and terrestrial habitats; two genera.

- *Bryometopus* Kahl, 1932
- *Thylakidium* Schewiakoff, 1893

Family JAROSCHIIDAE Foissner, 1993

Size, small to medium; shape, elongate, ovoid; free-swimming; somatic ciliation, holotrichous; **oral region anterior with several differently structured kinetidal elements – at least including right kinetofragments, seemingly derived as extensions of somatic kineties, a right paroral that may be developed as a polykinetid, and several oral polykinetids along the left**; macronucleus, globular; micronucleus, present; contractile vacuole, present; cytoproct (?); feeding on small flagellates and ciliates; in terrestrial habitats; two genera.

- *Jaroschia* Foissner, 1993
- *Pentahymena* Foissner, 1994

Family KREYELLIDAE Foissner, 1979

Size, very small; shape, ovoid, somewhat flattened; free-swimming; somatic ciliation, reduced or absent on left and dorsal sides; oral region, large relative to body size; **paroral typically extending only along the anterior half of the right border of the oral cavity**; **left serial oral polykinetids, inconspicuous, typically fewer than six, each may be reduced to one row of kinetosomes**; macronucleus, globular to ellipsoid; micronucleus, present; contractile vacuole, present; cytoproct, present in some species; bacterivorous; in freshwater and terrestrial habitats; three genera.

- *Kreyella* Kahl, 1931
- *Microdiaphanosoma* Wenzel, 1953
- *Orthokreyella* Foissner, 1984

Family TRIHYMENIDAE Foissner, 1988

Size, small; shape, elongated, ovoid; free-swimming; somatic ciliation, holotrichous; oral region, small relative to body size; **paroral short, slightly curved, in anterior right of oral region, appearing to be an extension of somatic Kinety 1**; **two rectangular left oral polykinetids, disposed in a "Λ" pattern**; macronucleus, globular to ellipsoid; micronucleus, large; contractile vacuole, present; cytoproct (?); bacterivorous (?); in terrestrial habitats; one genus.

- *Trihymena* Foissner, 1988

Incertae sedis in order Bryometopida

Family TECTOHYMENIDAE Foissner, 1993

Size, small; shape, ovoid; free-swimming; **somatic ciliation, holotrichous but sparse and with somatic Kinety No. 2 interrupted**; **oral structures including a simple or compound, U-shaped paroral, extending on right and over to posterior left, and typically five rectangular oral polykinetids on left**; division in reproductive cysts; macronucleus, globular to ellipsoid; micronucleus, present; contractile vacuole, present; cytoproct, present; bacterivorous; in terrestrial habitats; two genera.

- *Pseudokreyella* Foissner, 1985
- *Tectohymena* Foissner, 1993

Order Bryophryida de Puytorac, Perez-Paniagua, & Perez-Silva, 1979
(syn. Bryophryina)

Size, small to large; shape, elongate ovoid; free-swimming; somatic ciliation, holotrichous; **right oral kinetids at least including a series of radially oriented kinetosomal rows along the right border of the oral region, sometimes extending to almost encircle it** (except *Notoxoma* in which these are presumed to have been reduced to one kinetosome); left oral polykinetids, ranging from one to many that may extend out into the preoral suture (e.g., *Puytoraciella*); stomatogenesis, pattern not known; division in reproductive cysts; macronucleus, globular to ellipsoid; micronucleus, may be multiple; contractile vacuole, present, sometimes with collecting canals; cytoproct (?); feeding on bacteria and cyanobacteria; in temporary freshwater ponds and terrestrial habitats; one family.

Family BRYOPHRYIDAE de Puytorac, Perez-Paniagua, & Perez-Silva, 1979

With characteristics of the order; four genera and one genus *incertae sedis*.

- *Bryophrya* Wenzel, 1953
- *Notoxoma* Foissner, 1993
- *Parabryophrya* Foissner, 1985
- *Puytoraciella* Njiné, 1979

Incertae sedis in Family Bryophryidae

- *Telostomatella* Foissner, 1985

Order Bursariomorphida Fernández-Galiano, 1978
(syns. Bursaridida, Bursariida, Bursari(i)na)

Size, medium to large; shape, broadly ovoid; somatic ciliation, holotrichous, often very dense; oral cavity, funnel-like or cup-shaped, often expansive, and opening on the anterior and ventral surfaces; **left oral polykinetids, many, composed of three long rows, extending as an adoral zone along the left side of the expansive, deep anterior oral cavity**; stomatogenesis, pleurotelokinetal; carnivorous; resting cyst, may be heavy-walled with micropyle; in freshwater habitats, such as ponds and small lakes; two families.

Family BURSARIDIIDAE Foissner, 1993

Size, medium; shape, broad, ovoid, barrel- to tube-shaped; free-swimming; somatic ciliation, holotrichous, dense; **posterior end of oral cavity with serial left oral polykinetids curved to right or straight; paroral (?), a series of more densely ciliated kinetofragments, possibly extensions of somatic kineties, which surround the oral cavity opening**; macronucleus, globular to elongate ellipsoid and reniform; micronucleus, may be present; contractile vacuole, present; cytoproct (?); feeding on smaller protists; in freshwater habitats, typically in the plankton of small lakes and ponds; two genera.

- *Bursaridium* Lauterborn, 1894
- *Paracondylostoma* Foissner, 1980

Family BURSARIIDAE Bory de St. Vincent, 1826 (syn. Archiastomatidae)

Size, large; shape, broadly ovoid, with rounded posterior end (though tailed in one species) and truncate anterior end; free-swimming; somatic ciliation, holotrichous, dense; oral cavity prominent, funnel-like, opening at apical end of organism and remaining open for some distance onto ventral surface; **posterior end of oral cavity with serial left oral polykinetids in sigmoid curve to the left; paroral as series of oral polykinetids on the right wall of the oral cavity, separated from right somatic kineties by a non-ciliated band**; division while free-swimming; macronucleus, elongate, rod-like to vermiform; micronucleus, multiple, up to 35; conjugation, temporary, and only colpodean reported to conjugate; contractile vacuole, up to several hundred; cytoproct (?); feeding on smaller protists, other ciliates, and even metazoans, depending upon relative size; in freshwater lakes and temporary ponds; one genus.

- *Bursaria* O.F. Müller, 1773

Order Colpodida de Puytorac et al., 1974 (syns. Colpodina, Grossglockner(i)ida, Grossglocknerina)

Size, small to large; shape of many species highly asymmetrical, but detorsion before binary or palintomic fission; **right oral structure as paroral, associated with a few to many somewhat ordered or disordered rows to its right, sometimes forming a polykinetid, but reduced to a single row of dikinetids in some genera; left oral polykinetid composed of several to many well-ordered monokinetidal rows**; stomatogenesis, merotelokinetal; division, typically palintomic, in reproductive cysts; resting cysts common; in freshwater and terrestrial habitats; six families.

Family BARDELIELLIDAE Foissner, 1984

Size, small; shape, ovoid; free-swimming; somatic ciliation, holotrichous; **oral cavity in posterior half of cell with left oral polykinetid, very much longer than right oral polykinetid and extending out onto the cell surface to the anterior pole so that the oral region occupies the anterior two thirds of the ventral surface**; macronucleus, globular to ellipsoid; micronucleus, prominent; contractile vacuole, present; feeding (?); in terrestrial habitats; one genus.

- *Bardeliella* Foissner, 1984

Family COLPODIDAE Bory de St. Vincent, 1826 (syns. Exocolpodidae, Paracolpodidae)

Size, small to large; shape, typically kidney-shaped; free-swimming; somatic ciliation, holotrichous, except in smaller species, with somatic kineties curving to converge on an anterior ventral keel; somatic kineties on left postoral region may be distributed in a well-developed groove, whose ciliature directs food to the oral cavity; oral cavity in anterior to mid-half of body, a shallow depression to deeper tube, dependent on cell size; **right and left oral polykinetids of about equal length**; stomatogenesis, merotelokinetal, typically preceded both by complete dedifferentiation of the parental oral structures and by fission; division, palintomic, typically within reproductive cysts, except in *Exocolpoda*, which divides while swimming; macronucleus, globular to elongate ellipsoid; micronucleus, may be multiple; contractile vacuole, present, may have collecting canals in larger species; cytoproct, none (or poorly visible, impermanent?); feeding on bacteria and other smaller protists, including *Colpoda* species; in terrestrial habitats, widely distributed with one *Colpoda* species as an accidental (?) report from a skink (lizard); ten genera.

- *Apocolpoda* Foissner, 1993
- *Bresslaua* Kahl, 1931
- *Colpoda* O.F. Müller, 1773
- *Corticocolpoda* Foissner, 1993
- *Cosmocolpoda* Foissner, 1993
- *Exocolpoda* Foissner, Agatha, & Berger, 2002*
- *Idiocolpoda* Foissner, 1993
- *Krassniggia* Foissner, 1987
- *Kuehneltiella* Foissner, 1990
- *Pseudomaryna* Foissner, 2003*

Family GRANDORIIDAE Corliss, 1960

Size, small; shape, ovoid; free-swimming, but temporarily sessile; **somatic ciliation, holotrichous, with conspicuous bundle of caudal cilia (?) with which the ciliate attaches to the substrate**; **oral cavity, slightly subequatorial, with transversely elongated opening**; macronucleus, globular; micronucleus (?); contractile vacuole, present; cytoproct (?); bacterivorous (?); in terrestrial habitats, reported in soil of sewage-irrigated field; one genus.

NOTE: This family and genus are in need of redescription.

- *Grandoria* Corliss, 1960

Family GROSSGLOCKNERIIDAE Foissner, 1980
(syn. Grossglockneridae)

Size, small; shape, ovoid, elongate; free-swimming; somatic ciliation, holotrichous, but reduced in smaller species; **oral structures, near anterior pole, as a right anterior paroral of monokinetids and a single, small, left posterior polykinetid**; **cytopharynx everted as a small, microtubule-lined feeding tube used for puncturing fungi and yeasts, whose cytoplasm is ingested**; division, palintomic in reproductive cyst; macronucleus, globular to ellipsoid; micronucleus, present; contractile vacuole, present; cytoproct, present; feeding on the cytoplasm of fungal hyphae and solitary yeasts; in terrestrial habitats; five genera and two genera *incertae sedis*.

NOTE: Foissner (1993a, 1994b) separated this family into the Order Grossglockneriida on the basis of the novel feeding tube. However, preliminary molecular genetic evidence suggests that this is only an unusual colpodid feeding adaptation (Stechmann, Schlegel, & Lynn, 1998).

- *Fungiphrya* Foissner, 1999
- *Grossglockneria* Foissner, 1980
- *Mykophagophrys* Foissner, 1995
- *Nivaliella* Foissner, 1980
- *Pseudoplatyophrya* Foissner, 1980

Incertae sedis in Family Grossglockneriidae

- *Pseudoglaucoma* Wenzel, 1953
- *Rigchostoma* Vuxanovici, 1963

Family HAUSMANNIELLIDAE Foissner, 1987
(syn. Kalometopiidae)

Size, small to large; shape, kidney bean-like; free-swimming; somatic ciliation, holotrichous, with right kineties tending to curve anterior to the oral region; **oral region, equatorial, may be very cavernous, with right oral polykinetid tapering at its anterior end as it curves along the anterior border of the oral cavity, distinctly longer than left oral polykinetid**; division, palintomic within reproductive cyst; macronucleus, globular to ellipsoid, sometimes reniform; micronucleus, may be multiple; contractile vacuole, present; cytoproct, present; feeding on bacteria, fungi, and other protists, particularly *Colpoda* species; in terrestrial habitats; six genera.

- *Anictostoma* Foissner, 1993 (subj. syn. *Corallocolpoda*)
- *Avestina* Jankowski, 1980
- *Bresslauides* Blatterer & Foissner, 1988
- *Corallocolpoda* Alekperov, 1991
- *Hausmanniella* Foissner, 1984
- *Kalometopia* Bramy, 1962

Family MARYNIDAE Poche, 1913

Size, small to large; **shape, ovoid, with larger preoral lobe and smaller postoral lobe**; **free-swimming, but typically sessile, enclosed in a tubular or cup-shaped lorica or gelatinous sheath**; solitary or colonial; somatic ciliation, holotrichous, with leftward spiral, forming highly ordered oblique and longitudinal pattern along a midventral suture; **oral apparatus in posterior half of body, opening at base of posterior lobe**; right and left oral polykinetids of about equal length; fission, palintomic within reproductive cyst; macronucleus, globular to ellipsoid; micronucleus, present; contractile vacuole, present; cytoproct, present; bacterivorous; in small, temporary freshwater ponds and sometimes terrestrial habitats; four genera.

- *Ilsiella* Foissner, 1987
- *Maryna* Gruber, 1879
- *Mycterothrix* Lauterborn, 1898
- *Opisthostomatella* Corliss, 1960

Incertae sedis in Order Colpodida

- *Balantiophorus* Schewiakoff, 1889 (subj. syn. *Cyrtolophosis*)
- *Dragescozoon* Foissner, Agatha, & Berger, 2002*

Order Cyrtolophosidida Foissner, 1978
(syns. Cyrtolophosida, Cyrtolophosidina, Platyophryida, Platyophryina)

Size, small to large; shape, elongate, ovoid, sometimes laterally flattened; oral region, shallow, apical to subapical; **paroral as a file of dikinetids on right side of oral region; serial left oral polykinetids, few to many; micronucleus enclosed within the perinuclear space of the macronucleus (with some exceptions)**; stomatogenesis, pleurotelokinetal; fission, may occur in reproductive cyst; in freshwater and terrestrial habitats with some marine species; four families and one family *incertae sedis*.

Family CYRTOLOPHOSIDIDAE Stokes, 1888
(syns. Cyrtolophosidae, Cyrtolophosiidae)

Size, small; shape, narrow-ovoid; **free-swimming, but some species residing in a transparent gelatinous or mucous lorica or tube**; somatic ciliation, holotrichous, but can be sparse; oral region in anterior 1/3 of cell; **paroral typically in two segments, an anterior, conspicuously ciliated segment and a more posterior segment with inconspicuous cilia**; left oral polykinetids, three to five; stomatogenesis, pleurotelokinetal with parental paroral reorganized during the process; macronucleus, globular to ellipsoid; micronucleus, in perinuclear space; contractile vacuole, present; cytoproct, may be present; feeding on bacteria, microalgae, and smaller protists; in freshwater and terrestrial habitats; three genera and one genus *incertae sedis*.

- *Aristerostoma* Kahl, 1926
- *Cyrtolophosis* Stokes, 1885
- *Pseudocyrtolophosis* Foissner, 1980

Incertae sedis in Family Cyrtolophosididae

- *Plesiocaryon* Foissner, Agatha, & Berger, 2002*

Family PLATYOPHRYIDAE de Puytorac, Perez-Paniagua, & Perez-Silva, 1979
(syn. Reticulowoodruffiidae)

Size, small to large; shape, elongate, ovoid; free-swimming; somatic ciliation, holotrichous, with anterior ends of left somatic kineties sometimes forming several paratene-like rows bordering the serial left oral polykinetids; oral region, near obliquely truncate anterior end, with paroral as an uninterrupted file along the right margin of the anterior oral region; **serial left oral polykinetids, numerous, with adoral zone approximately the same length as paroral**; stomatogenesis, pleurotelokinetal with parental paroral maintained during the process; macronucleus, globular to ellipsoid, rarely in two nodules; micronucleus, in perinuclear space; contractile vacuole, present; cytoproct, present; feeding on bacteria, microalgae, other smaller protists, and in larger species, even metazoans, like nematode worms; in freshwater and terrestrial habitats; five genera and one genus *incertae sedis*.

- *Cirrophrya* Géllert, 1950
- *Platyophrya* Kahl, 1926
- *Platyophryides* Foissner, 1987
- *Reticulowoodruffia* Foissner, 1993
- *Semiplatyophrya* Wilbert & Kahan, 1986

Incertae sedis in Family Platyophryidae

- *Ottowphrya* Foissner, Agatha, & Berger, 2002*

Family SAGITTARIIDAE Grandori & Grandori, 1935
(for Proshymenidae)

Size, small; shape, ovoid; free-swimming, but some species in mucous sheaths; somatic ciliation, holotrichous, but sparse and with caudal cilium; **oral region on rounded anterior end of cell with paroral in one segment and serial left oral polykinetid zone of the same length**; macronucleus, globular to ellipsoid; micronucleus, in perinuclear space (?); contractile vacuole, present; cytoproct (?); bacterivorous; in terrestrial habitats; one genus.

- *Sagittaria* Grandori & Grandori, 1934

Family WOODRUFFIIDAE von Gelei, 1954 (syn. Woodruffidae)

Size, small to large; shape, broadly ovoid; free-swimming; **somatic ciliation, holotrichous, with anterior ends of left somatic kineties typically forming several paratene-like rows bordering the serial left oral polykinetids**; oral region, subapical, slanted, on right side of cell; right paroral dikinetid not segmented; **serial left oral polykinetids, in a more or less distinctive adoral zone extending into preoral suture, thus longer than paroral**; stomatogenesis, pleurotelokinetal with parental paroral probably reorganized during the process; macronucleus, globular to ellipsoid; micronucleus, in perinuclear space; contractile vacuole, present; cytoproct, present; feeding on cyanobacteria, fungal spores, and smaller protists, including other ciliates; in marine, freshwater, and terrestrial habitats; six genera.

- *Etoschophrya* Foissner, Agatha, & Berger, 2002*
- *Kuklikophrya* Njiné, 1979
- *Rostrophrya* Foissner, 1993
- *Rostrophryides* Foissner, 1987
- *Woodruffia* Kahl, 1931
- *Woodruffides* Foissner, 1987

Incertae sedis in Order Cyrtolophosidida

Family PSEUDOCHLAMYDONELLIDAE Buitkamp, Song, & Wilbert, 1989

Size, small; shape, ovoid; free-swimming; **somatic ciliation as kineties of dikinetids, restricted to right (= ventral) surface with a single somatic kinety of dikinetids on the anterior left**; oral region, equatorial, with an uninterrupted paroral on right and several rectangular oral polykinetids on left; **distinct cyrtos-like cytopharyngeal apparatus**; macronucleus, globular, may be in two nodules; micronucleus, in perinuclear space (?); contractile vacuole, present; cytoproct (?); feeding on microalgae, especially diatoms; in freshwater habitats, particularly spring-fed margins of rivers; two genera.

- *Hackenbergia* Foissner, 1997
- *Pseudochlamydonella* Buitkamp, Song, & Wilbert, 1989

Order Sorogenida Foissner, 1985

Size, small; shape, elongate, ovoid; free-swimming; somatic ciliation, holotrichous; **oral region at anterior pole of cell with paroral of dikinetids on right and several serial left oral polykinetids, together forming almost a closed circle around the cytostome**; stomatogenesis, pleurotelokinetal with parental structures retained during the process; macronucleus, globular to slightly ellipsoid; micronucleus, present; contractile vacuole, present; cytoproct, present; feeding on smaller *Colpoda* species; **only in terrestrial habitats, especially on submerged leaves, where it may aggregate to form aerial sorocarps, like some myxomycetes, enclosing "spores" (= resting cysts)**; one family.

Family SOROGENIDAE Bradbury & Olive, 1980

With characteristics of the order; one genus.

- *Sorogena* Bradbury & Olive, 1980

Incertae sedis in Class COLPODEA

- *Rhyposophrya* Kahl, 1933

Class **PROSTOMATEA** Schewiakoff, 1896 (syns. Apicostomata *p.p.*, Paramastigina *p.p.*, Prostomata *p.p.*, Prostomina *p.p.*, Telostomata *p.p.*)

Size, small to large; shape, ovoid to cylindroid; free-swimming with some sessile forms in loricae; alveoli, often well-developed; somatic ciliation, holotrichous, but with reduced ciliation in posterior in some taxa; **somatic kineties often arranged in complete circumferential paratenes with dikinetids at their anterior extremities**; somatic monokinetids usual, with radial transverse ribbon, slightly convergent postciliary ribbon, and anteriorly directed kinetodesmal fibril that does not overlap those of other kinetids; somatic dikinetid, when present, with posterior kinetosome derived from monokinetid and anterior kinetosome bearing only a tangential transverse ribbon; **oral region, apical, subapical, or lateral with cytostome at or near body surface; circumoral ciliation with oral dikinetids (a paroral homologue?), radial to tangential to perimeter of oral area; oral dikinetid postciliary ribbons extending laterally from each dikinetid, overlapping one another, and, in some species, forming a circular microtubular band that supports the walls of a shallow atrium**; cytopharyngeal apparatus, cyrtos-like, with nematodesmata originating from the bases of circumoral dikinetids; stomatogenesis, merotelokinetal, with migration of ventral anlagen subapically to apically; macronucleus, homomerous; bacterivorous, algivorous, carnivorous, and histophagous, including *Cryptocaryon* as a parasite of marine fishes; in freshwater, terrestrial, and marine habitats, widely distributed with some planktonic species; two orders and one family *incertae sedis*.

Order Prostomatida Schewiakoff, 1896
(syn. Vasicolina *p.p.*)

Size, small to medium; shape, cylindroid; free-swimming; loricae produced by several species; **somatic ciliation, holotrichous, clearly "radially symmetrical"**; paratenes, typically conspicuous; oral region, apical, surrounded by circumoral dikinetids; **brosse, absent**; **toxicysts, absent**; stomatogenesis, not described; microphagous; two families.

Family APSIKTRATIDAE Foissner, Berger, & Kohmann, 1994
(syn. Enchelyidae *p.p.*, Holophryidae *p.p.*)

Size, small to medium; shape, cylindroid; free-swimming; **somatic kineties, mostly bipolar, but some inserted on an indistinct suture as short segments between bipolar ones**; paratenes, present but not conspicuous; oral region, apical, surrounded by simple oral dikinetids; macronucleus, globular; micronucleus, present; contractile vacuole, present; cytoproct, present; feeding on cyanobacteria and microalgae; in freshwater habitats; one genus.

NOTE: Foissner, Berger, and Kohmann (1994) discovered problems with the nomenclature of *Holophrya*, *Prorodon*, and *Pseudoprorodon*, which necessitated the establishment of the new genus *Apsiktrata* with the type species *Urotricha gracilis* Penard, 1922. This genus presents the form of what was previously known as *Holophrya* (e.g. in Corliss, 1979; Small & Lynn, 1985).

- *Apsiktrata* Foissner, Berger, & Kohmann, 1994

Family METACYSTIDAE Kahl, 1926

Size, small to large; shape, cylindroid; **free-swimming, but living in a pseudochitinous or gelatinous lorica**; **somatic kineties, bipolar**; **paratenes conspicuous**; caudal cilia, present; **oral region, apical, with perioral ciliature composed of several rings of ciliature, surrounding the circumoral ciliature**; macronucleus, globular to ellipsoid; micronucleus, present; contractile vacuole, present; cytoproct (?); bacterivorous (?); in marine, brackish, and freshwater habitats; three genera.

- *Metacystis* Cohn, 1866
- *Pelatractus* Kahl, 1930
- *Vasicola* Tatem, 1869

Order Prorodontida Corliss, 1974
(syns. Bursellopsida *p.p.*, Colepina, Placina)

Size, small to large; shape, ovoid to cylindroid; free-swimming; alveoli, well-developed, including calcium carbonate concretions as skeletal plates in the Family Colepidae; somatic ciliation, holotrichous, but may be reduced in posterior half of cell, which typically bears one to many caudal cilia; somatic extrusomes as mucocysts; **oral extrusomes as toxicysts, may be in oral palps or extra-oral, near kinetids of "brosse"**; oral region, apical to subapical, surrounded by circumoral dikinetids; **brosse (homologous to the oral polykinetids of oligohymenophoreans?), typically of three or more dikinetidal rows bearing clavate cilia, varying from parallel to perpendicular to body axis, and developing on parental ventral surface**; cytostome, round or elliptical, sometimes in shallow atrium, which is lined by oral ridges supported by two unequal rows of microtubules (homologous to the oral ribs of oligohymenophoreans?); most species carnivores or scavengers, a few algivorous; widely distributed in freshwater and marine habitats; eight families and one family. *incertae sedis*.

Family BALANIONIDAE Small & Lynn, 1985

Size, small; **shape, ovoid with truncate, flattened anterior pole**; free-swimming; **somatic ciliation, restricted to an apparent girdle encircling the anterior 1/2–3/4 of the cell**; caudal cilium, present; **brosse, inconspicuous, with its units internal to oral dikinetids of circumoral ciliature**; oral region, on flattened anterior end, with very elongate oral "palps" bearing toxicysts, surrounding cytostome; macronucleus, globular to ellipsoid; micronucleus, present; contractile vacuole, present; cytoproct (?); feeding on bacteria, small algae, and heterotrophic flagellates; in marine and freshwater habitats, in plankton; one genus.

- *Balanion* Wulff, 1919

Family COLEPIDAE Ehrenberg, 1838

Size, small to large; shape, barrel-like, with prominent anterior and caudal spines often present; free-swimming; **alveoli as cuirass of longitudinal rows of armored calcium carbonate plates with small lateral teeth**; somatic ciliation, holotrichous, with caudal cilium, typically long; oral extrusomes as toxicysts; brosse as three, short, inconspicuous files of dikinetids; oral region, apical, surrounded by oral dikinetids whose circle may be broken by the intrusion of the brosse kinetids; macronucleus, globular to elongate ellipsoid; micronucleus, present; contractile vacuole, present; cytoproct (?); scavengers and histophages; in marine and freshwater habitats; eight genera.

- *Baikalocoleps* Obolkina, 1995
- *Coleps* Nitzsch, 1827
- *Kotinia* Obolkina in Aescht, 2001
- *Macrocoleps* Obolkina, 1995
- *Plagiopogon* Stein, 1859
- *Planicoleps* Dragesco & Dragesco-Kernéis, 1991
- *Tiarina* Bergh, 1881
- *Tiarinella* Obolkina, 1995

Family HOLOPHRYIDAE Perty, 1852
(syn. Cryptocaryonidae)

Size, small to large; shape, ovoid to cylindroid; free-swimming; somatic ciliation, holotrichous, often dense, sometimes with caudal cilia; oral region, apical or subapical, surrounded by circumoral ciliature; somatic extrusomes as mucocysts and oral extrusomes as toxicysts; **brosse as several to many kinetofragments, with somatic kineties parallel to the brosse (i.e., aklitoloph) or abutting against the brosse region (i.e., euklitoloph)**; oral region, apical to subapical, surrounded by circumoral dikinetids; macronucleus, globular to ellipsoid to elongate and even ribbon-like; micronucleus, present; contractile vacuole, present; cytoproct (?); scavengers and histophages with *Cryptocaryon* as a parasite of marine fishes; in marine and freshwater habitats; five genera and one genus *incertae sedis*.

NOTE: Foissner, Berger, and Kohmann (1994) discovered problems with the nomenclature of *Holophrya*, *Prorodon*, and *Pseudoprorodon*, which necessitated the redescription of the genus *Holophrya* with *Holophrya ovum* Ehrenberg, 1831 as type. This genus presents the form of what was previously known as *Prorodon* (e.g. in Corliss, 1979; Small & Lynn, 1985).

- *Cryptocaryon* Brown, 1951
- *Holophrya* Ehrenberg, 1831
- *Paraprorodon* Foissner, 1983
- *Pelagothrix* Foissner, Berger, & Schaumberg, 1999
- *Pleurofragma* Jankowski, 1976

Incertae sedis in Family Holophryidae

- *Fundenia* Vuxanovici, 1962

Family LAGYNIDAE Sola, Guinea, Longas, & Fernández-Galiano, 1990

Size, small; shape, pyriform when alive; free-swimming; **somatic ciliation as perioral ciliature or a girdle, completely encircling cell apex as "trikinetids"; oral region, apical, with brosse as 3–4 inconspicuous rows between perioral ciliature and circumoral dikinetids**; macronucleus, globular to reniform; micronucleus, may be multiple; contractile vacuole, present; cytoproct (?); feeding on flagellates and other smaller protists; in freshwater habitats; one genus.

- *Lagynus* Quennerstedt, 1867

Family PLACIDAE Small & Lynn, 1985

Size, small to medium; shape, ovoid, slightly flattened; free-swimming; **somatic ciliation, holotrichous, having slightly spiralling kineties with striae between**; somatic extrusomes as mucocysts; **"oral" extrusomes as toxicysts dispersed along the brosse or in a lateral pocket in *Spathidiopsis*; brosse as a single dikinetid file, extending posteriorly from the circumoral dikinetid ring**; oral region, subapical, a slightly elongate groove; macronucleus, ellipsoid to elongate ellipsoid; micronucleus, present; contractile vacuole, present; cytoproct (?); feeding on flagellates and smaller ciliates; in marine and brackish habitats; two genera.

- *Placus* Cohn, 1866
- *Spathidiopsis* Fabre-Domergue, 1889

Family PLAGIOCAMPIDAE Kahl, 1926

Size, small; shape, ovoid; free-swimming; somatic ciliation, holotrichous; caudal cilia, common; oral extrusomes as toxicysts; brosse of three units, on posterior right of oral area, opposite oral dikinetids; **oral region, subapical with oral dikinetids on dorsal-right, as semicircle accompanied by extensible "palps" or lappets in which toxicysts reside**; macronucleus, globular to ellipsoid; micronucleus, may be very small; contractile vacuole, present; cytoproct (?); feeding on bacteria, dinoflagellates, microalgae, and other ciliates; in marine, freshwater, and terrestrial habitats; three genera and one genus *incertae sedis*.

- *Chilophrya* Kahl, 1930
- *Paraurotricha* Foissner, 1983
- *Plagiocampa* Schewiakoff, 1893

Incertae sedis in Family Plagiocampidae

- *Plagiocampides* Foissner, Agatha, & Berger, 2002*

Family UROTRICHIDAE Small & Lynn, 1985 (syn. Bursellopsidae)

Size, small to large; shape, ovoid to cylindroid; free-swimming; **somatic ciliation, with posterior 1/4–1/3 of body non-ciliated and rest of body evenly ciliated**; caudal cilia, typical; brosse, two or three units, outside circumoral dikinetids, and quite elongate in *Longifragma*; oral region, apical to subapical, can be quite expansive, almost cavity-like; macronucleus, globular to elongate ellipsoid; micronucleus, present; contractile vacuole, present; cytoproct (?); feeding on bacteria, microalgae, and other smaller protists; in marine and freshwater habitats, often planktonic; six genera.

- *Bursellopsis* Corliss, 1960
- *Dissothigma* Jankowski, 1976
- *Longifragma* Foissner, 1984
- *Longitricha* Gajewskaja, 1933
- *Rhagadostoma* Kahl, 1926
- *Urotricha* Claparède & Lachmann, 1859

Incertae sedis in Class PROSTOMATEA

- *Amphibothrella* Grandori & Grandori, 1934
- *Peridion* Vuxanovici, 1962
- *Peridionella* Vuxanovici, 1963

Incertae sedis in Class PROSTOMATEA

Family PRORODONTIDAE Kent, 1881 (syn. Amphibot[h]rellidae [for Amphibothridae])

Size, medium to large; shape, ovoid; free-swimming; somatic ciliation, holotrichous; "**brosse" as an extension of the unclosed circumoral ciliature**; oral region, subapical, as an elongate groove; macronucleus, ribbon-like; micronucleus, may be multiple; contractile vacuole, present; cytoproct (?); feeding on other protists, even ciliates and small metazoans; in freshwater habitats; one genus.

NOTE: Foissner, Berger, and Kohmann (1994) discovered problems with the nomenclature of *Holophrya*, *Prorodon*, and *Pseudoprorodon*, which necessitated applying the name *Prorodon* to ciliates similar to the type *Prorodon niveus* Ehrenberg, 1834. This genus presents the form of what was previously known as *Pseudoprorodon* (e.g. in Corliss, 1979; Small & Lynn, 1985), which is an objective synonym of *Prorodon*.

- *Prorodon* Ehrenberg, 1834

Family MALACOPHRYIDAE Foissner, 1980

Size, small; shape, ovoid; free-swimming; **alveoli, regularly patterned, almost quadrangular**; somatic ciliation, holotrichous, with kinetids of bipolar kineties disposed in paratenes, and posterior partially non-ciliated; caudal cilia, present; oral region, subapical; **oral dikinetid along right side of oral area; two elongate "brosse" units along left side of oral area, similar (?) to oral polykinetids of oligohymenophoreans**; cyrtos with fine nematodesmata; macronucleus, globular; micronucleus, present; contractile vacuole, present; cytoproct (?); feeding (?); in terrestrial habitats; one genus.

NOTE: Deroux (1994) placed this family as *incertae sedis* in the Subclass Nassulia. We are placing it here for two reasons: the development of parateny, a feature characteristic of the prostomes; and a presumption that its oral polykinetids are homologous to "brosse" units. However, this placement must be confirmed by ultrastructural study and molecular genetic data.

- *Malacophrys* Kahl, 1926

Class **PLAGIOPYLEA** Small & Lynn, 1985
(syns. Contofragmea, Contofragmina *p.p.*, Plagiopylia, Spirotrichophora *p.p.*)

Size, small to large; shape, variable, but often flattened; free- swimming; alveoli, well-developed, often filled with dense material; somatic ciliation, holotrichous, very dense in large forms, but extremely reduced in smaller forms; somatic monokinetid with divergent postciliary ribbon, well-developed anteriorly-directed kinetodesmal fibril, and a transverse ribbon arising from dense material near triplets 2 and 3, extending laterally in Trimyemidae and anteriorly at least in Plagiopylidae, but if odontostomatids are correctly placed here, they typically have dikinetids; somatic extrusomes as mucocysts, which may be elongate and rod-shaped; oral region, variable, from subapical to post-equatorial, may be deeply invaginated; **cytostome partially encircled by one or two files of dikinetids (?), but if odontostomatids are correctly placed here, oral ciliature can include polykinetids**; stomatogenesis, holotelokinetal, but may be apokinetal in odontostomatids (?); macronucleus, homomerous; conjugation, temporary; mitochondria may be replaced by hydrogenosomes, which in many species are associated with endosymbiotic methanogens; bacterivorous and algivorous; in marine and freshwater habitats, especially common in anaerobic salt-marsh and interstitial biotopes, and sometimes as endocommensals in the digestive tracts of echinoids and hippopotami; one order and one order *incertae sedis*.

NOTE: Lynn (2004) characterized this class as a "riboclass" because there is strong support from small subunit rRNA gene sequences for uniting the included families, but no strong morphological synapomorphies. Ultrastructural observations (Lynn, 1991) reveal circumcytostomal dikinetid units at the base of the oral cavity of plagiopylids. Research is needed to confirm these as homologues of the trimyemid oral dikinetids. Stoeck et al. (2007) have demonstrated that the odontostomatid genus *Epalxella* strongly associates with trimyemids and plagiopylids, providing preliminary support for our transfer of the Order Odontostomatida to this class.

Order Plagiopylida Jankowski, 1978
(syn. Perikinetida, Trimyemida *p.p.*)

Size, small to large; **typically with sandwich-like arrangement of the hydrogenosome-methanogen assemblages;** three families.

Family PLAGIOPYLIDAE Schewiakoff, 1896
(syn. Paraplagiopylida, Paraplagiopylidae)

Size, small to medium; **shape, somewhat bean-shaped, flattened, with oral region at indented part**; free-swimming; **somatic ciliation, holotrichous, dense, with somatic kineties extending from dorsal surface over the anterior end to terminate on the oral region;** with striated band on right surface arising near right margin of oral cavity; **oral cavity, deep, transverse, opening ventrally with a more internal tubular part preceding the cytostome**; oral ciliature as extensions of somatic kineties, more densely packed with kinetosomes, lining the tubular part and terminating at the cytostome; macronucleus, globular to ellipsoid; micronucleus, present; contractile vacuole, present; cytoproct (?); feeding on bacteria and cyanobacteria; in marine, brackish, and freshwater habitats, particularly anaerobic biotopes with several species endocommensal in the digestive tracts of sea urchins and hippopotami; four genera.

- *Lechriopyla* Lynch, 1930
- *Paraplagiopyla* Thurston & Grain, 1971
- *Plagiopyla* Stein, 1860
- *Pseudoplagiopyla* Small & Lynn, 1985

Family SONDERIIDAE Small & Lynn, 1985
(syn. Parasonderiidae)

Size, small to large; shape, ovoid, somewhat flattened dorsoventrally; free-swimming; somatic ciliation, holotrichous, dense; with striated band on right surface arising near right margin of oral cavity; **oral cavity flattened, conical, deep, opening apically to subapically**; oral ciliature as extensions of somatic kineties, with densely packed kinetids, lining surfaces of cavity; macronucleus, globular to ellipsoid; micronucleus, present; contractile vacuole, present; cytoproct, present; feeding on bacteria, cyanobacteria, and microalgae, such as diatoms; in marine and brackish water habitats, rarely freshwater; four genera.

- *Oncosonderia* Jankowski, 1980
- *Parasonderia* Jankowski, 2007
- *Sonderia* Kahl, 1928
- *Sonderiella* Kahl, 1928

Family TRIMYEMIDAE Kahl, 1933
(syn. Sciadostom[at]idea)

Size, small; shape, ovoid, tapered at both ends; free-swimming; **somatic ciliation, holotrichous, but kineties with kinetosomes much reduced in numbers, appearing as several spirals around body**; caudal cilium, prominent; **without striated band**; oral cavity, a shallow depression, opening to the right; **oral ciliature as two semicircular files of dikinetids on the left and several (2–4) dikinetid units at the right anterior end of the semicircular files**; macronucleus, globular; micronucleus, present; contractile vacuole, present; cytoproct (?); bacterivorous; in marine and freshwater anaerobic habitats, sometimes occurring as endocommensals in the intestines of echinoids; one genus.

– *Trimyema* Lackey, 1925

Incertae sedis in Class PLAGIOPYLEA

Order Odontostomatida Sawaya, 1940
(for Ctenostomata, Ctenostom[at]ida, Ctenostomina; syn. Odontostomata)

Size, small; **shape, discoid, laterally compressed, wedge- or helmet-shaped, typically nearly as wide as long, with armour-like cuirass and often short posterior spines**; somatic ciliature, reduced, typically as dikinetids or occasionally cirrus-like, with somatic kineties often separated into anterior and posterior segments; **oral polykinetids, inconspicuous, typically <10 in number**; paroral, sometimes present, typically inconspicuous; stomatogenesis, possibly apokinetal; macronucleus, globular, one to several; cytoproct (?); absent; feeding (?); in marine, brackish, and freshwater anaerobic habitats; three families.

NOTE: The Order Odontostomatida with its three included families is tentatively transferred to the Class PLAGIOPYLEA based only on the results of the small subunit rRNA gene sequence of *Epalxella*, which is unambiguously associated with plagiopylids and trimyemids (Stoeck et al., 2007). It will be necessary to at least obtain gene sequences of representatives of the other two families to resolve this uncertainty.

Family DISCOMORPHELLIDAE Corliss, 1960
(for Discomorphidae [for Ctenostom [at] idae])

Size, small; **shape, discoidal, laterally compressed, smooth in outline except for two prominent anterior spines and one posterior spine**; free-swimming; somatic ciliature, sparse and fragmented into several cirrus-like clusters posteriorly; **a preoral band of cilia on a conspicuous transverse ridge anterior to the oral cavity**; oral cavity in posterior half with several oral polykinetids and a paroral; macronucleus, globular; micronucleus (?); contractile vacuole, present; cytoproct (?); feeding (?); in freshwater anaerobic habitats; one genus.

– *Discomorphella* Corliss, 1960

Family EPALXELLIDAE Corliss, 1960
(for Epalcidae and Epalxidae)

Size, small; shape, box-like, generally with short posterior spines, and some species with well-developed spine overhanging hidden oral cavity; free-swimming; **somatic ciliature, relatively dense in its short anterior and posterior linear kinetofragments, with those at the anterior end parallel to and anterior of the oral opening on both left and right sides of the body**; oral cavity, roughly equatorial, with several oral polykinetids; paroral, may be absent; macronucleus, globular to ellipsoid; micronucleus, present; contractile vacuole, present; cytoproct (?); feeding (?); in marine and freshwater anaerobic habitats, principally freshwater, but a few *Epalxella* species also in marine biotopes; three genera.

– *Epalxella* Corliss, 1960
– *Pelodinium* Lauterborn, 1908
– *Saprodinium* Lauterborn, 1908

Family MYLESTOMATIDAE Kahl in Doflein & Reichenow, 1929
(syns. Atopodiniidae, Mylestomidae)

Size, small; shape, discoid; free-swimming; **somatic ciliation very sparse, with anterior kinetofragments restricted to the ventral (= oral) surface, extending only slightly onto the left and right surface, and with longer cilia at posterior,**

cirrus-like in *Myelostoma*; oral cavity, in posterior half, with several oral polykinetids; paroral, may be absent; macronucleus, globular to ellipsoid; micronucleus, present; contractile vacuole, present; cytoproct (?); feeding (?); in marine and freshwater anaerobic habitats, predominantly in freshwater but a few *Mylestoma* species are marine; two genera.

– *Atopodinium* Kahl, 1932
– *Mylestoma* Kahl, 1928

Class **OLIGOHYMENOPHOREA** de Puytorac et al., 1974
(syns. Aspirigera *p.p.*, Aspirotricha *p.p.*, Cyrtostomata *p.p.*, Holotricha [Holotrichasina, Holotrichia] *p.p.*, Homoiotricha *p.p.*, Kinetodesmatophora *p.p.*, Membranellophora [Membranellata] *p.p.*, Stomatea *p.p.*, Axotrichidea + Peritrichidea; Hymenotricha (*sensu* Raabe) + Peritricha; Tetrahymenophora + Cyclohymenophora)

Size, small to medium, rarely large; shape, typically ovoid to elongate ovoid; free-swimming, but sessile and sedentary in Subclass Peritrichia; alveoli, well-developed; somatic ciliation, holotrichous; somatic monokinetids with anteriorly directed, distinct, overlapping kinetodesmal fibrils, divergent postciliary ribbons, and radial transverse ribbons (except in Subclass Peniculia, which has tangential transverse ribbons); posterior kinetosome of somatic dikinetids similar to that of a monokinetid unit while anterior kinetosome bears only a tangential transverse ribbon; parasomal sacs, typically to left or anterior of kinetosomes; somatic extrusomes as mucocysts, but trichocysts common in Subclass Peniculia in which mucocysts are rare; **oral apparatus with a distinct right paroral of dikinetids (i.e. stichodyad) and typically three left oral polykinetids, but oral apparatus absent in Subclass Astomatia, and highly modified in the Subclass Apostomatia**; stomatogenesis varies with subclass, of buccokinetal or parakinetal types; division while free-swimming, but typically in cyst in parasitic forms; macronucleus, homomerous, typically single; micronucleus, one to many; conjugation, usually temporary, but total in Subclass Peritrichia; cytoproct, typically located in the director meridian; feeding on bacteria and microalgae, but occasionally carnivorous in larger forms with the endocommensal astomes entirely osmotrophic (?) in marine, freshwater, and terrestrial habitats, distributed widely as free-living forms with many species of the Subclass Peritrichia as symphorionts and the entire Subclass Astomatia endocommensalistic; six subclasses.

Subclass Peniculia Fauré-Fremiet in Corliss, 1956
(syns. Trichohymenostomata and Vestibulata *sensu* von Gelei)

Size, medium; shape, ovoid; free-swimming; alveoli, well-developed; somatic ciliation, holotrichous, typically dense, with distinct pre- and postoral sutures, but Order Urocentrida with girdle-like ciliature; caudal cilia, often conspicuous; **somatic kinetids with tangential, not radial, transverse ribbons, but other fibrillar associates as for class**; somatic extrusomes as trichocysts, but mucocysts in Order Urocentrida; **oral structures, typically three left oral polykinetids with the long axis of the polykinetid (peniculus) parallel to the long axis of the oral cavity, and with alveoli between kinetosomal rows of oral polykinetids**; paroral, reduced, but present throughout interphase; no cyrtos, but nematodesmata may be associated with oral and perioral kinetosomes, sometimes loosely basket-like; **stomatogenesis, ophryobuccokinetal**; many species with endosymbiotic algae or bacteria; microphagous bacterivores, algivores, and some species carnivorous (e.g. *Neobursaridium*); cysts, rare; distributed widely, predominantly in freshwater habitats, but some marine species; two orders.

NOTE: Lynn and Small (2002) presented a single Order Peniculida divided into the Suborders Frontoniina and Parameciina. However, small subunit rRNA gene sequences do not strongly support this subdivision (Fokin, Andreoli, Verni, & Petroni, 2006; Strüder-Kypke, Wright, Fokin, & Lynn, 2000b), but do provide support, corroborated by morphology, for separation of the urocentrids at a higher level (Didier & de Puytorac, 1994).

Order Peniculida Fauré-Fremiet in Corliss, 1956
(syns. Frontoniina; Frontoniina + Quadrulina *p.p.*; Lembadionina, Parameciina, Peniculina)

Size, small to large; **somatic ciliation, holotrichous; somatic kinetids, predominantly dikinetids; somatic extrusomes as trichocysts**; seven families.

Family CLATHROSTOMATIDAE Kahl, 1926
(syn. Clathrostomidae)

Size, medium; shape, ovoid; free-swimming; somatic ciliation, holotrichous, with preoral and postoral sutures; ophryokineties, absent; oral region in anterior 1/4 of body; **oral apparatus with six oral polykinetids, each one a file of dikinetids; oral nematodesmata, forming a ring around cytopharynx**; macronucleus, elongated ellipsoid; micronucleus, present; contractile vacuole, present; cytoproct, present; feeding on bacteria and microalgae; in freshwater habitats; one genus.

- *Clathrostoma* Penard, 1922

Family FRONTONIIDAE Kahl, 1926
(for Chiliferidae)

Size, medium to large; shape, ovoid to bluntly ovoid; free-swimming; somatic ciliation, holotrichous, with postoral suture only; **ophryokineties, often many, to right of oral region**; **oral region, subapical, elongate in anterior 1/2–1/3 of body; prebuccal area, shallow or absent**; oral apparatus, typically three long, large polykinetids in oral cavity with the paroral along its right border; **cytostome expansible along postoral suture**; oral nematodesmata not forming a ring around the cytopharynx, but more prominent to its side and posterior; macronucleus, ellipsoid to elongate ellipsoid, sometimes band-form; micronucleus, present; contractile vacuole, single; cytoproct, present; often containing zoochlorellae; feeding on bacteria and microalgae; in marine, freshwater, and terrestrial habitats, typically free-living but one species commensal (?) on gills of *Amphioxus*; seven genera and two genera *incertae sedis*.

- *Apofrontonia* Foissner & Song, 2002*
- *Didieria* Small & Lynn, 1985
- *Disematostoma* Lauterborn, 1894
- *Frontonia* Ehrenberg, 1838
- *Frontoniella* Wetzel, 1927 (subj. syn. *Frontonia*)
- *Paraclathrostoma* Small & Lynn, 1985
- *Wenrichia* Jankowski, 1967 (subj. syn. *Disematostoma*)

Incertae sedis in Family Frontoniidae

- *Schistophrya* Kahl, 1933
- *Sigmostomum* Gulati, 1925 (subj. syn. *Frontonia*)

Family LEMBADIONIDAE Jankowski in Corliss, 1979

Size, small to medium; shape, broadly ovoid; free-swimming; somatic ciliation, holotrichous, with postoral suture; caudal cilia, long, forming a tuft; **oral region, expansive, occupying nearly entire ventral surface; left oral cilia appearing as one long polykinetid (probably longitudinally fused peniculi)**; **paroral, long, accompanied on its right by a long file of dikinetids (an ophryokinety?)**; macronucleus, elongate, ellipsoid; micronucleus, present; contractile vacuole, single; cytoproct (?); feeding on flagellates and other ciliates; in freshwater habitats, often planktonic; one genus.

- *Lembadion* Perty, 1849

Family MARITUJIDAE Jankowski in Small & Lynn, 1985

Size, medium; **shape, subspheroid, barrel-shaped;** free-swimming; **somatic ciliation, holotrichous, forming distinct transverse paratenes over the body; ophryokineties, numerous, encircling "right" of oral region; oral region, on anterior surface of body**; oral structures as a paroral and three long polykinetids (i.e. peniculi); macronucleus, ribbon-like; micronucleus, present; contractile vacuole, present; cytoproct (?); feeding on bacteria and microalgae (?); in freshwater habitats, planktonic; one genus.

- *Marituja* Gajewskaja, 1928

Family NEOBURSARIDIIDAE Dragesco & Tuffrau, 1967

Size, large, to over 700 μm in length; shape, elongate ovoid, typically rounded and broader at anterior end; free-swimming; **somatic ciliation, holotrichous, dense, with "heterotrich-looking", extensive, false adoral zone of polykinetids, actually formed by transverse paratenes of somatic kineties, lying in a prebuccal area as a much expanded preoral groove occupying the anterior half of the body**; oral region, midventral, expansive, with inner or right-most oral polykinetid of four widely spaced rows (i.e., a quadrulus); macronucleus, dumbbell-shaped; micronuclei, multiple; contractile vacuoles, two; cytoproct, present;

carnivorous on other ciliates; **in freshwater habitats, only pan-tropical**; one genus.

– *Neobursaridium* Balech, 1941

Family PARAMECIIDAE Dujardin, 1840

Size, medium; **shape, elongate ovoid, rounded and/or pointed at either or both ends (the "slipper-shaped animalcules," so long and well known);** free-swimming; **somatic ciliation, holotrichous, dense, with a prebuccal area as a preoral groove (formerly called a "vestibulum") not so much expanded, covered by paratenes, and leading to oral region**; oral cavity in anterior half to equatorial, with inner or right-most oral polykinetid of four widely spaced rows (i.e., a quadrulus) on dorsal wall of oral cavity; macronucleus, ellipsoid to elongate ellipsoid; micronucleus, may be multiple; contractile vacuoles, typically two; cytoproct, present; some species containing zoochlorellae; feeding on bacteria and microalgae; in brackish and freshwater habitats; one genus and one genus *incertae sedis*.

– *Paramecium* O.F. Müller, 1773

Incertae sedis in the Family Parameciidae

– *Physanter* Jankowski, 1975

Family STOKESIIDAE Roque, 1961

Size, medium; **shape, distinctly cone- or heart-shaped, with flattened ventral surface and humped dorsal surface**; free-swimming; somatic ciliation, holotrichous; **oral region, relatively large, as a V-shaped ventral depression on a somewhat flattened ventral body surface**; oral region, midventral, with inner or right-most oral polykinetid of four widely spaced rows (i.e., a quadrulus); macronucleus, ellipsoid to lenticular; micronucleus, may be multiple; contractile vacuole, present; cytoproct, present; feeding on flagellates and microalgae, such as diatoms; in freshwater habitats, typically planktonic; two genera.

– *Parastokesia* Jankowski, 1967 (subj. syn. *Disematostoma*)
– *Stokesia* Wenrich, 1929

Order Urocentrida Jankowski, 1980
(syn. Urocentrina)

Size, medium; **shape, short, cylindrical, with larger, rounded anterior half**; free-swimming, but may be temporarily attached to the substratum by a mucous thread; **somatic ciliation as a distinct equatorial girdle; caudal cilia, forming a conspicuous tuft that is used for temporary attachment to substrates by a mucous thread; somatic kinetids only as monokinetids with broad, tangential transverse ribbon; somatic extrusomes as mucocysts**; no depressed preoral area; oral region, equatorial to subequatorial; oral structures as a paroral along the right margin of the oral opening and three oral polykinetids of three rows each along the dorsal-left wall; macronucleus, band-like; micronucleus, present; contractile vacuole, single, with multiple collecting canals; cytoproct (?); feeding on bacteria and smaller protists; in freshwater habitats, typically in ponds where it attaches to the substrate and rotates on its mucous thread; one family.

Family UROCENTRIDAE Claparède & Lachmann, 1858
(syn. Calceolidae)

With characteristics of the order; one genus.

– *Urocentrum* Nitzsch, 1827

Subclass Scuticociliatia Small, 1967
(syns. Scuticostomata, Stichostomata *p.p.*)

Size, small to medium; shape, ovoid to elongate ovoid; mostly free-swimming, rarely restricted to a secreted lorica; alveoli, well-developed; somatic ciliation, holotrichous, though sometimes sparse, with thigmotactic field of somatic cilia in some symbiotic species; caudal cilia, often one or more; somatic dikinetids, predominant, both kinetosomes ciliated over much of the body; extrusomes as somatic mucocysts; mitochondria, large, elongate, cortically located, often-fused as a single cortical chondriome; **oral region, quite variable in shape and extent, with right paroral as a file of dikinetids divided into *a*, *b*, and *c* segments, especially conspicuous during stomatogenesis, and typically three oral polykinetids, often as membranoids, on left; stomatogenesis, scuticobuccokinetal, involving proliferation of**

kinetosomes at least from paroral *c* segment or a scutica or scuticovestige, posterior or parallel to paroral *a* and *b* segments; macronucleus, typically single but fragmented in some species; micronuclei, one to many; commonly bacterivorous, but facultatively (?) parasitic species, often histophagous on crustaceans and fish; cysts widespread; abundant in marine habitats, but also in some freshwater and terrestrial habitats, as free-living forms or in symbiotic association primarily with invertebrates, such as molluscs, echinoids, and annelids; three orders and one family *incertae sedis*.

NOTE: Lynn and Strüder-Kypke (2005) have demonstrated that this subclass is basically monophyletic using small subunit rRNA (SSUrRNA) gene sequences. However, the familial assignments of genera proposed in the classification below, which is based primarily on oral features, both morphostatic and stomatogenetic, receives little support from SSUrRNA gene sequences (Lynn & Strüder-Kypke, 2005; Shang, Song, & Warren, 2003). We are currently maintaining this morphological classification for the purposes of stability, and until there are more gene sequence data on this subclass, both from more genes and a wider diversity of genera and species.

Order Philasterida Small, 1967
(syns. Cinetochilina *p.p.*, Deuterostomatina *p.p.*, Loxocephalida *p.p.*, Loxocephalina *p.p.*, Pseudocohnilembina *p.p.*, Schizocaryumida *p.p.*, Urozonina *p.p.*, Thigmophryina *p.p.*)

Size, small to large; shape, ovoid to elongate ovoid, often flattened laterally, especially in symbiotic forms; somatic ciliation, holotrichous; **oral region, typically in anterior half, with paroral dikinetid shorter than other oral structures, typically by reduction of paroral *a* and *c* segments; no ribbed wall from paroral towards cytostome except in Cinetochilidae and Loxocephalidae (see NOTE); scutica typically present, often in anterior part of distinct director-meridian**; most commonly in brackish or marine habitats, including sand, with numerous species free-living, but some occurring as endocommensals in sea urchins, molluscs, coelenterates, annelids, sipunculids, and even the sea horse while others can be facultative parasites; 16 families.

NOTE: Li et al. (2006) have suggested establishment of the order Loxocephalida based on the sequences of the small subunit rRNA genes for *Dexiotrichides* and *Cardiostomatella*, two of the at least 14 genera that they suggest should be included in this order. We await additional sequence information before recognizing this order, but note that Jankowski (1964) has already suggested establishing the Suborder Loxocephalina.

Family CINETOCHILIDAE Perty 1852

Size, very small to small; **shape, ovoid to ellipsoid, usually flattened, sometimes looking very much like a baseball catcher's mitt!**; free-swimming; somatic ciliation, sparse, limited to ventral surface with one or more long caudal cilia; **oral area relatively large, midventral, with pronounced ribbed wall**; **in stomatogenesis, oral polykinetids 2 and 3 originating from scutica or scuticovestige**; macronucleus, globular to ellipsoid; micronucleus, present; contractile vacuole, present; cytoproct occupying all of director-meridian area on foreshortened ventral surface of some species; feeding on bacteria and smaller protists; in freshwater and terrestrial habitats, widespread with fewer species from brackish or marine habitats; seven genera.

- *Cinetochilum* Perty, 1849
- *Cinetozona* Olmo, Tellez, & Esteban, 1998
- *Platynematum* Foissner, Berger, & Kohmann, 1994
- *Pseudocinetochilum* Obolkina, 1995
- *Pseudoplatynematum* Bock, 1952
- *Sathrophilus* Corliss, 1960
- *Sphenostomella* Jankowski, 1980

Family COHNILEMBIDAE Kahl, 1933
(for Lembidae)

Size, medium; **shape, slender, finger-shaped, tapering to point anteriorly**; free-swimming; somatic ciliation, holotrichous, with one or more long caudal cilia; **oral region extending along tapered anterior, with oral polykinetid 1, very long relative to the other two oral polykinetids and with dense dikinetids of kinety *n* alongside it, forming a conspicuous, false "double-membrane"**; scutica, composed of several kinetosomes in a triangular arrangement with some trailing posteriorly as a file in the director meridian, providing the origin of kinetsomes for all oral polykinetids; macronucleus, globular to ellipsoid to ribbon-like; micronucleus, present; contractile vacuole, present; cytoproct, present; bacterivorous; in marine and saline terrestrial habitats, including Great Salt Lake; one genus.

- *Cohnilembus* Kahl, 1933

Family CRYPTOCHILIDAE Berger in Corliss, 1979

Size, medium to large; **shape, usually laterally compressed and tapered, anteriorly and posteriorly, commonly with caudal projection bearing one or more longer cilia**; free-swimming; somatic ciliation, holotrichous, dense; **oral region, usually posterior to equator with oral polykinetid 2 relatively well developed**; macronucleus, globular to ellipsoid; micronucleus, present; contractile vacuole, present; cytoproct, present; bacterivorous; in marine habitats as endocommensals in the intestines of sea urchins, and a few species in wood-boring molluscs; eight genera.

- *Biggaria* Aescht, 2001
- *Biggariella* Profant in Corliss, 1979 [nomen nudum]
- *Cryptochilum* Maupas, 1883
- *Metoikos* Berger & Thompson in Corliss, 1979 [nomen nudum]
- *Tanystomium* Berger in Corliss, 1979 [nomen nudum]
- *Thigmozoon* Santhakumari & Nair, 1973
- *Velistoma* Jankowski, 1980
- *Yagiua* Profant in Corliss, 1979 [nomen nudum]

Family ENTODISCIDAE Jankowski, 1973

Size, medium; **shape, ovoid, laterally flattened (= disc-like), with narrow, truncated anterior end and, in some species, with small caudal projection bearing single caudal cilium**; free-swimming; somatic ciliation, holotrichous; oral region, anterior, inconspicuous, with oral polykinetid 1 smaller than other two oral polykinetids; macronucleus, globular to ellipsoid; micronucleus, present; contractile vacuole, present; cytoproct, present; bacterivorous and detritivorous; in marine habitats as endocommensals in the intestines of sea urchins (*Entodiscus*), in the mantle cavity of bivalve molluscs (*Pectenita*) or in the esophagus of sipunculids (*Cryptochilidium*); four genera.

- *Cryptochilidium* Schouteden, 1906
- *Entodiscoides* Song, Warren, & Wilbert, 1996
- *Entodiscus* Madsen, 1931
- *Pectenita* Jankowski, 1973

Family ENTORHIPIDIIDAE Madsen, 1931

Size, medium to large; **shape, flattened laterally, with a prominent suture on the broad anterior end and with the posterior end tapered to tail**; somatic ciliation, holotrichous, dense, with single caudal cilium; **oral region, an anterior, small, inconspicuous cavity, overhung by the frontal lobe of the body, containing a short paroral and three small oral polykinetids**; macronucleus, ellipsoid to elongate ellipsoid; micronucleus, may be multiple; contractile vacuole, present; cytoproct, present; bacterivorous and detritivorous; in marine habitats as endocommensals in the intestines of sea urchins; one genus.

- *Entorhipidium* Lynch, 1929

Family LOXOCEPHALIDAE Jankowski, 1964 (syn. Cardiostomatellidae)

Size, small to large; shape, elongate-ovoid, with naked apical end in some genera, and darkly-appearing cytoplasm due to mineral inclusions; free-swimming; somatic ciliation, holotrichous; **somatic kineties with pronounced parateny near anterior end, appearing as perizonal kineties especially to the right of the oral region**; caudal cilium, one or more; **oral region, a small anterior cavity, with rectangular oral polykinetids arranged in a *Tetrahymena*-like pattern, and a pronounced ribbed wall extending from a short paroral**; postoral suture, conspicuous, replacing the director-meridian in some species; in stomatogenesis, all oral polykinetids originating from scutica or scuticovestige; macronucleus, globular to ellipsoid, rarely as multiple nodules; micronucleus, may be multiple; contractile vacuole, present, often at level of macronucleus; cytoproct, present, large and band-like; microphagous, typically on bacteria; in freshwater and occasionally brackish water and terrestrial habitats, preferring polysaprobic or sapropelic environments; eight genera.

- *Balanonema* Kahl, 1931
- *Cardiostomatella* Corliss, 1960
- *Dexiotricha* Stokes, 1885
- *Dexiotrichides* Kahl, 1931
- *Loxocephalus* Eberhard, 1862
- *Paradexiotricha* Grolière, 1975
- *Paraloxocephalus* Small & Lynn, 1985
- *Paratetrahymena* Thompson, 1963

Family ORCHITOPHRYIDAE Cépède, 1910
(syn. Paranophryidae)

Size, small to medium; shape, ovoid; free-swimming; somatic ciliation, holotrichous; caudal cilium, often present; oral region in anterior 1/3–1/2 body; **paroral *c* segment (= scuticovestige) aligned along midventral postoral region**; macronucleus, globular to ellipsoid; micronucleus, present; contractile vacuole, present; cytoproct, present; bacterivorous and histophagous; in marine habitats, especially significant as facultative parasites of crustaceans, asteroids, and fish; five genera.

- *Anophryoides* de Puytorac & Grolière, 1979 (subj. syn. *Paranophrys*)
- *Mesanophrys* Small & Lynn in Aescht, 2001
- *Metanophrys* de Puytorac, Grolière, Roque, & Detcheva, 1974
- *Orchitophrya* Cépède, 1907
- *Paranophrys* Thompson & Berger, 1965

Family PARALEMBIDAE Corliss & de Puytorac in Small & Lynn, 1985
(syn. Anophryidae)

Size, small; **shape, ovoid with anterior and posterior ends naked**; free-swimming; somatic ciliation, holotrichous; caudal cilium, long; **oral region, may be extensive, with long paroral beginning at level of oral polykinetid 2, which is more than three times longer than oral polykinetid 3, and with oral polykinetid 3 typically oriented transverse to previous two**; scutica, a large patch of kinetosomes; macronucleus, globular to ellipsoid, rarely elongate; micronucleus, present; contractile vacuole, present; cytoproct, present; bacterivorous and histophagous; in marine habitats with some species in marine detritus or "snow"; six genera.

- *Anophrys* Cohn, 1866
- *Cryptolembus* Gunderson, 1985
- *Magnalembus* Small & Lynn, 1985
- *Mesolembus* Small & Lynn, 1985
- *Ovolembus* Small & Lynn, 1985
- *Paralembus* Jankowski, 2007

Family PARAURONEMATIDAE Small & Lynn, 1985

Size, small; shape, pyriform to ovoid; free-swimming; somatic ciliation, holotrichous, not dense; caudal cilium, prominent; **oral region in anterior half of body, as shallow cavity with posterior segments of the paroral segment skewed to left of midventral postoral region**; **scutica often as linear file, in middle to left side of director meridian, sometimes extending anteriorly to the posterior end of the paroral, but can be "Y"-shaped**; macronucleus, globular to ellipsoid; micronucleus, present; contractile vacuole, present; cytoproct, present; life cycle of several genera including microstome-macrostome transformation; bacterivorous or histophagous; in marine habitats; four genera.

- *Glauconema* Thompson, 1966
- *Miamiensis* Thompson & Moewus, 1964
- *Parauronema* Thompson, 1967
- *Potomacus* Thompson, 1966

Family PHILASTERIDAE Kahl, 1931
(syn. Frontoniidae *p.p.*, Porpostomatidae *p.p.*)

Size, small to large; shape, elongate to finger-shaped, though ovoid in smaller genera, with anterior end bluntly tapered; free-swimming; somatic ciliation, holotrichous, dense; caudal cilium, single, may be inconspicuous; **oral region, an anterior cavity or depression, usually shallow, rarely extending to equator**; **oral polykinetid 1 triangular, equal to or smaller than oral polykinetid 2**; **scutica, elongated as posterior extension of paroral into director meridian**; macronucleus, globular to ellipsoid; micronucleus, present; contractile vacuole, present; cytoproct, present; bacterivorous, but sometimes histophagous; in marine habitats, predominantly as free-living forms, although several species are facultative parasites, being histophagous on or in invertebrates and vertebrates; seven genera.

- *Helicostoma* Cohn, 1866
- *Kahlilembus* Grolière & Couteaux, 1984
- *Madsenia* Kahl, 1934
- *Paraphilaster* Grolière, de Puytorac, & Grain, 1980
- *Philaster* Fabre-Domergue, 1885
- *Philasterides* Kahl, 1931
- *Porpostoma* Moebius, 1888 (subj. syn. *Helicostoma*)

Family PSEUDOCOHNILEMBIDAE Evans & Thompson, 1964

Size, small; shape, elongate-pyriform; free-swimming; somatic ciliation, holotrichous, sparse; caudal cilium, present; **oral region, long and shallow; oral polykinetids with long axes aligned more or less with long axis of body and oral region; paroral extending to middle of oral polykinetid 1, which itself extends to the anterior end and in line with the paroral, oral polykinetid 2 as one kinetosomal file parallel to these, and oral polykinetid 3 as tiny left posterior-lateral extension of oral polykinetid 2**; macronucleus, globular to ellipsoid; micronucleus, present; contractile vacuole, present; cytoproct, present; bacterivorous or scavengers; in marine, brackish, and saline habitats as free-living forms with occasional species endocommensal in the intestines of echinoids, but freshwater and coprozoic strains known in one species; one genus.

- *Pseudocohnilembus* Evans & Thomspon, 1964

Family SCHIZOCARYIDAE Jankowski, 1979
(syn. Schizocary[um]idae)

Size, medium; **shape, ovoid and somewhat dorsoventrally flattened**; free-swimming, but highly thigmotactic; somatic ciliation, holotrichous, dense; **somatic kinetids, cirrus-like, as unique polykinetids with the posterior rightmost kinetosome bearing the typical three fibrillar associates of the class and 2–12 or more kinetosomes clustered anteriorly to it; oral region in anterior half as an elongate groove with oral ciliature apparently a series of transverse rows of kinetosomes**; division by palintomy within a reproductive cyst, involving dedifferentiation of trophont oral structures prior to cell divisions; **macronucleus, fragmented, up to eight or more irregular pieces**; micronucleus, present; contractile vacuole, present; cytoproct (?); bacterivorous and detritivorous; in marine habitats as endocommensals in the esophagus of echinoderms, so far recorded only from the Pacific Ocean basin; one genus.

NOTE: Lynn and Strüder-Kypke (2002) have used small subunit rRNA gene sequences to relate *Schizocaryum* to philasterine scuticociliates, close to the genus *Anophryoides*. Careful examination of division morphogenesis may uncover developmental characters that corroborate this placement.

- *Schizocaryum* Poljansky & Golikova, 1957

Family THIGMOPHRYIDAE Chatton & Lwoff, 1926
(syns. Cochliodomidae, Conchophyllidae, Myxophyllidae)

Size, medium; shape, elongate ovoid, laterally flattened; free-swimming; **somatic ciliation, holotrichous, dense, with very dense thigmotactic ciliature, on the anterior left surface of the body; oral region, a cavity in the posterior 1/4 of the body, with oral ciliature as a reduced and inconspicuous paroral and a single oral polykinetid**; macronucleus, globular to ellipsoid, but sometimes band-form and nodular; micronucleus, present; contractile vacuole, present; cytoproct (?); bacterivorous(?); in marine and terrestrial habitats as commensals in the mantle cavity and occasionally the slime of terrestrial pulmonates and especially marine bivalve molluscs, and one species endocommensal in a nemertine worm, which itself lives in the mantle cavity of a bivalve mollusc; six genera.

- *Cochliodomus* Raabe, 1971
- *Cochliophilus* Kozloff, 1945
- *Conchophyllum* Raabe, 1936
- *Myxophthirus* Da Silva Neto, 1992
- *Myxophyllum* Raabe, 1934
- *Thigmophrya* Chatton & Lwoff, 1923

Family THYROPHYLACIDAE Berger in Corliss, 1961
(syn. Thyrophylaxidae)

Size, large; shape, ovoid, laterally compressed, with prominent anterodorsal suture and minute caudal projection; free-swimming; somatic ciliation, holotrichous, dense; **oral region as large, deep oral cavity with right somatic kineties lining its right wall and a large oral polykinetid 2 lining the entire left wall**; macronucleus, ellipsoid to elongate ellipsoid; micronuclei, numerous; contractile vacuoles, multiple; cytoproct, present; **carnivorous on other ciliates**; in marine habitats as endocommensals in the intestines of echinoids, so far recorded only from the Pacific Ocean basin; two genera.

- *Plagiopyliella* Poljansky, 1951
- *Thyrophylax* Lynn & Berger, 1973

Family URONEMATIDAE Thompson, 1964

Size, small; **shape, ovoid to elongate-ovoid, with anterior end slightly flattened and anterior pole conspicuously naked**; free-swimming; somatic ciliation, holotrichous, sparse; caudal cilium, one or more; oral region, anterior, shallow depression; **oral polykinetids reduced, with oral polykinetid 1 typically non-ciliated and with relatively inconspicuous oral ciliature on the other two oral polykinetids**; **scutica, a triangular arrangement of ciliated kinetosomes in the director meridian**; macronucleus, globular to ellipsoid; micronucleus, present; contractile vacuole, present; cytoproct, present; bacterivorous; in marine, freshwater, and occasionally terrestrial habitats; seven genera.

- *Homalogastra* Kahl, 1926
- *Pseuduronema* Hoare, 1927
- *Urocyclon* Song & Wilbert, 2000
- *Uronema* Dujardin, 1841
- *Uronemella* Song & Wilbert, 2002*
- *Uronemopsis* Kahl, 1931
- *Uropedalium* Kahl, 1928

Family UROZONIDAE Grolière, 1975
(syn. Urozonatidae)

Size, very small; **shape, ovoid, but well rounded at anterior and posterior ends**; free-swimming; **somatic ciliation as a series of short kineties, forming a single equatorial belt of somatic ciliature;** caudal cilium, single, long; **oral region as deep equatorial cavity with two oral polykinetids of two rows of kinetosomes, transverse to the longitudinal axis of the oral region**; scutica of several kinetosomes in "V" configuration; macronucleus, globular; micronucleus, present; contractile vacuole, present; cytoproct, present; bacterivorous; in freshwater polysaprobic habitats; one genus.

- *Urozona* Schewiakoff, 1889

Incertae sedis in Order Philasterida

- *Andreula* Kahl, 1934
- *Bizonula* Corliss, 1960
- *Cryptostomina* Fedele, 1938
- *Eurychilum* André, 1910
- *Lembadionella* Kahl, 1933
- *Ptyssostoma* Hentschel, 1927
- *Pusilloburius* Corliss, 1979
- *Rhinodisculus* Mansfeld, 1923
- *Sertumia* Tucolesco, 1962

Order Pleuronematida Fauré-Fremiet in Corliss, 1956
(syns. Conchophthiriina *p.p.*, Deuterostomatina *p.p.*, Pleuronematorina)

Size, very small to medium, with occasional striking exceptions; shape, ovoid; free-swimming, but with some forms restricted to loricae; somatic ciliation, holotrichous, but often more sparse posteriorly, with thigmotactic ciliation often well developed in some taxa; caudal cilia, typically one or more, conspicuous in many species; extrusomes as somatic mucocysts, sometimes of two types, one being rod-like; **oral region, often expansive, with paroral often prominent with long cilia, forming a curtain or velum as the organism feeds; paroral infraciliary base with a short *a* and an elongate *b* segment and with *c* segment as a permanent scutica or scuticovestige; ribbed wall conspicuous, may be in two fields**; cytostome, equatorial or subequatorial in anterior 3/4 of body, rarely leaving room for a director-meridian; **stomatogenesis of opisthe oral structures derived from paroral of proter and scutica**; microphagous, predominantly bacterivorous, but some species algivorous; in marine, freshwater, and terrestrial habitats, widely distributed as free-living forms, sometimes in sands, but with some species commensalistic in molluscs and other invertebrates, and some species coprozoic; nine families.

Family CALYPTOTRICHIDAE Small & Lynn, 1985

Size, small; shape, elongate ovoid; **free-swimming, but residing in a tubular lorica**; somatic ciliation, holotrichous, dense; caudal cilium, present; **oral region occupying most of ventral surface with conspicuous velum, extending around posterior of oral region and onto its left posterior margin, and with oral polykinetid 1 relatively longer and narrower than oral polykinetid 2**; cytostome, postequatorial; macronucleus, globular; micronucleus (?); contractile vacuole, present; cytoproct (?); bacterivorous; in brackish and freshwater habitats; one genus.

- *Calyptotricha* Phillips, 1882

Family CONCHOPHTHIRIDAE Kahl in Doflein & Reichenow, 1929
(syn. Conchophthiriidae)

Size, small to medium; shape, generally ellipsoidal to broadly reniform, laterally compressed; free-swimming; somatic ciliation, holotrichous, dense; caudal cilia, may be present, sometimes long; **thigmotactic ciliature on right anterior as field of structurally differentiated ciliated somatic dikinetids; oral region, nearly equatorial, as relatively small cavity into which vestibular kineties may extend on its right or left side; oral polykinetids reduced, obliquely oriented along left anterior wall of oral cavity; germinal row of stomatogenic kinetosomes deep in oral cavity**; macronucleus, globular to ellipsoid, rarely nodular; micronucleus, present; contractile vacuole, present; cytoproct, present; bacterivorous and detritivorous; in freshwater habitats as endocommensals in the mantle cavity (on gills) of bivalve molluscs; two genera.

- *Conchophthirus* Stein, 1861
- *Conchoscutum* Raabe, 1947

Family CTEDOCTEMATIDAE Small & Lynn, 1985

Size, small; shape, ovoid; free-swimming; somatic ciliation, holotrichous, sparse, typically of ciliated dikinetids; caudal cilia, prominent; **oral region, midventral with cytostome posterior to equator; velum of paroral segments *a*, *b*, and *c*, comb-like, as an open "C", not extending to the left of the cytostome; ribbed wall, conspicuous; oral polykinetid 3 at right angle to longitudinal axis of oral region**; macronucleus, globular to ellipsoid, rarely nodular; micronucleus, present; contractile vacuole, present; cytoproct, present; bacterivorous; in marine habitats; five genera.

- *Compsosomella* Small & Lynn, 1985
- *Ctedoctema* Stokes, 1884
- *Hippocomos* Czapik & Jordan, 1977
- *Paractedectema* Small & Lynn, 1985
- *Paractedoctema* Song & Wilbert, 2000

Family CYCLIDIIDAE Ehrenberg, 1838

Size, very small to small; shape, ovoid to elongate-ovoid, often with glabrous anterior and posterior zones; free-swimming; **somatic ciliation, holotrichous, sparse, but denser in the anterior half of the body in some genera; caudal cilium, distinctive, one to several**; oral region, not prominent, with cytostome variable in position and oral ciliature not conspicuous; **paroral dikinetid, often inconspicuous, typically with its postcytostomal curve not extending anterior and left of cytostome**; oral polykinetids, often highly fragmented; macronucleus, globular to ellipsoid; micronucleus, large, often located in anterior third of body; contractile vacuole, present; cytoproct, present; bacterivorous; in marine, brackish, freshwater, and terrestrial habitats, widely distributed with interstitial, anaerobic, and coprozoic species; ten genera.

- *Apocyclidium* Foissner, Agatha, & Berger, 2002*
- *Caspionella* Jankowski, 1980
- *Cristigera* Roux, 1899
- *Cyclidium* O.F. Müller, 1773
- *Echinocyclidium* Jankowski, 1980
- *Isocyclidium* Esteban, Finlay, & Embley, 1993
- *Mesogymnus* Berger in Corliss, 1979 [*nomen nudum*]
- *Paracyclidium* Grolière, de Puytorac, & Grain, 1980
- *Protocyclidium* Alekperov, 1993
- *Pseudocyclidium* Small & Lynn, 1985

Family DRAGESCOIDAE Jankowski, 1980

Size, small; shape, ovoid, flattened; free-swimming; **somatic ciliation, uneven, with**

several close-set thigmotactic kineties along right-ventral anterolateral margin; oral region, midventral, with single oral polykinetid, which runs adjacent to the inside curve of the paroral and which is presumably a fusion of three oral polykinetids; macronucleus, ellipsoid; micronucleus, present; contractile vacuole, present; cytoproct, present; bacterivorous(?); in marine habitats as ectosymbionts on certain strongylocentrid echinoids; one genus.

- *Dragescoa* Jankowski, 1974

Family HISTIOBALANTIIDAE de Puytorac & Corliss in Corliss, 1979
(syns. Sulciferiidae, Sulcigeridae)

Size, medium to large; shape, elliptical in outline, with right side slightly concave and anterior end a little narrower than posterior; free-swimming; **somatic ciliation, holotrichous, dense, with longer bristle-like cilia interspersed between regular cilia, and kineties having prominent preoral and postoral secant systems; oral region, an expansive and deep groove, with the posterior end of the paroral as an enlarged kinetosomal field, almost "polykinetid"-like, and oral polykinetids only on anterior 1/3–1/2 of oral region, and oblique to long axis of oral area**; macronucleus, globular to elongate ellipsoid, sometimes multiple; micronucleus, present; contractile vacuole, present; cytoproct, present; microphagous on bacteria and algae; in marine and freshwater habitats, often planktonic; one genus and one genus *incertae sedis*.

NOTE: The description of this family will need to be modified if *Sulcigera* is confirmed to be a member.

- *Histiobalantium* Stokes, 1886

Incertae sedis in Family Histiobalantiidae

- *Sulcigera* Gajewskaja, 1928

Family PENICULISTOMATIDAE Fenchel, 1965

Size, medium; shape, reniform, strongly flattened laterally; free-swimming; somatic ciliation, holotrichous, dense, with preoral and postoral secant systems; **oral region, on ventral margin with long paroral, the anterior segment of the long, linear oral polykinetid 2 of >2 files of kinetosomes, and the posterior segment of oral polykinetid 2 oriented transversely to the longitudinal axis of the oral cavity**; macronucleus, ellipsoid, relatively large; micronucleus, may be multiple; contractile vacuole, present; cytoproct, present; bacterivorous (?); in marine and freshwater habitats as endocommensals in the mantle cavity of bivalve molluscs and in the intestines of certain sea urchins; three genera.

- *Echinosociella* Berger in Small & Lynn, 1985
- *Mytilophilus* Antipa & Dolan, 1985
- *Peniculistoma* Jankowski, 1964

Family PLEURONEMATIDAE Kent, 1881
(for Aphthoniidae; syn. Larvulinidae (?))

Size, small to large; shape, ovoid; typically free-swimming; somatic ciliation, holotrichous; caudal cilia or bristles, long, stiff; somatic extrusomes as prominent, rod-shaped mucocysts; **oral region, shallow groove, occupying much of ventral surface, dominated by paroral cilia, present as a stiff velum and distinctly curling around the subequatorial cytostome; oral polykinetid 2, typically as two distinct segments or derived from two segments – a long, linear, anterior segment with no more than two files of zig-zag kinetosomes, and a posterior segment, typically "V"-shaped**; macronucleus, globular to ellipsoid; micronucleus, present; contractile vacuole, present; cytoproct, present; algivorous and bacterivorous; in marine and freshwater habitats, occasionally associated with some invertebrates as an ectocommensal (e.g., *Pleurocoptes* on hydractinian coelenterates); four genera.

- *Gajewskiella* Obolkina, 1989
- *Pleurocoptes* Wallengren, 1896
- *Pleuronema* Dujardin, 1841
- *Schizocalyptra* Dragesco, 1968

Family THIGMOCOMIDAE Kazubski, 1958

Size, small; shape, ovoid, flattened laterally, bluntly tapered posteriorly; free-swimming; **somatic ciliation, reduced, as relatively densely ciliated anterior kinetofragments with clearly delineated area of thigmotactic ciliature on the concave lateral left surface and posterior half**

of body sparsely ciliated with incomplete left lateral somatic kineties; oral region, equatorial, with short paroral and a large oral polykinetid, broader anteriorly and tapering posteriorly; macronucleus, globular; micronucleus, present; contractile vacuole, present; cytoproct (?); feeding (?); in terrestrial habitats as endoparasites (?) in the renal organ of snails, such as *Oxychilus*; three genera.

- *Baikalothigma* Jankowski, 1982
- *Cotensicoma* Jankowski, 1982
- *Thigmocoma* Kazubski, 1958

Incertae sedis in Order Pleuronematida

- *Larvulina* Penard, 1922

Order Thigmotrichida Chatton & Lwoff, 1922
(syns. Arhynchodina, Diplohymenina, Hemispeirina *p.p.*, Hysterocinetia *p.p.*, Hysterocinetida *p.p.*, Hysterocinetina *p.p.*, Parastomatida, Parastomatina, Protoptychostomatina *p.p.*, Stomatina, Stomodea, Thigmotricha, Thigmotrichina)

Size, small to medium; shape, ovoid to elongate ovoid, laterally compressed in many species and with an anterior sucker in some species; free-swimming, but highly thigmotactic, attached to host tissues by thigmotactic cilia; somatic ciliation, uniform, frequently dense; **thigmotactic cilia as anterior differentiations of somatic kineties, sometimes in a separate field**; **oral ciliature mostly subequatorial in location, often spiraled around the posterior pole of the body, where the cytostome is located, or at the posterior pole but in reduced form**; paroral not forming a prominent velum; ribbed wall, inconspicuous or absent; **oral polykinetid 3 reduced or absent**; stomatogenesis of opisthe oral structures involving kinetosomes of proter's paroral and scutica; bacterivorous (?) or detritivorous (?); in marine and freshwater habitats as commensals with one major group widely occurring in lamellibranch molluscs, and another mainly in oligochaete annelids, although other hosts occasionally involved; four families and two families *incertae sedis*.

NOTE: Chatton and Lwoff (1949, 1950) remain the classical works on this group. However, the monographic papers of Raabe (1967, 1970a, 1971a, 1971b, 1972), the latter published posthumously, should be consulted as the most recent revisionary works.

Family ANCISTRIDAE Issel, 1903
(syns. Ancistrumidae, Protophryidae)

Size, small to medium; shape, ovoid, occasionally elongate; free-swimming, but typically attached to the host; **somatic ciliation, holotrichous, with thigmotactic ciliature, anterior dorso-lateral left, not set apart from other somatic kineties; oral region, extending nearly length of body, with cytostome presumed to have moved progressively posterior-poleward as genera diversified; oral ciliature, conspicuous, winding in arc of >360° around antapical pole in some species, with oral polykinetid 2 long and terminating anterior to the cytostome**; macronucleus, globular to ellipsoid; micronucleus, present; contractile vacuole, present; cytoproct (?); bacterivorous (?); in marine and freshwater habitats as commensals in the mantle cavity and less often, intestine of molluscs, such as prosobranch limpets, pulmonates, and lamellibranchs; eight genera.

- *Ancistrella* Cheissin, 1931
- *Ancistrum* Maupas, 1883
- *Ancistrumina* Raabe, 1959
- *Fenchelia* Raabe, 1970
- *Protophyra* Kofoid, 1903
- *Protophyropsis* Raabe, 1959
- *Semiboveria* Raabe, 1970
- *Syncilancistrumina* Knight & Thorne, 1982

Family HEMISPEIRIDAE König, 1894
(syns. Boveriidae)

Size, small; shape, ovoid, occasionally elongate; free-swimming, but typically attached to the host; **somatic ciliation, holotrichous, but reduced to a smaller number of spiraled kineties in many species, becoming oblique and even almost horizontal in some; thigmotactic area, distinct, of reduced dorsal kineties enclosed in a secant system, very pronounced in certain genera**; **oral region in posterior 1/3 of body, with its ciliature, often reduced, with oral polykinetid 2, at least, hook-like to strongly curved, looping behind the cytostome and, in the extreme, curving to well over half-way around the posterior pole**; macronucleus, globular to ellipsoid; micronucleus, present; contractile vacuole, present; cytoproct (?);

bacterivorous (?) and detritivorous; in marine and freshwater habitats as commensals in the mantle cavity of molluscs, on the integument of certain echinoderms or in the respiratory trees of holothurian echinoderms; seven genera.

- *Ancistrospira* Chatton & Lwoff, 1926
- *Boveria* Stevens, 1901
- *Cheissinia* Chatton & Lwoff, 1949
- *Hemispeira* Fabre-Domergue, 1888
- *Plagiospira* Issel, 1903
- *Proboveria* Chatton & Lwoff, 1936
- *Protospirella* Aescht, 2001

Family HYSTEROCINETIDAE Diesing, 1866
(syns. Ladidae, Protoptychostomatinae, Ptychostom[at]idea, Raabellocinetinae)

Size, medium to large; shape, elongate ovoid, somewhat flattened laterally; free-swimming, but typically attached to the host; somatic ciliation, holotrichous, dense; **thigmotactic sucker, prominent, essentially at apical end of left lateral surface, comprised of segments of anterior kineties surrounded by a non-ciliated strip or field, often in a horseshoe shape and generally strengthened by fibers or other skeletal structures; oral region at posterior pole with paroral and three oral polykinetids, often reduced in size, with oral polykinetid 2 being two kinetosomes wide and in two segments – a peristomal segment around the posterior pole and an infundibular segment in the oral cavity proper**; reproductive methods include posterior budding or catenulation in some species; macronucleus, globular to ribbon-like; micronucleus, may be multiple; contractile vacuole, may be multiple; cytoproct (?); bacterivorous (?) and detritivorous; in freshwater and terrestrial habitats as commensals in the intestines of oligochaete annelids, with a few species of two genera (*Hysterocineta*, *Ptychostomum*) in the gut of certain freshwater snails; 21 genera.

NOTE: De Puytorac (1994f) and Ngassam, de Puytorac, and Grain (1994) have included this family and the Family Paraptychostomidae in the subclass Hysterocinetia. These two families are distinguished from other thigmotrich scuticociliates by some astome-like features: an anterior sucker occurs in these two families and also in haptophryid astomes; and division by chain-formation occurs in hysterocinetids (e.g., *Kozloffia*) and some astomes (e.g., Family Haptophryidae, Family Radiophryidae). Gene sequences will help to determine whether the hysterocinetids and paraptychostomids should be separated at this high taxonomic level or remain as families within the Order Thigmotrichida.

- *Amieta* Ngassam & Grain, 1998
- *Coelothigma* de Puytorac, 1969
- *Coronthigma* Jankowski, 1980
- *Cotylothigma* Raabe, 1949
- *Craticuloscuta* Kozloff, 1965
- *Drilocineta* Raabe, 1972
- *Elliptothigma* Meier, 1954
- *Eminothigma* Jankowski in Corliss, 1979
- *Epicharocotyle* Kozloff, 1965
- *Hysterocineta* Diesing, 1866
- *Hysterocinetoides* Ngassam & de Puytorac, 1994
- *Kozloffia* de Puytorac, 1969
- *Kysthothigma* Raabe, 1949
- *Metaptychostomum* Ngassam & Grain, 1997
- *Preptychostomum* de Puytorac, 1969
- *Proptychostomum* Ngassam & Grain, 1997
- *Protoptychostomum* Raabe, 1949
- *Ptychostomum* Stein, 1860
- *Raabellocineta* de Puytorac, Grolière, & Grain, 1979
- *Taeniocineta* Raabe, 1972
- *Thurstonia* de Puytorac, 1969

Family PARAPTYCHOSTOMIDAE Ngassam, de Puytorac, & Grain, 1994
(syn. Paraptychostomatidae)

Size, medium; shape, elongate ovoid, with extreme lateral flattening; free-swimming, but typically attached to host tissues; somatic ciliation, holotrichous, dense, of monokinetids; **thigmotactic sucker, prominent, essentially at apical end of left lateral surface, comprised of 7–9 short, isolated kineties and supported by cytoskeletal bundles; oral region at posterior pole, extending as a cavity, inwards and anteriorly to the level of the equator with oral infraciliature as a paroral and three oral polykinetids of which oral polykinetid 2 is continuous and of more than two files of kinetosomes wide**; macronucleus, elongate ellipsoid, orthogonal to the longitudinal axis of the cell; micronucleus, present; contractile vacuole, present; cytoproct (?); carnivorous on other ciliates; in freshwater and terrestrial habitats as endocommensals in the digestive tracts of oligochaete annelids (e.g., *Alma*); one genus.

– *Paraptychostomum* Ngassam, de Puytorac, & Grain, 1994

Incertae sedis in Order Thigmotrichida

Family NUCLEOCORBULIDAE Santhakumari & Nair, 1970

Size, large, up to 500 μm; shape, cylindroid, with posterior cellular extension of variable size; free-swimming, but highly thigmotactic; **anterior fixation sucker with its many kineties running nearly at right angles to longitudinal axis of the body and other somatic kineties**; oral ciliature appears to be at posterior pole, surrounded by long oral(?) cilia; **macronucleus, huge, branching**; micronucleus (?); contractile vacuole (?); cytoproct (?); carnivorous on other ciliates; in marine habitats in the mantle cavity of shipworms, species of eulamellibranch molluscs of the genera *Nausitora* and *Teredo*; one genus.

NOTE: This intriguing genus needs careful redescription.

– *Nucleocorbula* Santhakumari & Nair, 1970

Family PROTANOPLOPHRYIDAE Miyashita, 1929
(syn. Protoanoplophryidae)

Size, medium to large, up to 1,500 μm; shape, often elongate, laterally compressed and highly astomatid-like in appearance; free-swimming; pellicle, thickened; somatic ciliation, holotrichous, dense; **anterior secant system in the form of an asymmetrical "Λ" onto which somatic kineties insert with the suture area supported by fibers that condense to form a mucron at the tip of the "Λ"**; **presumed oral region, located a short distance from the anterior pole, with no paroral, but bordered by two parallel "adoral" kineties**; division, sometimes by budding; macronucleus, ribbon-like; micronucleus, present; **contractile vacuoles, numerous, in two rows**; cytoproct (?); osmotrophic (?), via a short canal that ends in a digestive vacuole (?); in freshwater habitats as endocommensals in the intestines of prosobranch snails, such as the genus *Bythinia*; one genus.

NOTE: Jankowski (2007) proposed the new order Parastomatida Jankowski, 2007 to include this unusual family.

– *Protanoplophrya* Miyashita, 1929

Incertae sedis in the Subclass Scuticociliatia

Family AZERIDAE Alekperov, 1985

Size, small; shape, ovoid; free-swimming; somatic ciliation, holotrichous; **oral region, broad, displaced posteriorly, apparently with three oral polykinetids of which oral polykinetid 1 is triangular and oral polykinetids 2 and 3 are rectangles oriented transversely to the longitudinal axis of the oral region**; macronucleus, ellipsoid; micronucleus, present; contractile vacuole (?); cytoproct (?); feeding on bacteria and smaller protists; in freshwater habitats, but reported only once from a reservoir in Azerbaijan; one genus

– *Azerella* Alekperov, 1985

Subclass Hymenostomatia Delage & Hérouard, 1896
(syns. Homoiotricha *p.p.*, Hymenostom[at]orida, Hymenostomida, Hymenostomina, Stichostomata *p.p.*, Tetrahymenophora *s.s.*)

Size, small to medium; shape, typically ovoid to elongate ovoid; somatic ciliation, typically holotrichous, often heavy, with preoral suture, but no postoral one; caudal cilium, rarely found; somatic kinetids mostly monokinetids with dikinetids only at anterior tip of body; somatic extrusomes as mucocysts; oral region in anterior 1/4 of cell, when not absent altogether, and usually inconspicuous; **oral structures as a paroral dikinetid (i.e., haplokinety, stichodyad, undulating membrane) of only the *b* segment, which may be unciliated and reduced, and typically three oral polykinetids as membranelles**; **stomatogenesis parakinetal**; division, free-swimming or in cyst; contractile vacuole(s), present; cytoproct, present; bacterivorous with some carnivorous and some histophagous (e.g., *Ophryoglena*) or parasitic (e.g., *Ichthyophthirius*, *Ophryoglena*) forms; polymorphic life cycles especially characteristic of carnivorous and parasitic forms; predominantly in freshwater habitats; two orders.

Order Tetrahymenida Fauré-Fremiet in Corliss, 1956
(syns. Apohymenida *p.p.*, Dishymenida *p.p.*, Tetrahymenina, Tetrahymenorina)

Size, small to medium; shape, typically ovoid; somatic ciliation, holotrichous, with the first anterior somatic kinetid in each kinety as a dikinetid; oral region, inconspicuous, except in species that undergo microstome-to-macrostome transformation; **oral structures with right oral *b* segment of paroral (haplokinety, undulating membrane) and three left oral polykinetids (membranelles) in oral cavity**; **stomatogenesis, monoparakinetal, typically involving the rightmost postoral somatic kinety**; microphagous forms, primarily bacterivorous, but some histophagous and several polymorphic forms with carnivorous macrostome stage; complex life cycle in histophagous and parasitic species; in freshwater habitats, sometimes terrestrial, and others as facultative or obligate parasites associated mainly with invertebrate hosts, but occasionally also vertebrate hosts, often fishes but one report from a mammal; five families with one family *incertae sedis*.

Family CURIMOSTOMATIDAE Jankowski, 1968
(syn. Curinostomatidae)

Size, small; shape, pyriform to ovoid; free-swimming; **somatic ciliation, holotrichous, with the silverline system similar to that of *Tetrahymena* (Family Tetrahymenidae)**; **oral region, absent, hence astomatous**; macronucleus, globular; micronucleus, present; contractile vacuole, present; cytoproct, present; osmotrophic (?); **in freshwater and terrestrial habitats as obligate parasites of molluscs and turbellarians**; two genera.

NOTE: Rasmussen and Orias (1975) have described an astomatous mutant of *Tetrahymena thermophila*. It is entirely possible that the genera in this family represent such "natural" mutants, and thus should not be separated at this level from genera in the Family Tetrahymenidae. Gene sequence data should easily test this hypothesis.

- *Curimostoma* Kozloff, 1954
- *Dogielella* Poljansky, 1925

Family GLAUCOMIDAE Corliss, 1971
(syns. Bromeliophryidae, Bursostom[at]idae, Espejoiidae, Frontoniidae *p.p.*, Tetrahymenidae *p.p.*)

Size, small to medium; shape, ovoid to ellipsoid; free-swimming; **somatic ciliation, holotrichous, with right ventral kineties curving left, but rarely twisting abruptly anteriorly to run parallel to the anterior suture (except in *Glaucomella*)**; **oral cavity, relatively large, with at least posterior portion of paroral, non-ciliated, and completely resorbed in *Bursostoma*, and with either or both oral polykinetids 2 and 3 having >3 kinetosomal rows**; **a small group of kinetosomes, the so-called X group, typically anterior to the enlarged oral polykinetid 2**; macronucleus, globular to ellipsoid, sometimes reniform; micronucleus, present; contractile vacuole, present; cytoproct, present; microphagous, but several species carnivorous on other ciliates, with some undergoing a microstome-to-macrostome transformation to become carnivores; resting cysts in some species; in freshwater, and occasionally terrestrial, habitats; 12 genera and three genera *incertae sedis*.

- *Bromeliophrya* Foissner, 2003*
- *Bursostoma* Vörösváry, 1950
- *Chasmatostoma* Engelmann, 1862 (subj. syn. *Colpoda*)
- *Dapedophrya* Foissner, 1995
- *Dichilum* Schewiakoff, 1893
- *Epenardia* Corliss, 1971
- *Espejoia* Bürger, 1908
- *Glaucoma* Ehrenberg, 1830
- *Glaucomella* Grolière, 1977
- *Jaocorlissia* Small & Lynn, 1985
- *Monochilum* Schewiakoff, 1893
- *Physalophrya* Kahl, 1931

Incertae sedis in the Family Glaucomidae

- *Discozoon* Vuxanovici, 1960
- *Pinchatia* Shibuya, 1931
- *Pleurochilidium* Stein, 1860

Family SPIROZONIDAE Kahl, 1926

Size, small; shape, elongate ovoid; free-swimming; **somatic ciliation, holotrichous, with somatic kineties on left side and dorsal left twisting slightly; caudal cilia, forming a posterior ring; oral structures with paraoral having isolated dikinetids at its anterior end (= an *a* segment?) and oral polykinetid 1 with its first kinetosomal row as long as oral polykinetid 2, but its kinetosomal rows 2 and 3 shorter**; macronucleus, globular to ellipsoid; micronucleus, present; contractile vacuole, present; cytoproct (?);

feeding on bacteria, flagellates, and microalgae; in freshwater saprobic habitats, very rare; two genera.

- *Spirozona* Kahl, 1926
- *Stegochilum* Schewiakoff, 1893

Family TETRAHYMENIDAE Corliss, 1952 (syns. Deltopylidae, Frontoniidae *p.p.*, Leucophry[i]dae)

Size, small; shape, pyriform to elongate-ovoid to cylindrical; free-swimming; somatic ciliation, holotrichous, with one and up to nine postoral kineties; caudal cilium in some species; **oral structures with paroral dikinetid ciliated along its entire length, not covered by pellicular structures, and three oral polykinetids, each of equal number of rows of kinetosomes, although not always of the same length**; macronucleus, globular to ellipsoid to ribbon-like, rarely nodular; micronucleus, present, but some amicronucleate "species" recognized; contractile vacuole, present; cytoproct, present; bacterivorous, but with some species of *Tetrahymena* exhibiting microstome-to-macrostome transformation; cysts, both resting and reproductive, in some species; in freshwater or terrestrial habitats as free-living forms, but others as facultative and obligate parasites in variety of hosts, such as slugs, snails, clams, enchytraeid worms, midges, mosquitoes, tadpoles, and fishes, and one species found in the urinary tract of a dog; four genera.

NOTE: Foissner, Strüder-Kypke, van der Staay, Moon-van der Staay, & Hackstein (2003) have described some unusual tetrahymenids from bromeliad "tanks" and assigned these to the genus *Lambornella* and suggest that there may be several other undescribed genera. However, these all group well within the clade of the genus *Tetrahymena* using the small subunit rRNA gene sequences. Thus, it is likely that *Lambornella* and these other "undescribed genera" can be considered junior synonyms of the genus *Tetrahymena*. What they represent is the developmental plasticity of the oral features of this clade as has been demonstrated by investigations of developmental mutants of *Tetrahymena* (Frankel, 1991 and references therein).

- *Deltopylum* Fauré-Fremiet & Mugard, 1946
- *Lambornella* Keilin, 1921
- *Paraglaucoma* Kahl, 1926
- *Tetrahymena* Furgason, 1940

Family TURANIELLIDAE Didier, 1971

Size, small to large; shape, elongate-ovoid, sometimes tapering posteriorly; free-swimming; **somatic ciliation, holotrichous, with right ventral kineties curving left, twisting anterior of the oral region, sometimes abruptly, to run parallel to the anterior suture**; **one or more somatic kineties interrupted by left edge of oral cavity**; **oral structures, in anterior half of cell, with paroral dikinetid, whose anterior part is ciliated and whose posterior part is of non-ciliated kinetosomes lying beneath finger-like extensions of the ribbed wall pellicle, and three elongate oral polykinetids**; macronucleus, globular to elongate ellipsoid; micronucleus, present; contractile vacuole, present; cytoproct, present; bacterivorous, but with dimorphic life cycle in *Turaniella*, which has a macrostome form carnivorous on other ciliates; in freshwater habitats; four genera.

- *Colpidium* Stein, 1860
- *Dexiostoma* Jankowski, 1967
- *Paracolpidium* Ganner & Foissner, 1989
- *Turaniella* Corliss, 1960

Incertae sedis in Order Tetrahymenida

Family TRICHOSPIRIDAE Kahl, 1926

Size, small; shape, ovoid; free-swimming; **somatic ciliation, holotrichous, except for a special band of cilia associated with a pellicular ridge that spirals dextrally posteriorly, ending in a transverse ring of cilia; caudal cilia, forming a tuft**; oral region, small, in anterior 1/4 of cell; **oral structures as anterior extensions of a number of somatic kineties invaginating into a shallow cavity**; macronucleus, globular; micronucleus, present; contractile vacuole, present; cytoproct (?); bacterivorous; in freshwater saprobic habitats; one genus.

NOTE: The oral structures of this ciliate need careful investigation to determine if there is a "cryptic" tetrahymenine oral configuration. This family might be better placed in the Class PLAGIOPYLEA.

- *Trichospira* Roux, 1899

Order Ophryoglenida Canella, 1964
(syn. Ophryoglenina)

Size, small to large; shape, elongate ovoid to spherical; somatic ciliation, holotrichous, very dense, with preoral suture; **oral region, inconspicuous, with paroral and three oral polykinetids and its wall "supported" by the organelle of Lieberkühn in at least one stage in the life cycle**; **stomatogenesis teloparakinetal, with dedifferentiation and replacement of parental oral structures, accompanied by marked regression of the paroral in the differentiated oral apparatus**; division, free-swimming or by palintomy in a cyst; histophagous forms, generally feeding on moribund or wounded invertebrates, though several species attack healthy fishes; in freshwater habitats; polymorphic life cycle, including resting cysts; two families.

Family ICHTHYOPHTHIRIIDAE Kent, 1881
(syn. Ichthyophthiridae)

Size, small to large; shape, variable, with the *Tetrahymena*-like theront, elongate ovoid, and with the encysted tomont, spherical; free-swimming, but moving within the epithelial tissues in the parasitic phase; somatic ciliation, holotrichous, dense; caudal cilium, present in theront stage; **oral structures, inconspicuous, with *Tetrahymena*-like microstome oral apparatus in theront and reduced (?) oral ciliature in trophont**; reproduction by palintomy in a cyst away from fish host, producing up to 2,000 tomites; macronucleus, globular to reniform; micronucleus, may be multiple; contractile vacuoles, may be multiple; cytoproct (?); feeding on cells and body fluids of hosts; **in freshwater habitats, widespread in distribution with trophonts invading epithelial tissues of gills and integument of fishes, causing white spot disease**; two genera.

- *Ichthyophthirioides* Roque & de Puytorac, 1968 (subj. syn. *Ichthyophthirius*)
- *Ichthyophthirius* Fouquet, 1876

Family OPHRYOGLENIDAE Kent, 1881

Size, medium to large; shape, elongate ovoid; free-swimming; somatic ciliation, holotrichous, dense; caudal cilium, may be present; **oral region, with inconspicuous opening, bordered on the right by conspicuous vestibular kineties and with oral apparatus in a deep cavity with three oral polykinetids, of which the posterior end of polykinetid 2 is enlarged, its cilia beating like a small brush**; reproduction by palintomy, typically in a cyst and producing 4–128 tomites that develop into small, slender theronts; macronucleus, ellipsoid to elongate ellipsoid to ribbon-like; micronucleus, may be multiple; contractile vacuole, may be multiple; cytoproct, present; histophagous on dying or dead invertebrates, but some species may be facultatively (?) parasitic, for example, in bivalve molluscs; in freshwater habitats; two genera.

- *Ophryoglena* Ehrenberg, 1831
- *Protophryoglena* Mugard, 1949 (subj. syn. *Ophryoglena*)

Incertae sedis in Order Hymenostomatida

- *Blepharostoma* Schewiakoff, 1893 [nomen nudum]
- *Neoichthyophthirius* Bauer & Yunchis, 2001

Subclass Apostomatia Chatton & Lwoff, 1928
(syns. Apohymenida, Apostomata, Apostomea, Apostomina)

Size, small to medium; shape, variable during the polymorphic life cycle, from ovoid to very elongate; free-swimming; somatic ciliation, not dense, holotrichous in mature forms, with kineties, often spiralled and typically numbering <22; **oral structures, highly modified, with a short paroral and three small oral polykinetids sometimes present, and an additional "oral or sensory (?)" structure of unknown function, the rosette, accompanied by three short kineties, designated the *x*, *y*, and *z* kineties**; cytostome, variable, from broad region on cortex to inconspicuous or absent in certain stages; stomatogenesis, possibly mixokinetal, often with involvement of three or four specialized kinetofragments; reproduction may involve palintomy and catenulation; contractile vacuole, present; cytoproct, absent; macronucleus, homomerous in trophonts, and heteromerous in tomites of many species; symbiotic (parasitic) in or on hosts from various invertebrate groups; in marine, rarely freshwater, habitats with only one possible terrestrial host – edaphic acari – reported; **complex polymorphic life cycles, involving phoront, which is encysted on host, trophont, tomont, and tomite stages**; three orders.

NOTE: The classic work of Chatton and Lwoff (1935a) still stands as the authoritative monograph on this group. De Puytorac (1994h) subdivides this group up considerably, establishing many new subfamilies. In our opinion, these subdivisions need confirmation by molecular genetic data.

Order Apostomatida Chatton & Lwoff, 1928
(syns. Cyrtostomatina, Foettingeriida *p.p.*, Gemmotomida, Gemmotomina, Incitophorina, Sanguicolida *p.p.*, Sanguicolina *p.p.*)

Size, small to medium; shape, ovoid to spherical; **somatic ciliation, holotrichous, not dense, with *x*, *y*, and *z* kineties that can be associated with an *a* kinety or an *a*, *b*, and *c* kineties**; oral apparatus, as for subclass, and with rosette; tomites formed by multiple fission, either by palintomy in a cyst or by catenulation; **trophonts, sanguicolous or exuviotrophic**; in marine, occasionally freshwater, habitats in crustacean hosts, such as hermit crabs, shrimps, and a copepod (with sea anemones as alternating host, for species with such an obligate cycle), but members of one genus (*Phtorophrya*, Family Foettingeriidae) hyperparasites of other apostomatids and an atypical family (Family Cyrtocaryidae) found in polychaete annelids; three families.

Family COLLINIIDAE Cépède, 1910
(syns. Colliniida, Colliniinae)

Size, small; shape, roughly pyriform with tapered end as posterior; free-swimming; **somatic ciliation, holotrichous, except for broad bare band medially coursing down dorsal surface**; tomite with nine kineties; no ogival field nor lateral canal, but small rosette near *x* kinety; macronucleus, ellipsoid, relatively large; micronucleus, present; contractile vacuole, may be in multiple rows; **in marine habitats as sanguicolous forms in the hemocoelomic fluid of amphipods, isopods, and euphausiids, sometimes causing mass mortality**; incompletely known life cycle; three genera.

- *Collinia* Cépède, 1910
- *Metacollinia* Jankowski, 1980
- *Paracollinia* Jankowski, 1980

Family CYRTOCARYIDAE Corliss, 1979
(syn. Cyrtocaryumidae)

Size of tomite very small, but trophont to medium size; **shape of trophont, pyriform**; free-swimming; **somatic ciliation, holotrichous, with up to 60 spiraled kineties in the trophont, with an area of strong thigmotactic cilia; caudal cilium in tomites; neither cytostome nor rosette**; division of trophonts in host, by unequal binary fission, and outside host, by catenulation; macronucleus, elongated, large, partially coiled; micronucleus (?); contractile vacuole, present; **in marine habitats with tomites becoming phoronts on a crustacean before infecting the lateral caeca of the digestive tract of polychaete annelids;** one genus.

- *Cyrtocaryum* Fauré-Fremiet & Mugard, 1949

Family FOETTINGERIIDAE Chatton, 1911
(syns. Foettingeriinae, Gymnodinioidae, Gymnodinioinae, Gymnodinioididae, Phtorophrydae, Phtorophryida, Phtorophryinae, Polyspiridae, Spirophryinae, Synophryinae, Terebrospirinae, and possibly Kofoidellidae and Perezellidae)

Size, small in tomite stage, to large in feeding trophont; shape, ovoid to spherical; free-swimming; **somatic kineties, right-spiralled, ranging from about 9 to <22; oral apparatus, as for subclass and with rosette**; macronucleus, globular to extremely elongate and ribbon-like, depending upon the life cycle stage; micronucleus, present; contractile vacuole, present; exuviotrophic, but one genus (*Phtorophrya*) hyperparasite on *Gymnodinioides* species and another (*Synophrya*) with sanguicolous stage; in marine, rarely brackish and freshwater (?), habitats as symbionts (parasites?) on marine crustacea, such as hermit crabs, shrimp, and copepods; 17 genera and three genera *incertae sedis*.

- *Calospira* Chatton & Lwoff, 1935
- *Foettingeria* Caullery & Mesnil, 1903
- *Gymnodinioides* Minkiewicz, 1912
- *Hyalophysa* Bradbury, 1966
- *Hyalospira* Miyashita, 1933 (subj. syn. *Gymnodinioides*)
- *Metaphrya* Ikeda, 1917
- *Ophiuraespira* Chatton & Lwoff, 1930
- *Pericaryon* Chatton, 1911
- *Phoretophrya* Chatton, A. & M. Lwoff, 1930
- *Phtorophrya* Chatton, A. & M. Lwoff, 1930
- *Polyspira* Minkiewicz, 1912
- *Rosea* de Puytorac, 1994 [nomen nudum]

- *Spirophrya* Chatton & Lwoff, 1924
- *Synophrya* Chatton & Lwoff, 1926
- *Terebrospira* Debaisieux, 1960
- *Traumatiophtora* Chatton & Lwoff, 1931
- *Vampyrophrya* Chatton & Lwoff, 1931

Incertae sedis in Family Foettingeriidae

- *Jeppsia* Corliss, 1960
- *Kofoidella* Cépède, 1910
- *Perezella* Cépède, 1910

Order Astomatophorida Jankowski, 1966
(syns. Astomophorina, Nephrocolida, Nephrocolina)

Size, small to medium; shape of trophont, long, vermiform; free-swimming, but trophont attached by its anterior end to host tissue; somatic ciliation, holotrichous, kineties much spiralled and somatic ciliature markedly thigmotactic; **no cytostome (in stages of life cycles known to date), but remnants of oral ciliature; fission of tomont-trophont by sequential formation of tomites (catenulation) or by multiple transverse fission with tomites remaining connected**; macronucleus, very variable, from fragmented nodules to an irregular network; micronucleus, present but obscure; contractile vacuole (?); **in marine habitats as parasites, for example, in the internal organs, such as liver, kidney, and gonad of cephalopods**; one family.

Family OPALINOPSIDAE Hartog, 1906
(syns. Chromidinida, Chromidinidae)

With characteristics of the order; two genera.

- *Chromidina* Gonder, 1905
- *Opalinopsis* Foettinger, 1881

Order Pilisuctorida Jankowski, 1966

Size, small to large; shape, ovoid to elongate; free-swimming but attached to host in the feeding state; **body with ventral adhesive organelle**; **species of most genera permanently in so-called "neotenic" tomite stage; somatic kineties of tomite arched, following rim of flattened ventral surface**; **mature trophonts (e.g., *Conidophrys*), nonciliated, immobile, characteristically attached to seta or cuticle of host, and a migrating tomite, which is ciliated but lacks a cytostome; tomites produced by synchrony or strobilation**; macronucleus, elongate ellipsoid, irregularly shaped to band-form and ribbon-like; micronucleus, present; contractile vacuole, present; feeding on tissue fluids (i.e., hemolymph); in marine habitats on the cuticular processes of amphipods, isopods, decapods, and cirripeds, and possibly a terrestrial mite; single host life cycle; one family.

NOTE: De Puytorac (1994h) recognized three monotypic families in this order, each including one of the three genera here included in the single family of this monotypic order. Perhaps gene sequence data will confirm that the genetic diversity of these forms warrants this higher order taxonomy.

Family CONIDOPHRYIDAE Kirby, 1941
(for Pilisuctoridae; syns. Ascophryidae, Askoellida, Askoellidae, Conidophryidae)

With characteristics of the order; three genera.

- *Ascophrys* Campillo & Deroux, 1974
- *Askoella* Fenchel, 1965
- *Conidophrys* Chatton & Lwoff, 1934

Subclass Peritrichia Stein, 1859
(syns. Cyclohymenophora, Dexiotricha, Peritrichasina, Peritrichidea, Peritrichorida, Stomatoda)

Size of zooids, small to medium, rarely large, but colonial forms can be macroscopic; shape, characteristically inverted bell- or goblet-shaped or conical-cylindrical; sessile and sedentary except as dispersive telotrochs or swarmers, although several taxa are always free-swimming; **prominent holdfast derivatives or scopula, which secretes the stalk of sessile species and includes a field of thigmotactic cilia, at aboral pole**; alveoli, well-developed, with pellicle perforated by pores (= parasomal sacs?); **somatic ciliature, reduced to subequatorial locomotor fringe, the trochal band, which is permanently ciliated on mobile species and temporarily ciliated on the dispersal stage or telotroch of sessile species**; oral region, often retractable, encircling apical pole as peristome, bordered by a more or less prominent collarette; **oral ciliature, conspicuous, winding counterclockwise around the border of the prominent peristome, from its outer terminus**

on an elevated central, often extensible epistomial disk, with peristomial part of the paroral (haplokinety, stichodyad) on the outside and oral polykinetid (polykinety) 1 on the inside, but both actually originating in an oral cavity, the infundibulum, at the terminus of which is the cytostome; oral ciliature of infundibulum including inner parts of paroral and oral polykinetid 1, which is accompanied at its inner terminus by infundibular polykinetids 2 and 3 (formerly called peniculi); **stomatogenesis, ophryobuccokinetal, with involvement of germinal field or row of kinetosomes**; fission, with its plane parallel to the major body axis, suggesting that the apical-antapical axis is a secondary adaptation to sessility; **conjugation, invariably total, typically involving fusion of a migratory microconjugant with a stationary macroconjugant; contractile vacuole and cytoproct, emptying into infundibulum**; bacterivorous and microphagous, with symbiotic species ingesting detritus and tissue debris of host; in marine, freshwater, and rarely terrestrial habitats, very widespread with species generally free-living but many occurring as commensals or even parasites on or in diverse hosts, ranging from protozoa to vertebrates; two orders.

NOTE: Gong et al. (2006) demonstrated that the mobilid peritrichs are a separate lineage from the sessilid peritrichs based on small subunit rRNA gene sequences, but the two lineages are apparently not sister clades. We remain conservative in our assignment of both sessilids and mobilids to the Subclass Peritrichia until sampling of additional genes and taxa refutes this assignment.

Order Sessilida Kahl, 1933
(syns. Aloricata + Loricata, Astylozo(on)ina *p.p.*, Cothurniina *p.p.*, Epistylina *p.p.*, Fibrodiscida *p.p.*, Lagenophryina *p.p.*, Loricina *p.p.*, Natantina *p.p.*, Operculariina *p.p.*, Opisthonectina *p.p.*, Ophrydiina *p.p.*, Rovinjellina *p.p.*, Scyphidiina *p.p.*, Sedentaria, Sessilia, Sessili[i]da, Stylophorina, Syncyathellina *p.p.*, Thigmodiscina, Vorticellina *p.p.*)

Size of zooids, small to medium, rarely large; shape, inverted bell- or goblet-shaped or conical-cylindrical; **zooids, dimorphic, with mature zooids or trophonts, sedentary or sessile, commonly stalked or with inconspicuous adhesive disc, attached to substrate by scopula, but a few species presumed to be secondarily mobile, and dispersal stage as migratory telotroch**; oral region and oral ciliature as for subclass; fission, isotomic or anisotomic, followed in many species by development into arboroid colonies; resting cysts; feeding on bacteria, microalgae, and sometimes detritus; free-living or ectosymbionts, with a few species as endosymbiotic forms; 14 families.

NOTE: Recent molecular genetic research has demonstrated that the morphology of peritrichs may be misleading us and suggests we are a long way from achieving a natural classification of this order (Clamp & Williams, 2006; Gong et al., 2006).

Family ASTYLOZOIDAE Kahl, 1935
(syns. Astylozooidae, Astylozoonidae, Hastatellidae)

Size, small to medium; shape, bell-shaped; **trophont, free-swimming, a secondarily(?) stalkless form, which swims with oral end forward using oral ciliature; telotrochs, lacking somatic ciliature; trophonts, with one or two rigid caudal bristles or short caudal cilia arising from scopula**; oral ciliature as for subclass; macronucleus, band-shaped; micronucleus, present; contractile vacuole, present; cytoproct, not observed; cysts, in some species; in freshwater habitats, especially the plankton of temporary or small ponds; two genera.

- *Astylozoon* Engelmann, 1862
- *Hastatella* Erlanger, 1890

Family ELLOBIOPHRYIDAE Chatton & Lwoff, 1929

Size, medium; **shape, ovoid-cylindrical, with a pair of elongate, cylindrical, and contractile aboral projections, encircling filamentous body parts of the host and cemented together by derivatives of the scopula at their tips to form a closed circle in firm attachment**; oral region as for subclass, but with epistomial disk slightly depressed in relation to prominent collarette; fission, anisotomic; macronucleus, ellipsoid; micronucleus, present; contractile vacuole, present; cytoproct, not observed; in marine habitats as ectocommensals on gills of lamellibranch molluscs and fishes, scale spines of fishes, and oral tentacles of bryozoans; two genera.

- *Caliperia* Laird, 1953
- *Ellobiophrya* Chatton & Lwoff, 1923

Family EPISTYLIDIDAE Kahl, 1933
(syns. Apiosomatidae, Epistylidae, Ichthyophyllinae, Nedulidae, Nidulidae, Nidulinae)

Size, from small to large with some species of *Campanella* and *Epistylis* having zooids up to 600 μm in length; shape, ovoid, cylindrical-conical or campanulate; **trophonts, contractile, on non-contractile stalk, which may be extremely difficult to resolve with light microscopy (e.g., *Apiosoma*)**; if stalkless, in lorica; solitary or colonial; oral region as for subclass, but with peristomial lip and with epistomial disk, only slightly projecting; oral ciliature, making from one to as many as five complete turns around the peristome; macronucleus, elongate, ellipsoid or band-shaped; micronucleus, present; contractile vacuole, present; cytoproct, not observed; in freshwater and occasionally marine habitats as free-living forms or as symphorionts associated with diverse hosts – from other peritrichs to molluscs, crustaceans, aquatic insects, and some vertebrates, such as freshwater fishes on whose integument *Apiosoma* is especially widely found; 11 genera, of which one is a fossil genus from the Lower Triassic.

- *Apiosoma* Blanchard, 1885
- *Campanella* Goldfuss, 1820
- *Epistylis* Ehrenberg, 1830
- *Foissnerella* Jankowski, 1986
- *Heteropolaria* Foissner & Schubert, 1977
- *Ichthyophyllum* Jankowski, 1976 [nomen nudum]
- *Nuchterleinella* Matthes, 1990
- *Opisthostyla* Stokes, 1886 (subj. syn. *Rhabdostyla*)
- *Rhabdostyla* Kent, 1881
- *Triadopercularia* Weitschat & Guhl, 1994 (fossil)
- *Uvelinus* Jankowski, 1985

Family LAGENOPHRYIDAE Bütschli, 1889
(syns. Lagenophryiidae, Lagenophryinae, Stylohedrinae)

Size, small to medium; **shape, ovoid, flattened, bilaterally symmetrical; trophont, completely enclosed in an ellipsoid, ovoid or hemispheroid lorica to which they are attached only at part of the interior margin of the aperture; lorica, closed by opposing lip-like folds, oriented parallel to the transverse axis of the lorica or with a flap-like operculum**; stalk, absent; scopula, greatly enlarged in area, apparently acting only during attachment of the telotroch; oral region as for subclass, but without peristomial lip and with an epistomial disk extensible by means of an elongate contractile base; **fission, anisotomic or asymmetric, with telotroch forming to left side of parent (when viewed from above)**; macronucleus, elongate, often band-shaped or compact ellipsoid; micronucleus, present; contractile vacuole, present; cytoproct, present; in marine, brackish, and freshwater habitats as symphorionts, most commonly on gills of freshwater amphipods and decapods, especially crayfish, but also shrimps and crabs, as well as on aquatic plants; six genera and one genus *incertae sedis*.

- *Clistolagenophrys* Clamp, 1991
- *Lagenophrys* Stein, 1852
- *Operculigera* Kane, 1969
- *Paralagenophrys* Clamp, 1987
- *Setonophrys* Jankowski, 1986
- *Stylohedra* Kellicott, 1884 (subj. syn. *Lagenophrys*)

Incertae sedis in Family Lagenophryidae

- *Eilymophrys* Corliss, 1979

Family OPERCULARIIDAE Fauré-Fremiet in Corliss, 1979
(syns. Bezedniellidae, Entziellidae, Entziellinae, Operculariinae)

Size, small to medium; shape, typically ovoid to elongate cylindrical; zooid, contractile, but peristomial lip does not fold outward on eversion; stalk, typically non-contractile; solitary or colonial, with highly developed theca in many species; **oral region as for subclass, but without distinct peristomial lip and with a prominent, extensible and contractile epistomial disk that is on a stalk with a furrow separating and elevating this disc from the margin of the peristome, but epistomial disk secondarily reduced in some endocommensal species**; oral cavity with an infundibulum, often dilated dorsally, into which contractile vacuole and cytoproct open; macronucleus, band-shaped or ellipsoid; micronuclues, present; contractile vacuole, present; cytoproct, present; in freshwater habitats, very commonly as epibionts on insects and other arthropods, but *Operculariella* is endocommensal in the esophagus of a beetle and *Orsomia* is associated with an oligochaete annelid; 14 genera and one genus *incertae sedis*.

- *Ballodora* Dogiel & Furssenko, 1921
- *Bezedniella* Stloukal & Matis, 1994
- *Cyathopercularia* Jankowski, 1980
- *Discophryson* Jankowski, 1985
- *Discotheca* Jankowski, 1967
- *Entziella* Stiller, 1951
- *Opercularia* Goldfuss, 1820
- *Operculariella* Stammer, 1948
- *Orbopercularia* Lust in Guhl, 1979
- *Orbopyxidiella* Lust in Guhl, 1979
- *Propyxidium* Corliss, 1979
- *Scyphidiella* Guhl, 1979
- *Spirocochlearia* Jankowski, 1976 [nomen nudum]
- *Syncyathella* Jankowski, 1976 [nomen nudum]

Incertae sedis in Family Operculariidae

- *Orsomia* Baer, 1952

Family OPHRYDIIDAE Ehrenberg, 1838

Size, medium to large; **shape, bottle-, vase-, or spindle-shaped, often with long and highly contractile oral end, neck-like**; zooids, solitary (*Gerda*) or forming gelatinous colonies, often up to 15 cm in diameter; green-colored from endosymbiotic zoochlorellae in the individual zooids; **stalk, atypical, with scopula producing long peduncular fibers**; oral region as for subclass; macronucleus, elongate; micronucleus (?); **contractile vacuole, aborally located, with a long canal connecting it to the infundibulum**; cytoproct (?); cysts; exclusively in freshwater habitats; two genera.

- *Gerda* Claparède & Lachmann, 1858
- *Ophrydium* Bory de St. Vincent, 1824

Family OPISTHONECTIDAE Foissner, 1976
(syns. Telotrochididae, Telotrochidiidae, Telotrochiidae)

Size, medium; shape, cylindrical to bell-shaped, narrowing toward oral end; **zooid, free-swimming with aboral end forward, secondarily (?) stalkless but with inactive scopula at the aboral pole and a permanently ciliated telotrochal band; oral region, as for subclass, but with a small separate group of rigid cilia prominent in one genus (i.e., *Opisthonecta*)**; macronucleus, bandform; micronucleus, present; contractile vacuole, present; cytoproct (?); in freshwater habitats, especially the plankton; two genera.

NOTE: Fauré-Fremiet (1950c) argued that these genera were probably not phylogenetically related and suggested that *Opisthonecta* derived from a stalked *Epistylis*-like ancestor while *Telotrochidium* derived from an *Opercularia*-like ancestor. While *Telotrochidium* has not yet been sequenced, *Opisthonecta* does appear to group with an *Epistylis* clade (e.g., Lynn & Strüder-Kypke, 2005).

- *Opisthonecta* Fauré-Fremiet, 1906
- *Telotrochidium* Kent, 1881

Family ROVINJELLIDAE Matthes, 1972

Size, medium; shape, elongate, cylindrical-conical; **zooids in two-membered colonies, sharing a stalk that is in two parts – in the lorica, a proximal part that folds into accordion-like pleats on contraction, and a non-contractile, distal part outside the lorica that is attached to the substrate; lorica, gaping widely at its upper (= oral) end**; oral region as for subclass, with a large, slightly protuberant epistomial disk; macronucleus, band-shaped; micronucleus (?); contractile vacuole, present; cytoproct (?); in marine and freshwater habitats on crustaceans; four genera and one genus *incertae sedis*.

- *Grainis* Jankowski, 1997
- *Rovinjella* Matthes, 1972
- *Shellositon* Jankowski, 1993
- *Tauriella* Naidenova, 1985

Incertae sedis in Family Rovinjellidae

- *Delamurea* Naidenova, 1978

Family SCYPHIDIIDAE Kahl, 1933
(syn. Corlissettidae)

Size, small to medium; shape, cylindroid to elongate bell-shaped; **zooid, solitary, stalkless**

- *Anoplophryopsis* de Puytorac, 1954
- *Buetschliella* Awerinzew, 1908
- *Herpinella* de Puytorac, 1954
- *Rhizocaryum* Caullery & Mesnil, 1907

Incertae sedis in Family Buetschliellidae

- *Hysterophrya* de Puytorac, Grolière & Grain, 1979

Family CLAUSILOCOLIDAE de Puytorac in Corliss, 1979
(syn. Proclausilocolidae)

Size, medium; shape, comma-shaped or broadly ellipsoidal, flattened, and not especially elongate; free-swimming; **somatic ciliation, holotrichous, dense, with anterior horseshoe-shaped secant system and with anterior thigmotactic area, variously developed**; macronucleus, globular to ellipsoid; micronucleus, present; contractile vacuoles, scattered; in terrestrial habitats in the digestive tracts of gastropods and African oligochaete annelids; three genera.

- *Clausilocola* Lom, 1959
- *Haptophryopsis* de Puytorac, 1971
- *Proclausilocola* Lom, 1959

Family CONTOPHRYIDAE de Puytorac, 1972

Size, medium; shape, elongate-ellipsoidal; free-swimming; **somatic ciliation, holotrichous, with anterior, sagittal secant system, just behind a circular nonciliated area**; **endoskeletal fibers associated with somatic kineties only in lower (= ventral) part of body, and attachment structures developed as a single median or two pairs of cytoskeletal hook(s), which may appear as a "Λ"-shape**; fission, subequatorial; macronucleus, elongate ellipsoid to ribbon-like; micronucleus, present; contractile vacuoles, in a single row; in terrestrial habitats in the digestive tracts of tropical oligochaete annelids of the Family Glossoscolecidae; two genera.

- *Contophyra* de Puytorac & Dragesco, 1969
- *Dicontophrya* de Puytorac & Dragesco, 1969

Family HAPTOPHRYIDAE Cépède, 1923
(for Discophryidae Cépède [*non* Collin]; syns. Cepediettinae, Haptophryinae, Lachmannellinae, Sieboldiellinae, Sieboldiellininae)

Size, large, up to 2,000 μm in length; shape, elongate; free-swimming; **somatic ciliation, holotrichous, dense, with kineties converging anteriorly onto horseshoe-shaped suture line, and with a thigmotactic region, which may be supported by rigid cortical armature**; **adhesive sucker, conspicuous, at apical end of body, sometimes provided with two or more hooks or spines**; fission, anisotomic, sometimes with catenulation; macronucleus, globular to elongate ellipsoid; micronucleus, present; **contractile vacuole as a long canal, emptied by several pores**; in marine and freshwater habitats in the digestive tracts of turbellarians and anuran and urodelean amphibians; five genera.

- *Annelophrya* Lom, 1959
- *Cepedietta* Kay, 1942
- *Haptophrya* Stein, 1867
- *Lachmannella* Cépède, 1910
- *Steinella* Cépède, 1910

Family HOPLITOPHRYIDAE Cheissin, 1930
(syns. Hoplitophryinae, Jirovecellinae, Juxtaradiophryinae, Mesnilellidae, Mesnilellinae, Mixtophryinae)

Size, small to medium; **shape, elongate, cylindrical, tapered posteriorly**; free-swimming; somatic ciliation, holotrichous, moderate to light; **ectoplasm thickened at the apical end, underlain by a fibrous cytoskeleton associated with a "Λ"-shaped attachment structure, which may be reduced or absent**; fission, anisotomic, often with catenulation; macronucleus, elongate, ribbon-like; micronucleus, present; contractile vacuoles, generally in a single row; in freshwater and terrestrial habitats in the digestive tracts of oligochaete annelids; 12 genera.

- *Akidodes* Lom in Aescht, 2001
- *Anglasia* Delphy, 1936
- *Buetschliellopsis* de Puytorac, 1954
- *Delphyella* de Puytorac, 1969
- *Hoplitophrya* Stein, 1860
- *Jirovecella* Lom, 1957
- *Juxtamesnilella* de Puytorac, 1969
- *Juxtaradiophrya* de Puytorac, 1954
- *Mesnilella* Cépède, 1910
- *Mixtophrya* de Puytorac, 1969
- *Protoradiophryopsis* Georgévitch, 1950
- *Radiophryoides* Lom, 1956

Family INTOSHELLINIDAE Cépède, 1910

Size, small to large; shape, cylindrical, elongate; free-swimming; somatic ciliation, holotrichous, dense, with kineties often loosely spiraled; **an anterior, non-ciliated attachment structure as an elaborate plate or unclosed ring with nodes and spines**; fission, anisotomic, with catenulation; macronucleus, elongate, ribbon-like; micronucleus, present; contractile vacuoles, typically in one row, which may be spiralled; in freshwater habitats in the digestive tract of oligochaete annelids; three genera.

- *Intoshellina* Cépède, 1910
- *Monodontophrya* Vejdovsky, 1892
- *Spirobuetschliella* Hovasse, 1950

Family MAUPASELLIDAE Cépède, 1910
(syns. Acanthophryinae, Maupasellinae)

Size, small to medium; shape, cylindrical, rounded at both ends; free-swimming; somatic ciliation, holotrichous, not dense; **cytoskeletal fibers, relatively short, ending in a small, anterior attachment spine, which may be fixed or mobile**; fission, isotomic or anisotomic, sometimes with catenulation; macronucleus, elongate, ribbon-like, rarely ramified; micronucleus, present; contractile vacuoles, in one or two rows; in freshwater and terrestrial habitats as endosymbionts in the intestines of oligochaete annelids and leeches; four genera.

- *Acanthophrya* Heidenreich, 1935
- *Buchneriella* Heidenreich, 1935
- *Georgevitchiella* de Puytorac, 1957
- *Maupasella* Cépède, 1910

Family RADIOPHRYIDAE de Puytorac, 1972
(syns. Acanthodiophryinae, Anthonyellinae, Durchoniellinae, Eudrilophryinae, Metarocoelophryinae, Metaradiophryinae, Radiophryinae)

Size, small to large; shape, generally ovoid, much flattened, occasionally elongate; free-swimming; **somatic ciliation, holotrichous, dense, with two, rarely three, anterior secant systems; cytoskeletal fibers, underlying much of lower (= ventral) surface and extending almost the entire length of somatic kineties, with, in some species, an apical end dominated by prominent "Λ"-shaped cytoskeletal organelle onto which numerous fibers converge**; hooks or spines or other attachment structures, often present; fission, anisotomic, sometimes extremely so, but catenulation not common; macronucleus, elongate ellipsoid to ribbon-like; micronucleus, present; contractile vacuoles, in one or two rows; in freshwater, marine, and terrestrial habitats as endosymbionts in the digestive tracts of oligochaete annelids, a few polychaete annelids, and occasionally a freshwater lamellibranch mollusc; 18 genera.

- *Acanthodiophrya* de Puytorac & Dragesco, 1969
- *Anthonyella* Delphy, 1936
- *Cheissiniophrya* de Puytorac & Dragesco, 1969
- *Coelophrya* de Puytorac & Dragesco, 1969
- *Desmophrya* Raabe, 1933
- *Dicoelophrya* de Puytorac & Dragesco, 1969
- *Durchoniella* de Puytorac, 1954
- *Eudrilophrya* de Puytorac, 1971
- *Helellophrya* Aescht, 2001
- *Hovasseiella* de Puytorac, 1955
- *Metaracoelophrya* de Puytorac & Dragesco, 1969
- *Metaradiophrya* Heidenreich, 1935 [nomen nudum]
- *Mimophrya* de Puytorac, 1969
- *Mrazekiella* Kijenskij, 1926
- *Ochridanus* Georgévitch, 1941
- *Paracoelophrya* de Puytorac, 1969 [nomen nudum]
- *Radiophrya* Rossolimo, 1926
- *Radiophryopsis* Georgévitch, 1941

Incertae sedis in Phylum Ciliophora

- *Antipia* Lepsi, 1927
- *Arachnidiopsis* Penard, 1918 [nomen dubium]
- *Benthontophrys* Foissner & Gschwind, 1988
- *Bipalmatum* Gajewskaja, 1924
- *Ceratospathula* Foissner, 2003*
- *Chanostoma* Daday, 1884
- *Conocladium* Schröder, 1914
- *Euploia* Lohmann, 1920 [nomen dubium]
- *Hyloplotes* Butschinsky, 1897
- *Isosticha* Kiesselbach, 1936
- *Litosolenius* Stokes, 1893
- *Macrocytopharynx* Li & Wang, 2002*
- *Orcavia* Tucolesco, 1962
- *Pachystomos* Rudzinska, 1952
- *Parablaste* Cragin, 1889
- *Pompholyxia* Fabre-Domergue, 1886
- *Rhynchodinium* Cunha & Penido, 1927
- *Sigalasia* Delphy, 1938 [nomen dubium]
- *Spirocytopharynx* Li & Wang, 2002*

Family COELOSOMIDIIDAE Corliss, 1961
(for Coelosom[at]idea; syns. Conchostomatidae, Orthostomatida *p.p.*)

Size, medium to large; **shape, cylindroid, elongate**; free-swimming; somatic ciliation, holotrichous, dense; **oral region as an anterior cavity lined by extensions of somatic kineties**; macronucleus, elongate ellipsoid to ribbon-like; micronucleus, present; contractile vacuole, may be multiple; cytoproct (?); feeding on bacteria and microalgae; in marine habitats; two genera and one genus *incertae sedis*.

- *Coelosomides* Strand, 1928
- *Conchostoma* Fauré-Fremiet, 1963

Incertae sedis in Family Coelosomidiidae

- *Epimecophrya* Kahl, 1933

ADDENDUM: While the draft of this 3rd edition was "in galley", Jankowski (2007) published a major revision of the Phylum Ciliophora. It is regrettable that we are unable to consider and to respond in detail to many of the changes proposed by him, due both to the time constraints of our publishing schedule and the need to keep "galley changes" to a minimum.

Jankowski (2007) has accepted the major classes proposed in **Chapter 17** with the exception of using Class CONTOFRAGMEA Jankowski, 1980 instead of Class PLAGIOPYLEA Small & Lynn, 1985 (Addendum. Table 1). However, he departs from our treatment in some cases at the subclass level and often below (cf. Table 4.1). A review of his large and significant "chapter" indicates that he has proposed at least 50 new genera, one new subfamily, six new families, three new suborders, and two new orders.

In addition, there are a number of differences in the assignment by Jankowski (2007) of genera to families compared to the assignments in **Chapter 17**. The majority of the differences occur in relation to assignments in four subclasses – the subclasses Stichotrichia, Haptoria, Suctoria, and Peritrichia. In at least three of these subclasses, we have already learned that morphology and molecules do not corroborate each other well: for the Stichotrichia, see Foissner et al. (2004); for the Haptoria, see Strüder-Kypke et al. (2006); and for the Peritrichia, see Clamp & Williams (2006). Thus, it is perhaps not surprising that there should be differences between the taxonomy proposed herein and that proposed by Jankowski (2007).

As noted in **Chapter 17**, we have decided to remain conservative in our treatments of the various taxa, not departing significantly from Corliss (1979) and Lynn & Small (2002) unless there is compelling evidence, ideally from both morphology and molecules, to do so. In relation to the many new taxa proposed by Jankowski (2007), we refrain from supporting them until there is the same kind of compelling new evidence that these taxa should be differentiated at the levels recommended by him. That is, we prefer to treat these new taxa as conjectural hypotheses about relationships that need the support of significant new evidence to test their robustness. Thus, the differing views presented herein and by Jankowski (2007) provide some competing hypotheses about relationships among ciliates that will provide research questions for the future.

ADDENDUM. TABLE 1. Classification of the Phylum Ciliophora Doflein, 1901 with authorships as proposed by Jankowski (2007)

Phylum CILIPHORA Doflein, 1901	Class **PHYLLOPHARYNGEA** de Puytorac et al., 1974
Subphylum *POSTCILIODESMATOPHORA* Gerassimova & Seravin, 1976	Subclass Hypostomatia Schewiakoff, 1896
	Order Gymnozoida Jankowski, 2007
	Order Chlamydodontida Deroux, 1976
Class **KARYORELICTEA** Corliss, 1974	Order Dysteriida Deroux, 1976
Order Trachelocercida Jankowski, 1978	Subclass Rhynchodia Chatton & Lwoff, 1939
Order Loxodida Jankowski, 1980	Order Rhynchodida Chatton & Lwoff, 1939
Order Protoheterotrichida Nouzarède, 1977	Suborder Ancistrocomina Jankowski, 1980
	Suborder Sphenophyrina Jankowski, 1980
Class **HETEROTRICHEA** Stein, 1859	Suborder Hypocomatina Deroux, 1976
Order Heterotrichida Stein, 1859	Subclass Chonotrichia Wallengren, 1895
Suborder Heterotrichina Stein, 1859	Order Chilodochonida Batisse, 1994
Suborder Coliphorina Jankowski, 1964	Order Exogemmida Jankowski, 1972
Order Peritromida Jankowski, 1978	Suborder Lobochonina Jankowski, 1967
Subphylum *INTRAMACRONUCLEATA* Lynn, 1996	Suborder Spirochonina Jankowski, 2007
Class **SPIROTRICHEA** Bütschli, 1889	Order Cryptogemmida Jankowski, 1978
Subclass Protocruziidia de Puytorac, Grain & Mignot, 1987	Subclass Suctoria Claparède & Lachmann, 1859
Order Protocruziida Jankowski, 1980	Order Vermigemmida Jankowski, 1980
Subclass Phacodiniidia Small & Lynn, 1985	Order Podophryida Jankowski, 1967
Order Phacodiniida Small & Lynn, 1985	Order Exogenida Collin, 1912
	Order Tachyblastonida Jankowski, 1978

(continued)

ADDENDUM. TABLE 1. (continued)

Subclass Hypotrichia Stein, 1859
Order Stichotrichida Fauré-Fremiet, 1961
Order Euplotida Jankowski, 1980
Suborder Kiitrichida Tuffrau & Fleury, 1994
Suborder Discocephalina Wicklow, 1982
Suborder Euplotina Jankowski, 1979
Order Plagiotomida Albaret, 1974
Subclass Licnophoria Lynn, 2003
Order Licnophorida Corliss, 1957
Subclass Oligotrichia Bütschli, 1887
Order Halteriida Petz & Foissner, 1992
Order Strombidiida Jankowski, 1980
Order Strobilidiida Jankowski, 1980
Order Tintinnida Kofoid & Campbell, 1929

Class **ARMOPHOREA** Lynn, 2004
Order Metopida Jankowski, 1980
Order Armophorida Jankowski, 1964
Order Odontostomatida Sawaya, 1940
Order Clevelandellida de Puytorac & Grain, 1976

Class **LITOSTOMATEA** Small & Lynn, 1981
Subclass Haptoria Corliss, 1974
Order Haptorida Corliss, 1974
Order Cyclotrichiida Jankowski, 1980
Order Pleurostomatida Schewiakoff, 1896
Subclass Trichostomatia Bütschli, 1889
Order Vestibuliferida de Puytorac et al., 1974
Order Archistomatida de Puytorac et al., 1974
Order Blepharocorythida Wolska, 1971
Order Entodiniomorphida Reichenow in Doflein & Reichenow, 1929
Order Reikostomatida Jankowski, 2007

Class **OLIGOHYMENOPHOREA** de Puytorac et al., 1974
Subclass Peniculia Fauré-Fremiet in Corliss, 1956
Order Peniculida Fauré-Fremiet in Corliss, 1956
Suborder Frontoniina Jankowski, 1980
Suborder Lembadionina Jankowski, 1980
Suborder Urocentrina Jankowski, 1980
Suborder Parameciina de Fromentel, 1874
Subclass Hymenostomatia Delage & Herouard, 1896
Order Tetrahymenida Fauré-Fremiet in Corliss, 1956
Order Ophryoglenida Canella, 1964
Order Scuticociliatida Small, 1967
Suborder Loxocephalina Jankowski, 1980
Suborder Philasterina Small, 1967
Suborder Pleuronematina Fauré-Fremiet in Corliss, 1956
Order Ephelotida Raabe, 1964
Order Endogenida Collin, 1912 (given as subclass, p. 705)
Order Evaginogenida Jankowski, 1978 (given as subclass, p. 721)
Order Neotenea Jankowski, 1978 (given as subclass, p. 728)

Class **NASSOPHOREA** Small & Lynn, 1981
Order Synhymeniida de Puytorac et al., 1974
Order Nassulida Jankowski, 1968
Suborder Nassulopsina de Puytorac, 1994
Suborder Parahymenostomatina Grain et al., 1976
Order Colpodidiida Foissner, Agatha & Berger, 2002
Order Microthoracida Jankowski, 1967

Class **COLPODEA** Small & Lynn, 1981
Subclass Colpodia Foissner, 1985
Order Colpodida de Puytorac et al., 1974
Order Bursariomorphida Fernández-Galiano, 1978
Order Sorogenida Foissner, 1985
Order Bryophryida de Puytorac, Perez-Paniagua & Perez-Silva, 1979
Order Cyrtolophosidida Foissner, 1978
Subclass Bryometopia Foissner, 1985
Order Bryometopida Foissner, 1985

Class **PROSTOMATEA** Small & Lynn, 1985
Order Prostomatida Schewiakoff, 1896
Order Prorodontida Corliss, 1974

Class **CONTOFRAGMEA** Jankowski, 1980
Order Plagiopylida Jankowski, 1978
Order Trimyemida Jankowski, 1980
Order Parastomatida Jankowski, 2007
Order Hysterocinetida Jankowski, 1973
Order Thigmotrichida Chatton & Lwoff, 1922
Subclass Apostomatia Chatton & Lwoff, 1928
Order Foettingeriida Jankowski, 1980
Order Colliniida Jankowski, 1980
Order Conidophryida Jankowski, 1980
Subclass Astomatia Schewiakoff, 1896
Order Hoplitophryida Jankowski, 1980
Order Anoplophryida Poche, 1913
Order Haptophryida Jankowski, 1980
Subclass Peritrichia Stein, 1859
Order Sessilida Kahl, 1933
Suborder Vorticellina de Fromentel, 1875
Suborder Operculariina Jankowski, 1980
Order Mobilida Kahl, 1933
Suborder Urceolariina Jankowski, 2007
Suborder Trichodinina Jankowski, 1980

References

Introductory Remarks

No paper or book is included in this bibliography that has not been cited directly in one or more of the preceding chapters. References are given uniformly and in full and have been carefully checked for accuracy against the original sources. Deliberately excluded, with rare exception, are abstracts, other very short notes, and unpublished Magisterial theses or Doctoral dissertations.

Having the proper date of publication of a paper is very important, particularly in the field of taxonomy. Unfortunately, certain numbers of some journals, on occasion, appear in the year following the supposed time of their appearance. When we have been able to determine that this has happened, we show two dates for the papers of authors so "trapped": the true date, following the author's name, plus the journal's earlier (incorrect) cover-date in parentheses after the volume number (just before the pagination). For multiple authorships with the same first or senior author, "et al." is used in the text unless there would be confusion because of an identical year of appearance of the two (or more) papers implicated (in the latter case, names of the first two authors are cited).

The date of a taxonomic name is not to be confused with date(s) of citation of paper(s) by the author of that name. There has not been space enough to include all original papers, often only notes or even abstracts, in which new taxa have first been described. For example, mention of a genus "X-us Smythe, 1973" does not mean that the publication so indicated appears in the bibliography. If it is to be included, then that particular "Smythe (1973)" or "(Smythe, 1973)" must also appear in the text elsewhere, separate from combination with the generic name.

For those readers interested in "statistics" – and to save anyone (including reviewers!) the time and labor of counting – there are, on the following pages, ca. 3,000 references, a number much greater than the entire bibliography of the First Edition and ca. 50% more than that of the Second Edition. Many of these are recent works that have appeared within the past decade. However, a large number are the "classical" references to significant works, conserved from the **References** chapters of BOTH the First and Second Editions.

AAAA

Abboud–Abi Saab, M. (1989). Distribution and ecology of tintinnids in the plankton of Lebanese coastal waters (eastern Mediterranean). *Journal of Plankton Research*, *11*, 203–222.

Acosta–Mercado, D., & Lynn, D. H. (2003). The edaphic quantitative protargol stain: A sampling protocol for assessing soil ciliate abundance and diversity. *Journal of Microbiological Methods*, *53*, 365–375.

Addicott, J. F. (1974). Predation and prey community structure: An experimental study of the effect of mosquito larvae on the protozoan communities in pitcher plants. *Ecology*, *55*, 475–492.

Adl, S. M. (2003). *The ecology of soil decomposition* (p. 352). Oxford: Oxford University Press.

Adl, S. M., Simpson, A. G. B., Farmer, M. A., Andersen, R. A., Anderson, O. R., Barta, J. R., et al. (2005). The new higher level classification of eukaryotes with emphasis on the taxonomy of protists. *Journal of Eukaryotic Microbiology*, *52*, 399–451.

Adl, S., & Berger, J. D. (2000). Timing of life cycle morphogenesis in synchronous samples of *Sterkiella*

Allen, R. D. (1984). *Paramecium* phagosome membrane: From oral region to cytoproct and back again. *Journal of Protozoology, 31*, 1–6.

Allen, R. D. (2000). The contractile vacuole and its membrane dynamics. *BioEssays, 22*, 1035–1042.

Allen, R. D., & Fok, A. K. (2000). Membrane trafficking and processing in *Paramecium. International Review of Cytology, 198*, 277–318.

Allen, R. D., & Naitoh, Y. (2002). Osmoregulation and contractile vacuoles of protozoa. *International Review of Cytology, 215*, 351–304.

Allen, R. D., & Wolf, R. W. (1974). The cytoproct of *Paramecium caudatum*: Structure and function, microtubules, and fate of food vacuole membranes. *Journal of Cell Science, 14*, 611–631.

Allen, R. D., & Wolf, R. W. (1979). Membrane recycling at the cytoproct of *Tetrahymena. Journal of Cell Science, 35*, 217–227.

Allen, R. D., Ueno, M. S., Pollard, L. W., & Fok, A. K. (1990). Monoclonal antibody study of the decorated spongiome of contractile vacuole complexes of *Paramecium. Journal of Cell Science, 96*, 469–475.

Allen, R. D., Schroeder, C. C., & Fok, A. K. (1992). Endosomal system of *Paramecium*: Coated pits to early endosomes. *Journal of Cell Science, 101*, 449–461.

Allen, S. L. (1967). Genomic exclusion: A rapid means of inducing homozygous diploid lines in *Tetrahymena pyriformis* syngen 1. *Science, 155*, 575–577.

Allen, S. L., & Li, C. I. (1974). Nucleotide sequence divergence among DNA fractions of different syngens of *Tetrahymena pyriformis. Biochemical Genetics, 12*, 213–233.

Allen, S. L., & Weremiuk, S. L. (1971). Intersyngenic variations in the esterases and acid phosphatases of *Tetrahymena pyriformis. Biochemical Genetics, 5*, 119–133.

Allen, S. L., Byrne, B. C., & Cronkite, D. L. (1971). Intersyngenic variations in the esterases of bacterized *Paramecium aurelia. Biochemical Genetics, 5*, 135–150.

Allen, S. L., Farrow, S. W., & Golembiewski, P. A. (1973). Esterase variations between 14 syngens of *Paramecium aurelia* under axenic growth. *Genetics, 73*, 561–573.

Allen, S. L., Nerad, T. A., & Rushford, C. L. (1983a). Comparison of the esterases and acid phosphatases in *Paramecium multimicronucleatum*, syngens 1-5, *P. jenningsi*, *P. caudatum*, and the *P. aurelia* complex. *Journal of Protozoology, 30*, 148–154.

Allen, S. L., Rushford, C. L., Nerad, T. A., & Lau, E. T. (1983b). Intraspecies variability in the esterases and acid phosphatases of *Paramecium jenningsi* and *Paramecium multimicronucleatum*: Assignment of unidentified paramecia; Comparison with the *P. aurelia* complex. *Journal of Protozoology, 30*, 155–163.

Almagor, M., Ron, A., & Bar–Tana, J. (1981). Chemotaxis in *Tetrahymena thermophila. Cell Motility, 1*, 261–268.

Alonso, P. (1978). Les chromosomes polytènes des ciliés. *Année Biologique, 17*, 369–386.

Al–Rasheid, K. A. S. (1996). Records of free–living ciliates in Saudia Arabia. II. Freshwater benthic ciliates of Al–Hassa Oasis, Eastern Region. *Arab Gulf Journal of Scientific Research, 15*, 187–205.

Al–Rasheid, K. A. S. (1999a). Free–living marine interstitial hypotrichid ciliates from Jubail Marine Wildlife Sanctuary in the Arabian Gulf. *Journal of Agricultural Science, 4*, 53–62.

Al–Rasheid, K. A. S. (1999b). Plagiopylid, oligotrichid and choreotrichid marine ciliates from the Jubail Marine Wildlife Sanctuary, the Arabian Gulf, Saudi Arabia. *Kuwait Journal of Science and Engineering, 26*, 173–179.

Al–Rasheid, K. A. S. (1999c). Marine interstitial pleurostomatid ciliates from Jubail Marine Wildlife Sanctuary, the Arabian Gulf, Saudi Arabia. *Saudi Journal Biological Sciences, 6*, 65–75.

Al–Rasheid, K. A. S. (1999d). A review of marine and brackish water interstitial ciliates from the Arabian Gulf, its offshore islands and Al Hassa Oasis with notes on their ecological status and recovery after the 1991 Gulf War oil spill. *Arab Gulf Journal Scientific Research, 17*, 336–368.

Al–Rasheid, K. A. S. (2001). Redescription of the rare heterotrichid ciliate, *Copemetopus subsalsus* Villeneuve–Brachon, 1940. *Journal of Eukaryotic Microbiology, 48*, 188–193.

Al–Rasheid, K. A. S., & Foissner, W. (1999). Apical feeding in the karyorelictids (Protozoa, Ciliophora) *Sultanophrys* and *Tracheloraphis* sp. *Journal of Eukaryotic Microbiology, 46*, 458–463.

Al–Rasheid, K. A. S., Ali, M. A., Sakran, T., Baki, A. A. A., & Ghaffar, F. A. A. (2000). Trichodinid ectoparasites (Ciliophora: Peritrichida) of some River Nile fish, Egypt. *International Journal of Parasitology, 49*, 131–137.

Al–Rasheid, K. A. S., Nilsson, J. R., & Larsen, H. F. (2001). *Blepharisma intermedium* Padmavathi, 1959 (Ciliophora: Heterotrichida) from Al–Hassa inland hypersaline oasis in Saudi Arabia. *Acta Protozoologica, 40*, 63–69.

Altschuler, M. I., & Yao, M. C. (1985). Macronuclear DNA of *Tetrahymena thermophila* exists as defined subchromosomal–sized molecules. *Nucleic Acids Research, 13*, 5817–5831.

Alvarez–Pellitero, P., Palenzuela, O., Padros, F., Sitja–Bobadilla, A., Riaza, A., Silva, A., et al. (2004). Histophagous scuticociliatids (Ciliophora) parasitizing turbot *Scophthalmus maximus*: Morphology, in vitro culture and virulence. *Folia Parasitologica, 51*, 177–187.

Amblard, C., Sime–Ngando, T., Rachiq, S., & Bourdier, G. (1993). Importance of ciliated protozoa in relation to the bacterial and phytoplankton biomass in an oligo–mesotrophic lake, during the spring diatom bloom. *Aquatic Sciences*, *55*, 1–9.

Amiet, J. L., & Affa'a, F. M. (1985). À propos des stratégies d'infestations chez les protozoaires parasites ou endocommensaux des amphibiens anoures du Cameroun. *Revue Écologie (Terre et Vie)*, 40, 389–398.

Ammermann, D. (1968). Die Kernverhältnisse des Ciliaten *Protocruzia depressa* n. sp. *Archiv für Protistenkunde*, *110*, 434–438.

Ammermann, D. (1971). Morphology and development of the macronuclei of the ciliates *Stylonychia mytilus* and *Euplotes aediculatus*. *Chromosoma*, *33*, 309–238.

Ammermann, D. (1982). Mating types in *Stylonychia mytilus* Ehrbg. *Archiv für Protistenkunde*, *126*, 373–381.

Ammermann, D., & Schlegel, M. (1983). Characterization of two sibling species of the genus *Stylonychia* (Ciliata, Hypotrichia): *S. mytilus* Ehrenberg, 1838 and *S. lemnae* n. sp. I. Morphology and reproductive behavior. *Journal of Protozoology*, *30*, 290–294.

Ammermann, D., Steinbrück, G., von Berger, L., & Hennig, W. (1974). The development of the macronucleus in the ciliated protozoan, *Stylonychia mytilus*. *Chromosoma*, *45*, 401–429.

Ammermann, D., Schlegel, M., & Hellmer, K. H. (1989). North American and Eurasian strains of *Stylonychia lemnae* (Ciliophora, Hypotrichida) have a high genetic identity, but differ in the nuclear apparatus and in their mating behavior. *European Journal of Protistology*, *25*, 67–74.

Amos, W. B., Routledge, L. M., & Yew, F. F. (1975). Calcium–binding proteins in a vorticellid contractile organelle. *Journal of Cell Science*, *19*, 203–213.

Anargyrou, K., Petrikkos, G. L., Suller, M. T. E., Skiada, A., Siakantaris, M. P., Osuntoyinbo, R. T., et al. (2003). Pulmonary *Balantidium Coli* infection in a leukemic patient. *American Journal of Hematology*, *73*, 180–183.

André, J., & Fauré–Fremiet, E. (1967). Formation et structure des concrétions calcaires chez *Prorodon morgani* Kahl. *Journal de Microscopie (Paris)*, *6*, 391–398.

André, J., & Fauré–Fremiet, E. (1984). Mitochondries. In P. P. Grassé (Ed.), *Traité de zoologie* (Vol. 2, Fasc. 1, pp. 207–217). Paris: Masson.

Antipa, G. A. (1972). Structural differentiation in the somatic cortex of a ciliated protozoan, *Conchophthirus curtus* Engelmann 1862. *Protistologica*, *7*(year 1971), 472–501.

Antipa, G. A., & Dolan, J. (1985). *Mytilophilus pacificae* n. g., n. sp.: A new mytilid endocommensal ciliate (Scuticociliatida). *Transactions of the American Microscopical Society*, *104*, 360–368.

Antipa, G. A., & Hatzidimitriou, G. (1981). Morphogenesis of *Conchophthirus curtus*: A study of the morphological events associated with binary fission. *Journal of Protozoology*, *28*, 206–214.

Antipa, G. A., & Small, E. B. (1971). The occurrence of thigmotrichous ciliated protozoa inhabiting the mantle cavity of unionid molluscs of Illinois. *Transactions of the American Microscopical Society*, *90*, 463–472.

Antipa, G. A., Martin, K., & Rintz, M. T. (1983). A note on the possible ecological significance of chemotaxis in certain ciliated protozoa. *Journal of Protozoology*, *30*, 55–57.

Antonius, A. (1999). *Halofolliculina corallasia*, a new coral–killer ciliate on Indo–Pacific reefs. *Coral Reefs*, *18*, 300.

Archbold, J. H. G., & Berger, J. (1985). A qualitative assessment of some metazoan predators of *Halteria grandinella*, a common freshwater ciliate. *Hydrobiologia*, *126*, 97–102.

Arslanyolu, M., & Doerder, F. P. (2000). Genetic and environmental factors affecting mating type frequency in natural isolates of *Tetrahymena thermophila*. *Journal of Eukaryotic Microbiology*, *47*, 412–418.

Arthur, J. R., & Lom, J. (1984a). Trichodinid protozoa (Ciliophora: Peritrichida) from freshwater fishes of Rybinsk Reservoir, USSR. *Journal of Protozoology*, *31*, 82–91.

Arthur, J. R., & Lom, J. (1984b). Some trichodinid ciliates (Protozoa: Peritrichida) from Cuban fishes, with a description of *Trichodina cubanensis* n. sp. from the skin of *Cichlasoma tetracantha*. *Transactions of the American Microscopical Society*, *103*, 172–184.

Arthur, J. R., & Margolis, L. (1984). *Trichodina trutta* Mueller, 1937 (Ciliophora: Peritrichida), a common pathogenic ectoparasite of cultured juvenile salmonid fishes in British Columbia: Redescription and examination by scanning electron microscopy. *Canadian Journal of Zoology*, *62*, 1842–1848.

Arthur, J. R., Cone, D. K., Cusack, R. R., Barker, D. E., & Burt, M. D. B. (2004). Two species of *Trichodina* (Ciliophora: Peritrichida) from cultured flatfishes (Pleuronectiformes) in Atlantic Canada. *Comparative Parasitology*, *71*, 247–250.

Asai, D. J., & Forney, J. D. (Eds.). (2000). *Tetrahymena thermophila*. *Methods in cell biology* (Vol. 62). New York: Academic Press.

Asai, H., Ninomiya, T., Kono, R. I., & Moriyama, Y. (1998). Spasmin and a putative spasmin binding protein(s) isolated from solubilized spasmonemes. *Journal of Eukaryotic Microbiology*, *45*, 33–39.

Ascaso, C., Wierzchos, J., Speranza, M., Gutiérrez, J. C., Gonzalez, A. M., los Rios, A., et al. (2005). Fossil protists and fungi in amber and rock substrates. *Micropaleontology*, *51*, 59–72.

Astrovsky, K. M., Schech, J. M., Sheppard, B. J., Obenschain, C. A., Chin, A. M., Kacergis, M. C., et al. (2002). High mortality due to *Tetrahymena* sp. infection in laboratory–maintained zebrafish (*Brachydanio rerio*). *Comparative Medicine*, *52*, 363–367.

Aufderheide, K., Frankel, J., & Williams, N. E. (1980). Formation and positioning of surface related structures in protozoa. *Microbiological Reviews, 44,* 252–302.

Aufderheide, K. J. (1983). Mitochondrial associations with specific components of the cortex of *Tetrahymena thermophila* Nanney & McCoy, 1976: Microtubules and the epiplasm. *Journal of Protozoology*, *30*, 457–460.

Aufderheide, K. J., Daggett, P. M., & Nerad, T. A. (1983). *Paramecium sonneborni* n. sp., a new member of the *Paramecium aurelia* species–complex. *Journal of Protozoology*, *30*, 128–131.

Aury, J. M., Jaillon, O., Duret, L., Noel, B., Jubin, C., Porcel, B. M., et al. (2006). Global trends of whole–genome duplications revealed by the ciliate *Paramecium tetraurelia. Nature*, *4444*, 171–178.

Austerberry, C. F., Snyder, R. O., & Yao, M. C. (1989). Sequence microheterogeneity is generated at junctions of programmed DNA deletions in *Tetrahymena thermophila. Nucleic Acids Research*, *17*, 7263–7272.

Ax, P., & Ax, R. (1960). Experimentelle Untersuchungen über die Salzgehaltstoleranz von Ciliaten aus dem Brackwasser und Süsswasser. *Biologisches Zentralblatt*, *79*, 7–31.

Azad, I. S., AL Marzouk, A., James, C. M., Almatar, S., & AL Gharabally, H. (2007). Scuticociliatosis–associated mortalities and histopathology of natural infection in cultured silver pomfret (*Pampus argenteus* Euphrasen) in Kuwait. *Aquaculture*, *262*, 202–210.

Azam, F., Fenchel, T., Field, F. G., Gray, J. S., Meyer–Reil, L. A., & Thingstad, F. (1983). The ecological role of water–column microbes in the sea. *Marine Ecology– Progress Series*, *10*, 257–263.

BBBB

Baird, S. E., & Klobutcher, L. A. (1988). Genetic characterization and use of a restriction fragment length variant in the hypotrichous ciliate *Euplotes crassus. Journal of Protozoology*, *35*, 459–465.

Baird, S. E., & Klobutcher, L. A. (1989). Characterization of chromosome fragmentation in two protozoans and identification of a candidate fragmentation sequence in *Euplotes crassus. Genes and Development*, *3*, 585–597.

Baird, S. E., & Klobutcher, L. A. (1991). Differential amplification and copy number control in the hypo–trichous ciliate *Euplotes crassus. Journal of Protozoology*, *38*, 136–140.

Baird, S. E., Fino, G. M., Tausta, S. L., & Klobutcher, L. A. (1989). Micronuclear genome organization in *Euplotes crassus. Molecular and Cellular Biology*, *9*, 3793–3807.

Bakowska, J., & Jerka–Dziadosz, M. (1980). Ultrastructural aspect of size dependent regulation of surface pattern of complex ciliary organelle in a protozoan ciliate. *Journal of Embryology and Experimental Morphology*, *59*, 355–375.

Bakowska, J., Nelsen, E. M., & Frankel, J. (1982a). Regulation of the pattern of basal bodies within the oral apparatus of *Tetrahymena thermophila. Journal of Embryology and Experimental Morphology*, *69*, 83–105.

Bakowska, J., Nelsen, E. M., & Frankel, J. (1982b). Development of the ciliary pattern of the oral apparatus of *Tetrahymena thermophila. Journal of Protozoology*, *29*, 366–382.

Balazi, P., & Matis, D. (2002). Composition, seasonal dynamics and feeding groups of ciliated protozoa in the pelagial of the Morava river. *Biologia*, *57*, 153–160.

Balczon, J. M., & Pratt, J. R. (1996). The functional responses of two benthic algivorous ciliated protozoa with differing feeding strategies. *Microbial Ecology*, *31*, 209–224.

Ball, G. H. (1969). Organisms living on and in Protozoa. In T. T. Chen (Ed.), *Research in Protozoology* (Vol. 3, pp. 565–718). London/New York: Pergamon Press.

Ball, I. R., & Fernando, C. H. (1968). On *Urceolaria mitra* (Protozoa, Peritricha) epizoic on North American flatworms. *Canadian Journal of Zoology*, *46*, 981–985.

Bamforth, S. S. (1963). Limnetic protozoa of southeastern Louisiana. *Proceedings of the Louisiana Academy of Sciences*, *26*, 120–134.

Bamforth, S. S. (1973). Population dynamics of soil and vegetation protozoa. *American Zoologist*, *13*, 171–176.

Bamforth, S. S. (1980). Terrestrial protozoa. *Journal of Protozoology*, *27*, 33–36.

Bamforth, S. S. (1981). Protist biogeography. *Journal of Protozoology*, *28*, 2–9.

Banchetti, R., Erra, F., Ricci, N., & Dini, F. (2003). Ethogram of *Aspidisca sedigita. Canadian Journal of Zoology*, *81*, 14–20.

Bank, M., van der, Basson, L., & Van As, J. G. (1989). A new species of *Trichodina* (Ciliophora: Peritrichia) from the urinary bladder of *Bufo gutturalis* Power, 1927. *South African Journal of Zoology*, *24*, 270–272.

Barbaree, J. M., Fields, B. S., Feeley, J. C., Gorman, G. W., & Martin, W. T. (1986). Isolation of protozoa from water associated with a legionellosis outbreak and demonstration of intracellular multiplication of *Legionella pneumophila. Applied and Environmental Microbiology*, *51*, 422–424.

Barbieri, S. M., & Orlandi, M. J. L. G. (1989). Ecological studies on the planktonic protozoa of a eutrophic reservoir (Rio Grande Reservoir – Brazil). *Hydrobiologia, 183*, 1–10.

Barcelo, J. A., & Calkins, J. (1980). The relative importance of various environmental factors on the vertical distribution of the aquatic protozoan *Coleps spiralis. Photochemistry and Photobiology, 31*, 67–73.

Bardele, C. F. (1968). *Acineta tuberosa*. I. Der Feinbau des adulten Suktors. *Archiv für Protistenkunde, 110*, 403–421.

Bardele, C. F. (1970). Budding and metamorphosis in *Acineta tuberosa*. An electron microscopic study on morphogenesis in Suctoria. *Journal of Protozoology, 17*, 51–70.

Bardele, C. F. (1974). Transport of materials in the suctorian tentacle. In Symposium on *Transport at the cellular level* (Vol. 28, pp. 191–208). Cambridge: Symposia of the Society of Experimental Biology, Cambridge University Press.

Bardele, C. F. (1981). Functional and phylogenetic aspects of the ciliary membrane: A comparative freeze–fracture study. *BioSystems, 14*, 403–421.

Bardele, C. F. (1983). Mapping of highly ordered membrane domains in the plasma membrane of the ciliate *Cyclidium. Journal of Cell Science, 61*, 1–30.

Bardele, C. F. (1987). Auf neuen Wegen zu alten Zielen – moderne Ansätze zur Verwandtschaftsanalyse der Ciliaten. *Verhandlungen der Deutschen Zoologisehen Gesellschaft, 80*, 59–75.

Bardele, C. F. (1989). From ciliate ontogeny to ciliate phylogeny: A program. *Bollettino di Zoologia, 56*, 235–243.

Bardele, C. F. (1999). Ultrastructure, morphogenesis, and a discussion of the phylogeny of the prostome ciliate *Balanion planctonicum. European Journal of Protistology, 35*, 17–33.

Bardele, C. F., & Grell, K. G. (1967). Elektronenmikroskopische Beobachtungen zur Nahrungsaufnahme bei dem Suktor *Acineta tuberosa* Ehrenberg. *Zeitschrift für Zellforschung, 80*, 108–123.

Bardele, C. F., & Klindworth, T. (1996). Stomatogenesis in the karyorelictean ciliate *Loxodes striatus*: A light and scanning microscopical study. *Acta Protozoologica, 35*, 29–40.

Bardele, C. F., & Kurth, T. (2001). Light and scanning electron microscopical study of stomatogenesis in the cyrtophorid ciliate *Chlamydodon mnemosyne* Ehrenberg, 1837. *Acta Protozoologica, 40*, 49–61.

Bardele, C. F., Foissner, W., & Blanton, R. L. (1991). Morphology, morphogenesis and systematic position of the sorocarp forming ciliate *Sorogena stoianovitchae* Bradbury & Olive, 1980. *Journal of Protozoology, 38*, 7–17.

Barker, H. A., & Taylor, C. V. (1931). A study of the conditions of encystment of *Colpoda cucullus. Physiological Zoology, 4*, 620–634.

Barker, J., & Brown, M. R. W. (1994). Trojan Horses of the microbial world: Protozoa and survival of bacterial pathogens in the environment. *Microbiology, 140*, 1253–1259.

Barlow, B. M., & Finley, H. E. (1976). Comparative studies on four species of *Vorticella* by conventional microscopy. *Transactions of the American Microscopical Society, 95*, 346–351.

Barnett, A. (1966). A circadian rhythm of mating type reversals in *Paramecium multimicronucleatum*, syngen 2, and its genetic control. *Journal of Cellular Physiology, 67*, 239–270.

Baroin, A., Perasso, R., Qu, L. H., Brugerolle, G., Bachellerie, J. P., & Adoutte, A. (1988). Partial phylogeny of the unicellular eukaryotes based on rapid sequencing of a portion of 28S ribosomal RNA. *Proceedings of the National Academy Sciences of the United States of America, 85*, 3474–3478.

Baroin–Tourancheau, A., Delgado, P., Perasso, R., & Adoutte, A. (1992). A broad molecular phylogeny of ciliates: Identification of major evolutionary trends and radiations within the phylum. *Proceedings of the National Academy of Sciencesof the United States of America, 89*, 9764–9768.

Baroin–Tourancheau, A., Tsao, N., Klobutcher, L. A., Pearlman, R. E., & Adoutte, A. (1995). Genetic code deviations in the ciliates: Evidence for multiple and independent events. *EMBO Journal, 14*, 3262–3267.

Baroin–Tourancheau, A., Villalobo, E., Tsao, N., Torres, A., & Pearlman, R. E. (1998). Protein coding gene trees in ciliates: Comparison with rRNA–based phylogenies. *Molecular Phylogenetics and Evolution, 10*, 299–309.

Barría de Cao, M. S. (1992). Abundance and species composition of Tintinnina (Ciliophora) in Bahía Blanca Estuary, Argentina. *Estuarine Coastal and Shelf Science, 34*, 295–303.

Barron, G. L., & Szijarto, E. (1982). A new hyphomycete parasitic on the ciliated protozoans *Vorticella* and *Opercularia. Canadian Journal of Botany, 60*, 1031–1034.

Barros, F. S. M., Vasconcelos, S. D., Arruda, M. E., Confalonieri, U. E. C., Luitgards–Moura, J. F., & Honorio, N. A. (2006). Tetrahymenidae infection in mosquito populations in a malaria–endemic region of the Amazon. *Journal Invertebrate Pathology, 91*, 199–201.

Barth, D., Krenek, S., Fokin, S. I., & Berendonk, T. U. (2006). Intraspecific genetic variation in *Paramecium* revealed by mitochondrial cytochrome *c* oxidase 1 sequences. *Journal of Eukaryotic Microbiology, 53*, 20–25.

Basson, L. (1989). An endoparasitic trichodinid (Ciliophora: Peritrichia) from the urinary system of *Barbus trimaculatus* Peters, 1852. *Suid–Afrikaanse Tydskrif für Dierkunde*, *24*, 260–262.

Basson, L., & Van As, J. G. (1989). Differential diagnosis of the genera in the family Trichodinidae (Ciliophora: Peritrichida) with the description of a new genus ectoparasitic on freshwater fish from southern Africa. *Systematic Parasitology*, *13*, 153–160.

Basson, L., & Van As, J. G. (1991). Trichodinids (Ciliophora: Peritrichia) from a calanoid copepod and catfish from South Africa with notes on host specificity. *Systematic Parasitology*, *18*, 147–158.

Basson, L., & Van As, J. G. (1994). Trichodinid ectoparasites (Ciliophora: Peritrichida) of wild and cultured freshwater fishes in Taiwan, with notes on their origin. *Systematic Parasitology*, *28*, 197–222.

Basson, L., Van As, J. G., & Fishelson, L. (1990). A new species of *Trichodina* (Ciliophora: Peritrichia) from the intestine of the surgeonfish *Acanthurus xanthopterus*. *International Journal of Parasitology*, *20*, 785–787.

Batisse, A. (1967a). Ultrastructure de la loge et du style chez *Acineta tuberosa* Ehrenberg et *Paracineta homari* Sand. *Protistologica*, *2*(3) (year 1966), 29–41.

Batisse, A. (1967b). Ultrastructure de la coque et du disque chez le genre *Metacineta* Butschli. *Protistologica*, *3*, 395–411.

Batisse, A. (1968). Quelques aspects de l'ultrastructure de *Pseudogemma pachystyla* Collin. *Protistologica*, *4*, 271–282.

Batisse, A. (1969). Les ultrastructures squelettiques chez certains Thecacinetidae. *Protistologica*, *4*(year 1968), 477–492.

Batisse, A. (1970). Les structures pédonculaires dans les genres *Tokophrya* Bütschli et *Choanophrya* Hartog (Ciliata, Suctorida). *Protistologica*, *5*(year 1969), 387–412.

Batisse, A. (1973). Premières observations sur l'ultrastructure de *Trematosoma bocqueti* (Guilcher), Batisse (Ciliata, Suctorida). *Protistologica*, *8*(year 1972), 477–496.

Batisse, A. (1975). Propositions pour une nouvelle systématique des acinétiens (Ciliophora, Kineto–fragmophora, Suctorida). *Comptes Rendus de l' Académie des Sciences, Paris*, *280*, 1797–1800, 2121–2124. (published in two parts, as indicated by the pagination, but with the identical title)

Batisse, A. (1986). Acinétiens nouveaux ou mal connus des côtes méditerranéennes françaises: II. *Dendro–somides grassei* n. sp. (Suctorida, Ophryodendrina). *Protistologica*, *22*, 11–21.

Batisse, A. (1994a). Sous–classe des Chonotrichia Wallengren, 1895. In P. de Puytorac (Ed.), *Traité de zoologie, infusoires ciliés* (Vol. 2, pp. 433–473). Paris: Masson.

Batisse, A. (1994b). Sous–classe des Suctoria Claparède et Lachmann, 1858. In P. de Puytorac (Ed.), *Traité de zoologie, infusoires ciliés* (Vol. 2, pp. 493–563). Paris: Masson.

Batisse, A., & Crumeyrolle, J. B. (1988). Sur deux chonotriches nouveaux de l'archipel Kerguelen (Ciliophora, Chonotrichida). *Comptes Rendus de l'Académie des Sciences, Paris*, *307*, 635–640.

Batson, B. S. (1983). *Tetrahymena dimorpha* sp. nov. (Hymenostomatida: Tetrahymenidae), a new ciliate parasite of Simuliidae (Diptera) with potential as a model for the study of ciliate morphogenesis. *Philosophical Transactions of the Royal Society of London, Series B*, *301*, 345–363.

Batson, B. S. (1985). A paradigm for the study of insect–ciliate relationships: *Tetrahymena sialidos* sp. nov. (Hymenostomatida: Tetrahymenidae), parasite of larval *Sialis lutaria* (Linn.) (Megaloptera: Sialidae). *Philosophical Transactions of the Royal Society of London, Series B*, *310*, 123–144.

Bauchop, T. (1979). The rumen ciliate *Epidinium* in primary degradation of plant tissues. *Applied and Environmental Microbiology*, *37*, 1217–1223.

Bauer–Nebelsick, M., Bardele, C. F., & Ott, J. A. (1996). Electron microscopic studies on *Zoothamnium niveum* (Hemprich & Ehrenberg, 1831) Ehrenberg, 1838 (Oligohymenophora, Peritrichida), a ciliate with ectosymbiotic, chemoautotrophic bacteria. *European Journal of Protistology*, *32*, 202–215.

Baumgartner, M., Stetter, K. O., & Foissner, W. (2002). Morphological, small subunit rRNA, and physiological characterization of *Trimyema minuta* (Kahl, 1931), an anaerobic ciliate from submarine hydrothermal vents growing from 28°C to 52°C. *Journal of Eukaryotic Microbiology*, *49*, 227–238.

Beale, G. H. (1954). *The genetics of Paramecium aurelia* (179 pp.). Cambridge: Cambridge University Press.

Beams, H. W., & Kessel, R. G. (1973). Studies on the fine structure of the protozoan *Cyclidium*, with special reference to the mitochondria, pellicle and surface–associated bacteria. *Zeitschrift für Zellforschung*, *139*, 303–310.

Beaver, J. R., & Crisman, T. L. (1981). Acid precipitation and the response of ciliated protozoans in Florida lakes. *Verhandlungen der Internationalen Vereinigung für Theoretische und Angewandte Limnologie*, *21*, 353–358.

Beaver, J. R., & Crisman, T. L. (1982). The trophic response of ciliated protozoans in freshwater lakes. *Limnology and Oceanography*, *27*, 246–253.

Beaver, J. R., & Crisman, T. L. (1989a). The role of ciliated protozoa in pelagic freshwater ecosystems. *Microbial Ecology*, *17*, 111–136.

Beaver, J. R., & Crisman, T. L. (1989b). Analysis of the community structure of planktonic ciliated protozoa relative to trophic state in Florida lakes. *Hydrobiologia, 174*, 177–184.

Beech, C. D., & Landers, S. C. (2002). Ciliated protozoan colonization of substrates from Dauphin Island, Alabama. *European Journal of Protistology, 38*, 83–89.

Beers, C. D. (1927). The relation between hydrogen–ion concentration and encystment in *Didinium nasutum. Journal of Morphology and Physiology, 44*, 21–28.

Beers, C. D. (1938). Structure and division in the astomatous ciliate *Metaradiophrya asymmetrica* n. sp. *Journal of the Elisha Mitchell Scientific Society, 54*, 111–125.

Beers, C. D. (1944). The maintenance of vitality in pure lines of the ciliate *Tillina magna. American Naturalist, 78*, 68–76.

Beers, C. D. (1945). Some factors affecting excystment of *Tillina magna. Physiological Zoology, 18*, 80–99.

Beers, C. D. (1946a). History of the nuclei of *Tillina magna* during division and encystment. *Journal of Morphology, 78*, 181–200.

Beers, C. D. (1946b). *Tillina magna:* Micronuclear number, encystment and vitality in diverse clones; capabilities of amicronucleate races. *Biological Bulletin, 91*, 256–271.

Beers, C. D. (1948). Excystment in the ciliate *Bursaria truncatella. Biological Bulletin, 94*, 86–98.

Beers, C. D. (1966a). The excystment process in the ciliate *Nassula ornata* Ehrbg. *Journal of Protozoology, 13*, 79–83.

Beers, C. D. (1966b). Distribution of *Urceolaria spinicola* (Ciliata, Peritrichida) on the spines of the sea urchin *Strongylocentrotus droebachiensis. Biological Bulletin, 131*, 219–229.

Beier, C. L., Horn, M., Michel, R., Schweikert, M., Görtz, H. D., & Wagner, M. (2002). The genus *Caedibacter* comprises endosymbionts of *Paramecium* spp. Related to the Rickettsiales (Alphaproteobacteria) and to *Francisella tularensis* (Gammaproteobacteria). *Applied and Environmental Microbiology, 68*, 6043–6050.

Bell, T. A., Arume, C. S., & Lightner, D. V. (1987). Efficacy of formalin in reducing the levels of peritrichous ciliates on cultured marine shrimp. *Journal of Fish Diseases, 10*, 45–51.

Belzecki, G., & Michalowski, T. (2006). The ability of the rumen ciliate, *Diploplastron affine*, to digest polysaccharides from fungal and bacterial cell walls. *Journal of Animal and Feed Sciences, 15*, 7–10.

Beijerinck, M. W. (1913). *De infusies en de ontdekking der bakteriën. Jaarboeck van de Koninklijke Akademie v. Wetenschappen*. Amsterdam: Müller.

Beisson, J., & Sonneborn, T. M. (1965). Cytoplasmic inheritance of the organization of the cell cortex in *Paramecium aurelia. Proceedings of the National Academy of Sciences of the United States of America, 53*, 275–282.

Beisson, J., Cohen, J., Lefort–Tran, M., Pouphile, M., & Rossignol, M. (1980). Control of membrane fusion in exocytosis. Physiological studies on a *Paramecium* mutant blocked in the final step of the trichocyst extrusion process. *Journal of Cell Biology, 85*, 213–227.

Benitez, L., & Gutiérrez, J. C. (1997). Encystment–specific mRNA is accumulated in the resting cyst of the ciliate *Colpoda inflata. Biochemistry and Molecular Biology International, 41*, 1137–1141.

Benyahya, M., Senaud, J., & Bohatier, J. (1992). Colonisation et dégradation de fragments de paille de blé par les ciliés du rumen *Epidinium*, *Entodinium* et *Isotricha*: étude en microscopie électronique. *Annales des Sciences Naturelles–Zoologie et Biologie Animale, Paris, Série 13, 13*, 103–119.

Benwitz, G. (1982). Entwicklung der Haptocysten von *Ephelota* (Suctoria, Ciliata). *Protistologica, 18*, 459–472.

Benwitz, G. (1984). Die Entladung der Haptocysten von *Ephelota gemmipara* (Suctoria, Ciliata). *Zeitschrift für Naturforschung, 39c*, 812–817.

Beran, A. (1990). *Mikromorphologische Untersuchungen zur Morphogenese der peniculinen Ciliaten Frontonia atra und Paramecium polycaryum. Dissertation* (p. 134). Tübingen, Germany: Eberhard–Karls–Universität.

Bereczky, M. C. s. (1990). Interspecific relationship of some suctoria species in the Danube. *Archiv für Protistenkunde, 138*, 251–255.

Berger, H. (1999). Monograph of the Oxytrichidae (Ciliophora, Hypotrichia). *Monographiae Biologicae, 78*, 1–1080.

Berger, H. (2001). *Catalogue of ciliate names. 1. Hypotrichs* (p. 206). Salzburg: Verlag Helmut Berger.

Berger, H. (2006a). *Bibliography of hypotrichs and euplotids (Ciliophora)* (p. 425). Salzburg: Verlag Helmut Berger.

Berger, H. (2006b). Monograph of the Urostyloidea (Ciliophora, Hypotricha). *Monographiae Biologicae, 85*, 1304.

Berger, H., & Foissner, W. (1987). Morphology and biometry of some soil hypotrichs (Protozoa: Ciliophora). *Zoologische Jahrbücher Systematik, 114*, 193–239.

Berger, H., & Foissner, W. (1988). The morphogenesis of *Kahliella franzi* (Foissner, 1982) nov. comb. and *Oxytricha gigantea* Horváth, 1933 (Ciliophora, Hypotrichida). *Archiv für Protistenkunde, 136*, 65–77.

Berger, H., & Foissner, W. (1989a). Morphology and biometry of some soil hypotrichs (Protozoa, Ciliophora) from Europe and Japan. *Bulletin of the British Museum of Natural History (Zoology), 55*, 19–46.

Berger, H., & Foissner, W. (1989b). Morphology and morphogenesis of *Parakahliella haideri* nov. spec. (Ciliophora, Hypotrichida). *Bulletin of the British Museum of Natural History (Zoology), 55*, 11–17.

Berger, H., & Foissner, W. (1997). Cladistic relationships and generic characterization of oxytrichid hypotrichs (Protozoa, Ciliophora). *Archiv für Protistenkunde*, *148*, 125–155.

Berger, H., Foissner, W., & Adam, H. (1983). Morphology and morphogenesis of *Fuscheria terricola* n. sp. and *Spathidium muscorum* (Ciliophora: Kinetofragminophora). *Journal of Protozoology*, *30*, 529–535.

Berger, H., Foissner, W., & Adam, H. (1984). Taxonomie, Biometrie und Morphogenese einiger terricoler Ciliaten (Protozoa: Ciliophora). *Zoologische Jahrbücher Systematik*, *111*, 339–367.

Berger, H., Foissner, W., & Adam, H. (1985). Morphological variation and comparative analysis of morphogenesis in *Parakahliella macrostoma* (Foissner, 1982) nov. gen. and *Histriculus muscorum* (Kahl, 1932) (Ciliophora, Hypotrichida). *Protistologica*, *21*, 295–311.

Berger, H., Foissner, W., & Adam, H. (1997). Morphological variation and comparative analysis of morphogenesis in *Parakahliella macrostoma* (Foissner, 1982) nov. gen. and *Histriculus muscorum* (Kahl, 1932) (Ciliophora, Hypotrichida). *Protistologica*, *21*, 295–311.

Berger, J. (1965). The infraciliary morphology of *Euplotes tuffraui* n. sp., a commensal in strongylocentrotid echinoids, with comments on echinophilous populations of *Euplotes balteatus* (Dujardin) (Ciliata: Hypotrichida), *Protistologica, 1*, 17–31.

Berger, J. (1978). Quantification of ciliophoran species descriptions: An appeal to reason. *Transactions of the American Microscopical Society*, *97*, 121–126.

Berger, J. (1979). The feeding behavior of *Didinium nasutum* on an atypical prey ciliate (*Colpidium campylum*). *Transactions of the American Microscopical Society*, *98*, 487–494.

Berger, J. (1980). Feeding behaviour of *Didinium nasutum* on *Paramecium bursaria* with normal or apochlorotic zoochlorellae. *Journal of General Microbiology*, *118*, 397–404.

Berger, J., & Hatzidimitriou, G. (1978). Multivariate morphometric analyses of demic variation in *Ancistrum mytili* (Ciliophora: Scuticociliatida) commensal in two mytilid pelecypods. *Protistologica*, *14*, 133–153.

Berger, J., & Lynn, D. H. (1984). Somatic and oral cortical ultrastructure of the plagiopylid ciliates *Lechriopyla mystax* and *Plagiopyla minuta* Powers, 1933. *Journal of Protozoology, 31*, 433–443.

Berger, J., & Lynn, D. H. (1992). Hydrogenosome–methanogen assemblages in the echinoid endocommensal plagiopylid ciliates, *Lechriopyla mystax* Lynch, 1930 and *Plagiopyla minuta* Powers, 1933. *Journal of Protozoology*, *39*, 4–8.

Berger, J. D. (2001). Riding the ciliate cell cycle – a thirty–five–year prospective. *Journal of Eukaryotic Microbiology*, *48*, 505–518.

Bergquist, D. C., Eckner, J. T., Urcuyo, I. A., Cordes, E. E., Hourdez, S., Macko, S. A., et al. (2007). Using stable isotopes and quantitative community characteristics to determine a local hydrothermal vent food web. *Marine Ecology and Progressive Series*, *330*, 49–65.

Berk, S. G., Parks, L. H., & Ting, R. S. (1991). Photo–adaptation alters the ingestion rate of *Paramecium bursaria*, a mixotrophic ciliate. *Applied and Environmental Microbiology*, *57*, 2312–2316.

Berk, S. G., Mills, A. L., Hendricks, D. L., & Colwell, R. R. (1978). Effects of ingesting mercury containing bacteria on mercury tolerance and growth rates of ciliates. *Microbial Ecology, 4*, 319–330.

Bernard, C., & Fenchel, T. (1994). Chemosensory behaviour of *Strombidium purpureum*, an anaerobic oligotrich with endosymbiotic purple non–sulphur bacteria. *Journal of Eukaryotic Microbiology*, *41*, 391–396.

Bernard, C., & Rassoulzadegan, F. (1990). Bacteria or microflagellates as a major food source for marine ciliates: Possible implications for the microzooplankton. *Marine Ecology and Progressive Series*, *64*, 147–155.

Bernard, C., & Rassoulzadegan, F. (1993). The role of picoplankton (cyanobacteria and plastidic picoflagellates) in the diet of tintinnids. *Journal Plankton Research*, *15*, 361–373.

Bernard, C., & Rassoulzadegan, F. (1994). Seasonal variations of mixotrophic ciliates in the northwest Mediterranean Sea. *Marine Ecology and Progressive Series*, *108*, 295–301.

Bernard, F., & Bohatier, J. (1981). Ultrastructure et mise en place des organelles buccaux au cours de la régénération orale chez *Stentor coeruleus* (cilié hétérotriche). *Canadian Journal of Zoology*, *59*, 2306–2318.

Bernatzky, G., Foissner, W., & Schubert, G. (1981). Rasterelektronenmikroskopische und biometrische Untersuchungen über die Variabilität der Form, Struktur und Grösse des Gehäuses einiger limnischer Tintinnina (Protozoa, Ciliophora). *Zoologica Scripta*, *10*, 81–90.

Bernhard, D., & Schlegel, M. (1998). Evolution of histone H4 and H3 genes in different ciliate lineages. *Journal of Molecular Evolution*, *46*, 344–354.

Bernhard, D., Leipe, D. D., Sogin, M. L., & Schlegel, K. M. (1995). Phylogenetic relationships of the Nassulida within the phylum Ciliophora inferred from the complete small subunit rRNA gene sequences of *Furgasonia blochmanni*, *Obertrumia georgiana*, and *Pseudomicrothorax dubius*. *Journal of Eukaryotic Microbiology*, *42*, 126–131.

Bernhard, D., Stechmann, A., Foissner, W., Ammerman, D., Hehn, M., & Schlegel, M. (2001). Phylogenetic relationships within the class Spirotrichea (Ciliophora) inferred from small subunit rRNA gene sequences. *Molecular Phylogenetics and Evolution*, *21*, 86–92.

Berninger, U., Finlay, B. J., & Canter, H. M. (1986). The spatial distribution and ecology of zoochlorellae-bearing ciliates in a productive pond. *Journal of Protozoology*, *33*, 557–563.

Biagini, G. A., Finlay, B. J., & Lloyd, D. (1998). Protozoan stimulation of anaerobic microbial activity: Enhancement of the rate of terminal decomposition of organic matter. *FEMS Microbiology Ecology*, *27*, 1–8.

Bick, H. (1972). *Ciliated protozoa: An illustrated guide to the species used as biological indicators in freshwater biology* (p. 198). Geneva: World Health Organization.

Biggar, R. B., & Wenrich, D. H. (1932). Studies on ciliates from Bermuda sea urchins. *Journal of Parasitology*, *18*, 252–257.

Bihn, J. P., Lilly, D. M., & Napolitano, R. L. (1974). Synthetic medium for the axenic cultivation of four species of *Paramecium*. *Journal of Protozoology*, *21*, 421.

Blackbourn, D. J., Taylor, F. J. R., & Blackbourn, J. (1973). Foreign organelle retention by ciliates. *Journal of Protozoology*, *20*, 286–288.

Blackburn, E. H. (1986). Telomeres. In J. G. Gall (Ed.), *The molecular biology of ciliated protozoa* (chap. 6, pp. 155–178). New York: Academic Press.

Blackburn, E. H. (1992). Telomerase. *Annual Review of Biochemistry*, *61*, 113–129.

Blackburn, E. H., & Gall, J. G. (1978). A tandemly repeated sequence at the termini of the extrachromosomal ribosomal RNA genes in *Tetrahymena*. *Journal of Molecular Biology*, *120*, 33–53.

Blanton, R. L., & Olive, L. S. (1983a). Ultrastructure of aerial stalk formation by the ciliated protozoan *Sorogena stoianovitchae*. *Protoplasma*, *116*, 125–135.

Blanton, R. L., & Olive, L. S. (1983b). Stalk function during sorogenesis by the ciliated protozoan *Sorogena stoianovitchae*. *Protoplasma*, *116*, 136–144.

Blanton, R. L., Warner, S. A., & Olive, L. S. (1983). The structure and composition of the stalk of the ciliated protozoan *Sorogena stoianovitchae*. *Journal of Protozoology*, *30*, 617–624.

Blatterer, H., & Foissner, W. (1988). Beitrag zur terricolen Ciliatenfauna (Protozoa: Ciliophora) Australiens. *Stapfia*, *17*, 1–84.

Bleyman, L. K. (1996). Ciliate genetics. In K. Hausmann & P. C. Bradbury (Eds.), *Ciliates. Cells as organisms* (pp. 291–324). Stuttgart: Gustav Fischer Verlag.

Bobyleva, N. N., Kudriavtsev, B. N., & Raikov, I. B. (1980). Changes of the DNA content of differentiating and adult macronuclei of the ciliate *Loxodes magnus* (Karyorelictida). *Journal of Cell Science*, *44*, 374–394.

Bockstahler, K. R., & Coats, D. W. (1993). Grazing of the mixotrophic dinoflagellate *Gymnodinium sanguineum* on ciliate populations in Chesapeake Bay. *Marine Biology*, *116*, 477–487.

Bohatier, J. (1972). Étude cytologique de la régénération de la région buccale chez le cilié *Lacrymaria olor* (O.F.M. 1786). *Protistologica*, *8*, 439–459.

Bohatier, J. (1981). Macromolecular synthesis during oral regeneration in the ciliate *Condylostoma magnum*: Effects of different inhibitors (II). Protein syntheis and microtubule assembly. *Journal of Experimental Zoology*, *218*, 203–217.

Bohatier, J. (1995). Morphogenèse de régénération chez les protozoaires ciliés: données cytologiques et biochimiques. *Annales des Sciences Naturelles–Zoologie et Biologie Animale, Paris, Série 13*, *16*, 59077.

Bohatier, J., & Kink, J. (1977). Étude des synthèses protéiques au cours des processus morphogénétiques de division et de régénération chez *Dileptus anser*: Action de la cycloheximide. *Protistologica*, *13*, 509–528.

Bohatier, J., & Njiné, T. (1973). Observations ultrastructurales sur le cilié holotriche gymnostome *Litonotus quadrinucleatus* Dragesco et Njiné. *Protistologica*, *9*, 359–372.

Bohatier, J., Iftode, F., Didier, P., & Fryd–Versavel, G. (1978). Sur l'ultrastructure des genres *Spathidium* et *Bryophyllum*, ciliés Kinetofragmophora (de Puytorac et al., 1974). *Protistologica*, *14*, 189–200.

Bohatier, J., Senaud, J., & Benyahya, M. (1990). In situ degradation of cellulose fibres by the entodiniomorph rumen ciliate *Polyplastron multivesiculatum*. *Protoplasma*, *154*, 122–131.

Böhm, P., & Hausmann, K. (1981). Cytochemical investigations of the alveolar plates of the Euplotidae (Ciliophora, Hypotrichida). *Protoplasma*, *106*, 309–316.

Boltovsky, D., Vivequin, S. M., & Swanberg, N. R. (1991). Vertical distribution of tintinnids and associated microplankton in the upper layers of the Barents Sea. *Sarsia*, *76*, 141–151.

Boltovsky, D., Vivequin, S. M., & Swanberg, N. R. (1995). Tintinnids and other microplankton from the Greenland Sea: Abundance and distribution in the marginal ice zone (May–June 1989). *P. S. Z. N. I. Marine Ecology*, *16*, 117–131.

Bonet, F. (1956). Zonificación microfaunística de las Calizas Cretácicas del Este de México. 20th International Geological Congress, 1956 (pp. 3–102). Mexico City, Mexico.

Bonhomme, A. (1990). Rumen ciliates: Their metabolism and relationships with bacteria and their hosts. *Animal Feed Science and Technology*, *30*, 203–266.

Bonhomme, A., Quintana, C., & Durand, M. (1980). Electron microprobe analysis of zinc incorporation into rumen protozoa. *Journal of Protozoology*, *27*, 491–497.

Bonhomme, A., Fonty, G., & Senaud, J. (1982). Essai d'obtention et de survie de ciliés entodiniomorphes du rumen en culture axéniques. *Annals of Microbiology*, *133B*, 335–341.

Bonhomme, A., Grain, J., & Collet, J. Y. (1989). Étude ultrastructurale de *Troglodytella gorillae*, cilié de l'intestin des gorilles. Considérations sur la systématique des ciliés entodiniomorphes. *European Journal of Protistology*, *24*, 225–237.

Bonhomme–Florentin, A. (1985). Attachement des ciliés du caecum de cheval aux fragments végétaux. Dégradation des chloroplastes. Attachement des bactéries aux ciliés du caecum. *Reproduction Nutrition Dévelopment*, *25*, 127–139.

Bonhomme–Florentin, A. (1994). Ciliés du rumen – métabolisme et relations avec les bactéries et leur hôte. In P. de Puytorac (Ed.), *Traité de zoologie, infusoires ciliés* (Vol. 2, pp. 381–394). Paris: Masson..

Bonhomme–Florentin, A., Blancou, J., & Latteur, B. (1978). Étude des variations saisonnières de la microfaune du rumen de zébus. *Protistologica*, *14*, 283–289.

Bonifaz, E. L., & Plaut, W. (1974). The synthesis and distribution of DNA in the macronucleus of *Euplotes eurystomus*. *Chromosoma*, *46*, 261–278.

Borchsenius, S. N., & Sergejeva, G. I. (1979). Characterization of DNA from the ciliate *Bursaria truncatella* O. F. Müll. vegetative cells. *Tsitologiya*, *51*, 327–333.

Borden, D., Whitt, G. S., & Nanney, D. L. (1973a). Isozymic heterogeneity in *Tetrahymena* strains. *Science*, *181*, 279–280.

Borden, D., Whitt, G. S., & Nanney, D. L. (1973b). Electrophoretic characterization of classical *Tetrahymena pyriformis* strains. *Journal of Protozoology*, *20*, 693–700.

Borden, D., Miller, E. T., Whitt, G. S., & Nanney, D. L. (1977). Electrophoretic analysis of evolutionary relationships in *Tetrahymena*. *Evolution*, *31*, 91–102.

Borror, A. C. (1963). Morphology and ecology of the benthic ciliated protozoa of Alligator Harbor, Florida. *Archiv für Protistenkunde*, *106*, 465–534.

Borror, A. C. (1972). Tidal marsh ciliates (Protozoa): morphology, ecology, systematics. *Acta Protozoologica, 10,* 29–71.

Borror, A. C. (1979). Redefinition of the Urostylidae (Ciliophora, Hypotrichida) on the basis of morphogenetic characters. *Journal of Protozoology*, *26*, 544–550.

Borror, A. C., & Hill, B. F. (1995). The order Euplotida (Ciliophora): Taxonomy, with division of *Euplotes* into several genera. *Journal of Eukaryotic Microbiology, 42,* 457–466.

Botes, H., Basson, L., & Van As, L. L. (2001a). Two new species of *Mantoscyphidia* Jankowski, 1980 (Ciliophora: Peritrichia), gill symbionts of *Haliotis* Linnaeus, 1758 (Mollusca: Archaeogastropoda) from the south coast of South Africa. *Acta Protozoologica*, *40*, 131–140.

Botes, H., Van As, L. L., Basson, L., & Van As, J. G. (2001b). Reproductive processes of *Ellobiophrya* sp., hypersymbiont of scyphidiid peritrichs. *Proceedings of the Microscopy Society of Southern Africa*, *31*, 76.

Bouland, C., de Puytorac, P., & Bricourt, E. (1987). *Orchitophrya stellarum*, cilié prétendu astome, est un scuticocilié. *Annales des Sciences Naturelles–Zoologie et Biologie Animale, Paris, Serie 13*, *8*(years 1986–1987), 249–257.

Bower, S. M., & Meyer, G. R. (1993). *Stegotricha enterikos* gen. nov., sp. nov. (Class Phyllopharyngea, Order Rhynchodida), a parasitic ciliate in the digestive gland of Pacific oysters (*Crassostrea gigas*), and its distribution in British Columbia. *Canadian Journal of Zoology*, *71*, 2005–2017.

Bowers, N., & Pratt, J. R. (1995). Estimation of genetic variation among soil isolates of *Colpoda inflata* (Stokes) (Protozoa: Ciliophora) using the polymerase chain reaction and restriction fragment length polymorphism analysis. *Archiv für Protistenkunde*, *145*, 29–36.

Bowers, N., Kroll, T. T., & Pratt, J. R. (1998). Diversity and geographic distribution of riboprints from three cosmopolitan species of *Colpoda* Müller (Ciliophora: Colpodea). *European Journal of Protistology*, *34*, 341–347.

Boxma, B., de Graaf, R. M., van der Staay, G. W. M., van Alen, T. A., Ricard, G., Gabaldon, T., et al. (2005). An anaerobic mitochondrion that produces hydrogen. *Nature*, *434*, 74–79.

Braband, A., Faafeng, B. A., Källqvist, T., & Nilssen, J. P. (1983). Biological control of undesirable cyanobacteria in culturally eutrophic lakes. *Oecologia (Berlin)*, *60*, 1–5.

Bradbury, P. C. (1965). The infraciliature and argyrome of *Opisthonecta benneguyi* Fauré–Fremiet. *Journal of Protozoology*, *12*, 345–363.

Bradbury, P. C. (1966a). The life cycle and morphology of the apostomatous ciliate, *Hyalophysa chattoni* n. g., n. sp. *Journal of Protozoology*, *13*, 209–225.

Bradbury, P. C. (1966b). The fine structure of the mature tomite of *Hyalophysa chattoni. Journal of Protozoology*, *13*, 591–607.

Bradbury, P. C. (1973). The fine structure of the cytostome of the apostomatous ciliate *Hyalophysa chattoni. Journal of Protozoology*, *20*, 405–414.

Bradbury, P. C. (1974). The fine structure of the phoront of the apostomatous ciliate, *Hyalophysa chattoni. Journal of Protozoology*, *21*, 112–120.

Bradbury, P. C. (1975a). *Conidophrys pitelkae*, a new species of pilisuctorian from cuticular hairs of *Crangon crangon* (Linneus). *Acta Protozoologica*, *14*, 161–170.

Bradbury, P. C. (1975b). Stored membranes associated with feeding in apostome trophonts with different diets. *Protistologica*, *10*(year 1974), 533–542.

Bradbury, P. C. (1982). The fine structure of *Conidophrys pitelkae* Bardbury related to its life cycle and taxonomic position in the Apostomatida. *Journal of Protozoology, 29*, 184–194.

Bradbury, P. C. (1989). Evidence for hymenostome affinities in an apostome ciliate. *Journal of Protozoology, 36*, 95–103.

Bradbury, P. C. (1996). Pathogenic ciliates. In K. Hausmann & P. C. Bradbury (Eds.), *Ciliates: Cells as organisms* (pp. 463–477). Stuttgart: Gustav Fischer Verlag.

Bradbury, P. C., & Clamp, J. C. (1991). The fine structure of the cinctum of *Ellobiophrya conviva* (Ciliophora: Peritricha). *Journal of Protozoology, 38*, 312–320.

Bradbury, P. C., & Goyal, V. (1976). The fine structure of a parasitic ciliate *Terebrospira* during ingestion of the exoskeleton of a shrimp *Palaemonetes. Tissue & Cell, 8*, 573–582.

Bradbury, P. C., & Olive, L. S. (1980). Fine structure of the feeding stage of a sorogenic ciliate, *Sorogena stoianovitchae* gen. n., sp. n. *Journal of Protozoology, 27*, 267–277.

Bradbury, P. C., & Outka, D. E. (1967). The structure of *Colpoda elliotti* n. sp. *Journal of Protozoology, 14*, 344–348.

Bradbury, P. C., & Trager, W. (1967). Excystation of apostome ciliates in relation to molting of their crustacean hosts. II. Effect of glycogen. *Biological Bulletin, 133*, 310–316.

Bradbury, P. C., & Tyson, G. E. (1982). The fine structure of *Conidophrys pitelkae* Bradbury related to its life cycle and taxonomic position in the Apostomatida. *Journal of Protozoology, 29*, 184–194.

Bradbury, P. C., Clamp, J. C., & Lyon, J. T. III. (1974). *Terebrospira chattoni* sp. n., a parasite of the endocuticle of the shrimp *Palaemonetes pugio* Holthuis. *Journal of Protozoology, 21*, 678–686.

Bradbury, P. C., Deroux, G., & Campillo, A. (1987). The feeding apparatus of a chitinivorous ciliate. *Tissue & Cell, 19*, 351–363.

Bradbury, P. C., Song, W., & Zhang, L. (1997). Stomatogenesis during the formation of the tomite of *Hyalophysa chattoni* (Hymenostomatida: Ciliophora). *European Journal of Protistology, 33*, 409–419.

Bräucker, R., Machemer–Röhnisch, S., & Machemer, H. (1994). Graviresponses in *Paramecium caudatum* and *Didinium nasutum* examined under varied hypergravity conditions. *Journal of Experimental Biology, 197*, 271–294.

Braun, V., & Miyake, A. (1975). Composition of blepharmone, a conjugation–inducing glycoprotein of the ciliate *Blepharisma. FEBS Letters, 53*, 131–134.

Bretschneider, L. H. (1959). Die submikroskopische Struktur der Pellikula von *Epidinium ecaudatum* (Ophryoscolecidae). *Proceedings of the National Academy of Sciences Amsterdam (Series. C), 62*, 542–555.

Bretschneider, L. H. (1962). Das Paralabialorgan der Ophryoscoleciden. *Proceedings of the National Academy of Sciences Amsterdam (Series C), 65*, 423–437.

Broers, C. A. M., Molhuizen, H. O. F., Stumm, C. K., & Vogels, G. D. (1992). An electromigration technique, applied for the concentration of anaerobic protozoa from mass cultures. *Journal of Microbiological Methods, 14*, 217–220.

Broglio, E., Johansson, M., & Jonsson, P. R. (2001). Trophic interaction between copepods and ciliates: Effects of prey swimming behavior on predation risk. *Marine Ecology Progress Series, 220*, 179–186.

Bronowski, J. (1970). New concepts in the evolution of complexity. *Synthese, 21*, 228–246.

Brunk, C. F., Kahn, R. W., & Sadler, L. A. (1990). Phylogenetic relationships among *Tetrahymena* species determined using the polymerase chain reaction. *Journal of Molecular Evolution, 30*, 290–297.

Brunk, C. F., Lee, L. C., Tran, A. B., & Li, J. (2003). Complete sequence of the mitochondrial genome of *Tetrahymena thermophila* and comparative methods for identifying highly divergent genes. *Nucleic Acids Research, 31*, 1673–1682.

Bruns, P. J. (1986). Genetic organization of *Tetrahymena*. In J. G. Gall (Ed.), *The molecular biology of ciliated protozoa* (chap. 2, pp. 27–44). New York: Academic Press.

Bruns, P. J., & Brussard, T. T. (1974). Pair formation in *Tetrahymena*, an inducible developmental system. *Journal of Experimental Zoology, 188*, 337–344.

Buonanno, F., Saltalamacchia, P., & Miyake, A. (2005). Defence function of pigmentocysts in the karyorelictid ciliate *Loxodes striatus. European Jorunal Of Protistology, 41*, 151–158.

Buchmann, K., & Nielsen, M. E. (1999). Chemoattraction of *Ichthyophthirius multifiliis* (Ciliophora) theronts to host molecules. *International Journal of Parasitology, 29*, 1415–1423.

Buchmann, K., Sigh, J., Nielsen, C. V., & Dalgaard, M. (2001). Host responses against the fish parasitizing ciliate *Ichthyophthirius multifiliis. Veterinary Parasitology, 100*, 105–116.

Buchmann, K., Jensen, P. B., & Kruse, K. D. (2003). Effects of sodium percarbonate and garlic extract on *Ichthyophthirius multifiliis* theronts and tomocysts: In vitro experiments. *North American Journal of Aquaculture, 65*, 21–24.

Budin, K., & Philippe, H. (1998). New insights into the phylogeny of eukaryotes based on ciliate Hsp70 sequences. *Molecular Biology and Evolution, 15*, 943–956.

Buhse, H. E., Jr. (1966). Oral morphogenesis during transformation from microstome to macrostome and macrostome to microstome in *Tetrahymena vorax* strain V_2 type S. *Transactions of the American Microscopical Society*, *85*, 305–313.

Buhse, H. E., Jr. (1967). Microstome–macrostome transformation in *Tetrahymena vorax* strain V_2 type S induced by a transforming principle, stomatin. *Journal of Protozoology*, *14*, 608–613.

Buhse, H. E., Jr., & Williams, N. E. (1982). A comparison of cortical proteins in *Tetrahymena vorax* microstomes and macrostomes. *Journal of Protozoology*, *29*, 222–226.

Buitkamp, U. (1977). Die Ciliatenfauna der Savanne von Lamto (Elfenbeinküste). *Acta Protozoologica*, *16*, 249–276.

Bulit, C., Díaz–Avalos, C., Signoret, M., & Montagnes, D. J. S. (2003). Spatial structure of planktonic ciliate patches in a tropical coastal lagoon: An application of geostatistical methods. *Aquatic Microbial Ecology*, *30*, 185–196.

Bullington, W. E. (1940). Some ciliates from Tortugas. *Papers from the Tortugas Laboratory*, *32*, 179–221.

Burchill, B. R., George, S. B., Lindberg, R. E., & Sims, J. P. (1974). Conjugation in *Stentor muliformis*. *Transactions of the American Microscopical Society*, *93*, 134–135.

Burgess, P. J., & Matthews, R. A. (1994a). A standardized method for the in vivo maintenance of *Cryptocaryon irritans* (Ciliophora) using the grey mullet *Chelon labrosus* as an experimental host. *Journal of Parasitology*, *80*, 288–292.

Burgess, P. J., & Matthews, R. A. (1994b). *Cryptocaryon irritans* (Ciliophora): Photoperiod and transmission in marine fish. *Journal of the Marine Biological Association of the United Kingdom*, *74*, 535–542.

Burgess, P. J., & Matthews, R. A. (1995a). *Cryptocaryon irritans* (Ciliophora): Acquired protective immunity in the thick–lipped mullet, *Chelon labrosus*. *Fish Shellfish Immunology*, *5*, 459–468.

Burgess, P. J., & Matthews, R. A. (1995b). Fish host range of seven isolates of *Cryptocaryon irritans* (Ciliophora). *Journal of Fish Biology*, *46*, 727–729.

Burkill, P. H., Mantoura, R. F. C., Llewellyn, C. A., & Owens, N. J. P. (1987). Microzooplankton grazing and selectivity of phytoplankton in coastal waters. *Marine Biology*, *93*, 581–590.

Burkovsky, I. V. (1970). The ciliates of the mesopsammon of the Kandalaksha Gulf (White Sea). II. *Acta Protozoologica*, *8*, 47–65. (in Russian with English summary)

Burbanck, W. D., & Eisen, J. D. (1960). The inadequacy of monobacterially–fed *Paramecium aurelia* as food for *Didinium nasutum*. *Journal of Protozoology*, *7*, 201–206.

Burlakova, L. E., Karatayev, A. Y., & Molloy, D. P. (1998). Field and laboratory studies of zebra mussel (*Dreissena polymorpha*) infection by the ciliate *Conchophthirus acuminatus* in the Republic of Belarus. *Journal of Invertebrate Pathology*, *71*, 251–257.

Burlakova, L. E., Padilla, D. K., Karatayev, A. Y., & Minchin, D. (2006). Endosymbionts of *Dreissena polymorpha* in Ireland: Evidence for the introduction of adult mussels. *Journal of Molluscan Studies*, *72*, 207–210.

Burns, C. W., & Gilbert, J. J. (1993). Predation on ciliates by freshwater calanoid copepods: Rates of predation and relative vulnerabilities of prey. *Freshwater Biology*, *30*, 377–393.

Burreson, E. M. (1973). Symbiotic ciliates from solitary ascidians in the Pacific northwest, with a description of *Parahypocoma rhamphisokarya* n. sp. *Transactions of the American Microscopical Society*, *92*, 517–522.

Burt, R. L. (1940). Specific analysis of the genus *Colpoda* with special reference to the standardization of experimental material. *Transactions of the American Microscopical Society*, *59*, 414–432.

Burt, R. L., Kidder, G. W., & Claff, C. L. (1941). Nuclear reorganization in the Colpodidae. *Journal of Morphology*, *69*, 537–561.

Bush, M. (1934). The morphology of *Haptophrya michiganensis* Woodhead, an astomatous ciliate from the intestinal tract of *Hemidactylium scutatum* (Schlegel). *University of California Publications in Zoology*, *39*, 251–276.

Buskey, E. J., & Stoecker, D. K. (1989). Behavioral responses of the marine tintinnid *Favella* sp. to phytoplankton: Influence of chemical, mechanical and photic stimuli. *Journal of Experimental Marine Biology and Ecology*, *132*, 1–16.

Bussers, J. C. (1976). Structure et composition du kyste de résistance de 4 protozoaires ciliés. *Protistologica*, *12*, 87–100.

Bussers, J. C. (1984). Les appareils protecteurs des infusoires ciliés. In P. P. Grassé (Ed.), *Traité de zoologie: Vol. 2, Fasc. 1. Infusoires ciliés* (pp. 439–455). Paris: Masson.

Bütschli, O. (1887–1889). Protozoa. Abt. III. Infusoria und System der Radiolaria. In H. G. Bronn (Ed.), *Klassen und Ordnung des Thiers–Reichs* (Vol. 1, pp. 1098–2035). Leipzig: C. F. Winter.

CCCC

Callieri, C., Karjalainen, S. M., & Passoni, S. (2002). Grazing by ciliates and heterotrophic nanoflagellates on picocyanobacteria in Lago Maggiore, Italy. *Journal of Plankton Research*, *24*, 785–796.

Calvo, P., & De Miguel, M. (1995/1996). Study of *Histriculus cavicola* cyst wall using different lectins. *Archiv für Protistenkunde*, *146*, 329–339.

Calvo, P., Fernandez–Aliseda, M. C., Garrido, J., & Torres, A. (2003). Ultrastructure, encystment and cyst wall composition of the resting cyst of the peritrich ciliate *Opisthonecta henneguyi*. *Journal of Eukaryotic Microbiology*, *50*, 49–56.

Cameron, S. L. (2002). Taxonomy and phylogeny of endosymbiotic ciliates (Ciliophora: Litostomatea) associated with Australian herbivorous marsupials. *International Journal of Parasitology*, *33*, 347–355.

Cameron, S. L. (2003). Taxonomy and phylogeny of endosymbiotic ciliates (Ciliophora: Litostomatea) associated with Australian herbivorous marsupials. *International Journal of Parasitology*, *33*, 347–355.

Cameron, S. L., & O'Donoghue, P. J. (2001). Stomatogenesis in the ciliate genus *Macropodinium* Dehority, 1996 (Litostomatea: Macropodiniidae). *European Journal of Protistology*, *37*, 199–206.

Cameron, S. L., & O'Donoghue, P. J. (2002a). The ultrastructure of *Amylovorax dehorityi* comb. nov. and erection of the Amylovoracidae fam. nov. (Ciliophora: Trichostomatia). *European Journal of Protistology*, *38*, 29–44.

Cameron, S. L., & O'Donoghue, P. J. (2002b). The ultrastructure of *Macropodinium moiri* and revised diagnosis of the Macropodiniidae (Litostomatea: Trichostomatia). *European Journal Protistology*, *38*, 179–194.

Cameron, S. L., & O'Donoghue, P. J. (2003a). Trichostome ciliates from Australian marsupials. II. *Polycosta* gen. nov. (Litostomatea: Polycostidae fam. nov.). *European Journal Protistology*, *39*, 83–99.

Cameron, S. L., & O'Donoghue, P. J. (2003b). Trichostome ciliates from Australian marsupials. IV. Distribution of the ciliate fauna. *European Journal Protistology*, *39*, 139–147.

Cameron, S. L., & O'Donoghue, P. J. (2004a). Phylogeny and biogeography of the "Australian" trichostomes (Ciliophora: Litostomata). *Protist*, *155*, 215–235.

Cameron, S. L., & O'Donoghue, P. J. (2004b). Morphometric and cladistic analyses of the phylogeny of *Macropodinium* (Ciliophora: Litostomatea: Macropodiniidae). *Acta Protozoologica.*, *43*, 43–53.

Cameron, S. L., O'Donoghue, P. J., & Adlard, R. D. (2000). Novel isotrichid ciliates endosymbiotic in Australian macropodid marsupials. *Systematic Parasitology*, *46*, 45–57.

Cameron, S. L., Adlard, R. D., & O'Donoghue, P. J. (2001a). Evidence for an independent radiation of endosymbiotic litostome ciliates within Australian marsupial herbivores. *Molecular Phylogenetics and Evolution*, *20*, 302–310.

Cameron, S. L., O'Donoghue, P. J., & Adlard, R. D. (2001b). Four new species of *Macropodinium* (Ciliophora: Litostomatea) from Australian wallabies and pademelons. *Journal of Eukaryotic Microbiology*, *48*, 542–555.

Cameron, S. L., Wright, A. D. G., & O'Donoghue, P. J. (2003). An expanded phylogeny of the Entodiniomorphida (Ciliophora: Litostomatea). *Acta Protozoologica*, *42*, 1–6.

Canella, M. F. (1951). Gimnostomi dei generi *Holophrya, Amphileptus* e *Lionotus* predatori di *Carchesium polypinum* e di altri peritrichi fissi. *Annali dell'Università di Ferrara (Nuova Serie, Sect. III)*, *1*, 1–80.

Canella, M. F. (1954). Ricerche sulla microfauna delle acque interne ferraresi. Introduzione allo studio dei ciliate e dei rotiferi. *Pubblicazioni del Museo Civico di Storia Naturale Ferrara*, *4*, 1–154.

Canella, M. F. (1957). Studi e ricerche sui tentaculiferi nel quadro della biologia generale. *Annali dell'Università di Ferrara (Nuova. Serie, Sect. III)*, *1*, 259–716.

Canella, M. F., & Rocchi–Canella, I. (1964). Morfo–fisiologia e morfogenesi dell'apparato orale di *Ophryoglena* (Ciliata, Holotricha, Hymenostomatida). *Annali dell'Università di Ferrara (Nuova Serie., Sect. III)*, *2*, 189–292.

Canella, M. F., & Rocchi–Canella, I. (1976). Biologie des Ophryoglenina (ciliés hyménostomes histophages). *Annali dell'Universtità di Ferrara (Nuova Serie, Sect. III)*, *3*(Suppl. 2), 1–510.

Canter, H. M., & Dick, M. W. (1994). *Eurychasmopsis multisecunda* gen. et sp. nov., a parasite of the suctorian ciliate *Podophrya*. *Mycological Research*, *98*, 105–117.

Canter, H. M., Heaney, S. I., & Lund, J. W. G. (1990). The ecological significance of grazing on planktonic populations of cyanobacteria by the ciliate *Nassula*. *New Phytologist*, *114*, 247–263.

Capdeville, Y., Cardoso de Almeida, M. L., & Deregnacourt, C. (1987). The membrane–anchor of *Paramecium* temperature–specific surface antigens is a glycosylinositol phospholipid. *Biochemical and Biophyscial Research Communications*, *147*, 1219–1225.

Caprette, C. L., & Gates, M. A. (1994). Quantitative analyses of interbreeding in populations of *vannus*–morphotype *Euplotes*, with special attention to the nominal species *E. vannus* and *E. crassus*. *Journal of Eukaryotic Microbiology*, *41*, 316–324.

Capriulo, G. M., & Carpenter, E. J. (1983). Abundance, species composition and feeding impact of tintinnid micro–zooplankton in central Long Island Sound. *Marine Ecology–Progress Series*, *10*, 277–288.

Capriulo, G. M., & Small, E. B. (1986). Discovery of an apostome ciliate (*Collinia beringensis* n. sp.) endoparasitic in the Bering Sea euphausiid *Thysanoessa inermis*. *Diseases of Aquatic Organisms*, *1*, 141–146.

Capriulo, G. M., Gold, K., & Okubo, A. (1982). Evolution of the lorica in tintinnids: A possible selective advantage. *Annales de l'Institut Océanographique, Paris, 58*, 319–323.

Carey, P. G., & Warren, A. (1983). The role of surface topography in the taxonomy of peritrich ciliates. *Protistologica, 19*, 73–89.

Carl, G. R., & Brown, R. D. (1983). Protozoa in the forestomach of the collared peccary (*Tayassu tajacu*). *Journal of Mammology, 64*, 709.

Cariou, J. B., Dolan, J. R., & Dallot, S. (1999). A preliminary study of tintinnid diversity in the NW Mediterranean Sea. *Journal of Plankton Research, 21*, 1065–1075.

Caron, F. (1990). Eukaryotic codes. *Experientia, 46*, 1106–1107.

Caron, F., & Meyer, E. (1985). Does *Paramecium primaurelia* use a different genetic code in its macronucleus? *Nature, 314*, 185–188.

Carrias, J. F., Amblard, C., & Bourdier, G. (1994). Vertical and temporal heterogeneity of planktonic ciliated protozoa in a humic. *Journal of Plankton Research, 16*, 471–485.

Carrias, J. F., Amblard, C., & Bourdier, G. (1998). Seasonal dynamics and vertical distribution of planktonic ciliates and their relationship to microbial food resources in the oligomesotrophic Lake Pavin. *Archiv für Hydrobiologie, 143*, 227–255.

Carver, C. E., Mallet, A. L., Warnock, R., & Douglas, D. (1996). Red–coloured digestive glands in cultured mussels and scallops: The implication of *Mesodinium rubrum. Journal of Shellfish Research, 15*, 191–201.

Cavalier–Smith, T. (1998). A revised six–kingdom system of life. *Biological Reviews, 73*, 203–266.

Cavalier–Smith, T. (1991). Cell diversification in heterotrophic flagellates. In D. J. Patterson & J. Larsen (Eds.), *The Biology of Free-living Heterorophic Flagellates* (pp. 113–131). Oxford: Clarendon Press.

Cawthorn, R. J., Lynn, D. H., Despres, B., MacMillan, R., Maloney, R., Loughlin, M., et al. (1996). Description of *Anophryoides haemophila* n. sp. (Scuticociliatida: Orchitophryidae), a pathogen of American lobsters *Homarus americanus. Diseases of Aquatic Organisms, 24*, 143–148.

Cépède, C. (1910). Recherches sur les infusoires astomes. Anatomie, biologie, éthologie parasitaire, systématique. *Archives de Zoologie Expérimentale et Générale (Série 5), 3*, 341–609.

Chantangsi, C., Lynn, D. H., Brandl, M. T., Cole, J. C., Hetrick, N., & Ikonomi, P. (2007). Barcoding ciliates: A comprehensive study of 75 isolates of the genus *Tetrahymena. International Journal of Systematic and Evolutionary Microbiology, 57*, 2412–2425.

Chao, A., Li, P. C., Agatha, S., & Foissner, W. (2006). A statistical approach to estimate soil ciliate diversity and distribution based on data from five continents. *Oikos, 114*, 479–493.

Chapman, G. B., & Kern, R. C. (1983). Ultrastructural aspects of the somatic cortex and contractile vacuole of the ciliate, *Ichthyophthirius multifiliis* Fouquet. *Journal of Protozoology, 30*, 481–490.

Chardez, D. (1983). Sur la variabilité de deux espèces d'Ophryscolecidae (Ciliophora Entodiniomorphida). *Bulletin des Recherches Agronomiques de Gembloux, 18*, 257–264.

Chatton, E., & Lwoff, A. (1927). *Pottsia infusoriorum* n. g., n. sp., acinétien parasite des folliculines et des cothurnies. *Bulletin de l'Institut Océanographique, 489*, 1–12.

Chatton, E., & Lwoff, A. (1929). Contribution à l'étude de l'adaptation: *Ellobiophrya donacis* Ch. et Lw., péritriche vivant sur les branchies de l'acéphale *Donax vittatus* da Costa. *Bulletin Biologique de la France et de la Belgique, 63*, 321–349.

Chatton, E., & Lwoff, A. (1930). Imprégnation, par diffusion argentique, de l'infraciliature des ciliés marins et d'eau douce, après fixation cytologique et sans dessiccation. *Comptes Rendus des Séances de la Société de Biologie et de ses Filiales, 104*, 834–836.

Chatton, E., & Lwoff, A. (1935a). Les ciliés apostomes. Morphologie, cytologie, éthologie, évolution, systématique. Première partie. Aperçu historique et général. Etude monographique des genres et des espèces. *Archives de Zoologie Experimentale et Générale, 77*, 1–453.

Chatton, E., & Lwoff, A. (1935b). La constitution primitive de la strie ciliaire des infusoires. La desmodexie. *Comptes Rendus des Séances de la Société de Biologieet de ses Filiales, 118*, 1068–1072.

Chatton, E., & Lwoff, A. (1939a). Sur le suçoir des infusoires thigmotriches rhynchoidés (Hypocomidae et Sphenophryidae) et sa genèse. *Comptes Rendus de l'Académie des Sciences, 209*, 333–336.

Chatton, E., & Lwoff, A. (1939b). Sur la systématique de la tribu des thigmotriches rhynchoïdes. Les deux familles des Hypocomidae Bütschli et des Ancistrocomidae n. fam. Les deux genres nouveaux, *Heterocoma* et *Parahypocoma. Comptes Rendus de l'Académie des Sciences, 209*, 429–431.

Chatton, E., & Lwoff, A. (1949). Recherches sur les ciliés thigmotriches. I. *Archives de Zoologie Expérimentaleet Générale, 86*, 169–253. (Chatton passed away in 1947.)

Chatton, E., & Lwoff, A. (1950). Recherches sur les ciliés thigmotriches. II. *Archives de Zoologie Expérimentale et Générale, 86*, 393–485. (Chatton passed away in 1947.)

Chatton, E., & Séguéla, J. (1936). Un hypotriche de la branchie de *Ciona intestinalis* L., intermédialre entre les Euplotidae et les Aspidiscidae: *Euplotaspis cio-*

naecola n. g., n. sp. *Bulletin de la Société Zoologique de France*, *61*, 332–340.

Chen, Z. G., & Song, W. B. (2001). Phylogenetic positions of *Uronychia transfuga* and *Diophrys appendiculata* (Euplotida, Hypotrichia, Ciliophora) within hypotrichous ciliates inferred from the small subunit ribosomal RNA gene sequences. *European Journal of Protistology*, *37*, 291–301.

Chen, Z. G., & Song, W. B. (2002). Phylogenetic positions of *Aspidisca steini* and *Euplotes vannus* within the order Euplotida (Hypotrichia: Ciliophora) inferred from complete small subunit ribosomal RNA gene sequences. *Acta Protozoologica*, *41*, 1–9.

Chen, Z. G., Song, W. B., & Warren, A. (2000). Studies on six *Euplotes* spp. (Ciliophora: Hypotrichida) using RAPD fingerprinting, including a comparison with morphometric analyses. *Acta Protozoologica*, *39*, 209–216.

Chen, Z. G., Song, W. B., & Warren, A. (2003). Species separation and identification of *Uronychia* spp. (Hypotrichia: Ciliophora) using RAPD fingerprinting and ARDRA riboprinting. *Acta Protozoologica*, *42*, 83–90.

Chen–Shan, L. (1969). Cortical morphogenesis in *Paramecium aurelia* following amputation of the posterior region. *Journal of Experimental Zoology*, *170*, 205–228.

Chernyshova, N. B. (1976). Morphological peculiarities and certain biological problems of the genus *Apiosoma* members (Ciliata, Peritricha) from the young of predatory fishes. *Parazitologia*, *10*, 170–177.

Chessa, M. G., Gallus, L., Tiano, L., Trielli, F., & Corrado, M. U. D. (2001). Variations in macronuclear chromatin structure and chromatin extrusion in excystment from resting cysts of *Colpoda inflata*. *European Journal of Protistology*, *37*, 281–290.

Chessa, M. G., Largana, I., Trielli, F., Rosati, G., Politi, H., Angelini, C., et al. (2002). Changes in the ultrastructure and glycoproteins of the cyst wall of *Colpoda cucullus* during resting encystment. *European Journal of Protistology*, *38*, 373–381.

Cheung, P. J., Nigrelli, R. F., & Ruggieri, G. D. (1980). Studies on the morphology of *Uronema marinum* Dujardin (Ciliatea: Uronematidae) with a description of the histopathology of the infection in marine fishes. *Journal of Fish Diseases*, *3*, 295–303.

Cho, P. L. (1971). Cortical patterns in two syngens of *Glaucoma*. *Journal of Protozoology*, *18*, 180–183.

Choi, J. K., Coats, D. W., Brownlee, D. C., & Small, E. B. (1992). Morphology and infraciliature of three species of *Eutintinnus* (Ciliophora: Tintinnia) with guidelines for interpreting protargol–stained tintinnine ciliates. *Journal of Protozoology*, *39*, 80–92.

Christaki, U., Dolan, J. R., Pelegri, S., & Rassoulzadegan, F. (1998). Consumption of picoplankton–size particles by marine ciliates: Effects of physiological state of the ciliate and particle quality. *Limnology and Oceanography*, *43*, 458–464.

Christaki, U., Jacquet, S., Dolan, J. R., Vaulot, D., & Rassoulzadegan, F. (1999). Growth and grazing on *Prochlorococcus* and *Synechococcus* by two marine ciliates. *Limnology and Oceanography*, *44*, 52–61.

Christner, B. C., Kvitko, B. H., II, & Reeve, J. N. (2003). Molecular identification of Bacteria and Eukarya inhabiting an Antarctic cryoconite hole. *Extremophiles*, *7*, 177–183.

Christopher, M. H., & Patterson, D. J. (1983). *Coleps hirtus*, a ciliate illustrating facultative symbioses between protozoa and algae. *Annales de la Station Biologique de Besse-en-Chandesse, 17,* 278–296.

Claff, C. L., Dewey, V. C., & Kidder, G. W. (1941). Feeding mechanisms and nutrition in three species of *Bresslaua*. *Biological Bulletin*, *81*, 221–234.

Clamp, J. C. (1973). Observations on the host–symbiont relationships of *Lagenophrys lunatus* Imamura. *Journal of Protozoology*, *20*, 558–561.

Clamp, J. C. (1982). *Ellobiophrya conviva* n. sp., a commensal of marine ectoprocts, and a revision of the family Ellobiophryidae (Ciliophora, Peritricha). *Journal of Protozoology*, *29*, 149–156.

Clamp, J. C. (1987). Five new species of *Lagenophrys* (Ciliophora, Peritricha, Lagenophryidae) from the United States with observations on their developmental stages. *Journal of Protozoology*, *34*, 382–392.

Clamp, J. C. (1990). Redescription of three species of *Lagenophrys* (Ciliophora: Peritricha: Lagenophryidae) and a new North American species of *Lagenophrys* from hypogean amphipods. *Transactions of the American Microscopical Society*, *109*, 1–31.

Clamp, J. C. (1991). Revision of the family Lagenophryidae Bütschli, 1889 and description of the family Usconophryidae n. fam. (Ciliophora, Peritricha). *Journal of Protozoology*, *38*, 355–377.

Clamp, J. C. (1992). Three new species of lagenophryid peritrichs (Ciliophora) ectocommensal on freshwater decapod crustaceans from Madagascar. *Journal of Protozoology*, *39*, 732–740.

Clamp, J. C., & Coats, D. W. (2000). *Planeticovorticella finleyi* n. g., n. sp. (Peritrichia, Vorticellidae), a planktonic ciliate with a polymorphic life cycle. *Invertebrate Biology*, *119*, 1–16.

Clamp, J. C., & Williams, D. (2006). A molecular phylogenetic investigation of *Zoothamnium* (Ciliophora, Peritrichia, Sessilida). *Journal of Eukaryotic Microbiology*, *53*, 494–498.

Clamp, J. C., Bradbury, P. C., Strüder–Kypke, M. C., & Lynn, D. H. (2008). Phylogenetic position of the apostome ciliates (Phylum Ciliophora, Subclass Apostomatia) tested using small subunit rRNA gene sequences. *Denisia*, *23*, 395–402.

Clark, T. B., & Brandl, D. G. (1976). Observations on the infection of *Aedes sierrensis* by a tetrahymenine ciliate. *Journal of Invertebrate Pathology*, *28*, 341–349.

Clark, T. G., Gao, Y., Gaertig, J., Wang, X. T., & Cheng, G. (2001). The i–antigens of *Ichthyophthirius multifiliis* are GPI–anchored proteins. *Journal of Eukaryotic Microbiology*, *48*, 332–337.

Clarke, K. J., Finlay, B. J., Esteban, G., Guhl, B. E., & Embley, T. M. (1993). *Cyclidium porcatum* n. sp.: A free–living anaerobic scuticociliate containing a stable complex of hydrogenosomes, Eubacteria and Archaeobacteria. *European Journal of Protistology*, *29*, 262–270.

Clément–Iftode, F. (1967). Sur unne *Sphaerophrya* ectoparasite de *Nassula elegans*. *Protistologica*, *3*, 127–132.

Clerot, J. C., Iftode, F., Budin, K., Jeanmaire–Wolf, R., Coffe, G., & Fleury–Aubusson, A. (2000). Fine oral filaments in *Paramecium*: A biochemical and immunological analysis. *Journal of Eukaryotic Microbiology*, *48*, 234–245.

Cleven, E. J. (2004). Seasonal and spatial distribution of ciliates in the sandy hyporheic zone of a lowland stream. *European Journal Protistology*, *40*, 71–84.

Coats, W. D. (1988). *Dubosquella cachoni* n. sp., a parasitic dinoflagellate lethal to its tintinnine host *Eutintinnus pectinis*. *Journal of Protozoology*, 35, 607–617.

Coats, W. D., & Heisler, J. J. (1989). Spatial and temporal occurrence of the parasitic dinoflagellate *Duboscquella cachoni* and its tintinnine host *Eutintinnus pectinis* in Chesapeake Bay. *Marine Biology*, *101*, 401–409.

Coats, D. W., & Small, E. B. (1976). Stomatogenesis in the philasterine scuticociliate *Philaster hiatti* Thompson, 1969. *Transactions of the American Microscopical Society*, *95*, 718–725.

Coats, W. D., Bockstahler, K. R., Berg, G. M., & Sniezek, J. H. (1994). Dinoflagellate infections of *Favella panamensis* from two North American estuaries. *Marine Biology*, *119*, 105–113.

Cochran–Stafira, D., & von Ende, C. N. (1998). Integrating bacteria into food webs: Studies with *Sarracenia purpurea* inquilines. *Ecology*, *79*, 880–898.

Coffe, G., Le Caer, J. P., Lima, O., & Adoutte, A. (1996). Purification, in vitro assembly, and preliminary sequence analysis of epiplasmins, the major constituent of the membrane skeleton of *Paramecium*. *Cell Motility and the Cytoskeleton*, *34*, 137–151.

Cohen, J., Beisson, J., & Tucker, J. B. (1980). Abnormal microtubule deployment during defective macronuclear division in a *Paramecium* mutant. *Journal of Cell Science*, *44*, 153–167.

Cohen, J., Adoutte, A., Grandchamp, S., Houdebine, L. H., & Beisson, J. (1982). Immunocytochemical study of microtubular structures throughout the cell cycle of *Paramecium*. *Biology of the Cell*, *44*, 35–44.

Cohen, J., Garreau de Loubresse, N., & Beisson, J. (1984). Actin microfilaments in *Paramecium*: Localization and role in intracellular movements. *Cell Motility*, *4*, 443–468.

Coleman, A. W. (2005). *Paramecium aurelia* revisited. *Journal of Eukaryotic Microbiology*, *52*, 68–77.

Coleman, G. S. (1989). Protozoal–bacterial interaction in the rumen. In J. V. Nolan, R. A. Leng, & D. I. Demeyer (Eds.), *The roles of protozoa and fungi in ruminant digetsion* (pp. 13–27). Armidale, Australia: Penambul Books.

Coleman, G. S., Laurie, J. I., & Bailey, J. E. (1977). Cultivation of rumen ciliate *Entodinium bursa* in presence of *Entodinium caudatum*. *Journal of General Microbiology*, *101*, 253–258.

Colgin–Bukovsan, L. (1976). The genetics of mating type in the suctorian *Tokophrya lemnarum*. *Genetical Research, Cambridge*, *27*, 303–314.

Colgin–Bukovsan, L. (1977). The process of conjugation in the suctorian *Tokophrya lemnarum*. *Archiv für Protistenkunde*, *119*, 247–263.

Colgin–Bukovsan, L. (1979). Life cycles and conditions for conjugation in the suctorian *Tokophrya lemnarum*. *Archiv für Protistenkunde*, *121*, 223–237.

Collin, B. (1909). La conjugaison d'*Anoplophrya branchiarum* (Stein) (*A. circulans* Balbiani). *Archives de Zoologie Expérimentale et Générale, Paris (Série 5)*, *1*, 345–388.

Collin, B. (1912). Étude monographique sur les acinétiens. II. Morphologie, Physiologie, Systematique. *Archives de Zoologie Expérimentale et Générale*, *51*, 1–457.

Collins, T., Baker, R. L., Wilhelm, J. M., & Olmsted, J. B. (1980). A cortical scaffold in the ciliate *Tetrahymena*. *Journal of Ultrastructure Research*, *70*, 92–103.

Colom, G. (1965). Essais sur la biologie, la distribution géographique et stratigraphique des tintinnoïdiens fossiles. *Eclogae Geologicae Helvetiae.*, *58*, 319–334.

Colorni, A., & Diamant, A. (1993). Ultrastructural features of *Cryptocaryon irritans*, a ciliate parasite of marine fish. *European Journal of Prostistology*, *29*, 425–434.

Committee on Cultures, Society of Protozoologists. (1958). A catalogue of laboratory strains of free–living and parasitic protozoa. *Journal of Protozoology*, *5*, 1–38.

Conover, R. K., & Brunk, C. F. (1986). Macronuclear DNA molecules of *Tetrahymena thermophila*. *Molecular and Cellular Biology*, *6*, 900–905.

Cook, J. A., Chubb, J. C., & Veltkamp, C. J. (1998). Epibionts of *Asellus aquaticus* (L.) (Crustacea, Isopoda): An SEM study. *Freshwater Biology*, *39*, 423–438.

Coppellotti, O. (1990). Description of *Uronema marinum* (Ciliophora, Scuticociliatida) from the Antarctica and observations on the nuclear events in conjugation. *Polar Biology*, *10*, 365–371.

Corliss, J. O. (1952). Le cycle autogamique de *Tetrahymena rostrata*. *Comptes Rendus de l'Académie des Sciences*, *235*, 399–402.

Corliss, J. O. (1955). The opalinid infusorians: Flagellates or ciliates? *Journal of Protozoology*, *2*, 107–114.

Corliss, J. O. (1956). On the evolution and systematics of ciliated protozoa. *Systematic Zoology*, *5*, 68–91; 5, 121–140.

Corliss, J. O. (1958a). The phylogenetic significance of the genus *Pseudomicrothorax* in the evolution of holotrichous ciliates. *Acta Biologica Hungarica*, *8*, 367–388.

Corliss, J. O. (1958b). The systematic position of *Pseudomicrothorax dubius*, ciliate with a unique combination of anatomical features. *Journal of Protozoology*, *5*, 184–193.

Corliss, J. O. (1959). An illustrated key to the higher groups of the ciliated protozoa, with definition of terms. *Journal of Protozoology, 6*, 265–281.

Corliss, J. O. (1960a). Comments on the systematics and phylogeny of the protozoa. *Systematic Zoology*, *8*(year 1959), 169–190.

Corliss, J. O. (1960b). *Tetrahymena chironomi* sp. nov., a ciliate from midge larvae, and the current status of facultative parasitism in the genus *Tetrahymena*. *Parasitology*, *50*, 111–153.

Corliss, J. O. (1961). *The ciliated protozoa: Characterization, classification, and guide to the literature* (310 pp.). London/New York: Pergamon Press.

Corliss, J. O. (1962a). Taxonomic procedures in classification of protozoa. *Symposium of the Society for General Microbiology*, *12*, 37–67.

Corliss, J. O. (1962b). Taxonomic–nomenclatural practices in protozoology and the new International Code of Zoological Nomenclature. *Journal of Protozoology*, *9*, 307–324.

Corliss, J. O. (1965). L'autogamie et la sénescence du cilié hyménostome *Tetrahymena rostrata* (Kahl). *Année Biologique*, *4*, 49–69.

Corliss, J. O. (1967). An aspect of morphogenesis in the ciliate Protozoa. *Journal of Protozoology*, *14*, 1–8.

Corliss, J. O. (1968). The value of ontogenetic data in reconstructing protozoan phylogenies. *Transactions of the American Microscopical Society*, *87*, 1–20.

Corliss, J. O. (1970). The comparative systematics of species comprising the hymenostome ciliate genus *Tetrahymena*. *Journal of Protozoology*, *17*, 198–209.

Corliss, J. O. (1972a). Common sense and courtesy in nomenclatural taxonomy. *Systematic Zoology*, *21*, 117–122.

Corliss, J. O. (1972b). Current status of the International Collection of Ciliate Type–Specimens and guidelines for future contributors. *Transactions of the American Microscopical Society*, *91*, 221–235.

Corliss, J. O. (1972c). *Tetrahymena* and some thoughts on the evolutionary origin of endoparasitism. *Transactions of the American Microscopical Society*, *91*, 566–573.

Corliss, J. O. (1973). History, taxonomy, ecology, and evolution of species of *Tetrahymena*. In A. M. Elliott (Ed.), *Biology of Tetrahymena, Dowden* (pp. 1–55). Stroudsburg, PA: Hutchinson & Ross.

Corliss, J. O. (1974a). The changing world of ciliate systematics: Historical analysis of past efforts and a newly proposed phylogenetic scheme of classification for the protistan phylum Ciliophora. *Systematic Zoology*, *23*, 91–138.

Corliss, J. O. (1974b). Remarks on the composition of the large ciliate class Kinetofragmophora de Puytorac et al., 1974, and recognition of several new taxa therein, with emphasis on the primitive order Primociliatida n. ord. *Journal of Protozoology*, *21*, 207–220.

Corliss, J. O. (1975a). Three centuries of protozoology: A brief tribute to its founding father, A. van Leeuwenhoek of Delft. *Journal of Protozoology*, *22*, 3–7.

Corliss, J. O. (1975b). Nuclear characteristics and phylogeny in the protistan phylum Ciliophora. *BioSystems*, *7*, 338–349.

Corliss, J. O. (1976). On lumpers and splitters of higher taxa in ciliate systematics. *Transactions of the American Microscopical Society*, *95*, 430–442.

Corliss, J. O. (1977). Annotated assignment of families and genera to the orders and classes currently comprising the Corlissian scheme of higher classification for the phylum Ciliophora. *Transactions of the American Microscopical Society*, *96*, 104–140.

Corliss, J. O. (1979). *The ciliated protozoa: Characterization, classification, and guide to the literature* (2nd ed., 455 pp.). London/New York: Pergamon Press.

Corliss, J. O. (1980). Some practical considerations with respect to the taxonomy of ciliated protozoa. *Monitore Zoologico Italiano, Nuova Serie.*, *14*, 173–189.

Corliss, J. O. (1995). The ambiregnal protists and the Codes of nomenclature: A brief review of the problem and of proposed solutions. *Bulletin of Zoological Nomenclature*, *52*, 11–17.

Corliss, J. O. (1997). Biodiversity, classification, and numbers of species of protists. In P. H. Raven & T. Williams (Eds.), *Nature and human society. The quest for a sustainable world* (pp. 130–155). Washington, DC: National Academy Press.

Corliss, J. O. (2004). Why the world needs protists. *Journal of Eukaryotic Microbiology*, *51*, 8–22.

Corliss, J. O., & Coats, D. W. (1976). A new cuticular cyst–producing tetrahymenid ciliate, *Lambornella clarki* n. sp., and the current status of ciliatosis in culicine mosquitoes. *Transactions of the American Microscopical Society*, *95*, 725–739.

Corliss, J. O., & Daggett, P. M. (1983). "*Paramecium aurelia*" and "*Tetrahymena pyriformis*": Current status of the taxonomy and nomenclature of these popularly known and widely used ciliates. *Protistologica, 19*, 307–322.

Corliss, J. O., & Dougherty, E. C. (1967). An appeal for stabilization of certain names in the protozoan family Tetrahymenidae (subphylum Ciliophora, order Hymenostomatida), with special reference to the generic name *Tetrahymena* Furgason, 1940. *Bulletin of Zoological Nomenclature, 24*, 155–185.

Corliss, J. O., & Esser, S. C. (1974). Comments on the role of the cyst in the life cycle and survival of free–living protozoa. *Transactions of the American Microscopical Society, 93*, 578–593.

Corliss, J. O., & Foissner, W. (1997). Case 3035. *Trachelocerca* Ehrenberg (Ciliophora): Proposed conservation of authorship as Ehrenberg (1840), with fixation of *Vibrio sagitta* Müller, 1786 as the type species. *Bulletin of Zoological Nomenclature, 54*, 219–221.

Corliss, J. O., & Hartwig, E. (1977). The "primitive" interstitial ciliates: Their ecology, nuclear uniquenesses, and postulated place in the evolution and systematics of the phylum Ciliophora. Akad. Wiss. Lit. (Mainz) Math.–Naturwiss. Kl. *Mikrofauna Meeresboden, 61*, 59–82.

Corliss, J. O., & Snyder, R. A. (1986). A preliminary description of several new ciliates from the Antarctica, including *Cohnilembus grassei* n. sp. *Protistologica, 22*, 39–46.

Corliss, J. O., de Puytorac, P., & Lom, J. (1965). Resolution of persistent taxonomic and nomenclatural problems involving ciliate protozoa assignable to the astome family Haptophryidae Cépède, 1923. *Journal of Protozoology, 12*, 265–273.

Corliss, J. O., Berl, D., & Laird, M. (1979). A note on the occurrence of the ciliate *Tetrahymena*, potential biocontrol agent, in the blackfly vector of onchocerciasis from Ivory Coast. *Transactions of the American Microscopical Society, 98*, 587–591.

Couch, J. A. (1973). Ultrastructural and protargol studies of *Lagenophrys callinectes* (Ciliophora: Peritrichida). *Journal of Protozoology, 20*, 638–647.

Cox, F. E. G. (1963). The initiation of cultures as a means of diagnosis of *Balantidium coli*. Progress in Protozoology. *Proceedings of the First International Congress on Protozoology*, Prague, August 1961, pp. 524–526.

Cox, G. W. (1982). *Systylis hoffi* (Protozoa: Ciliata) in California vernal pools. *Transactions of the American Microscopical Society, 101*, 294–298.

Crawford, D. W. (1989). *Mesodinium rubrum*: The phytoplankter that wasn't. *Marine Ecology–Progress Series, 58*, 161–174.

Crawford, D. W., & Purdie, D. A. (1992). Evidence for avoidance of flushing from an estuary by a planktonic, phototrophic ciliate. *Marine Ecology–Progress Series, 79*, 259–265.

Cremonte, F., & Figueras, A. (2004). Parasites and possible cause of mass mortalities of the critically endangered clam *Mesoderma mactroides* on the Atlantic coast of Argentina. *Bulletin of the European Association of Fish Pathologists, 24*, 166–171.

Creutz, C. E., Mohanty, S., Defalco, T., & Kretsinger, R. H. (2002). Purine composition of the crystalline cytoplasmic inclusions of *Paramecium tetraurelia. Protist, 153*, 39–45.

Crha, J., Stříž, J., Skřivánek, M., & Valach, Z. (1991). Rumen ciliates in lamb postnatal ontogenesis. *Acta Veterinaria Brno, 60*, 137–141.

Crippa Franceschi, T., Schieti Cavazza, B., & Boccardo Rinesi, M. (1967). La comparsa della senescenza in *Colpoda cucullus* O. F. M. *Atti dell'Accademia Ligure di Scienze e Lettere., 24*, 75–92.

Croft, K. E., Dalby, A. B., Hogan, D. J., Orr, K. E., Hewitt, E. A., Africa, R. J., et al. (2003). Macronuclear molecules encoding actins in spirotrichs. *Journal of Molecular Evolution, 56*, 341–350.

Cronkite, D. L. (1974). Genetics of chemical induction of conjugation in *Paramecium aurelia. Genetics, 76*, 703–714.

Cronkite, D. L., & Pierce, S. K. (1989). Free amino acids and cell volume regulation in the euryhaline ciliate *Paramecium calkinsi. Journal of Experimental Zoology, 251*, 275–284.

Cronkite, D., & Van den Brink, S. (1981). The role of oxygen and light in guiding "photoaccumulation" in the *Paramecium bursaria–Chlorella* symbiosis. *Journal of Experimental Zoology, 217*, 171–177.

Cróquer, A., Bastidas, C., Lipscomb, D., Rodríguez–Martínez, R. E., Jordan–Dahlgren, E., & Guzman, H. M. (2006). First report of folliculinid ciliates affecting Caribbean scleractinian corals. *Coral Reefs, 25*, 187–191.

Crosbie, P. B. B., & Munday, B. L. (1999). Environmental factors and chemical agents affecting the growth of the pathogenic marine ciliate *Uronema nigricans. Diseases of Aquatic Organisms, 36*, 213–219.

Culverhouse, P. F., Ellis, R., Simpson, R. G., Williams, R., Pierce, R. W., & Turner, J. T. (1994). Automatic categorization of 5 species of *Cymatocylis* (Protozoa, Tintinnida) by artificial neural-network. *Marine Ecology-Progress Series, 107*, 273–280.

Cunha, A. M. da, & Muniz, J. (1925). Contribution to the knowledge of Ciliata parasitic in Mammalia of Brazil. *Sciencia Medica, Rio de Janeiro, 3*, 732–747. (in Portuguese and English)

Cunha, A. M. da, & Muniz, J. (1927). Ciliés parasites de mammifères du Brésil. *Comptes Rendus des Séances de la Société de Biologie et de ses Filiales, 97*, 825–827.

Cuny, M., Milet, M., & Hayes, D. H. (1979). Strain–dependent variation of the protein composition of *Tetrahymena* ribosomes. *FEBS Letters, 101*, 77–84.

Curds, C. R. (1969). *An illustrated key to the British freshwater ciliated protozoa commonly found in activated sludge* (90 pp.). London: Ministry of Technology, H.M.S.O.

Curds, C. R. (1975a). A guide to the species of the genus *Euplotes* (Hypotrichida, Ciliatea). *Bulletin of the British Museum (Natural History) Zoology*, *28*, 1–61.

Curds, C. R. (1975b). Protozoa. In C. R. Curds & H. A. Hawkes (Eds.), *Ecological aspects of used–water treatment*: *Vol 1. The organisms and their ecology* (pp. 203–268). New York/London: Academic Press.

Curds, C. R. (1985). The species problem in ciliates – a taxonomist's view. *Atti della Societa Toscana di Scienza Naturali, Memorie Serie B*, *92*, 29–41.

Curds, C. R., & Cockburn, A. (1970a). Protozoa in biological sewage–treatment processes. I. A survey of the protozoan fauna of British percolating filters and activated–sludge plants. *Water Research*, *4*, 225–236.

Curds, C. R., & Cockburn, A. (1970b). Protozoa in biological sewage–treatment processes. II. Protozoa as indicators in the activated–sludge process. *Water Research*, *4*, 237–249.

Curds, C. R., & Vandyke, J. M. (1966). The feeding habits and growth rates of some fresh–water ciliates found in activated–sludge plants. *Journal of Applied Ecology*, *3*, 127–137.

Curry, A., & Butler, R. D. (1975). Bacteria–like endobionts of the suctorian, *Discophrya* sp. *Protoplasma*, *86*, 189–203.

Curry, A., & Butler, R. D. (1976). The ultrastructure, function and morphogenesis of the tentacle in *Discophrya* sp. (Suctorida) Cileatea. *Journal Ultrastracture Research*, *56*, 164–176.

Curry, A., & Butler, R. D. (1982). Asexual reproduction in the suctorian *Discophrya collini*. *Protoplasma*, *111*, 195–205.

Curtenaz, S., & Peck, R. K. (1992). A monoclonal antibody study of protein distribution in the membrane skeleton of the ciliate *Pseudomicrothorax*. *Journal of Cell Science*, *103*, 1117–1125.

Czapik, A. (1965). *Prorodon raabei* n. sp. et sa biologie. *Acta Protozoologica*, *3*, 21–26.

Czapik, A., & Jordan, A. (1976). Les observations sur les ciliés d'une mare. *Acta Protozoologica*, *15*, 277–288.

DDDD

Dale, T. (1987). Diel vertical distribution of planktonic ciliates in Lindåspollene, Western Norway. *Marine Microbial Food Webs*, *2*, 15–28.

Dale, T. (1988). Oil pollution and plantkon dynamics. VI. Controlled ecosystem experiments in Lindåspollene, Norway, June 1981: Effects on planktonic ciliates following nutrient additon to natural and oil–polluted enclosed water columns. *Sarsia*, *73*, 179–191.

Dale, T., & Dahl, E. (1987a). Mass occurrence of planktonic oligotrichous ciliates in a bay in southern Norway. *Journal Plankton Research*, *9*, 871–879.

Dale, T., & Dahl, E. (1987b). A red tide in southern Norway caused by mass occurrence of the planktonic ciliate *Tiarina fusus*. *Fauna*, *40*, 98–103. (in Norwegian with English summary)

Dale, T., & Lynn, D. H. (1998). Stomatogenesis of the ciliate genus *Strombidinopsis* with an improved description of *S. spiniferum* and *S. acuminatum*. *Journal of Eukaryotic Microbiology*, *45*, 210–217.

Dallai, R., & Luporini, P. (1989). Ultrastructural aspects of the development of the cytoplasmic connection between mating cells of the ciliate *Euplotes crassus*. *European Journal of Protistology*, *24*, 125–132.

Dan, X. M., Li, A. X., Lin, X. T., Teng, N., & Zhu, X. Q. (2006). A standardized method to propagate *Cryptocaryon irritans* on a susceptible host pompano *Trachinotus ovatus*. *Aquaculture*, *258*, 127–133.

Davis, C. C. (1973). A seasonal quantitative study of the plankton of Bauline Long Pond, a Newfoundland lake. *Le Naturaliste Canadien*, *100*, 85–105. (check)

Davis, C. C. (1981). Variations of lorica shape in the genus *Ptychocylis* (Protozoa: Tintinnina) in relation to species identification. *Journal of Plankton Research*, *3*, 433–443.

Dawson, J. A., & Hewitt, D. C. (1931). The longevity of encysted colpodas. *American Naturalist*, *65*, 181–186.

Dayeh, V. R., Chow, S. L., Schirmer, K., Lynn, D. H., & Bols, N. C. (2004). Evaluating the toxicity of Triton X-100 to protozoan, fish, and mammalian cells using fluorescent dyes as indicators of cell viability. *Ecotoxicology and Environmental Safety*, *57*, 375–382.

Debaisieux, P. (1960). Ciliates apostomes parasites de palaemon. *Cellule*, *60*, 331–352.

DeBiase, A. E., Sanders, R. W., & Porter, K. G. (1990). Relative nutritional value of ciliate protozoa and algae as food for *Daphnia*. *Microbial Ecology*, *19*, 199–210.

Deflandre, G., & Deunff, J. (1957). Sur la présence de ciliés fossiles de la famille des Folliculinidae dans un silex du Gabon. *Comptes Rendus de l'Académie des Sciences*, *244*, 3090–3093.

Dehority, B. A. (1979). Ciliate protozoa in the rumen of Brazilian water buffalo, *Bubalus bubalis* Linnaeus. *Journal of Protozoology*, *26*, 536–544.

Dehority, B. A. (1985). Rumen ciliates of musk–oxen (*Ovibos moschatus* Z.) from the Canadian Arctic. *Journal of Protozoology*, *32*, 246–250.

Dehority, B. A. (1986). Rumen ciliate fauna of some Brazilian cattle: Occurrence of several ciliates new to

the rumen, including the cycloposthid *Parentodinium africanum*. *Journal of Protozoology, 33*, 416–421.
Dehority, B. A. (1990). Rumen ciliate protozoa in Ohio white–tailed deer (*Odocoileus virginianus*). *Journal of Protozoology, 37*, 473–475.
Dehority, B. A. (1994). Rumen ciliate protozoa of the Blue Duiker (*Cephalophus monticola*), with observations on morphological variation lines within the species *Entodinium dubardi*. *Journal of Eukaryotic Microbiology, 41*, 103–111.
Dehority, B. A. (1995). Rumen ciliates of the pronghorn antelope (*Antilocapra americana*), mule deer (*Odocoileus hemionus*), white–tailed deer (*Odocoileus virginianus*) and elk (*Cervus canadensis*) in the northwestern United States. *Archiv für Protistenkunde, 146*, 29–36.
Dehority, B. A. (1996). A new family of entodiniomorphid protozoa from the marsupial forestomach, with descriptions of a new genus and five new species. *Journal of Eukaryotic Microbiology, 43*, 285–295.
Dehority, B. A. (2006). Studies on the variation in caudal spination of *Epidinium*. *Zootaxa, 1305*, 33–39.
Deitmer, J. W., Machemer, H., & Martinac, B. (1984). Motor control in three types of ciliary organelles in the ciliate *Stylonychia*. *Journal of Comparative Physiology A, 154*, 113–120.
Delgado, P., Calvo, P., & Torres, A. (1987). Encystment in the hypotrichous ciliate *Paraurostyla weissei*: Ultrastructure and cytochemistry. *Journal of Protozoology, 34*, 104–110.
Delgado, P., Romero, M. R., & Torres, A. (1990). Microtubular systems of *Paramecium* in division: Pattern of cytospindle assembly. *Journal of Protozoology, 37*, 182–186.
Delgery, C. C., Cragg, S. M., Busch, S., & Morgan, E. A. (2006). Effects of the epibiotic heterotrich ciliate *Mirofolliculina limnoriae* and of moulting on faecal pellet production by the wood–boring isopods, *Limnoria tripunctata* and *Limnoria quadripunctata*. *Journal of Experimental Marine Biology and Ecology, 334*, 165–173.
Delmonte Corrado, M. U., Chessa, M. G., & Pelli, P. (1996). Ultrastructural survey of mucocysts throughout the life cycle of *Colpoda cucullus* (Ciliophora, Colpodea). *Acta Protozoologica, 35*, 125–129.
Demar–Gervais, C. (1971). Quelques précisions sur le déterminisme de la conjugaison chez *Fabrea salina* Henneguy. *Protistologica, 7*, 177–195.
Demar–Gervais, C., & Génermont, J. (1971). Donnée expérimentales sur le mécanisme de l'éclosion des kystes de *Fabrea salina*. *Protistologica, 7*, 421–433.
Deroux, G. (1965). Origine des cinéties antérieures, gauches et buccales dans le genre *Dysteria* Huxley. *Comptes Rendus de l'Académie des Sciences, 260*, 6689–6691.
Deroux, G. (1970). La série "chlamydonellienne" chez les Chlamydodontidae (holotriches, Cyrtophorina Fauré–Fremiet). *Protistologica, 6*, 155–182.
Deroux, G. (1974). Quelques précisions sur *Strobilidium gyrans* Schewiakoff. *Cahiers de Biologie Marine, 15*, 571–588.
Deroux, G. (1975). Les dispositifs adhésifs ciliaires chez les Cyrtophorida et la famille des Hypocomidae. *Protistologica, 10*(year 1974), 379–396.
Deroux, G. (1976a). Le plan cortical des Cyrtophorida unité d'expression et marges de variabilité. I. Le cas des Plesiotrichopidae, fam. nov., dans la nouvelle systématique. *Protistologica, 12*, 469–481.
Deroux, G. (1976b). Le plan cortical des Cyrtophorida unité d'expression et marges de variabilité. II. Cyrtophorida a thigmotactisme ventral généralisé. *Protistologica, 12*, 483–500.
Deroux, G. (1977). Le plan cortical des Cyrtophorida. III. Structures differenciatrices chez les Dysteriina. *Protistologica, 12*(year 1976), 505–538.
Deroux, G. (1978). The hypostome ciliate order Synhymeniida: From *Chilodontopsis* of Blochmann to *Nassulopsis* of Fauré–Fremiet. *Transactions of the American Microscopical Society, 97*, 458–469.
Deroux, G. (1994a). Sous–classe des Cyrtophoria Fauré–Fremiet in Corliss, 1956. In P. de Puytorac (Ed.), *Traité de zoologie, infusoires ciliés* (Vol. 2, pp. 401–431). Paris: Masson.
Deroux, G. (1994b). Sous–classe des Nassulia Jankowski, 1967. In P. de Puytorac (Ed.), *Traité de zoologie, infusoires ciliés* (Vol. 2, pp. 585–600). Paris: Masson.
Deroux, G., & Dragesco, J. (1968). Nouvelles données sur quelques ciliés holotriches cyrtophores à ciliature ventrale. *Protistologica, 4*, 365–403.
Deroux, G., Iftode, F., & Fryd, G. (1974). Le genre *Nassulopsis* et les ciliés fondamentalement hypostomiens. *Comptes Rendus de l'Académie des Sciences, Paris, 278*, 2153–2156.
Deroux, G., Campillo, A., & Bradbury, P. C. (1975). *Ascophrys rodor* (Campillo et Deroux) parasite de la crevette rose *P. serratus* en élevage. *Revue des Travaux de l'Institut des Pêches Maritimes., 39*, 359–379.
Detcheva, R. (1973). Investigation of the infusorial fauna of the Black Sea region. *Bulletin of the Institute of the Zoological Museum, Academy of Bulgarian Sciences, 38*, 137–148. (in Russian with English summary)
Detcheva, R. (1976). Particularités ultrastructurales du cilié *Cyrtolophosis mucicola* Stokes, 1885. *Comptes Rendus des Séances de la Société de Biologieet de ses Filiales, 170*, 112–115.
Detcheva, R., de Puytorac, P., & Grolière, C. A. (1981). Some ultrastructural characteristics of the polysaprobic ciliate *Trimyema compressum*. *Transactions of the American Microscopical Society, 100*, 65–73.

De Terra, N. (1983). Cortical control over the direction of macronuclear elongation in the heterotrich ciliate *Stentor coeruleus*. *Journal of Protozoology*, *30*, 215–217.

De Terra, N. (1985). Does the oral apparatus of the ciliate *Stentor* inhibit oral development by release of a diffusible substance? *Journal of Embryology and Experimental Morphology*, *87*, 241–247.

Dias, N., Mortara, R. A., & Lima, N. (2003). Morphological and physiological changes in *Tetrahymena pyriformis* for the in vitro cytotoxicity assessment of Triton X-100. *Toxicology In Vitro, 17,* 357–366.

Dias, R. J. P., & D'Agosto, M. (2006). Feeding behavior of *Frontonia leucas* (Ehrenberg) (Protozoa, Ciliophora, Hymenostomatida) under different environmental conditions in a lotic system. *Revista Brasileira Zoologia*, *23*, 758–763.

Dive, D., Blaise, C., & Le Du, A. (1991). Standard protocol proposal for undertaking the *Colpidium campylum* ciliate protozoan growth inhibition test. *Zeitschrift für Angewandte Zoologie, 78,* 79–90.

Díaz, S., Martín–González, A., & Gutiérrez, J. C. (2006). Evaluation of heavy metal acute toxicity and bioaccumulation in soil ciliated protozoa. *Environment International*, *32*, 711–717.

Díaz, S., Martín–González, A., Borniquel, S., & Gutiérrez, J. C. (2000). *Cyrtolophosis elongata* (Colpodea Ciliophora): Some aspects of ciliary pattern, division, cortical and nuclear changes during encystment and resting cyst ultrastructure. *European Journal of Protistology*, *36*, 367–378.

Dickerson, H. W., & Dawe, D. L. (1995). *Ichthyophthirius multifiliis* and *Cryptocaryon irritans* (Phylum Ciliophora). In P. T. K. Woo (Ed.), *Fish diseases and disorders. Protozoan and metazoan infections* (Vol. 1, pp. 181–227). Wallingford, Washington: CAB International.

Dickerson, H. W., Clark, T. G., & Leff, A. A. (1993). Serotypic variation among isolates of *Ichthyophthirius multifiliis* based on immobilization. *Journal of Eukaryotic Microbiology*, *40*, 816–820.

Didier, P. (1971). Contribution à l'étude comparée des ultrastructures corticales et buccales des ciliés hyménostomes péniculiens. *Annales de la Station Biologique de Besse–en–Chandesse*, *5*(year 1970), 1–274.

Didier, P. (1976). Organisation ultrastructurale des cils des membranelles de quelques ciliés Oligohymenophora. Corps paraxonémiens et coalescence ciliaire. *Protistologica*, *12*, 37–47.

Didier, P., & Detcheva, R. (1974). Observations sur le cilié *Cohnilembus verminus* (O. F. Müller 1786): morphogenèse de bipartition et ultrastructure. *Protistologica*, *l0*, 159–174.

Didier, P., & Dragesco, J. (1979). Organisation ultrastructurale du cortex de *Phacodinium metchnicoffi* (cilié hétérotriche). *Protistologica*, *15*, 33–42.

Didier, P., & de Puytorac, P. (1969). Observations sur l'ultrastructure buccale du cilié péniculien: *Urocentrum turbo* M. *Annales de la Station Biologique de Besse–en–Chandesse*, *3*(year 1968), 243–246.

Didier, P., & de Puytorac, P. (1994). Sous–classe Peniculia Fauré–Fremiet in Corliss, 1956. In P. de Puytorac (Ed.), *Traité de zoologie, infusoires ciliés* (Vol. 2, pp. 603—;620.) Paris: Masson.

Didier, P., & Wilbert, N. (1976). Sur l'ultrastructure de *Paranophrys thompsoni* n. sp. (Oligohymenophora Scuticociliatida). *Protistologica*, *12*, 341–350.

Didier, P., Iftode, F., & Versavel, G. (1970). Morphologie, morphogenèse de bipartition et ultrastructures de *Turaniella vitrea* Brodsky (cilié hyménostome péniculien). II. Aspects de l'ultrastructure de *Turaniella vitrea* Brodsky. *Protistologica*, *6*, 21–30.

Didier, P., de Puytorac, P., Wilbert, N., & Detcheva, R. (1980). À propos d'observations sur l'ultrastructure du cilié *Cyrtolophosis mucicola* Stokes, 1885. *Journal of Protozoology*, *27*, 72–79.

Dieckmann, J. (1988). Infraciliature and morphogenesis of *Paraholosticha sterkii* (Garnjobst, 1934) n. comb. (Ciliophora, Hypotrichida). *European Journal of Protistology*, *23*, 218–228.

Diener, D. R., Burchill, B. R., & Burton, P. R. (1983). Microtubules and filaments beneath the fission furrow of *Stentor coeruleus. Journal of Protozoology*, *30*, 83–90.

Diggles, B. K. (1997). Some information on the morphology of *Cryptocaryon irritans* from south–east Queensland, Australia. *European Journal Protistology*, *33*, 200–210.

Diggles, B. K., & Adlard, R. D. (1995). Taxonomic affinities of *Cryptocaryon irritans* and *Ichthyophthirius multifiliis* inferred from ribosomal RNA sequence data. *Diseases of Aquatic Organisms*, *22*, 39–43.

Diggles, B. K., & Adlard, R. D. (1997). Intraspecific variation in *Cryptocaryon irritans. Journal of Eukaryotic Microbiology*, *44*, 25–32.

Diggles, B. K., & Lester, R. J. G. (1996). Infections of *Cryptocaryon irritans* on wild fish from south-east Queensland, Australia. *Diseases of Aquatic Organisms*, *25*, 159–167.

Diller, W. F. (1936). Nuclear reorganization processes in *Paramecium aurelia*, with descriptions of autogamy and "hemixis." *Journal of Morphology*, *59*, 11–67.

Dini, F. (1981). Relationship between breeding systems and resistance to mercury in *Euplotes crassus* (Ciliophora: Hypotrichida). *Marine Ecology–Progress Series*, *4*, 195–202.

Dini, F., & Gianni, A. (1985). Breeding systems in the *Euplotes vannus–crassus–minuta* groups. *Atti della Societa Toscana di Scienza Naturali, Memorie Serie B*, *92*, 75–93.

Dini, F., & Luporini, P. (1979). Multiple mating type system of the marine ciliate *Euplotes crassus* (Dujardin). *Archiv für Protistenkunde*, *121*, 238–245.

Dini, F., & Luporini, P. (1980). Genic determination of the autogamy trait in the hypotrich ciliate *Euplotes crassus*. *Genetical Research*, *35*, 107–119.

Dini, F., & Luporini, P. (1985). Mating–type polymorphic variation in *Euplotes minuta* (Ciliophora: Hypotrichida). *Journal of Protozoology*, *32*, 111–117.

Dini, F., & Nyberg, D. (1993). Sex in ciliates. In J. Gwynfryn Jones (Ed.), *Advances in microbial ecology* (Vol. 13, pp. 85–153). New York: Plenum Press.

Dini, F., & Nyberg, D. (1994). Adolescence and the reversibility of maturity in *Euplotes crassus*. *Journal of Eukaryotic Microbiology*, *41*, 373–380.

Dini, F., & Nyberg, D. (1999). Growth rates of marine ciliates on diverse organisms reveal ecological specializations within morphospecies. *Microbial Ecology*, *37*, 13–22.

Dini, F., Bracchi, P., & Gianní, A. (1987). Mating types in *Aspidisca* sp. (Ciliophora, Hypotrichida): A cluster of cryptic species. *Journal of Protozoology*, *34*, 236–243.

Dini, F., Bleyman, L. K., & Giubbilini, P. (1990). Non–Mendelian inheritance of early maturity in *Euplotes crassus*. *Journal Protozoology*, *37*, 475–478.

Dini, F., Guella, G., Giubbilini, P., Mancini, I., & Pietra, F. (1993). Control of interspecific relationships in marine ciliate protists by most evolved natural products. *Naturwissenschaften*, *80*, 84–86.

Dini, F., Raikov, I. B., & Bracchi, P. (1999). Nuclear phenomena during autogamy in the marine ciliate *Euplotes crassus*: A tangled cytogenetic process fostering evolutionary conservatism. *Acta Protozoologica*, *38*, 39–48.

Dioni, W. (1972). Un nuevo genero de Folliculinidae de agua dulce: *Botticula ringueleti* nov. gen. nov. sp. del Rio Parana medio. *Acta Zoologica Lilloana*, *29*, 309–315.

Dippell, R. V. (1968). The development of basal bodies in *Paramecium*. *Proceedings of the National Academy of Sciences of the United States of America*, *61*, 461–468.

Dive, D. G. (1973). La nutrition holozoïque des protozoaires ciliés. Ses conséquences dans l 'épuration naturelle et artificielle. *Année Biologique*, *12*, 343–380.

Do Huh, M., Thomas, C. D., Udomkusonsri, P., & Noga, E. J. (2005). Epidemic trichodinosis associated with severe epidermal hyperplasia in largemouth bass, *Micropterus salmoides*, from North Carolina, USA. *Journal of Wildlife Diseases*, *41*, 647–653.

Doak, T. G., Doerder, F. P., Jahn, C. L., & Herrick, G. (1994). A proposed superfamily of transposase genes: Transposon–like elements in ciliated protozoa and a common "D35E" motif. *Proceedings of the National Academy of Sciences of the United States of America*, *91*, 942–946.

Dobberstein, R. C., & Palm, H. W. (2000). Trichodinid ciliates (Peritrichia: Trichodiniidae) from the Bay of Kiel, with description of *Trichodina claviformis* sp. n. *Folia Parasitologica*, *47*, 81–90.

Dobrzańska, J. (1959). Investigations on ciliates living in lamellibranchiates of small water bodies. *Bulletin de l'Académie Polonaise des Sciences, Class II, Série des Sciences Biologique*, *6*, 113–118.

Dobrzańska, J. (1961). Further study on *Sphenophrya dreissenae* Dobrzańska, 1958 (Ciliata, Thigmotricha). *Acta Parasitologica Polonica*, *9*, 117–140.

Dobrzańska–Kaczanowska, J. (1963). Comparaison de la morphogenèse des ciliés: *Chilodonella uncinata* (Ehrbg.), *Allosphaerium paraconvexa* sp. n. et *Heliochona scheuteni* (Stein). *Acta Protozoologica*, *1*, 353–394.

Doerder, F. P., Gates, M. A., Eberhardt, F. P., & Arslanyolu, M. (1995). High frequency of sex and equal frequencies of mating types in natural populations of *Tetrahymena thermophila*. *Proceedings of the National Academy of Sciences of the United States of America*, *92*, 8715–8718.

Doerder, F. P., Arslanyolu, M., Saad, Y., Kaczmarek, M., Mendoza, M., & Mita, B. (1996). Ecological genetics of *Tetrahymena thermophila*: Mating types, i–antigens, multiple alleles and epistasis. *Journal of Eukaryotic Microbiology*, *43*, 95–100.

Dogiel, V. A. (1927). Monographie der Familie Ophryoscolecidae. I. *Archiv für Protistenkunde*, *59*, 1–288.

Dogiel, V. A. (1946). Phylogeny of the family Ophryoscolecidae from a palaeontological and parasitological standpoint. *Zoologichesky Zhurnal*, *25*, 395–402. (in Russian with English summary)

Dohi, K. (1982). Seasonal change of tintinnid community in Funka Bay. *Bulletin of the Plankton Society of Japan*, *29*, 77–87. (in Japanese with English summary)

Dohme, F., Machmuller, A., Estermann, B. L., Pfister, P., Wasserfallen, A., & Kreuzer, M. (1999). The role of the rumen ciliate protozoa for methane suppression caused by coconut oil. *Letters in Applied Microbiology*, *29*, 187–192.

Dolan, J. R. (1991). Guilds of ciliate microzooplankton in the Chesapeake Bay. *Estuarine Coastal and Shelf Science*, *33*, 137–152.

Dolan, J. R. (1997). Phosphorus and ammonia excretion by planktonic protists. *Marine Geology*, *139*, 109–122.

Dolan, J. R. (2000). Tintinnid ciliate diversity in the Mediterranean Sea: Longitudinal patterns related to water column structure in late spring–early summer. *Aquatic Microbial Ecology*, *22*, 69–78.

Dolan, J., & Antipa, G. A. (1985). Comparative stomatogenesis of two endocommensal scuticociliates, *Peniculistoma mytili* and *Mytilophilus pacificae* from marine mytilid mussels. *Protistologica*, *21*, 323–332.

Dolan, J. R., & Coats, D. W. (1991a). Preliminary prey digestion in a predacious estuarine ciliate and the use of digestion data to estimate ingestion. *Limnology and Oceanography*, *36*, 558–565.

Dolan, J. R., & Coats, D. W. (1991b). A study of feeding in predaceous ciliates using prey ciliates labeled with fluorescent microspheres. *Journal of Plankton Research*, *13*, 609–627.

Dolan, J. R., & Pérez, M. T. (2000). Costs, benefits and characteristics of mixotrophy in marine oligotrichs. *Freshwater Biology*, *45*, 227–238.

Dolan, J. R., Sall, N., Metcalfe, A., & Gasser, B. (2003). Effects of turbulence on the feeding and growth of a marine oligotrich ciliate. *Aquatic Microbial Ecology*, *31*, 183–192.

Domenech, R., Gaudes, A., Lopez-Doval, J. C., Salvado, H., & Munoz, I. (2006). Effects of short–term nutrient addition on microfauna diversity in a Mediterranean stream. *Hydrobiologia*, *568*, 207–215.

Dorothy, K. P., Satyanarayana, B., Kalavati, C., Raman, A. V., & Dehairs, F. (2003). Protozoa associated with leaf litter degradation in Coringa mangrove forest, Kakinada Bay, east coast of India. *Indian Journal of Marine Sciences*, *32*, 45–51.

Doucet, C. M., & Maly, E. J. (1990). Effect of copper on the interaction between the predator *Didinium nasutum* and its prey *Paramecium caudatum*. *Canadian Journal of Fisheries and Aquatic Sciences*, *47*, 1122–1127.

Dovgal, I. (1996). Keys for identification of tentaculous infusoria (Ciliophora, Suctoria) of the Ukrainian fauna. *Vestnik Zoologii*, *2*(Suppl.), 1–42. (in Russian with English summary)

Dovgal, I. (2002). Evolution, phylogeny and classification of Suctorea (Ciliophora). *Protistology*, *2*, 194–270.

Dragesco, A., Dragesco, J., Coste, F., Gasc, C., Romestand, B., Raymon, J. C., et al. (1995). *Philasterides dicentrarchi*, n. sp., (Ciliophora, Scuticociliatida), a histophagous opportunistic parasite of *Dicentrarchus labrax* (Linnaeus, 1758), a reared marine fish. *European Journal of Protistology*, *31*, 327–340.

Dragesco, J. (1962). Capture et ingestion des proies chez les infusoires ciliés. *Bulletin Biologique de la France et de la Belgique*, *46*, 123–167.

Dragesco, J. (1963). Compléments à la connaissance des ciliés mésopsammiques de Roscoff. I. Holotriches. *Cahiers de Biologie Marine*, *4*, 91–119.

Dragesco, J. (1965). Ciliés mésopsammiques d'Afrique Noire. *Cahiers Biologie Marine*, *6*, 357–399.

Dragesco, J. (1968). *Metopus jankowskii* n. sp., *Sonderia sinuata* Kahl et *Discocephalus minimus* n. sp., ciliés nouveaux ou mal connus. *Annales de la Faculté des Sciences du Cameroun*, *1*, 77–88.

Dragesco, J. (1970). *Ciliés libres du Cameroun*. Annales de la Faculté des Sciences (Numéro hors série, 141 pp.). Yaoundé: Université Fédérale du Cameroun.

Dragesco, J. (1972). Ciliés libres de l'Ouganda. *Annales de la Faculté des Sciences de Cameroun*, *9*, 87–126.

Dragesco, J. (1984a). Organites sous–cuticulaires éjectables éjectisomes (trichocystes). In P. P. Grassé (Ed.), *Traité de zoologie*: *Vol. 2, Fasc. 1. Infusoires ciliés* (pp. 181–199). Paris: Masson.

Dragesco, J. (1984b). Écologie des ciliés. In P. P. Grassé (Ed.), *Traité de zoologie*: *Vol. 2, Fasc. 1. Infusoires ciliés* (pp. 761–800). Paris: Masson.

Dragesco, J. (1999). Revision des geléiides (Ciliophora, Karyorelictea). *Stapfia*, *66*, 1–91.

Dragesco, J. (2002). Infraciliature de quinze especes de ciliés mesopsammiques marins comprenant *Trachelocerca stephani* comb. nova, *T. bodiani* comb. nova, *Tracheloraphis filiformis* spec. nova, *T. exilis* spec. nova et *Sathrophilus arenicolus* spec. nova. *Linzer Biologische Beiträge*, *34*, 1545–1626.

Dragesco, J., & Dragesco–Kernéïs, D. (1986). Ciliés libres de l'Afrique intertropicale. *Collection Faune Tropicale, Paris*, *26*, 1–559.

Dragesco, J., & Dragesco–Kernéïs, D. (1991). Free–living ciliates from the coastal are of Lake Tanganyika (Africa). *European Jorunal of Protistology*, *26*, 216–235.

Dragesco, J., & Tuffrau, M. (1967). *Neobursaridium gigas* Balech, 1941, cilié holotriche hymenostome pan–tropical. *Protistologica*, *3*, 133–146.

Dragesco, J., Auderset, G., & Baumann, M. (1965). Observations sur la structure et la genèse des trichocystes toxiques et des protrichocystes de *Dileptus* (ciliés holotriches). *Protistologica*, *1*, 81–90.

Dragesco, J., Iftode, F., & Fryd–Versavel, G. (1974). Contribution à la connaissance de quelques ciliés holotriches rhabdophores; I. Prostomiens. *Protistologica*, *10*, 59–75.

Dragesco, J., Fryd–Versavel, G., Iftode, F., & Didier, P. (1977). Le cilié *Platyophrya spumacola* Kahl, 1926: morphologie, stomatogenèse et ultrastructure. *Protistologica*, *13*, 419–434.

Drake, J. F., & Tsuchiya, H. M. (1977). Growth kinetics of *Colpoda steinii* on *Escherichia coli*. *Applied and Environmental Microbiology*, *34*, 18–22.

Drzewi ska, J., & Goli ska, K. (1987). Relationship between the size of cell and the number of its ciliary rows in the ciliate *Dileptus*. *Acta Protozoologica*, *26*, 19–10.

Duquette, S. L., Altwegg, R., & Anholt, B. R. (2005). Factors affecting the expression of inducible defences in *Euplotes*: Genotype, predator density and experience. *Functional Ecology*, *19*, 648–655.

Dute, R., & Kung, C. (1978). Ultrastructure of the proximal region of somatic cilia in *Paramecium tetraurelia*. *Journal of Cell Biology*, *78*, 451–464.

Duval, B., & Margulis, L. (1995). The microbial community of *Ophrydium versatile* colonies – endosymbionts, residents, and tenants. *Symbiosis*, *18*, 181–210.

Dworakowska, I. (1966). The structures of the nuclear apparatus in *Plagiotoma lumbrici* Dujardin and their transformations in division cycle. *Acta Protozoologica*, *4*, 109–124.

Dyer, B. D. (1989). *Metopus*, *Cyclidium*, and *Sonderia*: Ciliates enriched and cultured from sulfureta of a microbial mat community. *BioSystems*, *23*, 41–51.

EEEE

Eadie, J. M. (1967). Studies on the ecology of certain rumen ciliate protozoa. *Journal of General Microbiology*, *49*, 175–194.

Eadie, J. M. (1979). A study of variation in size in the rumen ciliate protozoa *Polyplastron multivesiculatum* (Dogiel et Fedorowa) and *Eudiplodinium maggii* (Fiorentini). *Protistologica*, *15*, 293–305.

Eagles, D. A., Gregg, R. A., & Spoon, D. M. (1980). Electrical events correlated with contractile vacuole activity in the suctorian, *Heliophrya. Journal of Protozoology*, *27*, 426–431.

Earl, P. R. (1991). Taxonomic adjustments in the Spirotrichophora, new name for Class III of the Ciliophora, including a checklist of *Nycthotheroides* and note of *Ozakia* n. gen. *Publicaciones Biologicas Facultad de Ciencias Biologicas Universidad Autonoma de Nuevo Leon, Mexico, 5,* 62–69.

Ederington, M. C., McManus, G. B., & Harvey, H. R. (1995). Trophic transfer of fatty acids, sterols, and a triterpenoid alcohol between bacteria, a ciliate, and the copepod *Acartia tonsa. Limnology and Oceanography*, *40*, 860–867.

Edwards, E. S., & Burkill, P. H. (1995). Abundance, biomass and distribution of microzooplankton in the Irish Sea. *Journal of Plankton Research*, *17*, 771–782.

Egerter, D. E., & Anderson, J. R. (1985). Infection of the western treehole mosquito, *Aedes sierrensis* (Diptera: Culicidae), with *Lambornella clarki* (Ciliophora: Tetrahymenidae). *Journal of Invertebrate Pathology*, *46*, 296–304.

Ehret, C. F. (1960). Organelle systems and biological organization. *Science*, *132*, 115–123.

Ehret, C. F. (1967). Paratene theory of the shapes of cells. *Journal of Theoretical Biology*, *15*, 263–272.

Ehret, C. F., & deHaller, G. (1963). Origin, development, and maturation of organelles and organelle systems of the cell surface in *Paramecium. Journal of Ultrastructure Research*, *6*(Suppl.), 1–42.

Ehret, C. F., & McArdle, E. W. (1974). The structure of *Paramecium* as viewed from its constituent levels of organization. In W. J. Wagtendonk van (Ed.), *Paramecium – A current survey* (pp. 263–338). Amsterdam/London/New York: Elsevier.

Eichenlaub–Ritter, U. (1985). Spatiotemporal control of functional specification and distribution of spindle microtubules with 13, 14, and 15 protofilaments during mitosis in the ciliate *Nyctotherus. Journal of Cell Science*, *76*, 337–355.

Eichenlaub–Ritter, U., & Ruthmann, A. (1982a). Holokinetic composite chromosomes with "diffuse" kinetochores in the micronuclear mitosis of a heterotrichous ciliate. *Chromosoma*, *84*, 701–716.

Eichenlaub–Ritter, U., & Ruthmann, A. (1982b). Evidence for three "classes" of microtubules in the interpolar space of the mitotic micronucleus of a ciliate and the participation of the nuclear envelope in conferring stability to microtubules. *Chromosoma*, *85*, 687–706.

Eichenlaub–Ritter, U., & Ruthmann, A. (1983). The oral apparatus of the ciliate *Nyctotherus ovalis* Leidy: Possible involvement of microtubules in the structural support of a cortical domain and the translocation of vesicles. *Differentiation*, *24*, 97–110.

Eichenlaub–Ritter, U., & Tucker, J. B. (1984). Microtubules with more than 13 protofilaments in the dividing nuclei of ciliates. *Nature*, *307*, 60–62.

Eigner, P. (1995). Divisional morphogenesis in *Deviata abbrevescens* nov. gen., nov. spec., *Neogeneia hortualis* nov. gen., nov. spec., and *Kahliella simplex* (Horváth) Corliss and redefinition of the Kahliellidae (Ciliophora, Hypotrichida). *European Journal of Protistology*, *31*, 341–366.

Eigner, P. (1997). Evolution of morphogenetic processes in the Orthoamphisiellidae n. fam., Oxytrichidae, and Parakahliellidae n. fam., and their depiction using a computer method (Ciliophora, Hypotrichida). *Journal of Eukaryotic Microbiology*, *44*, 553–573.

Eigner, P. (1999). Comparison of divisional morphogenesis in four morphologically different clones of the genus *Gonostomum* and update of the natural hypotrich system (Ciliophora, Hypotrichida). *European Journal of Protistology*, *35*, 34–48.

Eigner, P. (2001). Divisional morphogenesis in *Uroleptus caudatus* (Stokes, 1886), and the relationship between the Urostylidae and the Parakahliellidae, Oxytrichidae, and Orthoamphisiellidae on the basis of morphogenetic processes (Ciliophora, Hypotrichida). *Journal of Eukaryotic Microbiology*, *48*, 70–79.

Eigner, P., & Foissner, W. (1992). Divisional morphogenesis in *Bakuella pampinaria* nov. spec. and reevaluation of the classification of the urostylids (Ciliophora, Hypotrichida). *European Journal of Protistology*, *28*, 460–470.

Eigner, P., & Foissner, W. (1994). Divisional morphogenesis in *Amphisiellides illuvialis* n. sp., *Paramphisiella caudata* (Hemberger) and *Hemiamphisiella terricola* Foissner, and redefinition of the Amphisiellidae (Ciliophora, Hypotrichida). *Journal of Eukaryotic Microbiology*, *41*, 243–261.

Eisen, J. A., Coyne, R. S., Wu, M., Wu, D., Thiagarajan, M., Wortman, J. R., et al. (2006). Macronuclear genome sequence of the ciliate *Tetrahymena thermophila*, a model eukaryote. *PLoS Biology*, *4*, e286. DOI: 10.1371/journal.pbio.0040286

Eisler, K. (1988). Electron microscopical observations on the ciliate *Furgasonia blochmanni* Fauré–Fremiet, 1967. Part I: An update to morphology. *European Journal of Protistology*, *24*, 75–93.

Eisler, K. (1989). Electron microscopical observations on the ciliate *Furgasonia blochmanni* Fauré–Fremiet, 1967. Part II: Morphogenesis and phylogenetic conclusions. *European Journal Protistology*, *24*, 181–199.

Eisler, K. (1992). Somatic kineties or paroral membrane: Which came first in ciliate evolution? *BioSystems*, *26*, 239–254.

Eisler, K., & Bardele, C. F. (1983). The alveolocysts of the Nassulida: Ultrastructure and some phylogenetic considerations. *Protistologica*, *19*, 95–102.

Eisler, K., & Bardele, C. F. (1986). Cortical morphology and morphogenesis of the nassulid ciliates *Furgasonia blochmanni* Fauré–Fremiet, 1967 and *Nassula citrea* Kahl, 1930. *Protistologica*, *22*, 461–476.

Eisler, K., & Peck, R. K. (1998). Ultrastructural analysis of the cortical modifications during trichocyst secretory granule docking in the ciliate *Pseudomicrothorax dubius*. *European Journal of Protistology*, *34*, 39–50.

Ejercito, M., & Wolfe, J. (2003). Caspase–like activity is required for programmed nuclear elimination during conjugation in *Tetrahymena*. *Journal of Eukaryotic Microbiology*, *50*, 427–429.

Ekanem, A. P., Obiekezie, A., Kloas, A., & Knopf, K. (2004). Effects of crude extracts of *Mucuna pruriens* (Fabaceae) and *Carica papaya* (Caricaceae) against the protozoan fish parasite *Ichthyophthirius multifiliis*. *Parasitology Research*, *92*, 361–366.

Elliott, A. M. (Ed.). (1973a). *Biology of Tetrahymena*. Stroudsburg, PA: Dowden, Hutchinson & Ross.

Elliott, A. M. (1973b). Life cycle and distribution of *Tetrahymena*. In A. M. Elliott (Ed.), *Biology of Tetrahymena* (pp. 259–286). Stroudsburg, PA: Dowden, Hutchinson & Ross.

Elliott, A. M., & Nanney, D. L. (1952). Conjugation in *Tetrahymena*. *Science*, *116*, 33–34.

Elloumi, J., Carrias, J. F., Ayadi, H., Sime–Ngando, T., Boukhris, M., & Bouain, A. (2006). Composition and distribution of planktonic ciliates from ponds of different salinity in the solar saltwork of Sfax, Tunisia. *Estuarine Coastal and Shelf Science*, *67*, 21–29.

Eloff, A. K., & Van Hoven, W. (1979). Intestinal protozoa of the African elephant *Loxodonta africana* (Blumenbach). *South African Journal of Zoology*, *15*, 83–90.

El Serehy, H. A. H., & Sleigh, M. A. (1993). Ciliates in the plankton of the River Itchen Estuary, England. *Acta Protozoologica*, *32*, 183–190.

Elston, R. A., Cheney, D., Frelier, P., & Lynn, D. (1999). Invasive orchitophryid ciliate infections in juvenile Pacific and Kumomoto oysters, *Crassostrea gigas* and *Crassostrea sikamea*. *Aquaculture*, *174*, 1–14.

Elwood, H. J., Olsen, G. J., & Sogin, M. L. (1985). The small–subunit ribosomal RNA gene sequences from the hypotrichous ciliates *Oxytricha nova* and *Stylonychia pustulata*. *Molecular Biology and Evolution*, *2*, 399–410.

Embley, T. M., & Finlay, B. J. (1993). Systematic and morphological diversity of endosymbiotic methanogens in anaerobic ciliates. *Antonie van Leeuwenhoek*, *64*, 261–271.

Embley, T. M., & Finlay, B. J. (1994). The use of small subunit rRNA sequences to unravel the relationships between anaerobic ciliates and their methanogen endosymbionts. *Microbiology*, *140*, 225–235.

Embley, T. M., Finlay, B. J., Thomas, R. H., & Dyal, P. L. (1992). The use of rRNA sequences and fluorescent probes to investigate the phylogenetic positions of the anaerobic ciliate *Metopus palaeformis* and its archaeobacterial endosymbiont. *Journal of General Microbiolgy*, *138*, 1479–1487.

Embley, T. M., Finlay, B. J., Dyal, P. L., Hirt, R. P., Wilkinson, M., & Williams, A. G. (1995). Multiple origins of anaerobic ciliates with hydrogenosomes within the radiation of aerobic ciliates. *Proceedings of the Royal Society of London, B. Biological Science*, *262*, 87–93.

Endoh, H., & Kobayashi, T. (2006). Death and harmony played by nucleus and mitochondria – Nuclear apoptosis during conjugation of *Tetrahymena*. *Autophagy*, *2*, 129–131.

Eperon, S. (1980). Sur la stomatogenèse et les relations phylogénétiques du cilié péritrichie *Thuricola folliculata* (O. F. Müller, 1786). *Protistologica*, *16*, 549–564.

Eperon, S. (1985). Cytokinesis of *Thuricola folliculata* (Ciliophora, Peritrichida). *Journal of Protozoology*, *32*, 296–305.

Eperon, S., & Grain, J. (1983). Sur les fibres postciliaires de la polycinétie de *Thuricola folliculata* (O. F. Müller, 1786) (Ciliophora, Peritrichida) en bipartition. *Protistologica*, *19*, 123–129.

Eperon, S., & Peck, R. K. (1993). Immunological similarities among trichocyst secretory proteins of *Paramecium caudatum*, *Paramecium tetraurelia* and *Pseudomicrothorax dubius*. *European Journal of Protistology*, *29*, 435–445.

Epstein, M., & Eckert, R. (1973). Membrane control of ciliary activity in the protozoan *Euplotes*. *Journal of Experimental Biology*, *58*, 437–462.

Epstein, S. S., & Shiaris, M. P. (1992). Rates of microbenthic and meiobenthic bacterivory in a temperate muddy tidal flat community. *Applied and Environmental Microbiology*, *58*, 2426–2431.

Epstein, S. S., Bazylinski, D. A., & Fowle, W. H. (1998). Epibiotic bacteria on several ciliates from marine sediments. *Journal of Eukaryotic Microbiology*, *45*, 64–70.

Erdos, G. W., & Olive, L. S. (1971). *Endemosarca*: A new genus with proteomyxid affinities. *Mycologia*, *63*, 877–883.

Erra, F., Iervasi, A., Ricci, N., & Banchetti, R. (2001). Movement of the cirri during the creeping of *Euplotes crassus* (Ciliata, Hypotrichida). *Canadian Journal of Zoology*, *79*, 1353–1362.

Esposito, F., Ricci, N., & Nobili, R. (1976). Mating–type–specific soluble factors (gamones) in cell interaction of conjugation in the ciliate *Oxytricha bifaria*. *Journal of Experimental Zoology*, *197*, 275–281.

Esteban, G., & Fernández–Galiano, D. (1989). Morphology and morphogenesis in *Carchesium polypinum* (Ciliophora: Peritrichida). *Transactions of the American Microscopical Society*, *108*, 345–353.

Esteban, G., & Finlay, B. J. (1994). A new genus of anaerobic scuticociliate with endosymbiotic methanogens and ectobiotic bacteria. *Archiv für Protistenkunde*, *144*, 350–356.

Esteban, G., & Finlay, B. J. (2004). Marine ciliates (Protozoa) in Central Spain. *Ophelia*, *58*, 13–22.

Esteban, G., & Olmo, J. L. (1997). *Cristigera pleuronemoides* Roux, 1901: A cosmopolitan ciliate that embraces at least three nominal species. *European Journal of Protistology.*, *33*, 426–431.

Esteban, G., Finlay, B. J., & Embley, T. M. (1993a). New species double the diversity of anaerobic ciliates in a Spanish lake. *FEMS Microbiology Letters*, *109*, 93–100.

Esteban, G., Guhl, B. E., Clarke, K. J., Embley, T. M., & Finlay, B. J. (1993b). *Cyclidium porcatum* n. sp.: A free–living anaerobic scuticociliate containing a stable complex of hydrogenosomes, Eubacteria and Archaeobacteria. *European Journal of Protistology*, *29*, 262–270.

Esteban, G., Fenchel, T., & Finlay, B. (1995). Diversity of free–living morphospecies in the ciliate genus *Metopus*. *Archiv für Protistenkunde*, *146*, 137–164.

Esteban, G. F., Finlay, B. J., Olmo, J. L., & Tyler, P. A. (2000). Ciliated protozoa from a volcanic crater–lake in Victoria, Australia. *Journal of Natural History*, *34*, 159–189.

Esteban, G. F., Finlay, B. J., Charubhun, N., & Charubhun, B. (2001). On the geographic distribution of *Loxodes rex* (Protozoa, Ciliophora) and other alleged endemic species of ciliates. *Journal of Zoology London*, *255*, 139–143.

Estève, J. C. (1970). Distribution of acid phosphatase in *Paramecium caudatum*: Its relation with the process of digestion. *Journal of Protozoology*, *17*, 24–35.

Estève, J. C. (1972). L'appareil de Golgi des ciliés. Ultrastructure, particulièrement chez *Paramecium*. *Journal of Protozoology*, *19*, 609–618.

Estève, J. C. (1982). Défaillances de controle du comportement alimentaire chez le ciliés gymnostomes *Dileptus* et *Lacrymaria*. *Protistologica*, *18*, 517–526.

Estève, J. C. (1984a). La vacuole pulsatile des ciliés. In P. P. Grassé (Ed.), *Traité de zoologie*: *Vol. 2, Fasc. 1. Infusoires ciliés* (pp. 477–502). Paris: Masson.

Estève, J. C. (1984b). Calcium, calmoduline, sites glucidiques et ingestion chez le cilié: *Dileptus* sp. *Protistologica*, *20*, 15–25.

Ettienne, E. M., & Selitsky, M. (1974). The antagonistic effects of antimitotic agents on contraction and re–extension in the ciliate *Spirostomum ambiguum*. *Journal of Cell Science*, *16*, 377–383.

Evans, F. R., & Corliss, J. O. (1964). Morphogenesis in the hymenostome ciliate *Pseudocohnilembus persalinus* and its taxonomic and phylogenetic implications. *Journal of Protozoology*, *11*, 353–370.

Evans, F. R., & Thompson, J. C., Jr. (1964). Pseudocohnilembidae n. fam., a hymenostome ciliate family containing one genus, *Pseudocohnilembus* n. g., with three new species. *Journal of Protozoology*, *11*, 344–352.

Evans, M. S., Sicko–Goad, L. M., & Omair, M. (1979). Seasonal occurrence of *Tokophrya quadripartita* (Suctoria) as epibionts on adult *Limnocalanus macrurus* (Copepoda: Calanoida) in southeastern Lake Michigan. *Transactions of the American Microscopical Society*, *98*, 102–109.

Evans, M. S., Sell, D. W., & Beeton, A. M. (1981). *Tokophrya quadripartita* and *Tokophrya* sp. (Suctoria) associations with crustacean zooplankton in the Great Lakes Region. *Transactions of the American Microscopical Society*, *100*, 384–391.

Ewing, M. S., & Kocan, K. M. (1992). Invasion and development strategies of *Ichthyophthirius multifiliis*, a parasitic ciliate of fish. *Parasitology Today*, *8*, 204–208.

Ewing, M. S., Kocan, K. M., & Ewing, S. A. (1983). *Ichthyophthirius multifiliis*: Morphology of the cyst wall. *Transactions of the American Microscopical Society*, *102*, 122–128.

Ewing, M. S., Kocan, K. M., & Ewing, S. A. (1985). *Ichthyophthirius multifiliis* (Ciliophora) invasion of gill epithelium. *Journal of Protozoology.*, *32*, 305–310.

FFFF

Fabczak, H., Fabczak, S., & Song, P. S. (1993a). Photosensory transduction in ciliates. III. The temporal relation between membrane potentials and photomotile responses in *Blepharisma japonicum. Photochemistry Photobiology*, *57*, 872–876.

Fabczak, H., Fabczak, S., Song, P. S., Checcucci, G., Ghetti, F., & Lenci, F. (1993b). Photosensory transduction in ciliates. Role of intracellular pH and comparison between *Stentor coeruleus* and *Blepharisma japonicum. Journal of Photochemistry and Photobiology B, Biology*, *21*, 47–52.

Fabczak, H., Sobierajska, K., & Fabczak, S. (2004). Identification of possible phosphoducins in the ciliate *Blepharisma japonicum. Protist*, *155*, 181–192.

Fahrni, J. F. (1982). Morphologie et ultrastructure de *Spirochona gemmipara* Stein, 1852 (Ciliophora, Chonotrichida). I. Structures corticales et buccales de l'adulte. *Journal of Protozoology*, *29*, 170–184.

Fahrni, J. F. (1983). Morphologie et ultrastructure de *Spirochona gemmipara* Stein, 1852 (Ciliophora, Chonotrichida). II. Appareil cytoproctal et système excréteur de l'adulte. *Journal of Protozoology*, *30*, 65–73.

Fahrni, J. F. (1984). Morphologie et ultrastructure de *Spirochona gemmipara* Stein, 1852 (Ciliophora, Chonotrichida). III. Appareil de fixation chez la larve et chez l'adulte. *Journal of Protozoology*, *31*, 221–232.

Fahrni, J. F. (1985). Morphogenesis in the heterotrich ciliate *Climacostomum virens*. II. Oral development during regeneration. *Journal of Protozoology*, *32*, 460–473.

Fauré–Fremiet, E. (1910). La fixation chez les infusoires ciliés. *Bulletin Scientifique de la France et de la Belgique*, *44*, 27–50.

Fauré–Fremiet, E. (1930). Growth and differentiation of the colonies of *Zoothamnium alternans* (Clap. and Lachm.). *Biological Bulletin*, *58*, 28–51.

Fauré–Fremiet, E. (1932). Division et morphogenèse chez *Folliculina ampulla* O. F. Müller. *Bulletin Biologique de la France et de la Belgique*, *66*, 78–110.

Fauré–Fremiet, E. (1937). *Licnophora lyngbycola*, a new species of infusorian from Woods Hole. *Biological Bulletin*, *72*, 212–216.

Fauré–Fremiet, E. (1945). *Podophrya parasitica* nov. sp. *Bulletin Biologique de la France et de la Belgique*, *79*, 85–97.

Fauré–Fremiet, E. (1948a). Les mécanismes de la morphogenèse chez les ciliés. *Folia Biotheoretica*, *3*, 25–58.

Fauré–Fremiet, E. (1948b). Le rythme de marée de *Strombidium oculatum* Gruber. *Bulletin Biologique de la France et de la Belgique*, *82*, 3–23.

Fauré–Fremiet, E. (1950a). Morphologie comparée et systématique des ciliés. *Bulletin Société de Zoologique de France*, *75*, 109–122.

Fauré–Fremiet, E. (1950b). Morphologie comparée des ciliés holotriches trichostomes. *Anais da Academia Brasileira de Ciencias*, *22*, 257–261.

Fauré–Fremiet, E. (1950c). Une nouvelle vorticellide libre, *Telotrochidium johanninaen*. sp. *Bulletin Société de Zoologique de France*, *75*, 148–150.

Fauré–Fremiet, E. (1953). La bipartition énantiotrope chez les ciliés oligotriches. *Archives d'Anatomie Microscopique et de Morphologie Experimentale*, *42*, 209–225.

Fauré–Fremiet, E. (1955). La position systématique du genre *Balantidium. Journal of Protozoology*, *2*, 54–58.

Fauré–Fremiet, E. (1957). Le macronucléus hétéromère de quelques ciliés. *Journal of Protozoology*, *4*, 7–17.

Fauré–Fremiet, E. (1962a). Le genre *Paranassula* Kahl (Ciliata Cyrtophorina). *Cahiers de Biologie Marine*, *3*, 61–77.

Fauré–Fremiet, E. (1962b). Le genre *Trimyema* Lackey et les caractéristiques structurales des ciliés Trichostomatida. *Journal of Protozoology*, *9*, 146–149.

Fauré–Fremiet, E. (1965). Morphologie des Dysteriidae (Ciliata Cyrtophorina). *Comptes Rendus de l'Académie des Science*, *260*, 6679–6684.

Fauré–Fremiet, E. (1967a). La frange ciliaire des Nassulidae (Ciliata Cyrtophorina) et ses possibilités évolutives. *Comptes Rendus de l'Académie des Science*, *264*, 68–72.

Fauré–Fremiet, E. (1967b). Le genre *Cyclogramma* Perty 1852. *Journal of Protozoology*, *14*, 456–464.

Fauré–Fremiet, E. (1968). Les genres *Dexiotricha* Stokes et *Loxocephalus* Eberhard dans leur relations auxomorphiques. *Protistologica*, *4*, 115–125.

Fauré–Fremiet, E. (1970). Remarques sur la systématique des ciliés Oligotrichida. *Protistologica*, *5*(year 1969), 345–352.

Fauré–Fremiet, E. (1973). Le genre *Parasonderia* n. g. *Comptes Rendus de l'Académie des Sciences*, *276*, 2011–2014. (posthumous publication)

Fauré–Fremiet, E. (1984). Organites fixateurs. In P. P. Grassé (Ed.), *Traité de zoologie*: *Vol. 2, Fasc. 1. Infusoires ciliés* (pp. 405–437). Paris: Masson.

Fauré–Fremiet, E., & André, J. (1965a). Étude au microscope électronique de *Tillina praestans* Penard (cilié Trichostomatida). *Archives de Zoologie Expérimentale et Générale*, *105*, 345–353.

Fauré–Fremiet, E., & André, J. (1965b). L'organisation du cilié gymnostome *Plagiocampa ovata* Gelei. *Archives de Zoologie Expérimentale et Générale*, *105*, 361–367.

Fauré–Fremiet, E., & André, J. (1967). Étude au microscope électronique du cilié *Pseudomicrothorax dubius* Maupas. *Journal of Protozoology*, *14*, 464–473.

Fauré–Fremiet, E., & André, J. (1968). L'organisation corticale des Ciliata. *Comptes Rendus de l'Académie des Sciences*, *266*, 487–490.

Fauré–Fremiet, E., & Mugard, H. (1946). Sur un infusoire holotriche histiophage *Deltopylum rhabdoides*,

n. gen., n. sp. *Bulletin Société de Zoologique de France*, *71*, 161–164.

Fauré–Fremiet, E., & Mugard, H. (1949a). Le dimorphisme de *Espejoia mucicola*. *Hydrobiologia*, *1*, 379–389.

Fauré–Fremiet, E., & Mugard, H. (1949b). Un infusoire apostome parasite d'un polychète: *Cyrtocaryum halosydnae*, n. gen., n. sp. *Comptes Rendus de l'Académie des Sciences*, *228*, 1753–1755.

Fauré–Fremiet, E., & Tuffrau, M. (1955). *Sonderia labiata*, n. sp., cilié trichostome psammobie. *Hydrobiologia*, *7*, 210–218.

Fauré–Fremiet, E., André, J., & Ganier, M. C. (1968a). Structure fine de l'organite fixateur des Dysteriidae (Ciliata Cyrtophorina). *Comptes Rendus de l'Académie des Sciences, Paris*, *267*, 954–957.

Fauré–Fremiet, E., André, J., & Ganier, M. C. (1968b). Calcification tégumentaire chez les ciliés du genre *Coleps* Nitzsch. *Journal de Microscopie (Paris)*, *7*, 693–704.

Favard, P., Carasso, N., & Fauré–Fremict, E. (1963). Ultrastructure de l'appareil adhésif des urcéolaires (ciliés péritriches). *Journal de Microscopie (Paris)*, *2*, 337–368.

Felsenstein, J. (1978). Cases in which parsimony or compatibility methods will be positively misleading. *Systematic Zoology*, *27*, 401–410.

Felsenstein, J. (2004). *PHYLIP (Phylogeny Inference Package) version 3.6a2, distributed by the author*. Seattle: Department of Genetics, University of Washington.

Fenchel, T. (1965a). Ciliates from Scandinavian molluscs. *Ophelia*, *2*, 71–174.

Fenchel, T. (1965b). On the ciliate fauna associated with the marine species of the amphipod genus *Gammarus* J. G. Fabricius. *Ophelia*, *2*, 281–303.

Fenchel, T. (1980a). Suspension feeding in ciliated protozoa: Structure and function of feeding organelles. *Archiv für Protistenkunde*, *123*, 239–260.

Fenchel, T. (1980b). Suspension feeding in ciliated protozoa: Functional response and particle size selection. *Microbial Ecology*, *6*, 1–11.

Fenchel, T. (1980c). Relation between particle size selection and clearance in suspension–feeding ciliates. *Limnology and Oceanography*, *25*, 733–738.

Fenchel, T. (1980d). The protozoan fauna from the gut of the green turtle, *Chelonia mydas* L. with a description of *Balantidium bacteriophorus* sp. nov. *Archiv für Protistenkunde*, *123*, 22–26.

Fenchel, T. (1987). *Ecology of protozoa* (pp. 197). Berlin: Springer

Fenchel, T. (1990). Adaptive significance of polymorphic life cycles in protozoa: Responses to starvation and refeeding in two species of marine ciliates. *Journal of Experimental Marine Biology and Ecology*, *136*, 159–177.

Fenchel, T. (1993). Methanogenesis in marine shallow water sediments: The quantitative role of anaerobic protozoa with endosymbiotic methanogenic bacteria. *Ophelia*, *37*, 67–82.

Fenchel, T. (2004). Orientation in two dimensions: Chemosensory motile behaviour of *Euplotes vannus*. *European Journal of Protistology*, *40*, 49–55.

Fenchel, T., & Bernard, C. (1993). Endosymbiotic purple non–sulphur bacteria in an anaerobic ciliated protozoon. *FEMS Microbiology Letters*, *110*, 21–25.

Fenchel, T., & Bernard, C. (1996). Behavioural responses in oxygen gradients of ciliates from microbial mats. *European Journal of Protistology*, *32*, 55–63.

Fenchel, T., & Finlay, B. J. (1984). Geotaxis in the ciliated protozoan *Loxodes*. *Journal of Experimental Biology*, *110*, 17–33.

Fenchel, T., & Finlay, B. J. (1986a). The structure and function of Müller vesicles in loxodid ciliates. *Journal of Protozoology*, *33*, 69–76.

Fenchel, T., & Finlay, B. J. (1986b). Photobehavior of the ciliated protozoon *Loxodes*: taxic, transient, and kinetic responses in the presence and absence of oxygen. *Journal of Protozoology*, *33*, 139–145.

Fenchel, T., & Finlay, B. J. (1990a). Oxygen toxicity, respiration and behavioural responses to oxygen in free–living anaerobic ciliates. *Journal of General Microbiology*, *136*, 1953–1959.

Fenchel, T., & Finlay, B. J. (1990b). Anaerobic free–living protozoa: Growth efficiencies and the structure of anaerobic communities. *FEMS Microbiology Ecology*, *74*, 269–275.

Fenchel, T., & Finlay, B. J. (1991a). The biology of free–living anaerobic ciliates. *European Journal Protistology*, *26*, 210–215.

Fenchel, T., & Finlay, B. J. (1991b). Endosymbiotic methanogenic bacteria in anaerobic ciliates: Significance for the growth efficiency of the host. *Journal of Protozoology*, *38*, 18–22.

Fenchel, T., & Finlay, B. J. (1991c). Synchronous division of an endosymbiotic methanogenic bacterium in the anaerobic ciliate *Plagiopyla frontata* Kahl. *Journal of Protozoology*, *38*, 22–28.

Fenchel, T., & Finlay, B. J. (1992). Production of methane and hydrogen by anaerobic ciliates containing symbiotic methanogens. *Archives of Microbiology*, *157*, 475–480.

Fenchel, T., & Hansen, P. J. (2006). Motile behaviour of the bloom–forming ciliate *Mesodinium rubrum*. *Marine Biology Research*, *2*, 33–40.

Fenchel, T., & Jonsson, P. R. (1988). The functional biology of *Strombidium sulcatum*, a marine oligotrich ciliate (Ciliophora, Oligotrichina). *Marine Ecology Progress Series*, *48*, 1–15.

Fenchel, T., & Small, E. B. (1980). Structure and function of the oral cavity and its organelles in the

hymenostome ciliate, *Glaucoma. Transactions of the American Microscopical Society, 99*, 52–60.
Fenchel, T., Perry, T., & Thane, A. (1977). Anaerobiosis and symbiosis with bacteria in free–living ciliates. *Journal of Protozoology, 24*, 154–163.
Fenchel, T., Kristensen, L. D., & Rasmussen, L. (1990). Water column anoxia: Vertical zonation of planktonic protozoa. *Marine Ecology Progress Series, 62*, 1–10.
Fenchel, T., Bernard, C., Esteban, G., Finlay, B. J., Hansen, P. J., & Iversen, N. (1995). Microbial diversity and activity in a Danish fjord with anoxic deep water. *Ophelia, 43*, 45–100.
Ferguson, H. W., Hicks, B. D., Lynn, D. H., Ostland, V. E., & Bailey, J. (1987). Cranial ulceration in Atlantic salmon *Salmo salar* associated with *Tetrahymena* sp. *Diseases of Aquatic Organisms, 2*, 191–195.
Fernández–Galiano, D. (1979). Transfer of the widely known "spirotrich" ciliate *Bursaria truncatella* O.F.M. to the Vestibulifera as a separate order there, the Bursariomorphida. *Transactions of the American Microscopical Society, 98*, 447–454.
Fernández–Galiano, D. (1986). Sur les fibres catétodesmales de *Nyctotheroides cordiformis* (protozoaires, cilié). *Protistologica, 22*, 61–63.
Fernández–Galiano, D., & Calvo, P. (1992). Redescription of *Phacodinium metchnikoffi* (Ciliophora, Hypotrichida): General morphology and taxonomic position. *Journal of Protozoology, 39*, 443–448.
Fernández–Galiano, D., & Campos, M. I. (1992). *Ophryoscolex monoacanthus* n.sp., an entodiniomorphid ciliate in the rumen of the chamois *Rupicapra rupicapra. Transactions of the American Microscopical Society, 111*, 44–49.
Fernández–Galiano, D., & Fernández–Leborans, G. (1980). *Caenomorpha medusula* Perty, 1852 (Heterotrichida, Armophorina): Nouvelles données sur la ciliature et l'infraciliature. *Protistologica, 16*, 5–10.
Fernández–Galiano, D., Esteban, G., & Munoz, A. (1988). The stomatogenic process in *Opercularia coarctata* (Ciliophora, Peritrichida). *Journal of Protozoology, 35*, 1–4.
Fernández–Galiano, M. T., Fernández–Galiano, D., & Madrigal–Sesma, M. J. (1986). Presencia de quistes viables de *Colpoda* (Ciliophora, Colpodida) en el tubo digestivo de reptiles y anfibios. *Revista Ibérica de Parasitologia, 46*, 223–228.
Fernández–Galiano, T., Serrano, S., & Fernández–Galiano, D. (1985). General morphology and stomatogenesis of two species of the genus *Entodinium* (Ciliophora, Entodiniomorphida). *Acta Protozoologica, 24*, 181–186.
Fernández–Leborans, G. (2001). A review of the species of protozoan epibionts on crustaceans. III. Chonotrich ciliates. *Crustaceana, 74*, 581–607.
Fernández–Leborans, G. (2003). Comparative distribution of protozoan epibionts on *Mysis relicta* Loven, 1869 (Mysidacea) from three lakes in northern Europe. *Crustaceana, 76*, 1037–1054.
Fernández–Leborans, G., & Córdoba, M. J. H. (1997). Epibiosis of two sessile protozoan ciliate species on decapod crustaceans. *Cytobios, 91*, 143–154.
Fernández–Leborans, G., & Fernández–Fernández, D. (1999). Seasonal characterization of a marine sublittoral area using epibenthic protist communities. *Bulletin of Marine Science, 65*, 631–644.
Fernández–Leborans, G., & Gabilondo, R. (2006). Taxonomy and distribution of the hydrozoan and protozoan epibionts on *Pagurus bernhardus* (Linnaeus, 1758) (Crustacea, Decapoda) from Scotland. *Acta Zoologica, 87*, 33–48.
Fernández–Leborans, G., & Sorbe, J. C. (1999). A new chonotrichid, protozoan, epibiont of a bathyal mysid from the Capbreton canyon (SE Bay of Biscay). *Cytobios, 100*, 73–83.
Fernández–Leborans, G., & Tato–Porto, M. L. (2000a). A review of the species of protozoan epibionts on crustaceans. II. Suctorian ciliates. *Crustaceana, 73*, 1205–1237.
Fernández–Leborans, G., & Tato–Porto, M. L. (2000b). Presence of two suctorian ciliate species of the genera *Acineta* and *Pelagacineta*, epibionts on marine copepod crustaceans. *Cytobios, 103*, 139–148.
Fernández–Leborans, G., & Tato–Porto, M. L. (2000c). A review of the species of protozoan epibionts on crustaceans. I. Peritrich ciliates. *Crustaceana, 73*, 643–683.
Fernández–Leborans, G., & Tato–Porto, M. L. (2002). Distribution of the protozoan epibiont *Ophryodendron mysidacii* (Ciliophora, Suctoria) on the mysid *Schistomysis parkeri* (Crustacea). *Journal of Natural History, 36*, 505–513.
Fernández–Leborans, G., & Zaldumbide, M. C. (1983). *Chlorella* sp. (Chlorophyta) endosymbiotic of *Stentor coeruleus* (Protozoa, Ciliophora). *Archiv für Protistenkunde, 127*, 193–200.
Fernández–Leborans, G., Tato–Porto, M. L., & Sorbe, J. C. (1996). The morphology and life cycle of *Ophryodendron mysidacii* sp. nov. a marine suctorian epibiont on a mysid crustacean. *Journal of Zoology, London, 238*, 97–112.
Fernández–Leborans, G., Córdoba, M. J., & Gómez del Arco, P. (1997). Distribution of ciliate epibionts on the portunid crab *Liocarinus depurator* (Decapoda: Brachyura). *Invertebrate Biology, 116*, 171–177.
Fernández–Leborans, G., Valgañon, B., & Castro de Zaldumbide, M. (1999). Characterization of a marine sublittoral area facing the open sea, using epibenthic protists. *Bulletin of Marine Science, 65*, 725–743.

Fernández–Leborans, G., Hanamura, Y., & Nagasaki, K. (2002). A new suctorian, *Flectacineta isopodensis* (Protozoa: Ciliophora) epibiont on marine isopods from Hokkaido (Northern Japan). *Acta Protozoologica*, *41*, 79–84.

Fernández–Leborans, G., Arndt, C. E., & Gabilondo, R. (2006). Protozoan epibionts and their distribution on the Arctic ice–amphipod *Gammarus wilkitzkii* from Spitzbergen, Norway*Arctic Antarctic Alpine Research*, *38*, 343–356.

Ferry, T., Bouhour, D., De Monbrison, F., Laurent, F., Dumouchel–Champagne, H., Picot, S., et al. (2004). Severe peritonitis due to *Balantidium coli* acquired in France. *European Journal of Clinical Microbiology and Infectious Diseases*, *23*, 393–395.

Fidler, S. J., Jayaraman, S., & Kloetzel, J. A. (1985). Nuclear roles in the post–conjugant development of the ciliate *Euplotes aediculatus*. III. Roles of old macronuclear fragments in nuclear and cortical development. *Journal of Protozoology*, *32*, 429–436.

Finlay, B. J. (1982). Effects of seasonal anoxia on the community of benthic ciliated protozoa in a productive lake. *Archiv für Protistenkunde*, *125*, 215–222.

Finlay, B. J., & Berninger, U. G. (1984). Coexistence of congeneric ciliates (Karyorelictida: *Loxodes*) in relation to food resources in two freshwater lakes. *Journal of Animal Ecology*, *53*, 929–943.

Finlay, B. J., & Clarke, K. J. (1999). Ubiquitous dispersal of microbial species. *Nature*, *400*, 828.

Finlay, B. J., & Fenchel, T. (1989). Hydrogenosomes in some anaerobic protozoa resemble mitochondria. *FEMS Microbiology*, *65*, 311–314.

Finlay, B. J., & Fenchel, T. (1991). An anaerobic protozoon, with symbiotic methanogens, living in municipal landfill material. *FEMS Microbiology Ecology*, *85*, 169–180.

Finlay, B. J., & Fenchel, T. (1992). Methanogens and other bacteria as symbionts of free–living anaerobic ciliates. *Symbiosis*, *14*, 375–390.

Finlay, B. J., & Fenchel, T. (1996). Ecology: Role of ciliates in the natural environment. In K. Hausmann & P. C. Bradbury (Eds.), *Ciliates: Cells as organisms* (pp. 417–440). Stuttgart: Gustav Fischer Verlag.

Finlay, B. J., & Fenchel, T. (1999). Divergent perspectives on protist species richness. *Protist*, *150*, 229–233.

Finlay, B. J., & Maberly, S. C. (2000). *Microbial diversity in Priest Pot: A productive pond in the English Lake District* (pp. 73). Ambleside, Cumbria: Freshwater Biological Association.

Finlay, B. J., Bannister, P., & Stewart, J. (1979). Temporal variation in benthic ciliates and the application of association analysis. *Freshwater Biology*, *9*, 45–53.

Finlay, B. J., Fenchel, T., & Gardener, S. (1986). Oxygen perception and O2toxicity in the freshwater ciliated protozoon *Loxodes*. *Journal of Protozoology*, *33*, 157–165.

Finlay, B. J., Clarke, K. J., Cowling, A. J., Hindle, R. M., Rogerson, A., & Berninger, U. G. (1988). On the abundance and distribution of protozoa and their food in a productive freshwate pond. *European Journal Protistology*, *23*, 205–217.

Finlay, B. J., Embley, T. M., & Fenchel, T. (1993). A new polymorphic methanogen, closely related to *Methanocorpusculum parvum*, living in stable symbiosis within the anaerobic ciliate *Trimyema* sp. *Journal of General Microbiology*, *139*, 371–378.

Finlay, B. J., Esteban, G., Clarke, K. J., Williams, A. G., Embley, T. M., & Hirt, R. P. (1994). Some rumen ciliates have endosymbiotic methanogens. *FEMS Microbiology Letters*, *117*, 157–162.

Finlay, B. J., Corliss, J. O., Esteban, G., & Fenchel, T. (1996). Biodiversity at the microbial level: The number of free–living ciliates in the biosphere. *Quarterly Review of Biology*, *71*, 221–237.

Finlay, B. J., Esteban, G. F., Clarke, K. J., & Olmo, J. L. (2001). Biodiversity of terrestrial protozoa appears homogeneous across local and global spatial scales. *Protist*, *152*, 355–366.

Finlay, B. J., Esteban, G. F., & Fenchel, T. (2004). Protist diversity is different? *Protist*, *155*, 15–22.

Finlay, B. J., Span, A. S. W. & Harman, J. M. P. (1983). Nitrate respiration in primitive eukaryotes. Nature, *303*, 333–336.

Finley, H. E. (1936). A method for inducing conjugation within *Vorticella* cultures. *Transactions of the American Microscopical Society*, *60*, 323–326.

Finley, H. E., & Bacon, A. L. (1965). The morphology and biology of *Pyxicola nolandi* n. sp. (Ciliata, Peritrichida, Vaginicolidae). *Journal of Protozoology*, *12*, 123–131.

Finley, H. E., & McLaughlin, D. (1965). Cultivation of the peritrich *Telotrochidium henneguyi* in axenic and non–axenic media. *Journal of Protozoology*, *12*, 41–47.

Finley, H. E., McLaughlin, D., & Harrison, D. M. (1959). Non–axenic and axenic growth of *Vorticella microstoma*. *Journal of Protozoology*, *6*, 201–205.

Finley, M. J., & Bruns, P. J. (1980). Costimulation in *Tetrahymena*. II. A nonspecific response to heterotypic cell–cell interactions. *Developmental Biology*, *79*, 81–94.

Fischer–Defoy, D., & Hausmann, K. (1981). Microtubules, microfilaments, and membranes in phagocytosis: Structure and function of the oral apparatus of the ciliate *Climacostomum virens*. *Differentiation*, *20*, 141–151.

Fischer–Defoy, D., & Hausmann, K. (1982). Ultrastructural characteristics of algal digestion by *Climacostomum virens* (Ciliata) (Ehrenberg) Stein. *Zoomorphology*, *100*, 121–130.

Fitzgerald, M. E. C., Simco, B. A., & Coons, L. B. (1982). Ultrastructure of the peritrich ciliate *Ambiphrya ameiuri* and its attachment to the gills of the catfish *Ictalurus punctatus*. *Journal of Protozoology*, *29*, 213–217.

Fleury, A. (1983). Exemple d'expression et de potentialités morphogénètiques chez des unicellulaires: Le cas des hypotriches (protozoaires ciliés). *Annales de la Station Biologique de Besse–en–Chandesse*, *17*, 124–228.

Fleury, A. (1988). The use of correlated ultrastructural and morphogenetic characters in evolutionary taxonomy of hypotrich ciliates. *BioSystems*, *21*, 309–316.

Fleury, A. (1991a). Dynamics of the cytoskeleton during morphogenesis in the ciliate *Euplotes*. I. Basal bodies related microtubular system. *European. Journal of Protistology*, *27*, 99–114.

Fleury, A. (1991b). Dynamics of the cytoskeleton during morphogenesis in the ciliate *Euplotes*. II. Cortex and continuous microtubular systems. *European Journal of Protistology*, *27*, 220–237.

Fleury, A., & Fryd–Versavel, G. (1982). Aspects de la morphogenèse chez *Kahliella* (cilié hypotriche). *Protistologica*, *18*, 135–145.

Fleury, A., & Laurent, M. (1994). Transmission of surface pattern through a dedifferentiated stage in the ciliate *Paraurostyla*. Evidence from the analysis of microtubule and basal body development. *Journal of Eukaryotic Microbiology*, *41*, 276–291.

Fleury, A., Iftode, F., Deroux, G., & Fryd–Versavel, G. (1985a). Unité de diversité chez les hypotriches (protozoaires, ciliés): II – Éléments d'ultrastructure comparée chez divers représentants du sous–ordre des Euhypotrichina. *Protistologica*, *21*, 505–524.

Fleury, A., Iftode, F., Deroux, G., Fryd–Versavel, G., & Génermont, J. (1985b). Relation entre caractères ultrastructuraux et morphogénétiques chez les hypotriches (Protozoa, Ciliata): Définition des Euhypotrichina n. s. o. et des Pseudohypotrichina n. s. o. *Comptes Rendus de l'Académie des Sciences, Paris*, *300*(Series 3), 567–570.

Fleury, A., Iftode, F., Deroux, G., & Fryd–Versavel, G. (1986). Unité de diversité chez les hypotriches (protozoaires, ciliés): II – Éléments d'ultrastructure comparée chez divers représentants du sous–ordre des Pseudohypotrichina et remarques générales. *Protistologica*, *21*, 65–87.

Fleury, A., Delgado, P., Iftode, F., & Adoutte, A. (1992). A molecular phylogeny of ciliates: What does it tell us about the evolution of the cytoskeleton and of developmental strategies? *Developmental Genetics*, *13*, 247–254.

Fleury, A., Le Guyader, H., Iftode, F., Laurent, M., & Bornens, M. (1993). A scaffold for basal body patterning revealed by a monoclonal antibody in the hypotrich ciliate *Paraurostyla weissei*. *Developmental Genetics Biology*, *157*, 285–302.

Foissner, I., & Foissner, W. (1986). *Ciliomyces spectabilis*, nov. gen., nov. spec., a zoosporic fungus which parasitizes cysts of the ciliate *Kahliella simplex*. II. Asexual reproduction, life cycle and systematic account. *Zeitschrift für Parasitenkunde*, *72*, 43–55.

Foissner, I., & Foissner, W. (1987). The fine structure of the resting cysts of *Kahliella simplex* (Ciliata, Hypotrichida). *Zoologischer Anzeiger*, *218*, 65–74.

Foissner, I., & Foissner, W. (1995). *Ciliatosporidium platyophryae* nov. gen., nov. spec. (Microspora incerta sedis), a parasite of *Platyophrya terricola* (Ciliophora, Colpodea). *European Journal of Protistology*, *31*, 248–259.

Foissner, W. (1975). Der elektronenmikroskopische Nachweis der fibrillären Natur des Silberliniensystems bei peritrichen Ciliaten. *Zeitschrift für Naturforschung*, *30*, 818–822.

Foissner, W. (1977). Elektronenmikroskopische Untersuchungen über die Lage und Natur des Silberliniensystems von *Paramecium*. *Mikroskopie*, *33*, 260–276.

Foissner, W. (1978). Das Silberliniensystem und die Infraciliatur der Gattungen *Platyophrya* Kahl, 1926, *Cyrtolophosis* Stokes, 1885 und *Colpoda* O.F.M., 1786: Ein Beitrag zur Systematik der Colpodida (Ciliata, Vestibulifera). *Acta Protozoologica*, *17*, 215–231.

Foissner, W. (1979a). Taxonomische Studien über die Ciliaten des Grossglocknergebietes (Hohe Tauern, Österreich). Familien Microthoracidae, Chilodonellidae und Furgasoniidae. *Sitzungsberichte der Akademie der Wissenschaften Wien*, *188*, 27–43.

Foissner, W. (1979b). Morphologie, Infraciliatur und Silberliniensystem von *Phascolodon vorticella* Stein, *Chlamydonella alpestris* nov. spec. und *Trochilia minuta* (Roux) (Ciliophora, Cyrtophorida). *Protistologica*, *15*, 557–563.

Foissner, W. (1979c). Morphologie, Infraciliatur und Silberliniensystem von *Plagiocampa* Kahl, 1926 (Prostomatida, Plagiocampidae) und *Balanonema sapropelica* nov. spec. (Philasterina, Loxocephalidae). *Protistologica*, *14*(year 1978), 381–389.

Foissner, W. (1979d). Über ein Massenauftreten von *Ophrydium eutrophicum* nov. spec. (Ciliophora, Peritrichida) und *Cristatella mucedo* Cuvier (Bryozoa, Cristatellidae) in zwei Voralpenseen (Wallersee, Fuschlsee). *Berichte der Naturwissenschaftlich – Medizinischen Vereinigung in Salzburg*, *Vol. 3–4*, 95–100.

Foissner, W. (1979e). Ökologische und systematische Studien über das Neuston alpiner Klein–gewässer, mit besonderer Berücksichtigung der Ciliaten. *International Review of Hydrobiology*, *64*, 99–140.

Foissner, W. (1980a). Taxonomische studien über die Ciliaten des Grossglocknergebietes (Hohe Tauern, Österreich). IX. Ordnungen Heterotrichida und

Hypotrichida. *Berichte der Naturwissenschaftlich – Medizinischen Vereinigung in Salzburg, 5*, 71–117.

Foissner, W. (1980b). Artenbestand und Struktur der Ciliatenzönose in alpinen Kleingewässern (Hohe Tauern, Österreich). *Archiv für Protistenkunde, 123*, 99–126.

Foissner, W. (1980c). *Malacophrys viridis* sp. n. (Malacophryidae fam. n.), ein Ciliat (Protozoa, Ciliophora) mit Merkmalen der Kinetofragminophora und Oligohymenophora. *Zoologica Scripta, 9*, 81–88.

Foissner, W. (1980d). Colpodide Ciliaten (Protozoa: Ciliophora) aus alpinen Böden. *Zoologische Jahrbücher Systematik, 107*, 391–432.

Foissner, W. (1981a). Die Gemeinschaftsstruktur der Ciliatenzönose in alpinen Böden (Hohe Tauern, Österreich) und Grundlagen für eine Synökologie der terricolen Ciliaten (Protozoa, Ciliophora). *Veröff. Österr. MaB–Programms, 4*, 7–52.

Foissner, W. (1981b). Das Silberliniensystem der Ciliaten: Tatsachen, Hypothesen, Probleme. *Mikroskopie (Wien), 38*, 16–26.

Foissner, W. (1982). Ökologie und Taxonomie der Hypotrichida (Protozoa: Ciliophora) einiger österreichischer Böden. *Archiv für Protistenkunde, 126*, 19–143.

Foissner, W. (1983a). Die Morphogenese von *Urosoma macrostyla* (Wrzesniowski, 1870) (Ciliophora: Oxytrichidae). *Archiv für Protistenkunde, 127*, 413–428.

Foissner, W. (1983b). Morphologie und Infraciliatur zweier ectocommensaler Ciliaten (Protozoa: Ciliophora) von *Cyprinus carpio* L. (Pisces: Cypriniformes): *Heteropolaria lwoffi* (Fauré–Fremiet, 1943) (Peritrichida: Epistylididae) und ihr Predator *Pseudoamphileptus macrostoma* (Chen, 1955) nov. gen. (Pleurostomatida: Amphileptidae). *Zoologische Jahrbücher Systematik, 110*, 399–418.

Foissner, W. (1984a). Infraciliatur, Silberliniensystem und Biometrie einiger neuer und wenig bekannter terrestrischer, limnischer und mariner Ciliaten (Protozoa: Ciliophora) aus den Klassen Kinetofragminophora, Colpodea und Polyhymenophora. *Stapfia, 12*, 1–165.

Foissner, W. (1984b). Taxonomische Studien über die Ciliaten des Grossglocknergebietes (Hohe Tauern, Österreich). I. Familien Holophryidae, Prorodontidae, Plagiocampidae, Colepidae, Enchelyidae und Lacrymariidae nov. fam. *Annalen des Naturhistorischen Museums in Wien, 84B*, 49–85.

Foissner, W. (1985a). Morphologie und Infraciliatur der Genera *Microthorax* und *Stammeridium* und Klassifikation der Microthoracina Jankowski, 1967 (Protozoa: Ciliophora). *Zoologischer Anzeiger, Jena, 214*, 33–53.

Foissner, W. (1985b). Klassifikation und Phylogenie der Colpodea (Protozoa: Ciliophora). *Archiv für Protistenkunde, 129*, 239–290.

Foissner, W. (1985c). Die Morphologie und die Infraciliatur einiger Ciliaten (Protozoa: Ciliophora) aus dem Darm der Seeigel (Euechinoidea) *Paracentrotus lividus* und *Arbacia lixula*. *Archiv für Protistenkunde, 130*, 355–366.

Foissner, W. (1987). Soil protozoa: Fundamental problems, ecological significance, adaptations in ciliates and testaceans, bioindicators, and guide to the literature. *Progress in Protistology, 2*, 69–212.

Foissner, W. (1988a). Taxonomic and nomenclatural revision of Sládežek's lists of ciliates (Protozoa: Ciliophora) as indicators of water quality. *Hydrobiologia, 166*, 1–64.

Foissner, W. (1988b). Taxonomie und Ökologie einiger Ciliaten (Protozoa, Ciliophora) des Saprobiensystems. II. Familie Chilodonellidae. *Hydrobiologia, 162*, 21–45.

Foissner, W. (1988c). Gemeinsame Arten in der terricolen Ciliatenfauna (Protozoa: Ciliophora) von Australien und Afrika. *Stapfia, 17*, 85–133.

Foissner, W. (1990). *Kuehneltiella terricola* gen. nov., sp. nov. – a carnivorous ciliate (Protozoa, Ciliophora) from a sandy soil in Australia. *Biology and Fertility of Soils, 9*, 110–118.

Foissner, W. (1991). Basic light and electron microscopic methods for taxonomic studies of ciliated protozoa. *European Journal of Protistology, 27*, 313–330.

Foissner, W. (1993a). *Colpodea (Ciliophora)* (pp. 798). Stuttgart: Gustav Fischer Verlag.

Foissner, W. (1993b). *Corticocolpoda kaneshiroae* n. g., n. sp., a new colpodid ciliate (Protozoa, Ciliophora) from the bark of ohia trees in Hawaii. *Journal of Eukaryotic Microbiology, 40*, 764–775.

Foissner, W. (1994a). Morphology and morphogenesis of *Circinella arenicola* nov. gen., nov. spec., a cephalized hypotrich (Ciliophora, Hypotrichida) from sand dunes in Utah, USA. *European Journal of Protistology, 30*, 156–170.

Foissner, W. (1994b). Classe des Colpodea Small and Lynn, 1981. In P. de Puytorac (Ed.), *Traité de zoologie, infusoires ciliés* (Vol. 2, pp. 223–258). Paris: Masson.

Foissner, W. (1994c). *Bryometopus hawaiiensis* sp. n., a new colpodid ciliate from a terrestrial biotope of the Hawaiian Archipelago. *Annales des Naturhistorischen Museums in Wien, 96B*, 19–27.

Foissner, W. (1995a). *Kentrophoros* (Ciliophora, Karyorelictea) has oral vestiges: A reinvestigation of *K. fistulosus* (Fauré–Fremiet, 1950) using protargol impregnation. *Archiv für Protistenkunde, 146*, 165–179.

Foissner, W. (1995b). Tropical protozoan diversity: 80 ciliate species (Protozoa, Ciliophora) in a soil sample from a tropical dry forest of Costa Rica, with descriptions of four new genera and seven new species. *Archiv für Protistenkunde, 145*, 37–79.

Foissner, W. (1995/1996). The infraciliature of *Cryptopharynx setigerus* Kahl, 1928 and

Apocryptopharynx hippocampodes nov. gen., nov. spec. (Ciliophora, Karyorelictea), with an account on evolution of loxodid ciliates. *Archiv für Protistenkunde*, *146*, 309–327.

Foissner, W. (1996a). A redescription of *Remanella multinucleata* (Kahl, 1933) nov. gen., nov. comb. (Ciliophora, Karyorelictea), emphasizing the infraciliature and extrusomes. *European Journal of Protistology*, *32*, 234–250.

Foissner, W. (1996b). Ontogenesis in ciliated protozoa with emphasis on stomatogenesis. In K. Hausmann & P. C. Bradbury (Eds.), *Ciliates. Cells as organsisms* (pp. 95–177). Stuttgart: Gustav Fischer Verlag.

Foissner, W. (1996c). Updating the trachelocercids (Ciliophora, Karyorelictea). II. *Prototrachelocerca* nov. gen. (Prototrachelocercidae nov. fam.), with redescription of *P. fasciolata* (Sauerbrey, 1928) nov. comb. and *P. caudata* (Dragesco & Raikov, 1966) nov. comb. *European Journal of Protistology*, *32*, 336–355.

Foissner, W. (1997a). Infraciliature and systematic position of the marine interstitial ciliates (Protozoa, Ciliophora) *Lopezoterenia torpens* (Kahl, 1931) nov. gen., nov. comb., *Discotricha papillifera* Tuffrau, 1954, and *Paraspathidium fuscum* (Kahl, 1928) Fjeld, 1955. *Revista de la Sociedad Mexicana de Historia Natural*, *47*, 41–63.

Foissner, W. (1997b). Faunistic and taxonomic studies on ciliates (Protozoa, Ciliophora) from clean rivers in Bavaria (Germany), with descriptions of new species and ecological notes. *Limnologica*, *27*, 179–238.

Foissner, W. (1997c). Soil ciliates (Protozoa: Ciliophora) from evergreen rain forests of Australia, South America and Costa Rica: Diversity and description of new species. *Biology and Fertility of Soils*, *25*, 317–339.

Foissner, W. (1997d). Global soil ciliate (Protozoa, Ciliophora) diversity: A probability–based approach using large sample collections from Africa, Australia and Antarctica. *Biodiversity and Conservation*, *6*, 1627–1638.

Foissner, W. (1997e). Protozoa as bioindicators in agroecosystems, with emphasis on farming practices, biocides, and biodiversity. *Agriculture Ecosystems and Environment*, *62*, 93–103.

Foissner, W. (1997f). Updating the trachelocercids (Ciliophora, Karyorelictea). IV. Transfer of *Trachelocerca entzi* Kahl, 1927 to the Gymnostomatea as a new genus, *Trachelotractus* gen. n. (Helicoprorodontidae). *Acta Protozoologica*, *36*, 63–74.

Foissner, W. (1997g). Updating the trachelocercids (Ciliophora, Karyorelictea). V. Redescription of *Kovalevaia sulcata* (Kovaleva, 1966) gen. n., comb. n. and *Trachelocerca incaudata* Kahl, 1933. *Acta Protozoologica*, *36*, 197–219.

Foissner, W. (1998a). An updated compilation of world soil ciliates (Protozoa, Ciliophora), with ecological notes, new records, and descriptions of new species. *European Journal of Protistology*, *34*, 195–235.

Foissner, W. (1998b). The karyorelictids (Protozoa: Ciliophora), a unique and enigmatic assemblage of marine, interstitial ciliates: A review emphasizing ciliary patterns and evolution. In G. H. Coombs, K. Vickerman, M. A. Sleigh, & A. Warren (Eds.), *Evolutionary relationships among protozoa* (chap. 17, pp. 305–325). London: Chapman & Hall.

Foissner, W. (1999a). Notes on the soil ciliate biota (Protozoa, Ciliophora) from the Shimba Hills in Kenya (Africa): Diversity and description of three new genera and ten new species. *Biodiversity and Conservation*, *8*, 319–389.

Foissner, W. (1999b). Description of two new, mycophagous soil ciliates (Ciliophora: Colpodea): *Fungiphrya strobli* n. g., n. sp. and *Grossglockneria ovata* n. sp. *Journal of Eukaryotic Microbiology*, *46*, 34–42.

Foissner, W. (1999c). Protist diversity: Estimates of the near–imponderable. *Protist*, *150*, 363–368.

Foissner, W. (2003). The Myriokaryonidae nov. fam., a new family of spathidiid ciliates (Ciliophora, Gymnostomatea). *Acta Protozoologica*, *42*, 113–143.

Foissner, W. (2005a). Two new "flagship" ciliates (Protozoa, Ciliophora) from Venezuela: *Sleighophrys pustulata* and *Luporinophrys micelae. European Journal Protistology*, *41*, 99–117.

Foissner, W. (2005b). The unusual, lepidosome–coated resting cyst of *Meseres corlissi* (Ciliophora: Oligotrichea): Transmission electron microscopy and phylogeny. *Acta Protozoologica*, *44*, 217–230.

Foissner, W. (2006). Biogeography and dispersal of microorganisms: A review emphasizing protists. *Acta Protozoologica*, *45*, 111–136.

Foissner, W., & Adam, H. (1981). Morphologie und Infraciliatur von *Parafurgasonia sorex* (Penard, 1922) nov. gen. und *Obertrumia georgiana* (Dragesco, 1972) nov. gen. (Protozoa: Ciliophora). *Zoologischer Anzeiger, Jena*, *207*, 303–319.

Foissner, W., & Agatha, S. (1999). Morphology and morphogenesis of *Metopus hasei* Sondheim, 1929 and *M. inversus* (Jankowski, 1964) nov. comb. (Ciliophora, Metopida). *Journal of Eukaryotic Microbiology*, *46*, 174–193.

Foissner, W., & Al–Rasheid, K. A. S. (1999). Ontogenesis in a trachelocercid ciliate (Ciliophora: Karyorelictea), *Sultanophrys arabica*, with an account of evolution at the base of the ciliate tree. *Acta Protozoologica*, *38*, 273–290.

Foissner, W., & Al–Rasheid, K. (2006). A unified organization of the stichotrichine oral apparatus, including a description of the buccal seal (Ciliophora: Spirotrichea). *Acta Protozoologica*, *45*, 1–16.

Foissner, W., & Berger, H. (1999). Identification and ontogenesis of the *nomen nudum* hypotrichs

(Protozoa: Ciliophora) *Oxytricha nova* (= *Sterkiella nova* sp. n.) and *O. trifallax* (= *S. histriomuscorum*). *Acta Protozoologica*, *38*, 215–248.

Foissner, W., & Didier, P. (1983). Nahrungsaufnahme, Lebenszyklus und Morphogenese von *Pseudoplatyophrya nana* (Kahl, 1926) (Ciliophora, Colpodida). *Protistologica*, *19*, 103–109.

Foissner, W., & Dragesco, J. (1996a). Updating the trachelocercids (Ciliophora, Karyorelictea). I. A detailed description of the infraciliature of *Trachelolophos gigas* n. g., n. sp. and *T. filum*. (Dragesco & Dragesco–Kernéis, 1986) n. comb. *Journal of Eukaryotic Microbiology*, *43*, 12–25.

Foissner, W., & Dragesco, J. (1996b). Updating the trachelocercids (Ciliophora, Karyorelictea). III. Redefinition of the genera *Trachelocerca* Ehrenberg and *Tracheloraphis* Dragesco, and evolution in trachelocercid ciliates*Archiv Protistenkunde*, *147*, 43–91.

Foissner, W., & Foissner, I. (1984). First record of an ectoparasitic flagellate on ciliates: An ultrastructural investigation of the morphology and mode of attachment of *Spiromonas gonderi* nov. spec. (Zoomastigophora, Spiromonadidae) invading the pellicle of ciliates of the genus *Colpoda* (Ciliophora, Colpodidae). *Protistologica*, *20*, 635–648.

Foissner, W., & Foissner, I. (1985). Oral monokinetids in the free–living haptorid ciliate *Enchelydium polynucleatum* (Ciliophora, Enchelyidae): Ultrastructural evidence and phylogenetic implications. *Journal of Protozoology*, *32*, 712–722.

Foissner, W., & Foissner, I. (1988). The fine structure of *Fuscheria terricola* Berger et al., 1983 and a proposed new classification of the subclass Haptoria Corliss, 1974 (Ciliophora, Litostomatea). *Archiv für Protistenkunde*, *135*, 213–235.

Foissner, W., & Foissner, I. (1994). Fine structure of *Cosmocolpoda naschbergeri* (Ciliophora, Colpodida). *Archiv für Protistenkunde*, *144*, 129–136.

Foissner, W., & Foissner, I. (1995). Fine structure and systematic position of *Enchelyomorpha vermicularis* (Smith, 1899) Kahl, 1930, an anaerobic ciliate (Protozoa, Ciliophora) from domestic sewage. *Acta Protozoologica*, *34*, 21–34.

Foissner, W., & Kreutz, M. (1998). Systematic position and phylogenetic relationships of the genera *Bursaridium*, *Paracondylostoma*, *Thylakidium*, *Bryometopus*, and *Bursaria* (Ciliophora: Colpodea). *Acta Protozoologica*, *37*, 227–240.

Foissner, W., & Leipe, D. (1995). Morphology and ecology of *Siroloxophyllum utriculariae* (Penard, 1922) n. g., n. comb. (Ciliophora, Pleurostomatida) and an improved classification of pleurostomatid ciliates. *Journal of Eukaryotic Microbiology*, *42*, 476–490.

Foissner, W., & Pfister, G. (1997). Taxonomic and ecologic revision of urotrichs (Ciliophora, Prostomatida) with three or more caudal cilia, including a user–friendly key. *Limnologica*, *27*, 311–347.

Foissner, W., & Pichler, M. (2006). The unusual, lepidosome–coated resting cyst of *Meseres corlissi* (Ciliophora: Oligotrichea): Genesis of four complex types of wall precursors and assemblage of the cyst wall. *Acta Protozoologica*, *45*, 339–366.

Foissner, W., & Schiffmann, H. (1975). Vergleichende Studien an argyrophilen Strukturen von vierzehn peritrichen Ciliaten. *Protistologica*, *10*(year 1974), 489–508.

F oissner, W., & Schiffmann, H. (1976). Biometrische und morphologische Untersuchungen über die Variabilität von argyrophilen Strukturen bei peritrichen Ciliaten. *Protistologica*, *11*(year 1975), 415–428.

Foissner, W., & Schiffmann, H. (1978). Taxonomie und Phylogenie der Gattung *Colpidium* (Ciliophora, Tetrahymenidae) und Neubeschreibung von *Colpidium truncatum* Stokes, 1885. *Naturkundliches Jahrbuch der Stadt Linz*, *24*, 21–41.

Foissner, W., & Schubert, G. (1983). Morphologische und diskriminanzanalytische Trennung von *Colpoda aspera* Kahl, 1926 und *Colpoda elliotti* Bradbury et Outka, 1967 (Ciliophora: Colpodidae). *Acta Protozoologica*, *22*, 127–138.

Foissner, W., & Simonsberger, P. (1975a). Elektronenmikroskopischer Nachweis der subpelliculären Lage des Silberliniensystems bei *Colpidium colpoda* (Ciliata, Tetrahymenidae). *Protoplasma*, *86*, 65–82.

Foissner, W., & Simonsberger, P. (1975b). Vergleichende licht– und rasterelektronenmikroskopische Untersuchungen an trocken präparierten Silberliniensystemen von Ciliaten (Protozoa). *Mikroskopie*, *31*, 193–205.

Foissner, W., & Wilbert, N. (1979). Morphologie, infraciliatur und ökologie der limnischen Tintinnina: *Tintinnidium fluviatile* Stein, *Tintinnidium pusillum* Entz, *Tintinnopsis cylindrata* Daday and *Codonella cratera* (Leidy) (Ciliophora, Polyhymenophora). *Journal of Protozoology*, *26*, 90–103.

Foissner, W., & Wilbert, N. (1981). A comparative study of the infraciliature and silverline system of the fresh–water scuticociliates *Pseudocohnilembus putrinus* (Kahl, 1928) nov. comb., *P. pusillus* (Quennerstedt, 1869) nov. comb., and the marine form *P. marinus* Thompson, 1966. *Journal of Protozoology*, *28*, 291–297.

Foissner, W., & Wölfl, S. (1994). Revision of the genus *Stentor* Oken (Protozoa, Ciliophora) and description of *S. araucanus* nov. spec. from South American lakes. *Journal of Plankton Research*, *16*, 255–289.

Foissner, W., Agatha, S., & Berger, H. (2002). Soil ciliates (Protozoa, Ciliophora) from Namibia (Southwest Africa), with emphasis on two contrasting environments, the Etosha Region and the Namib Desert. Part I: Text and Line Drawings. *Denisia, 5*, 1–1063; Part II: Photographs. *Denisia*, *5*, 1064–1459.

Foissner, W., Adam, H., & Foissner, I. (1982). Morphologie und Infraciliatur von *Bryometopus pseudochilodon* Kahl, 1932, *Balantidioides dragescoi* nov.

spec. und *Kahliella marina* nov. spec. und Revision der Genus *Balantidioides* Penard, 1930 (Protozoa, Ciliophora). *Protistologica, 18*, 211–225.

Foissner, W., Oleksiv, I., & Müller, H. (1990). Morphologie und Infraciliatur einiger Ciliaten (Protozoa: Ciliophora) aus stagnierenden Gewässern. *Archive für Protistenkunde, 138,* 191–206.

Foissner, W., Schlegel, M., & Prescott, D. M. (1987). Morphology and morphogenesis of *Onychodromus quadricornutus* n. sp. (Ciliophora, Hypotrichida), an extraordinarily large ciliate with dorsal horns. *Journal of Protozoology, 34*, 150–159.

Foissner, W., Blatterer, H., Berger, H., & Kohmann, F. (1991). Taxonomische und ökologische Revision der Ciliaten des Saprobiensystems – Band I: Cyrtophorida, Oligotrichida, Hypotrichia, Colpodea. Informationsberichte des Bayer. Landesamtes für Wasserwirtschaft. Heft 1/91, 471 pp.

Foissner, W., Berger, H., & Kohmann, F. (1994). Taxonomische und ökologische Revision der Ciliaten des Saprobiensystems – Band III: Hymenostomata, Prostomatida, Nassulida. Informationsberichte des Bayer. Landesamtes für Wasserwirtschaft. Heft 1/94, 548 pp.

Foissner, W., Berger, H., & Schaumburg, J. (1999). *Identification and ecology of limnetic plankton ciliates* (pp. 793). Munich: Bavarian State Office of Water Management.

Foissner, W., Stoeck, T., Schmidt, H., & Berger, H. (2001). Biogeographical differences in a common soil ciliate, *Gonostomum affine* (Stein), as revealed by morphological and RAPD–fingerprint analysis. *Acta Protozoologica, 40*, 83–97.

Foissner, W., Strüder–Kypke, M., van der Staay, G. W. M., Moon–van der Staay, S. Y., & Hackstein, J. H. P. (2003). Endemic ciliates (Protozoa, Ciliophora) from tank bromeliads (Bromeliaceae): A combined morphological, molecular, and ecological study. *European Journal Protistology, 39*, 365–372.

Foissner, W., Moon–van der Staay, S. Y., van der Staay, G. W. M., Hackstein, J. H. P., Krautgartner, W. D., & Berger, H. (2004). Reconciling classical and molecular phylogenies in the stichotrichines (Ciliophora, Spirotrichea), including new sequences from some rare species. *European Journal of Protistology, 40*, 265–281.

Foissner, W., Müller, H., & Weisse, T. (2005a). The unusual, lepidosome–coated resting cyst of *Meseres corlissi* (Ciliophora: Oligotrichea): Light and scanning electron microscopy, cytochemistry. *Acta Protozoologica, 44*, 201–215.

Foissner, W., Berger, H., Xu, K., & Zechmeister–Boltenstern, S. (2005b). A huge, undescribed soil ciliate (Protozoa: Ciliophora) diversity in natural forest stands of Central Europe. *Biodiversity and Conservation, 14*, 617–701.

Fok, A. K., & Allen, R. D. (1975). Cytochemical localization of peroxisomes in *Tetrahymena pyriformis. Journal of Histochemistry& Cytochemistry, 23*, 599–606.

Fok, A. K., & Allen, R. D. (1979). Axenic *Paramecium caudatum*. I. Mass culture and structure. *Journal of Protozoology, 26*, 463–470.

Fokin, S. I. (1993a). Stability of the type of mitochondrial cristae in ciliates (Protista, Ciliophora). *Tsitologiya, 35*, 3–10. (in Russian with English summary)

Fokin, S. I. (1993b). Morphology of the ciliate *Protophrya ovicola* (Scuticociliatida, Ancistridae). *Tsitologiya, 35*, 11–14. (in Russian with English summary)

Fokin, S. I. (1994). Morphology of the transitional zone of the cilia in Ciliophora. I. Class Oligohymenophora. *Tsitologiya, 36*, 345–352. (in Russian with English summary)

Fokin, S. I. (1995). Morphology of the transitional zone of the cilia in Ciliophora. II. Classes Kinetofragminophora and Polyhymenophora. *Tsitologiya, 37*, 145–153. (in Russian with English summary)

Fokin, S. (1997). Morphological diversity of the micronuclei in *Paramecium. Archiv für Protistenkunde, 148*, 375–387.

Fokin, S. I. (1998). Strategies of the macronuclear endocytobionts of *Paramecium* during the sexual process of the host. *Symbiosis, 25*, 323–342.

Fokin, S. I., & Skovorodkin, I. N. (1997). Experimental analysis of the resistance of *Paramecium caudatum* (Ciliophora) against infection by bacterium *Holospora undulata. European Journal of Protistology, 33*, 214–218.

Fokin, S. I., & Chivilev, S. M. (2000). *Paramecium* morphometric analysis and taxonomy. *Acta Protozoologica, 39*, 1–14.

Fokin, S., & Görtz, H. D. (1993). *Caedibacter macronucleorum* sp. nov., a bacterium inhabiting the macronucleus of *Paramecium duboscqui. Archiv für Protistenkunde, 143*, 319–324.

Fokin, S., Brigger, T., Brenner, J., & Görtz, H. D. (1996). *Holospora* species infecting the nuclei of *Paramecium* appear to belong into two groups of bacteria. *European Journal of Protistology, 32*(Suppl. I), 19–24.

Fokin, S., Stoeck, T., & Schmidt, H. (1999). *Paramecium duboscqui* Chatton, Brachon, 1933. Distribution, ecology and taxonomy. *European Journal of Protistology, 35*, 161–167.

Fokin, S., Andreoli, I., Verni, F., & Petroni, G. (2006). *Apofrontonia dohrni* n. sp. and the phylogenetic relationships within Peniculia (Protista, Ciliophora, Oligohymenophorea). *Zoologica Scripta, 35*, 289–300.

Fokin, S. I., Sabaneyeva, E. V., Borkhsenius, O. N., Schweikert, M., & Görtz, H. D. (2000). *Paramecium calkinsi* and *P. putrinum* (Ciliophora, Protista) harboring alpha–subgroup bacteria in the cytoplasm. *Protoplasma, 213*, 176–183.

Fokin, S. I., Giamberini, L., Molloy, D. P., & de Vaate, A. B. (2003). Bacterial endocytobionts within endosymbiotic ciliates in *Dreissena polymorpha* (Lamellibranchia: Mollusca). *Acta Protozoologica*, *42*, 31–39.

Fokin, S. I., Skovorodkin, I. N., Schweikert, M., & Görtz, H. D. (2004). Co–infection of the macronucleus of *Paramecium caudatum* by free–living bacteria together with the infectious *Holospora obtusa*. *Journal of Eukaryotic Microbiology*, *51*, 417–424.

Fokin, S. I., Schweikert, M., Brummer, F., & Görtz, H. D. (2005). *Spirostomum* spp. (Ciliophora, Protista), a suitable system for endocytobiosis research. *Protoplasma*, *225*, 93–102.

Forer, A., Nilsson, J. R., & Zeuthen, E. (1970). Studies on the oral apparatus of *Tetrahymena pyriformis* GL. *Comptes Rendus des Travaux du Laboratoires Carlsberg*, *38*, 67–86.

Forge, T. A., Berrow, M. L., Darbyshire, J. F., & Warren, A. (1993). Protozoan bioassays of soil amended with sewage sludge and heavy metals, using the common soil ciliate *Colpoda steinii*. *Biology and Fertility of Soils*, *16*, 282–286.

Frankel, J. (1960). Morphogenesis in *Glaucoma chattoni*. *Journal of Protozoology*, *7*, 362–376.

Frankel, J. (1961). Spontaneous astomy: Loss of oral areas in *Glaucoma chattoni*. *Journal of Protozoology*, *8*, 250–256.

Frankel, J. (1975). An analysis of the spatial distribution of ciliary units in a ciliate, *Euplotes minuta*. *Journal of Embryology and Experimental Morphology*, *33*, 553–580.

Frankel, J. (1989). *Pattern formation. Ciliate studies and models* (pp. 314). Oxford: Oxford University Press.

Frankel, J. (1991). The patterning of ciliates. *Journal of Protozoology*, *38*, 519–525.

Frankel, J., Nelsen, E. M., & Jenkins, L. M. (1977). Mutations affecting cell division in *Tetrahymena pyriformis* syngen 1. II. Phenotypes of single and double homozygotes. *Developmental Biology*, *58*, 255–275.

Frankel, J., Jenkins, L. M., Bakowska, J., & Nelsen, E. M. (1984a). Mutational analysis of patterning of oral structures in *Tetrahymena*. 1. Effects of increased size on organization. *Journal of Embryology and Experimental Morphology*, *82*, 41–66.

Frankel, J., Nelsen, E. M., Bakowska, J., & Jenkins, L. M. (1984b). Mutational analysis of patterning of oral structures in *Tetrahymena*. 2. A graded basis for the individuality of intracellular structural arrays. *Journal of Embryology and Experimental Morphology*, *82*, 67–96.

Freeman, A. S., Kinsella, J. M., Cipolletta, C., Deem, S. L., & Karesh, W. B. (2004). Endoparasites of western lowland gorillas (*Gorilla gorilla gorilla*) at Bai Hokou, Central African Republic. *Journal of Wildlife Diseases*, *40*, 775–781.

Frenkel, M. A. (1975). Regulation of macronuclear DNA content in the holotrichous ciliate *Tillina magna* (Trichostomatida, Colpodidae). *Protistologica*, *11*, 41–48.

Frenkel, M. A. (1978). The DNA contents in micronuclei and chromatin aggregates of dividing macronucleus within reproduction cysts of ciliate *Colpoda steini*. *Tsitologiya*, *20*, 465–469. (in Russian with English summary)

Frenkel, M. A. (1980). Fine structure of the macronucleus of active and encysted (dividing) ciliates *Colpoda steini*. *Protistologica*, *16*, 339–351.

Frenkel, M. A. (1982). *Tillina magna* (Ciliophora, Colpodidae): Structural changes of the macronucleus correlated with its division. *Protistologica*, *18*, 259–272.

Frenkel, M. A. (1992). Fine structure of the macronucleus in the resting cysts of the ciliate *Tillina magna*. *Archiv für Protistenkunde*, *141*, 27–40.

Frenkel, M. A., Kudryavtsev, B. N., & Kudryavtseva, M. V. (1973). The DNA contents in macro– and micronuclei of *Colpoda steini* Maupas before and after the division of ciliates within reproduction cysts. *Tsitologiya*, *15*, 353–357. (in Russian with English summary)

Fried, J., & Lemmer, H. (2003). On the dynamics and function of ciliates in sequencing batch biofilm reactors (SBBR). *Water Science and Technology*, *47*, 189–196.

Fried, J., Mayr, G., Berger, H., Transpurger, W., Psenner, R., & Lemmer, H. (2000). Monitoring protozoa and metazoa biofilm communities for assessing wastewater quality impact and reactor up–scaling effects. *Water Science and Technology*, *41*, 309–316.

Fried, J., Ludwig, W., Psenner, R., & Schleifer, K. H. (2002). Improvement of ciliate identification and quantification: A new protocol for fluorescence in situ hybridization (FISH) in combination with silver stain techniques. *Systematic and Applied Microbiology*, *25*, 555–571.

Friedrich, J., Dandekar, T., Wolf, M., & Müller, T. (2005). ProfDist: A tool for the construction of large phylogenetic trees based on profile distances. *Bioinformatics*, *21*, 2108–2109.

Froissard, M., Kissmehl, R., Dedieu, J. C., Gulik–Krzywicki, T., Plattner, H., & Cohen, J. (2002). N–ethylmaleimide–sensitive factor is required to organize functional exocytotic microdomains in *Paramecium*. *Genetics*, *161*, 643–650.

Frontczak–Baniewicz, M., & Jerka–Dziadosz, M. (1992). Conjugation–specific cortical reorganization during zygocyst formation in the hypotrich ciliate *Paraurostyla weissei*. *Acta Protozoologica*, *31*, 59–67.

Fryd–Versavel, G., & Tuffrau, M. (1978). Compléments à la connaissance du genre *Transitella* Gellert, 1950. *Protistologica*, *14*, 91–98.

Fryd–Versavel, G., Iftode, F., & Wilbert, N. (1975). Le genre *Espejoia* Burger, 1908: morphologie et morphogenèse dans deux espèces, *E. mucicola* Penard et *E. culex* Smith 1897. *Protistologica, 10*(year 1974), 301–309.

Fryd–Versavel, G., Iftode, F., & Dragesco, J. (1976). Contribution à la connaissance de quelques ciliés gymnostomes. II. Prostomiens, pleurostomiens: morphologies stomatogenèse. *Protistologica, 11*(year 1975), 509–530.

Fu, C. J., & Miao, W. (2006). Cloning and characterization of a new multi–stress inducible metallothionein gene in *Tetrahymena pyriformis. Protist, 157*, 193–203.

Fujishima, M. (1988). Conjugation. In H. D. Görtz (Ed.), *Paramecium* (chap. 5, pp. 70–84). Berlin: Springer.

Fujishima, M., & Heckmann, K. (1984). Intra– and interspecies transfers of endosymbionts in *Euplotes. Journal of Experimental Zoology, 230*, 339–345.

Fujishima, M., Sawabe, H., & Iwatsuki, K. (1990). Scanning electron microscopic observations of differentiation from the reproductive short form to the infectious long form of *Holospora obtusa. Journal of Protozoology, 37*, 123–128.

Furgason, W. H. (1940). The significant cytostomal pattern of the "*Glaucoma–Colpidium* group," and a proposed new genus and species, *Tetrahymena geleii. Archiv für Protistenkunde, 94*, 224–266.

Furness, D. N., & Butler, R. D. (1983). The cytology of sheep rumen ciliates. I. Ultrastructure of *Epidinium caudatum* Crawley. *Journal of Protozoology, 30*, 676–687.

Furness, D. N., & Butler, R. D. (1985). The cytology of sheep rumen ciliates. II. Ultrastructure of *Eudiplodinium maggii. Journal of Protozoology, 32*, 205–214.

Furness, D. N., & Butler, R. D. (1986). Morphogenesis of infraciliature in the entodiniomorphid ciliate *Eudiplodinium maggii. Journal of Ultrastructure and Molecular Structure Research, 94*, 16–29.

Furness, D. N., & Butler, R. D. (1988). The functional and evolutionary significance of the ultrastructure of the Ophryscolecidae (Order Entodiniomorphida). *Journal of Protozoology, 35*, 34–38.

Furuya, H., Ota, M., Kimura, R., & Tsuneki, K. (2004). Renal organs of cephalopods: A habitat for dicyemids and chromidinids. *Journal of Morphology, 262*, 629–643.

Fyda, J. (1998). Predator–induced morphological changes in the ciliate *Colpidium* (Protozoa, Ciliophora). *European Journal of Protistology, 34*, 111–117.

Fyda, J., Warren, A., & Wolinska, J. (2005). An investigation of predator–induced defence responses in ciliated protozoa. *Journal of Natural History, 39*, 1431–1442.

Fyda, J., Kennaway, G., Adamus, K., & Warren, A. (2006). Ultrastructural events in the predator–induced defence response of *Colpidium kleini* (Ciliophora: Hymenostomatia). *Acta Protozoologica, 45*, 461–464.

GGGG

Gachter, E., & Weisse, T. (2006). Local adaptation among geographically distinct clones of the cosmopolitan freshwater ciliate *Meseres corlissi*. I. Temperature response. *Aquatic Microbial Ecology, 45*, 291–300.

Gaertig, J., Gao, Y., Tishgarten, T., Clark, T. G., & Dickerson, H. W. (1999). Surface display of a parasite antigen in the ciliate *Tetrahymena thermophila. Nature Biotechnology, 17,* 462–465.

Gaino, E., & Rebora, M. (2000). *Ophryoglena* sp. (Ciliata: Oligohymenophora) in *Caenis luctuosa* (Ephemeroptera: Caenidae). *Acta Protozoologica, 39*, 225–231.

Galil, B. S., & Hülsmann, N. (1997). Protist transport via ballast water – Biological classification of ballast tanks by food web interactions. *European Journal of Protistology, 33*, 244–253.

Gall, J. G. (1959). Macronuclear duplication in the ciliated protozoan *Euplotes. Journal of Biophysical and Biochemical Cytology, 5*, 295–308.

Gall, J. G. (Ed.). (1986). *The molecular biology of ciliated protozoa* (pp. 352). New York: Academic Press.

Ganner, B., Foissner, W., & Adam, H. (1986/1987). Morphogenetic and biometric comparison of four populations of *Urosomoida agiliformis* (Ciliophora, Hypotrichida). *Annales des Sciences Naturelles–Zoologie et Biologie Animale Paris, Série 13, 8*, 199–207.

García, C. M., & Niell, F. X. (1993). Seasonal changes in a saline temporary lake (Fuenta de Piedra, southern Spain). *Hydrobiologia, 267*, 211–223.

Garcia–Rodriguez, T., Perez–Paniagua, F., & Perez–Silva, J. (1981). Morphogenèse de division et ultrastructures corticale et buccale chez *Bresslaua vorax* Kahl (cilié Colpodidae). *Protistologica, 17*, 533–542.

Garreau de Loubresse, N., Keryer, G., Viguès, B., & Beisson, J. (1988). A contractile cytoskeletal network of *Paramecium*: The infraciliary lattice. *Journal of Cell Science, 90*, 351–364.

Garrison, D. L., Gibson, A., Coale, S. L., Gowing, M. M., Okolodkov, Y. B., Fritsen, C. H., et al. (2005). Sea–ice microbial communities in the Ross Sea: Autumn and summer biota. *Marine Ecology Progress Series, 300*, 39–52.

Gast, V., & Horstmann, U. (1983). N–remineralization of phyto– and bacterioplankton by the marine ciliate *Euplotes vannus. Marine Ecology Progress Series, 13*, 55–60.

Gates, M. A. (1977). Application of the analysis of positional information to the cirral patterns of the ciliate genus *Euplotes. Nature, 268*, 362–364.

Gates, M. A. (1978a). An essay on the principles of ciliate systematics. *Transactions of the American Microscopical Society, 97*, 221–235.

Gates, M. A. (1978b). Morphometric variation in the hypotrich ciliate genus *Euplotes*. *Journal of Protozoology, 25*, 338–350.

Gates, M. A. (1979). Pattern analysis in biology: A simple method for morphogenetically constrained systems. *American Naturalist, 114*, 344–349.

Gates, M. A. (1988). Dorsal kinetosomal distributions in the ciliate genus *Euplotes*. *Journal of Protozoology, 35*, 506–510.

Gates, M. A. (1990). Morphological drift accompanying nascent population differentiation in the ciliate *Euplotes vannus*. *Journal of Protozoology, 37*, 78–86.

Gates, M. A., & Berger, J. (1974). A biometric study of three strains of *Tetrahymena pyriformis* (Ciliatea: Hymenostomatida). *Canadian Journal of Zoology, 52*, 1167–1183.

Gates, M. A., & Berger, J. (1976a). Morphological stability in *Tetrahymena pyriformis*. *Transactions of the American Microscopical Society, 95*, 11–22.

Gates, M. A., & Berger, J. (1976b). Morphometric inseparability of *Paramecium primaurelia* and *P. pentaurelia*. *Transactions of the American Microscopical Society, 95*, 507–514.

Gates, M. A., & Ramphal, C. (1991). Daily rhythm of mating in the ciliated protist *Euplotes vannus*. *Transactions of the American Microscopical Society, 110*, 128–143.

Gates, M. A., Powelson, E. E., & Berger, J. (1975). Syngenic ascertainment in *Paramecium aurelia*. *Systematic Zoology, 23*(year 1974), 482–489.

Gaze, W. H., & Wootten, R. (1998). Ectoparasitic species of the genus *Trichodina* (Ciliophora: Peritrichida) parasitising British freshwater fish. *Folia Parasitologica, 45*, 177–190.

Gaze, W. H., & Wootten, R. (1999). An SEM study of adhesive disk skeletal structures isolated from trichodinids (Ciliophora: Peritrichida) of the genera *Trichodina* Ehrenberg, 1838 and *Paratrichodina* Lom, 1963. *Systematic Parasitology, 43*, 167–174.

Geiman, Q. M., & Wichterman, R. (1937). Intestinal protozoa from Galapagos tortoises (with descriptions of three new species). *Journal of Parasitology, 23*, 331–347.

Gelei, J. von. (1932). Die reizleitenden Elemente der Ciliaten in nass hergestellten Silberbzw. Goldpräparaten. *Archiv für Protistenkunde, 77*, 152–174.

Gelei, J. von. (1934a). Der feinere Bau des Cytopharynx von *Paramecium* und seine systematische Bedeutung. *Archiv für Protistenkunde, 82*, 331–362.

Gelei, J. von. (1934b). Eine mikrotechnische Studie über die Färbung der subpelliculären Elemente der Ciliaten. *Zeitschrift für Wissenschaftliche Mikroskopie und für Mikroskopische Technik, 51*, 103–178.

Gelei, J. von. (1950). Die Morphogenese der Einzeller mit Rücksicht auf die morphogenetischen Prinzipien von Sewertzoff. *Acta Biologica Hungarica, 1*, 69–134.

Gelei, J. von. (1954). Über die Lebensgemeinschaft einiger temporärer Tümpel auf einer Bergwiese im Börzsönygebirge (Oberungarn). III. Ciliaten. *Acta Biologica Academiae Scientiarum Hungaricae, 5*, 259–343.

Gelei, J. von, & Horváth, P. (1931). Eine nasse Silber–bzw. Goldmethode für die Herstellung der reizleitenden Elemente bei den Ciliaten. *Zeitschrift für Wissenschaftliche Mikroskopie und für Mikroskopische Tecknik, 48*, 9–29.

Génermont, J. (1984). L'appareil nucléaire. In P. P. Grassé (Ed.), *Traité de zoologie*: *Vol. 2, Fasc. 1. Infusoires ciliés* (pp. 269–310). Paris: Masson.

Génermont, J., Machelon, V., & Demar, C. (1985). The *vannus* group of genus *Euplotes*. Sibling species and related forms; evolutionary significance and taxonomical implications. *Atti della Societa Toscana di Scienza Naturali, Memorie Serie B, 92*, 53–65.

Génermont, J., Demar, C., Fryd–Versavel, G., Tuffrau, H., & Tuffrau, M. (1992). Polygenic control of an all–or–none morphological trait in *Euplotes* (Ciliata, Hypotrichea). Evolutionary significance of naturally occurring morphological variation in ciliates. *Genetics Selection Evolution, 24*, 89–105.

Gerassimova, Z. P. (1976). The ultrastructure of cortical fibrillar systems in the ciliates *Colpoda steini* and *Tillina magna*. *Tsitologiya, 18*, 255–260.

Gerassimova, Z. P., & Seravin, L. N. (1976). Ectoplasmic fibrillar system of Infusoria and its role for the understanding of their phylogeny. *Zoologichesky Zhurnal, 55*, 645–656. (in Russian with English summary)

Gerassimova, Z. P., Sergejeva, G. I., & Seravin, L. N. (1979). Ciliary and fibrillar structures of the ciliate *Bursaria truncatella* and its systematic position. *Acta Protozoologica, 18*, 355–370. (in Russian with English summary)

Gerber, C. A., Lopez, A. B., Shook, S. J., & Doerder, F. P. (2002). Polymorphism and selection at the SerH immobilization antigen locus in natural populations of *Tetrahymena thermophila*. *Genetics, 160*, 1469–1479.

Geyer, J. J., & Kloetzel, J. A. (1987a). Cellular dynamics of conjugation in the ciliate *Euplotes aediculatus*. I. Cytoskeletal elements. *Journal of Morphology, 192*, 27–42.

Geyer, J. J., & Kloetzel, J. A. (1987b). Cellular dynamics of conjugation in the ciliate *Euplotes aediculatus*. II. Cellular membranes. *Journal of Morphology, 192*, 43–61.

Giese, A. C. (1973). *Blepharisma: The biology of a light–sensitive protozoan* (366 pp). Stanford, CA: Stanford University Press.

Gifford, D. J. (1985). Laboratory culture of marine planktonic oligotrichs (Ciliophora, Oligotrichida). *Marine Ecology Progress Series, 23*, 257–267.

Gijzen, H. J., & Barugahare, M. (1992). Contribution of anaerobic protozoa and methanogens to hindgut metabolic activities of the American cockroach, *Periplaneta americana. Applied and Environmental Microbiology, 58*, 2565–2570.

Gil, R. (1984). Further ultrastructural observations on *Frontonia leucas* (Ciliata: Holotricha). *Transactions of the American Microscopical Society, 103*, 353–363.

Gilbert, J. J. (1994). Jumping behavior in the oligotrich ciliates *Strobilidium velox* and *Halteria grandinella*, and its significance as a defense against rotifer predators. *Microbial Ecology, 27*, 189–200.

Gilbert, J. J., & Jack, J. D. (1993). Rotifers as predators on small ciliates. *Hydrobiologia, 255/256*, 247–253.

Gilbert, J. J., & Schröder, T. (2003). The ciliate epibiont *Epistylis pygmaeum*: Selection for zooplankton hosts, reproduction and effect on two rotifers. *Freshwater Biology, 48*, 878–893.

Gilchrist, F. M. C., Van Hoven, W., & Stenson, M. O. (1994). Five new species of Trichostomatida (ciliated protozoa) from the colon of wild African rhinoceroses. *Systematic Parasitology, 28*, 187–196.

Gillies, C., & Hanson, E. D. (1963). A new species of *Leptomonas* parasitizing the macronucleus of *Paramecium trichium. Journal of Protozoology, 10*, 467–473.

Gilligan, D. M., & Satir, B. H. (1983). Stimulation and inhibition of secretion in *Paramecium*: Role of divalent cations. *Journal of Cell Biology, 97*, 224–234.

Gilron, G. L., Lynn, D. H., & Roff, J. C. (1991). The annual cycle of biomass and production of tintinnine ciliates in a tropical neritic region near Kingston, Jamaica. *Marine Microbial Food Webs, 5*, 95–113.

Gilron, G. L., Gransden, S. G., Lynn, D. H., Broadfoot, J., & Scroggins, R. (1999). A behavioural toxicity test using the ciliated protozoan *Tetrahymena thermophila*. I. Method description. *Environmental Toxicology and Chemistry, 18*, 1813–1816.

Gismervik, I. (2004). Podite carrying ciliates dominate the benthic ciliate community in the kelp forest. *Aquatic Microbial Ecology, 36*, 305–310.

Gismervik, I. (2006). Top–down impact by copepods on ciliate numbers and persistence of copepods depends on copepod and ciliate species composition. *Journal of Plankton Research, 28*, 499–507.

Gocmen, B., Dehority, B. A., Talu, G. H., & Rastgeldy, S. (2001). The rumen ciliate fauna of domestic sheep (*Ovis ammon aires*) from the Turkish Republic of Northern Cyprus. *Journal of Eukaryotic Microbiology, 48*, 455–459.

Goff, L. J., & Stein, J. R. (1981). Digestion in the peritrich ciliate *Ophrydium versatile. Protoplasma, 107*, 235–254.

Goggin, C. L., & Murphy, N. E. (2000). Conservation of sequence in the internal transcribed spacers and 5.8 S ribosomal RNA among geographically separated isolates of parasitic scuticociliates (Ciliophora, Orchitophryidae). *Diseases of Aquatic Organisms, 40*, 79–83.

Gold, K. (1971). Growth characteristics of the mass–reared tintinnid *Tintinnopsis beroidea. Marine Biology, 8*, 105–108.

Gold, K. (1973). Methods for growing Tintinnida in continuous culture. *American Zoologist, 13*, 203–208.

Gold, K. (1979). Scanning electron microscopy of *Tintinnopsis parva*: Studies on particle accumulation and the striae. *Journal of Protozoology, 26*, 415–419.

Gold, K., & Morales, E. A. (1975). Seasonal changes in lorica sizes and the species of Tintinnina in the New York Bight. *Journal of Protozoology, 22*, 520–528.

Gold, K., & Morales, E. A. (1976). Studies on the sizes, shapes, and the development of the lorica of agglutinated Tintinnida. *Biological Bulletin, 150*, 377–392.

Gold, K., & Pollingher, U. (1971). Microgamete formation and growth rate of *Tintinnopsis beroidea. Marine Biology, 11*, 324–329.

Golder, T. K. (1974). *Woodruffia metabolica*: An exception to the rule of desmodexy. *Science, 185*, 57–59.

Golder, T. K. (1976). The macro– micronuclear complex of *Woodruffia metabolica. Journal of Ultrastructure Research, 54*, 169–175.

Golder, T. K., & Lynn, D. H. (1980). *Woodruffia metabolica*: The systematic implications of its somatic and oral ultrastructures. *Journal of Protozoology, 27*, 160–169.

Golembiewska, M., & Radzikowski, S. (1980). Ultrastructure and some properties of the surface coat of *Chilodonella steini* and *Chilodonella cucullulus. Archiv für Protistenkunde, 123*, 367–381.

Golikova, M. N. (1965). Der Aufbau des Kernapparates und die Verteilung der Nukleinsäuren und Proteine bei *Nyctotherus cordiformis* Stein. *Archiv für Protistenkunde, 108*, 191–216.

Golini, V. I., & Corliss, J. O. (1981). A note on the occurrence of the hymenostome ciliate *Tetrahymena* in chironomid larvae (Diptera: Chironomidae) from the Laurentian Great Lakes. *Transactions of the American Microscopical Society, 100*, 89–93.

Golińska, K. (1979). Assessment of cell proportions during regeneration of *Dileptus anser* (Ciliata). *Wilhelm Roux's Archives, 187*, 307–321.

Golińska, K. (1982a). Regulation of ciliary pattern in *Dileptus* (Ciliata). I. Sensory cilia and their conversion into locomotor cilia. *Journal of Embryology and Experimental Morphology, 68*, 99–114.

Golińska, K. (1982b). Regulation of ciliary pattern in *Dileptus* (Ciliata). II. Formation of a cortical domain of sensory cilia from a domain of locomotor cilia. *Journal of Cell Science*, *62*, 459–475.
Golińska, K. (1984). Diminution of microtubular organelles after experimental reduction in cell size in the ciliate, *Dileptus*. *Journal of Cell Science*, *70*, 25–39.
Golińska, K. (1986). Modifications of size and pattern of microtubular organelles in overfed cells of a ciliate *Dileptus*. *Journal of Embryology and Experimental Morphology*, *93*, 85–104.
Golińska, K. (1987). Temperature–induced modifications in size and pattern of microtubular organelles in the ciliate, *Dileptus*. I. Supernumeray microtubules in axonemes of sensory cilia. *Journal of Cell Science*, *87*, 349–356.
Golińska, K. (1988). Temperature–induced modifications in size and pattern of microtubular organelles in a ciliate, *Dileptus*. II. Formation and spatial arrangement of the microtubular skeleton of the oral parts. *Protoplasma*, *147*, 125–134.
Golińska, K. (1991). Cortical organellar complexes, their structure, formation, and bearing upon cell shape in a ciliate, *Dileptus*. *Protoplasma*, *162*, 160–174.
Golińska, K. (1995). Formation and orientation of skeletal elements during development of oral territory in a ciliate, *Dileptus*. *Acta Protozoologica*, *34*, 101–113.
Golińska, K., & Afon'kin, S. Y. u. (1993). Preparatory changes and the development of the conjugation junction in a ciliate, *Dileptus*. *Protoplasma*, *173*, 144–157.
Golińska, K., & Bohatier, J. (1975). Action of actinomycin D upon regenerative and divisional stomatogenesis in *Dileptus*. *Acta Protozoologica*, *14*, 1–15.
Golińska, K., & Kink, J. (1977). Proportional regulation of body form and cortical organelle pattern in the ciliate *Dileptus*. *Journal of Cell Science*, *24*, 11–29.
Gómez–Gutiérrez, J., Peterson, W. T., & Morado, J. F. (2006). Discovery of a ciliate parasitoid of euphausiids off Oregon, USA: *Collinia oregonensis* n. sp. (Apostomatida: Colliniidae). *Diseases of Aquatic Organisms*, *71*, 33–49.
Gómez–Gutiérrez, J., Peterson, W. T., De Robertis, A., & Brodeur, R. D. (2003). Mass mortality of krill caused by parasitoid ciliates. *Science*, *301*, 339.
Gonda, K., & Numata, O. (2002). p85 binds to G–actin in a Ca^{2+} /calmodulin–dependent manner, thus regulating the initiation of cytokinesis in *Tetrahymena*. *Biochemical and Biophysical Research Communications*, *292*, 1098–1103.
Gong, Y. C., Yu, Y. H., Feng, W. S., & Shen, Y. F. (2005). Phylogenetic relationships among Trichodinidae (Ciliophora: Peritricha) derived from the characteristic values of denticles. *Acta Protozoologica*, *44*, 237–243.
Gong, Y. C., Yu, Y. H., Villalobo, E., Zhu, F. Y., & Miao, W. (2006). Reevaluation of the phylogenetic relationship between mobilid and sessilid peritrichs (Ciliophora, Oligohymenophorea) based on small subunit rRNA gene sequences. *Journal of Eukaryotic Microbiology*, *53*, 397–403.
González, J. (1979). La loge du cilié péritriche *Thuricola folliculata* (O. Müller, 1786): Structure, nature chimique et formation. *Protistologica*, *15*, 487–493.
Goodey, T. (1915). Note on the remarkable retention of vitality by protozoa from old stored soils. *Annals of Applied Biology*, *1*, 395–399.
Goodrich, J. P., & Jahn, T. L. (1943). Epizoic Suctoria (Protozoa) from turtles. *Transactions of the American Microscopical Society*, *62*, 245–253.
Goosen, N. K., Wagener, S., & Stumm, C. K. (1990). A comparison of two strains of the anaerobic ciliate *Trimyema compressum*. *Archives of Microbiology*, *153*, 187–192.
Gordo, F. P., Herrera, S., Castro, A. T., Buran, B. G., & Diaz, R. A. M. (2002). Parasites from farmed ostriches (*Struthio camelus*) and rheas (*Rhea americana*) in Europe. *Veterinary Parasitology*, *107*, 137–160.
Gorovsky, M. A., Yao, M. C., Keevert, J. B., & Pleger, G. L. (1975). Isolation of micro– and macronuclei of *Tetrahymena pyriformis*. *Methods in Cell Biology*, *9*, 311–327.
Görtz, H. D. (1982a). The behaviour and fine structure of the dorsal bristles of *Euplotes minuta*, *E. aediculatus*, and *Stylonychia mytilus* (Ciliata, Hypotrichida). *Journal of Protozoology*, *29*, 353–359.
Görtz, H. D. (1982b). Discharge of cortical ampules in *Euplotes aediculatus* Pierson, 1943 (Ciliophora, Hypotrichida). *Archiv für Protistenkunde*, *125*, 31–40.
Görtz, H. D. (1983). Endonuclear symbionts in ciliates. *International Review of Cytology Supplement*, *14*, pp. 145–176.
Görtz, H. D. (Ed.). (1988a). *Paramecium*. Berlin: Springer.
Görtz, H. D. (1988b). Endocytobiosis. In H. D. Görtz (Ed.), *Paramecium* (chap. 22, pp. 393–405). Berlin: Springer.
Görtz, H. D. (1996). Symbiosis in ciliates. In K. Hausmann & P. C. Bradbury (Eds.), *Ciliates. Cells as organsisms* (pp. 441–462). Stuttgart: Gustav Fischer Verlag.
Görtz, H. D., & Dieckmann, J. (1980). Life cycle and infectivity of *Holospora elegans* Haffkine, a micronucleus–specific symbiont of *Paramecium caudatum* (Ehrenberg). *Protistologica*, *16*, 591–603.
Görtz, H. D., & Dieckmann, J. (1987). *Leptomonas ciliatorum* n. sp. (Kinetoplastida, Trypanosomatidae) in the macronucleus of a hypotrichous ciliate. *Journal of Protozoology*, *34*, 259–263.
Görtz, H. D., & Maier, G. (1991). A bacterial infection in a ciliate from sewage sludge. *Endocytobiosis and Cell Research*. *8*, 45–52.

Görtz, H. D., & Wiemann, M. (1987). Colonization of the ciliate *Stentor multiformis* by three different endocytobionts. *Endocytobiosis and Cell Research*, *4*, 177–184.

Gould, S. J. (1977). *Ontogeny and phylogeny*. Cambridge, MA: Harvard University Press.

Goulder, R. (1980). The ecology of two species of primitive ciliated protozoa commonly found in standing freshwaters (*Loxodes magnus* Stokes and *L. striatus* Penard). *Hydrobiologia*, *72*, 131–158.

Goussard, B., Collet, J. Y., Garin, Y., Tutin, C. E. G., & Fernandez, M. (1983). The intestinal entodiniomorph ciliates of wild lowland gorillas (*Gorilla gorilla gorilla*) in Gabon, West Africa. *Journal of Medical Primatology*, *12*, 239–249.

Graham, L. E., & Graham, J. M. (1980). Endosymbiotic *Chlorella* (Chlorophyta) in a species of *Vorticella* (Ciliophora). *Transactions of the American Microscopical Society*, *99*, 160–166.

Grain, J. (1966a). Etude cytologique de quelques ciliés holotriches endocommensaux des ruminants et des équidés. *Protistologica*, *2*, 59–141; 2, 5–51.

Grain, J. (1966b). Confirmation, par la microscopie électronique, de la place des *Balantidium* dans l'ordre des Trichostomatida. *Comptes Rendus de l'Académie des Sciences.*, *263*, 1864–1867.

Grain, J. (1968). La genèse des mucocystes dans les formes prékystiques du cilié *Balantidium elongatum*. *Journal de Microscopie (Paris)*, *7*, 993–1006.

Grain, J. (1969). Le cinétosome et ses dérivés chez les ciliés. *Année Biologie, 8*, 53–97.

Grain, J. (1970). Structure et ultrastructure de *Lagynophrya fusidens* Kahl, 1927. *Protistologica*, *6*, 37–51.

Grain, J. (1972). Étude ultrastructurale d'*Halteria grandinella* O.F.M., (cilié oligotriche) et considérations phylogénétiques. *Protistologica*, *8*, 179–197.

Grain, J. (1984). Cinétosome, cil, systèmes fibrillaires en relation avec le cinétosome. In P. P. Grassé (Ed.), *Traité de zoologie*: *Vol. 2, Fasc. 1. Infusoires ciliés* (pp. 35–179). Paris: Masson.

Grain, J. (1994). Classe Vestibuliferea de Puytorac et al., 1974. In P. de Puytorac (Ed.), *Traité de zoologie, infusoires ciliés* (Vol. 2, pp. 311–379). Paris: Masson.

Grain, J., & Batisse, A. (1974). Étude ultrastructurale du cilié chonotriche *Chilodochona quennerstedti* Wallengren, 1895. I. Cortex et structures buccales. *Journal of Protozoology*, *21*, 95–111.

Grain, J., & Golińska, K. (1969). Structure et ultrastructure de *Dileptus cygnus* Claparède et Lachmann, 1859, cilié holotriche gymnostome. *Protistologica*, *5*, 269–291.

Grain, J., & de Puytorac, P. (1974). Particularités ultrastructurales des cinétosomes et de leurs annexes fibrillaires chez certains ciliés astomes Hoplitophryidae. *Journal de Microscopie (Paris)*, *19*, 231–246.

Grain, J., & Senaud, J. (1985). Dégradation de fragments du luzerne fraîche par le cilié d rumen *Epidinium ecaudatum*: Attachment, ingestion et digestion. *Protistologica*, *21*, 447–466.

Grain, J., de Puytorac, P., & Bohatier, J. (1973). Essai de systématique des ciliés gymnostomes fondée sur les caractéristiques de l'infraciliature circumorale. *Comptes Rendus de l'Académie des Sciences, Paris*, *277*, 69–72.

Grain, J., Peck, R. K., Didier, P., & Rodrigues de Santa Rosa, M. (1976). Importance de la microscopie électronique dans les études de systématique chez les unicellulaires; un exemple: les ciliés du genre *Cyclogramma. Comptes Rendus de l'Académie des Sciences, Paris*, *282*, 735–738.

Grain, J., de Puytorac, P., Detcheva, R., Grolière, C. A., Iftode, F., & Fryd–Versavel, G. (1979). Étude ultrastructurale d'un cilié Kinetofragmophora: *Placus striatus* Cohn, 1866. *Protistologica*, *14*(year 1978), 391–404.

Grain, J., de Puytorac, P., & Grolière, C. A. (1982). Quelques précisions sur l'ultrastructure et la position systématique du cilié *Mesodinium rubrum*, et sur la constitution de ses symbiontes chloroplastiques. *Protistologica*, *18*, 7–21.

Granados, L. C., & Chinchilla, M. (1990). Distribución de ciliados epibiontes en *Macrobrachium rosenbergii* (Decapoda: Natantia). *Revista de Biologia Tropical*, *38*, 1–5.

Grassmick, R. A., & Rowley, W. A. (1973). Larval mortality of *Culex tarsalis* and *Aedes aegypti* when reared with different concentrations of *Tetrahymena pyriformis*. *Journal of Invertebrate Pathology*, *22*, 86–93.

Greco, N., Bussers, J. C., van Daele, Y., & Goffinet, G. (1990). Ultrastructural localization of chitin in the cystic wall of *Euplotes muscicola* (Ciliata, Hypotrichia). *European Journal Protistology*, *26*, 75–80.

Green, J. (1974). Parasites and epibionts of Cladocera. *Transactions of the Zoological Society of London*, *32*, 417–515.

Greenwood, S. J., Sogin, M. L., & Lynn, D. H. (1991a). Phylogenetic relationships within the class Oligohymenophorea, phylum Ciliophora, inferred from the complete small subunit rRNA gene sequences of *Colpidium campylum*, *Glaucoma chattoni*, and *Opisthonecta henneguyi*. *Journal of Molecular Evolution*, *33*, 163–174.

Greenwood, S. J., Schlegel, M., Sogin, M. L., & Lynn, D. H. (1991b). Phylogenetic relationships of *Blepharisma americanum* and *Colpoda inflata* with the phylum Ciliophora inferred from the complete small subunit rRNA gene sequences. *Journal of Protozoology*, *38*, 1–6.

Greider, C. W., & Blackburn, E. H. (1987). The terminal transferase of *Tetrahymena* is a ribonucleoprotein with two kinds of primer specificity. *Cell, 51*, 887–898.

Grell, K. G. (1950). Der Generationswechsel des parasitischen Suktors *Tachyblaston ephelotensis* Martin. *Zeitschrif für Parasitenkunde, 14*, 499–534.

Grell, K. G. (1962). Morphologie und Fortpflanzung der Protozoen (einschliesslich Entwicklungsphysiologie und Genetik). *Fortschritte der Zoologie, 14*, 1–85.

Grell, K. G. (1973). *Protozoology* (554 pp.). Berlin, Heidelberg/New York: Springer.

Grell, K. G., & Benwitz, G. (1984). Die Ultrastruktur von *Ephelota gemmipara* Hertwig und *E. plana* Wailes (Suctoria): Ein Vergleich I. Die adulte Form. *Protistologica, 20*, 205–233.

Grell, K. G., & Meister, A. (1982a). Über die taxonomische Stellung der Hypocomidae (Ciliata). *Zeitschrift für Naturforschung, 37c*, 1050–1052.

Grell, K. G., & Meister, A. (1982b). Die Ultrastruktur von *Acinetopsis rara* Robin (Suctoria) II. Zellbau und Schwärmerbildung. *Protistologica, 18*, 403–421.

Grell, K. G., & Meister, A. (1983). Fine structure of *Hypocoma acinetarum* Collin. *Protistologica, 19*, 51–72.

Grell, K. G., & Meister, A. (1984). Beiträge zur Ultrastruktur der Konjugation von *Ephelota gemmipara* R. Hertwig (Suctoria). *Protistologica, 20*, 65–86.

Greslin, A. F., Prescott, D. M., Oka, Y., Loukin, S. H., & Chappell, J. C. (1989). Reordering of nine exons is necessary to form a functional actin gene in *Oxytricha nova. Proceedings of the National Academy of Sciences of the United States of America, 86*, 6264–6268.

Greuet, C., Gayol, P., Salvano, P., & Laval–Peuto, M. (1986). Preliminary report on the ultrastructural organization of the contractile appendix of *Tontonia appendiculariformis* (Ciliophora Oligotrichina). *Cell Motility and the Cytoskeleton, 6*, 217–224.

Griffiths, B. S. (1986). Mineralization of nitrogen and phosphorus by mixed cultures of the ciliate protozoan *Colpoda steinii*, the nematode *Rhabditis* sp. and the bacterium *Pseudomonas fluorescens. Soil Biology & Biochemistry, 18*, 637–641.

Griffiths, B. S. (2002). Spatial distribution of soil protozoa in an upland grassland. *European Journal of Protistology, 37*, 371–373.

Griffiths, H. I., & Evans, J. G. (1994). Infestation of the freshwater ostracod *Cypria ophthalmica* (Jurine) by the peritrich *Nuchterleinella corneliae* Matthes. *Archiv für Protistenkunde, 144*, 315–317.

Grigorovich, I. A., Dovgal, I. V., MacIsaac, H. J., & Monchenko, V. (2001). *Acineta nitocrae*: A new suctorian epizooic on nonindigenous harpacticoid copepods, *Nitocra hibernica* and *N. incerta*, in the Laurentian Great Lakes. *Archiv für Hydrobiologie, 152*, 161–176.

Grim, J. N. (1982). Subpellicular microtubules of *Euplotes eurystomus*: Their geometry relative to cell form, surface contours and ciliary organelles. *Journal of Cell Science, 56*, 471–484.

Grim, J. N. (1987). The kinetid structure of the choreotrichous ciliate *Strobilidium velox* and an assessment of its evolutionary lineage. *Journal of Protozoology, 34*, 117–123.

Grim, J. N. (1988). A somatic kinetid study of the pycnotrichid ciliate *Vestibulongum corlissi*, n. g., n. sp. (Class: Litostomatea), symbiont in the intestines of the surgeonfish, *Acanthurus xanthopterus. Journal of Protozoology, 35*, 227–230.

Grim, J. N. (1989). The vestibuliferan ciliate *Balantidium acanthuri* n. sp. from two species of the surgeonfish, genus *Acanthurus. Archiv für Protistenkunde, 137*, 157–160.

Grim, J. N. (1992). Descriptions of two sympatric and phylogenetically diverse ciliated protozoa, *Balantidium zebrascopi* n. sp. and *Paracichlidotherus leeuwenhoeki* n. gen., n. sp., symbionts in the intestines of the surgeonfish, *Zebrasoma scopas. Transactions of the American Microscopical Society, 111*, 149–157.

Grim, J. N. (1993a). Description of somatic kineties and vestibular organization of *Balantidium jocularum* sp. n., and possible taxonomic implications for the Class Litostomatea and the genus *Balantidium. Acta Protozoologica, 32*, 37–45.

Grim, J. N. (1993b). Endonuclear symbionts within a symbiont: The surgeonfish intestinal symbiont, *Balantidium jocularum* (Ciliophora) is host to a Gram–positive macronuclear inhabiting bacterium. *Endocytobiosis and Cell Research, 9*, 209–214.

Grim, J. N. (1998). A comparison of three populations of the ciliate genus, *Paracichlidotherus* Grim, 1992. New fish hosts, and biogeography; Revised genus description. *Journal of Eukaryotic Microbiology, 45*, 40–44.

Grim, J. N. (2006). Food vacuole contents in the ciliate, *Balantidium jocularum* (Balantidiidae), a symbiont in the intestine of the surgeonfish, *Naso tonganus* (Acanthuridae). *Journal of Eukaryotic Microbiology, 53*, 269–274.

Grim, J. N., & Halcrow, K. R. (1979). The oligotrich ciliate *Strobilidium gyrans*: Its fresh–water environment, laboratory culture conditions, and stalking behavior. *Transactions of the American Microscopical Society, 98*, 529–536.

Grim, J. N., & Monganaro, C. A. (1985). Form of the extrusomes and secreted material of the ciliated protozoon *Pseudourostyla cristata*, with some phylogenetic interpretations: A light, scanning electron, and transmission electron microscopic study. *Transactions of the American Microscopical Society, 104*, 350–359.

Grim, J. N., Reed, K. S., & Fishelson, L. (1995/1996). Description of a new species of ciliates, *Metanyctotherus paulini* n. sp., symbiont from the intestines of the angelfish *Apolemichthys xanthotis*. Assessment of species status for extant metanyctotherans. *Archiv für Protistenkunde*, *146*, 341–347.

Grim, J. N., Clements, K. D., & Byfield, T. (2002). New species of *Balantidium* and *Paracichlidotherus* (Ciliophora) inhabiting the intestines of four surgeonfish species from the Tuvalu Islands, Pacific Ocean. *Journal of Eukaryotic Microbiology*, *49*, 146–153.

Grimes, B. H. (1976). Notes on the distribution of *Hyalophysa* and *Gymnodinioides* on crustacean hosts in coastal North Carolina and a description of *Hyalophysa trageri* sp. n. *Journal of Protozoology*, *23*, 246–251.

Grimes, B. H., & Bradbury, P. C. (1992). The biology of *Vampyrophrya pelagica* (Chatton & Lwoff, 1930), a histophagous apostome ciliate associated with marine calanoid copepods. *Journal of Protozoology*, *39*, 65–79.

Grimes, G. W. (1972). Cortical structure in nondividing and cortical morphogenesis in dividing *Oxytricha fallax*. *Journal of Protozoology*, *19*, 428–445.

Grimes, G. W. (1973). Morphological discontinuity of kinetosomes during the life cycle of *Oxytricha fallax*. *Journal of Cell Biology*, *57*, 229–232.

Grimes, G. W. (1982). Pattern determination in hypotrich ciliates. *American Zoologist*, *22*, 35–46.

Grimes, G. W., & Adler, J. A. (1976). The structure and development of the dorsal bristle complex of *Oxytricha fallax* and *Stylonychia pustulata*. *Journal of Protozoology*, *23*, 135–143.

Grimes, G. W., McKenna, M. E., Goldsmith–Spoegler, C. M., & Knaupp, E. A. (1980). Patterning and assembly of ciliature are independent processes in hypotrich ciliates. *Science*, *209*, 281–283.

Grimm, M., Brunen–Nieweler, C., Junker, V., Heckmann, K., & Beier, H. (1998). The hypotrichous ciliate *Euplotes octocarinatus* has only one type of tRNACys with GCA anticodon encoded on a single macronuclear DNA molecule. *Nucleic Acids Research*, *26*, 4557–4565.

Grimstone, A. V. (1959). Cytology, homology and phylogeny – a note on "organic design." *American Naturalist*, *93*, 273–282.

Grolière, C. A. (1975a). Etude comparée de la stomatogenèse chez quelques ciliés hymenostomes des genres *Paralembus* Kahl, 1933, *Philaster* Fabre-Domergue, 1885, *Paraurónema* Thompson, 1967, *Tetrahymena* Furgason, 1940. *Protistologica*, *10*(year 1974), 319–331.

Grolière, C. A. (1975b). La stomatogenèse du cilié *Platyophrya spumacola* Kahl 1927; son intérêt pour la compréhension de la diversification buissonnante des Kinetophragmophora de Puytorac et coll. *Comptes Rendus de l'Académie des Sciences*, *280*, 861–864.

Grolière, C. A. (1975c). Etude de quelques ciliés hyménostomes des eaux acides de la region de Besse–en–Chandesse. *Annales de la Station Biologique de Besse–en–Chandesse*, *9*(years 1974–1975), 79–109.

Grolière, C. A. (1977). Contribution à l'étude des ciliés des sphaignes et des étendues d'eau acides. I. Description de quelques espèces de gymnostomes, hypostomes, hyménostomes et hétérotriches. *Annales de la Station Biologique de Besse–en–Chandesse*, *10*(year 1975–1976), 265–297.

Grolière, C. A. (1980). Morphologie et stomatogenèse chez deux ciliés Scuticociliatida des genres *Philasterides* Kahl, 1926 et *Cyclidium* O. F. Müller, 1786. *Acta Protozoologica*, *19*, 195–206.

Grolière, C. A., & Detcheva, R. (1974). Description et stomatogenèse de *Pleuronema puytoraci* n. sp. (Ciliata, Holotricha). *Protistologica*, *10*, 91–99.

Grolière, C. A., & Njiné, T. (1973). Etude comparée de la dynamique des populations de ciliés dans différents biotopes d'une mare de forêt pendant une année. *Protistologica*, *9*, 5–16.

Grolière, C. A., de Puytorac, P., & Detcheva, R. (1980a). À propos d'observations sur la stomatogenèse et l'ultrastructure du cilié *Protocruzia tuzeti* Villeneuve–Brachon, 1940. *Protistologica*, *16*, 453–466.

Grolière, C. A., de Puytorac, P., & Grain, J. (1980b). Observations de quelques espèces de ciliés endocommensaux d'échinides du Golfe du Mexique et de la mer des Antilles. *Protistologica*, *16*, 233–239.

Grønlien, H. K., Berg, T., & Løvlie, A. M. (2002). In the polymorphic ciliate *Tetrahymena vorax*, the non–selective phagocytosis seen in microstomes changes to a highly selective process in macrostomes. *Journal of Experimental Biology*, *205*, 2089–2097.

Gruby, D., & Delafond, O. (1843). Recherches sur des animalcules se développant en grande nombre dans l'estomac et dans les intestins, pendant la digestion des animaux herbivores et carnivores. *Comptes Rendus de l'Académie des Sciences, Paris*, *17*, 1304–1308.

Gruchy, D. G. (1955). The breeding system and distribution of *Tetrahymena pyriformis*. *Journal of Protozoology*, *2*, 178–185.

Gu, F., Chen, L., Ni, B., & Zhang, X. (2002). A comparative study of the electron microscopic enzymo–cytochemistry of *Paramecium bursaria* from light and dark cultures. *European Journal of Protistology*, *38*, 267–278.

Guella, G., Dini, F., & Pietra, F. (1995). 128. From epiraikovenal, an instrumental niche–exploitation sesquiterpenoid of some strains of the marine ciliate protist *Euplotes raikovi*, to an unusual intramolecular *tele*–dienone–olefin [2 + 2]: photocycloaddition. *Helvetica Chimica Acta*, *78*, 1747–1754.

Guella, G., Dini, F., & Pietra, F. (1996). 39. Epoxyfocardin and its putative biogenetic precursor, focardin, bioactive, new–skeleton diterpenoids of the marine ciliate *Euplotes focardii* from Antarctica. *Helvetica Chimica Acta*, *79*, 439–448.

Guhl, B. E., & Finlay, B. J. (1993). Anaerobic predatory ciliates track seasonal migrations of planktonic photosynthetic bacteria. *FEMS Microbiology Letters*, *107*, 313–316.

Guhl, B. E., Finlay, B. J., & Schink, B. (1996). Comparison of ciliate communities in the anoxic hypolimnia of three lakes: General features and the influence of lake characteristics. *Journal of Plankton Research*, *18*, 335–353.

Guhl, W. (1979). Bietrag zur Systematik, Biologie und Morphologie der Epistylidae (Ciliata, Peritricha). *Archiv für Protistenkunde*, *121*, 417–483.

Guhl, W., & Haider, G. (1988). Urceolariidae (Peritricha). In D. Matthes, W. Guhl, & G. Haider (Eds.), *Suctoria und Urceolariidae (Peritricha)* (pp. 227–309). Stuttgart: Gustav Fischer Verlag.

Guilcher, Y. (1951). Contribution à l'étude des ciliés gemmipares, chonotriches et tentaculifères. *Annales des Sciences Naturelle–Zoologie et Biologie Animale (Série 11)*, *13*, 33–132.

Guinea, A., Gil, R., & Fernández–Galiano, D. (1987). Structure of the buccal apparatus of *Urocentrum turbo* (Ciliophora: Hymenostomatida). *Transactions of the American Microscopical Society*, *106*, 53–62.

Guinea, A., Sola, A., Rueda, M. C., & Fernández–Galiano, D. (1988). Infraciliature, système myonémique, argyrome et bipartition de l'*Astylozoon pyriforme* (Ciliophora, Peritrichida). *Canadian Journal of Zoology*, *66*, 2104–2109.

Guinea, A., Gil, R., Serrano, S., & Sola, A. (1990). Ultrastructural data on the infraciliature of *Astylozoon pyriforme* (Ciliophora: Peritrichida). *Transactions of the American Microscopical Society*, *109*, 247–253.

Guinea, A., Anadón, R., & Fernández–Galiano, D. (1992). Light and electron microscopic study of *Balantidium entozoon* (Ciliophora, Vestibuliferida): Somatic cortex and vestibular cavity. *Archiv für Protistenkunde*, *142*, 41–50.

Guirao, B., & Joanny, J. F. (2007). Spontaneous creation of macroscopic flow and metachronal waves in an array of cilia. *Biophysical Journal*, *92*, 1900–1917.

Guirong Su, N. R., Hua, Z. X., Zhu, S., & Imai, S. (2000). Rumen ciliated protozoan fauna of the yak (*Bos grunniens*) in China with the description of *Entodinium monuo* n. sp. *Journal of Eukaryotic Microbiology*, *47*, 178–182.

Gunderson, J. (1984). Chonotrichs associated with *Nebalia pugettensis* [Abstract No. 11]. *Journal of Protozoology*, *31*, 3A–4A.

Gutiérrez, J. C., & Martín–González, A. (1990). Evidence for protein degradation and synthesis during encystment of the ciliate *Colpoda inflata*. *Microbios Letters*, *43*, 57–63.

Gutiérrez, J. C., Torres, A., & Perez–Silva, J. (1983). Fine structure of the cyst wall of *Laurentiella acuminata* (Hypotrichida: Oxytrichidae). *Transactions of the American Microscopical Society*, *102*, 55–59.

Gutiérrez, J. C., Martín–González, A., & Matsusaka, T. (1990). Towards a generalized model of encystment (cryptobiosis) in ciliates: A review and a hypothesis. *BioSystems*, *24*, 17–24.

Gutiérrez, J. C., Martín–González, A., & Callejas, S. (1998a). Nuclear changes, macronuclear chromatin reorganization and DNA modifications during ciliate encystment. *European Journal of Protistology*, *34*, 97–103.

Gutiérrez, J. C., Izquierdo, A., Martín–González, A., & Callejas, S. (1998b). Cryptobiosis of colpodid ciliates: A microbial eukaryotic cell differentiation model. *Recent Research Developments in Microbiology*, *2*, 1–15.

HHHH

Haacke-Bell, B., Hohenberger-Bregger, R., & Plattner, H. (1990). Trichocysts of *Paramecium*: Secretory organelles in search of their function. *European Journal of Protistology*, *25*, 289–305.

Haas, W., Haberl, B., Hofmann, M., Kerschensteiner, S., & Ketzer, U. (1999). *Ichthyophthirium multifiliis* invasive stages find their fish hosts with complex behavior patterns and in response to different chemical signals. *European Journal of Protistology*, *35*, 129–135.

Hackney, C. M., & Butler, R. D. (1981a). Electrically induced tentacle retraction in the suctorian protozoon *Discophrya collini* (Root). *Journal of Protozoology*, *28*, 151–157.

Hackney, C. M., & Butler, R. D. (1981b). Tentacle retraction in glycerinated *Discophrya collini* and the localization of HMM–binding filaments. *Journal of Cell Science*, 65–75.

Hackney, C. M., & Butler, R. D. (1981c). Distribution of calcium in the suctorian *Discophrya collini*: An X–ray microanalytical study. *Tissue & Cell*, *13*, 453–459.

Hackney, C. M., AL–Khazzar, A. R., & Butler, R. D. (1982). Tentacle contraction and ultrastructure in *Discophrya collini*: The response to cations. *Protoplasma*, *112*, 92–100.

Hackstein, J. H. P., & Stumm, C. K. (1994). Methane production by terrestrial arthropods. *Proceedings of the National Academy of Sciences of the United States of America*, *91*, 5441–5445.

Hackstein, J. H. P., Van Hoek, A. H. A. M., Leunissen, J. A. M., & Huynen, M. (2001). Anaerobic ciliates and their methanogenic endosymbionts. In J.

Seckbach (Ed.), *Symbiosis* (pp. 451–464). Dordrecht, The Netherlands: Kluwer.

Hadas, O., & Berman, T. (1998). Seasonal abundance and vertical distribution of protozoa (flagellates, ciliates) and bacteria in Lake Kinneret, Israel. *Aquatic Microbial Ecology*, *14*, 161–170.

Hadas, O., Malinsky–Rushansky, N., Pinkas, R., & Cappenberg, T. E. (1998). Grazing on autotrophic and heterotrophic picoplankton by ciliates isolated from Lake Kinneret, Israel. *Journal of Plankton Research*, *20*, 1435–1448.

Hadži, J. (1951). Studien über Follikuliniden. *Dela. Razred za Prirodoslovne in Medicinske Vede. Slovenska Akademija Znanosti in Umetnosti, 4*, 1–390.

Haentzsch, M., Schmidt, S. L., Bernhard, D., Ammerman, D., Berendonk, T. U., & Schlegel, M. (2006). A PCR–based method to distinguish the sibling species *Stylonychia mytilus* and *Stylonychia lemnae* (Ciliophora, Spirotrichea) using isocitrate dehydrogenase gene sequences. *Journal of Eukaryotic Microbiology*, *53*, 343–347.

Haider, G. (1964). Monographie der Famille Urceolariidae (Ciliata, Peritricha, Mobilia) mit besonderer Berücksichtigung der im süddeutschen Raum vorkommenden Arten. *Parasitologische Schriftenreihe*, *17*, 1–251.

Hajibabaei, M., Janzen, D. H., Burns, J. M., Hallwachs, W., & Hebert, P. D. N. (2006). DNA barcodes distinguish species of tropical Lepidoptera. *Proceedings of the National Academy of Sciences of the United States of America*, *103*, 968–971.

Haller, G. de. (1984a). Les armatures squelettiques. In P. P. Grassé (Ed.), *Traité de zoologie*: *Vol. 2, Fasc. 1. Infusoires ciliés* (pp. 359–367). Paris: Masson.

Haller, G. de. (1984b). Les myonèmes. In P. P. Grassé (Ed.), *Traité de zoologie*: *Vol. 2, Fasc. 1. Infusoires ciliés* (pp. 369–380). Paris: Masson.

Haller, G. de. (1984c). L'appareil alimentaire. In P. P. Grassé (Ed.), *Traité de zoologie*: *Vol. 2, Fasc. 1. Infusoires ciliés* (pp. 457–476). Paris: Masson.

Hamelmann, U., Eichenlaub–Ritter, U., & Ruthmann, A. (1986). The effect of trifluoroperazine on microtubules, nuclear division and the nuclear membranes of the ciliate *Nyctotherus ovalis*. *Chromosoma*, *93*, 453–459.

Hamels, I., Moens, T., Muylaert, K., & Vyverman, W. (2001). Trophic interactions between ciliates and nematodes from an intertidal flat. *Aquatic Microbial Ecology*, *26*, 61–72.

Hamilton, E. P., Williamson, S., Dunn, S., Merriam, V., Lin, C., Vong, L., et al. (2006). The highly conserved family of *Tetrahymena thermophila* chromosome breakage elements contains an invariant 10–base–pair core. *Eukaryotic Cell*, *5*, 771–780.

Hamilton, J. M. (1952). Studies on loricate Ciliophora. II. *Folliculina boltoni* Kent. *Proceedings of the Iowa Academy Science*, *58*, 469–476.

Hamilton, R. D., & Preslan, J. E. (1969). Cultural characteristics of a pelagic marine hymenostome ciliate, *Uronema* sp. *Journal of Experimental Marine Biology and Ecology*, *4*, 90–99.

Hammerschmidt, B., Schlegel, M., Lynn, D. H., Leipe, D. D., Sogin, M. L., & Raikov, I. B. (1996). Insights into the evolution of nuclear dualism in the ciliates revealed by phylogenetic analysis of rRNA sequences. *Journal of Eukaryotic Microbiology*, *43*, 225–230.

Hanke–Bucker, G., Lanners, H. N., & Hauser, M. (2000). Nuclear phenomena during conjugation of *Heliophrya erhardi* (Ciliata, Suctoria). II. The orientation of postmeiotic and metagamic division spindles is organized by the cytoskeleton. *European Journal of Protistology*, *36*, 211–228.

Hansen, P. J. (1991). *Dinophysis* – a planktonic dinoflagellate genus which can act both as a prey and a predator of a ciliate. *Marine Ecology–Progress Series*, *69*, 201–204.

Hansen, P. J. (1995). Growth and grazing response of a ciliate feeding on the red tide dinoflagellate *Gyrodinium aureolum* in monoculture and in mixture with a non–toxic alga. *Marine Ecology–Progress Series*, *121*, 65–72.

Hansen, P. J., & Fenchel, T. (2006). The bloom–forming ciliate *Mesodinium rubrum* harbours a single permanent endosymbiont. *Marine Biology Research*, *2*, 169–177.

Hansen, P. J., Cembella, A. D., & Moestrup, Ø. (1992). The marine dinoflagellate *Alexandrium ostenfeldii*: Paralytic shellfish toxin concentration, composition, and toxicity to a tintinnid ciliate. *Journal of Phycology*, *28*, 597–603.

Hanson, E. D. (1962). Morphogenesis and regeneration of oral structures in *Paramecium aurelia*: An analysis of intracellular development. *Journal of Experimental Zoology*, *150*, 45–68.

Hanyu, N., Kuchino, Y., & Nishimura, S. (1986). Dramatic events in ciliate evolution: Alteration of UAA and UAG termination codons to glutamine codons due to anticodon mutations in two *Tetrahymena* tRNAsGln. *EMBO Journal*, *5*, 1307–1311.

Harmsworth, G. C., & Sleigh, M. A. (1992). The abundance of different peritrich ciliates on stone surfaces in contrasting lowland streams throughout the year. *Journal of Protozoology*, *39*, 58–65.

Harmsworth, G. C., & Sleigh, M. A. (1993). Colonization of non–living surfaces in streams by peritrich ciliates. *European Journal of Protistology*, *29*, 294–301.

Harper, D. S., & Jahn, C. L. (1989). Differential use of termination codons in ciliated protozoa. *Proceedings* of the National Academy of Sciences of the United States of America, 86, 3252–3256.

Harrison, G. W. (1995). Comparing predator–prey models to Luckinbill's experiment with *Didinium* and *Paramecium*. *Ecology*, *76*, 357–374.

Harry, O. G. (1980). Damage to the eyes of the bivalve *Chlamys opercularis* caused by the ciliate *Licnophora auerbachii*. *Journal of Invertebrate Pathology*, *36*, 283–291.

Hartman, H. (1993). Evolution of the eukaryotic flagellum. *Nanobiology*, *3*, 37–39.

Hartmann, H. J., Taleb, H., Aleya, L., & Lair, N. (1993). Predation on ciliates by the suspension–feeding calanoid copepod *Acanthodiaptomus denticornis*. *Canadian Journal of Fisheries and Aquatic Sciences*, *50*, 1382–1393.

Harumoto, T. (1994). The role of trichocyst discharge and backward swimming in escaping behavior of *Paramecium* from *Dileptus margaritifer*. *Journal of Eukaryotic Microbiology*, *41*, 560–564.

Harumoto, T., & Miyake, A. (1991). Defensive function of trichocysts in *Paramecium*. *Journal of Experimental Zoology*, *260*, 84–92.

Harumoto, T., Miyake, A., Ishikawa, N., Sugibayashi, R., Zenfuku, K., & Iio, H. (1998). Chemical defense by means of pigmented extrusomes in the ciliate *Blepharisma japonicum*. *European Journal of Protistology*, *34*, 458–470.

Hascall, G. K. (1973). The stalk of the suctorian *Tokophrya infusionum*: Histochemistry, biochemistry, and physiology. *Journal Protozoology*, *20*, 701–704.

Hascall, G. K., & Rudzinska, M. A. (1970). Metamorphosis in *Tokophrya infusionum*; an electron–microscope study. *Journal of Protozoology*, *17*, 311–323.

Hashimoto, K. (1966). Stomatogenesis in resting cysts of Colpodidae. *Journal of Protozoology*, *13*, 383–390.

Hatai, K., Chukanhom, K., Lawhavinit, O. A., Hanjavanit, C., Kunitsune, M., & Imai, S. (2001). Some biological characteristics of *Tetrahymena corlissi* isolated from guppy in Thailand. *Fish Pathology*, *36*, 195–199.

Hattori, R., & Hattori, T. (1993). Soil aggregates as microcosms of bacteria–protozoa biota. *Geoderma*, *56*, 493–501.

Hatzidimitriou, G., & Berger, J. (1978). Morphology and morphogenesis of *Ancistrum mytili* (Scuticociliatida: Thigmotrichina), a commensal ciliate of mytilid pelecypods. *Protistologica*, *13*(year 1977): 477–495.

Hauser, M., & Hausmann, K. (1982). Electron microscopic localization of ATPase activity in the cytopharyngeal basket of the ciliate *Pseudomicrothorax dubius*. *Differentiation*, *22*, 67–72.

Hauser, M., & van Eys, H. (1976). Microtubules and associated microfilaments in the tentacles of the suctorian *Heliophrya erhardi* Matthes. *Journal of Cell Science*, *20*, 589–617.

Hauser, M., Beinbrech, G., Gröschel–Stewart, U., & Jockusch, B. M. (1975). Localisation by immunological techniques of myosin in nuclei of lower eukaryotes. *Experimental Cell Research*, *95*, 127–135.

Hauser, M., Hausmann, K., & Jockusch, B. M. (1980). Demonstration of tubulin, actin and –actinin by immunofluorescence in the microtubule–microfilament complex of the cytopharyngeal basket of the ciliate *Pseudomicrothorax dubius*. *Experimental Cell Research*, *125*, 265–274.

Hausmann, K. (1977). Elektronenmikroskopische Untersuchungen an den Konocysten von *Loxophyllum meleagris* Dujardin (Euciliata, Holotricha). *Archiv für Protistenkunde*, *119*, 233–238.

Hausmann, K. (1978). Extrusive organelles in protists. *International Review of Cytology*, *52*, 197–276.

Hausmann, K. (1979). Characteristics of the membranes of the pellicle of the ciliate *Pseudomicrothorax dubius*. *Protoplasma*, *100*, 199–213.

Hausmann, K. (1980). Zur Digestion bei *Pseudomicrothorax dubius* Mermod (Ciliophora). Nahrungsvakuolen–Vesikulation im Anschluss an die Phagocytose. *Zoomorphology*, *96*, 231–241.

Hausmann, K. (1983). Zur Organisation der kontraktilen Vakuole von *Pseudomicrothorax dubius* (Ciliophora, Kinetofragminophorea, Hypostomatia). *Archiv für Protistenkunde*, *127*, 319–325.

Hausmann, K., & Hausmann, E. (1981a). Structural studies on *Trichodina pediculus* (Ciliophora, Peritricha). I. The locomotor fringe and the oral apparatus. *Journal of Ultrastructure Research*, *74*, 131–143.

Hausmann, K., & Hausmann, E. (1981b). Structural studies on *Trichodina pediculus* (Ciliophora, Peritricha). II. The adhesive disc. *Journal of Ultrastructure Research*, *74*, 144–155.

Hausmann, K., & Kaiser, J. (1979). Arrangement and structure of plates in the cortical alveoli of the hypotrich ciliate, *Euplotes vannus*. *Journal of Ultrastructure Research*, *67*, 15–22.

Hausmann, K., & Peck, R. K. (1978). Microtubules and microfilaments as major components of a phagocytic apparatus: The cytopharyngeal basket of the ciliate *Pseudomicrothorax dubius*. *Differentiation*, *11*, 157–167.

Hausmann, K., & Peck, R. K. (1979). The mode of function of the cytopharyngeal basket of the ciliate *Pseudomicrothorax dubius*. *Differentiation*, *14*, 147–158.

Hausmann, K., & Rüskens, A. (1984). Untersuchungen zur Verdauung beim Ciliaten *Nassula aurea* Ehrenberg. *Archiv für Protistenkunde*, *128*, 77–87.

Hausmann, K., & Walz, B. (1979). Feinstrukturelle und mikroanalytische Untersuchungen an den Kristallen und Lithosomen des Ciliaten *Euplotes vannus*. *Protoplasma*, *99*, 67–77.

Hausmann, K., Hülsmann, N., Polianski, I., Schade, S., & Weitere, M. (2002). Composition of benthic protozoan communities along a depth transect in the eastern Mediterranean Sea. *Deep–Sea Research*, *49*, 1959–1970.

Hebert, P. D. N., Cywinska, A., Ball, S. L., & DeWaard, J. R. (2003). Biological identifications through DNA barcodes. *Proceedings of the Royal Society of London, Biological Sciences, Series B*, *270*, 313–321.

Hebert, P. D. N., Stoeckle, M. Y., Zemlak, T. S., & Francis, C. M. (2004). Identification of birds through DNA barcodes. *PLOS Biology*, *2*, e312. DOI: 10.1371/journal.pbio.0020312

Heckmann, K. (1963). Paarungssystem und genabhängige Paarungstypdifferenzierung bei dem hypotrichen Ciliaten *Euplotes vannus* O. F. Müller. *Archiv für Protistenkunde*, *106*, 393–421.

Heckmann, K. (1964). Experimentelle Untersuchungen an *Euplotes crassus*. I. Paarungssystem, Konjugation und Determination der Paarungstypen. *Zeitschrift für Verebungslehre*, *95*, 114–124.

Heckmann, K. (1965). Totale Konjugation bei *Urostyla hologama* n. sp. *Archiv für Protistenkunde*, *108*, 55–62.

Heckmann, K. (1975). *Omikron*, ein essentieller Endosymbiont von *Euplotes aediculatus*. *Journal of Protozoology*, *22*, 97–104.

Heckmann, K. (1983). Endosymbionts of *Euplotes*. *International Review of Cytology*, *14*(Suppl.), 111–144.

Heckmann, K., & Frankel, J. (1968). Genic control of cortical pattern in *Euplotes*. *Journal of Experimental Zoology*, *168*, 11–37.

Heckmann, K., & Kuhlmann, H. W. (1986). Mating types and mating inducing substances in *Euplotes octocarinatus*. *Journal of Experimental Zoology*, *237*, 87–96.

Heckmann, K., & Schmidt, H. J. (1987). *Polynucleobacter necessarius* gen nov., sp. nov., an obligately endosymbiotic bacterium living in the cytoplasm of *Euplotes aediculatus*. *International Journal of Systematic Bacteriology*, *37*, 456–457.

Heckmann, K., & Siegel, R. W. (1964). Evidence for the induction of mating type substances by cell to cell contact. *Experimental Cell Research*, *36*, 688–691.

Heckmann, K., Ten Hagen, R., & Görtz, H. D. (1983). Freshwater *Euplotes* species with a 9 type 1 cirrus pattern depend upon endosymbionts. *Journal of Protozoology*, *30*, 284–289.

Hegarty, R. S. (1999). Reducing rumen methane emissions through elimination of rumen protozoa. *Australian Journal of Agricultural Research*, *50*, 1321–1327.

Helftenbein, E. (1985). Nucleotide sequence of a macronuclear DNA molecule coding for α–tubulin from the ciliate *Stylonychia lemnae*. Special codon usage: TAA is not a translation termination codon. *Nucleic Acids Research*, *13*, 415–433.

Hellung–Larsen, P., Leick, V., & Tommerup, N. (1986). Chemoattraction in *Tetrahymena*: On the role of chemokinesis. *Biological Bulletin*, *170*, 357–367.

Hemberger, H., & Wilbert, N. (1982). Revision der Familie Keronidae Dujardin, 1840 (Ciliophora, Hypotrichida) mit einer Beschreibung der Morphogenese von *Kerona polyporum* Ehrenberg, 1835. *Archiv für Protistenkunde*, *125*, 261–270.

Henebry, M. S., & Ridgeway, B. T. (1979). Epizoic ciliated protozoa of planktonic copepods and cladocerans and their possible use as indicators of organic water pollution. *Transactions of the American Microscopical Society*, *98*, 495–508.

Henk, W. G. (1979). Ruthenium red staining of surface structures of *Discophrya collini* (Root). *European Journal of Cell Biology*, *19*, 83–88.

Henk, W. G., & Paulin, J. J. (1977). Scanning electron microscopy of budding and metamorphosis in *Discophrya collini* (Root). *Journal of Protozoology*, *24*, 134–139.

Herrick, G. (1994). Germline–soma relationships in ciliate protozoa: The inception and evolution of nuclear dimorphism in one–celled animals. *Seminars in Developmental Biology*, *5*, 3–12.

Herrick, G., Cartinhour, S., Dawson, D., Ang, D., Sheets, R., Lee, A., et al. (1985). Mobile elements bounded by C4A4 telomeric repeats in *Oxytricha fallax*. *Cell*, *43*, 759–768.

Herrick, G., Hunter, D., Williams, K., & Kotter, K. (1987). Alternative processing during development of a macronuclear chromosome family in *Oxytricha*. *Genes and Development*, *1*, 1047–1058.

Hertwig, R. (1889). Über die Konjugation der Infusorien. *Abhandlungen der Bayerische Akademie Wissenschaften*, *17*, 150–233.

Hewett, S. W. (1980a). Prey–dependent cell size in a protozoan predator. *Journal of Protozoology*, *27*, 311–313.

Hewett, S. W. (1980b). The effect of prey size on the functional and numerical responses of a protozoan predator to its prey. *Ecology*, *61*, 1075–1081.

Hewett, S. W. (1987). Prey size and survivorship in *Didinium nasutum* (Ciliophora: Gymnostomatida). *Transactions of the American Microscopical Society*, *106*, 134–138.

Hewett, S. W. (1988). Predation by *Didinium nasutum*: Effects of predator and prey size. *Ecology*, *69*, 135–145.

Hewitt, E. A., Müller, K. M., Cannone, J., Hogan, D. J., Gutell, R., & Prescott, D. M. (2003). Phylogenetic relationships among 28 spirotrichous ciliates documented by rDNA. *Molecular Phylogenetics and Evolution*, *29*, 258–267.

Hibberd, D. J. (1977). Observations on the ultrastructure of the cryptomonad endosymbiont of the red–water ciliate *Mesodinium rubrum*. *Journal of the Marine Biological Association of the United Kingdom, 57*, 45–61.

Hidayat, C., Hillman, K., Newbold, C. J., & Stewart, C. S. (1993). The contributions of bacteria and protozoa to ruminal forage fermentation in vitro, as determined by microbial gas production. *Animal Feed Science and Technology, 42*, 193–208.

Hill, B. F. (1981). The cortical morphogenetic cycle associated with cell division in *Diophrys* Dujardin, 1841 (Ciliophora, Hypotrichida). *Journal of Protozoology, 28*, 215–221.

Hill, B. F. (1990). *Uronychia transfuga* (O. F. Müller, 1786) Stein, 1850 (Ciliophora, Hypotrichida, Uronychiidae): Cortical structure and morphogenesis during division. *Journal of Protozoology, 37*, 99–107.

Hill, D. L. (1972). *The biochemistry and physiology of Tetrahymena* (230 pp.). New York/London: Academic Press.

Hill, J. (1752). *An history of animals, including the several classes of Animalcula visible only by the assistance of microscopes*: *Vol. 3. Compleat body of natural history* (584 pp.). London: Osborne.

Hiller, S. A. (1991). *Bursellopsis spaniopogon* n. spec. (Ciliophora: Prostomatida). I. Infraciliature and ultrastructure of the somatic and oral cortex. *European Journal of Protistology, 27*, 1–16.

Hiller, S. A. (1992). *Bursellopsis spaniopogon* (Ciliophora: Prostomatida). II. Stomatogenesis as revealed by light microscopy and scanning electron microscopy and some phylogenetic implications concerning prostome ciliates. *European Journal of Protistology, 28*, 102–119.

Hiller, S. A. (1993a). Ultrastructure of *Prorodon* (Ciliophora, Prostomatida). I. Somatic cortex and some implications concerning kinetid evolution in prostomatid and colpodid ciliates. *Journal of Eukaryotic Microbiology, 40*, 467–486.

Hiller, S. A. (1993b). Ultrastructure of *Prorodon* (Ciliophora, Prostomatida). II. Oral cortex and phylogenetic conclusions. *Journal of Eukaryotic Microbiology, 40*, 486–501.

Hiller, S., & Bardele, C. F. (1988). *Prorodon aklitolophon* n. spec. and the "dorsal brush" as a character to identify subgroups in the genus *Prorodon*. *Archiv für Protistenkunde, 136*, 213–236.

Hillman, K., Lloyd, D., & Williams, A. G. (1988). Interactions between the methanogen *Methanosarcina barkeri* and rumen holotrich ciliate protozoa. *Letters in Applied Microbiology, 7*, 49–53.

Hillman, K., Williams, A. G., & Lloyd, D. (1991). Evaluation of matrices in the rumen simulation technique (RUSITEC) for the maintenance of ciliate protozoa. *Letters in Applied Microbiology, 12*, 129–132.

Hino, T., Kametaka, M., & Kandatsu, M. (1973). The cultivation of rumen oligotrich protozoa. III. White clover factors which stimulate the growth of entodinia. *Journal of General and Applied Microbiology, 19*, 397–413.

Hirazawa, N., Oshima, S. I., Hara, T., Mitsuboshi, T., & Hata, K. (2001). Antiparasitic effect of medium–chain fatty acids against the ciliate *Cryptocaryon irritans* infestation in the red sea bream *Pagrus major*. *Aquaculture, 198*, 219–228.

Hirt, R. P., Dyal, P. L., Wilkinson, M., Finlay, B. J., Roberts, D. McL., & Embley, T. M. (1995). Phylogenetic relationships among karyorelictids and heterotrichs inferred from small subunit rRNA sequences: Resolution at the base of the ciliate tree. *Molecular Phylogenetics and Evolution, 4*, 77–87.

Hitchen, E. T., & Butler, R. D. (1972). A redescription of *Rhyncheta cyclopum* Zenker (Ciliatea, Suctorida). *Journal of Protozoology, 19*, 597–601.

Hitchen, E. T., & Butler, R. D. (1973a). Ultrastructural studies of the commensal suctorian, *Choanophrya infundibulifera* Hartog. I. Tentacle structure, movement and feeding. *Zeitschrift für Zellforschung, 144*, 37–57.

Hitchen, E. T., & Butler, R. D. (1973b). Ultrastructural studies of the commensal suctorian, *Choanophrya infundibulifera* Hartog. II. Tentacle morphogenesis. *Zeitschrift für Zellforschung, 144*, 59–73.

Hiwatashi, K. (1981). Sexual interactions of the cell surface in *Paramecium*. In D. H. O'Day & P. A. Horgen (Eds.), *Sexual interactions in eukaryotic microbes* (pp. 351–378). New York: Academic Press.

Hofer, R., Salvenmoser, W., & Fried, J. (2005). Population dynamics of *Capriniana piscium* (Ciliophora, Suctoria) on the gill surface of Arctic char (*Salvelinus alpinus*) from high mountain lakes. *Archiv für Hydrobiologie, 162*, 99–109.

Hofmann, A. H. (1987). Stomatogenesis in cyrtophorid ciliates. II. *Chilodonella cyprini* (Moroff, 1902): The kinetofragment as an anlagen–complex. *European Journal of Protistology, 23*, 165–184.

Hofmann, A. H., & Bardele, C. F. (1987). Stomatogenesis in cyrtophorid ciliates. I. *Trithigmostoma steini* (Blochmann, 1895): From somatic kineties to oral kineties. *European Journal of Protistology, 23*, 2–17.

Hoffman, D. C., & Prescott, D. M. (1997). Phylogenetic relationships among hypotrichous ciliates determined with the macronuclear gene encoding the large, catalytic subunit of DNA polymerase α. *Journal of Molecular Evolution, 45*, 301–310.

Hoffman, G. L. (1988). Ciliates of freshwater fishes. In P. Kreier (Ed.), *Parasitic protozoa* (Vol. 2, pp. 583–632). New York: Academic Press.

Hoffman, G. L., Landolt, M., Camper, J. E., Coats, D. W., Stookey, J. L., & Burek, J. D. (1975). A disease of freshwater fishes caused by *Tetrahymena corlissi* Thompson, 1955, and a key for identification of holotrich ciliates of freshwater fishes. *Journal of Parasitology*, *61*, 217–223.

Hofmann–Münz, A. H. (1991). The oral apparatus of *Colpoda variabilis* (Ciliophora, Colpodidae). II. Ultrastructure of the oral ciliature and its implications on ciliate phylogeny. *European Journal of Protistology*, *26*, 288–302.

Hogan, D. J., Hewitt, E. A., Orr, K. E., Prescott, D. M., & Müller, K. (2001). Evolution of IESs and scrambling in the actin I gene in hypotrichous ciliates. *Proceedings of the National Academy of Sciences of the United States of America*, *98*, 15101–15106.

Holler, S., Gälle, A., & Pfennig, N. (1994). Degradation of food compounds and growth response on different food quality by the anaerobic ciliate *Trimyema compressum*. *Archives of Microbiology*, *161*, 94–98.

Holt, P. A. (1972). An electron microscope study of the rhabdophorine ciliate *Didinium nasutum* during excystment. *Transactions of the American Microscopical Society*, *91*, 144–168.

Holt, P. A., & Chapman, G. B. (1971). The fine structure of the cyst wall of the ciliated protozoon *Didinium nasutum*. *Journal of Protozoology*, *18*, 604–614.

Holt, P. A., Lynn, D. H., & Corliss, J. O. (1974). An ultrastructural study of the tentacle–bearing ciliate *Actinobolina smalli* n. sp. and its systematic and phylogenetic implications. *Protistologica*, *9*(year 1973), 521–541.

Hori, M., Tomikawa, I., Przybos, E., & Fujishima, M. (2006). Comparison of the evolutionary distances among syngens and sibling species of *Paramecium*. *Molecular Phylogenetics and Evolution*, *38*, 697–704.

Horowitz, S., & Gorovsky, M. A. (1985). An unusual genetic code in nuclear genes of *Tetrahymena*. *Proceedings of the National Academy of Sciences of the United States of America*, *82*, 2452–2455.

Horwath, L., Lang, M., & Tamas, G. (1978). The use of copperoxychloride during larval growth as a preventative measure against the spread of Ciliata–exoparasites. *Bulletin of Fish Culture in Israel*, *30*, 80–84.

Hosein, R. E., Williams, S. A., & Gavin, R. H. (2005). Directed motility of phagosomes in *Tetrahymena thermophila* requires actin and Myo1p, a novel unconventional myosin. *Cell Motility and the Cytoskeleton*, *61*, 49–60.

Hoshino, R., Hayashi, S., & Imamura, N. (2006). Intraspecific genetic divergence of *Paramecium bursaria* and reconstruction of the paramecian phylogenetic tree. *Acta Protozoologica*, *45*, 377–386.

Hovasse, R. (1950). *Spirobütschliella chattoni* nov. gen., nov. sp., cilié astome parasite en Méditerranée de serpulien *Potamoceros triqueter* L., et parasité par la microsporidie *Gurleya nova*, sp. nov. *Bulletin de l'Institut Océanographique*, *962*, 1–10.

Hovasse, R. (1984a). Les associations symbiotiques. In P. P. Grassé (Ed.), *Traité de zoologie*: *Vol. 2, Fasc. 1. Infusoires ciliés* (pp. 311–343). Paris: Masson.

Hovasse, R. (1984b). Les associations parasitaires. In P. P. Grassé (Ed.), *Traité de zoologie*: *Vol. 2, Fasc. 1. Infusoires ciliés* (pp. 345–357). Paris: Masson.

Hu, X., & Song, W. (2000). Morphology and morphogenesis of a marine ciliate, *Gastrostyla pulchra* (Perejaslawzewa, 1885) Kahl, 1932 (Ciliophora, Hypotrichida). *European Journal of Protistology*, *36*, 201–210.

Hu, X., & Song, W. (2001a). Morphology and morphogenesis of *Holosticha heterofoissneri* nov. spec. from the Yellow Sea (Ciliophora, Hypotrichida). *Hydrobiologia*, *448*, 171–179.

Hu, X., & Song, W. (2001b). Morphological redescription and morphogenesis of the marine ciliate, *Pseudokeronopsis rubra* (Ciliophora: Hypotrichida). *Acta Protozoologica*, *40*, 107–115.

Hu, X., & Song, W. (2001c). Description of *Zoothamnium chlamydis* sp. n. (Protozoa: Ciliophora: Peritrichida), an ectocommensal peritrichous ciliate from cultured scallop in North China. *Acta Protozoologica*, *40*, 215–220.

Hu, X., & Song, W. (2002). Studies on the ectocommensal ciliate, *Trachelostyla tani* nov. spec. (Protozoa: Ciliophora: Hypotrichida) from the mantle cavity of the scallop *Chlamys farrei*. *Hydrobiologia*, *481*, 173–179.

Hu, X., Warren, A., & Song, W. (2004). Observations on the morphology and morphogenesis of a new marine hypotrich ciliate (Ciliophora, Hypotrichida) from China. *Journal of Natural History*, *38*, 1059–1069.

Huang, B., & Mazia, D. (1975). Microtubules and filaments in ciliate contractility. In S. lnoué & R. E. Stephens (Eds.), *Molecules and cell movement* (pp. 389–409). New York: Raven Press.

Huang, B., & Pitelka, D. R. (1973). The contractile process in the ciliate, *Stentor coeruleus*. I. The role of microtubules and filaments. *Journal of Cell Biology*, *57*, 704–728.

Huang, B., Piperno, G., & Luck, D. J. L. (1979). Paralyzed flagella mutants of *Chlamydomonas reinhardtii*. Defective for axonemal doublet microtubule arms. *Journal of Biological Chemistry*, *254*, 3091–3099.

Hudson, D. A., & Lester, R. J. G. (1992). Relationships between water quality parameters and ectocommensal ciliates on prawns (*Penaeus japonicus* Bate) in aquaculture. *Aquaculture*, *105*, 269–280.

Hufnagel, L. A. (1992). Cortical ultrastructure and chemoreception in ciliated protists (Ciliophora). *Microscopy Research and Technique*, *22*, 225–264.

Hufschmid, J. D. (1983). Étude de l'appareil nucléaire du cilié hétérotriche *Climacostomum virens. Revue Suisse de Zoologie*, *90*, 817–828.

Hufschmid, J. D. (1984). Bacterial endosymbionts 'theta' of the heterotrich ciliate *Climacostomum virens. Experientia*, *40*, 1161–1163.

Hull, R. W. (1954). Feeding processes in *Solenophrya micraster* Penard, 1914. *Journal of Protozoology*, *1*, 178–182.

Hunt, G. W., & Chein, S. M. (1983). Seasonal distribution, composition and abundance of the planktonic Ciliata and Testacea of Cayuga Lake. *Hydrobiologia*, *98*, 257–266.

Huttenlauch, I. (1985). SEM study of the skeletal plates of *Coleps nolandi* Kahl, 1930. *Prostistologica*, *21*, 499–503.

Huttenlauch, I. (1987). Ultrastructural aspects of the somatic and buccal infraciliature of *Coleps amphacanthus* Ehrenberg, 1833. *Protoplasma*, *136*, 191–198.

Huttenlauch, I., & Bardele, C. F. (1987). Light and electron microscopical observations on the stomatogenesis of the ciliate *Coleps amphacanthus* Ehrenberg, 1833. *Journal of Protozoology*, *34*, 183–192.

Huttenlauch, I., & Peck, R. K. (1991). The membrane skeleton of *Pseudomicrothorax*. II. Biochemical and immunological characterization of the glycosylated cytoskeletal proteins. *Journal of Cell Science*, *100*, 707–715.

Huttenlauch, I., & Stick, R. (2003). Occurrence of articulins and epiplasmins in protists. *Journal of Eukaryotic Microbiology*, *50*, 15–18.

Huttenlauch, I., Peck, R. K., & Stick, R. (1998a). Articulins and epiplasmins: Two distinct classes of cytoskeletal proteins of the membrane skeleton in protists. *Journal of Cell Science*, *111*, 3367–3378.

Huttenlauch, I., Peck, R. K., Plessmann, U., Weber, K., & Stick, R. (1998b). Characterization of two articulins, the major epiplasmic proteins comprising the membrane skeleton of the ciliate *Pseudomicrothorax. Journal of Cell Science*, *111*, 1909–1919.

Huttenlauch, I., Geisler, N., Plessmann, U., Peck, R. K., Weber, K., & Stick, R. (1995). Major epiplasmic proteins of ciliates are articulins: Cloning, recombinant expression, and structural characterization. *Journal of Cell Biology*, *130*, 1401–1412.

IIII

Iftode, F., & Fleury–Aubusson, A. (2003). Structural inheritance in *Paramecium*: Ultrastructural evidence for basal body and associated rootlets polarity transmission through binary fission. *Biology of the Cell*, *95*, 39–51.

Iftode, F., & Grain, J. (1975). L'appareil fibrillaire buccal isolé de *Turaniella vitrea* et son évolution au cours de la stomatogenèse. *Journal of Protozoology*, *22*, 88–96.

Iftode, F., Versavel, G., & Didier, P. (1970). Morphologie, morphogenèse de bipartition et ultrastructures de *Turaniella vitrea* Brodsky (cilié hyménostome péniculien). I. Structures infraciliaires et morphogenèse. *Protistologica*, *5*(year 1969), 523–533.

Iftode, F., Fryd–Versavel, G., Wicklow, B. J., & Tuffrau, M. (1983). Le genre *Transitella*: stomatogenèse, ultrastructure. Affinités de la famille des Transitellidae. *Protistologica*, *19*, 21–39.

Iftode, F., Fryd–Versavel, G., & Lynn, D. H. (1984). New details of the oral structures of *Colpidium* and *Turaniella* and transfer of the genus *Colpidium* to the Turaniellidae Didier, 1971 (Tetrahymenina, Hymenostomatida). *Protistologica*, *20*, 463–474.

Iftode, F., Cohen, J., Ruiz, F., Torres, A., Chen–Shan, L., Adoutte, A., et al. (1989). Development of surface pattern during division of *Paramecium*. I. Mapping of duplication and reorganization of cortical cytoskeletal structures in wild–type. *Development*, *105*, 191–211.

Iglesias, R., Paramá, A., Alvarez, M. F., Leiro, J., Fernández, J., & Sanmartín, M. L. (2001). *Philasterides dicentrarchi* (Ciliophora, Scuticociliatida) as the causative agent of scuticociliatosis in farmed turbot *Scophthalmus maximus* in Galicia (NW Spain). *Diseases of Aquatic Organisms*, *46*, 47–55.

Iglesias, R., Paramá, A., Alvarez, M. F., Leiro, J., & Sanmartín, M. L. (2002). Antiprotozoals effective in vitro against the scuticociliate fish pathogen *Philasterides dicentrarchi. Diseases of Aquatic Organisms*, *49*, 191–197.

Iglesias, R., Paramá, A., Alvarez, M. F., Leiro, J., Aja, C., & Sanmartín, M. L. (2003). In vitro growth requirements for the fish pathogen *Philasterides dicentrarchi* (Ciliophora, Scuticociliatida). *Veterinary Parasitology*, *111*, 19–30.

Ike, K., Imai, S., & Ishii, T. (1985). Establishment of intestinal ciliates in new–born horses. *Japanese Journal of Veterinary Science*, *47*, 39–43.

Imai, S. (1979). A new suctorian ciliate *Allantosoma japonensis* sp. nov., from the large intestine of the horse. *Proceedings of the Japanese Society of Systematic Zoology*, *16*, 1–4.

Imai, S. (1998). Phylogenetic taxonomy of rumen ciliate protozoa based on their morphology and distribution. *Journal of Applied Animal Research*, *13*, 17–36.

Imai, S., & Ogimoto, K. (1978). Scanning electron and fluorescent microscopic studies on the attachment of spherical bacteria to ciliate protozoa in the ovine rumen. *Japanese Journal of Veterinary Science*, *40*, 9–19.

Imai, S., & Rung, G. (1990a). Rumen ciliates from the Mongolian gazelle, *Procarpa gutturosa*. *Japanese Journal of Veterinary Science*, *52*, 1063–1067.

Imai, S., & Rung, G. (1990b). Ciliate protozoa in the forestomach of the Bactrian camel in Inner–Mongolia, China. *Japan Journal of Veterinary Science*, *52*, 1069–1075.

Imai, S., & Yamazaki, Y. (1988). Scanning electron microscopy of the adoral zone of *Cycloposthium* Bundle (Ciliophora, Entodiniomorphida). *Journal of Protozoology*, *35*, 578–583.

Imai, S., Katsuno, M., & Ogimoto, K. (1979). Type of the pattern of the rumen ciliate composition of the domestic ruminants and the predator–prey interaction of the ciliates. *Japanese Journal of Zootechnology and Science*, *50*, 79–87. (in Japanese with English summary)

Imai, S., Tashiro, H., & Ishii, T. (1983). Scanning electron microscopy of the adoral ciliary zone of *Entodinium* Stein (Ciliophora, Entodiniomorphida). *Journal of Protozoology*, *30*, 466–472.

Imai, S., Han, S. S., Cheng, K. J., & Kudo, H. (1989). Composition of the rumen ciliate population in experimental herds of cattle and sheep in Lethbridge, Alberta, Western Canada. *Canadian Journal of Microbiology*, *35*, 686–690.

Imai, S., Ikeda, S. I., Collet, J. Y., & Bonhomme, A. (1991). Entodiniomorphid ciliates from the wild lowland gorilla with the description of a new genus and three new species. *European Journal of Protistology*, *26*, 270–278.

Imai, S., Kudo, H., Fukuta, K., Abudullah, N., Ho, Y. W., & Onodera, R. (1995). *Isotricha jalaludinii* n. sp. found from the rumen of lesser mouse deer, *Tragulus javanicus*, in Malaysia. *Journal of Eukaryotic Microbiology*, *42*, 75–77.

Imai, S., Tsurimaki, S., Goto, E., Wakita, K., & Hatai, K. (2000). *Tetrahymena* infection in guppy, *Poecilia reticulata*. *Fish Pathology*, *35*, 67–72.

Inaba, F., Imamoto, K., & Suganuma, Y. (1966). Electron–microscopic observations on nuclear exchange during conjugation in *Paramecium multimicronucleatum*. *Proceedings of the Japan Academy*, *42*, 394–398.

Inagaki, Y., & Doolittle, W. F. (2001). Class I release factors in ciliates with variant genetic codes. *Nucleic Acids Research*, *29*, 921–927.

International Commission on Zoological Nomenclature. (1970). Opinion 915. *Tetrahymena* Furgason, 1940 (Ciliophora, Hymenostomatida): Preserved under the plenary powers and related matters. *Bulletin of Zoological Nomenclature*, *27*, 33–38.

International Commission on Zoological Nomenclature. (1999a). *International Code of Zoological Nomenclature* (4th ed., 306 p.), adopted by the International Union of Biological Sciences. London: International Trust for Zoological Nomenclature.

International Commission on Zoological Nomenclature. (1999b). Opinion 1923. *Trachelocerca* Ehrenberg (Ciliophora): Authorship conserved as Ehrenberg (1840), and *Vibrio sagitta* Müller, 1786 fixed as the type species. *Bulletin of Zoological Nomenclature*, *56*, 150–151.

Iriberri, J., Ayo, B., Santamaria, E., Barcina, I., & Egea, L. (1995). Influence of bacterial density and water temperature on the grazing activity of two freshwater ciliates. *Freshwater Biology*, *33*, 223–231.

Ishida, M., Suzaki, T., & Shigenaka, Y. (1991a). Studies on the mechanism of cell elongation in *Blepharisma japonicum*. 4. Three dimensional construction of the kinetosomal complex and its functional role on cell elongation. *E uropean Journal of Protistology*, *27*, 46–54.

Ishida, M., Shigenaka, Y., Suzaki, T., & Sugiyama, M. (1991b). Studies on the mechanism of cell elongation in *Blepharisma japonicum*. 3. Cytoplasmic calcium ions may correlate to cell elongation in calmodulin–dependent manner. *European Journal of Protistology*, *26*, 365–369.

Ishida, M., Suzaki, T., Shigenaka, Y., & Sugiyama, M. (1992). Ultrastructure and calcium–dependent contraction of the myonemal network in a heterotrich ciliate, *Blepharisma japonicum*. *European Journal of Protistology*, *28*, 279–287.

Ishida, M., Allen, R. D., & Fok, A. K. (2001). Phagosome formation in *Paramecium*: Roles of somatic and oral cilia and of solid particles as revealed by video microscopy. *Journal of Eukaryotic Microbiology*, *48*, 640–646.

Israel, R. L., Pond, S. L. K., Muse, S. V., & Katz, L. A. (2002). Evolution of duplicated alpha–tubulin genes in ciliates. *Evolution*, *56*, 1110–1122.

Itabashi, T., Mikami, K., Fang, J., & Asai, H. (2002). Phylogenetic relationships between *Vorticella convallaria* and other species inferred from small subunit rRNA gene sequences. *Zoological Science*, *19*, 931–937.

Ito, A., & Imai, S. (1998). Infraciliary bands in the rumen ophryoscolecid ciliate *Ostracodinium gracile* (Dogiel, 1925), observed by light microscopy. *Journal of Eukaryotic Microbiology*, *45*, 628–636.

Ito, A., & Imai, S. (2000a). Ciliates from the cecum of capybara (*Hydrocheorus hydrochaeris*) in Bolivia. 1. The families Hydrochoerellidae n. fam., Protohalliidae, and Pcynotrichidae. *European Journal of Protistology*, *36*, 53–84.

Ito, A., & Imai, S. (2000b). Ciliates from the cecum of capybara (*Hydrocheorus hydrochaeris*) in Bolivia.

2. The family Cycloposthiidae. *European Journal of Protistology*, *36*, 169–200.

Ito, A., & Imai, S. (2005). Infraciliature and morphogenesis in three rumen *Diplodinium* ciliates, *Diplodinium polygonale*, *Diplodinium leche*, and *Diplodinium nanum*, observed by light microscopy. *Journal of Eukaryotic Microbiology*, *52*, 44–51.

Ito, A., Imai, S., & Ogimoto, K. (1993). Rumen ciliates of ezo deer (*Cervus nippon yesoensis*) with the morphological comparison with those of cattle. *Journal of Veterinary Medical Science, 55,* 93–98.

Ito, A., Miyazaki, Y., & Imai, S. (2001). Light microscopic observations of infraciliature and morphogenesis in six species of rumen *Ostracodinium* ciliates. *Journal of Eukaryotic Microbiology*, *48*, 440–448.

Ivan, M., Neill, L., & Entz, T. (2000). Ruminal fermentation and duodenal flow following progressive inoculations of fauna–free wethers with major individual species of ciliate protozoa or total fauna. *Journal of Animal Science*, *78*, 750–759.

Iwadate, Y., Katoh, K., Kikuyama, M., & Asai, H. (1999). Ca^{2+} triggers toxicyst discharge in *Didinium nasutum*. *Protoplasma*, *206*, 20–26.

Izquierdo, A., Martín–González, A., Diaz, S., & Gutiérrez, J. C. (1999). Resting cyst wall proteins and glycoproteins of four colpodid ciliates: A comparative analysis. *Microbios*, *99*, 27–43.

JJJJ

Jack, J. D., & Gilbert, J. J. (1993). Susceptibilities of different–sized ciliates to direct suppression by small and large cladocerans. *Freshwater Biology*, *29*, 19–29.

Jackson, K., & Berger, J. (1985a). Life history attributes of some ciliated protozoa. *Transactions of the American Microscopical Society*, *104*, 52–63.

Jackson, K., & Berger, J. (1985b). Survivorship curves of ciliate protozoa under starvation conditions and at low bacterial levels. *Protistologica*, *21*, 17–24.

Jacobson, P. M., & Lynn, D. H. (1992). Ultrastructure and cell size–dependent regulation of the adoral zone of membranelles in the heterotrich ciliate *Stentor coeruleus*. *Developmental Genetics*, *13*, 203–209.

Jahn, C. L., & Klobutcher, L. A. (2002). Genome remodelling in ciliated protozoa. *Annual Review of Microbiology*, *56*, 489–520.

Jahn, C. L., Nilles, L. A., & Krikau, M. F. (1988a). Organization of the *Euplotes crassus* micronuclear genome. *Journal of Protozoology*, *35*, 590–601.

Jahn, C. L., Prescott, K. E., & Waggener, M. W. (1988b). Organization of the micronuclear genome of *Oxytricha nova*. *Genetics*, *120*, 123–134.

Jahn, C. L., Krikau, M. F., & Shyman, S. (1989). Developmentally coordinated en masse excision of a highly repetitive element in *E. crassus*. *Cell*, *59*, 1009–1018.

Jakobsen, H. H. (2001). Escape response of planktonic protists to fluid mechanical signals. *Marine Ecology–Progress Series*, *214*, 67–78.

Jakobsen, H. H. (2002). Escape of protists in predator–generated feeding currents. *Aquatic Microbial Ecology*, *26*, 271–281.

Jakobsen, H. H., & Montagnes, D. J. S. (1999). A redescription of *Balanion comatum* Wulff, 1919 (Prorodontida, Ciliophora), with notes on its cultivation and behaviour. *Journal of Eukaryotic Microbiology*, *46*, 198–205.

Jakobsen, H. H., Everett, L. M., & Strom, S. L. (2006). Hydromechanical signalling between the ciliate *Mesodinium pulex* and motile protist prey. *Aquatic Microbial Ecology*, *44*, 197–206.

James, M. R., Burns, C. W., & Forsyth, D. J. (1995). Pelagic ciliated protozoa in two monomictic, southern temperate lakes of contrasting trophic state: Seasonal distribution and abundance. *Journal Plankton Research*, *17*, 1479–1500.

Janisch, R. (1980). A freeze–etch study of the ultrastructure of *Colpoda cucullus* protective cysts. *Acta Protozoologica*, *19*, 239–246.

Jankowski, A. W. (1963). Pathology of Ciliophora. II. Life cycles of suctoria parasiting in *Urostyla* and *Paramecium*. *Tsitologiya*, *5*, 428–439. (in Russian with English summary)

Jankowski, A. W. (1964a). Morphology and evolution of Ciliophora. I. The new system of sapropelebiotic Heterotrichida. *Zoologichesky Zhurnal*, *43*, 503–517. (in Russian with English summary)

Jankowski, A. W. (1964b). Morphology and evolution of Ciliophora. III. Diagnoses and phylogenesis of 53 sapropelebionts, mainly of the order Heterotrichida. *Archiv für Protistenkunde*, *107*, 185–294.

Jankowski, A. W. (1966a). Morphology and evolution of the Ciliophora. VI. Morphology and development cycle of *Conidophrys enkystotrophos* sp. nov. in connection with phylogenetic and systematic problems of the family Conidophryidae Guilcher 1951. *Materialy IV Konferencii uchenyh Moldavii Zoologia Kishinev*, 123–129. (in Russian)

Jankowski, A. W. (1966b). The origin of the order Apostomatida Chatton and Lwoff. In *Problems of sea biology: Abstracts of symposium of young scientists* (pp. 137–140). Kiev: Naukova Dumka. (in Russian)

Jankowski, A. W. (1967a). A new system of ciliate Protozoa (Ciliophora). *Akademii Nauk SSSR, Trudy Zoologicheskogo Instituta*, *43*, 3–54. (in Russian)

Jankowski, A. W. (1967b). The boundaries and composition of the genera *Tetrahymena* and *Colpidium*. *Zoologichesky Zhurnal*, *46*, 17–23. (in Russian with English summary)

Jankowski, A. W. (1967c). Ciliates from the mantle cavity of *Balanus* in Barents Sea. *Parazitologia, 1*, 82–93. (in Russian with English summary)

Jankowski, A. W. (1968a). Taxonomy of the suborder Nassulina Jank., 1967 (Ciliophora, Ambihymenida). *Zoologichesky Zhurnal, 47*, 990–1001. (in Russian with English summary)

Jankowski, A. W. (1968b). Morphology and phylogeny of *Nyctotherus cordiformis*. *Parazitologia, 2*, 261–271. (in Russian with English summary)

Jankowski, A. W. (1972a). Cytogenetics of *Paramecium putrinum* C. et L., 1858. *Acta Protozoologica, 10*, 285–394.

Jankowski, A. W. (1972b). Commensological sketches. 3. New genera of Chonotricha Exogemmina symbiotic with Isopoda, Amphipoda and Decapoda. *Zoologichesky Zhurnal, 51*, 1617–1624. (in Russian with English summary)

Jankowski, A. W. (1973a). Commensological sketches. 4. New genera of Chonotricha Endogemmina symbiotic with Leptostraca. *Zoologichesky Zhurnal, 52*, 15–24. (in Russian with English summary)

Jankowski, A. W. (1973b). *Fauna of the USSR: Infusoria Subclass Chonotricha*. (Vol. 2, No. 1, 355 pp.). Leningrad: Akademii Nauk SSSR, Nauka. (in Russian)

Jankowski, A. W. (1973c). Taxonomic revision of subphylum Ciliophora Doflein, 1901. *Zoologichesky Zhurnal, 52*, 165–175. (in Russian with English summary)

Jankowski, A. W. (1973d). Suctorian infection in *Bursaria truncatella* O. F. M. *Parazitologia, 7*, 401–407. (in Russian with English summary)

Jankowski, A. W. (1974a). Parasitic Ciliophora. 2. *Ichthyonyctus* gen. n. from the intestine of tropical fishes. *Zoologichesky Zhurnal, 53*, 278–279. (in Russian with English summary)

Jankowski, A. W. (1974b). Commensological sketches. 6. Ectocommensals of *Strongylocentrotus intermedius* in Busse Lagoon, Southern Sakhalin. *Hydrobiologichesky Zhurnal, 10*, 60–68. (in Russian with English summary)

Jankowski, A. W. (1975). A conspectus of the new system of subphylum Ciliophora Doflein, 1901 [Abstract]. In U. S. Balashov (Ed.), *Account of scientific sessions on results of scientific work, Year 1974: Abstracts of reports* (pp. 26–27). Leningrad: Akademii Nauk SSSR, Zoologicheskogo Instituta. (in Russian)

Jankowski, A. W. (1978). [Revision of a system of class Polyhymenophora (Spirotricha)]. *Tezisky Doklady Zoologischeski Institut Akademia NAUK SSSR, 1978*, 39–40. (in Russian)

Jankowski, A. W. (1979). Revision of the order Hypotrichida Stein, 1859 (Protozoa, Ciliophora). Generic catalogue, phylogeny, taxonomy. *Trudy Zoologischeski Institute, Leningrad, 86*, 48–85. (in Russian with English summary)

Jankowski, A. W. (1980). Conspectus of a new system of the phylum Ciliophora. *Trudy Zoologicheskogo Instituta Leningrad, 94*, 103–121. (in Russian)

Jankowski, A. W. (1981). New species, genera and families of tentacled Infusoria (Class Suctoria). *Trudy Zoologicheskogo Instituta, Leningrad, 107*, 80–115. (in Russian)

Jankowski, A. W. (1985). Life cycles and taxonomy of generic groups *Scyphidia*, *Heteropolaria*, *Zoothamnium* and *Cothurnia* (class Peritricha). *Trudy Zoologicheskogo Instituta Leningrad, 129*, 74–100, 104. (in Russian with English summary)

Jankowski, A. W. (1986). New and little known genera of ciliated protozoa (Phylum Ciliophora). *Trudy Zoologicheskogo Instituta Leningrad, 144*, 72–88.

Jankowski, A. W. (1997). *Grainis* gen. n. allied to *Ruthiella*, a symbiont of endemic gammarids in Lake Baikal (Ciliophora: Peritricha). *Zoosystematica Rossica, 6*, 1–5.

Jankowski, A. W. (2007). [Phylum Ciliophora Doflein, 1901.] *In*: Alimov, A. F. (Ed.), *Protista. Part 2, Handbook on Zoology* (pp. 415–993). St. Petersburg: Russian Academy of Sciences, Zoological Institute. (in Russian with English summary).

Janssen, M. P. M., Oosterhoff, C., Heijmans, G. J. S. M., & Van der Voet, H. (1995). The toxicity of metal salts and the population growth of the ciliate *Colpoda cucullus*. *Bulletin of Environmental Contamination and Toxicology, 54*, 597–605.

Jareño, M. A. (1976). Une technique simple pour obtenir des montages permanents de chromosomes polytènes de *Stylonychia mytilus*. *Protistologica, 12*, 593–597.

Jee, B. Y., Kim, K. H., Park, S. I., & Kim, Y. C. (2000). A new strain of *Cryptocaryon irritans* from the cultured olive flounder *Paralichthys olivaceus*. *Diseases of Aquatic Organisms, 43*, 211–215.

Jee, B. Y., Kim, Y. C., & Park, M. S. (2001). Morphology and biology of parasite responsible for scuticociliatosis of cultured olive flounder *Paralichthys olivaceus*. *Diseases of Aquatic Organisms, 47*, 49–55.

Jenkins, R. A. (1973). Fine structure. Ch. 3, pp. 39–93. In A. C. Giese (Ed.), *Blepharisma. The biology of a light-sensitive protozoan* (chap. 3, pp. 39–93). Stanford: Stanford University Press.

Jennings, H. S. (1938). Sex reaction types and their interrelation in *Paramecium bursaria*. I. *Proceedings of the National Academy of Sciences of the United States of America, 24*, 112–117.

Jeong, H. J., Yoon, J. Y., Kim, J. S., Yoo, Y. D., & Seong, K. A. (2002). Growth and grazing rates of the prostomatid ciliate *Tiarina fusus* on red-tide and toxic algae. *Aquatic Microbial Ecology, 28*, 289–297.

Jeong, H. J., Yoo, Y. D., Kim, J. S., Kang, N. S., Kim, T. H., & Kim, J. H. (2004). Feeding by the marine protozoan *Tokophrya infusionum*. *Mechanisms of Ageing and Development, 26,* 217–229.

Karatayev, A. Y., Burlakova, L. E., Molloy, D. E., Volkova, L. K., & Volosyuk, V. V. (2002). Field and laboratory studies of *Ophryoglena* sp. (Ciliata: Ophryoglenidae) infection in zebra mussels, *Dreissena polymorpha* (Bivalvia: Dreissenidae). *Journal of Invertebrate Pathology, 79,* 80–85.

Karpenko, A. A., Railkin, A. I., & Seravin, L. N. (1977). Feeding behaviour of unicellular animals. II. The role of prey mobility in the feeding behaviour of protozoa. *Acta Protozoologica, 16,* 333–344.

Karpov, A. S., Goodkov, A. V., & Marinich, M. A. (1991). The variety of algae–bearing invertebrates from inner lakes of the Valamo Island. *Zoologichesky Zhurnal, 70,* 5–11. (in Russian with English summary)

Kasai, H., Osawa, S., Kobayashi, T., & Yoshimizu, M. (2002). Prevention of scuticociliatosis in Japanese flounder by treatment of water–supply with a high quality UV lamp. *Fish Pathology, 37,* 199–200.

Katashima, R. (1959). Mating types in *Euplotes eurystomus*. *Journal of Protozoology, 6,* 75–83.

Katashima, R. (1960). Correlations between nuclear behavior and disposition of the nuclei in cytoplasm during conjugatin of *Euplotes patella*. *Journal of Science of the Hiroshima University, Series B, Division 1, 18,* 239–263.

Kato, S., & Taniguchi, A. (1993). Tintinnid ciliates as indicator species of different water masses in the western North Pacific Polar Front. *Fisheries Oceanography, 2,* 166–174.

Katz, L. A. (2001). Evolution of nuclear dualism in ciliates: A reanalysis in light of recent molecular data. *International Journal of Systematic and Evolutionary Microbiology, 51,* 1587–1592.

Katz, L. A., Lasek–Nesselquist, E., & Snoeyenbos–West, O. L. O. (2003). Structure of the micronuclear (α–tubulin gene in the phyllopharyngean ciliate *Chilodonella uncinata*: Implications for the evolution of chromosomal processing. *Gene, 315,* 15–19.

Katz, L. A., Bornstein, J. G., Lasek–Nesselquist, E., & Muse, S. V. (2004). Dramatic diversity of ciliate histone H4 genes revealed by comparisons of patterns of substitutions and paralog divergences among eukaryotes. *Molecular Biology and Evolution, 21,* 555–562.

Katz, L. A., McManus, G. B., Snoeyenbos–West, O. L. O., Griffin, A., Pirog, K., Costas, B., et al. (2005). Reframing the 'Everything is everywhere' debate: Evidence for high gene flow and diversity in ciliate morphospecies. *Aquatic Microbial Ecology, 41,* 55–65.

Katz, L. A., Snoeyenbos–West, O., & Doerder, F. P. (2006). Patterns of protein evolution in *Tetrahymena thermophila*: Implications for estimates of effective population size. *Molecular Biology and Evolution, 23,* 608–614.

Katzen, A. L., Cann, G. M., & Blackburn, E. H. (1981). Sequence specific fragmentation of macronuclear DNA in a holotrichous ciliate. *Cell, 24,* 313–320.

Kaul, N., & Sapra, G. R. (1983). Feeding behaviour and digestion pattern in *Stylonychia mytilus* Ehrenberg. *Archiv für Protistenkunde, 127,* 167–180.

Kawakami, H. (1984). Ultrastructural study of an endosymbiotic alga and its host ciliate *Stentor niger*. *Journal of Protozoology, 31,* 247–253.

Kawakami, H. (1991). An endosymbiotic *Chlorella*–bearing ciliate: *Platyophrya chlorelligera* Kawakami, 1989. *European Journal of Protistology, 26,* 245–255.

Kawakami, H., & Yagiu, R. (1964). The electron microscopical study of the change of fine structures in the ciliate, *Colpoda cucullus*, during its life cycle. VI. The abandonment of the third layer of resting cyst membrane during the stage of excystment. *Zoological Magazine, Tokyo, 73,* 112–117. (in Japanese with English summary)

Kazubski, S. L. (1958). *Thigmocoma acuminata* gen. nov., sp. nov. (Thigmotricha–Thigmocomidae fam. nov.) a parasite of the renal organ of *Schistophallus orientalis* Cless. (Pulmonata–Zonitidae). *Bulletin de l'Académie Polonaise des Sciences, 6,* 167–172.

Kazubski, S. L. (1963). Studies on the parasitic ciliate *Thigmocoma acuminata* Kazubski (Thigmotricha–Thigmocomidae). *Acta Protozoologica, 1,* 237–278.

Kazubski, S. L. (1964). Contribution to the study of the fauna of parasitic ciliates in the land snail *Poiretia algira* (Brug.) from Yugoslavia. *Acta Protozoologica, 2,* 159–162.

Kazubski, S. L. (1967). Study on the growth of skeletal elements in *Trichodina pediculus* Ehrbg. *Acta Protozoologica, 5,* 37–48.

Kazubski, S. L. (1979). Morphological variability of *Trichodina vesicularum* Fauré–Fremiet and *T. faurefremieti* Kazubski, parasites of newts from Poland and France. *Acta Protozoologica, 18,* 385–400.

Kazubski, S. L. (1980). *Trichodina ranae* (Ciliata Peritrichida) a parasite of *Rana esculenta* and its morphological variability. *Acta Protozoologica, 19,* 207–224.

Kazubski, S. L. (1981). Further investigation on morphological variability of *Semitrichodina sphaeronuclea* f. *macrodentata* (Lom) (Ciliata, Peritrichida), a parasite of land snails. *Acta Protozoologica, 20,* 385–392.

Kazubski, S. L. (1988). Morphological variation in a ciliate, *Trichodina reticulata* Hirschmann et Partsch, 1955 (Peritrichida), in tadpoles from small ponds. *Acta Protozoologica, 27,* 259–269.

Kazubski, S. L. (1991). Morphological variation of the ciliate *Trichodina pediculus* Ehrenberg, 1838. III. Parasitizing on Crucian carp (*Carassius carassius* (L.)) from small ponds in Kortowo (Olsztyn). *Acta Protozoologica, 30,* 187–192.

Kazubski, S. L., & Migala, K. (1968). Urceolariidae from breeding carp – *Cyprinus cyprio* L . in Zabieniec and remarks on the seasonal variability of trichodinids. *Acta Protozoologica*, *6*, 137–160.

Kazubski, S. L., & Migala, K. (1974). Studies on the distinctness of *Chilodonella cyprini* (Moroff) and *Ch. hexasticha* (Kiernik) (Chlamydodontidae, Gymnostomatida), ciliate parasites of fishes. *Acta Protozoologica*, *13*, 9–40.

Kent, E. B. (1980). Effect of food level on reproduction and metamorphosis in the suctorian protozoon *Tokophrya lemnarum* Stein. *Microbial Ecology*, *6*, 153–159.

Kent, E. B. (1981). Life history responses to resource variation in a sessile predator, the ciliate protozoan *Tokophrya lemnarum* Stein. *Ecology*, *62*, 296–302.

Kent, W. S. (1880–1882). *A manual of the Infusoria* (Vols. 1–3, 913 pp.). London: David Bogue.

Kepner, R. L., Jr., Wharton, R. A., Jr., & Coats, D. W. (1999). Ciliated protozoa in two antarctic lakes: Analysis by quantitative protargol staining and examination of artificial substrates. *Polar Biology*, *21*, 285–294.

Kerr, S. J. (1983). Colonization of blue–green algae by *Vorticella* (Ciliata: Peritrichida). *Transactions of the American Microscopical Society*, *102*, 38–47.

Kersken, H., Tiggemann, R., Westphal, C., & Plattner, H. (1984). The secretory contents of *Paramecium tetraurelia* trichocysts: Ultrastructural-cytochemical characterization. *Journal of Histochemistry and Cytochemistry, 32,* 179–192.

Keryer, G., Adoutte, A., Ng, S. F., Cohen, J., Garreau de Loubresse, N., Rossignol, M., et al. (1990). Purification of the surface membrane–cytoskeleton complex (cortex) of *Paramecium* and identification of several of its protein constituents. *European Journal of Protistology*, *25*, 209–225.

Khan, M. M., & Ip, Y. K. (1986). Parasites of toads from Singapore, with a description of *Balantidium singaporensis* sp. n. (Ciliophora: Balantidiidae). *Zoological Science*, *3*, 543–546.

Kidder, G. W., & Dewey, V. C. (1951). The biochemistry of ciliates in pure culture. In A. Lwoff (Ed.), *Biochemistry and physiology of protozoa* (Vol. 1, pp. 323–400). New York: Academic Press..

Kiersnowksa, M., Peck, R. K., & de Haller, G. (1988). Cell to cell recognition between the ciliate *Pseudomicrothorax dubius* and its food organisms: The role of surface charges. *Protoplasma*, *143*, 93–100.

Kim, O. T. P., Yura, K., Go, N., & Harumoto, T. (2004). Highly divergent actins from karyorelictean, heterotrich, and litostome ciliates. *Journal of Eukaryotic Microbiology*, *51*, 227–233.

Kim, Y. O., & Taniguchi, A. (1995). Excystment of the oligotrich ciliate *Strombidium conicum. Aquatic Microbial Ecology*, *9*, 149–156.

Kim, Y. O., & Taniguchi, A. (1997). Seasonal variation of excytment pattern of the planktonic oligotrich ciliate *Strombidium conicum. Marine Biology*, *128*, 207–212.

Kim, Y. O., Suzuki, T., & Taniguchi, A. (2002). A new species in the genus *Cyrtostrombidium* (Ciliophora, Oligotrichia, Oligotrichida): Its morphology, seasonal cycle and resting stage. *Journal of Eukaryotic Microbiology*, *49*, 338–343.

Kimball, R. F. (1942). The nature and inheritance of mating types in *Euplotes patella. Genetics*, *27*, 269–285.

Kink, J. (1972). Observations on morphology and oral morphogenesis during regeneration of ciliate *Lacrymaria olor* (O.F.M. 1786) (Holotricha, Gymnostomatida). *Acta Protozoologica*, *10*, 205–214.

Kink, J. (1976). A localized region of basal body proliferation in growing cells of *Dileptus visscheri* (Ciliata, Gymnostomata). *Journal of Cell Science*, *20*, 115–133.

Kink, J. (1978). Dedifferentiation of fibrillar structures during encystment of *Dileptus visscheri* (Gymnostomatida). *Acta Protozoologica*, *17*, 31–45.

Kirby, H. (1950). *Materials and methods in the study of protozoa* (72 pp.). Berkeley, CA: University of California Press.

Kitajima, Y., & Thompson, G. A., Jr. (1977). *Tetrahymena* strives to maintain the fluidity interrelationships of all its membranes constant. Electron microscope evidence. *Journal of Cell Biology*, *72*, 744–755.

Kitching, J. A. (1967). Contractile vacuoles, ionic regulation, and excretion. In T. T. Chen (Ed.), *Research in protozoology* Vol. 1, pp. 307–336). London/New York: Pergamon Press.

Kivimaki, K. L., Riordan, G. P., & Lipscomb, D. (1997). The ultrastructure of *Zosterodasys agamalievi* (Ciliophora: Synhymeniida). *Journal of Eukaryotic Microbiology*, *44*, 226–236.

Klaas, J., II. (1974). Two new gastric mucin cultivation media and a chemically defined maintenance medium for *Balantidium coli. Journal of Parasitology*, *60*, 907–910.

Klaveness, D. (1984). Studies on the morphology, food selection and growth of two planktonic freshwater strains of *Coleps* sp. *Protistologica*, *20*, 335–349.

Klein, B. M. (1926). Ergebnisse mit einer Silbermethode bei Ciliaten. *Archiv für Protistenkunde*, *56*, 243–279.

Klein, B. M. (1928). Die Silberliniensysteme der Ciliaten. Weitere Resultate. *Archiv für Protistenkunde*, *62*, 177–260.

Klein, B. M. (1929). Weitere Beiträge zur Kenntnis des Silberliniensystems der Ciliaten. *Archiv für Protistenkunde*, *65*, 183–257.

Kleynhans, C. J. (1982). The rumen ciliates of the greater kudu *Tragelaphus strepsiceros* (Pallas) from South Africa and Zimbabwe with a description of one new species. *South African Journal of Zoology*, *17*, 11–14.

Kleynhans, C. J., & Van Hoven, W. (1976). Rumen protozoa of the giraffe with a description of two new species. *East Afrcian Wildlife Journal*, *14*, 203–214.

Klieve, A. V., & Hegarty, R. S. (1999). Opportunities for biological control of ruminal methanogenesis. *Australian Journal of Agricultural Research*, *50*, 1315–1319.

Klindworth, T., & Bardele, C. F. (1996). The ultrastructure of the somatic and oral cortex of the karyorelictean ciliate *Loxodes striatus*. *Acta Protozoologica*, *35*, 13–28.

Klobutcher, L. A. (1987). Micronuclear organization of macronuclear genes in the hypotrichous ciliate *Oxytricha nova*. *Journal of Protozoology*, *34*, 424–428.

Klobutcher, L. A., & Herrick, G. (1995). Consensus inverted terminal repeat sequence of *Paramecium* IESs: Resemblance to termini of Tc1–related and *Euplotes* Tec transposons. *Nucleic Acids Research*, *23*, 2006–2013.

Klobutcher, L. A., & Herrick, G. (1997). Developmental genome reorganization in ciliated protozoa: The transposon link. In W. E. Cohn & K. Moldave (Ed.), *Progress in nucleic acid research and molecular biology* (Vol. 56, pp. 1–62). New York: Academic Press.

Klobutcher, L. A., Swanton, M. A., Donini, P., & Prescott, D. M. (1981). All gene–sized DNA molecules in four species of hypotrichs have the same terminal sequence and an unusual 3' terminus. *Proceedings of the National Academy of Sciences of the United States of America*, *78*, 3015–3019.

Klobutcher, L. A., Jahn, C. L., & Prescott, D. M. (1984). Internal sequences are eliminated from genes during macronuclear development in the ciliated protozoan *Oxytricha nova*. *Cell*, *36*, 1045–1055.

Kloetzel, J. A. (1974). Feeding in ciliated protozoa. I. Pharyngeal disks in *Euplotes*: A source of membrane for food vacuole formation? *Journal of Cell Science*, *15*, 379–401.

Kloetzel, J. A. (1991). Identification and properties of *plateins*, major cortical proteins in the cortical alveolar plates of *Euplotes*. *Journal of Protozoology*, *38*, 392–401.

Kloetzel, J. A., Hill, B. F., & Kosaka, T. (1992). Genetic variants of plateins (alveolar plate proteins) among and within species of *Euplotes*. *Journal of Protozoology*, *39*, 92–101.

Kloetzel, J. A., Baroin–Tourancheau, A., Miceli, C., Barchetta, S., Farmar, J., Banerjee, D., et al. (2003). Cytoskeletal proteins with N–terminal signal peptides: Plateins in the ciliate *Euplotes* define a new family of articulins. *Journal of Cell Science*, *116*, 1291–1303.

Knochel, M., Kissmehl, R., Wissmann, J. D., Momayezi, M., Hentschel, J., Plattner, H., et al. (1996). Annexins in *Paramecium* cells. Involvement in site–specific positioning of secretory organelles. *Histochemistry and Cell Biology*, *105*, 269–281.

Ko, Y. G., & Thompson, G. A., Jr. (1992). Immobilization antigens from *Tetrahymena thermophila* are glycosylphosphatidylinositol–linked proteins. *Journal of Protozoology*, *39*, 719–723.

Kobayashi, T., & Endoh, H. (2003). Caspase–like activity in programmed nuclear death during conjugation of *Tetrahymena thermophila*. *Cell Death and Differentiation*, *10*, 634–640.

Koenig, K. M., Newbold, C. J., McIntosh, F. M., & Rode, L. M. (2000). Effects of protozoa on bacterial nitrogen recycling in the rumen. *Journal of Animal Science*, *78*, 2431–2445.

Kofoid, C. A., & Campbell, A. S. (1929). A conspectus of the marine and freshwater Ciliata belonging to the suborder Tintinnoinea, with descriptions of new species principally from the Agassiz expedition to the eastern tropical Pacific, 1904–1905. *University of California Publications in Zoology*, *34*, 1–403.

Kofoid, C. A., & Campbell, A. S. (1939). The Ciliata: The Tintinnoinea. Reports on the scientific results of the expedition to the eastern tropical Pacific…1904–1905. *Bulletin of the Museum Comparative Zoology Harvard*, *84*, 1–473.

Komala, Z., & Przybos, E. (1990). Studies on the distribution of species of the *Paramecium aurelia* species complex in the Western Sudetes of Poland. *Folia Biologica (Krakow)*, *38*, 35–42.

Komala, Z., & Przybos, E. (1994). *Paramecium triaurelia* and other species of the *aurelia* complex in the Kaczawskie Mts and Plateau (Western Sudeten). *Folia Biologica (Krakow)*, *42*, 27–33.

Komori, R., Sato, H., Harumoto, T., & Takagi, Y. (2005). A new mutation in the timing of autogamy in *Paramecium tetraurelia*. *Mechanisms of Ageing and Development*, *126*, 752–759.

Kopp, M., & Tollrian, R. (2003). Trophic size polyphenism in *Lembadion bullinum*: Costs and benefits of an inducible offense. *Ecology*, *84*, 641–651.

Kormos, J., & Kormos, K. (1957a). Die Entwicklungsge schichtlichen Grundlagen des Systems der Suctorien. *Acta Zoologica Hungarica*, *3*, 147–162.

Kormos, J., & Kormos, K. (1957b). Determination in der Entwicklung der Suctorien. I. Die Determination der Stelle der Embryoorganisierung. *Acta Biologica Hungarica*, *7*, 365–383.

Kormos, J., & Kormos, K. (1958). Determination in der Entwicklung der Suctorien. II. Neue Untersuchungen über determinativen Zusammenhang zwischen der Schwärmerbildung und der Metamorphose. *Acta Biologica Hungarica*, *9*, 25–45.

Kosaka, T. (1982). Inbreeding depression in marine *Euplotes woodruffi* (Ciliophora). *Journal of Science of the Hiroshima University, Series B, Division 1, 30*, 101–110.

Kosaka, T. (1990). Methods of inducing selfing, selfing and its role in the life cycle of *Euplotes woodruffi* syngen 3 (Ciliophora). *Journal of Protozoology, 37*, 33–39.

Kosaka, T. (1991a). Mature cells attracting cells of the complementary mating type in *Euplotes woodruffi* syngen 3 (Ciliophora, Hypotrichida). *Zoological Science, 8*, 681–692.

Kosaka, T. (1991b). Life cycle of *Paramecium bursaria* syngen 1 in nature. *Journal of Protozoology, 38*, 140–148.

Kosaka, T. (1994). Life cycle of *Paramecium bursaria* syngen 1 in a natural pond. *Zoological Science, 11*, 517–526.

Kosciuszko, H., & Prajer, M. (1991). New habitats of species of the *Paramecium aurelia* complex in Scandinavia (southern Sweden). *Folia Biologica (Krakow), 39*, 25–27.

Kovaleva, V. G., & Golemansky, V. G. (1979). Psammobiotic ciliates of the Bulgarian coast of the Black Sea. *Acta Protozoologica, 18*, 265–284. (in Russian with English summary)

Kovaleva, V. G., & Raikov, I. B. (1978). Diminution and re–synthesis of DNA during development and senescence of the "diploid" macronuclei of the ciliate *Trachelonema sulcata* (Gymnostomata, Karyorelictida). *Chromosoma, 67*, 177–192.

Kowalczyk, C. A., Anderson, A. M., Arce–Larreta, M., & Chalker, D. L. (2006). The germ line limited M element of *Tetrahymena* is targeted for elimination from the somatic genome by a homology–dependent mechanism. *Nucleic Acids Research, 34*, 5778–5789.

Kozel, T. R. (1986). Scanning electron microscopy of theronts of *Ichthyophthirius multifiliis*: Their penetration into host tissues. *Transactions of the American Microscopical Society, 105*, 357–364.

Kozloff, E. N. (1946). Studies on ciliates of the family Ancistrocomidae Chatton and Lwoff (order Holotricha, suborder Thigmotricha). IV. *Heterocineta janickii* Jarocki, *Heterocineta goniobasidis* sp. nov., *Heterocineta fluminicolae* sp. nov., and *Enerthecoma properans* Jarocki. *Biological Bulletin, 91*, 200–209.

Kozloff, E. N. (1954). Studies on an astomatous ciliate from a fresh–water limpet, *Ferrissia peninsulae. Journal of Protozoology, 1*, 200–206.

Kozloff, E. N. (1955). *Lwoffia cilifera* gen. nov., sp. nov., a ciliated member of the family Sphenophryidae (Holotricha: Thigmotricha). *Biological Bulletin, 108*, 283–289.

Kozloff, E. N. (1956). *Tetrahymena limacis* from the terrestrial pulmonate gastropods *Monadenia fidelis* and *Prophysaon andersoni. Journal of Protozoology, 3*, 204–208.

Kozloff, E. N. (1961). A new genus and two new species of ancistrocomid ciliates (Holotricha: Thigmotricha) from sabellid polychaetes and from a chiton. *Journal of Protozoology, 8*, 60–63.

Kozloff, E. N. (1965a). *Colligocineta furax* gen. nov., sp. nov., an ancistrocomid ciliate (Holotricha: Thigmotricha) from the sabellid polychaete *Laonome kröyeri* Malmgren. *Journal of Protozoology, 12*, 333–334.

Kozloff, E. N. (1965b). Morphological studies on holotrichous ciliates of the family Hysterocinetidae. II. *Craticuloscuta escobari* gen. nov., sp. nov., and *Epicharocotyle kyburzi* gen. nov., sp. nov. *Journal of Protozoology, 12*, 335–339.

Kozloff, E. N. (1966). *Phalacrocleptes verruciformis* gen. nov., sp. nov., an unciliated ciliate from the sabellid polychaete *Schizobranchia insignis* Bush. *Biological Bulletin, 130*, 202–210.

Kozloff, E. N. (1976). New ancistrocomid ciliates from polychaete annelids. *Transactions of the American Microscopical Society, 95*, 622–627.

Krainer, K. H., & Foissner, W. (1990). Revision of the genus *Askenasia* Blochmann, 1895, with proposal of two new species, and description of *Rhabdoaskenasia minima* n. g., n. sp. (Ciliophora, Cyclotrichida). *Journal of Protozoology, 37*, 414–427.

Krsinic, F., & Grbec, B. (2006). Horizontal distribution of tintinnids in the open waters of the South Adriatic (Eastern Mediterranean). *Scientia Marina, 70*, 77–88.

Kruger, J., Basson, L., & Van As, J. G. (1993). On the ultrastructure of the adhesive disc of *Trichodina xenopodos* Fantham, 1924 and *T. heterodentata* Duncan, 1977 (Ciliophora: Peritrichida). *Acta Protozoologica, 32*, 245–253.

Krumholz, L. R., Forsberg, C. W., & Veira, D. M. (1983). Association of methanogenic bacteria with rumen protozoa. *Canadian Journal of Microbiology, 29*, 676–680.

Kubota, T., Tokoroyama, T., Tsukuda, Y., Koyama, H., & Miyake, A. (1973). Isolation and structure determination of blepharismin, a conjugation initiating gamone in the ciliate *Blepharisma. Science, 179*, 400–402.

Kuck, A., & Ruthmann, A. (1983). Conical kinetochores with single microtubules in the micronuclear mitosis of the ciliate *Colpoda steinii. Zeitschrift für Naturforschung, 38 C*, 672–674.

Kuck, A., & Ruthmann, A. (1985). Micronuclear division and microtubular stability in the ciliate *Colpoda steinii. Chromosoma, 93*, 95–103.

Kuhlmann, H. W. (1993). Life cycle dependent phototactic orientation in *Ophryoglena catenula. European Journal of Protistology, 29*, 344–352.

Kuhlmann, H. W. (1994). Escape response of *Euplotes octocarinatus* to turbellarian predators. *Archiv für Protistenkunde*, *144*, 163–171.

Kuhlmann, H. W. (1998). Do phototactic ciliates make use of directional antennas to track the direction of light? *European Journal of Protistology*, *34*, 244–253.

Kuhlmann, H. W., & Heckmann, K. (1985). Interspecific morphogens regulating prey–predator relationships in Protozoa. *Science*, *227*, 1347–1349.

Kuhlmann, H. W., & Heckmann, K. (1989). Adolescence in *Euplotes octocarinatus*. *Journal of Experimental Zoology*, *251*, 316–328.

Kuhlmann, H. W., & Heckmann, K. (1991). Nuclear processes in *Euplotes octocarinatus* during conjugation. *European Journal of Protistology*, *26*, 370–386.

Kuhlmann, H. W., & Heckmann, K. (1994). Predation risk of typical ovoid and 'winged' morphs of *Euplotes* (Protozoa, Ciliophora). *Hydrobiologia*, *284*, 219–227.

Kuhlmann, H. W., & Hemmersbach–Krause, R. (1993a). Phototaxis in an "eyespot"–exposing ciliate. *Naturwissenschaften*, *80*, 139–141.

Kuhlmann, H. W., & Hemmersbach–Krause, R. (1993b). Phototaxis in the "stigma"–forming ciliate *Nassula citrea*. *Journal of Photochemistry and Photobiology B, Biology*, *21*, 191–195.

Kuhlmann, H. W., Bräucker, R., & Schepers, A. G. (1997). Phototaxis in *Porpostoma notatum*, a marine scuticociliate with a composed crystalline organelle. *European Journal of Protistology*, *33*, 295–304.

Kuhlmann, S., & Hausmann, K. (1980). Untersuchungen zu Nahrungserwerb und Nahrungsaufnahme bei *Homalozoon vermiculare* Stokes, 1887. 2. Mucocysten, Konocysten, Toxicysten. *Protistologica*, *16*, 125–134.

Kuhlmann, S., & Hausmann, K. (1983). Untersuchungen zur Nahrungsaufnahme bei *Homalozoon vermiculare* Stokes, 1887. *Protistologica*, *16*, 111–122.

Kuhlmann, S., Patterson, D. J., & Hausmann, K. (1980). Untersuchungen zu Nahrungserwerb und Nahrungsaufnahme bei *Homalozoon vermiculare* Stokes, 1887. 1. Nahrungserwerb und Feinstruktur der Oral Region. *Protistologica*, *16*, 39–55.

Kuhlmann, S., Walz, B., & Hausmann, K. (1983). Characterization of the parapharyngeal mass of the ciliate *Homalozoon vermiculare*. *Protoplasma*, *116*, 209–212.

Kumar, R. (2003). Effect of different food types on the postembryonic developmental rates and demographic parameters of *Phyllodiaptomus blanci* (Copepoda: Calanoida). *Archiv für Hydrobiologie*, *157*, 351–377.

Kuriyama, R., Besse, C., Gèze, M., Omoto, C. K., & Schrével, J. (2005). Dynamic organization of microtubules and microtubule–organizing centers during the sexual phase of a parasitic protozoan, *Lecudina tuzetae* (Gregarine, Apicomplexa). *Cell Motility and the Cytoskeleton*, *62*, 195–209.

Kurth, T., & Bardele, C. F. (2001). Fine structure of the cyrtophorid ciliate *Chlamydodon mnemosyne* Ehrenberg, 1837. *Acta Protozoologica*, *40*, 33–47.

Kurz, S., & Tiedtke, A. (1993). The Golgi apparatus of *Tetrahymena thermophila*. *Journal of Eukaryotic Microbiology*, *40*, 10–13.

Kusch, J. (1995). Adaptation of inducible defense in *Euplotes daidaleos* (Ciliophora) to predation risks by various predators. *Microbial Ecology*, *30*, 79–88.

Kusch, J. (1998). Local and temporal distribution of different genotypes of pond–dwelling *Stentor coeruleus*. *Protist*, *149*, 147–154.

Kusch, J., & Heckmann, K. (1996). Population structure of *Euplotes* ciliates revealed by RAPD fingerprinting. *Ecoscience*, *3*, 378–384.

Kusch, J., Welter, H., Stremmel, M., & Schmidt, H. J. (2000). Genetic diversity in populations of a freshwater ciliate. *Hydrobiologia*, *431*, 185–192.

Kusuoka, Y., & Watanabe, Y. (1987). Growth and survival of peritrich ciliates in an urban stream. *Oecologia*, *73*, 16–20.

Kusuoka, Y., & Watanabe, Y. (1989). Distinction of emigration by telotroch formation and death by predation in peritrich ciliates: SEM observations on the remaining stalk ends. *FEMS Microbiology Ecology*, *62*, 7–12.

LLLL

LaFountain, J. R., & Davidson, L. A. (1980). An analysis of spindle ultrastructure during anaphase of micronuclear division in *Tetrahymena*. *Cell Motility*, *1*, 41–61.

Lahlafi, T., & Metenier, G. (1991). Low–molecular–weight DNA in the heteromeric macronuclei of two cyrtophorid ciliates. *Biology of the Cell*, *73*, 79–88.

Lalpotu, P. A. (1980a). Studies on ciliates of the genus *Nyctotherus* Leidy, 1840. I. Parasites of Orthoptera. *Archiv für Protistenkunde*, *123*, 31–43.

Lalpotu, P. A. (1980b). Studies on ciliates of the genus *Nyctotherus* Leidy, 1840. II. Parasites of beetle. *Archiv für Protistenkunde*, *123*, 162–165.

Lalpotu, P. A. (1980c). Studies on ciliates of the genus *Nyctotherus* Leidy, 1840. III. Parasites of millipedes. *Archiv für Protistenkunde*, *123*, 261–266.

Landers, S. C. (1986). Studies on the phoront of *Hyalophysa chattoni* (Ciliophora, Apostomatida) encysted on grass shrimp. *Journal of Protozoology*, *33*, 546–552.

Landers, S. C. (1991a). The fine structure of secretion in *Hyalophysa chattoni*: Formation of the attachment peduncle and the chitinous phoretic cyst wall. *Journal of Protozoology*, *38*, 148–157.

Landers, S. C. (1991b). Secretion of the reproductive cyst wall by the apostome ciliate *Hyalphysa chattoni*. *European Journal of Protistology*, *27*, 160–167.

Landers, S. C. (1991c). Trichocyst digestion in *Hyalophysa chattoni* (Ciliophora: Apostomatida). *Transactions of the American Microscopical Society*, *110*, 118–127.

Landers, S. C., & Phipps, S. W. (2003). Ciliated protozoan colonization of substrates from Weeks Bay, Alabama. *Gulf of Mexico Science*, *21*, 79–85.

Landers, S. C., Confusione, A., & Defee, D. (1996). *Hyalophysa bradburyae* sp. n., a new species of apostome ciliate from the grass shrimp *Palaemonetes kadiakensis*. *European Journal of Protistology*, *32*, 372–379.

Landers, S. C., Zimlich, M. A., & Coate, T. (1999). Variations in the ventral ciliature of the crustacean symbiont *Hyalophysa* (Ciliophora, Apostomatida) from Mobile Bay and Dauphin Island, Alabama. *Gulf Research Reports*, *11*, 57–63.

Landers, S. C., Treadaway, R. A., Johnson, J. W., & Luckie, R. N. (2001). Food plaquette digestion in the ciliated protozoan *Hyalophysa chattoni*. *Invertebrate Biology*, *120*, 97–103.

Landis, W. G. (1981). The ecology, role of the killer trait, and interactions of five species of the *Paramecium aurelia* complex inhabiting the littoral zone. *Canadian Journal of Zoology*, *59*, 1734–1743.

Landis, W. G. (1986). The interplay among ecology, breeding systems, and genetics in the *Paramecium aurelia* and *Paramecium bursaria* complexes. *Progress in Protistology*, *1*, 287–307.

Landweber, L. F., & Kari, L. (1999). The evolution of cellular computing: Nature's solution to a computational problem. *BioSystems*, *52*, 3–13.

Landweber, L. F., Kuo, T. C., & Curtis, E. A. (2000). Evolution and assembly of an extremely scrambled gene. *Proceedings of National Academy of Sciences of the United States of America*, *97*, 3298–3303.

Lanners, H. N. (1973). Beobachtungen zur Konjugation von *Heliophrya (Cyclophrya) erhardi* (Rieder) Matthes, Ciliata, Suctoria. *Archiv für Protistenkunde*, *115*, 370–385.

Lanners, H. N. (1978). Ultrastructure of the fusion area during conjugation in the suctorian *Heliophrya erhardi* (Rieder) Matthes. *Journal of Protozoology*, *25*, 453–460.

Lanners, H. N., & Rudzinska, M. A. (1986). Exchange of pronuclei during conjugation in the suctorian ciliate *Heliophrya erhardi* is mediated by a microtubule–filament network. *Protistologica*, *22*, 89–96.

Lasek-Nesselquist, E., & Katz, L. A. (2001). Phylogenetic position of *Sorogena stoianovitchae* and relationships within the class Colpodea (Ciliophora) based on SSU rDNA sequences. *Journal of Eukaryotic Microbiology*, *48*, 604–607.

Latteur, B. (1968/1969). Révision systématique de la famille des Ophryoscolescidae Stein, 1858, sous–famille des Entodiniinae Lubinsky, 1957, genre *Entodinium* Stein, 1858. *Annales de la Société Royale Zoologiqe de Belgique*, *98*, 1–41; *99*: 3–25.

Latteur, B. (1970). Révision systématique de la famille des Ophryoscolescidae Stein, 1858, sous–famille des Diplodiniinae Lubinsky, 1957, genre *Diplodinium* (Schuberg, 1888) sensu novo. *Annales de la Société Royale Zoologique Belgique*, *100*, 275–312.

Laughlin, T. J., Henry, J. M., Phares, E. F., Long, M. V., & Olins, D. E. (1983). Methods for the large–scale cultivation of an *Oxytricha* (Ciliophora: Hypotrichida). *Journal of Protozoology*, *30*, 63–64.

Lauterborn, R. (1908). Protozoen–Studien. V. Teil. Zur Kenntnis einiger Rhizopoden und Infusorien aus dem Gebiete des Oberrheins. *Zeitschrift für Wissenschaftliche Zoologie*, *90*, 645–669.

Laval, M., & Tuffrau, M. (1973). Les ciliés endocommensaux d'une taret de Côte d'Ivoire, *Teredo adami*, mollusque Teredinidae. I. Infraciliature et polymorphisme de *Metanyctotherus rancureli*, sp. nov. (hétérotriche). *Protistologica*, *9*, 149–157.

Laval–Peuto, M. (1975). Cortex, périlemme et réticulum vésiculeux de *Cyttarocylis brandtii* (cilié tintinnide). Les ciliés à périlemme. *Protistologica*, *11*, 83–98.

Laval–Peuto, M. (1977). Reconstruction d'une lorica de forme *Coxiella* par le trophonte nu de *Favella ehrenbergii* (Ciliata, Tintinnina). *Comptes Rendus de l'Académie des Sciences*, *284*, 547–550.

Laval–Peuto, M. (1983). Sexual reproduction in *Favella ehrenbergii* (Ciliophora, Tintinnina). Taxonomical implications. *Protistologica*, *19*, 503–512.

Laval–Peuto, M. (1994). Classe des Oligotrichea Bütschli, 1887. Ordres des Tintinnida Kofoid et Campbell, 1929. In P. de Puytorac (Ed.), *Traité de zoologie, infusoires ciliés* (Vol. 2, pp. 182–219). Paris: Masson.

Laval–Peuto, M., & Brownlee, D. C. (1986). Identification and systematics of the Tintinnina (Ciliophora): Evaluation and suggestions for improvement. *Annales de l'Institut Océanographiqe, Paris*, *62*, 69–84.

Laval–Peuto, M., & Febvre, M. (1986). On plastid symbiosis in *Tontonia appendiculariformis* (Ciliophora, Oligotrichina). *BioSystems*, *19*, 137–158.

Laval–Peuto, M., & Rassoulzadegan, F. (1988). Autofluorescence of marine planktonic Oligotrichina and other ciliates. *Hydrobiologia*, *159*, 99–110.

Laval–Peuto, M., Gold, K., & Storm, E. R. (1979). The ultrastructure of *Tintinnopsis parva*. *Transactions of the American Microscopical Society*, *98*, 204–212.

Laval–Peuto, M., Grain, J., & Deroux, G. (1994). Classe des Oligotrichea Bütschli, 1887. Ordres des Oligotrichida Bütschli, 1887 et des Choreotrichida

Small & Lynn, 1985. In P. de Puytorac (Ed.), *Traité de zoologie, infusoires ciliés* (Vol. 2, pp. 153–179). Paris: Masson.

Lavrentyev, P. J., McCarthy, M. J., Klarer, D. M., Jochem, F., & Gardner, W. S. (2004). Estuarine microbial food web patterns in a Lake Erie coastal wetland. *Microbial Ecology*, *48*, 567–577.

Lawn, R. M., Heumann, J. M., Herrick, G., & Prescott, D. M. (1978). The gene–size DNA molecules in *Oxytricha*. *Cold Spring Harbor Symposia on Quantitave Biology*, *42*, 483–492.

Lawrence, J. R., & Snyder, R. A. (1998). Feeding behaviour and grazing impacts of a *Euplotes* sp. on attached bacteria. *Canadian Journal of Microbiology*, *44*, 623–629.

Laybourn, J. (1976). Respiratory energy losses in *Podophrya fixa* Müller in relation to temperature and nutritional status. *Journal of General Microbiology*, *96*, 203–208.

Laybourn–Parry, J., Quayle, W., & Henshaw, T. (2002). The biology and evolution of Antarctic saline lakes in relation to salinity and trophy. *Polar Biology*, *25*, 542–552.

Laybourn–Parry, J., Perriss, S. J., Seaton, G. G. R., & Rohozinski, J. (1997). A mixotrophic ciliate as a major contributor to plankton photosynthesis in Australian lakes. *Limnology and Oceanography*, *42*, 1463–1467.

Leakey, R. J. G., Burkill, P. H., & Sleigh, M. A. (1993). Planktonic ciliates in Southampton water: Quantitative taxonomic studies. *Journal of the Marine Biological Association of the United Kingdom*, *73*, 579–594.

Leakey, R. J. G., Fenton, N., & Clarke, A. (1994). The annual cycle of planktonic ciliates in nearshore waters at Signy Island, Antarctica. *Journal of Plankton Research*, *16*, 841–856.

Leander, B. S., & Keeling, P. J. (2003). Morphostasis in alveolate evolution. *Trends in Ecology and Evolution*, *18*, 395–402.

Lee, C. C., & Fenchel, T. (1972). Studies on ciliates associated with sea ice from Antarctica. II. Temperature responses and tolerances of ciliates from antarctic, temperate and tropical habitats. *Archiv für Protistenkunde*, *114*, 237–244.

Lee, J. J. (2002). An annotated glossary of protozoological terms. In J. J. Lee, G. F. Leedale, & P. C. Bradbury (Eds.) *An Illustrated Guide to the Protozoa* (Vol. 2 (2000), pp. 1346–1385). Lawrence, Kansas: Society of Protozoologists.

Lee, J. J., & Soldo, A. T. (1992). *Protocols in protozoology*. Kansas, USA: Society of Protozoologists, Lawrence.

Lee, S., Basu, S., Tyler, C. W., & Wei, I. W. (2004). Ciliate populations as bio–indicators at Deer Island Treatment Plant. *Advances in Environmental Research*, *8*, 371–378.

Leeuwenhock, A. van. (1674). More observations from Mr. Leewenhook, in a letter of 7 September 1674, sent to the publisher. *Philosophical Transactions of the Royal Society*, *9*, 178–182.

Legrand, B. (1971). Recherches expérimentales sur le déterminisme de la contraction et les structures contractiles chez le spirostome. *Protistologica*, *6*(year 1970), 283–300.

Lei, Y., Choi, J. K., & Xu, K. 2002. Morphology and infraciliature of a new marine ciliate *Euplotidium smalli* n. sp. with description of a new genus, *Paraeuplotidium* n. g. (Ciliophora, Euplotida). *Journal of Eukaryotic Microbiology*, *49*, 402–406.

Leighton, B. J., Boom, J. D. G., Bouland, C., Hartwick, E. B., & Smith, M. J. (1991). Castration and mortality in *Pisaster ochraceus* parasitized by *Orchitophrya stellarum* (Ciliophora). *Diseases of Aquatic Organisms*, *10*, 71–73.

Leipe, D. (1989). *Prorodon spirogyrophagus* nov. spec. (Ciliophora, Prostomatea): Morphology, infraciliature and food–uptake. *European Journal of Protistology*, *24*, 392–401.

Leipe, D. D., & Hausmann, K. (1989). Somatic infraciliature in the haptorid ciliate *Homalozoon vermiculare* (Kinetophragminophora, Gymnostomata) Ditransversalia n. subcl. and phylogenetic implications. *Journal of Protozoology*, *36*, 280–289.

Leipe, D. D., & Hausmann, K. (1992). The nuclear apparatus of the ditransversal ciliate *Homalozoon vermiculare* (Ciliophora, Rhabdophora) during interphase and division. I. The macronucleus. *Journal of Protozoology*, *39*, 27–39.

Leipe, D. D., & Hausmann, K. (1993). The nuclear apparatus of the ditransversal ciliate *Homalozoon vermiculare* (Ciliophora, Rhabdophora) during interphase and division. I. The micronuclei and some observations on conjugation. *Journal of Eukaryotic Microbiology*, *40*, 447–458.

Leipe, D. D., Oppelt, A., Hausmann, K., & Foissner, W. (1992). Stomatogenesis in the ditransversal ciliate *Homalozoon vermiculare* (Ciliophora, Rhabdophora). *European Journal of Protistology*, *28*, 198–213.

Leipe, D. D., Bernhard, D., Schlegel, M., & Sogin, M. L. (1994). Evolution of 16S–like ribosomal RNA genes in the ciliophoran taxa Litostomatea and Phyllopharyngea. *European Journal of Protistology*, *30*, 354–361.

Leitner, A. R., & Foissner, W. (1997a). Morphology and infraciliature of *Microthorax pusillus* Engelmann 1862 and *Spathidium deforme* Kahl 1928, two ciliates (Protozoa, Ciliophora) from activated sludge. *Linzer Biologische Beiträge*, *29*, 349–368.

Leitner, A. R., & Foissner, W. (1997b). Taxonomic characterization of *Epicarchesium granulatum* (Kellicott, 1887) Jankowski, 1985 and *Pseudovorticella elongata* (Fromentel, 1876) nov. comb., two peritrichs (Protozoa, Ciliophora) from activated sludge. *European Journal of Protistology*, *33*, 13–29.

Le Mouel, A., Butler, A., Caron, F., & Meyer, E. (2003). Developmentally regulated chromosome fragmentation linked to imprecise elimination of repeated sequences in paramecia. *Eukaryotic Cell, 2*, 1076–1090.

Lenk, S. E., Small, E. B., & Gunderson, J. (1984). Preliminary observations of feeding in *Tracheloraphis* spp. *Origins of Life, 13*, 229–234.

Lenk, S. E., Hollander, B. A., & Small, E. B. (1989). Ultrastructure of feeding in the karyorelictean ciliate *Tracheloraphis* examined by scanning electron microscopy. *Tissue & Cell, 21*, 189–194.

Lenzi, P., & Rosati, G. (1993). Ultrastructural study of *Euplotidium itoi* (Ciliata Hypotrichida). *European Journal of Protistology, 29*, 453–461.

Leonildi, A., Erra, F., Banchetti, R., & Ricci, N. (1998). The ethograms of *Uronychia transfuga* and *Uronychia setigera* (Ciliata, Hypotrichida): A comparative approach for new insights into the behaviour of protozoa. *European Journal of Protistology, 34*, 426–435.

Leppänen, J. M., & Bruun, J. E. (1986). The role of pelagic ciliates including the autotrophic *Mesodinium rubrum* during the spring bloom of 1982 in the open northern Baltic proper. *Ophelia, 4*(Suppl.), 147–157.

Leiro, J., Arranz, J. A., Parama, A., Alvarez, M. F., & Sanmartin, M. L. (2004). In vitro effects of the polyphenols resveratrol, mangiferin and (-)–epigall ocatechin–3–gallate on the scuticociliate fish pathogen *Philasterides dicentrarchi. Diseases of Aquatic Organisms, 59*, 171–174.

Levandowsky, M., Cheng, T., Kehr, A., Kim, J., Gardner, L., Silvern, L., et al. (1984). Chemosensory responses to amino acids and certain amines by the ciliate *Tetrahymena*: A flat capillary assay. *Biological Bulletin, 167*, 322–330.

Levine, N. D. (1972). Relationships between certain protozoa and other animals. In T. T. Chen (Ed.), *Research in protozoology* (Vol. 4, pp. 291–350). London/New York: Pergamon Press.

Lewis, L. M., Witkus, E. R., & Vernon, G. M. (1976). The role of microtubules and microfilaments in the micronucleus of *Paramecium bursaria* during mitosis. *Protoplasma, 89*, 203–219.

Li, L., & Desser, S. S. (1983). *Trichodina algonquinensis*, a new species of peritrich ciliate from Ontario freshwater fish, and observations on its transmission. *Canadian Journal of Zoology, 61*, 1159–1164.

Li, L., & Song, W. (2006a). Phylogenetic position of the marine ciliate, *Certesia quadrinucleata* (Ciliophora; Hypotrichia; Hypotrichida) inferred from the complete small subunit ribosomal RNA gene sequence. *European Journal of Protistology, 42*, 55–61.

Li, L., & Song, W. (2006b). Phylogenetic position of two cyrtophorid ciliates, *Dysteria procera* and *Hartmannula derouxi* (Ciliophora: Phyllopharyngea: Dysteriida) inferred from the complete small subunit ribosmal RNA gene sequences. *Acta Protozoologica, 45*, 265–270.

Li, L., Song, W., Warren, A., Wang, Y., Ma, H., Hu, X., et al. (2006). Phylogenetic position of the marine ciliate, *Cardiostomatella vermiforme* (Kahl, 1928) Corliss, 1960 inferred from the complete SSUrRNA gene sequence, with establishment of a new order Loxocephalida n. ord. (Ciliophora, Oligohymenophorea). *European Journal of Protistology, 42*, 107–114.

Liang, A., & Heckmann, K. (1993). *Blepharisma* uses UAA as a termination codon. *Naturwissenschaften, 80*, 225–226.

Liebsch, H. (1976). Untersuchungen zur Nahrungsselektion von *Stentor coeruleus* (Ehrenberg). *Zoologischer Anzeiger, Jena, 196*, 1–8.

Lilly, D. M., & Klosek, R. C. (1961). A protein factor in the nutrition of *Paramecium caudatum. Journal of General Microbiology, 24*, 327–334.

Lin, M., & Prescott, D. M. (1985). Electron microscope autoradiography of DNA synthesis in the replication of two hypotrichous ciliates. *Journal of Protozoology, 32*, 144–149.

Lin, X., Song, W., & Warren, A. (2004). Redescription of the rare marine ciliate, *Prodiscocephalus borrori* (Wicklow, 1982) from shrimp–culturing waters near Qingdao, China, with redefinitions of the genera *Discocephalus*, *Prodiscocephalus* and *Marginotricha* (Ciliophora, Hypotrichida, Discocephalidae). *European Journal of Protistology, 40*, 137–146.

Lin, Y., Lin, T. L., Wang, C. C., Wang, X., Stieger, K., Klopfleisch, R., et al. (2002). Variation in primary sequence and tandem repeat copy number among i–antigens of *Ichthyophthirius multifiliis. Molecular and Biochemical Parasitology, 120*, 93–106.

Lindholm, T. (1985). *Mesodinium rubrum* – a unique photosynthetic ciliate. *Advances in Aquatic Microbiology, 3*, 1–48.

Lindholm, T., Lindroos, P., & Mörk, A. C. (1988). Ultrastructure of the photosynthetic ciliate *Mesodinium rubrum. BioSystems, 21*, 141–149.

Lindley, J. A. (1978). Continuous plankton records: The occurrence of apostome ciliates (Protozoa) on Euphausiacea in the North Atlantic Ocean and North Sea. *Marine Biology, 46*, 131–136.

Lindstrom, E. S., Weisse, T., & Stadler, P. (2002). Enumeration of small ciliates in culture by flow cytometry and nucleic acid staining. *Journal of Microbiological Methods, 49*, 173–182.

Linnaeus, C. (1758, 1767). *Systema Naturae* (Vol. 1, 10th ed., 823 pp.; 12th ed., 1327 pp.). Holmiae: Salvii. (latter includes *Vorticella*)

Lipscomb, D. L., & Riordan, G. P. (1990). The ultrastructure of *Chaenea teres* and an analysis of the phylogeny of the haptorid ciliates. *Journal of Protozoology, 37*, 287–300.

Lipscomb, D. L., & Riordan, G. P. (1991). Ultrastructure of *Helicoprorodon multinucleatum* (Ciliophora).

Transactions of the American Microscopical Society, 11, 93–117.

Lipscomb, D. L., & Riordan, G. P. (1992). A reexamination of the ultrastructure of *Didinium nasutum* and a reanalysis of the phylogeny of the haptorid ciliates. *Journal of Protozoology, 39*, 110–121.

Lloyd, D., Hillman, K., Yarlett, N., & Williams, A. G. (1989). Hydrogen production by rumen holotrich protozoa: Effects of oxygen and implications for metabolic control by in situ conditions. *Journal of Protozoology, 36*, 205–213.

Lobban, C. S., Hallam, S. J., Mukherjee, P., & Petrich, J. W. (2007). Photophysics and multifunctionality of hypericin-like pigments in heterotrich ciliates: A phylogenetic perspective. *Photochemistry and Photobiology, 83*, 1–21.

Lobban, C. S., Schefter, M., Simpson, A. G. B., Pochon, X., Pawlowski, J., & Foissner, W. (2002). *Maristentor dinoferus* n. gen., n. sp., a giant heterotrich ciliate (Spirotrichea: Heterotrichida) with zooxanthellae, from coral reefs on Guam, Mariana Islands. *Marine Biology, 140*, 411–423 and *141*, 207–208.

Lobo da Cunha, A., & Azevedo, C. (1990). Ultrastructural localization of phosphatases with cerium in a ciliated protozoan. *Acta Histochemica and Cytochemica, 23*, 467–473.

Lobo da Cunha, A., & Azevedo, C. (1993). Ultrastructural and cytochemical identification of peroxisomes in the ciliate *Ichthyophthirius multifiliis*. *Journal of Eukaryotic Microbiology, 40*, 169–171.

Lobo da Cunha, A., & Azevedo, C. (1994). The Golgi apparatus of the ciliated protozoon *Ichthyophthirius multifiliis*. A cytochemical study. *European Journal of Protistology, 30*, 97–103.

Loeblich, A. R., Jr., & Tappan, H. (1968). Annotated index to genera, subgenera and suprageneric taxa of the ciliate order Tintinnida. *Journal of Protozoology, 15*, 185–192.

Lohse, K., Gutierrez, A., & Kaltz, O. (2006). Experimental evolution of resistance in *Paramecium caudatum* against the bacterial parasite *Holospora undulata*. *Evolution, 60*, 1177–1186.

Loidl, J., & Scherthan, H. (2004). Organization and pairing of meiotic chromosomes in the ciliate *Tetrahymena thermophila*. *Journal of Cell Science, 117*, 5791–5801.

Lom, J. (1958). A contribution to the systematics and morphology of endoparasitic trichodinids from amphibians, with a proposal of uniform specific characteristics. *Journal of Protozoology, 5*, 251–263.

Lom, J. (1964). The morphology and morphogenesis of the buccal ciliary organelles in some peritrichous ciliates. *Archiv für Protistenkunde, 107*, 131–162.

Lom, J. (1966). Sessiline peritrichs from the surface of some freshwater fishes. *Folia Parasitologica (Praha), 13*, 36–56.

Lom, J. (1973). The mode of attachment and relation to the host in *Apiosoma piscicola* Blanchard and *Epistylis lwoffi* Fauré–Fremiet, ectocommensals of freshwater fish. *Folia Parasitologica (Praha), 20*, 105–112.

Lom, J. (1994). Sous–classe des Peritrichia Stein, 1859. In P. de Puytorac (Ed.), *Traité de zoologie, infusoires ciliés* (Vol. 2, pp. 681–737). Paris: Masson.

Lom, J. (1995). Trichodinidae and other ciliates (Phylum Ciliophora). In P. T. K. Woo (Ed.), *Fish diseases and disorders. Protozoan and metazoan infections* (Vol. 1, pp. 229–262). Wallingford: CAB International.

Lom, J., & Corliss, J. O. (1968). Observations on the fine structure of two species of the peritrich ciliate genus *Scyphidia* and on their mode of attachment to their host. *Transactions of the American Microscopical Society, 87*, 493–509.

Lom, J., & Corliss, J. O. (1971). Morphogenesis and cortical ultrastructure of *Brooklynella hostilis*, a dysteriid ciliate ectoparasitic on marine fishes. *Journal of Protozoology, 18*, 261–281.

Lom, J., & Kozloff, E. N. (1967). The ultrastructure of *Phalacrocleptes verruciformis*, an unciliated ciliate parasitizing the polychaete *Schizobranchia insignis*. *Journal of Cell Biology, 33*, 355–364.

Lom, J., & Kozloff, E. N. (1968). Observations on the ultrastructure of the suctorial tube of ancistrocomid ciliates. *Folia Parasitologica (Praha), 15*, 291–308.

Lom, J., & Kozloff, E. N. (1970). Ultrastructure of the cortical regions of ancistrocomid ciliates. *Protistologica, 5*(year 1969), 173–192.

Lom, J., & Laird, M. (1969). Parasitic protozoa from marine and euryhaline fish of Newfoundland and New Brunswick. I. Peritrichous ciliates. *Canadian Journal of Zoology, 47*, 1367–1380.

Lom, J., & Nigrelli, R. F. (1970). *Brooklynella hostilis* n. g., n. sp., a pathogenic cyrtophorine ciliate in marine fishes. *Journal of Protozoology, 17*, 224–232.

Lom, J., Corliss, J. O., & Noirot–Timothée, C. (1968). Observations on the ultrastructure of the buccal apparatus in thigmotrich ciliates and their bearing on thigmotrich–peritrich affinities. *Journal of Protozoology, 15*, 824–840.

Long, Z. T., & Karel, I. (2002). Resource specialization determines whether history influences community structure. *Oikos, 96*, 62–69.

López–Ochoterena, E. (1966). Ciliados mesosaprobicos de Chapultepec (sistematica, morfologia, ecologia). *Revista de la Sociedad Mexicana de Historia Natural, 26*(year 1965), 115–247.

Loret, P., Le Gall, S., Dupuy, C., Blanchot, J., Pastoureaud, A., Delesalle, B., et al. (2000). Heterotrophic protists as a trophic link between picocyanobacteria and the pearl oyster *Pinctada margaritifera* in the Takapoto lagoon (Tuamotu

Archipelago, French Polynesia). *Aquatic Microbial Ecology*, *22*, 215–226.

Lozupone, C. A., Knight, R. D., & Landweber, L. F. (2001). The molecular basis of nuclear genetic code change in ciliates. *Current Biology*, *11*, 65–74.

Lu, L., Shi, X. B., & Ng, S. F. (1991). Nuclear control of cortical reorganization during sexual reproduction of *Stylonychia mytilus*: A study with amicronucleates. *European Journal of Protistology*, *26*, 330–339.

Lubinsky, G. (1955a). On some parasites of parasitic protozoa. I. *Sphaerita hoari* sp. n. – a chytrid parasitizing *Eremoplastron bovis*. *Canadian Journal of Microbiology*, *1*, 440–450.

Lubinsky, G. (1955b). On some parasites of parasitic protozoa. II. *Sagittospora cameroni* gen. n., sp. n. – a phycomycete parasitizing Ophryoscolecidae. *Canadian Journal of Microbiology*, *1*, 675–684.

Lubinsky, G. (1957a). Studies on the evolution of the Ophryoscolecidae (Ciliata: Oligotricha). II. On the origin of the higher Ophryoscolecidae. *Canadian Journal of Zoology*, *35*, 135–140.

Lubinsky, G. (1957b). Studies on the evolution of the Ophryoscolecidae (Ciliata: Oligotricha). III. Phylogeny of the Ophryoscolecidae based on their comparative morphology. *Canadian Journal of Zoology*, *35*, 141–159.

Lucchesi, P., & Santangelo, G. (2004). How often does conjugation in ciliates occur? Clues from a seven–year study on marine sandy shores. *Aquatic Microbial Ecology*, *36*, 195–200.

Luckinbill, L. S. (1973). Coexistence in laboratory populations of *Paramecium aurelia* and its predator *Didinium nasutum*, *Ecology*, *54*, 1320–1327.

Lueken, W., Ricci, N., & Krüppel, T. (1996). Rhythmic spontaneous depolarizations determine a slow–and–fast rhythm in walking of the marine hypotrich *Euplotes vannus*. *European Journal of Protistology*, *32*, 47–54.

Lüftenegger, G., Foissner, W., & Adam, H. (1985). *r*– and *K*–selection in soil ciliates: A field and experimental approach. *Oecologia*, *66*, 574–579.

Lund, E. E. (1941). The feeding mechanisms of various ciliated protozoa. *Journal of Morphology*, *69*, 563–573.

Luporini, P., & Dini, F. (1977). The breeding system and the genetic relationship between autogamous and non–autogamous sympatric populations of *Euplotes crassus* (Dujardin) (Ciliata Hypotrichida). *Monitore Zoologico Italiano Nuova Serie*, *11*, 119–154.

Luporini, P., & Miceli, C. (1986). Mating pheromones. In J. G. Gall (Ed.), *Molecular biology of ciliated protozoa* (pp. 263–299). New York: Academic Press.

Luporini, P., Miceli, C., & Ortenzi, C. (1983). Evidence that the ciliate *Euplotes raikovi* releases mating inducing factors (gamones). *Journal of Experimental Zoology*, *226*, 1–9.

Luporini, P., Vallesi, A., Miceli, C., & Bradshaw, R. A. (1995). Chemical signalling in ciliates. *Journal of Eukaryotic Microbiology*, *42*, 208–212.

Luporini, P., Alimenti, C., Ortenzi, C., & Vallesi, A. (2005). Ciliate mating types and their specific protein pheromones. *Acta Protozoologica*, *44*, 89–101.

Lwoff, A. (1923). Sur la nutrition des infusoires. *Comptes Rendus de l'Académie des Sciences*, *176*, 928–930.

Lwoff, A. (1950). *Problems of morphogenesis in ciliates* (103 pp.). New York: Wiley.

Lynch, J. E. (1929). Studies on the ciliates from the intestine of *Strongylocentrotus*. I. *Entorhipidium* gen. nov. *University of California Publications in Zoology*, *33*, 27–56.

Lynch, J. E. (1930). Studies on the ciliates from the intestine of *Strongylocentrotus*. II. *Lechriopyla mystax*, gen. nov., sp. nov. *University of California Publications in Zoology*, *33*, 307–350.

Lynn, D. H. (1975). *Woodruffia metabolica*: Exception to the rule of desmodexy questioned. *Science*, *188*, 1040–1041.

Lynn, D. H. (1976a). Comparative ultrastructure and systematics of the Colpodida: Structural conservatism hypothesis and a description of *Colpoda steinii* Maupas, 1883. *Journal of Protozoology*, *23*, 302–314.

Lynn, D. H. (1976b). Comparative ultrastructure and systematics of the Colpodida: An ultrastructural description of *Colpoda maupasi* Enriquez, 1908. *Canadian Journal of Zoology*, *54*, 405–420.

Lynn, D. H. (1976c). Comparative ultrastructure and systematics of the Colpodida (Ciliophora): Structural differentiation in the cortex of *Colpoda simulans*. *Transactions of the American Microscopical Society*, *95*, 581–599.

Lynn, D. H. (1977a). Comparative ultrastructure and systematics of the Colpodida (Ciliophora). Fine structural specialization associated with large body size in *Tillina magna* Gruber, 1880. *Protistologica*, *12*(year 1976), 629–648.

Lynn, D. H. (1977b). Proportional control of organelle position by a mechanism which similarly monitors cell size in wild type and conical form–mutant Tetrahymena. *Journal of Embryology and Experimental Morphology*, *42*, 261–274.

Lynn, D. H. (1978). Size increase and form allometry during evolution of ciliate species in the genera *Colpoda* and *Tillina* (Ciliophora: Colpodida). *BioSystems*, *10*, 201–211.

Lynn, D. H. (1979a). Changes in the classification of several major taxa of Ciliophora on the basis of a numerical phenetic analysis. *Journal of Protozoology*, *26*, 359–366.

Lynn, D. H. (1979b). Fine structural specializations and evolution of carnivory in *Bresslaua* (Ciliophora:

Colpodida). *Transactions of the American Microscopical Society*, *98*, 353–368.

Lynn, D. H. (1980). The somatic cortical ultrastructure of *Bursaria truncatella* (Ciliophora, Colpodida). *Transactions of the American Microscopical Society*, *99*, 349–359.

Lynn, D. H. (1981). The organization and evolution of microtubular organelles in ciliated protozoa. *Biological Reviews of the Cambridge Philosophical Society*, *56*, 243–292.

Lynn, D. H. (1982). Dimensionality and contractile vacuole function in ciliated protozoa. *Journal of Experimental Zoology*, *223*, 219–229.

Lynn, D. H. (1985). Cortical ultrastructure of *Coleps bicuspis* Noland, 1925 and the phylogeny of the Class Prostomatea (Ciliophora). *BioSystems*, 18, 387–397.

Lynn, D. H. (1991). The implications of recent descriptions of kinetid structure to the systematics of the ciliated protists. *Protoplasma*, *164*, 123–142.

Lynn, D. H. (1996a). My journey in ciliate systematics. *Journal of Eukaryotic Microbiology*, *43*, 253–260.

Lynn, D. H. (1996b). Systematics of ciliates. In K. Hausmann & P. C. Bradbury (Eds.), *Ciliates. Cells as organsisms* (pp. 51–72). Stuttgart: Gustav Fischer Verlag.

Lynn, D. H. (2004). Morphology or molelcules: How do we identify the major lineages of ciliates (Phylum Ciliophora)? *European Journal of Protistology*, *39*(Year 2003), 356–364.

Lynn, D. H., & Berger, J. (1972). Morphology, systematics, and demic variation of *Plagiopyliella pacifica* Poljansky, 1951 (Ciliatea: Philasterina), an entocommensal of strongylocentrotid echinoids. *Transactions of the American Microscopical Society*, *91*, 310–336.

Lynn, D. H., & Berger, J. (1973). The Thyrophylacidae, a family of carnivorous philasterine ciliates entocommensal in strongylocentrotid echinoids. *Transactions of the American Microscopical Society*, *92*, 533–557.

Lynn, D. H., & Corliss, J. O. (1991). Ciliophora. In F. W. Harrison & J. O. Corliss (Eds.), *Microscopic anatomy of invertebrates* (Vol. 1, chap. 5, pp. 333–467). New York: Wiley–Liss.

Lynn, D. H., & Didier, P. (1978). Caractéristiques ultrastructurales du cortex somatique et buccal du cilié *Colpidium campylum* (Oligohymenophora, Tetrahymenina) quant à la position systématique de *Turaniella*. *Canadian Journal of Zoology*, *56*, 2336–2343.

Lynn, D. H., & Foissner, W. (1994). The feet of *Pseudochlamydonellopsis plurivacuolata* (Ciliophora, Cyrtophorida) and a brief review of tentacle–like structures of ciliates. *European Journal of Protistology*, *30*, 423–430.

Lynn, D. H., & Frombach, S. (1987). The ultrastructure and systematics of *Schizocaryum dogieli*, a ciliate endocommensal in Pacific echinoids. *Canadian Journal of Zoology*, *65*, 3133–3143.

Lynn, D. H., & Gilron, G. L. (1993). Strombidiid ciliates from coastal waters near Kingston Harbour, Jamaica (Ciliophora, Oligotrichia, Strombidiidae). *Journal of Marine Biological Association of the United Kingdom*, *73*, 47–65.

Lynn, D. H., & Malcolm, J. R. (1983). A multivariate study of morphometric variation in species of the ciliate genus *Colpoda* (Ciliophora: Colpodida). *Canadian Journal of Zoology*, *61*, 307–316.

Lynn, D. H., & Montagnes, D. J. S. (1988). Taxonomic descriptions of some conspicuous species of strobilidiine ciliates (Ciliophora: Choreotrichida) from the Isles of Shoals, Gulf of Maine. *Journal of Marine Biological Association of the United Kingdom*, *68*, 639–658.

Lynn, D. H., & Montagnes, D. J. S. (1991). Global production of heterotrophic marine planktonic ciliates. In P. C. Reid, C. M. Turley, & P. H. Burkhill (Eds.), *Protozoa and their role in marine processes*, NATO Publications (Vol. G25, pp. 281–307). Berlin: Springer.

Lynn, D. H., & Munawar, M. (1999). Abundance, biomass, and diversity of planktonic ciliates (Ciliophora) in Lake Erie. In M. Munawar, T. Edsall, & I. F. Munawar (Eds.), *State of Lake Erie (SOLE) – Past, present and future*, Ecovision World Monograph Series (pp. 155–168). Leiden:, Backhuys.

Lynn, D. H., & Nicholls, K. H. (1985). Cortical microtubular structures of the ciliate *Lepidotrachelophyllum fornicis* Nicholls & Lynn, 1984 and phylogeny of the litostomate ciliates. *Canadian Journal of Zoology*, *63*, 1835–1845.

Lynn, D. H., & Small, E. B. (1981). Protist kinetids: Structural conservatism, kinetid structure, and ancestral states. *BioSystems*, *14*, 317–322.

Lynn, D. H., & Small, E. B. (1990). Phylum Ciliophora. In L. Margulis, J. O. Corliss, M. Melkonian, & D. J. Chapman (Eds.), *Handbook of protoctists* (pp. 498–523). Boston: Jones & Bartlett.

Lynn, D. H., & Small, E. B. (1997). A revised classification of the phylum Ciliophora Doflein, 1901. *Revista de la Sociedad Mexicana de Historia Natural*, *47*, 65–78.

Lynn, D. H., & Small, E. B. (2002). Phylum Ciliophora, Doflein, 1901. In J. J. Lee, G. F. Leedale, & P. C. Bradbury (Eds.), *An illustrated guide to the protozoa* (Vol. 1 (2000), pp. 371–656). Kansas, USA: Society of Protozoologists, Lawrence.

Lynn, D. H., & Sogin, M. L. (1988). Assessment of phylogenetic relationships among ciliated protists using partial ribosomal RNA sequences derived from reverse transcripts. *BioSystems*, *21*, 249–254.

Lynn, D. H., & Strüder–Kypke, M. (2002). Phylogenetic position of *Licnophora*, *Lechriopyla*, and *Schizocaryum*, three unusual ciliates (Phylum Ciliophora) endosymbiotic in echinoderms (Phylum Echinodermata). *Journal of Eukaryotic Microbiology*, *49*, 460–468.

Lynn, D. H., & Strüder–Kypke, M. (2005). Phylogenetic relationships among species in the genera *Entodiscus*, *Plagiopyliella*, *Thyrophylax*, and *Entorhipidium*, scuticociliate endosymbionts of echinoids. *Journal of Parasitology*, *91*, 1190–1199.

Lynn, D. H., & Strüder-Kypke, M. (2006). Species of *Tetrahymena* identical by small subunit rRNA gene sequences are discriminated by mitochondrial cytochrome *c* oxidase 1 sequences. *Journal of Eukaryotic Microbiology*, *53*, 385–387.

Lynn, D. H., & Tucker, J. B. (1976). Cell size and proportional distance assessment during determination of organelle position in the cortex of the ciliate *Tetrahymena*. *Journal of Cell Science*, *21*, 35–46.

Lynn, D. H., & Zimmerman, A. M. (1981). Cytoskeletal function of cortical microtubules in the ciliate *Colpoda*. *Protoplasma*, *108*, 29–38.

Lynn, D. H., Molloy, D., & Lebrun, R. (1981). *Tetrahymena rotunda* n. sp. (Hymenostomatida, Tetrahymenidae), a ciliate parasite of the hemolymph of *Simulium* (Diptera, Simuliidae). *Transactions of the American Microscopical Society*, *100*, 134–141.

Lynn, D. H., Montagnes, D. J. S., & Riggs, W. (1987). Divider size and the cell cycle after prolonged starvation of *Tetrahymena corlissi*. *Microbial Ecology*, *13*, 115–127.

Lynn, D. H., Montagnes, D. J. S., & Small, E. B. (1988). Taxonomic descriptions of some conspicuous species in the family Strombidiidae (Ciliophora: Oligotrichida) from the Isles of Shoals, Gulf of Maine. *Journal of the Marine Biological Association of the United Kingdom*, *68*, 259–276.

Lynn, D. H., Roff, J. C., & Hopcroft, R. R. (1991a). Annual abundance and biomass of aloricate ciliates in tropical neritic waters off Kingston, Jamaica. *Marine Biology*, *110*, 437–448.

Lynn, D. H., Frombach, S., Ewing, M. S., & Kocan, K. M. (1991b). The organelle of Lieberkühn as a synapomorphy for the Ophryoglenina (Ciliophora: Hymenostomatida). *Transactions of the American Microscopical Society*, *110*, 1–11.

Lynn, D. H., Montagnes, D. J. S., Dale, T., Gilron, G. L., & Strom, S. L. (1991c). A reassessment of the genus *Strombidinopsis* Kent, 1881 (Ciliophora, Choreotrichida) with descriptions of four new planktonic species and remarks on its taxonomy and phylogeny. *Journal of the Marine Biological Association of the United Kingdom*, *71*, 597–612.

Lynn, D. H., Wright, A. D. G., Schlegel, M., & Foissner, W. (1999). Phylogenetic relationships of orders within the class Colpodea (Phylum Ciliophora) inferred from small subunit rRNA gene sequences. *Journal of Molecular Evolution*, *48*, 605–614.

Lynn, D. H., Gransden, S. G., Wright, A. D. G., & Josephson, G. (2000). Characterization of a new species of the ciliate *Tetrahymena* (Ciliophora: Oligohymenophorea) isolated from the urine of a dog: First report of *Tetrahymena* from a mammal. *Acta Protozoologica*, *39*, 289–294.

MMMM

Ma, H., & Song, W. (2003). Stomatogenesis of the marine ciliate *Metanophrys sinensis* (Protozoa: Ciliophora: Scuticociliatida). *Journal of the Marine Biological Association of the United Kingdom*, *83*, 407–409.

Ma, H., Song, W., & Hu, X. (2001). Stomatogenesis of the marine ciliate *Paranophrys magna* (Protozoa: Ciliophora: Scuticociliatida) from Qingdao, China. *Journal of the Marine Biological Association of the United Kingdom*, *81*, 377–382.

Ma, H., Song, W., & Ma, H. (2002). Morphogenetic processes in division of *Uronemella filificum* (Ciliophora, Scuticociliatida). *Journal of Ocean University of Qingdao*, *1*, 140–144.

Ma, H., Gong, J., & Song, W. (2003a). Stomatogenesis of the marine ciliate *Pleuronema coronatum* Kent, 1881 (Ciliophora, Scuticociliatida). *Acta Zoologica Sinica*, *49*, 829–834.

Ma, H., Song, W., Hu, X., & Warren, A. (2003b). Morphology and stomatogenesis of *Pseudocohnilembus hargisi* (Ciliophora: Scuticociliatida). *Journal of the Marine Biological Association of the United Kingdom*, *83*, 399–405.

Ma, H., Song, W., Gong, J., & Warren, A. (2004). Reconsideration of stomatogenesis in *Uronema marinum* Dujardin, 1841 during asexual division (Protozoa: Ciliophora: Scuticociliatida). *Acta Zoologica Sinica*, *50*, 823–827.

MacDougall, M. S. (1935). Cytological studies of the genus *Chilodonella* Strand, 1926 (*Chilodon* Ehrbg., 1838). I. The conjugation of *Chilodonella* sp. *Archiv für Protistenkunde*, *84*, 199–206.

Macek, M., Šimek, K., Pernthaler, J., Vyhnálek, V., & Psenner, R. (1996). Growth rates of dominant planktonic ciliates in two freshwater bodies of different trophic degree. *Journal of Plankton Research*, *18*, 463–481.

Macek, M., Callieri, C., Simek, K., & Vazquez, A. L. (2006). Seasonal dynamics, composition and feeding patterns of ciliate assemblages in oligotrophic lakes covering a wide pH range. *Archiv für Hydrobiologie*, *166*, 261–287.

Machelon, V. (1986). Incompatibilités sexuelles et conjugaisons intraclonales chez une espèce du genre *Euplotes* (ciliés hypotriches). *Protistologica, 22*, 235–240.

Machelon, V., & Demar, C. (1984). Electrophoretic variations among the genus *Euplotes* (Ciliata: Hypotrichida): Comparative data for the sibling species complex *Euplotes vannus* and survey of infrageneric variability. *Journal of Protozoology, 31*, 74–82.

Machelon, V., & Génermont, J. (1984). A biometrical analysis of morphologica variation within a section of a genus *Euplotes* (Ciliata, Hypotrichida), with special reference to the *E. vannus* complex of sibling species. *Origins of Life, 13*, 249–267.

Machemer, H., & Sugino, K. (1989). Electrophysiological control of ciliary beating: A basis of motile behaviour in ciliated protozoa. *Comparative Biochemistry And Physiology, 94A*, 365–374.

Machemer, H., & Teunis, P. F. M. (1996). Sensory–motor coupling and motor responses. In K. Hausmann & P. C. Bradbury (Eds.), *Ciliates: Cells as organisms* (pp. 379–402). Stuttgart: Gustav Fischer Verlag.

Maciejewska, A. (2007). Molecular phylogenetics of representative *Paramecium* species. *Folia Biologica, 55*, 1–8.

MacKeen, P. C., & Mitchell, R. B. (1977). Prey–capture capacity in the suctorian protozoon *Tokophrya lemnarum* Stein. *Transactions of the American Microscopical Society, 96*, 68–75.

Madoni, P. (1986). Abundance, productivity and energy flow of a population of *Coleps hirtus* Nitzch (Protozoa, Ciliophora) during the first stage of colonization in an experimental ricefield. *Archiv für Protistenkunde, 132*, 43–51.

Madoni, P. (1990). The ciliated protozoa of the monomictic Lake Kinneret (Israel): Species composition and distribution during stratification. *Hydrobiologia, 190*, 111–120.

Madoni, P. (1996). The contribution of ciliated protozoa to plankton and benthos biomass in a European ricefield. *Journal of Eukaryotic Microbiology, 43*, 193–198.

Madoni, P. (2000). The acute toxicity of nickel to freshwater ciliates. *Environmental Pollution, 109*, 53–59.

Madoni, P. (2003). Protozoa as indicators of wastewater treatment efficiency. In *The Handbook of water and wastewater microbiology* (chap. 22, pp. 361–371). Amsterdam: Elsevier.

Madoni, P., & Ghetti, P. F. (1980). Étude de la dynamique des populations de ciliés d'un torrent expérimental pendant deux années. *Hydrobiologia, 74*, 273–282.

Madoni, P., & Romeo, M. G. (2006). Acute toxicity of heavy metals towards freshwater ciliated protists. *Environmental Pollution, 141*, 1–7.

Madoni, P., & Sartore, F. (2003). Long–term changes in the structure of ciliate communities in a small isolated pond. *Italian Journal of Zoology, 70*, 313–320.

Madoni, P., Berman, T., Hadas, O., & Pinkas, R. (1990). Food selection and growth of the planktonic ciliate *Coleps hirtus* isolated from a monomictic subtropical lake. *Journal of Plankton Research, 12*, 735–741.

Madoni, P., Davoli, D., Gorbi, G., & Vescovi, L. (1996). Toxic effect of heavy metals on the activated sludge protozoan community. *Water Research, 30*, 135–141.

Maguire, B., Jr. (1963a). The exclusion of *Colpoda* (Ciliata) from superficially favorable habitats. *Ecology, 44*, 781–794.

Maguire, B., Jr. (1963b). The passive dispersal of small aquatic organisms and their colonization of isolated bodies of water. *Ecological Monographs, 33*, 161–185.

Maguire, B., Jr., & Belk, D. (1967). *Paramecium* transport by land snails. *Journal of Protozoology, 14*, 445–447.

Malvin, G. M., & Wood, S. C. (1992). Behavioral hypothermia and survival of hypoxic protozoans *Paramecium caudatum. Science, 255*, 1423–1425.

Maloney, M. S., McDaniel, W. S., Locknar, S. A., & Torlina, H. M. (2005). Identification and localization of a protein immunologically related to caltractin (centrin) in the myonemes of membranelles of the heterotrich ciliate *Stentor coeruleus. Journal of Eukaryotic Microbiology, 52*, 328–338.

Maly, E. J. (1975). Interactions among the predatory rotifer *Asplanchna* and two prey, *Paramecium* and *Euglena. Ecology, 56*, 346–358.

Maly, E. J. (1978). Stability of the interaction between *Didinium* and *Paramecium*: Effects of dispersal and predator time lag. *Ecology, 59*, 733–741.

Mandal, A. K., & Nair, K. N. (1975). *Plagiotoma dichogasteri* sp. nov. (Plagiotomidae) a new heterotrichous ciliate from the earthworm *Dichogaster bolaui* (Michaelsen) of Orissa, India. *Archiv für Protistenkunde, 117*, 38–40.

Martín, A., Palacios, G., Olmo, A., Martín–González, A., Ruiz–Perez, L. M., & Gutiérrez, J. C. (1997). Karyotypic variability in ribosomal DNA subchromosome size among colpodid ciliates, a possible tool to differentiate colpodid species. *Applied and Environmental Microbiology, 63*, 1602–1605.

Martin, A. J., & Montagnes, D. J. S. (1993). Winter ciliates in a British Columbian fjord: Six new species and an analysis of ciliate putative prey. *Journal of Eukaryotic Microbiology, 40*, 535–549.

Martin, J. (1982). Évolution des patrons morphogénétiques et phylogenèse dans le sour–ordre des Sporadotrichina (Ciliophora, Hypotrichida). *Protistologica, 18*, 431–447.

Martin, J., Fedriani, C., & Perez–Silva, J. (1983). Morphogenetic pattern in *Laurentiella acuminata* (Ciliophora, Hypotrichida): Its significance for the comprehension of ontogeny and phylogeny of hypotrichous ciliates. *Journal of Protozoology, 30*, 519–529.

Martin–Cereceda, M., Serrano, S., & Guinea, A. (2001a). Biofilm communities and operational monitoring of a rotating biological contactor system. *Water Air and Soil Pollution*, *126*, 193–206.

Martin–Cereceda, M., Alvarez, A. M., Serrano, S., & Guinea, A. (2001b). Confocal and light microscope examination of protozoa and other microorganisms in the biofilms from a rotating biological contactor wastewater treatment plant. *Acta Protozoologica*, *40*, 263–272.

Martín–González, A., Serrano, S., & Fernández–Galiano, D. (1984). Fenomenos nucleares y corticales durante la conjugacion de *Glaucoma chattoni* (Ciliophora, Glaucomidae). *Microbiologia Espanola*, *37*, 89–98.

Martín–González, A., Serrano, S., Guinea, A., & Fernández–Galiano, D. (1986). Further studies on *Urocentrum turbo* O. F. M. (Ciliata): Morphology and morphogenesis. *Archiv für Protistenkunde*, *132*, 11–21.

Martín–González, A., Serrano, S., & Fernández–Galiano, D. (1987). Cortical morphogenesis and conjugation process in *Caenomorpha medusula* (Ciliophora, Heterotrichida). *European Journal of Protistology*, *23*, 111–121.

Martín–González, A., Benitez, L., & Gutiérrez, J. C. (1992a). Ultrastructural analysis of resting cysts and encystment in *Colpoda inflata*. 2. Encystment process and a review of ciliate resting cyst classification. *Cytobios*, *72*, 93–106.

Martín–González, A., Benitez, L., Palacios, G., & Gutiérrez, J. C. (1992b). Ultrastructural analysis of resting cysts and encystment in *Colpoda inflata*. 1. Normal and abnormal resting cysts. *Cytobios*, *72*, 7–18.

Masaki, M. E., Harumoto, T., Terazima, M. N., Miyake, A., Usuki, Y., & Iio, H. (1999). Climacostol, a defense toxin of the heterotrich ciliate *Climacostomum virens* against predators. *Tetrahedron Letters*, *40*, 8227–8229.

Maslin–Leny, Y., & Bohatier, J. (1984). Cytologie ultrastructurale de *Trichodina* et *Tripartiella* (ciliés péritriches). *Protistologica*, *20*, 113–132.

Massana, R., & Pedrós–Alió, C. (1994). Role of anaerobic ciliates in planktonic food webs: Abundance, feeding, and impact on bacteria in the field. *Applied and Environmental Microbiology*, *60*, 1325–1334.

Massana, R., Stumm, C. K., & Pedrós–Alió, C. (1994). Effects of temperature, sulfide, and food abundance on growth and feeding of anaerobic ciliates. *Applied and Environmental Microbiology*, *60*, 1317–1324.

Matsuda, A., Mayer, K. M., & Forney, J. D. (2004). Identification of single nucleotide mutations that prevent developmentally programmed DNA elimination in *Paramecium tetraurelia*. *Journal of Eukaryotic Microbiology*, *51*, 664–669.

Matsuoka, T., & Kotsuki, H. (2001). Photosensory transduction in unicellular eukaryote *Blepharisma*. *Journal of Experimental Zoology*, *289*, 467–471.

Matsusaka, T., & Hongo, F. (1984). Cytochemical and electrophoretic studies on the cyst wall of a ciliate, *Histriculus muscorum* Kahl. *Journal of Protozoology*, *31*, 471–475.

Matsusaka, T., Nakamura, T., & Nagata, K. (1984). Ultrastructure, disintegration and formation of a cirrus in the vegetative, encysting and excysting ciliate, *Histriculus muscorum*. *Journal of Electron Microscopy, Japan*, *33*, 217–229.

Matthes, D. (1971). Parasitische Suktorien. *Zoologischer Anzeiger*, *186*, 272–291.

Matthes, D. (1973). Zur Bakteriensymbiose in der Gattung *Discophrya* (Ciliata, Suctoria). *Zoologischer Anzeiger*, *190*, 149–153.

Matthes, D. (1974). Sessile Ciliaten und ihre Adaptationen an die symphorionte Lebensweise. *Zoologischer Anzeiger*, *192*, 153–164.

Matthes, D. (1988). Suctoria. In D. Matthes, W. Guhl, & G. Haider (Eds.), *Suctoria und Urceolariidae (Peritricha)* (pp. 1–226). Stuttgart: Gustav Fischer Verlag.

Matthes, D. (1990). *Nüchterleinella corneliae* n. g., n. sp. – ein Peritrich auf *Cypria ophthalmica* (Ostracoda). *Archiv für Protistenkunde*, *138*, 139–141.

Matthes, D., & Gräf, W. (1967). Ein Flavobacterium (*Flavobacterium buchneri* n. sp.) als Endosymbiont zweier Sauginfusorien. *Zeitschrift für Morphologie und Ökologie der Tiere*, *58*, 381–395.

Matthes, D., & Guhl, W. (1975). Systematik, Anpassungen und Raumparasitismus auf Hydrophiliden lebender operculariformer Epistyliden. *Archiv für Protistenkunde*, *117*, 110–186.

Matthews, B. F., Matthews, R. A., & Burgess, P. J. (1993). *Cryptocaryon irritans* Brown, 1951 (Ichthyophthiriidae): The ultrastructure of the somatic cortex throughout the life cycle. *Journal of Fish Diseases*, *16*, 339–349.

Matthews, D. C. (1968). The folliculinids (Protozoa) of Ago Bay, Japan, and their relation to the epifauna of the pearl oyster (*Pinctada martensii*). *Pacific Science*, *22*, 232–250.

Matthews, R. A. (2005). *Ichthyophthirius multifiliis* Fouquet and ichthyophthiriosis in freshwater teleosts. *Advances in Parasitology*, *59*, 159–241.

Maupas, E. (1889). La rajeunissement karyogamique chez les ciliés. *Archives de Zoologie Experimentale et Générale (Série 2)*, *7*, 149–517.

Mayén–Estrada, R., & Aladro–Lubel, M. A. (2002). Distribution and prevalence of 15 species of epibiont peritrich ciliates on the crayfish *Cambarellus patzcuarensis* Villalobos, 1943 in Lake Pátzcuaro, Michoacán, Mexico. *Crustaceana*, *74*, 1213–1224.

Mayén–Estrada, R., & Aladro–Lubel, M. A. (2004). First record of *Conidophrys pilisuctor* (Ciliophora: Pilisuctorida) as ectosymbiont of *Hyalella azteca* from Mexico. *Hydrobiologia*, *529*, 19–26.

Mayer, K. M., Mikami, K., & Forney, J. D. (1998). A mutation in *Paramecium tetraurelia* reveals functional and structural features of developmentally excised DNA elements. *Genetics, 148*, 139–149.

Mayr, E. (1970). *Populations, species and evolution* (453 pp.). Cambridge, MA: Harvard University Press.

Mazei, Y. A., & Burkovsky, I. V. (2003). Vertical structure of the interstitial ciliate community in the Chernaya River estuary (the White Sea). *Protistology, 3*, 107–120.

McAllister, T. A., Bae, H. D., Jones, G. A., & Cheng, K. J. (1994). Microbial attachment and feed digestion in the rumen. *Journal of Animal Science, 72*, 3004–3018.

McArdle, E. W., Bergquist, B. L., & Ehret, C. F. (1980). Structural changes in *Tetrahymena rostrata* during induced encystment. *Journal of Protozoology, 27*, 388–397.

McCoy, J. W. (1974a). Biology and systematics of the ciliate genus *Cyrtolophosis* Stokes, 1885. *Acta Protozoologica, 13*, 41–52.

McCoy, J. W. (1974b). Updating the tetrahymenids. I. General considerations and application to the genus *Colpidium. Acta Protozoologica, 13*, 161–176.

McCoy, J. W. (1977). *Cyrtolophosis*: A "lower" ciliate with a buccal cavity. *Protistologica, 13*, 497–502.

McEwan, N. R., Gatherer, D., Eschenlauer, S. C. P., McIntosh, F. M., Calza, R. E., Wallace, R. J., & Newbold, C. J. (2000). An unusual codon usage pattern in the ciliate family Ophryscolecidae and its implications for determining the source of cloned DNA. *Anaerobe, 6*, 21–28.

McFarland, C. P., Chang, W. J., Kuo, S. Y., & Landweber, L. F. (2006). Conserved linkage of two genes on the same chromosome in spirotrichous ciliates. *Chromosoma, 115*, 129–138.

McKanna, J. A. (1973a). Fine structure of the contractile vacuole pore in *Paramecium. Journal of Protozoology, 20*, 631–638.

McKanna, J. A. (1973b). Cyclic membrane flow in the ingestive–digestive system of peritrich protozoans. I. Vesicular fusion at the cytopharynx. II. Cup–shaped coated vesicles. *Journal of Cell Science, 13*, 663–686.

McLaughlin, D., Johnson, G. R., & Bradley, C. J. (1974). Growth of the peritrich *Opercularia coarctata* in axenic culture. *Journal of Protozoology, 21*, 561–564.

McLaughlin, N. B., & Buhse, H. E., Jr. (2004). Localization by indirect immunofluorescence of tetrin, actin, and centrin to the oral apparatus and buccal cavity of the macrostomal form of *Tetrahymena vorax. Journal of Eukaryotic Microbiology, 51*, 253–257.

McNair, J. N. (1979). A model of tentacle function in certain suctorians. *Journal of Theoretical Biology, 78*, 593–610.

McTavish, C., & Sommerville, J. (1980). Macronuclear DNA organization and transcription in *Paramecium aurelia. Chromosoma, 78*, 147–164.

Medlin, L., Elwood, H. J., Stickel, S., & Sogin, M. L. (1988). The characterization of enzymatically amplified eukaryote 16S–like rRNA coding regions. *Gene, 71*, 491–499.

Meier, R., Lefort–Tran, M., Pouphile, M., Reisser, W., & Reisser, W. (1984). Comparative freeze–fracture study of the perialgal and digestive vacuoles in *Paramecium bursaria. Journal of Cell Science, 71*, 121–140.

Menzies, E., Das, N., & Wood, D. C. (2004). Behaviors producing photodispersal in *Stentor coeruleus. Photochemistry and Photobiology, 80*, 401–407.

Messick, G. A., & Small, E. B. (1996). *Mesanophrys chesapeakensis* n. sp., a histophagous ciliate in the blue crab, *Callinectes sapidus*, and associated histopathology. *Invertebrate Biology, 115*, 1–12.

Metcalf, M. M. (1923). The opalinid ciliate infusorians. *Bulletin of the United States National Museum, 120*, 1–484.

Metcalf, M. M. (1940). Further studies on the opalinid ciliate infusorians and their hosts. *Proceedings of the United States National Museum, 87*, 465–635.

Méténier, G. (1977). Effects of cell density, culture age and "stomatin" on the yield of the microstome–macrostome transformation in *Tetrahymena paravorax. Protistologica, 12*, (year 1976): 389–398.

Méténier, G. (1984a). Quelques aspects biochimiques du cycle cellulaire des ciliés. In P. P. Grassé (Ed.), *Traité de zoologie: Vol. 2, Fasc. 1. Infusoires ciliés* (pp. 503–535). Paris: Masson.

Méténier, G. (1984b). Actin in *Tetrahymena paravorax*: Ultrastructural localization of HMM–binding filaments in glycerinated cells. *Journal of Protozoology, 31*, 205–215.

Méténier, G., & Grolière, C. A. (1979). Précisions temporelles sur les processus de la stomatogenèse de *Tetrahymena paravorax* analysés en microscopie photonique. *Journal of Protozoology, 26*, 75–82.

Meyer, E., & Duharcourt, S. (1996). Epigenetic regulation of programmed genomic rearrangements in *Paramecium aurelia. Journal of Eukaryotic Microbiology, 43*, 453–461.

Meyer, E. B., & Nanney, D. L. (1987). Isozymes in the ciliated protozoan *Tetrahymena. Isozymes: Current Topics in Biological and Medical Research, 13*, 61–101.

Meyer, E., Butler, A., Dubrana, K., Duharcourt, S., & Caron, F. (1997). Sequence–specific epigenetic effects of the maternal somatic genome on developmental rearrangements of the zygotic genome in *Paramecium aurelia. Molecular and Cellular Biology, 17*, 3589–3599.

Meyer, F., Schmidt, H. J., Plümper, E., Hasilik, A., Mersmann, G., Meyer, H. E., Engström, A., & Heckmann, K. (1991). UGA is translated as cysteine in pheromone 3 of *Euplotes octocarinatus. Proceedings of the National Academy of Sciences of the United States of America, 88*, 3758–3761.

Miao, W., Yu, Y. H., & Shen, Y. F. (2001). Phylogenetic relationships of the subclass Peritrichia (Oligohymenophorea, Ciliophora) with emphasis on the genus *Epistylis*, inferred from small subunit rRNA gene sequences. *Journal of Eukaryotic Microbiology*, *48*, 583–587.

Miao, W., Yu, Y. H., Shen, Y. F., & Zhang, X. Y. (2004a). Intraspecific phylogeography of *Carchesium polypinum* (Peritrichia, Ciliophora) from China, inferred from 18S–ITS1–5.8S ribosomal DNA. *Science in China, Series C., Life Sciences*, *47*, 11–17.

Miao, W., Fen, W. S., Yu, Y. H., Zhang, X. Y., & Shen, Y. F. (2004b). Phylogenetic relationships of the subclass Peritrichia (Oligohymenophorea, Ciliophora) inferred from small subunit rRNA gene sequences. *Journal of Eukaryotic Microbiology*, *51*, 180–186.

Miao, W., Simpson, A. G. B., Fu, C., & Lobban, C. S. (2005). The giant zooxanthellae–bearing ciliate *Maristentor dinoferus* (Heterotrichea) is closely related to Folliculinidae. *Journal of Eukaryotic Microbiology*, *52*, 11–16.

Middlebrook, K., Emerson, C. W., Roff, J. C., & Lynn, D. H. (1987). Distribution and abundance of tintinnids in the Quoddy Region of the Bay of Fundy. *Canadian Journal of Zoology*, *65*, 594–601.

Miceli, C., La Terza, A., & Melli, M. (1989). Isolation and structural characterization of cDNA clones encoding the mating pheromone Er–1 secreted by the ciliate *Euplotes raikovi. Proceedings of the National Academy of Sciences of the United States of America*, *86*, 3016–3020.

Miceli, C., La Terza, A., Bradshaw, R. A., & Luporini, P. (1992). Identification and structural characterization of a cDNA clone encoding a membrane–bound form of the polypeptide pheromone E*r*–1 in the ciliated protozoan *Euplotes raikovi. Proceedings of the National Academy of Sciences of the United States of America*, *89*, 1988–1992.

Michalowski, T., Muszynski, P., & Landa, I. (1991). Factors influencing the growth of rumen ciliates *Eudiplodinium maggii* in vitro. *Acta Protozoologica*, *30*, 115–120.

Michalowski, T., Belzecki, G., Kwiatkowska, E., & Pajak, J. J. (2003). The effect of selected rumen fauna on fibrolytic enzyme activities, bacterial mass, fibre disappearance and fermentation pattern in sheep. *Journal of Animal and Feed Sciences*, *12*, 45–64.

Millecchia, L. L., & Rudzinska, M. A. (1970). The ultrastructure of brood pouch formation in *Tokophrya infusionum. Journal of Protozoology*, *17*, 574–583.

Millecchia, L. L., & Rudzinska, M. A. (1971). The ultrastructure of nuclear division in a suctorian, *Tokophrya infusionum. Zeitschrift für Zellforschung*, *115*, 149–164.

Miller, S. (1968). The predatory behavior of *Dileptus anser. Journal of Protozoology*, *15*, 313–319.

Miltko, R., Michalowski, T., Pristas, P., Javorsky, P., & Hackstein, J. H. P. (2006). Factors influencing morphological variability of rumen ciliates from the genus *Ophryscolex. Journal of Animal and Feed Sciences*, *15*, 35–38.

Minkiewicz, R. (1912). Un cas de reproduction extraordinaire chez un protiste, *Polyspira delagei* Minkiew. *Comptes Rendus de l'Académie des Sciences, Paris*, *155*, 733–737.

Mislan, T. W., & Smith–Somerville, H. E. (1986). Food vacuole morphology and membrane retrieval in the microstomal form of *Tetrahymena vorax. Journal of Protozoology*, *33*, 172–179.

Mitchell, D. R. (2004). Speculations on the evolution of 9+2 organelles and the role of central pair microtubules. *Biology of the Cell, 96,* 691–696.

Mitchell, E. A. D., & Meisterfeld, R. (2005). Taxonomic confusion blurs the debate on cosmopolitanism versus local endemism of free–living protists. *Protist*, *156*, 263–267.

Miyake, A. (1964). Induction of conjugation by cell–free preparations of *Paramecium multimicronucleatum. Science*, *146*, 1583–1585.

Miyake, A. (1981). Cell interaction by gamones in *Blepharisma*. In D. H. O'Day & P. A. Horgen (Eds.), *Sexual interactions in eukaryotic microbes* (pp. 95–129). New York: Academic Press.

Miyake, A. (1996). Fertilization and sexuality in ciliates. In K. Hausmann & P. C. Bradbury (Eds.), *Ciliates. Cells as organsisms* (pp. 243–290). Stuttgart: Gustav Fischer Verlag.

Miyake, A., & Beyer, J. (1974). Blepharmone: A conjugation–inducing glycoprotein in the ciliate *Blepharisma. Science*, *185*, 621–623.

Miyake, A., & Bleyman, L. K. (1976). Gamones and mating types in the genus *Blepharisma* and their possible taxonomic classification. *Genetical Research, Cambridge*, *27*, 267–275.

Miyake, A., & Harumoto, T. (1990). Asymmetrical cell division in *Blepharisma japonicum*: Difference between daughter cells in mating–type expression. *Experimental Cell Research*, *190*, 65–68.

Miyake, A., & Harumoto, T. (1996). Defensive function of trichocyts in *Paramecium* against the predatory ciliate *Monodinium balbiani. European Journal of Protistology*, *32*, 128–133.

Miyake, A., & Nobili, R. (1974). Mating reaction and its daily rhythm in *Euplotes crassus. Journal of Protozoology*, *21*, 584–587.

Miyake, A., Harumoto, T., Salvi, B., & Rivola, V. (1990). Defensive function of pigment granules in *Blepharisma japonicum. European Journal of Protistology*, *25*, 310–315.

Miyake, A., Harumoto, T., & Iio, H. (2001). Defence function of pigment granules in *Stentor coeruleus*. *European Journal of Protistology*, *37*, 77–88.

Miyoshi, N. T., Kawano, T., Tanaka, M., Kadono, T., Kosaka, T., Kunimoto, M., Takahashi, T., & Hosoya, H. (2003). Use of *Paramecium* species in bioassays for environmental risk management: Determination of IC50 values for water pollutants. *Journal of Health Science*, *49*, 429–435.

Mizobuchi, N., Yokoigawa, K., Harumoto, T., Fujisawa, H., & Takagi, Y. (2003). Catalase is the bacteria–derived detoxifying substance against paramecia–killing toxin in wheat grass powder infusion. *Journal of Eukaryotic Microbiology*, *50*, 299–303.

Modenutti, B. E., & Balseiro, E. G. (2002). Mixotrophic ciliates in an Andean lake: Dependence on light and prey of an *Ophrydium naumanni* population. *Freshwater Biology*, *47*, 121–128.

Modeo, L., Petroni, G., Bonaldi, M., & Rosati, G. (2001). Trichites of *Strombidium* (Ciliophora, Oligotrichida) are extrusomes. *Journal of Eukaryotic Microbiology*, *48*, 95–101.

Modeo, L., Petroni, G., Rosati, G., & Montagnes, D. J. S. (2003). A multidisciplinary approach to describe protists: Redescriptions of *Novistrombidium testaceum* Anigstein, 1914 and *Strombidium inclinatum* Montagnes, Taylor, and Lynn, 1990 (Ciliophora, Oligotrichia). *Journal of Eukaryotic Microbiology*, *50*, 175–189.

Modeo, L., Rosati, G., Andreoli, I., Gabrielli, S., Verni, F., & Petroni, G. (2006). Molecular systematics and ultrastructural characterization of a forgotten species: *Chattonidium setense* (Ciliophora, Heterotrichea). *Proceedings of the Japan Academy Series B*, *82*, 359–374.

Modigh, M. (2001). Seasonal variations of photosynthetic ciliates at a Mediterranean coastal site. *Aquatic Microbial Ecology*, *23*, 163–175.

Mogensen, M. M., & Butler, R. D. (1984). Cytological studies of *Trichophrya rotunda* (Hentschel). *Journal of Protozoology*, *31*, 101–111.

Mohamed, I., Husser, M., Sehring, I., Hentschel, J., Hentschel, C., & Plattner, H. (2003). Refilling of cortical calcium stores in *Paramecium*: In situ analysis in correlation with store–operated calcium influx. *Cell Calcium*, *34*, 87–96.

Mohr, J. L. (1966). On the age of the ciliate group Chonotricha. In H. Barnes (Ed.), *Some contemporary studies in marine science* (pp. 535–543). London: G. Allen & Unwin.

Mohr, J. L., Matsudo, H., & Leung, Y. M. (1970). The ciliate taxon Chonotricha. *Oceanography and Marine Biology. An Annual Review*, *8*, 415–456.

Mohr, S., Gerten, D., & Adrian, R. (2002). Minimal algal food requirements in the presence of protozoan prey for the rotifer *Brachionus calyciflorus*. *Journal of Plankton Research*, *24*, 723–728.

Mollenbeck, M. (1999). Genetic relationship of 32 cell lines of the *Euplotes octocarinatus* species complex revealed by random amplified polymorphic DNA (RAPD) fingerprinting. *Molecular Ecology*, *8*, 1971–1979.

Molloy, D. P., Karatayev, A. Y., Burlakova, L. E., Kurandina, D. P., & Laruelle, F. (1997). Natural enemies of zebra mussels: Predators, parasites, and ecological competitors. *Revues in Fisheries Science*, *5*, 27–97.

Molloy, D. P., Lynn, D. H., & Giamberini, L. (2005). *Ophryoglena hemophaga* n. sp. (Ciliophora: Ophryoglenidae), a parasite of the digestive gland of Zebra Mussels (*Dreissena polymorpha*). *Diseases of Aquatic Organisms*, *65*, 237–244.

Montagnes, D. J. S. (1996). Growth responses of planktonic ciliates in the genera *Strobilidium* and *Strombidium*. *Marine Ecology–Progress Series*, *130*, 241–254.

Montagnes, D. J. S. (1999). Population dynamics of the marine planktonic ciliate *Strombidinopsis multiauris*: Its potential to control phytoplankton blooms. *Aquatic Microbial Ecology*, *20*, 167–181.

Montagnes, D. J. S., & Lynn, D. H. (1989). The annual cycle of *Mesodinium rubrum* in the waters surrounding the Isles of Shoals, Gulf of Maine. *Journal Plankton Research*, *11*, 193–201.

Montagnes, D. J. S., & Lynn, D. H. (1991). Taxonomy of choreotrichs, the major marine planktonic ciliates, with emphasis on the aloricate forms. *Marine Microbial Food Webs*, *5*, 59–74.

Montagnes, D. J. S., & Lynn, D. H. (1993). A quantitative protargol stain (QPS) for ciliates and other protists. In P. F. Kemp, B. F. Sherr, E. B. Sherr, & J. J. Cole (Eds.), *Aquatic microbial ecology* (pp. 229–240). Boca Raton, FL: Lewis Publishers.

Montagnes, D. J. S., & Weisse, T. (2000). Fluctuating temperatures affect growth and production rates of planktonic ciliates. *Aquatic Microbial Ecology, 21*, 97–102.

Montagnes, D. J. S., Lynn, D. H., Roff, J. C., & Taylor, W. D. (1988). The annual cycle of heterotrophic planktonic ciliates in the waters surrounding the Isles of Shoals, Gulf of Maine: An assessment of their trophic role. *Marine Biology*, *99*, 21–30.

Montagnes, D. J. S., Wilson, D., Brooks, S. J., Lowe, C., & Campey, M. (2002). Cyclical behaviour of the tide–pool ciliate *Strombidium oculatum*. *Aquatic Microbial Ecology*, *28*, 55–68.

Moore, B. E., & Dehority, B. A. (1993). Effects of diet and hindgut defaunation on diet digestibility and microbial concentrations in the cecum and colon of the horse. *Journal of Animal Science*, *71*, 3350–3358.

Morado, J. F., & Small, E. B. (1994). Morphology and stomatogenesis of *Mesanophrys pugettensis* n. sp. (Scuticociliatida: Orchitophryidae), a facultative parasitic ciliate of the Dungeness Crab, *Cancer magister* (Crustacea: Decapoda). *Transactions of the American Microscopical Society, 113*, 343–364.

Morado, J. F., & Small, E. B. (1995). Ciliate parasites and related diseases of Crustacea: A review. *Revues in Fisheries Science, 3*, 275–354.

Morado, J. F., Giesecke, R. H., & Syrjala, S. E. (1999). Molt related mortalities of the Dungeness crab *Cancer magister* caused by a marine facultative ciliate *Mesanophrys pugettensis. Diseases of Aquatic Organisms, 38*, 143–150.

Morat, G. (1982). Variations quantitatives des teneurs en AND nucléaire lors de la reproduction végétative chez le cilié *Ophryoglena catenula* Savoie, 1966. *Archiv für Protistenkunde, 125*, 137–161.

Morat, G., Chessa, M. G., & Crippa–Franceschi, T. (1981). Étude de la régulation des teneurs en ADN nucléaire au cours de la reproduction végétative et de l'enkystement d'attente chez le cilié *Colpoda cucullus. Protistologica, 17*, 313–329.

Moreira, D., Le Guyader, H., & Philippe, H. (1999). Unusually high evolutionary rate of the elongation factor 1α genes from the Ciliophora and its impact on the phylogeny of eukarytoes. *Molecular Biology and Evolution, 16*, 234–245.

Moreira, D., Kervestin, S., Jean–Jean, O., & Philippe, H. (2002). Evolution of eukaryotic translation elongation and termination factors: Variations in evolutionary rate and genetic code deviations. *Molecular Biology and Evolution, 19*, 189–200.

Morelli, A., Ricci, N., & Verni, F. (1999). Orthokinetic and klinokinetic reactions in the behaviour of *Litonotus lamella* predating on *Euplotes crassus. European Journal of Protistology, 35*, 168–174.

Morin, G. B., & Cech, T. R. (1988). Phylogenetic relationships and altered genome structures among *Tetrahymena* mitochondrial DNAs. *Nucleic Acids Research, 16*, 327–346.

Morin, L. (2000). Long branch attraction effects and the status of "basal eukaryotes": Phylogeny and structural analysis of the ribosomal RNA gene cluster of the free–living diplomonad *Trepomonas agilis. Journal of Eukaryotic Microbiology, 47*, 167–177.

Moritz, C. M., Montagnes, D., Carleton, J. H., Wilson, D., & McKinnon, A. D. (2006). The potential role of microzooplankton in a northwestern Australian pelagic food web. *Marine Biology Research, 2*, 1–13.

Moss, A. G., Estes, A. M., Muellner, L. A., & Morgan, D. D. (2001). Protistan epibionts of the ctenophore *Mnemiopsis mccradyi* Mayer. *Hydrobiologia, 451*, 295–304.

Moss, A. R., Jouany, J. P., & Newbold, J. (2000). Methane production by ruminants: Its contribution to global warming. *Annales de Zootechnie, 49*, 231–253.

Mugard, H. (1949). Contribution à l'étude des infusoires hyménostomes histiophages. *Annales des Sciences Naturelles–Zoologie et Biologie Animale (Série 11), 10*(year 1948), 171–268.

Mugard, H., & Lorsignol, L. (1956). Étude de la division et de la régénération chez deux Ophryoglenidae d'eau douce: *Ophryoglena pectans* and *Deltopylum rhabdoides. Bulletin Biologique de la France et de la Belgique, 90*, 446–464.

Mulisch, M. (1988). Perinuclear organization in the heterotrich ciliate *Stentor coeruleus. Protoplasma, 143*, 170–175.

Mulisch, M., & Hausmann, K. (1983). Lorica construction in *Eufolliculina* sp. (Ciliophora, Heterotrichida). *Journal of Protozoology, 30*, 97–104.

Mulisch, M., & Hausmann, K. (1984). Structure and ultrastructure of the oral apparatus of *Eufolliculina uhligi* Mulisch and Patterson, 1983. *Protistologica, 20*, 415–429.

Mulisch, M., & Hausmann, K. (1988). Transmission electron microscopical observations on stomatogenesis during metamorphosis of *Eufolliculina uhligi* (Ciliophora: Heterotrichida). *Journal of Protozoology, 35*, 450–458.

Mulisch, M., & Hausmann, K. (1989). Localization of chitin in ultrathin sections of cysts of two ciliated protozoa, *Blepharisma undulans* and *Pseudomicrothorax dubius*, using colloidal gold conjugated wheat germ agglutinin. *Protoplasma, 152*, 77–86.

Mulisch, M., Barthlott, W., & Hausmann, K. (1981). Struktur und Ultrastruktur von *Eufolliculina* spec. Schwärmer und sessiles Stadium. *Protistologica, 17*, 285–312.

Mulisch, M., Herth, W., Zugenmaier, P., & Hausmann, K. (1983). Chitin fibrils in the lorica of the ciliate *Eufolliculina uhligi*: Ultrastructure, extracellular assembly and experimental inhibition. *Biology of the Cell, 49*, 169–178.

Mulisch, M., Heep, T., Sturm, W., & Borcherding, J. (1998). Redescription of *Ascobius lentus*, a rare freshwater folliculinid (Ciliophora: Heterotrichida) from a pond in Germany. *Acta Protozoologica, 37*, 29–39.

Müller, H. (1989). The relative importance of different ciliate taxa in the pelagic food web of Lake Constance. *Microbial Ecology, 18*, 261–273.

Müller, H. (1991). *Pseudobalanion planctonicum* (Ciliophora, Prostomatida): Ecological significance of an algivorous nanociliate in a deep mesoeutrophic lake. *Journal of Plankton Research, 13*, 247–262.

Müller, H. (2002). Laboratory study of the life cycle of a freshwater strombidiid ciliate. *Aquatic Microbial Ecology, 29*, 189–197.

Müller, H., & Schlegel, A. (1999). Responses of three freshwater planktonic ciliates with different feeding modes to cryptophyte and diatom prey. *Aquatic Microbial Ecology, 17*, 49–60.

Müller, H., & Weiss, T. (1994). Laboratory and field observations on the scuticociliate *Histiobalantium* from the pelagic zone of Lake Constance, FRG. *Journal of Plankton Research, 16*, 391–401.

Müller, H., & Wunsch, C. (1999). Seasonal dynamics of cyst formation of pelagic strombidiid ciliates in a deep prealpine lake. *Aquatic Microbial Ecology, 17*, 37–47.

Müller, H., Foissner, W., & Weiss, T. (2006). Role of soil in the life cycle of *Meseres corlissi* (Ciliophora: Oligotrichea): Experiments with two clonal strains from the type locality, an astatic meadow pond. *Aquatic Microbial Ecology, 42*, 199–208.

Müller–Graf, C. D. M., Collins, D. A., & Woolhouse, M. E. J. (1996). Intestinal parasite burden in five troops of olive baboons (*Papio cynocephalus anubis*) in Gombe Stream National Park, Tanzania. *Parasitology, 112*, 489–497.

Müller, M. (1993). The hydrogenosome. *Journal of General Microbiology, 139*, 2879–2889.

Müller, O. F. (1773). *Vermium Terrestrium et Fluviatilium, seu Animalium Infusorium, Helminthicorum et Testaceorum, non Marinorum, Succincta Historia* (Vol. 1, 135 pp.). Havniae et Lipsiae:, Heineck & Faber.

Mullis, K. B., & Faloona, F. A. (1987). Specific synthesis of DNA in vitro via a polymerase catalyzed chain reaction. *Methods in Enzymology, 255*, 335–350.

Munderle, M., Sures, B., & Taraschewski, H. (2004). Influence of *Anguillicola crassus* (Nematoda) and *Ichthyophthirius multifiliis* (Ciliophora) on swimming activity of European eel *Anguilla anguilla. Diseases of Aquatic Organisms, 60*, 133–139.

Muñoz, A., Téllez, C., & Fernández–Galiano, D. (1989). Description of the infraciliature and morphogenesis in the ciliate *Urotricha ondina* n. sp. (Prorodontida, Urotrichidae). *Journal of Protozoology, 36*, 104–109.

Myung, G., Yih, W., Kim, H. S., Park, J. S., & Cho, B. C. (2006). Ingestion of bacterial cells by the marine photosynthetic ciliate *Myrionecta rubra. Aquatic Microbial Ecology, 44*, 175–180.

NNNN

Nagano, N., Iwatsuki, Y., Kamiyama, T., & Nakata, H. (2000). Effects of marine ciliates on the survivability of the first–feeding larval surgeonfish, *Paracanthurus hepatus*: Laboratory rearing experiments. *Hydrobiologia, 432*, 149–157.

Nagasawa, S. (1988). The copepod *Centropages abdominalis* as a carrier of the stalked ciliate *Zoothamnium. Hydrobiologia, 167/168*, 255–258.

Naitoh, Y., Tominaga, T., Ishida, M., Fok, A. K., Aihara, M. S., & Allen, R. D. (1997). How does the contractile vacuole of *Paramecium* expel fluid? Modelling the expulsion mechanism. *Journal of Experimental Biology, 200*, 713–721.

Nakagawa, Y., Ota, T., Endo, Y., Taki, K., & Sugisaki, H. (2004). Importance of ciliates as prey of the euphausiid *Euphausia pacifica* in the NW North Pacific. *Marine Ecology–Progress Series, 271*, 261–266.

Nakai, Y., & Imai, S. (1989). Cytochemical identification of reserve polysaccharides in rumen ciliates by microspectrophotometry. *Japanese Journal of Parasitology, 38*, 216–220.

Nakajima, Y., Ishida, M., & Mikami, K. (2002). Microtubules mediate germ–nuclear behavior after meiosis in conjugation of *Paramecium caudatum. Journal of Eukaryotic Microbiology, 49*, 74–81.

Nakamura, Y., & Hirata, A. (2006). Plankton community structure and trophic interactions in a shallow and eutrophic estuarine system, Ariake Sound, Japan. *Aquatic Microbial Ecology, 44*, 45–57.

Nakaoka, Y., Kinugawa, K., & Kurotani, T. (1987). Ca^{2+} –dependent photoreceptor potential in *Paramecium bursaria. Journal of Experimental Biology, 131*, 107–115.

Nalecz–Jawecki, G. (2004). Spirotox–*Spirostomum ambiguum* acute toxicity test – 10 years of experience. *Environmental Toxicology, 19*, 359–364.

Nalecz–Jawecki, G., & Sawicki, J. (2002). A comparison of sensitivity of Spirotox biotest with standard toxicity tests. *Archives of Environmental Contamination and Toxicology, 42*, 389–395.

Nanney, D. L. (1966). Corticotypes in *Tetrahymena. American Naturalist, 100*, 303–318.

Nanney, D. L. (1967). Cortical slippage in *Tetrahymena. Journal of Experimental Zoology, 166*, 163–170.

Nanney, D. L. (1968). Patterns of cortical stability in *Tetrahymena. Journal of Protozoology, 15*, 109–112.

Nanney, D. L. (1974). Aging and long–term temporal regulation in ciliated protozoa. A critical review. *Mechanisms of Ageing and Development, 3*, 81–105.

Nanney, D. L. (1980). *Experimental ciliatology* (p. 304). New York: Wiley.

Nanney, D. L., & McCoy, J. W. (1976). Characterization of the species of the *Tetrahymena pyriformis* complex. *Transactions of the American Microscopical Society, 95*, 664–682.

Nanney, D. L., Chen, S. S., & Meyer, E. B. (1978). Scalar constraints in *Tetrahymena* evolution. Quantitative basal body variations within and between species. *Journal of Cell Biology, 79*, 727–736.

Nanney, D. L., Cooper, L. E., Simon, E. M., & Whitt, G. S. (1980a). Isozymic characterization of three mating groups of the *Tetrahymena pyriformis* complex. *Journal of Protozoology*, *27*, 451–459.

Nanney, D. L., Nyberg, D., Chen, S. S., & Meyer, E. B. (1980b). Cytogeometric constraints in *Tetrahymena* evolution: Contractile vacuole pore positions in nineteen species of the *Tetrahymena pyriformis* complex. *American Naturalist*, *115*, 705–717.

Nanney, D. L., Meyer, E. B., Simon, E. M., & Preparata, R. M. (1989). Comparison of ribosomal and isozymic phylogenies of tetrahymenine ciliates. *Journal of Protozoology*, *36*, 1–8.

Nanney, D. L., Park, C., Preparata, R., & Simon, E. M. (1998). Comparison of the sequence differences in the variable 23S domain among sets of cryptic species of ciliated protozoa. *Journal of Eukaryotic Microbiology*, *45*, 91–100.

Nelsen, E. M. (1981). The undulating membrane of *Tetrahymena*: Formation and reconstruction. *Transactions of the American Microscopical Society*, *100*, 285–295.

Nelsen, E. M., & Debault, L. E. (1978). Transformation in *Tetrahymena pyriformis*: Description of an inducible phenotype. *Journal of Protozoology*, 113–119.

Nematbakhsh, S., & Bergquist, B. L. (1993). Periodicity and the influence of temperature and cellular size in contractile vacuole contraction intervals. *Transactions of American Microscopical Society*, *112*, 292–305.

Nenninger, U. (1948). Die Peritrichen der Umgebung von Erlangen mit besonderer Berücksichtigung ihrer Wirtsspezifität. *Zoologische Jahrbücher Systematik*, *77*, 169–266.

Nerad, T. A., Schaffer, S. A., Small, E. B., & Mangold, L. A. (1995). *Trimyema shoalsia* sp. n., an anaerobic, microaerotolerant marine ciliate from Appledore Island, Gulf of Maine, USA. *Acta Protozoologica*, *34*, 289–298.

Ng, S. F. (1990). Developmental heterochrony in ciliated protozoa: Overlap of asexual and sexual cycles during conjugation. *Biological Reviews*, *65*, 19–101.

Ng, S. F., & Williams, R. J. (1977). An ultrastructural investigation of 180°–rotated ciliary meridians in *Tetrahymena pyriformis*. *Journal of Protozoology*, *24*, 257–263.

Ngassam, P. (1983). Trois espèces nouvelles de ciliés astomes des genres *Almophrya* de Puytorac et Dragesco, 1968, *Maupasella* Cépède, 1910 et *Njinella* nov. g. endocommensaux d'annelides oligochètes de la région de Yaoundé. *Protistologica*, *19*, 131–135.

Ngassam, P., & Grain, J. (1997). Six espèces nouvelles d'Hysterocinetidae appartenant aux deux genres nouveaux *Metaptychostomum* et *Proptychostomum*, ciliés endocommensaux d'un oligochète terricole du Cameroun. *Annales des Sciences Naturelles–Zoologie et Biologie Animale, Paris, Series 13*, *18*, 41–49.

Ngassam, P., & Grain, J. (2002). Description of the new genus *Hysterocinetoides* (Ciliophora, Hysterocinetidae) and revision of the systematics of the subclass Hysterocinetia. *European Journal of Protistology*, *38*, 11–17.

Ngassam, P., de Puytorac, P., & Grain, J. (1994). On *Paraptychostomum almae* n. g., n. sp., a commensal ciliate from the digestive tract of oligochaetes of the Cameroons, in a new subclass Hysterocinetia. *Journal of Eukaryotic Microbiology*, *41*, 155–162.

Nicholls, K. H., & Lynn, D. H. (1984). *Lepidotrachelophyllum fornicis* n. gen., n. sp., a ciliate with an external layer of organic scales (Ciliophora, Litostomatea, Haptoria). *Journal of Protozoology*, *31*, 413–419.

Nicol, S. (1984). *Ephelota* sp., a suctorian found on the euphausiid *Meganyctiphanes norvegica*. *Canadian Journal of Zoology*, *62*, 744–745.

Nielsen, C. V., & Buchmann, K. (2000). Prolonged in vitro cultivation of *Ichthyophthirius multifiliis* using an EPC cell line as substrate. *Diseases of Aquatic Organisms*, *42*, 215–219.

Nielsen, T., & Villadsen, I. S. (1985). Pleiotropic effects of a temperature–sensitive mouthless mutation in *Tetrahymena thermophila* on the excretion of extracellular enzymes. *Journal of Protozoology*, *32*, 634–639.

Niess, D., Reisser, W., & Wiessner, W. (1982). Photobehaviour of *Paramecium bursaria* infected with different symbiotic and aposymbiotic species of *Chlorella*. *Planta*, *156*, 475–480.

Nikolaev, S. I., Berny, C., Fahrni, J., Bolivar, I., Polet, S., Myl'nikov, A. P., et al. (2004). The twilight of Heliozoa and rise of Rhizaria, an emerging supergroup of amoeboid eukaryotes. *Proceedings of the National Academy of Sciences of the United States of America*, *101*, 8066–8071.

Nilsson, J. R. (1969). The fine structure of *Neobursaridium gigas* (Balech). *Comptes Rendus des Travaux du Laboratoire Carlsberg*, *37*, 49–76.

Nilsson, J. R. (1976). Physiological and structural studies on *Tetrahymena pyriformis* GL. *Comptes Rendus des Travaux du Laboratoire Carlsberg*, *40*, 215–355.

Nilsson, J. R. (1979). Phagotrophy in *Tetrahymena*. In M. Levandowsky & S. H. Hutner (Eds.), *Biochemistry and physiology of protozoa* (2nd. ed., chap. 8, pp. 339–379). New York: Academic Press.

Nilsson, J. R. (1984). On starvation–induced autophagy in *Tetrahymena*. *Carlsberg Research Communications*, *49*, 323–340.

Nilsson, J. R. (1987). Structural aspects of digestion of *Escherichia coli* in *Tetrahymena*. *Journal of Protozoology*, *34*, 1–6.

Nilsson, J. R. (2000). On macronuclear "subnuclei", or genome segregation, and "amitosis" in amicronucleate *Tetrahymena pyriformis* GL. A study utilizing the effect of vanadate on nuclear division. *European Journal of Protistology*, *36*, 423–436.

Nilsson, J. R., & Coleman, J. R. (1977). Calcium–rich, refractile granules in *Tetrahymena pyriformis* and their possible role in the intracellular ion–regulation. *Journal of Cell Science*, *24*, 311–325.

Nilsson, J. R., & van Deurs, P. (1983). Coated pits and pinocytosis in *Tetrahymena*. *Journal of Cell Science*, *63*, 209–222.

Njiné, T. (1972). La transformation microstome–macrostome et macrostome–microstome chez *Tetrahymena paravorax* Corliss 1957. *Annales de la Faculté des Sciences de Cameroun*, *10*, 69–84.

Njiné, T. (1979). Compléments a l'étude des ciliés libres du Cameroun. *Protistologica*, *15*, 343–354.

Njiné, T. (1980). Structure et morphogenèse buccales du cilié *Leptopharynx* (Mermod, 1914). *Protistologica*, *15*(year 1979), 459–465.

Njiné, T., & Didier, P. (1980). Étude ultrastructurale des ciliés du genre *Leptopharynx* Mermod, 1914. *Protistologica*, *16*, 155–166.

Njiné, T., & Ngassam, P. (1993). Morphogenesis of bipartition in two species of the genus *Ptychostomum* (Ciliophora, Hysterocinetidae). *European Journal of Protistology*, *29*, 396–406.

Nobili, R. (1966). Mating types and mating type inheritance in *Euplotes minuta* Yocom (Ciliata, Hypotrichida). *Journal of Protozoology*, *13*, 38–41.

Noble, A. E. (1932). On *Tokophrya lemnarum* Stein (Suctoria) with an account of its budding and conjugation. *University of California Publications in Zoology*, *37*, 477–520.

Noguchi, M., Kurahashi, S., Kamachi, H., & Inoue, H. (2004). Control of the ciliary beat by cyclic nucleotides in intact cortical sheets from *Paramecium*. *Zoological Science*, *21*, 1167–1175.

Noirot–Timothée, C. (1960). Etude d'une famille de ciliés: les Ophryoscolecidae. Structures et ultrastructures. *Annales des Sciences Naturelles–Zoologie et Biologie Animale (Série 12)*, *2*, 527–718.

Noirot–Timothée, C., & Lom, J. (1965). L'ultrastructure de l'haplocinétie des ciliés péritriches. Comparaison avec la membrane ondulante des hyménostomes. *Protistologica*, *1*, 33–40.

Noland, L. E. (1927). Conjugation in the ciliate *Metopus sigmoides* C. and L. *Journal of Morphology*, *44*, 341–361.

Novotny, R. T., Lynn, D. H., & Evans, F. R. (1977). *Colpoda spiralis* n. sp., a colpodid ciliate found inhabiting treeholes (Colpodida, Ciliophora). *Journal of Protozoology*, *24*, 364–369.

Nozawa, Y., Iida, H., Fukushima, H., Ohki, K., & Ohnishi, S. (1974). Studies on *Tetrahymena* membranes: Temperature–induced alterations in fatty acid composition of various membrane fractions in *Tetrahymena pyriformis* and its effect on membrane fluidity as inferred by spin–label study. *Biochimica et Biophysica Acta*, *367*, 134–147.

Nyberg, D. (1974). Breeding systems and resistance to environmental stress in ciliates. *Evolution*, *28*, 367–380.

Nyberg, D. (1981a). Three new "biological" species of *Tetrahymena* (*T. hegewischi* n. sp., *T. sonneborni* n. sp., *T. nipissingi* n. sp.) And temperature tolerance of members of the "pyriformis" complex. *Journal of Protozoology*, *28*, 65–69.

Nyberg, D. (1981b). Fertility is not a function of geographic distance in *Tetrahymena*. *Journal of Heredity*, *72*, 94–96.

Nyberg, D. (1988). The species concept and breeding systems. In H. D. Görtz (Ed.), *Paramecium* (chap. 3, pp. 41–58). Berlin: Springer.

OOOO

Obolkina, L. A. (1995). New species of the family Colepidae (Prostomatida, Ciliophora) from Lake Baikal. *Zoologichesky Zhurnal*, *74*, 3–19. (in Russian with English summary)

Obolkina, L. A. (2006). Planktonic ciliates of Lake Baikal. *Hydrobiologia*, *568*(S), 193–199.

Ohman, M. D., & Snyder, R. A. (1991). Growth kinetics of the omnivorous oligotrich ciliate *Strombidium* sp. *Limnology and Oceanography*, *36*, 922–935.

Ólafsdóttir, S. H., & Svavarsson, J. (2002). Ciliate (Protozoa) epibionts of deep–water asellote isopods (Crustacea): Pattern and diversity. *Journal of Crustacean Biology*, *22*, 607–618.

Olins, D. E., & Olins, A. L. (1994). The replication band of ciliated protozoa. *International Review of Cytology*, *153*, 137–170.

Olins, A. L., Olins, D. E., Derenzini, M., Hernandez–Verdun, D., Gounon, P., Robert–Nicoud, M., et al. (1988). Replication bands and nucleoli in the macronucleus of *Euplotes eurystomus*: An ultrastructural and cytochemical study. *Biology of the Cell*, *62*, 83–93.

Olive, L. S. (1978). Sorocarp development by a newly discovered ciliate. *Science*, *202*, 530–532.

Olive, L. S., & Blanton, R. L. (1980). Aerial sorocarp development by the aggregative ciliate, *Sorogena stoianovitchae*. *Journal of Protozoology*, *27*, 293–299.

Olivotto, I., Zenobi, A., Rollo, A., Migliarini, B., Avella, M., & Carnevali, O. (2005). Breeding, rearing and feeding studies in the cleaner goby *Gobiosoma evelynae*. *Aquaculture*, *250*, 175–182.

Olmo, J. L., Tellez, C., & Esteban, G. F. (1998). *Cinetozona pyriformis* n. g., n. sp.: A relative of the ciliate genera *Urozona* and *Cinetochilum* (Ciliophora, Scuticociliatida). *Journal of Eukaryotic Microbiology, 45*, 448–451.

Olsson, P., Granéli, E., Carlsson, P., & Abreu, P. (1992). Structuring of a postspring phytoplankton community by manipulation of trophic interactions. *Journal of Experimental Marine Biology and Ecology, 158*, 249–266.

Omoto, C. K., & Kung, C. (1980). Rotation and twist of the central–pair microtubules in the cilia of *Paramecium. Journal of Cell Biology, 87*, 33–46.

Orias, E. O. (1963). Mating–type determination in variety 8, *Tetrahymena pyriformis. Genetics, 48*, 1509–1518.

Orias, E. (1976). Derivation of ciliate architecture from a simple flagellate: An evolutionary model. *Transactions of the American Microscopical Society, 95*, 415–429.

Orias, E. (1981). Probable somatic DNA rearrangements in mating type determination in *Tetrahymena thermophila*: A review and a model. *Developmental Genetics, 2*, 185–202.

Orias, E. O. (1991a). Evolution of amitosis of the ciliate macronucleus: Gain of the capacity to divide. *Journal of Protozoology, 38*, 217–221.

Orias, E. O. (1991b). On the evolution of the karyorelict ciliate life cycle: Heterophasic ciliates and the origin of ciliate binary fission. *BioSystems, 25*, 67–73.

Orias, E. O., & Pollock, N. A. (1975). Heat–sensitive development of the phagocytotic organelle in a *Tetrahymena* mutant. *Experimental Cell Research, 90*, 345–356.

Orias, E., Hamilton, E. P., & Orias, J. D. (2000). *Tetrahymena* as a laboratory organism: Useful strains, cell culture, and cell line maintenance. In D. J. Asai & J. D. Forney (Eds.), *Tetrahymena thermophila. Methods in cell biology* (Vol. 62, pp. 189–211). New York: Academic Press.

Orias, J., Hamilton, E. P., & Orias, E. (1983). A microtubule meshwork associated with gametic pronucleus transfer across a cell–cell junction. *Science, 222*, 181–184.

Ortenzi, C., Alimenti, C., Vallesi, A., De Pretoro, B., La Terza, A., & Luporini, P. (2000). The autocrine mitogenic loop of the ciliate *Euplotes raikovi*: The pheromone membrane–bound forms are the cell binding sites and potential signalling receptors of soluble pheromones. *Molecular Biology of the Cell, 11*, 1445–1455.

Ossipov, D. V., Borchsenius, O. N., & Podlipaev, S. A. (1980). Some peculiarities in organization of the nuclear apparatus in a ciliate, *Paramecium bursaria*, infected with symbiotic bacteria, *Holospora acuminata. Acta Protozoologica, 19*, 315–326. (in Russian with English summary)

Owen, R. W., Gianesella–Galvão, S. F., & Kutner, M. B. B. (1992). Discrete, subsurface layers of the autotrophic ciliate *Mesodinium rubrum* off Brazil. *Journal of Plankton Research, 14*, 97–105.

Ozaki, Y., & Yagiu, R. (1941). Studies on the marine ciliates of Japan, mainly from the Setonaikai (the inland Sea of Japan). II. *Journal of Science of the Hiroshima University, Series B, Division 1, 9*, 159–180.

Özer, A. (2000). The occurrence of three species of *Trichodina* (Ciliophora: Peritrichia) on *Cyprinus carpio* in relation to culture conditions, seasonality and host characteristics. *Acta Protozoologica, 39*, 61–66.

PPPP

Packroff, G. (2000). Protozooplankton in acidic mining lakes with special respect to ciliates. *Hydrobiologia, 433*, 157–166.

Packroff, G., & Wöfl, S. (2000). A review on the occurrence and taxonomy of heterotrophic protists in extreme acidic environments of pH values < = 3. *Hydrobiologia, 433*, 153–156.

Padnos, M., & Nigrelli, R. F. (1947). *Endosphaera engelmanni* endoparasitic in *Trichodina spheroidesi* infecting the puffer, *Sphaeroides maculatus. Zoologica, 32*, 169–172.

Pantin, C. F. A. (1951). Organic design. *Advancement of Science, London, 8*, 138–150.

Pantin, C. F. A. (1966). Homology, analogy and chemical identity in the Cnidaria. In W. J. Rees (Ed.), *The Cnidaria and their evolution. Symposia of the zoological society of London* (Vol. 16, pp. 1–17). London: Academic Press.

Paramá, A., Iglesias, R., Álvarez, M. F., Leiro, J., Aja, C., & Sanmartín, M. L. (2003). *Philasterides dicentrarchi* (Ciliophora, Scuticociliatida): Experimental infection and possible routes of entry in farmed turbot (*Scophthalmus maximus*). *Aquaculture, 217*, 73–80.

Paramá, A., Iglesias, R., Álvarez, M. F., Sanmartín, M. L., & Leiro, J. (2004). Chemotactic responses of the fish–parasitic scuticociliate *Philasterides dicentrarchi* to blood and blood components of the turbot *Scophthalmus maximus*, evaluated using a new microplate multiassay. *Journal of Microbiological Methods, 58*, 361–366.

Paramá, A., Arranz, J. A., Álvarez, M. F., Sanmartín, M. L., & Leiro, J. (2005). Ultrastructure and phylogeny of *Philasterides dicentrarchi* (Ciliophora, Scuticociliatia) from farmed turbot in NW Spain. *Parasitology*. DOI: 10.1017/S0031182005009534.

Paranjape, M. A. (1980). Occurrence and significance of resting cysts in a hyaline tintinnid, *Helicostomella subulata* (Ehre.) Jorgensen. *Journal of Experimental Marine Biology and Ecology, 48*, 23–33.

Parfenova, E. V., Afon'kin, S. Yu., Yudin, A. l. L., & Etingof, R. N. (1989). Characterization and partial purification of mating pheromone excreted by mating type II cells of the ciliate *Dileptus anser. Acta Protozoologica*, *28*, 11–21.

Parker, J. G. (1976). Cultural characteristics of the marine ciliated protozoan, *Uronema marinum* Dujardin. *Journal of Experimental Marine Biology and Ecology*, *24*, 213–226.

Patterson, D. J. (1976). Observations on the contractile vacuole complex of *Blepharisma americanum* Suzuki, 1954 (Ciliophora, Heterotrichida). *Archiv für Protistenkunde*, *118*, 235–242.

Patterson, D. J. (1980). Contractile vacuoles and associated structures: Their organization and function. *Biological Reviews*, *55*, 1–46.

Paul, R. G., Butler, R. D., & Williams, A. G. (1989). Ultrastructure of the rumen ciliate *Dasytricha ruminantium. European Journal of Protistology*, *24*, 205–215.

Paul, R. G., Williams, A. G., & Butler, R. D. (1990). Hydrogenosomes in the rumen entodiniomorphid ciliate *Polyplastron multivesiculatum. Journal of General Microbiology*, *136*, 388–396.

Paulin, J. J. (1967). The fine structure of *Nyctotherus cordiformis* (Ehrenberg). *Journal of Protozoology*, *14*, 183–196.

Paulin, J. J. (1996). Morphology and cytology of ciliates. In K. Hausmann & P. C. Bradbury (Eds.), *Ciliates: Cells as organisms* (pp. 1–40). Stuttgart: Gustav Fischer Verlag.

Paulin, J. J., & Corliss, J. O. (1964). The somatic and oral infraciliature of the enigmatic ciliate *Cyathodinium piriforme. Journal Protozoology*, *11*, 438–444.

Paulin, J. J., & Corliss, J. O. (1969). Ultrastructural and other observations which suggest suctorian affinities for the taxonomically enigmatic ciliate *Cyathodinium. Journal of Protozoology*, *16*, 216–223.

Paulin, J. J., & Krascheninnikow, S. (1973). An electron microscopic study of *Balantidium caviae. Acta Protozoologica*, *12*, 97–103.

Pearson, P. J., & Tucker, J. B. (1977). Control of shape and pattern during the assembly of a large microtubule bundle. Evidence for a microtubule–nucleating–template. *Cell and Tissue Research*, *180*, 241–252.

Peck, R. K. (1975). Morphology and morphogenesis of *Pseudomicrothorax, Glaucoma* and *Dexiotricha*, with emphasis on the types of stomatogenesis in holotrichous ciliates. *Protistologica*, *10*(year 1974), 333–369.

Peck, R. K. (1977a). Cortical ultrastructure of the scuticociliates *Dexiotricha media* and *Dexiotricha colpidiopsis* (Hymenostomata). *Journal of Protozoology*, *24*, 122–134.

Peck, R. K. (1977b). The ultrastructure of the somatic cortex of *Pseudomicrothorax dubius*: Structure and function of the epiplasm in ciliated protozoa. *Journal of Cell Science*, *25*, 367–385.

Peck, R. K. (1978). Ultrastructure of the somatic and buccal cortex of the tetrahymenine hymenostome *Glaucoma chattoni. Journal of Protozoology*, *25*, 122–134.

Peck, R. K. (1985). Feeding behavior in the ciliate *Pseudomicrothorax dubius* is a series of morphologically and physiologically distinct events. *Journal of Protozoology*, *32*, 492–501.

Peck, R. K., & Duborgel, F. (1985). Effects of cations on phagocytosis in the ciliate *Pseudomicrothorax dubius. Journal of Protozoology*, *32*, 501–508.

Peck, R. K., & Hausmann, K. (1980). Primary lysosomes of the ciliate *Pseudomicrothorax dubius*: Cytochemical identification and role in phagocytosis. *Journal of Protozoology*, *27*, 401–409.

Peck, R. K., Pelvat, B., Bolivar, I., & Haller, G. de. (1975). Light and electron microscopic observations on the heterotrich ciliate *Climacostomum virens. Journal of Protozoology*, *22*, 368–385.

Peck, R. K., Duborgel, F., Huttenlauch, I., & Haller, G. de. (1991). The membrane skeleton of *Pseudomicrothorax*. I. Isolation, structure and composition. *Journal of Cell Science*, *100*, 693–706.

Peck, R. K., Swiderski, B., & Tourmel, A. M. (1993a). Involvement of the trans–Golgi network, coated vesicles, vesicle fusion, and secretory product condensation in the biogenesis of *Pseudomicrothorax* trichocysts. In H. Plattner (Ed.), *Membrane trafficking in protozoa* (Vol. 2, pp. 1–26). Greenwich, CT: JAI Press. *Advances in Cellular Molecular Biology of Membranes*, *2A*, 1–25.

Peck, R. K., Swiderski, B., & Tourmel, A. M. (1993b). Secretory organelle biogenesis: Trichocyst formation in a ciliated protozoan. *Biology of the Cell*, *77*, 277–287.

Pedersen, M. F., & Hansen, P. J. (2003). Effects of high pH on the growth and survival of six marine heterotrophic protists. *Marine Ecology-Progress Series, 260*, 33–41.

Pedrós–Alió, C., Massana, R., Latasa, M., Garciá–Cantizano, J., & Gasol, J. M. (1995). Predation by ciliates on a metalimnetic *Cryptomonas* population: Feeding rates, impact and effects of vertical migration. *Journal of Plankton Research*, *17*, 2131–2154.

Pelvat, B., & Haller, G. de. (1976). Macronuclear DNA in *Stentor coeruleus*: A first approach to its characterization. *Genetical Research*, *27*, 277–289.

Penard, E. (1919). On *Folliculina boltoni* (S. Kent). *Journal Royal Microscopical Society, London*, *1919*, 305–319.

Perasso, R., & Beisson, J. (1978). Temporal pattern of mitochondrial multiplication during the cell cycle of *Paramecium. Biology of the Cell*, *32*, 275–290.

Pérez, M. T., Dolan, J. R., & Fukai, E. (1997). Planktonic oligotrich ciliates in the NW Mediterranean: Growth rates and consumption by copepods. *Marine Ecology–Progress Series*, *155*, 89–101.

Perez–Paniagua, F., & Perez–Silva, J. (1978). Réorganisation buccale au cours de la division chez *Tillina* sp. (cilié Colpodida). *Protistologica*, *14*, 405–411.

Perez–Paniagua, F., Perez–Silva, J., & de Puytorac, P. (1979a). Observation, en microscope électronique, de certains stades du développement d l'infraciliature buccale, au cours de la morphogenèse de division du cilié *Tillina* sp. *Protistologica*, *15*, 531–539.

Perez–Paniagua, F., Perez–Silva, J., & de Puytorac, P. (1979b). Étude structurale et ultrastructurale de la stomatogenèse de bipartition du cilié *Colpoda steinii* Maupas, 1883. *Protistologica*, *15*, 151–161.

Perez–Paniagua, F., de Puytorac, P., & Savoie, A. (1980). Caractéristiques de la stomatogenèse et des ultrastructures corticales et buccale du cilié Colpodidea *Bursaria truncatella* O. F. Müller, 1773. *Journal of Protozoology*, *27*, 300–308.

Pérez–Uz, B. (1995). Growth rate variability in geographically diverse clones of *Uronema* (Ciliophora: Scuticociliatida). *FEMS Microbiology Ecology*, *16*, 193–204.

Pérez–Uz, B. (1996). Bacterial preferences and growth kinetic variation in *Uronema marinum* and *Uronema nigricans* (Ciliophora: Scuticociliatida). *Microbial Ecology*, *31*, 189–198.

Pérez–Uz, B., Song, W., & Warren, A. (1996). Morphogenetic processes in division of *Uronema filificum* Kahl 1931 (Ciliophora: Scuticociliatida) and implications for its systematic position in the family Uronematidae. *European Journal of Protistology*, *32*, 262–270.

Perez–Uz, B., Franco, C., Martin–Cereceda, M., Arregui, L., Campos, I., Serrano, S., et al. (1998). Biofilm characterization of several wastewater treatment plants with rotating biological contactors in Madrid (Spain). *Water Science and Technology*, *37*, 215–218.

Pérez Reyes, R., & López–Ochoterena, E. (1963). *Sphaerophyra sol* (Ciliata: Suctoria) parasitic in some Mexican ciliates. *Journal of Parasitology*, *49*, 697.

Pernberg, J., & Machemer, H. (1995). Voltage–dependence of ciliary activity in the ciliate *Didinium nasutum*. *Journal of Experimental Biology*, *198*, 2537–2545.

Perret, E., Albert, M., Bordes, N., & Soyer–Gobillard, M. O. (1991). Microtubule spindle and centrosome structure during the cell cycle in a dinoflagellate *Crypthecodinium cohnii* B.: An immunocytochemical study. *BioSystems*, *25*, 53–65.

Perriss, S. J., Laybourn–Parry, J., & Jones, R. I. (1994). Chlorophyll contents and photosynthetic rates of the freshwater mixotrophic ciliate *Strombidium viride* (Ciliophora: Oligotrichida). *Archiv für Hydrobiologie*, *130*, 473–483.

Perriss, S. J., Laybourn–Parry, J., & Marchant, H. J. (1995). Widespread occurrence of populations of the unique autotrophic ciliate *Mesodinium rubrum* (Ciliophora: Haptorida) in brackish and saline lakes of the Vestfold Hills (eastern Antarctica). *Polar Biology*, *15*, 423–428.

Peters, H., Van As, L. L., Basson, L., & Van As, J. G. (2004). A new species of *Ellobiophrya* Chatton et Lwoff, 1923 (Ciliophora: Peritrichia) attached to *Mantoscyphidia* Jankowski, 1980 (Ciliophora: Peritrichia) species. *Acta Protozoologica*, *43*, 163–172.

Peters–Regehr, T., Kusch, J., & Heckmann, K. (1997). Primary structure and origin of a predator released protein that induces defensive morphological changes in *Euplotes*. *European Journal of Protistology*, *33*, 389–395.

Peterson, D. S., Gao, Y., Asokan, Y., & Gaertig, J. (2002). The circumsporozoite protein of *Plasmodium falciparum* is expressed and localized to the cell surface in the free-living ciliate *Tetrahymena thermophila*. *Molecular and Biochemical Parasitology*, *122*, 119–126.

Petroni, G., Spring, S., Schleifer, K. H., Verni, F., & Rosati, G. (2000). Defensive extrusive ectosymbionts of *Euplotidium* (Ciliophora) that contain microtubule–like structures are bacteria related to *Verrucomicrobia*. *Proceedings of the National Academy of Sciences of the United States of America*, *97*, 1813–1817.

Petroni, G., Dini, F., Verni, F., & Rosati, G. (2002). A molecular approach to the tangled intrageneric relationships underlying phylogeny in *Euplotes* (Ciliophora, Spirotrichea). *Molecular Phylogenetics and Evolution*, *22*, 118–130.

Pettigrosso, R. E., Barria de Cao, M. S., & Popovich, C. A. (1997). Planktonic ciliates during a diatom bloom in Bahia Blanca estuary, Argentina. I. Aloricate ciliates. *Oebalia*, *23*, 3–19.

Petz, W. (1994). Morphology and morphogenesis of *Strombidium kryalis* nov. spec. (Ciliophora; Strombidiidae) from Antarctic sea ice. *Archiv für Protistenkunde*, *144*, 185–195.

Petz, W., & Foissner, W. (1992). Morphology and morphogenesis of *Strobilidium caudatum* (Fromentel), *Meseres corlissi* n. sp., *Halteria grandinella* (Müller), and *Strombidium rehwaldi* n. sp., and a proposed phylogenetic system for oligotrich ciliates (Protozoa, Ciliophora). *Journal of Protozoology*, *39*, 159–176.

Petz, W., & Foissner, W. (1993). Morphogenesis in some freshwater tintinnids (Ciliophora, Oligotrichida). *European Journal of Protistology*, *29*, 106–120.

Petz, W., & Foissner, W. (1996). Morphology and morphogenesis of *Lamtostyla edaphoni* Berger and Foissner and *Onychodromopsis flexilis* Stokes, two hypotrichs (Protozoa: Ciliophora) from Antarctic soils. *Acta Protozoologica, 35*, 257–280.

Petz, W., & Foissner, W. (1997). Morphology and infraciliature of some soil ciliates (Protozoa, Ciliophora) from continental Antarctica, with notes on the morphogenesis of *Sterkiella histriomuscorum. Polar Record, 33*, 307–326.

Petz, W., Foissner, W., & Adam, H. (1993). Culture, food selection and growth rate in the mycophagous ciliate *Grossglockneria acuta* Foissner, 1980: First evidence of autochthonous soil ciliates. *Soil Biology & Biochemistry, 17*, 871–875.

Petz, W., Song, W., & Wilbert, N. (1995). Taxonomy and ecology of the ciliate fauna (Protozoa, Ciliophora) in the endopelagial and pelagial of the Weddell Sea, Antarctica. *Stapfia, 40*, 223.

Pfister, G. B., Auer, H., & Arndt, H. (2002). Pelagic ciliates (Protozoa, Ciliophora) of different brackish and freshwater lakes – a community analysis at the species level. *Limnologica, 32*, 147–168.

Philippe, H., & Adoutte, A. (1998). The molecular phylogeny of Eukaryota: Solid facts and uncertainties. In G. H. Coombs, K. Vickerman, M. A. Sleigh, & A. Warren (Eds.), *Evolutionary relationships among protozoa* (chap. 2, pp. 25–56). London: Chapman & Hall.

Philippe, H., Chenuil, A., & Adoutte, A. (1994). Can the Cambrian explosion be inferred from molecular phylogeny? *Development, 120*(Suppl.), 15–25.

Piccinni, E., Irato, P., Cavallini, L., & Ammermann, D. (1992). Effects of cadmium in *Stylonychia lemnae, Stylonychia notophora* and *Oxytricha granulifera*: Isolation of a cadmium-binding protein. *Journal of Protozoology, 39*, 589–593.

Piekarski, G. (1939). Cytologische Untersuchungen an einem normalen und einem mikronucleuslosen Stamm von *Colpoda steinii* Maupas. *Archiv für Protistenkunde, 92*, 117–130.

Pierce, E., Isquith, I. R., & Repak, A. J. (1978). Quantitative study of cannibal–giantism in *Blepharisma. Acta Protozoologica, 17*, 493–501.

Pierce, R. W., & Coats, D. W. (1999). The feeding ecology of *Actinophrys sol* (Sarcodina: Heliozoa) in Chesapeake Bay. *Journal of Eukaryotic Microbiology, 46*, 451–457.

Pierce, R. W., & Turner, J. T. (1992). Ecology of planktonic ciliates in marine food webs. *Reviews in Aquatic Science, 6*, 139–181.

Pierce, R. W., & Turner, J. T. (1993). Global biogeography of marine tintinnids. *Marine Ecology–Progress Series, 94*, 11–26.

Pierce, R. W., Carlton, J. T., Carlton, D. A., & Geller, J. B. (1997). Ballast water as a vector for tintinnid transport. *Marine Ecology–Progress Series., 149*, 295–297.

Pinheiro, M. D. O., Power, M. E., Butler, B. J., Dayeh, V. R., Slawson, R., Lee, L. E. J., et al. (2007). Use of *Tetrahymena thermophila* to study the role of protozoa in inactivation of viruses in water. *Applied and Environmental Microbiology, 73*, 643–649.

Pitelka, D. R. (1969). Fibrillar systems in protozoa. In T. T. Chen (Ed.), *Research in protozoology* (Vol. 3, pp. 279–388). London/New York: Pergamon Press.

Plachter, H. (1979). Licht- und rasterelektronenmikroskopische Untersuchung des Schwärmers von *Discophrya guilcherae* Matthes 1954 (Suctoria, Ciliata). *Archiv für Protistenkunde, 121*, 372–380.

Platt, L. D., & Hausmann, K. (1993). On the ultrastructure of *Bresslauides australis* Blatterer & Foissner, 1988 (Ciliophora, Colpodea). *Archiv für Protistenkunde, 143*, 297–310.

Plattner, H. (1975). Ciliary granule plaques: Membrane–intercalated particle aggregates associated with Ca^{2+} –binding sites in *Paramecium. Journal of Cell Science, 18*, 257–269.

Plattner, H., & Kissmehl, R. (2003). Molecular aspects of membrane trafficking in *Paramecium. International Review of Cytology, 232*, 185–216.

Plattner, H., Miller, F., & Bachmann, L. (1973). Membrane specializations in the form of regular membrane–to–membrane attachment sites in *Paramecium*. A correlated freeze–etching and ultrathin–sectioning analysis. *Journal of Cell Science, 13*, 687–719.

Plattner, H., Reichel, K., Matt, H., Beisson, J., Lefort–Tran, M., & Pouphile, M. (1980). Genetic dissection of the final exocytosis steps in *Paramecium tetraurelia* cells: Cytochemical determination of Ca^{2+}–ATPase activity over preformed exocytosis sites. *Journal of Cell Science, 46*, 17–40.

Plattner, H., Diehl, S., Husser, M. R., & Hentschel, J. (2006). Sub–second calcium coupling between outside medium and subplasmalemmal stores during overstimulation/depolarisation–induced ciliary beat reversal in *Paramecium. Cell Calcium, 39*, 509–516.

Poljansky, G. I. (1934). Geschlechtsprozesse bei *Bursaria truncatella* O. F. Müll. *Archiv für Protistenkunde, 81*, 420–546.

Poljansky, G. I., & Golikova, M. N. (1959). On the infusoria dwelling in the intestine of sea urchins. Part III. Infusoria of sea urchins of the Barents Sea. *Zoologichesky Zhurnal, 38*, 1138–1145. (in Russian with English summary)

Poljansky, G. I., & Raikov, I. B. (1976). Polymerization and oligomerization phenomena in protozoan evolution. *Transactions of the American Microscopical Society, 95*, 314–326.

Poljansky, G. I., & Sergejeva, G. I. (1981). The autoradiographic study of DNA synthesis in the ciliate *Bursaria truncatella* during the new macronucleus development. *Tsitologiya, 23*, 666–673. (in Russian with English summary)

Poljansky, G. I., & Strelkow, A. (1938). Etude expérimentale sur la variabilité de quelques ophryoscolécidés. *Archives de Zoologie Expérimentale et Générale, 80*, 1–123.

Pollack, S. (1974). Mutations affecting the trichocysts in *Paramecium aurelia*. I. Morphology and description of the mutants. *Journal of Protozoology, 21*, 352–362.

Pomel, S., Diogon, M., Bouchard, P., Pradel, L., Ravet, V., Coffe, G., et al. (2006). The membrane skeleton in *Paramecium*: Molecular characterization of a novel epiplasmin family and preliminary GFP expression results. *Protist, 157*, 61–75.

Pomeroy, L. R. (1974). The ocean's food web, a changing paradigm. *BioScience, 24*, 499–504.

Pond, F. R., Gibson, I., Lalucat, J., & Quackenbush, R. L. (1989). R–body producing bacteria. *Microbiological Reviews, 53*, 25–67.

Popenko, V. I., Cherny, N. E., Ivanova, J. L., & Yakovleva, M. G. (1998a). Ultrastructure of the macronucleus in the resting cysts of the ciliate *Bursaria truncatella*. *European Journal of Protistology, 34*, 18–28.

Popenko, V. I., Novikova, E. G., Churikov, N. A., Viktorova, L. S., & Cherni, N. E. (1998b). DNA size and chromatin structural organization in the macronucleus of a ciliate *Bursaria truncatella*. *Molecular Biology, 32*, 285–292.

Posch, T., & Arndt, H. (1996). Uptake of sub–micrometre and micrometre–sized detrital particles by bacterivorous and omnivorous ciliates. *Aquatic Microbial Ecology, 10*, 45–53.

Posch, T., Jezbera, J., Vrba, J., Šimek, K., Pernthaler, J., Andreatta, S., et al. (2001). Size selective feeding in *Cyclidium glaucoma* (Ciliophora, Scuticociliatida) and its effect on bacterial community structure: A study from a continuous cultivation system. *Microbial Ecology, 42*, 217–227.

Post, F. J., Borowitzka, L. J., Borowitzka, M. A., Mackay, B., & Moulton, T. (1983). The protozoa of a Western Australian hypersaline lagoon. *Hydrobiologia, 105*, 95–113.

Postberg, J., Alexandrova, O., & Lipps, H. J. (2006). Synthesis of pre–rRNA and mRNA is directed to a chromatin–poor compartment in the macronucleus of the spirotrichous ciliate *Stylonychia lemnae*. *Chromosome Research, 14*, 161–175.

Pourriot, R., Iltis, A., & Leveque–Duwat, S. (1967). Le plancton des mares natronées du Tchad. *Internationale Revue der Gesamten Hydrobiologie und Hydrographie (Leipzig), 52*, 535–543.

Powders, V. N. (1970). Altitudinal distribution of the protozoan *Cepedietta michiganensis* in the salamanders *Plethodon glutinosus* and *Plethodon jordani* in eastern Tennessee. *American Midland Naturalist, 83*, 393–403.

Powelson, E. E., Gates, M. A., & Berger, J. (1975). A biometrical analysis of 22 stocks of four syngens of *Paramecium aurelia*. *Canadian Journal of Zoology, 53*, 19–32.

Powers, P. B. A. (1933). Studies on the ciliates from sea urchins. I. General taxonomy. *Biological Bulletin, 65*, 106–121.

Pratt, J. R., & Rosen, B. H. (1983). Associations of species of *Vorticella* (Peritrichida) and planktonic algae. *Transactions of the American Microscopical Society, 102*, 48–54.

Preer, J. R., Jr. (2000). Epigenetic mechanisms affecting macronuclear development in *Paramecium* and *Tetrahymena*. *Journal of Eukaryotic Microbiology, 47*, 515–524.

Preer, J. R., Jr., Preer, L. B., & Jurand, A. (1974). Kappa and other endosymbionts in *Paramecium aurelia*. *Bacteriological Reviews, 38*, 113–163.

Preer, J. R., Jr., Preer, L. B., Rudman, B. M., & Barnett, A. J. (1985). Deviation from the universal code shown by the gene for surface protein 51A in *Paramecium*. *Nature, 314*, 188–190.

Prelle, A. (1963). Bipartition et morphogenèse chez le cilié holotriche *Woodruffia metabolica* (Johnson et Larson, 1938). *Comptes Rendus de l'Académie des Sciences, Paris, 257*, 2148–2151.

Prelle, A. (1966). Quelques aspects de l'ultrastructure du cilié *Leptopharynx costatus* Mermod. *Protistologica, 1*(year 1965), 23–28.

Premke, K., & Arndt, H. (2000). Predation on heterotrophic flagellates by protists: Food selectivity determined using a live–staining technique. *Archiv für Hydrobiologie, 150*, 17–28.

Preparata, R. M., Meyer, E. B., Preparata, F. P., Simon, E. M., Vossbrinck, C. R., & Nanney, D. L. (1989). Ciliate evolution: The ribosomal phylogenies of the tetrahymenine ciliates. *Journal of Molecular Evolution, 28*, 427–441.

Prescott, D. M. (1994). The DNA of ciliated protozoa. *Microbiological Reviews, 58*, 233–267.

Prescott, D. M. (1998). Invention and mystery in hypotrich DNA. *Journal of Eukaryotic Microbiology, 45*, 575–581.

Prescott, D. M. (1999). The evolutionary scrambling and developmental unscrambling of germline genes in hypotrichous ciliates. *Nucleic Acids Research, 27*, 1243–1250.

Prescott, D. M. (2000). Genome gymnastics: Unique modes of DNA evolution and processing in ciliates. *Nature Reviews. Genetics, 1*, 191–198.

Prescott, D. M., Murti, K. G., & Bostock, C. J. (1973). Genetic apparatus of *Stylonychia* sp. *Nature, 242*, 596–600.

Price, C. M., Adams, A. K., & Vermeesch, J. R. (1994). Accumulation of telomerase RNA and telomere protein transcripts during telomere synthesis in *Euplotes*. *Journal of Eukaryotic Microbiology*, *41*, 267–275.

Przybos, E. (1993). Species of the *Paramecium aurelia* complex in Spain. *Microbiologia SEM*, *9*, 113–117.

Przybos, E. (1995). Intraspecies differentiation of *Paramecium triaurelia*. I. Strains from Spain. *Folia Biologica*, *43*, 5–7.

Przybos, E., & Fokin, S. (1997). Species of the *Paramecium aurelia* complex Sonneborn in Germany. *Archiv für Protistenkunde*, *148*, 167–172.

Przybos, E., & Fokin, S. (2000). Data on the occurrence of species of the *Paramecium aurelia* complex world–wide. *Protistology*, *1*, 179–184.

Przybos, E., Fokin, S., Stoeck, T., & Schmidt, H. (1999). Occurrence and ecology of *Paramecium jenningsi* strains. *Folia Biologica (Krakow)*, *47*, 53–59.

Pucciarelli, S., Marziale, F., Di Giuseppe, G., Barchetta, S., & Miceli, C. (2005). Ribosomal cold adaptation: Characterization of the genes encoding the acidic ribosomal P0 and P2 proteins from the Antarctic ciliate *Euplotes focardii*. *Gene*, *360*, 103–110.

Puytorac, P. de. (1954). Contribution à l'étude cytologique et taxonomique des infusoires astomes. *Annales des Sciences Naturelles–Zoologie et Biologie Animale (Série 11)*, *16*, 85–270.

Puytorac, P. de. (1957). Nouvelles donées sur les ciliés Hoplitophryidae. I. Nécessité d'une mise au point de la classification. *Archives de Zoologie Experimentale et Générale*, *94*(Notes et Rev.), 89–120.

Puytorac, P. de. (1960). Observations sur quelques ciliés astomes des oligochètes du lac d'Ohrid (Yougoslavie). II. Familles des Hoplitophryidae (Hoplitophryinae et Maupasellinae), des Intoshellinidae et des Anoplophryidae. *Journal of Protozoology*, *7*, 278–289.

Puytorac, P. de. (1961). Les *Lomielia* nouveau genre de cilié astome, parasite des vers Megascolecidae, reposent le problème des relations phylétiques entre les deux grands groupes d'infusoires astomes: Anoplophryidae et Hoplitophryidae. *Comptes Rendus de l'Académie des Sciences, Paris*, *252*, 2939–2941.

Puytorac, P. de. (1964). Quelques aspects de l'ultrastructure du cilié: *Prorodon viridis* Ehrbg. Kahl. *Acta Protozoologica*, *2*, 147–152.

Puytorac, P. de. (1970). Les ciliés astomes Hoplitophryidae. I. Description de nouvelles espèces. *Protistologica*, *5*(year 1969), 255–268.

Puytorac, P. de. (1972). Les ciliés astomes Hoplitophryidae. II. Révision de la systématique de ce groupe. *Protistologica*, *8*, 5–42.

Puytorac, P. de. (1984a). Le cortex. In P. P. Grassé (Ed.), *Traité de zoologie*: *Vol. 2, Fasc. 1. Infusoires ciliés* (pp. 9–33). Paris: Masson.

Puytorac, P. de. (1984b). Le polymorphisme. In P. P. Grassé (Ed.), *Traité de zoologie*: *Vol. 2, Fasc. 1. Infusoires ciliés* (pp. 581–620). Paris: Masson.

Puytorac, P. de. (1994a). Phylum Ciliophora Doflein, 1901. In P. de Puytorac (Ed.), *Traité de zoologie, infusoires ciliés* (Vol. 2, pp. 1–15). Paris: Masson.

Puytorac, P. de. (1994b). Sous–classe des Rhynchodia Chatton and Lwoff, 1939. In P. de Puytorac (Ed.), *Traité de zoologie, infusoires ciliés* (Vol. 2, pp. 475–492). Paris: Masson.

Puytorac, P. de. (1994c). Sous–classe des Prostomatia Schewiakoff, 1896. In P. de Puytorac (Ed.), *Traité de zoologie, infusoires ciliés* (Vol. 2, pp. 569–584). Paris: Masson.

Puytorac, P. de. (1994d). Classe des Plagiopylea Small et Lynn, 1985. In P. de Puytorac (Ed.), *Traité de zoologie, infusoires ciliés* (Vol. 2, pp. 259–263). Paris: Masson.

Puytorac, P. de. (1994e). Sous–classe des Scuticociliatia Small, 1967. In P. de Puytorac (Ed.), *Traité de zoologie, infusoires ciliés* (Vol. 2, pp. 621–679). Paris: Masson.

Puytorac, P. de. (1994f). Sous–classe des Hysterocinetia Diesing, 1866. In P. de Puytorac (Ed.), *Traité de zoologie, infusoires ciliés* (Vol. 2, pp. 739–750). Paris: Masson.

Puytorac, P. de. (1994g). Sous–classe des Astomatia Schewiakoff, 1896. In P. de Puytorac (Ed.), *Traité de zoologie, infusoires ciliés* (Vol. 2, pp. 751–787). Paris: Masson.

Puytorac, P. de. (1994h). Sous–classe des Apostomatia Chatton et Lwoff, 1928. In P. de Puytorac (Ed.), *Traité de zoologie, infusoires ciliés* (Vol. 2, pp. 823–846). Paris: Masson.

Puytorac, P. de., & Dragesco, J. (1969a). Quatre espèces nouvelles de ciliés astomes chez les *Alma emini* (Mchlsn) (ver Criodrilinae) du Gabon. *Annales de la Station Biologique de Besse–en–Chandesse*, *3*(year 1968), 259–266.

Puytorac, P. de., & Dragesco, J. (1969b). Description de six genres nouveaux de ciliés astomes Hoplitophryidae endoparasites de vers Glossoscolecidae au Gabon. *Biologia Gabonica*, *5*, 5–27.

Puytorac, P. de., & Grain, J. (1969). Structure et ultrastructure de *Sicuophora xenopi* n. gen., n. sp., cilié hétérotriche parasite du batracien *Xenopus fraseri* Boul. *Protistologica*, *4*(year 1968), 405–414.

Puytorac, P. de., & Grain, J. (1972). Bactéries intramitochondriales et particularités de l'ultrastructure cytostomo–pharyngienne chez le cilié *Urotricha ovata* Kahl. *Comptes Rendus des Séances de la Société de Biologie et de ses Filiales*, *166*, 604–611.

Puytorac, P. de., & Grain, J. (1975). Etude de la tomitogenèse et de l'ultrastructure de *Collinia orchestiae*, cilié apostome sanguicole, endoparasite du crustacé *Orchestia gammarella* Pallas. *Protistologica*, *11*, 61–74.

Puytorac, P. de., & Grain, J. (1976). Ultrastructure du cortex buccal et évolution chez les ciliés. *Protistologica, 12*, 49–67.

Puytorac, P. de., & Mauret, P. (1956). Détermination de certaines des conditions écologiques propres aux différents ciliés parasites du tube digestif d'*Allolobophrya savignyi* G. et H. (Oligochète). *Bulletin Biologique de la France et de la Belgique, 90*, 122–141.

Puytorac, P. de., & Njiné, T. (1980). À propos des ultrastructures corticale et buccale du cilié hypostome *Nassula tumida* Maskell, 1887. *Protistologica, 16*, 315–327.

Puytorac, P. de., & Oktem, N. (1967). Observations cytologiques sur les nyctothères des genres *Nyctotherus* Leidy et *Prosicuophora* n. gen., ciliés parasites de batraciens anoures du Gabon. *Biologia Gabonica, 3*, 223–243.

Puytorac, P. de., & Perez–Paniagua, F. (1979). Une application de l'hypothèse du conservatisme structural du cortex chez les ciliés: les Colpodidea de Puytorac et coll., 1979, avec inclusion du genre *Bursaria* O.F.M., 1786. *Comptes Rendus de l'Académie des Sciences, Paris, 289*, 1163–1165.

Puytorac, P. de., & Savoie, A. (1968). Observations cytologiques et biologiques sur *Prorodon palustris* nov. sp. *Protistologica, 4*, 53–60.

Puytorac, P. de., Raikov, I. B., & Nouzarède, M. (1973a). Particularités des ultrastructures corticale et buccale du cilié marin *Geleia nigriceps* Kahl. *Comptes Rendus des Séances de la Société de Biologie et de ses Filiales, 167*, 982–985.

Puytorac, P. de., Savoie, A., & Roque, M. (1973b). Observations cytologiques et biologiques sur le cilié polymorphe *Glaucoma ferox* Savoie et de Puytorac 1971. *Protistologica, 9*, 45–63.

Puytorac, P. de., Didier, P., Detcheva, R., & Grolière, C. A. (1974a). Sur la morphogenèse de bipartition et l'ultrastructure du cilié *Cinetochilum margaritaceum* Perry. *Protistologica, 10*, 223–238.

Puytorac, P. de., Batisse, A., Bohatier, J., Corliss, J. O., Deroux, G., Didier, P., et al. (1974b). Proposition d'une classification du phylum Ciliophora Doflein, 1901 (réunion de systématique, Clermont–Ferrand). *Comptes Rendus de l'Académie des Sciences, 278*, 2799–2802.

Puytorac, P. de., Grain, J., & Rodriguez de Santa Rosa, M. (1976). À propos de l'ultrastructure corticale du cilié hypotriche *Stylonychia mytilus* Ehrbg, 1838: les caractéristiques du cortex buccal adoral et paroral des Polyhymenophora Jankowski, 1967. *Transactions of the American Microscopical Society, 95*, 327–345.

Puytorac, P. de., Grain, J., Grolière, C. A., & Lopez–Ochoterena, E. (1978). Sur l'ultrastructure du cilié *Proboveria rangiae* sp. nov. Endocommensal du lamellibranche *Rangia cuneata. Protistologica, 14*, 503–512.

Puytorac, P. de., Grolière, C. A., & Grain, J. (1979a). Nouveaux arguments (*Hysterophrya* gen. nov., *Raabellocineta* gen. nov.) En faveur de liaisons phylétiques possibles entre ciliés Astomatida et Scuticociliatida. *Comptes Rendus de l'Académie des Sciences, Paris, 289*, 857–859.

Puytorac, P. de., Perez–Paniagua, F., & Perez–Silva, J. (1979b). À propos d'observations sur la stomatogènese et l'ultrastructure du cilié *Woodruffia metabolica* (Johnson et Larson, 1938). *Protistologica, 15*, 231–243.

Puytorac, P. de., Didier, P., Detcheva, R., & Foissner, W. (1983a). Sur l'ultrastructure du cilié Colpodida *Pseudoplatyophrya nana* (Kahl, 1926). *Protistologica, 19*, 423–434.

Puytorac, P. de., Perez–Paniagua, F., Garcia–Rodriguez, T., Detcheva, R., & Savoie, A. (1983b). Observations sur la stomatogenèse du cilié Oligohymenophora *Ophryoglena mucifera* Mugard, 1948. *Journal of Protozoology, 30*, 234–247.

Puytorac, P. de., Grain, J., Legendre, P., & Devaux, J. (1984). Essai d'application de l'analyse phénétique à la classification du phylum des Ciliophora. *Journal of Protozoology, 31*, 496–507.

Puytorac, P. de., Grolière, J. C., Detcheva, R., & Grain, J. (1985). Ultrastructural observations of the vestibuliferan ciliate *Plagiopyla nasuta* Stein, 1860. *Annales des Sciences Naturelles-Zoologie et Biologie Animale, 7*, 189–198. (in French)

Puytorac, P. de., Grain, J., & Mignot, J. P. (1987). *Précis de Protistologie* (pp. 581). Paris: Boubée.

Puytorac, P. de., Batisse, A., Deroux, G., Fleury, A., Grain, J., Laval–Peuto, M., & Tuffrau, M. (1993). Proposition d'une nouvelle classification du phylum des protozoaires Ciliophora Doflein, 1901. *Comptes Rendus de l'Académie des Sciences, Life Sciences, 316*, 716–720.

Puytorac, P. de., Grain, J., & Legendre, P. (1994). An attempt at reconstructing a phylogenetic tree of the Ciliophora using parsimony methods. *European Journal of Protistology, 30*, 1–17.

Pyne, C. K. (1978). Electron microscopic studies of macronuclear development in the ciliate *Chilodonella uncinata. Cytobiologie, 18*, 145–160.

Pyne, C. K., & Tuffrau, M. (1970). Structure et ultrastructure de l'appareil cytopharyngien et des tubules complexes en relation avec celui–ci chez le cilié gymnostome *Chilodonella uncinata* Ehrbg. *Journal de Microscopie (Paris), 9*, 503–516.

Pyne, C. K., Ruch, F., Leeman, U., & Schneider, S. (1974). Development of the macronuclear anlage in the ciliate *Chilodonella uncinata*. I. Morphological and cytophotometric studies on the evolution of DNA. *Chromosoma, 48*, 225–238.

QQQQ

Quackenbush, R. L. (1988). Endosymbionts of killer paramecia. In H. D. Görtz (Ed.), *Paramecium* (chap. 23, pp. 406–418). Berlin: Springer.

RRRR

Raabe, J., & Raabe, Z. (1961). Urceolariidae from fresh–water and terrestrial molluscs in Poland. *Acta Parasitologica Polonica*, *9*, 141–152.

Raabe, Z. (1964). The taxonomic position and rank of Peritricha. *Acta Protozoologica, 2,* 19–32.

Raabe, Z. (1967). Ordo Thigmotricha (Ciliata–Holotricha). I. *Acta Protozoologica*, *5*, 1–36.

Raabe, Z. (1970a). Ordo Thigmotricha (Ciliata–Holotricha). II. Familia Hemispeiridae. *Acta Protozoologica*, *7*, 117–180.

Raabe, Z. (1970b). Ordo Thigmotricha (Ciliata–Holotricha). III. Familiae Ancistrocomidae et Sphenophryidae. *Acta Protozoologica*, *7*, 385–463.

Raabe, Z. (1971a). The morphogenetic principles of Sewertzoff, their extension and application to Protozoa. *Acta Protozoologica*, *9*, 1–22.

Raabe, Z. (1971b). Ordo thigmotricha (ciliata–holotricha). IV. Familia thigmophryidae. *Acta Protozoologica*, *9*, 121–170.

Raabe, Z. (1972). Ordo Thigmotricha (Ciliata–Holotricha). V. Familiae Hysterocinetidae et Protoanoplophryidae. *Acta Protozoologica*, *10*, 115–184. (posthumous publication)

Radek, R., & Hausmann, K. (1996). Phagotrophy of ciliates. In K. Hausmann & P. C. Bradbury (Eds.), *Ciliates: Cells as organisms* (pp. 197–219). Stuttgart: Gustav Fischer Verlag.

Radzikowski, S. (1973). Die Entwicklung des Kernapparates und die Nukleinsäuresynthese während der Konjugation von *Chilodonella cucullulus* O. F. Müller. *Archiv für Protistenkunde*, *115*, 419–428.

Radzikowski, S. (1979). Asynchronous replication of polytene chromosome segments of the new macronucleus anlage in *Chilodonella cucullulus* O. F. Müller. *Protistologica*, *15*, 521–526.

Radzikowski, S. (1985). Replication, division and mechanisms controlling the variable DNA content in the heteromeric macronucleus of *Chilodonella steini* (Ciliata). *Archiv für Protistenkunde*, *130*, 381–396.

Radzikowski, S., & Golembiewska, M. (1977). *Chilodonella steini*. Morphology and culture method. *Protistologica*, *13*, 381–389.

Radzikowski, S., & Golembiewska–Skoczylas, M. (1999). Kinety shifts and kinety number regulation in *Trithigmostoma steini*. *European Journal of Protistology*, *35*, 136–145.

Radzikowski, S., & Steinbrück, G. (1990). Location of rdna in the heteromeric macronucleus of *Chilodonella steini*. *European Journal of Protistology*, *25*, 249–254.

Raffioni, S., Luporini, P., & Bradshaw, R. A. (1989). Purification, characterization, and amino acid sequence of the mating pheromone E*r*–10 of the ciliate *Euplotes raikovi*. *Biochemistry*, *28*, 5250–5256.

Raikov, I. B. (1969). The macronucleus of ciliates. In T. T. Chen (Ed.), *Research in protozoology* (Vol. 3, pp. 1–128). London/New York: Pergamon Press.

Raikov, I. B. (1972). Nuclear phenomena during conjugation and autogamy in ciliates. In T. T. Chen (Ed.), *Research in protozoology* (Vol. 4, pp. 147–289). London/New York: Pergamon Press.

Raikov, I. B. (1974a). Three new types of extrusive organoids in lower marine ciliates. *Tsitologia*, *16*, 626–630. (in Russian with English summary)

Raikov, I. B. (1974b). Etude ultrastructurale des bactéries epizoïques et endozoïques de *Kentrophoros latum* Raikov, cilié holotriche mésopsammique. *Cahiers de Biologie Marine*, *15*, 379–393.

Raikov, I. B. (1978). Ultrastructure du cytoplasme et des nématocystes du cilié psammophile *Remanella multinucleata* (Gymnostomata, Loxodidae). Existence des nématocystes chez les ciliés. *Protistologica*, *14*, 413–432.

Raikov, I. B. (1982). *The protozoan nucleus. Morphology and evolution* (pp. 474). Vienna: Springer.

Raikov, I. B. (1985). Primitive never–dividing macronuclei of some lower ciliates. *International Review of Cytology*, *95*, 267–325.

Raikov, I. B. (1992). Unusual extrusive organelles in karyorelictid ciliates: An argument for the ancient origin of this group. *BioSystems*, *28*, 195–201.

Raikov, I. B. (1993). The orthonematocyst, a new type of extrusome found in *Remanella rugosa* and *Remanella brunnea* (Ciliophora: Karyorelictida). *European Journal of Protistology*, *29*, 81–87.

Raikov, I. B. (1994a). The nuclear apparatus of some primitive ciliates, the karyorelictids: Structure and divisional reorganization. *Bolletino di Zoologia*, *61*, 19–28.

Raikov, I. B. (1994b). Somatic cortical and endoplasmic fine structure of *Remanella granulosa* Kahl (Ciliophora, Karyorelictida). *Archiv für Protistenkunde*, *144*, 7–16.

Raikov, I. B. (1995). Structure and genetic organization of the polyploid macronucleus of ciliates: A comparative review. *Acta Protozoologica*, *34*, 151–171.

Raikov, I. B. (1996). Nuclei of ciliates. In K. Hausmann & P. C. Bradbury (Eds.), *Ciliates. Cells as organsisms* (pp. 221–242). Stuttgart: Gustav Fischer Verlag.

Raikov, I. B., & Kovaleva, V. G. (1995). Comparative ultrastructure of the cytoplasm in species of the genus *Tracheloraphis* (Ciliophora: Karyorelictida) I. Somatic cortex. *Archiv für Protistenkunde, 145*, 80–93.

Raikov, I. B., Gerassimova–Matvejeva, Z. P., & Puytorac, P. de. (1976). Cytoplasmic fine structure of the marine psammobiotic ciliate *Tracheloraphis dogieli* Raikov. I. Somatic infraciliature and cortical organelles. *Acta Protozoologica, 14*(year 1975), 17–42.

Ramoino, P., Fronte, P., Fato, M., Beltrame, F., Robello, M., & Diaspro, A. (2001). Fluid phase and receptor–mediated endocytosis in *Paramecium primaurelia* by fluorescence confocal laser scanning microscopy. *European Biophysics Journal, 30*, 305–312.

Ramsey, J. M., Brownlee, D. C., & Small, E. B. (1980). *Potomacus pottsi* Thompson, 1967 (Scuticociliatida, Ciliophora): Stomatogenesis with redescription of the microstome trophont. *Transactions of the American Microscopical Society, 99*, 43–51.

Rapport, D. J., Berger, J., & Reid, D. B. W. (1972). Determination of food preference of *Stentor coeruleus. Biological Bulletin, 142*, 103–106.

Rapport, E. W., Rapport, D. J., Berger, J., & Kupers, V. (1976). Induction of conjugation in *Stentor coeruleus. Transactions of the American Microscopical Society, 95*, 220–224.

Rasmussen, L., & Orias, E. (1975). *Tetrahymena*: Growth without phagocytosis. *Science, 190*, 464–465.

Rassoulzadegan, F. (1980). Granulometric analysis of the particles used by a tintinnid *Stenosemella ventricosa* (Clap. & Lachm.) Jörg., during lorica building. *Protistologica, 16*, 507–510.

Rassoulzadegan, F. (1982). Dependence of grazing rate, gross growth efficiency and food size range on temperature in a pelagic oligotrichous ciliate *Lohmanniella spiralis* Leeg., fed on naturally occurring particulate matter. *Annales de l'Institut Océanographique, Paris, 58*, 177–184.

Rassoulzadegan, F., & Etienne, M. (1981). Grazing rate of the tintinnid *Stenosemella ventricosa* (Clap. & Lachm.) Jörg. On the spectrum of the naturally occurring particulate matter from a Mediterranean neritic area. *Limnology and Oceanography, 26*, 258–270.

Rassoulzadegan, F., Laval–Peuto, M., & Sheldon, R. W. (1988). Partitioning of the food ration of marine ciliates between pico– and nanoplankton. *Hydrobiologia, 159*, 75–88.

Rautian, M. S., & Potekhin, A. A. (2002). Electrokaryotypes of macronuclei of several *Paramecium* species. *Journal of Eukaryotic Microbiology, 49*, 296–304.

Rawlinson, N. G., & Gates, M. A. (1985). The excystment process in the ciliate *Euplotes muscicola*: An integrated light and scanning electron microscopy study. *Journal of Protozoology, 32*, 729–735.

Ray, C., Jr. (1956). Meiosis and nuclear behavior in *Tetrahymena pyriformis. Journal of Protozoology, 3*, 88–96.

Regali–Seleghim, M. H., & Godinho, M. J. L. (2004). Peritrich epibiont protozoans in the zooplankton of a subtropical shallow aquatic ecosystem (Monjolinho Reservoir, Sao Carlos, Brazil). *Journal of Plankton Research, 26*, 501–508.

Reid, P. C. (1987). Mass encystment of a planktonic oligotrich ciliate. *Marine Biology, 95*, 221–230.

Reid, P. C., & John, A. W. G. (1978). Tintinnid cysts. *Journal of Marine Biological Association of the United Kingdom, 58*, 551–557.

Reid, P. C., & John, A. W. G. (1981). A possible relationship between chitinozoa and tintinnids. *Review of Paleobotany Palynology, 34*, 251–262.

Reid, P. C., & John, A. W. G. (1983). Resting cysts in the ciliate class Polyhymenophorea: Phylogenetic implications. *Journal of Protozoology, 30*, 710–713.

Reisser, W. (1984). The taxonomy of green algae endosymbiotic in ciliates and a sponge. *British Phycological Journal, 19*, 309–318.

Reisser, W. (1986). Endosymbiotic associations of freshwater protozoa and algae. *Progress in Protistology, 1*, 195–214.

Reisser, W., & Häder, D. P. (1984). Role of endosymbiotic algae in photokinesis and photophobic responses of ciliates. *Photochemistry and Photobiology, 39*, 673–678.

Reisser, W., Vietze, S., & Widowski, M. (1988). Taxonomic studies on endocytobiotic chlorophycean algae isolated from different American and European strains of *Paramecium bursaria. Symbiosis, 6*, 253–270.

Repak, A. J. (1968). Encystment and excystment of the heterotrichous ciliate *Blepharisma stoltei* Isquith. *Journal of Protozoology, 15*, 407–412.

Repak, A. J. (1983). Suitability of selected marine algae for growing the marine heterotrich ciliate *Fabrea salina. Journal of Protozoology, 30*, 52–54.

Repak, A. J. (1986). Suitability of selected bacteria and yeasts for growing the estuarine heterotrich ciliate *Fabrea salina* (Henneguy). *Journal of Protozoology, 33*, 219–222.

Repak, A. J., & Isquith, I. R. (1974). The systematics of the genus *Spirostomum* Ehrenberg, 1838. *Acta Protozoologica, 12*, 325–333.

Repak, A. J., Isquith, I. R., & Nabel, M. (1977). A numerical taxonomic study of the heterotrich ciliate genus *Blepharisma. Transactions of the American Microscopical Society, 96*, 204–218.

Reynolds, B. D. (1936). *Colpoda steini*, a facultative parasite of the land slug, *Agriolimax agrestis. Journal of Parasitology, 22*, 48–53.

Reynoldson, T. B. (1955). Factors influencing population fluctuations of *Urceolaria mitra* (Peritricha) epizoic on freshwater triclads. *Journal of Animal Ecology, 24*, 57–83.

Reynoldson, T. B. (1956). The population dynamics of host specificity in *Urceolaria mitra* (Peritricha) epizoic on freshwater triclads. *Journal of Animal Ecology, 25*, 127–143.

Ricci, N. (1989). Microhabitats of ciliates: Specific adaptations to different substrates. *Limnology and Oceanography, 34*, 1089–1097.

Ricci, N. (1990). The behaviour of ciliated protozoa. *Animal Behaviour, 40*, 1048–1069.

Ricci, N. (1996). Ethology of ciliates. In K. Hausmann & P. C. Bradbury (Eds.), *Ciliates: Cells as organisms* (pp. 403–416). Stuttgart: Gustav Fischer Verlag.

Ricci, N., & Esposito, F. (1981). Mating–type specific soluble factors mediating the preconjugant cell interactions in a marine species of *Blepharisma* (Protozoa, Ciliata). *Monitore Zoologico Italiano, 15*, 107–115.

Ricci, N., Cionini, K., Banchetti, R., & Erra, F. (1999). Rheotaxis in *Uronychia setigera* (Ciliata, Hypotrichida). *Journal of Eukaryotic Microbiology, 46*, 268–277.

Rickards, J. C., & Lynn, D. H. (1985). Can ciliates adjust their intermembranellar spacing to prey size? *Transactions of the American Microscopical Society, 104*, 333–340.

Rieder, N. (1971). Elektronenoptische Untersuchungen an *Didinium nasutum* O. F. Müller (Ciliata, Gymnostomata) in Interphase und Teilung. *Forma et Functio, 4*, 46–86.

Rieder, N. (1973). Elektronenoptische und histochemische Untersuchungen an der Cystenhülle von *Didinium nasutum* O. F. Müller (Ciliata, Holotricha). *Archiv für Protistenkunde, 115*, 125–131.

Rieder, N., Ott, H. A., Pfundstein, P., & Schoch, R. (1982). X–ray microanalysis of the mineral contents of some protozoa. *Journal of Protozoology, 29*, 15–18.

Riggio, D. C., Ricci, N., Banchetti, R., & Seyfert, H. M. (1987). Cannibals of *Oxytricha bifaria* (Ciliata, Hypotrichida). Macro– and micro–nuclear DNA content. *Canadian Jorunal of Zoology, 65*, 847–851.

Rigos, G., Pavlidis, M., & Divanach, P. (2001). Host susceptibility to *Cryptocaryon* sp. infection of Mediterranean marine broodfish held under intensive culture conditions: A case report. *Bulletin of the European Association Fish Pathologists, 21*, 33–36.

Riley, J. L., & Katz, L. A. (2001). Widespread distribution of extensive chromosomal fragmentation in ciliates. *Molecular Biology and Evolution, 18*, 1372–1377.

Rivera, F., Castro, F., Moreno, G., Lugo, A., Gallegos, E., & Norouzian, M. (1988). Protozoa of a rotating biological contactor treatment plant in Mexico. *Water, Air, and Soil Pollution, 42*, 281–301.

Rivera, F., Lugo, A., Ramirez, E., Bonilla, P., Calderon, A., Rodriguez, S., et al. (1992). Seasonal distribution of air–borne protozoa in Mexico City and its suburbs. *Water, Air, and Soil Pollution, 61*, 17–36.

Roberts, D. McL., Warren, A., & Curds, C. R. (1983). Morphometric analysis of outline shape applied to the peritrich genus *Vorticella*. *Systematic Zoology, 32*, 377–388.

Rodrigues de Santa Rosa, M. (1976). Observations sur l'ultrastructure du cilié *Coleps hirtus* Nitzsch, 1817. *Protistologica, 12*, 205–216.

Rojo–Herguedas, I., & Olmo, J. L. (1999). The ciliated protozoa of the pitcher plant *Sarracenia purpurea*. *Acta Protozoologica, 38*, 155–159.

Roque, M. (1961). Recherches sur les infusoires ciliés: les hyménostomes péniculiens. *Bulletin Biologique de la France et de la Belgique, 95*, 432–519.

Roque, M., Puytorac, P. de, & Savoie, A. (1971). Caractéristiques morphologiques et biologlques de *Tetrahymena bergeri* sp. nov., cilié hyménostome tetrahyménien. *Protistologica, 6*(year 1970), 343–351.

Rosati, G., & Modeo, L. (2003). Extrusomes of ciliates: Diversification, distribution, and phylogenetic implications. *Journal of Eukaryotic Microbiology, 50*, 383–402.

Rosati, G., Verni, F., & Ricci, N. (1984). The cyst wall of *Oxytricha bifaria* (Ciliata Hypotrichida). III. Cytochemical investigation. *Protistologica, 20*, 197–204.

Rosati, G., Verni, F., Bracchi, P., & Dini, F. (1987). An ultrastructural analysis of the ciliated protozoon *Aspidisca* sp. *Transactions of the American Microscopical Society, 106*, 31–52.

Rosati, G., Verni, F., & Dini, F. (1998). Mating by conjugation in two species of the genus *Aspidisca* (Ciliata, Hypotrichida): An electron microscopic study. *Zoomorphology, 118*, 1–12.

Rosati, G., Modeo, L., Melai, M., Petroni, G., & Verni, F. (2004). A multidisciplinary approach to describe protists: A morphological, ultrastructural, and molecular study of *Peritromus kahli* Villeneuve–Brachon, 1940 (Ciliophora, Heterotrichea). *Journal of Eukaryotic Microbiology, 51*, 49–59.

Rose, C. R., & Finley, H. E. (1976). A method for mass cultivation of sessile peritrich protozoa. *Transactions of the American Microscopical Society, 95*, 541–544.

Roth, M., Lin, M., & Prescott, D. M. (1985). Large scale synchronous mating and the study of macronuclear development in *Euplotes crassus*. *Journal of Cell Biology, 101*, 79–84.

Routledge, L. M. (1978). Calcium–binding proteins in the vorticellid spasmoneme. Extraction and characterization by gel electrophoresis. *Journal of Cell Biology, 77*, 358–370.

Rowland, S. J., Mifsud, C., Nixon, M., & Boyd, P. (2006). Effects of stocking density on the performance of the Australian freshwater silver perch (*Bidyanus bidyanus*) in cages. *Aquaculture, 253*, 301–308.

Rudberg, P., & Sand, O. (2000). Bistable membrane potential of the ciliate *Coleps hirtus*. *Journal of Experimental Biology, 203*, 757–764.

Rudzinska, M. A. (1962). The use of a protozoan for studies on aging. III. Similarities between young overfed and old normally fed *Tokophrya infusionum*: A light and electron microscope study. *Gerontologica, 6*, 206–226.

Rudzinska, M. A. (1965). The fine structure and function of the tentacle in *Tokophrya infusionum*. *Journal of Cell Biology, 25*, 459–477.

Rudzinska, M. A. (1970). The mechanism of food intake in *Tokophrya infusionum* and ultrastructural changes in food vacuoles during digestion. *Journal of Protozoology, 17*, 626–641.

Rudzinska, M. A. (1973). Do Suctoria really feed by suction? *BioScience, 23*, 87–94.

Rudzinska, M. A. (1974). Ultrastructural localization of acid phosphatase in starved *Tokophrya infusionum*. *Journal of Protozoology, 21*, 721–728.

Rudzinska, M. A. (1980). Internalization of macromolecules from the medium in Suctoria. *Journal of Cell Biology, 84*, 172–183.

Rudzinska, M. A., Jackson, G. J., & Tuffrau, M. (1966). The fine structure of *Colpoda maupasi* with special emphasis on food vacuoles. *Journal of Protozoology, 13*, 440–459.

Ruffolo, J. J., Jr. (1976a). Fine structure of the dorsal bristle complex and pellicle of *Euplotes*. *Journal of Morphology, 148*, 469–487.

Ruffolo, J. J., Jr. (1976b). Cortical morphogenesis during the cell division cycle in *Euplotes*: An integrated study using light optical, scanning electron and transmission electron microscopy. *Journal of Morphology, 148*, 489–527.

Ruffolo, J. J., Jr. (1978). Intracellular calculi of the ciliate protozoon *Euplotes eurystomus*: Morphology, localization, and possible stages in formation. *Transactions of the American Microscopical Society, 97*, 381–386.

Ruiz, F., Garreau de Loubresse, N., & Beisson, J. (1987). A mutation affecting basal body duplication and cell shape in *Paramecium*. *Journal of Cell Biology, 104*, 417–430.

Ruthmann, A., & Hauser, M. (1974). Mitosis–like macronuclear division in a ciliate. *Chromosoma, 45*, 261–272.

Ruthmann, A., & Kuck, A. (1985). Formation of the cyst wall of the ciliate *Colpoda steinii*. *Journal of Protozoology, 32*, 677–682.

Ryan, P. G., Watkins, B. P., Lewis Smith, R. I., Dastych, H., Eicker, A., Foissner, W., et al. (1989). Biological survey of Robertskollen, western Dronning Maud Land: Area description and preliminary species lists. *South African Journal of Antarctic Research, 19*, 10–20.

SSSS

Saad, Y. M., & Doerder, F. P. (1995). Immobilization antigen variation in natural isolates of *Tetrahymena thermophila*. *European Journal of Protistology, 31*, 45–53.

Sadler, L. A., & Brunk, C. F. (1992). Phylogenetic relationships and unusual diversity in histone H4 proteins within the *Tetrahymena pyriformis* complex. *Molecular Biology and Evolution, 9*, 70–84.

Salvano, P. (1975). Comparaison du fonctionnement des bandes de réorganisation d'*Euplotes crassus* (Dujardin) et de *Strombidium* sulcatum Claparède et Linné après analyse microspectrographique en UV. *Journal of Protozoology, 22*, 230–232.

Salt, G. W. (1967). Predation in an experimental protozoa population (*Woodruffia–Paramecium*). *Ecological Monographs, 37*, 113–144.

Salt, G. W. (1974). Predator and prey densities as controls of the rate of capture by the predator *Didinium nasutum*. *Ecology, 55*, 434–439.

Salt, G. W. (1975). Changes in the cell volume of *Didinium nasutum* during population increase. *Journal of Protozoology, 22*, 112–115.

Salt, G. W. (1979). Density, starvation, and swimming rate in *Didinium* populations. *American Naturalist, 113*, 135–143.

Sánchez–Silva, R., Villalobo, E., Morin, L., & Torres, A. (2003). A new noncanonical nuclear genetic code: Translation of UAA into glutamate. *Current Biology, 13*, 442–447.

Sanders, R. W. (1988). Feeding by *Cyclidium* sp. (Ciliophora, Scuticociliatida) on particles of different sizes and surface properties. *Bulletin of Marine Science, 43*, 446–457.

Sanders, R. W. (1991). Mixotrophic protists in marine and freshwater ecosystems. *Journal of Protozoology, 38*, 76–81.

Sanders, R. W. (1995). Seasonal distributions of the photosynthesizing ciliates *Laboea strobila* and *Myrionecta rubra* (=*Mesodinium rubrum*) in an estuary of the Gulf of Maine. *Aquatic Microbial Ecology, 9*, 237–242.

Sanders, R. W., & Wickham, S. A. (1993). Planktonic protozoa and metazoa: Predation, food quality and population control. *Marine Microbial Food Webs, 7*, 197–223.

Sand–Jensen, K., Pedersen, O., & Geertz–Hansen, O. (1997). Regulation and role of photosynthesis in the colonial symbiotic ciliate *Ophrydium versatile*. *Limnology and Oceanography*, *42*, 866–873.

Sandon, H. (1941a). Studies on South African endozoic ciliates. I. *Paranyctotherus (Balantidium) kirbyi* (Rodriguez) emend., gen. nov. from the rectum of the clawed toad, *Xenopus laevis*. *South African Journal of Medical Sciences*, *6*, 116–127.

Sandon, H. (1941b). Studies on South African endozoic ciliates. II. *Meiostoma georhychi* gen. nov., sp. nov., from the caecum of the *Georhychus capensis*. *South African Journal of Medical Sciences*, *6*, 128–135.

Sanger, F., Nicklen, S., & Coulson, A. R. (1977). DNA sequencing with chain–terminating inhibitors. *Proceedings of the National Academy of Sciences of the United States of America*, *74*, 5463–5467.

Santos, S. M., Guinea, A., & Fernández–Galiano, D. (1986). Étude de l'infraciliature et du processus de bipartition chez *Nyctotherus ovalis* Leidy, 1850 (Ciliophora, Heterotrichida). *Protistologica*, *22*, 351–358.

Santovito, G., Irato, P., Palermo, S., Boldrin, F., Hunziker, P., & Piccinni, E. (2001). Identification, cloning and characterisation of a novel copper–metallothionein in *Tetrahymena pigmentosa*. Sequencing of cDNA and expression. *Protist*, *152*, 219–229.

Sapra, G. R., & Kloetzel, J. A. (1975). Programmed autophagocytosis accompanying conjugation in the ciliate *Stylonychia mytilus*. *Development Biology*, *42*, 84–94.

Sasaki, T., Konoha, Y., Toyoda, T., Yasaka, Y., Przybos, E., & Nakaoka, Y. (2006). Correlation between thermotolerance and membrane properties in *Paramecium aurelia*. *Journal of Experimental Biology*, *209*, 3580–3586.

Satir, B. H. (1989). Signal transduction events associated with exocytosis in ciliates. *Journal of Protozoology*, *36*, 382–389.

Satir, B. H., & Wissig, S. L. (1982). Alveolar sacs of *Tetrahymena*: Ultrastructural characteristics and similarities to subsurface cisterns of muscle and nerve. *Journal of Cell Science*, *55*, 13–33.

Satir, B. H., Schooley, C., & Satir, P. (1973). Membrane fusion in a model system. Mucocyst secretion in *Tetrahymena*. *Journal of Cell Biology*, *56*, 153–176.

Satir, P., & Barkalow, K. L. (1996). Cilia: Structure and molecular biology. In K. Hausmann & P. C. Bradbury (Eds.), *Ciliates: Cells as organisms* (pp. 355–377). Stuttgart: Gustav Fischer Verlag.

Sattler, C. A., & Staehelin, L. A. (1974). Ciliary membrane differentiation in *Tetrahymena pyriformis*. *Tetrahymena* has four types of cilia. *Journal of Cell Biology*, *62*, 473–490.

Sattler, C. A., & Staehelin, L. A. (1979). Oral cavity of *Tetrahymena pyriformis*. A freeze–fracture and high–voltage electron microscopy study of the oral ribs, cytostome, and forming food vacuole. *Journal of Ultrastructure Research*, *66*, 132–150.

Sawicka, K., Kaczankowski, A., & Kaczanowska, J. (1983). Kinetics of ingestion and egestion of food vacuoles during cell cycle of *Chilodonella steini*. *Acta Protozoologica*, *22*, 157–167.

Schlegel, M., & Eisler, K. (1996). Evolution of ciliates. In K. Hausmann & P. C. Bradbury (Eds.), *Ciliates. Cells as organsisms* (pp. 73–94). Stuttgart: Gustav Fischer Verlag.

Schlegel, M., & Steinbrück, G. (1986). Interspezifische Variabilität bei hypotrichen Ciliaten. *Zeitschrift für Zoologische Systematik und Evolutionsforschung*, *24*, 247–266.

Schlegel, M., Kramer, M., & Hahn, K. (1988). Taxonomy and phylogenetic relationship of eight species of the genus *Euplotes* (Hypotrichida, Ciliophora) as revealed by enzyme electrophoresis. *European Journal of Protistology*, *24*, 22–29.

Schlott–Idl, K. (1984). Qualitative und quantitative Untersuchungen der pelagischen Ciliaten des Piburger Sees (Tirol, Österreich). *Limnologica*, *15*, 43–54.

Schmahl, G., Mehlhorn, H., & Taraschewski, H. (1989). Treatment of fish parasites. 5. The effects of sym. Triazinone (Toltrazuril) on fish parasitic Ciliophora (*Ichthyophthirius multifiliis* Fouquet, 1876, *Apiosoma amoebae* Grenfell, 1884, *Trichodina* sp. Ehrenberg, 1831). *European Journal of Protistology*, *24*, 152–161.

Schmidt, A. R., von Eynatten, H., & Wagreich, M. (2001). The Mesozoic amber of Schliersee (southern Germany) is Cretaceous in age. *Cretaceous Research*, *22*, 423–428.

Schmidt, H. J. (1982). New methods for cultivating, harvesting, and purifying mass cultures of the hypotrich ciliate *Euplotes aediculatus*. *Journal of Protozoology*, *29*, 132–135.

Schmidt, H. J. (1996). Molecular biology of ciliates. In K. Hausmann & P. C. Bradbury (Eds.), *Ciliates. Cells as organsisms* (pp. 325–353). Stuttgart: Gustav Fischer Verlag.

Schmidt, S. L., Ammerman, D., Schlegel, M., & Bernhard, D. (2006a). *Stylonychia lemnae* strains from a North American site differ in a single nucleotide within the small subunit rDNA from Eurasian strains. *Journal of Eukaryotic Microbiology*, *53*, 308–309.

Schmidt, S. L., Bernhard, D., Schlegel, M., & Fried, J. (2006b). Fluorescence in situ hybridzation with specific oligonucleotide rRNA probes distinguishes the sibling species *Stylonychia lemnae* and *Stylonychia mytilus* (Ciliophora, Spirotrichea). *Protist*, *157*, 21–30.

Schmidt, S. L., Bernhard, D., Schlegel, M., & Foissner, W. (2007). Phylogeny of the Stichotrichia (Ciliophora; Spirotrichea) reconstructed with nuclear small subunit rRNA gene sequences. Discrepancies and accord-

ances with morphological data. *Journal of Eukaryotic Microbiology, 54*, 201–209.
Schmidt, S. L., Foissner, W., Schlegel, M., & Bernhard, D. (2007). Molecular phylogeny of the Heterotrichea (Ciliophora, Postciliodesmatophora) based on small subunit rRNA gene sequences. *Journal of Eukaryotic Microbiology, 54,* 358–363.
Schnittler, M., Unterseher, M., & Tesmer, J. (2006). Species richness and ecological characterization of myxomycetes and myxomycete–like organisms in the canopy of a temperate deciduous forest. *Mycologia, 98*, 223–232.
Schönborn, W., Dörfelt, H., Foissner, W., Krientiz, L., & Schäfer, U. (1999). A fossilized microcenosis in Triassic amber. *Journal of Eukaryotic Microbiology, 46*, 571–584.
Schrenk, H. G., & Bardele, C. F. (1987). The fine structure of the paralabial organelle in the rumen ciliate *Ophryoscolex purkinjei* Stein, 1858. *Journal of Protozoology, 34*, 97–104.
Schrenk, H. G., & Bardele, C. F. (1991). The fine structure of *Saprodinium dentatum* Lauterborn, 1908 as a representative of the Odontostomatida (Ciliophora). *Journal of Protozoology, 38*, 278–293.
Schulz, S., Wagener, S., & Pfennig, N. (1990). Utilization of various chemotrophic and phototrophic bacteria as food by the anaerobic ciliate *Trimyema compressum. European Journal of Protistology, 26*, 122–131.
Schwarz, M. V. J., & Frenzel, P. (2005). Methanogenic symbionts of anaerobic ciliates and their contribution to methanogenesis in an anoxic rice field soil. *FEMS Microbiology Ecology, 52*, 93–99.
Schwartz, V. (1978). Struktur und Entwicklung des Makronucleus von *Paramecium bursaria. Archiv für Protistenkunde, 120*, 255–277.
Sehring, I. M., Reiner, C., Mansfeld, J., Plattner, H., & Kissmehl, R. (2007). A broad spectrum of actin paralogs in *Paramecium tetraurelia* cells display differential localization and function. *Journal of Cell Science, 120*, 177–190.
Selbach, M., & Kuhlmann, H. W. (1999). Structure, fluorescent properties and proposed function in phototaxis of the stigma apparatus in the ciliate *Chlamydodon mnemosyne. Journal of Experimental Biology, 202*, 919–927.
Selbach, M., Hader, D. P., & Kuhlmann, H. W. (1999). Phototaxis in *Chlamydodon mnemosyne*: Determination of the illuminance–response curve and the action spectrum. *Journal of Photochemistry and Photobiology B, Biology, 49*, 35–40.
Seravin, L. N., & Gerassimova, Z. P. (1978). A new macrosystem of ciliates. *Acta Protozoologica, 17*, 399–418.
Seravin, L. N., & Orlovskaja, E. E. (1977). Feeding behaviour of unicellular animals. I. The main role of chemoreception in the food choice of carnivorous protozoa. *Acta Protozoologica, 16*, 309–332.
Sergejeva, G. I., & Bobyleva, N. N. (1995). Polynemic structures in the differentiated macronucleus of the ciliate *Bursaria ovata* Beers 1952. *Acta Protozoologica, 35*, 115–124.
Sergejeva, G. I., Livolant, F., Conceicao, M., Da., Samoshkin, A. A., Bolshakova, N. N., et al. (1995). Ciliates of the genus *Bursaria* in the state of cryptobiosis. *Tsitologiya, 37*, 1189–1206. (in Russian with English summary)
Serrano, S., Martín–González, A., & Fernández–Galiano, D. (1985). Cytological observations on the ciliate *Coleps hirtus* Nitzch, 1817: Vegetative cell, binary fission and conjugation. *Acta Protozoologica, 24*, 77–84.
Serrano, S., Martín–González, A., & Fernández–Galiano, D. (1988). *Trimyema compressum* Lackey, 1925: Morphology, morphogenesis and systematic implications. *Journal of Protozoology, 35*, 315–320.
Serrano, S., Martín–González, A., & Fernández–Galiano, D. (1990). General morphology and cytological events during the conjugation process in *Acaryophrya collaris* (Kahl, 1926) (Ciliophora, Haptorina). *Archiv für Protistenkunde, 138*, 65–74.
Sewertzoff, A. N. (1931). *Die Morphologische Gesetzmässigkeiten der Evolution* (371 pp.). Jena: Fischer–Verlag.
Shang, H., & Song, W. (2005). Separation and relationship of ten marine scuticociliates (Protozoa, Ciliophora) using RAPD fingerprinting method. *Acta Oceanologica Sinica, 24*, 78–85.
Shang, H., Chen, Z. G., and Song, W. (2002). Species separation among seven *Euplotes* spp. (Protozoa: Ciliophora: Hypotrichida) using PCR/RFLP analysis of nuclear ribosomal DNA. *Journal of Zoology, London, 258*, 375–379.
Shang, H., Song, W., & Warren, A. (2003). Phylogenetic positions of two ciliates, *Paranophrys magna* and *Mesanophrys carcini* (Ciliophora: Oligohymenophorea), within the subclass Scuticociliatia inferred from complete small subunit rRNA gene sequences. *Acta Protozoologica, 42*, 171–181.
Shao, C., Song, W., Hu, Z., Ma, H., Zhu, M., & Wang, M. (2006). Cell division and morphology of the marine ciliate, *Condylostoma spatiosum* Ozaki and Yagiu (Ciliophora, Heterotrichida) based on a Chinese population. *European Journal of Protistology, 42*, 9–19.
Sherr, E., & Sherr, B. (1988). Role of microbes in pelagic food webs: A revised concept. *Limnology and Oceanography, 35*, 1225–1227.
Shi, X. (1980). The morphology and morphogenesis of the buccal apparatus of *Paramecium* and their phylogenetic implications. II. Stomatogenesis. *Acta Zoologica Sinica, 26*, 289–300 (in Chinese with English summary)

Shin, M. K., Hwang, U. W., Kim, W., Wright, A. D. G., Krawczyk, C., & Lynn, D. H. (2000). Phylogenetic position of the ciliates *Phacodinium* (Order Phacodiniida) and *Protocruzia* (Subclass Protocruziidia) and systematics of the spirotrich ciliates examined by small subunit ribosomal RNA gene sequences. *European Journal of Protistology, 36*, 293–302.

Shippen–Lentz, D., & Blackburn, E. (1989). Telomere terminal transferase activity from *Euplotes crassus* adds large numbers of TTTTGGGG repeats onto telomeric primers. *Molecular and Cellular Biology, 9*, 2761–2764.

Siegel, R. W. (1963). New results on the genetics of mating types in *Paramecium bursaria. Genetical Research, Cambridge, 4*, 132–142.

Silva Neto, I. D. Da. (1992). Observations sur la structure et l'ultrastructure du cilié *Myxophthirus anomalocardiae* gen. nov., sp. nov. (Scuticociliatida, Thigmophryidae), parasite du bivalve *Anomalocardia brasiliana* Gmelin, 1791. *European Journal of Protistology*, 421–429.

Silva Neto, I. D. Da. (1993a). Structural and ultrastructural observations of the ciliate *Phacodinium metchnicoffi* Certes, 1891 (Heterotrichea, Phacodiniida). *European Journal of Protistology, 29*, 209–218.

Silva Neto, I. D. Da. (1993b). Observations sur l'ultrastructure du cilié *Fabrea salina* Henneguy, 1890 (Heterotrichida). *Annales des Sciences Naturelles–Zoologie et Biologie Animale, Paris, Série 13, 14*, 1–12.

Silva Neto, I. D. Da. (1994a). Observations sur l'ultrastructure du cilié hétérotriche, *Licnophora auerbachi* Cohn, 1866, épibionte de l'étoile de mer. *Annales des Sciences Naturelles–Zoologie et Biologie Animale Paris, Série 13, 15*, 49–62.

Silva Neto, I. D. Da. (1994b). Morphologie et ultrastructure du cilié *Condylostomides grolieri* gen. n., sp. n. (Ciliophora: Heterotrichida). *Acta Protozoologica, 33*, 149–158.

Silva Neto, I. D. Da., & Grolière, C. A. (1993). Observations sur l'ultrastructure de cilié *Fabrea salina* Henneguy, 1890 (Heterotrichida). *Annales des Sciences Naturelles–Zoologie, Paris, Série 13, 14*, 1–12.

Silver, M. W., Gowing, M. M., Brownlee, D. C., & Corliss, J. O. (1984). Ciliated protozoa associated with oceanic sinking detritus. *Nature, 309*, 246–248.

Šimek, K., Vrba, J., & Hartman, P. (1994). Size–selective feeding by *Cyclidium* sp. on bacterioplankton and various sizes of cultured bacteria. *FEMS Microbiology Ecology, 14*, 157–168.

Šimek, K., Bobková, J., Macek, M., Nedoma, J., & Psenner, R. (1995). Ciliate grazing on picoplankton in a eutrophic reservoir during the summer phytoplankton maximum: A study at the species and community level. *Limnology and Oceanography, 40*, 1077–1090.

Šimek, K., Macek, M., Pernthaler, J., Straškrabová, V., & Psenner, R. (1996). Can freshwater planktonic ciliates survive on a diet of picoplankton? *Journal of Plankton Research, 18*, 597–613.

Sime–Ngando, T., Demers, S., & Juniper, S. K. (1999). Protozoan bacterivory in the ice and the water column of a cold temperate lagoon. *Microbial Ecology, 37*, 95–106.

Simon, E. M., & Orias, E. (1987). Genetic instability in the mating type system of *Tetrahymena pigmentosa. Genetics, 117*, 43–58.

Simon, E. M., & Meyer, E. B. (1992). Suicide is not the inevitable outcome of "perpetual" selfing in tetrahymenines collected from natural habitats. *Developmental Genetics, 13*, 47–52.

Simon, E. M., Meyer, E. B., & Preparata, R. M. (1985). New wild *Tetrahymena* from Southeast Asia, China, and North America, including *T. malaccensis*, *T. asiatica*, *T. nanneyi*, *T. caudata*, and *T. silvana* n. spp. *Journal of Protozoology, 32*, 183–189.

Simpson, G. G. (1961). *Principles of animal taxonomy* (p. 247). New York: Columbia University Press.

Singh, S., & Dash, S. C. (1992). *Paramecium* colonizing urinary tract of a patient on dialysis: A rare entity. *Nephron, 62*, 243–244.

Sirgel, W. F. (1983). A new ciliate genus *Trichodoxa* n. g. (Ciliata, Peritricha, Mobilina, Trichodinidae) with two new species from the genital system of terrestrial pulmonates. *Journal of Protozoology, 30*, 118–125.

Skogstad, A., Granskog, L., & Klaveness, D. (1987). Growth of freshwater ciliates offered planktonic algae as food. *Journal of Plankton Research, 9*, 503–512.

Skotarczak, B., Przybos, E., Wodecka, B., & Maciejewska, A. (2004). Random amplified polymorphic DNA fingerprinting as a marker for *Paramecium jenningsi* strains. *Folia Biologica, 52*, 117–124.

Sládeček, V. (1973). System of water quality from the biological point of view. *Archiv für Hydrobiologie (Beihaft Ergebnisse der Limnologie), 7*, 1–218.

Sleigh, M. A. (1984). The integrated activity of cilia: Function and coordination. *Journal of Protozoology, 31*, 16–21.

Sleigh, M. A. (1989). Ciliary propulsion in protozoa. *Scientific Progress, Oxford, 73*, 317–331.

Sleigh, M. A., & Barlow, D. (1976). Collection of food by *Vorticella. Transactions of the American Microscopical Society, 95*, 482–486.

Sleigh, M. A., & Barlow, D. I. (1982). The use of cilia by Protozoa. Progress in Protozoology. Proceedings of the 6th International Congress of Protozoology, *Acta Protozoologica*, Pt. I, 141–147.

Sleigh, M. A., & Silvester, N. R. (1983). Anchorage functions of the basal apparatus of cilia. *Journal of Submicroscopic Cytology, 15*, 101–104.

Small, E. B. (1967). The Scuticociliatida, a new order of the class Ciliatea (phylum Protozoa, subphylum Ciliophora). *Transactions of the American Microscopical Society*, *86*, 345–370.

Small, E. B. (1973). A study of ciliate Protozoa from a small polluted stream in east–central Illinois. *American Zoologist*, *13*, 225–230.

Small, E. B. (1976). A proposed subphyletic division of the phylum Ciliophora Doflein, 1901. *Transactions of the American Microscopical Society*, *95*, 739–751.

Small, E. B. (1984). An essay on the evolution of ciliophoran oral cytoarchitecture based on descent from within a karyorelictean ancestry. *Origins Life*, *13*, 217–228.

Small, E. B., & Gross, M. E. (1985). Preliminary observations of protistan organisms, especially ciliates, from the 21°N hydrothermal vent site. *Bulletin of the Biological Society of Washington*, *6*, 401–410.

Small, E. B., & Lynn, D. H. (1981). A new macrosystem for the Phylum Ciliophora Doflein, 1901. *BioSystems*, *14*, 387–401.

Small, E. B., & Lynn, D. H. (1985). Phylum Ciliophora, Doflein, 1901. In J. J. Lee, S. H. Hutner, & E. C. Bovee (Eds.), *An illustrated guide to the protozoa*. (pp. 393–575). Kansas, USA: *Society of Protozoologists*, Lawrence.

Small, E. B., Antipa, G. A., & Marszaek, D. S. (1972). The cytological sequence of events in the binary fission of *Didinium nasutum* (0. F. M.). *Acta Protozoologica*, *9*, 275–282.

Small, E. B., Heisler, J., Sniezek, J., & Iliffe, T. M. (1986). *Glauconema bermudense* n. sp. (Scuticociliatida, Oligohymenophorea), a troglobitic ciliophoran from Bermudian marine caves. *Stygologia*, *2*, 167–179.

Small, H. J., Neil, D. M., Taylor, A. C., Bateman, K., & Coombs, G. H. (2005). A parasitic scuticociliate infection in the Norway lobster (*Nephrops norvegicus*). *Journal of Invertebrate Pathology*, *90*, 108–117.

Smalley, G. W., & Coats, D. W. (2002). Ecology of the red–tide dinoflagellate *Ceratium fusca*: Distribution, mixotrophy, and grazing impact on ciliate populations in Chesapeake Bay. *Journal of Eukaryotic Microbiology*, *49*, 63–73.

Smalley, G. W., Coats, D. W., & Adam, E. J. (1999). A new method using fluorescent microspheres to determine grazing on ciliates by the mixotrophic dinoflagellate *Ceratium furca*. *Aquatic Microbial Ecology*, *17*, 167–179.

Smith, H. E. (1982). Oral apparatus structure in the carnivorous macrostomal form of *Tetrahymena vorax*. *Journal of Protozoology*, *29*, 616–627.

Smith, W. O., Jr., & Barber, R. T. (1979). A carbon budget for the autotrophic ciliate *Mesodinium rubrum*. *Journal of Phycology*, *15*, 27–33.

Smith–Somerville, H. E., & Buhse, H. E., Jr. (1984). Changes in oral apparatus structure accompanying vacuolar formation in the macrostomal form of *Tetrahymena vorax*. A model for the formation of food vacuoles in *Tetrahymena*. *Journal of Protozoology*, *31*, 373–380.

Smith–Sonneborn, J. (1981). Genetics and aging in protozoa. *International Review of Cytology*, *73*, 319–354.

Sniezek, J. H., & Coats, D. W. (1996). Stomatogenesis in *Kyaroikeus cetarius* (Dysteriina, Kyaroikeidae, n. fam.): Clues to its systematic placement. *Journal of Eukaryotic Microbiology*, *43*, 113–119.

Sniezek, J. H., Capriulo, G. M., Small, E. B., & Russo, A. (1991). *Nolaclusilis hudsonicus* n. sp. (Nolaclusiliidae n. fam.) A bilaterally symmetrical tintinnine ciliate from the lower Hudson River estuary. *Journal of Protozoology*, *38*, 589–594.

Sniezek, J. H., Coats, D. W., & Small, E. B. (1995). *Kyaroikeus cetarius* n. g., n. sp.: A parasitic ciliate from the respiratory tract of odonticete Cetacea. *Journal of Eukaryotic Microbiology*, *42*, 260–268.

Snoeyenbos–West, O. L. O., Salcedo, T., McManus, G. B., & Katz, L. A. (2002). Insights into the diversity of choreotrich and oligotrich ciliates (Cl: Spirotrichea) based on genealogical analyses of multiple loci. *International Journal of Systematic and Evolutionary Microbiology*, *52*, 1901–1913.

Snoeyenbos–West, O. L. O., Cole, J., Campbell, A., Coats, D. W., & Katz, L. A. (2004). Molecular phylogeny of phyllopharyngean ciliates and their Group I introns. *Journal of Eukaryotic Microbiology*, *51*, 441–450.

Snoke, M. S., Berendon, T. U., Barth, D., & Lynch, M. (2006). Large global effective population sizes in *Paramecium*. *Molecular Biology and Evolution*, *23*, 2474–2479.

Snyder, R. A., & Brownlee, D. C. (1991). *Nolaclusilis bicornis* n. g., n. sp. (T intinnina: Tintinnidiidae): A tintinnine ciliate with novel lorica and cell morphology from the Chesapeake Bay estuary. *Journal of Protozoology*, *38*, 583–589.

Snyers, L., Hellings, P., Bovy–Kesler, C., & Thines–Sempoux, D. (1982). Occurrence of hydrogenosomes in the rumen ciliates Ophryoscolecidae. *FEBS Letters*, *137*, 35–39.

Sobierajska, K., Fabczak, H., & Fabczak, S. (2006). Photosensory transduction in unicellular eukaryotes: A comparison between related ciliates *Blepharisma japonicum* and *Stentor coeruleus* and photoreceptor cells of higher organisms. *Journal of Photochemistry and Photobiology B, Biology*, *83*, 163–171.

Sogin, M. L., & Elwood, H. J. (1986). Primary structure of the *Paramecium tetraurelia* small–subunit rRNA coding region: Phylogenetic relationships within the Ciliophora. *Journal of Molecular Evolution*, *23*, 53–60.

Sogin, M. L., Ingold, A., Karlok, M., Nielsen, H., & Engberg, J. (1986a). Phylogenetic evidence for the acquisition of ribosomal RNA introns subsequent to the divergences of some of the major *Tetrahymena* groups. *EMBO Journal*, *5*, 3625–3630.

Sogin, M. L., Swanton, M. T., Gunderson, J. H., & Elwood, H. J. (1986b). Sequence for the small subunit ribosomal RNA gene from the hypotrichous ciliate *Euplotes aediculatus*. *Journal of Protozoology*, *33*, 26–29.

Sola, A., Guinea, A., & Fernández–Galiano, D. (1985). Étude du processus de bipartition chez *Opisthonecta henneguyi* Fauré–Fremiet, 1906 (Ciliophora, Peritrichida). *Protistologica*, *21*, 333–338.

Sola, A., Guinea, A., Longás, J. F., & Fernández–Galiano, D. (1988). Observations sur l'infraciliature de *Plagiopyla nasuta* Stein, 1860. *Acta Protozoologica*, *27*, 279–286.

Sola, A., Guinea, A., & Fernández–Galiano, D. (1989a). Light microscopical study of the conjugation process in *Opisthonecta henneguyi* Fauré–Fremiet (Ciliophora: Peritrichida). *Acta Protozoologica*, *28*, 239–244.

Sola, A., Guinea, A., Longás, J. F., & Fernández–Galiano, D. (1989b). Quelques observations sur le cilié *Sonderia sinuata* Kahl, 1931 (Ciliophora, Trichostomatida). *Archiv für Protistenkunde*, *137*, 1–8.

Sola, A., Guinea, A., Fernández–Galiano, D., Longás, J. F., & Corliss, J. O. (1990a). Description of two species of nassophorean ("hypostome") ciliates, *Nassulopsis elegans* (Ehrenberg, 1833) and *Zosterodasys derouxi* n. sp., with comments on their systematic and phylogenetic affinities. *European Journal of Protistology*, *26*, 15–24.

Sola, A., Guinea, A., Longás, J. F., & Fernández–Galiano, D. (1990b). Buccal and somatic infraciliature of *Lagynus elegans* (Engelmann, 1862) Quennersted, 1867 (Ciliophora, Prostomatea): Remarks on its systematic position. *Journal of Protozoology*, *37*, 328–333.

Sola, A., Serrano, S., Guinea, A., & Longás, J. F. (1992). The infraciliature of two sapropelic heterotrich ciliates, *Cirranter mobilis* and *Saprodinium dentatum* (Ciliophora, Heterotrichida). *Archiv für Protistenkunde*, *142*, 51–57.

Soldo, A. T., & Godoy, G. A. (1972). The kinetic complexity of *Paramecium* macronuclear deoxyribonucleic acid. *Journal of Protozoology*, *19*, 673–678.

Soldo, A. T., & Merlin, E. J. (1972). The cultivation of symbiote–free marine ciliates in axenic medium. *Journal of Protozoology*, *19*, 519–524.

Soldo, A. T., & Merlin, E. J. (1977). The nutrition of *Parauronema acutum*. *Journal of Protozoology*, *24*, 556–562.

Soldo, A. T., Godoy, G. A., & van Wagtendonk, W. J. (1966). Growth of particle–bearing and particle–free *Paramecium* in axenic culture. *Journal of Protozoology*, *13*, 492–497.

Soldo, A. T., Godoy, G. A., & Brickson, S. A. (1974). Infectious particles in a marine ciliate. *Nature*, *249*, 284–286.

Soldo, A. T., Godoy, G. A., & Larin, F. (1978). Purine–excretory nature of refractile bodies in the marine ciliate *Parauronema acutum*. *Journal of Protozoology*, *25*, 416–418.

Soldo, A. T., Brickson, S. A., & Vazquez, D. (1992). The molecular biology of a bacterial endosymbiont. *Journal of Protozoology*, *39*, 196–198.

Soltynska, M. S. (1971). Morphology and fine structure of *Chilodonella cucullulus* (O. F. M.). Cortex and cytopharyngeal apparatus. *Acta Protozoologica*, *9*, 49–82.

Song, B. (2000). A comparative study on planktonic ciliates in two shallow mesotrophic lakes (China): Species composition, distribution and quantitative importance. *Hydrobiologia*, *427*,143–153.

Song, W. (1990a). A comparative analysis of the morphology and morphogenesis of *Gonostomum strenua* (Engelmann, 1862) (Ciliophora, Hypotrichida) and related species. *Journal of Protozoology*, *37*, 249–257.

Song, W. (1990b). Infraciliature and silverline system of the ciliate *Pseudotrachelocerca trepida* (Kahl, 1928), nov. gen., nov. comb., and establishment of a new family, the Pseudotrachelocercidae nov. fam. (Ciliophora, Haptoria). *European Journal of Protistology*, *26*, 160–166.

Song, W. (1993). Studies on cortical morphogenesis during cell division in *Halteria grandinella* (Müller, 1773) (Ciliophora, Oligotrichida). *Chinese Journal of Oceanology and Limnology*, *11*, 120–129.

Song, W. (1995). Studies on the stomatogenesis of *Frontonia acuminata* (Ciliophora, Hymenostomatida. *Acta Hydrobiologica Sinica*, *19*, 257–262 (in Chinese with English summary)

Song, W. (1997). Morphological and comparative studies on *Mesodinium* – red–tidal ciliates in the marine biotope (Protozoa, Ciliophora). *Chinese Journal of Applied and Environmental Biology*, *3*, 163–167. (in Chinese with English Summary)

Song, W. (2000). Morphological and taxonomical studies on some marine scuticociliates from China Sea, with description of two new species, *Philasterides armatalis* sp. n. and *Cyclidium varibonneti* sp. n. (Protozoa: Ciliophora: Scuticociliatida). *Acta Protozoologica*, *39*, 295–322.

Song, W. (2003). Reconsideration of the morphogenesis in the marine hypotrichous ciliate, *Aspidisca leptaspis* Fresenius, 1865 (Protozoa, Ciliophora). *European Journal of Protistology*, *39*, 53–61.

Song, W. (ed.) (2003). *Pathogenic Protozoa in Mariculture*. (483 p.) Beijing: Science Press. (in Chinese)

Song, W., & Bradbury, P. C. (1998). Studies on some new and rare reported marine planktonic ciliates (Ciliophora: Oligotrichia) from coastal waters in North China. *Journal of Marine Biological Association of the United Kingdom*, *78*, 767–794.

Song, W., & Hu, X. (1999). Divisional morphogenesis in the marine ciliate, *Hemigastrostyla enigmatica* (Dragesco & Dragesco–Kernéis, 1986) and redefinition of the genus *Hemigastrostyla* Song & Wilbert, 1997 (Protozoa, Ciliophora). *Hydrobiologia, 391*, 249–257.

Song, W., & Wang, M. (1996). Morphogenesis of *Strombidium sulcatum* during asexual binary division. *Progress in Natural Science, China, 6*, 741–746.

Song, W., & Wei, J. (1998). Morphological studies of three marine pathogenetic ciliates. *Acta Hydrobiologica Sinica, 22*, 361–366.

Song, W., & Wilbert, N. (1994). Morphogenesis of the marine ciliate *Diophrys oligothrix* Borror, 1965 during the cell division (Protozoa, Ciliophora, Hypotrichida). *European Journal of Protistology, 30*, 38–44.

Song, W., & Wilbert, N. (1997). Morphological investigations on some free living ciliates (Protozoa, Ciliophora) from China Sea with description of a new hypotrichous genus, *Hemigastrostyla* nov. gen. *Archiv für Protistenkunde, 148*, 413–444.

Song, W., & Wilbert, N. (2000a). Redefinition and redescription of some marine scuticociliates from China, with report of a new species, *Metanophrys sinensis* nov. spec. (Ciliophora, Scuticociliatida). *Zoologischer Anzeiger, 239*, 45–74.

Song, W., & Wilbert, N. (2000b). Ciliates from Antarctic sea ice. *Polar Biology, 23*, 212–222.

Song, W., & Wilbert, N. (2002). Reinvestigations of three "well–known" marine scuticociliates: *Uronemella filificum* (Kahl, 1931) nov. gen., nov. comb., *Pseudocohnilembus hargisi* Evans & Thompson, 1964 and *Cyclidium citrullus* Cohn, 1865, with description of the new genus *Uronemella* (Protozoa, Ciliophora, Scuticociliatida). *Zoologischer Anzeiger, 241*, 317–331.

Song, W., Wilbert, N., & Warren, A. (1999). Three new entocommensal ciliates from the digestive tract of sea urchins of the Weddell Sea, Antarctica (Protozoa, Ciliophora). *Polar Biology, 22*, 232–240.

Song, W., Wang, M., & Warren, A. (2000). Redescriptions of three marine ciliates, *Strombidium elegans* Florentin, 1901, *Strombidium sulcatum* Claparède & Lachmann, 1859 and *Heterostrombidium paracalkinsi* Lei, Xu & Song, 1999 (Ciliophora, Oligotrichida). *European Journal of Protistology, 36*, 327–342.

Song, W., Petz, W., & Warren, A. (2001). Morphology and morphogenesis of the poorly–known marine urostylid ciliate, *Metaurostylopsis marina* (Kahl, 1932) nov. gen., nov. comb. (Protozoa, Ciliophora, Hypotrichida). *European Journal of Protistology, 37*, 63–76.

Song, W., Warren, A., Ji, D., Wang, M., & Al–Rasheid, K. A. S. (2003). New contributions to two heterotrichous ciliates, *Folliculina simplex* (Dons, 1917), *Condylostoma curva* Burkovsky, 1970 and one licnophorid, *Licnophora lyngbycola* Fauré–Fremiet, 1937 (Protozoa, Ciliophora): Descriptions of morphology and infraciliature. *Journal of Eukaryotic Microbiology, 50*, 449–462.

Song, W., Wilbert, N., Chen, Z. G., & Shi, X. (2004). Considerations on the systematic position of *Uronychia* and related euplotids based on the data of ontogeny and 18S rRNA gene sequence analyses, with morphogenetic redescription of *Uronychia setigera* Calkins, 1902 (Ciliophora: Euplotida). *Acta Protozoologica, 43*, 313–328.

Sonneborn, T. M. (1937). Sex, sex inheritance and sex determination in *Paramecium aurelia. Proceedings of the National Academy of Sciences, 23*, 378–385.

Sonneborn, T. M. (1938). Mating types in *Paramecium aurelia*: Diverse conditions for mating in different stocks; occurrence, number and interrelations of the types. *Proceedings of the American Philosophical Society, 79*, 411–434.

Sonneborn, T. M. (1947). Recent advances in the genetics of *Paramecium* and *Euplotes. Advances in Genetics, 1*, 263–358.

Sonneborn, T. M. (1954). The relation of autogamy to senescence and rejuvenescence in *Paramecium aurelia. Journal of Protozoology, 1*, 38–53.

Sonneborn, T. M. (1957). Breeding systems, reproductive methods, and species problems in Protozoa. In E. Mayr (Ed.), *The species problem* (pp. 155–324). Washington, DC: AAAS Publication.

Sonneborn, T. M. (1964). The differentiation of cells. *Proceedings of the National Academy of Sciences, 51*, 915–919.

Sonneborn, T. M. (1970). Methods in *Paramecium* research. In D. M. Prescott (Ed.), *Methods in Cell Physiology* (Vol. 4, chap. 12, pp. 241–339). New York/London: Academic Press.

Sonneborn, T. M. (1975). The *Paramecium aurelia* complex of fourteen sibling species. *Transactions of the American Microscopical Society, 94*, 155–178.

Sonneborn, T. M. (1977). Genetics of cellular differentiation: Stable nuclear differentiation in eucaryotic unicells. *Annual Review of Genetics, 11*, 349–367.

Sonntag, B., Posch, T., Klammer, S., Teubner, K., & Psenner, R. (2006). Phagotrophic ciliates and flagellates in an oligotrophic, deep, alpine lake: Contrasting variability with seasons and depths. *Aquatic Microbial Ecology, 43*, 193–207.

Sperling, L., Tardieu, A., & Gulik–Krzywicki, T. (1987). The crystal lattice of *Paramecium* trichocysts before and after exocytosis by X–ray diffraction and freeze–fracture electron microscopy. *Journal of Cell Biology, 105*, 1649–1662.

Spoon, D. M., Chapman, G. B., Cheng, R. S., & Zane, S. F. (1976). Observations on the behavior and feeding mechanisms of the suctorian *Heliophrya erhardi* (Rieder) Matthes preying on *Paramecium.*

Transactions of the American Microscopical Society, *95*, 443–462.

Sprague, V., & Couch, J. (1971). An annotated list of protozoan parasites, hyperparasites, and commensals of decapod Crustacea. *Journal of Protozoology*, *18*, 526–537.

Stabell, T., Andersen, T., & Klaveness, D. (2002). Ecological significance of endosymbionts in a mixotrophic ciliate – an experimental test of a simple model of growth coordination between host and symbiont. *Journal of Plankton Research*, *24*, 889–899.

Stackebrandt, E., & Woese, C. R. (1981). Evolution of prokaryotes. In M. J. Carlisle, J. R. Collins, & B. E. B. Moseley (Eds.), *Molecular and cellular aspects of microbial evolution* (pp. 1–31). Cambridge: Cambridge University Press.

Stan, K., Belzecki, G., Kasperowicz, A., Kwiatkowska, E., & Michalowski, T. (2006). The ability of the rumen ciliate *Anoplodinium denticulatum* to utilize hemicellulosic material for in vitro growth. *Journal of Animal and Feed Science*, *15*, 39–42.

Stebbing, A. R. D., Soria, S., Burt, G. R., & Cleary, J. J. (1990). Water quality bioassays in two Bermudan harbours using the ciliate *Euplotes vannus*, in relation to tributylin distribution. *Journal of Experimental Marine Biology and Ecology*, *138*, 159–166.

Stechmann, A., Schlegel, M., & Lynn, D. H. (1998). Phylogenetic relationships between prostome and colpodean ciliates tested by small subunit rRNA sequences. *Molecular Phylogenetics and Evolution*, *9*, 48–54.

Steele, C. J., Barkovy–Gallagher, G. A., Preer, L. B., & Preer, J. R., Jr. (1994). Developmentally excised sequences in macronuclear DNA of *Paramecium*. *Proceedings of the National Academy of Sciences of the United States of America*, *91*, 2255–2259.

Steele, T. W., & McLennan, A. M. (1996). Infection of *Tetrahymena pyriformis* by *Legionella longbeachae* and other *Legionella* species found in potting mixes. *Applied and Environmental Microbiology*, *62*, 1081–1083.

Steffens, S., & Wilbert, N. (2002). Ciliates (Protozoa: Ciliophora) of selected ephemeral flushes on tropical inselbergs. *International Review of Hydrobiology*, *87*, 401–410.

Stein, F. (1851). Neue Beiträge zur Kenntniss der Entwicklungsgeschichte und des feineren Baues der Infusionsthiere. *Zeitschrift für Wissenschaftliche Zoologie*, *3*, 475–509.

Stein, F. (1854). *Die Infusionsthiere auf ihre Entwicklungsgeschichte untersucht* (265 pp.). Leipzig: W. Engelmann.

Stein, F. (1859). *Die Organismus der Infusionsthiere nach eigenen Forschungen in systematischer Reihenfolge bearbeitet. I* (pp. 206). Leipzig: W. Engelmann.

Stein, G. A. (1979). New data on parasitic ciliates (Peritricha, Urceolariidae) of fishes of the basin of Pacific Ocean. *Acta Protozoologica*, *18*, 531–552.

Steinbrück, G. (1983). Overamplification of genes in macronuclei of hypotrichous ciliates. *Chromosoma*, *88*, 156–163.

Steinbrück, G. (1990). Recent advances in the study of ciliate genes. *European Journal of Protistology*, *26*, 2–14.

Steinbrück, G., Radzikowski, S., Golembiewska–Skoczylas, M., & Sapetto–Rebow, B. (1995). Characterization of low and high molecular weight DNA in the macronucleus of the ciliate *Chilodonella steini*. *Acta Protozoologica*, *34*, 125–134.

Stelfox–Widdicombe, C. E., Archer, S. D., Burkill, P. H., & Stefels, J. (2004). Microzooplankton grazing in *Phaeocystis* and diatom–dominated waters in the southern North Sea in spring. *Journal of Sea Research*, *51*, 37–51.

Stelly, N., Balmefrezol, M., & Adoutte, A. (1975). Diamniobenzidine reactivity of mitochondria and peroxisomes in *Tetrahymena* and in wild–type and cytochrome oxidase–deficient *Paramecium*. *Journal of Histochemistry and Cytochemistry*, *23*, 686–696.

Stelly, N., Halpern, S., Nicolas, G., Fragu, P., & Adoutte, A. (1995). Direct visualization of a vast cortical calcium compartment in *Paramecium* by secondary ion mass spectrometry (SIMS) microscopy: Possible involvement in exocytosis. *Journal of Cell Science*, *108*, 1895–1909.

Stensdotter–Blomberg, U. (1998). Factors controlling pelagic populations of ciliates and helizoans – late summer investigations in an acidic lake before and after liming. *Journal of Plankton Research*, *20*, 423–442.

Stenson, J. A. E. (1984). Interactions between pelagic metazoan and protozoan zooplankton, an experimental study. *Hydrobiologia*, *111*, 107–112.

Stevens, N. M. (1901). Studies on ciliate infusoria. *Proceedings of the California Academy of Sciences Zoology (Series 3)*, *3*, 1–42.

Stickle, W. B., Weidner, E. H., & Kozloff, E. N. (2001). Parasitism of *Leptasterias* spp. (Echinodermata: Asteroidea) by the ciliated protozoan *Orchitophrya stellarum* (Scuticociliata). *Invertebrate Biology*, *120*, 88–95.

Stiller, J. (1941). Epizoische Peritrichen aus dem Balaton. *Arbeiten der I. Abteilung des Ungarischen Biologischen Forschungsinstitutes*, *13*, 211–223.

Stiller, J. (1971). Szájkoszorús csillósok – Peritricha. *Fauna Hung*, *105*, 245. (in Hungarian)

Stloukal, E., & Matis, D. (1993). Discovery of epizoic ciliate *Spirochona gemipara* Stein, 1852 (Ciliophora, Chonotrichida) in Slovakia. *Acta Zoologica University Comenianae*, *37*, 109–112.

Stock, C., Kruppel, T., Key, G., & Lueken, W. (1999). Sexual behaviour of *Euplotes raikovi* is accompanied by pheromone–induced modifications of ionic currents. *Journal of Experimental Biology*, *202*, 475–483.

Stock, C., Allen, R. D., & Naitoh, Y. (2001). How external osmolarity affects the activity of the contractile vacuole complex, the cytosolic osmolarity and the water permeability of the plasma membrane in *Paramecium multimicronucleatum*. *Journal of Experimental Biology*, *204*, 291–304.

Stock, C., Gronlien, H. K., Allen, R. D., & Naitoh, Y. (2002). Osmoregulation in *Paramecium*: In situ ion gradients permit water to cascade through the cytosol to the contractile vacuole. *Journal of Cell Science*, *115*, 2339–2348.

Stoeck, T., & Schmidt, H. J. (1998). Fast and accurate identification of European species of the *Paramecium aurelia* complex by RAPD–fingerprints. *Microbial Ecology*, *35*, 311–317.

Stoeck, T., Przybos, E., & Schmidt, H. J. (1998). A combination of genetics with inter– and intra–strain crosses and RAPD–fingerprints reveals different population structures within the *Paramecium aurelia* species complex. *European Journal of Protistology*, *34*, 348–355.

Stoeck, T., Przybos, E., Kusch, J., & Schmidt, H. J. (2000a). Intra–species differentiation and level of inbreeding of different sibling species of the *Paramecium aurelia* complex. *Acta Protozoologica*, *39*, 15–22.

Stoeck, T., Welter, H., Seitz–Bender, D., Kusch, J., & Schmidt, H. J. (2000b). ARDRA and RAPD–fingerprinting reject the sibling species concept for the ciliate *Paramecium caudatum* (Ciliophora, Protoctista). *Zoologica Scripta*, *29*, 75–82.

Stoeck, T., Fowle, W. H., & Epstein, S. S. (2003). Methodology of protistan discovery: From rRNA detection to quality scanning electron microscope images. *Applied and Environmental Microbiology*, *69*, 6856–6863.

Stoeck, T., Foissner, W., & Lynn, D. H. (2007). Small-subunit rRNA phylogenies suggest that *Epalxella antiquorum* (Penard, 1922) Corliss, 1960 (Ciliophora, Odontostomatida) is a member of the Plagiopylea. *Journal of Eukaryotic Microbiology, 54*, 436–442.

Stoecker, D. K. (1984). Particle production by planktonic ciliates. *Limnology and Oceanography*, *29*, 930–940.

Stoecker, D. K. (1988). Are marine planktonic ciliates suspension–feeders? *Journal of Protozoology*, *35*, 252–255.

Stoecker, D. K., & Evans, G. T. (1985). Effects of protozoan herbivory and carnivory in a microplankton food web. *Marine Ecology–Progress Series*, *25*, 159–167.

Stoecker, D. K., & Govoni, J. J. (1984). Food selection by young larval Gulf menhaden (*Brevoortia patronus*). *Marine Biology*, *80*, 299–306.

Stoecker, D. K., & Sanders, N. K. (1985). Differential grazing by *Acartia tonsa* on a dinoflagellate and a tintinnid. *Journal of Plankton Research*, *7*, 85–100.

Stoecker, D. K., Davis, L. H., & Provan, A. (1983). Growth of *Favella* sp. (Ciliata: Tintinnina) and other microzooplankters in cages incubated in situ and comparison to growth in vitro. *Marine Biology*, *75*, 293–302.

Stoecker, D. K., Davis, L. H., & Anderson, D. M. (1984). Fine scale spatial correlations between planktonic ciliates and dinoflagellates. *Journal of Plankton Research*, *6*, 829–842.

Stoecker, D. K., Michaels, A. E., & Davis, L. H. (1987a). Large proportion of marine planktonic ciliates found to contain functional chloroplasts. *Nature*, *326*, 790–792.

Stoecker, D. K., Michaels, A. E., & Davis, L. H. (1987b). Grazing by the jellyfish, *Aurelia aurita*, on microzooplankton. *Journal of Plankton Research*, *9*, 910–915.

Stoecker, D. K., Verity, P. G., Michaels, A. E., & Davis, L. H. (1987c). Feeding by larval and post–larval ctenophores on microzooplankton. *Journal Plankton Research*, *9*, 667–683.

Stoecker, D. K., Silver, M. W., Michaels, A. E., & Davis, L. H. (1988). Obligate mixotrophy in *Laboea strobila*, a ciliate which retains chloroplasts. *Marine Biology*, *99*, 415–423.

Stoecker, D. K., Silver, M. W., Michaels, A. E., & Davis, L. H. (1988/1989). Enslavement of algal chloroplasts by four *Strombidium* spp. (Ciliophora, Oligotrichida). *Marine Microbial Food Webs*, *3*, 79–100.

Stoecker, D. K., Taniguchi, A., & Michaels, A. E. (1989). Abundance of autotrophic, mixotrophic and heterotrophic planktonic ciliates in shelf and slope waters. *Marine Ecology–Progress Series*, *50*, 241–254.

Stoecker, D. K., Putt, M., Davis, L. H., & Michaels, A. E. (1991). Photosynthesis in *Mesodinium rubrum*: Species–specific measurements and comparison to community rates. *Marine Ecology–Progress Series*, *73*, 245–252.

Stoecker, D. K., Gallager, S. M., Langdon, C. J., & Davis, L. H. (1995). Particle capture by *Favella* sp. (Ciliata, Tintinnida). *Journal of Plankton Research*, *17*, 1105–1124.

Stoecker, D. K., Parrow, M. W., Burkholder, J. M., & Glasgow, H. B., Jr. (2002). Grazing by microzooplankton on *Pfiesteria piscicida* cultures with different histories of toxicity. *Aquatic Microbial Ecology*, *28*, 79–85.

Stott, R., May, E., Matsushita, E., & Warren, A. (2001). Protozoan predation as a mechanism for the removal of *Cryptosporidium* oocysts from wastewaters in constructed wetlands. *Water Science and Technology*, *44*, 191–198.

Stout, J. D. (1960). Morphogenesis in the ciliate *Bresslaua vorax* Kahl and the phylogeny of the Colpodidae. *Journal of Protozoology*, *7*, 26–35.

Strom, S. L. (2001). Light–aided digestion, grazing and growth in herbivorous protists. *Aquatic Microbial Ecology*, *23*, 253–261.

Strom, S. L., Postel, J. R., & Booth, B. C. (1993). Abundance, variability, and potential grazing impact of planktonic ciliates in the open subarctic Pacific Ocean. *Progress in Oceanography*, *32*, 185–203.

Strüder–Kypke, M. C., & Lynn, D. H. (2003). Sequence analyses of the small subunit rRNA gene confirm the paraphyly of oligotrich ciliates sensu lato and support the monophyly of the subclasses Oligotrichia and Choreotrichia (Ciliophora, Spirotrichea). *Journal of Zoology, London*, *260*, 87–97.

Strüder-Kypke, M. C., Kornilova, O. A., & Lynn, D. H. (2007). Phylogeny of trichostome ciliates (Ciliophora, Litostomatea) endosymbiotic in the Yakut Horse (*Equus caballus*). *European Journal of Protistology*, *43*, 319–328.

Strüder–Kypke, M. C., Wright, A. D. G., Fokin, S. I., & Lynn, D. H. (2000a). Phylogenetic relationships of the genus *Paramecium* inferred from small subunit rRNA gene sequences. *Molecular Phylogenetics and Evolution*, *14*, 122–130.

Strüder–Kypke, M. C., Wright, A. D. G., Fokin, S. I., & Lynn, D. H. (2000b). Phylogenetic relationships of the subclass Peniculia (Oligohymenophorea, Ciliophora) inferred from small subunit rRNA gene sequences. *Journal of Eukaryotic Microbiology*, *47*, 419–429.

Strüder–Kypke, M. C., Wright, A. D. G., Jerome, C. A., & Lynn, D. H. (2001). Parallel evolution of histophagy in ciliates of the genus *Tetrahymena*. *BMC Evolutionary Biology*, *1*, 5. Retrieved from http://www.biomedcentral.com/1471–2148/1/5

Strüder–Kypke, M. C., Wright, A. D. G., Foissner, W., Chatzinotas, A., & Lynn, D. H. (2006). Molecular phylogeny of litostome ciliates (Ciliophora, Litostomatea) with emphasis on free–living haptorian genera. *Protist*, *157*, 261–278.

Stuart, C. A., Kidder, G. W., & Griffin, A. M. (1939). Growth studies on ciliates. III. Experimental alteration of the method of reproduction in *Colpoda*. *Physiological Zoology*, *12*, 348–362.

Stumm, C. K., Gijzen, H. J., & Vogels, G. D. (1982). Association of methanogenic bacteria with ovine rumen ciliates. *British Journal of Nutrition*, *47*, 95–100.

Suárez, J., Guinea, A., & Fernández–Galiano, D. (1987a). Observations on the swarmer of *Discophrya collini* Root, 1914 (Ciliophora, Suctorida). *Archiv für Protistenkunde*, *133*, 251–255.

Suárez, J., Guinea, A., & Fernández–Galiano, D. (1987b). Morfología e infraciliación del embrión de *Tokophrya infusionum* Stein 1851 (Ciliophora, Suctorida). *Boletín de la Real Sociedad Española de Historia Natural (Biológica)*, *83*, 213–219.

Suchard, S. J., & Goode, D. (1982). Microtubule–dependent transport of secretory granules during stalk secretion in a peritrich ciliate. *Cell Motility*, *2*, 47–71.

Suganuma, Y., & Yamamoto, H. (1992). Conjugation in *Tetrahymena*: Ultrastructure of the meiotic prophase of the micronucleus. *European Journal of Protistology*, *28*, 434–441.

Sugibayashi, R., & Harumoto, T. (2000). Defensive function of trichocysts in *Paramecium tetraurelia* against heterotrich ciliate *Climacostumum virens*. *European Journal of Protistology*, *36*, 415–422.

Sugimoto, H., & Endoh, H. (2006). Analysis of fruiting body development in the aggregative ciliate *Sorogena stoianovitchae* (Ciliophora, Colpodea). *Journal of Eukaryotic Microbiology*, *53*, 96–102.

Sugiura, M., Kawahara, S., Iio, H., & Harumoto, T. (2005). Developmentally and environmentally regulated expression of gamone 1: The trigger molecule for sexual reproduction in *Blepharisma japonicum*. *Journal of Cell Science*, *118*, 2735–2741.

Suhama, M. (1969). The fine structure of ectoplasmic organelles of *Tillina canalifera* (Ciliata, Trichostomatida) from the trophic stage to the division cyst stage. *Journal of Science of the Hiroshima University, (Series B, Division 1)*, *22*, 103–118.

Summers, F. M. (1938). Some aspects of normal development in the colonial ciliate *Zoothamnium alternans*. *Biological Bulletin*, *74*, 117–129.

Sundararaman, V., & Hanson, E. D. (1976). Longitudinal microtubules and their functions during asexual reproduction in *Paramecium*. *Genetical Research Cambridge*, *27*, 205–211.

Sundermann, C. A., & Paulin, J. J. (1981). Ultrastructural features of *Allantosoma intestinalis*, a suctorian ciliated isolated from the large intestine of the horse. *Journal of Protozoology*, *28*, 400–405.

Sundermann, C. A., & Paulin, J. J. (1985). An ultrastructural and cytochemical study of the cell surface of a suctorian ciliate. *Journal of Protozoology*, *32*, 723–729.

Swofford, D. L. (2002). *PAUP*. Phylogenetic analysis using parsimony (*and other methods)* [Version 4]. Sunderland, MA: Sinauer.

Székely, C., & Bereczky, M. C. (1992). An unusual case of disease in pet fish stocks caused by *Coleps* sp. (Protozoa: Kinetofragminophorea). *Diseases of Aquatic Organisms*, *13*, 143–145.

TTTT

Tait, A. (1970). Enzyme variation between syngens in *Paramecium aurelia*. *Biochemical Genetics*, *4*, 461–470.

Tait, A. (1978). Species identification in protozoa: Glucosephosphate isomerase variation in the *Paramecium aurelia* group. *Biochemical Genetics*, *16*, 945–955.

Takagui, R. R., & Silveira, M. (1999). Studies of the ossicles from the ciliate protozoan *Spirostomum ambiguum. Japanese Journal of Applied Physics, Part 1, 38*, 5303–5304.

Takagi, Y. (1988). Aging. In: H. D. Görtz (Ed.), *Paramecium* (chap. 9, pp. 131–140). Berlin: Springer.

Takagi, Y. (1999). Clonal life cycle of *Paramecium* in the context of evolutionarily acquired mortality. In A. Macieira–Coelho (Ed.), *Progress in molecular and subcellular biology* (Vol. 24, pp. 81–101). Berlin, Heidelberg/New York: Springer.

Takagi, Y., Suzuki, T., & Shimada, C. (1987). Isolation of a *Paramecium tetraurelia* mutant with short clonal life–span and with novel life–cycle features. *Zoological Science, 4*, 73–80.

Takahashi, E. I., & Imai, S. (1989). Light and scanning electron microscopy of *Nyctotherus kyphodes* (Ciliophora, Plagiotomidae) from the Galapagos giant tortoise, *Testudo* sp. *Bulletin of the Nippon Veterinary and Zootechnical College, 38*, 9–15. (In Japanese with English summary)

Takahashi, T. (1974). Nuclear behavior in total and temporary conjugation of *Pseudourostyla levis* (Hypotrichida, Ciliata). *Journal of Science of the Hiroshima University (Series B, Division 1), 25*, 157–171.

Takeda, H., Sekiguchi, T., Nunokawa, S., & Usuki, I. (1998). Species–specificity of *Chlorella* for establishment of symbiotic association with *Paramecium bursaria* – Does infectivity depend upon sugar components of the cell wall? *European Journal of Protistology, 34*, 133–137.

Tamar, H. (1979). The movements of jumping ciliates. *Archiv für Protistenkunde, 122*, 290–327.

Tamar, H. (1992). Four marine species of *Mesodinium* (Ciliophora: Mesodiniidae). II. *Mesodinium pulex* Clap. & Lachm., 1858. *Archiv für Protistenkunde, 149*, 284–303.

Tamm, S. L., Sonneborn, T. M., & Dippell, R. V. (1975). The role of cortical orientation in the control of the direction of ciliary beat in *Paramecium. Journal of Cell Biology, 64*, 98–112.

Taniguchi, A., & Takeda, Y. (1988). Feeding rate and behavior of the tintinnid ciliate *Favella taraikaensis* observed with a high speed VTR system. *Marine Microbial Food Webs, 3*, 21–34.

Tannreuther, G. W. (1926). Life history of *Prorodon griseus. Biological Bulletin, 51*, 303–320.

Tappan, H., & Loeblich, A. R., Jr. (1968). Lorica composition of modern and fossil Tintinnida (ciliate Protozoa), systematics, geologic distribution, and some new tertiary taxa. *Journal of Paleontology, 42*, 1378–1394.

Tappan, H., & Loeblich, A. R., Jr. (1973). Evolution of the oceanic plankton. *Earth–Science Reviews, 9*, 207–240.

Tarcz, S., Przybos, E., Prajer, M., & Greczek–Stachura, M. (2006). Intraspecific variation of diagnostic rDNA genes in *Paramecium dodecaurelia, P. tredecaurelia*, and *P. quadecaurelia* (Ciliophora: Oligohymenophorea). *Acta Protozoologica, 45*, 255–263.

Tartar, V. (1961). *The biology of Stentor* (413 pp.). New York/London: Pergamon Press.

Tatchell, E. C. (1980). An ultrastructural study of extension and contraction in *Lacrymaria olor. Protistologica, 16,* 167–176.

Taylor, C. V., & Garnjobst, L. (1939). Reorganization of the "silverline system" in the reproductive cysts of *Colpoda duodenaria. Archiv für Protistenkunde, 92*, 73–90.

Taylor, C. V., & Strickland, A. G. R. (1939). Reactions of *Colpoda duodenaria* to environmental factors. II. Factors influencing the formation of resting cysts. *Physiological Zoology, 12*, 219–230.

Taylor, D. M., Lynn, D. H., & Gransden, S. G. (1995). *Vasichona opiliophila* n. sp., an ectosymbiotic ciliate (Chonotrichia; Exogemmida) on the maxillae of the snow crab, *Chionoecetes opilio. Canadian Journal of Zoology, 73*, 166–172.

Taylor, W. D. (1978). Growth responses of ciliate protozoa to the abundance of their bacterial prey. *Microbial Ecology, 4*, 207–214.

Taylor, W. D. (1979). Overlap among cohabiting ciliates in their growth responses to various prey bacteria. *Canadian Journal of Zoology, 57*, 949–951.

Taylor, W. D. (1983a). A comparative study of the sessile, filter–feeding ciliates of several small streams. *Hydrobiologia, 98*, 125–133.

Taylor, W. D. (1983b). Rates of population increase and mortality for sessile ciliates on artificial substrates in a stream riffle. *Canadian Journal of Zoology, 61*, 2023–2028.

Taylor, W. D. (1984). Phosphorus flux through epilimnetic zooplankton from Lake Ontario: Relationship with body size and significance to phytoplankton. *Canadian Journal of Fisheries and Aquatic Sciences, 41*, 1702–1712.

Taylor, W. D., & Berger, J. (1976). Growth responses of cohabiting ciliate protozoa to various prey bacteria. *Canadian Journal of Zoology, 54*, 1111–1114.

Taylor, W. D., & Heynen, M. L. (1987). Seasonal and vertical distribution of Ciliophora in Lake Ontario. *Canadian Journal of Fisheries and Aquatic Sciences, 44*, 2185–2191.

Taylor, W. D., Gates, M. A., & Berger, J. (1976). Morphological changes during the growth cycle of axenic and monoxenic *Tetrahymena pyriformis. Canadian Journal of Zoology, 54*, 2011–2018.

Tellez, C., Small, E. B., Corliss, J. O., & Maugel, T. K. (1982). The ultrastructure of specimens of *Paramecium multimicronucleatum* impregnated with silver by the Fernandez–Galiano method. *Journal of Protozoology, 29*, 627–634.

Terazima, M. N., & Harumoto, T. (2004). Defense function of pigment granules in the ciliate *Blepharisma japonicum* against two predatory protists, *Amoeba proteus* (Rhizopodea) and *Climacostomum virens* (Ciliata). *Zoological Science, 21*, 823–828.

Thilakaratne, I. D. S. I. P., Rajapaksha, G., Hewakopara, A., Rajapakse, R. P. V. J., & Faizal, A. C. M. (2003). Parasitic infections in freshwater ornamental fish in Sri Lanka. *Diseases of Aquatic Organisms, 54*, 157–162.

Thompson, J. C., Jr. (1958). Experimental infections of various animals with strains of the genus *Tetrahymena*. *Journal of Protozoology, 5*, 203–205.

Thompson, J. C., Jr. (1963). The generic significance of the buccal infraciliature in the family Tetrahymenidae and a proposed new genus and species, *Paratetrahymena wassi*. *Virginia Journal of Science, 14*, 126–135.

Thompson, J. C., Jr. (1964). A redescription of *Uronema marinum*, and a proposed new family Uronematidae. *Virginia Journal of Science, 15*, 80–87.

Thompson, J. C., Jr. (1965). *Pseudocohnilembus longisetus*, a hymenostome ciliate from Antarctica. *Virginia Journal of Science, 16*, 165–169.

Thompson, J. C., Jr. (1966a). *Potomacus pottsi* n. g., n. sp., a hymenostome ciliate from the Potomac River. *Journal of Protozoology, 13*, 459–462.

Thompson, J. C., Jr. (1966b). *Pseudocohnilembus marinus* n. sp., a hymenostome ciliate from the Virginia coast. *Journal of Protozoology, 13*, 463–465.

Thompson, J. C., Jr. (1967). *Parauronema virginianum* n. g., n. sp., a marine hymenostome ciliate. *Journal of Protozoology, 14*, 731–734.

Thompson, J. C., Jr. (1968). A description of *Cohnilembus verminus* from Eniwetok Atoll. *Journal of Protozoology, 15*, 396–399.

Thompson, J. C., Jr. (1969). *Philaster hiatti* n. sp., a holotrichous ciliate from Hawaii. *Journal of Protozoology, 16*, 81–83.

Thompson, J. C., Jr. (1972). Ciliated protozoa of the Antarctic Peninsula. In G. A. Llano (Ed.), *Antarctic terrestrial biology* (Vol. 20, pp. 261–288). Washington, DC: American Geophysical Union.

Thompson, J. C., Jr., & Corliss, J. 0. (1958). A redescription of the holotrichous ciliate *Pseudomicrothorax dubius* with particular attention to its morphogenesis. *Journal of Protozoology, 5*, 175–184.

Thurston, J. P., & Grain, J. (1971). Holotrich ciliates from the stomach of *Hippopotamus amphibius*, with descriptions of two new genera and four new species. *Journal of Protozoology, 18*, 133–141.

Thurston, J. P., & Noirot–Timothée, C. (1973). Entodiniomorph ciliates from the stomach of *Hippopotamus amphibius*, with descriptions of two new genera and three new species. *Journal of Protozoology, 20*, 562–565.

Tibbs, J. (1966). The cyst wall of *Colpoda steinii*. A substance rich in glutamic acid residues. *Biochemical Journal, 98*, 645–651.

Tibbs, J. (1968). Fine structure of *Colpoda steinii* during encystment and excystment. *Journal of Protozoology, 15*, 725–732.

Tibbs, J., & Marshall, B. J. (1970). Cyst wall protein synthesis and some changes on starvation and encystment in *Colpoda steinii*. *Journal of Protozoology, 17*, 125–128.

Timoshenko, O., & Imai, S. (1997). Three new intestinal protozoan species of the genus *Latteuria* n. g. (Ciliophora: Trichostomatida) from Asian and African elephants. *Parasitology International, 46*, 297–303.

Tirjaková, E., & Matis, D. (1985). To the knowledge of the soil ciliates (Ciliophora) from the territory of Mongolia. *Acta Facultatis Rerum Naturalism Universitatis Comenianae, Zoologica, 28*, 77–87.

Tiselius, P., Jonsson, P. R., & Verity, P. G. (1993). A model evaluation of the impact of food patchiness on foraging strategy and predation risk in zooplankton. *Bulletin of Marine Science, 53*, 247–264.

Tokura, M., Chagan, I., Ushida, K., & Kojima, Y. (1999). Phylogenetic study of methanogens associated with rumen ciliates. *Current Microbiology, 39*, 123–128.

Tooczko, B. (1980). Effect of colchicine on food ingestion in *Dileptus anser*. *Acta Protozoologica, 19*, 111–120.

Towne, G., & Nagaraja, T. G. (1990). Omasal ciliated protozoa in cattle, bison, and sheep. *Applied and Environmental Microbiology, 56*, 409–412.

Towne, G., Nagaraja, T. G., & Kemp, K. K. (1988). Ruminal ciliated protozoa in bison. *Applied and Environmental Microbiology, 54*, 2733–2736.

Tso, S. F., & Taghon, G. L. (1999). Factors affecting predation by *Cyclidium* sp. and *Euplotes* sp. on PAH–degrading and nondegrading bacteria. *Microbial Ecology, 37*, 3–12.

Tucker, J. B. (1967). Changes in nuclear structure during binary fission in the ciliate *Nassula*. *Journal of Cell Science, 2*, 481–498.

Tucker, J. B. (1968). Fine structure and function of the cytopharyngeal basket in the ciliate *Nassula*. *Journal of Cell Science, 3*, 493–514.

Tucker, J. B. (1970a). Morphogenesis of a large microtubular organelle and its association with basal bodies in the ciliate *Nassula*. *Journal of Cell Science, 6*, 385–429.

Tucker, J. B. (1970b). Initiation and differentiation of microtubule patterns in the ciliate *Nassula*. *Journal of Cell Science, 7*, 793–821.

Tucker, J. B. (1971a). Development and deployment of cilia, basal bodies, and other microtubular organelles in the cortex of the ciliate *Nassula*. *Journal of Cell Science, 9*, 539–567.

Tucker, J. B. (1971b). Microtubules and a contractile ring of microfilaments associated with a cleavage furrow. *Journal of Cell Science*, *8*, 557–571.

Tucker, J. B. (1972). Microtubule–arms and propulsion of food particles inside a large feeding organelle in the ciliate *Phascolodon vorticella*. *Journal of Cell Science*, *10*, 883–903.

Tucker, J. B. (1974). Microtubule–arms and cytoplasmic streaming and microtubule bending and stretching of intertubule links in the feeding tentacle of the suctorian ciliate *Tokophrya*. *Journal of Cell Biology*, *62*, 424–437.

Tucker, J. B. (1978). Endocytosis and streaming of highly gelated cytoplasm alongside rows of arm–bearing microtubules in the ciliate *Nassula*. *Journal of Cell Science*, *29*, 213–232.

Tucker, J. B., & Mackie, J. B. (1975). Configurational changes in helical microtubule frameworks in feeding tentacles of the suctorian ciliate *Tokophrya*. *Tissue & Cell*, *7*, 601–612.

Tucker, J. B., Dunn, M., & Pattisson, J. B. (1975). Control of microtubule pattern during the development of a large organelle in the ciliate *Nassula*. *Developmental Biology*, *47*, 439–453.

Tucker, J. B., Beisson, J., Roche, D. L. J., & Cohen, J. (1980). Microtubules and control of macronuclear 'amitosis' in *Paramecium*. *Journal of Cell Science*, *44*, 135–151.

Tuffrau, M. (1952). La morphogenèse de division chez les Colpodidae. *Bulletin Biologique de la France et de la Belgique*, *86*, 309–320.

Tuffrau, M. (1953). Les processus cytologiques de la conjugaison chez *Spirochona gemmipara* Stein. *Bulletin Biologique de la France de la Belgique*, *87*, 314–322.

Tuffrau, M. (1954). *Discotricha papillifera*, n. g., n. sp. Cilié psammobie de la famille des Trichopelmidae. *Journal of Protozoology*, *1*, 183–186.

Tuffrau, M. (1960). Révision du genre *Euplotes*, fondée sur la comparaison des structures superficielles. *Hydrobiologia*, *15*, 1–77.

Tuffrau, M. (1967). Perfectionnements et pratique de la technique d'imprégnation au protargol des infusoires ciliés. *Protistologica*, *3*, 91–98.

Tuffrau, M. (1968). Les structures fibrillaires somatiques et buccales chez les ciliés hétérotriches. *Protistologica*, *3*(year 1967), 369–394.

Tuffrau, M. (1984). Les aspects généraux de la morphogenèse chez les ciliés. In P. P. Grassé (Ed.), *Traité de zoologie*: *(Vol. 2, Fasc. 1., Infusoires ciliés* (pp. 537–580). Paris: Masson.

Tuffrau, M. (1987). Proposition d'une classification nouvelle de l'ordre Hypotrichida (Protozoa, Ciliophora), fondée sur quelques données récentes. *Annales des Sciences Naturelles–Zoologie et Biologie Animale, Paris, Séries 13*, *8*(1986–1987), 111–117.

Tuffrau, M. (1992). Observations on a rare ciliate, *Discomorphella pectinata* (Levander, 1894) Corliss, 1960. *Journal of Protozoology*, *39*, 122–125.

Tuffrau, M., & Fleury, A. (1977). Un nouveau type de morphogenèse chez les ciliés hypotriches: la bipartition sous kyste de *Paraholosticha muscicola* Kahl, 1935. *Protistologica*, *13*, 321–328.

Tuffrau, M., & Fleury, A. (1994). Classe des Hypotrichea Stein, 1859. In P. de Puytorac (Ed.), *Traité de zoologie, infusoires ciliés* (Vol. 2, pp. 83–151). Paris: Masson.

Tuffrau, M., & Fryd-Versavel, G. (1977). A new type of morphogenesis in the Ciliata Hypotricha division inside a cyst of *Paraholosticha muscicola*. *Protistologica, 13*, 321–328. (in French)

Tuffrau, M., & Laval–Peuto, M. (1978). Les ciliés endocommensaux d'un taret de Côte d'Ivoire, *Teredo adami*, mollusque Teredinidae. II. Infraciliature et polymorphisme de *Biggaria caryoselaginelloides* Tchang, 1958 (Hymenostome). *Protistologica*, *14*, 217–224.

Tuffrau, M., & de Puytorac, P. (1994). Classe des Heterotrichea Stein, 1859. In P. de Puytorac (Ed.), *Traité de zoologie, infusoires ciliés* (Vol. 2, pp. 35–79). Paris: Masson.

Tuffrau, M., Tuffrau, H., & Genermont, J. (1976). La réorganisation infraciliaire au cours de la conjugaison et l'origine du primordium buccal dans le genre *Euplotes*. *Journal of Protozoology*, *23*, 517–523.

Tuffrau, M., Fryd–Versavel, G., & Tuffrau, H. (1981). La réorganisation infraciliaire au cours de la conjugaison chez *Stylonychia mytilus*. *Protistologica*, *17*, 387–396.

Turkewitz, A. P. (2004). Out with a bang! *Tetrahymena* as a model system to study secretory granule biogenesis. *Traffic*, *5*, 63–68.

Turner, J. P. (1937). Studies on the ciliate *Tillina canalifera*. *Transactions of the American Microscopical Society*, *61*, 447–456.

Turner, J. T., Levinsen, H., Nielsen, T. G., & Hansen, B. W. (2001). Zooplankton feeding ecology: Grazing on phytoplankton and predation on protozoans by copepod and barnacle nauplii in Disko Bay, West Greenland. *Marine Ecological–Progress Series*, *221*, 209–219.

Twagilimana, L., Bohatier, J., Grolière, C. A., Bonnemoy, F., & Sargos, D. (1998). A new low–cost microbiotest with the protozoan *Spirostomum teres*: Culture conditions and assessment of sensitivity of the ciliate to 14 pure chemicals. *Ecotoxicology and Environmental Safety*, *41*, 231–244.

UUUU

Uhlig, G. (1960). Entwicklungsphysiologische Untersuchungen zur Morphogenese von *Stentor coeruleus* Ehrbg. *Archiv für Protistenkunde*, *105*, 1–109.

Urawa, S. (1992). Host range and geographical distribution of the ectoparasitic protozoans *Ichthyobodo neca-*

tor, *Trichodina truttae* and *Chilodonella piscicola* on hatchery–reared salmonids. *Scientific Reports of the Hokkaido Salmon Hatchery, 46*, 175–203.

Uspenskaya, Z. I., & Yudin, A. L. (1992). Clonal and temperature induced differences in serotypes in the ciliate *Dileptus anser. European Journal of Protistology, 28*, 85–93.

Usuki, I., & Irie, T. (1983). Inter– and intrasyngenic variation of *Paramecium* hemoglobin. II. *Paramecium caudatum, Paramecium jenningsi* and *Paramecium multimicronucleatum. Comparative Biochemistry and Physiology, Series B, 75*, 421–424.

VVVV

Vacchiano, E. J., Kut, J. L., Wyatt, M. L., & Buhse, H. E., Jr. (1991). A novel method for mass–culturing *Vorticella. Journal of Protozoology, 38*, 608–613.

Vacchiano, E., Dreisbach, A., Locascio, D., Castaneda, L., Vivian, T., & Buhse, H. E., Jr. (1992). Morphogenetic transformations and cytoskeletal elements of the stalked zooid and the telotroch stages in the peritrich ciliate *Vorticella convallaria. Journal of Protozoology, 39*, 101–106.

Valbonesi, A., & Guglielmo, L. (1988). Infestation of a lagoon zooplanktonic community with the epizoic peritrich *Zoothamnium intermedium* Precht (Peritrichia, Zoothamniidae). *Bolletino di Zoologia, 3*, 179–183.

Valbonesi, A., & Luporini, P. (1995). *Euplotes bisulcatus*, a key species for a better understanding of the organization and evolution of *Euplotes* argyrome patterns. *European Journal of Protistology, 31*, 32–37.

Valbonesi, A., Ortenzi, C., & Luporini, P. (1985). Electrophoretic comparison between *Euplotes vannus* and *E. crassus* strains. *Atti della Societa Toscana di Scienza Naturali, Memorie Serie B, 92*, 67–74.

Valbonesi, A., Ortenzi, C., & Luporini, P. (1988). An integrated study of the species problem in the *Euplotes crassus–minuta–vannus* group. *Journal of Protozoology, 35*, 38–45.

Valbonesi, A., Ortenzi, C., & Luporini, P. (1992). The species problem in a ciliate with a high multiple mating type system, *Euplotes crassus. Journal of Protozoology, 39*, 45–54.

Van As, L. L., & Basson, L. (1989). A further contribution to the taxonomy of the Trichodinidae (Ciliophora: Peritrichia) and a review of the taxonomic status of some fish ectoparasitic trichodinids. *Systematic Parasitology, 14*, 157–179.

Van As, L. L., & Basson, L. (1992). Trichodinid ectoparasites (Ciliophora: Peritrichida) of freshwater fishes of the Zambesi River system, with a reappraisal of host specificity. *Systematic Parasitology, 22*, 81–109.

Van As, L. L., & Basson, L. (1993). On the biology of *Pallitrichodina rogenae* gen. n., sp. n. and *P. stephani* sp. n. (Ciliophora: Peritrichida), mantle cavity symbionts of the giant African snail *Achatina* in Mauritius and Taiwan. *Acta Protozoologica, 32*, 47–62.

Van As, L. L., Van As, J. G., & Basson, L. (1999). *Licnophora limpetae* sp. n. (Ciliophora: Heterotrichea) ectosymbiont of South African limpets. *Acta Protozoologica, 38*, 249–253.

Van Bell, C. T. (1985). The 5S and 5.8S ribosomal RNA sequences of *Tetrahymena thermophila* and *T. pyriformis. Journal of Protozoology, 32*, 640–644.

van Bruggen, J. J. A., Stumm, C. K., & Vogels, G. D. (1983). Symbiosis of methanogenic bacteria and sapropelic protozoa. *Archives of Microbiology, 136*, 89–95.

van Bruggen, J. J. A., Zwart, K. B., van Assema, R. M., Stumm, C. K., & Vogels, G. D. (1984). *Methanobacterium formicicum*, an endosymbiont of the anaerobic cilaite *Metopus striatus* McMurrich. *Archives of Microbiology, 139*, 1–7.

van Bruggen, J. J. A., Zwart, K. B., Hermans, J. G. F., van Hove, E. M., Stumm, C. K., & Vogels, G. D. (1986). Isolation and characterization of *Methanoplanus endosymbiosus* sp. nov., an endosymbiont of the marine sapropelic ciliate *Metopus contortus* Quennerstedt. *Archives of Microbiology, 144*, 367–374.

Van de Peer, Y., Van der Auwera, G., & De Wachter, R. (1996). The evolution of stramenopiles and alveolates as derived by "substitution rate calibration" of small ribosomal subunit RNA. *Journal of Molecular Evolution, 42*, 201–210.

Van Den Bussche, R. A., Hoofer, S. A., Drew, C. P., & Ewing, M. S. (2000). Characterization of histone H3/H4 gene region and phylogenetic affinity of *Ichthyophthirius multifiliis* based on H4 DNA sequence variation. *Molecular Phylogenetics and Evolution, 14*, 461–468.

van Hoek, A. H. A. M., Sprakel, V. S. I., van Alen, T. A., Theuvenet, A. P. R., Vogels, G. D., & Hackstein, J. H. P. (1999). Voltage–dependent reversal of the anodic galvanotaxis in *Nyctotherus ovalis. Journal of Eukaryotic Microbiology, 46*, 427–433.

van Hoek, A. H. A. M., van Alen, T. A., Sprakel, V. S. I., Hackstein, J. H. P., & Vogels, G. D. (1998). Evolution of anaerobic ciliates from the gastrointestinal tract: Phylogenetic analysis of the ribosomal repeat from *Nyctotherus ovalis* and its relatives. *Molecular Biology and Evolution, 15*, 1195–1206.

van Hoek, A. H. A. M., Akhmanova, A. S., Huynen, M. A., & Hackstein, J. H. P. (2000a). A mitochondrial ancestry of the hydrogenosomes of *Nyctotherus ovalis. Molecular Biology and Evolution, 17*, 251–258.

van Hoek, A. H. A. M., van Alen, T. A., Sprakel, V. S. I., Leunissen, J. A. M., Brigge, T., Vogels, G. D., et al. (2000b). Multiple acquisitions of methanogenic Archaeal symbionts by anaerobic ciliates. *Molecular Biology and Evolution*, *17*, 251–258.

Van Houten, J., Hansma, J., & Kung, C. (1975). Two quantitative assays for chemotaxis in *Paramecium*. *Journal of Comparative Physiology*, *104A*, 211–223.

Van Houten, J., Martel, E., & Kasch, T. (1982). Kinetic analysis of chemokinesis of *Paramecium*. *Journal of Protozoology*, *29*, 226–230.

Van Hoven, W. (1978). Development and seasonal changes in the rumen protozoan population in young blesbok (*Damaliscus dorcas phillipsi* Harper 1939). *South African Journal of Wildlife Research*, *8*, 127–130.

Van Hoven, W. (1983). Rumen ciliates with description of two new species from three African reedbuck species. *Journal of Protozoology*, *30*, 688–691.

Van Hoven, W., Hamilton–Attwell, V. L., & Grobler, J. H. (1978). Rumen ciliate protozoa of the sable antelope *Hippotragus niger*. *South African Journal of Zoology*, *14*, 37–42.

Van Hoven, W., Gilchrist, F. M. C., & Hamilton–Attwell, V. L. (1987). Intestinal ciliated protozoa of African rhinoceros: Two new genera and five new species from the white rhino (*Ceratotherium simum* Burchell, 1817). *Journal of Protozoology*, *34*, 338–342.

Van Hoven, W., Gilchrist, F. M. C., & Hamilton–Attwell, V. L. (1988). A new family, genus, and seven new species of Entodiniomorphida (Protozoa) from the gut of African rhinoceros. *Journal of Protozoology*, *35*, 92–97.

Vannini, C., Schena, A., Verni, F., & Rosati, G. (2004). *Euplotes magnicirratus* (Ciliophora, Hypotrichia) depends on its bacterial endosymbiont '*Candidatus* Devosia euplotis' for food digestion. *Aquatic Microbial Ecology*, *36*, 19–28.

Vargas, R., & Hattori, T. (1990). The distribution of protozoa among soil aggregates. *FEMS Microbiology Ecology*, *74*, 73–78.

Vaudaux, P. E., Williams, N. E., Frankel, J., & Vaudaux, C. (1977). Interstrain variability of structural proteins in *Tetrahymena*. *Journal of Protozoology*, *24*, 453–458.

Vayssie, L., Garreau de Loubresse, N., & Sperling, L. (2001). Growth and form of secretory granules involves stepwise assembly but not differential sorting of a family of secretory proteins in *Paramecium*. *Journal of Cell Science*, *114*, 875–886.

Verhoeven, R. (2002). Ciliates in coastal dune soils of different stages of development. *European Journal of Soil Biology*, *38*, 187–191.

Veira, D. M. (1986). The role of ciliate protozoa in nutrition in the ruminant. *Journal of Animal Science*, *63*, 1547–1560.

Verity, P. G. (1985). Grazing, respiration, excretion, and growth rates of tintinnids. *Limnology and Oceanography*, *30*, 1268–1282.

Verity, P. G. (1987). Abundance, community composition, size distribution, and production rates of tintinnids in Narragansett Bay, Rhode Island. *Estuarine Coastal and Shelf Science*, *24*, 671–690.

Verity, P. G. (1991). Feeding in planktonic protozoans: Evidence for non–random acquisition of prey. *Journal of Protozoology*, *38*, 69–76.

Verity, P. G., & Villareal, T. A. (1986). The relative food value of diatoms, dinoflagellates, flagellates, and cyanobacteria for tintinnid ciliates. *Archiv für Protistenkunde*, *131*, 71–84.

Verni, F., & Gualtieri, P. (1997). Digestive process of the raptorial feeder ciliate *Litonotus lamella* (Rhabdophora, Litostomatea) visualized by fluorescence microscopy. *Micron*, *28*, 447–451.

Verni, F., & Rosati, G. (1990). Peculiar epibionts in *Euplotidium itoi* (Ciliata, Hypotrichida). *Journal of Protozoology*, *37*, 337–343.

Verni, F., Rosati, G., & Nobili, R. (1977). Infection mechanism of *eta* killing particles in sensitive cells of the ciliate *Euplotes crassus*. *Transactions of the American Microscopical Society*, *96*, 363–369.

Verni, F., Rosati, G., & Ricci, N. (1990). The cyst of *Oxytricha bifaria* (Ciliata Hypotrichida). II. The ultrastructure. *Protistologica*, *20*, 87–95.

Vigues, B., & David, D. (1989). Demonstration of a 58kd protein as a major component of the plasma membrane skeleton in the ciliate *Entodinium bursa*. *Protoplasma*, *149*, 11–19.

Vigues, B., & Grolière, C. A. (1985). Evidence for a Ca^{2+}– binding protein associated to non–actin microfilamentous systems in two ciliated protozoans. *Experimental Cell Research*, *159*, 366–376.

Vigues, B., Méténier, G., & Grolière, C. A. (1984a). Biochemical and immunological characterization of the microfibrillar ecto–endoplasmic boundary in the ciliate *Isotricha prostoma*. *Biology of the Cell*, *51*, 67–78.

Vigues, B., Méténier, G., & Sénaud, J. (1984b). The sub–surface cytoskeleton of the ciliate *Polyplastron multivesiculatum*: Isolation and major protein components. *European Journal of Cell Biology*, *35*, 336–342.

Vigues, B., Bricheux, G., Metivier, C., Brugerolle, G., & Peck, R. K. (1987). Evidence for common epitopes among proteins of the membrane skeleton of a ciliate, an euglenoid and a dinoflagellate. *European Journal of Protistology*, *23*, 101–110.

Villeneuve–Brachon, S. (1940). Recherches sur les ciliés hétérotriches: cinétome, argyrome, myonèmes, formes nouvelles ou peu connues. *Archives de Zoologie Expérimental et Générale*, *82*, 1–180.

Vinnikova, N. V., & Golikova, M. N. (1978). The ultrastructure of developed polytene chromosomes in macronuclear anlagen of *Nyctotherus cordiformis* (Protozoa, Ciliata). *Tsitologiya, 20*, 731–735. (in Russian with English summary)

Vivier, E. (1984). Conjugaison et phénomènes apparantés. In P. P. Grassé (Ed.), *Traité de zoologie: Vol. 2, Fasc. 1. Infusoires ciliés* (pp. 621–670). Paris: Masson.

Vogelbein, W. K., & Thune, R. L. (1988). Ultrastructural features of three ectocommensal protozoa attached to the gills of the red swamp crawfish, *Procambarus clarkii* (Crustacea: Decapoda). *Journal of Protozoology, 35*, 341–348.

Voncken, F. G. J., Boxma, B., van Hoek, A. H. A. M., Akhmanova, A. S., Vogels, G. D., Huynen, M., et al. (2002). A hydrogenosomal [Fe]–hydrogenase from the anaerobic chytrid *Neocallimastix* sp. *L2. Gene, 284*, 103–112.

Vopel, K., Reick, C. H., Arlt, G., Pohn, M., & Ott, J. A. (2002). Flow microenvironment of two marine peritrich ciliates with ectobiotic chemoautotrophic bacteria. *Aquatic Microbial Ecology, 29*, 19–28.

Vopel, K., Thistle, D., Ott, J. A., Bright, M., & Roy, H. (2004). Wave–induced H_2S flux sustains a chemoautotrophic symbiosis. *Limnology and Oceanography, 50*, 128–133.

Voss, H. J. (1992). Morphogenesis of *Amphisiella australis* Blatterer and Foissner, 1988 (Ciliophora, Hypotrichida). *European Journal of Protistology, 28*, 405–414.

Voss, H. J., & Foissner, W. (1996). Divisional morphogenesis in *Steinia sphagnicola* (Ciliophora, Hypotrichida): A comparative light and scanning electron microscopic study. *European Journal of Protistology, 32*, 31–46.

WWWW

Wagener, S., & Pfennig, N. (1987). Monoxenic culture of the anaerobic ciliate *Trimyema compressum* Lackey. *Archives of Microbiology, 149*, 4–11.

Wagener, S., Stumm, C. K., & Vogels, G. D. (1986). Electromigration, a tool for studies on anaerobic ciliates. *FEMS Microbiology Ecology, 38*, 197–203.

Wagener, S., Bardele, C. F., & Pfennig, N. (1990). Functional integration of *Methanobacterium formicicum* into the anaerobic ciliate *Trimyema compressum. Archives of Microbiology, 153*, 496–501.

Wagtendonk, W. J. van, (1974). *Paramecium – A current survey*. Amsterdam: Elsevier.

Wakita, M., & Hoshino, S. (1980). Physicochemical properties of a reserve polysaccharide from sheep rumen ciliates genus *Entodinium. Comparative Biochemistry and Physiology, 65B*, 571–574.

Walker, G. K. (1976). The fine structure of micronuclear division in the hypotrich ciliate *Gastrostyla steinii. Protistologica, 12*, 271–278.

Walker, G. K., & Goode, D. (1976). The role of microtubules in macronuclear division in the hypotrich ciliates *Gastrostyla steinii* and *Stylonychia mytilus. Cytobiologie, 14*, 18–37.

Walker, G. K., & Grim, J. N. (1973). Morphogenesis and polymorphism in *Gastrostyla steinii. Journal of Protozoology, 20*, 566–573.

Walker, G. K., & Maugel, T. K. (1980). Encystment and excystment in hypotrich ciliates. II. *Diophrys scutum* and remarks on comparative features. *Protistologica, 16*, 525–531.

Walker, G. K., Maugel, T. K., & Goode, D. (1980). Encystment and excystment in hypotrich ciliates. I. *Gastrostyla steinii. Protistologica, 16*, 511–524.

Walker, G. K., Edwards, C. A., & Suchard, S. J. (1989). Encystment in the peritrich ciliate *Telotrochidium henneguyi. Cytobios, 59*, 7–18.

Walker, M. H., & Roberts, E. (1982). The protozoan epizooites found on the gills of *Gammarus pulex. Hydrobiologia, 88*, 171–176.

Walker, M. H., & Roberts, E. (1988). A study of tomite structure and metamorphosis in *Dendrocometes paradoxus* Stein (Suctoria, Ciliophora) using silver staining and scanning electron microscopy. *European Journal of Protistology, 24*, 2–10.

Walker, M. H., Roberts, E. M. Usher, M. L. (1986). The fine structure of the trophont and stages in telotroch formation in *Circolagenophrys ampulla* (Ciliophora, Peritrichida). *Journal of Protozoology, 33*, 246–255.

Wallengren, H. (1897). Studier öfver Ciliata infusorier. III. Bidrag till Kännedomen om Fam. Urceolarina Stein. *Acta Universitatis Lundensis, 33*, 1–48.

Wallengren, H. (1900). Zur Kenntnis der vergleichenden Morphologie der hypotrichen Infusorien. *Bihang till K. Svenska Vetenskapsakademiens handlingar, 26*, 1–31.

Wang, J. (1977). Protozoa from some districts of Tibetan plateau. *Acta Zoologica Sinica, 23*, 131–160 (in Chinese with English summary)

Wang, X., & Dickerson, H. W. (2002). Surface immobilization antigen of the parasitic ciliate *Ichthyophthirius multifiliis* elicits protective immunity in channel catfish (*Ictalurus punctatus*). *Clinical Diagnostic Laboratory Immunology, 9*, 176–181.

Warren, A. (1986). A revision of the genus *Vorticella* (Ciliophora: Peritrichida). *Bulletin of the British Museum (Natural History), 50*, 1–57.

Warren, A. (1987). A revision of the genus *Pseudovorticella* Foissner & Schiffman, 1974 (Ciliophora: Peritrichida). *Bulletin of the British Museum (Natural History), 52*, 1–12.

Warren, A. (1988). A revision of the genus *Haplocaulus* Precht, 1935 (Ciliophora: Peritrichida). *Bulletin of the Britsih Museum (Natural History), 54*, 127–152.

Warren, A., & Carey, P. G. (1983). Lorica structure of the freshwater ciliate *Platycola decumbens* Ehrenberg, 1830 (Peritrichida, Vaginicolidae). *Protistologica, 19*, 5–20.

Washburn, J. O., Gross, M. E., Mercer, D. R., & Anderson, J. R. (1988). Predator–induced trophic shift of a free–living ciliate: Parasitism of mosquito larvae by their prey. *Science, 240*, 1193–1195.

Watanabe, T. (1978). A scanning electron–microscopic study of the local degeneration of cilia during sexual reproduction in *Paramecium caudatum. Journal of Cell Science, 32*, 55–66.

Watanabe, T. (1983). Local degeneration of cilia and nuclear activation during sexual reproduction in *Paramecium. Development Growth & Differentiation, 25*, 113–120.

Weide, T., Herrmann, L., Bockau, U., Niebur, N., Aldag, I., Laroy, W., et al. (2006). Secretion of functional human enzymes by *Tetrahymena thermophila. BMC Biotechnology, 6*, 19. DOI: 10.1186/1472–6750–6–19

Weis, D. S. (1974). Sparing effect of light on bacterial consumption of *Paramecium bursaria. Transactions of the American Microscopical Society, 93*, 135–140.

Weis, D. S. (1979). Correlation of sugar release and concanavalin A agglutinability with infectivity of symbiotic algae from *Paramecium bursaria* for aposymbiotic *P. bursaria. Journal of Protozoology, 26*, 117–119.

Weiske–Benner, A., & Eckert, W. A. (1985). Differentiation of nuclear structure during the sexual cycle in *Tetrahymena thermophila* I. Development and transcriptional activity of macronuclear anlagen. *Differentiation, 28*, 225–236.

Weisse, T. (2006). Freshwater ciliates as ecophysiological model organisms – lessons from *Daphnia*, major achievements, and future perspectives. *Archive Hydrobioloiga, 167*, 371–402.

Weisse, T., & Montagnes, D. J. S. (1998). Effect of temperature on inter– and intraspecific isolates of *Urotricha* (Prostomatida, Ciliophora). *Aquatic Microbial Ecology, 15*, 285–291.

Weisse, T., & Rammer, S. (2006). Pronounced ecophysiological clonal differences of two common freshwater ciliates, *Coleps spetai* (Prostomatida) and *Rimostrombidium lacustris* (Oligotrichida), challenge the morphospecies concept. *Journal of Plankton Research, 28*, 55–63.

Weisse, T., & Stadler, P. (2006). Effect of pH on growth, cell volume, and production of freshwater ciliates, and implications for their distribution. *Limnology and Oceanography, 51*, 1708–1715.

Weisse, T., Karstens, N., Meyer, V. C. L., Janke, L., Lettner, S., & Teichgräber, K. (2001). Niche separation in common prostome freshwater ciliates: The effect of food and temperature. *Aquatic Microbial Ecology, 26*, 167–179.

Weissman, P., Lonsdale, D. J., & Yen, J. (1993). The effect of peritrich ciliates on the production of *Acartia hudsonica* in Long Island Sound. *Limnology and Oceanography, 38*, 613–622.

Weitschat, W., & Guhl, W. (1994). Erster Nachweis fossiler Ciliaten. *Paläontologische Zeitschrift, 68*, 17–31.

Wenzel, F., & Meier–Tackmann, D. (1975). Untersuchungen über den Einfluss einiger Bakterienarten auf die Excystierung von *Colpoda cucullus*. O. F. Müller (Holotricha, Ciliata) aus Hunger–Dauercysten. *Protistologica, 11*, 125–133.

Wenzel, F., & Winkler, A. (1984). Quantitative Untersuchungen über die Nahrungsaufnahme von *Colpoda cucullus* O. F. Müller (Holotricha, Ciliata). *Zoologischer Anzeiger, Jena, 212*, 1–16.

Wessenberg, H., & Antipa, G. A. (1969). Studies on *Didinium nasutum*. I. Structure and ultrastructure. *Protistologica, 4*(year 1968): 427–447.

Wessenberg, H., & Antipa, G. A. (1970). Capture and ingestion of *Paramecium* by *Didinium nasutum. Journal of Protozoology, 17*, 250–270.

White, A. W., Sheath, R. G., & Hellebust, J. A. (1977). A red tide caused by the marine ciliate *Mesodinium rubrum* in Passamaquoddy Bay, including pigment and ultrastructure studies of the endosymbiont. *Journal of the Fisheries Research Board of Cananda, 34*, 413–416.

Wiackowski, K. (1988). Phenetic and cladistic numerical estimates of phylogenetic relationships in Urostylina (Ciliophora: Hypotrichida). *Acta Protozoologica, 27*, 1–20.

Wiackowski, K., Brett, M. T., & Goldman, C. R. (1994). Differential effects of zooplankton species on ciliate community structure. *Limnology and Oceanography, 39*, 486–492.

Wiackowski, K., Fyda, J., Pajdak–Stós, A., & Adamus, K. (2003). Predator–induced morphological defence in ciliates: Interclonal variation in sensitivity to the inducing factors. *Oikos, 100*, 534–540.

Wichterman, R. (1936). Division and conjugation in *Nyctotherus cordiformis* (Ehr.) Stein (Protozoa, Ciliata) with special reference to the nuclear phenomena. *Journal of Morphology, 60*, 563–611.

Wichterman, R. (1986). *The biology of Paramecium* (2nd ed., 559). New York: Plenum Press.

Wickham, S. A. (1995). *Cyclops* predation on ciliates: Species–specific differences and functional responses. *Journal of Plankton Research, 17*, 1633–1646.

Wickham, S. A., & Lynn, D. H. (1990). Relations between growth rate, cell size, and DNA content in colpodean ciliates (Ciliophora: Colpodea). *European Journal of Protistology*, *25*, 345–352.

Wickham, S. A., & Gilbert, J. J. (1991). Relative vulnerabilities of natural rotifer and ciliate communities to cladocerans: Laboratory and field experiments. *Freshwater Biology*, *26*, 77–86.

Wickham, S. A., & Gilbert, J. J. (1993). The comparative importance of competition and predation by *Daphnia* on ciliated protists. *Archiv für Hydrobiologie*, *126*, 289–313.

Wickham, S. A., Gilbert, J. J., & Berninger, U. G. (1993). Effects of rotifers and ciliates on the growth and survival of *Daphnia*. *Journal of Plankton Research*, *15*, 317–334.

Wicklow, B. J. (1981). Evolution with the order Hypotrichida (Ciliophora, Protozoa): Ultrastructure and morphogenesis of *Thigmokeronopsis jahodai* (n. gen., n. sp.); phylogeny in the Urostylina (Jankowski, 1979). *Protistologica*, *17*, 331–351.

Wicklow, B. J. (1982). The Discocephalina (n. subord.): Ultrastructure, morphogenesis and evolutionary implications of a group of endemic marine interstitial hypotrichs (Ciliophora, Protozoa). *Protistologica*, *18*, 299–330.

Wicklow, B. J. (1983). Ultrastructure and cortical morphogenesis in the euplotine hypotrich *Certesia quadrinucleata* Fabre–Domergue, 1885 (Ciliophora, Protozoa). *Journal of Protozoology*, *30*, 256–266.

Wicklow, B. J. (1988). Developmental polymorphism induced by intraspecific predation in the ciliated protozoon *Onychodromus quadricornutus*. *Journal of Protozoology*, *35*, 137–141.

Wicklow, B. J., & Borror, A. C. (1977). *Discotricha papillifera*: Structure and morphogenesis of a marine interstitial ciliate. *Journal of Protozoology*, *24*, 99–108.

Wilbert, N. (1986). Beitrag zur Morphologie und Ökologie einiger planktischer Ciliaten aus dem Poppelsdorfer Weiher in Bonn (BRD). *Archiv für Protistenkunde*, *131*, 59–69.

Wilbert, N., & Foissner, W. (1980). Eine Neubeschreibung von *Calyptotricha lanuginosum* Penard, 1922 (Ciliata, Scuticociliatida). *Archiv für Protistenkunde*, *123*, 12–21.

Wilbert, N., & Schmall, G. (1976). Morphologie und Infraciliatur von *Coleps nolandi* Kahl, 1930. *Protistologica*, *12*, 193–197.

Wilbert, N., & Schmeier, U. (1982). Survey of the intestinal opalines and ciliates in central European amphibians. *Archiv für Protistenkunde*, *125*, 271–285.

Wilks, S. A., & Sleigh, M. A. (1998). Grazing rates in *Euplotes mutabilis*: Relationship between particle size and concentration. *Microbial Ecology*, *36*, 165–174.

Wilks, S. A., & Sleigh, M. A. (2004). Lectin binding sites on *Euplotes mutabilis* (Tuffrau, 1960) and the implications for food particle selection. *European Journal of Protistology*, *40*, 153–162.

Wille, J. J., Jr., Weidner, E., & Steffens, W. L. (1981). Intranuclear parasitism of the ciliate *Euplotes* by a trypanosomatid flagellate. *Journal of Protozoology*, *28*, 223–227.

Willey, R. L., & Walkosz, R. A. (1975). The scopula of *Ophrydium versatile* (Ciliophora: Peritrichida). *Transactions of the American Microscopical Society*, *94*, 427–432.

Williams, A. G. (1986). Rumen holotrich ciliate protozoa. *Microbiological Reviews*, *50*, 25–49.

Williams, A. G., Joblin, K. N., Butler, R. D., Fonty, G., & Bernalier, A. (1993). Interactions bactéries–protistes dans le rumen. *Année Biologique*, *32*, 13–30.

Williams, D. B. (1980). Clonal aging in two species of *Spathidium* (Ciliophora: Gymnostomatida). *Journal of Protozoology*, *27*, 212–215.

Williams, D. B., Williams, B. D., & Hogan, B. K. (1981). Ultrastructure of the somatic cortex of the gymnostome ciliate *Spathidium spathula* (O. F. M.). *Journal of Protozoology*, *28*, 90–99.

Williams, N. E. (1960). The polymorphic life history of *Tetrahymena patula*. *Journal of Protozoology*, *7*, 10–17.

Williams, N. E. (1961). Polymorphism in *Tetrahymena vorax*. *Journal of Protozoology*, *8*, 403–410.

Williams, N. E. (1984). An apparent disjunction between the evolution of form and substance in the genus *Tetrahymena*. *Evolution*, *38*, 25–33.

Williams, N. E. (1986). The assembly of ciliates. *Journal of Protozoology*, *33*, 313–316.

Williams, N. E. (2004). The epiplasm gene *EPC1* influences cell shape and cortical pattern in *Tetrahymena thermophila*. *Journal of Eukaryotic Microbiology*, *51*, 201–206.

Williams, N. E., & Bakowska, J. (1982). Scanning electron microscopy of cytoskeletal elements in the oral apparatus of *Tetrahymena*. *Journal of Protozoology*, *29*, 382–389.

Williams, N. E., & Frankel, J. (1973). Regulation of microtubules in *Tetrahymena*. I. Electron microscopy of oral replacement. *Journal of Cell Biology*, *56*, 441–457.

Williams, N. E., & Williams, R. J. (1976). Macronuclear division with and without microtubules in *Tetrahymena*. *Journal of Cell Science*, *20*, 61–77.

Williams, N. E., Van Bell, C., & Newlon, M. (1980). Variation in surface proteins in *Tetrahymena*. *Journal of Protozoology*, *27*, 345–350.

Williams, N. E., Buhse, H. E., Jr., & Smith, M. G. (1984). Protein similarities in the genus *Tetrahymena* and a description of *Tetrahymena leucophrys* n. sp. *Journalof Protozoology*, *31*, 313–321.

Williams, N. E., Honts, J. E., & Dress, V. M. (1992). Protein polymorphism and evolution in the genus *Tetrahymena*. *Journal of Protozoology*, *39*, 54–58.

Williams, N. E., Tsao, C. C., Bowen, J., Hehman, G. L., Williams, R. J., & Frankel, J. (2006). The actin gene

ACT1 is required for phagocytosis, motility, and cell separation of *Tetrahymena thermophila. Eukaryotic Cell, 5*, 555–567.

Williams, R., McCall, H., Pierce, R. W., & Turner, J. T. (1994). Speciation of the tintinnid genus *Cymatocylis* by morphometric analysis of the loricae. *Marine Ecology–Progress Series, 107*, 263–272.

Winkler, R. H., & Corliss, J. O. (1965). Notes on the rarely described, green colonial protozoon *Ophrydium versatile* (O. F. M.) (Ciliophora, Peritrichida). *Transactions of the American Microscopical Society, 84*, 127–137.

Wirnsberger, E., Foissner, W., & Adam, H. (1985a). Cortical pattern in non–dividers, dividers and reorganizers of an Austrian population of *Paraurostyla weissei* (Ciliophora: Hypotrichida): A comparative morphological and biometrical study. *Zoologica Scripta, 14*, 1–10.

Wirnsberger, E., Foissner, W., & Adam, H. (1985b). Morphogenesis, fine structure, and phylogenetic relationships of the "heterotrich" ciliate *Bryometopus atypicus* (Protozoa, Colpodea). *Annales des Sciences Naturelles–Zoologie et Biologie Animale, Paris, Série 13, 7*, 113–128.

Wirnsberger, E., Foissner, W., & Adam, H. (1986). Biometric and morphogenetic comparison of the sibling species *Stylonychia mytilus* and *S. lemnae*, including a phylogenetic system for the oxytrichids (Ciliophora, Hypotrichida). *Archiv für Protistenkunde, 132*, 167–185.

Wirnsberger–Aescht, E., Foissner, W., & Foissner, I. (1989). Morphogenesis and ultrastructure of the soil ciliate *Engelmanniella mobilis* (Ciliophora, Hypotrichida). *European Journal of Protistology, 24*, 354–368.

Wise, B. N. (1965). The morphogenetic cycle in *Euplotes eurystomus* and its bearing on problems of ciliate morphogenesis. *Journal of Protozoology, 12*, 626–648.

Woelfl, S., & Geller, W. (2002). *Chlorella*–bearing ciliates dominate in an oligotrophic North Patagonian lake (Lake Pirehueico, Chile): Abundance, biomass and symbiotic photosynthesis. *Freshwater Biology, 47*, 231–242.

Wolfe, J. (1985). Cytoskeletal reorganization and plasma membrane fusion in conjugating *Tetrahymena. Journal of Cell Science, 73*, 69–85.

Wolska, M. (1962). A rare case of *Balantidium coli* in the caecum of a horse. *Bulletin de l'Académie Polonaise des Sciences, 10*, 425–429.

Wolfe, J., & Grimes, G. W. (1979). Tip transformation in *Tetrahymena*: A morphogenetic response to interactions between mating types. *Journal of Protozoology, 26*, 82–89.

Wolska, M. (1965). Studies on the representatives of the family Paraisotrichidae Da Cunha (Ciliata, Trichostomata). IV. General discussion. *Acta Protozoologica, 3*, 115–131.

Wolska, M. (1966). Study on the family Blepharocorythidae Hsiung. I. Preliminary remarks. *Acta Protozoologica, 4*, 97–103.

Wolska, M. (1971). Studies on the family Blepharocorythidae Hsiung. VI. Phylogenesis of the family and the description of the new genus *Circodinium* gen. n. with the species *C. minimum* (Gassovsky, 1918). *Acta Protozoologica, 9*, 171–194.

Wolska, M. (1981). Studies on the genus *Triadinium* Fior. (Ciliata, Entodiniomorphida). Comparison of *Triadinium galea* Gass. And *Triadinium caudatum* Fior. *Acta Protozoologica, 20*, 357–365.

Wolters, J. (1991). The troublesome parasites – molecular and morphological evidence that Apicomplexa belong to the dinoflagellate–ciliate clade. *BioSystems, 25*, 75–83.

Wong, L. C., & Landweber, L. F. (2006). Evolution of programmed DNA rearrangements in a scrambled gene. *Molecular Biology and Evolution, 23*, 756–763.

Wood, D. C. (1989). Localization of mechanoreceptors in the protozoan, *Stentor coeruleus. Journal of Comparative Physiology A, 165*, 229–235.

Woodruff, L. L. (1941). Endomixis. In G. N. Calkins & F. M. Summers (Eds.), *Protozoa in biological research* (pp. 646–665). New York: Columbia University Press.

Wright, A. D. G. (1999). Analysis of intraspecific sequence variation among eight isolates of the rumen symbiont, *Isotricha prostoma* (Ciliophora), from two continents. *Journal of Eukaryotic Microbiology, 46*, 445–446.

Wright, A. D. G., & Colorni, A. (2002). Taxonomic re–assignment of *Cryptocaryon irritans*, a marine fish parasite. *European Journal of Protistology, 37*, 375–378.

Wright, A. D. G., & Lynn, D. H. (1995). Phylogeny of the fish parasite *Ichthyophthirius* and its relatives *Ophryoglena* and *Tetrahymena* (Ciliophora, Hymenostomatia) inferred from 18S ribosomal RNA sequences. *Molecular Biology and Evolution, 12*, 285–290.

Wright, A. D. G., & Lynn, D. H. (1997a). Phylogenetic analysis of the rumen ciliate family Ophryoscolecidae based on 18S ribosomal RNA sequences, with new sequences from *Diplodinium*, *Eudiplodinium*, and *Ophryoscolex. Canadian Journal of Zoology, 75*, 963–970.

Wright, A. D. G., & Lynn, D. H. (1997b). Monophyly of the trichostome ciliates (Phylum Ciliophora: Class Litostomatea) tested using new 18S RNA sequences from the vestibuliferids, *Isotricha intestinalis* and *Dasytricha ruminantium*, and the haptorian *Didinium nasutum. European Journal of Protistology, 33*, 305–315.

Wright, A. D. G., & Lynn, D. H. (1997c). Maximum ages of ciliate lineages estimated using a small subunit rRNA molecular clock: Crown eukaryotes date back to the Paleoproterozoic. *Archiv für Protistenkunde, 148*, 329–341.

Wright, A. D. G., Dehority, B. A., & Lynn, D. H. (1997). Phylogeny of the rumen ciliates *Entodinium*, *Epidinium* and *Polyplastron* (Litostomatea: Entodiniomorphida) inferred from small subunit ribosomal RNA sequences. *Journal of Eukaryotic Microbiology*, *44*, 61–67.

Wright, J. F. (1981). *Tetrahymena pyriformis* (Ehrenberg) and *T. corlissi* Thompson parasitic in stream–dwelling triclads (Platyhelminthes: Turbellaria). *Journal of Parasitology*, *67*, 131–133.

Wunderlich, F., Speth, V., Batz, W., & Kleinig, H. (1973). Membranes of *Tetrahymena*. III. The effect of temperature on membrane core structures and fatty acid composition of *Tetrahymena* cells. *Biochimica et Biophysica Acta*, *298*, 39–49.

Wurtsbaugh, W. A., & Tapia, R. A. (1988). Mass mortality of fishes in Lake Titicaca (Peru–Bolivia) associated with the protozoan parasite *Ichthyophthirius multifiliis*. *Transactions of the American Fisheries Society*, *117*, 213–217.

XXXX

Xie, T., Sanderson, B., Frost, T., & Magnuson, J. J. (2001). Manipulation of host density affects infestation of a peritrich ciliate (*Epistylis lacustris*) on a calanoid copepod (*Leptodiaptomus minutus*) in Crystal Lake, Wisconsin, USA. *Journal Freshwater Ecology*, *16*, 557–564.

Xu, D., Song, W., Sun, P., & Chen, X. (2006). Morphology and infraciliature of the oligotrich *Strombidium rapulum* (Yagiu, 1933) Kahl, 1934 (Protozoa, Ciliophora, Oligotrichida) from the intestine of the sea urchin *Hemicentrotus pulcherrimus* Agassiz. *Zootaxa*, *1113*, 33–40.

Xu, D. H., Klesius, P. H., & Panangala, V. S. (2006). Induced cross–protection in channel catfish, *Ictalurus punctatus* (Rafinesque), against different immobilization serotypes of *Ichthyophthirius multifiliis*. *Journal of Fish Diseases*, *29*, 131–138.

Xu, K., & Foissner, W. (2004). Body, nuclear, and ciliary changes during conjugation of *Protospathidium serpens* (Ciliophora, Haptoria). *Journal of Eukaryotic Microbiology*, *51*, 605–617.

Xu, K., & Foissner, W. (2005). Morphology, ontogenesis and encystment of a soil ciliate (Ciliophora, Haptorida), *Arcuospathidium cultriforme* (Penard, 1922), with models for the formation of the oral bulge, the ciliary patterns, and the evolution of the spathidiids. *Protistology*, *4*, 5–55.

Xu, K., Song, W., & Warren, A. (2000). Observations on trichodinid ectoparasites (Ciliophora: Peritricha) from the gills of maricultured molluscs in China, with descriptions of three new species of *Trichodina* Ehrenberg, 1838. *Systematic Parasitology*, *45*, 17–24.

Xu, K. D., Song, W., & Warren, A. (2002). Taxonomy of trichodinids from the gills of marine fishes in coastal regions of the Yellow Sea, with descriptions of two new species of *Trichodina* Ehrenberg, 1830 (Protozoa: Ciliophora: Peritrichia). *Systematic Parasitology*, *51*, 107–120.

Xu, X., Kumakura, M., Kaku, E., & Takahashi, M. (2000). Odd mating–type substances may work as precursor molecules of even mating–type substances in *Paramecium caudatum*. *Journal of Eukaryotic Microbiology*, *48*, 683–689.

Xu, Z., Bowers, N., & Pratt, J. R. (1997). Variation in morphology, ecology, and toxicological responses of *Colpoda inflata* (Stokes) collected from five biogeographic realms. *European Journal of Protistology*, *33*, 136–144.

YYYY

Yamaoka, M., Watoh, T., & Matsuoka, T. (2004). Effects of salt concentration and bacteria on encystment induction in ciliated protozoan *Colpoda* sp. *Acta Protozoologica*, *43*, 93–98.

Yambot, A. V., & Song, Y. L. (2004). Short term in vitro culture of *Cryptocaryon irritans*, a protozoan parasite of marine fishes. *Fish Pathology*, *39*, 175–181.

Yambot, A. V., Song, Y. L., & Sung, H. H. (2003). Characterization of *Cryptocaryon irritans*, a parasite isolated from marine fishes in Taiwan. *Diseases of Aquatic Organisms*, *54*, 147–156.

Yamada, K., Kamagata, Y., Nakamura, K., Inamori, Y., & Nakamura, I. (1994). Selectivity of food bacteria for the growth of anaerobic ciliate *Trimyema compressum*. *Archives of Microbiology*, *161*, 229–233.

Yao, M. C., & Gall, J. G. (1979). Alteration of the *Tetrahymena* genome during nuclear differentiation. *Journal of Protozoology*, *26*, 10–13.

Yao, M. C., & Gorovsky, M. (1974). Comparison of the sequences of macronuclear and micronuclear DNA of *Tetrahymena pyriformis*. *Chromosoma*, *48*, 1–18.

Yao, M. C., & Yao, C. H. (1981). The repeated hexanucleotide C–C–C–C–A–A is present near the free ends of macronuclear DNA of *Tetrahymena*. *Proceedings of the National Academy of Sciences of the United States of America*, *78*, 7436–7439.

Yao, M. C., Duhacourt, S., & Chalker, D. L. (2002). Genome–wide rearrangements of DNA in ciliates. In R. Craigie, M. Gellert, & A. M. Labbowitz (Eds.), *Mobile DNA II* (pp. 730–758). Washington, DC: American Society for Microbiology.

Yarlett, N., Hann, A. C., Lloyd, D., & Williams, A. G. (1981). Hydrogenosomes in the rumen protozoon *Dasytricha ruminantium* Schuberg. *Biochemical Journal, 200*, 365–372.

Yarlett, N., Coleman, G. S., Williams, A. G., & Lloyd, D. (1984). Hydrogenosomes in known species of rumen entodiniomorphid protozoa. *FEMS Microbiology Letters, 21*, 15–19.

Yasindi, A. W., Lynn, D. H., & Taylor, W. D. (2002). Ciliated protozoa in Lake Nakuru, a shallow alkaline–saline lake in Kenya: Seasonal variation, potential production and role in the food web. *Archiv für Hydrobiologie, 154*, 311–325.

Yasuda, T., Numata, O., Ohnishi, K., & Watanabe, Y. (1980). A contractile ring and cortical changes found in the dividing *Tetrahymena pyriformis. Experimental Cell Research, 128*, 407–417.

Ye, A. J., & Romero, D. P. (2002). Phylogenetic relationships amongst tetrahymenine ciliates inferred by a comparison of telomerase rRNAs.*International Journal of Systematic and Evolutionary Microbiology, 52*, 2297–2302.

Yoshinaga, T. (2001). Effects of high temperature and dissolved oxygen concentration on the development of *Cryptocaryon irritans* (Ciliophora) with a comment on the autumn outbreaks of cryptocaryoniasis. *Fish Pathology, 36*, 231–235.

Yoshinaga, T., & Nakazoe, J. (1997). Acquired protection and production of immobilization antibody against *Cryptocaryon irritans* (Ciliophora, Hymenostomatida) in mummichog (*Fundulus heteroclitus*). *Fish Pathology, 32*, 229–230.

Yoshinaga, T., & Nakazoe, J. I. (2001). Acquired protection and production of immobilization antibody against *Cryptocaryon irritans* (Ciliophora, Hymenostomatida) in mummichog (*Fundulus heteroclitus*). *Fish Pathology, 32*, 229–230.

Yogosawa–Ohara, R., & Shigenaka, Y. (1985). Twisting contraction mechanism of a heterotrichous ciliate, *Spirostomum ambiguum*. 1. Role of the myoneme. *Cytobios, 44*, 7–17.

Yogosawa–Ohara, R., Suzaki, T., & Shigenaka, Y. (1985). Twisting contraction mechanism of a heterotrichous ciliate, *Spirostomum ambiguum*. 2. Role of longitudinal microtubular sheet. *Cytobios, 44*, 215–230.

Yusa, A. (1957). The morphology and morphogenesis of the buccal organelles in *Paramecium* with particular reference to their systematic significance. *Journal of Protozoology, 4*, 128–142.

ZZZZ

Zackroff, R. V., & Hufnagel, L. A. (1998). Relative potencies of different cytochalasins for the inhibition of phagocytosis in ciliates. *Journal of Eukaryotic Microbiology, 45*, 397–403.

Zackroff, R. V., & Hufnagel, L. A. (2002). Induction of anti–actin drug resistance in *Tetrahymena. Journal of Eukaryotic Microbiology, 49*, 475–477.

Zagon, I. S. (1970). *Carchesium polypinum:* Cytostructure after protargol silver deposition. *Transactions of the American Microscopical Society, 89*, 450–468.

Zaman, V. (1978). *Balantidium coli*. In P. Kreier (Ed.), *Parasitic protozoa* (Vol. 2, pp. 633–653). New York: Academic Press.

Zaritsky, A., Ben–Dov, E., Zalkinder, V., & Barak, Z. (1992). Digestibility by and pathogenicity of the protozoa *Tetrahymena pyriformis* to larvae of *Aedes aegypti. Journal of Invertebrate Pathology, 59*, 332–334.

Zeliff, C. C. (1933). A new protozoan from the larva of the beetle *Osmoderma scabra. Proceedings of the United States National Museum, 82*, 23, 1–3.

Zhang, W., & Wang, R. (2000). Summertime ciliate and copepod nauplii distributions and microzooplankton herbivorous activity in the Laizhou Bay, Bohai Sea, China. *Estuarine Coastal and Shelf Science, 51*, 103–114.

Zhang, W. J., Yang, J., Yu, Y. H., Shu, S. W., & Shen, Y. F. (2006). Population genetic structure of *Carchesium polypinum* (Ciliophora: Peritrichia) in four Chinese lakes inferred from ISSR fingerprinting: High diversity but low differentiation. *Journal of Eukaryotic Microbiology, 53*, 358–363.

Zilberg, D., & Sinai, T. (2006). Optimization and validation of a colorimetric assay for *Tetrahymena* sp. survival. *Research in Microbiology, 157*, 355–359.

Zingel, P., & Ott, I. (2000). Vertical distribution of planktonic ciliates in strongly stratified temperate lakes. *Hydrobiologia, 435*, 19–26.

Zingel, P., Huitu, E., Makela, S., & Arvola, L. (2002). The abundance and diversity of planktonic ciliates in 12 boreal lakes of varying trophic state. *Archiv für Hydrobiologie, 155*, 315–322.

Zou, S. F., & Ng, S. F. (1991). Commitment to the first cortical reorganization during conjugation in *Stylonychia mytilus*: An argument for homology with cortical development during binary fission. *Journal of Protozoology, 38*, 192–200.

Zufall, R. A., McGrath, C. L., Muse, S. V., & Katz, L. A. (2006). Genome architecture drives protein evolution in ciliates. *Molecular Biology and Evolution, 23*, 1681–1687.

Zwart, K. B., Goosen, N. K., Van Schijndel, M. W., Broers, C. A. M., Stumm, C. K., & Vogels, G. D. (1988). Cytochemical localization of hydrogenase activity in the anaerobic protozoa *Trichomonas vaginalis, Plagiopyla nasuta* and *Trimyema compressum. Journal of General Microbiology, 134*, 2165–2170.

Subject Index

D

J

K

S

Systematic Index

D

F

O

T